BIOLOGICAL CONTROL OF WEEDS IN AUSTRALIA

Sponsors

The following weed societies provided financial support so that this book could be presented in full colour.

- The Weed Society of Queensland Inc. www.wsq.org.au
- The Weeds Society of Western Australia Inc.

This has greatly improved the presentation of maps and photos. The foresight of the executives of both societies is acknowledged and will be appreciated by all readers of this book.

BIOLOGICAL CONTROL OF WEEDS IN AUSTRALIA

Editors:

Mic Julien
CSIRO Ecosystem Sciences

Rachel McFadyen
Weed Society of Queensland

Jim Cullen
CSIRO Ecosystem Sciences

CSIRO
PUBLISHING

National Library of Australia Cataloguing-in-Publication entry

Biological control of weeds in Australia/edited by M.H. Julien, R.E.C. McFadyen and J.M. Cullen.

9780643099937 (hbk.)
9780643104204 (epdf)
9780643104211 (epub)

Includes bibliographical references and index.

Weeds – Biological control – Australia.

Julien, M. H.
McFadyen, R. E. C.
Cullen, J. M.

632.5

Published by

CSIRO PUBLISHING
36 Gardiner Road, Clayton VIC 3168
Private Bag 10, Clayton South VIC 3169
Australia

Telephone: [+613] 9545 8555
Local call: 1300 788 000 (Australia only)
Fax: +61 3 9662 7555
Email: csiropublishing@csiro.au
Web site: www.publishing.csiro.au

Front cover (clockwise from top left): *Salvinia molesta* in Kakadu National Park (photo Mic Julien, CSIRO), *Opuntia stricta* in flower and fruit (photo JR Hosking, NSW Government), skeleton weed rust on *Chondrilla juncea* (photo JM Cullen, CSIRO), *Cactoblastis cactorum* larvae hollow out cactus segments (photo GR Sainty, consultant).

Back cover (left to right): Flowers and climbing habit of sollya *Billardiera heterophylla* (photo A Williams), Arum lily *Zantedeschia aethiopica* (photo JK Scott, CSIRO), biological control agent *Dialectica scalariella* released against Paterson's curse (photo CSIRO).

Set in 10/13 Adobe Minion Pro and ITC Stone Sans
Edited by Adrienne de Kretser (Righting Writing)
Cover and text design by James Kelly
Typeset by Desktop Concepts Pty Ltd, Melbourne
Index by Bruce Gillespie
Printed by Ingram Lightning Source

Preferred method of citation

Full volume

Julien M, McFadyen R and Cullen J (Eds) (2012) *Biological Control of Weeds in Australia.* CSIRO Publishing, Melbourne.

Individual chapters (example)

Heard TA and Segura R (2012) *Argenome mexicana* L. and *Argemone ochroleuca* Sweet – Mexican poppy. In *Biological Control of Weeds in Australia.* (Eds M Julien, R McFadyen and J Cullen) pp. xx–xx. CSIRO Publishing, Melbourne.

Feb26_RP_ILS

Foreword

Almost since the adoption of biological control as a means of alleviating issues caused by problematic plant species, Australia has been at the forefront of developing the process. It remains prominent nearly 100 years later.

The outstanding successes against *Opuntia stricta* in Australia confirmed the exceptional potential of a fledgling technology in the early part of the 20th century. The achievement has become the classic example of what is possible with biological control. What has been forgotten is the amount of effort that it took to achieve this early success: nearly two decades of exploration, screening and mass rearing on a massive scale. Several insect species were tried – and failed or were rejected – before *Cactoblastis* and cochineal became the champions that solved the problem.

This early success set the bar very high in terms of creating expectations that biological control will always be able to reduce problem plants to very low levels. The international community has still to fully appreciate that this type of miracle is the exception rather than the rule. Nevertheless, there is a growing awareness that seemingly less-successful programs can be extremely beneficial in checking the invasiveness of weeds and making them more manageable with other interventions, and that biological control brings outstanding economic benefits.

Australia continues to develop and use biological control against many different types of weeds. In doing so, entomologists, botanists and pathologists have accumulated a wealth of scientific expertise which continues to guide and enhance biological control research internationally. Australia has promoted biological control by training scientists from countries where biological control is not commonplace. Australian scientists have also set up research stations in other countries and have lent aid to scientists from abroad who need to explore for agents on plants that originate from Australia but are now problematic elsewhere.

This compilation provides a comprehensive overview of all these activities. The benefits that have been derived, both in Australia and internationally, continue to make biological control a plausible option for controlling the invasive species that bedevil so many parts of the world. There is no doubt that the historical accounts contained in these chapters will serve as a valuable reference source for many years to come. They should inspire future generations of students, scientists, agriculturalists, conservationists and urban planners to promote biological control as a preferred means of dealing with the invasive plant species they confront.

Dr John H. Hoffmann
Associate Professor and Head of Department
Department of Zoology, University of Cape Town
South Africa
May 2011

Contents

List of contributors

RJ Adair Department of Primary Industries, Bioprotection Research Platform, PO Box 48, Frankston, Vic 3199, Australia

Freda E Anderson CERZOS-UNS, Camino La Carrindanga Km 7, 8000, Bahía Blanca, Argentina

Bruce A Auld Charles Sturt University, Orange, NSW 2800, Australia

Peter Bailey South Australian Research & Development Institute, Entomology Unit, Waite Campus, GPO Box 397, Adelaide, SA 5001, Australia

J Baker Australian Government Land and Coasts, GPO Box 858, Canberra, ACT 2601 Australia

Jane Barton Contractor to Landcare Research, Private Bag 92170, Auckland, New Zealand

Faiz Bebawi Department of Employment, Economic Development and Innovation, Biosecurity Queensland, Tropical Weeds Research Centre, PO Box 187, Charters Towers, Qld 4820, Australia

Karen L Bell CSIRO Sustainable Ecosystems, GPO Box 2583, Brisbane, Qld 4001, Australia

DT Briese CSIRO Ecosystem Sciences, GPO Box 1700, Canberra, ACT 2601, Australia

Willie Cabrera-Walsh USDA-ARS South American Biological Control Laboratory, Bolivar 1559, Hurlingham, Buenos Aires, Argentina

Richard Chan CSIRO Ecosystem Sciences, GPO Box 2583, Brisbane, Qld 4001, Australia

Rachel Cruttwell McFadyen PO Box 88, Mt Ommaney, Qld 4074, Australia

Jim Cullen CSIRO Ecosystem Sciences, GPO Box 1700, Canberra, ACT 2601, Australia

JT Davies Department of Primary Industries, Parks, Water and Environment, 13 St John's Ave, New Town, Tas 7008, Australia

Michael Day Department of Employment, Economic Development & Innovation, Biosecurity Queensland, GPO Box 267, Brisbane, Qld 4001, Australia

K Dhileepan Department of Employment, Economic Development and Innovation, Biosecurity Queensland, GPO Box 267, Brisbane, Qld 4001, Australia

Katherine J Evans Perennial Horticulture Centre, Tasmanian Institute of Agricultural Research, University of Tasmania, 13 St Johns Ave, New Town, Tas 7008, Australia

Gio Fichera CSIRO Ecosystem Sciences, GPO Box 2583, Brisbane, Qld 4001, Australia

R Fogliani Department of Agriculture and Food, Locked Bag 4, Bentley Delivery Centre, WA, Australia

AJ Gordon ARC-Plant Protection Research Institute, Private Bag X5017, Stellenbosch 7599, South Africa

RH Groves CSIRO Plant Industry, GPO Box 1600, Canberra, ACT 2601, Australia

Tim A Heard CSIRO Ecosystem Sciences, GPO Box 2583, Brisbane, Qld 4001, Australia

Ken Henry SARDI Sustainable Systems, Entomology, GPO Box 397, Adelaide, SA 5001, Australia

Royce H Holtkamp Industry & Investment New South Wales, 4 Marsden Park Rd, Calala, NSW 2340, Australia

John R Hosking Industry & Investment New South Wales, 4 Marsden Park Rd, Calala, NSW 2340, Australia

FAC Impson Zoology Department, University of Cape Town, Rondebosch 7701, South Africa; and ARC-Plant Protection Research Institute, Private Bag X5017, Stellenbosch 7599, South Africa

JE Ireson Tasmanian Institute of Agricultural Research/University of Tasmania, 13 St John's Ave, New Town, Tas 7008, Australia

Mic Julien CSIRO Ecosystem Sciences, GPO Box 2583, Brisbane, Qld 4001, Australia

Raelene M Kwong Department of Primary Industries, Biosciences Research Division, PO Box 48, Frankston, Vic 3199, Australia

Catherine J Lockett Department of Employment, Economic Development & Innovation, Biosecurity Queensland, GPO Box 267, Brisbane, Qld 4001, Australia

Kathryn McCarren CSIRO Ecosystem Sciences, Private Bag 5, PO Wembley, WA 6913, Australia; and Entomology, WA Department of Agriculture & Food, Locked Bag 4, Bentley Delivery Centre, WA 6983, Australia

David McLaren Department of Primary Industries Victoria, BioSciences Research Division/La Trobe University, PO Box 48, Frankston, Vic 3199, Australia

Louise Morin CSIRO Ecosystem Sciences, GPO Box 1700, Canberra, ACT 2601, Australia

Tom Morley Department of Primary Industries, Biosciences Research Division, PO Box 48, Frankston, Vic 3199, Australia

Bill Palmer Department of Employment, Economic Development & Innovation, Biosecurity Queensland, GPO Box 276, Brisbane, Qld 4001, Australia

MF Purcell USDA-ARS Australian Biological Control Laboratory, GPO Box 2583, Brisbane, Qld 4001, Australia

Michelle Rafter School of Biological Sciences, University of Queensland, St Lucia, Qld 4072, Australia

Jean-Louis Sagliocco Department of Primary Industries, PO Box 48, Frankston, Vic 3199, Australia

Shon Schooler CSIRO Ecosystem Sciences, GPO Box 2583, Brisbane, Qld 4001, Australia

John K Scott CSIRO Ecosystem Sciences, Private Bag 5, PO Wembley, WA 6913, Australia

Ricardo Segura CSIRO Ecosystem Sciences, Mexican Field Station, A. Carlón No. 5, Col. Ejido 1°de Mayo, Boca del Río, C.P. 94297, Veracruz, Mexico

Wilmot Senaratne Department of Employment, Economic Development & Innovation, Biosecurity Queensland, GPO Box 276, Brisbane, Qld 4001, Australia

Andy W Sheppard CSIRO Ecosystem Sciences, GPO Box 1700, Canberra, ACT 2601, Australia

Nikki Sims-Chilton School of Biological Sciences, University of Queensland, St Lucia, Qld 4072, Australia

Matthew Smyth DAFF, Biosecurity Services Group, 7 London Circuit, Canberra, ACT 2601, Australia

Alejandro Sosa USDA-ARS South American Biological Control Laboratory, Bolivar 1559 (B1686EFA), Hurlingham, Buenos Aires, Argentina

David Sparks Department of Employment, Economic Development & Innovation, Level 5, 80 Ann St, Brisbane, Qld 4001, Australia

GR Strickland Ord Irrigation Cooperative, PO Box 573, Kununurra, WA 6743, Australia

Guadalupe Traversa Universidad Nacional del Sur, Bahía Blanca, Argentina

Rieks D van Klinken CSIRO Ecosystem Sciences, GPO Box 2583, Brisbane, Qld 4001, Australia

Wayne Vogler Department of Employment, Economic Development & Innovation, Tropical Weeds Research Centre, PO Box 187, Charters Towers, Qld 4820, Australia

John Weiss Department of Primary Industries, Biosciences Research Division, PO Box 48, Frankston, Vic 3199, Australia

Anna Williams MS 181, 109 Ruhle Rd, Pittsworth, Qld 4356, Australia

Paul B Yeoh CSIRO Ecosystem Sciences, Private Bag 5, PO Wembley, WA 6913, Australia

Introduction

The last full review of biological control of weeds in Australia was made by Frank Wilson, Division of Entomology, CSIRO (Wilson 1960). It included 10 sections, each covering a weed species, e.g. *Lantana camara*, or group of species, e.g. the *Opuntias*, *Eupatorium adenophorum* and *E. riparium* (now *Ageratina adenophora* and *A. riparia*), and *Xanthium pungens* (now *X. strumarium*) and *X. spinosum*. Since 1960, work has continued on all except one species listed by Wilson, nutgrass, *Cyperus rotundus*. Consequently all except nutgrass are again covered in the following chapters.

We achieved our aim of encouraging biological control researchers involved with particular weed species to review the biological control research up to late 2010. This has resulted in 73 chapters covering over 90 weed species, written by 47 authors. We thank the authors, especially the lead authors, for finding time to prepare their reviews.

The chapters are presented A to Z by scientific name with the exception of the first chapter, which is about exploration in Australia and biological control of Australian plants in South Africa and the USA. Since some chapters deal with multiple weed species an index of weeds by scientific name is included, along with an index of weed common names. A key word index includes the scientific names of many of the principal agents used.

Of the 73 chapters, one deals with exploration in Australia to control Australian plants that are weedy in the USA and South Africa, two are about the selection and redistribution of Australian native insects to control weedy Australian native shrubs, another covers surveys in WA conducted on an Australian native plant, *Billardiera heterophylla*, looking for suitable agents to control the plant in eastern Australia, and another considers *Tribulus terrestris* for which confirmation about its native status in Australian is required. Of the remaining chapters, 44 discuss weeds for which agents (242 arthropod and fungi species) have been released in Australia, and the rest cover weeds that were or are being studied, but for which no agents have yet been released. The agents released in Australia have been sourced from over 30 countries, indicating the truly international cooperation that has supported biological control of weeds research in Australia.

Many of the weed projects have resulted in very good control of the target weed and provided extraordinary benefits to Australian agriculture and environment. The oft-cited classical example of the cactus moth, *Cactoblastis cactorum*, controlling prickly pear, *Opuntia* spp., is the earliest example. There are many others, such as the projects against the weeds *Chondrilla juncea*, *Salvinia molesta*, *Asparagus asparagoides* and *Sida* spp., that have provided very significant positive returns on investment to the Australian taxpayer. The overall benefit:cost ratio for Australia's investment in biological control of weeds to 2005 was calculated to be 23:1 in direct economic returns to agriculture and health alone, not including environmental benefits (Page and Lacey 2006). Out of 29 programs assessed, 14 returned positive net benefits in dollar terms. A further three programs, not included in the analysis because of problems with data, also produced significant economic benefits.

We anticipate that the chapters in this book will be useful to weed researchers and managers, and to students of biological control. Once again we thank the authors for the efforts in making this compilation of their research available.

Mic Julien
Rachel McFadyen
Jim Cullen

REFERENCES

Page AR and Lacey KL (2006) Economic impact assessment of Australian weed biological control. CRC for Australian Weed Management. Technical Series No. 10.

Wilson F (1960) A review of the biological control of insects and weeds in Australia and Australian New Guinea. Commonwealth Agricultural Bureaux, Technical Communication No. 1. Commonwealth Institute of Biological Control, Ottawa, Canada.

Acknowledgements

Below are acknowledgements of the financial support provided for biological control of weeds research in Australia. This research is in the public interest, is not commercial and therefore is dependent largely on government support and rural industry levies.

Project (weed)	Funders
Acacia nilotica, prickly acacia	AWI, Qld Government, Qld Land Protection Fund, MLA
Ageratina adenophora, Crofton weed	Qld Government
Ageratina riparia, mistflower	NSW Government, Qld Government, Qld Land Protection Fund
Alternanthera philoxeroides, alligator weed	Australian Government, CRCAWM, CSIRO
Ambrosia artemisiifolia, annual ragweed	Qld Government, Qld Land Protection Fund
Anredera cordifolia, Madeira vine	NSW Government, Qld Government, Qld Land Protection Fund
Argemone mexicana and *Argemone ochroleuca*, Mexican poppy	Australian Government
Asparagus asparagoides, bridal creeper	Australian Government, Australian and New Zealand Environment and Conservation Council, CRCAWM, CRCWMS, CSIRO
Asphodelus fistulosus, onion weed	CSIRO
Baccharis halimifolia, groundsel bush	CRCAWM, Qld Government, Qld Lands Protection Fund
Billardiera heterophylla, sollya	CRCAWM
Bryophyllum delagoense, mother-of-millions	Qld Government, Qld Land Protection Fund
Cabomba caroliniana, cabomba	Australian Government, Burnet Mary Regional Group, CSIRO, 11 local community stakeholders administered by Lake Macdonald Catchment Care Group
Carduus nutans, nodding thistle	AWI, CSIRO
Carduus pycnocephalus and *Carduus tenuiflorus*, slender thistles	AWI, CRCWMS, CSIRO, Tas Government, Vic Government
Carthamus lanatus, saffron thistle	AWI, CRCWMS, MLA
Cassinia spp., cassinia	NSW Government
Chondrilla juncea, skeleton weed	CSIRO, GRDC
Chromolaena odorata, chromolaena	ACIAR, Qld Government
Chrysanthemoides monilifera, bitou bush and boneseed	ACT Government, Australian Government, CRCWMS, CSIRO, NSW Government, Qld Government, Qld Land Protection Fund, SA Government, Tas Government, Vic Government, WA Government
Cirsium vulgare, spear thistle	AWI, MLA, Vic Government
Cryptostegia grandiflora, rubber vine	MLA, Qld Government, Qld Land Protection Fund
Cylindropuntia imbricata, rope pear, and *Cylindropuntia rosea*, Hudson pear	Australian Government, Qld Government, NSW Government, NSW Western CMA
Cytisus scoparius, broom, Scotch broom or English broom	Australian Government, CSIRO, Goulburn-Murray Water, Hunter Pastoral Company, NSW Government, Vic Government
Echium plantagineum, Paterson's curse	Australian Government, CSIRO, MLA
Eichhornia crassipes, water hyacinth	ACIAR, AusAID, CSIRO

Project (weed)	Funders
Emex australis, doublegee, and *Emex spinosa*, lesser jack	AWI, CSIRO, Dried Fruits Research Council, GRDC, Rural Credits Development Fund, WA Government
Eremophila mitchellii, budda, *Eremophila sturtii*, turpentine bush, *Dodonaea viscosa*, narrow-leaf hopbush, and *Myoporum deserti*, poison-bush	Cuttagoa Catchment Landcare Group
Euphorbia paralias, sea spurge	CSIRO, Tas Government
Fumaria spp., fumitory	CRCAWM, CSIRO
Genista monspessulana, Cape broom	Australian Government, CRCWMS, Californian Dept of Forestry and Fire Protection, California Dept of Food and Agriculture, SA Government
Harrisia spp., Harrisia cactus	Qld Government, Qld Land Protection Fund
Heliotropium amplexicaule, blue heliotrope	AWI, CRCAWM, CSIRO, MLA, RIRDC
Heliotropium europaeum, common heliotrope	Australian Government, CSIRO, MLA
Hypericum perforatum, St John's wort	AWI, CRCAWM, CSIRO, MLA, RIRDC
Hyptis suaveolens, hyptis	CSIRO, NT Government
Jacobaea vulgaris, ragwort	Australian Government, CSIRO, Dairy Australia, Tas Government, Vic Government
Jatropha gossypiifolia, bellyache bush	CSIRO, Qld Government, Qld Land Protection Fund, MLA, NT Government
Lantana camara, lantana	ACIAR, Australian Government, CRCAWM, CSIRO, NSW Lantana Taskforce, Qld Government, Qld Land Protection Fund, Wet Tropics Management Authority, University of Qld
Lantana montevidensis, creeping lantana	Qld Government, Qld Land Protection Fund, North Burnett Landcare Group
Macfadyena unguis-cati, cat's claw creeper	Qld Government, Qld Land Protection Fund
Marrubium vulgare, horehound	AWI, CRCWMS, SA Government, Vic Government
Mikania micrantha, mile-a-minute	ACIAR, Qld Government
Mimosa diplotricha, giant sensitive plant	Qld Government
Mimosa pigra, mimosa	ACIAR, Australian Government, CSIRO, CTPM, NT Government
Moraea flaccida, one-leaf Cape tulip, and *Moraea miniata*, two-leaf Cape tulip	CSIRO, WA Government
Nassella spp., serrated tussock and Chilean needle grass	Australian Government, Vic Government, CSIRO, National collective of regional councils and the Department of Conservation of New Zealand, MLA, RIRDC, a consortium of 24 shires and councils from NSW and Vic, NSW Government and Vic Serrated Tussock Working Party
Onopordum acanthium, Scotch thistle, *Onopordum illyricum*, Illyrian thistle, and hybrids	AWI, CRCAWM, CSIRO, MLA
Opuntia robusta, wheel cactus	CRCAWM, University of Adelaide
Opuntia spp.	Australian Government, Qld Government, Qld Land Protection Fund, NSW Government
Parkinsonia aculeata, parkinsonia	Australian Government, CSIRO, Qld Government, Qld Land Protection Fund, MLA
Parthenium hysterophorus, parthenium	Qld Government, Qld Land Protection Fund, MLA, CTPM
Phyla canescens, lippia	Australian Government, CSIRO, NSW Government, Murray-Darling Committee Inc.

Project (weed)	Funders
Pistia stratiotes, water lettuce	CSIRO
Prosopis spp., mesquite	Australian Government, CSIRO, Pilbara Mesquite Management Committee, NT Government, WA Government
Raphanus raphanistrum, wild radish	CRCAWM, CSIRO, GRDC
Reseda lutea, cutleaf mignonette	SA Grains Industry Trust Fund, University of Reading
Rubus fruticosus, blackberry	Australian Government, CRCAWM, CRCWMS, CSIRO, Riverina Highlands Weed Working Group, WA Government
Rumex spp., docks	CSIRO, MLA, WA Government
Salvinia molesta, salvinia	ACIAR, CSIRO
Senecio madagascariensis, fireweed	Qld Government, Qld Land Protection Fund, MLA
Senna obtusifolia, sicklepod	Qld Government, Qld Land Protection Fund
Sida acuta, spinyhead sida, *Sida rhombifolia*, paddy's lucerne, and *Sida cordifolia*, flannel weed	CRCAWM, CSIRO, NT Government
Silybum marianum, variegated thistle	AWI
Solanum elaeagnifolium, silverleaf nightshade	MLA, Wheat Research Council of Australia
Sonchus oleraceus, sowthistle	CRCAWM, CSIRO
Sporobolus spp., weedy sporobolus grasses	CRCAWM, Qld Government, Qld Land Protection Fund, MLA
Tribulus terrestris, caltrop	Australian Dried Fruit Research and Development Council, CSIRO
Ulex europaeus, gorse	Australian Government
Xanthium occidentale, Noogoora burr	Australian Government, CSIRO
Xanthium spinosum, Bathurst burr	AWI, CRCWMS, Sandoz Corporation (Switzerland)
Zantedeschia aethiopica, arum lily	Australian Government, CSIRO

Abbreviations

STATES OF AUSTRALIA

NSW	New South Wales
NT	Northern Territory
Qld	Queensland
SA	South Australia
Tas	Tasmania
Vic	Victoria
WA	Western Australia

ORGANISATIONS

ACIAR	Australian Centre for International Agricultural Research
CABI	CABI International
CIBC	Commonwealth Institute for Biological Control (now CABI)
CIE	Commonwealth Institute of Entomology
CRCAWM	Cooperative Research Centre for Australian Weed Management
CRCWMS	Cooperative Research Centre for Weed Management Systems
CSIRO	Commonwealth Scientific and Industrial Research Organisation, Australia
CTPM	Cooperative Research Centre for Tropical Pest Management
USDA-ARS	United States Department of Agriculture – Agricultural Research Service
USNM	United States National Museum

FUNDING AGENCIES

AWI	Australian Wool Innovation
MLA	Meat and Livestock Australia
GRDC	Grains Research and Development Corporation
RIRDC	Rural Industries Research and Development Corporation

SPECIFIC TERMS

Biocontrol	biological control

Biological control of Australian plants: a South African and US perspective

FAC Impson, MF Purcell and AJ Gordon

ABSTRACT

A number of Australian plants are considered to be weedy in various parts of the world. Most of these are characterised by woody trees and shrubs which have been deliberately introduced into their new environments for specific purposes. To date only two countries; South Africa and the USA, are actively involved in biological control programs against these plants using agents sourced within Australia. The biological control program against weeds of Australian origin was initiated in South Africa in the 1960s and has targeted 15 species to date. The programme in the USA against four weed species began in the 1980s. The history of plant introductions and the biological control endeavours are reviewed here.

Key words: weed biological control, *Acacia*, *Hakea*, *Melaleuca*, *Hydrilla*, *Lygodium*, *Casuarina*.

INTRODUCTION

Australian plants are represented in almost every continent. Some are commonly traded and grown as ornamentals, but have no impact in their non-native habitats. Other species have become naturalised in introduced areas but are not seen to be a problem. However, others have adapted well to their new environments and have subsequently become highly invasive and troublesome weeds. The factors which determine which species become invasive have been discussed by many weed practitioners and ecologists (Crawley 1987; Drake *et al.* 1989; Daehler and Strong 1993; Blossey and Nötzold 1995; Blossey and Kamil 1996), and with increasing travel and trade between countries there is continued potential for the introduction and establishment of new weed problems.

Of those Australian plants which have achieved weedy status around the world, the majority are characterised by woody shrubs and trees. They include species from a number of genera, e.g. *Acacia* (Mimosaceae), *Atriplex* (Chenopodiaceae), *Callistemon* (Myrtaceae), *Casuarina* (Casuarinaceae), *Eucalyptus* (Myrtaceae), *Grevillia* (Proteaceae), *Hakea* (Proteaceae), *Leptospermum* (Myrtaceae), *Melaleuca* (Myrtaceae), *Myoporum* (Myoporaceae), *Paraserianthes* (Fabaceae), *Pittosporum* (Pittosporaceae) and *Syzygium* (Myrtaceae). The submerged aquatic plant, *Hydrilla verticillata* (L.F.) Royle (Hydrocharitaceae), and a climbing fern, *Lygodium mycrophyllum* (Cav.) R.Br. (Lygodiaceae), which have origins in Africa, Asia and the Pacific (*L. microphyllum* only) as well as Australia, are also invasive overseas.

The Australian plants that are problematic in other countries originated mostly from deliberate introductions. Commercial and agroforestry, dune stabilisation, the aquarium trade and horticulture have all played major roles in the distribution of such plants. Only relatively recently have risk assessments and guidelines for the safe introduction of new species been considered globally (Hughes 1995; Tucker and Richardson 1995; Walton *et al.* 1998; Wittenberg and Cock 2001). These have often been developed together with national strategies for the management of invasive alien plants. Such guidelines need to go hand in hand with strong permitting regulations (with monitoring procedures and containment requirements), good public awareness tools and early detection and rapid response systems to deal with new or potential invasions (i.e. http://invader.dbs.umt.edu/, http://www.sanbi.org.za), followed by appropriate forms of managing the invasions.

Despite the fact that the abovementioned plants are considered to be weedy in numerous countries and are controlled either chemically or mechanically in many, only two countries, South Africa and the USA, are actively involved in biological control programs to control these weeds. The potential for biological control of Australian *Acacia* species is also being investigated in Portugal.

This chapter documents the invasion pathways of Australian plants into South Africa and the USA and their impact as weeds in their non-native habitats. Attention is also focused on the history of the biological control programs and the various strategies employed to implement exploration in Australia for potential agents. Current biocontrol initiatives in both countries are also discussed.

SOUTH AFRICAN SITUATION

History of plant introductions

The earliest records of exotic plant introductions coincide with European settlement into the Cape region in the mid 1600s, when a provisioning station was set up on the shipping route between Europe and the East Indies (Shaughnessy 1980). However, it was not until the 1800s in the Western Cape that there were accounts of any Australian plants (Burchell 1822; Latrobe 1821 cited in Shaughnessy 1980). Many of the early records refer to species within private collections and government gardens. Although some are believed to have come direct from Botany Bay (Latrobe 1821), much of the seed of Australian acacias, for example, is known to have come via England (Bowie 1832 in Shaughnessy 1980), where Australian plants were popular horticultural species at the time. During 1835–65 genera such as *Eucalyptus*, *Melaleuca*, *Hakea*, *Leptospermum*, *Paraserianathes* (formerly *Albizia*), *Banksia*, *Casuarina*, *Callitris* and *Callistemon* were recorded in addition to the *Acacia* species, with further introductions between 1881 and 1896, probably due to demands from the forestry industry. With the exception of *Paraserianthes lophantha* (Willd.) Nielsen, *Hakea suaveolens* R.Br. and *Leptospermum laevigatum* (Gaertn.) F.Muell., there is evidence of multiple introductions of species from Europe and Australia, thus there is little likelihood that any species is present as a single genetic strain (Shaughnessy 1980).

The period from the mid to late 1800s saw many of the exotic species being spread widely in the Cape region and elsewhere in South Africa. The woody species proved to be fast growers and, through planting and experimentation, they rapidly became exploited for uses such as stabilisation of drift sands, building of shelters, hedges and livestock enclosures, firewood, timber and tannin production. It was believed that the development of forests would lead to the benefits of increased rainfall and improved soil fertility. The discovery of gold and diamonds in the interior of South Africa led to a population boom, and thus increased demands for wood and fodder species (Deacon 1986).

Between 1904 and 1908 thousands of seed samples (particularly of fodder species) were imported from all over the world, and distributed to farmers (Burtt-Davy 1904 in Wells *et al.* 1986). The Australian plants *Atriplex semibaccata* R.Br. and *Chenopodium carinatum* R.Br (both now considered invasive) were included; there are records of some 500 000 trees (including *Acacia dealbata* Link, *A. melanoxylon* R.Br., *H. drupacea* (C.F.Gaertn.) Roem. & Schult. (= *H. suaveolens*) and *Eucalyptus* species), being distributed annually by the Conservator of Forests for the Transvaal (Legat 1904–08; Wells *et al.* 1986). Some of the species, e.g. *Eucalyptus* and *Acacia*, were also planted as fire breaks between pine plantations, a practice which was still recommended as late as 1931.

Although many of the introduced Australian plants now have weedy status in South Africa, species within the genera *Acacia*, *Paraserianthes*, *Hakea*, *Leptospermum* and *Eucalyptus* are considered most problematic. Although some of these were widely planted, which may account in part for their abundance, there are some notable exceptions. There were limited plantings of *Hakea sericea* Schrad. & J.C.Wendl. and *Acacia longifolia* (Andr.) Willd., but they spread rapidly (Macdonald and Jarman 1984; Shaughnessy 1986). The climate, occurrence of fires and low-nutrient soils in South Africa were ideal for Australian plants (Macdonald 1985) and the scarcity of indigenous trees in the 'fynbos' vegetation of the Cape meant that there was little competition for the fast-growing introduced tree species (Oliver 1977; Hall *et al.* 1980). These factors, together with efficient dispersal mechanisms, high levels of seed production and, in some cases, the accumulation of large and long-lived seed banks, contributed to the success of these plants as invader species (Fugler 1979; Macdonald and Jarman 1984; Milton and Hall 1981; Holmes *et al.* 1987; Richardson *et al.* 1987).

Despite continued planting of many exotic forestry species to the early 1900s, there was a definite change in public attitude towards exotic plants from 1917, when several species were recognised as being troublesome. However, official attitudes remained uncertain, except towards *Hakea* species. As early as 1863, farmers in the Eastern Cape had met to discuss their concern regarding the spread of *H. sericea* (Phillips 1938). By 1937 forestry research began looking into different methods of controlling *Hakea* species; a year later the Department of Agriculture proclaimed them to be noxious weeds and began a campaign to educate landowners regarding the threats they posed (Shaughnessy 1980). Many of the introduced Australian plants are no longer recognised for the positive traits for which they were first introduced. There is more interest in their severe impacts on ecosystem functioning, biodiversity, water resources and agriculture. The campaign against alien invasive plants gained momentum with the advent of the Working for Water program in 1996. This program not only funds biological control efforts, but employs large numbers of unskilled workers to physically remove alien plants from water catchments and environmentally sensitive areas.

History of biological control

Biological control of invasive Australian plants dates back 50 years in South Africa with the initiation of a program against *H. sericea* in 1962. The early exploratory work, detailed by Kluge and Neser (1991), is synthesised in Table 1. Seed-attacking agents were the initial targets. The reasoning behind this was threefold:

- the prolific seed production and its high levels of accumulation on plants in South Africa were believed to be responsible for the plants' aggressiveness;
- due to the high seed production and lack of vegetative reproduction, it was felt that hakea could be effec-

Table 1: Exploratory work carried out in Australia for biological control agents to be used in South Africa against Australian plants

Name	Date	Objective	Target weed(s)	Department
D van V Webb	July 1962–Nov 1964	Survey	*H. sericea, H. gibbosa, H. drupacea*	Agric. Tech. Services
KM Moore	1962–1964 (opportunistic)	Survey	*H. sericea, H. gibbosa*	Forestry Commission, NSW
S Neser	July 1964–Dec 1968	Surveys, host-specificity studies	*H. sericea, H. gibbosa*	Agric. Tech. Services
MA van den Berg	Jan 1975–Dec 1976	Surveys of seed-attackers	*Acacia* spp.	PPRI
RL Kluge	Apr–May 1978	Survey	*H. sericea*	PPRI
S Neser	Jan 1981–Dec 1982	Survey, collecting	*H. sericea, H. gibbosa, A. longifolia, L. laevigatum*	PPRI
MJ Morris	Dec 1981–Dec 1984	Survey, host-specificity studies	*Acacia* spp., *A. saligna P. lophantha*	PPRI
AJ Gordon	July 1997, June 1998	Host-range	*H. sericea*	PPRI
RA Adair	1998–2003	Survey, host-range, impact	*Acacia* spp.	PPRI
FAC Impson, C Kleinjan	Nov 2005	Survey	*A. pycnantha, A. podalyriifolia*	PPRI
J Post, A Wood	Oct 2006	Survey	*P. lophantha*	PPRI
AJ Gordon	Feb 2008	Survey	*L. laevigatum*	PPRI
FAC Impson, J Post	Oct 2009	Survey	*A. elata*	PPRI

PPRI = Plant Protection Research Institute.

Figure 1: The gall wasp *Trichilogaster signiventris* (insert) and its galls on *Acacia pycnantha*. Photos: J Hoffmann, University of Cape Town.

tively suppressed by reducing the amount of seed produced (Neser 1968);

- insects attacking the vegetative parts of *Hakea* species were believed to be potentially oligophagous and thus posed possible risks to South African Proteaceae (Webb 1964).

By the mid 1970s the biological control initiative shifted to include certain *Acacia* species (Table 1). From the outset, although thorough surveys were carried out, the economic importance of several of the Australian *Acacia* species created a conflict of interest, which became pivotal in the consideration of agent selection (Dennill and Donnelly 1991; Impson and Moran 2004). Despite assurances that only seed-reducing agents would be considered and that there would be good consultation (Annecke 1975), strong concerns by the wattle industry (Stubbings 1977) delayed progress until the mid 1980s. During that period the wattle industry was earning approximately $US30 million per year; it was felt that the problems being caused by wattle invasions were largely unsubstantiated. A number of different seed-reducing agents (Figs 1 and 2) were subsequently introduced against 10 of the most invasive Australian *Acacia* species in South Africa, in addition to the closely related species, *Paraserianthes lophantha* (Table 2).

Unlike North American colleagues, the South African biocontrol researchers have not established an official research station in Australia from which to base surveys, preliminary host-specificity testing and collecting of potential agents, although some researchers such as Webb, Neser and Morris were based in Australia for lengthy periods in the 1960s and 1970s (Table 1). For the most part, researchers have undertaken relatively short (two- to five-week) survey and collecting trips to Australia, timed to coincide with the required life-stage of the target biocontrol agent. Since 1995, one or two such trips have been carried out annually. South Africa has benefited from collaboration with the USDA laboratory in Brisbane, e.g. with a shipment of *Bagous hydrillae* Hustache (Curculionidae) for *H. verticillata* from cultures maintained in Australia, which circumvented the need for South African researchers to survey and collect in Australia.

South African researchers have gained valuable insights into invasive Australian plants through first-hand observations of these plants in their native environments. In

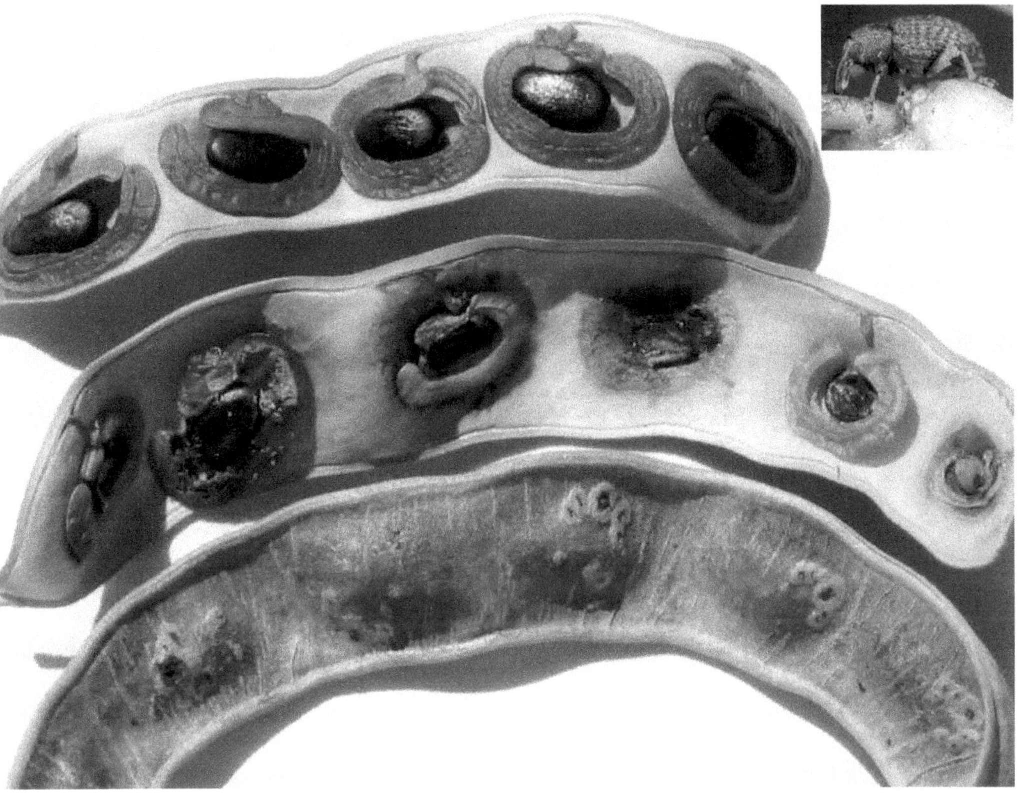

Figure 2: *Acacia cyclops* pods showing healthy seeds (top) and seeds damaged (middle) by the seed-feeding weevil, *Melanterius servulus*. Seed pod photo: F Impson, Plant Protection Research Institute, South Africa. Insert photo: J Hoffmann, University of Cape Town.

addition, targeting seed-reducing agents has greatly narrowed the time during which surveys are most beneficial. Other factors that allow South African scientists to work effectively during relatively short visits are that access into Australia for both exploratory work and collecting purposes is relatively easy, the permitting system in place for both collection and exportation of agents from Australia is straightforward and efficient, travel within Australia is easy and there are no language barriers. The initiation of the South African Working for Water program in 1996 led to substantial financial support for biological control and this, together with donations from the Drakenstein Trust, has greatly facilitated collecting trips for biological control agents from Australia. These trips have resulted in the testing and ultimate release of 16 agents against 14 target weed species (Table 2).

Various aspects of biological control programs against Australian weeds in South Africa have been well publicised and reviewed (Kluge and Neser 1991; Dennill and Donnelly 1991; Gordon 1999a, b; Dennill *et al.* 1999; Morris 1999; Impson *et al.* 2009). The impact of each biological control agent on its target weed is summarised in Table 2. With the exception of *Cydmaea binotata* Lea. (Curculionidae), released against *H. sericea* and *Erytenna consputa* Pascoe (Curculionidae) against *H. gibbosa* (Sm.) Cav., the remaining agents have caused moderate to extensive damage to their host plants. Only two of the agents which were tested in South Africa were never released. The first, a leaf-spotting bug, *Rayieria* sp. (Miridae), which was proposed for *A. longifolia*, was rejected at the time due to its potential damage to commercial wattle (Donnelly 1986). The second, a seed-feeding wasp, *Bruchophagus orarius* Neser and Prinsloo (Eurytomidae), although host-specific to *A. longifolia* (Kluge 1990) was never released, probably because two other agents were successfully established and sufficiently damaging to the host plant (Impson *et al.* 2009). *Bruchophagus acaciae* (Cameron) and *B. hydrillae* are currently in quarantine for host-specificity testing.

All the agents released in South Africa against Australian weeds target the flowers or fruit of their hosts, with the exception of *Uromycladium tepperianum* (Sacc.) McAlp. (Pileolariaceae) and *Aphanasium australe* (Boisduval) (Cerambycidae), which can cause mortality to their host

Table 2: Biological control agents imported into South Africa for which host-testing was, or is, being carried out

	Biocontrol agent	Family	Feeding guild	Origin	Date released	Damage to weed
Hakea sericea Schrad. & J.C.Wendl.	*Erytenna consputa* Pascoe	Curculionidae	Seed-feeder	NSW	1970	Extensive
	Carposina autologa Meyrick	Carposinidae	Seed-feeder	NSW	1970	Considerable
	Cydmaea binotata Lea	Curculionidae	Seed-feeder	NSW, Vic	1979	Trivial
	Aphanasium australe (Boisduval)	Cerambycidae	Stem-borer	NSW	2001	?
	Dicomada rufa Blackburn	Curculionidae	Flower bud-feeder	NSW	2006	?
Hakea gibbosa (Sm.) Cav.	*Erytenna consputa* Pascoe	Curculionidae	Seed-feeder	NSW	1975	Trivial
	Aphanasium australe (Boisduval)	Cerambycidae	Stem-borer	NSW	2001	?
Leptospermum laevigatum (Gaertn.) F.Muell.	*Aristaea thalassias* (Meyrick)	Gracillaridae	Leaf-miner	NSW	1996	Moderate
	Dasineura strobila Dorchin	Cecidomyiidae	Bud-galler	NSW	1994	Considerable
Acacia longifolia (Andr.) Willd.	*Trichilogaster acaciaelongifoliae* Froggatt	Pteromalidae	Bud-galler	NSW	1982	Extensive
	Melanterius ventralis Lea	Curculionidae	Seed-feeder	NSW	1985	Extensive
	Rayieria sp.	Miridae	Leaf-sucker	ACT	Not released	
	Bruchophagus orarius sp.n	Eurytomidae	Seed-feeder	NSW, Vic	Not released	
Acacia melanoxylon R.Br.	*Melanterius acaciae* Lea	Curculionidae	Seed-feeder	NSW	1986	Extensive
Acacia cyclops A.Cunn. ex G.Don	*Melanterius servulus* Pascoe	Curculionidae	Seed-feeder	WA	1994	Considerable
	Dasineura dielsi Rübsaamen	Cecidomyiidae	Flower-galler	WA	2001	Extensive
Acacia mearnsii De Wild.	*Melanterius maculatus* Lea	Curculionidae	Seed-feeder	NSW, ACT	1994	Considerable
	Dasineura rubiformis Kolesik	Cecidomyiidae	Flower-galler	NSW	2006	Considerable
Acacia dealbata Link.	*Melanterius maculatus* Lea	Curculionidae	Seed-feeder	ACT, NSW, Vic	1998	Moderate
	Bruchophagus acaciae (Cameron)	Eurytomidae	Seed-feeder	ACT, Vic, NZ	In quarantine	
Acacia decurrens (Wendl.)	*Melanterius maculatus* Lea	Curculionidae	Seed-feeder	ACT, NSW, Vic	2001	Moderate
Acacia saligna (Labill.) H.L.Wendl.	*Uromycladium tepperianum* (Sacc.) McAlp.	Pileolariaceae	Gall-forming rust fungus	WA	1987	Extensive

	Biocontrol agent	Family	Feeding guild	Origin	Date released	Damage to weed
	Melanterius compactus Lea	Curculionidae	Seed-feeder	WA	2001	Extensive
Acacia pycnantha Benth.	*Trichilogaster signiventris* (Girault)	Pteromalidae	Bud-galler	Vic	1987	Extensive
	Melanterius maculatus Lea	Curculionidae	Seed-feeder	SA, Vic	2003	?
Acacia baileyana F.Muell.	*Melanterius maculatus* Lea	Curculionidae	Seed-feeder	ACT, NSW	2006	?
Acacia podalyriifolia A.Cunn. ex G.Don	*Melanterius maculatus* Lea	Curculionidae	Seed-feeder	Qld	2008	?
	Bruchophagus acaciae (Cameron)	Eurytomidae	Seed-feeder	Qld	In quarantine	
Paraserianthes lophantha (Willd.) Nielsen	*Melanterius servulus* (Pascoe)	Curculionidae	Seed-feeder	WA	1989	Considerable
	Uromycladium tepperianum (Sacc.) McAlp.	Pileolariaceae	Gall-forming rust fungus	WA	In quarantine	
Hydrilla verticillata (L.F.) Royle	*Bagous hydrillae*	Curculionidae	Stem-boring weevil	Qld	In quarantine	

plants, *Acacia saligna* (Labill.) H. Wendl. and *H. gibbosa*, respectively. Seed-reducing agents have limitations, particularly where seed banks are large and long-lived. The outcome is never likely to be a reduction in population density of the plant; in these cases it is important to see the biological control program in the context of an integrated approach, together with mechanical or chemical control operations. These agents have an important role, however, in reducing spread or rates of spread and are suitable for areas which are inaccessible for mechanical clearing. Most importantly, as has been demonstrated several times in South Africa, they allow a compromise where conflicts of interest are an issue.

US SITUATION

History of plant introductions

Native American cultures imported cultivated crops from the Caribbean more than 10 000 years before European settlement (Winters 1967). However, the arrival of Europeans in the 1500s increased the importation of plant species, including fruits and ornamentals (Austin 1978; Knott 1976). Importation of species increased greatly in the 1900s with the increase in population (Simberloff *et al.* 1997). The USDA began overseas plant expeditions in 1897 and was ultimately responsible for introducing over 300 000 plant accessions by 1967 (Winters 1967) including forage and grain crops (Simberloff *et al.* 1997) as well as ornamentals (Ryerson 1967). Some species became aggressive and problematic invaders. Federal legislation to restrict importation of non-indigenous plant species that threaten natural or agricultural systems has increased since 1900 and mostly pertains to agricultural (Simberloff *et al.* 1997) rather than environmental weeds. For example, it was not until 1967 that laws were introduced in Florida preventing the release and propagation of aquatic plant species in natural ecosystems for recruitment and supply by commercial interests (Schmitz *et al.* 1991).

Plants from Australia are well represented in the USA as valued ornamentals though some were introduced for commercial purposes; e.g. *Melaleuca alternifolia* (Maiden & Betche) Cheel was grown in California for tea-tree oil production. Others are becoming invasive, but conflicts of interest exist. *Eucalyptus globulus* Labill. was introduced into California in the mid 1800s as an ornamental tree. It is invading large natural areas along the coast where it suppresses native plants, possibly due to allelopathic effects. However, it is a popular suburban tree and there is resistance to the implementation of control measures, although its contribution to inhibiting fire management and associated property losses may alter this view. Similarly, on the east coast, the popular bottlebrush *Callistemon viminalis* (Gaertn.) is becoming invasive in wetland areas.

Although introductions have occurred across the USA, Australian plants seem to be particularly adapted to the ecosystems of Florida. A well documented example is the broad-leaved paperbark tree, *Melaleuca quinquenervia* (Cav.) S.T. Blake (Myrtaceae), introduced into southern Florida in 1886 (Dray *et al.* 2006), to Orlando in the early 1900s (Meskimen 1962) and to the west coast of Florida in 1912 (Meskimen 1962; Pritchard 1976). Seeds were spread by aircraft over the eastern Everglades (Meskimen 1962) in land reclamation efforts and trees were planted for erosion control (Stocker and Sanders 1981). *M. quinquenervia* now infests over 200 000 ha of wetlands in south Florida (Bodle *et al.* 1994), of which up to 10% are virtual monocultures (Ferriter 1994). *M. quinquenervia* is highly invasive in forested and non-forested wetlands of south Florida (Fig. 3) where it is economically and environmentally detrimental (Turner *et al.* 1998). Because of high concentrations of essential oils in the leaves, canopy fires are extremely hot and difficult to control. It flowers several times a year and the large amounts of pollen produced may induce mild allergic reactions (Morton 1962; Lockey *et al.* 1981; Stanaland *et al.* 1986). *M. quinquenervia* has severely damaged wetland ecosystems and caused large reductions in biodiversity.

Old world climbing fern, *Lygodium microphyllum* (Lygodiaceae), is native to Africa, Asia, Australia and the Pacific Islands (Pemberton 1998). There is circumstantial evidence that it was introduced into Florida from north Qld in the late 1800s (Goolsby *et al.* 2006). It was naturalised in Florida by 1968 (Beckner 1968) and spread rapidly into wetland habitats in southern and central Florida (Boughton *et al.* 2009). Ferriter and Pernas (2006) reported that *L. microphyllum* inhabits almost 50 000 ha in southern Florida; this may be an underestimate (Volin *et al.* 2004). Natural areas in south-central Florida are believed to be vulnerable to invasion by *L. microphyllum* (Goolsby 2004; Volin *et al.* 2004) where this fern threatens native ecosystems (Pemberton and Ferriter 1998). The fern smothers other vegetation (Fig. 4) because of its growth habit (Nauman and Austin 1978; Pemberton and Ferriter 1998). Within infestations, large amounts of dried and combustible fern material accumulate, leading to a change in the fire ecology of wetland habitats (Pemberton and Ferriter 1998).

Australian pine/sheoak species, *Casuarina* (Casuarinaceae), were introduced as ornamental shade trees and

Figure 3: The broad-leaved paperbark tree *Melaleuca quinquenervia* infesting wetlands in southern Florida. Photo: F Laroche, South Florida Water Management District.

for windbreaks, but are problematic in natural ecosystems in Florida where they reduce biodiversity, particularly along coastlines. However, these trees are also highly valued as ornamentals and one, *C. cunninghamiana* Miq., is being considered for use in windbreaks to impede the spread of citrus canker. *Casuarina equisetifolia* L. was introduced into Hawaii and Florida in the late 1800s (Wheeler *et al.* 2011) while both *C. glauca* Sieb. ex Spreng. and *C. cunninghamiana* were introduced into Florida before 1924 (Morton 1980). *C. equisetifolia* is the most widespread; it colonises beaches, hammocks, flatwoods, marshes, swamps and tidal marshes (Wheeler *et al.* 2011). *C. glauca* is also widespread, though it prefers upland coastal areas (Gann *et al.* 2008). *C. cunninghamiana* is less widespread and is mainly restricted to riparian zones, reflecting its native habitat in Australia. *Casuarina* is estimated to have invaded over 160 000 ha in Florida (Cox 1999), including the Everglades (Doren and Jones 1997) and extensive areas of the coastline and barrier islands (Glisson 1997; Johnson and Barbour 1990).

Acacia species are commonly planted in California as ornamentals; two species have escaped cultivation and become weedy in coastal regions (DiTomaso and Healy

Figure 4: The climbing fern *Lygodium microphyllum* smothering native vegetation in southern Florida. Photo: M Purcell, USDA ARS Australian Biological Control Laboratory.

2007). *Acacia melanoxylon* grows to 15 m and is native to eastern Australia while *A. paradoxa* is a spiny shrub growing to 3 m, native to southern and eastern Australia (DiTomaso and Healy 2007). There are at least 11 other *Acacia* species that occasionally escape cultivation in California.

The submersed aquatic macrophyte hydrilla, *Hydrilla verticillata* (Hydrocharitaceae), is native to Asia, Australia, Europe and Africa (Buckingham and Bennett 1996). In the USA it occurs from Washington state (north-west coast) and Pennsylvania (north-east coast) to Texas and Florida in the south. It has monoecious and dioecious forms and was likely introduced to the USA from the Indian subcontinent and South Korea (Madeira *et al.* 1997). Hydrilla damages ecosystems by dominating aquatic habitats, altering plant communities and water chemistry. It also has economic impacts by restricting irrigation flow, hydroelectric power generation and navigation of waterways (Balciunas *et al.* 2002).

Though several of the plant species listed above are not uniquely Australian, the focus for exploration of these plants has been in Australia, with the exception of hydrilla. The reason relates largely to infrastructure and the establishment of the Australian Biological Control Laboratory by the US Department of Agriculture, Agricultural Research Service in Brisbane.

History of biological control

Unlike the South African experience, exploration for biological control agents for US weeds of Australian origin has a short history. During worldwide surveys to locate biological control agents for the submerged aquatic weed *H. verticillata*, a permanent laboratory was established in Townsville at the CSIRO Davies Laboratory in 1985, later moved to the nearby James Cook University with Joe Baciunas as the leader. This lab was jointly administered by the University of Florida and the USDA-ARS. A satellite laboratory was established at the CSIRO Long Pocket Laboratories in Brisbane in mid 1985. In 1989 the laboratory became an official USDA-ARS facility known as the Australian Biological Control Laboratory (ABCL) administered through USDA-ARS specific cooperative agreements with James Cook University and CSIRO Entomology. In 1996 the new director, Charles Turner, consolidated ABCL at one location in Brisbane at CSIRO.

ABCL is one of four overseas biological control laboratories (OBCLs) administered by the USDA-ARS, the others being located in Argentina, China and France. Having an extensive overseas presence has given the USA the capability to readily locate biological control agents for weeds originating from most parts of the globe. Recently, ABCL has expanded its exploratory research into Asia and the Pacific, which has been extremely important for discovering new agents for *H. verticillata* and *L. microphyllum*. Such research is conducted through collaborative research partnerships with agencies throughout the Australasian region.

Having a permanent base in Australia has contributed to the success of biological control programs in the USA, particularly in Florida. Research in the native range of targets has allowed a thorough assessment of natural enemies including discovery, seasonality, biology, efficacy and risk assessment. It facilitated ready access to potential agents for shipment to US quarantine facilities for further testing and for support of field releases. Nine agents from Australia have been released in the USA for three weed species; five have established (Table 3). The *M. quinquenervia* project identified more than 450 insect herbivores that attack this tree across a broad geographic range. Four species have been released in Florida and this weed is being successfully controlled. No significant non-target effects have been recorded. Many other agents have been assessed and rejected for reasons of specificity, lack of impact or toxicity. Assessing agents before shipment to quarantine facilities in the USA is cost-effective and streamlines the evaluation process from discovery to field release. This project has been conducted at a significant level for 25 years, giving an insight into the resources needed to combat serious invasive weeds. Cooperative agreements with host agencies (James Cook University and CSIRO) have contributed to the quality of research, tapping the expertise of co-located and other local scientists specialising in areas such as biological control, taxonomy, genomics and statistical analysis.

Exploration for biological control agents of hydrilla was the first ABCL project in 1985. Surveys were conducted in northern and south-eastern Qld, in northern NSW and sporadically at more distant locations including southern NSW and around Darwin in the NT. The first agent assessed for biological control was the stem-boring weevil *Bagous hydrillae* O'Brien (Curculionidae). Larvae were found damaging the stems of hydrilla at reservoirs, dams and lakes in Australia; damage was not found where water was flowing. Eggs were laid in the stems, which larvae tunnelled and fed upon on emergence. Pupation occurred only when the material became stranded along the shore by wave action, wind or draw-downs in the water level. This weevil had several field and

Table 3: Biological control agents imported into the USA from Australia for which host-testing was, or is, being carried out

Biocontrol agent	Biocontrol agent	Family	Feeding guild	Origin	Date released	Damage to weed
Melaleuca quinquenervia S.T. Blake	*Oxyops vitiosa* Pascoe	Curculionidae	Leaf-feeder	Qld	1997	Extensive
Melaleuca quinquenervia S.T. Blake	*Boreioglycaspis melaleucae* Moore	Psyllidae	Sap-sucker	Qld	2002	Extensive
Melaleuca quinquenervia S.T. Blake	*Fergusonina turneri* Taylor	Fergusoninidae	Bud-galler	NSW/Qld	2005	Failed to establish
Melaleuca quinquenervia S.T. Blake	*Lophodiplosis trifida* Gagné	Cecidomyiidae	Stem-galler	Qld	2008	Extensive
Melaleuca quinquenervia S.T. Blake	*Haplonyx multicolor* Lea	Curculionidae	Foliage-feeder	Qld	In quarantine	
Melaleuca quinquenervia S.T. Blake	*'Sphaerococcus' ferrugineus* Froggatt	Eriococcidae	Foliage-galler	Qld	In quarantine	
Melaleuca quinquenervia S.T. Blake	*Eucerocoris suspectus* Distant	Miridae	Sap-sucker	Qld	Rejected – specificity	
Melaleuca quinquenervia S.T. Blake	*Lophyrotoma zonalis* Rohwer	Pergidae	Leaf-feeder	Qld	Rejected – toxic	
Lygodium microphyllum (Cav.) R.Br.	*Austromusotima camptozonale* (Hampson)	Pyralidae	Leaf-feeder	Qld	2005	Failed to establish
Lygodium microphyllum (Cav.) R.Br.	*Floracarus perrepae* Knihinicki and Boczek	Eriophyiidae	Leaf-roller	Qld	2007	Failed to establish
Lygodium microphyllum (Cav.) R.Br.	*Neomusotima conspurcatalis* (Warren)	Pyralidae	Leaf-feeder	NT	2008	Moderate
Hydrilla verticillata (L.F.) Royle	*Hydrellia balciunasi* Bock	Ephydridae	Leaf-miner	Qld	1989	Minor
Hydrilla verticillata (L.F.) Royle	*Bagous hydrillae* O'Brien	Curculionidae	Adult leaf-feeder/larval stem-borer	Qld	1991	Failed to establish
Hydrilla verticillata (L.F.) Royle	*Ambia ptolycusalia* Walker	Pyralidae	Leaf-feeder	Qld	Rejected – specificity	
Hydrilla verticillata (L.F.) Royle	*Theila siennata* Warren	Pyralidae	Leaf-feeder	Qld	Rejected – specificity	

See description of damage to weed in Olckers and Hill (1999).

laboratory hosts (Balciunas *et al.* 1996), but was approved for release in Florida in 1991 given that it was unlikely to complete development on other submerged aquatic plants. Unlike other hosts, tunnelling by *B. hydrillae* larvae in hydrilla was likely to facilitate stem fragmentation, allowing the material to reach the shore and immatures to complete development. Alternate plant hosts with immatures which became stranded through drawdowns in the water level were unlikely to survive desiccation. Despite massive releases in Florida, *B. hydrillae* never established (Table 3). The reasons are unclear but may relate to the unsuitability of release sites for the unique biology of this weevil.

Hydrellia balciunasi Bock (Ephydridae), a leaf-mining fly found throughout the range of hydrilla in Australia, is a highly specific insect released in Florida in 1989. It established but is often displaced by *H. pakistanae* Deonier, a close relative from India which appears to be better adapted to Florida hydrilla. No other agents from Australia were released. Several defoliating pyralid moths were evaluated but their host-ranges were too broad. The species included *Ambia ptolycusalia* Walker (prev. *Nymphula eromenalis*), *Margarosticha repetitalis* Warren (prev. *Strepsinoma repititalis* Walker) and *Theila siennata* Warren (prev. *Aulacodes sienatta*). Only *A. ptolycusalia* and *T. siennata* were sent to quarantine in Florida.

Figure 5: The weevil *Oxyops vitiosa* has contributed to the successful reduction of the broad-leaved paperbark tree, *Melaleuca quinquenervia*, in Florida. Photo: M Purcell, USDA-ARS Australian Biological Control Laboratory.

Exploration for biological control agents against *M. quinquenervia* began in 1986 and is still under way. Four insects have been released in Florida – a leaf-feeding weevil *Oxyops vitiosa* Pascoe (1997) (Figs 5 and 6), a sap-sucking psyllid *Boreioglycaspis melaleuca* Moore (2002), a bud-galling fergusoninid fly *Fergusonina turneri* Taylor (2005) and a stem-galling cecidomyiid fly *Lophodiplosis trifida* Gagné (2009) (Table 3). *F. turneri* is the only insect which has not established. The remaining species are causing significant damage to *M. quinquenervia* in south Florida (Fig. 7). The weevil and psyllid alone have reduced stand densities (Tipping *et al.* 2009) and curtailed flowering by up to 95% (Center *et al.* 2008). Two agents which were sent to quarantine in the USA were not released. The defoliating sawfly *Lophyrotoma zonalis* Rohwer (Pergidae) was approved for release but subsequently found to contain two sawfly toxins, lophyrotomin and pergidin, which could have threatened native fauna that consumed these insects. A sap-sucking bug, *Eucerocorus suspectus* Distant (Miridae) was rejected due to a broad host-range.

The biological control program against *L. microphyllum* began in 1996. Compared to plants such as *M. quinquenervia*, this fern had few insect herbivores – only two species of mite and 20 insect herbivores have been recorded (Goolsby *et al.* 2003). The first agent discovered, the defoliating Pyralidae moth *Austromusotima camptozonale* Hampson from south-eastern Qld, was the first released (2005) in Florida, but thus far it has failed to establish though releases are ongoing. A highly specific eriophyid mite from tropical north Qld, *Floracarus perrepae* Knihinicki and Boczek, also failed to widely establish after its release in 2007. This mite is haplotype-specific and was collected and tested from the haplotype of the fern in the native range which most closely matched that of the introduced range in Florida (Goolsby *et al.* 2006). The reasons for its poor establishment are unknown, although there may be more genetic variation of *L. microphyllum* in Florida than indicated from the specimens genetically characterised. A second defoliating pyralid moth, *Neomusotima conspurcatalis* Warren, from the NT, did establish after its release in 2008 and has caused spectacular damage to *L. microphyllum* (Fig. 8), but only in a

Figure 6: Larvae of *Oxyops vitiosa* and the psyllid *Boreioglycaspis melaleuca* (white, larval secretions) severely damage young foliage of *Melaleuca quinquenervia*. Photo: M Purcell, USDA-ARS Australian Biological Control Laboratory.

Figure 7: The leaf-feeding weevil *Oxyops vitiosa*, the sap-sucking psyllid *Boreioglycaspis melaleuca* and the stem-galling cecidomyiid fly *Lophodiplosis trifida* are causing significant damage to *Melaleuca quinquenervia* in south Florida. Photo: M Purcell, USDA-ARS Australian Biological Control Laboratory.

Figure 8: Spectacular defoliation of *Lygodium microphyllum* at Jonathon Dickinson Park in southern Florida caused by the pyralid moth, *Neomusotima conspurcatalis*. Photo: M Purcell, USDA-ARS Australian Biological Control Laboratory.

localised area within Jonathon Dickinson Park near Jupiter in southern Florida. Its native range extends from tropical northern Australia into south-east Asia and China (Boughton and Pemberton 2009).

Extensive surveys have been conducted for biological control agents of sheoaks/Australian pine, *Casuarina* spp., and several promising agents have been discovered including seed-feeding torymid wasps, defoliating caterpillars and weevils, leaf tip gall-forming cecidomyiid midges and sap-feeding psyllids (Wheeler *et al.* 2011) as well as gall-forming eulophid wasps. As yet, none have been imported for evaluation into US quarantine facilities.

DISCUSSION

The biological control programs against Australian plants discussed here are currently active and ongoing in South Africa or the USA.

In South Africa, although much of the focus over the last 10 years has been on post release evaluation, new agents are being tested in quarantine (Table 2) and evaluation studies have demonstrated that additional biological control measures are required for several of the species. Surveys have recently been carried out for potential new agents for *L. laevigatum* in Australia, and it is highly probable that new agents will be sought in the near future to supplement the biological control efforts against certain of the *Acacia* species.

Besides the species which are currently under some level of biological control, attention to potentially problematic Australian plants has been raised through the early detection program and the South African Plant Invaders Atlas. Species such as tall kangaroo paw *Anigozanthos flavidus* DC. (Haemodoraceae), Australian weeping bottlebrush *Callistemon viminalis* (Gaertn.) G. Don, Australian tree fern *Cyathea cooperi* (W.J. Hooker ex F.

von Mueller) Domin, red silky oak *Grevillia banksii* R.Br., silky oak *Grevillia robusta* A.Cunn. ex R.Br., broad-leaved paperbark *M. quinquenervia*, sweet pittosporum *Pittosporum undulatum* Vent. and Australian brush cherry *Syzigium paniculatum* Banks ex Gaertn. are among those earmarked for attention. Currently recorded as plants to be aware of, or for eradication, some may become targets for biological control in the future.

At ABCL, biological control programs for *M. quinquenervia* and *L. microphyllum* are likely to continue in Australia at least for the short term. The focus for exploration on *H. verticillata* is now in Asia, though collections in remote areas of northern Australia may be possible. Support for *Casuarina* exploration has been suspended due to conflict of interest issues. The river sheoak *C. cunninghamiana* is being considered for use as a windbreak around citrus orchards to inhibit the spread of citrus canker (*Xanthomonas campestris* (Hasse) Dye) in Florida. However, alternative non-invasive trees could be used as windbreaks, or species-specific agents could be used for the other problematic *Casuarina* species. Other Australian plants which have become invasive in the USA include tuckeroo tree/carrot wood *Cupaniopsis anacardioides* (A.Rich.) Radlk., umbrella tree *Schefflera actinophylla* (Endl.) Harm and the tree fern *Cyathea cooperi*, which is a significant weed in Hawaii.

Although biological control efforts in the two countries have, for the most part, targeted different weeds, they share some problems in terms of Australian invasive plants, e.g. *M. quinquenervia*, *Acacia* species and *H. verticillata*. Where well developed biological control programs are in place, sharing research expertise can be mutually beneficial and result in the saving of a great deal of time, money and effort.

REFERENCES

Annecke DP (1975) Biological control of Australian *Acacia* species. Official communication Ref. 13/3/417/6 dated 26 November 1975 from Director, Plant Protection Research Institute, Pretoria, to Director, Wattle Research Institute, University of Natal, Pietermaritzburg.

Austin DF (1978) Exotic plants and their effects in southeastern Florida. *Environmental Conservation* **5**, 25–34.

Balciunas JK, Burrows DW and Purcell MF (1996) 'Final report on the Australian Surveys (1985–1992) for insect biological control agents of *Hydrilla verticillata*'. Technical Report A-96-5. US Army Corps of Engineers, Waterways Experiment Station, Vicksburg, MS.

Balciunas JK, Grodowitz MJ, Cofrancesco AF and Shearer JF (2002) Hydrilla. In *Biological Control of Invasive Plants in the Eastern United States*. (Eds R Van Driesche, S Lyon, B Blossey, M Hoddle and R Reardon) pp. 139–147. Publication FHTET-2002-04. Forest Health Technology Enterprise Team, USDA Forest Service.

Beckner J (1968) *Lygodium microphyllum*, another fern escaped in Florida. *American Fern Journal* **58**, 93–94.

Blossey B and Kamil J (1996) What determines the increased competitive ability of invasive non-indigenous plants? In *Proceedings of the 9th International Symposium on Biological Control of Weeds*. 19–26 January 1996, Stellenbosch, South Africa. (Eds VC Moran and JH Hoffmann) pp. 3–9. University of Cape Town, South Africa.

Blossey B and Nötzold R (1995) Evolution of increased competitive ability in invasive non-indigenous plants: a hypothesis. *Journal of Ecology* **83**, 887–889.

Bodle JM, Ferriter AP and Thayer DD (1994) The biology, distribution and ecological consequences of *Melaleuca quinquenervia* in the Everglades. In *Everglades: The Ecosystem and its Restoration*. (Eds SM Davis and JC Ogden) pp. 341–355. St Lucie Press, Delray Beach, Florida.

Boughton AJ and Pemberton RW (2009) Establishment of an imported natural enemy, *Neomusotima conspurcatalis* (Lepidoptera; Crambidae) against an invasive weed, Old World climbing fern, *Lygodium microphyllum*, in Florida. *Biocontrol Science and Technology* **19**, 769–772.

Boughton AJ, Bennett CA, Goolsby JA and Pemberton RW (2009) Laboratory host range testing of *Neomusotima conspurcatalis* (Lepidoptera: Crambidae) a potential biological control agent of the invasive weed, Old World climbing fern, *Lygodium microphyllum* (Lygodiaceae). *Biocontrol Science and Technology* **19**, 369–390.

Bowie J (1832) Hints on raising the leguminous plants of Australia and the Cape of Good Hope from seeds, on acclimatizing them in Europe and on their native habitats. *Gardener's Magazine* **8**(36), 5–9.

Buckingham GR and Bennett CA (1996) Laboratory biology of an immigrant Asian moth, *Parapoynx diminutalis* (Lepidoptera: Pyralidae) on *Hydrilla verticillata* (Hydrocharitaceae). *Florida Entomologist* **79**, 353–363.

Burchell WJ (1822) *Travels in the Interior of Southern Africa*. Vol. 1. Longman, London.

Burtt-Davy J (1904) Alien plants spontaneous in the Transvaal. *Report of the South African Association for the Advancement of Science* **2**, 252–299.

Center TD, Pratt PD, Tipping PW, Rayamajhi MB, Winеriter SA and Purcell MF (2008) Biological control of *Melaleuca quinquenervia*: goal-based assessment of success. In *Proceedings of the 12th International Symposium on Biological Control of Weeds*. 22–27 April 2007, La Grande Motte, France. (Eds MH Julien, R Sforza, MC Bon, HC Evans, PE Hatcher, HL Hinz and BG Rector) pp. 655–664. CABI, Wallingford, UK.

Cox GW (1999) *Alien Species in North America and Hawaii*. Island Press, Washington, DC.

Crawley MJ (1987) What makes a community invisible? In *Colonization, Succession and Stability*. (Eds AJ Gray, MJ Crawley and PJ Edwards) pp. 429–453. Blackwell Scientific Publications, Oxford.

Daehler CC and Strong DR (1993) Prediction and biological invasions. *Trends in Ecology and Evolution* **8**, 380.

Deacon J (1986) Human settlement in South Africa and archaeological evidence for alien plants and animals. In *The Ecology and Management of Biological Invasions in Southern Africa*. (Eds IAW Macdonald, FJ Kruger and AA Ferrar) pp. 3–19. Oxford University Press, Cape Town.

Dennill GB and Donnelly D (1991) Biological control of *Acacia longifolia* and related weed species (Fabaceae) in South Africa. *Agriculture, Ecosystems and Environment* **37**, 115–135.

Dennill GB, Donnelly D, Stewart K and Impson FAC (1999) Insect agents used for the biological control of Australian *Acacia* species and *Paraserianthes lophantha* (Willd.) Nielsen (Fabaceae) in South Africa. *African Entomology*, Memoir No. 1, 45–54.

DiTomaso JM and Healy EA (2007) *Weeds of California and other Western States*. University of California, Agriculture and Natural Resources, Publication 3488, California.

Donnelly D (1986) *Rayieria* sp. (Heteroptera: Miridae): host specificity, conflicting interests, and rejection as a biological control agent against the weed *Acacia longifolia* (Andr.) Willd. in South Africa. *Journal of the Entomological Society of Southern Africa* **49**, 183–191.

Doren RF and Jones DT (1997) Plant management in Everglades National Park. In *Strangers in Paradise: Impact and Management of Non-indigenous Species in Florida*. (Eds D Simberloff, DC Schmitz and TC Brown) pp. 275–286. Island Press, Washington, DC.

Drake JA, Mooney HA, di Castri F, Groves RH, Kruger FJ, Rejmánek M and Williamson M (1989) *Biological Invasions: A Global Perspective*. John Wiley, Chichester.

Dray FA, Bennett BC and Center TD (2006) Invasion history of *Melaleuca quinquenervia* (Cav.) S.T. Blake in Florida. *Castanea* **71**, 210–225.

Ferriter AP (1994) Extent of melaleuca infestation in Florida. In *Melaleuca Management Plan for Florida*. (Ed FB Laroche) pp. 12–17. South Florida Water Management District, West Palm Beach, Florida.

Ferriter A and Pernas T (2006) An explosion in slow motion: tracking the spread of *Lygodium microphyllum* in Florida. *Wildland Weeds* **9**, 7–9.

Fugler SR (1979) Some aspects of the autoecology of three *Hakea* species in the Cape Province, South Africa. MSc thesis. University of Cape Town, South Africa.

Gann GD, Bradley K and Woodmansee SW (2008) *Floristic Inventory of South Florida Database*. Institute for Regional Conservation. <www.regionalconservation.org>.

Glisson MW (1997) Management of state lands. In *Strangers in Paradise: Impact and Management of Non-indigenous Species in Florida*. (Eds D Simberloff, DC Schmitz and TC Brown) pp. 287–295. Island Press, Washington, DC.

Goolsby JA (2004) Potential distribution of the invasive Old World climbing fern, *Lygodium microphyllum* in North and South America. *Natural Areas Journal* **24**, 351–353.

Goolsby JA, Wright AD and Pemberton RW (2003) Exploratory surveys in Australia and Asia for natural enemies of Old World climbing fern, *Lygodium microphyllum*: Lygodiaceae. *Biological Control* **28**, 33–46.

Goolsby JA, De Barro PJ, Makinson JR, Pemberton RW, Hartley DM and Frohlich DR (2006) Matching the origin of an invasive weed for selection of a herbivore haplotype for a biological control programme. *Molecular Ecology* **15**, 287–297.

Gordon AJ (1999a) A review of established and new insect agents for the biological control of *Hakea sericea* Schrader (Proteaceae) in South Africa. *African Entomology*, Memoir No. 1, 35–43.

Gordon AJ (1999b) Biological control of Australian myrtle, *Leptospermum laevigatum* (J. Gaertn.) F. Muell. (Myrtaceae), in South Africa. *African Entomology*, Memoir No.1, 139–144.

Hall AV, De Winter M, De Winter B and van Oosterhout SAM (1980) 'Threatened plants of southern Africa'. South African National Scientific Programmes Report No 45. CSIR, Pretoria.

Holmes PM, Dennill GB and Moll EJ (1987) Effects of feeding by native alydid insects on the seed viability of an alien invasive weed, *Acacia cyclops*. *South African Journal of Science* **83**, 580–581.

Hughes CE (1995) Protocols for plant introductions with particular reference to forestry: changing perspectives on risks to biodiversity and economic development. In *British Crop Protection Proceedings No. 64: Weeds in a Changing World*. (Ed CH Stirton) pp. 15–32. British Crop Protection Council, Farnham.

Impson FAC and Moran VC (2004) Thirty years of exploration for and selection of a succession of *Melanterius* weevil species for biological control of invasive Australian acacias in South Africa: should we have done anything differently? In *Proceedings of the 11th International Symposium on Biological Control of Weeds*. 27 April–2 May 2003, Canberra. (Eds JM Cullen, DT Briese, DJ Kriticos, WM Lonsdale, L Morin and JK Scott) pp. 127–134. CSIRO Entomology, Canberra.

Impson FAC, Hoffmann JH and Kleinjan C (2009) Australian *Acacia* species (Mimosaceae) in South Africa. In *Biological Control of Tropical Weeds Using Arthropods*. (Eds R Muniappan, GVP Reddy and A Raman) pp. 38–62. Cambridge University Press, Cambridge.

Johnson AF and Barbour MG (1990) Dunes and maritime forests. In *Ecosystems of Florida*. (Eds RL Myers and JJ Ewel) pp. 429–480. University of Central Florida Press, Orlando.

Kluge RL (1990) The seed-attacking wasp *Bruchophagus* sp. (Hymenoptera: Eurytomidae) and its potential for biological control of *Acacia longifolia* in South Africa. In *Proceedings of the 7th International Symposium on Biological Control of Weeds*. 6–11 March 1988, Rome. (Ed ES Delfosse) pp. 349–356. Istituto Sperimentale per la Patologia Vegetale, Ministero dell'Agricoltura e delle Foreste.

Kluge RL and Neser S (1991) Biological control of *Hakea sericea* (Proteacea) in South Africa. *Agriculture, Ecosystems and Environment* **37**, 91–114.

Knott R (1976) Letter to Mrs Alfred G Kay.

Latrobe CI (1821) *A Journal of a Visit to South Africa*. 2nd edn. Seeley, London.

Legat CE (1904–08) 'Report(s) of the conservator of forests'. Report(s) of the Transvaal Dept of Agriculture, Pretoria.

Lockey RF, Stablein JJ and Binford LRF (1981) Melaleuca tree and respiratory disease. Allergin or irritant effect of Melaleuca pollen and odor, respectively, in patients with allergic and respiratory disease. In *Proceedings of Melaleuca Symposium*. (Ed RK Geiger) pp. 101–115. Florida Dept of Agriculture and Consumer Services, Divn of Forestry.

Macdonald IAW (1985) The Australian contribution to South Africa's invasive alien flora: an ecological analysis. *Proceedings of the Ecological Society of Australia* **14**, 225–236.

Macdonald IAW and Jarman ML (Eds) (1984) 'Invasive alien organisms in the terrestrial ecosystems of the fynbos biome, South Africa'. South African National Scientific Programmes Report No. 85. CSIR, Pretoria.

Madeira PT, Van TK, Steward KK and Schnell RJ (1997) Random amplified polymorphic DNA analysis of the phenetic relationships among world-wide accessions of *Hydrilla verticillata*. *Aquatic Botany* **59**, 217–236.

Meskimen GF (1962) A silvical study of the melaleuca tree in south Florida. MSc thesis. University of Florida, Gainesville, USA.

Milton SJ and Hall AV (1981) Reproductive biology of Australian acacias in the south western Cape Province, South Africa. *Transactions of the Royal Society of South Africa* **44**, 465–487.

Morris MJ (1999) The contribution of the gall-forming rust fungus *Uromycladium tepperianum* (Sacc.) McAlp. to the biological control of *Acacia saligna* (Labill.) Wendl. (Fabaceae) in South Africa. *African Entomology*, Memoir No. 1, 125–128.

Morton JF (1962) Ornamental plants with toxic and/or irritant properties. II. *Proceedings of the Florida State Horticultural Society* **75**, 484–491.

Morton JF (1980) The Australian pine or beefwood (*Casuarina equisetifolia* L.) an invasive 'weed' tree in Florida. *Proceedings of the Florida State Horticultural Society* **93**, 87–95.

Nauman CE and Austin DF (1978) Spread of *Lygodium microphyllum* in Florida. *American Fern Journal* **68**, 65–66.

Neser S (1968) Studies on some potentially useful insect enemies of needle-bushes (*Hakea* spp. – Proteaceae). PhD thesis. Australian National University, Canberra.

Olckers T and Hill MP (1999) Biological control of weeds in South Africa (1990–1998). *African Entomology*, Memoir No. 1.

Oliver EGH (1977) An analysis of the Cape flora. In *Proceedings of the 2nd National Weeds Conference of South Africa*. 1977, Stellenbosch, South Africa. pp. 1–18. AA Balkema, Cape Town.

Pemberton RW (1998) The potential of biological control to manage old world climbing fern (*Lygodium micro-*

phyllum), an invasive weed in Florida. *American Fern Journal* **88**, 176–182.

Pemberton RW and Ferriter AP (1998) Old world climbing fern (*Lygodium microphyllum*), a dangerous invasive weed in Florida. *American Fern Journal* **88**, 165–175.

Phillips EP (1938) The naturalized species of hakea. *Farming in South Africa* **13**, 424.

Pritchard PCH (1976) Melaleuca. *Florida Naturalist* **49**, 7–11.

Richardson DM, van Wilgen BW and Mitchell DT (1987) Aspects of the reproductive ecology of four Australian *Hakea* species (Proteaceae) in South Africa. *Oecologia* **71**, 345–354.

Ryerson KA (1967) The history of plant exploration and introduction in the United States Department of Agriculture. In *Proceedings of the International Symposium on Plant Introduction*. pp. 1–19. Escuela Agricola Panamericana, Honduras.

Schmitz DC, Nelson BV, Nall LE and Schardt JD (1991) Exotic aquatic plants in Florida: a historical perspective and review of the present aquatic plant regulation program. In *Proceedings of the Symposium on Exotic Pest Plants*. (Eds TD Center, RF Doren, RL Hofstetter, RL Myers and LD Whiteaker) pp. 303–326. Technical Report NPS/NREVER/NRTR-91/06. US Dept of the Interior, National Park Service, Denver.

Shaughnessy GL (1980) Historical ecology of alien woody plants in the vicinity of Cape Town, South Africa. PhD thesis. University of Cape Town, South Africa.

Shaughnessy GL (1986) A case study of some woody plant introductions to the Cape Town area. In *The Ecology and Management of Biological Invasions in Southern Africa*. (Eds IAW Macdonald, FJ Kruger and AA Ferrar) pp. 37–43. Oxford University Press, Cape Town.

Simberloff D, Schmitz D and Brown T (1997) *Strangers in Paradise: Impact and Management of Non-indigenous Species in Florida*. Island Press, Washington, DC.

Stanaland BE, Gennaro RN, Klotz SD, Sweeney MJ and White RS (1986) Isolation and characterization of cross-reactive allergenic components in *Callistemon citrinis* and *Melaleuca quinquenervia* pollens. *International Archives of Allergy and Applied Immunology* **86**, 35–41.

Stocker RK and Sanders DR (1981) Chemical control of *Melaleuca quinquenervia*. In *Proceedings of Melaleuca Symposium*. (Ed RK Geiger) pp. 129–134. Florida Dept of Agriculture and Consumer Services, Divn of Forestry.

Stubbings JA (1977) A case against controlling invasive Acacias. In *Proceedings of the 2nd National Weeds Conference*, Stellenbosch, South Africa. pp. 89–107. AA Balkema, Cape Town.

Tipping PW, Martin MR, Nimmoa KR, Pierce RM, Smart MD, White E, Madeira PT and Center TD (2009) Invasion of a West Everglades wetland by *Melaleuca quinquenervia* countered by classical biological control. *Biological Control* **48**, 73–78.

Tucker KC and Richardson DM (1995) An expert system for screening potentially invasive alien plants in South African fynbos. *Journal of Environmental Management* **44**, 309–338.

Turner CE, Center TD, Burrows DW and Buckingham GR (1998) Ecology and management of *Melaleuca quinquenervia*, an invader of wetlands in Florida, U.S.A. *Wetlands Ecology and Management* **5**, 165–178.

Volin JC, Lott MS, Muss JD and Owen D (2004) Predicting rapid invasion of the Florida everglades by old world climbing fern (*Lygodium microphyllum*). *Diversity and Distributions* **10**, 439–446.

Walton C, Ellis N and Pheloung P (1998) 'A manual for using the Weed Risk Assessment system (WRA) to assess new plants'. Australian Quarantine and Inspection Service, Canberra.

Webb D van V (1964) 'Investigations into the biological control of hakea in Australia'. Part I–III. Unpublished report. Dept of Agricultural Technical Services, Pretoria.

Wells MJ, Poynton RJ, Balsinhas AA, Musil KJ, Joffe H, van Hoepen E and Abbott SK (1986) The history of introduction of invasive alien plants to southern Africa. In *The Ecology and Management of Biological Invasions in Southern Africa*. (Eds IAW Macdonald, FJ Kruger and AA Ferrar) pp. 21–35. Oxford University Press, Cape Town.

Wheeler GS, Taylor GS, Gaskin JF and Purcell MF (2011) Ecology and management of sheoak (Casuarina spp.), an invader of coastal Florida, USA. *Journal of Coastal Research* **27**, 485–492.

Winters HF (1967) The mechanics of plant introduction. In *Proceedings of the International Symposium on Plant Introduction*. pp. 49–53. Escuela Agricola Panamericana, Honduras.

Wittenberg R and Cock MJW (Eds) (2001) *Invasive Alien Species: A Toolkit of Best Prevention and Management Practices*. CABI, Wallingford, UK.

Acacia nilotica subsp. *indica* (Benth.) Brenan – prickly acacia

Bill Palmer, Catherine Lockett and K Dhileepan

ABSTRACT

Prickly acacia, *Acacia nilotica* subsp. *indica*, is one of the worst woody weeds of northern Australia and is a Weed of National Significance. Since its introduction as a shade and fodder tree it has infested over 7 million ha of the Mitchell grass downs of western Qld, where it has changed natural grassland into woody savannah. It also infests several more discrete coastal areas. Investigations into biological control commenced after dramatic increases in prickly acacia in the mid 1970s. Faunal studies of the insects associated with *A. nilotica* were undertaken in Pakistan, Kenya and South Africa from local field stations. Six insect species were released in Qld after 1980. The bruchid *Bruchidius sahlbergi* established throughout the range of the weed, but was ineffective. The geometrid *Chiasmia assimilis* became well established in coastal areas, where it caused significant total defoliations. The moth *Cuphodes profluens* may have temporarily established at a coastal site before all prickly acacia trees were removed from that site. The other three insects (*Homichloda barkeri, Chiasmia inconspicua* and *Cometaster pyrula*) did not establish. Further exploration is being undertaken in India, where two pathogens and three insect species are being investigated. A particular feature of this project, perhaps unique, is that climatic modelling has suggested that neither prickly acacia nor some of the released agents are very well adapted for the areas of most concern in western Qld.

Key words: *Chiasmia assimilis, Bruchidius sahlbergi*, CLIMEX modelling.

INTRODUCTION

Prickly acacia, *Acacia nilotica* subsp. *indica* (Benth.) Brenan (Mimosaceae), is described as a thorny shrub or small tree usually growing to 5 m but sometimes to 10 m (March 2004). It has a characteristic umbrella-shaped canopy (Fig. 1) and is single-stemmed unless it has been damaged by fire, frost or mechanical control efforts. The tree has a very deep tap root, and several laterals close to the soil surface. The bipinnate leaves are finely divided, with a pair of thorns at each leaf base. Flowers are ball-shaped and golden-yellow. The pods contain eight to 15 seeds and have very characteristic constrictions between the seeds. A mature tree might produce up to 175 000 seeds that remain viable for several years (Carter *et al.* 1989). The plant is largely dispersed by livestock (particularly cattle) ingesting mature seed (Harvey 1981; Kriticos *et al.* 1999a).

Although prickly acacia was recorded in Australia in the 1860s it was not until the early 20th century that it was recognised as a beneficial plant for the western areas; lambing percentages and wool production could be increased by providing shade trees on natural grassland. Deliberate efforts supported by departmental recommendations were made to plant this tree along bore drains

Figure 1: Line drawings of *Acacia nilotica* subsp. *indica* showing various identifying features. Illustration: Qld Government.

and other watercourses. The tree became abundant in the western areas but remained much less common in coastal areas, where infestations were often associated with livestock introductions from the west.

While sheep remained the major enterprise in the western areas of Qld, prickly acacia was not generally regarded a serious problem. Undoubtedly the major reason was that sheep grazed prickly acacia seedlings and ensured a very high mortality rate for juvenile plants. Further, sheep were much less likely to disperse germinable seed in their manure, although significant dispersal of viable seed occurred through cud regurgitation (Harvey 1981). When cattle production became the dominant enterprise, increasingly from the 1970s to today when there are few sheep remaining, the weed problem became more and more evident (Brown and Carter 1998).

Prickly acacia has become one of the worst woody weeds of northern Australia, where it infests over 7 million ha of the Mitchell grass downs of western Qld (Mackey 1997). Smaller infestations of the weed also exist in coastal areas of Qld between Bowen and Townsville (Fig. 2). These are significant because coastal areas are climatically more favourable for the plant (Fig. 2) than the more arid western areas (Kriticos *et al.* 1999b, 2003b). Modelling also suggests that potential areas of spread under current climatic conditions are far greater than the present distribution, while global climate change may significantly increase the area at risk of invasion (Kriticos *et al.* 1999b, 2003b).

Prickly acacia was recognised as a serious weed after a dramatic increase in its abundance in the mid 1970s, which coincided with a series of wet years and the change

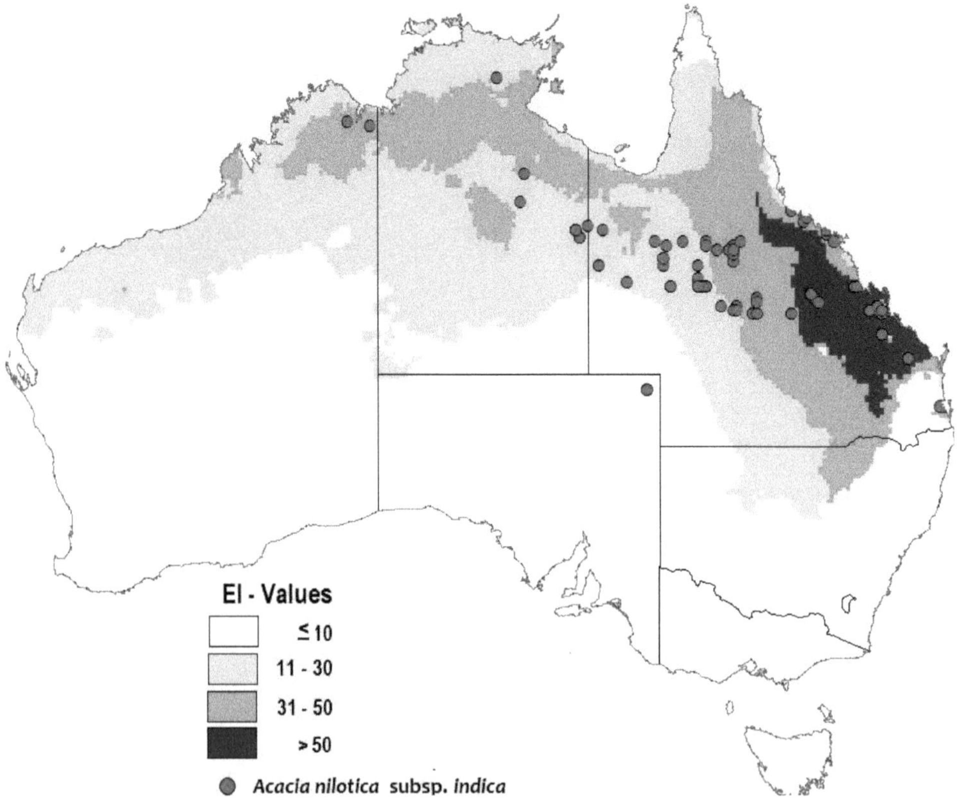

Figure 2: Distribution of *Acacia nilotica* subsp. *indica* in Australia based on herbarium records from the Australian Virtual Herbarium and its potential distribution, after Kriticos *et al.* (2003b), predicted by the climate modelling software CLIMEX (Sutherst *et al.* 2007). EI values >30 are considered climatically favourable while values ≤10 are considered unfavourable.

in enterprise from wool to cattle production. The detrimental effects of dense infestations have been described (Mackey 1996; Miller 1996a; March 2000): these include loss of pasture, soil erosion, mustering difficulties, loss of ecological values and injury to humans, animals and vehicles. The Mitchell grass downs is an area of particular concern as it is one of the last natural grasslands in Australia still in a reasonably pristine condition. This area is now in danger of conversion from grassland to thorny scrubland, similar to an African thornveldt (Fig. 3).

BIOLOGICAL CONTROL HISTORY

Australia is one of very few countries where prickly acacia is a concern. There are reports that it is weedy in Indonesia, but in most countries of its native range it is regarded as a very valuable natural resource. No other country has initiated biocontrol efforts for prickly acacia.

A biological control project was initiated by the Qld Department of Lands after the great proliferation of the weed in the 1970s. Since then, exploration has been undertaken in Pakistan, Kenya, South Africa and India with the result that six insect species were released after host testing at the Qld Government's Alan Fletcher Research Station, Sherwood, followed by mass rearing, release and evaluation, mainly by the Tropical Weeds Research Centre at Charters Towers. The project is still active, with efforts now concentrated in India.

During the project, attempts have been made to understand what types of agents might produce effective biocontrol. A plant population model (Kriticos *et al.* 1999a) indicated that seedlings and juvenile plants were the most susceptible life-stages of the plant to target for control and that flower, pod or seed feeders would be unlikely to be effective. Experimental evidence of the effects of artificial damage on juvenile plants was obtained by Dhileepan *et al.* (2009), who found that two defoliation events or a single damaging of the shoot could reduce plant vigour, while root damage had negligible effect.

PLANT TAXONOMY

Acacia is the largest genus (with over 950 endemic species) of flowering plants in Australia (Orchard and Wilson

(a)

Figure 3: A comparision of a) uninfested Mitchell grass downs and b) country severely infested with *Acacia nilotica* subsp. *indica* – the grasses have been lost and erosion is evident. Photos: Qld Government.

2001a) and the numerical and ecological importance of the genus presents particular issues for proposals to introduce biological control agents for prickly acacia. *Acacia* has been divided into three subgenera: *Acacia*, *Phyllodineae* and *Aculeiferum* (Maslin 2001). The subgenus *Acacia* is represented in Australia by nine endemic species and two introduced species (*A. nilotica* and *A. farnesiana* (L.) Willd.) (Maslin 2001). Except for two species in subgenus *Aculeiferum*, the remaining Australian *Acacia* spp. belong to subgenus *Phyllodineae* (Maslin 2001). Species of the subgenus *Acacia* have bipinnate leaves, whereas those of the *Phyllodinae* have either phyllodinous (seven sections of the subgenus) or bipinnate leaves (two sections, the *Botrycephalae* and *Pulchellae*).

Because the genus *Acacia sens. lat.* is polyphyletic (Maslin 2001) it has recently been split into into five genera with the name *Acacia* being conserved with a new type (Orchard and Maslin 2003). The old subgenus *Acacia* becomes the genus *Vachellia* Wight & Arn. and the Australian subgenus *Phyllodineae* becomes the new genus *Acacia*. New combinations for the New World (Seigler and Ebinger 2006) and Australian (Kodela and Wilson 2006) species of *Vachellia* are available, but not for African and Asian species. Because the transition is incomplete, the nomenclature of *Flora of Australia* (Orchard and Wilson 2001a, b) is used in this chapter.

Prickly acacia is one of nine subspecies of *A. nilotica* (L.) Willd. Ex Del., three being from the Indian subcontinent and six from Africa (Ross 1979; Brenan 1983). Although morphological (Brenan 1983) and leaf phenolic studies (Hannan-Jones 1999) suggested that the Australian populations of prickly acacia were from India, more detailed analysis was provided by molecular genetic distance studies using patterns of DNA sequence variation (Wardill *et al.* 2004, 2005). These studies confirmed that the Australian populations of *A. nilotica* mostly comprise subspecies *indica*, but a few individuals were genetically identical to an unidentified genotype from Pakistan not previously reported in Australia (Wardill *et al.* 2005).

EXPLORATION

The first initiative was a recommendation to explore Pakistan or India for suitable biocontrol agents (Girling 1980). Overseas studies, on Qld's behalf, for potential biocontrol agents began in 1980 with a team led by AI Mohyuddin based at the CABI institute at Rawalpindi in Pakistan (Mohyuddin 1981). Pakistan was chosen because the subspecies *Acacia nilotica* subsp. *indica* grew there and because its climate was considered similar to that of western Qld. Surveys were conducted throughout the plant's range in Pakistan. Some 43 phytophagous insect species were identified and 16 of these were considered stenophagous (Mohyuddin 1981). One weakness of this study was that it was not possible to extend the survey across the border into India.

Following preliminary exploration in Kenya by IAD Robertson in 1987 (Robertson 1987), a field station was established at the Kenya Forestry Research Institute at Maguga by J Marohasy in 1989, from which field surveys of *A. nilotica* subsp. *subalata* (Vatke) Brenan and *A. nilotica* subsp. *leiocarpa* Brenan were undertaken until 1992. The surveys found that over 90 phytophagous insect species were found to attack *A. nilotica*, but only a small number of them were considered possibilities for biological control (Marohasy 1993b, 1995). Palmer (1996) suggested that some implications drawn from the surveys in Pakistan and Kenya should be considered before further surveys were planned.

In the late 1990s attention switched to southern Africa. Surveys of the fauna on *A. nilotica* subsp. *kraussiana* (Benth.) Brenan were undertaken by several entomologists associated with the ARC-Plant Protection Research Institute in Pretoria. These surveys resulted in over 600 arthropod species being collected (Stals 1997; Stals *et al.* 2004). In 1999 the Qld Government's South African Field Station was established under the leadership of ABR Witt, who continued surveying for potential biocontrol agents.

Surveys of fauna on *A. nilotica* subsp. *indica* in India commenced in July 2008, through two collaborating research institutes, the Arid Forest Research Institute at Jodhpur, Rajasthan and the Institute of Forest Genetics & Tree Breeding at Coimbatore, Tamil Nadu. These areas were chosen after climate-matching that suggested that areas of western Qld (e.g. Winton, Barcaldine and Hughenden) matched most regions in India (Ecoclimatic Index = 50%) and that Rajasthan was climatically the most suitable region (Ecoclimatic Index = 70%) for exploration (Dhileepan *et al.* 2006).

So far, 77 insect species and 14 diseases have been recorded in Tamil Nadu and Karnataka in southern India, and 14 insect species and 11 diseases have been documented in Gujarat and Rajasthan in north-western India (Dhileepan *et al.* 2010).

CANDIDATES

Agents released

***Bruchidius sahlbergi* Schilsky (Coleoptera: Chrysomelidae).** The first insect introduced for prickly acacia was a seed-feeding bruchid from Pakistan. (There may be taxonomic issues with this species: African papers mentioning *B. sahlbergi*, e.g. Miller (1996b), may be referring to a different species.) *Bruchidius sahlberghi* is generally distributed in Pakistan and is particularly abundant in coastal areas to Peshawar and Rawalpindi (Mohyuddin 1986). It is also widespread in India. The beetles oviposit in ripening seed pods (Willson 1985) and larvae move between seed pods seeking unattacked seed. This insect was found to be sufficiently host-specific and was approved for release in Australia. Between 1982 and 1985, some 309 000 beetles were released across 96 sites mostly in the western areas of Qld (Willson 1985). The insect established readily and is now found in practically all areas where prickly acacia is found (Willson 1985; Palmer *et al.* 2005). However, it has not been an effective biocontrol because the seed predation before pod drop and removal by cattle is insufficient (Radford *et al.* 2001). The insect would possibly be more effective if cattle could be excluded from areas infested by prickly acacia while trees are bearing pods.

***Cuphodes profluens* Meyrick (Lepidoptera: Gracillariidae).** The larva of this species is a small tip-feeding caterpillar. The species was sourced from Pakistan (Mohyuddin 1986), where it is found fairly generally. It is multivoltine and common in summer, depending on new growth on the host plant. Host-range testing determined that the insect was host-specific and approval for its release in Australia was granted (Willson 1985). Over 3395 moths were released at 14 sites after November 1983, but indications of establishment occurred at only one coastal site near St Lawrence. Unfortunately, the prickly acacia at this site was later eradicated and the insect is now regarded as having failed to establish.

***Homichloda barkeri* (Jacoby) (Coleoptera: Chrysomelidae).** This leaf-feeding beetle was first found in Kenya on both *A. nilotica* subsp. *leiocarpa* and *A. nilotica* subsp. *subalata* (Marohasy 1993a). It was also found in South Africa on *A. nilotica* subsp. *kraussiana* (Stals 1997). The beetle oviposits batches of eggs in fissures in the bark. Larvae feed on the leaf pinnae before pupating in the soil. The insect has one or two generations before diapausing in the egg stage for the dry winter months. Host-specificity testing confirmed that it was specific to *A. nilotica* and it was approved for release in 1994 (Marohasy 1994). Great difficulty was experienced trying to rear this insect in sufficient numbers. New collections from Kenya were made in 1995 and 1997 and a satisfactory rearing method was ultimately devised. Over 10 000 beetles were released at 28 sites between 1996 and 1999 (Lockett and Palmer 2003). There has been no indication of establishment. Climatic modelling indicated that coastal areas of Qld were more suitable for *H. barkeri* than the western areas of Qld (Lockett and Palmer 2003).

***Chiasmia assimilis* (Warren) (Lepidoptera: Geometridae).** This leaf-feeder was originally sourced from Kenya, where it was collected on both *A. nilotica* subsp. *leiocarpa* and *A. nilotica* subsp. *subalata*. It was subsequently found in South Africa on *A. nilotica* subsp. *kraussiana*. Its biology, phenology, indicative host-range and climatic requirements have been described by Palmer *et al.* (2007). Females oviposit 400–500 eggs. Larvae feed on foliage for two to three weeks before pupating in leaf litter or soil. The life-cycle is completed in about one month and there can be several generations per year. The insect was tested against 74 plant species, mostly *Acacia* spp., and found to have a very limited host-range. Although the insect did complete its life-cycle on four Australian *Acacia* spp., the survival on these plants was insufficient to sustain a population and the insect was approved for release in 1999. *Acacia farnesiana* was not a host. About 74 000 were released, mostly in western Qld. Little indication of establishment was forthcoming. In 2002 a second population, sourced from South Africa, was imported and released. This time the insect successfully established. Very significant populations have been observed in coastal infestations since 2003 and have caused periodic defoliations. Experimental evidence indicated that the insect will attack juvenile plants, not just the established trees (C Lockett *et al.*, unpub.). The insect has also established at Hughenden and elsewhere on the Mitchell grass downs, but not in the high populations observed on the coast. Interestingly, climate modelling indicated that coastal areas of Qld would be more suitable than western areas for *C. assimilis* (Senaratne *et al.* 2006; Palmer *et al.* 2007).

***Chiasmia inconspicua* (Warren) (Lepidoptera: Geometridae).** This congener has attributes, host-range and history similar to those of *C. assimilis* (Palmer *et al.* 2007) and was also imported from Kenya. It was approved for release in 1998, after which 73 000 insects were released at 63 sites. There has been no indication of establishment and climate modelling has indicated that coastal areas of Qld would be more suitable than western areas for *C. inconspicua* (Senaratne *et al.* 2006; Palmer *et al.* 2007).

***Cometaster pyrula* (Hopffer) (Lepidoptera: Noctuidae).** This noctuid moth was the only insect released as a result of the South African surveys. Its biology, phenology, indicative host-range and climatic requirements were described by Palmer and Senaratne (2007). *Cometaster pyrula* is generally found across southern

Africa and has been recorded from Tanzania. Its only known host is *Acacia nilotica* subsp. *krausiiana* and it is rarely found in large numbers. Moths oviposit 300–400 eggs on the foliage. Larvae feed voraciously on foliage and eventually attain a length of over 6 cm before pupating in leaf litter. Some pupae enter a winter diapause from autumn onwards. The larvae are quite susceptible to disease when overcrowded; best rearing results were achieved if densities were no more than five caterpillars per potted plant. The insect was tested against 77 plant species, mostly *Acacia* spp., and found to be specific to *A. nilotica*. There were indications that survival of early instars was greater on subspecies *kraussiana* than on subspecies *indica*. The insect was approved for release in Australia and between 2004 and 2007 some 43 000 larvae and 1700 moths were released, mainly in coastal areas (Anon 2008). To date, there has been no evidence of establishment. Climate modelling has indicated that coastal areas of Qld would be more suitable than western areas for *C. pyrula* (Palmer and Senaratne 2007).

Agents considered but not released in Australia

Acacidiplosis spinosa **Gagné (Diptera: Cecidomyiidae).** Perhaps the most impressive insect agent seen in the field was this gall-forming midge found in Kenya. The galls formed by this insect are commonly seen and it appears to cause greatest damage in its termination of growth and inhibition of flowering. Culturing this species under quarantine conditions presented several difficulties, including that of getting potted plants to flower in the glasshouse. These difficulties were not overcome and the insect was not host-tested. This insect and several other Kenyan cecidomyiids from *A. nilotica* have been described by Gagné and Marohasy (1993).

Isturgia deerraria **(Walker) and** ***I. disputaria*** **(Guenée) (Lepidoptera: Geometridae).** The larvae of these moths are leaf-feeders. *Isturgia deerraria* was found on both *A. nilotica* subsp. *subalata* and *A. nilotica* subsp. *leiocarpa* in Kenya (Marohasy 1995). It was collected from Kenya, but host-testing in Brisbane confirmed field records (Kruger 2001) that the insect might attack several Australian native *Acacia* spp., though not the exotic *A. farnesiana*, and the insect was considered no further (Palmer and McLennan 2006). The second species, *I. disputaria*, was very interesting because it occurs on various subspecies of *A. nilotica* from India through to Africa. In Africa, it has also been collected on *Acacia* species native to Australia (Kruger 2001). Over the years, three different populations of this insect, from Pakistan, India and Kenya, were at least partially host-tested with similar results on each occasion. The insect could develop, but suboptimally, on some Australian native *Acacia* spp. though not the exotic *A. farnesiana*. Although a case might have been made for its release, the decision was made not to proceed (WA Palmer, unpub.). Both these *Isturgia* spp., being stenophagous, provided valuable information about the affinities of *A. nilotica* to the Australian flora (Palmer and McLennan 2006). These insects were more likely to utilise the bipinnate sections of subgenus *Phyllodineae* and the Australian species in subgenus *Acacia* than either the phyllodinous Australian *Acacia* spp. or *A. farnesiana*, also belonging to subgenus *Acacia* and thought to be of American origin (Orchard and Wilson 2001a).

Acizzia melanocepala **Burckhardt and Mifsud (Hemiptera: Psyllidae).** This psyllid feeds on the terminal shoots and is known from several subspecies of *A. nilotica* in Africa and the Arabian Peninsula (Burckhardt and Mifsud 1998). Host-testing of a Kenyan population in Brisbane indicated that *A. nilotica* subsp. *indica* was not a suitable host; this was supported by unsuccessful attempts to culture it on *A. nilotica* subsp. *indica* at the South African Field Station (Palmer and Witt 2006). Studies on this insect confirmed that differences between the subspecies of *A. nilotica* can be important for highly host-specific organisms.

Rhamphus **sp. (Coleoptera: Curculionidae).** This is a tiny weevil abundant on *A. nilotica* subsp. *kraussiana* in South Africa. Larvae mined the leaflets, causing them to be shed. The insect was imported into quarantine in Brisbane on several occasions, but it was not successfully cultured (Anon 2002).

Promising agents

In India, a leaf-webbing caterpillar (*Phycita* sp.), a scale insect (*Anomalococcus indicus* Ayyar), a leaf-feeding weevil (*Dereodus denticollis* Boheman) and a rust fungus (*Ravenelia acacia-arabicae* Mundkur & Thirumalachar) infecting leaves and inducing galls on the rachises and pods, have been prioritised for further studies (Dhileepan *et al.* 2010).

DISCUSSION

Prickly acacia, with its close affinity to the Australian native *Acacia* spp., its presence in a highly variable climate and its deep-rootedness, perennial habit and long-lived dispersible seeds, presents a formidable target for biological control. Indeed, adult prickly acacia plants exhibit five of the six traits listed by Crawley (1990) that describe a plant that is difficult to control with insect herbivores: it has a long growing period, a tough woody stem, high rates of regrowth, low food quality due to high tannin content and a large persistent seed bank (Kriticos *et al.* 1999a). This set of traits suggests that successful biological control against mature stands would be difficult.

That difficulty is compounded by the close affinity with some of Australia's native *Acacia* spp., particularly the bipinnate species, which limits possible biocontrol agents to those that are very host-specific. Climatic issues may also be important, particularly the fact that the areas in Australia with the worst weed problem have a very variable and different climate to that more suited to prickly acacia and possibly many of its insects.

The host-range testing of some potential insect agents of varying specificity has illustrated some general principles. The testing of *A. melanocephala* confirmed that some insects may have host-ranges limited to some subspecies of *A. nilotica*, and that insect species collected from subspecies other than *indica* may not utilise or prosper on the weed prickly acacia in Australia. The testing of the more stenophagous species such as the *Chiasmia* spp. and the *Isturgia* spp. indicated the likely affinities between *A. nilotica* and other *Acacia* groups. Stenophagous insects collected from *A. nilotica* were most likely to be able to develop on the bipinnate Australian *Acacia* spp., while the phyllodinous Australian *Acacia* spp. were not utilised as hosts to any extent.

Despite a well-funded project with 13 years of foreign exploration in four countries and commensurate Australian research and biocontrol activity, only six insect species have been approved for release and only two of these have established. This record may be improved by the present selection of potential insects and pathogens from India.

It is likely that useful biocontrol agents will have to be integrated with other management options including chemical and mechanical control, strict livestock movement control and grazing strategies with sheep, goats (Carter *et al.* 1990) or camels, to provide the best management strategies for this difficult weed (Jeffrey 1995; Kriticos *et al.* 2003a; Spies and March 2004). Feral camels are plentiful in Australia and their natural browse is *Acacia* spp. Graziers are now purchasing a few camels to run with their cattle. The camels do not compete with cattle for grasses and are large enough to break juvenile trees up to about 3 m in height. There has not been any formal assessment of their effect, but many graziers consider camels very useful and they may prove to be an effective biological control.

ACKNOWLEDGEMENTS

Scientists at the British Museum of Natural History, the International Institute of Entomology, the Biosystematics Division of the ARC-Plant Protection Research Institute, the Transvaal Museum and the Systematic Entomology Laboratory, Agricultural Research Service, USDA identified collected specimens. The CIBC Pakistan, the Kenya Forestry Research Institute, the ARC-Plant Protection Research Institute, the Arid Forest Research Institute and the Institute of Forest Genetics & Tree Breeding have hosted or are presently hosting survey teams. KADW Senaratne prepared Figure 2. Valuable comments on the manuscript were provided by colleagues at Biosecurity Queensland.

REFERENCES

Anon (2002) '2000/01 technical highlights: annual report on weed and pest animal research'. Qld Dept of Natural Resources and Mines, Brisbane.

Anon (2008) 'Technical highlights: invasive plant and animal research 2006–07'. Qld Dept of Primary Industries and Fisheries, Brisbane.

Brenan JPM (1983) *Manual on Taxonomy of Acacia Species: Present Taxonomy of Four Species of Acacia (A. albida, A. senegal, A. nilotica, A. tortilis).* FAO, Rome.

Brown JR and Carter J (1998) Spatial and temporal patterns of exotic shrub invasion in an Australian tropical grassland. *Landscape Ecology* **13**, 93–102.

Burckhardt D and Mifsud D (1998) Psylloidea (Insecta: Hemiptera) of the Arabian Peninsula. *Fauna of Arabia* **17**, 7–49.

Carter JO, Bolton MP and Cowan DC (1989) Prickly acacia: save dollars by early control measures. *Queensland Agricultural Journal* **115,** 121–126.

Carter JO, Newman P, Tindale P, Cowan D and Hodge PB (1990) Complementary grazing of sheep and goats on *Acacia nilotica. Proceedings of the Australian Rangelands Society* **6**, 271–272.

Crawley MJ (1990) Plant life-history and the success of weed biological control projects. In *Proceedings of the 7th International Symposium on the Biological Control of Weeds.* 6–11 March 1988, Rome, Italy. (Ed ES Delfosse) pp. 17–26. Instituto Sperimentale per la Patologia Vegetale. MAF, Rome.

Dhileepan K, Senaratne KADW and Raghu S (2006) A systematic approach to biological control agent exploration and prioritisation for prickly acacia (*Acacia nilotica* ssp. *indica*). *Australian Journal of Entomology* **45**, 303–307.

Dhileepan K, Lockett C, Robinson M and Pukallus K (2009) Prioritising potential guilds of specialist herbivores as biological control agents for prickly acacia through simulated herbivory. *Annals of Applied Biology* **154**, 97–105.

Dhileepan K, Balu A, Ahmed SI, Singh S, Srivastava KK, Senthilkumar M, Murugesan S, Senthilkumar P, Gorain M, Sharma A, Sharma N, Mahalakshmi R and Shivas R (2010) New biological control opportunities for prickly acacia: exploration in India. In *Proceedings of the 17th Australasian Weeds Conference*. 26–30 September 2010, Christchurch, New Zealand. (Ed SM Zydembos) pp. 231–234. New Zealand Plant Protection Society, Christchurch, New Zealand.

Gagné RJ and Marohasy JJ (1993) The gall midges (Diptera: Cecidomyiidae) of *Acacia* spp. (Mimosaceae) in Kenya. *Insecta Mundi* **7**, 77–124.

Girling DJ (1980) *The Possibilities for Biological Control of Acacia nilotica indica in Queensland: A Survey of the Literature*. Commonwealth Institute of Biological Control, London.

Hannan-Jones MA (1999) Variation in leaf phenolics in subspecies of *Acacia nilotica* (L.) Willd. ex Del. (Mimosaceae). In *Proceedings of the 12th Australian Weeds Conference*. 12–16 September 1999, Hobart, Tas. (Eds AC Bishop, M Boersma and CD Barnes) pp. 601–604. Tasmanian Weed Society, Hobart.

Harvey GJ (1981) Recovery and viability of prickly acacia (*Acacia nilotica* ssp. *indica*) seed ingested by sheep and cattle. In *Proceedings of the 6th Australian Weeds Conference*. 13–18 September 1981, Gold Coast, Qld. (Eds BJ Wilson and JT Swarbrick) pp. 197–201. Weed Society of Queensland, Brisbane.

Jeffrey PL (1995) Prickly acacia. In *Exotic Weeds and Their Control in North-west Queensland*. (Ed N March) pp. 3–9. Qld Dept of Lands, Brisbane.

Kodela PG and Wilson PG (2006) New combinations in the genus *Vachellia* (Fabaceae: Mimosoideae) from Australia. *Telopea* **11**, 233–244.

Kriticos D, Brown J, Radford I and Nicholas M (1999a) Plant population ecology and biological control: *Acacia nilotica* as a case study. *Biological Control* **16**, 230–239.

Kriticos DJ, Brown JR, Maywald GF, Sutherst RW and Adkins SW (1999b) A population model of *Acacia nilotica*: a tool for exploring weed management and the effects of climate change. In *Proceedings of the 6th International Rangeland Congress*. 19–23 July 1999, Townsville. (Eds D Eldridge and D Freudenberger) pp. 870–872. 6th International Rangeland Congress Inc., Aitkenvale, Qld.

Kriticos DJ, Brown JR, Maywald GF, Radford ID, Nicholas DM, Sutherst RW and Adkins SW (2003a) SPAnDX: a process-based population dynamics model to explore management and climate change impacts on an invasive alien plant, *Acacia nilotica*. *Ecological Modelling* **163**, 187–208.

Kriticos DJ, Sutherst RW, Brown JR, Adkins SW and Maywald GF (2003b) Climate change and the potential distribution of an invasive alien plant: *Acacia nilotica* ssp. *indica* in Australia. *Journal of Applied Ecology* **40**, 111–124.

Kruger M (2001) A revision of the tribe Macariini (Lepidoptera: Geometridae: Ennominae) of Africa, Madagascar and Arabia. *Bulletin of the Natural History Museum (Entomology)* **70**, 1–502.

Lockett CJ and Palmer WA (2003) Rearing and release of *Homichloda barkeri* (Jacoby) (Coleoptera: Chrysomelidae: Alticinae) for the biological control of prickly acacia, *Acacia nilotica* ssp. *indica* (Mimosaceae) in Australia. *Australian Journal of Entomology* **42**, 287–293.

Mackey AP (Ed) (1996) *Prickly Acacia* (Acacia nilotica) *in Queensland*. Pest status review series. Land Protection Branch, Qld Dept of Natural Resources, Brisbane.

Mackey AP (1997) The biology of Australian weeds 29. *Acacia nilotica* ssp. *indica* (Benth.) Brenan. *Plant Protection Quarterly* **12**, 7–17.

March N (2000) *Prickly Acacia Best Practice Manual*. Qld Dept of Natural Resources, Brisbane.

March N (2004) Prickly acacia: ecology and threat. In *Prickly Acacia National Case Studies Manual*. (Eds P Spies and N March) pp. 1–11. Qld Dept of Natural Resources, Mines and Energy, Brisbane.

Marohasy J (1993a) *Biocontrol of* Acacia nilotica *using Insects from Kenya*. Alan Fletcher Research Station, Qld Dept of Lands.

Marohasy J (1993b) Constraints to the introduction of biocontrol agents for prickly acacia (*Acacia nilotica*). In *Proceedings of the 10th Australian and 14th Asian-Pacific Weed Conference*. 6–10 September 1993, Brisbane. (Ed JT Swarbrick) pp. 33–38. Weed Society of Queensland, Brisbane.

Marohasy J (1994) Biology and host specificity of *Weiseana barkeri* (Col.: Chrysomelidae): a biological control agent for *Acacia nilotica* (Mimosaceae). *Entomophaga* **39**, 335–340.

Marohasy J (1995) Prospects for the biological control of prickly acacia, *Acacia nilotica* (L.) Willd. ex Del. (Mimosaceae) in Australia. *Plant Protection Quarterly* **10**, 24–31.

Maslin BR (2001) Introduction to *Acacia*. *Flora of Australia* **11A**, 3–13.

Miller EN (1996a) A property level economic assessment of prickly acacia (*Acacia nilotica*) on the Mitchell grass downs of Queensland. In *Proceedings of the 11th Australian Weeds Conference*. 30 September–3 October 1996, University of Melbourne, Melbourne. (Ed RCH Shepherd) pp. 227–230. Weed Science Society of Victoria, Melbourne.

Miller MF (1996b) *Acacia* seed predation by bruchids in an African savanna ecosystem. *Journal of Applied Ecology* **33**, 1137–1144.

Mohyuddin AI (1981) Phytophages associated with *Acacia nilotica* in Pakistan and possibilities of their introduction into Australia. In *Proceedings of the 5th International Symposium on Biological Control of Weeds*. 22–29 July 1980, Brisbane. (Ed ES Delfosse) pp. 161–166. CSIRO, Melbourne.

Mohyuddin AI (1986) *Investigations on the Natural Enemies of* Acacia nilotica *in Pakistan*. Commonwealth Institute of Biological Control, Rawalpindi.

Orchard AE and Maslin BR (2003) (1584) Proposal to conserve the name *Acacia* (Leguminosae: Mimosoideae) with a conserved type. *Taxon* **52**, 362–363.

Orchard AE and Wilson AJG (Eds) (2001a) *Flora of Australia. Vol. 11A, Mimosaceae, Acacia. Part 1*. ABRS/CSIRO Publishing, Melbourne.

Orchard AE and Wilson AJG (Eds) (2001b) *Flora of Australia. Vol. 11B, Mimosaceae, Acacia. Part 2*. ABRS/CSIRO Publishing, Melbourne.

Palmer WA (1996) Biological control of prickly acacia in Australia. In *Proceedings of the 11th Australian Weeds Conference*. 30 September–3 October 1996, University of Melbourne, Melbourne. (Ed RCH Shepherd) pp. 239–241. Weed Science Society of Victoria, Melbourne.

Palmer WA and McLennan A (2006) The host range of *Isturgia deerraria*, an insect considered for the biological control of *Acacia nilotica* in Australia. *African Entomology* **14**, 141–145.

Palmer WA and Senaratne KADW (2007) The host range and biology of *Cometaster pyrula*: a biocontrol agent for *Acacia nilotica* subsp. *indica* in Australia. *Biocontrol* **52**, 129–143.

Palmer WA and Witt ABR (2006) On the host range and biology of *Acizzia melanocephala* (Hemiptera: Psyllidae), an insect rejected for the biological control of *Acacia nilotica* subsp. *indica* (Mimosaceae) in Australia. *African Entomology* **14**, 387–390.

Palmer WA, Vitelli MP and Donnelly GP (2005) The phytophagous insect fauna associated with *Acacia nilotica* ssp. *indica* (Mimosaceae) in Australia. *Australian Entomologist* **32**, 173–180.

Palmer WA, Lockett CJ, Senaratne KADW and McLennan A (2007) The introduction and release of *Chiasmia inconspicua* and *C. assimilis* (Lepidoptera: Geometridae) for the biological control of *Acacia nilotica* in Australia. *Biological Control* **41**, 368–378.

Radford I, Nicholas DM and Brown JR (2001) Assessment of the biological control impact of seed predators on the invasive shrub *Acacia nilotica* (prickly acacia) in Australia. *Biological Control* **20**, 261–268.

Robertson IAD (1987) *A Preliminary Survey of the Natural Enemies of* Acacia nilotica *in Kenya*. CAB International Institute of Biological Control, Nairobi.

Ross JH (1979) A conspectus of the African acacia species. *Memoirs of the Botanical Survey of South Africa* **44**, 1–155.

Seigler DS and Ebinger JE (2006) New combinations in the genus *Vachellia* (Fabaceae: Mimosoideae) from the New World. *Phytologia* **87**, 139–178.

Senaratne KADW, Palmer WA and Sutherst RW (2006) Use of CLIMEX modelling to identify prospective areas for exploration to find new biological control agents for prickly acacia. *Australian Journal of Entomology* **45**, 298–302.

Spies P and March N (Eds) (2004) *Prickly Acacia National Case Studies Manual: Approaches to the Management of Prickly Acacia* (Acacia nilotica *subsp.* indica) *in Australia*. Qld Dept of Natural Resources, Mines and Energy, Brisbane.

Stals R (1997) *A Survey of Phytophagous Organisms associated with* Acacia nilotica *in South Africa*. ARC-Plant Protection Research Institute, Pretoria.

Stals R, Grobbelaar E, Neser S and Palmer WA (2004) The value of using taxonomists to survey for potential biological control agents of weeds. In *Proceedings of the 11th International Symposium on Biological Control of Weeds*. 27 April–2 May 2003, Canberra. (Eds JM

Cullen, DT Briese, DJ Kriticos, WM Lonsdale, L Morin and JK Scott) p. 64. CSIRO Entomology, Canberra.

Sutherst RW, Maywald GF and Kriticos D (2007) *CLIMEX Version 3 Users' Guide*. Hearne Scientific Software Pty Ltd, Melbourne.

Wardill TJ, Scott KD, Graham GC and Zalucki MP (2004) Isolation and characterization of microsatellite loci from *Acacia nilotica* ssp. *indica* (Mimosaceae). *Molecular Ecology Notes* **4**, 361–363.

Wardill TJ, Graham GC, Zalucki MP, Palmer WA, Playford J and Scott KD (2005) The importance of species identity in the biocontrol process: identifying the subspecies of *Acacia nilotica* (Leguminosae: Mimosoideae) by genetic distance and their implications for biological control. *Journal of Biogeography* **32**, 2145–2159.

Willson BW (1985) The biological control of *Acacia nilotica indica* in Australia. In *Proceedings of the 6th International Symposium on Biological Control of Weeds*. 19–25 August 1984, Vancouver. (Ed ES Delfosse) pp. 849–853. Agriculture Canada, Ottawa.

Ageratina adenophora (Spreng.) King & Robinson – crofton weed

Rachel Cruttwell McFadyen

ABSTRACT

Crofton weed, *Ageratina (Eupatorium) adenophora*, native to Mexico, is a perennial weed of pastures in coastal parts of Qld and NSW from Gympie to south of Sydney. A tephritid gall fly *Procecidochares utilis* from Mexico had given good control in Hawaii and was introduced into Qld in 1952. It established and spread, but its impact was greatly reduced by native parasitoids. A leaf-spot disease *Phaeoramularia eupatorii-odorati* (previously misidentified as *Cercospora eupatorii*), accidentally introduced at the same time, has had a significant impact on the weed. Crofton weed is no longer invasive into pastures and is now regarded as a minor weed in Australia.

Key words: *Eupatorium*, *Procecidochares utilis*, cercospora leaf-spot.

INTRODUCTION

Crofton weed *Ageratina (Eupatorium) adenophora* (Spreng.) King & Robinson is a shrubby perennial with many stems 1–2 m tall from a shallow rootstock (Fig. 1). It reproduces by seed, producing numerous small black achenes which are wind-dispersed on a short pappus of hairs. Masses of white flowers are produced in spring. The plant is unpalatable to stock and poisonous to horses. It is not very invasive in well-managed pastures, but invades bare ground readily then persists for many years, especially on steep slopes where machinery cannot be used. The plant is native to Central America and was introduced to Australia as an ornamental in the late 1800s. It was first recorded as naturalised near Sydney in 1904 and has since spread north along the east coast to the Mary River in Qld, with a few records from further north (Fig. 2). It is also naturalised in small patches around Adelaide and along the Murray River in SA (Parsons and Cuthbertson 2001). From the 1930s to the 1950s, crofton weed was highly invasive in pastures and small crops in the high-rainfall, high-fertility soils of the Border Ranges in Qld and NSW (Dodd 1961).

Crofton weed is a serious weed in many subtropical and high-altitude (up to 2000 m) tropical zones in the Old World, including Hawaii, West Africa, South and South-east Asia including southern China (Muniappan *et al.* 2009).

BIOLOGICAL CONTROL HISTORY

The first biological control program was undertaken by the Hawaiian Government, with surveys in 1944–45 by NLH Krauss in Mexico where the plant is native (Dodd 1961). Two insects were released on Maui Island in 1945 and 1955 respectively, the stem-galling tephritid *Procecidochares utilis* Stone and the flower-feeding tephritid *Xanthaciura connexionis* Benjamin (Julien and Griffiths 1998). *Xanthaciura connexionis* did not establish, but *P. utilis* established rapidly and gave good control in the drier areas of Maui.

Figure 1: Crofton weed growing on roadside in south-east Queensland. Photo: R McFadyen.

As a result of this success, the gall fly was progressively introduced to countries where crofton weed is a problem – Australia, New Zealand, India, South Africa and China – and successfully established in all. Control was, however, always limited by the impacts of native parasitoids (Julien and Griffiths 1998).

Further surveys in Mexico were undertaken by the South African Government in 2009 and 2010, and several insects have been imported into quarantine in Pretoria for detailed host-testing (F Heystek, *pers. comm.* 2010).

Agents released

***Procecidochares utilis* Stone (Diptera: Tephritidae).** This gall fly was the only agent deliberately released in Australia. In 1951 AP Dodd, employed by the Queensland Government, visited Hawaii, where the gall fly had spread widely since its release in 1945. Dodd undertook detailed host-tests on the fly and, as no other plants were attacked, in February 1952 a shipment of 2455 newly emerged adult gall flies was air-freighted from Hawaii to Qld (Dodd 1961). At the time, there was no requirement for a mandatory generation in quarantine and 1282 flies were directly released onto crofton weed infestations at four sites in the Border Ranges area of southern Qld. The remaining flies were kept to establish a breeding colony at the Brisbane laboratory.

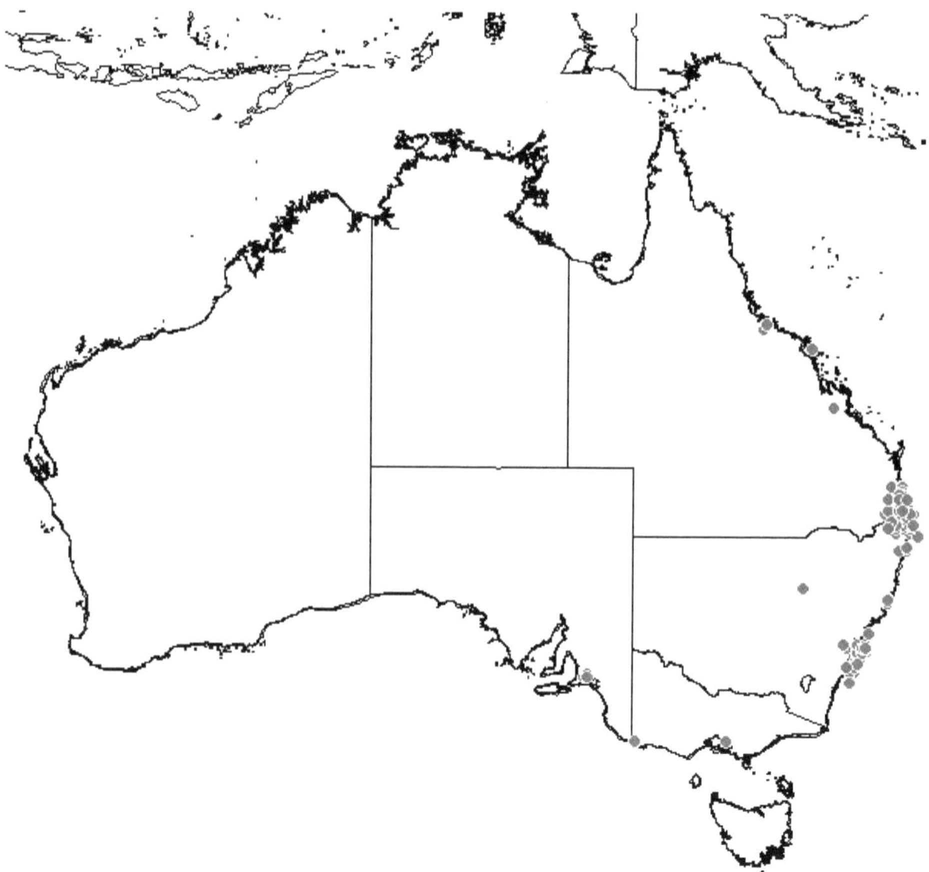

Figure 2: Distribution of crofton weed in Australia. Specimen data reproduced from Australia's Virtual Herbarium with permission of the Council of Heads of Australasian Herbaria Inc.

Adults are 3–4 mm long and strikingly patterned in black and grey. They live for about two weeks and are active during the day. Mating occurs on the host plant and the white narrow oval eggs are laid in twos or small groups between the paired bud leaves in stem tips. Up to 70 eggs can be laid by each female. Eggs hatch in about five days; larval feeding within the tip induces gall formation by the plant. Each gall may contain two to four larvae, and in high numbers galls can coalesce (Fig. 3). Pupation occurs within the gall after about 40 days and the pupal stage lasts about 20 days, or more at lower temperatures. Adults emerge via a clear window in the side of the gall (Dodd 1961; Haseler 1963).

Initial results with the gall fly were impressive, with immediate establishment and rapid spread. In the first two years, 'every plant supported scores of galls' and 'many stems died and the density of the plant's growth was greatly reduced' (Dodd 1961). However, attack by a suite of native hymenopterous parasitoids occurred rapidly; within two years of the initial release gall numbers fell. They never recovered to the initial high levels (Dodd 1961). This has remained the situation – galls are common and easily found throughout the Australian distribution of crofton weed, but parasitism remains high and gall numbers never exceed a few galls per plant, or one per stem. At these levels, the impact on the plant is minor. Stem dieback does not occur, nor is flowering prevented.

Phaeoramularia eupatorii-odorati **(Yen) Liu and Guo** (previously referred to as *Cercospora eupatorii* Peck). This leaf-spot fungus was never deliberately released, but was noticed in crofton weed in Qld in 1954 close to a site where adult gall flies from Hawaii had been released early in 1952 (Dodd 1961). The fungus, which is native to the Americas, was not present prior to 1952 but by 1957 was widespread in all infestations of crofton weed in Qld. In 1956 its presence was confirmed in Hawaii, and it is highly likely that it was introduced into Australia through spores carried on the bodies of the original flies imported from Hawaii (Dodd 1961). This leaf-spot fungus is highly host-specific, with *A. adenophora* the only known host. It is now widespread on crofton weed in Australia and elsewhere (Muniappan *et al.* 2009; Zhang *et al.* 2008).

The fungus causes small, brown spot lesions on the leaves, followed by general yellowing of the leaves and premature leaf-fall, leaving stems leafless except for the three or four youngest leaves. The initial attack in the 1950s was very severe, with widespread death of stems and seedling plants (Dodd 1961). Since then, the leaf-spot has been present in all crofton weed infestations and it continues to damage plants.

Figure 3: Crofton weed heavily galled by *Procecidochares utilis*. Photo: Qld Government.

DISCUSSION

No proper evaluation of the impact of the biocontrol agents has been undertaken, at the time of the introductions or since. No estimate could be made of the benefit:cost ratio of the program; however, it was believed to be positive (Page and Lacey 2006). It is clear that crofton weed is a much less invasive plant than it was in the 1940s, and the impact of the biocontrol agents is the likeliest cause. There have been significant changes in land use but much of the affected country is still used for cattle or horse pasture with very little weed control, yet crofton weed is no longer a problem in these pastures. For example, it is now impossible to find areas with 'a few hundred acres of dense crofton weed' as described by Dodd for the release sites in northern NSW (Dodd 1961), nor are patches of dense seedlings observed even in disturbed ground. Crofton weed is no longer considered a major weed in Australia; for example, in 1998 it was not listed as one of the 71 nationally worst weeds (Thorp and Lynch 2000) and there is no pressure for further biocontrol efforts. The biocontrol program must therefore be considered a success, even though the major impact probably came from the unintentionally introduced leaf-spot fungus rather than from the gall fly.

Crofton weed is still a major weed in southern China and in South Africa, where there is interest in additional biocontrol including the introduction of new agents (Zhang *et al.* 2008). If any of the potential new agents found by South African scientists in Mexico lead to improved control in these countries, importation into Australia might be considered.

ACKNOWLEDGEMENTS

My thanks to John Hosking, who supplied the map for Figure 2.

REFERENCES

Dodd AP (1961) Biological control of *Eupatorium adenophorum* in Queensland. *Australian Journal of Science* **23**, 356–366.

Haseler WH (1963) Biological studies of the crofton weed gall fly *Procecidochares utilis* Stone. MSc thesis. University of Queensland.

Julien MH and Griffiths MW (1998) *Biological Control of Weeds: A World Catalogue of Agents and their Target Weeds.* 4th edn. CABI, Wallingford, UK.

Muniappan R, Raman A and Reddy GVP (2009) *Ageratina adenophora* (Sprengel) King & Robinson (Asteraceae). In *Biological Control of Tropical Weeds using Arthropods.* (Eds R Muniappan, GVP Reddy and A Raman) pp. 63–73. Cambridge University Press, Cambridge.

Page AR and Lacey KL (2006) *Economic Impact Assessment of Australian Weed Biological Control: Technical Series 10.* CRC for Australian Weed Management, Adelaide.

Parsons WT and Cuthbertson EG (2001) *Noxious Weeds of Australia.* 2nd edn. CSIRO Publishing, Melbourne.

Thorp JR and Lynch R (2000) *The Determination of Weeds of National Significance.* National Weeds Strategy Executive Committee, Launceston.

Zhang F, Liu W-X, Wan F-H and Ellison CA (2008) Sustainable management based on biological control and ecological restoration of an alien invasive weed, *Ageratina adenophora* (Asteraceae) in China. In *Proceedings of the 12th International Symposium on Biological Control of Weeds.* 22–27 April 2007, La Grande Motte, France. (Eds MH Julien, R Sforza, MC Bon, HC Evans, PE Hatcher, HL Hinz and BG Rector) pp. 699–703. CABI, Wallingford, UK.

Ageratina riparia (Regel) K. & R. – mistflower

Shon Schooler, Bill Palmer and Louise Morin

ABSTRACT

Mistflower, *Ageratina riparia*, is an environmental weed in Australia, New Zealand, South Africa and Hawaii. The plant originates from Mexico and Central America. The State Department of Hawaii began the first surveys for biological control agents of mistflower in the 1950s in Mexico, and the surveys continued through the 1970s. These efforts resulted in the release in Hawaii of three agents – the moth *Hellinsia beneficus*, the gall fly *Procecidochares alani* and the fungus *Entyloma ageratinae* – which resulted in successful control in many areas. New Zealand later released the fungus and gall fly, which resulted in successful control. In 1985, the Qld Department of Lands brought the two insect species from Hawaii into quarantine to test host-specificity. The moth was not successfully reared in quarantine and was therefore not tested. Host-specificity testing of the gall fly indicated that oviposition did not occur on any of the 50 native or agricultural species tested; the insect was released in Australia in 1985 and on Norfolk Island in 1988. The gall fly has become widely established in south-eastern Qld but is not considered effective in reducing mistflower populations, possibly due to parasitism from several native wasp species. The mistflower fungus was not tested in Australia because in 1985 suitable quarantine facilities were not available for use by the Qld Department of Lands. Surprisingly, the fungus was found in south-eastern Qld in October 2010, near Lamington National Park. Its method of arrival is unknown, but it was probably an accidental introduction on hiking gear that had previously been used in northern New Zealand. Studies are underway to examine the distribution, spread and potential risk of the fungus to native Australian plants. Given the observations of its host-specificity and effectiveness in Hawaii and New Zealand, it is expected to benefit native plant diversity in Australia.

Key words: invasive riparian plant, rainforest, stream, predation, parasitism, *Eupatorium riparium*, *Oidaematophorus beneficus*.

INTRODUCTION

Mistflower, *Ageratina riparia* (Regel) K. & R. (= *Eupatorium riparium* Regel) Asteraceae, also known as hamakua pa-makani in Hawaii and as mistweed in parts of Australia, is native to Mexico and Central America. It is a stoloniferous perennial herb that invades wet habitats, particularly riparian areas and moist cliff faces (Figs 1, 2). In subtropical habitats, it grows through most of the year, flowering in late winter. Each seed has a pappus which, along with its small seed size, allows the plant to disperse into upstream and upslope habitats on prevailing winds. Part of its Hawaiian name, pa-makani, means 'windblown' (Neal 1965). In late spring, the flowering stalks die back to a stolon, particularly during dry periods. New shoots begin to grow in early summer.

Mistflower is considered a serious environmental weed in Hawaii, New Zealand, South Africa and Aus-

Figure 1: Illustration of mistflower. a) Entire plant with stoloniferous growth habit. b) Stem with gall made by the gall fly *Procecidochares alani*, showing emergence holes. c) Flowers. d) Seed with pappus. Drawing: S Wragg. © CSIRO.

tralia. It is primarily a problem in mid- to high-elevation rainforest areas where it creates a canopy over headwater streams and displaces native riparian plant species (Barton *et al.* 2007). It is also a problem in wet meadows, where it reduces forage quality for livestock (Trujillo 2005). Mistflower has no commercial feed value, although it is occasionally fed on by native red-necked pademelons, *Thylogale thetis* (Macropodidae), in Australia (Zancola *et al.* 2000).

Mistflower was introduced to NSW in 1875 as an ornamental garden plant and was first recorded in Queensland in the 1930s. It is currently present in eastern Australia from Wollongong to Cairns and in Western Australia south of Perth (Fig. 3). Unlike some other environmental weeds in Australia, mistflower has the ability to spread along riparian corridors into pristine catchment headwaters and negatively affect biotic diversity in ecologically important areas such as the World Heritage listed Lamington National Park and Border Ranges National Park. Mistflower adversely affects a number of threatened and endangered species in NSW including the plants *Brachyscome ascendens* (Asteraceae), *Daphnandra* sp. *C Illawarra* (Atherospermataceae), *Doryanthes palmeri* (Doryanthaceae), *Euphrasia bella* (Scrophulariaceae), *Irenepharsus trypherus* (Brassicaceae), *Rapanea* sp. (Myrsinaceae) and *Sarcochilus hartmannii* (Orchidaceae), and a native bird species, *Dasyornis brachypterus* (Pardatotidae) (Coutts-Smith and Downey 2006). Several native plants in Qld have been observed to be negatively affected by mistflower, including *Gaultheria* sp. Merino (Ericaceae), *Cyperus semifertilis* (Cyperaceae), *Isoglossa eranthemoides* (Acanthaceae), *Ozothamnus vagans* (Asteraceae), *Plectranthus*

Figure 2: a) Mistflower inflorescence. b) Mistflower invasion of stream and slope habitats at Springbrook National Park in Qld, 2005. Photos: S Schooler, CSIRO.

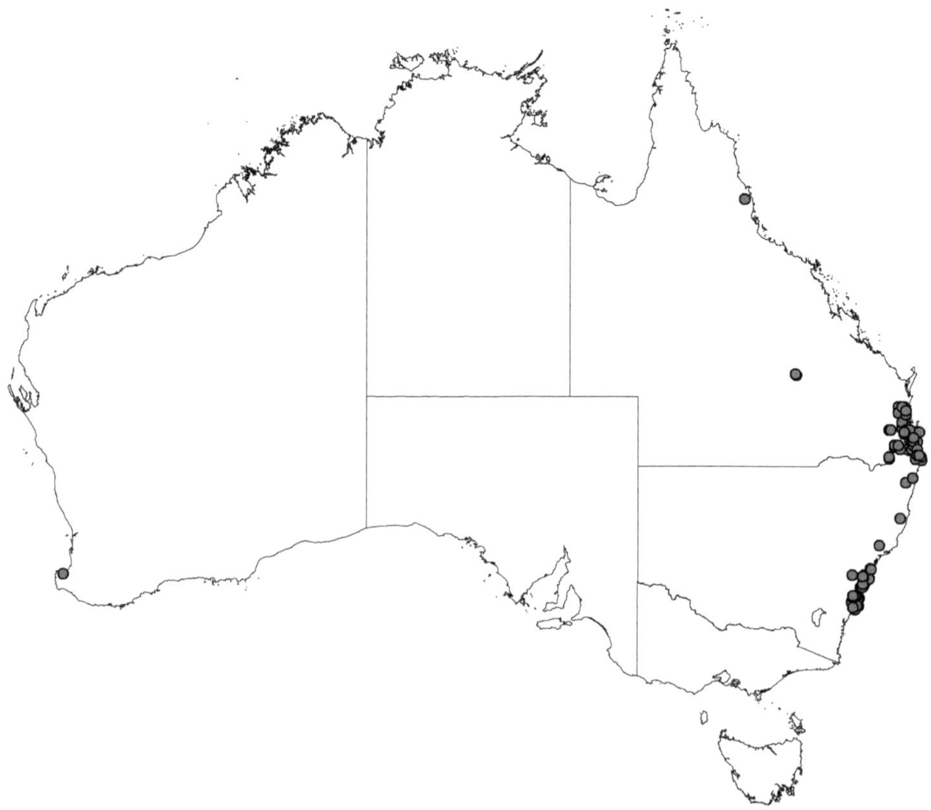

Figure 3: Distribution of *Ageratina riparia* in Australia, based on herbarium records extracted from the Australian Virtual Herbarium. Mistflower is also present on Norfolk Island (not shown). Specimen data reproduced from Australia's Virtual Herbarium with permission of the Council of Heads of Australasian Herbaria Inc.

nitidus (Lamiaceae), *Sarcochilus fitzgeraldii* (Orchidaceae) and *Wahlenbergia scopulicola* (Campanulaceae) (B Macdonald, Qld Herbarium, *pers. comm.* 2009).

Current management operations for mistflower are restricted to hand-pulling and the application of herbicides. Physical removal is labour-intensive and the weed often infests areas that are difficult to access (headwater riparian corridors and cliff faces). Herbicide application is increasingly expensive, often destroys desirable vegetation along with the weed (creating disturbed habitat for weed seed germination) and is known to harm fish and amphibian populations (Relyea 2005). Therefore, biological control has been considered the only viable option for control of mistflower in Australia.

BIOLOGICAL CONTROL HISTORY

Mistflower was introduced to Hawaii in about 1925 (Neal 1965). By 1973 it occupied over 62 500 ha of rangeland (Trujillo 1985). In 1973–75 the Hawaii Department of Agriculture and the University of Hawaii introduced three classical biological control agents – the plume moth *Hellinsia beneficus* (Yano and Heppner), formerly known as *Oidaematophorus beneficus*, (Lepidoptera: Pterophoridae) whose larvae feed on leaves, the gall fly *Procecidochares alani* Steyskal (Diptera: Tephritidae) that oviposits in the growing tips of plants, eventually leading to stem galls produced by the developing larvae (Fig. 4), and the white smut fungus *Entyloma ageratinae* sp. nov. Barreto and Evans (Ustilagnales), originally identified as a *Cercosporella* sp., that causes leaf damage (Fig. 5) leading to extensive defoliation (Trujillo 1985; Julien and Griffiths 1998). The program was successful, with the fungus being the most effective agent. It achieved successful control of the plant in wet areas within eight months and in dry areas within three to eight years (Morin *et al.* 1997; Trujillo 2005). The fungus was released in South Africa in 1989 and was observed to have spread and defoliated plants the next year (Morris 1991). In South Africa, *Entyloma ageratinae* is considered to have prevented mistflower from emerging as a weed of any concern (AR Wood, South African Agricultural Research Council, *pers. comm.* 2010).

Mistflower was recognised as a potential problem in New Zealand in 1933. The weed has become a serious

Figure 4: Mistflower gall interior showing multiple pupal cases of the gall fly, *Procecidochares alani*. Photo: S Schooler, CSIRO.

problem in forest margins, streams, riparian zones and pastures, where it forms dense mats of semi-woody stems that smother and displace native vegetation and inhibit regeneration. The fungus and gall fly were chosen as biocontrol agents due to their effectiveness in Hawaii (Fröhlich *et al.* 2000), tested for host-specificity and released in New Zealand in 1998 and 2001 respectively (Barton *et al.* 2007). The fungus spread readily and caused major epidemics among mistflower populations, leading to dramatic declines in plant abundance. The gall fly has been slow to disperse, has caused limited damage to plants and has thus probably not provided much control (Barton *et al.* 2007). The decline of mistflower populations has been associated with an increase in diversity of riparian plant communities (Barton *et al.* 2007).

Mistflower is a problem in northern NSW and south-eastern Qld. In 1983, the Qld Department of Lands obtained approval from the Australian Weeds Committee for mistflower to become a target for biological control. Host-specificity testing of agents resulted in the release of the gall fly in Queensland in 1986 and on Norfolk Island in 1988.

Figure 5: Leaf damage caused by the the white smut fungus *Entyloma ageratinae*. Photo: S Schooler, CSIRO.

EXPLORATION

Exploration in the native range was undertaken as part of the Hawaiian biological control program for mistflower (see Morin *et al.* 1997 for review). Hawaii sent an entomologist (most probably N Krauss) to explore Mexico and tropical America for candidate agents for mistflower in 1965; 31 insect species were reported (C Davis, State of Hawaii, unpub. report, 1966). Two weevils and two lepidopterans were selected for further study. A second team, comprising the plant pathologist E Trujillo and the entomologist W Rose, returned to Mexico in 1972–73 to survey further for potential agents. They collected and shipped two promising insect agents to Hawaii. In 1974, E Trujillo made an exploration trip to Jamaica, where a defoliating fungus had been reported on mistflower. Plant material infected by the fungus was shipped to Hawaii for host-specificity testing.

CANDIDATES

Following the 1983 approval of mistflower as a target for classical biological control in Australia, efforts were made to import the gall fly (*Procecidochares alani*) and the plume moth (*Hellinsia beneficus*) into quarantine facilities at the Alan Fletcher Research Station (Brisbane) for host-specificity testing.

Gall fly, *Procecidochares alani*. The gall fly was initially collected in Veracruz, Mexico, in 1973 and studied in Hawaii before its release there (H Nakao and SH Au, Hawaiian Dept of Agriculture, unpub. report, 1974). These flies are diurnally active and have boldly patterned wings. They do not feed, live for about two weeks and oviposit into the top or auxillary buds of mistflower. The resulting larvae stimulate the development of a tip gall. Communal galls are common and can contain up to 15 larvae, producing a gall of 25 mm length and 10 mm diameter. Larvae prepare emergence holes in the cuticle of the gall shortly before pupation and about four weeks after oviposition. The flies emerge through these transparent 'windows' after a further two weeks (CH Wild, Qld Dept of Lands, unpub. report, 1986).

In 1985, two shipments of field-collected gall flies were sent from Hawaii to the Alan Fletcher Research Station and additional host-specificity testing was undertaken. Fifty plant species approved by the then

Australian Department of Primary Industry as appropriate to evaluate the specificity of this agent were used in multiple-choice tests of oviposition. The tests indicated that the gall fly was restricted to mistflower; no oviposition or larval development was observed on any other plant (CH Wild, Qld Dept of Lands, unpub. report, 1986) (Table 1).

The gall fly was released in Queensland in 1986. During the next two years, over 18 500 flies were released at 25 sites in south-eastern Qld (Table 2). Surveys in 1987 documented establishment at 21 of the sites (84%) (G Diatloff, Qld Dept of Lands, unpub. report, 1987). In 1988, 400 flies were released on Norfolk Island.

The gall fly spread and established widely in south-eastern Qld but is not considered effective in reducing mistflower abundance, possibly because it is heavily parasitised by native parasitoids. From October 2009 to January 2010, 17 sites were surveyed within the centre of the mistflower distribution in Australia (north-eastern NSW and south-eastern Qld) (S Schooler, unpub. data). Mistflower galls were present at all the surveyed sites. The mean number of galls across sites was 0.37 per stem (SD = 0.20) and the mean percent parasitism rate was 54% (SD = 30%). Hymenopteran wasps from four families (Torymidae, Braconidae, Eupelmidae and Pteromalidae) emerged from galls. The most abundant parasitoid was a *Megastigmus* sp. (Torymidae), followed by a *Torymoides* sp. (Torymidae). Many galls were found to have been scraped or chewed (16%, SD = 14%), exposing the interior chamber, indicating that predators may have identified the larval or pupal gall flies as a source of protein.

Plume moth, *Hellinsia beneficus*. The plume moth was collected in Contreras, Mexico, in 1973. It was imported into Hawaii and tested for host-specificity, with related studies on biology, before its release there the same year. Female moths oviposit minute, elliptical translucent eggs singly onto the lower surface of young leaves and stem terminals. Eggs eclose in five to seven days with the first instar feeding on the underside of the leaf on which it was oviposited, leaving the upper epidermis of the leaf intact. Larvae complete development in 30–35 days and move first to the young growth of the terminal stem tips, then later to more mature leaves. The prepupae attach by silken threads to dried leaves or other litter at the base of the plant; adult moths emerge 10 days after pupation (H Nakao *et al.*, Hawaiian Dept of Agriculture, unpub. report, 1973).

Several shipments, mainly of eggs and neonate larvae, of the plume moth were made from Hawaii to the Alan Fletcher Research Station between 1985 and 1988. On each occasion, larvae were transferred to mistflower plants and they completed their development. However, on each occasion the resulting moths did not oviposit. It was thought that the moths did not mate satisfactorily in the laboratory. A laboratory colony could not be established in quarantine and efforts on this insect were eventually discontinued. One reason for the lack of enthusiasm in further exploring this insect was the knowledge that it had not been very effective in Hawaii (Morin *et al.* 1997).

White smut fungus, *Entyloma ageratinae*. Although reported to be extremely damaging to mistflower in Hawaii (Trujillo 1985), the white smut fungus was not seriously considered for introduction to Australia in the mid 1980s because quarantine facilities at the Qld Department of Lands were unsuitable for the testing of pathogens. Nonetheless, the Qld Department of Lands commissioned a study to clarify the taxonomy of the fungus (Barreto and Evans 1988), in preparation for possible future introductions. The fungus produces fruiting bodies that emerge through stomata on the lower surface of leaves seven to 10 days after infection. Copious amount of wind-borne spores are produced, infection sites become necrotic and lesions appear on the upper leaf surface. Leaves eventually die, and fall off once lesions have coalesced. Additional host-specificity tests conducted prior to the release of the fungus in New Zealand in 1998 (Fröhlich *et al.* 2000) provided further evidence of the high host-specificity of the fungus, leading to renewed interest in considering it for introduction in Australia. In light of previous host-specificity testing, it is anticipated that limited supplementary testing would have been required to obtain permission to release the fungus in Australia. However, in October 2010, prior to acquiring funding to complete the testing, the fungus was found near Lamington National Park in south-eastern Qld. There was no indication of its arrival method, but given its proximity to the park it is likely to have been accidentally introduced via hiking gear that was previously used in northern New Zealand.

DISCUSSION

Mistflower's ability to disperse, establish and reduce biotic diversity in relatively undisturbed areas that are ecologically important is one of the primary reasons it should be actively managed. However, there are currently no effective methods to control mistflower in Australia. The difficulty in accessing sites and the ecological sensitivity of many of the habitats invaded by mistflower make

Table 1: Host-specificity testing results for gall fly *Procecidochares alani* in Qld

Plant species	No of replications	Mean no. of galls/plant
Anacardiaceae		
Mangifera indica L.	2	0
Annonaceae		
Annona reticulata L.	2	0
Apiaceae		
Apium graveolens L.	2	0
Daucus carota L.	2	0
Asteraceae		
Ageratina adenophora (Spreng.) R.M.King & H.Rob.	2	0
A. riparia	24	48.3±3.2
Carthamus tinctorius L.	2	0
Chrysanthemum cinerariifolium (Trevir.) Vis.	2	0
Dahlia sp.	2	0
Eclipta prostrate (L.) L.	2	0
Helianthus annuus L.	2	0
Helichrysum ramosissimum Hook.	2	0
Lactuca sativa L.	2	0
Brassicaceae		
Brassica oleracea L.	2	0
B. rapa L.	1	0
Bromeliaceae		
Ananas comosus (L.) Merr.	2	0
Caricaceae		
Carica papaya L.	2	0
Chenopodiaceae		
Beta vulgaris L.	2	0
Convolvulaceae		
Ipomoea batatas (L.) Lam.	2	0
Cucurbitaceae		
Cucurbita maxima Duch.	2	0
Fabaceae		
Arachis hypogaea L.	2	0
Centrosema pubescens Bentham	2	0
Glycine max (L.) Merr.	2	0
Macroptilium atropurpureum (DC.) Urb.	2	0
Phaseolus vulgaris L.	2	0
Lauraceae		
Persea americana Mill.	2	0
Liliaceae		
Linum usitatissimum L.	2	0

Plant species	No of replications	Mean no. of galls/plant
Malvaceae		
Gossypium sp.	2	0
Mimosaceae		
Acacia concurrens Pedley		
Musaceae		
Musa sapientum L.	2	0
Myrtaceae		
Eucalyptus curtisii Blakely & C.T.White	2	0
Passifloraceae		
Passiflora edulis Sims	2	0
Pinaceae		
Pinus elliotti Engelm.	2	0
Poaceae		
Cenchrus ciliaris L.	2	0
Saccharum officinarum L.	2	0
Sorghum vulgare Pers.	2	0
Triticum sp.	2	0
Zea mays L.	2	0
Proteaceae		
Macadamia integrifolia Maiden & Betche	2	0
Rosaceae		
Fragaria vesca L.	2	0
Malus sylvestris Mill.	2	0
Prunus persica (L.) Batsch	2	0
Rosa sp.	2	0
Rutaceae		
Citrus sinensis (L.) Osbeck	2	0
Solanaceae		
Lycopersicon lycopersicum (L.) H.Karst. ex Farw.	2	0
Nicotiana tabacum L.	2	0
Solanum tuberosum L.	2	0
Theaceae		
Camellia sinensis (L.) Kuntze	2	0
Vitaceae		
Vitis vinifera L.	2	0
Zingiberaceae		
Zingiber officinale Roscoe	2	0

In these tests the fly was exposed to the above plant species in a multiple choice design. (C Wild, Department of Lands, unpublished report, 1986).

Table 2: Initial release locations and number of gall fly ***Procecidochares alani*** released in Qld in 1986 and persistence of those populations in 1987

Release site	No. released	Persisted 1987
Mapleton Falls	500	yes
Maleny	500	yes
Kureelpa Falls	500	yes
Palmwoods	500	yes
Kondalilla Falls	500	yes
Bellthorpe	500	yes
Cunninghams Gap	500	yes
Peachester	500	yes
Upper Ormeau	500	yes
Running Creek	500	yes
Mt Tenison Wood	500	yes
Forestry Rd Highvale	500	yes
Repeater Station Rd	500	no
Natural Arch	1500	yes
O'Reillys	500	yes
Beechmont	500	yes
Binna Burra	500	yes
Burnett Creek Rd	500	no
Teviot Creek Rd	500	no
Mt Tamborine	1000	yes
Lost World	500	yes
Upper Tallebudgera Creek	500	yes
Mt Mee State Forest (1)	500	yes
Mt Mee State Forest (2)	1000	yes
Purling Brook Falls	1000	no

physical and chemical management methods untenable. Although the gall fly *Procecidochares alani* has successfully established and spread widely, it is not damaging mistflower enough to reduce its competitive ability and thus facilitate the recovery of native plant communities. Parasitism from several native wasp species is partially responsible for this lack of control. However, damage from the gall fly may be reducing biomass and seed abundance to some extent, and therefore limiting the potential impacts and spread capability of mistflower. These effects should be measured in future research.

Evidence from Hawaii and New Zealand suggests that the white smut fungus *Entyloma ageratinae* would significantly reduce the abundance of mistflower and facilitate the return of a diverse native plant community. As the fungus has appeared in Australia, studies are planned to retrospectively test its host-specificity to assess risk to native plants, measure the reduction in mistflower abundance after exposure to the fungus and examine the expected benefits to native plant communities.

Further work on the plume moth *Hellinsia beneficus* might also be considered. Although all attempts to rear it in Australian quarantine facilities failed, it was successfully reared in the laboratory in Hawaii where host-specificity testing was undertaken. Further attempts may be made in the new quarantine facilities of the Ecosciences Precinct in Brisbane. Limited host-testing would be needed before the insect could be cleared for release in Australia.

ACKNOWLEDGEMENTS

The cooperation of the State Department of Hawaii in providing the biological control agents and advice on their rearing is acknowledged. Simon Benateau assisted with field and laboratory work examining the parasitism of mistflower gall flies. Many thanks to J La Salle of the Australian National Insect Collection for his rapid identification of the parasitoid wasps. Thanks to R McFadyen for spotting the mistflower fungus.

REFERENCES

Barreto RW and Evans HC (1988) Taxonomy of a fungus introduced into Hawaii for biological control of *Ageratina riparia* (Eupatorieae; Compositae), with observations on related weed pathogens. *Transactions of the British Mycological Society* **91**, 81–97.

Barton J, Fowler SV, Gianotti AF, Winks CJ, de Beurs M, Arnold GC and Forrester G (2007) Successful biological control of mistflower (*Ageratina riparia*) in New Zealand: agent establishment, impact and benefits to the native flora. *Biological Control* **40**, 370–385.

Coutts-Smith AJ and Downey PO (2006) *Impact of Weeds on Threatened Biodiversity in New South Wales*. Technical Series no. 11. CRC for Australian Weed Management, Adelaide.

Fröhlich J, Fowler SV, Gianotti A, Hill RL, Killgore E, Morin L, Sugiyama L and Winks CJ (2000) Biological control of mistflower (*Ageratina riparia*, Asteraceae): transferring a successful program from Hawaii to New Zealand. In *Proceedings of the 10th International Symposium on Biological Control of Weeds*. 4–14 July 1999,

Bozeman, Montana. (Ed. NR Spencer) pp. 51–57. Montana State University, Bozeman, USA.

Julien MH and Griffiths MW (1998) *Biological Control of Weeds: A World Catalogue of Agents and their Target Weeds*. 4th edn. CABI, Wallingford, UK.

Morin L, Hill RL and Matayoshi S (1997) Hawaii's successful biological control strategy for mist flower (*Ageratina riparia*): can it be transferred to New Zealand? *Biocontrol News and Information* **18**, 77–87.

Morris MJ (1991) The use of plant pathogens for biological weed control in South Africa. *Agriculture, Ecosystems and Environment* **37**, 239–255.

Neal MC (1965) *In Gardens of Hawaii*. Special Publication 50. Bishop Museum Press, Honolulu.

Relyea RA (2005) The impact of insecticides and herbicides on the biodiversity and productivity of aquatic communities. *Ecological Applications* **15**, 618–627.

Trujillo EE (1985) Biological control of Hamakua PaMakani with Cercosporella sp. in Hawaii. In *Proceedings of the 6th International Symposium on Biological Control of Weeds*. 19–25 August 1984, Vancouver. (Ed. ES Delfosse) pp. 661–671. Agriculture Canada, Vancouver.

Trujillo EE (2005) History and success of plant pathogens for biological control of introduced weeds in Hawaii. *Biological Control* **33**, 113–122.

Zancola BJ, Wild C and Hero J (2000) Inhibition of *Ageratina riparia* (Asteraceae) by native Australian flora and fauna. *Austral Ecology* **25**, 563–569.

Alternanthera philoxeroides (Martius) Grisebach – alligator weed

Mic Julien, Alejandro Sosa, Richard Chan, Shon Schooler and Guadalupe Traversa

ABSTRACT

Biological control of *Alternanthera philoxeroides*, alligator weed, began when George Vogt, USDA, conducted several surveys in South America during the 1960s. Three agents were released in the USA and two of them – the flea beetle *Agasicles hygrophila* and the moth *Arcola malloi* – were released in Australia in the 1970s. The flea beetle provides excellent control of alligator weed in aquatic habitats in warm temperate areas. However, alligator weed also grows in semi-aquatic and terrestrial situations and additional control agents are needed for those habitats. New surveys were conducted in South America from 2001 to 2005. Four insects have been tested but were not sufficiently host-specific for release in Australia as they completed their life-cycles on native *Alternanthera* species. Two others – the flea beetle *Phenrica* sp. and the fly *Ophiomyia marelli* – are being tested in quarantine. Several others – the stem-mining fly *Ophiomyia buski* and leaf-mining flies *Ophiomyia alternantherae* and *Pegomya* sp. – are candidates for future testing. The fungus *Uredo pacensis*, that has only been found in Bolivia, may also have potential.

Key words: biological control, native range surveys, host-testing, *Agasicles hygrophila, Arcola malloi, Disonycha argentinensis, Amynothrips andersoni, Uredo pacensis*, potential agents.

INTRODUCTION

Alternanthera philoxeroides (Martius) Grisebach (Amaranthaceae), alligator weed (Fig. 1), is a perennial herb from South America that spread to Australia, possibly in cargo ships, but was also deliberately introduced as an ornamental plant (Julien and Bourne 1988). It has been spread within and between catchments in Australia as a substitute culinary plant for *Alternanthera sessilis* (L.) R. Br. ex DC by the Sri Lankan community (Gunasekera 1999), on earth-moving equipment, in lawn turf and by boats (van Oosterhout 2007). In its native range of Argentina it occurs as an emergent species in shallow and ephemeral water bodies, especially roadside drains and low areas of floodplains. In Australia it occurs in similar situations and is particularly prevalent through the seasonally inundated floodplains of the lower Hunter River Valley of coastal NSW and associated drains (Fig. 2). It is entrenched along many rivers of the coastal plains and urban waterways, where it invades banks and may form floating mats (Julien *et al.* 1992; Julien 1995).

Alligator weed populations in Australia appear to be sterile hybrids (Sosa *et al.* 2008) and spread by clonal growth and fragmentation. Alligator weed grows as a terrestrial (Fig. 2), semi-terrestrial and aquatic plant (Fig. 5). It is highly competitive (Julien and Bourne 1988), although its competitive ability depends on duration of inundation (Schooler *et al.* 2010). Aquatic growth may be emergent (rooted to the substrate underwater) or free-floating over deep water, though mats are usually rooted to banks. Its biology and ecology are described in Julien *et al.* (1992) and Julien (1995).

Alligator weed occurs mostly on the eastern coastal strip (Fig. 3) and it covers only an estimated 4000 ha in Australia (van Oosterhout 2007), yet it was selected as a

Figure 1: *Alternanthera philoxeroides*, alligator weed. a) In aquatic habitats it has filamentous roots. b) In terrestrial habitats it has both filamentous and tap roots. c) Papery whitish floral head. d) Individual flower. Illustrations: S Fiscke, CSIRO.

Weed of National Significance (Thorp and Lynch 2000). It was nominated because of its potential for spread (Julien *et al.* 1995) that may be increased by use of physical control methods, a lack of suitable herbicides (Sainty *et al.* 1998) and its impact on agriculture and the environment (van Oosterhout 2007).

BIOLOGICAL CONTROL HISTORY

The first surveys for biological control agents in South America were made by G Vogt, USDA, from January to May 1960 (Vogt 1960) and January to June 1961 (Vogt 1961). Vogt collected over 30 insect species and a mite and suggested that five insects were 'major biotic suppressants' of alligator weed. Three of these were later studied in Argentina and the USA and subsequently released in the USA. The leaf-eating flea beetle *Agasicles hygrophila* Selman and Vogt (Coleoptera: Chrysomelidae) was released in 1964, the tip-distorting thrips *Amynothrips andersoni* O'Neill (Thysanoptera: Phlaeothripidae) in 1967 and the stem-boring moth *Arcola malloi* (Pastrana) (Lepidoptera: Pyralidae) in 1971. Releases of the flea beetle *A. hygrophila* and the moth *A. malloi* were made in Australia in 1977 and New Zealand in 1980, and of the flea beetle in Thailand (1981) and China (1986) (Julien and Griffiths 1998). These agents controlled aquatic mats of alligator weed in the USA (Coulson 1977), Australia (Julien 1981) and China (J Ding, *pers. comm.* 2010), but only in warmer regions where population development was not limited by winter cold. Cool temperatures contributed to lack of control in New Zealand (Stewart *et al.* 1999a, b). In Australia, although the moth *A. malloi* established widely, the key agent controlling the weed was *A. hygrophila*. Climate modelling for Australia and

Figure 2: Alligator weed growing in low-lying pastures in the lower Hunter Valley, NSW. Photo: G Pritchard, Port Stephens Shire Council.

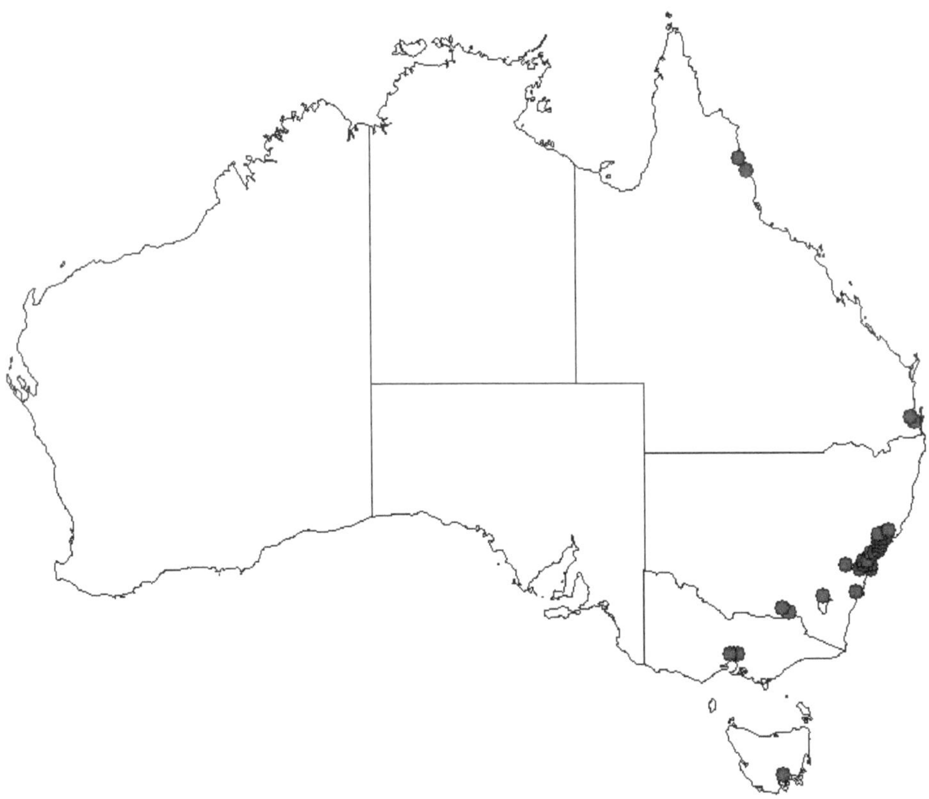

Figure 3: The distribution of alligator weed in Australia. Data source AVH. Specimen data reproduced from Australia's Virtual Herbarium with permission of the Council of Heads of Australasian Herbaria Inc.

elsewhere indicated that alligator weed had a much greater climatic tolerance than *Agasicles* (Julien *et al.* 1995).

Since the biological control agents were not able to control the weed across all habitats (semi-aquatic, seasonally flooded floodplains, terrestrial situations) nor in cooler climates (Julien and Bourne 1988; Julien and Stanley 1999; Julien *et al.* 1995), additional or better-adapted agents were required. Attempts to introduce a cold hardy biotype of *A. hygrophila* to the USA in 1979 were unsuccessful (Buckingham *et al.* 1983). A terrestrial insect, the flea beetle *Disonycha argentinensis* Jacoby (Coleoptera: Chrysomelidae), was released in Australia (Julien and Chan 1992) and New Zealand (Stewart *et al.* 1999a) but failed to establish in either country.

No systematic surveys had been conducted for plant pathogen natural enemies of alligator weed. Barreto and Torres (1999) reported the occurrence of *Nimbya alternantherae* Holcomb & Antonopoulus and *Cercospora alternantherae* Ellis & Langlois in Brazil. *C. alternantherae* failed to cause disease symptoms *in vitro.* However, *N. alternantherae*, which occurs in the USA (Holcomb and Antonopoulos 1976) and in China (Xiang *et al.* 1998), has potential as a mycoherbicide (Pomella *et al.* 2007).

PLANT TAXONOMY

Alligator weed has three morphotypes in Argentina – *A. philoxeroides* f. *philoxeroides* (Mart.) Griseb., *A. philoxeroides* f. *angustifolia* Süssenguth (Pedersen 1999) and a third, provisionally named *Santa Fe* (Sosa *et al.* 2004). In Argentina, there are small populations of fertile tetraploids and fertile hexaploids but most populations there are sterile hexaploids. The fertile forms produce viable seeds and seedlings in the field (Sosa *et al.* 2008). Cytogenetic analyses using GISH (genomic *in situ* hybridisation) indicated that the Argentine alligator weed populations sampled (including the three forms mentioned above and populations with different ploidy levels) and its close relative *Alternanthera aquatica* (D Parodi) Chodat share at least a third of their chromosomes, and are therefore likely part of a complex of hybrids (E Greizerstein *et al.*, unpub.).

Alligator weed populations in Australia comprise sterile hexaploid forms (Sosa *et al.* 2008). Current genetic analyses using ISSR indicate that a number of different genotypes exist in Australia with similarities to samples from Argentina of the forms *A. ph* f. angustifolia and *A. ph.* f. philoxeroides, suggesting multiple introductions from Argentina (D Gopurenko *et al., pers. comm.* 2010).

EXPLORATION

Exploration for new biological control agents, especially for cold-tolerant and terrestrially oriented agents, was carried out from October 2001 to April 2005. Surveys for insects were conducted in central and northern Argentina, southern Uruguay and southern Paraguay by A Sosa and H Cordo, USDA-ARS South American Biological Control Laboratory (SABCL) at Hurlingham, Buenos Aires, and occasionally with M Julien, CSIRO Entomology, Brisbane.

Surveys for plant pathogens were conducted throughout the native range of alligator weed in Argentina from December 2004 to May 2005 by G Traversa, Universidad Nacional del Sur (UNS), Bahia Blanca, Argentina, together with entomological surveys by A Sosa. A survey was also conducted in Bolivia by Traversa and Sosa in August 2007, specifically to search for the rust *Uredo pacensis.* At least two surveys per growing season were conducted in Argentina to look for this rust in the northwestern provinces during the summers of 2007 to 2009.

CANDIDATES

The most important biological control agent in use in Australia and elsewhere is the flea beetle *Agasicles hygrophila* (Fig. 4). However, this insect has aquatic orientation, for reasons unknown, thus high and damaging populations occur only on floating mats of the weed. Such mats, that used to extend to 70 m over water (Fig. 5), no longer occur in the active range of this beetle in Australia (Fig. 6) (Julien 1981). Although patches may extend several metres over water at the beginning of the growing season, the flea beetles achieve high enough populations to reduce them by mid summer. Maddox (1968) and Stewart *et al.* (1999a, b) describe the biology of *A. hygrophila,* while host-specificity tests on 14 plant species in the USA determined host-specificity (Maddox, *pers. comm.* 1974; Maddox and Resnik 1969), which was further confirmed by tests in Australia using the test plant list shown in Sands *et al.* (1982).

Host-testing of the moth *A. malloi* (Fig. 7), using the test plant list in Sands *et al.* (1982), determined that it was safe for release in Australia. It was released in 1977 and has become established (Julien *et al.* 1979), but does not overtly contribute to control. Its biology was reported by Maddox (1970) and Brown and Spencer (1973).

***Disonycha argentinensis* Jacoby (Coleoptera: Chrysomelidae).** This flea beetle was host-tested in Australia in 1979 and released in 1980 (Sands *et al.* 1982). It

Figure 4: The alligator weed flea beetle *Agasicles hygrophila*. Photo: J Greene, CSIRO

failed to establish in the field (Julien and Chan 1992) and a satisfactory explanation for this was never determined. Its biology was studied by Sands *et al.* (1982) and Cordo *et al.* (1984). Because this insect was terrestrially oriented and occurred widely in the native range of alligator weed in Argentina, it was decided to make another attempt to establish it in Australia. Host-testing in the 1970s did not include closely related native plants and hence further host-testing was conducted, including several *Alternanthera* species from New Zealand on behalf of Landcare Research. A colony of adults from Buenos Aires and Tucuman provinces, Argentina, were imported to a quarantine facility at Brisbane in March 2005. In subsequent tests *D. argentinensis* fed and developed on native *Alternanthera* species

Figure 6: The reduction in alligator weed as a result of damage by *Agasicles hygrophila*, July 1979. Photo: M Julien, CSIRO.

including *A. angustifolia* R. Br., *A. denticulata* R. Br. (three different forms), *A. nana* R. Br., *A. nodiflora* R. Br., *A. sessilis* (L.) R. Br. ex DC. (Australian material), *A. sessilis* (New Zealand material, but not native to NZ), *A. ficoidea* (L.) P. Beauv., *A. pungens* Kunth and a taxon of South American origin referred to as '*A. sissoo*' (this name does not appear to have been formally published). As a consequence, *D. argentinensis* was not released.

***Amynothrips andersoni* O'Neill (Thysanoptera: Phaeothripidae).** The host-specificity of this insect was studied in Argentina and Uruguay (Maddox 1973) and its biology was studied in the USA (Maddox and Mayfield 1979). During surveys from 2001 to 2005, this thrips was a very common and widespread natural enemy wherever alligator weed was found in Argentina. It was present almost year-round and as an adult in winter (Maddox and Mayfield 1979).

A. andersoni has both brachypterous (short-winged, flightless) and macropterous (long-winged, able to fly)

Figure 5: Alligator weed growing on Georges River at Chipping Norton, Sydney, March 1978. Photo: M Julien, CSIRO.

Figure 7: The alligator weed moth *Arcola malloi*. Photo: CSIRO.

forms (O'Neill 1968). Adults are black and about 2 mm long, immature stages are red. Adults and larvae suck plant juices from the tender developing tips of alligator weed. The host-range appears to be restricted to plants in the genus *Alternanthera*, specifically alligator weed and its close relative *A. aquatica* (Maddox 1973).

In March 2005 *A. andersoni* was sent to Australia in two shipments, with 229 adults and 240 nymphs in the first and 20 adults and 126 nymphs in the second. The material, from Buenos Aires in central Argentina and Tucuman in north-western Argentina, comprised mainly the macropterous form (79%). In host tests conducted in Brisbane, feeding and oviposition occurred on *A. nodiflora, A. denticulata* and *Ptilotus polystachyus* (Gaudich.) F.Muell. Although new hatchlings died on *P. polystachyus*, populations were sustained on the other two Australian natives.

The short-winged form of this thrips had been released in the USA in 1967 (O'Neill 1968); identification was confirmed from museum specimens held there (L Mound, *pers. comm.* 2005). To determine whether the short-winged form had different host-specificity, a shipment of 11 adults and 34 nymphs of *A. andersoni* from Georgia, USA, was received into quarantine in Brisbane during July 2005. Host-testing confirmed that this short-winged form had the same host-specificity as the Argentine long-winged material, therefore *A. andersoni* was considered unsuitable for release in Australia.

***Systena* spp. (Chrysomelidae: Alticinae).** Five species of *Systena* were collected from alligator weed. *S. testaceovittata* Clark, *S. s-littera colligata* Weise and *S. tenuis* Bechyné fed on other plant species as well as alligator weed in the field. *S. silvestrii* Bechyné was found only in northern Argentina in Chaco, Salta and Misiones provinces and was collected on *A. aquatica* as well as alligator weed, but as very few specimens were found rearing was not possible. *S. nitentula* Bechyné was the most abundant species (Cabrera *et al.* 2005) and was found only on the *angustifolia* form of alligator weed and on *A. aquatica* in Brazil. It was common in northern Argentina, southern Paraguay and southern Brazil, mostly in terrestrial habitats.

Adults of *S. nitentula* are 34 mm long and feed on leaves. Larvae feed on and inside stems, crown and roots; pupation is in the galleries made by larvae in the stems and roots and in the substrate. The life-cycle was about 40 days in the laboratory (Cabrera *et al.* 2005). Field observations in Argentina indicate that the insects are mainly associated with alligator weed growing in dry conditions. Adults collected from Orán, Salta province, Argentina, were sent to Australia, 45 adults on 14 January and 22 on 7 April 2009. In quarantine tests in Brisbane *S. nitentula* fed and developed colonies on *A. angustifolia, A. denticulata* (two colour forms and one form from New Zealand), *A. nana, A. nodiflora* and *A. sessilis* (one form from Australia and one from New Zealand). Therefore, this insect was not considered suitable for release in Australia.

***Clinodiplosis alternantherae* Gagné (Diptera: Cecidomyiidae).** Feeding by the larvae of this midge in the new stem tip tissues causes tip galls and tip deformation. Pupation occurs in the gall; adults lived for just a day in the laboratory. The life-cycle was three to four weeks. The midge was the most common insect found throughout the range of alligator weed in Argentina, Uruguay, Paraguay and Brazil, attacking all forms of alligator weed. It also attacked *A. aquatica* in Argentina, Paraguay and Brazil. No other plant species were observed to be attacked in the native range. Colonies collected near Buenos Aires in 2006 and 2007 were sent to Australia for host-specificity testing. The midge was found able to complete its life-cycle on several Australian native and some exotic *Alternanthera* species, so was not considered suitable for release.

***Phenrica* possibly *litoralis* (Bechyné).** This flea beetle has been found only in Misiones province in northern Argentina on alligator weed growing in terrestrial situations. Adults are about 2 mm long, feed on leaves and lay eggs under leaves and on stems near the soil surface. Larvae feed on older leaves; pupation occurs in the substrate. The life-cycle appears to be about six weeks. Adults collected from San Ignacio, Misiones province, were imported to Australia in January 2010. Host-testing determined that this insect is not specific and can feed and develop on a range of *Alteranthera* species including Australian natives and on plants from several other families.

***Ophiomya marellii* (Bréthes) (Diptera: Agromyzidae).** This node galling fly occurs widely across the range of alligator weed in Argentina on the various forms. Galls were found on stem nodes in the substrate or older sections of above-ground stems. The insect has mostly been collected from terrestrial habitats but appears to withstand temporary submergence (Sosa *et al.* 2004). A number of parasitoids collected from galls have been submitted for identification. The biology and life-history of *O. marellii* was provided by Marelli (1926) and Spencer (1963). During January 2010, collections of galls from Santa Fé and Entre Rios provinces and five locations in Buenos Aires province were imported into quarantine in Brisbane. Adults and parasites emerged, and adults were placed on Australian alligator weed to try to develop a colony for host-testing.

Other leaf- and stem-mining flies

***Ophiomyia alternantherae* (Spencer) (Diptera: Agromyzidae).** This leaf-mining fly, described by Spencer (1963, 1990) was found on all forms of alligator weed in Argentina (Sosa *et al.* 2004). Larvae feed inside the leaf, forming galleries in which pupation occurs. Several unidentified hymenopteran parasitoids emerged from harvested leaves.

***Ophiomyia buski* (Frost) (Diptera: Agromyzidae).** This stem-mining fly (Spencer 1963) is uncommon and has been collected at only a few sites both from the *angustifolia* and *philoxeroides* forms of alligator weed in Argentina. Larvae tunnel in the epidermis of internodes, causing linear mines about 1 mm wide and up to 2 cm long.

***Pegomya* sp. (Diptera: Anthomyiidae).** Another leafminer appears to be a new species of *Pegomya*. The larvae feed on the mesophyll portion of the leaf, leaving the epidermis intact. Pupae are larger than those of *O. alternantherae* and pupation occurs on the ground, suggesting that this insect favors terrestrial environments. It has been found on two occasions at different sites in Buenos Aires province. The biology, life-histories and host-ranges of these insects have not yet been studied.

Fungi

Currently, the only fungus identified with potential for classical biological control of alligator weed is the rust *Uredo pacensis* Lindquist (1957), which was found to be abundant at Coroico, La Paz province, Bolivia. This rust has not been found in Argentina, southern Paraguay, southern Brazil or Uruguay. We are seeking ways to obtain material and conduct host-specificity tests in Australia. The first test will determine if the rust will accept the Australian populations of alligator weed. Another rust, *Uredo nitidula* Arthur (1920), has been recorded on alligator weed from Guatemala but there have been no surveys to seek it.

Various other fungi have been isolated from alligator weed in Argentina but none are suitable as classical biological control agents (G Traversa, *pers. comm.* 2009). One, *Nimbya* sp., was found in Argentina and has been isolated in Australia. It is under study as a bioherbicide (Gilbert *et al.* 2008).

DISCUSSION

Excellent control of alligator weed was achieved with the flea beetle *A. hygrophila* in aquatic habitats and warm temperate regions, particularly central coastal NSW (Julien *et al.* 1979). However, much of the alligator weed in Australia grows in riparian and floodplain terrestrial habitats that are only occasionally covered with water, where the flea beetle cannot thrive. Although the moth *V. malloi* is relatively common, in Australia it has not reached densities capable of adequately controlling the weed. Consequently, recent studies aimed to provide agents that would attack alligator weed across a broader range of habitats. Two known agents not established in Australia, the flea beetle *D. argentinensis* and the thrips *A. andersoni*, and two new potential agents, *S. nitentula* and *C. alternantherae*, all failed host-specificity tests as they attacked native *Alternanthera* species. Several species, a node-galler and another flea beetle, are in quarantine in preparation for host-specificity testing. Two others, a stem- and a leaf-miner, could be tested in the future.

None of the insects yet to be tested are particularly exciting; they do not form large and damaging populations in Argentina and it is not obvious that they are restricted there by parasitism, predation or disease, though these controls cannot be ruled out. One other potential biological control agent remains to be assessed, the rust *U. pacensis*. The fact that it has been found only in Bolivia suggests that it might be a *formae specialis* on the Bolivian form of alligator weed. In addition to completing host-testing of the four insects, future effort should be directed to obtaining the rust from Bolivia and determining if it will attack Australian populations of alligator weed. If it does and is shown to be host-specific, it may be the best hope as a significant biological control agent for Australia.

Due to the difficulty of containing and managing alligator weed infestations, its environmental and economic impacts and consequently its status as a Weed of National Significance, it is likely that adequate funding will be available to complete the testing of the candidate agents listed above. However, the presence of closely related native species means that finding an effective and suitably host-specific agent remains the primary issue in determining the success of biological control of alligator weed in the temperate climates and terrestrial habitats in Australia.

ACKNOWLEDGEMENTS

We acknowledge the help of Nora Cabrera (Chrysomelidae taxonomist), Graciela Valladares (Agromyzidae taxonomist), Jimena Dorado, Eduardo Martinez Mullo, Victoria Cardo and Marta Telesnicki (laboratory and field work) in Argentina, Gio Fichera and Dalio Mira (laboratory work) in Australia and Justine Murray for the map (Fig. 3).

REFERENCES

Arthur JC (1920) New Species of Uredineae XII. *Bulletin of the Torrey Botanical Club* **47**, 465–480.

Barreto RW and Torres ANL (1999) *Nimbya alternantherae* and *Cercospora alternantherae*: two new records of fungal pathogens on *Alternanthera philoxeroides* (alligatorweed) in Brazil. *Australasian Plant Pathology* **28**, 103–107.

Brown JL and Spencer NR (1973) *Vogtia malloi*, a newly introduced Phycitine moth (Lepidoptera: Pyralidae) to control alligatorweed. *Environmental Entomology* **2**, 519–523.

Buckingham GR, Boucias D and Theriot RF (1983) Reintroduction of the alligatorweed flea beetle (*Agasicles hygrophila* Selman and Vogt) into the United States from Argentina. *Journal of Aquatic Plant Management* **21**, 101–102.

Cabrera N, Sosa AJ, Dorado J and Julien M (2005) Redescription of *Systena nitentula* Bechyné (Coleoptera: Chrysomelidae), a flea beetle injurious to alligator weed (*Alternanthera philoxeroides*). *Annals of the American Society of Entomology* **98**, 643–652.

Cordo HA, DeLoach CJ and Ferrer M (1984) Biology and larval host range of the flea beetle *Disonycha argentinensis* (Coleoptera: Chrysomeldiae) on alligatorweed in Argentina. *Annals of the Entomological Society of America* **77**, 134–141.

Coulson JR (1977) Biological control of alligatorweed, 1959–1972. A review and evaluation. *ARS, USDA Technical Report No. 1547.*

Gilbert R, Gurr GM, Priest MJ and Mitchell A (2008) Towards a bioherbicide for alligator weed. In *Proceedings of the 16th Australian Weeds Conference.* 18–22 May 2008, Cairns, Australia. (Eds RD van Klinken, VA Osten, FD Panetta and JC Scanlan) p. 283. Queensland Weeds Society, Brisbane.

Gunasekera L (1999) Alligator weed: an aquatic weed present in Australian backyards. *Plant Protection Quarterly* **14**, 77–78.

Holcomb GE and Antonopoulos AA (1976) *Alternaria alternantherae*: a new species found on alligatorweed. *Mycologia* **68**, 1125–1129.

Julien MH (1981) Control of aquatic *Alternanthera philoxeroides* in Australia: another success for *Agasicles hygrophila*. In *Proceedings of the 5th International Symposium on Biological Control of Weeds.* 22–29 July 1980, Brisbane, Australia. (Ed. ES Delfosse) pp. 583588. CSIRO, Melbourne.

Julien MH (1995) *Alternanthera philoxeroides* (Mart.) Griseb. In *The Biology of Australian Weeds.* (Eds RH Groves, RCH Shepherd and RG Richardson), pp. 112. R.G. and F.J. Richardson, Frankston.

Julien MH and Bourne AS (1988) Alligator weed is spreading in Australia. *Plant Protection Quarterly* **3**, 91–96.

Julien MH and RR Chan (1992) Biological control of alligator weed: unsuccessful attempts to control terrestrial growth using the flea beetle *Disonycha argentinensis* (Col: Chrysomelidae). *Entomophaga* **37,** 215–221.

Julien MH and Griffiths MW (1999) *Biological Control of Weeds: A World Catalogue of Agents and their Target Weeds.* 4th edn. CABI, Wallingford, UK.

Julien MH and Stanley JN (1999) The management of alligator weed: a challenge for the new millennium. In *Proceedings of the 10th Biennial Noxious Weeds Conference.* 20–22 July 1999, Ballina, Australia. (Eds R Ensbey, P Blackmore and A Simpson) pp. 213. NSW Agriculture, Australia.

Julien MH, Broadbent JE and Harley KLS (1979) The current status of biological control of *Alternanthera philoxeroides* in Australia. In *Proceedings of the 7th AsianPacific Weed Science Society Conference.* 26–30 November 1979, Sydney. (Eds RW Medd and BA Auld) pp. 231–235. Council of the Australian Weed Science Societies.

Julien MH, Bourne AS and Low VHK (1992) Growth of the weed *Alternanthera philoxeroides* (Martius) Grisebach (alligator weed) in aquatic and terrestrial habitats in Australia. *Plant Protection Quarterly* **7**, 102–108.

Julien MH, Skarratt B and Maywald GF (1995) Potential geographical distribution of alligator weed and its biological control by *Agasicles hygrophila. Journal of Aquatic Plant Management* **33**, 55–60.

Lindquist JC (1957) Notas Uredinológicas IV: *Uredo pacensis. Revista Facultad de Agronomía* **33**, 80.

Maddox D M (1968) Bionomics of an alligatorweed flea beetle, *Agasicles* sp. in Argentina. *Annals of the Entomological Society of America* **61**, 1299–1305.

Maddox DM (1970) The bionomics of a stem borer, *Vogtia malloi* (Lepidoptera: Phycitidae) on alligatorweed in Argentina. *Annals of the Entomological Society of America* **63**, 1267–1273.

Maddox DM (1973) *Amynothrips andersoni* (Thysanoptera: Phlaeothripidae), a thrips for the biological control of alligatorweed. I. Host specificity studies. *Environmental Entomology* **2**, 30–37.

Maddox DM and Mayfield A (1979) Biology and life history of *Amynothrips andersoni*, a thrips for the biological control of alligatorweed. *Annals of the Entomological Society of America* **72**, 136–140.

Maddox DM and Resnik ME (1969) Determination of host specificity of the alligatorweed flea beetle, *Agasicles* n. sp., with radioisotopes. *Journal of Economic Entomology* **62**, 996–999.

Marelli C (1926) La agromyza productora de agallas nodicolas en la lagunilla y sus parasitos e hiperparasitos. *Revista de la S.E.A.* **2**, 14–21.

O'Neill K (1968) Amynothrips andersoni: a new genus and species injurious to alligatorweed (Thysanoptera: Phlaeothripidae). *Proceedings of the Entomological Society of Washington* **70**, 175–183.

Pomella AWV, Barreto RWand Charudattan C (2007) *Nimbya alternantherae* a potential biocontrol agent for alligatorweed, *Alternanthera philoxeroides. BioControl* **52**, 271–288.

Sainty G, McCorkelle G and Julien M (1998) Control and spread of alligator weed, *Alternanthera philoxeroides*, in Australia: lessons from other regions. *Wetlands Ecology and Management* **5**, 195–201.

Sands DPA, Kassulke RC and Harley KLS (1982) Host specificity of *Disonycha argentinensis* [Col.: Chrysomelidae]: an agent for the biological control of *Alternanthera philoxeroides* (alligator weed) in Australia. *Entomophaga* **27**, 163–172.

Schooler SS, Cook T, Prichard G and Yeates A (2010) Disturbance-mediated competition: the interacting roles of inundation regime and physical and herbicidal control in determining native and invasive plant abundance. *Biological Invasions* DOI 10.1007/s10530-010-9722-y.

Sosa AJ, Julien MH and Cordo HA (2004) New research on *Alternanthera philoxeroides* (alligator weed) in its South American native range. In *Proceedings of the 11th International Symposium on Biological Control of Weeds.* 27 April–2 May 2003, Canberra. (Eds JM Cullen, DT Briese, DJ Kriticos, WM Lonsdale, L Morin and JK Scott) pp. 180–185. CSIRO Entomology, Canberra.

Sosa AJ, Greizerstein E, Cardo MV, Telesnicki MC and Julien MH (2008) The evolutionary history of an invasive species: alligator weed, *Alternanthera philoxeroides*. In *Proceedings of the 12th International Symposium on Biological Control of Weeds.* 22–27 April 2007, La Grande Motte, France. (Eds MH Julien, R Sforza, MC Bon, HC Evans, PE Hatcher, HL Hinz and BG Rector) pp. 435–442. CABI, Wallingford, UK.

Spencer KA (1963) A synopsis of the neotropical Agromyzidae (Diptera). *Transactions of the Royal Entomological Society of London* **115**, 291–389.

Spencer KA (1990) *Host Specialization in the World Agromyzidae (Diptera).* Kluwer Academic, Netherlands.

Stewart CA, Chapman RB and Emberson RM (1999a) The effect of temperature on the development and survival of *Agasicles hygrophila* Selman & Vogt (Coleoptera: Chrysomeldiae), a biological control agent for alligator weed (*Alternanthera philoxeroides*). *New Zealand Journal of Zoology* **26**, 11–20.

Stewart CA, Chapman RB, Barrington AM and Frampton CMA (1999b) Influence of temperature on adult longevity, oviposition and fertility of *Agasicles hygrophila* Selman & Vogt (Coleoptera: Chrysomelidae). *New Zealand Journal of Zoology* **26**, 191–197.

Thorp JR and Lynch R (2000) *The Determination of Weeds of National Significance.* National Weeds Strategy Executive Committee, Launceston.

van Oosterhout E (2007) *Alligator Weed Control Manual. Eradication and Suppression of Alligator Weed* (Alternanthera philoxeroides) *in Australia.* NSW Dept of Primary Industries.

Vogt GB (1960) *Exploration for Natural Enemies of Alligator Weed and Related Plants in South America.* USDA, ARS Special Report PI4, October 1960.

Vogt G B (1961) *Exploration for Natural Enemies for Alligator Weed and Related Plants in South America.* USDA, ARS Special Report PI5, October 1961.

Vogt GB (1973) *Exploration for Natural Enemies of Alligator Weed and Related Plants in South America.* Technical Report no. 3, Biological control of alligator weed. Aquatic Plant Control Program, US Army Engineers, Vicksburg, Mississippi, pp. B3–B66.

Xiang M, Liu R and Yeng Y (1998) *Nimbya alternantherae*, a new record of the genus *Nimbya* from China. *Mycosystema* **17**, 283.

Ambrosia artemisiifolia L. – annual ragweed

Bill Palmer and Rachel Cruttwell McFadyen

ABSTRACT

Ambrosia artemisiifolia, annual ragweed, is an annual herb native to subtropical and temperate North America that is capable of causing very serious allergenic health issues in humans. In Australia, this weed has increased in abundance since the mid 1900s, particularly in south-eastern Qld and northern NSW. A biological control project was initiated by the Qld Department of Lands in 1984, utilising three insect species, *Epiblema strenuana*, *Zygogramma bicolorata* and *Stobaera concinna* imported primarily for parthenium. Further exploration in North America led to the release of *Zygogramma suturalis*, which did not establish in Australia. Two other possible insects, *Ophraella communa* and *Asphondylia ambrosiae*, were partially investigated, but not progressed to introduction. An Argentinian *Liothrips* sp. was host-tested in quarantine, but rejected for release. Annual ragweed is now thought to be under good biological control in most areas in eastern Australia. An economic evaluation of this project in 2006 indicated that project costs of $625 000 resulted in an estimated net present value of $52 million and a benefit:cost ratio of 103:7. These figures place annual ragweed as one of Australia's most successful projects.

Key words: *Epiblema strenuana*, *Zygogramma bicolorata*, *Stobaera concinna*, *Zygogramma suturalis*, *Parthenium hysterophorus*.

INTRODUCTION

Annual ragweed (*Ambrosia artemisiifolia* L.) (Asteraceae) is an annual herb (Fig. 1) native to subtropical and temperate North America. It was introduced into Australia early last century and by 1940 was recorded in several localities. There was a gradual increase in both the density of the weed and the area affected, until by 1984 it was a common and abundant weed of suitable land throughout the coastal region of south-eastern Qld and northern and central NSW (McFadyen 1984).

It is not known how annual ragweed was introduced into Australia, but it was first recorded as naturalised in Murwillumbah in northern NSW in 1908 and was known as 'Dunbible weed', which suggests an origin in that area. The first record in Qld was at Currumbin, just north of the NSW border, in 1915. Between 1940 and 1947, it was recorded as common at Kyogle in NSW and there seems no doubt that the main areas prior to 1940 were in northern NSW. During the next 20 years, infestations established and increased in the Albert and Coomera River valleys in the Gold Coast hinterland, and at several points along the highways north into Brisbane. Isolated infestations also occurred at Toogoolawah in the Brisbane Valley (possibly originating from a wartime American air force base in the region) and at Stanthorpe. By 1970 the weed was widespread in the Albert, Coomera and other minor river systems. Numerous small infestations were present along roads and in creeks in several Brisbane suburbs, but these were very restricted in extent. The 1974 floods resulted in an enormous increase in the area affected, and much of the floodplain of the Brisbane River and associated creeks became infested (McFadyen 1984). The current distribution of annual ragweed in Australia and the areas climatically favourable for annual ragweed are given in Figure 2.

Figure 1: Line drawing of *Ambrosia artemisiifolia* (from a painting by S Monteith) showing various identifying features.

Annual ragweed is an example of a plant that is a very serious problem within its native range. The plant is endemic to North America, presently found in the eastern portions from Canada to Texas, and is a major cause of severe seasonal allergic rhinitis in humans. It is so serious that in many areas of the USA, ragweed pollen counts are given with the weather forecasts in the flowering season. Annual ragweed is also a weed of cultivation and is often seen growing abundantly in fallowed fields. Other ragweeds are also problematic, in particular giant ragweed, *A. trifida* L.

Annual ragweed has also become a serious weed in many European and Asian countries, including Russia, Croatia, Hungary and China. The weed is spreading rapidly in southern and eastern Europe (Coutinot *et al.* 2008), and by 2000 up to 80% of pollen-sensitive patients showed positive reactions to ragweed (Rybncek and Jäger 2001).

In Australia, annual ragweed is a weed of waste land and neglected pastures (Fig. 3) in coastal eastern Australia between Gympie (26°S) in the north and Coffs Harbour (31°S) in the south. It is not a major agricultural or environmental weed, but allergic rhinitis due to ragweed pollen was already noted as a problem in ragweed areas of NSW in the 1980s (Morrison 1984). In 1997, 30–40% of people sampled in the coastal city of Casino, NSW, most of whom suffered from allergic symptoms during the ragweed flowering season, showed a positive skin test reaction to ragweed pollen (Bass *et al.* 2000).

BIOLOGICAL CONTROL HISTORY

Perhaps because of its abundance and weed status, annual ragweed is unusual in that most of the faunistic studies were undertaken by scientists from its native range rather than by foreign 'explorers' working for countries where the plant is exotic. The arthropod fauna of annual ragweed, mainly in its northern distribution, was studied by Harris and Piper (1970), with some supplementary collecting mainly in the southern USA contributed by WA Palmer in the 1980s. The faunas of the ragweeds of the western USA were studied by RD Goeden and his team from the University of California, Riverside (Goeden and Ricker 1974, 1975, 1976). Over 70 species of phytophagous arthropods are now known from annual ragweed in its native range (Goeden and Palmer 1995).

The then USSR, through O Kovalev of the Academy of Sciences of the USSR, initiated a biological control program for annual ragweed in 1966. Kovalev established collaborations with American and Canadian scientists, particularly RD Goeden, LA Andres and P Harris, that led to several insect species being released in the USSR. The noctuid *Tarachidia candefacta* (Hübner) was the first insect released, in 1968 (Kovalev and Runeva 1970). Another insect, the leaf-feeding chrysomelid *Zygogramma suturalis* (F.), established well and in high populations produced a spectacular wave-like defoliation through stands of annual ragweed (Kovalev and Vechernin 1986). A second point of interest about this introduction was that *Z. suturalis* was imported as a flightless form with reduced wing muscles and flight articulations, but developed flight muscles five generations after its release (Kovalev 1995). The former Yugoslavia also introduced *Z. suturalis* from the USA to release in Croatia in 1985 (Igrc *et al.* 1995).

In 1975 and 1976, RE McFadyen, then of the CIBC station in north-western Argentina, undertook a survey of insects on *Ambrosia elatior* L. (previously misidentified

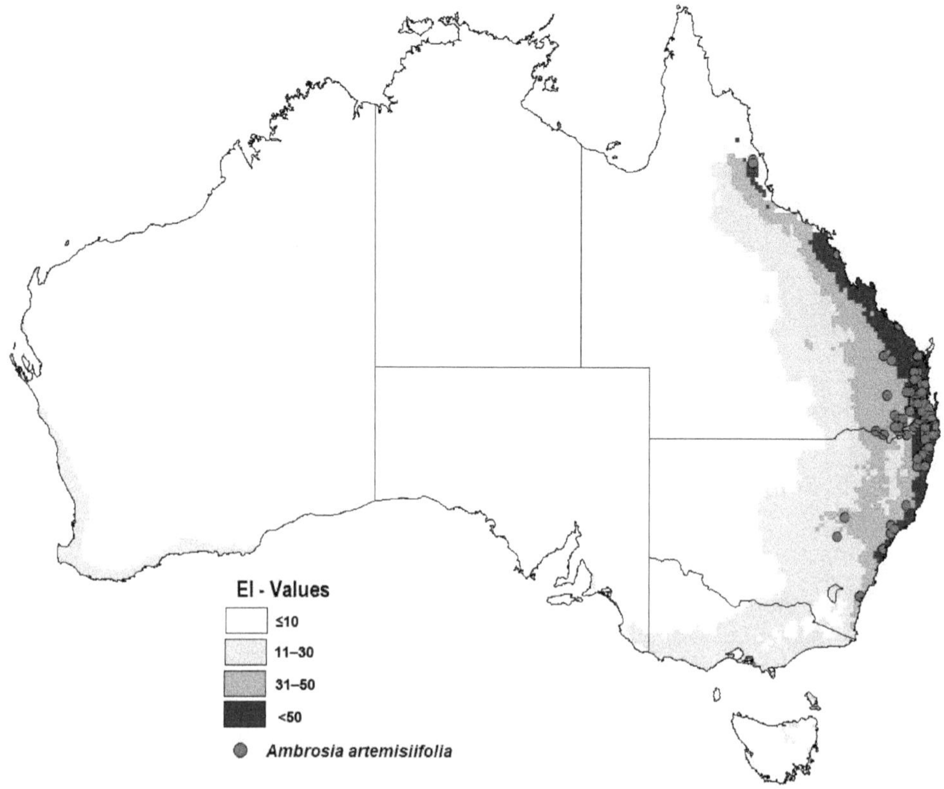

Figure 2: Distribution of *Ambrosia artemisiifolia* in Australia and its potential distribution predicted using the climate modelling software CLIMEX. EI values >30 are considered favourable while values ≤10 are considered climatically unfavourable. Data based largely on herbarium records from the Australian Virtual Herbarium.

as *A. tenuifolia* Sprengel) for Canada. The tip-feeding *Liothrips* sp. was sent to Canada for host-testing, but was rejected because of feeding on sunflower (P Harris, *pers. comm.* 1992).

China developed a biological control program in the 1980s which resulted in the release and establishment of *Z. suturalis*, supplied by Canada and the USSR, in 1987 (Wan *et al.* 1995). The stem-galling moth *Epiblema strenuana* (Walker), obtained from Australia, was released in 1991 and has established (Wan *et al.* 2003). More recently, there has been considerable interest in the leaf-feeding beetle *Ophraella communa* LeSage (Zhang *et al.* 2005; Wang and Meng 2009).

In Australia, Qld commenced a small project in 1984, mainly because of overlap with the larger project against parthenium *Parthenium hysterophorus* L. and with the establishment of its North American Field Station in Temple, Texas. Surveying on annual and perennial ragweed was undertaken from this base and one insect species was shipped back to Australia. Further surveys were undertaken from Tucuman in north-west Argentina from 1987 to 1992, also as an adjunct to surveys on parthenium.

PLANT TAXONOMY

The taxonomy and phylogeny of the ragweeds are well understood (Wodehouse 1971). From a biological control perspective, the interesting aspect is the phylogenetic relationship of the ragweeds to other taxa in the Asteraceae. The subtribe *Ambrosiinae* consists of several genera (Stuessy 1977) largely regarded as weedy. In Australia, the subtribe is represented by only four genera – *Ambrosia*, *Parthenium*, *Xanthium* and *Iva* – the first three of which include serious weeds and are subjects of chapters in this book. The *Ambrosiineae* is placed in the tribe Heliantheae. Australia's native flora has very few representatives of the Heliantheae, but the tribe does include the commercial crop sunflower *Helianthus annuus* L.

EXPLORATION

From the studies of Harris and Piper (1970), Goeden and Ricker (1974, 1975, 1976) plus the surveys by WA Palmer in Texas and Florida between 1984 and 1990, over 70 species of phytophagous arthropods are now known from annual ragweed in its North American native range

Figure 3: An infestation of *Ambrosia artemisiifolia* taken in the Brisbane suburb of Moggill with J Cummings providing an indication of its height. Photo: Qld Government.

(Goeden and Palmer 1995). Surveys were also undertaken in north-western Argentina from 1987 to 1992 by C Garcia, employed by Qld and based in Tucuman. His prime aim was to locate biocontrol agents against parthenium but annual ragweed *A. elatior*, common in the area, was included in all surveys, as many host-specific insects feed on both species. No specific search for pathogens was undertaken in the Australian surveys, but a form of *Puccinia xanthii* Schwein was found on *A. artemisiifolia* near Austin, Texas.

CANDIDATES

As previously stated, the insects attacking *A. artemisiifolia* in its native range have been extensively studied in the USA and there is an extensive literature published. Less is known on the diseases though *P. xanthii* has been considered (Kiss 2007).

Insect species investigated for Australia but not released

***Tarachidia candefacta* (Hübner) (Lepidoptera: Noctuidae).** This moth had been successfully established on annual ragweed in Russia (Kovalev and Runeva 1970). One small shipment was sent from Tennessee to Australia in 1989. Unfortunately the larvae in the shipment entered a diapause and were not successfully cultured. No further attempts were made with this species.

Gall midges (Diptera: Cecidomyiidae). Three species were studied in the USA: *Contarinia partheniicola* (Cockerell), *Rhopalomyia ambrosiae* Gagné and *Asphondylia ambrosiae* Gagné. It was difficult to locate sizeable populations for the first two species, and no shipments were made to Australia or laboratory colonies established. The third, *A. ambrosiae*, causes a characteristic globular stem gall and was found in both central Texas and Florida. Several shipments were sent to Australia in 1987 and adult emergence, mating and oviposition occurred in quarantine, but the galls did not develop. The most probable reason for this failure was that *Asphondylia* spp. generally require fungal symbionts for successful gall formation (Borkent and Bissett 1985; Goeden and Palmer 1995). However, no further attempts have been made to rear *A. ambrosiae* in Australia.

***Ophraella communa* LeSage (Coleoptera: Chrysomelidae).** This leaf-feeder has a wide distribution in North America, from Canada to Mexico and right across the USA (LeSage 1986), and was early recognised as a potential agent (Harris and Piper 1970). Preliminary host-testing was conducted at the North American Field Station, but in these tests the insect also fed and bred on commercial sunflower (Palmer and Goeden 1991). As a result the insect was considered unsuitable for introduction to Australia. Interestingly, some time after this experimental work, *O. communa* was reported from a *Helianthus* sp. (Futuyma 1990), perhaps justifying the decision and increasing confidence in the use of laboratory tests as a predictor of field hosts. Although *O. communa* was not progressed for Australia, there remains interest in this insect for Europe and Asia and even for augmentative release in Canada (Dernovici *et al.* 2006). The insect has been released in Asia and is now found in Japan (Yamazaki *et al.* 2000), China (Hu and Meng 2007) and Taiwan (Wang and Chiang 1998). A second species, *O. slobodkini* Futuyma, known only from annual ragweed in the USA, has since been found and might warrant further investigation (Futuyma 1991).

***Liothrips* sp. (Thysanoptera: Phlaeothripidae).** This was the only agent to be reared and host-tested in Australia, but not released. It was first collected from *A. elatior* in Tucuman, Argentina, in 1976 by RE McFadyen for Canada. In quarantine in Canada, it developed equally well on *A. artemisiifolia* and was host-tested, but rejected due to significant feeding on sunflower. Because the thrips causes severe damage to the plant, and because sunflower is never attacked in Argentina where it is a common crop in the same areas, the thrips was brought into quarantine in Brisbane in 1990 for further detailed host-testing (McFadyen and Weggler-Beaton 2000).

Adult thrips are black, 2.5–3 mm long, and feed in young shoots and buds before the leaves have unfurled. Feeding causes necrotic lesions, followed by cessation of bud and shoot growth, with reduced internode distances. Eggs are laid in the base of emerging leaves or in the bud. Nymphs are red with black patches increasing as they mature and feed in the shortened buds and shoots, which are often killed from the tip down. Pupation occurs in leaf litter and adults overwinter in the soil or leaf litter, emerging to oviposit in spring. All stages are then present on plants until ragweed dies back in autumn.

In Argentina, *Liothrips* sp. is attacked by two unidentified anthocorids and by the predatory larvae of the cecidomyiid *Lestodiplosis* sp. Nymphs were parasitised by *Tetrastichus gentilei* Del Guercio (Hymenoptera: Eulophidae), a frequent parasite of the genus *Liothrips*.

Detailed host-testing was undertaken on a range of plants in the tribe Heliantheae. Some adult feeding occurred on a number of plants, but nymphal development was only possible on ragweed, *Xanthium* spp. and very young sunflower seedlings. Sunflower seedlings with more than two pairs of true leaves (usually >23 days from sowing) were not accepted for oviposition and did not support nymphal development. Sunflower seedlings <20 days from sowing with one or two pairs of true leaves supported complete nymphal development to fertile adults, although development was much slower and survival much poorer than on ragweed. However, given the importance of sunflower as a crop in Australia, the risk was regarded as unacceptable and the quarantine colony destroyed (McFadyen and Weggler-Beaton 2000).

Species released in Australia

***Zygogramma suturalis* (F.) (Coleoptera: Chrysomelidae).** This leaf-feeder is the only species released in Australia solely for control of annual ragweed. It is well documented in the literature following its release in Russia and Croatia in 1979 (Kovalev and Vechernin 1986). Adult beetles collected in Tennessee were sent to Australia, where the species was host-tested prior to its release in 1990. However, rearing was not very successful because there was a long daylight requirement for adults to initiate oviposition, and summer day length in southern Qld at 27°S does not meet this until full summer (December). As a result, oviposition was poor and only one or two very small releases of about 300 adults each were made in the Stanthorpe district (28°40'S). The insect did not establish.

***Zygogramma bicolorata* Pallister (Coleoptera: Chrysomelidae).** This species was collected from *Parthenium hysterophorus* in Mexico and released in Qld in 1980 as a biocontrol agent for parthenium (McFadyen and McClay 1981). Mass rearing in cages in the open took place in Brisbane from 1980 to 1983 and adults from these cages rapidly established colonies on annual ragweed growing in horse pastures and waste land in the area. The beetles continued to spread naturally and from 1984 to 1985 large numbers of adults were collected for release into ragweed infestations in Qld south of Brisbane and in northern NSW. The beetle is now present on ragweed throughout its range in eastern Australia, though abundance varies considerably in different localities and seasons.

Adult beetles are about 6 mm long, and easily recognisable by the elytral pattern of dark brown or black lines on a cream background. The yellow eggs are laid singly on the stems and leaves and young larvae feed on bud leaves, moving out as they grow to feed on larger leaves. Mature larvae pupate in the soil, and adults re-enter the soil to overwinter, emerging with rainfall in spring or early summer. Damage to the plants can be significant and large populations can completely defoliate ragweed infestations. In Australia, larvae and newly emerged adults are heavily predated by a pentatomid bug, and eggs are parasitised by the introduced pteromalid parasitoid *Erixestus zygogrammae* Cave and Grissell (Withers *et al.* 1998). In wet years, there can also be very high mortality in pupae and adults in the soil from the entomopathogenic fungus, *Beauvaria bassiana* (Bals.-Criv.) Vuill. and other soil-dwelling fungi.

***Epiblema strenuana* Walker (Lepidoptera: Torticidae).** This stem-galling moth was collected from parthenium in Mexico and released in Qld in 1982 as a biocontrol agent for this weed. It established rapidly and spread, and from 1984 was released on annual ragweed in southern Qld and northern NSW (McFadyen 1985). Adults are 6–10 mm long, and lay eggs singly on the underside of leaves. Larvae bore into stem tips or axillary buds, causing the plant to develop into a fusiform gall in which the larva feeds and pupates. All stages of the plant, from seedlings to flowering plants, are attacked and single galls may kill seedlings, while larger plants require 30 or more galls to prevent growth and seed production. In summer, the duration of the life-cycle is around five weeks and in the ragweed areas there are three to five generations over summer (McFadyen 1987). Females lay 400–500 eggs, with very large females laying over 1000 eggs; as a result, very rapid increases can occur in favourable conditions. In autumn, larvae enter pre-pupal diapause to overwinter within the galls. Emergence occurs in spring in response to rising temperature, and the first spring generation can suffer very high mortality if, due to seasonally dry conditions, there is little or no green ragweed available for oviposition. Several parasitoids have been recorded in Australia, but parasitism is generally low (McFadyen 1985).

***Stobaera concinna* (Stäl) (Hemiptera: Delphacidae).** This sap-sucker was collected from parthenium weed in Mexico and was first released in Qld in 1983 as a biocontrol agent for parthenium (McFadyen 1992). It established and has persisted on annual ragweed in south-eastern Qld ever since, but at such low levels that there is no significant impact on the weed.

EVALUATION

There has been no formal study to determine the impact of biological control on annual ragweed. However, the two agents *E. strenuana* and *Z. bicolorata* are known to be widespread and exerting a degree of control in most affected areas from Brisbane to the north coast of NSW. Certainly there is much less annual ragweed around Brisbane than there was in the 1980s; the plant is now relatively rare and no longer causes significant allergic symptoms in the flowering season. Annual ragweed is more abundant in areas of the north coast of NSW, such as around Lismore, though there, too, populations are far below those observed in the 1980s. For example, the Northern Rivers Catchment Management Authority does not list annual ragweed as a serious weed in the current Invasive Plants Strategy for the region, and there has been no publication on control or management of annual ragweed in Australian weed conferences or journals since 1984.

An economic evaluation of this very small project (Page and Lacey 2006) estimated that the project cost a total of $625 000, resulting in a net present value of $52 million and a benefit:cost ratio of 103:7 at a discount rate of 8%. These figures place the annual ragweed project as one of Australia's most successful, in part because of the unusually low research costs since three of the four insects had been previously tested and released against parthenium weed.

DISCUSSION

Only a small proportion of the insects known to feed on annual ragweeds in their native range have ever been considered for use as biocontrol agents, or subjected to detailed host-specificity testing. Given that annual ragweed is no longer considered a serious weed in eastern Australia and that the incidence of allergic rhinitis due to ragweed pollen is falling steadily in the affected regions (McFadyen, unpub.), it is unlikely that any further work will be undertaken by Australian authorities. However, there is continuing interest from Europe, where annual ragweed is spreading rapidly, largely through the movement of contaminated seed sold as wild bird food. Some of the agents used in Australia may be tested for use in Europe, as may be other yet-untested agents (Coutinot *et al.* 2008).

This project is perhaps unique in that effective biocontrol has been achieved with insects that occur naturally on another plant in a different genus, *Parthenium*. Some reasons why insects from parthenium were utilised have been given. There is probably another. Most of the ragweed in North America occurs in areas that experience much severer winter conditions than are found in Australia, especially the areas where ragweed has become abundant, such as south-eastern Qld and northern NSW. The experience with *Z. suturalis* is an example of a temperate insect not doing well in subtropical areas. Without resorting to climatic modelling, workers were aware that insects from northern USA and Canada would be unlikely to do well in Australia. Parthenium, on the other hand, grows in areas climatically much more similar to areas infested in Australia.

ACKNOWLEDGEMENTS

Several American and Canadian scientists such as RD Goeden, LA Andres and P Harris assisted the project in various ways. The project was supported by Rural Lands Protection Trust funds. KADW Senaratne kindly developed the CLIMEX model.

REFERENCES

Bass DJ, Delpech V, Beard J, Bass P and Walls RS (2000) Ragweed in Australia. *Aerobiologia* **16**, 107–111.

Borkent A and Bissett J (1985) Gall midges (Diptera: Cecidomyiidae) are vectors for their fungal symbionts. *Symbiosis* **1**, 185–194.

Coutinot D, Starfinger U, McFadyen RC, Volkovitsh MG, Kiss L, Cristofaro M and Ehret P (2008) Feasibility of biological control of common ragweed *Ambrosia artemisiifolia*, a noxious and highly allergenic weed in Europe. In *Proceedings of the 12th International Symposium on Biological Control of Weeds.* 22–27 April

2007, La Grande Motte, France. (Eds MH Julien, R Sforza, MC Bon, HC Evans, PE Hatcher, HL Hinz and BG Rector) pp. 717–719. CABI, Wallingford, UK.

Dernovici SA, Teshler MP and Watson AK (2006) Is sunflower (*Helianthus annuus*) at risk to damage from *Ophraella communa*, a natural enemy of common ragweed (*Ambrosia artemisiifolia*)? *Biocontrol Science and Technology* **16**, 669–686.

Futuyma DJ (1990) Observations on the taxonomy and natural history of *Ophraella* Wilcox (Coleoptera: Chrysomelidae), with a description of a new species. *Journal of the New York Entomological Society* **98**, 163–186.

Futuyma DJ (1991) A new species of *Ophraella* Wilcox (Coleoptera: Chrysomelidae) from the south-eastern United States. *Journal of the New York Entomological Society* **99**, 643–653.

Goeden RD and Palmer WA (1995) Lessons learned from studies of the insects associated with Ambrosiinae in North America in relation to the biological control of weedy members of this group. In *Proceedings of the 8th International Symposium on Biological Control of Weeds.* 2–7 February 1992, Canterbury, New Zealand. (Eds ES Delfosse and RR Scott) pp. 565–573. DSIR/CSIRO, Melbourne.

Goeden RD and Ricker DW (1974) The phytophagous insect fauna of the ragweed, *Ambrosia acanthicarpa*, in southern California. *Environmental Entomology* **3**, 827–834.

Goeden RD and Ricker DW (1975) The phytophagous insect fauna of the ragweed, *Ambrosia confertiflora*, in southern California. *Environmental Entomology* **4**, 301–306.

Goeden RD and Ricker DW (1976) The phytophagous fauna of the ragweed, *Ambrosia psilostachya*, in southern California. *Environmental Entomology* **5**, 1169–1177.

Harris P and Piper GL (1970) Ragweed (*Ambrosia* spp.: Compositae): its North American insects and the possibilities for its biological control. *Commonwealth Institute of Biological Control Technical Bulletin* **13**, 117–140.

Hu YP and Meng L (2007) Potential impacts of alien herbivorous insect *Ophraella communa* (Coleoptera: Chrysomelidae) on non-target plants in mainland China. *Chinese Journal of Ecology* **26**, 56–60.

Igrc J, Deloach CJ and Zlof V (1995) Release and establishment of *Zygogramma suturalis* F. (Coleoptera, Chrysomelidae) in Croatia for control of common ragweed (*Ambrosia artemisiifolia* L.). *Biological Control* **5**, 203–208.

Kiss L (2007) Is *Puccinia xanthii* a suitable biological control agent of *Ambrosia artemisiifolia*? *Biocontrol Science and Technology* **17**, 535–539.

Kovalev OV (1995) The role of phylogenetically young taxa in formation of new weeds: coenotic regulation of the co-evolution of phytophages. In *Proceedings of the 8th International Symposium on Biological Control of Weeds.* 2–7 February 1992, Canterbury, New Zealand. (Eds ES Delfosse and RR Scott) pp. 133–135. DSIR/CSIRO, Melbourne.

Kovalev OV and Runeva TD (1970) *Tarachidia candefacta* Hübn. (Lepidoptera, Noctuidae), a promising phytophage in the biological control of weeds of the genus *Ambrosia* L. *Entomology Review* **49**, 816.

Kovalev OV and Vechernin VV (1986) Description of a new wave process in populations with reference to introduction and spread of the leaf beetle *Zygogramma suturalis* F. (Coleoptera, Chrysomelidae). *Entomologicheskoye Obozreniye* **1**, 21–38.

LeSage L (1986) A taxonomic monograph of the nearctic galurucine genus *Ophraella* Wilcox (Coleoptera: Chrysomelidae). *Memoirs of the Entomological Society of Canada* **133**, 11–33.

McFadyen RE (1984) Annual ragweed in Queensland. In *Proceedings of the 7th Australian Weeds Conference.* 17–24 September 1984, Perth, WA. (Ed. RW Madden) pp. 205–209. Weed Society of Western Australia, Perth.

McFadyen RE (1985) The biological control programme against *Parthenium hysterophorus* in Queensland. In *Proceedings of the 6th International Symposium on Biological Control of Weeds.* 19–25 August 1984, Vancouver. (Ed. ES Delfosse) pp. 789–796. Agriculture Canada, Ottawa.

McFadyen RE (1987) The effect of climate on the stem-galling moth, *Epiblema strenuana*, in eastern Australia. In *Proceedings of the 8th Australin Weeds Conference.* 21–25 September 1987, Sydney. (Eds D Lemerle and AR Leys) pp. 97–99. Weed Society of New South Wales, Sydney.

McFadyen REC (1992) Biological control of parthenium weed in Australia. *Crop Protection* **24**, 400–407.

McFadyen RE and McClay AS (1981) Two new insects for the biological control of parthenium weed in Queensland. In *Proceedings of the 6th Australian Weeds Conference.* 13–18 September 1981, Gold Coast, Qld. (Eds

BJ Wilson and JT Swarbrick) pp. 145–149. Weed Society of Qld, Australia.

McFadyen RE and Weggler-Beaton K (2000) The biology and host specificity of *Liothrips* sp (Thysanoptera : phlaeothripidae), an agent rejected for biocontrol of annual ragweed. *Biological Control* **19**, 105–111.

Morrison IC (1984) The changing face of pollen allergy in Queensland. *Medical Journal of Australia Special Supplement*, 12–23.

Page AR and Lacey KL (2006) *Economic Impact Assessment of Australian Weed Biological Control.* CRC for Australian Weed Management, Adelaide.

Palmer WA and Goeden RD (1991) The host range of *Ophraella communa* LeSage (Coleoptera: Chrysomelidae). *Coleopterists Bulletin* **45**, 115–120.

Rybncek O and Jäger S (2001) Ambrosia (ragweed) in Europe. *Allergy and Clinical Immunology International: Journal of the World Allergy Organisation* **13**, 60–66.

Stuessy TF (1977) Heliantheae: systematic review. In *The Biology and Chemistry of the Compositae.* (Eds. VH Heywood, JB Harbone and BL Turner) pp. 621–672. Academic Press, London.

Wan FH, Wang R and Ding J (1995) Biological control of *Ambrosia artemisiifolia* with introduced insect agents, *Zygogramma suturalis* and *Epiblema strenuana*, in China. In *Proceedings of the 8th International Symposium on Biological Control of Weeds.* 2–7 February 1992, Canterbury, New Zealand. (Eds ES Delfosse and RR Scott) pp. 193–200. DSIR/CSIRO, Melbourne.

Wan F-H, Ma J, Guo J-Y and You L-S (2003) Integrated control effects of *Epiblema strenuana* (Lepidoptera: Tortricidae) and *Ostrinia orientalis* (Lepidoptera: Pyralidae) against ragweed, *Ambrosia artemisiifolia* (Compositae). *Acta Entomologica Sinica* **46**, 473–478.

Wang C-L and Chiang M-Y (1998) New record of a fastidious chrysomelid, *Ophraella communa* LeSage (Coleoptera: Chrysomelidae), in Taiwan. *Plant Protection Bulletin (Taichung)* **40**, 185–188.

Wang H, Li B and Meng L (2009) Host-selection behavior of exotic *Ophraella communa* (Chrysomelidae, Coleoptera) for non-target plants. *Chinese Journal of Biological Control* **25**, 120–124.

Withers TM, Christensen JA and Burwell CJ (1998) *Erixestus zygogrammae* Cave and Grissell (Hymenoptera : Pteromalidae): an exotic egg parasitoid of *Zygogramma bicolorata* Pallister (Coleoptera : Chrysomelidae) in Australia. *Australian Journal of Entomology* **37**, 83–84.

Wodehouse RP (1971) *Hayfever Plants.* Hafner Publishing, New York.

Yamazaki K, Imai C and Natahara Y (2000) Rapid population growth and food-plant exploitation pattern in an exotic leaf beetle, *Ophraella communa* LeSage (Coleoptera: Chrysomelidae), in western Japan. *Applied Entomology and Zoology* **35**, 215–223.

Zhang L, Yang X, Li W and Cui J (2005) A new record of *Ophraella communa* of mainland China. *Chinese Bulletin of Entomology* **42**, 227–228.

Anredera cordifolia (Ten.) Steenis – Madeira vine

Bill Palmer and Wilmot Senaratne

ABSTRACT

Madeira vine, *Anredera cordifolia*, is a serious environmental weed in Australia, New Zealand, Sri Lanka and subtropical areas of South Africa. It is also listed as an invasive plant in parts of the USA, including Hawaii, and in some Pacific Island countries. Two countries (South Africa and Australia) are cooperating to effect biological control. Exploration in Brazil and Argentina has been conducted by South African and Argentinean scientists and two chrysomelid leaf-feeding species have been selected for further study. *Plectonycha correntina* was sourced in Argentina, where its biology was elucidated and preliminary host-testing undertaken. It was then shipped to South Africa and from there to Australia, where comprehensive host-range testing was undertaken. The insect was found to be sufficiently host-specific and approved for release in 2011. The second chrysomelid, *Phenrica* sp., was studied in South Africa and remains a prospect. However, it has not been easy to rear and cultures in both South Africa and Australia have been lost. This is a relatively new biocontrol project and requires greater inputs, particularly in foreign exploration, because it is anticipated that Madeira vine will become an increasingly serious weed in Australia.

Key words: *Plectonycha correntina*, *Phenrica* sp., South America, Basellaceae.

INTRODUCTION

The vine *Anredera cordifolia* (Ten.) Steenis (family Basellaceae), which is known by various common names including Madeira vine, potato vine, lambs' tail vine and sweet mignonette, is a serious environmental weed in Australia (Vivian-Smith *et al.* 2007), New Zealand, Sri Lanka and subtropical areas of South Africa. It is also listed as an invasive plant in parts of the USA, including Hawaii, and in some Pacific Island countries.

Madeira vine is a vigorous perennial climber or scrambling shrub, forming dense mats that cover trees and shrubs. Stems are up to 30 m in length with succulent heart-shaped leaves and flowers on axillary drooping racemes that resemble lambs' tails in the summer and autumn (Fig. 1). The dense mats of Madeira vine cover the canopy stratum (Fig. 3), ultimately smothering supporting vegetation and threatening the biodiversity of riparian and rainforest communities (Floyd 1989). Reproduction is predominantly by aerial and subterranean tubers (Fig. 1) and seed production is rarely observed outside the native range (Swarbrick 1999; Vivian-Smith *et al.* 2007). Tuber density can be as high as 1500 m^{-2} and tubers can remain viable for up to 15 years. The persistence of these tubers may well present a difficulty for successful biocontrol.

It is thought that Madeira vine was introduced into Australia as an ornamental (Floyd 1989). It is found principally in coastal areas of Qld and NSW though it is now known from all states and territories except the NT (Fig.

Figure 1: Line drawings of *Anredera cordifolia* showing various identifying features. Drawing by W Roux, reprinted from Henderson (2001) with permission.

2). Eco-climatic modelling (Sutherst *et al.* 2007) suggests that much of coastal Australia is suitable habitat for Madeira vine (Fig. 2).

BIOLOGICAL CONTROL HISTORY

Madeira vine is a relatively recent target for biological control, probably because it is regarded as an environmental weed that has not yet caused significant economic losses to agricultural or pastoral activities. South Africa, through the Agricultural Research Council – Plant Protection Research Institute (ARC-PPRI), and Australia, through Biosecurity Qld and the NSW Department of Environment and Conservation, both recognised the need to find a biological control for this growing weed problem. The organisations have approached the problem cooperatively and have shared resources, knowledge and potential agent species. Investigations began in 2001 and continue to the present.

Following submissions from Qld, Madeira vine was approved as a target for biological control in Australia by the Natural Resource Management Standing Committee in 2006. This approval process is the mechanism for ensuring that any conflicts of interest about the plant are identified and it must be undertaken before agents are approved for release in Australia.

In December 2009, application was made to release *Plectonycha correntina* Lacordaire (Coleoptera: Chrysomelidae) (see below) in Australia. This application is believed to be the first biological control agent to be processed by the new procedures of Biosecurity Australia, which involve assessment through an import risk analysis.

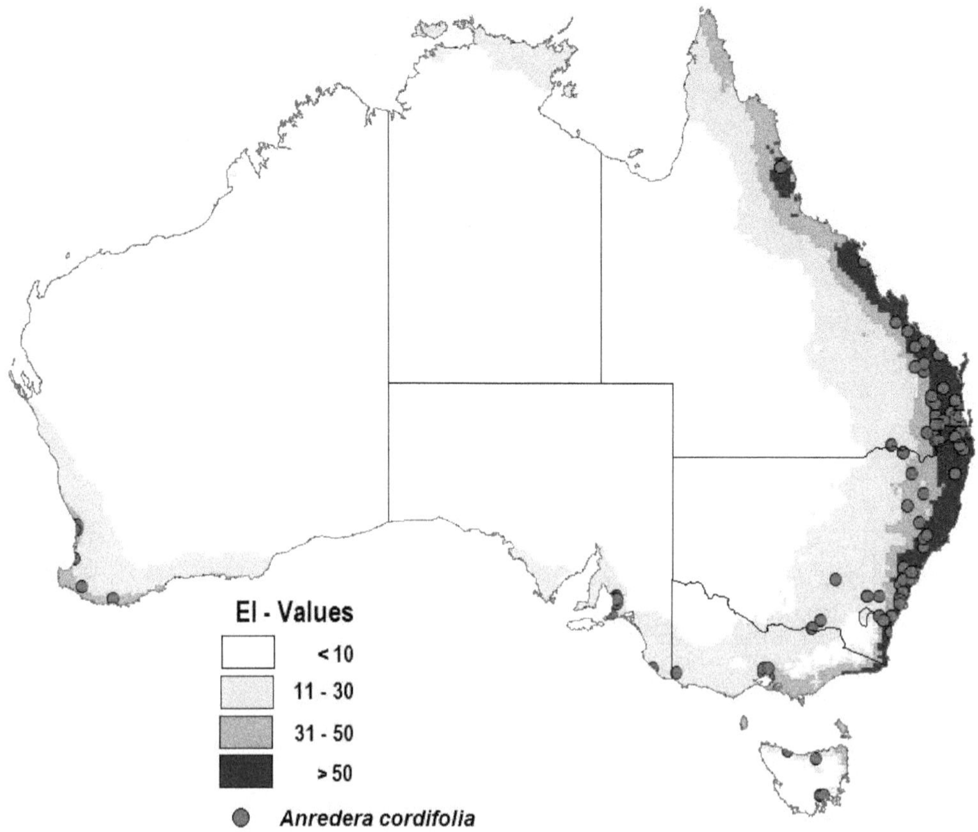

Figure 2: Distribution of *Anredera cordifolia* in Australia and its potential distribution predicted by the climate modelling software CLIMEX. EI values >30 are considered very favourable while values ≤10 are considered climatically unfavourable. Data based on herbarium records from the Australian Virtual Herbarium.

PLANT TAXONOMY

Madeira vine is a member of the family Basellaceae, which is a small family of mainly South American origin (Sperling 1987; Eriksson 2007). No member of the Basellaceae is native or endemic to Australia; Madeira vine is the only species in the family to be naturalised here. Only one other species of Basellaceae is presently under cultivation in Australia: *Basella alba* L. (Ceylon spinach) is grown as a leaf vegetable, particularly by the Asian community, but it is not grown commercially to any extent.

Two subspecies of *A. cordifolia* have been recognised (Xifreda *et al.* 1999). *Anredera cordifolia* subsp. *cordifolia* is autotriploid and virtually sterile while *A. cordifolia* subsp. *gracilis* (Meirs) Xifreda & Argimón produces viable pollen and fruit (Vivian-Smith *et al.* 2007). Both subspecies are thought to be in Australia but there have been no definitive studies conducted.

The fact that Madeira vine is the only basellaceaeous species naturalised in Australia should increase the opportunity to find suitable safe agents for release. A wider host range for a potential biocontrol agent might be acceptable since there are no closely related native or economically important species.

EXPLORATION

Madeira vine is native to South America and is known from Brazil and Argentina. Surveys for potential biocontrol agents have been largely opportunistic and have comprised short-duration visits by the South Africans, S Neser and colleagues, to South America, where agents for several weed targets were sought simultaneously. The search area included parts of Argentina and Brazil. These surveys have been supported by the resources of the USDA-ARS South American Biological Control Laboratory at Hurlingham, Argentina, where preliminary studies and shipment to South Africa were undertaken.

CANDIDATES

Two insect species have been selected for further study, in the belief that they are stenophagous and damaging.

Figure 3: A severe infestation of *Anredera cordifolia* showing the vine gaining a stranglehold on a small tree. Photo: Qld Government.

Plectonycha correntina **Lacordaire (Coleoptera: Chrysomelidae).** This beetle is regarded as the better prospect. This insect was known from Argentina, Paraguay and Uruguay (Monrós 1947) and the recent collections used for biological control investigations were made from both Argentina and Brazil.

Biology and preliminary host-range studies were undertaken at the South American Biological Control Laboratory, USDA-ARS, Hurlingham (Cagnotti *et al.* 2007). Further observations on the biology were made by Biosecurity Qld in Brisbane. Eggs are cylindrical and yellowish in colour. Groups of eight to 15 eggs are oviposited in two oblique rows, usually on the underside of leaves. There are four instars, the first three of which are gregarious. Newly hatched larvae have a white egg-shaped body and a dark brown head and pronotum. Once they begin feeding they become covered in a transparent gelatinous substance, which later incorporates frass and exuviae. Fourth instars disband and migrate to the lower sections of the host, where their gelatinous cover is shed before burrowing into the soil. Adults emerge after 20 days and are black with a reddish brown pronotum and elytra. Black spots may or may not occur on the elytra. As with larvae, adults are found on the underside of leaves. After a pre-oviposition period of six days, females lay over 500 eggs.

Comprehensive host-testing in Brisbane confirmed the results of Cagnotti *et al.* (2007). The insect was tested on 37 plant species selected by the centrifugal phylogenetic method; only *B. alba* supported, albeit poorly, the insect through its life-cycle (WA Palmer, unpub.). The insect was approved for release in Australia in February 2011 and the first releases made around Brisbane in April that year. It is too early to assess the establishment of this insect.

Phenrica **sp. (Coleoptera: Chrysomelidae).** The biology of this second chrysomelid was studied in South Africa (van der Westhuizen 2006). Eggs are cylindrical in shape. Initially orange-red with a pale tip, they become progressively paler from the tip as they mature. Eggs are generally oviposited in groups on the underside of leaves and eclose in six days. The larval stage takes approximately 14 days and consists of three instars. All instars are solitary and feed on leaves and young growth. Pupation occurs just below ground level and adults emerge after 17 days. Initially adults have a pale yellow head, pronotum and abdomen, and black forewing with yellowish white markings. After eight to 12 days males change to red-orange and females to red. Like the larvae, adults feed on leaves and young growth. Adults jump readily when disturbed. Adult females oviposit after 14 days, following which they are reproductive for an average of 188 days. On average, six eggs are oviposited per female per day. Under laboratory conditions the generation period (egg to egg) is seven to eight weeks, so three generations are possible during warmer months.

Although *Phenrica* sp. was cultured through several generations in South Africa, problems eventually arose with the culture and it was lost. Two attempts were made to import the South African culture into quarantine in Australia. Although the culture 'spluttered' through some generations, it also was eventually lost. An attempt to recollect the insect was made on a subsequent trip to South America, but it was not found.

DISCUSSION

This project is only in its early stages and clearly requires much more work before it will be successful. Further exploration within the native range is required to find more potential agents. Ideally, long-term study of the plant and its fauna in the native range is highly desirable. More information about the two subspecies of Madeira vine, particularly as they occur in Australia, would also be very useful.

The first insect for release, *P. correntina*, has many attributes of a good biocontrol agent. It is a chrysomelid beetle, is highly fecund and multivoltine, feeds as both larvae and adult, is very damaging in the laboratory and has a protective mechanism for the early instars. Rapid increases in population might therefore be expected, accompanied by commensurate damage.

ACKNOWLEDGEMENTS

Biological control efforts in Australia have been largely dependent on the initiatives first taken by, and the close cooperation with, the ARC-Plant Protection Research Institute and particularly S Neser, who explored in the native range and selected the species as potential agents, and L van der Westuizen who provided cultures for Australia. The CLIMEX model was developed by KADW Senaratne. Valuable comments on the manuscript were provided by colleagues at Biosecurity Queensland.

REFERENCES

Cagnotti C, Mc Kay F and Gandolfo D (2007) Biology and host specificity of *Plectonycha correntina* Lacordaire (Chrysomelidae), a candidate for the biological control of *Anredera cordifolia* (Tenore) Steenis (Basellaceae). *African Entomology* **15**, 300–309.

Eriksson R (2007) A synopsis of Basellaceae. *Kew Bulletin* **62**, 297–320.

Floyd AG (1989) The vine weeds of coastal rainforest. In *Proceedings of the 5th Biennial Noxious Plant Conference*. (Ed. P Gorham) pp. 109–115. NSW Agriculture & Fisheries, Sydney.

Henderson L (2001) *Alien Weeds and Invasive Plants: A Complete Guide to Declared Weeds and Invaders in South Africa*. Plant Protection Research Institute, Agricultural Research Council, South Africa.

Monrós F (1947) Revisión del género *Plectonycha* Lac. *Anales de la Sociedad Científica Argentina* **144**, 46–64.

Sperling CR (1987) Systematics of the Basellaceae. PhD thesis. Harvard University, Cambridge, Mass.

Sutherst RW, Maywald GF and Kriticos D (2007) *CLIMEX Version 3 User's Guide*. Hearne Scientific Software, Melbourne.

Swarbrick JT (1999) Seedling production by Madeira vine (*Anredera cordifolia*). *Plant Protection Quarterly* **14**, 38–39.

van der Westhuizen L (2006) The evaluation of *Phenrica* sp. 2 (Coleoptera: Chrysomelidae: Alticinae), as a possible biological control agent for Madeira vine, *Anredera cordifolia* (Ten.) Steenis in South Africa. Rhodes University, Grahamstown, South Africa.

Vivian-Smith G, Lawson BE, Turnbull I and Downey PO (2007) The biology of Australian weeds. 46. *Anredera cordifolia* (Ten.) Steenis. *Plant Protection Quarterly* **22**, 2–10.

Xifreda CC, Argimon S and Wulff AF (1999) Infraspecific characterization and chromosome numbers in *Anredera cordifolia* (Basellaceae). *Thaiszia* **9**, 99–108.

Argemone mexicana L. and *Argemone ochroleuca* Sweet – Mexican poppy

Tim A Heard and Ricardo Segura

ABSTRACT

Mexican poppy is the collective common name in Australia for the closely related species *Argemone mexicana* and *Argemone ochroleuca* (Papaveraceae). Three potential prospects for biocontrol were identified. Two species of Languriidae (*Languria sanguinicollis* and *Languria convexicollisi*) are common, damaging and probably specific on *Argemone* spp. in Mexico. The root weevil *Conotrachelus leucophaeatus* is also very common, widespread and damaging especially on *A. ochroleuca* in Jalisco state. Further effort in the evaluation of these insects has been suspended as these weeds are considered to be low priority in Australia. However, they remain serious weeds in warmer parts of the world and biocontrol efforts could lead to useful outcomes there.

Key words: Papaveraceae, invasive weeds, agent selection.

INTRODUCTION

In Australia, Mexican poppy is the collective common name for the closely related species *Argemone mexicana* L. and *Argemone ochroleuca* Sweet (Papaveraceae). *Argemone mexicana* is now a pan-tropical invader, but *A. ochroleuca* is the most common species in Australia (Fig. 1). They are annual prickly herbs to 1 m tall. In the native range of North America, centred in Mexico, they grow through the dry winter season but elsewhere, including Australia, they may grow year-round provided there is soil moisture.

Members of the genus *Argemone* are readily identified by the presence of yellow sap. The two species are distinguished by the colour of their petals – bright yellow in *Argemone mexicana* and pale yellow in *A. ochroleuca* (Fig. 2). In Australia *Argemone mexicana* can also be confused with *A. subfusiformis* G.B.Ownbey, American poppy. Both have bright yellow petals, but those of *A. mexicana* are 1.7–2.5 cm wide while those of *A. subfusiformis* are 2.8–3.3 cm wide. Mexican poppy produces large numbers of small seeds, most of which fall beneath the parent plant, resulting in a locally dense population. Other seeds are spread by water, animals or machinery (Parsons and Cuthbertson 1992). The majority of seeds normally do not germinate during their first season; they enter the seed bank and continue to produce seedlings for several years (Karlsson *et al.* 2003).

These two ubiquitous tropical weed species have no obvious restriction to particular agronomic or environmental situations (Karlsson *et al.* 2003). They are serious weeds of river channels and crops in many parts of Australia. They are not particularly aggressive competitors of crops or established pastures as they persist only until the crop or perennial species become established, then gradually disappear (Parsons and Cuthbertson 1992). However, they have become a problem weed of various crops in parts of Africa, Asia and South America. They are highly toxic to humans and have resulted in many deaths, particularly associated with consumption of contaminated mustard oil (Babu *et al.* 2006). Plants are unpalatable and

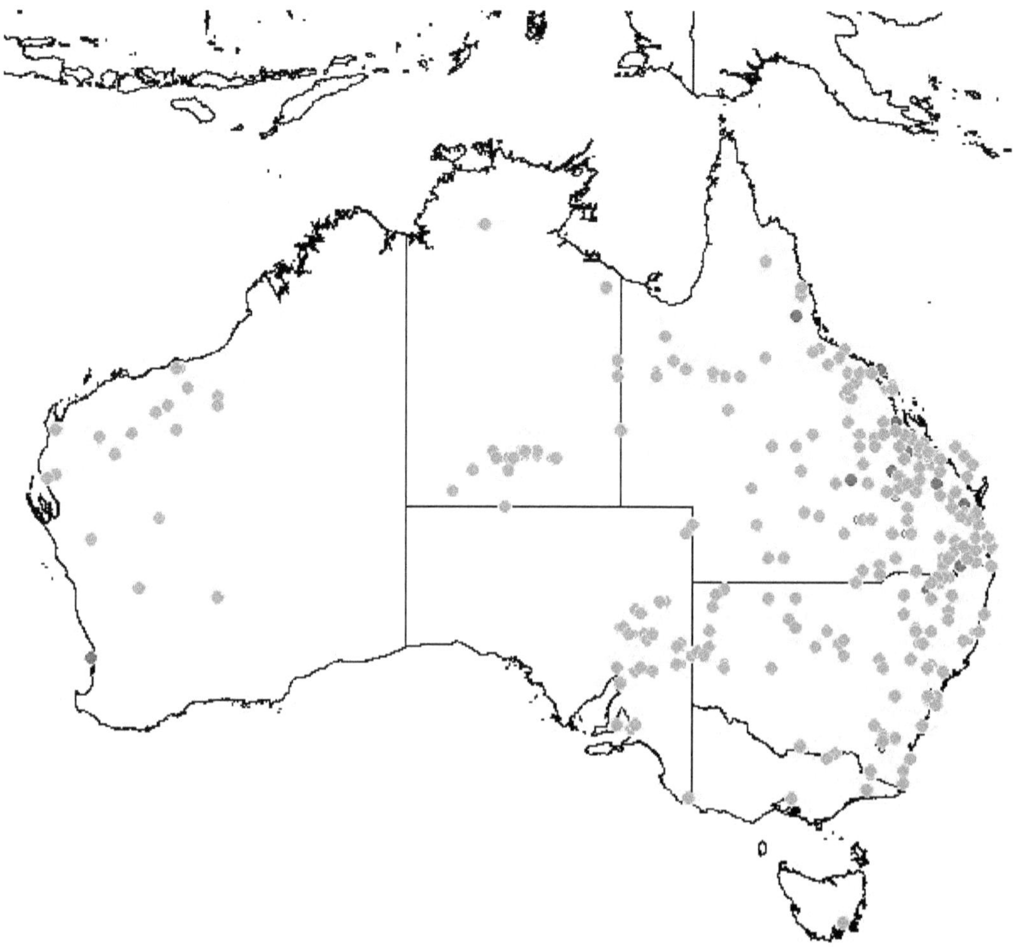

Figure 1: The distribution of *Argemone mexicana* (red dots) and *Argemone ochroleuca* (green dots) in Australia. Specimen data reproduced from Australia's Virtual Herbarium with permission of the Council of Heads of Australasian Herbaria Inc.

thus rarely eaten by stock, but plant material mixed with hay or seed mixed with grain may poison animals eating such products (Holm *et al.* 1977).

BIOLOGICAL CONTROL HISTORY

Exploration in the native range for natural enemies was undertaken from 1998 to 2002. The project was terminated prior to fully testing any agents as there were doubts that this weed would cause sufficient impact to warrant a biocontrol program (Grace *et al.* 2006).

South Africa has more recently developed a biocontrol program against this weed and has begun to import agents into quarantine for testing. Two species of curculionid beetles, preliminarily identified as *Conotrachelus* cf. *leucophaeatus* Fahraeus and *Sirocalodes* cf. *wickhami* (Champion), were collected in Mexico and brought into South Africa for evaluation. Preliminary host-specificity trials seem to indicate a clear preference for the target weed (van der Westhuizen 2011).

PLANT TAXONOMY

Argemone is a monophyletic genus with a centre of distribution in south-west USA and northern Mexico, which has diversified into other parts of the Americas (Schwarzbach and Kadereit 1999). Three species occur in Australia – *A. mexicana*, *A. ochroleuca* and *A. subfusiformis*. *A. mexicana* was determined as diploid (n=14), *A. ochroleuca* as hexaploid (n=42) with a regular meiosis and *A. subfusiformis* as a triploid (n=18–20) with an irregular meiosis (Lozzia *et al.* 2006).

EXPLORATION

Native range

A total of 23 specimens of *Argemone* collected during exploration work in Mexico were identified by Jerzy Rzedowski, of the Instituto de Ecología, Pátzcuaro, Michoacán, Mexico. Most were *A. ochroleuca* and *A. mexicana*. *A. mexicana* occurs on the Gulf and Pacific coasts. *A. ochroleuca* is distributed more centrally on the Mexican

Figure 2: *Argemone mexicana* (left) and *Argemone ochroleuca* (right) growing together in experimental garden of the CSIRO Mexican Field Station. Photo: R Segura, CSIRO.

altiplano and Trans-Mexican volcanic belt. Other species encountered include *A. grandiflora* Sweet subsp. *grandiflora*, *A. arida* Rose and *A. echinata* G.B.Ownbey.

A total of 115 sites were surveyed for agents. Some sites were visited multiple times, resulting in a total of 126 collections (a collection refers to stopping at a site and collecting all insects and fungi found). Most sites were in Mexico with fewer in Guatemala, Venezuela, Puerto Rico and the Dominican Republic (Fig. 3). Mexican sites were in the states of Veracruz, Tabasco, Morelos, Puebla, Tlaxcala, Estado de Mexico, Hidalgo, Queretaro, Guanajuato, Michoacan, Jalisco, Colima, Oaxaca and Chiapas.

Insects were collected directly from the plant, and plant material including seeds, flowers and stems was collected to rear internally feeding insects. Where possible, the plant part that an insect was using was noted. Species were defined as phytophagous based on field observations, rearing from plant material and information in the literature. Surveys for natural enemies were primarily conducted by staff at the CSIRO Mexican Field Station, particularly Ricardo Segura, Moisés Martínez and Manuel Juárez. Occasionally, Australian-based scientist Tim Heard participated in the surveys.

The majority of species were identified to species by taxonomic specialists (see acknowledgements for details), although some could only be identified to genus or higher. Unidentified species were grouped into morphospecies.

CANDIDATES

Natural enemies (native range)

A total of 698 specimens were collected. Of these, 274 were sent for identification and the remainder were held at the Mexican Field Station. The specimens were identified into 36 taxa. Thirteen of these species were excluded as they were not herbivores, leaving a total of 23 natural enemies (Table 1). The few insect species collected from

Figure 3: Sites surveyed for natural enemies of *Argemone mexicana* and *Argemone ochroleuca*.

Table 1: Natural enemies of *A. mexicana* and *A. ochroleuca* collected in surveys, 1998–2002

Order	Family	Species	Abundance	Larval feeding site	Adult feeding site	Host-specificity
Coleoptera						
	Chrysomelidae					
		Omophoita octomaculata Crotch	Rare	Leaf	Leaf	Polyphagous
	Curculionidae					
		Conotrachelus leucophaeatus Fahrs.	Abundant	Roots	All plant parts	Poss. specific
		Conotrachelus sp.	Abundant	Roots	All plant parts	Unknown
		Sirocalodes wickhami (Ch.)	Abundant	Fruits, immature seeds	All plant parts	Poss. specific
		Curculionidae unidentified sp.	Common	Stem	Stem	Unknown
Erotylidae						
		Erotylidae unidentified sp.	Common	Stem	Stem	Polyphagous
	Languriidae					
		Languria convexicollis Horn	Abundant	All plant parts	All plant parts	Poss. specific
		Languria sanguinicollis Chrevrolat	Abundant	All plant parts	All plant parts	Poss. specific
		Languria poss. *mozardi*	Rare	All plant parts	All plant parts	Unknown
	Tenebrionidae					
		Blapstinus sp.	Rare	Stem	Stem	Unknown
		Lobometopon sp.	Rare	Stem	Stem	Unknown
Hemiptera						
	Cicadellidae					
		Agrosoma bispinella Medler	Rare	Leaf	Leaf	Unknown
		Agrosoma syklis Medler	Common	Leaf	Leaf	Unknown
		Scaphytopius sp.	Common	All plant parts	All plant parts	Polyphagous
	Coreidae					
		Largus cinctus Herrich-Schaeffer	Common	Leaf	Leaf	Polyphagous
	Miridae					
		Lygus mexicanus Kelton	Rare	All plant parts	All plant parts	Polyphagous

Order	Family	Species	Abundance	Larval feeding site	Adult feeding site	Host-specificity
	Psyllidae					
		Poss. *Mitrapsylla* sp.	Common	Apical growth	Apical growth	Poss. specific
	Rhyparochromidae					
		Atrazonotus umbrosus Distant	Rare	Fruit	Fruit	Unknown
Lepidoptera						
	Noctuidae					
		Copitarsia sp.	Abundant	Leaf		Unknown
	Tortricidae					
		Amorbia sp.	Rare	Leaf		Unknown
		Argyrotaenia sp.	Rare	Leaf		Unknown
		Clepsis sp.	Rare	Leaf		Prob. polyphagous
Hyphomycete						
		Cercosporidium guanicense (F. Stevens) Deighton	Abundant	Leaf		Poss. specific

Rare=1–2 specimens collected; common=2–4 specimens collected; abundant = >4 specimens collected.

species other than *A. mexicana* and *A. ochroleuca* are not included in Table 1. For example, *Neoterpes ephelidaria* (Hulst) was collected only from *A. arida*.

A pathogen identified by H Evans as *Cercosporidium guanicence* (F. Stevens) Deighton (synonym = *Passalora guanicensis* (F.Stevens) U.Braun & R.F.Castañeda) was found on one collection. It is reported from other countries in tropical America, often in association with the pathogen *Deightoniella argemonensis* Pollack and Matthews (Pollack and Matthews 1976); both pathogens may have potential as agents although, in our observations, they attacked older and senescent leaves.

Agents released

No agents have been released in Australia or any other country. In Australia, the project ceased before it reached this stage (Grace *et al.* 2006). The continuing project in South Africa may result in agent releases.

Promising agents

Langunia sanguinicollis **Chevrolat (Coleoptera: Languriidae).** *Languria sanguinicollis* was very common in our collections, with 95 specimens collected in 18 collections in the Mexican states of Veracruz, Oaxaca, Chiapas, Puebla and Queretaro. All collections were made on *A. mexicana* with the exception of one individual from *A. grandiflora* subsp. *grandiflora* Sweet. *Languria sanguinicollis* breeds in the roots and adults eat all plant parts including pollen (Fig. 4). It has one or maybe two generations per year. This species has a more southern and/or eastern distribution than *L. convexicollis*, the other common species on *Argemone* in Mexico (see below).

Figure 4: An adult of *Languria sanguinicollis* on a flower of *Argemone mexicana* in Mexico. Photo: R Segura, CSIRO.

A preliminary host-specificity study was conducted at a field site called Actopan between Veracruz and Xalapa where *A. mexicana* occurs with the related plant *Bocconia frutescens* L. This test examined adult feeding and survival in a no-choice situation. Adults were put into bags on a leaf of a living plant for three days. The adults were then removed and counted, and the area of damage estimated. Three replications of each treatment were done. Adults of *L. sanguinicollis* survived on *A. mexicana*, but died quickly on the test plant. Adult feeding was restricted to *A. mexicana*.

There are 14 members of the genus *Languria* in North America and another four species in Central America. They are generally bicoloured in North America (red thorax, dark elytra) but unicoloured in Central America (Vaurie 1948). The larvae of the Languriidae of North America are stem-borers. The adults feed on pollen and leaves of the host plant. The only species to become an economic pest in North America is the clover stem-borer *Languria mozardi* Latrielle, which infests red clover and alfalfa (and many other species in various families) throughout the USA, southern Canada and northern Mexico. Feeding by the larvae interferes with the vitality of the plant and sometimes the plant breaks off above the oviposition hole (Vaurie 1948).

Languria convexicollis **Horn (Coleoptera: Languriidae).** *Languria convexicollis* was less common than *L. sanguinicollis* in our collections, with 24 specimens collected, all in the state of Jalisco. All collections were made on *A. ochroleuca* with the exception of one individual, from *A. mexicana*. In the literature, *L. convexicollis* is also known from North America, recorded in British Columbia (Canada) and California, Arizona and Utah. At the US National Museum there are 21 specimens from California labelled 'borer in *Argemone platyceros* var. *hispida*'. The males of this species have a very convex thorax (Vaurie 1948). *Languria convexicollis*, like *L. sanguinicollis*, breeds in the roots and the adults eat all plant parts.

Conotrachelus leucophaeatus **Fahraeus (Coleoptera: Curculionidae).** The root weevil *Conotrachelus leucophaeatus* breeds on roots and stems and the adults eat all plant parts including flowers (Fig. 5). Feeding by the larvae can seriously damage the host plant, leading to premature collapse and death. *C. leucophaeatus* is very common; during our surveys we collected over 100 specimens in 27 collections in the Mexican states of Jalisco, Guanajuato, Hidalgo, Oaxaca and Veracruz. It was recorded on the host species *A. ochroleuca*, *A. mexicana* and *A. echinata*. It is especially common on *A. ochroleuca*

Figure 5: Adults of *Conotrachelus leucophaeatus* feeding on a fruit of *Argemone* sp. in Mexico. Photo: R Segura, CSIRO.

in Jalisco. *C. leucophaeatus* has one or maybe two generations per year. It was successfully reared on Mexican poppy plants grown at the La Aguada field plot of the Mexican Field Station in 1999.

A preliminary host-specificity study was conducted at a field site called Actopan between Veracruz and Xalapa, where *A. mexicana* occurs with related plant *B. frutescens*. This test examined adult feeding and survival in a no-choice situation. Adults were put into bags on a leaf of a living plant for three days. The adults were then removed and counted, and the area of damage estimated. Three replications of each treatment were done. The results show that adult feeding by *C. leucophaeatus* is restricted to *A. mexicana*.

An artificial diet has been useful for rearing *C. leucophaeatus* and *Languria* spp. *C. leucophaeatus* can be reared on diet for multiple generations. However, the adults often emerge deformed. The pupae appear very healthy and it seems to be a physical problem related to eclosion rather than a nutritional deficiency.

***Sirocalodes wickhami* (Champion) (Coleoptera: Curculionidae).** The weevil *Sirocalodes wickhami* breeds in flowers, fruits and seeds and the adults eat the same. Feeding by the larvae can destroy a large proportion of the seeds. It is known only from Puebla and Oaxaca states in Mexico. It is univoltine. It was less common in our collections than *C. leucophaeatus* and the *Languria* spp. *Sirocalodes tescorum* (Fall) was locally common on *Argemone munita* Dur. and Hilg. at high elevations in California (Goeden and Ricker 1985).

DISCUSSION

The fauna on *Argemone* appears to be depauperate with only 23 species of herbivores found. Of these, four species (17%) are likely to be highly specialised. Our 23 species are comparable with the 15 species on *Argemone corymbosa* Greene and *A. munita* in southern California (Goeden and Ricker 1985).

The four specialised species hold considerable promise as agents, but they are unlikely to be further tested in Australia unless Mexican poppy increases in importance. However, these agents may be beneficial to other countries where Mexican poppy is more important. Researchers in South Africa are pursuing two of these agents, or closely related ones.

To prove the host-specificity of the promising agents, they need to be imported into quarantine. It is difficult to do host-specificity studies in Mexico as there are few other species of Papaveraceae there. The one species found in the field in Mexico, *Bocconia frutescens*, was useful in delineating the host-range of the candidates, but more tests on other species of Papaveraceae and related families will be necessary.

ACKNOWLEDGEMENTS

Tania Yonow (CSIRO) designed the collection database. Karen Bell drew Figure 3. J Rzedowski, Instituto de Ecología, Pátzcuaro, Michocan, México identified the plant species. Harry C Evans identified the pathogen. The following taxonomists kindly identified insect specimens: Douglass R Miller, Jens Prena, John W Brown, Michael G Pogue, Natalia J Vandenberg, Norman E Woodley, Steven W Lingafelter, Stuart H McKamey, Thomas J Henry, Piotr Wegrzynowicz and Robert Anderson. Field and laboratory work were conducted by Moises Martinez and Manuel Juarez (CSIRO-MFS).

REFERENCES

Babu CK, Khannaa SK and Das M (2006) Safety evaluation studies on argemone oil through dietary exposure for 90 days in rats. *Food and Chemical Toxicology* **44**, 1151–1157.

Goeden RD and Ricker DW (1985) Prickly poppies, *Argemone corymbosa* and *Argemone munita*, in southern California: native weeds attacked by few insects. *Annals of the Entomological Society of America* **78,** 214–216.

Grace B, Paynter J, van Klinken R and Wirf L (2006) Biocontrol of weeds in the Northern Territory: what has it achieved? In *Proceedings of the 15th Australian Weeds Conference.* 24–28 September 2006, Adelaide. (Eds C Preston, JH Watts and ND Crossman) pp. 553–556. Weed Management Society of South Australia, Adelaide.

Holm L, Plucknett D, Pancho J and Herberger J (1977) *The World's Worst Weeds: Distribution and Biology.* University of Hawaii Press, Honolulu.

Karlsson LM, Tamado T and Milberg P (2003) Seed dormancy pattern of the annuals *Argemone ochroleuca* and *A. mexicana* (Papaveraceae). *Flora* **198**, 329–339.

Lozzia ME, Andrada AR and Cristobal ME (2006) Cytotaxonomic studies in species of Argemone. *Lilloa* **43**, 99–106 [only abstract seen].

Parsons WT and Cuthbertson EG (1992) *Noxious Weeds of Australia.* Inkata Press, Melbourne.

Pollack FG and Matthews FD (1976) *Deightoniella argemonensis*, a new fungus on Mexican prickle poppy, associated with *Cercosporidium guanicense. Mycologia* **68**, 1093–1097.

Schwarzbach AE and Kadereit JW (1999) Phylogeny of prickly poppies, *Argemone* (Papaveraceae), and the evolution of morphological and alkaloid characters based on ITS nrDNA sequence variation. *Plant Systematics and Evolution* **218**, 257–279.

van der Westhuizen L and Mpedi P (2011) The initiation of a biological control programme against *Argemone mexicana* L. and *Argemone ochroleuca* Sweet subsp. *ochroleuca* (Papaveraceae) in South Africa. *African Entomology* **19**, 223–229.

Vaurie P (1948). A review of the North American Languriidae. *Bulletin of the American Museum of Natural History* **92,** 119–155.

Asparagus asparagoides (L.) Druce – bridal creeper

Louise Morin and John K Scott

ABSTRACT

Bridal creeper (*Asparagus asparagoides*, Asparagaceae), a scrambling vine of South African origin, invades a range of natural habitats in warm temperate climates of Australia and is a weed of citrus orchards. It can quickly dominate and smother understorey vegetation and directly threatens several native plant species. It produces an extensive mat of below-ground storage organs that ensure persistence over many years. A biological control program was initiated in the late 1980s. Research initially clarified the identity of the target plant and its relatives, and delimited its distribution in South Africa. Surveys for natural enemies principally targeted the Western Cape province and were followed by in-depth research at selected sites to gather data on the phenology, impact and host-range of candidate biological control agents. An undescribed Erythroneurini leaf-hopper (formerly referred to as *Zygina* sp.), the rust fungus *Puccinia myrsiphylli* and a *Crioceris* sp. leaf beetle were imported into Australian containment facilities for comprehensive host-specificity testing. All three agents were approved for release between 1999 and 2002. A *Eurytoma* sp. seed wasp was also considered a promising candidate agent, but host-specificity testing could not be performed due to rearing difficulties and concerns were raised about the conflict of interest with producers of cultivated asparagus seed. Large-scale releases of the leaf-hopper and rust fungus were carried out in partnership with community groups, land managers and schools. The rust fungus has had a major impact in reducing bridal creeper populations, particularly in moist coastal areas where climatic conditions are conducive to rust epidemics. The leaf-hopper can also adversely affect bridal creeper, but its populations have a tendency to fluctuate from year to year, hence limiting its impact. Establishment of the leaf beetle has been extremely poor despite considerable release efforts and this agent is not currently contributing to the program.

Key words: *Myrsiphyllum*, rust fungus *Puccinia myrsiphylli*, Erythroneurini leaf-hopper, leaf beetle *Crioceris* sp.

INTRODUCTION

Bridal creeper (*Asparagus asparagoides* (L.) Druce) (Asparagaceae) is an introduced, dense, scrambling vine that is capable of invading a range of habitats in warm temperate climates of Australia, ranging from coastal heath on sandy dunes to eucalyptus and banksia woodlands or forests (Morin *et al.* 2006a). It is naturalised in all states, but not in the two territories (Fig. 1). It is most prevalent in temperate and Mediterranean regions of southern Australia, although it has been found from time to time in cooler areas of Qld (e.g. vicinity of Toowoomba). Bridal creeper has been the target of an eradication campaign for several years in Tas (Cooper and Warren 2006), but new small infestations keep emerging. The species has reached most of its predicted distribution

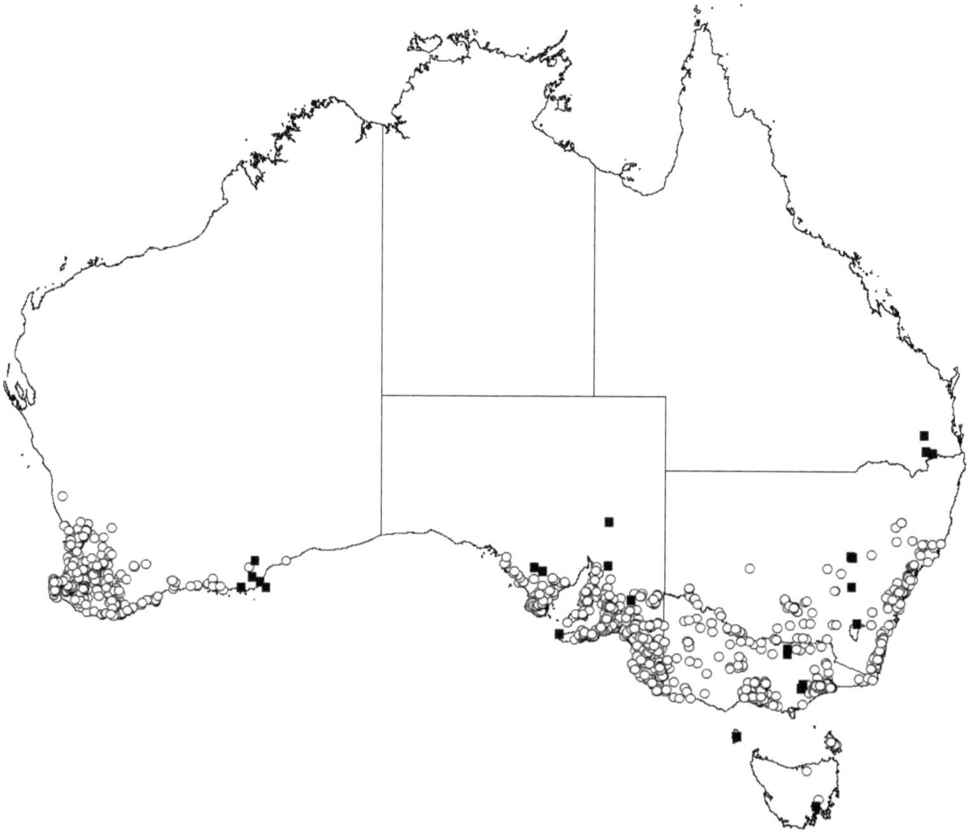

Figure 1: Distribution of bridal creeper (*Asparagus asparagoides*) in Australia. Open circles are based on information obtained during the release program of the biological control agents up to 2006. Solid squares correspond to verified herbarium records from Australia's Virtual Herbarium (Council of Heads of Australasian Herbaria 2006). Reproduced from Morin *et al.* (2006a).

based on climate modelling, except in non-coastal parts of south-east Qld, throughout Tas and along the west coast and immediate hinterland of WA from Perth to Kalbarri (Scott and Batchelor 2006).

Bridal creeper was ranked in the top 20 Weeds of National Significance in the late 1990s (Thorpe and Lynch 2000). It can invade both disturbed and undisturbed natural ecosystems, where it can quickly dominate and smother understorey vegetation and change the structure, floristic composition and ecology of the system (Humphries *et al.* 1991; Turner and Virtue 2006) (Fig. 2a). In WA, for example, Turner *et al.* (2008b) found that areas infested by bridal creeper had only half the number of native plant species present in nearby non-infested areas. At least 16 native plant species in NSW and SA are directly threatened by bridal creeper (Downey 2006), including the orchids *Pterostylis bryophila* Jones, *Pterostylis arenicola* M.A.Clem. & J.Stewart (Orchidaceae) and *Pterostylis* sp. Halbury (Sorensen and Jusaitis 1995; Quarmby 2010) and the small shrub *Pimelea spicata* R.Br. (Thymelaeaceae) (Willis *et al.* 2003). Bridal creeper is also recognised as a threat to four endangered ecological communities in NSW: littoral rainforest, river-flat eucalypt forest on coastal floodplains, swamp-oak floodplain forest and subtropical coastal floodplain forest (Coutts-Smith and Downey 2006).

Bridal creeper is not a weed of agricultural systems, except for citrus orchards in irrigation areas of NSW, Vic and SA where it can smother trees and displace citrus roots, leading to reduced fruit production (Kwong and Holland-Clift 2004). The cost of controlling bridal creeper in orchards, using slashing and sometimes herbicides, can be as high as $A2000 per ha per year.

Bridal creeper is a geophyte with extensive below-ground storage organs that produce above-ground foliage consisting of thin, wiry, twisting shoots and cladodes (leaf-like stems) (Morin *et al.* 2006a). Shoots can grow as long as 6 m while climbing understorey shrubs and small trees. The below-ground biomass of mature plants consists of a mat of branching rhizomes bearing numerous small fleshy tubers, often tapering into a small root, and can represent up to 90% of the plant's total biomass (Raymond 1999) (Fig. 2b). Sweetly scented flowers produced in the axils of cladodes, generally in late winter to early

Figure 2: a) Typical infestations of bridal creeper (*Asparagus asparagoides*) in a natural ecosystem. b) Bridal creeper foliage and below-ground biomass, consisting of fleshy tubers attached to a network of rhizomes. c) Adult of the undescribed Erythroneurini leaf-hopper (formerly referred to as *Zygina* sp.) on bridal creeper. d) Rust fungus *Puccinia myrsiphylli*. Photos a, c and d: CSIRO. Photo b: J Hosking, NSW Government.

spring in Australia, develop into round red berries that are primarily dispersed by frugivorous birds (Raymond 1999; Stansbury 2001). Above-ground parts die back annually, usually at the onset of higher temperatures and drier conditions, but below-ground organs persist to the next growing season. New shoots emerge from adventitious buds on rhizomes either slightly before or with the onset of late summer or autumn rains.

Herbicides, hand-weeding and grazing have been used to manage bridal creeper in natural ecosystems (Morin *et al.* 2006a). However, the costs and potential off-target damage associated with these control methods and the inaccessibility of some infested sites preclude their widespread use. The extensive distribution of bridal creeper in Australia, its environmental impact and limitations of traditional control methods have justified the substantial investment made in a classical biological control program, which was estimated in 2002 to have reached approximately $A6 million (Morin *et al.* 2006b; Morin and Edwards 2006).

BIOLOGICAL CONTROL HISTORY

So far, there has been no attempt to establish biological control programs for bridal creeper in any countries other than Australia. Bridal creeper was proposed as a target species for biological control in Australia by the SA Animal and Plant Control Commission and was approved in 1991 by the Standing Committee on Agriculture (equivalent to the current Natural Resource Management Standing Committee) (Australian Weeds Committee 2011).

The biological control program for bridal creeper was initiated by CSIRO in the late 1980s. Research first focused on clarifying the identity of the target plant and its relatives and determining its precise distribution in South Africa, the country of origin (Kleinjan and Edwards 1999). Once appropriate areas were located, field surveys of natural enemies were undertaken, followed by in-depth research at selected sites to gather data on the phenology, impact and host-range of candidate biological control agents (Scott and Kleinjan 1991; Witt and Edwards 2000, 2002; Kleinjan *et al.* 2004 a, b; Kleinjan and

Edwards 2006). Subsequent research conducted in Australian containment facilities focused on the most promising candidate agents and culminated with the release in Australia of an undescribed Erythroneurini leaf-hopper (Fig. 2c), the rust fungus *Puccinia myrsiphylli* (Thuem) Wint. (Fig. 2d), and a *Crioceris* sp. leaf beetle between 1999 and 2002 (Batchelor *et al.* 1999; Morin *et al.* 2001, 2006b; KL Batchelor and TL Woodburn, unpub. data).

PLANT TAXONOMY

In their latest Australian treatment of the Asparagaceae, Clifford and Conran (1987) included bridal creeper in the genus *Myrsiphyllum*, based on Obermeyer's (1984) revision of the family, which divided it into three genera, *Asparagus, Protasparagus* and *Myrsiphyllum*. In a subsequent revision, Malcomber and Demissew (1993) concluded that the Asparagaceae contains only one genus, *Asparagus*, with two subgenera, *Asparagus* (which includes Obermeyer's *Protasparagus*) and *Myrsiphyllum*. The decision that the family should comprise a single genus was upheld by Fellingham and Meyer (1995), but they argued that the subgeneric status of *Asparagus* and *Myrsiphyllum* is not warranted. A recent molecular phylogeny of *Asparagus* found evidence, however, that southern African species are potentially paraphyletic, supporting subdivision of *Asparagus* into more than three groups (Fukuda *et al.* 2005).

Following an extensive examination of herbarium specimens and field surveys in South Africa, Kleinjan and Edwards (1999) identified two distinct forms of *A. asparagoides*: a widespread form that corresponds to bridal creeper found in Australia and another form that is restricted to the south-western part of the Western Cape province of South Africa. The two forms may warrant separation at the species level. The south-western Cape form is differentiated from bridal creeper by its rhizomes that grow primarily upwards towards the soil surface, larger tubers and more waxy and elongated cladodes. It was not reported from Australia until 2004, when a few small populations were located in south-eastern SA and western Vic (Coles *et al.* 2006). Unless otherwise specified, the following sections only refer to the widespread form, bridal creeper.

EXPLORATION

The native range of bridal creeper is southern Africa (Obermeyer 1984; Kleinjan and Edwards 1999). Besides Australia, bridal creeper has naturalised in California, the Azores, Canary Islands, Portugal and Sicily, and is considered a weed in New Zealand (see Morin *et al.* 2006a for references).

In South Africa, bridal creeper is widely distributed and occurs in the three main rainfall zones (winter, summer and even rainfall), but is not considered a weed (Kleinjan and Edwards 1999). The winter-rainfall region of the Western Cape province climatically matches southern Australia, where bridal creeper is most prevalent. The region was identified as the likely geographical source of founder populations in Australia by comparing sequences of non-coding DNA regions from the chloroplast (*trn*L-*trn*F) and nuclear ribosomal (ITS1 and ITS2) genomes of South African and Australian bridal creeper accessions (L Morin, unpub. data).

In the late 1980s, CSIRO began explorations for natural enemies of bridal creeper in South Africa, specifically targeting areas in the Western Cape (Scott and Kleinjan 1991). Searches were carried out by JK Scott, followed by PB Edwards with the assistance of locally engaged staff CA Kleinjan and ABR Witt, and concentrated on organisms that attack either above-ground or below-ground organs of bridal creeper. Other members of the Asparagaceae, formerly in the genus *Myrsiphyllum*, were also examined for presence of natural enemies when they were encountered. Six study sites throughout the region were established and visited at regular intervals to gather data on the seasonal phenology of the plant and its associated natural enemies.

CANDIDATES

Natural enemies

In Australia, surveys of 34 bridal creeper infested sites carried out in WA in 1996–97 prior to the release of biological control agents showed very little leaf damage and no fruits or seed damage (KL Batchelor and JK Scott, unpub. data). In Vic, the weevil *Phlyctinus callosus* Boheman and larvae of an unidentified Lepidoptera were observed feeding on bridal creeper foliage in the mid 1990s, while light brown apple moth (*Epiphyas postvittana* Walker) larvae were seen eating ripe fruit pulp but not damaging seeds (Raymond 1999). The Victoria Plant Pathology Herbarium holds a culture of the facultative parasitic fungus *Phoma* sp. isolated from a diseased specimen of *bridal creeper* collected at Keith, SA in 1990 (VPRI 16978). Damage by snails (*Helix aspersa* Müller) has sometimes been observed on bridal creeper in NSW (L Morin, *pers. obs.*).

The following insect natural enemies were found feeding on bridal creeper foliage during surveys in South

Africa: larvae of the moths *Agrotis* sp., *Euplexia augens* Felder & Rogenhoven, *Lycophotia oliveata* (Hampson) and *Cucullia terrensis* Felder & Rogenhoven, an unidentified cecidomyiid gall fly, a *Hespera* sp. flea beetle, an undescribed Erythroneurini leaf-hopper (formerly referred to as *Zygina* sp.) (M Fletcher, *pers. comm.* 2009), a *Crioceris* sp. leaf beetle and adults of the weevil *Brachycerus parilis* Haaf. (Kleinjan and Edwards 2006). Larvae of the moth *Zalaca snelleni* (Wallengren) and the wasp *Eurytoma* sp. were also found feeding on fruits or immature seeds. No insects were observed feeding on belowground organs of bridal creeper despite the extensive excavations undertaken (Kleinjan and Edwards 2006). The rust fungus *P. myrsiphylli* was the sole pathogen observed attacking foliage of bridal creeper (Kleinjan *et al.* 2004b).

The south-western Cape form of *A. asparagoides* was also examined for natural enemies during surveys. Kleinjan and Edwards (2006) observed the Erythroneurini leaf-hopper, *Z. snelleni* and *Eurytoma* sp. on above-ground parts as well as weevil larvae, probably of a *Brachycerus* species, feeding on tubers. Infection by the rust fungus was never seen on this form of *A. asparagoides*.

Agents released

The undescribed Erythroneurini leaf-hopper, rust fungus and *Crioceris* sp. leaf beetle were identified as having the greatest potential for biological control of bridal creeper based on field observations in South Africa and preliminary host-specificity laboratory tests (Witt and Edwards 2000, 2002; Kleinjan *et al.* 2004b). Recent examination of the leaf-hopper (voucher ASCTHE034290) by Murray Fletcher of NSW Agricultural Scientific Collections Unit (Orange, NSW) and Christopher Dietrich of Illinois Natural History Survey (Champaign, Illinois) revealed that it probably belongs to an undescribed genus; consequently, extensive taxonomic work would be required to obtain a complete identification. The *Crioceris* sp. was examined by E. Grobbelaar, a specialist in southern African Chrysomelidae from the National Collection of Insects, Pretoria and found to be an undescribed species as well. These three candidate agents were imported into Australian quarantine facilities and underwent detailed host-specificity testing on native and exotic plant species selected following Wapshere's (1974) centrifugal phylogenetic sequence approach (Batchelor *et al.* 1999; Morin *et al.* 2001, 2006b; KL Batchelor and TL Woodburn, unpub. data). All three agents were tested on a range of species in Australia that belong to the Asparagaceae (all introduced except for the native *Asparagus racemosus* Willd.) and other closely related families within the Order Asparagales, and species fitting in the safeguard categories outlined by Wapshere (1974).

Leaf-hopper, undescribed Erythroneurini. The leaf-hopper feeds on mesophyll cells of cladodes and young stems of bridal creeper. Feeding damage initially appears as small chlorotic flecks that develop into zigzag patterns, which coalesce and completely discolour cladodes (Witt and Edwards 2000). The insect is easy to rear due to its multivoltine life-cycle and consequently host-specificity testing was straightforward. The leaf-hopper fed, oviposited and developed some nymphs on a wide range of plant species in non-choice tests, but not in choice tests, in which only bridal creeper and closely related species were affected (Batchelor *et al.* 1999). A series of additional field experiments were performed in South Africa to further assess the risk posed by the leaf-hopper to non-target plants. While some feeding occurred on other species such as cherry, peach, almond and sunflower, exposed to leaf-hoppers originating from senescing bridal creeper plants, neither eggs nor nymphs were found on these non-target plants (Batchelor *et al.* 1999). On the basis of the evidence gathered, the agent was approved for release in Australia in 1999.

Only 40% of release sites monitored in the early 2000s reported that the leaf-hopper had established one year after release (L Morin, unpub. data). Establishment failures were scattered across the bridal creeper range. Although the leaf-hopper has been seen causing impressive damage at some sites, its populations and thus impact have a tendency to fluctuate from year to year, probably due to parasitism or poor survival during summer (Batchelor and Woodburn 2002b; Joder *et al.* 2002). Up to 95% of leaf-hopper eggs examined at field sites in WA were found to be parasitised by a *Stethynium* sp. (Hymenoptera: Mymaridae) (Joder *et al.* 2002). Another unidentified ectoparasitic larva of the Dryinidae family (Hymenoptera) was also detected in the field in WA (Batchelor and Woodburn 2002b), but so far has not been investigated. Nonetheless, the leaf-hopper may play a role in regulating bridal creeper populations at drier inland sites, where conditions are often less conducive for rust epidemics.

Rust fungus *Puccinia myrsiphylli*. Like other rust fungi, *P. myrsiphylli* parasitises living plant cells of its host and acts as a resource sink through continuous absorption of plant nutrients. Initial tests revealed that isolates of the rust fungus collected from sites in the winter-rainfall region of the Western Cape province were highly pathogenic towards Australian bridal creeper accessions (Morin *et al.* 2001). In contrast, isolates from the aseasonal-rainfall region produced only a low number of fruiting bodies and limited sporulation. Isolates from the two regions differed not only in pathogenicity but in their banding patterns from random amplification of polymorphic DNA polymerase chain reaction (RAPD-PCR) and ITS sequences. Host-specificity testing was performed using a single rust isolate (ex. Joubertsdal Farm near Swellendam). Only bridal creeper, and none of the other plant species tested including other *Asparagus* spp., was infected by the rust (Morin *et al.* 2001). The rust fungus was approved for release in 2000 and the isolate used in host-specificity tests was the only one released.

The rust fungus was confirmed established one year after release at 81.5% of sites monitored in the early 2000s (L Morin, unpub. data). Establishment failures occurred mainly in the inland regions of NSW and WA and were scattered across the bridal creeper range in SA. The impact of the rust fungus on bridal creeper has been highest in moist coastal areas where climatic conditions are conducive to epidemics (Morin *et al.* 2006b).

Glasshouse and field experiments showed that damage caused by the rust fungus and leaf-hopper, alone or in combination, not only severely affects bridal creeper's foliage but can reduce its below-ground biomass, thus affecting the plant's ability to produce new shoots the following growing season (Batchelor and Woodburn 2002b; Morin *et al.* 2002, 2006b, 2006c; Spafford Jacob *et al.* 2007; Turner *et al.* 2008a, 2010; Morin and Reid 2009). For example, Turner *et al.* (2010) reported that potted plants exposed to both agents in the field for a period of 15 weeks had above-ground biomass, rhizome length, tuber number and total tuber biomass reduced on average by 37%, 29%, 39% and 51%, respectively. In another study on naturally occurring bridal creeper infestations, the rust fungus was reported to have reduced bridal creeper cover by more than 75% in a 15 month period (Turner *et al.* 2008a). In longer-term field experiments, including comparisons between and after release and fungicide exclusion trials, consecutive seasonal epidemics of the rust fungus resulted in a drastic reduction of bridal creeper populations, which was followed by natural recovery of native plant species at some sites (Morin *et al.* 2006b; Stansbury *et al.* 2007; Morin and Reid 2009). These experiments highlighted the importance of several consecutive years of rust epidemic in fully exhausting bridal creeper's below-ground biomass and the need to incorporate activities that facilitate the recovery of native plant communities, such as follow-up control of other weeds or adding local native seeds or seedlings, to restore the ecosystem (Turner *et al.* 2006, 2008b).

Leaf beetle *Crioceris* sp. (Coleoptera: Chrysomelidae). Both adults and larvae of this leaf beetle feed on young expanding shoots and cladodes of bridal creeper when plants are actively growing (Witt and Edwards 2002). Damage inflicted by larvae is the most severe and can reduce bridal creeper's ability to climb. Results from host-specificity tests showed that some oviposition occurred on most *Asparagus* spp. and at very low levels on the two *Allium* spp. (Alliaceae) tested (KL Batchelor and TL Woodburn, unpub. data). Larval development to pupae, however, was recorded only on the introduced weed *Asparagus densiflorus* (Kunth) Jessop and to a much lesser extent on cultivated asparagus *Asparagus officinalis* L., with only 4% of eggs developing through to pupae. The only native *Asparagus* sp. in Australia, *A. racemosus*, supported limited oviposition and no larval development beyond the first instar. Based on these results and those of previous tests conducted in South Africa (Witt and Edwards 2002), the leaf beetle was approved for release in 2002.

Releases of the leaf beetle were carried out by CSIRO and the Victorian Department of Primary Industries, with the aim of establishing nursery sites from which the community could source insects for further releases in their region. By 2008, however, the leaf beetle was confirmed established at only one site in SA and two in WA, although it had been released since 2002 at 82 sites in all southern states except Tas (L Morin, unpub. data). Predation or parasitism may have played a role in this poor establishment rate (Reilly *et al.* 2004; Morin *et al.* 2006b). At this stage it is unclear what contribution the leaf beetle may make to the biological control program because it has established poorly, has only one to two generations per year, and feeds on very young shoots which are in limited supply when the other agents are active (Witt and Edwards 2002; Morin *et al.* 2006b).

Release program. Following the initial release of the leaf-hopper and rust fungus, simple protocols were designed to assist the community in making releases of the agents across the range of bridal creeper (Batchelor and Woodburn 2002a; Woodburn *et al.* 2002; Batchelor *et al.* 2004). The rearing protocol developed for the leaf-hopper was embraced by many primary schools and became one of the keystones for the Australian Weed Warriors program (Kwong 2002). Workshops were organised in various regions to demonstrate the release techniques and to encourage land managers and community groups to participate (Woodburn *et al.* 2002). By the end of the official release program in partnership with the community in July 2008, the leaf-hopper and rust fungus had been released at 898 and 2122 sites across southern Australia, respectively (L Morin, unpub. data). The dramatic visual impact of the rust fungus stimulated some stakeholders to use conventional spray equipment to apply a suspension of rust spores, obtained by soaking field-collected infected material in water and straining, onto large infestations to speed up spread (Overton and Overton 2006).

Other promising agents

The seed wasp *Eurytoma* sp. is considered a promising candidate agent for bridal creeper and other invasive *Asparagus* species based on several years of monitoring conducted at four sites in South Africa in the early 1990s (Kleinjan and Edwards 2006). It has a complementary mode of action to the other agents introduced in Australia that damage bridal creeper's foliage, which could play a role in preventing some fruit production (Morin and Edwards 2006). Field observations revealed that the wasp's host-range is most likely restricted to the Asparagaceae family (Kleinjan and Edwards 2006). It is, however, recorded from cultivated asparagus in South Africa; a conflict of interest with asparagus seed producers would arise should it be proposed for introduction in Australia (Kleinjan and Edwards 2006). The wasp was imported into an Australian containment facility for further study in 1999, but host-specificity testing could not be performed due to rearing difficulties, because of the short life-span of adults and a lack of a continuous supply of fruiting *Asparagus* plants (Kleinjan and Edwards 2006; Morin *et al.* 2006b). The fruit moth *Z. snelleni*, cecidomyiid stem-tip galler and weevil *B. parilis* appeared promising in the field in South Africa, but knowledge of these species is still extremely limited.

DISCUSSION

The high profile of the bridal creeper biological control program in Australia is primarily due to the impressive damage caused by the rust fungus and the active engagement of the community in the redistribution of the agents. The benefit:cost ratio of 2:1 estimated for the program in 2004–05 was based only on the benefits to the Murray Valley citrus industry; it did not include the significant environmental and social benefits derived from the biological control of bridal creeper (Page and Lacey 2006). Morin and Edwards (2006) concluded that the program benefited from the initial attention given to clarifying the identity of the target weed and locating the most suitable areas for searches of natural enemies in South Africa. No attempts were made, however, to predict possible levels of parasitism following the release of the insect agents in Australia and no formal assessment was undertaken to decide in what order prioritised agents should be released. They also commented that the selection process for suitable rust isolates could have been streamlined if the molecular analysis to pinpoint the origin of Australian bridal creeper populations had been carried out at the onset of the program. The genetic uniformity observed in Australian bridal creeper was a bonus for the program as it meant less cost and time were required to find and test a single rust isolate rather than a suite of isolates to target different genotypes.

The south-western Cape form of *A. asparagoides* discovered in Australia in 2004 could jeopardise benefits gained from this biological control program if it invades areas previously occupied by bridal creeper. Field observations and glasshouse experiments revealed that all three agents released for bridal creeper can survive and reproduce on young growth of the south-western Cape form, but cause considerably less damage than on bridal creeper (Coles *et al.* 2006; M Neave and L Morin, unpub. data). Biological control of this form may need to be contemplated if current eradication efforts are unsuccessful and it becomes highly invasive. Other invasive *Asparagus* species in Australia, such as *A. densiflorus/Asparagus aethiopicus* L., *Asparagus scandens* Thunb. and *Asparagus declinatus* L. are becoming increasingly important weeds and may also warrant consideration as potential targets for biological control.

ACKNOWLEDGEMENTS

We acknowledge the collaboration from the Departments of Primary Industries of NSW and Vic, and Water, Land & Biodiversity Conservation of SA and hundreds of community members.

REFERENCES

Australian Weeds Committee (2011) *Target Species for Biological Control.* <http://www.weeds.org.au/management.htm> (accessed 3.8.11).

Batchelor KL and Woodburn TL (2002a) Community involvement in distributing the bridal creeper leaf-hopper *Zygina* sp., a biocontrol agent for bridal creeper, *Asparagus asparagoides*. In *Proceedings of the 13th Australian Weeds Conference*. 8–13 September 2002, Perth. (Eds H Spafford Jacob, J Dodd and JH Moore) pp. 453–455. Plant Protection Society of WA, Perth.

Batchelor KL and Woodburn TL (2002b) Population development and impact of the bridal creeper leafhopper *Zygina* sp. in Western Australia. In *Proceedings of the 13th Australian Weeds Conference*. 8–13 September 2002, Perth. (Eds H Spafford Jacob, J Dodd and JH Moore) pp. 381–384. Plant Protection Society of WA, Perth.

Batchelor K, Scott JK, Witt ABR and Woodburn TL (1999) Oversummering and host specificity of *Zygina* sp. (Cicadellidae), a potential agent for the control of bridal creeper, *Asparagus asparagoides* (Asparagaceae). In *Proceedings of the 12th Australian Weeds Conference*. 12–16 September 1999, Hobart. (Eds AC Bishop, M Boersma and CD Barnes) p. 246. Tasmanian Weed Society, Devonport.

Batchelor KL, Morin L, Swirepik A and Woodburn TL (2004) Community involvement in the distribution of the biological control agents for bridal creeper, *Asparagus asparagoides*. In *Proceedings of the 11th International Symposium on Biological Control of Weeds*. 27 April–2 May 2003, Canberra. (Eds JM Cullen, DT Briese, DJ Kriticos, WM Lonsdale, L Morin and JK Scott) pp. 381–385. CSIRO Entomology, Canberra.

Clifford HT and Conran JG (1987) 2. *Asparagus*, 3. *Protasparagus*, 4. *Myrsiphyllum*. *Flora Australia* **45**, 159–165.

Coles RB, Willing KL, Conran JG and Gannaway D (2006) The identification and distribution of Western Cape form of bridal creeper (*Asparagus asparagoides* (L.) Druce) in the south-east of South Australia and western Victoria. *Plant Protection Quarterly* **21**, 104–108.

Cooper J and Warren W (2006) Tasmania and asparagus weeds: hanging in the balance. *Plant Protection Quarterly* **21**, 117.

Council of Heads of Australasian Herbaria (2006) Australia's Virtual Herbarium. <http://www.chah.gov.au/avh/public_query.jsp> (accessed 3.8.11).

Coutts-Smith AJ and Downey PO (2006) The impact of weeds on threatened biodiversity in New South Wales. CRC for Australian Weed Management, Technical Series 11, Glen Osmond, SA. < http://www.southwest-nrm.org.au/sites/default/files/uploads/ihub/coutts-smith-aj-downey-po-2006-impact-weeds-threatened-biodiversity-new-south-wales.pdf> (accessed 3.8.11).

Downey PO (2006) The weed impact to native species (WINS) assessment tool: results from a trial for bridal creeper (*Asparagus asparagoides* (L.) Druce) and ground asparagus (*Asparagus aethiopicus* L.) in southern New South Wales. *Plant Protection Quarterly* **21**, 109–116.

Fellingham AC and Meyer NL (1995) New combinations and a complete list of *Asparagus* species in southern Africa (Asparagaceae). *Bothalia* **25**, 205–209.

Fukuda T, Ashizawa H, Suzuki R, Ochiai T, Nakamura T, Kanno A, Kameya T and Yokoyama J (2005) Molecular phylogeny of the genus *Asparagus* (Asparagaceae) inferred from plastid *petB* intron and *petD-rpoA* intergenic spacer sequences. *Plant Species Biology* **20**, 121–132.

Humphries SE, Groves RH and Mitchell DS (1991) Plant invasions of Australian ecosystems: a status review and management directions. *Kowari* **2**, 1–134.

Joder A, Spafford Jacob H, Batchelor K and Woodburn T (2002) Biology of *Stethynium* sp. nov. (Hymenoptera: Mymaridae), an egg parasitoid of *Zygina* sp. (Hemiptera: Cicadellidae). In *Proceedings of the 13th Australian Weeds Conference*. 8–13 September 2002, Perth. (Eds H Spafford Jacob, J Dodd and JH Moore) p. 412. Plant Protection Society of WA, Perth.

Kleinjan CA and Edwards PB (1999) A reappraisal of the identification and distribution of *Asparagus asparagoides* in southern Africa. *South African Journal of Botany* **65**, 23–31.

Kleinjan CA and Edwards PB (2006) *Asparagus* weeds in Australia: a South African perspective with emphasis on biological control prospects. *Plant Protection Quarterly* **21**, 63–68.

Kleinjan CA, Edwards PB and Hoffmann JH (2004a) Impact of foliage feeding by *Zygina* sp. on tuber biomass and reproduction of *Asparagus asparagoides* (L.): relevance to biological control in Australia. *Biological Control* **30**, 36–41.

Kleinjan CA, Morin L, Edwards PB and Wood AR (2004b) Distribution, host range and phenology of the rust fungus *Puccinia myrsiphylli* in South Africa. *Australasian Plant Pathology* **33**, 263–271.

Kwong R (2002) Weed Warriors: increasing awareness of weeds in schools. In *Proceedings of the 13th Australian Weeds Conference*. 8–13 September 2002, Perth. (Eds H Spafford Jacob, J Dodd and JH Moore) pp. 442–445. Plant Protection Society of WA, Perth.

Kwong R and Holland-Clift S (2004) Biological control of bridal creeper, *Asparagus asparagoides* (L.) W.Wight, in citrus orchards. In *Proceedings of the 14th Australian Weeds Conference*. 6–9 September 2004, Wagga Wagga, NSW. (Eds BM Sindel and SB Johnson) pp. 329–332. Weed Society of New South Wales, Sydney.

Malcomber ST and Demissew S (1993) The status of *Protasparagus* and *Myrsiphyllum* in Asparagaceae. *Kew Bulletin* **48**, 63–78.

Morin L and Edwards PB (2006) Selection of biological control agents for bridal creeper: a retrospective review. *Australian Journal of Entomology* **45**, 287–291.

Morin L and Reid AM (2009) Effectiveness of the rust *Puccinia myrsiphylli* in reducing populations of the invasive plant bridal creeper in Australia. In *Conference Handbook: 17th Biennial Conference of the Australasian Plant Pathology Society*. 29 September–1 October 2009, Newcastle, NSW. p. 115. Australasian Plant Pathology Society.

Morin L, Armstrong J and Driver F (2001) The science underpinning the release of the bridal creeper rust in Australia. In *Conference Handbook: 13th Biennial Conference of the Australasian Plant Pathology Society*. 24–27 September 2001, Cairns, Qld. p. 78. Australasian Plant Pathology Society.

Morin L, Willis AJ, Armstrong J and Kriticos D (2002) Spread, epidemic development and impacts of the bridal creeper rust in Australia: summary of results. In *Proceedings of the 13th Australian Weeds Conference*. 8–13 September 2002, Perth. (Eds H Spafford Jacob, J Dodd and JH Moore) pp. 385–388. Plant Protection Society of WA, Perth.

Morin L, Batchelor KL and Scott JK (2006a) The biology of Australian weeds. 44 *Asparagus asparagoides* (L.) Druce. *Plant Protection Quarterly* **21**, 46–62.

Morin L, Neave M, Batchelor K and Reid A (2006b) Biological control: a promising tool for managing bridal creeper, *Asparagus asparagoides* (L.) Druce, in Australia. *Plant Protection Quarterly* **21**, 69–77.

Morin L, Reid A and Willis AJ (2006c) Impact of the rust fungus *Puccinia myrsiphylli* on the below-ground biomass of bridal creeper. In *Proceedings of the 15th Australian Weeds Conference*. 24–28 September 2006, Adelaide. (Eds C Preston, JH Watts and ND Crossman) p. 608. Weed Management Society of South Australia, Adelaide.

Obermeyer AA (1984) Revision of the genus *Myrsiphyllum* Willd. *Bothalia* **15**, 77–88.

Overton D and Overton B (2006) The development of 'spore water' on Kangaroo Island for rapid spread of bridal creeper rust fungus. *Plant Protection Quarterly* **21**, 77–78.

Page AR and Lacey KL (2006) *Economic Impact Assessment of Australian Weed Biological Control*. CRC for Australian Weed Management, Technical Series No. 10, Glen Osmond, SA. <http://www.southwestnrm.org.au/sites/default/files/uploads/ihub/page-ar-lacey-kl-2006-economic-impact-assessment-australian-weed-biological-control.pdf> (accessed 3.8.11).

Quarmby JP (2010) Recovery plan for twelve threatened orchids in the Lofty Block region of South Australia 2010. SA Dept of Environment and Natural Resources, Adelaide. <http://155.187.2.69/biodiversity/threatened/publications/recovery/pubs/lofty-block-orchids-recovery-plan.pdf > (accessed 3.8.11).

Raymond KL (1999) Ecology of *Asparagus asparagoides* (bridal creeper), an environmental weed of southern Australia. PhD thesis. Monash University, Victoria.

Reilly T, Spafford Jacob H and Batchelor K (2004) The effectiveness of ant baiting to reduce predation of *Crioceris* sp. (Coleoptera: Chrysomelidae), a biological control agent for bridal creeper. In *Proceedings of the 14th Australian Weeds Conference*. 6–9 September 2004, Wagga Wagga, NSW. (Eds BM Sindel and SB Johnson) pp. 115–118. Weed Society of New South Wales, Sydney.

Scott JK and Batchelor KL (2006) Climate-based prediction of potential distribution of introduced *Asparagus* species in Australia. *Plant Protection Quarterly* **21**, 91–98.

Scott JK and Kleinjan CA (1991) Bridal creeper (*Myrsiphyllum asparagoides*) in Australia and developments towards its biological control. *Plant Protection Quarterly* **6**, 116–119.

Sorensen B and Jusaitis M (1995) The impact of bridal creeper on an endangered orchid. In *Weeds of Conservation Concern: Seminar and Workshop Papers*. 5–6 April 1995, Adelaide. (Eds D Cooke and J Choate) pp. 27–31. Dept of Environmental and Natural Resources and Animal and Plant Control Commission, Adelaide.

Spafford Jacob H, Reilly TE and Batchelor KL (2007) The presence of *Zygina* sp. and *Puccinia myrsiphylli* reduces survival and influences oviposition of *Crioceris* sp. *BioControl* **52**, 113–127.

Stansbury CD (2001) Dispersal of the environmental weed bridal creeper, *Asparagus asparagoides*, by silvereyes, *Zosterops lateralis*, in south-western Australia. *Emu* **101**, 39–45.

Stansbury CD, Batchelor KL, Morin L, Woodburn TL and Scott JK (2007) Standardized support to measure biomass and fruit production by the invasive climber (*Asparagus asparagoides*). *Weed Technology* **21**, 820–824.

Thorp JR and Lynch R (2000) *The Determination of Weeds of National Significance*. National Weeds Strategy Executive Committee, Launceston.

Turner PJ and Virtue JG (2006) An eight-year removal experiment measuring the impact of bridal creeper (*Asparagus asparagoides* (L.) Druce) and the potential benefit from its control. *Plant Protection Quarterly* **21**, 79–84.

Turner PJ, Scott JK and Spafford Jacob H (2006) Barrier to restoration: the decomposition of bridal creeper's root system. In *Proceedings of the 15th Australian Weeds Conference*. 24–28 September 2006, Adelaide. (Eds C Preston, JH Watts and ND Crossman) pp. 827–830. Weed Management Society of South Australia, Adelaide.

Turner PJ, Morin L, Williams D and Kriticos DJ (2010) Interactions between a leafhopper and rust fungus on the invasive plant *Asparagus asparagoides* in Australia: a case of two agents being better than one for biological control. *Biological Control* **54**, 322–330.

Turner PJ, Scott JK and Spafford H (2008a) Implications of successful biological control of bridal creeper (*Asparagus asparagoides* (L.) Druce) in south-west Australia. In *Proceedings of the 16th Australian Weeds Conference*. 18–22 May 2008, Cairns, Qld. (Eds RD Van Klinken, VA Osten, FD Panetta and JC Scanlan) pp. 390–392. Queensland Weeds Society, Brisbane.

Turner PJ, Spafford H and Scott JK (2008b) The ecological barriers to the recovery of bridal creeper (*Asparagus asparagoides* (L.) Druce) infested sites: impacts on vegetation and the potential increase in other exotic species. *Austral Ecology* **33**, 713–722.

Wapshere AJ (1974) A strategy for evaluating the safety of organisms for biological weed control. *Annals of Applied Biology* **77**, 201–211.

Willis AJ, McKay R, Vranjic JA, Kilby MJ and Groves RH (2003) Comparative seed ecology of the endangered shrub, *Pimelea spicata* and a threatening weed, bridal creeper: smoke, heat and other fire-related germination cues. *Ecological Management and Restoration* **4**, 55–65.

Witt ABR and Edwards PB (2000) Biology, distribution, and host range of *Zygina* sp. (Hemiptera: Cicadellidae), a potential biological control agent for *Asparagus asparagoides*. *Biological Control* **18**, 101–109.

Witt ABR and Edwards PB (2002) Aspects of the biology, distribution, and host range of *Crioceris* sp. (Col.: Chrysomelidae: Criocerinae), a potential biological control agent for *Asparagus asparagoides* in Australia. *Biological Control* **23**, 56–63.

Woodburn TL, Morin L, Batchelor KL, Swirepik A and Armstrong J (2002) Enhancement of biological control of bridal creeper by using community as a resource. In *Proceedings of the 13th Australian Weeds Conference*. 8–13 September 2002, Perth. (Eds H Spafford Jacob, J Dodd and JH Moore) pp. 701–703. Plant Protection Society of WA, Perth.

Asphodelus fistulosus L. – onion weed

Jim Cullen

ABSTRACT

Onion weed, *Asphodelus fistulosus*, is an annual to short-lived perennial herb, reproducing and spreading by seed, that has invaded extensive areas of the drier sandy soils of south-eastern Australia and, to a lesser extent, WA. It invades both cropping and grazing systems, but is particularly prevalent in disturbed or overgrazed land, where farmers have pushed strongly for biological control. Preliminary surveys in onion weed's Mediterranean and North African native habitat found a rust fungus *Puccinia barbeyi* that severely damages the plant. Initial host-specificity tests demonstrated a very high level of specificity, even within the genus *Asphodelus*, and this rust would be good candidate for introduction. However, the importance of onion weed as a target for biological control has been debatable and, although demand for biological control remains, recent work has failed to demonstrate significant losses in either productivity or biodiversity from the presence of the weed. There was therefore little prospect of onion weed being declared a target for biological control and work was terminated.

INTRODUCTION

Asphodelus fistulosus L. (Liliaceae) is an annual, biennial or short-lived perennial herb of Mediterranean–Eurasian origin that grows up to 75 cm high with several upright branched flowering stems in spring–summer. The flowers are whitepink and up to 13 200 seeds can be produced per plant (Parsons and Cuthbertson 2001; Roark 1955) (Fig. 1). Reproduction is by seed and to some extent by branching at the base, yielding larger and more spreading clumps. Seed can germinate at any time, but predominantly in summer–autumn. Plants do not usually flower until the second summer and there can be quite high mortality of young plants in the first summer (Parsons and Cuthbertson 2001).

The plant favours sandy, neutral to alkaline soils and has spread widely through the drier areas of SA and Vic and the semi-arid grazing lands of NSW (Fig. 2). It is not toxic but is unpalatable to stock, thriving in overgrazed pastures where it can form dense infestations, reducing the availability of productive pasture to stock (Fig. 3). It also invades cropping systems, though the availability of effective herbicides in combination with cultivation and competitive pastures has facilitated control in such systems (Pitt *et al.* 2006). It was almost certainly introduced as an ornamental in the 19th century and was first proclaimed a weed in Vic in 1895. It has also spread to North America and New Zealand, but appears to be weedy only in Australia (Parsons and Cuthbertson 2001).

BIOLOGICAL CONTROL HISTORY

There has been no interest in biological control of onion weed apart from in Australia. Starting in 1971, the CSIRO Montpellier laboratory in France undertook a series of surveys of several important weeds of Mediterranean

Figure 1: Onion weed, *Asphodelus fistulosus*. Photo: J Cullen, CSIRO.

origin, including *Echium plantagineum* L. (Boraginaceae), *Heliotropium europaeum* L. (Boraginaceae), *Hypericum perforatum* L. (Clusiaceae) and various thistles. As a result of continuous pressure from the rural community for biological control of onion weed, this weed was included in some of the surveys.

A promising pathogen, the rust *Puccinia barbeyi* (Roum.) Magnus, was discovered and research carried out on the biology and host-specificity of this species (Hasan 1991). While the level of specificity seemed high and the rust had obvious potential, the priority for onion weed as a biological control project declined. No case was put forward for approval of the species as a target for biological control and no introductions were made into Australia.

EXPLORATION

The surveys of onion weed from the Montpellier laboratory were not comprehensive and did not yield any arthropods that merited further work. However, during surveys in southern Spain and North Africa, the rust *P. barbeyi* was found to severely attack the weed. An isolate of this rust, collected near Almeria in southern Spain, was imported to the Montpellier laboratory, where it was cultured for several years (Hasan 1991).

CANDIDATES

Agents tested but not released

***Puccinia barbeyi* (Roum.) Magnus.** When found in southern Spain, this rust had been reported only on *Asphodelus* spp. as a monoecious microcyclic rust, multiplying by aecial and telial stages, but it was possible that

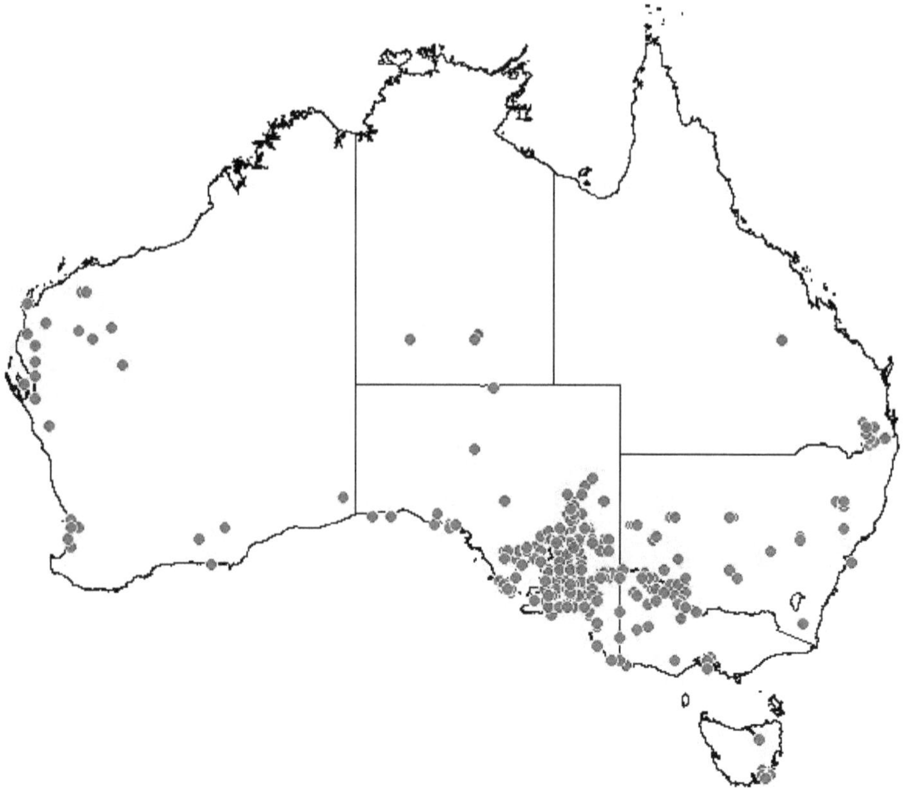

Figure 2: Distribution of onion weed, *Asphodelus fistulosus*, in Australia. Specimen data reproduced from Australia's Virtual Herbarium with permission of the Council of Heads of Australasian Herbaria Inc.

Figure 3: Infestation of onion weed, *Asphodelus fistulosus*, in the Victorian mallee. Photo: J Cullen, CSIRO.

the other spore stages had occurred on another host and had not been recorded. Hasan (1991) was able to demonstrate that the rust is an autoecious species with all its life-cycle on the one host plant. The teliospores germinate on *A. fistulosus* to produce basidiospores that infect the plant. The resulting aeciospores can reinfect the plant, producing aecia and more aeciospores. Both aeciospores and teliospores are capable of germinating irrespective of light or darkness and without any resting period. The rust therefore cycles continuously on the host, producing aeciospores and teliospores, the latter particularly as the infected plant parts begin to dry. Both spore stages can produce serious damage to the plant, destroying or severely reducing the growth of established plants and often killing small plants and seedlings.

In preliminary host-specificity tests the following related species, all from the Liliaceae, were exposed to aeciospores of *P. barbeyi*: onion (*Allium cepa* L.), leek (*A. porrum* L.), garlic (*A. sativum* L.), two wild species of *Allium* (*A. vineale* L. and *A. paniculatum* L.), tulip (*Tulipa* sp.) and the wild *Asphodelus cerasifer* Gay. None of the test plants, including *A. cerasifer*, showed any signs of infection, but in each test *A. fistulosus* showed normal development of aecia and later, telia (Hasan 1991).

The culture was terminated when it became apparent that onion weed was not going to be declared a target for biological control.

DISCUSSION

This was always a minor project, using the opportunity and resources provided by the Montpellier base and the enthusiasm for investigating the potential for the use of plant pathogens for weed control resulting from the successful control of skeleton weed *Chondrilla juncea* L. (this volume) (Hasan 1980). The project was essentially exploratory and never reached the stage of seeking industry funding. The rust found in southern Spain and North Africa was a very promising species and there was public demand for biological control. However, more objective examination of the impact of onion weed failed to show any significant differences in stocking rates at low, medium or high densities of onion weed over a five-year period, nor any evidence of a negative correlation with biodiversity in terms of numbers of native plant species. Overall, there was no evidence of significant detrimental effect from onion weed in the SA rangelands, though there is still demand for biological control from WA and NSW (Pitt *et al.* 2006). However, it is doubtful whether the weed's impact will ever be great enough to attract external funding for a project.

ACKNOWLEDGEMENTS

The author would like to thank Louise Morin for assistance in locating some of the material used in writing this section.

REFERENCES

Hasan S (1980) Plant pathogens and biological control of weeds. *Review of Plant Pathology* **59**, 349–356.

Hasan S (1991) The biology and host specificity of the onion weed rust, *Puccinia barbeyi*, a potentially useful agent for biological control in Australia. *Annals of applied Biology* **118**, 19–25.

Parsons WT and Cuthbertson EG (2001) *Noxious Weeds of Australia*. 2nd edn. CSIRO Publishing, Melbourne.

Pitt JL, Virtue JG and Feuerherdt LJ (2006) Onion weed: pest or perception? In *Proceedings of the 15th Australian Weeds Conference*. 24–28 September 2006, Adelaide. (Eds C Preston, JH Watts and ND Crossman) pp. 454–457. Weed Management Society of South Australia, Adelaide.

Roark B (1955) The autecology of *Asphodelus fistulosus* L. PhD thesis. University of Adelaide.

Baccharis halimifolia L. – groundsel bush

Bill Palmer and Nikki Sims-Chilton

ABSTRACT

The North American dioecious shrub *Baccharis halimifolia* (Asteraceae) became a very serious weed of south-eastern Qld and northern NSW approximately 50 years after its introduction into Australia in the late 1800s. Dairying and forestry interests were particularly affected. One of Australia's largest biocontrol projects was initiated in 1960 and has continued, almost uninterrupted, until the present. Exploration for potential agents was undertaken from bases in Florida and Texas and from Curitiba, Brazil. Ultimately the entire range of *B. halimifolia*, from New York to Texas, was searched and extensive surveying was undertaken on several congeners found in western USA, Mexico and Brazil. Altogether 13 insects and one rust were released in Australia and a further two insect species were approved for release. Very few weed biocontrol projects have released so many agents. Six agents established after release, four of these collected from *B. halimifolia* itself. Three agents – the gall midge *Rhopalomyia californica*, the stem-borer *Hellinsia balanotes* and the rust *Puccinia evadens* – are thought to give a measure of control over most of the weed's range, while the leaf-feeder *Trirhabda bacharidis* and the stem-borer *Megacyllene mellyi* give some control in limited areas. Recent analyses have suggested that changing climate and land utilisation patterns, as well as biocontrol, may have contributed to the reduced abundance of the weed in recent years.

Key words: *Rhopalomyia californica*, *Hellinsia balanotes*, *Puccinia evadens*, *Trirhabda bacharidis*, *Megacyllene mellyi*, south-eastern Qld, evaluation.

INTRODUCTION

The shrub *Baccharis halimifolia* L. (Asteraceae: Astereae: Baccharidinae) is known in Australia as groundsel bush and in its native range of the eastern USA as sea myrtle, saltbush, groundsel tree or consumption weed. 'Groundsel' refers to the similarity of the flowering head (Fig. 1) to those of the *Senecio* genus which have groundsel as a common name (Parsons and Cuthbertson 2001).

Groundsel bush was brought to Australia in the late 19th century as an ornamental plant. It soon naturalised around towns and was recognised as a potential weed by the turn of the century (Bailey 1900). It then invaded rural areas and became a significant weed in cattle pastures. By 1958 it had been declared a noxious weed in Qld and in many shires in northern NSW (McFadyen 1979) due to its ability to invade pastures and native *Melaleuca* wetlands (McFadyen 1973). Groundsel bush could form dense stands in pastures (Fig. 3), creating impenetrable thickets and even limiting stock movement. The plant is generally not eaten by stock and there are laboratory studies suggesting its toxicity. Ultimately productive pasture is replaced by dense, woody, unpalatable and perennial vegetation. Groundsel bush also became a problem for forestry plantations, where it was capable of overgrowing the recently planted pine seedlings. By 1971 the cost of control of groundsel bush in Qld was estimated at \$500 000, increasing to over \$1 million by 1977. Initially, chemical and mechanical control methods were used. However, due to the extent of the invasion and the cost of

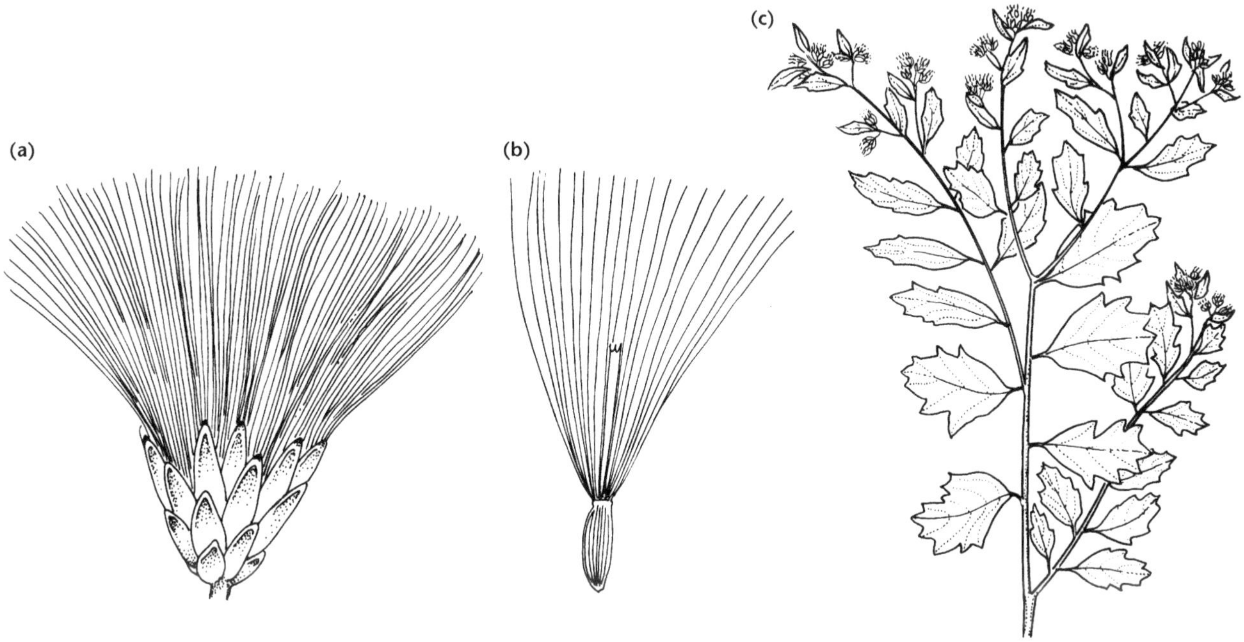

Figure 1: Reproductive structures of *Baccharis halimifolia*. a) Capitulum from female plant. b) Seed plus pappus. c) Flowering branch of male plant. Illustration: Nicola Oram, Botanic Gardens Trust.

control, groundsel bush became an increasingly good candidate for biological control.

Groundsel bush is an erect, dioecious perennial shrub growing to 6 m high (Mahler and Waterfall 1964) with stem bases to 18 cm in diameter and alternate leaf arrangement (Fig. 1). It generally flowers after one year and thereafter in the autumn, with male plants preceding the female plants. Sex of plants is easily determined during the flowering period – male flowers are yellow due to the large amount of pollen and female flowers white due to the pappus attached to each achene (Fig. 1). Female flowers are wind- and sometimes insect-pollinated (Westman *et al.* 1975; Panetta 1979). Groundsel bush is a very prolific seeder, producing up to 1.5 million seeds per plant (Westman *et al.* 1975; Panetta 1979). Seeds are attached to a pappus and are readily dispersed by wind (Boldt 1989a) and water (Panetta 1977). They may drift up to 140 m from a 2 m high plant (Diatloff 1964) but most seeds fall within a few metres of the parent plant (Parsons and Cuthbertson 2001). Seedlings establish readily in pastures.

Groundsel bush has been found on a wide range of soil types, from dry infertile forest soils to rich volcanic loams and low-lying clay soils with high moisture content (Winders 1937). It commonly grows in moist soils with high organic content (Egler 1952) and is often found in soil covered by brackish water with a salt content ranging from 0.5% to 2% (Boldt 1989a). Groundsel bush also has a wide tolerance to soil pH and has been found in soils with a pH ranging from 3.8 to 9 (Westman *et al.* 1975) and chlorinity from 0% to 2%. In Australia it is found in coastal areas along the eastern seaboard from Gladstone, Qld to Coffs Harbour, NSW (Fig. 2). The areas of Australia climatically suitable for groundsel bush are also shown in Figure 2. Interestingly, this projection indicated that much of southern Australia would be climatically very favourable for this weed.

Although Australia has been most seriously affected by groundsel bush, the species has also become a significant weed in southern France, the Iberian peninsula and around the Black Sea in Russia (Westman *et al.* 1975).

BIOLOGICAL CONTROL HISTORY

Groundsel bush was targeted for biological control by the Qld Government in the early 1960s. Short-term exploratory surveys indicated that several insect species were potential biocontrol agents. A base was established at the Archbold Biological Station in central Florida to undertake a three-year research program and export selected insect species to Australia. From 1973 interest switched to the centre of origin of the Baccharidinae, Brazil, with a base in Curitiba. In 1982 the North American Field Station was established by the first author at Temple, Texas to undertake faunal surveys throughout the USA and Mexico and to study and ship potential agents.

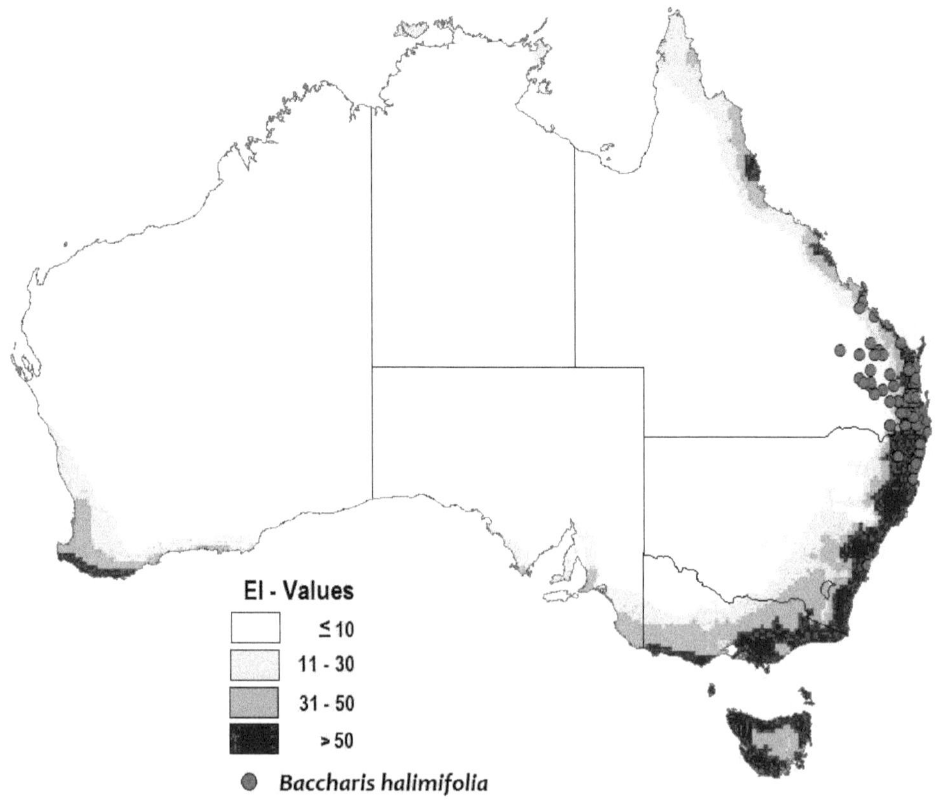

Figure 2: Distribution of *Baccharis halimifolia* in Australia and its potential distribution as predicted after Sims-Chilton (2009) using the climate modelling software CLIMEX (Sutherst *et al.* 2007). EI values >30 are considered climatically favourable while values ≤10 are considered unfavourable. Data based on records from the Australian Virtual Herbarium.

The establishment of the North American Field Station coincided with an interest by the US Department of Agriculture (USDA) in biological control of several native plants including *B. neglecta* Britt., *B. halimifolia* and *B. salicifolia* (R & P.) Pers. (Boldt 1989a). The USDA conducted pre-release faunal surveys of these plants and detailed studies of several Brazilian insects supplied by Qld. However, by the time studies were complete, there was a change in attitude about the value of native plants and no biocontrol was ultimately attempted. Three insects previously released in Australia were also sent to Russia for release but the outcome is not known.

Figure 3: A severe infestation of *Baccharis halimifolia*, showing the shrub suppressing pasture. Photo: Qld Government.

From 1968 to 2001, Qld Government scientists in Brisbane undertook extensive investigations of selected insects, to define their host-ranges in quarantine using Australian plant species, mass-rear approved insects and monitor successful establishments. All told, 35 agents were imported for testing, 14 were released and seven are established (Sims-Chilton *et al.* 2009). The biological control program against groundsel bush is estimated to have cost $9.6 million (Page and Lacey 2006).

Finally, the whole Australian program was evaluated via a PhD project (Sims-Chilton 2009) supported by the Cooperative Research Centre for Australian Weed Management.

PLANT TAXONOMY

Baccharis L. is an entirely American genus recently thought to encompass about 375 species and is currently the largest genus within the tribe Astereae (Giuliano 2005). Most species are South American, with only 20 occurring in the USA (Mahler and Waterfall 1964). *Baccharis* has been divided into 21 sections (Nesom 1990). *B. halimifolia* is the type species for section *Baccharis*, which is widespread in both North and South America. Other species in this section include *B. angustifolia* Michx., *B. conferta* Kunth., *B. dioica* Vahl, *B. glomeruliflora* Pers., *B. neglecta*, *B. pilularis* DC and *B. salicina* T. & G.

B. halimifolia is the only member of the Baccharidinae naturalised or cultivated in Australia. The tribe Astereae is represented in Australia by only a few native species, a few exotic ornamentals and no commercial crops or pasture plants. From a biocontrol perspective, this phylogenetic distance from important flora enhanced the prospects of finding good agents.

EXPLORATION

The first surveys for potential biocontrol agents for groundsel bush were made by FD Bennett of the Commonwealth Institute of Biological Control, with two six-week surveys of the USA in May and October 1960, and two surveys in Brazil in 1963 (Bennett 1963). These scoping studies and a faunal survey of the closely related *B. pilularis* in California (Tilden 1951a) indicated that several potential organisms were available for study. The project at the Archbold Biological Station concentrated more on the study of several species nominated by Bennett (1963) rather than on further surveys for new prospects. This work was undertaken by three entomologists (WH Haseler, G Diatloff and B Willson), each spending a spring–autumn period in Florida before returning to Australia.

Further exploration then switched to Brazil, the putative centre of origin for the genus *Baccharis*, where PJ McFadyen established the field station at Curitiba in 1973 (McFadyen 1979). This field station remained in use for several years after McFadyen returned to Australia in 1976, and was run by C Garcia.

Interest returned to North America with the establishment of the North American Field Station at Temple, Texas in 1982. Detailed faunistic surveys were undertaken of *B. halimifolia*, *B. angustifolia* and the closely related *B. neglecta* (Palmer 1987; Palmer and Bennett 1988). Later, closely related congeners, *B. dioica* and *B. conferta*, were surveyed in central Mexico (Palmer and Pullen 1994). Altogether, some 182 phytophagous species have been collected on groundsel bush in its native range and 8–10% of these were considered to be monophagous or narrowly stenophagous. Interestingly, while only a small proportion of the insect fauna (16%) was common to the area west of the Mississippi and the eastern seaboard, a much higher proportion (72%) of the narrowly stenophagous species were common to both areas (Palmer and Bennett 1988).

Faunal surveys of *B. neglecta* in Texas were undertaken by PE Boldt and TO Robbins (1987), as well as surveys of the less closely related *B. salicina*, *B. pteronioides* DC and *B. bigelovii* Gray (Boldt and Robbins 1994).

CANDIDATES

Agents released

Over the course of the program, 13 insect species (Julien and Griffiths 1998) and one rust fungus were introduced and released as biocontrol agents in Australia. Seven were of US origin and the remainder were sourced from Brazil.

***Trirhabda bacharidis* (Weber) (Coleoptera: Chrysomelidae).** This leaf-feeder is found on several species (*B. halimifolia*, *B. angustifolia*, *B. neglecta* and *B. glomeruliflora*). Its native range extends from near Corpus Christi on the Gulf Coast of Texas along the eastern seaboard to Maine in the north. *T. bacharidis* is the most obvious phytophage of these *Baccharis* spp. and on occasion defoliation is obvious even from a moving car. *T. bacharidis* was found to be sufficiently host-specific (Boldt 1989b; Palmer and Haseler 1992b). It was released in Australia in 1969 but established in only six sites. In its native range, it is univoltine with eggs being the overwintering stage. However, in Australia, its phenology is quite different: egg and larval stages are completed before winter and the insect overwinters as pupae in the soil. Perhaps this change in phenology contributed to the very limited areas of establishment for an insect with such a large and diverse native range. In 1983 further introductions of this insect, this time collected in Texas, were made but these populations also changed phenology after release and there was no improvement in the insect's range or effect. Recent surveying (Sims-Chilton 2009) indicated that the distribution of *T. bacharidis* is still limited, being observed at only two sites in south-eastern Qld.

***Rhopalomyia californica* (Felt) (Diptera: Cecidomyiidae).** This gall fly was the first agent to produce widespread and obvious damage to groundsel bush in Australia. The fly is native to California where its host is *B. pilularis*. Eggs are laid in shoot tips and larval feeding causes swollen galls in stem tips. Heavy attack slows growth and may kill seedling plants. It was early recognised as a major phytophage of *B. pilularis* (Tilden 1951b) although usually heavily parasitised. More recently, its potential effect on *B. pilularis* was demonstrated in elegant studies (Ehler *et al.* 1984; Ehler 1989) showing that, when parasitoid populations were suppressed by broad-scale malathion baiting for medfly eradication, populations of *R. californica* increased dramatically and were damaging to the host plant. Host-specificity testing in Florida demonstrated that the insect was specific to some *Baccharis* spp. and it was approved for release in Australia in 1968 (Palmer *et al.* 1993). A few small releases were made in Australia the next year but there was no establishment. In 1982 another attempt was made to establish the insect in Australia, starting with about 100 galls from one plant in a residential garden in Berkeley, California. Fortuitously, these galls were not heavily parasitised and 654 flies emerged from them in Brisbane and a colony was successfully established (Palmer *et al.* 1993). Subsequent shipments to Brisbane all failed to produce cultures because parasitism rates were >99% and insufficient flies emerged. A well-funded mass-rearing program was undertaken for three years, resulting in the widespread establishment of the insect. For some years results were spectacular (McFadyen 1985) but in later years attack was less pronounced (Palmer *et al.* 1993). Recent surveys (Sims-Chilton 2009) indicate that *R. californica* is widespread throughout the distribution of groundsel bush and may offer some level of growth and fecundity control when galling is very abundant.

Hellinsia balanotes **(Meyrick) (Lepidoptera: Pterophoridae).** This stem borer is widely distributed in the eastern USA on *B. halimifolia*, *B. glomeruliflora*, *B. neglecta* and *B. angustifolia*. The insect was found to be host-specific and it was released in Australia in 1969 (Palmer and Haseler 1992a). However, it did not establish, possibly because too few were released. The insect was recollected in 1982–83, mostly from the Gainesville area of Florida. Collecting was a very onerous job and very high rates of parasitism were encountered. Eventually, sufficient were received to establish a culture in Australia. Mass-rearing was also difficult, requiring considerable human and physical resources. The insect was successfully established throughout the weed's range in Australia, and surveys (Sims-Chilton *et al.* 2009) indicate that *H. balanotes* is visually very damaging. However, it may offer little in the way of control of growth or fecundity as groundsel bush recovers rapidly from stem-boring damage.

Several leaf-feeding beetles (Coleoptera: Chrysomelidae). Several beetles from Brazil were sufficiently host-specific for release in Australia. These were *Lioplacis elliptica* (Stål), *Metallactus patagonicus* Suffrian, *Metallactus nigrofasciatus* Suffrian, *Stolas fuscata* (Klug) and *Anacassis phaeopoda* Buzzi. All were collected from species other than *B. halimifolia* but were reared for several generations on *B. halimifolia* in Australia during the host-testing and mass-rearing phases. None established in Australia. *S. fuscata* was also investigated in the USA for the biocontrol of *B. salicifolia* (R. & P.) Pers., found to be sufficiently host-specific but ultimately not released (Boldt 1989c).

Aristotelia ivae **Busck (Gelechiidae),** ***Bucculatrix ivella*** **Busck (Bucculatricidae) and** ***Lorita baccharivora*** **Pogue (Tortricidae).** These three microlepidopterons, collected from *B. halimifolia* in the USA where they were found only at low densities, were studied as potential biocontrol agents. All three were sufficiently host-specific and were released in Australia (Palmer and Diatloff 1987; Diatloff and Palmer 1988). The leaf-feeder *A. ivae* and the leaf-miner *B. ivella* established and are now widespread, but have contributed little to control of the weed.

Megacyllene mellyi **(Chevrolat) (Coleoptera: Cerambycidae).** This stem-borer from Brazil was collected in *B. microdonta* AP DC and, relatively rarely, in *B. dracunculifolia* DC. It is also known from *B. spicata* (Lam.) Baillon, *B. tridentate* Vahl. and the closely related genus *Baccharidastrum* (McFadyen 1979, 1983). Host-specificity testing showed that the insect was sufficiently host-specific and it was released in Australia in 1978 (McFadyen 1983). Releases continued until 1984. It established only in limited areas with shallow saline soils close to the coast but in some of these, e.g. the small islands of southern Moreton Bay, it produced quite spectacular results (Tomley 1990). *M. mellyi* is the only insect from South America to become established in Australia. Subsequently, *M. mellyi* was investigated in the USA as a possible biocontrol agent for *B. neglecta* and *B. halimifolia*. Its host-specificity was confirmed but the insect was ultimately not released in the USA (Boldt 1987).

Helipodus intricatus **(Boheman) (Coleoptera: Curculionidae).** The stem-borer from Brazil was collected on several species including *B. microdonta*, *B. dracunculifolia* and *B. spicata*. Host-specificity testing showed that this univoltine insect was sufficiently host-specific (P McFadyen and C Garcia, unpub.) and it was approved for release in 1982. However, no mass-rearing was undertaken and only a few individuals were released in 1983 and 1984 as adults emerged from the laboratory colony (P McFadyen, unpub.). The insect did not establish in the field.

Puccinia evadens **Harkness (Uredinales: Pucciniaceae).** This rust fungus is known from *B. halimifolia* in Florida and from *B. pilularis* subsp. *consanguinea* in California. Host-specificity studies were undertaken on the Florida population and it was found to be highly host-specific (Verma *et al.* 1996). The rust was released in Australia in 1997. It quickly established over the weed's range and caused severe dieback in some areas (Tomley and Willsher 2002). However, later studies indicated it has minimal impact on the growth of groundsel bush and may even promote growth (Sims-Chilton 2009).

Agents considered but not released

Two insects were approved for release in Australia but then not released. In both cases, there was difficulty culturing the insect in quarantine and at the time leaf-feeding lepidopterans were not regarded as particularly useful. The leaf-feeder, *Itame varadaria* (Walker) (Lepidoptera: Geometridae) from *B. halimifolia* along the Gulf Coast of Texas, was clearly sufficiently host-specific but never observed to be damaging in its native range (Palmer 1989). Another leaf-feeder, *Prochoerodes truxaliata* (Guenée) (Lepidoptera: Geometridae), from *B. pilularis* around Davis, California, was on occasion observed to be damaging and was sufficiently host-specific (Palmer and Tilden 1988; Ehler *et al.* 1990) but was never released.

A very promising species, the crown-boring *Amniscus perplexus* Haldeman (Coleoptera: Cerambycidae), was found on *B. halimifolia* along the Gulf Coast of Texas where it was quite damaging. An application to release this insect was not pursued because the adult beetles fed on the bark of several plants, including citrus, in cage experiments. However, oviposition and larval development occurred only on *B. halimifolia* and this insect remains a potential biocontrol agent for *B. halimifolia* (Palmer and Tomley 1993).

Two other insects were investigated at the North American Field Station, where preliminary testing and field observations suggested they were host-specific. *Slaterocoris pallipes* (Knight) (Hemiptera: Miridae) is a univoltine species occurring from New York to North Carolina. In a multiple-choice test, *S. pallipes* fed only on *B. halimifolia* and *B. neglecta* (Palmer 1993a). *Stobaera pallida* Osborn (Hemiptera: Delphacidae) is widely distributed on various *Baccharis* spp., all of section *Baccharis*, in the USA and Mexico. In multiple-choice tests, *S. pallida* oviposited only on *B. halimifolia* and *B. neglecta* and not on 31 other plant species (Palmer 1993b). Both insects were shipped to Australia but neither was successfully cultured in quarantine.

EVALUATION

B. halimifolia is now generally considered by affected communities to have been brought under control, although some large populations still exist. Producer groups, such as the dairy industry, no longer call for further research effort. The weed is rarely mentioned in the media, local governments are spending less on its chemical control and from the highway it is clear that it is much less abundant than previously. As a result, most stakeholders regard groundsel bush as successfully controlled through biological means.

From 2005–09, the biological control of groundsel bush was selected in a PhD study to demonstrate effective strategies and quantitative tools for evaluating biological control programs (Sims-Chilton 2009). Groundsel bush was a useful case study of potentially variable success, as biological control began decades earlier and the plant still exists in large populations in Australia. Several techniques were used in a novel manner to assess groundsel bush biological control.

A large-scale survey determined the current distribution of the agents and groundsel bush in south-eastern Qld and northern NSW (Sims-Chilton *et al.* 2009). Results indicated that *H. balanotes*, *P. evadens*, *M. mellyi*, *R. californica*, *A. ivae*, *B. ivella* and *T. bacharidis* still exist in populations of groundsel bush but are patchy in their effectiveness due to factors such as rainfall, elevation, plant size and population density.

A combination of surveys, experiments and modelling techniques was used to assess the overall impact of herbivory on populations of groundsel bush, which depended on the agent's effect on individual plant vital rates (e.g. growth and fecundity) and the sensitivity of the population growth rate to changes in these vital rates (Schutzenhofer and Knight 2007). Individual plant level effects were examined through the use of observational damage data, insect exclusion experiments and statistical models. It was evident that the agents are damaging groundsel bush plants, but at the current levels of damage they do not significantly reduce individual plant growth or fecundity. Simulation models of high levels of damage indicated that the agents have the potential to reduce individual plant vital rates such as growth and fecundity.

The thesis also examined how agent damage at the individual plant level translated into population level impacts (Sims-Chilton 2009). Most models of plant population dynamics have been developed to assist with the selection of effective agents by identifying the most vulnerable stages of the weed's life-cycle, where damage would cause major adverse effects on population growth rates. Very few models have been developed after agent release, to determine which agents may be having a negative impact on population growth rate. A model presented in the thesis suggested that, at observed levels, the agents are not reducing groundsel bush population growth rate. When high levels of damage were simulated, it appeared that population growth rate reduction may be possible.

It is often assumed that a reduction in populations of an invasive plant following the introduction of biological control agents is a result of successful biological control. However, just as for the initial invasion, other factors, e.g. climate, may have an influence on invasive species populations. CLIMEX, a climate modelling tool (Skarratt *et al.* 1995; Sutherst *et al.* 2004), was used to explore the possibility that long-term changes in climate may have played a role in reducing the populations of groundsel bush since 1960, by predicting how the favourability of a location for the growth and persistence of groundsel bush has changed over time. Sims-Chilton *et al.* (2010) demonstrated that prior to the introduction of biological control of groundsel bush (pre 1960), temperature and rainfall parameters in areas of eastern Australia were optimal for groundsel bush growth. Climate simulations indicated that since the 1960s temperature and rainfall in these areas have become less favourable for growth of groundsel bush.

An economic evaluation of this very large long-term project was undertaken (Page and Lacey 2006). The study estimated that the project cost a total of $9.6 million, resulting in an estimated net present value of -$900 000 (i.e. loss) and a benefit:cost ratio of 0.7 at a discount rate of 8%.

Since the introduction of these biological control agents, the status of groundsel bush as a pest has decreased. A number of factors may have contributed to its decline, including mechanical and chemical control efforts by landholders and councils, the introduction of biological control, land use changes and development, and changes in climate.

One interesting observation from the biological project is that five of the seven agents that eventually established in Australia came from *B. halimifolia* itself. *R. californica* came from *B. pilularis*, also in section *Baccharis*, while only one of seven South American insects, from more distantly related species, established. This was a clear example when sourcing agents from the actual target and then from phylogenetically very close taxa was more successful than investigating the more distant associations.

The evaluation of the program (Sims-Chilton 2009) suggested a need for continued integrated control because the biological control may not be entirely responsible for the improved status of the weed and the agents are patchy in distribution. Currently, groundsel bush is a declared weed, which means that by law it has to be controlled. The established biological agents are damaging individual plants but average levels of damage in the field may not be sufficient to cause large population impact. The agents should not be considered the solution to large infestations but may assist in minimising localised populations in areas which are difficult to access.

The groundsel bush biological control program and evaluation has no further funding at this stage. Future work could focus on examining the effectiveness of herbicide and mechanical treatments in combination with biological control. Additionally, research could be conducted to test the models developed in the Sims-Chilton thesis. And, of course, *A. perplexus, I. varadaria* and *P. truxaliata* could be released.

DISCUSSION

This chapter reports one of Australia's largest biocontrol efforts, a project that spanned almost 40 years of continuous research, foreign exploration in several countries on two continents, and the release of 14 agents. Very few biocontrol projects result in more than 10 agents being ultimately released (Julien and Griffiths 1998). The large number released in this project is a testament to the effort, as well as the fact that *Baccharis* is relatively isolated phylogenetically from Australia's native flora and economic plants.

ACKNOWLEDGEMENTS

Special acknowledgement needs to be made of the generous identification services provided by the Systematic Entomology Laboratory (SEL), ARS-USDA and the associated Department of Entomology, National Museum of Natural History. Taxonomists from these institutions provided identifications, *gratis*, for literally hundreds of submitted series. The Archbold Biology Station in Florida, the Federal University of Parana, Brazil and the USDA-ARS in Texas very kindly hosted field stations and provided specialist facilities such as quarantine, glasshouses and temperature cabinets. KADW Senaratne prepared Figure 2. Valuable comments on the manuscript were provided by colleagues at the Alan Fletcher Research Station.

REFERENCES

Bailey FM (1900) *The Queensland Flora, Part 3.* Queensland Government, Brisbane.

Bennett FD (1963) *Final Report on Surveys of the Insects attacking Baccharis spp. in the South-East United States of America and Brazil.* Commonwealth Institute of Biological Control, Curepe, Trinidad.

Boldt PE (1987) Host-specificity and laboratory rearing studies of *Megacyllene mellyi* (Coleoptera, Cerambycidae), a potential biological control agent of *Baccharis neglecta* Britt. (Asteraceae). *Proceedings of the Entomological Society of Washington* **89**, 665–672.

Boldt PE (1989a) Baccharis *(Asteraceae): A Review of its Taxonomy, Phytochemistry, Ecology, Economic Status, Natural Enemies and the Potential for its Biological Control in the United States.* Texas Agricultural Experiment Station, College Station, Texas.

Boldt PE (1989b) Biology and host-specificity of *Trirhabda bacharidis* (Coleoptera: Chrysomelidae) on *Baccharis* (Asteraceae: Astereae). *Environmental Entomology* **18**, 78–84.

Boldt PE (1989c) Host-specificity studies of *Stolas fuscata* (Klug) (Coleoptera: Chrysomelidae) for the biological control of *Baccharis salicifolia* (R. & P.) Pers. (Asteraceae). *Proceedings of the Entomological Society of Washington* **91**, 502–508.

Boldt PE and Robbins TO (1987) Phytophagous and pollinating insect fauna of *Baccharis neglecta* (Compositae) in Texas. *Environmental Entomology* **16**, 887–895.

Boldt PE and Robbins TO (1994) Phytophagous insect faunas of *Baccharis salicina, B. pteronioides,* and *B. bigelovii* (Asteraceae) in the south-western United States and northern Mexico. *Environmental Entomology* **23**, 47–57.

Diatloff G (1964) How far does groundsel seed travel? *Queensland Agricultural Journal* **90**, 354–356.

Diatloff G and Palmer WA (1988) The host-specificity and biology of *Aristotelia ivae* Busck (Gelechiidae) and *Lorita baccharivora* (Tortricidae), two microlepidoptera selected as biological control agents for *Baccharis halimifolia* (Asteraceae) in Australia. *Proceedings of the Entomological Society of Washington* **90**, 458–461.

Egler FE (1952) South-east saline Everglades vegetation, Florida and its management. *Vegetation* **3**, 213–265.

Ehler LE (1989) Effect of malathion-bait sprays on the spatial structure of a parasite guild. *Entomologia Experimentalis et Applicata* **52**, 15–21.

Ehler LE, Endicott PC, Hertlein MB and Alvarado-Rodriguez B (1984) Medfly eradication in California: impact of malathion-bait sprays on an endemic gall midge and its parasitoids. *Entomologia Experimentalis et Applicata* **36**, 201–208.

Ehler LE, Kinsey MG and Palmer WA (1990) Further observations on the biology and host-specificity of *Prochoerodes truxaliata* (Guenee) (Lepidoptera: Geometridae), a biological control agent for *Baccharis halimifolia* L. in Australia. *Pan-Pacific Entomologist* **66**, 79–88.

Giuliano DA (2005) New infragenera in *Baccharis* (Asteraceae, Astereae). *Novon* **15**, 534–541.

Julien MH and Griffiths MH (1998) *Biological Control of Weeds: A World Catalogue of Agents and their Target Weeds.* 4th edn. CABI, Wallingford, UK.

Mahler WF and Waterfall UT (1964) Baccharis (Compositae) in Oklahoma, Texas and New Mexico. *Southwestern Naturalist* **9**, 189–202.

McFadyen PJ (1973) Insects for biological control. *Queensland Agricultural Journal* **99**, 607–611.

McFadyen PJ (1979) South American insects introduced into Australia for the control of the weed *Baccharis halimifolia* L. Ph D thesis. University of Queensland, Brisbane.

McFadyen PJ (1983) Host-specificity and biology of *Megacyllene mellyi* (Col.: Cerambycidae) introduced into Australia for the biocontrol of *Baccharis halimifolia* (Compositae). *Entomophaga* **28**, 65–72.

McFadyen PJ (1985) Introduction of the gall fly *Rhopalomyia californica* from the USA into Australia for the control of the weed *Baccharis halimifolia.* In *Proceedings of the 6th International Symposium on Biological Control of Weeds.* Vancouver. (Ed ES Delfosse) pp. 779–787. Agriculture Canada, Ottawa.

Nesom GL (1990) Infrageneric taxonomy of North and Central American *Baccharis* (Asteraceae: Astereae). *Phytologia* **68**, 40–46.

Page AR and Lacey KL (2006) *Economic Impact Assessment of Australian Weed Biological Control.* CRC for Australian Weed Management, Adelaide.

Palmer WA (1987) The phytophagous insect fauna associated with *Baccharis halimifolia* L. and *B. neglecta* Britton in Texas, Louisiana and northern Mexico. *Proceed-*

ings of the Entomological Society of Washington **89**, 185–199.

Palmer WA (1989) Food plant specificity and biology of *Itame varadaria* (Walker) (Geometridae), a North American moth introduced into Australia to control the weed *Baccharis halimifolia. Journal of the Lepidopterists' Society* **43**, 305–312.

Palmer WA (1993a) A note on the host-specificity of the mirid *Slaterocoris pallipes* (Knight). *Proceedings of the Entomological Society of Washington* **95**, 640–641.

Palmer WA (1993b) On the host-range of the delphacid planthopper *Stobaera pallida* Osborn (Homoptera). *Proceedings of the Entomological Society of Washington* **95**, 241–244.

Palmer WA and Bennett FD (1988) The phytophagous insect fauna associated with *Baccharis halimifolia* L. in the eastern United States. *Proceedings of the Entomological Society of Washington* **90**, 216–228.

Palmer WA and Diatloff G (1987) Host-specificity and biology of *Bucculatrix ivella* Busck, a potential biological control agent for *Baccharis halimifolia* in Australia. *Journal of the Lepidopterists' Society* **41**, 23–28.

Palmer WA and Haseler WH (1992a) The host-specificity and biology of *Oidaematophorus balanotes* (Meyrick) (Lepidoptera: Pterophoridae): a species introduced into Australia for the biological control of *Baccharis halimifolia* L. *Journal of the Lepidopterists' Society* **46**, 195–202.

Palmer WA and Haseler WH (1992b) The host-specificity and biology of *Trirhabda bacharidis* (Weber) (Coleoptera: Chrysomelidae): a species introduced into Australia for the biological control of *Baccharis halimifolia* L. *Coleopterists Bulletin* **46**, 61–66.

Palmer WA and Pullen KR (1994) Phytophagous insects associated with *Baccharis conferta* Kunth and *B. dioica* Vahl (Asteraceae: Astereae) in Mexico. *Proceedings of the Entomological Society of Washington* **96**, 757–763.

Palmer WA and Tilden JW (1988) Host-specificity and biology of *Prochoerodes truxaliata* (Guenée) (Geometridae), a potential biocontrol agent for the rangeland weed *Baccharis halimifolia* L. in Australia. *Journal of the Lepidopterists' Society* **41**, 199–208.

Palmer WA and Tomley AJ (1993) The host-range and biology of *Amniscus perplexus* Haldeman (Coleoptera: Cerambycidae), a candidate evaluated for the control of *Baccharis halimifolia* in Australia. *Coleopterists Bulletin* **47**, 27–34.

Palmer WA, Diatloff G and Melksham J (1993) The host-specificity of *Rhopalomyia californica* Felt (Diptera: Cecidomyiidae) and its importation into Australia as a biological control agent for *Baccharis halimifolia* L. *Proceedings of the Entomological Society of Washington* **95**, 1–6.

Panetta FD (1977) The effects of shade upon seedling growth in groundsel bush (*Baccharis halimifolia* L.). *Australian Journal of Agricultural Research* **28**, 681–690.

Panetta FD (1979) The effects of vegetation development upon achene production in the woody weed, groundsel bush (*Baccharis halimifolia* L.). *Australian Journal of Agricultural Research* **30**, 1053–1065.

Parsons WT and Cuthbertson EG (2001) *Noxious Weeds of Australia.* 2nd edn. CSIRO Publishing, Melbourne.

Schutzenhofer MR and Knight TM (2007) Population-level effects of augmented herbivory on *Lespedeza cuneata*: implications for biological control. *Ecological Applications* **17**, 965–971.

Sims-Chilton NM (2009) Evaluation of the biological control program of groundsel bush (*Baccharis halimifolia* L. Asteraceae). PhD thesis. University of Queensland, Brisbane.

Sims-Chilton NM, Zalucki MP and Buckley YM (2009) Patchy herbivore and pathogen damage throughout the introduced Australian range of groundsel bush, *Baccharis halimifolia*, is influenced by rainfall, elevation, temperature, plant density and size. *Biological Control* **50**, 13–20.

Sims-Chilton NM, Zalucki MP and Buckley YM (2010) Long-term climate effects are confounded with the biological control programme against the invasive weed *Baccharis halimifolia* in Australia. *Biological Invasions* **12**, 3145–3155.

Skarratt DB, Sutherst RW and Maywald GF (1995) *CLIMEX for Windows Version 1.0.* CSIRO and CRC for Tropical Pest Management, Brisbane.

Sutherst RW, Maywald GE, Bottomley W and Bourne A (2004) *CLIMEX v2.* Hearne Scientific Software, Melbourne.

Sutherst RW, Maywald GF and Kriticos D (2007) *CLIMEX Version 3 User's Guide.* Hearne Scientific Software, Melbourne.

Tilden JW (1951a) The insect associates of *Baccharis pilularis* De Candolle. *Microentomology* **16**, 149–188.

Tilden JW (1951b) Observations on *Rhopalomyia californica* Felt. *Pan-Pacific Entomologist* **27**, 20–27.

Tomley AJ (1990) *Megacyllene mellyi*: a biological control agent for groundsel bush, *Baccharis halimifolia*, in Queensland. In *Proceedings of the 9th Australian*

Weeds Conference. (Ed. JW Heap) pp. 513–515. Crop Science Society of South Australia, Adelaide.

Tomley AJ and Willsher L (2002) Field releases and initial impact of groundsel bush rust in Australia. In *Proceedings of the 13th Australian Weeds Conference*. 8–13 September 2002, Perth. (Eds H Spafford Jacob, J Dodd, JH Moore) pp. 391–393. Plant Protection Society of WA, Perth.

Verma U, Charudattan R, Devalerio JT and Tomley AJ (1996) *Puccinia evadens*, a biological control agent for *Baccharis halimifolia*. In *Proceedings of the 9th International Symposium on Biological Control of Weeds*. 19–26 January 1996, Stellenbosch, South Africa. (Eds VC Moran and JH Hoffman) p. 234. University of Cape Town, Rondebosch, South Africa.

Westman WE, Panetta FD and Stanley TD (1975) Ecological studies on reproduction and establishment of the woody weed, groundsel bush (*Baccharis halimifolia* L.: Asteraceae). *Australian Journal of Agricultural Research* **26**, 855–870.

Winders CW (1937) Groundsel bush in south-eastern Queensland. *Queensland Agricultural Journal* **63**, 656–664.

Billardiera heterophylla (Lindl.) L. Cayzer & Crisp – sollya

Anna Williams

ABSTRACT

Billardiera heterophylla, sollya, is native to south-western WA and has become weedy in south-eastern Australia. Surveys for natural enemies were conducted in 2003–06 in the native range and in south-eastern Australia. Over 229 taxa from 13 insect orders were collected but further investigations have not been made.

Key words: native weed, surveys.

INTRODUCTION

Billardiera heterophylla (Lindl.) L. Cayzer & Crisp (= *Sollya heterophylla* Lindl. (CHAH, 2007)) (Pittosporaceae) is a perennial, evergreen woody shrub with climbing branches. It bears blue bell-shaped flowers (Figs 1 and 2) from October to March. The flowers have five petals, are ~10 mm long and hang in clusters of up to 10. Cylindrical fleshy fruit, 20–30 mm long, ripen from green to purple, soften when mature in winter and contain numerous seeds. Leaves are alternate, mostly ovate and glabrescent on both surfaces. Leaf size is variable, from 30–50 mm × 10–20 mm (Cayzer *et al.* 2004). Common names include sollya, Australian bluebell and bluebell creeper.

Sollya has had a complicated nomenclatural history. The revision of *Billardiera* by Cayzer *et al.* (2004) incorporated *Sollya* within *Billardiera* rather than as a separate small genus within the family Pittosporaceae. Their analysis characterises *Sollya heterophylla* Lindl., *Sollya heterophylla* var. *pallida* Ser., *Sollya heterophylla* var. *angustifolia* Benth., *Sollya erecta* C.R.P. Andrews, and *Sollya fusiformis* (Labill.) Briq., which had been described between 1832 and 1921, as synonyms of a new species *Billardiera heterophylla* (Lindl.) L. Cayzer & Crisp. *Sollya salicifolia* Marnock, *Sollya heterophylla* var. *salicifolia* Ser., *Billardiera salicifolia* F. Cels, *Sollya linearis* Lindl., *Sollya heterophylla* var. *linearis* (Lindl.) Pax., *Sollya fusiformis* Payer, *Sollya heterophylla* var. *linearis* Otto, *Billardiera elongate* Schnizl. and *Billardiera hambruchiana* Seem. & J.A. Schmidt, described between 1839 and 1870, are listed as synonyms of *Billardiera fusiformis* Labill. which was originally described in 1805 and is cryptically similar to, and often confused with, *B. heterophylla*.

Figure 1: Flowers and climbing habit of sollya, *Billardiera heterophylla*. Photo: A Williams.

Figure 2: Sollya, *Billardiera heterophylla*, at Angahook-Lorne State Park, Vic. Photo: A Williams.

B. heterophylla is indigenous to southern WA but has been in cultivation (under the name *S. heterophylla)* since at least 1831 as Lindley (1831) includes a drawing of a specimen grown in England. In Australia, it has been listed in nursery catalogues since at least 1860 (Brookes and Barley 1992). It has been a very popular horticultural plant, prized in gardening books for its vigorous growth and wide tolerance of soil types and growing conditions including tolerance to frost and resistance to drought (Lord 1948; Wrigley and Fagg 1979; Canberra Botanic Gardens 1972). Plants can be propagated from seed or cuttings (Wrigley and Fagg 1979).

Sollya is now an invasive environmental weed in south-eastern Australia (Fig. 3), with infestations present in all southern states (NSW, Vic, SA and Tas). It is also naturalised in the ACT. The largest known infestations are in the lower south-east of SA. At long-established sites in this area *B. heterophylla* has greater than 80% cover in the understorey (Bachmann and Johnson, *pers. comm.* 2007). Physical and chemical control have had limited success in controlling invasive *B. heterophylla* populations. Small plants can be pulled out by hand and mature plants can be dug out relatively easily as *B. heterophylla* has a shallow root system. Chemical control can be achieved using foliar sprays, or by cutting the stem as close to the ground as possible and applying an appropriate herbicide. However, these methods are not practical for many invasive *B. heterophylla* populations in environmental reserves.

Figure 3: Sollya, *Billardiera heterophylla*, growing in state forest at Mt Burr, SA. Photo: A Williams.

BIOLOGICAL CONTROL HISTORY

B. heterophylla was selected as a model species for a postgraduate project of the CRC for Australian Weed Management to address the hypothesis that classical biological control is a possible management option for Australian plants that have become invasive elsewhere in Australia, outside their indigenous range. It was selected because it is well established as an environmental weed and in some areas is considered to be more serious than exotic species, being capable of destroying almost all indigenous understorey vegetation and associated fauna within invaded areas (Carr 2001).

The Weeds CRC project on *B. heterophylla* continued from 2003 to 2006. It included exploration for natural enemies both within the indigenous range in WA and for 'new association' agents in the invasive range. Comparisons of population ecology of the plant in the invasive and indigenous ranges, determination of germination requirements and molecular identification of invasive *Billardiera* genotypes were also conducted but results have not been published.

At least 229 different taxa, from 13 insect orders, were collected from *B. heterophylla* plants in its indigenous range. However, no formal identifications of collected specimens have been obtained. The most common damage observed on *B. heterophylla* plants within the indigenous range were bud galls and tip galls caused by dipterous larvae, and tip rolling caused by an unidentified larva. A witches broom formation (bunching) was also common and initially thought to be caused by eriophyid mites; however, mites were not found in samples sent for identification to DPI Vic, Knoxfield. On one occasion, coleopteran larvae were found boring in the stems of a mature plant growing on a roadside near Dunsborough, WA. These natural enemies appeared capable of stunting the growth of *B. heterophylla* and reducing reproductive capacity. However, due to lack of resources, none of these potential agents were successfully reared in the laboratory. No host-specificity testing was conducted and there was no assessment of damage caused.

A black blotchy discolouration on fruit and leaves was observed in a population at Stirling Ranges National Park, WA. The disease was investigated by CSIRO scientists in Perth but no causal agent was discovered.

Research on biological control of *B. heterophylla* ceased in 2007 and there is currently no prospect of any further efforts regarding this plant.

REFERENCES

Brookes M and Barley R (1992) *Plants Listed in Nursery Catalogues in Victoria 1855–1889.* Ornamental Plant Collections Association, Melbourne.

Canberra Botanic Gardens (1972) *Growing Australian Plants, vol. 2.* Australian Government Publishing Service, Canberra.

Carr GW (2001) Australian plants as weeds in Victoria. *Plant Protection Quarterly* **16**, 124–125.

Cayzer LW, Crisp MD and Telford IRH (2004) Cladistic analysis and revision of *Billardiera* (Pittosporaceae). *Australian Systematic Botany* **17**, 83–125.

Council of Heads of Australasian Herbaria (2007) *Australian Plant Census.* http://www.chah.gov.au/chah/apc/.

Lindley J (1831) *Sollya heterophylla. Edwards's Botanical Register* **17**, 14–66.

Lord EE (1948) *Shrubs and Trees for Australian Gardens.* Lothian Publishing, Melbourne.

Wrigley JW and Fagg M (1979) *Australian Native Plants: A Manual for their Propagation, Cultivation and Use in Landscaping.* Collins, Sydney.

Bryophyllum delagoense (Ecklon & Zeher) Schinz – mother-of-millions

Bill Palmer and Michelle Rafter

ABSTRACT

Since its escape as a garden ornament over half a century ago, *Bryophyllum delagoense* or mother-of-millions has become an increasing weed problem and is now a declared plant in Qld and NSW. This exotic plant from Madagascar is responsible for significant stock losses, limitations in stock management and trading opportunities and loss of ecological values. A biological control project began with exploration in Madagascar and southern Africa between 1999 and 2002. Four agents were found to be stenophagous, detrimental to the plant and to have other characters associated with effective biocontrol agents. These were the weevils *Osphilia tenuipes* and *Alcidodes sedi*, the wasp *Eurytoma bryophylli* and the leaf beetle *Rhembastus* sp. All four were brought to Qld for comprehensive host-testing. In the quarantine laboratory, *Osphilia tenuipes* and *A. sedi* developed on several *Bryophyllum* spp., on the commercial ornamental *Kalanchoe blossfeldiana* and on *K. spathulata*, a possibly native plant growing in the Kimberleys. *Eurytoma bryophylli* and *Rhembastus* sp. were not successfully cultured when shipped to Qld but preliminary host-testing in South Africa indicated that they had similar host-ranges to the two weevils. Because there are clear conflicts of interest associated with the release of any of these insects, the support of the appropriate Ministerial Council is required before authorities would consider approving them for release. This could be provided through a *Biological Control Act*. Because the actual risk to the commercial nursery industry is considered to be small, gazettal under this Act of the *Bryophyllum* spp. and the four agents is being considered. In 2002 the South African citrus thrips *Scirtothrips aurantii* appeared in Qld. The population does not attack citrus or other horticultural crops but does attack *Bryophyllum* spp. and other crassulaceous plants. This thrips appears to have given a significant measure of control of mother-of-millions in many areas of Qld. The final outcome for the biological control of mother-of-millions will depend on whether the selected potential agents can be approved for release, as no further exploration for additional agents is planned.

Key words: Madagascar, *Osphilia tenuipes*, *Alcidodes sedi*, *Eurytoma bryophylli*, *Rhembastus* sp., *Scirtothrips aurantii*, *Biological Control Act*.

INTRODUCTION

Bryophyllum delagoense (Eckl. & Zeyh.) Schinz (Crassulaceae) is a drought-tolerant garden ornamental of Madagascan origin that became naturalised in Australia, where it is known as mother-of-millions because of its tremendous reproductive capacity. Because it is toxic to stock, it poses a serious threat to pasture and stock routes. The first reported outbreak was at Chinchilla, Qld in 1940; the species is now widely distributed in Qld and NSW (Hannan-Jones and Playford 2002).

B. delagoense is an erect succulent (Fig. 1) that can grow to about 1.2 m. It has whorls made up of three phyllodes at

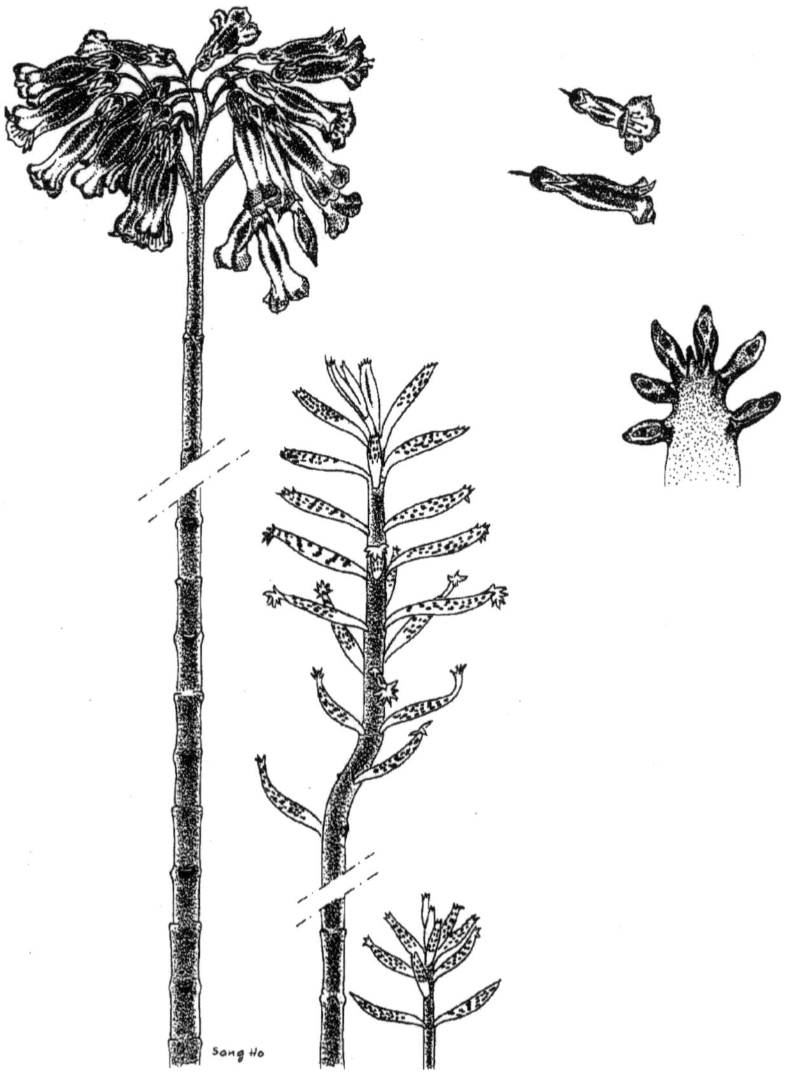

Figure 1: Drawings of mother-of-millions showing flowers, stem, phyllodes and plantlets. Illustration: Ho (1934–35). Used with permission.

each node, with each phyllode set at 120° to its neighbour and each successive whorl at 60° to the whorl beneath it. (Gomez-Campo 1974). The reproductive cycle is semelparous and is bi- or triannual depending on environmental conditions (Gomez-Campo 1974). Flowering, in early winter, is conspicuous, with a red flowering bell produced at the growing apex of mature plants. Seed production is prolific, with up to 20 000 seeds produced per inflorescence (Hannan-Jones, unpub.) (Fig. 3). The species is best known, however, for its vegetative reproduction. Six to eight plantlets are produced at the terminal end of each phyllode. These plantlets drop to the ground, where they put down roots to become new plants.

Mother-of-millions is most commonly found in disturbed areas such as tree-lined roadsides or degraded pastures. It prefers shallow stony soils and granite outcrops (Hannan-Jones and Playford 2002). In its introduced range, mother-of-millions is found on a broad range of soils from sand to heavy clay, as long as surface organic matter is present (Batianoff and Franks 1998). Its current distribution extends down the eastern coastline from Cairns, through much of Qld and NSW, and recently into Vic (Fig. 2). It has been reported in SA and on the Abrolhos Islands in WA (Longman *et al.* 2000). Mother-of-millions is increasing in distribution in western Qld as it is dispersed down river drainage systems with each flood event (Hannan-Jones and Playford 2002). In time, it is likely to occupy a much larger area of Australia (Fig. 2).

Mother-of-millions is successful in Australia, at least in part because it can withstand drought. Like other

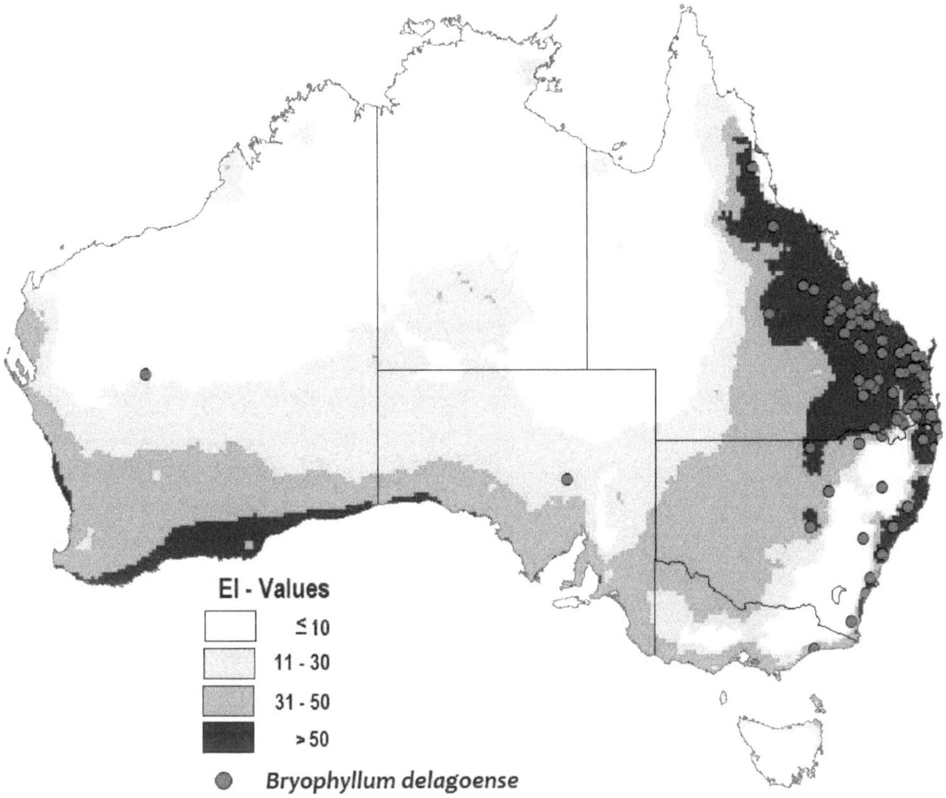

Figure 2: Distribution of *Bryophyllum delagoense* in Australia and its potential distribution as predicted using the climate modelling software CLIMEX (Sutherst *et al.* 2007). EI values >30 are considered very favourable while values ≤10 are considered climatically unfavourable. Data based on records from the Australian Virtual Herbarium.

species of Crassulaceae and most Cactaceae, these plants use crassulacean acid metabolism, which means they can take up CO_2 for photosynthesis at night. During the day, when water loss in plants is generally highest, their stomata are closed. Mother-of-millions is also able to translocate water from old to new phyllodes and plants can increase wax secretions on phyllodes to prevent desiccation when it is particularly dry (Ayerbe and Gomez-Campo 1977; Schäfer and Lüttge 1987). Its efficiency in water use is demonstrated by the fact that the plant occasionally grows in the crevices of trees.

Bryophyllum spp. are toxic to livestock because they contains bufadienolides, which are a type of cardiac glycoside (Oelrichs *et al.* 1992). Ingestion of flowers results in ruminal atony, diarrhoea, heart rate abnormalities, dyspnoea and death (McKenzie and Dunster 1986). Poisonings occur from May through to September, which coincides with the flowering period, as the flower heads are five times more toxic than stem and leaf material (McKenzie and Dunster 1986). Forty-one recorded poisoning incidents affecting 379 cattle in Qld occurred between 1960 and 1984 (McKenzie and Armstrong 1986). Poisoning incidents and stock losses are underreported; treatment is costly and is effective only if administered before respiratory system damage occurs (McKenzie and Dunster 1986). Mother-of-millions can compete with pasture as a result of growth-inhibiting root exudates, a feature called allelopathy (Groner 1975; McKenzie and Dunster 1986). This feature, combined with the plant's prolific reproductive capabilities, enables the species to be highly invasive. Chemical control is costly and not always highly effective. Time-consuming monitoring and follow-up chemical treatments are required for effective control.

A hybrid of mother-of-millions, *Bryophyllum* x *houghtonii* (D.B. Ward) P.I. Forst., is also naturalised and a serious weed in Australia. Both mother-of-millions and the hybrid are declared Class 2 pest plants in Qld and W2 or W3 in NSW.

BIOLOGICAL CONTROL HISTORY

Mother-of-millions has become naturalised in many countries, including the continental USA, Hawaii, Caribbean islands, South Africa and East Africa. However, it

Figure 3: A severe infestation of *Bryophyllum delagoense* in southern Qld. The attractive flowers are highly toxic to livestock. Photo: Qld Government.

has become a serious weed only in Australia and even here its weed status is relatively recent. Biological control has not been considered in any country other than Australia.

Interest in its biological control began in the mid 1980s as the weed continued to spread in various areas of Qld. Property owners and stakeholders, particularly in the western Darling Downs where prickly pear had been so successfully controlled decades earlier (Dodd 1940), lobbied heavily for a biocontrol program. The weed was declared noxious in 1996 and planning began for a biocontrol project.

The push for biocontrol coincided with plans for Qld to establish a field station in South Africa. The South African Field Station was eventually established, under the auspices of the ARC-Plant Protection Research Institute at Rietondale near Pretoria, to enable exploration and study of potential agents for mother-of-millions, prickly acacia and the weedy sporobolus grasses. Exploration in Madagascar and southern Africa and preliminary biology and host-range studies were undertaken there between 1999 and 2002. Prospective agents were then imported into the quarantine facilities in Brisbane for comprehensive host-testing.

PLANT TAXONOMY

The Madagascan plant genus, *Bryophyllum* Salsb. (family Crassulaceae), is now represented in Australia by six species and a hybrid which naturalised after their introduction as ornamentals (Hannan-Jones and Playford 2002). *B. delagoense*, known as mother-of-millions, and its hybrid *B.* x *houghtonii* (D. B. Ward) P. I. Forst., known as hybrid mother-of-millions (Forster 2006), are serious weeds in Qld and northern NSW. The remaining species, *B. daigremontianum* (Raym.-Hamet & H. Perrier) A. Berger, *B. beauverdii* (Raym.-Hamet) A. Berger, *B. fedtschenkoi* (Raym.-Hamet & H. Perrier) Lauz.-March., *B. pinnatum* (L. f.) Oken and *B. proliferum* Bowie, are regarded as both minor ornamentals and minor weeds in Australia. Another Madagascan genus, *Kalanchoe* Adans., is very closely related to *Bryophyllum*, such that some authors (Boiteau and Allorge-Boiteau 1995) consider all *Bryophyllum* spp. to fall within a broadly circumscribed *Kalanchoe* (Hannan-Jones and Playford 2002). Three species, *K. longiflora* Schltr. ex J.M. Wood, *K. lateritia* Engl. and *K. spathulata* DC (previously thought to be *K. crenata* (Andrews) Haw.(Kenneally 1983)), are also now naturalised in Australia. Several other *Kalanchoe* spp. are grown as ornamentals but have not naturalised. Six *Crassula* spp. are the only endemic representatives of Crassulaceae in Australia. These species are typically ephemeral and characterised by their very thin stems.

The understanding of the relationship of Crassulaceae to other plant families has also changed in recent years. Whereas Crassulaceae had been considered a component of the order Rosales (Anon 1981), it is now placed in the Saxifragales (Stevens 2001 onwards).

EXPLORATION

All *Bryophyllum* spp. are native only to Madagascar (Hannan-Jones and Playford 2002), while the most closely related *Kalanchoe* spp. are represented more widely in Madagascar and continental Africa (Baldwin 1938). From a biological control perspective, Madagascar is therefore the primary area of interest, with other areas of Africa a further possibility.

Preliminary surveys were undertaken in the 1980s by J Marohasy, who was stationed in Madagascar to undertake faunistic surveys for rubber vine. A further exploratory investigation in Madagascar was undertaken by the botanists L Dreyer and D Du Puy in 1997 to determine the distribution and taxonomy of *Bryophyllum* spp. They advised that *B. delagoense* was found south of 22°S and that it displayed a clear centre of diversity in central southern Madagascar. The plant appeared to be substrate-specific, with most collections made on open granite outcrops and no collections from basalt soils (D Du Puy and J Moat, Kew Royal Botanic Gardens, unpub. report).

Exploration in Madagascar, South Africa and Botswana was undertaken between 1999 and 2002 under the direction of ABR Witt of the South African Field Station.

Madagascar was the primary area of search and several trips were made to that island (Witt and Rajaonarison 2004). Typically, infestations consisted of a few adult plants with large numbers of juveniles and were found in drier rocky habitats at the southern end of the island. Some 23 insect species were found on mother-of-millions in these surveys. Three species, *Osphilia tenuipes* (Fairemaire) (Coleoptera: Curculionidae), the chrysomelid leaf beetle *Rhembastus* sp. (Coleoptera: Chrysomelidae) and the wasp *Eurytoma bryophylli* Neser (Hymenoptera: Eurytomidae), were found to be damaging and probably sufficiently host-specific to warrant further study.

Exploration was also made throughout South Africa and southern Botswana. Mother-of-millions derives its specific epithet, *delagoense*, from its early discovery by European botanists at Delgoa Bay, Mozambique (Tolken and Leistner 1986). Mother-of-millions naturalisation in southern Africa is therefore thought to predate European settlement, quite possibly occurring much earlier. Further, many species of closely related genera such as *Kalanchoe* are native to southern Africa. Some 10 insect species were found on mother-of-millions in southern Africa and *Alcidodes sedi* (Marshall) (Coleoptera: Curculionidae) was considered a potential agent.

Altogether, a fairly small arthropod fauna was found in these surveys. This would be consistent with the fauna expected on a specialist, small plant of relatively very small natural distribution (Strong *et al.* 1984).

CANDIDATES

Four insect species were identified from the surveys of Madagascar and southern Africa as potential biocontrol agents. These agents were investigated further at the South African Field Station and all were imported into quarantine in Brisbane for full evaluation. Unfortunately, investigations indicated that all four species had similar host-ranges that included other crassulaceous species. These host-ranges presented problems for obtaining permission to release in Australia.

The stem-boring conderine weevil, *O. tenuipes*, was found infesting *B. delagoense* and (on one occasion) *B. pubescens* Baker in southern Madagascar (Witt 2004). The insect was relatively common and damaging. Dead plants with hollowed stems and characteristic emergence holes indicative of *O. tenuipes* damage were found at several sites (Witt 2004). Preliminary host-testing was conducted in quarantine in South Africa; the weevil was found to be narrowly stenophagous but oviposition and larval development also occurred on other *Bryophyllum* spp., *Kalanchoe sexangularis* N. E. Br., *Crassula sarcocaulis* Ecklon & Zeher and an *Echeveria* sp. (Witt 2004). Comprehensive host-range testing in quarantine in Brisbane showed that the insect would not attack any Australian Crassulaceae but that some exotic species, particularly *K. blossfeldiana*, *K. spathulata* and ornamental *Echeveria* spp., could attract oviposition and support normal larval development.

The second stem-boring weevil, *A. sedi*, was found to cause damage similar to that caused by *O. tenuipes* but in southern Africa and not Madagascar (Witt *et al.* 2004b). *A. sedi* was collected from various *Bryophyllum*, *Kalanchoe* and *Crassula* species, so was evidently not monophagous. The insect was cultured and preliminary host-testing in South Africa confirmed its field host-range. Because the host-range was confined to some Crassulaceae and the large size of the insect would preclude it from attacking Australian natives (all very thin-stemmed *Crassula* spp.), the insect was sent to quarantine in Brisbane for full host-range testing. These tests showed that some exotic species, particularly *K. blossfeldiana*, *K. spathulata*, ornamental *Echeveria* spp. and *Saxifraga stolonifera* Curtis, could attract oviposition and support normal larval development. A particular characteristic of the insect was that it was very sedentary and very rarely seen to fly. In this respect it was quite the opposite of *O. tenuipes*, which is a strong and fast flier.

The phytophagous eurytomid wasp, *E. bryophylli*, infested *B. delagoense* at various times of the year in southern Madagascar in areas between Ampanihy and Fort Dauphin (Witt *et al.* 2004a). The larvae of this wasp tunnel in the cylindrical succulent leaves, causing extensive damage. Details of its morphology have been described (Neser 2008). It was imported into South Africa for life-history studies and preliminary host-range testing and proved to be a suitable candidate agent because it was easy to rear, had multiple generations per year and was damaging to its host plant in the field and under laboratory conditions (Witt *et al.* 2004a). However, preliminary host-range trials revealed that *E. bryophylli* is stenophagous, with larvae able to complete development on three other species in the Crassulaceae. It was imported into quarantine facilities in Brisbane on three occasions but was not successfully cultured. It was not pursued further as it was difficult to culture and because preliminary testing indicated a host-range similar to that of *O. tenuipes* and *A. sedi*.

The leaf-feeding beetle *Rhembastus* sp. was found feeding on new growth and bulbils of *B. delagoense*, but not on several other *Bryophyllum* spp. and *Kalanchoe* spp. growing in southern Madagascar (Witt *et al.* 2006). These beetles were more common in December and February than in the winter months and there were indications of multivoltinism. Preliminary host tests in South Africa indicated a similar host-range to that of *O. tenuipes* and *E. bryophylli*. It was imported into quarantine in Qld but was not successfully cultured and was progressed no further.

RESOLVING CONFLICTS OF INTEREST

All four potential agents for mother-of-millions are stenophagous rather than host-specific, with host-ranges spanning several *Bryophyllum* spp., some *Kalanchoe* spp. and possibly other Crassulaceae. Those tested did not attack the *Crassula* spp. native to Australia. The exotic ornamental *K. blossfeldiana*, which is sold commercially in some quantity, has been shown to be a host and might be attacked should any of the agents be released in Australia.

Although testing indicates attack could occur on *K. blossfeldiana*, other circumstances would limit this. Most commercially produced *K. blossfeldiana* is grown and sold as disposable 'pots of colour' during the winter months (when the insects are not very active) into areas far removed from mother-of-millions, such as Sydney and Melbourne. Any attack in wholesale nurseries in Qld, the most likely scenario, would be controlled by the routine insecticide treatments applied for other insect pests. Other susceptible species have little commercial value. In summary, there is little likelihood of significant detrimental damage to non-target species should the agents be released in Australia.

Australia has legislation to resolve conflicts of interest when the benefits of introducing a biological control clearly outweigh costs associated with either beneficial aspects of the target organism or non-target attack by the agent organism (Anon 1985; McLaren *et al.* 2006). This legislation is the *Biological Control Act 1984* (Cwlth) and mirror legislation enacted by the state jurisdictions. These Acts all require that both the target organism and the agent organisms be declared under the Act. Once agent organisms are declared, legal injunctions preventing release cannot be lodged and neither can those making the releases be held liable for anticipated non-target effects. Declaration of target organisms and agent organisms are made by unanimous decision of the appropriate Ministerial Council after a Biological Control Authority conducts adequate community and stakeholder consultation and develops a benefit:cost analysis or an ecological impact statement. Only three potential conflicts of interest (Paterson's curse, blackberry and rabbit calicivirus) have been considered under a *Biological Control Act*. In 2011, Qld indicated to the Primary Industries Ministerial Council that it intended to declare *Bryophyllum* spp. as target organisms and all four potential insects as agent organisms under the Qld *Biological Control Act 1987*. The Ministerial Council supported that decision. This case will therefore be the first use of a *Biological Control Act* for over 15 years.

In preparation for this application through the *Biological Control Act*, consultations with the nursery industry and economic analyses were undertaken. It is thought that the group most at risk from non-target effects is wholesale nurseries (but the risk to them is low) and that the benefit:cost ratio for releasing the biocontrol agents is better than 20:1 for Qld (Palmer *et al.*, unpub.).

SCIRTOTHRIPS AURANTII ISSUE

South African citrus thrips, *Scirtothrips aurantii* (Faure), is a well-known pest of citrus and mango in its native range, which encompasses a large part of Africa from Egypt to South Africa and the islands of Cape Verde and Mauritius (Faure 1929; CABI and EPPO 1997). Feeding induces superficial scarring on fruit, which renders damaged fruit commercially unacceptable and results in major economic losses for growers (Lewis 1997). The species is reported to be highly polyphagous, with reported hosts ranging from 70 to 100 plant species (Gilbert 1990).

S. aurantii was discovered on *B. delagoense* in Brisbane in 2002 (Palmer 2005). The discovery was a major cause for concern and eradication was considered. However, it was soon discovered that the insect was too widespread to eradicate. The insect is now widespread in southern Qld (Morris and Mound 2004) but has not displayed any of its reported polyphagous tendencies (Manners and Dhileepan 2005; Rafter *et al.* 2008). The Australian population has remained on plants of a single host plant family, Crassulaceae, primarily on *B. delagoense* and *B.x houghtonii* in the field.

The recent incursion of *S. aurantii* has had a visually noticeable impact on *B. delagoense* populations in south-eastern Qld and on the western Darling Downs (comments from several graziers at Qld Murray-Darling Committee field day, November 2007). The feeding damage caused by the thrips and the apparent suppression of *B. delagoense* populations in the area have caused landholders to view the insect as a potentially useful biocontrol agent (Rafter *et al.* 2011). Many landholders thought the insect was intentionally released and have been collecting thrips-infested *B. delagoense* and moving it to uninfected areas to aid dispersal. Therefore, resolution of the threat posed by the incursion of *S. aurantii* to Australian horticulture and verification of its impact on *Bryophyllum* populations are required.

Preliminary research on *S. aurantii* in Australia exposed to citrus and mango in choice tests has revealed that the insect's host-range is likely to be the genus *Bryophyllum* (Rafter *et al.* 2008). The thrips settled and reproduced on mango in two of five replicates, but only on flushing mango leaves. Only two of 19 nymphs produced on flushing mango foliage completed development. These two died without reproducing, so mango does not seem to be a host plant for this thrips. Field sampling of *S. aurantii* on *B. delagoense* in the field demonstrated the insect affects the weed, both structurally and reproductively, when compared to *B. delagoense* in non-*S. aurantii* infested areas. Specifically, feeding damage caused by thrips impairs flowering ability (>50% damage at growing apices) and reduces the number of viable plantlets produced (Rafter *et al.* 2008).

DISCUSSION

Mother-of-millions and its hybrid continue to be serious weeds in Qld and NSW. They have not reached their ultimate distribution in Australia and it is anticipated that they will spread further south. A naturalised population of mother-of-millions was recently discovered in Vic.

The prospects for biological control of these weeds appear to be quite good. Like the prickly pear (Dodd 1940), mother-of-millions is a specialist succulent plant. The potential biocontrol agents all appear quite damaging, at least in the laboratory, and have other characteristics of good biocontrol agents.

No further biocontrol work will be undertaken until the non-target issues are resolved through the *Biological Control Act*. The next step then would be the release of the two weevils, *O. tenuipes* and *A. sedi*. Any further progression of the other agents or more faunal survey would depend on the outcome with these two agents, together with that of *S. aurantii*.

A lesson from this project has been the difficulty of using a *Biological Control Act* which has not been used in the last 15 years. There is little remaining 'corporate memory' and senior executives in government agencies are now reluctant to recommend its use. The uncertainty and lack of decision at key points in this process led to two insects being cultured in quarantine facilities for over eight years after host-testing was completed.

ACKNOWLEDGEMENTS

This project was enabled by the kind hospitality of ARC - Plant Protection Research Institute in hosting the South African Field Station at Rietondale and the Government of Madagascar for allowing faunal survey and export of promising agents. KADW Senaratne developed the CLIMEX model for *B. delagoense* and prepared Figure 2.

REFERENCES

Anon (1981) *Flora of Australia. Vol. 1. Introduction*. Australian Government Publishing Service, Canberra.

Anon (1985) *Biological Control of* Echium *Species (including Paterson's Curse/Salvation Jane)*. Industries Assistance Commission, Canberra.

Ayerbe I and Gomez-Campo C (1977) Drought hardening in excised shoots of *Bryophyllum tubiflorum* Harv. *Annals of Botany* **41**, 649–651.

Baldwin JT Jr (1938) *Kalanchoe*: the genus and its chromosomes. *American Journal of Botany* **25**, 572–579.

Batianoff GN and Franks AJ (1998) Environmental weed invasions on south-east Queensland foredunes. *Proceedings of the Royal Society of Queensland* **107**, 15–34.

Boiteau P and Allorge-Boiteau L (1995) *Kalanchoe de Madagascar*. ICSN, Karthala.

CABI and EPPO (1997) Data sheets on quarantine pests *Scirtothrips aurantii*. In *Quarantine Pests for Europe*. (Eds IM Smith, DG McNamara, PR Scott and M Holderness) pp. 497–500. CABI, Cambridge.

Dodd AP (1940) *The Biological Campaign against the Prickly Pear*. Commonwealth Prickly Pear Board, Brisbane.

Faure JC (1929) The South African citrus thrips and five other new species of *Scirtothrips* Shull. *Bulletin of Transvaal University Collage* **18**, 1–18.

Forster PI (2006) *Bryophyllum* x *houghtonii* (D.B. Ward) P.I. Forst., a new combination in Crassulaceae for the hybrid mother-of-millions. *Austrobaileya* **7**, 383.

Gilbert MJ (1990) Relative population levels of citrus thrips *Scirtothrips aurantii* on commercial citrus and adjacent bush. *South African Journal of Zoology* **25**, 72–76.

Gomez-Campo C (1974) Phyllotactic patterns in *Bryophyllum tubiflorum* Harv. *Botanical Gazette* **135**, 73–79.

Groner MG (1975) Alleopathic influence of *Kalanchoe daigremontiana* on other species of plants. *Botanical Gazette* **136**, 207–211.

Hannan-Jones MA and Playford J (2002) The biology of Australian weeds 40. *Bryophyllum* Salisb. species. *Plant Protection Quarterly* **17**, 42–57.

Ho S (1934–35) Australian botanical illustration. In *Poisonous Plants of S.E. Australia: A Portfolio of Illustrations.* Australian National Botanic Gardens, Australian Government, Canberra.

Kenneally KF (1983) *Kalanchoe crenata* (R. Br.) Haw (Crassulaceae): a new record from Australia. *Australian Systematic Botany Society Newsletter* **34**, 4–5.

Lewis T (1997) *Thrips as Crop Pests.* CABI, New York.

Longman VM, Harvey JM and Keighery GJ (2000) *Bryophyllum delagoense* (Crassulaceae): a new weed for Western Australia and a potentially serious problem for the Abrolhos Islands. *Nuytsia* **13**, 399–400.

Manners AG and Dhileepan K (2005) Australian *Scirtothrips aurantii* Faure (Thysanoptera: Thripidae) only survived on mother-of-millions (*Bryophyllum delagoense*) in a no-choice trial. *Plant Protection Quarterly* **20**, 33–35.

McKenzie RA and Armstrong TA (1986) Poisoning of cattle by *Bryophyllum* plants. *Queensland Agricultural Journal* **10**, 105–108.

McKenzie RA and Dunster PJ (1986) Heart and flowers: *Bryophyllum* poisoning of cattle. *Australian Veterinary Journal* **63**, 222–227.

McLaren DA, Palmer WA and Morfe TA (2006) Costs associated with declaring organisms through the Biological Control Act when conflicts of interest threaten weed biological control projects. In *Proceedings of the 15th Australian Weeds Conference.* 24–28 September 2006, Adelaide. (Eds C Preston, JH Watts and ND Crossman) pp. 549–552. Weed Management Society of South Australia, Adelaide.

Morris DC and Mound LA (2004) Molecular relationships between populations of South African citrus thrips (*Scirtothrips aurantii* Faure) in South Africa and Queensland, Australia. *Australian Journal of Entomology* **43**, 353–358.

Neser OC (2008) *Eurytoma bryophylli* sp. n. (Hymenoptera: Eurytomidae), a leaf-borer of *Bryophyllum delagoense* (Crassulaceae) from Madagascar and a candidate for the biocontrol of the plant in Australia. *African Entomology* **16**, 60–67.

Oelrichs PB, MacLeod JK and Capon RJ (1992) Isolation and identification of the toxic principles in *Bryophyllum tubiflorum* (Kalanchoe). In *Proceedings of the 3rd International Symposium of Poisonous Plants.* (Ed. L James) pp. 288–292. Iowa State University Press, Ames.

Palmer WA (2005) Corrections to Morris and Mound (2004) 'Molecular relationships between populations of South African citrus thrips (*Scirtothrips aurantii* Faure) in South Africa and Queensland, Australia'. *Australian Journal of Entomology* **44**, 1.

Rafter MA, Gillions RM and Walter GH (2008) Generalist herbivores in weed biological control: a natural experiment with a reportedly polyphagous thrips. *Biological Control* **44**, 188–195.

Rafter MA, Palmer WA and Walter GH (2011) Assessing the biological control potential of an adventitiously-established 'pest', *Scirtothrips aurantii* (Faure), on a weed, *Bryophyllum delagoense* (Eckl. & Zeyh.), in Queensland. *Biological Control* **58**, 346–353.

Schäfer C and Lüttge U (1987) Water translocation in *Kalanchoe daigremontiana* during periods of drought. *Plant Cell and Environment* **10**, 761–766.

Stevens PF (2001 onwards) Angiosperm phylogeny website. Version 7, May 2006 [more or less continuously updated since]. <http://www.mobot.org/MOBOT/research/APweb/>.

Strong DR, Lawton JH and Southwood R (1984) *Insects on Plants.* Harvard University Press, Cambridge, Mass.

Sutherst RW, Maywald GF and Kriticos D (2007) *CLIMEX Version 3 User's Guide.* Hearne Scientific Software, Melbourne.

Tolken HR and Leistner OA (1986) *Bryophyllum delagoense. Flowering Plants of Africa* **49,** pl. 1938.

Witt ABR (2004) Initial screening of the stem-boring weevil *Osphilia tenuipes*, a candidate agent for the biological control of *Bryophyllum delagoense* in Australia. *Biocontrol* **49**, 197–209.

Witt ABR and Rajaonarison JH (2004) Insects associated with *Bryophyllum delagoense* (Crassulaceae) in Madagascar and prospects for biological control of this weed. *African Entomology* **12**, 1–7.

Witt ABR, McConnachie AJ and Docherty S (2004a) Distribution and aspects of the biology and host range of *Eurytoma* sp (Hymenoptera: Eurytomidae), a candidate agent for the biological control of *Bryophyllum delagoense* (Ecklon & Zeyher) Schinz (Crassulaceae) in Australia. *African Entomology* **12**, 201–207.

Witt ABR, McConnachie AJ and Stals R (2004b) *Alcidodes sedi* (Col.: Curculionidae), a natural enemy of *Bryophyllum delagoense* (Crassulaceae) in South Africa and a possible candidate agent for the biological control of this weed in Australia. *Biological Control* **31**, 380–387.

Witt ABR, McConnachie AJ, Palmer WA and Grobbelaar E (2006) Distribution, biology and host range of *Rhembastus* sp. (Coleoptera: Chrysomelidae), a candidate for the biological control of *Bryophyllum delagoense* (Crassulaceae) in Australia. *Biocontrol Science and Technology* **16**, 859–869.

Cabomba caroliniana Gray – cabomba

Shon Schooler, Willie Cabrera-Walsh and Mic Julien

ABSTRACT

Cabomba is a submerged aquatic plant that originates from two areas in the Americas – north-eastern Argentina and the south-eastern USA. Despite its restricted native distribution, it has been invading new habitats worldwide and has a large latitudinal range, from cold temperate lakes in Canada and Holland to tropical waters in Australia. It prefers the slow-moving waters of lakes and impoundments where it produces thick stands that reduce the abundance of native species, negatively affect irrigation and potable water treatment, and impede human activities such as boating, fishing and swimming. There are no sustainable methods to manage cabomba and a biological control research program was instigated by the CSIRO in 2003. Several potential biological control agents have been identified in the native range and on closely related plants in the Cabomba genus. The host-specificity of an aquatic weevil, *Hydrotimetes natans*, is being tested in quarantine.

Key words: invasive aquatic macrophyte, insect, herbivore, water fanwort, *Paracles* sp., *Paraponyx diminutalis*.

INTRODUCTION

Cabomba (*Cabomba caroliniana* Gray, Cabombaceae), or water fanwort, has a disjunct native range – in South America centred around northern Argentina and in the southern USA along the Atlantic and Gulf coasts (Orgaard 1991). The plant's tolerance of fragmentation and delicate appearance make it a desirable aquarium plant (Hiscock 2003) and consequently it has been introduced into many countries by the aquarium trade. Cabomba has subsequently been introduced into lakes and streams, both accidentally through the dumping of aquarium water and on purpose to enable cultivation for later collection and sale. It is primarily spread across catchments by watercraft, boat trailers and eel-trapping cages.

Cabomba is a fast-growing submerged aquatic plant that has the potential to infest permanent water bodies in a range of regions, from tropical to cool temperate, throughout the world. It is considered a serious pest in the USA north of the Carolinas and along the west coast, in Canada, the Netherlands, Japan, India, China and Australia, and is present in Hungary, South Africa, France and the UK. Cabomba grows well in slow-moving water bodies and is found mostly in areas of permanent standing water less than 4 m deep; it occurs at depths up to 6 m in Australia (Schooler and Julien 2006). The weed is recognised by its opposing pairs of finely dissected underwater leaves that are feathery or fan-like in appearance (Fig. 1). Small white flowers bearing three petals and three sepals (Fig. 2) extend above the water surface, making infestations more visible in summer months (Fig. 3). Reproduction is almost entirely vegetative throughout most of its introduced range and any fragment that includes a node can grow into a new plant (Sanders 1979).

Cabomba negatively affects the environment, recreational activities, public safety and water quality (Mackey

Figure 1: *Cabomba caroliniana.* a) Plant. b) Fruit. c) Flower. d) Non-dissected upper leaves. e) Dissected lower leaves. Illustration: S Wragg, CSIRO.

and Swarbrick 1997). The weed can smother native submerged plants and may also reduce germination of desirable native emergent plants. Alteration of the flora by cabomba is thought to have reduced populations of platypus (*Ornithorhynchus anatinus* Shaw, Ornithorhynchidae), water rats (*Hydromys chrysogaster* Geoffroy, Muridae) (Mackey and Swarbrick 1997) and populations of the endangered Mary River cod (*Maccullochella peelii mariensis Rowland,* Percichthyidae) (T Anderson, *pers. comm.* 2004). The long stems of cabomba impede boat movement and they can get tangled in propellers, paddles and fishing lines. They are also a potential danger to swimmers. In its native range, however, cabomba seldom covers large expanses or becomes a problematic plant.

It is difficult to assess the cost of the damage caused by cabomba, but some data on the cost of managing cabomba are available. Herbicides are largely ineffective and herbicide use is severely regulated in or around public water supplies (Anderson and Diatloff 1999). Eradication of cabomba from the Darwin River (12.521°S, 131.055°E) is currently being attempted using herbicides. However, supplying water to residents, water quality testing, public education and repeated applications of herbicide have cost over $360 000 in the 2006–07 financial year alone (S Wingrave, *pers. comm.* 2007). Physical control methods are also being used to manage cabomba. From 2000 to 2005, Caloundra Shire Council in Qld spent over $240 000 per year to physically remove cabomba around public areas of Ewen Maddock Reservoir (26.797°S, 152.990°E) using scuba divers and suction pumps (R Rainbird, *pers. comm.* 2007) and Noosa Shire Council spent in excess of $140 000 per year to mechanically remove cabomba from

Figure 2: Cabomba flower with three petals and three sepals. Note small non-dissected upper leaves floating on the water under the flower. Photo: S Schooler, CSIRO.

a small fraction of Lake Macdonald (26.385°S, 152.929°E) using a floating weed harvester (B McMullen, *pers. comm.* 2007). The only sustainable method of managing cabomba appears to be biological control (Schooler *et al.* 2006).

Cabomba has a wide potential distribution. In Australia it is found from Melbourne to Darwin (Fig. 4). Current latitudinal distribution is from monsoonal tropical (Darwin River, Darwin, 12°S) to cold temperate (Loosdrecht Lakes, Holland, 52°N) environments and it can persist under ice in continental temperate climates (Wilson *et al.* 2007). It is primarily a problem in lakes and reservoirs but it can also establish populations in streams, rivers and irrigation canals. It is often found near bridges, either because there are pools of slow-moving water and/or because it was planted for later collection and sale.

BIOLOGICAL CONTROL HISTORY

In 2003, CSIRO commenced a project to discover and test biological control agents from the native range in South America for possible release in the introduced range of

Figure 3: Cabomba infestation in a small farm dam in south-east Qld, 2006. The white flowers emerge above the surface, making the weed more visible during the summer months. Maximum depth is 4 m. Photo: S Schooler, CSIRO.

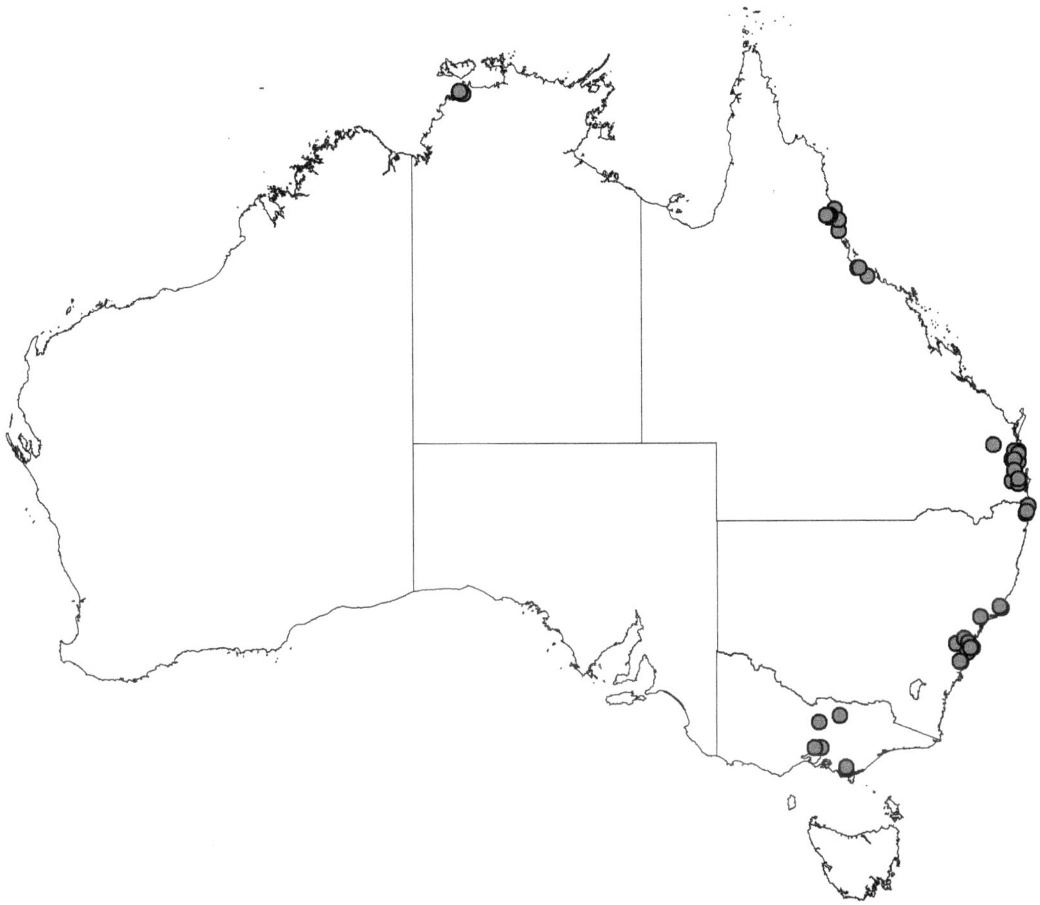

Figure 4: Distribution of *Cabomba caroliniana* in Australia based on herbarium records as of February 2010. Location data records extracted from the Australian Virtual Herbarium. Specimen data reproduced from Australia's Virtual Herbarium with permission of the Council of Heads of Australasian Herbaria Inc.

this weed. Surveys for herbivores in the native range are being completed and host-specificity tests conducted on selected agents in quarantine.

Hydrilla verticillata (L.f.) Royle (Hydrocharitaceae) is the only other submerged weed for which biological control has been attempted, with four agents released in the USA (Julien and Griffiths 1998). So far that project has not resulted in effective control. This may be because the specific life-cycle requirements of the agent are not met in the introduced range (Center *et al.* 2002).

PLANT TAXONOMY

Few plants in Australia are closely related to *Cabomba caroliniana*. Cabomba occurs within Nymphaeales, which consists of two families, Nymphaeaceae and Cabombaceae. One of only two genera in the Cabombaceae, *Cabomba* consists of *C. caroliniana* and four other species (*C. aquatica* Aublet, *C. palaeformis* Fassett, *C. furcata* Schultes and Schultes and *C. haynesii* Wiersema), all of which are endemic to Central and South America (Orgaard 1991).

In its native range in South America there are two varieties of cabomba: *C. caroliniana* Gray var. *caroliniana* Gray, a stout plant with white flowers, and the smaller and more delicate *C. caroliniana* var. *flavida* Orgaard that has yellow flowers. These two varieties do not grow in the same habitats. *C. c. flavida* occurs only in small streams, shallow ponds and irrigation ditches, whereas *C. c. caroliniana* occurs in larger and deeper water bodies. In North America there is another variety, *C. c.* var. *pulcherrima* Harper, known only from the south-eastern USA (North Carolina to Florida). Preliminary genetic analysis indicates that most of the populations of cabomba in Australia are a hybrid between *C. c. caroliniana* and *C. c.* var. *pulcherrima*. However, two localised populations are different; one is more similar to *C. c. caroliniana* and the second appears to be a hybrid between *C. c. caroliniana* and *C. aquatica* (A Weiss, *pers. comm.* 2008). This suggests that there have been at least three separate introductions of cabomba into Australia.

EXPLORATION

Northern and central Argentina and southern Paraguay and Uruguay were surveyed from 2003 to 2009 (W Cabrera-Walsh, unpub. data). In Brazil, surveys were conducted in the Pantanal, Mato Grosso do Sul, during 2004 and in the four south-eastern states during 2006 (C Wikler, *pers. comm.* 2006). Waterways and water bodies along more than 21 000 km of road were surveyed. Cabomba was mostly found in clear-water lakes and streams with low current velocity at 1–4 m depth. These environments were found in Argentina (in the province of Corrientes and in isolated locations along the Paraná-Uruguay basin in the provinces of Formosa, Chaco, Entre Ríos and Buenos Aires), in southern Paraguay and in a few isolated populations on the south-east coast of Brazil. Surveys were also conducted in tropical America (Venezuela, Costa Rica, Puerto Rico and Mexico) on three congeneric species (*C. palaeformis, C. furcata,* and *C. aquatica*) in 2006–08 by Ricardo Segura and Shon Schooler, CSIRO. Specimens of the herbivorous insects found during these surveys are currently housed at the CSIRO Mexican Field Station, where they are being identified.

Ad hoc surveys have been conducted in southern USA by various researchers during the last decade. In 2007–09 several surveys were conducted by Min Rayamajhi, USDA-ARS, looking specifically for plant pathogens. No potential cabomba biological control agents have yet been identified in the USA.

CANDIDATES

After six years of sampling, we believe we have identified most potential biological control agents in the native range of *C. caroliniana* in South America. The area with the greatest density of cabomba sites has been located and numerous lakes in this area sampled throughout the year, to ensure that we have identified potential agents with differing life-cycles. Surveys have focused on arthropods because pathogens are difficult to recognise when sampling submerged aquatic plants with highly dissected leaves. Three potential agents have been found and studies have examined their life-cycles and methods for laboratory rearing. Preliminary host-specificity testing has been carried out at the USDA-ARS South American Biological Control Laboratory in Buenos Aires, Argentina.

Among many obviously generalist herbivores, the three potentially specialist phytophagous insects (Table 1) include an aquatic weevil (*Hydrotimetes natans* Kolbe) and two moth species, *Paracles* sp. and *Paraponyx diminutalis* (Snellen).

***Hydrotimetes natans* Kolbe (Coleoptera: Curculionidae)** This weevil feeds on plant tips as an adult, and the larvae mine the plant stems (Schooler *et al.* 2009; Cabera-Walsh *et al.* 2011). Development from egg to adult requires about 40 days in the laboratory. At high densities, adults can cause extensive tip damage and the larvae can induce

Table 1: Species of herbivores associated with *Cabomba caroliniana* in Argentina

Species	Order: Family	Natural hosts
Hydrotimetes natans Kolbe	Coleoptera: Curculionidae	Cabomba
Ilyodytes lembulus Kuschel	Coleoptera: Curculionidae	*Egeria naias*
Oryzophagus oryzae Lima	Coleoptera: Curculionidae	Rice, others
Dryops sp	Coleoptera: Dryopidae	Cabomba, *Egeria* spp.
Paracles nr. *burmeisteri*	Lepidoptera: Arctiidae	Cabomba, *Eichhornia azurea*
Paraponyx diminutalis (Snellen)	Lepidoptera: Pyralidae	Cabomba, *E. naias*
Synclita nr. *obliteralis* Walker	Lepidoptera: Pyralidae	Cabomba, *Nymphoides indica*
Rhopalosiphum nymphaeae L.	Hemiptera: Aphididae	Generalist
Hedriodiscus sp.	Diptera: Stratyomyidae	Generalist
Ephydra riparia Fallen	Diptera: Ephydridae	Generalist
Midges	Diptera: Chironomidae	Detritivorous
Limpets, snails	Gastropoda	Generalists

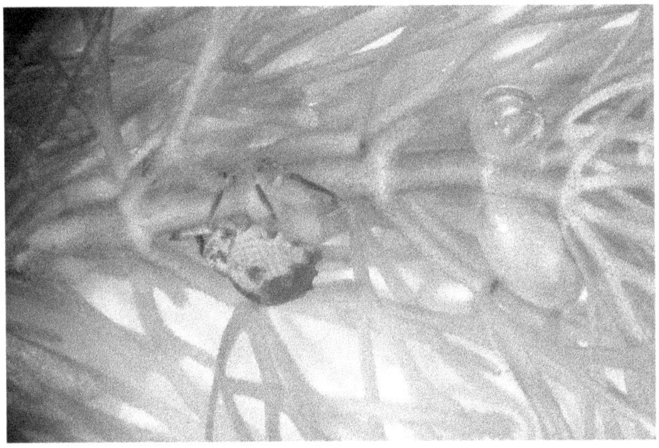

Figure 5: *Hydrotimetes natans* adult weevil on a cabomba stem. Note pupal case to the right of the weevil. Photo: R Chan, CSIRO.

stem decay. In the field, adults are present year-round. They survive for approximately one year in the laboratory. From December to February, adults are often observed alone or mating on the flowers, but they remain underwater most of the time (Fig. 5). Larvae have been found within plant stems from October to July with populations peaking at the beginning of summer (Cabera-Walsh *et al.* 2011). Tunnels have been found through the stem from base to tip down to depths of 3 m. The weevil can mine stems greater than 1.8 mm in diameter, depending on larval stage.

To determine host-range of the weevil in the field, samples of submerged plants occurring near cabomba were collected and suspended in cloth Berlese funnels. Thus far, *H. natans* larvae have only been extracted from cabomba samples (Table 2). Some adult weevils (<0.6%) were found on *Egeria naias* and *Utricularia platensis* when these plants were collected growing intertwined with cabomba in the field. However, we have found no evidence that this weevil uses them as a host.

Host-range tests have been conducted in shadehouses to confirm the field observations, but rearing *H. natans* in the laboratory in containers small enough to closely observe its development and behaviour has not yet been possible. In 2006, colonies were raised outdoors in Buenos Aires, in 1000 L glass tanks where stems with larvae, pupae and adults were observed. In these tanks, cabomba was grown alongside *Egeria densa* Planchon (Hydrocharitaceae), *Potamogeton illinoensis* Morong (Potomogetonaceae) and *P. pusillus* L., thus exposing these plants to *H. natans*. These are key test plant species as they are the plants in Argentina that are closely related to potential hosts in Australia – native *Potamogeton* species and native Australian Hydrocharitaceae (e.g. *H. verticillata*, *Maidenia rubra* Rendle, *Blyxa* spp.). During the 11 month trial period, no *H. natans* of any stage and no damage attributable to *H. natans* was observed on plants other than cabomba (Walsh *et al.* in press).

Table 2: Plant species sampled in the field for presence of *Hydrotimetes natans*

Species	No. of samples	*H. natans* adults	*H. natans* larvae
CABOMBACEAE			
C. caroliniana var. *caroliniana*	111	975	879
C. caroliniana var. *flavida*	9	0	0
Cabomba haynesii	5	0	0
HYDROCHARITACEAE			
Egeria naias	19	3	0
Egeria densa	2	0	0
MENYANTHACEAE			
Nymphoides indica	2	0	0
POTAMOGETONACEAE			
Potamogeton illinoensis	2	0	0
Potamogeton gayi	4	0	0
LENTIBULARIACEAE			
Utricularia platensis	5	2	0
Utricularia foliosa	3	0	0

One sample = 24 L tray (approx. 8 kg fresh weight).

Table 3: Field and no-choice laboratory host-range of *Paracles* sp. larvae

Host plant	Field feeding	Laboratory tests, first instars		Laboratory tests, third instars	
		Feeding	Development	Feeding	Development
Cabomba caroliniana	Yes	Yes	Normal	Yes	Full
Egeria densa	No	Yes	Normal	Yes	Full
Egeria naias	No	No	0	Yes	Partial
Eichhornia azurea	Yes	Yes	1-2 moults, poor development	Yes	Partial
Myriophyllum aquaticum	No	No	0	Yes	No
Ceratophyllum demersum	No	Yes	1 moult and death	Yes	Partial
Panicum sp.	Yes	No	1-2 moults, poor development	Yes	Full
Nymphoides indica	No	No	No	No	No
Potamogeton gayi	No	No	No	Yes	No
Potamogeton illinoensis	No	No	No	Yes	No
Hydrocleis nymphoides	Yes	No	No	Yes	No
Ludwigia peploides	No	No	No	No	No

***Paracles* sp. (Lepidoptera: Arctiidae)** The large aquatic larvae of this moth were collected in several locations in Corrientes province. They cause heavy defoliation of cabomba and are most often found on the leaves near the apical tips. They feed underwater, keeping air bubbles amid short hydrophobic hairs on the dorsum. The cocoon is also aquatic, resembling a canoe, and pupating larvae sometimes form a 'raft' of cocoons woven together in which only silk tufts at the ends remain dry and above the water surface. The adult is brown and has a 25 mm wing-span. It stretches its wings on the floating cocoon upon emergence. The life-cycle of *Paracles* sp. lasts about 40 days. Mating occurs within two days after emergence and the female lays 20–70 whitish eggs (in a mass covered with pale orange scales) on any vertical surface over the water. The eggs hatch in approximately eight days; the larval and pupal stages last 22 and 10 days respectively, at ambient temperature.

Laboratory no-choice and preference host-range tests were performed with this moth on 12 aquatic plants. Since cabomba has no related species at the family level in Argentina, test plants were selected based on damage observed in the field and their co-occurrence with cabomba. When first instars were transferred to test plants in no-choice trials, a narrow host-range was observed. However, feeding of mature larvae was observed in the field on some plants that had been rejected in the laboratory. The survival of more mature larvae on plants rejected by the first instars was therefore tested. More plants were accepted as hosts, and supported development, for second and third instars than for first instars. Mature larvae could develop into fertile adults on the aquatic grass *Panicum* sp. (Poaceae) as well as cabomba, and live for a considerable length of time on four other species (Table 3).

Further choice tests were conducted in three inter-connected 1000 L tanks under cages. In this mixed-choice/no-choice test, 10 gravid females were allowed to choose to oviposit over tanks with *Eichhornia azurea*, *Egeria densa* and cabomba, but the larvae were confined to feeding only on the plant species over which the female had laid the eggs. Adults oviposited indiscriminately on any vertical surface and eggs were laid in every tank. Complete development was obtained on all three plant species, though *E. densa* had been rejected when neonates had been transferred to it in previous experiments. This suggests that, in the first test, initial feeding on cabomba, however brief, precluded acceptance of the other test plant. However, when the larvae emerged directly onto these alternative hosts, development was possible (Tables 3 and 4). This imprinting process has been observed before for other Lepidoptera larvae (Renwick 2001).

It is still not known if *Paracles* sp. can accept these plants as permanent hosts in the field and *Paracles* has not been observed on *E. densa* in the field. However, it suggests plasticity in host acceptance. Australia has several native aquatic genera within the Hydrocharitaceae

Table 4: Mixed-choice test for larval development of *Paracles* sp.

Host	Larval survival (%)	Pupation time (d)	Adult emergence (%)
C. caroliniana	33.3	32±1	100
Egeria densa	26.7	35±1.5	100
Eichhornia azurea	20	40±4	100

Gravid females were allowed to oviposit over any of the three plant species, but larvae were allowed to feed only on the species upon which the female had laid her eggs.

(e.g. *Hydrilla*, *Blyxa*, *Vallisneria* and *Maidenia*), so the priority of this potential agent has been reduced.

***Paraponyx diminutalis* (Snellen) (Lepidoptera: Pyralidae).** The gilled larvae of this moth feed on the terminal shoots of cabomba stems, stunting stem growth. Because of their cryptic habits the larvae are hard to detect despite being >15 mm long in later instars. At the end of its larval stage (45 days), *P. diminutalis* spins an irregular cocoon attached to the plant tips and wrapped with live leaves. In the field it is quite common during spring and early summer but its presence is erratic on cabomba; it is more common on *Egeria naias*. *P. diminutalis* is a well-known natural enemy of several aquatic plants in its native range of south-east Asia (Indonesia to Pakistan) and East Africa (Balloch and Sana-Ullah 1974; Buckingham and Bennett 1989). Unfortunately, the literature (referenced above) and observations in Argentina (Schooler *et al.* 2009) indicate that this species is not host-specific enough to consider it a candidate for biological control of cabomba in Australia.

Surveys on other Cabomba species

The suite of specialist natural enemies found on cabomba in Argentina is relatively small, so surveys of populations of four closely related *Cabomba* species, for additional potential biological control agents, were begun in 2005. Since no closely related native or economically important plants occur in Australia, effective agents that are host-specific to the *Cabomba* genus may be found.

DISCUSSION

In addition to the biological control agent surveys, the effects of water quality and physical variables on cabomba populations and distribution in the native and introduced ranges were examined. There are four main differences between cabomba populations in their native and introduced ranges.

- The climatic variation in the native range is quite narrow (restricted to the Parana-Uruguay basin in eastern Argentina), while climatic variation in the introduced range is extreme (tropical to cold continental climates).
- Cabomba is rarely found in depths less than 1.5 m in the native range (usually 1.5–4 m) because floating vegetation usually occurs around the edges of lakes and ponds. This vegetation greatly reduces light penetration and consequently prevents the growth of cabomba. However, native floating vegetation is rare in most water bodies in subtropical Australia and cabomba grows up to the water's edge.
- Cabomba tends to grow in oligotrophic conditions in the native range, whereas it grows in both oligotrophic and eutrophic conditions in its introduced range.
- Populations are patchy in the native range and often grow mixed with other species (Table 5), but approach monospecific stands in the introduced range. This dearth of large dense stands in the native range is presumably due to the combination of herbivory by natural enemies and competition (and disturbance) caused by floating vegetation.

In Australia, a sampling program was developed to detect the environmental factors that produce variations in plant biomass. Apart from water depth and clarity, no seasonal related factors could be identified in subtropical habitats (Schooler *et al.* 2006). This is not likely to be the case in temperate climates, but it has yet to be assessed for Australia. A similar assessment in China determined that the plant presented maximum biomass in summer (Yu *et al.* 2004), suggesting that plant behaviour will vary according to climate throughout its introduced distribution. However, once established it seems to hold its ground.

In Australia, cabomba grows in dense and temporally stable monospecific stands, thus it is expected that an agent that is present year-round and is host-specific should fluctuate with its host population. The result should be a reduction of cabomba abundance, particularly in areas of deep water where disturbance will exacerbate the effects of herbivore damage and in shallow areas where damage to cabomba will increase the relative competitive ability of native plant species (Schooler *et al.* 2006).

Table 5: Submerged plant species observed to grow mixed with cabomba in its native range

Plant species	Locality	Province/state, country
Ceratophyllum demersum	Mburucuya Cross Creek	Corrientes, Argentina Florida, USA
Chara sp.	Many	Corrientes, Argentina
Egeria densa	Otamendi Iberá	Buenos Aires, Argentina Corrientes, Argentina
Egeria naias	Many	Corrientes, Argentina
Myriophyllum aquaticum	Otamendi	Buenos Aires, Argentina
Myriophyllum spicatum	San Marcos	Texas, USA
Hydrilla verticillata	San Marcos Orange Lake	Texas, USA Florida, USA
Utricularia spp.	Iberá Loch Loosa	Corrientes, Argentina Florida, USA
Vallisneria americana	San Marcos	Texas, USA

Host-specificity testing on the most promising agent, *H. natans*, is being conducted in quarantine in Brisbane and is expected to be completed in 2011. However, this may be delayed due to difficulties in obtaining export permits from Argentina and difficulties in rearing the weevil in quarantine. Given that *P. diminutalis* and *Paracles* spp. are not host-specific, future research will be concentrated on the organisms found in surveys of the other *Cabomba* species.

ACKNOWLEDGEMENTS

We thank Russell Rainbird, Peter Clifton, Colin and Kay Seibs, Tom Anderson, the Noosa Shire Council and the Lake Macdonald Catchment Care Group (LMCCG) for their support. Ricardo Segura (CSIRO Mexican Field Station), Min Rayamajhi (USDA-ARS) and Charles Wikler (UniCentro, Brazil) helped to organise and conduct surveys. Federico Mattioli and Joaquín Sacco assisted in the field and the laboratory. This work was supported by the Australian Department of Environment and Heritage, the Australian Department of Agriculture, Fisheries and Forestry, the Burnett Mary Regional Group and a number of stakeholder groups administrated by the LMCCG.

REFERENCES

Anderson T and Diatloff G (1999) Cabomba management attempts in Queensland. In *Proceedings of the 10th Biennial Noxious Weeds Conference*. 20–22 July 1999, Ballina, NSW. (Ed P Blackmore) pp. 42–45. NSW Agriculture.

Balloch GM and Sana-Ullah (1974) Insects and other organisms associated with *Hydrilla verticillata* (L.f.) L.C. (Hydrocharitaceae) in Pakistan. In *Proceedings of the 3rd International Symposium on the Biological Control of Weeds*. 10–14 September 1973, Montpellier, France. (Ed AJ Wapshere) pp. 61–66. Miscellaneous Publication 8. Commonwealth Institute of Biological Control, Farnham Royal, UK.

Buckingham GR and Bennett CA (1989) Laboratory host range of *Paraponyx diminutalis* (Lepidoptera: Pyralidae), an Asian aquatic moth adventive in Florida and Panama on *Hydrilla verticillata* (Hydrocharitaceae). *Environmental Entomology* **18**, 526–530.

Cabrera-Walsh G, Schooler SS and Julien MH (2011) Biology and preliminary host range of *Hydrotimetes natans* Kolbe (Coleoptera: Curculionidae), natural enemy candidate for biological control of *Cabomba caroliniana* Gray (Cabombaceae) in Australia. *Australian Journal of Entomology* **50**, 200–206.

Center TD, Dray FA, Jubinsky GP and Grodowitz MJ (2002) *Insects and other Arthropods that Feed on Aquatic and Wetland Plants*. Technical Bulletin 1870. US Dept of Agriculture, Agricultural Research Service, Springfield, VA.

Hiscock P (2003) *Encyclopedia of Aquarium Plants*. Barron's Educational Series, Hauppauge, New York.

Julien MH and Griffiths MW (1998) *Biological Control of Weeds: A World Catalogue of Agents and their Target Weeds*. 4th edn. CABI, Wallingford, UK.

Mackey AP and Swarbrick JT (1997) The biology of Australian weeds 32. *Cabomba caroliniana* Gray. *Plant Protection Quarterly* **12**, 154–165.

Orgaard M (1991) The Genus Cabomba (Cabombaceae): a taxonomic study. *Nordic Journal of Botany* **11**, 179–203.

Renwick JAA (2001) Variable diets and changing taste in plant–insect relationships. *Journal of Chemical Ecology* **27**, 1063–1076.

Sanders DR (1979) The ecology of *Cabomba caroliniana*. In *Weed Control Methods for Public Health Applications.* (Ed EO Gangstad) pp. 133–146. CRC Press, Boca Raton.

Schooler SS and Julien MH (2006) Effects of depth and season on the population dynamics of *Cabomba caroliniana* in SE Queensland. In *Proceedings of the 15th Australian Weed Conference.* 24–28 September 2006, Adelaide. (Eds C Preston, JH Watts and ND Crossman) pp. 768–771. Weed Management Society of South Australia, Adelaide.

Schooler SS, Julien MH and Walsh GC (2006) Predicting the response of *Cabomba caroliniana* to biological control agent damage. *Australian Journal of Entomology* **45**, 326–329.

Schooler S, Cabrera-Walsh W and Julien MH (2009) *Cabomba caroliniana* Gray (Cabombaceae). In *Biological Control of Tropical Weeds using Arthropods.* (Ed R Muniappan, GVP Reddy and A Raman) pp. 88–107. Cambridge University Press, Cambridge.

Wilson CE, Darbyshire SJ and Jones R (2007) The biology of invasive alien plants in Canada. 7. *Cabomba caroliniana* A. Gray. *Canadian Journal of Plant Science* **87**, 615–638.

Yu J, Ding B, Yu M, Jin X, Shen H and Jiang W (2004) The seasonal dynamics of the submerged plant communities invaded by *Cabomba caroliniana* Gray. *Acta Ecologica Sinica/Shengtai* **24**, 2149–2156.

Carduus nutans L. – nodding thistle

Jim Cullen and Andy W Sheppard

ABSTRACT

Nodding thistle, *Carduus nutans*, native to Europe and North Africa, is a serious weed of pastures in the USA, Canada, New Zealand and Argentina. In Australia it was introduced relatively recently and is still more or less confined to parts of the tablelands of NSW, but with the potential to spread much more widely. With the benefit of considerable work already carried out for, or by, the USA, Canada and NZ, the Australian program concentrated on assessing the relative merits of the agents available, based on field studies in areas of the native range in southern Europe climatically similar to the target regions of Australia. Two populations of the receptacle weevil *Rhinocyllus conicus* from two such areas, plus a third population from NZ, originally from northern-central France, were released in Australia and their performances compared. The NZ population significantly outperformed the other two populations, though apparently climatically less well suited. The oviposition period of this weevil did not coincide with the main period of seed production of nodding thistle in Australia; to try to cover this period, the seed fly *Urophora solstitialis* was introduced and released. This was more effective, but still not effective enough to reduce seed production sufficiently. In addition, fly emergence was initially asynchronous with flowering and it was largely outcompeted by *R. conicus* early in the season, a result not expected from the native range studies. The rosette weevil *Trichosirocalus horridus* was introduced from NZ and released, with significant effects on rosette vigour and thus seed production. These three agents together reduced seed production by 81%; seed banks and rosette density subsequently declined. The success of this program was calculated to yield a benefit:cost ratio of 6.9.

Key words: *Rhinocyllus conicus*, *Urophora solstitialis*, *Trichosirocalus horridus*, seed banks, climate matching, interspecific competition.

INTRODUCTION

Nodding thistle *Carduus nutans* L. (Asteraceae) is a monocarpic annual or biennial herb, germinating in Australia throughout the year to form a prickly rosette which may grow up to 1 m in diameter before producing a tall flowering stem approximately 1 m in height in the first or second summer after germination (Fig, 1). Its biology has been well described by Popay and Medd (1995). Rosettes from autumn- and winter-germinating seedlings may produce flowering stems in their first or second summer, depending mainly on size attained during winter and spring. The majority of those from spring- and summer-germinating cohorts produce stems in the following summer. Numerous, predominantly crimson-purple flower heads, 20–60 mm diameter, are produced on the multi-branched stem. Each head characteristically droops at a right angle or more to the stem and up to 20 000 seeds may be produced per plant (commonly averaging approximately 10 000). Although equipped with a pappus, most seeds fall near the parent plant (91% within 1–2 m in NZ)

Figure 1: Rosettes and flowering stems of nodding thistle, *Carduus nutans*. Photos: J Cullen, CSIRO.

(Kelly *et al.* 1988). The seed has a short innate dormancy, but thereafter may germinate as soon as moisture conditions are suitable. Germination may be delayed in low light or when buried; seed buried to 15–20 cm remains viable for up to 20 years. Rosettes are avoided by grazing livestock (except perhaps goats and donkeys), thereby reducing the productive area of pasture, but their competitive effect extends well beyond the area occupied, further reducing productivity for some distance around each plant (Thompson *et al.* 1987). The species was accidentally introduced into Australia from NZ as a contaminant of pasture or pea seed in the 1940s. It established and spread in pastures on moderate- to high-fertility soils in the cool and damp (500–900 mm annual rainfall) tablelands of NSW, where it occurs in fairly discrete northern, central and southern regions (Figs 2 and 3). Small infestations occur in Vic and Tas, but these are actively contained. Single infestations found in SA and WA have been eradicated (Popay and Medd 1995). Nodding thistle is of Eurasian and North African origin and has an international reputation as a serious weed of pastures, being considered an important weed in NZ, the USA, Canada and Argentina (Parsons and Cuthbertson 2001).

BIOLOGICAL CONTROL HISTORY

Nodding thistle had been the target of biological control in Canada, the USA and NZ, where infestations are much more extensive, before being considered in Australia. Six agents were released in North America (Gassmann and Kok 2002) and three in NZ. The distribution of this thistle in Australia is relatively limited because of its recent introduction and active control measures, but its potential has always been recognised and it was an obvious target once resources could be diverted to it. The program had the advantage of considerable knowledge of agents, and aspects of the earlier work had provided material for discussion concerning biological control. Apart from the aim of controlling nodding thistle, the Australian program was considered an opportunity to test ideas regarding procedure, particularly concerning climate matching and competition between natural enemies.

While populations of potential agents were available from previous work by other countries and had involved significant success, these were derived from the Rhine Valley (Alsace, France) and from Austria, in northern-central continental Europe, whereas the climate of Australian infestations more closely matched that of southern Europe. Biological control of *C. nutans* in Australia was first considered in 1976, based on encouraging reports from North America and NZ. The program started in 1977 when a proposed host-specificity test list of the receptacle weevil *Rhinocyllus conicus* Froehlich was approved by AQIS; testing against the list was carried out in 1978–80 in France. A full program did not commence until 1985 with studies of the field biology, interaction

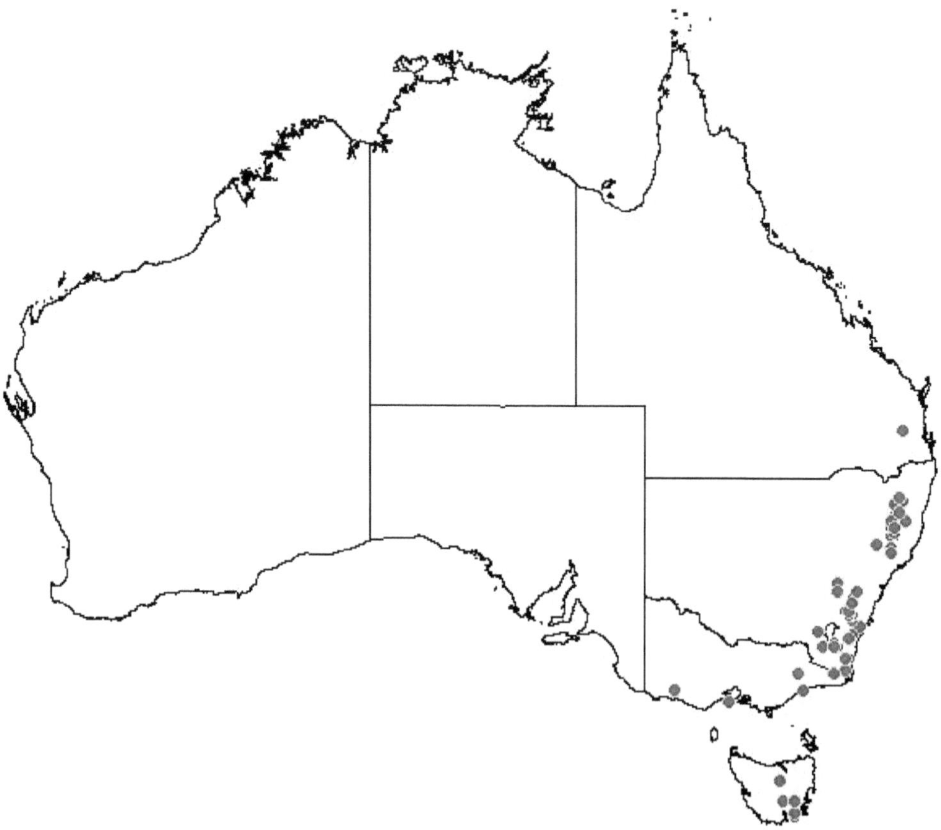

Figure 2: Distribution of nodding thistle in Australia. Specimen data reproduced from Australia's Virtual Herbarium with permission of the Council of Heads of Australasian Herbaria Inc.

and effectiveness of natural enemies on natural populations of *C. nutans* at sites north of Montpellier on the Larzac plateau of the southern Massif Central of France. Such studies showed that these native populations were limited by seed-feeding insects (Jongejans *et al.* 2006). After confirmatory host-specificity tests, populations of

Figure 3: Infestation of nodding thistle, *Carduus nutans*, at Kybeyan, NSW. Photo: J Cullen, CSIRO.

R. conicus and the seed gall fly *Urophora solstitialis* (L.) were imported from this region and released in Australia. *R. conicus* was also obtained from Scritto (central Italy) and from NZ (originally from Alsace). Later in the program, the rosette weevil *Trichosirocalus horridus* Panzer (= *Ceuthorhynchidius horridus* Panzer) was imported from NZ where it was having a significant impact (Jessep 1989), in contrast to its apparent lesser importance in the field in southern France.

All three species established and, in combination, reduced the density and importance of *C. nutans* in Australia (Swirepik and Smith 2002).

PLANT TAXONOMY

Carduus nutans L. (Asteraceae: Carduinae) is considered a complex of four subspecies, with *C. nutans* subsp. *nutans* the predominant form in Australia (Doing *et al.* 1969; Medd 1981). *C. nutans* subsp. *nutans* occurs in North America, but the dominant subspecies is *C. nutans* subsp. *leiophyllus*, which is considered synonymous with *C. thoermeri*. The latter occurs in Australia, but only as a ruderal of minor importance, confined to south-eastern Qld (Medd 1981).

C. nutans belongs to the subtribe Cardueae in the tribe Cynareae. There are only two species of Cardueae recorded as native to Australia – *Rhaponticum australe* (Gaudich.) Sojak (= *Stemmacantha australis* (Gaudichaud-Beaupre) Dittr.) and *Hemistepta lyrata* Bunge. The recorded range of *H. lyrata* is from northern NSW to Qld, and Asia. Searches of possible sites in the 1980s, based on herbarium records, failed to find any plants. This species could not be included in host-specificity studies.

EXPLORATION

Surveys in Europe over several years by CIBC on behalf of Canada and by the USDA discovered a wide range of natural enemies (Zwölfer 1965; Harris and Zwölfer 1971; Boldt 1978; Boldt and Campobasso 1978; Redfern 1983; Zwölfer and Harris 1984; Freese 1993). These studies found 33 oligophagous insects recorded on *C. nutans* in the native range (Gassmann and Kok 2002). The more promising of these included *R. conicus*, *U. solstitialis*, *T. horridus*, the rosette weevil *Ceuthorhynchus trimaculatus* (F.), the stem- and root-boring syrphid *Cheilosia corydon* (Harris), the seed flies *Urophora sibynata* (Rondani) and *Tephritis hyoscami* L., the chrysomelid *Psylliodes chalcomera* Illig., the moth *Epiblema pflugiana* (Hayworth) (Olethreutidae), the weevils *Larinus jaceae* F. and *Larinus* sp. and the rust *Puccinia carduorum* Jacky. Of these, *R. conicus* and *T. horridus* were introduced to Canada and the USA where they subsequently produced significant reductions in density of the weed (Harris 1984; Kok 2001).

The Australian program was based in Montpellier (France) and apart from the early host-specificity work by AJ (Tony) Wapshere (CSIRO) and the detailed ecological studies at the Larzac sites, referred to above, led by Jim Cullen and Andy Sheppard (CSIRO), it included surveys further afield in France and northern Italy by Jean-Paul Aeschlimann (CSIRO), in similar cooler areas whose climate matched that of the Australian distribution of *C. nutans*. The work by CSIRO added five flower-inhabiting species to the list compiled from the earlier surveys – the weevil *Larinus sturnus* Schaller, the phalacrid beetle *Olibrus bisignatus* Ménétriès, the Lepidoptera *Cochylis posterana* Zeller (Tortricidae), *Homeosoma nebulellum* D&S (Pyralidae) and *Epiblema scutulana* D&S (Olethreutidae), plus the smut fungus *Ustilago cardui* Fischer von Waldheim.

Five species were considered for introduction to Australia – *R. conicus*, *U. solstitialis*, *T. horridus*, *C. trimaculatus* and *C. corydon*.

CANDIDATES

Natural enemies

For the Australian program, considerations of possible interspecific competition, prior knowledge and experience elsewhere and observations in the field sites in Europe determined the order of priority for introduction among the five species initially considered.

The Larzac studies showed that survival of *R. conicus* and *U. solstitialis* was unaffected when the two species occurred together because the fly avoided capitula heavily attacked by *R. conicus* (A Sheppard, unpub. data 1989). Given the known success of *R. conicus* elsewhere, this species was introduced first. Following release, it became clear that *R. conicus* was not well synchronised with the phenology of *C. nutans* seed production in Australia, which led to a decision to import *U. solstitialis* from the Larzac. This was released in 1991. This insect had not been used in North America or NZ. *Trichosirocalus horridus* and *C. trimaculatus* were seen as more direct competitors in the same niche without a clear indication of which might be the more effective. Controlled pot trials, using five, 10 and 20 larvae per rosette, found little effect of treatment on receptacle and seed production or dry weight for either species (J-L Sagliocco, unpub. data 1989, 1991). However, from the literature, *C. trimaculatus* had a broader host-range, and given the results being obtained in NZ, a population of *T. horridus* (originally from Alsace, France via Canada to NZ) was imported and released in Australia in 1993. *Ceuthorhynchus trimaculatus* and other species might have been considered later if initial success in Australia had been limited and resources had continued to be available.

Agents tested and not released

***Cheilosia corydon* (Harris) (Diptera: Syrphidae).** The adults of this syrphid lay eggs on the young stems of *C. nutans* in early summer and the larvae bore into the stem, then progress down to the rootstock. They were observed to cause multi-stemming and significant damage in the field in France (A Sheppard, unpub. data 1989). However, it is extremely difficult to induce reproduction in adults under confined conditions and, following only limited success in the USA, this species was not released in Australia.

Agents released

Rhinocyllus conicus **Froehlich (Coleoptera: Curculionidae).** There is a significant amount of geographic differentiation in this species in the native range that also relates to host plant use (Klein and Seitz 1994). In this respect, *R. conicus* on *C. nutans* seems to be quite closely related throughout Europe. The adults of this weevil (Fig. 4) overwinter and become active in late spring, laying eggs on the outside of the developing thistle heads. The eggs hatch in approximately a week and the larvae chew their way into the receptacle, where they feed on receptacle tissue and developing achenes, forming a gall-like callus (Zwölfer and Harris 1984; Gassmann and Kok 2002). A single larva destroys on average 26 seeds (Sheppard *et al.* 1994). Fully developed larvae pupate in the heads and new adults emerge in summer. After feeding to a limited extent on foliage, they aestivate for the remainder of summer and enter diapause for winter. A partial second generation of weevils has been noted in some native regions, very early emerging adults laying eggs in the second half of the thistle head season, the subsequent larvae destroying some of the seed developing at this time and the adults emerging in late summerautumn (Gassmann and Kok 2002).

Figure 4: Adult of *Rhinocyllus conicus*. Photo: D McClenaghan, CSIRO.

The host-specificity testing of the Alsace (France) population of this agent carried out for North America and summarised in Zwölfer and Harris (1984) greatly assisted the risk assessment for Australia. This population was imported into Australia via NZ (see below) and required no further testing. For the two other southern European populations of this agent planned for release in Australia (Table 1), additional host-range testing was carried out at CSIRO Montpellier facility in 1977–80, with additional studies in 1988. The key test plant was cultivated artichoke, *Cynara scolymus* L. (Asteraceae), because a population of *R. conicus* had been reported in the European literature as a pest of this crop in Algeria (AJ Wapshere, *pers. comm.* 1977). Testing found only the occasional unprotected eggs laid on this plant and some unexpectedly high adult feeding on *Chysanthemum* sp., but no oviposition. Permission was granted for the release of all three populations of *R. conicus* in 1988.

The release and establishment of *R. conicus* in Australia has been described by Woodburn and Cullen (1995). The three populations introduced were from the Larzac and from Scritto (Italy), which have climates similar to those of the southern and northern tablelands of NSW respectively, plus the original one from Alsace, via Canada, where it had been successful in controlling *C. nutans*, and NZ. Both French and Italian populations were heavily infected with *Nosema*, a microsporidian parasite, and had to be carefully reared, with single pair matings and constant microscopic examination, to free the colonies of the disease. The population from NZ was disease-free. Each population was first released in separate sites in each of the three main regions of the NSW tablelands (northern, central and southern), the French and NZ populations in 1988 and the Italian population in 1989 and 1990. Establishment was variable, the NZ population establishing readily and spreading in all three regions. The French population established well in the southern region but struggled in the central and northern regions. This was due to late flowering of thistles, missing *R. conicus* activity, and/or sheep grazing the flower heads which contained developing larvae. The Italian population fared poorly, struggling to establish in the southern region site and not being recovered in the central and northern sites. A partial second generation was observed only in the NZ population in Australia, as observed previously in NZ and North America for this population.

Table 1: Adult feeding and oviposition host-specificity testing of *Rhinocyllus conicus* for Australia

Test plant	N	Adult feeding (rating[b] or mm² per pair ± SE)	% adult mortality during 1-week test	Oviposition on test plant capitula (eggs/week/female ± SE)
1978–80 trials southern French *R. conicus* from *C. nutans*				
Family Asteraceae; tribe Cardueae				
Carduus nutans L.	6	+++	8	Yes
Carthamus tinctorius L.	6	0	8	No
Cynara scolymus L.	6	+	84	No
tribe Cichorieae				
Lactuca sativa L.	6	0	45	No
tribe Heliantheae				
Helianthus annus L.	6	+	11	No
tribe Anthemideae				
Chrysanthemum sp.	6	++ to +++	8	No
Family Caricaceae				
Carica papaya L.[a]	6	0	50	No data
Family Poaceae				
Oryza sativa L.[a]	6	0	45	No data
Sorghum vulgare L.[a]	6	+	62	No data
Saccharum officinarum L.[a]	6	+	11	No data
Family Fabaceae				
Arachis hypogaea L.[a]	6	+	8	No data
Glycine max L.[a]	6	+	95	No data
Acacia sp.[a]	6	0	70	No data
Family Malvaceae				
Gossypium sp.[a]	6	+	92	No data
Family Musceae				
Musa sapientum L.[a]	6	+	56	No data
Family Myrtaceae				
Eucalyptus sp.[a]	6	0	100	No data
Family Proteaceae				
Macadamia ternifolia F. Muell.[a]	6	0	97	No data
1988 field trials a) southern French Larzac plateau *R. conicus* from *C. nutans*				
Carduus nutans L	13	No data	0	5.4±1.2
Cynara scolymus L. var Violet de Prov	13	0.8±0.8	10	0.07±0.07
1988 field trials b) Italian Larzac plateau *R. conicus* from *C. nutans*				
Carduus nutans L	7	No data	0	7.2±1.9
Cynara scolymus L. var Violet de Prov	7	7.8±2.3	8	0±0
1988 field trials b) French La Crau *R. conicus* from *S. marianum*				
Silybum marianum (L.) Gaertner	9	No data	0	14.3±1.7
Cynara scolymus L. var Violet de Prov.	9	5±2	33	0.1±0.1
Carduus pycnocephalus L.	9	0±0	57	0±0

a: Test consisted of one detached leaf each from six plant individuals – oviposition not recorded.
b: + = very slight feeding. ++ = definite feeding. +++ = same level as control.
Additional to Zwölfer and Harris (1984), adapted from AJ Wapshere (unpub. data 1980) and J-L Sagliocco and J Vitou (unpub. data 1988). The control plants are underlined. Unless stated, 197880 trials used 36 adults for each test, collected from *C. nutans* in active oviposition and exposed to one test plant for a week with the control run in parallel. 1988 trials used field test plants with one *R. conicus* pair per test, with control then test plant tested in sequence.

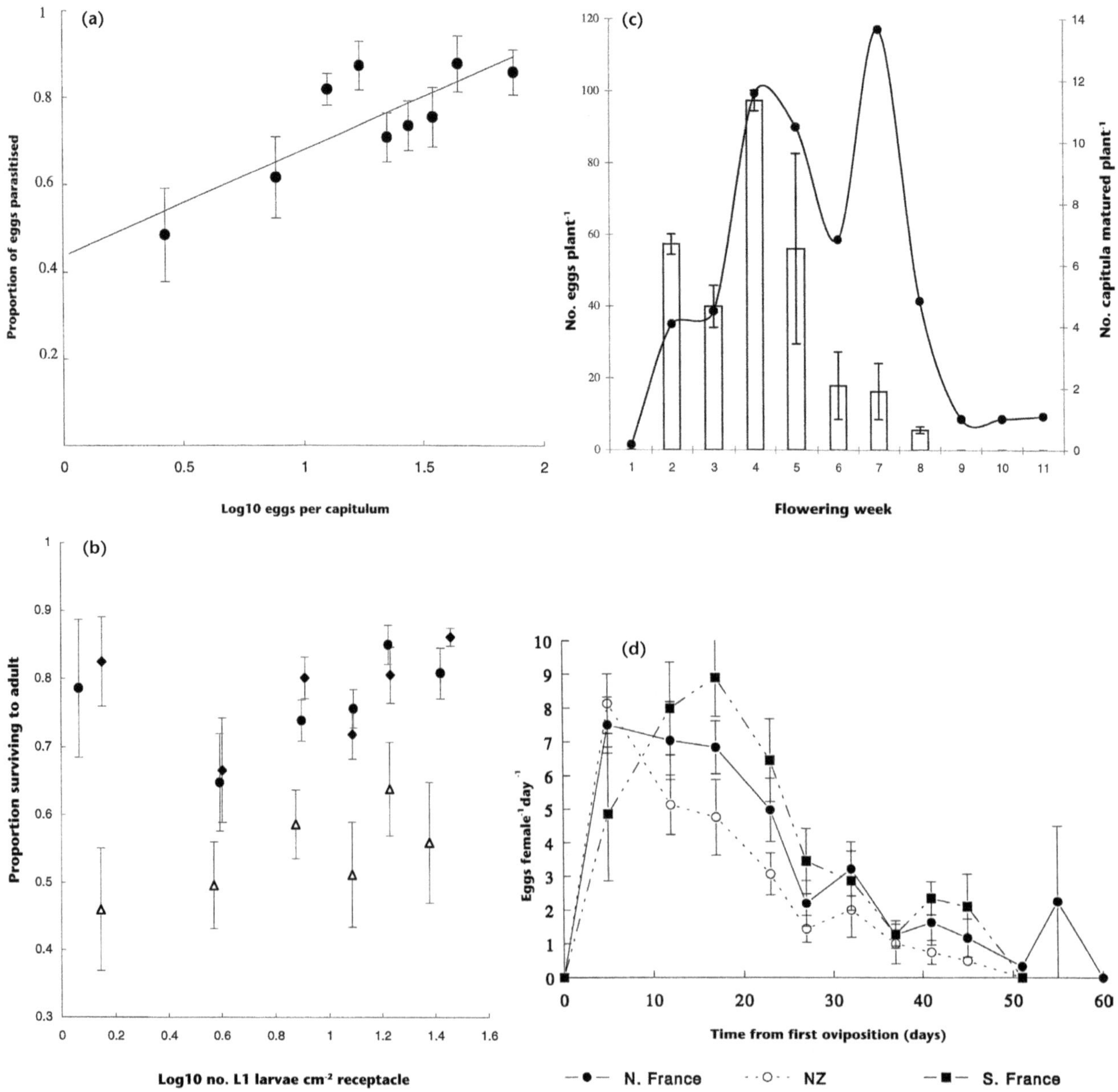

Figure 5: a) Oviposition over time with respect to capitulum availability in *Rhinocyllus conicus* in France (Larzac) for 50 random plants in 1989. b) *R. conicus* larval survival (±SE) in response to density per capitulum in three native thistle populations (different symbols). c) Egg parasitism in relation to egg density per capitulum. d) Oviposition patterns (± SE) of three populations of *R. conicus* – Larzac population (filled squares), Alsace population (filled circles) and NZ population originally collected from Alsace, but after 25 years in the exotic range without significant parasitism or interspecific competition (open circles). 10 caged pairs per population at 27°C were transferred to fresh suitable capitula every two to four days until oviposition ceased.

The high clumping of oviposition by *R. conicus* on the early capitula (Woodburn 1996b) caused high intraspecific larval mortality per capitulum in Australia (Woodburn 1996b), but not in the native range (Fig. 5a, b). In the native range, intraspecific competition was in part being offset by high levels of egg parasitism (Fig. 5c). High intraspecific competition in Australia should be a strong adaptive selection pressure for *R. conicus* to extend oviposition across more of the flowering period, especially given the near-absence of other capitulum feeders in much of the exotic range and the many unexploited capitula available later in the flowering cycle (Woodburn 1996b). To test this, *R. conicus* was collected from the original Alsace collection sites of the NZ population in

1989. The oviposition phenology of this native population was compared to that of the Larzac population and the NZ exotic population that had experienced at least 25 years in North America and NZ in the absence of interspecific competition or significant egg parasitism levels. The oviposition phenology (eggs per day over time) of all these populations of *R. conicus* were, however, identical (Fig. 5d).

Regardless of origin, it quickly became apparent that despite the success in Canada and Virginia, USA, the longer flowering season in Australia, assisted by intermittent summer rains, and poor synchronisation of adult weevil emergence with the onset of flower head production, meant that a large proportion of the seasonal seed production would not be affected by *R. conicus* activity (Woodburn and Cullen 1993). The maximum reduction in seasonal seed production recorded was 35.7%, which is insufficient to produce control in Australia, as is the 40% recorded in NZ for infestations (Kelly and McCallum 1995).

Urophora solstitialis **(L.) (Diptera: Tephritidae).** *Urophora solstitialis* overwinters as fully fed larvae in diapause within their galls in the thistle capitula, pupation and adult emergence taking place in spring (Fig. 6). Adult emergence is prolonged and coincides with the bolting of the thistle stems and production of the first buds. Oviposition occurs in these buds, eggs being laid singly, each into a developing floral tube. Newly hatched larvae eat their way down through the developing ovule into the receptacle, where they induce the development of a gall within which they feed. The multi-locus gall of up to 25 larvae lignifies and incorporates surrounding seed. Each larva consumes one seed and prevents the maturation of five or six others (Sheppard *et al.* 1994). A large proportion of the larvae developing early in the season pupate, and a second generation of adults emerges later in the season. The proportion of larvae pupating declines steadily until all later developing larvae enter diapause as fully fed larvae (Woodburn 1993).

The existence of a second generation of this insect was seen as a means of reducing seed production after *R. conicus* activity had ceased. Field studies in France confirmed this, and that there were no negative interactions between *R. conicus* and *U. solstitialis* in the thistle capitula (Sheppard *et al.* 1994) with adult flies avoiding capitula heavily attacked by *R. conicus* (A Sheppard, unpub. data 1989). It was imported to Australia from the Larzac in 1990. Host-specificity of this and subsequently considered agents required closer attention to Australian native species, with *R. australe* being a key test plant. Specificity was confirmed (Woodburn 1993) and this seed fly released at sites in all three regions in December 1991.

Figure 6: Adult of *Urophora solstitialis*. Photo: D McClenaghan, CSIRO.

The species established at all sites and by the second season, 1992–93, a reduction of 45% was recorded in seed production at the Berridale site in the southern region (Woodburn 1996a). However, synchronisation of adult emergence in spring with flower bud production was poor, with many flies emerging too early and dying before buds suitable for oviposition developed. The second generation was therefore small and late, with a distinct decrease in attack between the first and second generations when seed production was highest. Despite this, seed reduction by *U. solstitialis* was higher than by *R. conicus*.

Trichosirocalus horridus **Panzer (Coleoptera: Curculionidae).** The biology, host-specificity, release and establishment of *T. horridus* in Australia have been well described by Woodburn (1997). In Australia, adults (Fig. 7) become active in early autumn following summer aestivation, and eggs are laid on the lower surface of the rosette leaves through autumn, winter and spring. Newly hatched larvae crawl to the centre of the rosette, where they feed first on the unexpanded leaves surrounding the meristem and then on the meristem itself. Fully developed larvae pupate in the soil and new adults emerge in late spring–early summer. These feed briefly on cauline leaves before aestivating over summer. Larval feeding directly reduces the vigour of the plant and the damage to the meristem stimulates the formation of subcrowns, which are also attacked, as are sub-subcrowns that may be produced in turn. As well as a reduction in vigour and therefore seed production, the morphology of the rosette is altered by the larval attack, the leaves becoming spineless and softer, rendering them more prone to grazing.

Figure 7: Adult of *Trichosirocalus horridus*. Photo: D McClenaghan, CSIRO.

T. horridus was imported to Australia from NZ and was therefore originally of German origin, having first been established in Canada, from where the NZ population was obtained. The taxonomy of this agent has been the subject of some debate as *T. horridus s.l.* in Europe was considered to be two species – *T. horridus* and a newly described species *Trichosirocalus mortadelo* (Alonso-Zarazaga and Sanchez-Ruiz 2002). However, more recent examination of material from NZ suggests that there was only one species, which most resembled *T. horridus* (R Groenteman and R Oberprieler, *pers. comm.* 2006). After confirmation of its host-specificity, *T. horridus* was released initially at a site in the southern region of the NSW tablelands in 1993 and subsequently in the other regions.

Woodburn (1997) found that, in contrast to the USA where *T. horridus* had been studied by McAvoy and Kok (1985) and Cartwright and Kok (1985), larval development occurred continuously during the slightly milder winter in Australia, reduction of vigour was more severe and seed production was reduced significantly regardless of rosette size. Some rosettes were actually killed by *T. horridus* feeding and overall seed production was reduced by 72%. This weevil has therefore been the most effective agent established in Australia for control of *C. nutans*.

DISCUSSION

Apart from the main aim of controlling *C. nutans* in Australia, this program presented the opportunity to compare the performances of different climatically matched populations of at least one agent and to explore possible negative interspecific interactions between two different agents.

Climate matching

For *R. conicus*, the comparison was between a population from an area in Europe (Larzac, France) climatically similar to the southern tablelands of *C. nutans* distribution in Australia, one from a second area (Scritto, Italy) with a climate apparently similar to the northern tablelands and one from a region (Alsace, France) whose climate did not match the Australian *C. nutans* distribution very well, but which had succeeded in Canada and the USA and had established well in NZ. The two populations imported directly from France and Italy had to be cleaned of a *Nosema* infection and the benefit was clearly evident when the fecundity of infected and non-infected French *R. conicus* was compared in quarantine: a mean of 21.6 eggs over 10.4 days for infected females versus 97.9 eggs over 21.8 days for uninfected females (Woodburn and Cullen 1995). Disease was therefore not a point of difference, but the Italian population struggled to establish, particularly in the region to which it was supposed to be most climatically suited. The southern French population established, but had slow population increase; the NZ (Alsace) population proved to be far more vigorous, establishing readily in all regions and spreading far more rapidly. Interpretation of this result raises some interesting issues and suggestions: (a) climate matching is only important to a certain level depending on the species, (b) some native populations are inherently better colonisers or benefit better from conditions more favourable than their native habitat, (c) there had been genetic change in the NZ population as a consequence of periods of population explosion in Canada and NZ which meant it was now a more effective coloniser, (d) there were other factors unknown or in combination that produced this result. Woodburn and Cullen (1995) noted that the fecundity of the NZ population was significantly higher and that this population had a partial second generation, in contrast to the southern French population. It seems unusual that the NZ population originating from the most northern and coldest part in Europe was the only population to exhibit a partial second generation, as insect generation number is usually correlated with mean annual temperature. One explanation may be that a second generation in *R. conicus* is maladaptive in a more Mediterranean climate as the hot dry season curtails the flowering period of *C. nutans*. In the Australian distribution of *C. nutans* the climate is more akin to a Mediterranean climate in terms of average rainfall and temperature, but the distribution of the rainfall is different. Eastern Australia has much higher occurrences of summer rainfall than a true Mediterranean climate and this is the most likely reason why *C. nutans* has a flowering period on average nearly twice as long in Australia than in southern France (Woodburn and Sheppard 1996). This might be sufficient to give the NZ *R. conicus* with a partial second generation an adaptive advantage. Whether these properties were sufficient to explain the difference in population performance observed, however, is still unknown.

Interspecific interactions

The interaction between capitulum insects has interested previous researchers (Zwölfer 1979) and was of particular interest in this program, given that *R. conicus* and *U. solstitialis* were the two most important capitulum species in the southern France sites and likely candidates for introduction (Sheppard *et al.* 1994). Zwölfer (1973) had found that survival of *R. conicus* was markedly reduced in the presence of *U. solstitialis* and suggested that *R. conicus* alone should be introduced (to Canada). However, Harris (1991) suggested that *U. solstitialis* could be the more effective on its own. *U. solstitialis* was eventually released in North America against *C. nutans* in 1996 some 27 years after *R. conicus*, but its establishment has not been confirmed (Gassmann and Kok 2002).

The southern French field studies showed that within the capitulum there was little effect of one species on the other and that *U. solstitialis* tended to avoid heads already colonised by *R. conicus*, attacking the smaller secondary capitula (Woodburn and Sheppard 1995; A Sheppard, unpub. data 1989). At the levels of infestation and the observed insect synchrony in France this was not a major issue, tending to spread out attack by the two species. However, in Australia, *U. solstitialis* emerged so early that only the very first large capitula were available for oviposition. All these were so severely attacked by *R. conicus* as to cause abortion or malformation, often killing the resident *R. conicus* larvae and rendering the capitula unavailable to *U. solstitialis*. The fly's first-generation population was therefore much reduced, as was the second as a consequence (Woodburn 1996b). Despite this, *U. solstitialis* was still contributing more to seed reduction than *R. conicus*, so the latter's redistribution was not assisted further in Australia because of this possible negative interaction (Woodburn and Briese 1996). Thus, while the field data in the country of origin correctly established the importance of both insects as seed predators and that their interaction under the observed conditions was not negative, under the conditions of asynchrony with much heavier insect exploitation of the capitulum resources, without parasitoids, in Australia, this interaction became more significant. This had not been foreseen. Evidence from laboratory studies suggested that *U. solstitialis* was likely to resynchronise its spring adult emergence, more in line with *C. nutans* flowering phenology in Australia, after several generations (Woodburn 1996a; Woodburn and Cullen 1996). Nonetheless, extrapolation of species interactions from the native to the introduced range needs to be done with care. As it turned out, *T. horridus* attacking the overwintering rosettes provided the most effective long-term control of the weed.

Evaluation

A comparison of *C. nutans* demography in its native range with that in Australia showed that reduction of the soil seed bank would be the key to successful control. Seed banks in Australia were in the range of 3000–13 000 seeds m^{-2} compared with undetectable to 140 seeds m^{-2} in France. This was a result of a longer flowering season, higher seed production per plant (57–24 334 in Australia in contrast to 0–1116 in France) and a higher seed rain (171–38 500 seeds m^{-2} in Australia in contrast to 0.4–240 m^{-2} in France) (Woodburn and Sheppard 1996). Shea (1996) modelled *C. nutans* dynamics in NZ where microsites for germination are often abundant, as in Australia, and concluded that a reduction of at least 65% in seed production was required for populations to become seed-limited. The best result of the combined impact of *R. conicus* and *U. solstitialis* was 71% (Woodburn 1996b), but this was rare. However, the establishment of *T. horridus* produced a reduction of 72% due to that species alone, and 81% with the combined effect of all three agents. At two sites, the mean seed bank level declined from 9500 seeds m^{-2} in 1989 to 397seeds m^{-2} in 2001 (Swirepik and Smyth 2002) and density trends suggested that high- and medium-density infestations would disappear by 2005.

Using population modelling, Shea *et al.* (2010) have shown that the relative importance of different biological control agents can be predicted if the population dynamics of the target species are well understood in their exotic range. For example, they found that *C. nutans* population growth and spread rates were much higher in NZ than in Australia, where rosette growth and survival was much lower. Based on knowledge of the impacts of the various biocontrol agents, they predicted that the order of agent effectiveness would be *T. horridus*, *U. solstitialis* then *R. conicus* for Australia, but the order was predicted to be *U. solstitialis*, *R. conicus* then *T. horridus* in NZ, where seed production played a greater role in population growth (Shea *et al.* 2005, 2006).

The decline in density of *C. nutans* translates directly to a decrease in the problem caused by the weed, and therefore its costs. The replacement of *C. nutans* by productive pasture has been observed in several sites and increases productivity from grazing. At the same time, the declining population has meant decreased control costs and decreased contamination of wool from thistle fragments. A recent benefit:cost study of this program based on Young and Woodburn (2002), which used estimates of costs from the wool industry, estimated an NPV of $69.4 million in 2004–05 dollars and a benefit:cost ratio of 6.9 (Page and Lacey 2006).

Non-target effects

In North America, *R. conicus* has been found to attack other species in the Carduinae, with considerable concern over rare native North American *Cirsium* spp. (Turner *et al.* 1987; Louda *et al.* 1997). This has not been an issue in Australia as there are no native species sufficiently closely related.

ACKNOWLEDGEMENTS

A number of farmers permitted fenced or controlled-access experimental sites on their land to allow detailed follow-up work. Their cooperation is gratefully acknowledged.

REFERENCES

Alonso-Zarazaga MA and Sanchez-Ruiz M (2002) Revision of the *Trichosirocalus horridus* (Panzer) species complex, with description of two new species infesting thistles (Coleoptera: Curculionidae, Ceutorhynchinae). *Australian Journal of Entomology* **41**, 199–208.

Boldt PE (1978) Foreign exploration for the biological control of *Carduus* spp. In *Biological Control of Thistles in the Genus Carduus in the United States: A Progress Report.* (Ed KE Frick) pp. 11–17. USDA.

Boldt P and Campobasso G (1978) Phytophagous insects on *Carduus macrocephalus* in Italy. *Environmental Entomology* 7, 904–909.

Cartwright R and Kok LT (1985) Growth response of musk and plumeless thistles (*Carduus nutans* and *C. acanthoides*) to damage by *Trichosirocalus horridus* (Coleoptera: Curculionidae). *Weed Science* **33**, 57–62.

Doing H, Biddiscombe EF and Knedlhans S (1969) Ecology and distribution of the *Carduus nutans* group (nodding thistles) in Australia. *Vegetation* **17**, 313–351.

Freese G (1993) The insect complexes associated with the stems of seven thistle species. *Entomologia Generalis* **19**, 191–207.

Gassmann A and Kok LT (2002) Musk thistle (nodding thistle). In *Biological Control of Invasive Plants in the Eastern United States.* (Eds FV Driesche, B Blossey, M Hoodle, S Lyon and R Reardon) pp. 229–245. USDA-FS. Forest Health Technology Enterprise Team. Morgantown, West Virginia.

Harris P (1984) *Carduus nutans* L., nodding thistle and *C. acanthoides* L., plumeless thistle (Compositae). In *Biological Control Programmes against Insects and Weeds in Canada 1969–1980.* (Eds JS Kelleher and MA Hulme) pp. 115–126. CAB, Slough.

Harris P (1991) Classical biocontrol of weeds: its definition, selection of effective agents, and administrative-political problems. *Canadian Entomologist* **123**, 827–849.

Harris P and Zwölfer H (1971) *Carduus acanthoides* L., welted thistle, and *C. nutans* L., nodding thistle. In *Biological Control Programmes against Insects and Weeds in Canada 1959–1968. Technical Communication CIBC* **4**, 76–79.

Jessep CT (1989) Introduction of the crown weevil *Trichosirocalus horridus* as an additional biocontrol agent against nodding thistle. *Proceedings of the NZ Weed and Pest Control Conference* **42**, 8–10 August, New Plymouth, NZ. (Ed AJ Popay) pp. 52–54. NZ Weed and Pest Control Society, Palmerston North.

Jongejans E, Sheppard AW and Shea K (2006) What controls the population dynamics of the invasive thistle *Carduus nutans* in its native range? *Journal of Applied Ecology* **43**, 877–886.

Kelly D and McCallum K (1995) Evaluating the effectiveness of *Rhinocyllus conicus* on *Carduus nutans* in New Zealand. In *Proceedings of the 8th International Symposium on Biological Control of Weeds.* 2–7 February 1992, Canterbury, New Zealand. (Eds ES Delfosse and RR Scott) pp. 205–212. DSIR/CSIRO, Melbourne.

Kelly D, Cameron H and Alex J (1988) Wind dispersal of nodding thistle seeds and pappi. In *Proceedings of the New Zealand Weed and Pest Control Conference* **41**. 9–11 August, Auckland. (Ed AJ Popay) pp. 207–209. NZ Weed and Pest Control Society, Palmerston North.

Kelly D, McCallum K, Schmidt CJ and Scanlan PM (1990) Seed predation in nodding and slender winged thistles by nodding thistle receptacle weevil. In *Proceedings of the New Zealand Weed and Pest Control Conference* **43**. 14–16 August, Dunedin. (Ed AJ Popay) pp. 212–215. NZ Weed and Pest Control Society, Palmerston North.

Klein M and Seitz A (1994) Geographic differentiation between populations of *Rhinocyllus conicus* Frőlich (Coleoptera: Curculionidae): concordance of allozyme and morphometric analysis. *Zoological Journal of the Linnean Society* **110**, 181–191.

Kok LT (2001) Classical biological control of nodding and plumeless thistles. *Biological Control* **21**, 206–213.

Louda SM, Kendall D, Connor J and Simberloff D (1997) Ecological effects of an insect introduced for the biological control of weeds. *Science* **277**, 1088–1090.

McAvoy TJ and Kok LT (1985) Viability and development rate of overwintering eggs of *Trichosirocalus horridus* (Coleoptera: Curculionidae). *Environmental Entomology* **14**, 284–288.

Medd RW (1981) Taxonomy of *Carduus nutans* in north-eastern New South Wales, Australia. In *Proceedings of the 8th Asian Pacific Weed Science Society Conference.* (Ed BV Venkata Rao) pp. 171–173. APWSS, Bangalore.

Page AR and Lacey KL (2006) *Carduus nutans* (nodding thistle). In *Economic Impact Assessment of Australian Weed Biological Control.* pp. 59–60. CRC for Australian Weed Management, Adelaide.

Parsons WT and Cuthbertson EG (2001) *Noxious Weeds of Australia.* 2nd edn. CSIRO Publishing, Melbourne.

Popay AI and Medd RW (1995) The biology of Australian weeds 21. *Carduus nutans* L. ssp. *nutans. Plant Protection Quarterly* **5**, 3–13.

Redfern M (1983) *Insects and Thistles.* Naturalists' Handbooks 4. Richmond Publishing, Slough, UK.

Shea K (1996) Estimating the impact of control efforts: models of population dynamics. In *Proceedings of a Thistle Management Workshop.* (Eds TL Woodburn, DT Briese and S Corey) *Plant Protection Quarterly* **11**, 263–265.

Shea K, Kelly D, Sheppard AW and Woodburn TL (2005) Context-dependent biological control of an invasive thistle. *Ecology* **86**, 3174–3181.

Shea K, Sheppard A and Woodburn T (2006) Seasonal life-history models for the integrated management of the invasive weed nodding thistle *Carduus nutans* in Australia. *Journal of Applied Ecology* **43**, 517–526.

Shea K, Jongejans E, Skarpaas O, Kelly D and Sheppard AW (2010) Optimal management strategies to control local population growth or population. *Ecological Applications* **20**, 1148–1161.

Sheppard AW, Cullen JM and Aeschlimann J-P (1994) Predispersal seed predation in *Carduus nutans* L. (Asteraceae) populations in southern Europe. *Acta Oecologia* **15**, 529–541.

Swirepik A and Smyth M (2002) Biological control of broad-leafed pasture weeds (Paterson's curse, *Onopordum* and nodding thistles): what have we achieved and where to from here? In *Proceedings of the 13th Australian Weeds Conference.* 8–13 September 2002, Perth. (Eds Spafford Jacob H, Dodd J and Moore JH) pp. 373–376. Plant Protection Society of WA, Perth.

Thompson A, Saunders AE and Martin P (1987) The effect of nodding thistle (*Carduus nutans*) on pasture production. In *Proceedings of the New Zealand Weed and Pest Control Conference* **40**. 11–13 August 1987, Nelson, NZ. (Ed AJ Popay) pp. 222–225. NZ Weed and Pest Control Society, Palmerston North.

Turner CE, Pemberton RW and Rosenthal SS (1987) Host utilization of native *Cirsium* thistles (Asteraceae) by the introduced weevil *Rhinocyllus conicus* (Coleoptera: Curculionidae) in California. *Environmental Entomology* **16**, 111–115.

Woodburn TL (1993) Host-specificity testing, release and establishment of *Urophora solstitialis* (L.) (Diptera: Tephritidae), a potential biological control agent for *Carduus nutans* L. in Australia. *Biocontrol Science and Technology* **3**, 419–426.

Woodburn TL (1996a) Reduction of seed set in nodding thistle (*Carduus nutans*) by the seed fly *Urophora solstitialis*, in Australia. In *Fruit fly Pests: A World Assessment of their Biology and Management.* (Eds BA McPheron and GJ Steck) pp. 165–169. St Lucie Press, Florida.

Woodburn TL (1996b) Interspecific competition between *Rhinocyllus conicus* and *Urophora solstitialis*, two biocontrol agents released in Australia against *Carduus nutans*. In *Proceedings of the 9th International Symposium on Biological Control of Weeds.* 19–26 January 1996, Stellenbosch, South Africa. (Eds VC Moran and JH Hoffmann) pp. 409–415. University of Cape Town, South Africa.

Woodburn TL (1997) Establishment in Australia of *Trichosirocalus horridus*, a biological control agent for *Carduus nutans*, and preliminary assessment of its impact on plant growth and reproductive potential. *Biocontrol Science and Technology* **7**, 645–656.

Woodburn TL and Briese DT (1996) The contribution of biological control to the management of thistles. In *Proceedings of a Thistle Management Workshop* (Eds TL Woodburn, DT Briese and S Corey). *Plant Protection Quarterly* **11**, 250–253.

Woodburn TL and Cullen JM (1993) Effectiveness of *Rhinocyllus conicus*, a biological control agent for *Carduus nutans*, in Australia. In *Proceedings of the 10th Australian and 14th Asian-Pacific Weed Conference.* 6–10 September, Brisbane. pp. 99–103. Queensland Weed Society, Brisbane.

Woodburn TL and Cullen JM (1995) Release and establishment of the thistle-head weevil, *Rhinocyllus conicus* in Australia. In *Proceedings of the 8th International Symposium on Biological Control of Weeds.* 2–7 February 1992, Canterbury, NZ. (Eds ES Delfosse and RR Scott) pp. 411–414. DSIR/CSIRO, Melbourne.

Woodburn TL and Cullen JM (1996) Impact of *Rhinocyllus conicus* and *Urophora solstitialis* on achene set in *Carduus nutans* in Australia. In *Compositae: Biology and Utilization. Proceedings of the International Compositae Conference*, Kew, 1994 (Ed DJN Hind) pp. 307–319. Royal Botanic Gardens, Kew.

Woodburn TL and Sheppard AW (1995) *Urophora solstitialis*, a potential biological control agent for *Carduus nutans* in Australia. In *Proceedings of the 8th International Symposium on Biological Control of Weeds.* 2–7 February 1992, Canterbury, NZ. (Eds ES Delfosse and RR Scott) p. 415. DSIR/CSIRO, Melbourne.

Woodburn TL and Sheppard AW (1996) The demography of *Carduus nutans* as a native and an alien weed. In *Proceedings of a Thistle Management Workshop.* (Eds TL Woodburn, DT Briese and S Corey). *Plant Protection Quarterly* **11**, 236–238.

Young R and Woodburn TL (2002) Evaluation of research on the biological control of nodding thistle *C. nutans*. Unpublished report. CSIRO Entomology, CSIRO, Canberra.

Zwölfer H (1965) Preliminary list of phytophagous insects attacking wild Cynareae (Compositae) in Europe. *Technical Bulletin CIBC* **6**, 81–154.

Zwölfer H (1973) Competition and coexistence in phytophagous insects attacking the heads of *Carduus nutans* L. In *Proceedings of the 2nd International Symposium on Biological Control of Weeds.* 4–7 October 1971, Rome. (Ed PH Dunn) pp. 74–81. Commonwealth Agricultural Bureaux, Slough, England.

Zwölfer H (1979) Strategies and counterstrategies in insect population systems competing for space and food in flower heads and plant galls. *Fortschritte der Zoologie* **5**, 331–353.

Zwölfer H and Harris P (1984) Biology and host-specificity of *Rhinocyllus conicus* (Froel.) (Col., Curculionidae), a successful agent for biocontrol of the thistle, *Carduus nutans* L. *Journal of Applied Entomology* **97**, 36–62.

Carduus pycnocephalus L. – slender thistle
Carduus tenuiflorus Curt. – slender thistle

RH Groves and Andy W Sheppard

ABSTRACT

The two closely related slender thistles, *Carduus pycnocephalus* and *Carduus tenuiflorus*, occur as introduced annuals in southern Australian pasture and natural ecosystems, where they are capable of reducing pasture productivity and hence animal production. One strain of a rust fungus, *Pucccinia cardui-pycnocephali*, is known to have been introduced accidentally to Australia at least 50 years ago, but this strain was ineffective in reducing seed output. A classical program for the biological control of these two taxa of slender thistles for Australia began with research in southern France in 1986. The program included surveys and assessments of potential insect agents for biological control against these thistles, but none were considered suitable. Host-specificity testing was then undertaken of a further two strains of the rust collected in Mediterranean Europe, with their importation for additional testing in quarantine in Canberra and their eventual field release in southern Australia between 1993 and 1995. One rust strain was shown to be more effective at limiting seed production in one plant taxon and the second in the other taxon. Subsequent monitoring of releases in 1998 suggest that at two localities in Vic and Tas the mean number of inflorescences and dry weight of *C. tenuiflorus* plants decreased with rust infection. The monitoring program did not continue, however. We tentatively conclude that mycological control continues to reduce slender thistle densities in southern Australia and, if biological control can be integrated with pasture management involving competitive perennial plants, plant population densities will be decreased still further.

Key words: integrated control, pasture management, *Puccinia cardui-pycnocephali*, rust fungus, southern Australia.

INTRODUCTION

Two closely related plant species are referred to as slender thistles, namely *Carduus pycnocephalus* L. and *C. tenuiflorus* Curt. (Asteraceae). Both are annuals that may co-occur in pasture and natural ecosystems in southern Australia, to which they were introduced from Europe over 100 years ago (Parsons 1977; Parsons and Cuthbertson 2001). In 1992, industry losses to Australia resulting from reduced stock carrying capacities and wool contamination were estimated to exceed $A5 million per year (Burdon *et al.* 2000).

The two species are widely distributed across the winter-rainfall regions of southern Australia (Fig. 1), where they typically occupy disturbed sites or bare patches following summer drought. Reproduction of the species relies on dispersal of seeds, which usually germinate in autumn. Seedlings then pass winter as rosettes; the plants flower and seed in spring and early summer (Groves and Kaye 1989), depending on soil moisture and fertility levels as well as the effects of increasing daylength. Groves and Kaye (1989) showed that, of the seven thistle species

(a)

(b)

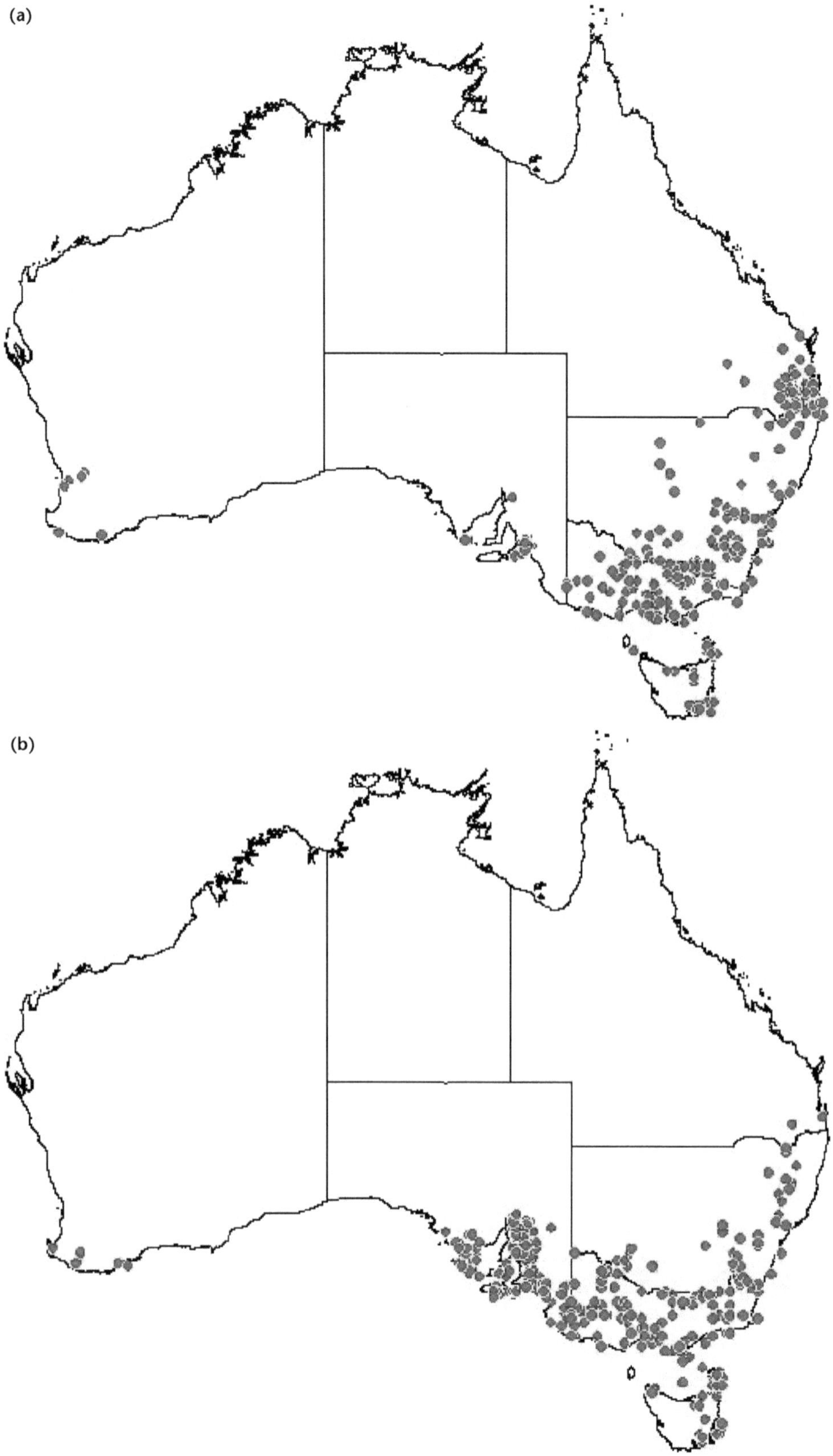

Figure 1: Distribution of the two slender thistles, a) *Carduus pycnocephalus* and b) *C. tenuiflorus*, in southern Australia. Specimen data reproduced from Australia's Virtual Herbarium with permission of the Council of Heads of Australasian Herbaria Inc.

they tested for responses of seed germination to a range of day/night temperatures, *C. pycnocephalus* germinated relatively more at low temperatures (15/5°C and 15/0°C), a result that may help to explain its distribution being limited to southern Australia. Slender thistle demography has been studied by Parsons (1977) and Groves *et al.* (1984) in Vic and NSW respectively, and in Mediterranean Europe by Olivieri (1984). All showed significant seedling mortality in late autumn/early winter at the seedling or young rosette stages.

One or both of these slender thistles occur as aliens in southern Australia, New Zealand and western USA where they are pasture weeds. They also occur in South America (Brazil and Argentina), South Africa, Afghanistan, Iran, south-western Asia, Greece, the former Soviet Union and Britain (Parsons and Cuthbertson 2001).

BIOLOGICAL CONTROL HISTORY

The first surveys were undertaken by CABI in the 1960s across Europe (Zwölfer 1965) and then by USDA in the 1970s (Goeden 1974; Batra *et al.* 1981) to initiate classical biological control of the slender thistles in the USA. Several attempts to use *Rhinocyllus conicus* Frölich (Coleoptera: Curculionidae) to control the weed in south-western USA were mostly ineffective because of asynchronous developmental differences between the insect and the host species *C. pycnocephalus* (Goeden and Ricker 1985). The subsequent Australian program was cooperative between CSIRO Plant Industry and Entomology, and commenced in 1986 in Canberra and in Montpellier, France. It concerned searches for both entomological (Table 1) and mycological candidates; the latter proved to be more effective in the longer term, when released in the field in southern Australia (Burdon *et al.* 2000).

PLANT TAXONOMY

The slender thistles are both in the tribe Cardueae and subtribe Carduinae which includes many other common invasive thistles in Australia, most notably *Carduus nutans* (nodding thistle) and *Cirsium vulgare* (spear thistle), with many common natural enemies. No Australian native species occur in this tribe. Parsons and Cuthbertson (2001) described the taxonomic differences between the two species. Briefly, *C. pycnocephalus* has rosette leaves that may have lighter-coloured areas along the veins of the leaves and at the base of the spines on the flowering stems (never in *C. tenuiflorus*), its flowers are in clusters of three or four (typically up to eight in *C. tenuiflorus*) and its inner involucral bracts are shorter than the adjacent florets (they are longer than the adjacent florets in *C. tenuiflorus*). Seeds of *C. pycnocephalus* are yellowish-fawn and slightly longer than those of *C. tenuiflorus*, which has grey-brown seeds. The single most distinguishing morphological feature between the two species, in the authors' experience, however, is that the stem of *C. tenuiflorus* has spiny wings along most of its length up to the base of the flower heads (Fig. 2); in *C. pycnocephalus* the wings are absent or poorly developed below the flower heads. Hybrids with morphological characteristics intermediate between the two species can occur in mixed populations, but are not common.

EXPLORATION

Both thistle species originate in southern Europe and northern Africa (Parsons and Cuthbertson 2001). The natural distribution of *C. pycnocephalus* extends further east through south-western Asia as far as Pakistan, while that of *C. tenuiflorus* extends further north in Europe to Scandinavia and Britain. Exploration for natural enemies for the Australian biological control program against slender thistles was built on previous surveys by Zwölfer (1964) and Goeden (1974). These surveys (for insects and plant pathogens) were centred on the region of southern France and Italy where the species are sympatric and there is a Mediterranean-type climate; they took place across southern Europe from 1986 to 1991 in parts of Spain, France, Italy, former Yugoslavia and Greece and were conducted by Jean-Paul Aeschlimann.

CANDIDATES

Arthropods

Sampling of 117 sites led to the identification of 27 herbivorous insects (25 species on *C. pycnocephalus* and 12 species on *C. tenuiflorus*; Table 1) that were typical of communities found attacking other thistles (Batra *et al.* 1981). They included root, stem and capitulum weevils, captulum-feeding tephritid flies and three species of fungal pathogen found using these two *Carduus* species. There was little evidence, however, that slender thistles had highly specific insect herbivores; nearly all species were stenophagous within the genus *Carduus* or the tribe Cardueae and often moved between thistle species at sites with mixtures of thistle species (Sheppard *et al.* 1991). The insect species considered most promising as biologi-

Figure 2: a) Close-up of *Carduus tenuiflorus* flowerheads. b) *Carduus pyncocephalus* plant. c) Infestation of *Carduus tenuiflorus*. Photos: Vic Dept of Primary Industries.

cal control candidates are listed in Table 2, with the reasons they were not studied further.

Populations of *C. pycnocephalus* were studied over three years at three separate sites in southern France during this period to assess the field impacts of natural enemies on this species (Sheppard *et al.* 1991).

Agents tested but not released

Ceutorhynchus trimaculatus F. To measure the potential impact of *C. trimaculatus*, rosette weevil, a field trial was established at Montpellier. On 1 November 1989 seedlings were transplanted into a plot, fertilised and watered as necessary. After one month an insecticide treatment (1.5 g L^{-1} Orthene monthly and Lannate 1 mL L^{-1} bimonthly) was given and zero, five and 10 first instar larvae transferred into notches on leaf midribs on plants, with six plants per treatment randomly arranged across three blocks. In mid March, five plants per treatment were randomly harvested to dissect and confirm attack rates. Total seed production per plant was estimated by multiplying up from a random subsample of 10–100% of capitula dissected per plant on which glue spray was

Table 1: Insects that were known thistle herbivores found on *C. pycnocephalus* and *C. tenuiflorus* during surveys and field studies in the western Mediterranean, 1986–91

Species	On *C. pycnocephalus*	On *C. tenuiflorus*	Region/country
Coleoptera: Curculionidae			
Rhinocyllus conicus (Frölich)	X		S. France, Italy
Ceutorhynchus trimaculatus F.	X		S. France
Trichosirocalus horridus Panzer	X		S. France
Larinus jaceae F.	X	X	S. France
Larinus planus F.	X	X	S. France, Spain
Lixus elongatus Goeze	X	X	S. France
Coleoptera: Apionidae			
Apion gibbirostre Gyll.	X		S. France
Coleoptera: Chrysomelidae			
Psylliodes chalcomera (Ill.)	X		S. France
Sphaeroderma testaceum Weise	X		S. France
Galeruca interrupta Ill.		X	S. France
Cassida rubiginosa Muell.	X	X	S. France, Spain
Coleoptera: Anobiidae			
Lasioderma haemorrhoidale (Ill.)	X		S. France
Coleoptera: Pharacridae			
Olibrus bisignatus Men.	X		S. France
Coleoptera: Bruchidae			
Spermophagus sericeus (Goeff.)	X	X	S. France
Diptera: Tephritidae			
Urophora stylata (F.)	X	X	Italy
Xyphosia miliaria (Schrank)		X	Italy
Terellia serratulae (L.)	X	X	S. France
Chaetostomella onotrophes Locis	X		S. France
Acanthiophilus helianthi Rossi	X		S. France
Diptera: Muscidae			
Phaomia trimaculata Bché	X		S. France
Diptera: Agromyzidae			
Chromatomyia sp.	X		Spain
Diptera: Chloropidae			
Chlorops sp.		X	S. France
Diptera: Stratiomyidae			
Chloromyia sp.	X		S. France
Lepidoptera: Tortricidae			
Cochylis posterana (Zeller)	X		S. France
Cnephasia stephensiana Doubleday	X		S. France
Lepidoptera: Cosmoterigidae			
Pyroderces argyrogrammos Zeller	X		
			S. France
Hemiptera: Tingidae	X		
Tingis cardui (L.)			S. France

Table 2: Insects considered to have potential as biological control agents for slender thistles, later discounted based on abundance and indirect assessment of their impact on thistles in the western Mediterranean

Order/Family	Species (niche)	Strengths	Weaknesses
Coleoptera			
Curculionidae	*Larinus planus* F (capitulum feeder)	Observed only on slender thistles in surveys	Low attack levels at sites where measured
	Ceutorhynchus trimaculatus F (root crown and leaves)	First species found attacking rosettes in autumn	Stenophagous on Cardueae and narrow distribution on slender thistles
Apionidae	*Ceratapion gibbirostre* Gyll. (roots and root crown)	Widespread and abundant and high attack rates per plant	Stenophagous on Cardueae and low attack level per plant
Phalacridae	*Olibrus bisignatus* Ménétriès (seed-feeder)	High attack levels per plant where present	*Carduus* specialist, narrow distribution and low attack rates per plant
Homoptera			
Aphididae	*Brachycaudus helichrysi* Kalt. (leaves and stems)	Widespread and abundant	Generalist

applied to mature capitula to prevent seed dispersal. The results for this experiment found 0%, 14%, 12% and 21% of rosettes with larvae over the four treatments. High rates of natural field attack in the control treatment led to use of the insecticide treatment as the attack-free treatment; compared to this, seed production was 60%, 90% and 67% lower in the control, five larvae and 10 larvae treatments respectively (Fig. 3). Despite these encouraging results, a decision was made not to continue with this agent as the literature indicated the species is stenophagous on the tribe Cardueae and not widely distributed on slender thistles (Table 2), being found most often in the presence of other thistles.

The results of detailed field studies showed that, from late autumn to mid winter, rosettes were attacked mainly by the weevil *Ceutorhynchus trimaculatus* F. with lesser damage from another weevil *Trichosirocalus horridus* Panzer and the chrysomelid beetle *Sphaeroderma testaceum* F. Other insect damage was mostly confined to the short flowering period. At this time there were high numbers of the apionid *Ceratapion gibbirostre* Gyll. in the roots, and damage within the capitula was from the

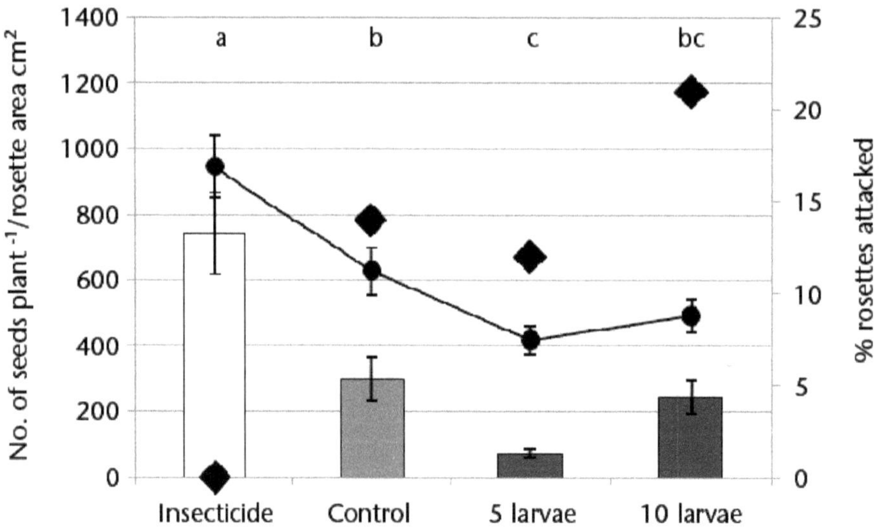

Figure 3: Maximum rosette area (line and filled circles), seed production (bars) per *Carduus pynocephalus* plant (± SE, treatments with a different letter above were significantly different in log. transformed ANOVA, $p < 0.05$) and percent of rosettes attacked by *C. trimaculatus* (right Y axis) in a field experiment on the impact of this weevil on plant productivity in the native range using treatments of insecticide, zero, five and 10 added first instar larvae at the early rosette stage. Adapted from internal CSIRO report by J-L Sagliocco 1990.

phalacrid beetle *Olibrus bisignatus* Menetries and the capitulum weevil *Rhinocyllus conicus* Frölich. Once the thistles had finished flowering, most flower-feeding insects moved to later-flowering thistle species nearby. The capitulum weevil *Larinus planus* F. was found only on *C. pynocephalus* at these sites, but rarely and in low numbers. The only measurable impact of insects came from seed counts where losses of 1–12% were attributed to insects (Sheppard *et al.* 1991).

Agents tested and released

Puccinia cardui-pycnocephali. Survey results showed that this autoecious rust fungus (Fig. 4) was widespread. The results of a separate study had shown it could significantly reduce flowering under glasshouse conditions (Olivieri 1984), so field experiments were undertaken to assess its impact. This rust was found to significantly limit seed production of plant material grown in southern France (at the CSIRO Biological Control Laboratory, Montpellier), with different rust strains shown to differentially affect the two species (Chaboudez *et al.* 1993). Isolates of the rust were collected from Yugoslavia, Italy, France and Spain as well as one isolate from Australia. The virulence and aggressiveness of two of these strains, one collected from Italy (IT2) and one from southern France (FR3), were tested on southern French populations of the two *Carduus* species and compared with the one rust strain already present in southern Australia. The two Mediterranean isolates were shown to be more aggressive than the Australian isolate (Chaboudez *et al.* 1993), with isolate IT2 more pathogenic on plants of *C. pycnocephalus* and FR3 more pathogenic on *C. tenuiflorus.*

Results of host-specificity testing at Montpellier, especially on species in the Asteraceae family, showed that the two strains collected from the Mediterranean were host-specific, with the exception that pustules of both strains developed on cotyledons and the first true leaves of *Cynara scolymus* L. (Asteraceae), globe artichoke, but leaves produced later were unaffected. The host plant material on which the rust was assessed originated from six sites in WA and one in NSW. Permission for field release of these two strains of rust was granted in 1993 and releases commenced in October 1993. Methods for the release of the pathogen are described more fully in Burdon *et al.* (2000).

Figure 4: Rust fungus, *Pucccinia cardui- pycnocephali*, on *Carduus tenuiflorus*. Photo: Vic Dept of Primary Industries.

Of some significance to biological control of these two carduine thistles is that in most years floral development of the related *Carduus nutans* L. (nodding thistle) occurs after that of *C. pycnocephalus* and *C. tenuiflorus.* One biotype of an insect released in Australia for the biological control of *C. nutans* (namely, *Rhinocyllus conicus*) may lay its eggs on the earlier-developing flower heads of the slender thistles and even limit subsequent seed production of the slender thistles if the species co-occur. Once bud development on *C. nutans* occurs, however, the weevil moves onto this species where it contributes to control of this thistle.

DISCUSSION

Two strains of the rust fungus *Puccinia cardui-pycnocephali* are capable of significantly reducing plant growth and seed production of the slender thistles in southern Australia (Burdon *et al.* 2000). The rosette weevil, *C. trimaculatus*, was shown to have a high impact on rosette growth in field trials in its native range, but was not released as part of the slender thistle program. One other biotype of the weevil *R. conicus* can reduce growth and seed production of *C. pycnocephalus* in southern California (Goeden and Ricker 1985) but was not pursued for Australia. The magnitude of reduction in seed production by the host is critical in both cases, as the slender thistles regenerate annually and only from seed. In each region, monitoring of host populations has been a single instance or short-term, thus any conclusion about longer-term effects of either agent must be qualified. Nevertheless, we conclude that biological control of slender thistle populations has the potential to be successful, and is likely to be a necessary component of an integrated management system for these pasture weeds.

Bendall (1973) showed that slender thistle growth could be limited in pastures by grazing management. If rosettes can be shaded by vigorous growth of competitive pasture species (e.g. perennial grasses and/or annual clovers) during the vegetative stages of the thistle life-cycle and if rust infection and/or the destruction of seeds can further reduce seed production, then a longer-term reduction in the soil seed store of these two thistles becomes a real possibility in theory. In practice, quantification of the effects of this combination of treatments awaits field documentation and is unlikely to be achieved. Such quantification will be long-term and expensive in research terms, but less expensive than the cost of these and other thistles to Australian agriculture.

ACKNOWLEDGEMENTS

We acknowledge the following colleagues for their contribution to this biological control program: Jean-Paul Aeschlimann, Jeremy Burdon, Pierre Chaboudez, Jim Cullen, Jean-Louis Sagliocco, Peter Thrall and Janine Vitou.

REFERENCES

Batra SWT, Coulson JR, Dunn PH and Boldt PG (1981) *Insects and Fungi associated with Carduus Thistles, Compositae.* Technical Bulletin USDA 1616. USDA, Beltsville, USA.

Bendall GM (1973) Control of slender thistle *Carduus pycnocephalus* L. and *C. tenuiflorus* Curt. (Compositae) in pasture by grazing management. *Australian Journal of Agricultural Research* **24**, 831–837.

Burdon JJ, Thrall PH, Groves RH and Chaboudez P (2000) Biological control of *Carduus pycnocephalus* and *C. tenuiflorus* using the rust fungus *Puccinia cardui-pycnocephali. Plant Protection Quarterly* **15**, 14–17.

Chaboudez P, Burdon JJ and Groves RH (1993) 'Application for field release of the slender thistle rust fungus (*Puccinia cardui-pycnocephali*): classical biological control agent for the noxious weeds *C. pycnocephalus* and *C. tenuiflorus*'. CSIRO Division of Plant Industry Report to the Australian Quarantine Inspection Service and Environment Australia, Canberra.

Goeden RD (1974) Comparative study of the phytophagous insect faunas of Italian thistle *Carduus pycnocephalus*, in southern California and southern Europe relative to biological weed control. *Environmental Entomology* **3**, 464–474.

Goeden RD and Ricker DW (1985) Seasonal asynchrony of Italian thistle, *Carduus pycnocephalus*, and the weevil, *Rhinocyllus conicus* (Coleoptera: Curculionidae), introduced for biological control in southern California. *Environmental Entomology* **118**, 433–436.

Groves RH and Kaye PE (1989) Germination and phenology of seven introduced thistle species in southern Australia. *Australian Journal of Botany* **37**, 351–359.

Groves RH, Olivieri I and Parsons WT (1984) Survivorship of *Carduus* spp. in France and Australia. In *Proceedings of the 4th International Conference on Mediterranean Ecosystems.* August 1984, Perth. (Ed B Dell) pp. 65–66. University of Western Australia, Nedlands, WA.

Olivieri I (1984) Effect of *Puccinia cardui-pycnocephali* on slender thistles (*Carduus pycnocephalus* and *C. tenuiflorus*). *Weed Science* **32**, 508–510.

Parsons WT (1977) The ecology and physiology of two species of *Carduus* as weeds of pastures in Victoria. PhD thesis. University of Melbourne, Australia.

Parsons WT and Cuthbertson EG (2001) *Noxious Weeds of Australia.* 2nd edn. CSIRO Publishing, Melbourne.

Sheppard AW, Aeschlimann J-P, Sagliocco J-L and Vitou J (1991) Natural enemies and population stability of the winter-annual *Carduus pycnocephalus* L. in Mediterranean Europe. *Acta Oecologica* **12**, 707–726.

Zwölfer H (1965) 'A list of phytophagous insects attacking wild Cynareae in Europe'.Weed Projects in Canada, Progress Report 12. CIBC, Dèlemont, Switzerland.

Carthamus lanatus L. – saffron thistle

Louise Morin and Andy W Sheppard

ABSTRACT

Saffron thistle, *Carthamus lanatus* subsp. *lanatus* and *baeticus* (Asteraceae), is the most abundant thistle in southern Australia, where it primarily invades cereal crops and grazing areas. A classical biological control program for saffron thistle began in the early 1990s, with a series of surveys in the Mediterranean regions of Europe and regular monitoring of study sites in France. Two fly species, *Urophora mauritanica* that attacks flower heads and *Botanophila turcica* that feeds on rosette crowns, were selected as the most promising insects for biological control and studied in more detail. The rust fungus *Puccinia sommieriana* and the leaf-spot fungus *Septoria centrophylli* were the only pathogens considered for preliminary host-specificity testing. The two fungi and the fly *U. mauritanica* were found to be capable of attacking safflower cultivars, a closely related crop species to saffron thistle, and were thus not pursued further. The rosette crown-feeding fly *B. turcica* was the only candidate agent that did not attack safflower. Field studies showed that this fly could kill saffron thistle rosettes but in most instances attacked plants were able to compensate and continue to grow, although considerably less seed was produced. Nonetheless, questions were raised about the ability of this candidate agent to reduce populations of saffron thistle. Consequently, comprehensive host-specificity testing was not performed and it was not released in Australia. Research has also been exploring the potential of the inundative approach (bioherbicide) for saffron thistle control using a *Phomosis* sp. fungus that already occurs in Australia, and is ongoing.

Key words: flowerhead fly *Urophora mauritanica*, rosette crown-feeding fly *Botanophila turcica*, rust fungus *Puccinia sommieriana*, leaf-spot fungus *Septoria centrophylli*, safflower.

INTRODUCTION

Saffron thistle *Carthamus lanatus* L. (Asteraceae) is native to the Mediterranean region of Europe and Western Asia and present in most temperate regions of the world, but is considered a serious weed only in Australia (Peirce 1995; Parsons and Cuthbertson 2001). It is found in all states and territories (Fig. 1) and is the most prevalent thistle in the southern regions where it invades both cropping, particularly cereals, and grazing areas (Briese 1988; Peirce 1995; Crump *et al.* 1996b) (Fig. 2a). Its impact on agriculture includes damage to livestock by eye and mouth injury, contamination of wool with vegetable matter, competition with pastures and reductions in grain yields of 50–70% (Peirce 1995). Briese (1988) listed saffron thistle as the most economically important thistle species in NSW. The cost of saffron thistle to the wool and meat industries in NSW, in terms of lost production, wool contamination and control, was estimated to be more than $A110 million per year in the early 2000s (Grace *et al.* 2004).

Saffron thistle is a spiny and lightly hairy winter annual herb that can grow to 40–90 cm tall (Peirce 1995)

Figure 1: Distribution of saffron thistle (*Carthamus lanatus*) in Australia. Reproduced from Peirce (1995) with permission from JR Peirce.

(Fig. 2b). It reproduces by large seeds, which are well adapted to persist in the soil, and does not require a cold period to trigger flowering (Grace *et al.* 2002b, 2004). Following seed germination, primarily after summer and autumn rains, plants remain as rosettes over winter (Fig. 2c) until a flowering stem elongates in mid spring and produces flowers in early summer. Pasture competition and grazing strongly influence saffron thistle's growth, fecundity and survival (Grace *et al.* 1999, 2002c). Cultivation and herbicide applications are effective control methods in crops (Parsons and Cuthbertson 2001). Improving fertility through superphosphate applications combined with grazing management is recommended to reduce infestations in pastures (Peirce 1996; Grace *et al.* 2002c). However, given the vast areas of grazing land infested, it is likely that reclamation of the bulk of saffron thistle infestations will remain beyond reach unless classical biological control is implemented. The most important challenge for biological control has been to find agents that cause severe damage to saffron thistle, but not on the closely related cultivated oil-seed crop safflower *Carthamus tinctorius* L. (Asteraceae) (Wapshere 1984).

BIOLOGICAL CONTROL HISTORY

Australia is the only country that has investigated options for the biological control of saffron thistle. From 1990 to 1995, CSIRO carried out a series of surveys of natural enemies of saffron thistle in southern Europe (Aeschlimann 1997). Regular observations were also made at a number of permanent study sites infested by saffron thistle within a 100 km radius of Montpellier, France. In addition, three field sites in France where safflower was grown were monitored to gather information on pests affecting this crop. In April 1995, the International Institute of Biological Control (now known as CABI) was commissioned by CSIRO to undertake a short survey in Greece specifically targeting fungal pathogens occurring on saffron thistle (Aeschlimann 1997; H Evans, unpub. data). Preliminary host-specificity tests were performed with two insects (the flowerhead fly *Urophora mauritan-*

Figure 2: a) Pasture infested by saffron thistle, *Carthamus lanatus*, at Leadville, NSW. b) Saffron thistle inflorescence. c) Saffron thistle rosette. Photos: a, b) J Hosking, NSW Government. c) M Jourdan, CSIRO.

ica Macquart and the rosette crown-feeding fly *Botanophila turcica* Hennig) and two fungal pathogens (the rust fungus *Puccinia sommieriana* Sacc. and leaf-spot fungus *Septoria centrophylli* Bubák & Ranójevic) identified as having potential for biological control (Aeschlimann 1997; Morin and Jourdan 2001; Vitou *et al.* 2001). The biology and impact of the rosette crown-feeding *B. turcica* on saffron thistle was also studied in the field (Sheppard and Vitou 2000; Vitou *et al.* 2001). None of these potential agents were investigated further nor released in Australia, either because of their low impact on saffron thistle or their ability to attack safflower.

Since the mid 1990s, biological control of saffron thistle using the inundative approach (development of bioherbicide) has been explored at Charles Sturt University, Wagga Wagga, NSW. Isolates of the leaf-blight fungus *Phomopsis* sp. recovered from diseased saffron thistle in NSW were shown to be capable of killing plants four to 10 days after exposure to optimum conditions for disease development (Crump *et al.* 1996a, b). Some of the isolates were also capable of killing other Asteraceae weeds such as Noogoora burr *Xanthium occidentale* Bertol., Bathurst burr *Xanthium spinosum* L. and bitou bush *Chrysanthemoides monilifera* (L.) Norl. subsp. *rotundata* (DC.) Norl., thus enhancing the commercial prospects of this potential bioherbicide (Crump *et al.* 1999). The isolates proposed for development as a bioherbicide were reported to be unrelated to *Phomopsis* species that cause disease on agricultural crops (Ash *et al.* 2010). Since these isolates do not readily produce spores on artificial media, hyphal fragments of the fungus were used as infective propagules in subsequent studies. Research to explore the potential of micro-encapsulation to protect propagules against dehydration was initiated through a partnership with a small research and development private company (Generex, Tamworth) in the late 1990s (Crump *et al.* 1999; Chittick *et al.* 2003). The ability of the fungus to produce phytotoxins is currently being examined through collaboration with the University of Naples, Italy (G Ash, *pers. comm.* 2010).

PLANT TAXONOMY

There are two recognised subspecies within saffron thistle – *lanatus* and *baeticus* (Hanelt 1976). In Australia, only the *baeticus* subspecies is recorded, although there is a range of morphological and phenological variants found in the field (Peirce 1990; Grace *et al.* 2002a). A study using intersimple sequence repeats (ISSR) fingerprinting to describe genetic variation in saffron thistle in NSW indicated that there were two distinct groups, each correlating with the northern and southern regions of the state (Ash *et al.* 2003).

EXPLORATION

Surveys for natural enemies of saffron thistle were performed by CSIRO staff in March–June over five consecutive years in the early 1990s in Mediterranean regions of Spain, France, Italy, Greece and the former Yugoslavia (Aeschlimann 1997). The south of Greece, where both subspecies of saffron thistle and their hybrids co-occur, was subsequently targeted to search for fungal plant pathogens (Aeschlimann 1997; H Evans, unpub. data).

CANDIDATES

Natural enemies

In their fungal database, Farr and Rossman (2011) list four rust and four powdery mildew fungi from saffron thistle in the European native range. These species, however, are also reported to occur on other Asteraceae hosts or to infect a range of unrelated plant species. The leaf-spot fungus *S. centrophylli* is the only species in the database that is solely reported from saffron thistle. In Australia, a range of fungi have been identified from diseased saffron thistle plants lodged in the NSW Plant Pathology Herbarium (DAR): *Alternaria carthami* Choudhury from leaf-spots and blight, *Phoma exigua* Desm. from dieback, *Phytophthora drechsleri* Tucker and *Rhizoctonia* sp. from root rot, *Sordaria fimicola* Ces.&de Not. from damaged seeds and *Phomopsis* species from canker and dieback.

In WA and NSW, the native moth *Helicoverpa punctigera* Wallengren, which is a pest of several crops in Australia, has been reported to attack saffron thistle's flowering heads (Peirce 1995; Grace *et al.* 2002a). Sap-sucking insects such as aphids and thrips have also been observed in NSW (Grace *et al.* 2002a).

Of the 42 arthropods and four fungal pathogens found on saffron thistle during surveys and field monitoring in the native range (Aeschlimann 1997; Grace *et al.* 2002a; Sheppard and Vitou 2000; H Evans, unpub. data), only six insects and two pathogens, which were damaging and appeared sufficiently host-specific based on observations, were investigated. One species of capitulum weevil, *Larinus flavescens* Germar, was abandoned once it was found in the literature to also attack safflower and related species in the genera *Carduus* and *Centaurea* (Campobasso *et al.* 1999). Among the four fly species, *Acantiophilus* sp., *Chaetorellia* sp., *Terellia* sp. and *Urophora* sp., found attacking saffron thistle flower heads, *U. mauritanica* was selected for preliminary testing as it had not been recorded from commercially grown safflower in Europe (Aeschlimann 1997). The other insect selected for further study was the rosette crown-feeding fly *B. turcica* on the basis of the damage it caused on saffron thistle in the field. The rust fungus *P. sommieriana* and the leaf-spot fungus *S. centrophylli* were the only pathogens that caused severe disease symptoms in the field and were thus studied in more detail (H Evans, unpub. data).

Natural enemies tested but not released

***Urophora mauritanica* Macquart (Diptera: Tephritidae).** Eggs of this univoltine flowerhead fly are laid on very young inflorescences of plants and each of the developing larvae induces the development of one gall into which it overwinters as prepupa. Five or more galls developing simultaneously on an inflorescence can almost totally suppress seed production (Aeschlimann 1997). In choice tests comprising saffron thistle and safflower, the flowerhead fly normally developed on both species, although attack levels were significantly higher on saffron thistle (Aeschlimann 1997). Similar results were obtained in subsequent field experiments.

***Botanophila turcica* Hennig (Diptera: Anthomyiidae).** This univoltine rosette crown-feeding fly lays eggs individually close to the collar of young plantlets. First-instar larvae burrow into and feed on the root meristem where they developed through three successive larval instars (Aeschliman 1997; Vitou *et al.* 2001) (Fig. 3). A detailed study of the effect of the fly at three sites in southern France showed that although the fly reduced plant relative growth rate, attacked plants were able to compensate at two of the sites and rosettes continued to grow (Sheppard and Vitou 2000). At the site with the lowest relative growth rate, the fly killed 12% of plants and attacked plants produced 71% less seed than those not attacked. In no-choice larval feeding tests, Vitou *et al.* (2001) demonstrated that the fly could not complete normal development on safflower. The suggested explanation was that safflower does not form a true rosette and therefore there was insufficient meristem tissue for larvae to complete development. While these results suggested the fly was highly specific, it was not pursued further due to its perceived inability to reduce populations of saffron thistle on its own.

***Puccinia sommieriana* Sacc.** This microcyclic rust fungus species does not have an alternate host and only produces telia with non-deciduous teliospores, which upon germination produce infective basidiospores that

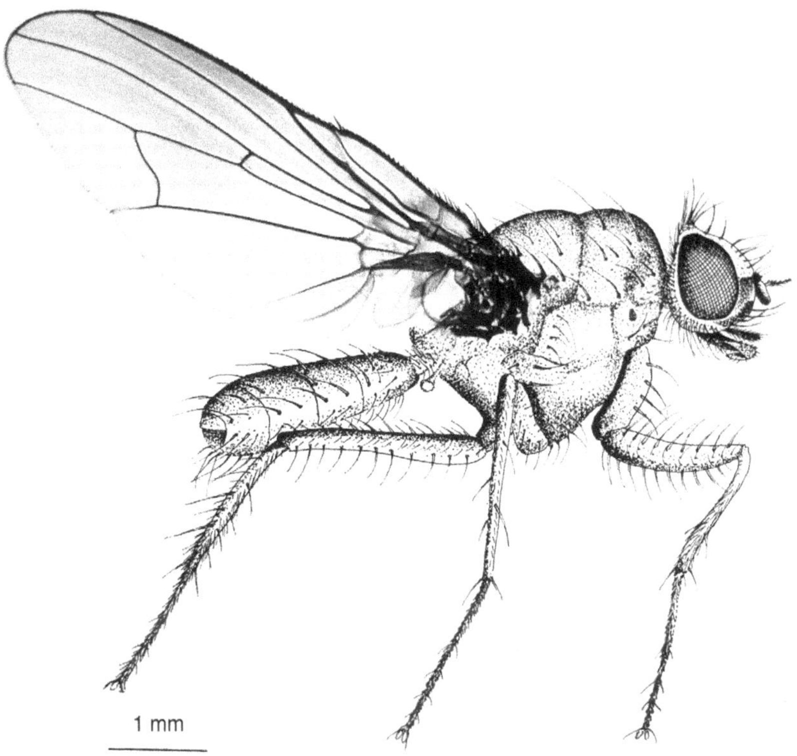

Figure 3: The rosette crown-feeding fly *Botanophila turcica.* Drawing: Michel Vitou.

are spread over short distances by wind. It was observed causing severe damage to young rosettes at one of the Greek sites surveyed in 1996 (H Evans, unpub. data). Prior to this survey, there were no records of this species on saffron thistle apart from the original description of a specimen from Malta (Saccardo 1913). In April 1998, fresh accessions of the rust were collected at three sites in the Peloponnese, Greece, where severe infection occurred (M Jourdan, unpub. data). These accessions were purified and three isolates cultured in a glasshouse to produce inoculum for testing. In pathogenicity tests, the three isolates successfully infected all Australian saffron thistle accessions inoculated. One of the isolates was then tested on four safflower cultivars and found to develop normal telia on leaves, bracts and stems of plants (Morin and Jourdan 2001).

Septoria centrophylli **Bubák & Ranójevic.** This leaf-spot fungus is a facultative parasite that can survive on dead infected tissue. When environmental conditions are conducive, spores are produced on necrotic leaf lesions or on infected residues on the soil surface and are disseminated by splashing or wind-borne rain droplets. Four isolates of the leaf-spot fungus, recovered from diseased material collected in Greece in 1998, were tested on two saffron thistle accessions from Australia and on four safflower cultivars (Morin and Jourdan 2001). All plants developed large necrotic lesions on leaves within 13 days of inoculation.

DISCUSSION

None of the four candidate classical biological control agents for saffron thistle investigated in the 1990s have been released in Australia. The ability of the flowerhead fly *U. mauritanica*, rust *P. sommieriana* and leaf-spot fungus *S. centrophylli* to attack safflower cultivars is the main reason why these candidates were not pursued further. A comprehensive analysis comparing the potential benefits to graziers from classical biological control of saffron thistle with the costs to the safflower industry of additional insect and disease damage, would be required to address the conflict of interest between these two parties. A quantitative comparison of the severity of the two diseases on saffron thistle and safflower, as has been done for the flowerhead fly (Aeschliman 1997), would be necessary to gather data to predict the damage that could occur on the crop. While there is not currently a major safflower industry in Australia, the high-quality oil produced from this crop could lead to a resurgence of this industry; this would need to be taken into account in such analysis. Current research into developing genetically modified safflower, to produce industrial fatty acids for high-value industrial biomaterial applications, may also lead to a considerable increase in the cropping of this species (CSIRO 2011).

In contrast, the rosette crown-feeding fly, *B. turcica*, was found in preliminary host-specificity testing to be highly specific towards saffron thistle. It would thus not present a conflict of interest with the safflower industry should it be proposed for release in Australia. Nonetheless, questions were raised about the ability of this candidate agent to reduce populations of saffron thistle in Australia. This, and a lack of further funding, meant that comprehensive host-specificity testing was not performed. Sheppard and Vitou (2000) identified that as a maximum of only three eggs per saffron thistle rosette was found, attack levels could be a key limitation to this insect having a severe impact on saffron thistle populations. A congeneric species of fly has been released against thistles of the genus *Onopordum* in Australia, but did not establish (Briese 2006).

It is difficult to predict what will eventuate from the research to develop a bioherbicide for saffron thistle based on a *Phomopsis* sp. The market potential of a biological control product that targets a single weed species of pastures and rangeland is limited, considering the low economic returns from such land. The demonstrated ability of some isolates of *Phomopsis* sp. to kill other Asteraceae weeds (Crump *et al.* 1999) may prove crucial to attracting a commercial partner to support the future development of a bioherbicide based on this fungus.

Saffron thistle remains the most common thistle in Australia, and the most costly to agriculture. Biological control of this winter annual, however, appears to have only a very slim chance whatever the investment, unless an acceptance of potential damage to safflower allows consideration of agents with broader specificity (including the capitulum weevil *L. flavescens* found commonly during surveys but never investigated as its host-range was known to include safflower). In the meantime, managing this weed will involve applying the outcomes of the significant amount of research into its integrated management (Grace *et al.* 1999, 2002a, b, c, 2004), as careful management of the perennial component of pasture is the most effective strategy against such annual weeds.

ACKNOWLEDGEMENTS

We acknowledge contributions by Mireille Jourdan, Thierry Thomann and Janine Vitou, based at the CSIRO European Laboratory in Montpellier. Thanks are also extended to Dr Gavin Ash, Charles Sturt University, for providing an update of the bioherbicide project targeting saffron thistle.

REFERENCES

Aeschlimann J-P (1997) Reappraising the potential of biological control against the weed *Carthamus lanatus*. *Entomophaga* **42**, 559–568.

Ash GJ, Raman R and Crump NS (2003) An investigation of genetic variation in *Carthamus lanatus* in New South Wales, Australia, using intersimple sequence repeats (ISSR) analysis. *Weed Research* **43**, 208–213.

Ash GJ, Stodart B, Sakuanrungsirikul S, Anschaw E and Crump N (2010) Genetic characterization of a novel *Phomopsis* sp., a putative biocontrol agent for *Carthamus lanatus*. *Mycologia* **102**, 54–61.

Briese DT (1988) Weed status of twelve thistle species in New South Wales. *Plant Protection Quarterly* **3**, 135–141.

Briese DT (2006) Can an *a priori* strategy be developed for biological control? The case of *Onopordum* spp. thistle in Australia. *Australian Journal of Entomology* **45**, 317–323.

Campobasso G, Colonnelli E, Knutson L, Terragitti G and Cristofaro M (1999) *Wild Plants and their Associated Insects in the Palearctic Region, Primarily Europe and the Middle East.* US Dept of Agriculture, Agricultural Research Service, ARS-147, Beltsville, USA.

Chittick AT, Ash GJ, Kennedy RA and Harper JDI (2003) Microencapsulation: an answer to the formulation quandary? In *Proceedings of the 6th International Bioherbicide Workshop – Bioherbicides: The Next Generation.* 27 April 2003, Canberra. International Bioherbicide Group. <http://ibg.ba.cnr.it/Workshops.htm> (accessed 3.8.11).

Crump NS, Ash GJ and Nikandrow A (1996a) *Phomopsis* spp. associated with die-back of saffron thistle (*Carthamus lanatus*). *Australasian Plant Pathology* **25**, 143.

Crump NS, Ash GJ and Nikandrow A (1996b) Potential of native pathogens for control of saffron thistle. *Plant Protection Quarterly* **11**, 254–255.

Crump NS, Ash GJ and Fagan RJ (1999) The development of an Australian bioherbicide. In *Proceedings of the 12th Australian Weeds Conference.* 12–16 September 1999, Hobart. (Eds AC Bishop, M Boersma and CD Barnes) pp. 235–237. Tasmanian Weed Society, Devonport.

CSIRO (2011) Crop biofactories. <http://www.csiro.au/files/files/pk18.pdf> (accessed 3.8.11).

Farr DF and Rossman AY (2011) Fungal databases, systematic mycology and microbiology laboratory. ARS, USDA. <http://nt.ars-grin.gov/fungaldatabases/> (accessed 3.8.11).

Grace BS, Sindel BM, Sheppard AW and Whalley RDB (1999) Ecology and management of saffron thistle (*Carthamus lanatus* L.) in New South Wales pastures. In *Proceedings of the 12th Australian Weeds Conference.* 12–16 September 1999, Hobart. (Eds AC Bishop, M Boersma and CD Barnes) pp. 13–15. Tasmanian Weed Society, Devonport.

Grace BS, Sheppard AW, Vitou J, Whalley RDB and Sindel BM (2002a) The ecology of saffron thistle in France and eastern Australia. In *Proceedings of the 13th Australian Weeds Conference.* 8–13 September 2002, Perth. (Eds H Spafford Jacob, J Dodd and JH Moore) pp. 529–532. Plant Protection Society of WA, Perth.

Grace BS, Sheppard AW, Whalley RDB and Sindel BM (2002b) Seedbanks and seedling emergence of saffron thistle (*Carthamus lanatus*) in eastern Australian pastures. *Australian Journal of Agricultural Research* **53**, 1327–1334.

Grace BS, Whalley RDB, Sheppard AW and Sindel BM (2002c) Managing saffron thistle in pastures with strategic grazing. *Rangeland Journal* **24**, 313–315.

Grace BS, Sheppard AW, Whalley RDB and Sindel BM (2004) Recent news about saffron thistle (*Carthamus lanatus* L.). *Plant Protection Quarterly* **19**, 36–39.

Hanelt P (1976) *Carthamus.* In *Flora Europaea*, vol. 4. (Eds TG Tutin, AD Chter, RA DeFilips and TBK Richardson) pp. 302–303. Cambridge University Press, Cambridge.

Morin L and Jourdan M (2001) Biological control of saffron thistle with fungi: are we facing a potential conflict of interest? In *Conference Handbook: 13th Biennial Conference of the Australasian Plant Pathology Society.* 24–27 September 2001, Cairns, Qld. p. 346.

Parsons WT and Cuthbertson EG (2001) *Noxious Weeds of Australia.* 2nd edn. CSIRO Publishing, Melbourne.

Peirce JR (1990) Morphological and phonological variation in three populations of saffron thistle (*Carthamus lanatus* L.) from Western Australia. *Australian Journal of Agricultural Research* **41**, 1191–1201.

Peirce JR (1995) *Carthamus lanatus* L. In *The Biology of Australian Weeds.* (Eds RCH Shepherd, RH Groves and RG Richardson) pp. 51–66. R.G. Richardson and F.J. Richardson, Melbourne.

Saccardo PA (1913) Notae Mycologiciae. VI. Fungi Melitenses. *Annili Mycologici* **11**, 546–573.

Sheppard AW and Vitou J (2000) The effect of a rosette-crown fly, *Botanophila turcica*, on growth, biomass allocation and reproduction of the thistle *Carthamus lanatus. Acta Oecologica* **21**, 337–347.

Vitou J, Briese DT, Sheppard AW and Thomann T (2001) Comparative biology of two rosette crown-feeding flies of the genus *Botanophila* (Dipt., Anthomyiidae) with potential for biological control of their thistle hosts. *Journal of Applied Entomology* **125**, 89–95.

Wapshere AJ (1984) The possibilities of biological control of thistles in Australia. In *Proceedings of the 4th Australian Applied Entomological Research Conference.* 24–28 September 1984, Adelaide. (Eds P Bailey and D Swincer) pp. 324–332. South Australia Government Press, Adelaide.

Cassinia spp. – cassinia

Royce H Holtkamp

ABSTRACT

Cassinia spp. are Australian native plants which have become weeds primarily in the tablelands and slopes regions of mainland Australia. Three species, *Cassinia arcuata*, *C. laevis* and *C. quinquefaria*, are causing problems, usually in response to soil or plant disturbance such as clearing or overgrazing. The potential of two native scale insects to control *Cassinia* spp. was first recognised in 1979 and a biological control program was formally commenced in 1988 with the deliberate transfer of *Austrotachardia* sp. for *C. arcuata*. This was followed in 1991 by the deliberate transfer of *Paratachardina* sp. for *C. laevis* and *C. quinquefaria*. This program differs from classical biological control programs reported elsewhere in this volume because native agents were used to control native plant species which had become weedy in their native ranges.

Key words: woody weeds, scale insects, native weeds, native agents for biological control.

INTRODUCTION

Cassinia spp. (Asteraceae) are native shrubs which are widespread in Qld, NSW, Vic (Campbell 1977, 1990) and SA (Fig. 1). They generally grow on infertile acid soils and invade pastures in response to soil or plant disturbance such as clearing or overgrazing and become weeds because they are unpalatable, competitive, unproductive and difficult and costly to control (Fig. 2) (Campbell *et al.* 1990). They also harbour noxious animals and make stock mustering difficult.

The three main problem species are *Cassinia arcuata* R.Br., *C. laevis* R.Br. (Fig. 3) and *C. quinquefaria* R.Br. *Cassinia arcuata* is present on 616 000 ha in central and southern NSW and is a declared noxious weed in 10 shires (Campbell 1990; Campbell *et al.* 1990). Preliminary estimates indicate *C. laevis* and *C. quinquefaria* are present on 250 000 ha in north-western NSW (Campbell *et al.* 1994). On arable land *Cassinia* spp. can be controlled by ploughing, sowing improved pastures and applying fertilisers to overcome nutrient deficiencies and soil acidity (McGowen *et al.* 1990). Conventional methods of control, e.g. herbicide spraying and aerial distribution of pasture seed and fertilisers, are impractical on non-arable land because herbicide treatments are uneconomic and pasture establishment is difficult on unploughed acid soils (Campbell 1990).

Native scale insects have killed small areas of *C. arcuata* in central NSW since 1979 (J Dellow, *pers. comm.* 1994) and patches of *C. laevis* and *C. quinquefaria* in north-western NSW since 1988 (L McCormick, *pers. comm.* 1994). However, between 1988 and 1993 relatively large areas of *Cassinia* spp. were killed by *Austrotachardia* sp. (Hemiptera: Kerriidae) in central NSW (Campbell and Wykes 1991, 1992) and by *Paratachardina* sp. (Hemiptera: Kerriidae) in north-western NSW (Holtkamp and Campbell 1995). Although similar scales were described by Froggatt (1899, 1903), no data on their biology was available.

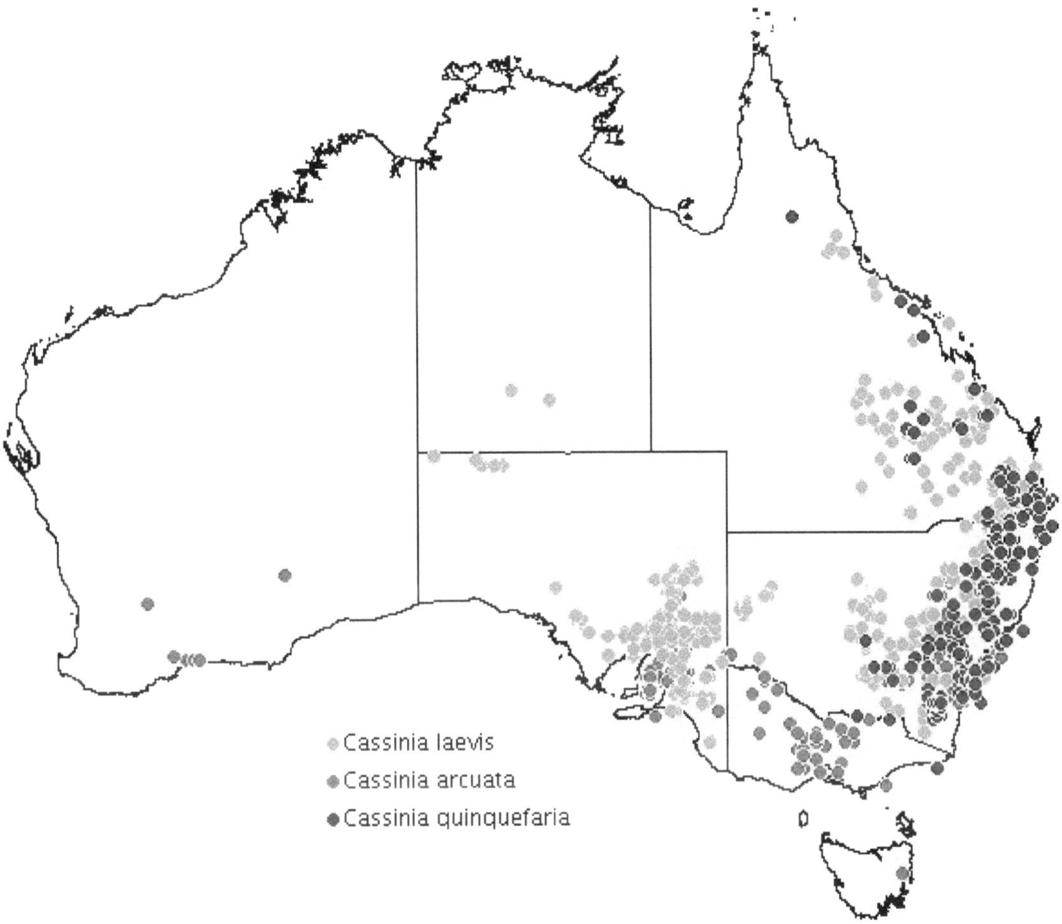

Figure 1: Distribution of *Cassinia arcuata, Cassinia laevis* and *Cassinia quinquefaria* in Australia. Specimen data reproduced from Australia's Virtual Herbarium with permission of the Council of Heads of Australasian Herbaria Inc, September 2010.

There are currently 45 species of *Cassinia* recognised in Australia (B Lepschi, *pers. comm.* 2010). However, this figure is uncertain due to ongoing revision of the genus. The majority of these species are not considered to be weedy.

BIOLOGICAL CONTROL HISTORY

Observations began on *Austrotachardia* sp. in 1988 and *Paratachardina* sp. in 1991 to record their biology and to ascertain whether they could be used for biological control of *Cassinia* spp.

Figure 2: Infestation of *Cassinia laevis* near Manilla, NSW. Photo: R Holtkamp, NSW Government.

Figure 3: *Cassinia laevis* in flower. Photo: J Hosking, NSW Government.

CANDIDATE SPECIES

Both insect species used as biological control agents for *Cassinia* spp. are Australian native scales and both are univoltine with first instar nymphs (crawlers), which are the only mobile stage, being produced in late spring to early summer (Holtkamp and Campbell 1995).

Insects

***Austrotachardia* sp. (Hemiptera: Kerriidae).** No formal surveys of the distributions of *Austrotachardia* sp. have been carried out but it has been observed between Gulgong and Uarbry, north of Coolah, throughout the Pilliga scrub, Bathurst, Hill End and in a 40 km arc north of Orange, all in central NSW.

Austrotachardia sp. was studied at 'Daydawn', Kerrs Creek near Orange, NSW and at the Orange Agricultural Research and Veterinary Centre, NSW from 1988 to 1992. By 1990 around 800 ha of *C. arcuata* had been killed by natural infestations of this scale (Holtkamp and Campbell 1995). In the field, *Austrotachardia* sp. crawlers were produced in December of each year. Cuttings from infected branches, taken in December, were transferred to new host plants on uninfected properties; within four years in excess of 250 ha of *C. arcuata* were killed and pasture had regenerated in these areas (Campbell *et al.* 1994).

***Paratachardina* sp. (Hemiptera: Kerriidae).** No formal surveys of the distributions of *Paratachardina* sp. (Fig. 4) have been conducted, but it has been observed in the Tamworth, Armidale, Inverell, Tenterfield, Glen Innes, Bingara, Barraba, Bundarra, Manilla and Merriwa districts of central NSW (Campbell *et al.* 1994). *Paratachardina* sp. was studied in the Tamworth and Manilla districts from 1991 to 1994. During this time substantial areas of *C. laevis* and *C. quinquefaria* were killed. Unfortunately, no accurate estimates of the area involved were obtained. It was noted that each year crawlers of *Paratachardina* sp. were produced in the field in November.

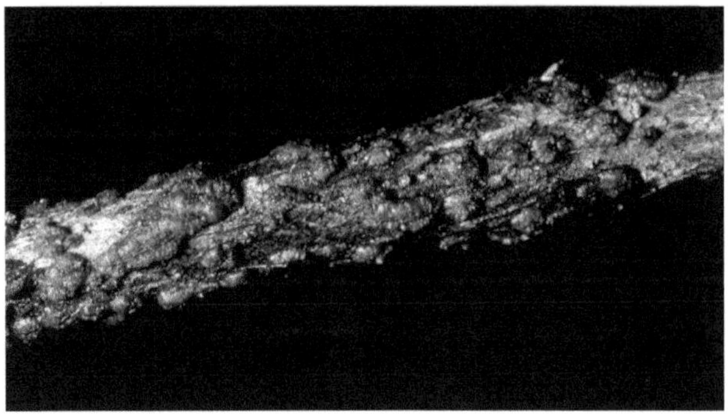

Figure 4: *Cassinia laevis* infested with *Paratachardina* sp. Photo: J Hosking, NSW Government.

Both *Austrotachardia* sp. and *Paratachardina* sp. are commonly covered by the black sooty mould, *Capnodium walteri* Sacc., (Holtkamp and Campbell 1995), which grows on the honeydew exudates produced by these scales. The mould appears to have no detrimental effect on either scales or plants.

DISCUSSION

This program differed from classical biological control because native agents were used to control plant species which had become weedy in their native range. There is only one other example of this type of biological control being implemented in Australia, where two native agents were used to control woody weeds in western NSW and south-western Qld (Sparks 2000). There are examples from other countries of native agents being used to control both native and exotic weed species (Julien and Griffiths 1998).

Human dispersal has also been used to augment the naturally slow spread of some classical biological control agents. Successful examples include *Dactylopius* spp. on *Opuntia* and *Cylindropuntia* spp. in Australia (Hosking *et al.* 1988), *Hypogeococcus festerianus* (Lizer y Trelles) on *Harrisia* spp. in Australia (Hosking *et al.* 1988) and *Dactylopius austrinus* De Lotto on *Opuntia aurantiaca* Lindley in South Africa (Moran and Zimmermann 1991).

Successful biological control of *Cassinia* spp. relies on the ability of these scale insects to colonise host plants and inflict damage. Because female scales were sessile, correct determination of the time of crawler production was critical to ensure establishment on uninfected plants as this was the only time when insects could be successfully relocated to new plant infestations. There was a relatively narrow window of opportunity – if cuttings were transferred too early the plants dried, resulting in death of the attached females. If transferred too late, the emerged crawlers died in the absence of suitable host plants.

The project determined that the critical times when crawlers were active – November and December for *Paratachardina* sp. and *Austrotachardia* sp., respectively. This enabled field days to be held where interested landholders could collect infected cuttings. Over 400 landholders were supplied with cuttings containing *Austrotachardia* sp. and over 200 landholders were supplied with cuttings containing *Paratachardina* sp. (Campbell *et al.* 1994).

Many landholders reported subsequent reductions in infestations of *Cassinia* spp. and resultant regeneration of pasture species (Campbell *et al.* 1994). The project resulted in an increased awareness of the possibility of biological control of *Cassinia* spp. among affected landholders and many continue to perform annual transfers of scale insects on their properties.

Successful biological control of *C. arcuata* was estimated to yield a saving of $2 million per year to NSW, in 1990 dollar terms (Holtkamp and Campbell 1995).

ACKNOWLEDGEMENTS

John Hosking provided comments on an earlier draft of this manuscript.

REFERENCES

Campbell MH (1977) Assessing the area and distribution of serrated tussock (*Nassella trichotoma*), St Johns wort (*Hypericum perforatum*) and sifton bush (*Cassinia arcuata*) in NSW. NSW Agriculture Technical Bulletin 18.

Campbell MH (1990) Distribution, ecology and control of *Cassinia arcuata* in New South Wales. *Australian Journal of Experimental Agriculture* **30**, 215–220.

Campbell MH and Wykes PJ (1991) Possible biocontrol of sifton bush by scale insects. In *Proceedings of the 6th Conference of the Grassland Society of New South Wales.* Orange, NSW. (Ed. DL Michalk) pp. 101–102. Grassland Society of New South Wales.

Campbell MH and Wykes PJ (1992) Possible biocontrol of native weeds by native insects. In *Proceedings of the 1st International Weed Control Congress.* 17–21 February, Monash University, Melbourne. (Eds JH Combellack *et al.*) pp. 106–108. Weed Science Society of Victoria, Melbourne.

Campbell MH, McGowen IJ, Milne BR and Vere DT (1990) The biology of Australian weeds. 22. *Cassinia arcuata* R.Br. *Plant Protection Quarterly* **5**, 162–168.

Campbell MH, Holtkamp RH, McCormick LH, Wykes PJ, Donaldson JF, Gullan PJ and Gillespie PS (1994) Biological control of the native shrubs *Cassinia* spp. using the native scale insects *Austrotachardia* sp. and *Paratachardina* sp. (Hemiptera: Kerriidae) in New South Wales. *Plant Protection Quarterly* **9**, 64–68.

Froggatt WW (1899) Scale insects that produce lac. *Agricultural Gazette of New South Wales* **10**, 1159–1161.

Froggatt WW (1903) *Australian Insects.* William Brooks and Co., Sydney.

Holtkamp RH and Campbell MH (1995) Biological control of *Cassinia* spp. (Asteraceae). In *Proceedings of the 8th International Symposium on Biological Control of Weeds.* 2–7 February 1992, Canterbury, NZ. (Eds ES Delfosse and RR Scott) pp. 447–449. CSIRO Publishing, Melbourne.

Hosking JR, McFadyen RE and Murray ND (1988) Distribution and biological control of cactus species in eastern Australia. *Plant Protection Quarterly* **3**, 115–123.

Julien MH and Griffiths MW (1998) *Biological Control of Weeds: A World Catalogue of Agents and their Target Weeds.* 4th edn. CABI, Wallingford, UK.

McGowen IJ, Campbell MH and Milne BR (1990) Sifton bush. NSW Agriculture Agfact P7.6.49.

Moran VC and Zimmermann HG (1991) Biological control of jointed cactus, *Opuntia aurantiaca* (Cactaceae), in South Africa. *Agriculture, Ecosystems and Environment* **37**, 5–27.

Sparks DR (2000) *Biological Control of Woody Weeds in Western New South Wales and within South-western Queensland.* Cuttagoa Catchment Landcare Group report.

Chondrilla juncea L. – skeleton weed

Jim Cullen

ABSTRACT

Chondrilla juncea, skeleton weed, is a herbaceous perennial of Eurasian origin, first recorded in Australia in 1917. It spread throughout the south-eastern Australian wheat belt from the 1920s to 1940s, where it dramatically reduced yields, increased production costs and proved difficult to control. By the 1950s it was considered Australia's worst weed. It has a deep perennial rootstock which produces rosettes from autumn to spring and a multi-branched, wiry, aerial stem during summer. This bears many capitula and the seeds, each bearing a pappus, are readily dispersed by wind. Once established in an area, density increases by regeneration from lateral roots and from segments of the rootstock when damaged, e.g. by cultivation. Densities in excess of 100 rosettes per m^2 were common. *C. juncea* is an apomict and exists as a number of genetically discrete forms, three of which occur in Australia. One of these was by far the most widespread and economically important. Detailed studies in Mediterranean Europe eastwards to Iran examined the impact of natural enemies as part of the weed's overall ecology and examined nine species in some detail. Two were discarded as insufficiently host-specific, three more because of unresolved issues over specificity and four were introduced, three of which established widely. Of these, the rust *Puccinia chondrillina*, the first plant pathogen deliberately introduced for weed control, was extremely successful, dramatically reducing the density of the weed over a few years. The eriophyid mite *Eriophyes chondrillae* and the gall midge *Cystiphora schmidti* also established widely and assisted in weakening the plant. The economic benefit of this program has been calculated as $1.426 billion in 2004–05 dollars with a benefit cost:ratio of 112:1. The strains of the rust and the mite initially introduced attacked only the common widespread form of the weed and the other two forms gradually spread over most of the weed's distribution. Later introductions of additional strains of *P. chondrillina* and *E. chondrillae* for the other two forms did not provide effective control, but the importance of those forms never reached that of the common form and skeleton weed is no longer considered a significant problem.

Key words: *Puccinia chondrillina*, *Eriophyes chondrillae*, *Aceria chondrillae*, *Cystiphora schmidti*, *Bradyrrhoa gilveolella*, *Oporopsamma wertheimsteini*.

INTRODUCTION

Skeleton weed *Chondrilla juncea* L. (Asteraceae) is a herbaceous perennial of Eurasian origin. It has an extremely long vertical rhizome or tap root (up to 4 m deep in light soil) with one or more annual top growths, commencing as rosettes in autumn, which produce vertical branched stems up to 1 m tall, with many capitula, in spring/summer (Fig. 1). These die back in late summer and new rosette growth regenerates from the rootstock. *Chondrilla juncea* is an obligate apomict, pollination not being required to produce viable seed, and occurs as a number

Figure 1: a) Rosettes. b) An infestation of *Chondrilla juncea* in summer, showing stems, 1971. Photos: E Cuthbertson, NSW, and J Green, CSIRO.

of genetically distinct forms, three of which occur in Australia, designated as forms A, B and C (Hull and Groves 1973) and commonly referred to as the narrow, intermediate and broad-leaf forms. Form A, the most widespread and common form, has been the main problem.

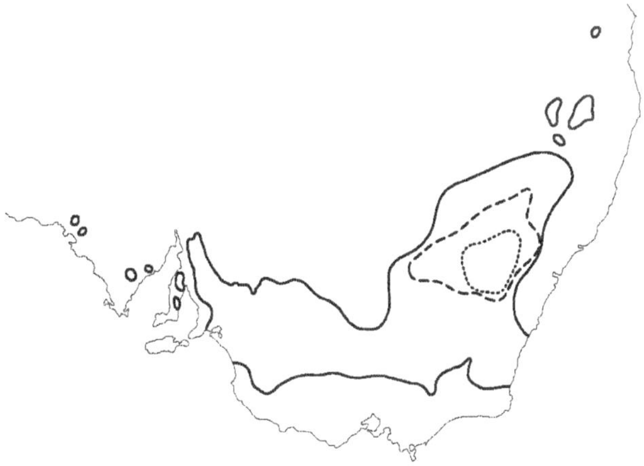

Figure 2: Distribution of skeleton weed *Chondrilla juncea* in Australia in the 1970s showing the more confined distributions of form B (dotted line) and form C (dashed line) at that time. Adapted from Hull and Groves (1973).

It was first recorded near Wagga Wagga, NSW in 1917, probably imported accidentally as a contaminant of grapevine rootstock, and spread rapidly through the NSW cereal belt in the 1920s and 1930s. It continued to spread into Vic and SA, with a small infestation occurring in southern Qld (Fig. 2). Outbreaks were later found in WA, where it continues to spread (Cullen and Groves 1978).

In cereal cropping in particular, cultivation cuts the rootstock into smaller pieces, each capable of establishing a new plant. Once established in a region, reproduction is mainly vegetative from lateral roots and cut pieces of rootstock, though spread to new areas is mainly by windblown seed. The plant is extremely competitive for nitrogen and water, particularly in dry seasons when yield can be reduced by 80% (Wells 1970). In addition, the wiry stems become entangled in harvesting machinery, making harvesting impossible in some situations. Wheat production was abandoned in several areas and by the 1950s *C. juncea* was considered Australia's worst weed.

BIOLOGICAL CONTROL HISTORY

Although skeleton weed is present in North and South America, the only interest in biological control was from Australia until late in the program. In Australia, biological control was first suggested in 1936 (Currie 1936) but not until 1966 did a program commence, with studies of its ecology in Australia and the establishment of the CSIRO Biological Control Unit at Montpellier in southern France. The work of this laboratory, led by AJ (Tony) Wapshere, examined the status of *C. juncea* in its native habitat. It concluded that, apart from some significant differences in agricultural practice, the presence of several natural enemies, particularly a rust fungus *Puccinia chondrillina* Bubak & Syd. (Fig. 3), was instrumental in maintaining populations of the weed 10–100 times lower than those observed in Australia (Wapshere 1970).

Despite their obvious effect on local *C. juncea* populations, accessions of the rust fungus and a second damaging agent, an eriophyid mite *Eriophyes (Aceria) chondrillae* Canestrini, collected at the beginning of the program, did not attack the common Australian form (form A), the main target of the program. It took another 18 months to find a strain of the fungus (in Italy) and a strain of the mite (in Greece) that attacked form A. Host-specificity testing was carried out with *P. chondrillina*, *E. chondrillae*, a gall midge *Cystiphora schmidti* Rubs. from Greece and local populations of the aphids *Uroleucon chondrillae* (Nevsk.) and *Chondrillobium blattnyi* (Pintera). The

Figure 3: Skeleton weed rust on rosettes of *Chondrilla juncae.* Photo: J Cullen, CSIRO.

proposal to introduce a rust fungus to Australia to operate in wheat crops initiated considerable discussion with plant quarantine authorities and advisers, resulting in an extensive host test list and an inspection of testing facilities in Montpellier by an Australian plant pathologist. Despite opposition from some pathologists, approval was given for importation and release in 1971 after three years of testing. This was the first time in the recent history of plant quarantine control that a plant pathogen has been approved for deliberate introduction to combat a weed. *Eriophyes chondrillae* and *C. schmidti* were also approved and imported, later in 1971.

All three species were released in 1971 and established readily. *P. chondrillina* spread rapidly and widely, causing a severe epidemic (Cullen *et al.* 1973) and having a dramatic impact on populations (Cullen 1978). Top growths were consistently heavily attacked (Fig. 4) and often killed; over nine to 10 years, as the reserves of *C. juncea* steadily declined and recruitment was virtually eliminated, populations declined dramatically (Fig. 5).

The strain of *P. chondrillina* introduced did not attack forms B and C of the weed. The population of *E. chondrillae* introduced persisted to a very limited extent on form B, but not at all on form C. *Cystiphora schmidti* attacked all three forms but was not sufficiently effective on its own. Forms B and C continued to spread through most of the former distribution of form A (Burdon *et al.* 1981); the program was continued for several years to try to find strains or other agents that would attack these forms. No subsequent introductions successfully attacked these forms, but they have not posed the same problem, possibly because they may be less competitive, and *C. juncea* is no longer a weed of serious concern. The introduction and success of *P. chondrillina* ignited interest in the use of pathogens for weed control in Australia and around the world and it remains a major success.

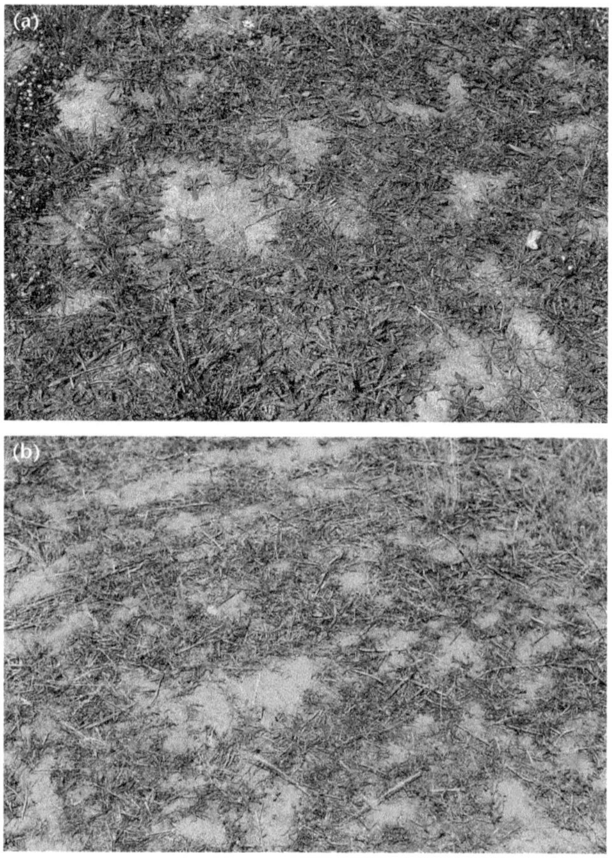

Figure 4: Impact of rust on rosettes on skeleton weed, *Chondrilla juncea*, spring 1972. a) Initial infection. b) Six weeks later. Photos: J Cullen, CSIRO.

The Australian success against *C. juncea* stimulated the initiation of a program in the USA and a small importation program in Argentina. *P. chondrillina* was imported to the USA from Europe and from the USA to Argentina. It established in both countries, but with variable impact on the different forms present in the USA. *E. chondrillae* was also imported to the USA from Europe and thence to Argentina, establishing widely in both countries. *C. schmidti* was imported to the USA and Argentina from Australia, but established only in the USA. *Bradyrrhoa gilveolella* (Tr.) was imported to Argentina from Greece, but did not establish (Julien and Griffiths 1998). Attempts are being made to establish *B. gilveolella* in the USA and a major investigation of the genetic differences in forms of *C. juncea* is ongoing (M Schwarzlaender, *pers. comm.* 2010).

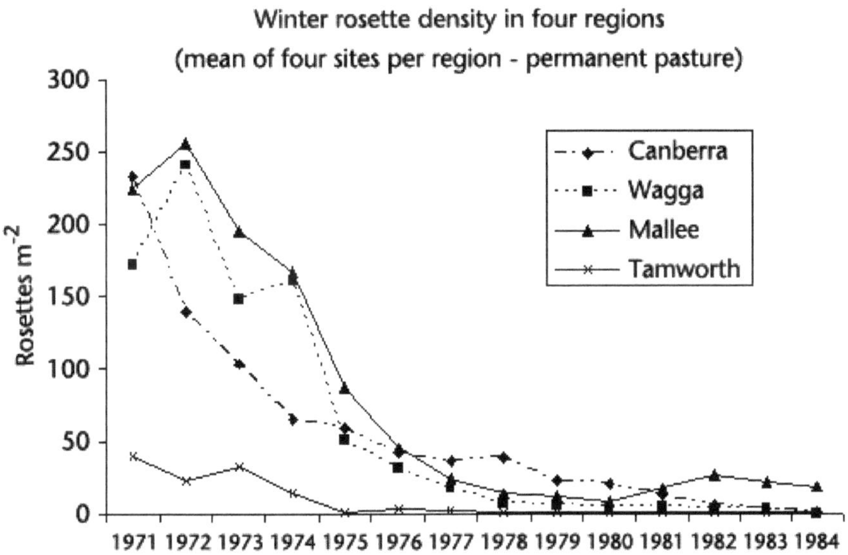

Figure 5: Density of rosettes of *Chondrilla juncea* in winter in permanent pasture in the four main regions of the distribution. Mean of four sites per region.

EXPLORATION

The CSIRO Biological Control Unit was established in Montpellier in 1966 specifically as a base for exploration in the native range of *C. juncea* in the Mediterranean region, in climatic regions similar to the regions infested in Australia. While initial local surveys established the importance of *P. chondrillina* and *E. chondrillae* (Wapshere 1970), further exploration to the east was necessary to find strains virulent against Australian *C. juncea* (form A) (Caresche 1973; Hasan 1972). It was clear from the literature that *C. juncea* was the species of *Chondrilla* that had spread furthest west from its probable centre of origin in south-western Russia and that exploration further east would be valuable.

Following location of a suitable strain of *P. chondrillina* in Italy, work in Greece located not only a suitable strain of *E. chondrillae*, but additional natural enemies whose spread further west had possibly been prevented by the cold mountains of the Balkans. These included the gall midge *C. schmidti* and the moth *B. gilveolella* (Wapshere 1973), whose larvae feed on the tap root of *C. juncea* below ground. Both species were imported to Australia during the program. Subsequent exploration further east located further possible agents and a temporary laboratory was established in Iran to study the moth *Oporopsamma wertheimsteini* (Rbl.), the subterranean scale insect *Neomargarodes chondrillae* Ark and the weevil *Sphenoptera foveola* Kerr. Of these species, *O. wertheimsteini* was later imported into quarantine in Australia for further testing. Exploration was led by AJ Wapshere, with L (Louis) Caresche carrying out a lot of the entomological work. Siraj Hasan carried out the plant pathology studies and was responsible for the work in Iran.

Later in the program, following the success of *P. chondrillina* on form A of *C. juncea* in Australia and therefore the increased emphasis on finding strains virulent against forms B and C, an intense electrophoretic study was made of the genetics of *C. juncea* to identify the forms sufficiently for close matching. It was also hoped to identify its possible centre(s) of variation. Six of the isozymes studied allowed sufficiently clear differentiation between different forms (Chaboudez 1989). Chaboudez discovered the existence of sexually reproducing diploid *C. juncea* in two regions of Turkey around Izmir, and a mechanism for the production of new triploid apomictic forms. These regions also showed considerable variation in forms of *C. juncea*. However, no exact matches were found for the Australian forms. Turkish forms were found with five of the isozymes matching those of form B, but only four out of six for form C. However, given the high probability of the existence of variation in strains of *P. chondrillina*, agreements were reached to establish trap gardens planted with Australian forms B and C in these regions to try to find suitable strains of *P. chondrillina* that would develop on them (Chaboudez *et al.* 1992). Three strains pathogenic against form B were subsequently imported to Australia, but none were found for form C. This work ceased in 1996 and although the possibility of searching further east, into south-west Russia, was considered, the lesser importance of forms B and C in Australia did not justify further work and no further exploration was carried out.

CANDIDATES

Natural enemies

Table 1 lists the natural enemies recorded as causing damage to *C. juncea* in its native range. The capitulum midge *Dasyneura* n. sp., the seed flies *Ensina sonchi* L. and *Tephritis nesii* Wied. and the leaf-miner *Melanoagromyza cunctata* Hend. were recorded in a very early limited survey (Currie 1936), but in the recent program, along with *Tephritis rasa* Seguy, were judged as causing insignificant damage and/or of being generalists and were not studied further (Wapshere 1970).

Agents tested and not released

***Erysiphe cichoracearum* DC and *Leveillula taurica* (Lev,) Arnaud (Erysiphaceae).** Both species of powdery mildew produce masses of powdery white mycelia on rosettes in autumn and spring, with conidiophores bearing conidia. The latter detach readily and disperse to establish new infections, only humid conditions being necessary for germination. The rosettes and flowering shoots can become completely covered in a mycelial coating. In late summer, cleistothecia are produced containing asci which bear ascospores. The ascospores ensure persistence over winter if temperatures are too low for conidia to survive.

The species occur together and their appearance on the plant is so similar that distinguishing which species is/are present can only be determined by microscopic examination of the conidia, cleistothecia, asci or

Table 1: Natural enemies causing damage to skeleton weed, *Chondrilla juncea*

Agent	Common name	Damage	Distribution	Comment
Dasyneura sp. N. (Diptera: Cecidomyidae)	Seed head midge	Reduces seed set	Europe*	Insufficient damage
Ensina sonchi L. (Diptera: Tephritidae)	Seed fly	Reduces seed set	Europe*	Insufficient damage
Tephritis nesii Wied. (Diptera: Tephritidae)	Seed fly	Reduces seed set	Europe*	Insufficient damage
Tephritis rasa Seguy (Diptera: Tephritidae)	Seed fly	Reduces seed set	Europe*	Insufficient damage
Melanagromyza cunctata Hend. (Diptera: Agromyzidae)	Leaf miner	Mines rosette leaves	Europe*	Insufficient damage
Puccinia chondrillina	Rust fungus	Attacks all aerial parts	Throughout distribution of *C. juncea*	Tested and released
Erysiphe cichoracearum	Powdery mildew	Attacks mainly rosettes	Throughout distribution	Tested but not released
Leveillula taurica	Powdery mildew	Attacks mainly stems	Throughout distribution	Testing not completed, not released
Eriophyes chondrillae	Gall mite	Galls stem growing points	Throughout distribution	Tested and released
Uroleucon chondrillae	Aphid		Throughout distribution	Testing not completed, not released
Chondrillobium blattnyi	Aphid		Throughout distribution	
Cystiphora schmidti	Gall midge	Galls rosettes and stems	Greece is western limit	Tested and released
Bradyrrhoa gilveolella	Moth	Larvae attack root	Greece is western limit	Tested and released
Oporopsamma wertheimsteini	Moth	Larvae attack rosette base and roots	Iran is western limit	Testing incomplete, not released
S. clarescens	Weevil		Iran is western limit	Preliminary testing only
N. chondrillae	Root scale insect	Feeds on root	Iran is western limit	Preliminary testing only

* Distribution not investigated further.

ascospores (Hasan 1973). Neither species was particularly damaging in southern France, but *E. cichoracearum* caused substantial rosette mortality in south-eastern Italy and *L. taurica* was extremely damaging in eastern Greece, particularly in late summer.

Both species occur as a number of specialised forms and there was considerable evidence that the forms found on *C. juncea* were extremely host-specific, at least to the genus *Chondrilla*. An extensive testing program, involving three varieties of each known host of *E. cichoracearum* in Australia, was undertaken for this species (Hasan 1974a). No infection of any other species was observed, but importation was refused on the grounds of insufficient knowledge of the genetics of host-specificity and its stability in powdery mildews. No further work was carried out on these species.

Uroleucon (Dactynotus) chondrillae **(Nevsk.) (Aphididae, Dactynotinae)**. This large brown aphid completes its life-cycle on all aerial parts of *C. juncea*, but requires extremely high populations to cause significant damage. Crowding leads to emigration and populations are generally low, particularly on rosettes in autumn and spring (Wapshere 1970). *Uroleucon chondrillae* is recorded only from *Chondrilla* spp., but in host-specificity tests some feeding and reproduction were observed on other species of Cichoriaceae (Caresche *et al.* 1974). While feeding on lettuce (*Lactuca sativa* L.) was limited, the difficulty of assessing the probability of the aphid acting as a vector of lettuce diseases, combined with apparent lack of potential for damage to *C. juncea*, meant that further work on this species was ceased.

Chondrillobium blattnyi **(Pintera) (Aphididae, Myzinae).** This aphid also completes its life-cycle on *C. juncea*, but in summer it is found only on the very small stem leaves and is common only on the rosettes. Under laboratory conditions it can kill rosettes at reasonably low populations, but this was never observed in the field. *Chondrillobium blattnyi* has been recorded only from *C. juncea*, but probing of all Cichoriaceae exposed to it was observed, including lettuce, where it stayed for six days (Caresche *et al.* 1974). As for *U. chondrillae*, the specificity difficulties this raised, together with its limited potential, prompted the cessation of further work on this species.

Oporopsamma wertheimsteini **Rebel (Lepidoptera, Tortricidae).** This is a univoltine moth whose larvae feed around the base of the rosette and top of the root, causing significant damage in spring. Summer is passed as pupae in the soil and adults are active in autumn. Eggs are laid on rosettes and these overwinter, hatching in spring.

Initial specificity testing in Iran showed a high level of specificity to *Chondrilla*, but partially developed larvae could feed and develop on the roots of lettuce, though eggs were not laid on this species and *O. wertheimsteini* was not recorded from it in the field (Hasan and Wapshere 1977). The moth was imported to quarantine in Australia for more detailed study of larval behaviour and experiments were carried out involving the establishment of first instar larvae on *C. juncea* plants at varying distances from lettuce plants. There was only bare soil surface separating the host and test species. The lettuce plants were later assessed for damage both when *C. juncea* plants remained alive and some time after *C. juncea* plants were killed, part way through larval development, forcing the larvae to starve or migrate to lettuce plants. Initial results (J Cullen, unpub.) demonstrated that lettuce was not at risk, especially given that lettuce is not normally grown in proximity to *C. juncea*. Unfortunately, the culture was lost due to rearing difficulties in quarantine and it was not possible to import further material from Iran. This species remains a promising agent if able to adjust to the milder climate characteristic of *C. juncea* in Australia. It could be worth considering in the current North American program, particularly for the more continental climatic regions.

Neomargarodes chondrillae **Ark (Hemiptera, Coccidae)** This root-feeding coccid was common in north-eastern Iran and is known from southern Russia. Adults emerge from cysts on the root in mid to late summer and lay their eggs in the soil close to the host plant in late summer. First instar larvae are active briefly in autumn before overwintering in the soil. They re-emerge in spring to feed on the young rosettes and seedlings before descending the root and attaching to it, where they complete development by late spring, becoming increasingly cyst-like. Several larvae at a time may be found on a root, which cracks and splits; the stress on the plant can reduce populations in the dry late summer.

This species was found on the commercially important lettuce and chicory *Cichorium intybus* L. in the field. It was concluded that it was not sufficiently host-specific and no further work was carried out (Hasan 1978).

Sphenoptera clarescens **Kerr (Coleoptera, Buprestidae).** The larvae of this species feed on the root of *C. juncea* over a long period and are most common in central northern Iran. Adults emerge from spring to autumn and lay eggs on the flowering stem of *C. juncea*; the larvae then bore into the stem and feed, before descending into the root. They exit the plant below the soil surface and feed externally on the root before re-entering and overwintering as mid instar larvae. They continue to feed internally, even though the root may die as a result of the feeding and generation time may be in excess of one year (Hasan 1978).

Damage is extensive, with many plants dying, but initial screening showed that the larvae feed on a number of wild and cultivated Asteraceae, including lettuce and chicory. They were also found on these species in the field and no further work was carried out due to the lack of specificity.

Agents released

Puccinia chondrillina **Bubak & Syd. (Uredinales).** *P. chondrillina* is a macrocyclic autoecious rust, recorded only from *Chondrilla* spp. and reproducing throughout most of its range by urediniospores. Teliospores are normally found only in winter in regions with a colder continental climate. They have been observed only once in Australia, in one of the highest populations of *C. juncea*, in the Snowy Mountains. Urediniospores infect *C. juncea* via the stomata, requiring surface water for a variable number of hours depending on the temperature (Hasan and Jenkins 1972). A field bioassay of overnight infection conditions in Canberra, following its introduction to Australia, showed that suitable conditions for infection would be common in most regions and most seasons except for summer in the drier areas and winter in the cooler regions (Cullen *et al.* 1973).

All above-ground parts of the plant are attacked, including the flower buds, leading to dieback and premature death of the rosettes and stems. Any top growths that regenerate as a result of initial infection are also attacked. Seeding is dramatically reduced in infected plants and root reserves steadily decline over time (Hasan and Wapshere 1973).

The strain imported (IT32) for form A was obtained from *C. juncea* morphologically similar to form A, discovered in Vieste, Italy. Later isozyme analysis of plants from this site showed that the source plants were not identical to form A, although IT32 was highly pathogenic against it (Chaboudez 1989).

A total of 57 species and varieties were tested against this strain over a period of three years, showing very narrow specificity to *C. juncea* and its forms (Hasan 1972). It was imported to Australia in 1971 (Hasan 1974b) and cultured through two generations in quarantine to ensure it was pure, in particular, free from its common fungal parasite *Sphaerellopsis filum* (Biv.: Fr.) B. Sutton. Inoculations in the field were carried out in May–June around Canberra and Wagga Wagga, both as a suspension of urediniospores in water sprayed onto plants in the field and covered with plastic overnight, and by transplanting infected plants into the field. Details of its colonisation and spread are given in Cullen *et al.* (1973). It was widespread throughout the range of *C. juncea* in Australia by March 1972, and spring–summer 197273 was marked by massive urediniospore production and dieback of the aerial parts of the plant. Farmers were reported as wondering why machinery used in infested land turned 'rusty' immediately. Though large-scale death of whole rootstocks with declining reserves took three to four seasons, the weed was immediately less competitive and farmers noted increased wheat yields. The effect of *P. chondrillina* on root growth of *C. juncea*, both alone and in competition with subclover *Trifolium subterraneum* L., a common management option for the weed, was demonstrated experimentally by Groves and Williams (1975).

Strains TU21 (1980), IT36 (1982) and TU788 (1996) were later introduced and colonised on form B (Julien and Griffiths 1998) after work in Europe demonstrated their pathogenicity against this form, but they were not as virulent as IT32 against form A. Initial observations showed no establishment of strain TU21, but there was establishment and spread of strain IT36. A recent survey by J Cullen and R Groves (unpub.) showed that *P. chondrillina* can now be found on form B throughout most of its range, apart from the hotter drier mallee. There is currently no reliable means of distinguishing between the strains, but it can be assumed that the rust found on form B is most likely to be strain IT36 as it was already established and widespread when strain TU788 was colonised. However, TU788 could also have established in the presence of strain IT36. The impact on form B has not been measured, but the severity of attack and impact on plants in the field has not approached that of strain IT32 on form A.

Eriophyes (Aceria) chondrillae **Canestrini (Eriophyidae).** *E. chondrillae* is a barely visible eriophyid mite that feeds on growing points on the stem, causing deformation and the production of complex leafy galls up to 4 cm diameter within which large mite populations develop (Fig. 6). Dispersal of mites from the galls is by wind or by crawling to new growing points. Stem production and therefore flower production can be completely aborted. As stems senesce in autumn, mites disperse to the new rosettes where they overwinter without reproducing, causing no obvious damage until the stems elongate the

Figure 6: Galls of *Eriophyes chondrillae* on skeleton weed stems. Photo: J Cullen, CSIRO.

following spring (Caresche and Wapshere 1974). Cullen *et al.* (1982) demonstrated that severe infestations not only reduce seeding, but weaken the roots and significantly reduce capacity for regeneration.

E. chondrillae occurs throughout the distribution of *C. juncea*, but in a number of different strains specific to different forms of its host. Strains from southern France and Italy did not attack any of the Australian forms, but a strain from Dionyssos in Greece attacked form A more strongly than the Greek form of *C. juncea* it was obtained from (Caresche 1973). *E. chondrillae* is known only from *Chondrilla* spp. and host-specificity tests with the Greek strain showed no gall development on any species other than *C. juncea*.

E. chondrillae was imported in 1971 and colonised throughout the Australian distribution of *C. juncea* from 1971 to 1974. Whole galls were placed on developing stems, the mites migrating to the stems as the galls dried. The species readily established and heavy infestations with severe stunting and premature death of stems were observed at several sites, but not uniformly. Higher populations were observed in drier sites with free-draining soils and limited vegetation cover (J Cullen, unpub.) while overwintering mite populations were influenced by the degree of overlap between living stems with galls and regenerating rosettes in autumn (Cullen 1974).

The strain of *E. chondrillae* imported and colonised was chosen for its severe effect on form A of *C. juncea*. A very limited number of galls could be produced on form B and none on form C (Caresche and Wapshere 1974). This was confirmed in Australia, where populations could be maintained quite successfully on form B in the laboratory. Galls were found on form B in the field at two sites where *E. chondrillae* had been colonised on form A two and five years previously. The extent to which there had been any selection for the ability to colonise form B since introduction was tested by Cullen and Moore (1983) and laboratory results showed some differences from the standard population maintained on form A, but it was concluded that the complexity of the relationship between growing conditions for the plant, its physiology and the complex interaction with *E. chondrillae* made definitive conclusions difficult.

Cystiphora schmidti **Rübsaamen (Diptera: Cecidomyidae).** The biology and specificity of *C. schmidti* were described in detail by Caresche and Wapshere (1975a). The adult is a delicate, short-lived midge that lays single eggs just below the epidermis of the rosette leaves, stems and stem leaves. The developing larva feeds on the underlying tissues and stimulates the production of gall tissue, forming a small gall approximately 3–4 mm long, which may be green or develop a deep purple colour (Fig. 7). Pupation occurs in the gall, from where the adult emerges a few days later. The total life-cycle takes 25–45 days depending on temperature and there may be four to six generations per year. Mature larvae, prepupae and pupae may overwinter in the galls remaining on dead dry tissue on the plant or in the litter. Large numbers of galls may be found on a single plant, stems often turning purple with galls. Rosettes can be severely weakened, with leaves dying back, and stem growth can be severely stunted, with significantly reduced seed production.

Cystiphora schmidti was imported in 1971, reared in quarantine to ensure the absence of parasites, and first colonised in the spring and early summer of 1971. It was then colonised throughout the range of *C. juncea* during 1972–74. The midge was distributed as pupae within galls on dry leaf tissue held in small containers that allowed adult dispersal, but provided protection from rain, sun

Figure 7: Galls of *Cystiphora schmidti* on stem of skeleton weed. Photo: J Green, CSIRO.

and ants. Starter colonies of 15, 50, 150 and 500 pupae were used to ascertain the influence of founder population size, but the major determinant of successful colonisation was the condition of host material available. Young fleshy growth, whether rosette leaves or stems, was readily colonised even by small numbers, but older drier stems in summer were less successful (J Cullen, unpub.).

Under favourable conditions, population increase can be rapid with large populations developing, particularly on stems in summer, leading to stem dieback. Dispersal of at least 16 km was recorded in one season and the midge's ability to readily recolonise areas where populations had died out due to prolonged lack of host plant material demonstrated its host-finding ability. Until 1978 no parasitism was observed, but an Australian wasp, *Tetrastichus* sp., was found in that year. It became very common in all populations of *C. schmidti*, reaching parasitism levels close to 100% in late summer. Despite this, damaging populations of *C. schmidti* still develop, particularly early in the season. With its ability to attack all forms of *C. juncea*, this species remains a useful component of *C. juncea* control.

Bradyrrhoa gilveolella **(Lepidoptera: Phycitidae).** The larvae of this moth feed on the root of *C. juncea*, in silken tubes of latex and soil particles. Older larvae cut deep grooves in or through the root, causing significant damage and destroying regenerating buds. Populations of this species were found in Greece, but no further west. It has two or three generations a year and in southern Greece is found commonly from mid summer through to autumn. It does not seem to persist through winter, with negligible numbers found in spring. In northern Greece, where it is cooler, it can be found year-round and it was speculated that southern populations might be re-established each spring by migration (Wapshere 1973). Its biology and specificity are described by Caresche and Wapshere (1975b).

B. gilveolella was first imported to Australia in 1973 as pupae to avoid the risk of diseased larvae, known to be attacked by a species of *Beauveria* fungus. It had not been reared for a full generation in the laboratory and initial work in quarantine encountered severe problems, with very low larval survival either on growing plants or on cut root pieces and very variable and often very prolonged development times, making synchronous emergence of sufficient adults extremely problematic. This was only overcome by developing an artificial diet and increasing the temperature for larval establishment and growth, but the founder population had been reduced to between three and nine adults. (Single pair matings had been deemed too risky and it was therefore difficult to determine how many adults contributed to the eggs recovered.) Subsequent lack of establishment in the field, despite the release of very large numbers of larvae and adults, plus experimental evidence, suggested the ability of larvae to establish on plants at field temperatures had been significantly reduced during the rearing program (Cullen 1985). Adult fecundity and egg fertility had also decreased, symptomatic of inbreeding.

A second population was imported in 1977 and a number of measures taken to limit the effect of laboratory rearing (Cullen 1985). Larvae of this colony were better able to establish on plants in the field and releases of first and mid instar larvae and adults were made at two sites around Canberra, in the coolest part of *C. juncea*'s range in Australia. Evidence of larval establishment was found at both sites and a small number of adults were recovered in virgin female baited traps two field generations later. This was promising and the colonies were supplemented with more releases, but no further recoveries were made.

Given the observations in Greece, warmer regions were not initially favoured for release, but there was the possibility that disappearance over winter might have been linked to fungal disease that was no longer present in the population. The prevalence of sandy soils, apparently favoured in Greece, in the mallee plus the occurrence of reasonable populations of healthy plants of forms B and C of *C. juncea* and warmer temperatures favouring rapid development, suggested it was worthwhile attempting establishment in that region. Large numbers of larvae were released at a site in the mallee region, but despite the thousands of larvae released on plants, no recoveries were made.

It was clear that establishment and persistence of the larvae on plants in the field was a critical factor. Apart from any behavioural or physiological interactions with the plant, predation by ants was known to be significant. Ants had been excluded from the release sites in Canberra in the first season. It is possible that continued work over a longer period might have established a population around Canberra, although if adult migration had been a factor the task would have been even harder. Work on this species ceased in 1985.

DISCUSSION

This program ran from late 1966 for more than 30 years, the first release (strain IT32 of *P. chondrillina*) being in 1971 and the last (strain TU788 of *P. chondrillina*) in

1996. The bulk of the program was finished by 1985, but attempts to find strains of *P. chondrillina* pathogenic against forms B and C of *C. juncea* prolonged the program at a lower level for another 12 years.

The program had a number of notable features:

- it was based on an extensive study of the weed's ecology in both its introduced and native ranges;
- it demonstrated the advantage of exploration as close as possible to the centre of origin of the weed;
- it was the first program to propose the deliberate importation and release of an exotic plant pathogen for weed control, which was spectacularly successful;
- it included detailed post-release field evaluation which allowed the demonstration of enormous economic benefits;
- it established the critical importance of characterising the identity of the weed and finding the appropriate strains of extremely specific agents.

The studies in Australia established benchmarks for population density and demography in relation to distribution and agricultural practice, while describing the variation in forms of the weed in Australia (Groves and Hull 1970; Hull and Groves 1973). The study of the ecology of *C. juncea* in the Mediterranean region (Wapshere *et al.* 1974, 1976) allowed the role of differing agricultural practices and soil types, compared with Australia, to be put in perspective. It also elucidated the role of natural enemies and allowed a comparison of their relative effectiveness. It was this work that discarded the species considered in the very early surveys and prompted the radical proposal to introduce a rust fungus, rather than an arthropod, as the top priority. This in itself was a historic decision and it was fortunate that the Director of Plant Quarantine at the time, a pathologist, was sufficiently objective, in the face of significant opposition, to allow the introduction, provided host-specificity could be demonstrated. It sparked the production of entirely new guidelines for the introduction of plant pathogens into Australia.

Though nothing like as dramatic, the introduction and release of an eriophyid mite *E. chondrillae* was also a first. It resulted from the studies carried out by Wapshere and his colleagues in the Mediterranean region.

As noted earlier, both *P. chondrillina* and *E. chondrillae* demonstrated the importance of finding the strain most appropriate for a particular form of the weed, work that led to the genetic analysis of the forms of *C. juncea*, their possible derivation and the use of trap gardens in centres of speciation discovered in Turkey. Accurate characterisation of the host weed is now more often the norm in biological control programs.

The increasing variety in forms of *C. juncea*, species of *Chondrilla* and species of natural enemies, with increasing proximity to the assumed centre of origin of *C. juncea*, is also well documented (Wapshere 1974) and therefore the increasing number of natural enemies to be considered.

The impact of *P. chondrillina*, *E. chondrillae* and *C. schmidti* on the ecology of *C. juncea* in Australia was followed quantitatively across its range for several years (Cullen and Groves 1978), which allowed calculation of quite accurate estimates of benefit to the grains industry, including an independent cost:benefit analysis by the then Industries Assistance Commission (Cullen 1978, 1985; Marsden *et al.* 1980). This demonstrated a benefit (NPV) of $290.2 million in 1975 dollars to the year 2000 and a benefit:cost ratio of 112 at a 10% discount rate. This analysis was updated in 2006, calculating a benefit of $1.426 billion in 2004–05 dollar terms (Page and Lacey 2006).

While this impact was obtained without recourse to considering integration with other control methods, as noted earlier, Groves and Williams (1975) demonstrated the additive influence of competition with subclover. The fact that this plant was the most common alternative management technique for *C. juncea* at the time means it almost certainly hastened the decline of the weed. However, Groves and Cullen (1980) noted that the interaction could be complex, with *P. chondrillina* infection declining under lush subclover growth, but increasing when growth was restricted, thus ensuring constant pressure on the plant.

Both forms B and C have spread throughout most of the former distribution of form A; in WA the sporadic outbreaks of forms A and C are sought out and eradication measures employed. While this approach is continued, there is no incentive to try to establish any biological control agents. Despite widespread occurrence of *P. chondrillina* (probably mainly strain IT36) on form B and occasional impact by *C. schmidti* on forms B and C of *C. juncea* in the eastern states, these forms cannot be considered as under control. However, their importance as weeds has been far less than that of form A and, apart from the ongoing eradication campaign in WA, *C. juncea* is only rarely considered an important weed in Australia.

ACKNOWLEDGEMENTS

The collaboration of several research organisations in Europe and Turkey, notably the CNRS (France), the

Regional Plant Protection Institute, Ankara and the University of Izmir, Turkey, plus the University of Montpellier, is gratefully acknowledged.

REFERENCES

Burdon JJ, Groves RH and Cullen JM (1981) The impact of biological control on the distribution and abundance of *Chondrilla juncea* in south-eastern Australia. *Journal of Applied Ecology* **18**, 957–966.

Caresche LA (1973) The host specialisation of three insects and a mite living on *Chondrilla juncea* L. In *Proceedings of 2nd International Symposium on Biological Control of Weeds.* 4–7 October 1971, Rome. (Ed PH Dunn) pp. 127–131. Commonwealth Agricultural Bureaux, Slough, UK.

Caresche LA and Wapshere AJ (1974) Biology and host specificity of the *Chondrilla* gall mite *Aceria chondrillae* (G. Can.) (Acarina, Eriophyidae). *Bulletin of Entomological Research* **64**, 183–192.

Caresche LA and Wapshere AJ (1975a) The *Chondrilla* gall midge *Cystiphora schmidti* Rübsaamen (Diptera: Cecidomyidae). II. Biology and host specificity. *Bulletin of Entomological Research* **65**, 55–64.

Caresche LA and Wapshere AJ (1975b) Biology and host specificity of the *Chondrilla* root moth *Bradyrrhoa gilveolella* (Treitschke) (Lepidoptera, Phycitidae). *Bulletin of Entomological Research* **65**, 171–185.

Caresche LA, Hasan S and Wapshere AJ (1974) The biology and host specificity of two aphids *Dactynotus chondrillae* (Nevsk.) and *Chondrillobium blattnyi* (Pintera) (Hemiptera) living on *Chondrilla juncea* L. *Bulletin of Entomological Research* **64**, 277–288.

Chaboudez P (1989) Modes de reproduction et variabilité génétique des populations de *Chondrilla juncea* L.: implications dans la lutte biologique contre cette mauvais herbe. PhD thesis. University of Montpellier, France.

Chaboudez P, Hasan S and Espiau C (1992) Exploiting the clonal variability of *Chondrilla juncea* to detect virulent strains of *Puccinia chondrillina* for use in Australia. In *Proceedings of 1st International Weed Control Congress.* 17–21 February 1992, Melbourne. (Ed RG Richardson) pp. 118–121. Weed Science Society Victoria, Melbourne.

Cullen JM (1974) Seasonal and regional variation in the success of organisms imported to combat skeleton weed *Chondrilla juncea* L. in Australia. In *Proceedings of 3rd International Symposium on Biological Control of Weeds.* 10–14 September 1973, Montpellier. (Ed AJ Wapshere) pp. 111–117. Commonwealth Agricultural Bureaux, Slough, UK.

Cullen JM (1978) Evaluating the success of the programme for the biological control of *Chondrilla juncea* L. In *Proceedings of 4th International Symposium on Biological Control of Weeds.* 30 August–2 September 1976, Gainesville, Florida. (Ed TE Freeman) pp. 117–121. University of Florida, Gainesville.

Cullen JM (1985) Bringing the cost benefit analysis of biological control of *Chondrilla juncea* up to date. In *Proceedings of 6th International Symposium on Biological Control of Weeds.* 19–25 August, Vancouver. (Ed ES Delfosse) pp. 145–152. Agriculture Canada, Ottawa.

Cullen JM and Groves RH (1978) The population biology of *Chondrilla juncea* L. in Australia. *Proceedings of the Ecological Society of Australia* **10**, 121–134.

Cullen JM and Moore AD (1983) The influence of three populations of *Aceria chondrillae* on three forms of *Chondrilla juncea. Journal of Applied Ecology* **20**, 235–243.

Cullen JM, Kable PF and Catt M (1973) Epidemic spread of a rust imported for biological control. *Nature* **244**, 462–464.

Cullen JM, Groves RH and Alex JF (1982) The influence of *Aceria chondrillae* on the growth and reproductive capacity of *Chondrilla juncea. Journal of Applied Ecology* **19**, 529–537.

Currie GA (1936) A report on a survey of weed problems in Australia. Pamphlet No. 60. Council for Scientific & Industrial Research Australia.

Groves RH and Cullen JM (1980) *Chondrilla juncea*: the ecological control of a weed. In *The Ecology of Pests: Some Australian Case Histories.* (Eds RL Kitching and RE Jones) pp. 6–17. CSIRO Publishing, Melbourne.

Groves RH and Hull VJ (1970) Variation in density and cover of *Chondrilla juncea* L. in south-eastern Australia. *Field Station Records of the Division of Plant Industry CSIRO (Australia)* **9**, 57–71.

Groves RH and Williams JD (1975) Growth of skeleton weed (*Chondrilla juncea* L.) as affected by growth of subterranean clover (*Trifolium subterraneum* L.) and infection by *Puccinia chondrillina* Bubak & Syd. *Australian Journal of Agricultural Research* **26**, 975–983.

Hasan S (1972) Specificity and host specialisation of *Puccinia chondrillina. Annals of Applied Biology* **72**, 257–263.

Hasan S (1973) The powdery mildews as potential biological control agents of skeleton weed (*Chondrilla juncea* L.). In *Proceedings of 2nd international Symposium on Biological Control of Weeds*. 4–7 October 1971, Rome. (Ed PH Dunn) pp. 114–119. Commonwealth Agricultural Bureaux, Slough, UK.

Hasan S (1974a) Host specialisation of powdery mildew from *Chondrilla juncea*. *Australian Journal of Agricultural Research* **25**, 459–465.

Hasan S (1974b) First introduction of a rust fungus in Australia for the biological control of skeleton weed. *Phytopathology* **64**, 253–254.

Hasan S (1978) Studies of new organisms in Iran for the biological control of skeleton weed (*Chondrilla juncea*). In *Proceedings of 4th International Symposium on Biological Control of Weeds*. 30 August–2 September 1976, Gainesville, Florida. (Ed TE Freeman) pp. 122–123. University of Florida, Gainesville.

Hasan S and Jenkins PT (1972) The effect of some climatic factors on infectivity of the skeleton weed rust, *Puccinia chondrillina*. *Plant Disease Reporter* **56**, 858–860.

Hasan S and Wapshere AJ (1973) The biology of *Puccinia chondrillina*, a potential biological control agent of skeleton weed. *Annals of Applied Biology* **74**, 325–332.

Hasan S and Wapshere AJ (1977) Biology and host specificity of the *Chondrilla* crown moth, *Oporopsamma wertheimsteini* (Rebel) (Lepidoptera: Tortricidae). *Bulletin of Entomological Research* **67**, 619–625.

Hull VJ and Groves RH (1973) Variation in *Chondrilla juncea* L. in south-eastern Australia. *Australian Journal of Botany* **21**, 113–135.

Julien MH and Griffiths MW (1998) *Biological Control of Weeds: A World Catalogue of Agents and their Target Weeds*. 4th edn. CABI, Wallingford, UK.

Marsden JS, Martin GE, Parham DJ, Ridsdill-Smith TJ and Johnston BJ (1980) *Returns on Agricultural Research*. CSIRO Publishing, Melbourne.

Page AR and Lacey KL (2006) *Chondrilla juncea* (skeleton weed). In *Economic Impact Assessment of Australian Weed Biological Control*. pp. 65–66. CRC for Australian Weed Management, Adelaide.

Wapshere AJ (1970) The assessment of biological control potential of the organisms attacking *Chondrilla juncea* L. In *Proceedings of 1st International Symposium on Biological Control of Weeds*. 6–8 March 1969, Delémont, Switzerland. (Ed FJ Simmonds) pp. 81–89. Commonwealth Agricultural Bureaux, Slough, UK.

Wapshere AJ (1973) Recent work on the assessment of the biological control potential of the *Chondrilla juncea* organisms. In *Proceedings of 2nd International Symposium on Biological Control of Weeds*. 4–7 October 1971, Rome. (Ed PH Dunn) pp. 95–100. Commonwealth Agricultural Bureaux, Slough, UK.

Wapshere AJ (1974) Host specificity of phytophagous organisms and the evolutionary centres of plant genera or sub-genera. *Entomophaga* **19**, 301–309.

Wapshere AJ, Hasan S, Wahba WK and Caresche L (1974) The ecology of *Chondrilla juncea* in the western Mediterranean. *Journal of Applied Ecology* **11**, 783–799.

Wapshere AJ, Caresche L and Hasan S (1976) The ecology of *Chondrilla* in the eastern Mediterranean. *Journal of Applied Ecology* **13**, 545–553.

Wells GJ (1970) Skeleton weed (*Chondrilla juncea*) in the Victorian mallee. 2. Effect of legumes on soil fertility and the subsequent wheat crop, and the resultant effect on the weed population. *Australian Journal of Experimental Agriculture and Animal Husbandry* **10**, 622.

Chromolaena odorata (L.) King and Robinson – chromolaena

Michael Day and Rachel Cruttwell McFadyen

ABSTRACT

Chromolaena odorata is a perennial shrub native to the tropical Americas. It is now a major weed in Africa, Asia and the western Pacific, affecting small crops, plantations and grazing lands. In Australia, chromolaena is present only in northern Qld, where it is the target of a national cost-share eradication program. Biocontrol worldwide began in 1966, with surveys for insects conducted by CABI in the Caribbean. Following host-specificity testing, the beetle *Apion brunneonigrum* and the moth *Pareuchaetes pseudoinsulata* were introduced into Nigeria, Malaysia, Ghana, India and Sri Lanka in the 1970s. *P. pseudoinsulata* established in Malaysia, India and Sri Lanka. In the 1980s, *P. pseudoinsulata* was introduced and established in Guam, the Federated States of Micronesia and the Northern Marianas. It then spread naturally into Palawan in the Philippines, and was reintroduced and established in West Africa. South Africa, which has a different chromolaena biotype from that occurring in Asia and West Africa, began a biocontrol project in 1988. Four agents were released, with two agents, *Pareuchaetes insulata* and *Calycomyza eupatoriae*, establishing. Surveys by CABI in 1988–89 targeting pathogens identified eight species but none was investigated further. In 1993, ACIAR funded a project in Indonesia and the Philippines, extended to include PNG in 1998 and East Timor in 2004, to control the weed in these countries and thereby reduce the risk of spread into Australia. *P. pseudoinsulata* and the gall fly *Cecidochares connexa* were introduced into Indonesia, the Philippines and PNG, and the butterflies *Actinote anteas* and *A. thalia pyrrha* into Indonesia. All agents established in their respective countries and are providing good control, with the butterflies also aiding control of *M. micrantha* in Sumatra. In 2005, *C. connexa* was introduced into East Timor, where it established and is providing some control. Effective agents used elsewhere will be introduced into Australia if the current eradication program fails and the emphasis shifts to containment of the weed.

Key words: biological control, *Eupatorium*, *Cecidochares connexa*, *Pareuchaetes pseudoinsulata*.

INTRODUCTION

Chromolaena (Eupatorium) odorata (L.) King and Robinson, (Asteraceae), chromolaena, is a woody shrub, originating from the tropical Americas (Gautier 1992) where it is found from Florida to Argentina and the Caribbean. It now has a pantropical distribution and is viewed as a weed in over 50 countries throughout Africa, Asia and the western Pacific (Zachariades *et al.* 2009). It was first reported in Australia in 1994 around Bingil Bay in north Qld, probably introduced via contaminated pasture seed (Waterhouse 1994a). New infestations were found in 2003 at Mossman and Townsville, probably as a result of seed being accidentally carried in from other areas, and the weed has since been found on the Atherton Tableland (DEEDI 2009) (Fig. 1). It is also present on Cocos and

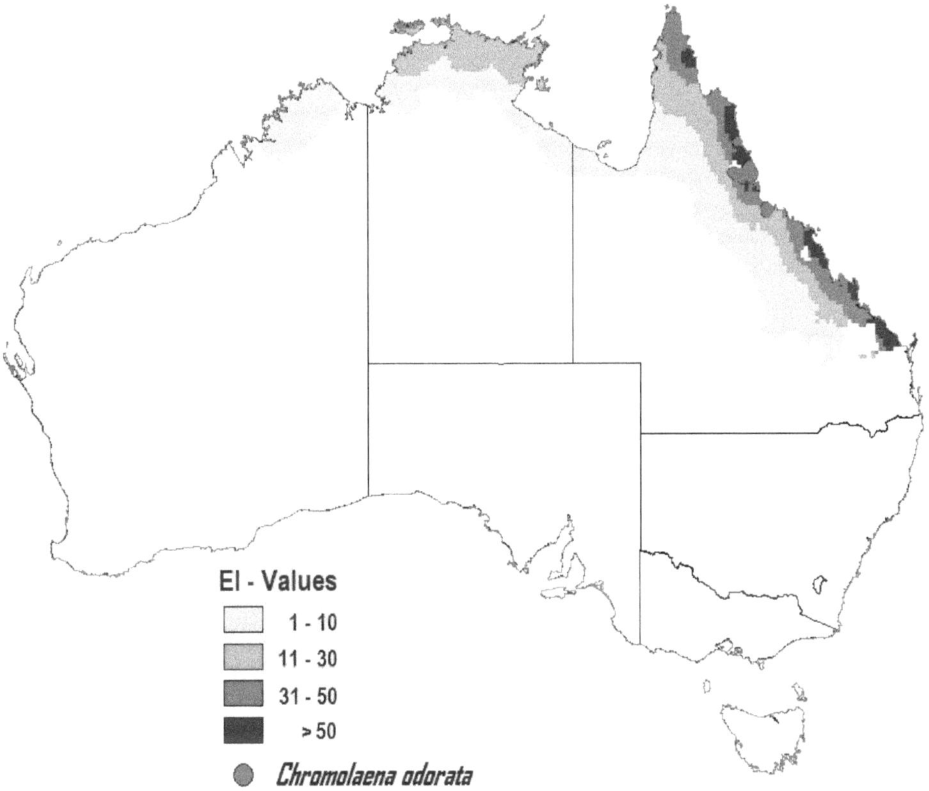

Figure 1: Distribution of *Chromolaena odorata* in Qld and its potential distribution based on a CLIMEX model. Source: Qld Government.

Figure 2: Flowering *Chromolaena odorata*. Photo: M Day, Qld Government.

Christmas Islands, which are Australian territories in the Indian Ocean.

Chromolaena is a fast-growing perennial shrub, with long branching stems which can clamber several metres high on supporting vegetation (Waterhouse 1994b) (Fig. 2). It flowers prolifically and seeds are dispersed by wind and on machinery, as well as by people, through the attachment of seeds to clothing (Day and Bofeng 2007; Zachariades *et al.* 2009). Chromolaena grows well in frost-free areas receiving over 1000 mm rainfall per year (Waterhouse 1994b) and can grow up to altitudes of 1500 m (Zachariades *et al.* 2009).

Chromolaena prefers open sunny areas where it can form dense monocultures, suppressing all other species and outcompeting preferred pasture species. It tolerates partial shade such as under plantation crops, where it can become the dominant species. It interferes with the harvesting of coconuts and oil palm and is a major weed in rubber and teak (Day and Bofeng 2007; Zachariades *et al.* 2009). In small crops, it can quickly establish in newly cleared land and smother crops such as taro and cassava (Waterhouse 1994b; Day and Bofeng 2007). During the dry season, it dies back to become a fire hazard. Fire does not usually kill mature plants, which reshoot from the crown (Waterhouse 1994b).

Chromolaena has been declared a Class 1 weed in Qld (i.e. all plants must be eradicated) and has been the target of a national cost-share eradication program since 1995, costing over $1.4 million per year (DEEDI 2009). If left uncontrolled in north Qld, chromolaena could invade agricultural and grazing areas, reducing productivity. It could also invade many natural ecosystems, becoming the dominant understorey species and reducing biodiversity. It has the ability to change fire regimes, which could greatly affect rainforest regeneration. A climate-matching model suggests it could spread west into coastal lands in the NT and south down the Qld coast to Sandy Cape (24.7°S 153.2°E) (Kriticos *et al.* 2005) (Fig. 1). Under the current eradication program, biocontrol is not an option for Australia, as all plants have to be destroyed when found. Should the eradication program fail, biocontrol could be initiated as part of a containment program, with agents effective elsewhere introduced into Australia (Day and Panetta 2010).

BIOLOGICAL CONTROL HISTORY

Biocontrol of chromolaena began in 1966, with surveys by CIBC (now CABI) in Trinidad, funded by the Nigerian Institute for Oil Palm Research (Cruttwell 1974). Over 240 phytophagous insects were found and, following host-specificity testing, four were recommended for release. These were the seed-feeding weevil *Apion brunneonigrum* Béguin-Billecocq (Coleoptera: Apionidae), the arctiid moth *Pareuchaetes pseudoinsulata* Rego Barros (Lepidoptera: Arctiidae), the tip-boring moth *Mescinia parvula* (Zeller) (Lepidoptera: Pyralidae) and the gall-forming eriophyid mite *Acalitus adoratus* Keifer (Acari: Eriophyidae) (McFadyen 1996).

Apion brunneonigrum and *P. pseudoinsulata* were released in Nigeria and Malaysia in 1970 and soon after in Ghana (Timbilla 1996), India and Sri Lanka (McFadyen 1996). *P. pseudoinsulata* established in Malaysia, India and Sri Lanka but not in Nigeria or Ghana (McFadyen 1996). It subsequently spread naturally from Sabah, Malaysia into Palawan in the Philippines, where it was first recorded in 1985 (Aterrado 1987). It was reintroduced into Ghana in 1991 from Guam and successfully established (Timbilla 1996). *A. brunneonigrum* failed to establish in every country in which it was released (Julien and Griffiths 1998). *Mescinia parvula* and a tip-boring fly *Melanagromyza eupatoriella* Spencer (Diptera: Agromyzidae) were released in India, Thailand and Guam, but there was no establishment (Chacko 1988; Napompeth *et al.* 1988).

In 1983, a biocontrol project was initiated in Guam and both *A. brunneonigrum* and *P. pseudoinsulata* were subsequently released. However, only *P. pseudoinsulata* established (Muniappan *et al.* 2004). The gall fly *C. connexa* (Macquart) was introduced into Guam from Indonesia in 1998 and established. Both *P. pseudoinsulata* and *C. connexa* have established in Micronesia and the Northern Mariana Islands (Muniappan *et al.* 2004).

Biocontrol of chromolaena in South Africa began in 1988, with surveys conducted in 15 countries throughout its native range, resulting in the field release of four agents. Of these, only the leaf-mining fly *Calycomyza eupatorivora* Spencer (Diptera: Agromyzidae) and the moth *Pareuchaetes insulata* (Walker) have established. Both *Pareuchaetes aurata aurata* (Butler) and *P. pseudoinsulata* failed to establish despite the release of large numbers. Numerous other species, including *C. connexa*, have been tested but either could not be reared on the South African biotype or were not sufficiently host-specific (Zachariades *et al.* 2009).

A survey of fungi attacking chromolaena was conducted by CABI in southern Brazil between 1988 and 1989. Eight species were found of which three, *Anhellia niger* (Viégas) Arx (Myriangiales: Myriangiaceae), *Mycovellosiella perfoliata* (Ellis and Everh.) Munt.-Cvetk. (Capnodiales: Mycosphaerellaceae) and *Septoria ekmaniana* Petrak and Cif., were considered potential biocontrol agents (Barreto and Evans 1994). However, none was investigated further.

A project funded by the European Community began in 1990 in Côte d'Ivoire and Indonesia but it ended in 1992 before any agent had been established. Further work in Africa funded by the African Development Bank was then blocked, as chromolaena was considered an important fallow crop by some scientists (McFadyen 1996).

An ACIAR-funded project managed by the Qld Government began in Indonesia and the Philippines in 1993 and was extended to include PNG in 1998 (Orapa *et al.* 2002), to control the weed in these countries and thereby reduce the risk of spread into Australia (Day and Panetta 2010). *P. pseudoinsulata* was released in Indonesia in 1993 and in PNG in 1998. *C. connexa* was released in Indonesia in 1995 and in the Philippines and PNG in 2001, while the butterflies *Actinote anteas* (Doubleday and Hewitson) (Lepidoptera: Nymphalidae) and *Actinote thalia pyrrha* Fabr. were released in Indonesia in 1999. All these agents established and contribute to the control of the weed

(Desmier de Chenon *et al.* 2002a; Day and Bofeng 2007; Zachariades *et al.* 2009; R Desmier de Chenon, *pers. comm.* 2010). The leaf-mining fly *C. eupatorivora* was also released in PNG but failed to establish (Day and Bofeng 2007).

Following the success of the project in Indonesia and PNG, a new ACIAR-funded project administered by Charles Darwin University, Darwin was initiated in East Timor in 2004. *C. connexa* was introduced into East Timor from PNG and established in several places (Zachariades *et al.* 2009). However, subsequent civil unrest in the country prevented further releases and monitoring, although the gall fly is now known to be widespread throughout East Timor (M Day, *pers. obs.* 2009).

The gall fly has been introduced into numerous other countries (Zachariades *et al.* 2009). Recently, it was released in Côte d'Ivoire, West Africa, where it has established (R Desmier de Chenon, *pers. comm.* 2010). It was reintroduced into Thailand in 2009, as it had failed to establish from field releases conducted in 2002 (Muniappan *et al.* 2005).

PLANT TAXONOMY

The genus *Chromolaena*, tribe Eupatorieae, contains 165 species, all of Central and South American origin (King and Robinson 1987). There are no crops or ornamentals in the tribe (Waterhouse 1994b). There are 20 other species in the Eupatorieae in Australia but only *Adenostemma lavenia* (L.) Kuntze and *A. macrophyllum* (Blume) DC. are native. The weeds *Ageratina adenophora* (Spreng.) RM King & H Rob. (crofton weed), *Ageratina riparia* (Regel) RM King & H Rob. (mistflower) and *Mikania micrantha* Kunth have been targets for biocontrol.

There are two invasive biotypes of *C. odorata*, based on their distribution and ecology – the Asian–West African biotype, which is the biotype found in Australia, and the South African biotype. The morphological differences are comparatively small. The South African biotype has glabrous stems and white flowers, while the Asian–West African biotype has hairy stems and pale blue to lilac flowers (Zachariades *et al.* 2009). However, there are significant differences with respect to climatic preferences (Kriticos *et al.* 2005), susceptibility to fire and attack by some biocontrol agents. The gall fly, *Cecidochares connexa*, which is controlling chromolaena in many parts of south-east Asia and the western Pacific, will not oviposit or develop on the South African biotype (Zachariades *et al.* 2009).

EXPLORATION

Early exploration was undertaken between 1966 and 1973 by the second author, who surveyed chromolaena (targeting the West African biotype) in Trinidad and the Americas while working for CIBC on the project funded by Nigeria (Cruttwell 1974). From 1993 to 1997, Cesar Garcia, working for the Qld Government, undertook surveys in northern Argentina, southern Brazil and Colombia, and made shipments of *C. connexa* and *Actinote* species to quarantine in Australia for transshipment to Indonesia.

From the 1990s, entomologists from South Africa made many surveys in South and Central America and the northern Caribbean; their target was the biotype found in South Africa which occupies a distinctly cooler (higher-altitude) climatic range (Kriticos *et al.* 2005).

CANDIDATES

No agent has been released in Australia for the biocontrol of chromolaena. However, several agents utilised elsewhere (see Zachariades *et al.* 2009) have the potential to control the weed and could be introduced if a biocontrol program is adopted.

Promising agents

***Cecidochares connexa* (Macquart) (Diptera: Tephritidae).** This stem-galling fly is one of the most damaging agents on chromolaena. Adults are 34 mm long and strikingly patterned in black and grey (Fig. 3). They live for five to 11 days and are diurnal. Mating occurs on the host plant and females oviposit white narrowly oval eggs in twos or small groups between the paired bud leaves in the terminal or axillary buds. Up to 70 eggs can be laid by each female. Eggs hatch in about five days and larval feeding within the tip induces gall formation by the plant. Each gall may contain two to four larvae but in high numbers galls can coalesce (Fig. 4). Pupation occurs within the gall after about 40 days and the pupal stage lasts about 20 days. Adults emerge via a clear window in the side of the gall (McFadyen *et al.* 2003). Adults fly well and can find even single isolated plants. Galling often causes stem dieback and sometimes kills plants (Day and Bofeng 2007; Zachariades *et al.* 2009).

C. connexa was collected from the Caribbean coastal region of Colombia and imported into Indonesia via quarantine in Brisbane in 1993, to avoid accidental importation of rubber rust into Indonesia. Following host-specificity testing at Marihat Research Station in

Figure 3: Adult stage of *Cecidochares connexa*. Photo: C Wilson, Kangaroo Island Natural Resources Management Board.

northern Sumatra, the gall fly was first released in Sumatra and Java in 1995 (McFadyen *et al.* 2003). It established rapidly and was subsequently released on all the main islands of Indonesia, in the Philippines and in PNG. It was released in West Timor in 1995 and in East Timor in 2005 and is now widespread throughout Timor (Zachariades *et al.* 2009). The gall fly has now been released and has established in eight countries in Asia and the western Pacific, where it is aiding control (Zachariades *et al.* 2009). It has recently been reported to have established in Côte d'Ivoire, West Africa (R Desmier de Chenon, *pers. comm.* 2010).

Figure 4: *Chromolaena odorata* heavily galled by *Cecidochares connexa*. Photo: M Day, Qld Government.

In PNG, where the gall fly is the only agent established in most parts of the country, it has controlled chromolaena in at least eight of the 12 provinces in which chromolaena occurs. The gall fly is very mobile, moving over 100 km in four years, which has aided redistribution efforts. However, there is still a need to release the gall fly in some of the more remote areas and islands (Day and Bofeng 2007).

In other countries, such as Indonesia and Guam, effective control of chromolaena has been achieved due to the combined efforts of *C. connexa* and one or both of *P. pseudoinsulata* and *A. anteas* (McFadyen *et al.* 2003; Zachariades *et al.* 2009; R Desmier de Chenon, *pers. comm.* 2010). If a biocontrol program against chromolaena is initiated in Australia, *C. connexa* should be considered first for introduction due to its host-specificity, host-finding ability and impact on chromolaena. Two unidentified parasitoid wasps were found in Indonesia, but parasitism levels were generally low (McFadyen *et al.* 2003).

***Pareuchaetes pseudoinsulata* Rego Barros (Lepidoptera: Arctiidae).** This creamy-coloured moth lays pale yellow spherical eggs in clusters on the underside of leaves. Eggs hatch after about four days and the larvae initially feed gregariously on the underside of the young leaves. In the third instar, larvae become solitary and nocturnal, moving down into the leaf litter during the day. There are five instars and pupation occurs in a loose silk cocoon spun in leaf litter on the soil surface. The total life span is 44–64 days. When present in large numbers, larvae can totally defoliate large areas of chromolaena (Timbilla 1996; Day and Bofeng 2007). Following its establishment in Malaysia, Sri Lanka and India in the 1970s, this moth was introduced into Guam (1985), Federated States of Micronesia (FSM) (1988), Ghana (1991), Indonesia (1993), the Philippines (1995) and PNG (1998) (Zachariades *et al.* 2009).

In Guam, Ghana, and FSM, *P. pseudoinsulata* established and quickly spread, causing substantial damage to chromolaena (McFadyen 1996; Timbilla 1996). It established in Indonesia, although only after very large releases, and now contributes to the overall control (R Desmier de Chenon, *pers. comm.* 2010). It is widespread in the Philippines, although generally at low levels. In PNG, it is confined to the Markham Valley, Morobe province and the lower slopes in Eastern Highlands province, where it causes seasonal damage (Day and Bofeng 2007). In most Asian and western Pacific countries, larvae are parasitised by *Exorista xanthaspis* Wiedemann (Diptera: Tachinidae). Parasitism can reach high levels and probably has a significant impact on large populations (McFadyen 1997).

***Acalitus adoratus* Keifer (Acari: Eriophyidae).** This mite was host-tested in 1971 and found to be highly host-specific (McFadyen 1995), but it was not deliberately released in any country. It was accidentally introduced into Sabah, Malaysia in the 1970s, probably in plant material with releases of the weevil *A. brunneonigrum* from Trinidad (McFadyen 1995). The mite subsequently spread throughout south-east Asia, being confirmed in Thailand (1984), the Philippines (1987), southern China (McFadyen 1995) and PNG (Zachariades *et al.* 2009). Feeding by the mite causes abnormal and excessive growth of trichomes on the leaves and stems, but its impact on chromolaena is minor (McFadyen 1995). Although highly host-specific, it would not be a high priority for introduction into Australia.

***Actinote anteas* (Doubleday and Hewitson) and *A. thalia pyrrha* Fabr. (Lepidoptera: Nymphalidae).** These butterflies were collected from Costa Rica, Colombia and north-east Brazil, and released in Indonesia in 1999 after host-testing (Desmier de Chenon *et al.* 2002b). Females lay small orange spherical eggs in batches of about 400 on the underside of leaves. Larvae first feed gregariously on young leaves, becoming more solitary and feeding on older leaves as they mature. There are six instars and pupation occurs in a chrysalis suspended on a branch or petiole away from the feeding sites. The life-cycle takes about 73–84 days during summer (Desmier de Chenon *et al.* 2002b). During host-specificity testing, 56 plant species representing 22 families, including 18 species from Asteraceae, were assessed. Both *Actinote* spp. were able to complete their life-cycle on both chromolaena and *M. micrantha* and were subsequently approved for release (Desmier de Chenon *et al.* 2002b). Both butterflies established in the field and were later found to cause significant damage to *M. micrantha*, especially in the lowlands of Sumatra (R Desmier de Chenon, *pers. comm.* 2006).

Releases of these butterflies have not been made elsewhere in the world, although they are currently under consideration for release in southern China (Zhang *et al.* 2006). Both were rejected for release in South Africa because of attack on native *Mikania* species (Zachariades *et al.* 2009). As Australia has no native *Mikania* species, *Actinote* spp. could be considered for introduction into Australia. However, there is field evidence that successful mating requires display behaviour by several adults (C Garcia, *pers. comm.* 1996; R Desmier de Chenon, *pers. comm.* 2010). Therefore, establishment would require relatively large populations which might not be achieved where very little chromolaena is present.

Other agents that have been utilised

Of the other known potential agents, some such as *A. brunneonigrum* failed to establish where released and are not being considered for further study. Others such as *C. eupatorivora* have only recently been released and their impact is still unknown. There are several other agents under study in South Africa that might be of value to Australia (Zachariades *et al.* 2009).

DISCUSSION

As chromolaena is the target of a national cost-share eradication program in Australia, biocontrol has not been seriously considered for Australia. However, overseas results suggest that several agents could be useful if the program shifts from eradication to containment. The gall fly *C. connexa* would be the first candidate as it is highly specific, easy to culture and establish and has a high searching efficiency, locating isolated plants away from main infestations. The gall fly has successfully controlled chromolaena in several countries (Zachariades *et al.* 2009) and it is hoped that similar results could occur in Australia if the agent were released.

For Australia, extra testing of *C. connexa* should be minimal, involving the two native *Adenostemma* species and possibly the weeds *A. adenophora*, *A. riparia* and *Ageratum conyzoides* L. subsp. *conyzoides* (Asteraceae). Successful establishment should be possible even in relatively small stands of chromolaena. However, parasitism might be a limiting factor. Parasitism by native hymenoptera of the closely related gall fly *Procecidochares utilis* Stone has prevented successful biocontrol of crofton weed in many Asian and western Pacific countries, including Australia (Muniappan *et al.* 2009). However, these parasitoids do not attack *C. connexa*; presumably they are not adapted to the different eco-climatic zone occupied by chromolaena (lower altitudes, therefore hotter and seasonally dry).

Of the other agents that have been released on chromolaena, the butterflies *Actinote* spp. may be the next highest priority, as they also attack *M. micrantha*, a Class

1 weed in Qld. The weevil *A. brunneonigrum* failed to establish in any country in which it was released. *P. pseudoinsulata*, although a damaging agent, has been difficult to establish in most countries where it has been released. The mite *A. adoratus* has only a minor impact on chromolaena and the leaf-miner *C. eupatorivora* is yet to be fully assessed.

If any of the above agents are released into Australia and control chromolaena, the benefit:cost ratio will be higher than for most successful biocontrol projects, as the exploration and most of the detailed host-specificity testing have been undertaken by other countries. Similar principles have been used to justify overseas biocontrol work on *M. micrantha* (see chapter on *M. micrantha*, this volume).

ACKNOWLEDGEMENTS

The authors wish to acknowledge work conducted by research staff at BIOTROP, Indonesia, Indonesian Oil Palm Research Institute, Marihat, Indonesia, National Agricultural Research Institute (PNG) and Dr Roch Desmier de Chenon, France.

REFERENCES

Aterrado ED (1987) *Pareuchaetes pseudoinsulata* caterpillar on *Chromolaena odorata*: a new Philippine record. *Philippine Journal of Coconut Studies* **11**, 59–60.

Barreto RW and Evans HC (1994) The mycobiota of the weed *Chromolaena odorata* in southern Brazil with particular reference to fungal pathogens for biological control. *Mycological Research* **98**, 1107–1116.

Chacko MJ (1988) Prospects for the biological control of *C. odorata*. In *Proceedings of the 1st International Workshop on Biological Control of* Chromolaena odorata. 29 February–4 March 1988, Bangkok. (Ed. R Muniappan) pp. 25–33. Agricultural Experiment Station, Magnilao, Guam.

Cruttwell RE (1974) Insects and mites attacking *Eupatorium odoratum* in the neotropics. 4. An annotated list of the insects and mites recorded from *Eupatorium odoratum* L., with a key to the types of damage found in Trinidad. *Technical Bulletin, Commonwealth Institute of Biological Control* **17**, 87–125.

Day MD and Bofeng I (2007) Biocontrol of *Chromolaena odorata* in Papua New Guinea. In *Proceedings of the 7th International Workshop on the Biological Control and Management of* Chromolaena odorata *and* Mikania micrantha. 12–15 September 2006, Taiwan. (Eds P-Y Lai, GVP Reddy and R Muniappan) pp. 53–67. National Pingtung University of Science and Technology, Taiwan.

Day MD and Panetta FD (2010) Preventing and managing incursions of Class 1 weeds in Queensland. In *Proceedings of the Global Biosecurity Conference*. 28 February–3 March 2010, Brisbane. pp. 69–70. CRC National Plant Biosecurity.

Department of Employment, Economic Development and Innovation (2009) *National Siam Weed,* Chromolaena odorata, *Eradication Program: Annual Report*. July 2008–June 2009. Brisbane.

Desmier de Chenon R, Sipayung A and Sudharto P (2002a) A decade of biological control against *Chromolaena odorata* at the Indonesian Oil Palm Research Institute in Marihat. In *Proceedings of the 5th International Workshop on Biological Control and Management of* Chromolaena odorata. 23–25 October 2000, Durban. (Eds C Zachariades, R Muniappan and LW Strathie) pp. 46–52. ARC-PPRI, Pretoria.

Desmier de Chenon R, Sipayung A and Sudharto P (2002b) A new biological agent, *Actinote anteas*, introduced into Indonesia from South America for the control of *Chromolaena odorata*. In *Proceedings of the 5th International Workshop on Biological Control and Management of* Chromolaena odorata. 23–25 October 2000, Durban. (Eds C Zachariades, R Muniappan and LW Strathie) pp. 170–176. ARC-PPRI, Pretoria.

Gautier L (1992) Taxonomy and distribution of a tropical weed, *Chromolaena odorata* (L.) R. King and H. Robinson. *Candollea* **47**, 645–662.

Julien MH and Griffiths MW (1998) *Biological Control of Weeds: A World Catalogue of Agents and their Target Weeds*. 4th edn. CABI, Wallingford, UK.

King RM and Robinson H (1987) The genera of the Eupatorieae (Asteraceae). *Monographs in Systematic Botany from the Missouri Botanical Garden* **22**, 15–81.

Kriticos DJ, Yonow T and McFadyen RE (2005) The potential distribution of *Chromolaena odorata* (Siam weed) in relation to climate. *Weed Research* **45**, 246–254.

McFadyen REC (1995) The accidental introduction of the Chromolaena mite, *Acalitus adoratus*, into South-East Asia. In *Proceedings of the 8th International Symposium on Biological Control of Weeds*. 2–7 February 1992, Canterbury, New Zealand. (Eds ES Delfosse and RR Scott) pp. 649–652. DSIR/CSIRO, Melbourne.

McFadyen RE (1996) Biocontrol of *Chromolaena odorata*: divided we fail. In *Proceedings of the 9th International Symposium on Biological Control of Weeds*. 19–26 January 1996, Stellenbosch, South Africa. (Eds VC Moran and JH Hoffmann) pp. 455–459. University of Cape Town, South Africa.

McFadyen RE (1997) Parasitoids of the arctiid moth *Pareuchaetes pseudoinsulata* (Lep.: Arctiidae), an introduced biocontrol agent against the weed *Chromolaena odorata* (Asteraceae) in Asia and Africa. *Entomophaga* **42**, 467–470.

McFadyen REC, de Chenon RD and Sipayung A (2003) Biology and host specificity of the chromolaena stem gall fly, *Cecidochares connexa* (Macquart) (Diptera: Tephritidae). *Australian Journal of Entomology* **42**, 294–297.

Muniappan R, Englberger K, Bamba J and Reddy GVP (2004) Biological control of chromolaena in Micronesia. *ACIAR Technical Reports Series*, 11–12.

Muniappan R, Reddy GVP and Lai P-Y (2005) Distribution and biological control of *Chromolaena odorata*. In *Invasive Plants: Ecological and Agricultural Aspects*. (Ed Inderjit) pp. 223–233. Birkhauser Verlag, Basel.

Muniappan R, Raman A and Reddy GVP (2009) *Ageratina adenophora* (Sprengel) King and Robinson (Asteraceae). In *Biological Control of Tropical Weeds using Arthropods*. (Eds R Muniappan, GVP Reddy and A Raman) pp. 63–73. Cambridge University Press, Cambridge.

Napompeth B, Thi Hai N and Winotai A (1988) Attempts on biological control of siam weed *Chromolaena odorata* in Thailand. In *Proceedings of the 1st International Workshop on Biological Control of* Chromolaena odorata. 29 February–4 March 1988, Bangkok. (Ed R Muniappan) pp. 57–62. Agricultural Experiment Station, Magnilao, Guam.

Orapa W, Donnelly GP and Bofeng I (2002) The distribution of Siam weed, *Chromolaena odorata*, in Papua New Guinea. In *Proceedings of the 5th International Workshop on Biological Control and Management of* Chromolaena odorata. 23–25 October 2000, Durban. (Eds C Zachariades, R Muniappan and LW Strathie) pp. 19–25. ARC-PPRI, Pretoria.

Timbilla JA (1996) Status of *Chromolaena odorata* biological control using *Pareuchaetes pseudoinsulata*, in Ghana. In *Proceedings of the 9th International Symposium on Biological Control of Weeds*. 19–26 January 1996, Stellenbosch, South Africa. (Eds VC Moran and JH Hoffmann) pp. 327–331. University of Cape Town, South Africa.

Waterhouse BM (1994a) Discovery of *Chromolaena odorata* in northern Queensland, Australia. *Chromolaena odorata Newsletter* **9**, 12.

Waterhouse DF (1994b) *Biological Control of Weeds: Southeast Asian Prospects*. ACIAR, Canberra.

Zachariades C, Day M, Muniappan R and Reddy GVP (2009) *Chromolaena odorata* (L.) King and Robinson (Asteraceae). In *Biological Control of Tropical Weeds using Arthropods*. (Eds R Muniappan, GVP Reddy and A Raman) pp. 130–162. Cambridge University Press, Cambridge.

Zhang L, Han S, Zhigang L, Liu N, Liying L, Luo L, Peng T and Liu W (2006) Effects of the infestation by *Actinote thalia pyrrha* (Fabricius) on the physiological indexes of *Mikania micrantha* leaves. *Acta Ecologica Sinica* **26**, 1330–1336.

Chrysanthemoides monilifera (L.) T. Norl. – bitou bush and boneseed

RJ Adair, Tom Morley and Louise Morin

ABSTRACT

Chrysanthemoides monilifera invades native vegetation of southern Australia and is a major threat to the integrity of natural ecosystems. Two subspecies have mostly allopatric distributions: boneseed (subsp. *monilifera*) occurs throughout south-eastern Australia and bitou bush (subsp. *rotundata*) is mainly restricted to coastal habitats in NSW. Twelve potential agents have been studied, and some released, as part of a biological control program for *C. monilifera*. Four agents are established on bitou bush (the bitou tip moth *Comostolopsis germana*, the leaf-roller moth *Tortrix* sp., the bitou seed fly *Mesoclanis polana* and the bitou tortoise beetle *Cassida* sp. 3) and affect certain plant growth parameters, but have not yet caused substantial decline of infestations. One other agent (the boneseed leaf buckle mite *Aceria* sp.) is possibly established on boneseed. Four agents have failed to establish (three leaf beetles *Chrysolina scotti*, *C.* sp. B and *C. picturata* and the lacy-winged seed fly *Mesoclanis magnipalpis*). The bitou leaf beetle *Ageniosa electoralis* was rejected. Testing of the systemic rust fungus *Endophyllum osteospermi* is in progress and testing of the boneseed seed fly *Mesoclanis dubia* is incomplete.

Key words: *Chrysolina*, *Mesoclanis*, *Comostolopsis*, *Tortrix*, *Aceria*, *Endophyllum*, *Cassida*, *Ageniosa*.

INTRODUCTION

Chrysanthemoides monilifera (L.) T. Norl. (Asteraceae: Calenduleae) is a perennial woody shrub indigenous to southern Africa. Two subspecies, *C. m.* subsp. *monilifera* (L.) T. Norl. (boneseed) and *C. m.* subsp. *rotundata* (DC.) T. Norl. (bitou bush), are naturalised in southern Australia (Fig. 1) where they invade natural ecosystems. Boneseed is widespread and occurs in a broad range of bioregions and vegetation types, but is most abundant on sandy or gravelly soils of coastal hills and plains. Infestations are extensive in the Mount Lofty Ranges, SA, the Mornington Peninsula and You Yang Range in Vic and around Hobart and Launceston in Tas. Bitou bush occupies over 80% of the NSW coastline and is the dominant plant over 400 km, particularly in the northern coastal regions (Thomas and Leys 2002). The most significant bitou bush infestations occur from Jervis Bay in the south to Tweed Heads in the north; these landmarks coincide with southern and northern containment lines of a national management strategy (Downey 2004; Cherry *et al.* 2008). Small isolated infestations are known beyond the containment lines on the southern NSW coast, Menindee (western NSW) and at Frankston and Mallacoota (Vic), and these are targets for eradication. Both subspecies are Weeds of National Significance (WoNS) in Australia. Infestations of boneseed occur in New Zealand, where it is listed under the NZ *Biosecurity Act 1993*, and in California, Sicily, St Helena and southern France (Weiss *et al.* 1998). The impact of *C. monilifera* on natural ecosystems in Australia is severe and has received considerable attention in recent times, particularly for bitou bush. Grazing is detrimental to *C. monilifera* (Scurr *et al.* 2008); this partially explains why it has negligible impacts on grazing industries.

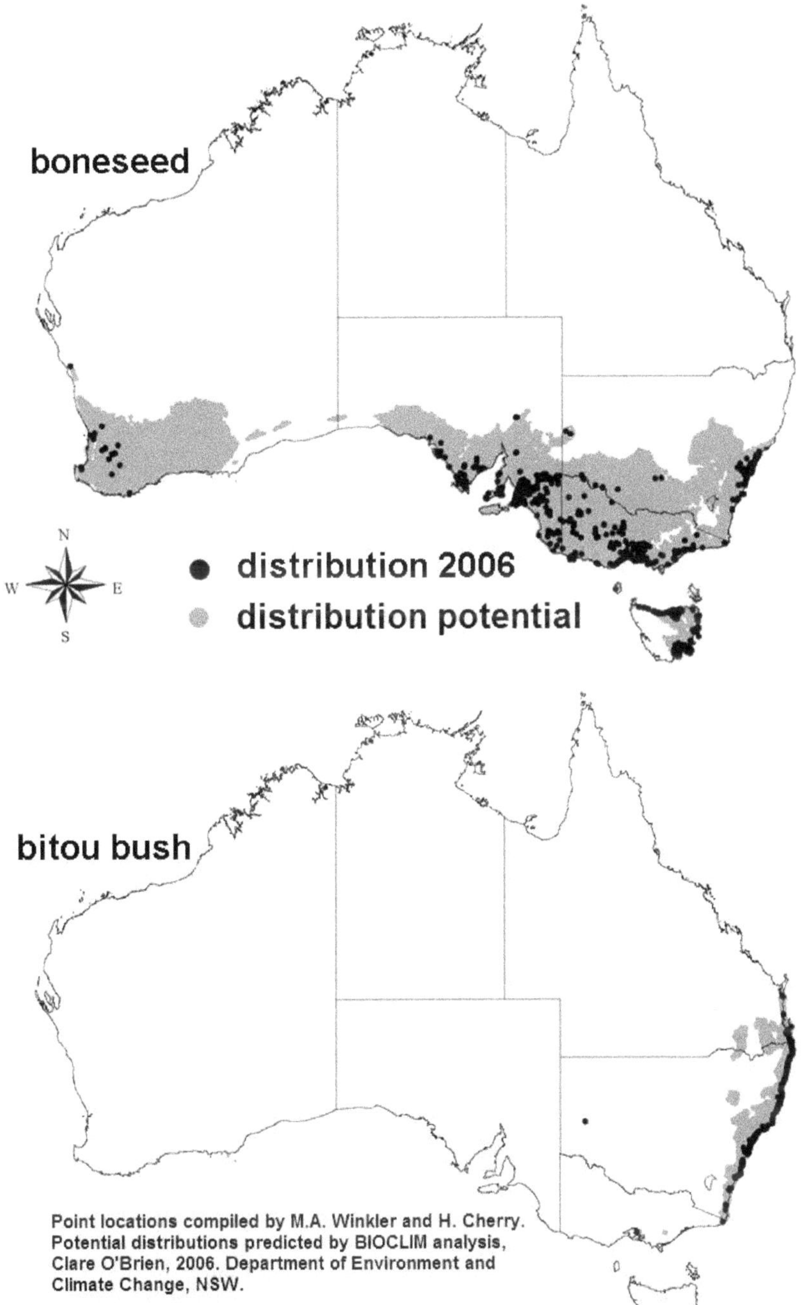

Figure 1: Australian distribution of *Chrysanthemoides monilifera* (boneseed and bitou bush). Information provided courtesy of the National Bitou Bush and Boneseed Management Group.

Impact (bitou bush)

Bitou bush invasion threatens at least 158 species, 26 ecological communities and three endangered plant populations in NSW (Downey *et al.* 2009). Bitou bush can dominate native vegetation and alter the characteristics and functioning of ecosystems in a variety of ways. These include cooler, moister and darker microhabitats, increased leaf litter decomposition rates and altered nutrient cycling (Lindsay and French 2004, 2005); suppression of indigenous plant germination, seedling growth and recruitment (Ens *et al.* 2009; French *et al.* 2008); reduced plant species richness in foredune shrublands and forested hind dunes (Mason and French 2008); increases in some moisture-loving invertebrate guilds, decreased beetle diversity and changes in abundance of ants, earwigs, centipedes and native cockroaches (French and Eardley 1997; Wilkie *et al.* 2007); negative effects on plant-foraging bird species, less abundant scavengers and less diverse raptors in bitou bush-invaded habitats (French and Zubovic 1997; Gosper 2004; Owers 1999).

Impact (boneseed)

In Australia, invasion of native vegetation by boneseed reduces floristic and soil seed bank diversity and emergence levels of native and exotic plant species (Thomas *et al.* 2005). Similarly, in NZ boneseed seedling abundance increases dramatically following canopy disturbance, while native seedling abundance is reduced (McAlpine *et al.* 2009). Grassy woodland, valley grassy forest and lowland forest vegetation communities in Vic are vulnerable to invasion by boneseed, where dense infestations eliminate most indigenous ground flora and prevent virtually all overstorey regeneration (Muyt 2001). Infestations in the You Yang Range directly threaten the endangered orchid *Pterostylis truncata* Fitzg. (brittle greenhood, Orchidaceae) (Bramwells 2003).

BIOLOGICAL CONTROL HISTORY

Groves (2008) provided details on the historical events associated with the biological control of *C. monilifera* in Australia; this review augments his account. In 1974, the Australian Weeds Committee declined a proposal for initiating research on biological control of *C. monilifera*. A decade later, a national conference on *Chrysanthemoides* (Love and Dyason 1985), hosted by two NSW Government agencies, decided that a biological control program should be developed. In 1986, the then Council of Nature Conservation Ministers (CONCOM) secured funding from state and federal government agencies to develop and implement the biological control program. A collaboration was formed between the CSIRO, responsible for overseas exploration, and the Keith Turnbull Research Institute (now Department of Primary Industries, Vic), which carried out host-testing and evaluation of potential agents in its quarantine facility in Melbourne. Other state agencies in NSW, Tas and SA were engaged to release approved agents and monitor their establishment and impact. In 1987, surveys and ecological studies commenced in South Africa to assess the phytophagous biota associated with *Chrysanthemoides* species. This phase ran for over a decade and was based mostly at Cape Town, South Africa. One hundred and thirteen species of arthropods and four species of pathogenic fungi were found to attack *C. monilifera* and the closely related *Chrysanthemoides incana* (Brum. f.) T. Norl. (Scott and Adair 1992). Forty-two of the arthropods and the four pathogens were identified as potential agents based on expected host-specificity, severity of damage to host, climate matching and level of predation; 12 of these have been fully or partially evaluated for release in Australia. Nine agents were approved for release from 1989 to 2008 – two moths, four beetles, two flies and one mite species (details below). One beetle was rejected, one rust fungus is being tested and tests on one fly are incomplete.

PLANT TAXONOMY

The genus *Chrysanthemoides* contains two species, *C. monilifera* and *C. incana*, and both are indigenous to southern Africa. The published taxonomy of *C. monilifera* recognises six subspecies, most of which have distinct habitat requirements and geographic distributions. Subspecies are distinguished by a combination of floral, vegetative and growth form characteristics (Norlindh 1943). Additional infraspecific entities are purported to occur in South Africa (Griffioen 1995), but critical examination of morphological and genetic characters is required to confirm this. Barker *et al.* (2009) offer some support for the recognition of infraspecific taxa based on DNA sequence data, but *Chrysanthemoides* remains unresolved as monophyletic. Hybridisation between taxa in South Africa is also suggested from molecular evidence (Barker *et al.* 2009), which partially explains why it is difficult to assign some plants to described taxonomic entities.

In Australia, analysis of morphological data indicates that bitou bush and boneseed are relatively homogeneous with no obvious patterns of phenotypic variation, although boneseed is recognised as having greater phenotypic variation than bitou bush (Simmons and Flint 1986). Putative hybridisation between bitou bush and boneseed is reported from several locations and is most recently documented at Daveys Bay (Vic), where a hybrid swarm occurs over approximately 20 ha (Adair and Butler 2010). Such hybridisation could complicate or jeopardise biological control if hybrids have a lower susceptibility to agents with particularly high levels of host-specificity.

CANDIDATES

Agents released

***Comostolopsis germana* Prout (Lepidoptera: Geometridae).** Larvae of *Comostolopsis germana* (bitou tip moth) feed in the apices of *C. monilifera* stems and branches.

Figure 2: The bitou tip moth *Comostolopsis germana*. Scale bar 10 mm. Photo: T Morley, Vic Government.

Individual larvae construct loose canopies with white silk and leaves and feed on young foliage, soft stems and buds. Pupation usually occurs within folded leaves on or near the host. The insect is multivoltine and several generations develop over a favourable season. Adults are green with white wavy lines on both fore and hind wings and rest appressed on foliage during the day (Fig. 2), taking flight for short distances when disturbed.

The bitou tip moth is specific to *C. monilifera* and *C. incana* (Adair and Scott 1989). In South Africa, it is most prevalent in warmer regions of the Eastern Cape and KwaZulu Natal and more abundant on bitou bush and *C. m.* subsp. *pisifera* (L.) T. Norl. than on other subspecies or *C. incana*. Between 1989 and 1992, the moth was released at 77 sites in NSW and establishment was confirmed at 44 sites (Downey *et al.* 2007). It has now colonised most of the range of bitou bush in NSW. Releases were also made on boneseed at 37 sites in Vic, SA and Tas, but the insect failed to establish. Predation pressure, poor climatic matching between the insect's origins in South Africa and release locations in south-east Australia, and host preferences have been implicated in lack of establishment (Adair and Edwards 1996; Weiss *et al.* 1998; Downey *et al.* 2007).

Using an insecticide exclusion technique (Adair and Holtkamp 1999), Holtkamp (2002) showed that densities of 100 bitou tip moth larvae m^{-2} can reduce seed production of bitou bush by more than 50% and sometimes by more than 80%. Larval densities up to 400 m^{-2} have been recorded and found to significantly reduce bitou bush flowering and fruiting (Holtkamp 1997).

***Chrysolina* spp. (Coleoptera: Chrysomelidae).** Of the eight Chrysomelinae known from *Chrysanthemoides* in South Africa (Adair and Scott 1991), three species of *Chrysolina* leaf beetles were selected for host-specificity testing (Fig. 3). The general biology of the three species is similar, although their distributions and host preferences in South Africa are distinctive. Elongate eggs are laid in clusters on the leaves of the host plant or in the soil around the plant base. The larvae and adults primarily feed on young foliage and stems in the upper canopy, but mature leaves are also eaten. Adults and larvae shelter on the underside of leaves, in leaf folds or in litter beneath the canopy. They drop from the host plant when disturbed, possibly as a predator avoidance strategy. All three species are specific to *C. monilifera* and are multivoltine with generation times ranging from 64 to 226 days at 22°C. Adults are mostly nocturnal. Larvae pupate in the soil within a cocoon constructed from soil particles and debris. There are four larval instars. In addition to the criteria mentioned above, their selection included consideration of the 'new association' hypothesis discussed by Hokkanen and Pimentel (1984).

***Chrysolina scotti* Daccordi (formerly referred to as *Chrysolina* sp. A).** The black boneseed leaf beetle is distinguished by shiny black elytra with a narrow white band along the lateroventral margin. The species is known from the Eastern Cape province of South Africa between Plettenberg Bay and Grahamstown, where it feeds on *C. monilifera* subsp. *pisifera*. Outbreak populations can cause severe defoliation and death of the host; the only parasitoid recorded in South Africa was a species of Tachinidae, which was rare (Adair and Scott 1991). The climate match index between the insect's natural distribution and areas of *C. monilifera* in south-eastern NSW is high. In host-specificity tests the black boneseed leaf beetle developed on both bitou bush and boneseed, with higher emergence rates on boneseed.

The black boneseed leaf beetle was first released at Studley Park, Melbourne in 1989 and subsequently released at 18 sites in Vic, one site in SA, five in NSW and 10 in Tas, but the species failed to establish at any site (R Adair, unpub. data). In southern Vic, arboreal-foraging ants and spiders are principal predators of eggs (Meggs 1995) and this undoubtedly adversely affected establishment of the beetle. In contrast, when chrysomelids under consideration for use as biological control agents were assessed for relative susceptibility to ant predation in South Africa using the canopy-foraging ant *Crematogaster peringueyi* Emery (Formicidae), tests indicated a range of behaviours or attributes would enable them to defend against predatory ants (J Scott, *pers. comm.* 1990).

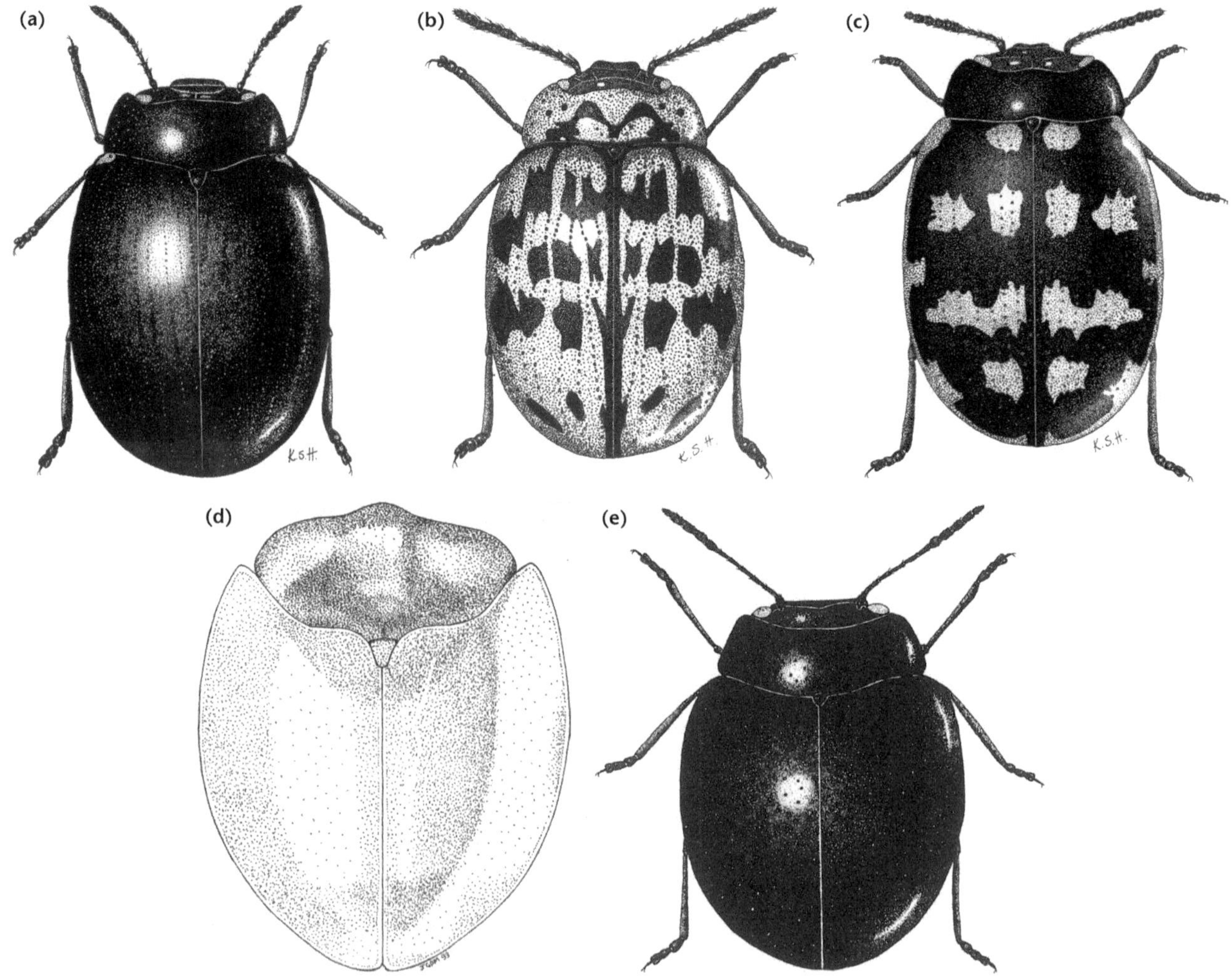

Figure 3: Chrysomelid beetles tested as biological control agents for *Chrysanthemoides monilifera*. a) *Chrysolina scotti*. b) *Chrysolina* sp. B. c) *Chrysolina picturata*. d) *Cassida* sp. 3. e) *Ageniosa electoralis*. Drawings: a–c, e) KS Harmer. d) JGD Morrison.

***Chrysolina* sp. B.** The elytra of this undescribed species (painted boneseed leaf beetle) are white with irregular and deeply dissected black blotches in four traverse bands. The thorax is mostly white with an M-shaped black dorsal band. The beetle is restricted to near-coastal areas on the west coast of the Western Cape province of South Africa in habitats with calcareous soils. The natural hosts of the painted boneseed leaf beetle are *C. m.* subsp. *pisifera* and *C. incana*. The species was approved for release in 1994 (Adair and Scott 1997) and distributed to four sites in Vic, two in Tas and eight in NSW (Downey *et al.* 2007). None were released in SA. The insect has not been recovered from any of these sites.

***Chrysolina picturata* (Clark).** In contrast to the painted boneseed leaf beetle, the blotched boneseed leaf beetle can be distinguished by four elytral bands of irregular white shapes on a black background. The thorax is wholly black (Adair and Scott 1997). It is known from the Western Cape province near Cape Agulhas in habitats with alkaline soils derived from calcareous sands, marine clays or limestone. The natural hosts in South Africa are boneseed and *C. m.* subsp. *pisifera*. The beetle was approved for release in 1992 and subsequently distributed to seven sites in Vic and two sites in SA. No establishment has been recorded (Downey *et al.* 2007).

***Cassida* sp. 3 (Coleoptera: Chrysomelidae).** One described and three undescribed species of *Cassida* are recorded from *Chrysanthemoides* in South Africa, and all are foliage feeders. The bitou tortoise beetle (Fig. 3d) was selected by matching the host subspecies and target climate in Australia (Kleinjan and Scott 1996). In South Africa, adults and larvae feed on bitou bush and are known from the Eastern Cape. Larvae form shallow feeding pits in the leaf surface and remain firmly attached.

Both larvae and adults are green and well camouflaged on the foliage of bitou bush. The bitou tortoise beetle was approved for release in 1995 and released at 12 sites in NSW (Downey *et al.* 2007). The tortoise beetle has established and is present at most of the initial release sites, but numbers remain low and impact on bitou bush appears to be negligible.

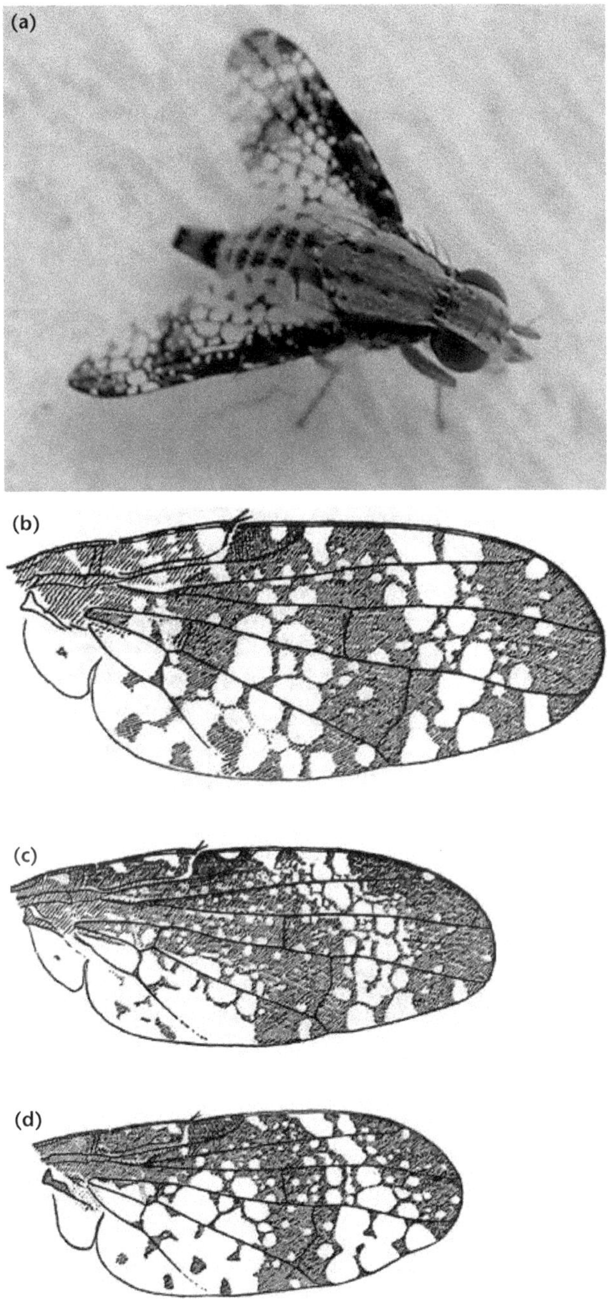

Figure 4: Seed flies of *Chrysanthemoides monilifera*. a) *Mesoclanis polana* female. Photo: T Morley, Vic Government. b–d) Wing pigmentation patterns of *Mesoclanis magnipalpis*, *M. dubia* and *M. polana*, respectively. Drawings adapted from Munro (1950).

***Mesoclanis* spp. (Diptera: Tephritidae).** The genus *Mesoclanis* is restricted to southern Africa and is associated only with *Chrysanthemoides* (Munro 1950). Species of *Mesoclanis* (seed flies) are distinguished principally by wing pigmentation patterns (Munro 1950) (Fig. 4). Three species were considered as biological control agents for *C. monilifera* in Australia (Scott and Adair 1992; Adair and Edwards 1996; Edwards and Brown 1997). The larvae of *Mesoclanis polana* Munro (bitou seed fly), *Mesoclanis magnipalpis* Bezzi (lacy-winged seed fly) and *Mesoclanis dubia* Walker (boneseed seed fly) feed in capitula of *C. monilifera*. White cigar-shaped eggs are inserted between buds and larvae bore through flowers, developing ovules and, to a lesser extent, receptacle tissue. Pupation mostly occurs within ovules (Adair and Bruzzese 2000) but may also occur among disc florets (*M. dubia*) or within the receptacle (*M. polana*) (Edwards and Brown 1997). *M. polana* is found on bitou bush in the Eastern Cape and KwaZula Natal and does not utilise boneseed in South Africa. Across the southern and the eastern seaboard, *M. magnipalpis* is known from bitou bush, boneseed, *C. m.* subsp. *pisifera* and *C. incana*, but is most prevalent in southern localities. *M. dubia* is known from bitou bush and boneseed (Edwards and Brown 1997; Munro 1950; T Morley, unpub. data).

***Mesoclanis polana*.** *Mesoclanis polana* was released in 1996. Following 10 releases of 50 to 800 flies each, by 1998 the bitou seed fly had dispersed over 1200 km, occupying virtually the entire range of bitou bush in Australia (Edwards *et al.* 1999). Although this fly can be reared on boneseed, emergence rates are low and it has not been recovered from boneseed in the field in Australia, even from regions where bitou bush and boneseed infestations are sympatric and *M. polana* is abundant (T Morley, unpub. data).

The impact of *M. polana* on seed production was monitored at eight latitudinally distinct sites in NSW between 1996 and 2004 (Edwards *et al.* 2009). High and persistent levels of seed reduction (58–86%) occurred in northern NSW, while considerably lower seed reduction (6–11%) occurred at southern sites. At most sites in this study, seed destruction levels were continuing to rise, suggesting higher levels of destruction may be achieved. In a separate study, Stuart *et al.* (2002) found reductions of 23–31% in seed production between 2001 and 2002.

***Mesoclanis magnipalpis*.** *Mesoclanis magnipalpis* was approved for release in Australia in 1997. Using flies derived from *C. m.* subsp. *pisifera* in South Africa, several releases totalling several hundred flies were made between 1998 and 2000 on boneseed and boneseed × bitou bush putative hybrids in Vic, and on boneseed in SA. No establishment was detected from these releases although the fly was twice recovered in the same season of release (A Bruzzese, *pers. comm.* 2004). In 2005, substantially larger numbers of the fly (also derived from *C. m.* subsp. *pisifera*) were released on boneseed in Vic and in the southern range of bitou bush in NSW (Morley 2005), where populations of *M. polana* are least abundant. Although some adults were recovered from boneseed sites during the same season as the releases, no flies were subsequently found. The reasons for this establishment failure are unknown, but some possibilities are discussed in Morley and Morin (2008). A further attempt to introduce *M. magnipalpis* was made in 2009 and comprised flies collected from boneseed in the Western Cape. Sixty-two flies were released in spring of that year at Frankston, Vic into a field cage containing 100 small flowering boneseed plants. At the time of writing (September 2010) no signs of *M. magnipalpis* had been observed for the 2010 flowering season, thus this introduction attempt seems to have failed (Morley 2010).

Further research is required to resolve survival strategies of *M. magnipalpis*. Larger releases of the fly collected from the boneseed in the Western Cape may be warranted. It would also be worthwhile to release on bitou bush in southern NSW flies that are sourced from bitou bush in its southern distribution in South Africa.

***Tortrix* Linnaeus sp. (Lepidoptera: Tortricidae).** *Tortrix* sp. (leaf-roller moth) is widespread in South Africa and is recorded from all six subspecies of *C. monilifera* (Scott and Brown 1992) with no clear preference for climate or *Chrysanthemoides* taxa. Larvae feed in shoot apices where they construct feeding shelters by using silk to join two or more neighbouring leaves. Leaves and stems within and around the shelter are consumed, often resulting in the death of terminal leaves and shoot tips. One to several larvae may occur in each shelter, particularly in the early stages of development. There are six larval instars and pupation occurs within leaf folds, and adpressed leaves or stems. Egg clusters are laid on leaf blades with a transparent waxy covering. The moth is multivoltine and usually has three generations per year, with oviposition peaks in July, November and late January.

In South Africa *Tortrix* sp. is the most damaging insect of *C. monilifera*. Outbreaks occur over many hectares, often significantly reducing seed production and killing whole bushes (Scott and Adair 1990). Host plant records and field observations in South Africa indicated that it was restricted to *Chrysanthemoides* (Scott and Brown 1992; Edwards 1997). However, in laboratory and cage tests, development was completed on 36 plant species from 11 families. To predict the likely field host-range, open-field studies were undertaken in South Africa using five different approaches, some of which were novel (Edwards 1997; R Adair, unpub. data). The results of these tests indicated that although the moth has a strong preference for oviposition and development on *C. monilifera*, there is a possibility that some oviposition could occur on plants located in very close proximity to *C. monilifera* and that limited larval development could be expected on *Calendula* and *Gazania* spp. if they occurred in close proximity to *Chrysanthemoides*. Approval for release of *Tortrix* sp. was granted on the basis of this risk assessment.

The first release was made at the You Yang Range, Vic in April 2000 and further releases were made at 112 sites across south-eastern Australia until 2004. In NSW, the leaf-roller moth is established on bitou bush at up to nine sites, with up to 70 larvae m^{-2} recorded (Downey *et al.* 2007; H Cherry, *pers. comm.* 2010). No establishment has occurred elsewhere. Low establishment success has been attributed to competition between neonates and predation, most likely by ants (Swirepik *et al.* 2004). In predator-exclusion experiments, up to 18 times more *Tortrix* sp. larvae and pupae were recorded when protected from predators (Ireson *et al.* 2002) and there were significantly more predatory arthropods and higher mortality of the moth at dune than at headland sites (Strakosch 2004). Although eggs of the moth are often heavily parasitised by *Trichogramma* sp. (Hymenoptera) in South Africa (P Edwards and P Müller, unpub. data) there has been no record of attack on eggs in Australia.

Tortrix sp. is the subject of a Weed Warriors program in NSW, which began in 2006. Weed Warriors engages students, weed officers, land managers and community groups to rear, release and monitor biological control agents (Schembri *et al.* 2008).

***Aceria* sp. (Acari: Eriophyidae)** *Aceria* sp. (boneseed leaf-buckle mite) induces formation of erinea (a type of gall) on the leaf blades of *C. monilifera*. Erinea are dense patches of hair-like structures whose formation disrupts normal leaf expansion, resulting in disfigurement by twisting or buckling. Erineum formation commences on shoot meristems before leaves become visible (Morley and Morin 2008). In South Africa, the mite has been recovered from boneseed, bitou bush, *C. m.* subsp. *pisifera* and *C. incana*. These mite populations probably represent host-specific biotypes.

Following extensive host-specificity testing of the mite in South Africa and quarantine in Australia, taxonomic problems with the eriophyids associated with *C. monilifera* erinea contributed to the delayed approval of its release in Australia. Smith Meyer (1981) attributed erineum formation on *C. monilifera* to *Aceria neseri* Meyer, but no mite that corresponded with that description was found from more than 1000 specimens sampled from a culture reared in quarantine in Australia and associated with boneseed leaf buckle mite erinea (T Morley and J Davies, unpub. data). Unfortunately, Smith Meyer's *A. neseri* holotype specimen is badly damaged and cannot be viewed for comparison with its description or other mites. *Aceria* sp. was released from quarantine in 2008, with more than 90 releases made up to autumn 2010 in Vic, SA and Tas. Small colonies have persisted 12 months after release at four sites in Vic and one site in Tas, and these are promising signs of establishment.

Agents tested but not released

***Ageniosa electoralis* (Vogel) (Coleoptera: Chrysomelidae).** The only agent rejected after completion of host-testing evaluation was the bitou leaf beetle *Ageniosa electoralis* (Fig. 3e). This beetle is restricted to coastal areas of KwaZulu Natal, South Africa, where it can cause extensive defoliation on bitou bush. Adults and larvae feed on young foliage and stems of the host and are mostly nocturnal. The adults are shiny dark green-black. Round dark eggs are laid in the leaf litter and soil around the base of the host plant. Pupation occurs in the soil in cocoons of cemented soil particles. In host-specificity tests, larvae were able to complete development and produce adults on seven species of plant including globe artichoke *Cynara scolymus* L. (Asteraceae) and carrot *Daucus carota* L. (Apiaceae). The bitou leaf beetle was rejected as a potential biological control agent pending additional information on the likelihood of attack on non-target hosts under field conditions (Adair and Scott 1993). Given failure of *Chrysolina* spp. released for biological control of *Chrysanthemoides*, further evaluation of the bitou leaf beetle is a low priority.

***Mesoclanis dubia*.** Of the *Chrysanthemoides* seed flies, least is known about the ecology and biology of boneseed seed fly *Mesoclanis dubia* (Fig. 4). Twenty-one percent (n=262) of seed flies from bitou bush fruit samples from St Francis Bay (Eastern Cape province, South Africa) in 1993 and 1994 were *M. dubia* (Edwards and Brown 1997), whereas this species comprised 55% (n=303) of flies from boneseed fruit samples collected from Stellenbosch and Paarl (Western Cape province) in 2009 (T Morley, unpub. data). All remaining flies in these samples were *M. magnipalpis*. Preliminary host-testing indicated that *M. dubia* oviposits readily on boneseed soon after emergence (R Adair, unpub. data). More comprehensive testing is necessary to fully assess the potential of this fly for biological control of boneseed in Australia.

Agents undergoing evaluation

***Endophyllum osteospermi* (Doige) A.R. Wood (Uredinales: Pucciniaceae).** The boneseed rust fungus is a systemic pathogen that is restricted to a small group of related plants in the Calendulae: *C. m.* subspp. *monilifera*, *rotundata*, *pisifera*, *canescens* (L.) T. Norl. and *subcanescens* (L.) T. Norl., *C. incana*, an undescribed taxon and three *Osteospermum* L. species (Wood 1998).

The fungus has a latent period of six months to three years after infection of a host before the development of symptoms manifesting as witches' brooms (Wood *et al.* 2004; Morin and Wood 2009). Witches' brooms are deformed branches with multiple swollen stems, short internodes and small chlorotic leaves that produce fewer flowers and drupes (Wood 2002, 2006). In a field study in South Africa, vegetative growth, number of buds, flower heads, fruiting heads and drupes of symptomless branches of infected bushes were 15–99% less than those of similar branches of uninfected bushes (Wood 2002), indicating that infected branches act as resource sinks. Plants severely infected over a few years died prematurely.

A three-tier system was developed to evaluate the host-specificity of the fungus. This was a novel approach, adopted because of the long latent period before symptoms appear (Wood 2006; Morin and Wood 2009). The first tier of tests involved inoculation of detached young leaves of test species and microscopic examination four days later. Germination of spores and penetration of epidermal cells by the fungus occurred on boneseed and bitou bush and three other closely related species, but also on four species outside the Calenduleae: *Gazania rigens* (L.) Gaertn., *Gerbera jamesonii* L., *Bedfordia arborescens* Hochr. (all Asteraceae) and *Eucalyptus cladocalyx* F.Muell. (Myrtaceae) (Wood 2006).

Second-tier tests utilising leaves still attached to whole potted plants were performed on six non-target species that were penetrated in first-tier tests. These second-tier tests involved microscopic examinations of leaves two weeks after inoculation to determine if the rust had successfully colonised. They were repeated several times in South Africa and under quarantine in Australia in 2007 and 2008. None of the tests were successful – the various leaf-clearing and staining techniques tried could not detect early signs of rust colonisation (Morin and Wood 2009). In light of these difficulties, a third tier of tests was initiated, whereby inoculated plants are kept for several years to allow symptoms to develop.

In South Africa during 2008 and 2009 and under quarantine in Australia in spring 2009 (one test species) whole potted plants were inoculated repeatedly with the fungus over a few weeks and, at the time of writing (September 2010), had been maintained for up to two years. Witches' brooms developed on some boneseed plants within one year of inoculation, but on none of the other species (A Wood and L Morin, unpub. data). An application for release in Australia will be prepared, when the third-tier tests are satisfactorily completed, if the risk to non-target species seems acceptable. Climate modelling predicts that the greatest impact of the rust fungus is likely to be on boneseed populations in Tas and southern Vic, areas where the plant is highly invasive (Wood *et al.* 2004).

Other promising agents

The beetle *Obereopsis pseudocapensis* Breuning (Coleoptera: Cerambycidae) attacks stems of *C. m.* subspp. *rotundata* and *pisifera* in the southern Cape coast and Eastern Cape (Neser and Morris 1985; Adair and Edwards 1996). Larvae bore internally within upper shoot tips, where they eat vascular tissue and cause stem death. Adults oviposit in leaf midveins and young larvae tunnel into nearby stem tissue. Preliminary testing and field observations indicate this insect is specific to *C. monilifera* and can cause considerable damage.

Two moth species are known to cause damage similar to that of *O. pseudocapensis*. An undetermined tip wilt moth (Gelechidae) feeds internally within the stem apices on *C. m.* subspp. *monilifera* and *pisifera*. (Adair and Edwards 1996) and larvae of an undetermined stem-girdling moth (Pyralidae) feed on the same hosts, but external feeding as well as internal feeding causes stems to wilt (Scott and Adair 1992, 1990). Their host-specificities are not known.

Several other damaging insect species feed internally on *C. monilifera* in South Africa. Investigation is warranted for the root- and crown-boring beetle *Sphenoptera* Dejean sp., (Coleoptera: Buprestidae) (Neser and Morris 1985), the shoot apex-galling cecidomyid *Cecidomyia deformans* Schiner (Diptera: Cecidomyiidae) and the peduncle-galling cecidomyid, *Lasioptera* Meigen sp. (Diptera: Cecidomyiidae).

Endemic natural enemies

While no endemic phytophagous arthropods cause significant herbivory damage to *C. monilifera* in Australia, there are at least 55 fungal species associated with bitou bush (Cother *et al.* 1996). The cosmopolitan fungus, *Sclerotinia sclerotiorum* (Lib.) de Bary is associated with bitou bush dieback in Australia; its potential for inundative biological control was investigated and rejected in the early 1990s (Cother *et al.* 1996; Cother 2000). The anamorphic fungus *Hendersonia osteospermi* Wakef. causes necrotic spots on the foliage of *C. monilifera* in South Africa (Crous *et al.* 2000) and its teleomorph was recently isolated from bitou bush in Australia and described as the new species *Austropleospora osteospermi* R.G. Shivas & L. Morin (Morin *et al.* 2010). *A. osteospermi* is implicated, with other fungi, in bitou bush dieback in Australia and further investigation is required (Morin *et al.* 2010).

DISCUSSION

Comostolopsis germana and *M. polana* have reduced the reproductive capacity of bitou bush in NSW (Holtkamp 2002; Edwards *et al.* 2009). However, further agents with potential to reduce canopy density and cause structural damage to bitou bush infestations are required. French *et al.* (2008) found that bitou bush seedling establishment success compared to that of native plants is a major contributor to bitou bush invasiveness, suggesting agents that target seedlings might effectively suppress bitou bush infestations. None of the agents so far released on boneseed have established, although there are early signs of establishment of the recently released mite, *Aceria* sp. The mite and *E. osteospermi*, which is under investigation, offer reasonable prospects that biological control agents will be established on boneseed in the near future.

Various factors are likely to have contributed to the failure of agents on *C. monilifera*. Predation by indigenous invertebrates appears to have hampered establishment or dispersal of *Chrysolina* spp. and *Tortrix* sp. These insects feed or shelter in the canopy and appear susceptible to Australian predators. Predatory mites are implicated in the decline of release stocks of *Tortrix* sp. (Ireson *et al.* 2002) and might hinder establishment of *Aceria* sp. The selection of future biological control agents for *C. monilifera* should avoid poorly concealed external feeders and focus closely on endophagous guilds. Incorrect climate and host matching appears to have contributed to establishment failure of some agents. In retrospect, the 'new association' hypothesis (Hokkanen and Pimentel 1984) that led to concentrated effort on the evaluation of leaf beetles failed to deliver effective agents. Greater emphasis on exploring a more diverse range of insect feeding guilds may have yielded better results.

Indirect effects on non-target invertebrate species of agents released for bitou bush have been extensively studied (Carvalheiro *et al.* 2007). Their introduction has led to changes in invertebrate community structure by increasing abundance of native parasitoids (Carvalheiro *et al.* 2007), although the magnitude of such perturbations will probably diminish if bitou bush infestations decline significantly.

The integration of biological control and other forms of management is likely to be required for adequate suppression of *C. monilifera* and this is recognised by many (e.g. Adair and Edwards 1996; Groves 2008). Control of bitou bush in NSW with low doses of aerially applied herbicides (Toth *et al.* 1993) enabled larger areas to be treated, and accelerated the containment of this invasive plant. Interactions between herbicide applications and biological control agents of bitou bush seem likely. Ainsworth and Holtkamp (1999) observed high numbers of *C. germana* on bitou bush seedlings following aerial spraying and noted that larvae on sprayed plants would perish due to lack of fresh foliage. They also found that *M. polana* avoided bitou bush sprayed with glyphosate.

Fire can be used to kill boneseed (Lane and Shaw 1978; Melland *et al.* 1999) and, depending on intensity, partially kill stands of bitou bush (Vranjic and Groves 1999). Canopy-dwelling agents of *C. monilifera* will be killed by fire and recolonisation will occur when regenerating hosts become phenologically acceptable, depending on proximity of viable agent populations in unburnt areas and agent dispersal capabilities. Otherwise, programs using wild or laboratory-reared agents may be required to hasten recolonisation.

Population modelling (Kriticos *et al.* 2003; Noble and Weiss 1989; Stuart 2002) suggests that pre-dispersal seed destruction by biological control agents might need to be greater than 95% to significantly affect bitou bush infestations, even when integrated with other control methods. A recent assessment of predicted bitou bush seed longevity (Schoeman *et al.* 2010), where the value of this parameter was suggested to be much lower than that assumed in models above, perhaps adds to uncertainty of such models' predictions.

Estimated costs to 2005 of the biological control program for *C. monilifera* were $7.1 million in 2005 dollar terms, with the potential to result in a net present value of $47.9 million and a benefit:cost ratio of 10.3 (Page and Lacey 2006). However, accrued benefits are 5% of those predicted, resulting in net present value of $2.5 million and a benefit:cost ratio of 0.5 at a 5% discount rate (Page and Lacey 2006). Nominal investment in this program between 2005 and 2010 was about $1.5 million, focused on the boneseed leaf-buckle mite, the boneseed rust fungus and the lacy-winged seed fly.

The program for biological control of *C. monilifera* received strong nationally coordinated financial support from 1987 to 2000, but support has declined since then. Although the search for and testing of new agents and mass-rearing and distribution of released agents has diminished, a broader national *C. monilifera* management program is facilitated through the WoNS program and associated biological control activities. We believe that introduction of the boneseed rust fungus, lacy-winged seed fly (possibly different biotypes for boneseed and bitou bush) and boneseed seed fly should be pursued and that other natural enemies of *C. monilifera*, particularly the endophages mentioned above, have potential as biological control agents in Australia. Further quantification of the impacts of *C. monilifera* invasions, particularly for boneseed, is warranted and we agree with Groves (2008) that documenting change in biodiversity metrics as a result of successful biological control will be a significant challenge.

ACKNOWLEDGEMENTS

We thank Petra Müller, Liesl Smith, Tony Gordon, Keith Appollis and Abraham Adonis for technical support and agent collections in South Africa. We thank Hillary Cherry and Kris French for reviewing and improving a draft of this manuscript.

REFERENCES

Adair RJ and Bruzzese AS (2000) Evaluation and host specificity of two seed flies *Mesoclanis polana* and *M. magnipalpis* (Diptera: Tephritidae): biological control agents for *Chrysanthemoides monilifera* (Asteraceae) in Australia. *Bulletin of Entomological Research* **90**, 467–474.

Adair RJ and Butler K (2010) Hybridisation within *Chrysanthemoides monilifera* and implications for the eradication of bitou bush from Victoria, Australia. In *Proceedings of the 17th Australian Weeds Conference.*

26–30 September 2010, Christchurch, New Zealand. (Ed SM Zydenbos) pp. 378–381. New Zealand Plant Protection Society.

Adair RJ and Edwards PB (1996) An attack strategy against *Chrysanthemoides monilifera*, a weed of native vegetation in Australia. In *Proceedings of the 9th International Symposium on Biological Control of Weeds.* 19–26 January 1996, Stellenbosch, South Africa. (Eds VC Moran and JH Hoffmann) pp. 429–434. University of Cape Town, South Africa.

Adair RJ and Holtkamp RH (1999) Development of a pesticide exclusion technique for assessing the impact of biological control agents for *Chrysanthemoides monilifera. Biocontrol Science and Technology* **9**, 383–390.

Adair RJ and Scott JK (1989) The life-history and host specificity of *Comostolopsis germana* Prout (Lepidoptera: Geometridae), a biological control agent of *Chrysanthemoides monilifera* (Compositae). *Bulletin of Entomological Research* **79**, 649–657.

Adair RJ and Scott JK (1991) Distribution, life-history, host specificity and suitability of an undescribed *Chrysolina* species (Coleoptera:Chrysomelidae) for the biological control of *Chrysanthemoides monilifera* (Compositae). *Bulletin of Entomological Research* **81**, 235–242.

Adair RJ and Scott JK (1993) Biology and host specificity of *Ageniosa electoralis* (Coleoptera: Chrysomelidae), a prospective biological control agent for *Chrysanthemoides monilifera* (Asteraceae). *Biological Control* **3**, 191–198.

Adair RJ and Scott JK (1997) Distribution, life history and host specificity of *Chrysolina picturata* and *Chrysolina* sp. B (Coleoptera: Chrysomelidae), two biological control agents for *Chrysanthemoides monilifera* (Compositae). *Bulletin of Entomological Research* **87**, 331–341.

Ainsworth N and Holtkamp R (1999) Integrating herbicide use with biological control of bitou bush. In *Proceedings of the 12th Australian Weeds Conference.*12–16 September 1999, Hobart. (Eds AC Bishop, M Boersma and CD Barnes) pp. 275–278. Tasmanian Weed Society, Devonport, Tas.

Barker NP, Howis S, Nordenstam B, Kallersjo M, Eldenas P, Griffioen C and Linder HP (2009) Nuclear and chloroplast DNA-based phylogenies of *Chrysanthemoides* Tourn. ex Medik. (Calenduleae; Asteraceae) reveal extensive incongruence and generic paraphyly, but support the recognition of intraspecific taxa in *C. monilifera. South African Journal of Botany* **75**, 560–572.

Bramwells H (2003) Flora and Fauna Guarantee Action Statement #63: Brittle Greenhood *Pterostylis truncata.* Vic Dept of Sustainability and Environment, Melbourne. <http://www.dse.vic.gov.au>.

Carvalheiro LG, Buckley YM, Ventim R and Memmott J (2007) Assessing indirect impacts of biological control agents on native biodiversity: a community-level approach. In *Proceedings of the 12th International Symposium on Biological Control of Weeds.* 22–27 April 2007, La Grande Motte, France (Eds MH Julien, R Sforza, MC Bon, HC Evans, PE Hatcher, HL Hinz and BG Rector), pp. 83–86. CABI, Wallingford, UK.

Cherry H, Willsher L and Whyte B (2008) Bitou bush and boneseed eradication and containment in Australia. *Plant Protection Quarterly* **23**, 38–40.

Cother EJ (2000) Pathogenicity of *Sclerotinia sclerotiorum* to *Chrysanthemoides monilifera* ssp. *rotundata* (bitou bush) and selected species of the coastal flora of eastern Australia. *Biological Control* **18**, 10–17.

Cother EJ, Nikandrow A and Gilbert RL (1996) *Sclerotinia sclerotiorum*, a potential biocontrol agent for *Chrysanthemoides monilifera* (bitou bush). In *Proceedings of the 9th International Symposium on Biological Control of Weeds.*19–26 January 1996, Stellenbosch, South Africa (Eds VC Moran and JH Hoffmann) pp. 529–530. University of Cape Town, South Africa.

Crous PW, Phillips AJL and Baxter AP (2000) *Phytopathogenic Fungi from South Africa.* Dept of Plant Pathology Press, University of Stellenbosch, South Africa.

Downey PO (2004) Bitou bush management and plant conservation: establishing priorities for control. In *Proceedings of the 14th Australian Weeds Conference.* 6–9 September 2004, Wagga Wagga, NSW. (Eds BS Sindel and SB Johnson) pp. 697–700. Weed Society of New South Wales.

Downey PO, Holtkamp RH, Ireson JE, Kwong RM and Swirepik AE (2007) A review of the *Chrysanthemoides monilifera* biological control program in Australia: 1987–2005. *Plant Protection Quarterly* **22**, 24–32.

Downey PO, Williams MC, Whiffen LK, Turner PJ, Burley AL and Hamilton MA (2009) Weeds and biodiversity conservation: a review of managing weeds under the New South Wales Threatened Species Conservation Act 1995. *Ecological Management and Restoration* **10**, s53–s58.

Edwards P (1997) Prospects for biological control of bitou bush and boneseed with '*Tortrix*' sp. In *Proceedings of the Bitou Bush Workshop.* 3–4 September 1997, Sydney. (Eds R Holtkamp, R Groves and S Corey) p. 18. NSW National Parks and Wildlife Service and CRC for Weed Management Systems, Adelaide.

Edwards PB and Brown EM (1997) *Mesoclanis* seed flies (Diptera: Tephritidae) on *Chrysanthemoides* (Asteraceae) in South Africa: distribution, attack strategy and impact on seed production. *Bulletin of Entomological Research* **87**, 127–135.

Edwards PB, Holtkamp RH and Adair RJ (1999) Establishment and rapid spread of the bitou seed fly *Mesoclanis polana* in eastern Australia. *Australian Journal of Entomology* **38**, 148–150.

Edwards PB, Adair RJ, Holtkamp RH, Wanjura WJ, Bruzzese AS and Forrester RI (2009) Impact of the biological control agent *Mesoclanis polana* (Tephritidae) on bitou bush (*Chrysanthemoides monilifera* subsp. *rotundata*) in eastern Australia. *Bulletin of Entomological Research* **99**, 51–63.

Ens EJ, Bremmer JB, French K and Korth J (2009) Identification of volatile compounds released by roots of an invasive plant, bitou bush (*Chrysanthemoides monilifera* ssp. *rotundata*), and their inhibition of native seedling growth. *Biological Invasions* **11**, 275–287.

French K and Eardley K (1997) The impact of weed infestations on litter invertebrates in coastal vegetation. In *Frontiers in Ecology: Building the Links.* (Ed I Lunt) p. 89. Elsevier Science, Amsterdam.

French K and Zubovic A (1997) Effect of the weed *Chrysanthemoides monilifera* (bitou bush) on bird communities. *Wildlife Research* **24**, 727–735.

French K, Ens E, Gosper CR, Lindsay E, Mason T, Owers B and Sullivan N (2008) Management implications of recent research into the effect of bitou bush invasion. *Plant Protection Quarterly* **23**, 24–28.

Gosper CR (2004) Consequences of weed invasion and control on plant–bird interactions and bird communities. PhD thesis. University of Wollongong, Australia.

Griffioen RC (1995) A taxonomic study of *Chrysanthemoides* Tourn. ex Medik. Compositae. MSc. thesis. University of Cape Town, South Africa.

Groves RH (2008) Research on bitou bush and boneseed: a work in progress. *Plant Protection Quarterly* **23**, 15–19.

Hokkanen H and Pimentel D (1984) New approach for selecting biological control agents. *Canadian Entomologist* **116**, 1109–1121.

Holtkamp RH (1997) The current status of bitou bush biological control in New South Wales. In *Proceedings of the Bitou Bush Workshop.* 3–4 September 1997, Sydney. (Eds R Holtkamp, R Groves and S Corey) p. 17. NSW National Parks and Wildlife Service and CRC for Weed Management Systems, Adelaide.

Holtkamp RH (2002) Impact of bitou tip moth, *Comostolopsis germana*, on bitou bush in New South Wales. In *Proceedings of the 13th Australian Weeds Conference.* 8–13 September 2002, Perth. (Eds H Spafford Jacob, J Dodd and JH Moore) pp. 405–406. Plant Protection Society of Western Australia, Perth.

Ireson JE, Davies JT, Holloway RJ and Chatterton WS (2002) Attempts to establish biological control agents for boneseed in Tasmania. In *Proceedings of the 13th Australian Weeds Conference.* 8–13 September 2002, Perth. (Eds H Spafford Jacob, J Dodd and JH Moore) pp. 407–411. Plant Protection Society of Western Australia, Perth.

Kleinjan CA and Scott JK (1996) Selection of *Cassida* spp. from southern Africa for the biological control of *Chrysanthemoides monilifera* in Australia. *Annals of Applied Biology* **128**, 373–385.

Kriticos DJ, Stuart RM and Ash JE (2003) Exploring interactions between cultural and biological control techniques: modelling bitou bush (*Chrysanthemoides monilifera* ssp. *rotundata*) and a seed fly (*Mesoclanis polana*). In *Proceedings of the 11th International Symposium on Biological Control of Weeds.* 27 April–2 May 2003, Canberra. (Eds JM Cullen, DT Briese, DJ Kriticos, WM Lonsdale, L Morin and JK Scott) pp. 559–566. CSIRO Entomology, Canberra.

Lane DW and Shaw K (1978) The role of fire in boneseed (*Chrysanthemoides monilifera* (L.) Norl.) control in bushland. In *Proceedings of the 1st Conference of the Council of Australian Weed Science Societies.* 12–14 April 1978, Melbourne. pp. 333–335. Weed Science Society of Victoria, Melbourne.

Lindsay EA and French K (2004) *Chrysanthemoides monilifera* ssp. *rotundata* invasion alters decomposition rates in coastal areas of south-eastern Australia. *Forest Ecology and Management* **198**, 387–399.

Lindsay EA and French K (2005) Litterfall and nitrogen cycling following invasion by *Chrysanthemoides monilifera* ssp. *rotundata* in coastal Australia. *Journal of Applied Ecology* **42**, 556–566.

Love A and Dyason R (1985) (Eds) *Bitou Bush and Boneseed: A National Conference on* Chrysanthemoides monilifera. 8–9 August 1984, Port Macquarie, Australia. NSW National Parks and Wildlife Service and NSW Dept of Agriculture.

Mason TJ and French K (2008) Impacts of a woody invader vary in different vegetation communities. *Diversity and Distributions* **14**, 829–838.

McAlpine KG, Timmins SM and Westbrooke I (2009) Bone-seed (*Chrysanthemoides monilifera* ssp. *monilifera*) invasion effects on native regeneration in New Zealand coastal plant communities. *New Zealand Journal of Ecology* **33**, 72–82.

Meggs J (1995) Biotic resistance to a chrysomelid beetle (*Chrysolina* sp.) introduced to Australia for the biological control of boneseed (*Chrysanthemoides monilifera* ssp. *monilifera*). Hons thesis. Monash University, Australia.

Melland R, Ainsworth N and Roush RT (1999) Reducing the fitness of the fittest: how to control a *Chrysanthemoides monilifera* ssp. *monilifera* (Norlindh) invasion. In *Proceedings of the 12th Australian Weeds Conference.* 12–16 September 1999, Hobart. (Eds AC Bishop, M Boersma and CD Barnes) pp. 294–297. Tasmanian Weed Society, Devonport.

Morin L and Wood A (2009) Boneseed rust: a highly promising candidate for biological control. In *Conference Handbook of the 17th Biennial Conference of the Australasian Plant Pathology Society.* 29 September–1 October 2009, Newcastle, Australia. p. 182. Australasian Plant Pathology Society.

Morin L, Shivas RG, Piper MC and Tan YP (2010) *Austropleospora osteospermi* gen. et sp. nov. and its host specificity and distribution on *Chrysanthemoides monilifera* ssp. *rotundata* in Australia. *Fungal Diversity* **40**, 65–74.

Morley T (2005) Lacy-winged seed fly releases for biological control of boneseed and bitou bush. *Under Control – Pest Plant and Animal News*, **31**, 19–20.

Morley T (2010) A new attempt to introduce the lacy-winged seed fly *Mesoclanis magnipalpis* to Australia for biological control of boneseed *Chrysanthemoides monilifera* ssp. *monilifera*. In *Proceedings of the 17th Australian Weeds Conference*, 26–30 September 2010, Christchurch, NZ. (Ed SM Zydenbos) pp. 243–244. New Zealand Plant Protection Society.

Morley TB and Morin L (2008) Progress on boneseed (*Chrysanthemoides monilifera* subsp. *monilifera*) (L.) Norlindh) biological control: the boneseed leaf buckle mite *Aceria* (Keifer) sp., the lacy-winged seed fly *Mesoclanis magnipalpis* Bezzi and the boneseed rust *Endophyllum osteospermi* (Doidge) A.R. Wood. *Plant Protection Quarterly* **23**, 29–31.

Munro HK (1950) Trypetid flies (Diptera) associated with the Calendulae, plants of the family Compositae in South Africa. I. A bio-taxonomic study of the genus *Mesoclanis. Journal of the Entomological Society of Southern Africa* **13**, 37–52.

Muyt A (2001) *Bush Invaders of South-east Australia. A Guide to the Identification of Environmental Weeds found in South-eastern Australia.* RG and FJ Richardson, Melbourne.

Neser S and Morris MJ (1985) Preliminary observations on natural enemies of *Chrysanthemoides monilifera* in South Africa. In *Proceedings of a Conference on* Chrysanthemoides monilifera. 8–9 August 1984, Port Macquarie, Australia. (Eds A Love and R Dyason) pp. 105–109. NSW National Parks and Wildlife Service and NSW Dept of Agriculture.

Noble IR and Weiss PW (1989) Movement and modelling of buried seed of the invasive perennial *Chrysanthemoides monilifera* in coastal dunes and biological control. *Australian Journal of Ecology* **14**, 55–64.

Norlindh T (1943) Studies in the Calenduleae. 1. Monograph on the genera *Dimorphotheca, Castalis, Osteospermum, Gibbaria* and *Chrysanthemoides*, p. 367. CWK Gleerup, Lund.

Owers B (1999) The effect of bitou bush infestation on the behaviour and nest predation of birds in coastal woodlands. Hons thesis. University of Wollongong, Australia.

Page AR and Lacey KL (2006) *Economic Impact Assessment of Australian Weed Biological Control.* CRC for Australian Weed Management. Technical Series No. 10.

Schembri A, Holtkamp R, Cherry H, Hauesler J and West J (2008) Community weed education initiatives in NSW: getting school children involved. *Plant Protection Quarterly* **23**, 42–43.

Schoeman J, Buckley YM, Cherry H, Long RL and Steadman KJ (2010) Inter-population variation in seed longevity for two invasive weeds: *Chrysanthemoides monilifera* ssp. *monilifera* (boneseed) and ssp. *rotundata* (bitou bush). *Weed Research* **50**, 67–75.

Scott JK and Adair RJA (1990) The commencement of biological control of bitou bush and boneseed (*Chrysanthemoides monilifera*). In *Proceedings of the 9th Australian Weeds Conference.* 6–10 August 1990, Adelaide. (Ed JW Heap) pp. 126–129. Crop Science Society of South Australia, Adelaide.

Scott JK and Adair RJ (1992) Selection of biological control agents for *Chrysanthemoides monilifera*, a weed of protected natural areas in Australia. In *Proceedings of the 8th International Symposium on Biological Control of Weed.* 2–7 February 1992, Lincoln, NZ. (Eds ES Delfosse and RR Scott) pp. 61–67. DSIR/CSIRO, Melbourne.

Scott JK and Brown EM (1992) Preliminary observations on the biology and host plants of *Tortrix* sp. (Lepidoptera: Tortricidae), a possible biological control agent for *Chrysanthemoides* spp. (Asteraceae). *Journal of the Entomological Society of South Africa* **55**, 245–253.

Scurr G, Kirkpatrick JB, Daniels GD and McQuillan P (2008) Biotic resistance to *Chrysanthemoides monilifera* ssp. *monilifera* in Tasmania. *Austral Ecology* **33**, 941–950.

Simmons DM and Flint PW (1986) Variation in *Chrysanthemoides monilifera* (Compositae) in eastern Australia. *Weed Research* **26**, 427–432.

Smith Meyer MKP (1981) South African Eriophyidae (Acari): the genus *Aceria* Keifer, 1944. *Phytophylactica* **13**, 117–126.

Strakosch N (2004) Does predation limit the establishment of the biological control agent *Tortrix* sp. in *Chrysanthemoides monilifera* ssp. *rotundata* infestations? Hons thesis. University of Wollongong, Australia.

Stuart RM (2002) A biological control model for *Chrysanthemoides monilifera* ssp. *rotundata* (Asteraceae) by *Mesoclanis polana* (Tephritidae). Hons thesis. Australian National University, Australia.

Stuart RM, Kriticos DJ and Ash JE (2002) Modelling the biological control of bitou bush (*Chrysanthemoides monilifera*: Asteraceae) by *Mesoclanis polana* (Tephritidae). In *Proceedings of the 13th Australian Weeds Conference*. 8–13 September 2002, Perth. (Eds H Spafford Jacob, J Dodd and JH Moore) pp. 591–594. Plant Protection Society of Western Australia, Perth.

Swirepik A, Aveyard R and Sheppard A (2004) The post-release larval mortality of the *Chrysanthemoides* leaf roller *Tortix* sp. in Australia. In *Proceedings of the 11th International Symposium on Biological Control of Weeds*. 27 April–2 May 2003, Canberra. (Eds JM Cullen, DT Briese, DJ Kriticos, WM Lonsdale, L Morin and JK Scott) p. 480. CSIRO Entomology, Canberra.

Thomas J and Leys A (2002) Strategic management of bitou bush (*Chrysanthemoides monilifera* ssp. *rotundata* (L.) T.Norl.). In *Proceedings of the 13th Australian Weeds Conference*. 8–13 September 2002, Perth. (Eds H Spafford Jacob, J Dodd and JH Moore) pp. 586–590. Plant Protection Society of Western Australia, Perth.

Thomas PB, Possingham H and Roush R (2005) Effects of boneseed (*Chrysanthemoides monilifera* (L.) Norl. ssp. *monilifera*) on the composition of the vegetation and the soil seed bank of an open eucalypt woodland. *Plant Protection Quarterly* **20**, 74–80.

Toth J, Milham PJ and Nazer C (1993) Aerially applied herbicides selectively control bitou bush in communities of indigenous plants on sand dunes. In *Proceedings of the 10th Australian and 14th Asian Pacific Weed Science Society Conference*. 6–10 September 1993, Brisbane. pp. 67–71. Weed Society of Queensland, Brisbane.

Vranjic JA and Groves RH (1999) 'Best-practice' management strategies for the South African weed, bitou bush (*Chrysanthemoides monilifera* subsp. *rotundata*). In *Proceedings of the 12th Australian Weeds Conference*. 12–16 September 1999, Hobart. (Eds AC Bishop, M Boersma and CD Barnes) pp. 288–293. University of Tasmania, Australia.

Weiss PW, Adair RJ and Edwards PB (1998) *Chrysanthemoides monilifera* (L.) T. Norl. In *The Biology of Australian Weeds*. (Eds FD Panetta, RH Groves and RCH Shepherd) pp. 49–61. RG and FJ Richardson, Melbourne.

Wilkie L, Cassis G and Gray M (2007) The effects on terrestrial arthropod communities of invasion of a coastal heath ecosystem by the exotic weed bitou bush (*Chrysanthemoides monilifera* ssp. *rotundata* L.). *Biological Invasions* **9**, 477–498.

Wood AR (1998) *Endophyllum osteospermi*, a new combination for *Aecidium osteospermi* (Basidiomycetes – Uredinales – Pucciniaceae). *South African Journal of Botany* **64**, 146.

Wood AR (2002) Infection of *Chrysanthemoides monilifera* ssp. *monilifera* by the rust fungus *Endophyllum osteospermi* is associated with a reduction in vegetative growth and reproduction. *Australian Plant Pathology* **31**, 409–415.

Wood AR (2006) Preliminary host specificity testing of *Endophyllum osteospermi* (Uredinales: Pucciniaceae), a biological control agent against *Chrysanthemoides monilifera* ssp. *monilifera*. *Biocontrol Science and Technology* **16**, 495–507.

Wood AR, Crous PW and Lennox CL (2004) Predicting the distribution of *Endophyllum osteospermi* (Uredinales: Pucciniaceae) in Australia based on its climatic requirements and distribution in South Africa. *Australasian Plant Pathology* **33**, 549–558.

Cirsium vulgare (Savi) Tenore – spear thistle

Jean-Louis Sagliocco, Raelene M Kwong and Tom Morley

ABSTRACT

Cirsium vulgare was identified as the most invasive thistle in Vic during the 1980s. A biological control program was initiated during the same decade and a receptacle weevil, a tephritid gall fly and a rosette weevil were released during the 1990s–2000s. Surveys were conducted on the establishment and impact of the fly but data are lacking on the survival of the receptacle and rosette weevils. It appears that the fly and the receptacle weevil have survived only at release sites with cool summers and cold winters. Detailed surveys are required to assess the overall success of this program.

Key words: *Urophora stylata*, *Rhinocyllus conicus*, *Trichosirocalus horridus*.

INTRODUCTION

Cirsium vulgare (Savi) Tenore (Asteraceae: Cardueae: Carduinae) (spear thistle) is a late-flowering monocarpic biennial, short-lived perennial or more rarely annual plant, growing 1–1.5 m tall, producing many capitula, each containing up to 200 achenes (Michaux 1989) (Fig. 1). *C. vulgare* originates in Eurasia and North Africa. In Europe, it is found in all countries except Crete, the Faroe Islands and Iceland (Tutin *et al.* 1976) and occurs in both cool temperate (oceanic to continental climates) and warm temperate zones (dry summer Mediterranean to dry summer steppe climates) (Klinkhamer and De Jong 1993). Its presence is generally linked to cultivation or land disturbance (Klinkhamer and De Jong 1993) and soil disturbance has been shown to trigger *C. vulgare* germination (Klinkhamer and De Jong 1988). This species has been introduced to several countries where it has also

Figure 1: Spear thistle, *Cirsium vulgare*. Photo: J-L Sagliocco, Vic Government.

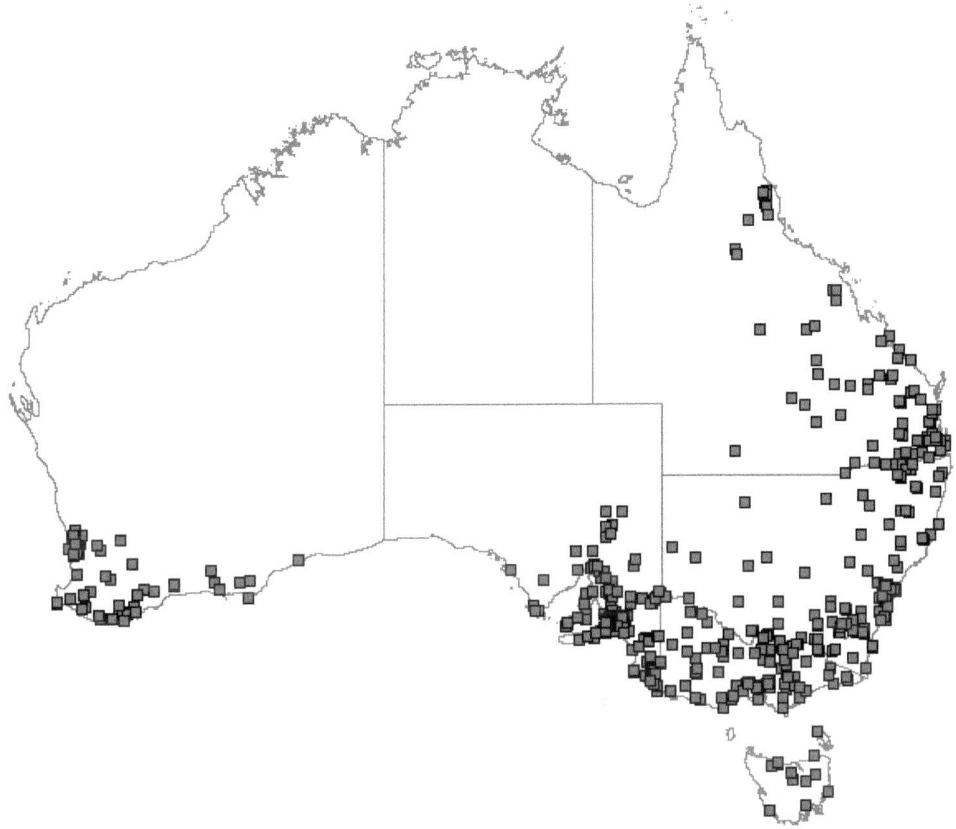

Figure 2: Distribution of *Cirsium vulgare* in Australia. Specimen data reproduced from Australia's Virtual Herbarium with permission of the Council of Heads of Australasian Herbaria Inc.

become invasive, in particular the USA and Canada, New Zealand and South Africa (Parsons and Cuthbertson 1992).

In Australia, *C. vulgare* occurs in all states except the NT (Fig. 2). It colonises habitats such as pastures, cereal cropping areas and, to a lesser extent, natural ecosystems. All parts of the plant are spiny and deter animals other than goats from grazing it. The cost of spear thistle to Victorian farmers alone was estimated in 1989 at over $58 million per year (Bruzzese and Heap 1989). The biological control program against spear thistle took place in the context of ecological studies conducted in Europe by CSIRO during the 1980s and 1990s to quantify the impact of natural enemies on thistles and to select biological control agents adapted to Australian situations.

BIOLOGICAL CONTROL HISTORY

The chronology of introduction of the three biological control agents against spear thistle in the different countries until 1994 is detailed in Julien and Griffiths (1998). Biological control of *C. vulgare* was first considered in Canada and the USA in the 1960s. European surveys identified the arthropod fauna associated with *C. vulgare*, among which the most promising agents were the receptacle weevil *Rhinocyllus conicus* (Frölich) and the gall-forming tephritid fly *Urophora stylata* (Fabricius) which could significantly reduce the number of seeds produced (Zwölfer 1965). An ecotype of *R. conicus* from spear thistle collected in eastern France was released in Canada in 1968. It established in British Columbia and Nova Scotia but its impact was unknown (Julien and Griffiths 1998). After the first release in 1969, different accessions of the weevil were introduced into the USA to control several species of the Eurasian thistles *Carduus*, *Cirsium* and *Silybum* (Pemberton 2000). Off-target damage on native thistles has been documented (Louda *et al.* 1997; Pemberton 2000) and *R. conicus sensu lato* has been reported developing successfully on 22 of the 90 native species of *Cirsium* in the USA (Pemberton 2000). In 1984 and 1986 South Africa released *R. conicus* collected from spear thistle in eastern France, but only the second introduction survived in the KwaZulu-Natal province. Later monitoring led to the recovery of weevil populations in

KwaZulu-Natal and Eastern Cape provinces with levels of damage to spear thistle capitula amounting to 41.5% and 12.6%, respectively (Hodson *et al.* 2003).

Several accessions of *Urophora stylata* imported from Germany, Switzerland, Austria and France were released in British Columbia and Quebec between 1973 and 1976 (Harris and Wilkinson 1981). Monitoring in British Columbia between 1973 and 1982 indicated that between 3% and 96% of capitula were galled depending on sites and years, with densities from two to over 30 larvae per capitulum (Harris and Wilkinson 1981). In the USA, *U. stylata* was released in Maryland and Washington states in 1983, followed by releases in Colorado, Oregon, Montana and California (Van Driesche *et al.* 2002). South Africa released *U. stylata* populations collected in western Germany and western France during 198389, but the insect survived at one site only (Zimmermann 1991).

The rosette weevil *Trichosirocalus horridus* (Panzer) from Italy (then referred to as *Ceuthorhynchidius horridus* (Panzer)) was introduced in the USA in 1974 to control Eurasian species *Carduus nutans* L. and *C. macrocephalus* Desf. (Julien and Griffiths 1998; Takahashi *et al.* 2009). Releases were conducted in 17 states and feeding on the native thistles *Cirsium discolor* (Muhl. ex Willd.) (McAvoy *et al.* 1987) and *C. altissimum* (Takahashi *et al.* 2009) was reported. In 1984, New Zealand released a population of *T. horridus* originating from Germany via Canada, where it had been introduced to control nodding thistle *Carduus nutans* (L.) ssp. *nutans*. The weevil was later found to have successfully established on spear thistle in NZ (Julien and Griffiths 1998). During the 1990s, a population of *T. horridus* from Italy was released in the USA (Wyoming) but its establishment was not confirmed (Julien and Griffiths 1998).

In Australia, *C. vulgare* was approved as a target for biological control in 1988. Due to farmers' concerns and surveys indicating that spear thistle was the most invasive of the thistles present in Vic with 9.7 million ha infested (Lane *et al.* 1980), the Victorian Government undertook surveys of its natural enemies in Vic (Bruzzese and Heap 1989) and investigated spear thistle biology (Bruzzese 1996). It was found that spear thistle was commonly infected by the rust fungus *Puccinia cnici* Mart. (Uredinales: Pucciniaceae), although only limited damage was caused. Despite its narrow host-range under laboratory conditions (Bruzzese *et al.* 1989), this fungus was not further investigated. Biological control efforts concentrated on the import and release of the agents *R. conicus* and *U. stylata* and the release of the rosette weevil *T. horridus*, previously imported to Australia to control *Carduus nutans* L.

CANDIDATES

Agents released in Australia

***Rhinocyllus conicus* (Frölich) (Coleoptera: Curculionidae).** The oligophagous receptacle weevil *R. conicus* originates in Eurasia and North Africa. Throughout its distribution range in western Europe, the insect has one or two generations per year, depending on the day length to which populations are exposed (Zwölfer and Harris 1984). Several insect ecotypes develop in flower heads of different thistle species: *Carduus*, *Cirsium*, *Silybum* and *Onopordum* spp. (Zwölfer and Harris 1984). In western France, the main and most common host plant of *R. conicus* is *C. vulgare* (Zwölfer 1983) while in the rest of its distribution range its main host is *Carduus nutans* (Zwölfer and Harris 1984). Females lay on average 200 eggs (Phillips 1982) and larvae feed in the receptacle on developing achenes (Fig. 3). During a detailed ecological study in southern France, it was observed that *R. conicus* larvae caused the destruction of 9–57% of the potential seed production of *C. nutans* (Sheppard *et al.* 1994; see chapter in this volume). However, no study was conducted in France on its damage to spear thistle.

Ecotypes of *R. conicus* collected from spear thistle in south-western France were imported to Vic between 1989 and 1995 and releases were conducted at infestations in the five major Victorian water catchments. By 1997, 17 releases had been conducted (Kwong and Bruzzese 1997). Due to termination of funding, no detailed survey to monitor the establishment or impact of this insect has

Figure 3: *Rhinocyllus conicus* damage and adult in spear thistle capitulum. Photo: J-L Sagliocco, Vic Government.

Figure 4: *Urophora stylata* galls in spear thistle capitulum. Photo: J-L Sagliocco, Vic Government.

been undertaken, thus little is known about its overall establishment, damage to spear thistle or dispersal. Observations on capitula containing larvae or diapausing adults collected in 2010 near Melbourne (Frankston) and in Gippsland suggest that isolated populations of this insect may still be present in some parts of Vic (T Morley, *pers. obs.*). The Gippsland observation was at a site located 22 km from the nearest release site, suggesting that natural dispersal had occurred.

Urophora stylata **Fabricius (Diptera: Tephritidae).** Adults *U. stylata* (spear thistle gall fly) live for about two months. Females lay eggs between bracts of unopened capitulum or between florets. Larvae hatch as second instar and feed on florets and developing achenes, causing the formation of multilocular galls (Redfern 1968) (Fig. 4). In Europe, *U. stylata* is reported to attack around 25% of spear thistle capitula (Zwölfer 1972). The importation of *U. stylata* from western France into Vic commenced in 1990 and, following approval to release from quarantine, the first release occurred in December 1993 at Yarram, Gippsland (Kwong and Bruzzese 1997). Between 1994 and 1997, 54 release sites were established in five Victorian water catchments and by 2001, 118 releases had been conducted (Kwong 2000). Monitoring in 2000–01 confirmed the establishment of fly populations at 12 of the 24 sites surveyed, representing a 50% establishment success (Swirepik and Smyth 2001). Monitoring in 2001–02 found *U. stylata* galls at 11 of the 29 sites surveyed (38% of the release sites) (Smyth 2002). Detailed impact assessment conducted at three of the 11 sites revealed that the percentage of capitula attacked ranged from 1% to 83%, depending on site and time of year, with the percentage of seeds destroyed ranging from 8% to 46% (Smyth 2002). A survey conducted in 2003 at 27 release sites found *U. stylata* at three sites only. The percentage of spear thistle capitula with galls was 58% (n=90, Gelantipy, Gippsland), 69% (n=300, Wilkin, south-west Vic) and 72% (n=300, Marysville, north-east Vic). At Wilkin, the seed reduction caused by *U. stylata* was 32%, from an average of 173 seeds per capitulum down to 119 (n= 40). At Marysville, it caused a 25% reduction in seed production, from an average of 145 seeds per capitulum down to 109 (n=200, T Morley, unpub. data). No systematic survey or impact study has been conducted since, but galls were recently collected at three other localities in north-eastern Vic, each at least 20 km from the nearest release site (T Morley and I Faithfull, *pers. obs.*).

Two releases of *U. stylata* were conducted in southern NSW in 2000 but, despite some early signs of establishment, no further monitoring has been conducted. It is suspected that very hot summers have caused high larval mortality (R Holtkamp, *pers. comm.*). *U. stylata* was released at two locations in Tas in 1997 but surveys in 2004 failed to recover the agent (Ireson *et al.* 2006). A new release was conducted in 2010 in the Tasmanian midlands (J Ireson, *pers. comm.* 2010). Parasitism has not been observed at any of the release sites in Australia.

Trichosirocalus horridus **(Panzer) (Coleoptera: Curculionidae).** Adults of the thistle rosette weevil *T. horridus* feed on leaves and larvae develop in veins of the rosette leaves and meristematic tissues at the centre of rosettes, causing necrosis and delayed flowering. An ecotype of the insect (previously referred to as *Ceuthorhynchidius horridus* (Panzer)), collected in Germany on *Carduus nutans*, was imported from NZ via Canada (Julien and Griffiths 1998) and was released for the control of this weed in eastern Australia (Julien and Griffiths 1998; Woodburn 1997). It was expected that, in the absence of its original host, the insect would establish on spear thistle and between 1997 and 1998, adult weevils from this importation were released on eight occasions on spear thistle at four geographically distinct localities in Vic. At one site near Strathdownie in south-western Vic, *T. horridus* was observed in 1999, two years after release (Kwong 2000), but no monitoring has been conducted since to confirm the survival of this population. Alonso-Zarazaga and Sánchez-Ruiz (2002) described *T. horridus* population introduced to Australia to control *C. nutans* under the name *T. mortadelo* Alonso-Zarazaga & Sánchez-Ruiz. However, the status of this new species is uncertain as recent molecular studies of Australian specimens strongly suggest it is a synonym of *T. horridus* (De Biase and Colonnelli, *pers. comm.* 2010).

DISCUSSION

Substantial efforts were made to distribute *R. conicus* and *U. stylata* throughout Vic during 1994-2000, with a lesser effort for *T. horridus*, of which eight releases were made. Recently, monitoring and impact surveys were conducted on *U. stylata* only. At the three sites where *U. stylata* was found in 2003 and where detailed monitoring was conducted, the percentage of capitula attacked varied from 58% to 72%. The Marysville site, where *U. stylata* had the greatest impact, was destroyed by bushfire in February 2009. Anecdotal evidence suggest that both *U. stylata* and *R. conicus* have survived and possibly colonised only locations with cool and wet climatic conditions, similar to the regions where they were originally collected. Surveys of the original release sites are required to assess the survival, dispersal, impact and overall success of *R. conicus* and *U. stylata*, together with assessment of survival of *U. stylata* in Tas and NSW. This is particularly desirable for *R. conicus*, which has been observed to destroy similar amounts of seeds in areas where it has been introduced for the biological control of *C. nutans*, as in its regions of origin (Sheppard *et al.* 1994). However, in South Africa, where *R. conicus* has been introduced to control spear thistle, the weevil did not disperse far from release sites and levels of damage were generally considered low (Hodson *et al.* 2003). Since spear thistle is the main host of *R. conicus* in western France, where this insect destroys a significant proportion of seeds (Zwölfer 1983), it would be advantageous to quantify its impact in Vic. In the case of *T. horridus*, the survival, impact and identity of the small populations released should also be revisited. Unfortunately, funding for thistle biological control ceased in the early 2000s, which made proper evaluation difficult.

ACKNOWLEDGEMENTS

We thank John Ireson and Royce Holtkamp for information on the status of *U. stylata* in Tas and NSW, and Drs Alessio De Biase and Enzo Colonnelli for information on the results of their molecular work on *Trichosirocalus* species.

REFERENCES

Alonso-Zarazaga MA and Sánchez-Ruiz M (2002) Revision of the *Trichosirocalus horridus* (Panzer) species complex, with description of two new species infesting thistles (Coleoptera: Curculionidae, Ceutorhynchinae). *Australian Journal of Entomology* **41**, 199–208.

Bruzzese E (1996) Ecology of *Cirsium vulgare* and *Silybum marianum* in relation to biological control. *Plant Protection Quarterly* **11**, Supplement 2, 245–249.

Bruzzese E and Heap J (1989) *Biological Control of Widespread Species of Thistles in Victoria*. Final Report to Wool Research and Development Fund, Vic Dept of Conservation, Forests and Lands, Melbourne.

Bruzzese E, Buxton RJ and Heap J (1989) Aspects of the biology of the spear thistle rust fungus in Victoria, Australia. In *Proceedings of the 7th International Symposium on the Biological Control of Weeds*. 6–11 March 1988, Rome. (Ed ES Delfosse) pp. 449–454. Ministero dell'Agricoltura e delle Foreste, Italy.

Harris P and Wilkinson ATS (1981) *Cirsium vulgare* (Savi) Ten., bull thistle (Compositae). In *Biological Control Programmes against Insects and Weeds in Canada 1969–1980*. (Eds JS Kelleher and MA Hulme) pp. 147–153. CAB, Slough, UK.

Hodson JL, Hoffmann JH and Zimmermann HG (2003) Biological control of spear thistle, *Cirsium vulgare* (Asteraceae), in South Africa: a modest start for *Rhinocyllus conicus* (Coleoptera: Curculionidae). *African Entomology* **11**, 15–20.

Ireson JE, Davies JT, Friend DA, Holloway RJ, Chatterton WS, Van Putten EI and McFadyen REC (2006) *Weeds of Pastures and Field Crops in Tasmania: Economic Impacts and Biological Control*. Technical Series no. 13. CRC for Australian Weed Management, Adelaide.

Julien MH and Griffiths MW (1998) *Biological Control of Weeds: A World Catalogue of Agents and their Target Weeds*. 4th edn. CABI, Wallingford, UK.

Klinkhamer PGL and De Jong TJ (1988) The importance of small-scale disturbance for seedling establishment in *Cirsium vulgare* and *Cynoglossum officinale*. *Journal of Ecology* **76**, 383–392.

Klinkhamer PGL and De Jong TJ (1993) *Cirsium vulgare* (Savi) Ten. *Journal of Ecology* **81**, 177–191.

Kwong R (2000) Biological control of thistles 1998–2000. Vic Dept of Natural Resources and Environment, Melbourne. Unpublished report.

Kwong R and Bruzzese E (1997) *Implementation of Biological Control of Thistles DVC 20*. Final Report to the Wool Research and Development Corporation. Vic Dept of Conservation and Natural Resources, Melbourne.

Lane D, Riches K and Combellack H (1980) A survey of the distribution of the noxious weeds in Victoria. Vic

Dept of Crown Lands and Survey, Melbourne. Unpublished report.

Louda SM, Kendall D, Connor J and Simberloff D (1997) Ecological effects of an insect introduced for the biological control of weeds. *Science* **277**, 1088–1090.

McAvoy TJ, Kok LT and Mays WT (1987) Dispersal of *Trichosirocalus horridus* (Panzer) (Coleoptera: Curculionidae) in southwest Virginia. *Journal of Entomological Science* **22**, 324–329.

Michaux B (1989) Reproductive and vegetative biology of *Cirsium vulgare* (Savi) Ten. (Compositae: Cynareae). *New Zealand Journal of Botany* **27**, 401–414.

Parsons WT and Cuthbertson EG (1992) *Noxious Weeds of Australia*. Inkata Press, Melbourne.

Pemberton RW (2000) Predictable risk to native plants in weed biological control. *Oecologia* **125**, 489–494.

Phillips D (1982) Weevils outperform chemicals in nodding thistle control. *New Zealand Journal of Agriculture* **145**, 16–17.

Redfern M (1968) The natural history of spear thistle-heads. *Field Studies* **2**, 669–717.

Sheppard AW, Cullen JM and Aeschlimann J-P (1994) Predispersal seed predation on *Carduus nutans* (Asteraceae) in southern Europe. *Acta Oecologica* **15**, 529–541.

Smyth M (2002) *Effective Biological Control of Paterson's Curse and Thistles*. Annual Report for Australian Wool Innovation and Meat and Livestock Australia.

Swirepik A and Smyth M (2001) *Effective Biological Control of Paterson's Curse and Thistles*. Annual Report for Australian Wool Innovation and Meat and Livestock Australia.

Takahashi M, Louda SM, Miller TEX and O'Brien CW (2009) Occurrence of *Trichosirocalus horridus* (Coleoptera: Curculionidae) on native *Cirsium altissimum* versus exotic *C. vulgare* in North American tallgrass prairie. *Environmental Entomology* **38**, 731–740.

Tutin TG, Heywood VH, Burges NA, Moore DM, Valentine DH, Walters SM and Webb DA (1976) *Flora Europaea*. Vol. 4. Cambridge University Press, Cambridge.

Van Driesche R, Lyon S, Blossey B, Hoddle M and Reardon R (Eds) (2002) *Biological Control of Invasive Plants in the Eastern United States*. USDA Forest Service.

Woodburn TL (1997) Establishment in Australia of *Trichosirocalus horridus*, a biological control agent for *Carduus nutans*, and preliminary assessment of its impact on plant growth and reproductive potential. *Biocontrol Science and Technology* **7**, 645–656.

Zimmermann HG (1991) Biological control of spear thistle, *Cirsium vulgare* (Asteraceae), in South Africa. *Agriculture, Ecosystems & Environment* **37**, 199–205.

Zwölfer H (1965) A preliminary list of phytophagous insects attacking wild Cynareae species in Europe. *Commonwealth Institute of Biological Control Technical Bulletin* **6**, 81–154.

Zwölfer H (1972) *Investigations on* Urophora stylata *Fabr., a Possible Agent for the Biological Control of* Cirsium vulgare *in Canada*. CIBC Progress Report no. 29.

Zwölfer H (1983) Host selection and oviposition behaviour in west-European ecotypes of *Rhinocyllus conicus* Froel. (Col.: Curculionidae). *Zeitschrift für Angewandte Entomologie* **95**, 113–122.

Zwölfer H and Harris P (1984) Biology and host specificity of *Rhinocyllus conicus* (Froel.) (Col., Curculionidae), a successful agent for the biocontrol of the thistle (*Carduus nutans*). *Zeitschrift für Angewandte Entomologie* **97**, 36–62.

Cryptostegia grandifolia (Roxb.) R. Br. – rubber vine

Bill Palmer and Wayne Vogler

ABSTRACT

Cryptostegia grandiflora, rubber vine, was introduced into Australia before 1875 and eventually became a major weed in northern Qld and a Weed of National Significance. It grows as a shrub or a scrambling vine capable of smothering tall trees and can form monocultures in riparian situations. Efforts to find biological control agents in the native range consisted of surveys and preliminary host-specificity testing performed from 1985–88 at a field station in Toliare, Madagascar. Although only a limited arthropod fauna was found, several species were selected for study. Two agents, the leaf-feeding moth *Euclasta whalleyi* and the autoecious rust fungus *Maravalia cryptostegiae*, were subsequently released in Australia. *E. whalleyi* established throughout the range of rubber vine, though it was not detected until nine years after release. On occasion it can be abundant on rubber vine, causing significant damage. *M. cryptostegiae* established very rapidly throughout most of the weed's range and is a very effective agent providing a high degree of control. An evaluation of the project estimated a benefit:cost ratio of $109 at a discount rate of 8%. Based on benefit:cost ratio, the rubber vine project was ranked third (behind those for prickly pear and skeleton weed) of all Australian weed biological control projects. The control of rubber vine is therefore one of Australia's best successes.

Key words: *Euclasta whalleyi*, *Maravalia cryptostegiae*, Madagascar, vine, Weed of National Significance.

INTRODUCTION

Rubber vine, *Cryptostegia grandiflora* (Roxb.) R. Brown (Apocynaceae), is a native of Madagascar that has become a serious problem in northern Australia to the extent that it is recognised as a Weed of National Significance (Thorp and Lynch 2000). It was introduced before 1875 for use as an ornamental and was considered especially useful for planting around mines and towns in northern Qld because of its hardiness (Mackey 1996). During World War II rubber vine was considered a potential source of latex after the fall of Singapore, and experimental crop plantings were made near Charters Towers (Howard 1964; Grice 1996).

Rubber vine (Fig. 1) is a woody perennial that grows as a shrub or as a sprawling vine capable of climbing and smothering tall trees. It will grow on most soil types from deep sands to alluvial floodplains and skeletal soils. It appears tolerant of salinity as it grows where sea water inundates, and adjacent to salt flats in coastal areas. It is commonly found in riparian areas, open grasslands and woodlands of northern Qld. At one time it densely occupied more than 700 000 ha in tropical Qld between the 400 mm and 1400 mm rainfall isohyets (Mackey 1996). Its distribution and favourable climatic range are given in Figure 2.

Rubber vine negatively affects the extensive grazing industry of northern Qld by reducing pasture yield and interfering with livestock management. It can cause stock losses due to poisoning and reduces land values (McGavin

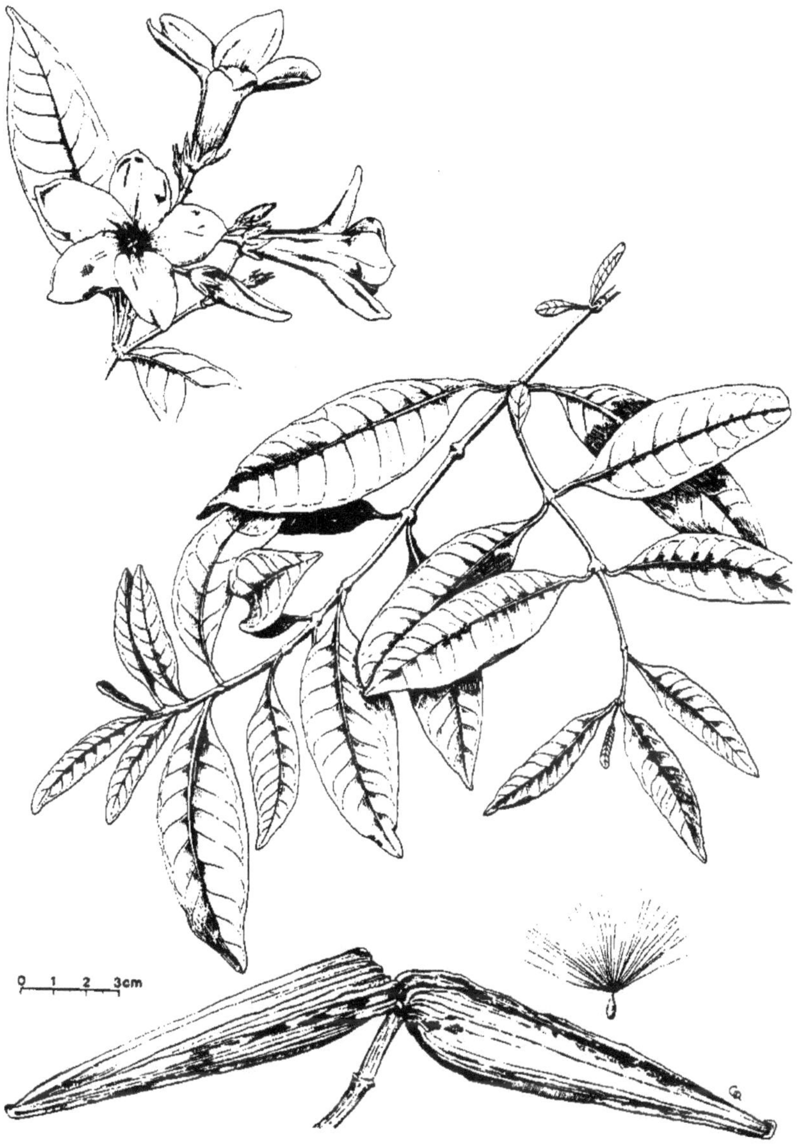

Figure 1: Line drawings of *Cryptostegia grandiflora* showing the inflorescence, seed pods and leaves. Qld Government.

1969; Chippendale 1991). The economic loss to the grazing industry caused by rubber vine was estimated at $18 million annually in 1995 (Mackey 1996). Environmental impacts, although difficult to quantify, are significant (Fig. 3). They include loss of wildlife habitat, reduced flora and fauna diversity and loss of amenity (Tomley 1995).

Rubber vine is a serious weed only in Australia, although there are reports of a *Cryptostegia* sp. becoming invasive and ecologically problematic in some of the semi-arid islands of the Caribbean. *C. grandiflora* has been confirmed as a weed invading national parks in Curacão (Tomley and Evans 2004).

BIOLOGICAL CONTROL HISTORY

Interest in the biological control of rubber vine began in the 1970s (McFadyen and Harvey 1990) as a consequence of the weed's spread and the recognition that other methods of control were largely uneconomic (Vitelli 1995). The initiative was driven by the then Rural Lands Protection Board and supported by the then Qld Department of Lands.

PLANT TAXONOMY

Asclepiadaceae has been subsumed within the family Apocynaceae Jussieu as subfamilies Periplocoideae R.

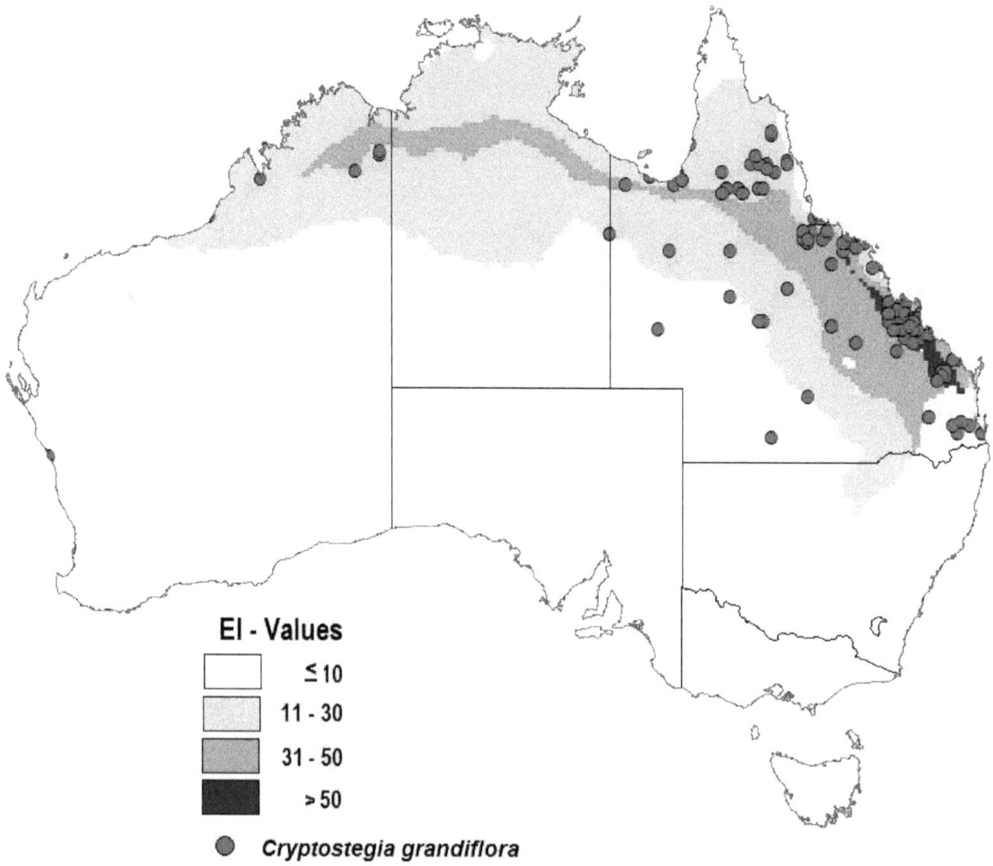

Figure 2: The distribution of *Cryptostegia grandiflora* in Australia, based on records from the Australian Virtual Herbarium, and its potential distribution predicted (after Kriticos *et al.* 2003) using the climate modelling software CLIMEX (Sutherst *et al.* 2007). EI values >30 are considered climatically favourable for growth while values ≤10 are considered unfavourable.

Brown ex Endl., Secamonoideae Endl. and Asclepiadoideae R. Brown ex Burnett (Endress and Bruyns 2000; Stevens 2001). Subfamily Periplocoideae, to which *Cryptostegia* belongs, remains essentially unchanged from previous treatments. Other genera in Periplocoideae in Australia include *Gymnanthera* R. Brown (with two native species), *Cryptolepis* R. Brown (one species) and *Finlaysonia* Wall. (one species) (Forster 1996). *Cryptostegia* is represented in Australia by two exotic species from Madagascar, *C. grandiflora* (rubber vine) and *C. madagascariensis* (Bojer) ex Decne. Two varieties of *C. madagascariensis* are sparingly naturalised in the NT but not considered weedy (Marohasy and Forster 1991). No plant species in the Periplocoideae is of economic significance in Australia.

From a biocontrol perspective, there were few closely related native or economic plant species and this fact improves the possibilities of finding acceptable agents. The native species most related to rubber vine were considered to be *Gymnanthera oblongata* (Burm. F.) P. S. Green (=*G. nitida* R. Br.), *G. cunninghamii* (Benth.) P. I. Forster (=*G. fruticosa* Wilson), *Cryptolepis grayi* P. I. Forster and *Finlaysonia obovata* Wall.

EXPLORATION

Exploration for potential biocontrol agents necessitated surveys in western Madagascar, the native range of rubber vine. Because of the very high rate of endemism associated with this island, Madagascar has not always permitted study of its flora and fauna by overseas scientists nor allowed the export of its live material. The initiation of exploration for a biocontrol project was therefore dependent on a relaxation of Madagascar's policy on these issues, which occurred in 1985 (McFadyen and Harvey 1990).

In 1985, RE McFadyen undertook preliminary investigations of rubber vine's natural enemies in Madagascar. She found good prospective agents and recommended to the Qld Department of Lands that a field station be established in Madagascar.

Figure 3: A severe infestation of *Cryptostegia grandiflora* showing both shrubs and vines growing over trees. Photo: Qld Government.

The following year, J Marohasy was sent to establish a field station at Toliara in south-western Madagascar. From this field station she surveyed the arthropod fauna and pathogens associated with *C. grandiflora* and *C. madagascariensis* during 1987 and 1988. She was joined for short intervals by HC Evans of CABI (UK), who concentrated on the pathogens. Details of the field station, working conditions and various adventures were described by Walton (2005).

The area of search was essentially the western side of Madagascar, where *C. grandiflora* occurs in the southern half and *C. madagascariensis* grows in the north (Marohasy and Forster 1991). These areas were approximately 1000 km north–south. Neither *Cryptostegia* spp. occurs on the eastern coast or in the plateau region, and both are limited to western regions with a rainfall below 1800 mm and an altitude below 600 m (Marohasy and Forster 1991).

CANDIDATES

***Euclasta whalleyi* Popescu-Gorj & Constantinescu (=*E. gigantalis* Viette) (Lepidoptera: Pyralidae).** This leaf-feeding moth was found along the entire western side of Madagascar, essentially the combined area of both its hosts, *C. grandiflora* in the south and *C. madagascariensis* in the north. After preliminary testing in Madagascar, it was imported into quarantine at Biosecurity Qld in Brisbane for host-specificity testing. There it was tested using a host test list of 52 plant species from 19 plant families selected by the centrifugal phylogenetic method. These studies indicated that *E. whalleyi* had a host-range restricted to species in the subfamily Periplocoideae, as some development was observed on the Australian natives *Gymnanthera oblonga* and *G. cunninghamii* (McFadyen and Marohasy 1990b). Larval development rates were 73% lower, mortality 300% greater and pupal weights 12% lighter for larvae reared on *G. oblongata* than for those reared on *C. grandiflora* (McFadyen *et al.* 2002). Despite indications of possible attack on the *Gymnanthera* spp. the insect was approved for release on the basis that rubber vine itself posed more risk to ecological values. Ten years after approval to release the insect in Australia was obtained and after it was widely established, field surveys indicated that there was no attack on *G. oblongata*, except when plants were growing in close proximity to infested rubber vine plants (McFadyen *et al.* 2002), thus vindicating the decision to release.

Between 1988 and 1991, over 115 000 adults and larvae of *E. whalleyi* were reared at the Tropical Weeds Research Centre, Charters Towers and released across northern Qld (Mo *et al.* 2000). For several years there was little indication of establishment, despite close monitoring and the use of field cages for releases. Establishment of the insect was not confirmed until 1995, when 'outbreak' populations were found around Charters Towers, nine years after the releases (Mo *et al.* 2000). Over the next few months, numbers built up spectacularly, with patches of rubber vine being completely defoliated and large numbers of moths seen around homes. In fact, the moths became a topic of conversation around the town, with people complaining about their numbers and general nuisance value. Surveys of release sites across Qld were made in 1995 and 1996; these confirmed that *E. whalleyi* was widely established throughout all major infestations of rubber vine (Mo *et al.* 2000). *E. whalleyi* is a good example of the delayed appearance of establishment of some introduced biocontrol agents.

***Maravalia cryptostegiae* (Cummins).** The primitive autoecious tropical forest rust fungus with an unusual life-cycle of two teliospore stages (Evans 1993), *Maravalia cryptostegiae* (Cummins) Ono was found on both *C. grandiflora* and *C. madagascariensis* in Madagascar. The life-cycle (Evans 1993) and infection mechanism (Evans and Fleureau 1993) were elucidated before host-specificity testing was undertaken. Seventy-one plant species selected using a centrifugal phylogenetic method were used for these tests (Evans and Tomley 1994). Results showed that *M. cryptostegiae* had a very narrow host-range but that it could produce fertile sori on *Cryptolepis grayi*, a newly discovered (Forster 1990) endemic species from northern Qld. Because *C. grayi* was geographically isolated from rubber vine in northern Qld and grew in an environment unsuitable for the rust (Evans and Tomley 1994), *M. cryptostegiae* was approved for release in Australia.

A rust accession originally collected from *C. madagascariensis*, which sporulated extensively and was shown in host-specificity tests to infect all species and varieties within the genus *Cryptostegia*, was released first (Evans and Tomley 1994). However, this rust accession did not cause severe disease symptoms on rubber vine in the field (Evans and Tomley 1996). Consequently, another accession was collected in 1994 specifically from rubber vine in Madagascar. This accession was screened in quarantine in the UK to assess pathogenicity and remove any associated mycoparasites, then shipped to Australia for release (Tomley and Hardwick 1996; Tomley and Evans 2004). Between 1995 and 1997, the rust accession (ex. rubber vine) was released during the summer or wet seasons at 69 sites spanning most of the infested range of the weed (Tomley and Evans 2004). Establishment was rapid and confirmed by 1996. Twenty months after release most of the rubber vine infestations of the Gulf and Peninsula regions of northern Qld were infected and evidence of significant damage was very noticeable. Significant infections were also found in other Qld regions.

***Nephele densoi* Keferstein (Lepidoptera: Sphingidae).** This hawk moth was found on *C. grandiflora* in Madagascar (Huwer and McFadyen 1999). After preliminary testing in Madagascar, it was imported into quarantine facilities at Biosecurity Qld in Brisbane for further testing. The insect was tested against 54 plant species selected from 18 families. It was shown that this insect had a host-range very similar to that of *E. whalleyi*, although it could develop better on *G. oblonga* than did *E. whalleyii* (Huwer and McFadyen 1999). Because *E. whalleyi* and *M. cryptostegiae* had been released and there were possibilities of attack on *G. oblonga*, application to release this insect was not made. If it is felt that new agents are needed, this decision might be worth revisiting.

***Steatococcus* n. sp. (Hemiptera: Margarodidae).** This soft scale was investigated after it was collected from stems of *C. grandiflora* in Madagascar and less frequently from four other genera, some within other subfamilies of Apocynaceae (McFadyen and Marohasy 1990a). Studies on its host-range performed in quarantine in Australia showed that colonies of the insect could survive up to six months on several genera in the subfamilies Rauvolfioideae and Apocynoideae, and indefinitely on several genera of Asclepiadoideae and Periplocoideae. It was not considered further for biological control. A diaspid scale *Hulaspis* n. sp. (Hemiptera: Diaspididae) was also rejected after it was later collected from *Byttneria* sp. (family Sterculiaceae), a plant unrelated to Apocynaceae (RE McFadyen, unpub.).

***Schizomyia cryptostegiae* Gagné (Diptera: Cecidomyiidae).** This new species of gall midge was also discovered on rubber vine in Madagascar (Gagné and Marohasy 1997). This insect galled flowers, causing irregular green to reddish succulent swellings on sepals, scales and bracts that ultimately caused infested flower buds to drop off. However, the galls were also collected from other genera in the subfamily Periplocoideae and from *Leptadenia madagascariensis* Decne (subfamily Asclepiadoideae) where that species was growing close to rubber vine. This insect was not investigated further.

EVALUATION

Evans (1993) reported that *M. cryptostegiae* infection caused defoliation in Madagascar and that this stress caused further damage by way of stem dieback. Tomley and Evans (2004) reported that the impact of *M. cryptostegiae* over the four years (1997–2001) following its release in Australia varied widely from site to site. Stem dieback and large-scale plant mortality occurred where infection was severe during the wet season, followed by high stress of infected plants induced by low water availability during the dry season. In contrast, rubber vine continued to grow and reproduce in areas where rust infection was low or where plants (even those severely infected by the rust during the wet season) were exposed to limited stress due to high water availability during the dry season.

In controlled studies on mature rubber vine (grown under natural temperature and humidity conditions) naturally infected with rust during 1998 at the Tropical Weeds Research Centre, severe rust infection did not affect stem basal diameter. It did, however, reduce plant height by almost 10%, leaf number by more than 50%, flower number by 47% and pod number by 80% (MP Vitelli and WD Vogler, unpub. data). A similar study on juvenile plants showed that severe infection reduced basal stem diameter and plant height by 34% and 37% respectively, considerably slowing plant growth rate and increasing the time to maturity (MP Vitelli and WD Vogler, unpub. data).

Detailed long-term field evaluation of *M. cryptostegiae* was conducted at six sites across the Australian range of rubber vine from 1997 to 2007, including a site near Hughenden where the climatic conditions were expected to be suboptimal for *M. cryptostegiae* development. Impact of the rust on plant population density, reproduction and soil seed banks was measured at least once per year. By 2001, density at sites other than Hughenden had declined by more than 40%, while seedling recruitment and pod production were negligible (Vogler and Lindsay 2002). This is in sharp contrast to pre-rust seedling densities of 53 675 ha^{-1} (IJ Dale, unpub.) and 2030 ha^{-1}, recorded soon after rust release in 1998 (Radford 2003), and pre-rust seed production of 190 000 ha^{-1} (Grice 1996) and 25 480 ha^{-1}, recorded soon after rust release in 1998 (Radford 2003). These data support the two-year observed decline in plant populations due to rapid depletion of soil seed banks (Bebawi *et al.* 2003), resulting in little seed to regenerate rubber vine populations.

By 2007 the density of rubber vine populations had declined by 50–98% at all sites except at Hughenden, where density had declined by 20%, largely due to a bushfire. Although *M. cryptostegiae* was present at Hughenden, it was at such low levels that damage was negligible. Seedling recruitment continued to be negligible except at Hughenden, where large numbers of seedlings were still appearing annually (WD Vogler, unpub. data). In northern Qld it appears that, where rust activity is relatively high, rubber vine populations are remaining relatively stable or are declining slowly and that, where rust activity is negligible, rubber vine populations are increasing slowly due to high levels of seed production and seedling recruitment.

E. whalleyi was also monitored at each field site; there was almost no damage due to *E. whalleyi* (WD Vogler, unpub. data). *E. whalleyi* is now considered a minor agent.

An economic evaluation of the project (Page and Lacey 2006) estimated that $3.6 million was spent between 1984 and 2004, returning benefits with a net present value of $232 million and a benefit:cost ratio of 108.8 at a discount rate of 8%. This analysis placed the rubber vine project third (behind those for prickly pear and skeleton weed) of all Australian weed biological control projects, when ranked by benefit:cost ratio.

DISCUSSION

Not only has biocontrol against rubber vine had substantial direct effects upon this weed, but leaf-fall induced by infection with *M. cryptostegiae* has enhanced the use of fire as a control technique by allowing grass to grow beneath plants (Bebawi and Campbell 2002). The higher grass biomass under a stand of defoliated rubber vine increases the available fuel for prescribed burning activities, thereby increasing fire intensity and efficacy as a control technique. The additional stress imposed on rubber vine by *M. cryptostegiae* infection also appears to improve the efficacy of fire as a control technique. However, *M. cryptostegiae* infection has a negative impact on herbicide control techniques by reducing herbicide efficacy (Vitelli and Madigan 1999). This has required some revision of herbicide recommendations although the overall negative impact has been minor.

Three interesting issues, from a biocontrol perspective, arose during this successful project and justify some additional comment. The subtleties of host plant–pathogen relationships were demonstrated with *M. cryptostegiae*. The rust accession originally released was first collected from *C. madagascariensis* in Madagascar and was cultured successfully for several generations on *C. grandiflora* while host-specificity testing was conducted. There was no indication that *C. grandiflora* was a suboptimal host. However, it soon became apparent that this accession of the rust was not causing severe disease symptoms on rubber vine after its release in the field. In contrast, the rust accession collected from *C. grandiflora* that was subsequently released, quickly established and caused damaging epidemics. The biocontrol agent was best collected from the actual target rather than from a closely related congener.

Very large numbers of *E. whalleyi* were released in northern Qld before 1991 and monitoring in subsequent years indicated that the insect had not established. Seven years after the initial releases however, high populations of *E. whalleyi* were found around Charters Tower and major defoliations occurred. A similar situation occurred with the parthenium beetle *Zygogramma bicolorata* Pallister (Dhileepan *et al.* 2000). These experiences confirm the principle that released agents can take considerable time before they become abundant in their new environment.

The case presented to the authorities to gain approval for the release of *E. whalleyi* in Australia was challenging, as this insect completed its life-cycle on the native *Gymnanthera* spp. Potential attack on an Australian native usually disqualifies an agent from further consideration, but in this case it was successfully argued that the weed itself or other methods for its control would endanger native species to a greater extent than would the agent. This argument might not prevail today, after passage of the *Environmental Protection and Biodiversity Conservation Act 1999*, because this Act is more prescriptive about damage to native flora.

ACKNOWLEDGEMENTS

This project could not have been undertaken without kind permission from the Government of Madagascar to conduct faunal surveys and ship potential agents to Australia and the UK. Very valuable help with the manuscript was provided by colleagues at Biosecurity Qld in Brisbane.

REFERENCES

Bebawi FF and Campbell SD (2002) Impact of early and late dry-season fires on plant mortality and seed banks within riparian and subriparian infestations of rubber vine (*Cryptostegia grandiflora*). *Australian Journal of Experimental Agriculture* **42**, 43–48.

Bebawi FF, Campbell SD and Lindsay AM (2003) Effects of burial and age on viability of rubber vine (*Cryptostegia grandiflora*) seeds. *Plant Protection Quarterly* **18**, 147–151.

Chippendale JF (1991) Potential returns to research on rubber vine (*Cryptostegia grandiflora* R.Br.) in north Queensland. MSc (Agric) thesis. University of Queensland, Brisbane.

Dhileepan K, Setter SD and McFadyen RE (2000) Impact of defoliation by the biocontrol agent *Zygogramma bicolorata* on the weed *Parthenium hysterophorus* in Australia. *Biocontrol* **45**, 501–512.

Endress ME and Bruyns PV (2000) A revised classification of the Apocynaceae s.l. *Botanical Review* **66**, 1–56.

Evans HC (1993) Studies on the rust *Maravalia cryptostegiae*, a potential biological control agent of rubber-vine weed (*Cryptostegia grandiflora*, Asclepiadaceae: Periplocoideae) in Australia. I: Life-cycle. *Mycopathologia* **124**, 163–174.

Evans HC and Fleureau L (1993) Studies on the rust, *Maravalia cryptostegiae*, a potential biological control agent of rubber-vine weed, *Cryptostegia grandiflora* (Asclepiadaceae: Periplocoideae), in Australia. II: Infection. *Mycopathologia* **124**, 175–184.

Evans HC and Tomley AJ (1994) Studies on the rust, *Maravalia cryptostegiae*, a potential biological control agent of rubber-vine weed, *Cryptostegia grandiflora* (Asclepiadaceae: Periplocoideae), in Australia. III: Host range. *Mycopathologia* **126**, 93–108.

Evans HC and Tomley AJ (1996) Greenhouse and field evaluations of the rubber vine rust, *Maravalia cryptostegiae*, on Madagascan and Australian Asclepiadaceae. In *Proceedings of the 9th International Symposium on Biological Control of Weeds.* (Eds VC Moran and JH Hoffman) pp. 165–169. University of Cape Town, South Africa.

Forster PI (1990) Notes on Asclepiadaceae, 2. *Austrobaileya* **3**, 273–289.

Forster PI (1996) Asclepiadaceae. *Flora of Australia* **28**, 197–307.

Gagné RJ and Marohasy J (1997) A new species of *Schizomyia kieffer* (Diptera: Cecidomyiidae) on rubber vine and other Asclepiadaceae in Madagascar. *Australian Journal of Entomology* **36**, 15–18.

Grice AC (1996) Seed production, dispersal and germination in *Cryptostegia grandiflora* and *Ziziphus mauritiana*, two invasive shrubs in tropical woodlands of northern Australia. *Australian Journal of Ecology* **21**, 324–331.

Howard RA (1964) The role of botanists during World War II in the Pacific Theatre. *Botanical Review* **60**, 197–257.

Huwer RK and McFadyen RE (1999) Biology and host range of the hawk moth *Nephele densoi* Keferstaein (Lep.: Sphingidae), a potential agent for the biological control of rubbervine *Cryptostegia grandiflora* R. Br. (Asclepiadaceae). *Biocontrol Science and Technology* **9**, 79–88.

Kriticos DJ, Sutherst RW, Brown JR, Adkins SW and Maywald GF (2003) Climate change and biotic invasions: a case history of a tropical woody vine. *Biological Invasions* **5**, 147–165.

Mackey AP (Ed) (1996) *Rubber Vine (*Cryptostegia grandiflora*) in Queensland.* Pest status review series. Land Protection Branch, Qld Dept of Natural Resources, Brisbane.

Marohasy J and Forster PI (1991) A taxonomic revision of *Cryptostegia* R. Br. (Asclepiadaceae: Periplocoideae). *Australian Systematic Botany* **4**, 571–577.

McFadyen RE and Harvey GJ (1990) Distribution and control of rubber vine, *Cryptostegia grandiflora*, a major weed in northern Australia. *Plant Protection Quarterly* **5**, 152–155.

McFadyen RE and Marohasy JJ (1990a) Biology and host plant range of the soft scale, *Steatococcus* new species (Hem,: Margarodidae) for the biological control of the weed *Cryptostegia grandiflora* (Asclepiadaceae). *Entomophaga* **35**, 437–439.

McFadyen RE and Marohasy JJ (1990b) A leaf feeding moth, *Euclasta whalleyi* (Lep.: Pyralidae) for the biological control of *Cryptostegia grandiflora* (Asclepiadaceae) in Queensland, Australia. *Entomophaga* **35**, 431–435.

McFadyen REC, Vitelli M and Setter C (2002) Host specificity of the rubber vine moth, *Euclasta whalleyi* Popescu-Gorj and Constantinescu (Lepidoptera: Crambidae: Pyraustinae): field host-range compared to that predicted by laboratory tests. *Australian Journal of Entomology* **41**, 321–323.

McGavin MD (1969) Rubber vine (*Cryptostegia grandiflora*) toxicity for ruminants. *Queensland Journal of Agriculture Animal Science* **26**, 9–19.

Mo J, Treviño M and Palmer WA (2000) Establishment and distribution of the rubber vine moth, *Euclasta whalleyi* Popescu-Gorj & Constantinescu (Lepidoptera: Pyralidae) following its release in Australia. *Australian Journal of Entomology* **39**, 344–350.

Page AR and Lacey KL (2006) *Economic Impact Assessment of Australian Weed Biological Control*. CRC for Australian Weed Management, Adelaide.

Radford IJ (2003) Demography, fecundity and natural enemy dynamics of rubber vine (*Cryptostegia grandiflora*) in riparian and upland sites of north Queensland. *Plant Protection Quarterly* **18**, 138–142.

Stevens PF (2001) Angiosperm phylogeny website. Version 7, May 2006 [fairly continuously updated]. Missouri Botanical Gardens. <http://www.mobot.org/MOBOT/research/APweb>.

Sutherst RW, Maywald GF and Kriticos D (2007) *CLIMEX Version 3 User's Guide*. Hearne Scientific Software, Melbourne.

Thorp JR and Lynch R (2000) *The Determination of Weeds of National Significance*. National Weeds Strategy Executive Committee, Launceston.

Tomley AJ (1995) The biology of Australian weeds 26. *Cryptostegia grandiflora* R. Br. *Plant Protection Quarterly* **10**, 122–130.

Tomley AJ and Evans HC (2004) Establishment of, and preliminary impact studies on, the rust, *Maravalia cryptostegiae*, of the invasive alien weed, *Cryptostegia grandiflora* in Queensland, Australia. *Plant Pathology* **53**, 475–484.

Tomley AJ and Hardwick G (1996) Bulking up, field distribution and establishment of rubber vine rust, *Maravalia cryptostegiae* in far north Queensland. In *Proceedings of the 11th Australian Weeds Conference*. 30 September–3 October 1996, University of Melbourne. (Ed RCH Shepherd) pp. 237–238. Weed Science Society of Victoria, Melbourne.

Vitelli JS (1995) Rubber vine. In *Exotic Weeds and their Control in North West Queensland*. (Ed N March) pp. 14–17. Qld Department of Lands, Mt Isa.

Vitelli J and Madigan B (1999) Effect of rust on the efficacy of foliar applied herbicides on rubber vine. In *Proceedings of the 12th Australian Weeds Conference*. 12–16 September 1999, Hobart. (Eds AC Bishop, M Boersma and CD Barnes) pp. 238–241. Tasmanian Weed Society, Hobart.

Vogler WD and Lindsay A (2002) The impact of the rust fungus *Maravalia cryptostegiae* on three rubber vine (*Cryptostegia grandiflora*) populations in tropical Queensland. In *Proceedings of the 13th Australian Weeds Conference*. 8–13 September 2002, Perth. (Eds H Spafford Jacob, J Dodd and JH Moore) pp. 180–181. Plant Protection Society of Western Australia, Perth.

Walton C (2005) *Reclaiming Lost Provinces: A Century of Weed Biological Control in Queensland*. Qld Dept of Natural Resources and Mines, Brisbane.

Cylindropuntia imbricata (Haw.) F.M. Knuth – rope pear *Cylindropuntia rosea* (DC.) Backeb.– Hudson pear

Royce H Holtkamp

ABSTRACT

Cylindropuntia spp. are plants of Mexican and southern USA origin which are commonly referred to as cholla in the USA. They have become weeds of arid and semi-arid regions of mainland Australia, with eight species currently recognised as naturalised. A biological control program against rope pear, *Cylindropuntia imbricata*, commenced in 1925 with the release of the cochineal insect *Dactylopius tomentosus*. This agent has been found to damage three additional species of *Cylindropuntia* in Australia. A biological control program is being conducted for control of Hudson pear, *Cylindropuntia rosea*.

Key words: *Cylindropuntia fulgida* var. *mamillata*, *Cylindropuntia kleiniae*, *Cylindropuntia leptocaulis*, *Cylindropuntia prolifera*, *Cylindropuntia spinosior*, *Cylindropuntia tunicata*, cochineal.

INTRODUCTION

Cylindropuntia spp. (Cactaceae) are present in every state of mainland Australia, with eight species recognised as naturalised. They are all shrubs which vary in height from 60 cm to 3 m. Stems are succulent, consisting of a number of cylindrical segments. These segments grow mostly end to end and possess areoles which contain glochids (very small, detachable barbed bristles) adjacent to longer spines. Glochids are found only in this genus and genera that were previously included in *Opuntia*. Long spines possess a papery detachable sheath. Leaves occur at the base of areoles and are usually not seen on mature segments as they are shed early. Flowers and fruit form within or near areoles on upper stem segments. Flowers vary through shades of yellow, pink and purple in Australian species, although *Cylindropuntia fulgida* var. *mamillata* (Schott ex Engelm.) Backeb. is not known to flower or fruit in Australia. Fruits contain pulp surrounding the seeds, which are of questionable viability in some species. Fruits of *C. rosea* (DC.) Backeb. have been regarded as not producing seed at all, but studies in late 2009 around the Lightning Ridge area (RJ Chinnock, *pers. comm.* 2009) discovered fruits containing seeds which have yet to be tested for viability. All species of *Cylindropuntia* spread via movement of detached segments which are capable of forming roots when they come into contact with the ground.

There are no Australian native members of the Cactaceae (Telford 1984). Members of the genus *Cylindropuntia* present in Australia are *C. fulgida* var. *mamillata*, *C. imbricata* (Haw.) FM Knuth, *C. kleiniae* (DC.) FM Knuth, *C. leptocaulis* (DC.) FM Knuth, *C. prolifera* (Engelm.) FM Knuth, *C. rosea*, *C. spinosior* (Engelm.) FM Knuth and *C. tunicata* (Lehm.) FM Knuth. These species were described in detail by Anderson (2001) and images of the species can be found on the PlantNET website (http://plantnet.rbgsyd.nsw.gov.au/).

Figure 1: *Cylindropuntia imbricata* in flower and fruit. Photo: J Hosking, NSW Government.

Cylindropuntia spp. occur naturally in the southern USA, Mexico, Chile, Ecuador and the Caribbean island of Hispaniola. There is conjecture, however, whether *C. tunicata* was deliberately carried by humans to areas outside North America for use as a medicinal herb and living fence (Anderson 2001).

Cylindropuntia was recognised as a subgenus of *Opuntia* by Engelmann in 1856 and raised to generic status by FM Knuth in 1935. This change has only been widely accepted after 2000. The name *Cylindropuntia* means cylindrical *Opuntia*, due to the shape of the stem segments. *C. imbricata*, rope pear, was referred to as *Opuntia imbricata* Haw. in Australian publications such as Dodd (1940), Mann (1970) and Hosking *et al.* (1988).

The major problem species of *Cylindropuntia* in Australia was *C. imbricata* (Fig. 1), for which a biological control program began in 1925. The earliest record of *C. imbricata* in Australia was as a garden plant in NSW; it was also growing wild at a number of sites in the Hunter Valley and north of Bathurst along the Turon River (Maiden 1911). No estimates of the area covered by any *Cylindropuntia* spp. have been made, but today the largest concentrations of *C. imbricata* are found around old mining areas and along a number of watercourses inland from the Great Dividing Range to south of Mt Isa, Qld (Hosking *et al.* 1988). The distribution of *C. imbricata* is shown in Figure 2 (AVH 2010). For additional detail on distribution of *C. imbricata* (as *O. imbricata*) in eastern Australia see Hosking *et al.* (1988). However, records in AVH (2010) are partial because cactus is rarely collected due to difficulties with preservation of specimens.

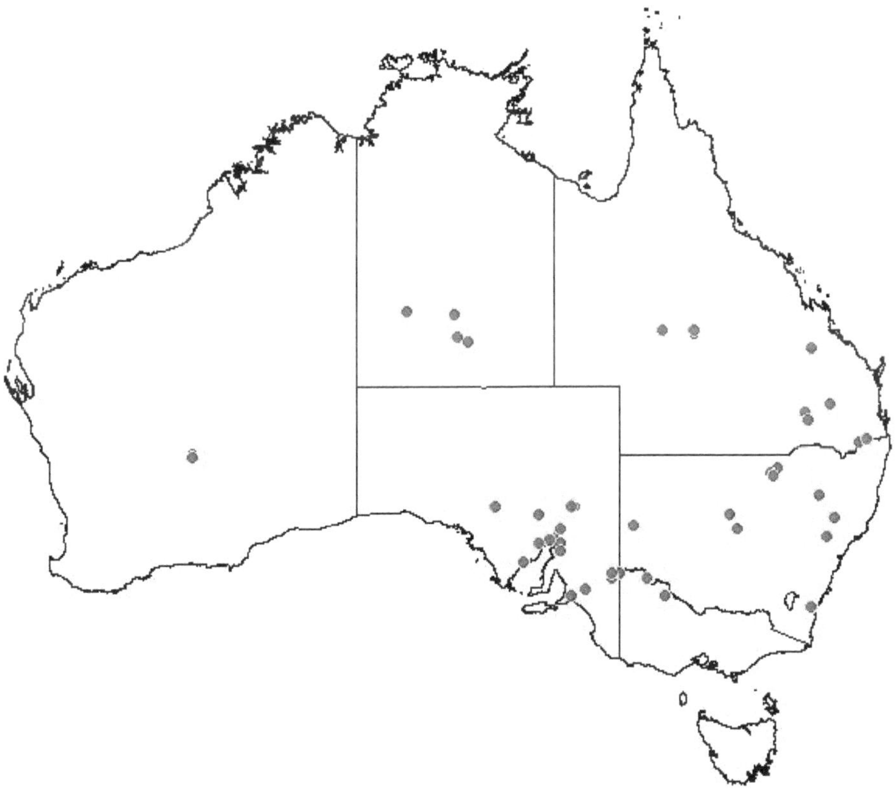

Figure 2: Distribution of *Cylindropuntia imbricata* in Australia. Specimen data reproduced from Australia's Virtual Herbarium with permission of the Council of Heads of Australasian Herbaria Inc and knowledge of the author.

BIOLOGICAL CONTROL HISTORY

The first deliberate use of insects to control a weed involved the use of a cochineal insect on an *Opuntia* sp. in India in 1863 (Tryon 1910). The biological control of *Opuntia* spp. has been discussed elsewhere in this book and in publications such as Dodd (1940), Mann (1970) and Hosking *et al.* (1988). The Commonwealth Prickly Pear Board (CPPB) was appointed in 1920 by the Commonwealth Government and the state governments of Queensland and NSW (Dodd 1940). The Board's charter included a search for prickly pear insects in the Americas, investigation of host restriction of prickly pear insects, selection and transport to Australia of suitable insects and large-scale rearing and distribution of the insects, followed by progress of established insects (Mann 1970). There was also to be a survey of plant disease organisms.

Studies in South Africa on *Dactylopius opuntiae* Cockerell (Hemiptera: Dactylopiidae) (Volchansky *et al.* 1999) found that different biotypes of *D. opuntiae* were responsible for the control of *Opuntia monocantha* Haw. and *O. stricta* (Haw.) Haw. The discovery of biotypes within species of *Dactylopius* led to the search for new biotypes of *Dactylopius tomentosus* (Lamarck) for the control of *Cylindropuntia* spp. in South Africa (Mathenge *et al.* 2009b). Misidentification of the target species as *C. rosea* led to the importation of a biotype of *D. tomentosus* specific to *C. rosea*. This agent's failure to cause damage led to a re-examination of the host plant, which was subsequently identified as *C. fulgida* var. *fulgida* (Engelmann) FM Knuth. Further research led to the discovery that *D. tomentosus* was capable of forming new associations within the genus *Cylindropuntia* (HG Zimmermann, *pers. comm.* 2009; Mathenge *et al.* 2009b), one of which has proven highly successful on *C. fulgida* var. *fulgida* (PPRI 2009).

CANDIDATES

The only insect introduced into Australia for biological control of *Cylindropuntia* spp. was *D. tomentosus*, which was introduced from Texas in 1925 by the CPPB (under the name *Dactylopius newsteadi* Cockerell) for the control of *C. imbricata*. For more detail see Dodd (1940), Mann (1970), Hosking *et al.* (1988) and Julien and Griffiths (1998). A number of other insects are known to attack *Cylindropuntia* spp. in their native range (Mann 1969; HG Zimmermann, *pers. comm.* 2009).

***Dactylopius tomentosus* (Lamarck) (Hemiptera: Dactylopiidae).** The family Dactylopiidae contains only one genus *Dactylopius* Costa with 10 currently recognised species (USDA 2009), all of which are restricted to the Cactaceae (De Lotto 1974, Mann 1970, USDA 2009). The native distribution of *D. tomentosus* appears to be parts of North America, namely Arizona, New Mexico and Texas in the USA and Mexico (Rodriguez *et al.* 2001). It was introduced into Australia in 1925 (Mann 1970) and South Africa in 1970 (De Lotto 1974; Moran and Zimmermann 1991) as a biological control agent for rope pear.

Figure 3: *Cylindropuntia rosea* infested with *Dactylopius tomentosus* in Mexico. Photo: H Zimmermann.

Adult female cochineal insects, including those of *D. tomentosus*, are sessile and remain attached to cladodes of their host plants (Fig. 3). Adult males are small free-flying insects that do not feed. Feeding by the immature stages and females of this sap-sucking insect causes stress to plants, ultimately leading to their death if cochineal insect populations are high. The impact of cochineal insects appears to increase when conditions are dry. A detailed study of the biology of *D. tomentosus* has been published (Mathenge *et al.* 2009a).

Diseases. Although not investigated by the CPPB, a number of pathogens are known to attack *Cylindropuntia*

Figure 4: *Cylindropuntia rosea* in flower and fruit. Photo: R Holtkamp, NSW Government.

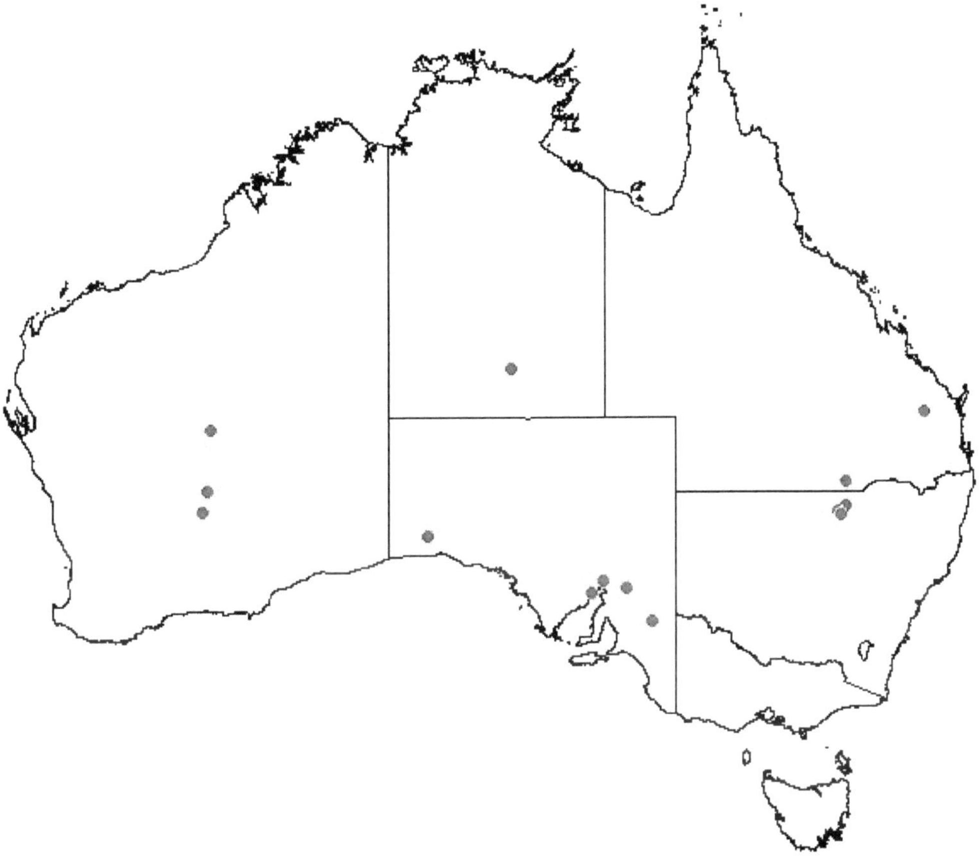

Figure 5: Distribution of *Cylindropuntia rosea* in Australia. Specimen data reproduced from Australia's Virtual Herbarium with permission of the Council of Heads of Australasian Herbaria Inc and knowledge of the author.

spp. in their native range (HG Zimmermann, *pers. comm.* 2009). Some might be suitable for further investigation as biological control agents.

DISCUSSION

Despite its introduction for biological control of *C. imbricata*, *D. tomentosus* has successfully established on three other *Cylindropuntia* species in Australia. These are *C. leptocaulis* and *C. tunicata*, on which it has a significant impact, and *C. rosea*, on which it is largely ineffective. *D. tomentosus* is highly effective as a control agent for *C. imbricata* when plants are small or regrowing, but is less effective on large plants. Studies in Australia (Hosking *et al.* 1988) and South Africa (Moran and Zimmermann 1991) have shown that cutting down large *C. imbricata* plants improves the level of biological control.

D. tomentosus was imported into South Africa by the Plant Protection Research Institute, following its success in Australia, where it has provided very successful biological control of *Cylindropuntia* spp. (Moran and Zimmermann 1991; Julien and Griffiths 1998). Recent research in South Africa has indicated the capacity for biotypes of *D. tomentosus* to form new associations within the genus *Cylindropuntia* (HG Zimmermann, *pers. comm.* 2009; Mathenge *et al.* 2009b, 2010; PPRI 2009).

D. tomentosus is one of the biological control agents introduced into Australia for *Opuntia*/*Cylindropuntia* spp. which have been discussed by Page and Lacey (2006). They calculated the benefit:cost ratio for the combined biological control program to be 312.3; in 2005 dollar terms, the investment of $21.1 million returned a net present value of $3110.3 million.

A current project in NSW is focusing on the introduction of a biotype of *D. tomentosus* which appears to be specific to *C. rosea* (Fig. 4), a species which is becoming a significant problem in the region around the opal fields of central northern NSW. This biotype of *D. tomentosus* was imported into Australian quarantine in August 2010 for host-specificity testing. The biology of *C. rosea* is discussed in detail in Johnson *et al.* (2009) and its distribution is shown in Figure 5 (AVH 2010).

ACKNOWLEDGEMENTS

Thanks to Drs John Hosking, Bob Chinnock, Helmuth Zimmermann, Catherine Mathenge and Carla Chavez-Moreno for their discussions on various aspects of *Cylindropuntia* spp. and their control. Dr John Hosking also provided comments on an earlier draft of this manuscript.

REFERENCES

Anderson EF (2001) *The Cactus Family.* Timber Press, Oregon.

AVH (2010) *Australia's Virtual Herbarium.* <http://www.chah.gov.au/avh/public_query.jsp> (accessed 28 May 2010).

De Lotto G (1974) On the status and identity of the cochineal insects (Homoptera: Coccoidea: Dactylopiidae). *Journal of the Entomological Society of Southern Africa* **37**, 167–193.

Dodd AP (1940) *The Biological Campaign against Prickly-pear.* Commonwealth Prickly Pear Board Bulletin. Government Printer, Brisbane.

Hosking JR, McFadyen RE and Murray ND (1988) Distribution and biological control of cactus species in eastern Australia. *Plant Protection Quarterly* **3**, 115–123.

Johnson SB, Hosking JR, Chinnock RJ and Holtkamp RH (2009) The biology of Australian weeds 53: *Cylindropuntia rosea* (DC.) Backeb. and *Cylindropuntia tunicata* (Lehm.) F.M.Knuth (Cactaceae). *Plant Protection Quarterly* **24**, 42–49.

Julien MH and Griffiths MW (1998) *Biological Control of Weeds: A World Catalogue of Agents and their Target Weeds.* 4th edn. CABI, Wallingford, UK.

Maiden JH (1911) The prickly pears of interest to Australians. II. *Agricultural Gazette of New South Wales* **22**, 696–698.

Mann J (1969) *Cactus-feeding Insects and Mites.* US National Museum Bulletin 256. Smithsonian Institution Press, Washington.

Mann J (1970) *Cacti Naturalised in Australia and their Control.* Government Printer, Brisbane.

Mathenge CW, Holford P, Hoffmann JH, Spooner-Hart R, Beattie GAC and Zimmermann HG (2009a) The biology of *Dactylopius tomentosus* (Hemiptera: Dactylopiidae). *Bulletin of Entomological Research* **99**, 551–559.

Mathenge CW, Holford P, Hoffmann JH, Zimmermann HG, Spooner-Hart R and Beattie GAC (2009b) Distinguishing suitable biotypes of *Dactylopius tomentosus* (Hemiptera: Dactylopiidae) for biological control of *Cylindropuntia fulgida* var. *fulgida* (Caryophyllales: Cactaceae) in South Africa. *Bulletin of Entomological Research* **99**, 619–627.

Mathenge CW, Holford P, Hoffmann JH, Zimmermann HG, Spooner-Hart R and Beattie GAC (2010) Determination of biotypes of *Dactylopius tomentosus* (Hemiptera: Dactylopiidae) and insights into the taxonomic relationships of their hosts *Cylindropuntia* spp. *Bulletin of Entomological Research* **100**, 347–358.

Moran VC and Zimmermann HG (1991) Biological control of cactus weeds of minor importance in South Africa. *Agriculture, Ecosystems and Environment* **37**, 37–55.

Page AR and Lacey KL (2006) *Economic Impact Assessment of Australian Weed Biological Control.* Technical Series 10. CRC for Australian Weed Management, Adelaide.

PPRI (Plant Protection Research Institute) 2009. Biocontrol of invasive cactus exceeds all expectations. *Plant Protection News* **79**, 1–2.

Rodriguez LC, Mendez MA and Niemeyer HM (2001) Direction of dispersal of cochineal (*Dactylopius coccus* Costa) within the Americas. *Antiquity* **75**, 73–77.

Telford IRH (1984) Cactaceae. In *Flora of Australia Vol. 4.* (Ed AS George). Commonwealth of Australia, Canberra.

Tryon H (1910) The 'wild cochineal,' with reference to its injurious action on prickly pear (*Opuntia* spp.) in India, etc., and to its availability for the subjugation of this plant in Queensland and elsewhere. *Queensland Agriculture Journal* **25**, 188–197.

USDA (2009) *ScaleNet* online checklist of the Coccoidea. <http://www.sel.barc.usda.gov/ catalogs/ dactylop/ DactylopiusAll.htm>.

Volchansky CR, Hoffmann JH and Zimmermann HG (1999) Host plant affinities of two biotypes of *Dactylopius opuntiae* (Homoptera: Dactylopiidae): enhanced prospects for biological control of *Opuntia stricta* (Cactaceae) in South Africa. *Journal of Applied Ecology* **36**, 85–91.

Cytisus scoparius (L.) Link – broom, Scotch broom or English broom

John R Hosking, Andy W Sheppard and Jean-Louis Sagliocco

ABSTRACT

Cytisus scoparius is a leguminous shrub that is a serious weed of cool climate areas in south-eastern Australia. Biological control for Australia was initiated because of large infestations in the Barrington Tops region of NSW. A biological control program against broom has been underway since 1993, carried out by several state and federal agencies, much of it in collaboration with other countries, particularly New Zealand. This program has resulted in the release of three insects and a mite in Australia. All the agents have established somewhere in Australia and a number are causing slight damage in southern NSW, Vic and Tas. No significant damage to broom has occurred at the Barrington Tops, but the mite was only released in this area in 2010. Biological control of broom has not been considered a success to date despite considerable ecological research suggesting natural enemies are important in the invasion dynamics of broom.

Key words: *Leucoptera spartifoliella, Arytainilla spartiophila, Bruchidius villosus, Aceria genistae, Exapion fuscirostre, Uromyces pisi-sativi, Parthenolecanium rufulum.*

INTRODUCTION

Cytisus scoparius (L.) Link (Fabaceae: Faboideae: Genisteae), broom, Scotch broom or English broom, is native to Europe. It has been introduced as an ornamental and has naturalised and is a weed in many countries including Australia, NZ, the USA, Canada, India, Iran, Japan, South Africa, Argentina and Chile (Parsons and Cuthbertson 1992; Zuloaga and Morrone 1999; Arroyo *et al.* 2000; Iwatsuki *et al.* 2001). Biological programs to control broom have been conducted in a number of countries, with the first agent released in 1960 in the USA.

Broom is a leguminous shrub lacking thorns and having several erect or ascendant stems which may later collapse to become prostrate. Plants grow to 4 m high, and often form dense thickets in cooler areas. The plant has large yellow pea-shaped flowers and trifoliate leaves, although some leaves near the inflorescence may have

Figure 1: *Cytisus scoparius* at the Barrington Tops, NSW. Note the distinctive pod with hairs only around the margins. Photo: JR Hosking, NSW Government.

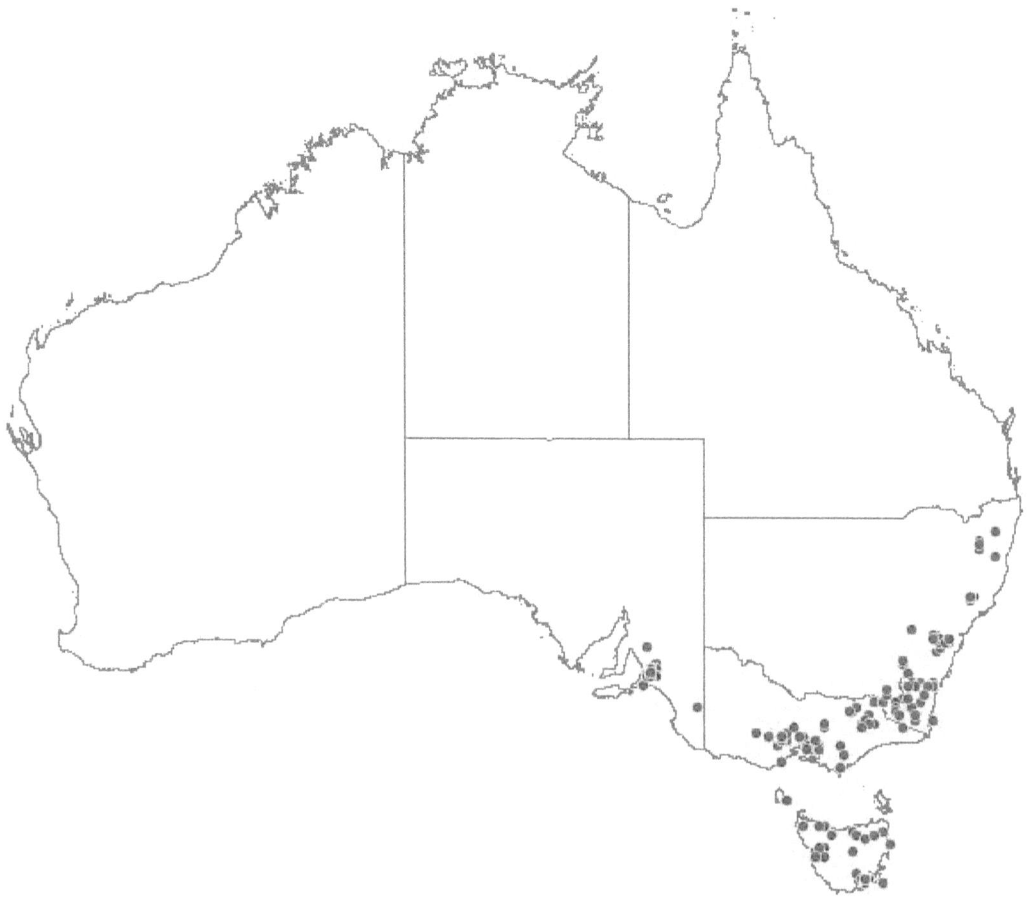

Figure 2: Distribution of broom, *Cytisus scoparius*, in Australia based on data in Australia's Virtual Herbarium. Specimen data reproduced from Australia's Virtual Herbarium (AVH 2011) with permission of the Council of Heads of Australasian Herbaria Inc.

single leaflets (Fig. 1). Two red- and yellow-flowered hybrids, known as *C. scoparius* cultivar 'Andreanus' and *C. scoparius* cultivar 'Andreanus aureus', have naturalised in Australia (Hosking *et al.* 1998).

Figure 3: Broom, *Cytisus scoparius*, infestation at the Barrington Tops, NSW. Photo: PR Sullivan, NSW Government.

The current distribution of broom in Australia, based on herbarium records, is shown in Figure 2. Some of the larger Australian broom infestations now occupy about 10 000 ha on the Barrington Tops, NSW (Waterhouse 1988) (Fig. 3). Broom also poses a serious threat to the Australian Alps national parks, with large infestations in the Central and Southern Tablelands of NSW, the Adelaide Hills and higher rainfall areas in Vic and Tas. It is mainly a weed of environmental and cattle-grazing areas (Hosking *et al.* 1998) but in Tas it also invades areas of *Eucalyptus nitens* H. Deane & Maiden plantation forestry (Barnes and Holz 2000). It is estimated that the area infested in Australia is over 230 000 ha (Hosking *et al.* 1998; Barnes and Holz 2000). A summary of information available on broom in Australia to 1998 can be found in Hosking *et al.* (1998). Since then there have been a number of published studies on broom in Australia (Downey and Smith 2000; Memmott *et al.* 2000; Sheppard and Hosking 2000; Sheppard *et al.* 2002; Paynter *et al.* 2003; Wearne and Morgan 2004, 2006; Stokes *et al.* 2006).

BIOLOGICAL CONTROL HISTORY

Biological control programs for broom have been conducted in the USA and NZ. The initial USA program, conducted for California by the US Department of Agriculture (USDA), resulted in the release of *Leucoptera spartifoliella* (Hübner) (Lepidoptera: Lyonetiidae), the broom twigminer, in 1960. At the time of release this insect was found to be already established in California, Oregon and Washington, probably due to accidental introduction on ornamental plants some time before 1940 (Julien and Griffiths 1998). *Exapion fuscirostre* (Fabricius) (Coleoptera: Brentidae) was released in California in 1964, Oregon in 1983 and Washington in 1989 (Syrett *et al.* 1999). This seed weevil established and causes seed destruction of up to 60% in California (Julien and Griffiths 1998) and 75–90% in Oregon (Syrett *et al.* 1999). In Oregon it is found in *C. scoparius* but not in sympatric populations of *Cytisus striatus* (Hill) Rothm. (D Isaacson, *pers. comm.* 1998). Following these releases the USA broom program ceased until 1997, partly because of increasing interest in shrubby legumes as ornamental and landscape plants (Syrett *et al.* 1999). In 1997 a proposal was prepared for movement of the broom seed bruchid, *Bruchidius villosus* Fabricius (Coleoptera: Chrysomelidae) to the western USA from the eastern USA where it had entered accidentally and spread widely (D Isaacson, *pers. comm.* 1997).

A NZ broom biological control program began in 1981. *Leucoptera spartifoliella* was found to have been accidentally introduced to NZ, with the first record of occurrence in 1950. This insect causes considerable damage at times on the South Island of NZ (Syrett and Harman 1995; Syrett *et al.* 1999). The first agent to be released in NZ was the broom seed bruchid, *B. villosus* in 1987 (Julien and Griffiths 1998). It is estimated that this bruchid will cause about 73% seed destruction once the insect has spread throughout NZ (Paynter *et al.* 2010). A broom psyllid, *Arytainilla spartiophila* (Förster) (Hemiptera: Psyllidae), was released in 1993 and has established in many areas. The gall-forming mite *Aceria genistae* (Nalepa) (Acari: Eriophyidae) was released in 2007 at Lincoln, near Christchurch, from a colony imported from southern France in 2006, and since then releases of *Gonioctena olivacea* (Förster) (Coleoptera: Chrysomelidae) have taken place (Q Paynter, *pers. comm.* 2009).

An Australian broom biological control program commenced in 1990. A study conducted by John Hosking in 1990 concluded that a broom biological control program was feasible and that it should operate through the International Institute for Biological Control (now called CABI) broom program being conducted from Silwood Park, UK, for NZ. Part of the funds was to be used for the program to be extended to Montpellier in southern France. It was thought that biological control agents sourced from southern Europe may be more suited climatically to the broom problem at the Barrington Tops as these areas were closer in latitude than the UK. Australian host-specificity studies were conducted by CSIRO and the NSW Government. The program resulted in the release of *L. spartifoliella*, *A. spartiophila*, *B. villosus* and *A. genistae*. The NSW program ceased when funding wound up in 2003. Details on agents released are covered below. Meanwhile, the Vic Department of Primary Industries (Vic DPI), which had been rearing and distributing insects since 1993, took over research with the re-importation of *A. genistae* from southern France in 2006, leading to its subsequent release. In 2010 Vic DPI plan to begin host-specificity testing of *G. olivacea* for Australia based on host-specificity test results, and material imported, from NZ.

From 1997 to 2002 Australia, NZ and the USA (Oregon and California) shared resources under an International Broom Initiative (which also worked on the biological control of other Genisteae). The three countries provided funding to CABI and initially focused on the evaluation and screening of the gall-forming mite *A. genistae*. This mite was found to be already present in California and had been recorded on gorse in NZ; subspecific taxonomic analysis became a requirement to allow further progress. Work was also initiated on other potential agents including *Hexomyza sarothamni* (Hendel) (Diptera: Agromyzidae) and *Asphondylia sarothamni* Loew (Diptera: Cecidomyiidae) (Syrett *et al.* 1999).

EXPLORATION

The USDA was the first organisation to investigate insects as potential biological control agents for *C. scoparius*. This work commenced in 1951 (Parker 1964).

CABI ran the native range research activities in Europe for NZ through the 1980s from Silwood Park in the UK. When Australia needed work done in Europe in the early 1990s, CABI also set up a research program based in Montpellier, France in collaboration with CSIRO at its European laboratory. From 1997 to 2002 Australia, NZ and the USA (Oregon and California) shared resources under the International Boom Initiative discussed above.

For more information on exploration and potential biological agents see Syrett *et al.* (1999).

CANDIDATES

Over 240 insects and mites and many fungi have been reported from *Cytisus* spp. (Syrett *et al.* 1996, 1999). Of these, 70 species relevant for biological control of broom and its close relatives are listed and discussed in Syrett *et al.* (1999). Three insects, *L. spartifoliella*, *A. spartiophila* and *B. villosus*, and the mite, *A. genistae*, have been released for biological control of broom in Australia.

Agent not released in Australia

Exapion fuscirostre. This broom seed weevil was brought into quarantine in Canberra, Australia, but could not be synchronised to the southern hemisphere under quarantine conditions. This problem also occurred in NZ (P Syrett, *pers. comm.* 1999). The weevil was released in the USA and records of 60–90% seed destruction in the USA have been recorded (Syrett *et al.* 1999).

Agents released in Australia

Leucoptera spartifoliella. The first agent to be tested in Australia was the multivoltine twig-mining moth *L. spartifoliella*. The testing program began late in 1990 with insects brought into quarantine in Canberra from NZ. Information from the USA host-specificity studies on *L. spartifoliella* and some additional host-specificity testing, particularly on Australian natives, found this insect to be specific to broom and to hybrids with broom as one of the parents. The moth was released in NSW in February 1993. It has established at 'Krawarree', a property near Braidwood and one of the original release sites, and it appeared to have established at the Barrington Tops, another of the original release sites, but has not been seen in the latter area in recent years. Other releases were made in the Blue Mountains and in other areas in southern NSW. The moth causes reasonable damage with multiple mines per stem at 'Krawarree'. While it is still too early to determine the level of damage likely to be caused by this moth across Australia, after 10 years at 'Krawarree' attack levels have levelled off without much long-term impact on the vigour of stands. The established population at 'Krawarree' led to *L. spartifoliella* redistribution across Vic. *L. spartifoliella* has also been released in Tas and SA. There is some evidence that it has established at low densities in Vic. Sampling conducted in 2007 at some release sites on 20 1 m broom branches showed average densities of 1.1 larval mines per metre near Omeo and 2.15 at Lake Dartmouth (J-L Sagliocco, unpub. data). However, insect densities appear to have declined since then. Releases were made in Tas in 1995 and 1998 with further releases since 2004; the only recovery to date is from a site near Oatlands in 2005, 12 months after release, but establishment still needs to be confirmed (J Ireson, *pers. comm.* 2010). The fate of *L. spartifoliella* in SA is not known. In Australia, parasitism by the pupal parasitoid *Megadicylus* sp. (near *M. dubius* (Girault)) (Hymenoptera: Pteromalidae) and an unidentified ichneumonid wasp appear to affect *L. spartifoliella* populations (Sheppard *et al.* 2006a).

Arytainilla spartiophila. The second agent was the univoltine Scotch broom psyllid, *A. spartiophila*, imported from France and tested against Australian natives and a few other species to add to test results from studies carried out by NZ (Syrett 1991). Permission to release *A. spartiophila* was obtained in 1994 and releases were made in December 1994 and January 1995 at a number of sites in NSW. Subsequent releases were made at two sites in Vic and at least one in SA. This psyllid is only known to have established at 'Krawarree' (but was not seen during a casual visit in 2010) and possibly in SA (Q Paynter, *pers. comm.* 2000). No recent comprehensive surveys have been carried out to verify long-term establishment.

Bruchidius villosus. A third agent, a univoltine broom seed bruchid, *B. villosus*, was also tested against Australian plants to complement NZ host-specificity testing. *B. villosus* was sourced from both NZ and France. Releases were made near Katoomba and at 'Krawarree' at the end of 1995. The difficulty of rearing this agent led to an application for direct release of adults shipped from NZ using an approved surface sterilisation protocol. On approval, this approach led to further releases of around 1000 bruchids at the end of 1996 at two locations in the Barrington Tops and at 'Krawarree', and in 1998 near Bright, Beechworth and Omeo in Vic. The bruchid established at 'Krawarree' and was still easy to find in early spring 2010 on the first opening flowers. It was also persisting at one site, near Omeo, in 2000 with occasional damage seen since at this site, but there was no evidence of establishment at the Barrington Tops. More effective monitoring of the impacts and spread of this agent at release sites where it has established would be timely. In NZ, where seed predation rates on broom up to 84% have been recorded (Paynter *et al.* 2010), this agent was found attacking seeds of tagasaste, *Chamaecytisus palmensis* (Christ) F.A. Bisby & K.W. Nicholls (Fabaceae: Faboideae: Genisteae) in 1999 contrary to the earlier testing results (Sheppard *et al.* 2006b). This exotic fodder plant is considered useful in WA so no further releases were made in Australia. Further work in NZ showed that egg-laying on *C. palmensis* seeds under laboratory conditions is much less than on *C. scoparius*

Figure 4: Galls formed on *Cytisus scoparius* by the broom gall mite, *Aceria genistae* (galled plant material from Tas). Photo: JR Hosking, NSW Government.

seeds and numbers of insects used were much lower in original tests, therefore the difference may have been the result of numbers involved in tests, or other minor differences in test procedures (Haines *et al.* 2004). Field surveys and choice tests in Europe also demonstrated that at the time of the original tests there was a poor understanding of the host range of *B. villosus* throughout its native range (Sheppard *et al.* 2006b).

Aceria genistae. The broom gall mite *A. genistae* was studied from 1993 but, due to taxonomic, rearing and testing complications, a release permit was not obtained until 2003. The mite infests dormant vegetative buds in autumn, causing them to become galls on bursting in spring (Fig. 4). *A. genistae sensu lato* was considered to have a wide host-range based on literature records, but the colonies tested from *C. scoparius* in France were shown to be restricted to *C. scoparius* and to hybrids with *C. scoparius* as one of the parents. Due to issues with obtaining a clean colony in quarantine (the existence of inquilines and predators in mite galls shipped from Europe required development of a technique to successfully transfer individual fecund mites to clean broom plants in quarantine) this agent was not released until 2008 in Vic and SA, by Vic DPI, and until 2009 in Tas. Releases were made at the Barrington Tops, Captains Flat and 'Krawarree' in NSW in October 2010; it is too early to determine if this mite will establish in NSW and, if so, what damage it will cause. In Vic, SA and Tas the first galls appeared a year after release and, in 2011, mite establishment was confirmed for these states (J-P Sagliocco, unpublished data). Monitoring of this mite and its damage is ongoing in all states.

Natural enemies not deliberately introduced to Australia

***Parthenolecanium rufulum* (Cockerell) (Hemiptera: Coccidae).** This scale insect is native to Europe. It occurs at the Barrington Tops and in southern NSW where it causes death of sections of broom bushes and can reduce flowering and seed production. On 'Krawarree' this insect was mostly responsible for a 33% reduction in seeds produced on control versus insecticide-treated bushes in their first year of seed production. This scale has a wide host-range in Europe, but in Australia it has been found only on broom (Hosking *et al.* 1998).

***Etiella behrii* (Zeller) (Lepidoptera: Pyralidae).** Broom seeds are eaten by this native moth in southern NSW (Hosking 1995). This moth is a pest of a number of leguminous crops in Australia (Common 1990).

Cerambycid beetles (Coleoptera: Cerambycidae). In the Braidwood area in southern NSW four cerambycids have been reared from girdled stems of broom. The larger *Uracanthus bivittata* Newman and *Strongylurus arduus* Elliott and McDonald are probably causing the girdling and the smaller *Pentacosmia scoparia* Newman and *Sybra acuta* Pascoe are probably secondary borers in the dead girdled stems (Wapshere and Hosking 1993). Such attack may be associated with large stem girth and it may therefore appear more widespread once stands mature.

***Acyrthosiphon pisum* (Harris) (Hemiptera: Aphididae).** This European pea aphid is variable in abundance, causing extensive foliar-marking to broom in some years. It also causes most insect-related damage to developing seedlings in Australia (Sheppard and P Hodge, unpub. data). This species may be the nominal subspecies that colonises herbaceous legumes, or it may belong to the nominal subspecies *spartii* Koch, which colonises woody Genisteae such as *Cytisus* and *Spartium*. Subspecies *spartii* has been recorded from Tas (M Carver, *pers. comm.* 1996) and may have spread to the mainland.

***Uromyces pisi-sativi* (Pers.) Liro (Pucciniaceae).** Broom rust was first observed on broom plants near the CSIRO quarantine facility in Canberra in May 2003 and was then found to be widespread throughout the range of broom in Australia, suggesting it had been in the country for some time. There is some confusion over the name of this rust and the implications this has for host-range (Morin *et al.* 2006), but this rust has not been reported on other legumes in Australia to date. This rust results in the death of some small broom plants but damage to larger plants is usually not significant (Morin *et al.* 2006).

Other insects and pathogens. Other insects have been found on broom in Australia (Syrett *et al.* 1999; Memmott *et al.* 2000). Fungal attack of mature broom plants, causing their premature death, is occasionally observed in Australia, but occurrences are isolated and the causal agent has not been identified. In some locations, once the initial invasive stand of broom has senesced subsequent plant survival appears lower, which might be explained by a build-up of antagonistic soil pathogens (A Sheppard, *pers. obs.* 1998). Post-dispersal seed predation has been measured in an experiment repeated in four countries including Australia. Such seed predation reached 80% per year in Australia but was higher in the native range and was considered to be largely caused by rodents (Paynter *et al.* 1996).

DISCUSSION

The Australian broom program commenced following concerns about a 10 000 ha broom infestation on the Barrington Tops, where broom was shown to affect native trees (Waterhouse 1988) and was thought to also be affecting other vegetation and animals. Recent research has demonstrated that, where broom grows in frost hollow areas on the Barrington Tops, it results in loss of most plant species (J Hosking, unpub. data). The Australian broom program was built on the programs already started in the USA and NZ. As a result of the NZ program, the involvement of IIBC and later the eastern states of the USA, the collaboration developed into a multi-country research agenda involving sharing of data, cross-testing of plant species in some of the host-specificity tests, joint funding of biological work on new potential agents and ecological studies relevant to biological control across three continents (Fowler *et al.* 1996; Rees and Paynter 1997; Paynter *et al.* 1998, 2000; Sheppard *et al.* 2000, 2002). Collaboration continues between Australia and NZ.

Despite this, the biological control of Scotch broom has not been successful in any country, although some high damage levels have been recorded. Nonetheless, the ecological research still suggests that natural enemies are the most important factor in determining the invasion dynamics of broom in its native range (Waloff and Richards 1977) and its exotic range (Paynter *et al.* 2003). Biological control efforts continue in the expectation of future success. The gall mite *A. genistae* has been recently observed killing plants in NZ (L Hayes, *pers. comm.* 2010), which may offer some hope.

At 'Krawarree', the only release site in Australia where at least two agents have established in large numbers for more than 10 years, the dense broom stands have not declined although seed production appears to have been lowered (J O'Brian, *pers. comm.* 2010).

ACKNOWLEDGEMENTS

The following significantly contributed to the program for Australia. In CSIRO: Jim Cullen, Tony Wapshere, Jean-Paul Aeschlimann, Louise Morin, Thierry Thomann, Janine Vitou and Mireille Jourdan. In CABI/IIBC (IIBC was merged with other organisations to form CAB International): Quentin Paynter and Simon Fowler. In NSW Agriculture: Peter Hodge. We acknowledge the international collaborative assistance of Landcare Research Ltd in NZ, CABI in the UK and the Oregon Department of Agriculture in the USA. We acknowledge the significant help of NSW national parks staff and of James and Mandy O'Brian at 'Krawarree' near Braidwood in NSW. Royce Holtkamp provided useful comments on this chapter.

REFERENCES

Arroyo MTK, Marticorena C, Matthei O and Cavieres L (2000) Plant invasions in Chile: present patterns and future predictions. In *Invasive Species in a Changing World.* (Eds HA Mooney and RJ Hobbs) pp. 385–421. Island Press, Washington, DC.

AVH (2011) *Australia's Virtual Herbarium.* <http://www.chah.gov.au/avh/public_query.jsp> (accessed 2.8.11).

Barnes CD and Holz GK (2000) Broom (*Cytisus scoparius* (L.) Link) competition and management in eucalypt tree farms. *Plant Protection Quarterly* **15**, 163–165.

Common IFB (1990) *Moths of Australia.* Melbourne University Press, Melbourne.

Downey PO and Smith JMB (2000) Demography of the invasive shrub Scotch broom (*Cytisus scoparius*) at Barrington Tops, New South Wales: insights for management. *Austral Ecology* **25**, 477–485.

Fowler SV, Harman HM, Memmott J, Paynter Q, Shaw R, Sheppard AW and Syrett P (1996) Comparing the population dynamics of broom, *Cytisus scoparius*, as a native plant in the United Kingdom and France and as an invasive alien weed in Australia and New Zealand. In *Proceedings of the 9th International Symposium on Biological Control of Weeds.* 19–26 January 1996, Stellenbosch, South Africa. (Eds VC Moran and JH Hoffmann) pp. 19–26. University of Capetown, South Africa.

Haines ML, Syrett P, Emberson RM, Withers TM, Fowler SV and Worner SP (2004) Ruling out a host-range expansion as the cause of the unpredicted non-target attack of tagasaste (*Chamaecytisus proliferus*) by *Bruchidius villosus*. In *Proceedings of the 11th International Symposium on Biological Control of Weeds*. 27 April–2 May 2003, Australian National University, Canberra (Eds JM Cullen, DT Briese, DJ Kriticos, WM Lonsdale, L Morin and JK Scott) pp. 271–276. CSIRO Entomology, Canberra.

Hosking JR (1995) The impact of seed and pod feeding insects on *Cytisus scoparius* (L.) Link. In *Proceedings of the 8th International Symposium on Biological Control of Weeds*. 2–7 February 1992, Canterbury, NZ (Eds ES Delfosse and RR Scott) pp. 45–51. DSIR/CSIRO, Melbourne.

Hosking JR, Smith JMB and Sheppard AW (1998) *Cytisus scoparius* (L.) Link ssp. *scoparius*. In *The Biology of Australian Weeds*. Vol. 2. (Eds FD Panetta, RH Groves and RCH Shepherd) pp. 77–88. RG and FJ Richardson, Melbourne.

Iwatsuki K, Boufford DE and Ohba H (Eds) (2001) *Flora of Japan* 2b. Kodansha, Tokyo.

Julien MH and Griffiths MW (1998) *Biological Control of Weeds: A World Catalogue of Agents and their Target Weeds*. 4th edn. CABI, Wallingford, UK.

Memmott J, Fowler SV, Paynter Q, Sheppard AW and Syrett P (2000) The invertebrate fauna on broom, *Cytisus scoparius*, in two native and two exotic habitats. *Acta Oecologica* **21**, 213–222.

Morin L, Sagliocco J-L, Hartley D, Hosking JR, Cramond P and Washington B (2006) Broom rust in Australia. In *Papers and Proceedings of the 15th Australian Weeds Conference*. 24–28 September 2006, Adelaide. (Eds C Preston, JH Watts and ND Crossman) pp. 569–572. Weed Management Society of South Australia, Adelaide.

Parker HR (1964) Life history of *Leucoptera spartifoliella* with results of host transfer tests conducted in France. *Journal of Economic Entomology* **57**, 566–569.

Parsons WT and Cuthbertson EG (1992) *Noxious Weeds of Australia*. Inkata Press, Melbourne.

Paynter Q, Fowler SV, Hinz HL, Memmott J, Shaw R, Sheppard AW and Syrett P (1996) Are seed-feeding insects of use for the biological control of broom? In *Proceedings of the 9th International Symposium on Biological Control of Weeds*. 19–26 January 1996, Stellenbosch, South Africa. (Eds VC Moran and JH Hoffmann) pp. 495–501. University of Capetown, South Africa.

Paynter Q, Fowler SV, Memmott J and Sheppard AW (1998) Factors affecting the establishment of *Cytisus scoparius* in southern France: implications for its control. *Journal of Applied Ecology* **35**, 582–595.

Paynter Q, Fowler SV, Memmott J, Shaw RH and Sheppard AW (2000) Determinants of broom (*Cytisus scoparius* (L.) Link) abundance in Europe. *Plant Protection Quarterly* **15**, 149–155.

Paynter Q, Downey PO and Sheppard AW (2003) Age structure and growth of the woody legume weed *Cytisus scoparius* in native and exotic habitats: implications for control. *Journal of Applied Ecology* **40**, 470–480.

Paynter Q, Main A, Gourlay AH, Peterson PG, Fowler SV and Buckley YM (2010) Disruption of an exotic mutualism can improve management of an invasive plant: varroa mite, honeybees and biological control of Scotch broom *Cytisus scoparius* in New Zealand. *Journal of Applied Ecology* **47**, 309–317.

Rees M and Paynter Q (1997) Biological control of Scotch broom: modelling the determinants of abundance and the potential impact of introduced insect herbivores. *Journal of Applied Ecology* **34**, 1203–1221.

Sheppard AW and Hosking JR (Eds) (2000) Broom management. Proceedings of a workshop held at Ellerston and Moonan on 16–17 November 1998. *Plant Protection Quarterly* **15**, 133–186.

Sheppard AW, Hodge P and Paynter Q (2000) Factors affecting broom regeneration in Australia and their management implications. *Plant Protection Quarterly* **15**, 156–161.

Sheppard AW, Hodge P, Paynter Q and Rees M (2002) Factors affecting invasion and persistence of broom *Cytisus scoparius* in Australia. *Journal of Applied Ecology* **39**, 721–734.

Sheppard A, Haines M and Thomann T (2006a) Native-range research assists risk analysis for non-targets in weed biological control: the cautionary tale of the broom seed beetle. *Australian Journal of Entomology* **45**, 292–297.

Sheppard AW, Hosking JR, Sagliocco J-L, Thomann T, Downey PO and Kwong RM (2006b) Biological control of brooms in Australia: an update. In *Papers and Proceedings of the 15th Australian Weeds Conference*. 24–28 September 2006, Adelaide. (Eds C Preston, JH Watts and ND Crossman) pp. 573–576. Weed Management Society of South Australia, Adelaide.

Stokes KE, Buckley YM and Sheppard AW (2006) A modelling approach to estimate the effect of exotic pollinators on exotic weed population dynamics: bumblebees and broom in Australia. *Diversity and Distributions* **12**, 593–600.

Syrett P (1991) Importation impact assessment: *Arytainilla spartiophila* for biological control of broom, *Cytisus scoparius*, in New Zealand. DSIR Plant Protection Research Report.

Syrett P and Harman HM (1995) *Leucoptera spartifoliella* Hübner as a biological control agent for broom in New Zealand. *Plant Protection Quarterly* **10**, 75–78.

Syrett P, Fowler SV and Emberson RM (1996) Are chrysomelid beetles effective agents for biological control of weeds? In *Proceedings of the 9th International Symposium on Biological Control of Weeds*. 19–26 January 1996, Stellenbosch, South Africa. (Eds VC Moran and JH Hoffmann) pp. 399–407. University of Capetown, South Africa.

Syrett P, Fowler SV, Coombs EM, Hosking JR, Markin GP, Paynter QE and Sheppard AW (1999) The potential for biological control of Scotch broom (*Cytisus scoparius*) (Fabaceae) and related weedy species. *Biocontrol News and Information* **20(1)**, 17N–34N.

Waloff N and Richards OW (1977) The effect of insect fauna on the growth, mortality and natality of broom, *Sarothamnus scoparius*. *Journal of Applied Ecology* **14**, 787–798.

Wapshere AJ and Hosking JR (1993) Biological control of broom, *Cytisus scoparius*, in Australia. In *Proceedings of the 10th Australian Weeds Conference and 14th Asian-Pacific Weed Science Society Conference*. Vol. 1. 6–10 September 1993, Brisbane. (Eds JT Swarbrick, CWL Henderson, RJ Jettner, L Streit and SR Walker) pp. 94–98. Weed Society of Queensland on behalf of the Council of Australian Weed Science Societies and the Asian-Pacific Weed Science Society, Brisbane.

Waterhouse BM (1988) Broom (*Cytisus scoparius*) at Barrington Tops, New South Wales. *Australian Geographical Studies* **26**, 239–248.

Wearne LJ and Morgan JW (2004) Community-level changes in Australian subalpine vegetation following invasion by the non-native shrub *Cytisus scoparius*. *Journal of Vegetation Science* **15**, 595–604.

Wearne LJ and Morgan JW (2006) Shrub invasion into subalpine vegetation: implications for restoration of the native ecosystem. *Plant Ecology* **183**, 361–376.

Zuloaga FO and Morrone O (Eds) (1999) *Catálogo de las plantas vasculares de la República Argentina II. Fabaceae–Zygophyllaceae* (*Dicotyledoneae*). Monographs in Systematic Botany 74. Missouri Botanical Garden Press, St Louis.

Echium plantagineum L. – Paterson's curse

Andy W Sheppard and Matthew Smyth

ABSTRACT

Paterson's curse, *Echium plantagineum* (Boraginaceae), called Salvation Jane, is a toxic winter annual pasture and cropping weed present in all states and territories of Australia, that is native to the Mediterranean region on acid soils. In southern Australia, where it often dominates grazed, fallow or disturbed land and roadsides from March to January (autumn to mid summer), it is an important and often dominant component of annual and perennial pastures. It is arguably the most widespread and costly broad-leaved pasture weed of southern Australia. When eaten by livestock it causes cumulative liver damage from pyrrolizidine alkaloid poisoning, often causing fatality, particularly of horses and other livestock with a non-ruminant digestive system. Paterson's curse competes directly with winter crops. A biological control program was started in 1972 with extensive surveys which identified 80 arthropods and six fungi as natural enemies of the genus *Echium*. Shortened to 32 species on *E. plantagineum*, 11 insects were eventually tested. The moth *Dialectica scalariella* was the first to be released in 1981, but failed to establish. This release led to a seven-year court injunction on the program following a challenge from two graziers and two apiarists, halted only after a major public inquiry and an independent cost:benefit analysis and followed by the passing of a *Biological Control Act* in 1984. In 1988 the moth was re-released and established, followed by the release and establishment of two root weevils, *Mogulones larvatus* and *M. geographicus*, two flea beetles, *Longitarsus echii* and *L. aeneus* (failed to establish), a cerambycid beetle *Phytoecia coerulescens* and a pollen beetle *Meligethes planiusculus* between 1988 and 1994. Field studies and economic analyses show that *M. larvatus* and *L. echii* had significantly contributed to a decline in the abundance of this weed. Both can kill the plant before flowering.

Key words: Boraginaceae, *Longitarsus echii, Longitarsus aeneus, Dialectica scalariella, Phytoecia coerulescens, Mogulones larvatus, Mogulones geographicus, Meligethes planiusculus, Biological Control Act*, cost:benefit analysis.

INTRODUCTION

Echium plantagineum L. (Boraginaceae), Paterson's curse or Salvation Jane, is a winter annual calcifuge herb native to the Mediterranean region, although probably endemic to the Iberian Peninsula. Introduced into all continents, it is a weed in South Africa and North and South America (Parsons and Cuthbertson 2001). *E. plantagineum* was first introduced into Australia as an ornamental, and deliberately spread. It is now common, and often abundant, throughout south-west WA, central and southern SA, south-east Qld, central and southern NSW, Vic and Tas (Fig. 1). At the limits of its range in the north it occurs as a roadside and disturbed area casual. In typical Mediterranean-type climates where annual rainfall is less than 1270 mm and temperatures range from 4°C in winter to 27°C in summer (Moore 1967) it can dominate pastures

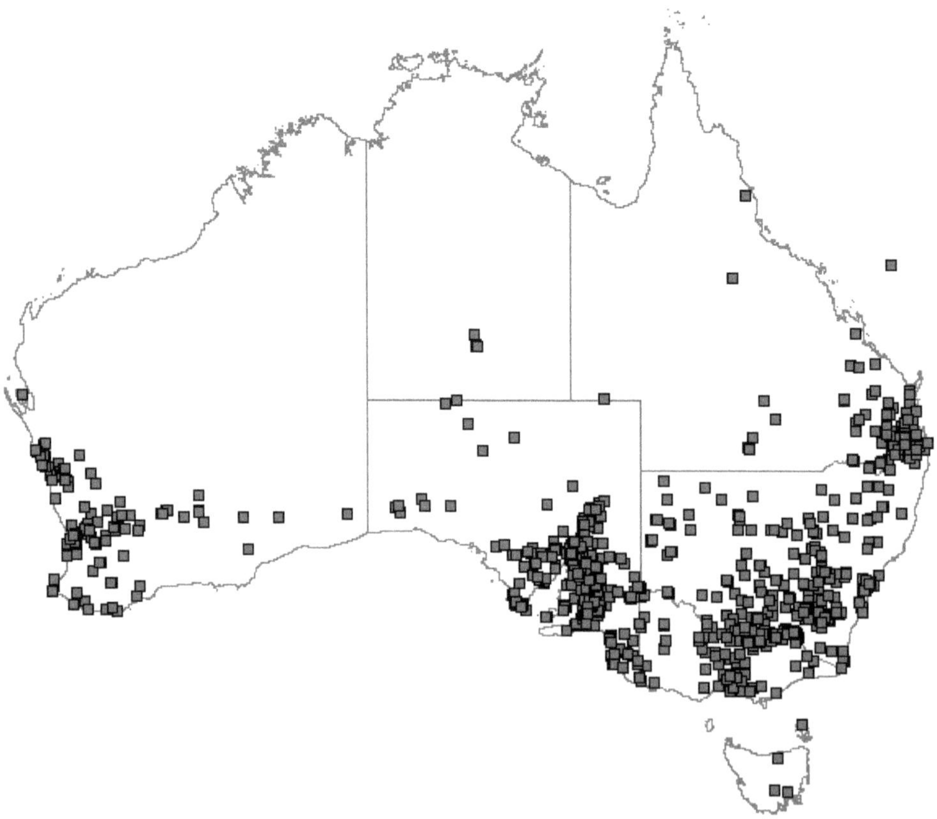

Figure 1: Distribution of Paterson's curse, *Echium plantagineum*, in Australia. Specimen data reproduced from Australia's Virtual Herbarium with permission of the Council of Heads of Australasian Herbaria Inc in May 2010.

and pasture-crop rotations, native pasture and marginal land, sometimes occurring in crops if sowing is early and/or land preparation poor.

E. plantagineum is perfectly adapted to germinate and grow rapidly in response to irregular rainfall from February to September (Fig. 2a). It has become abundant in Australia because of the predominance of sandy acid soils and a pastoral and rotational agriculture based on species-poor exotic communities (Fig. 2b). Australian native vegetation is poorly adapted to compete with *E. plantagineum* in such agro-ecosystems.

E. plantagineum contains pyrrolizidine alkaloids which cause cumulative chronic liver damage and fatality in grazing livestock, especially if substantial amounts of foliage are eaten over prolonged periods (Seaman *et al.* 1989). Monogastric livestock like horses are more susceptible than ruminants (Culvenor 1956; Culvenor *et al.* 1984), but even sheep show reduced weight gain and wool growth (Seaman *et al.* 1989). Despite this, opinions differ on almost every aspect of its weed status. While it may cause toxicity if dominant in pasture, its nutrient content and high productivity also make it a useful forage species in some seasons (Piggin and Sheppard 1995). In practice, *E. plantagineum* can be extensively grazed without severely harmful effects if it is subdominant in pastures (Moore 1971). Indeed, a key control measure called 'spray-graze' involves heavy or crash-grazing for a week or so after the application of a low rate of selective herbicide (Pearce 1972). Livestock consume large amounts of the plant over a short period, but the pasture is relatively free of the weed for the rest of the season. In crops, *E. plantagineum* reduces grain yield and contaminates grain harvests (penalties apply if seed contamination is above threshold levels; Grain Trade Australia 2010).

Paterson's curse remains one of Australia's most costly weeds because of the extent of its distribution and local abundance. In 1985 it was considered to infest an estimated 33 million ha at an annual cost to agriculture of c. $30 million (Industries Assistance Commission 1985), reducing pasture productivity and carrying capacity through competition with more beneficial pasture species. Infestations also decrease land values. More recently, the recoverable cost of *E. plantagineum* was estimated at $90 million per year based on the increased carrying capacity of livestock if

Figure 2: a) The plant. b) Paterson's curse (*Echium plantagineum*) infestation near Gundagai, NSW. Photos: a) A Sheppard, CSIRO. b) CSIRO.

biological control successfully removed the weed within a season (Nordblom *et al.* 2001).

Paterson's curse passes autumn and winter as a flat to semi-erect rosette of broadly ovate leaves with a well-developed tap root system, rising to 20–60 cm in height after the flowering stem(s) are produced in spring. Flowers on unfurling cymes are showy purple and broadly funnel-shaped and strongly attractive to pollinators (Fig. 2). The fruit has four nutlets with small brown to black subglobular seeds (Parsons and Cuthbertson 2001). Individual plants produce up to 10 000 seeds and create a persistent seed bank (up to 10 years) of up to 30 000 seeds m^{-2}. Summer rainfall can extend flowering but eventually plants senesce in summer.

In Australia, *E. plantagineum* has few competitors other than subclover in annual pastures (Piggin and Sheppard 1995) and poor suppression by natural enemies (Sheppard *et al.* 2001). In its native range, high-density stands are rarely observed. The abundance of *E. plantagineum* is influenced by strong interspecific competition from a diverse range of winter annual herbs in the same functional group in the native grazed systems it inhabits (Fernandez-Alés *et al.* 1993; Lavorel *et al.* 1999), maintaining much lower plant densities and resulting in much lower soil seed banks (Grigulis *et al.* 2001). These contrasting competitive situations in the native and exotic ranges also alter the effect of grazing on *E. plantagineum*. Grazing increases *E. plantagineum* abundance in Australia (Smyth *et al.* 1992) but decreases it in Mediterranean Europe (Forrester 1992), where highest abundance is found at moderate grazing intensities (Noy-Meir *et al.* 1989). Competition also reduces the chance of a seed in the seed bank successfully germinating and establishing in its native range (Grigulis *et al.* 2001).

Control options for *E. plantagineum* are well documented in Parsons and Cuthbertson (2001). They include herbicides, spray-grazing, crash-grazing by sheep (Smyth *et al.* 1997) and spray-topping to reduce seed production, but each of these options has significant economic, practical and animal health issues. The seed bank ensures that effective control in pasture can be achieved only through a dense competitive perennial grass cover (Sheppard and Smyth 2002), which can persist only in high-rainfall regions. On arable land, cropping is an effective method of control – most seed germinates when soil temperatures are high, allowing seedlings to be removed by cultivation, use of unselective herbicides prior to sowing (spray fallow) or selective herbicides within the crop.

BIOLOGICAL CONTROL HISTORY

Classical biological control was first suggested in 1928. It was listed as a priority of biological control by the Australian Weeds Committee and approval from the Australian Agricultural Council was granted in 1971 (Delfosse and Cullen 1981). Only Australia has undertaken a biological control program against *E. plantagineum*. CSIRO began work in 1972. A list of potential European biological control agents was proposed; specificity testing has led to nine of these being approved for release.

Figure 3: Biological control agents released against Paterson's curse. a) *Dialectica scalariella*. b) *Mogulones geographicus* and *M. larvatus*. c) *Phytoecia coerulescens*. d) *Longitarsus echii*. e) *Meligethes planiusculus*. Photos: CSIRO.

Following initial unsuccessful releases of *D. scalariella* (Fig. 3a), the biological control program was brought to a halt between 1981 and 1988 when a High Court injunction filed by two apiarists and two graziers from SA was granted against CSIRO (Cullen and Delfosse 1985). Key to the resolution of this conflict and the judicial consequences of the Australian Government releasing biological control agents was the passage of the *Biological Control Act* in 1984 (Commonwealth of Australia 1984). Under the provisions of this Act, the Industries Assistance Commission (IAC) conducted an independent inquiry into whether *Echium* spp. should be targeted for biological control (Delfosse and Cullen 1985; Industries Assistance Commission 1985). This inquiry included a very detailed cost:benefit analysis, probably the most comprehensive of any weed control project anywhere in the world at that time. The IAC recommended that the *Echium* biocontrol program would be in the nation's interest, with a benefit:cost ratio of 6.7–8.6 depending on the likely success of biological control reducing weed density to 'low' or 'negligible' levels (costs to bee-keeping and pollination and costs of research were estimated at $17 million, while benefits to livestock and cropping industries and public control costs were estimated at $114–149 million in 1985 prices), if the program were highly successful within 15 years. The IAC therefore initiated several actions leading to the declaration, on 3 March 1986, of all *Echium* spp. as 'target organisms' for biological control, and eight insects (see below) as 'agent organisms' under the Act. From 1986 to 1988, all seven Australian states and territories passed complementary biological control legislation and declared all *Echium* spp. and those eight insects for biological control. The Australian Weeds Committee, the Standing Committee on Agriculture and the Australian Agricultural Council all recommended that the program be restarted. Following the IAC inquiry, the plaintiffs challenged the validity of the *Biological Control Act*; a preliminary legal decision upheld the Act on 16 March. In the meantime, Australian agencies other than CSIRO were within their legal rights to re-initiate the biological control work. The Keith Turnbull Research Institute in Vic started collecting and shipping *D. scalariella* into Australia. The injunction on CSIRO was lifted on 17 November 1986.

The program then became a long-term collaborative venture between CSIRO and state agencies to complete host-range testing and obtain any remaining agent release permits. This led to a nationwide redistribution program for selected agents, with local community groups, which continued until the middle of the 2000s.

PLANT TAXONOMY

There are two other herbaceous species of *Echium* of European origin naturalised in Australia, *Echium vulgare* L. and *Echium italicum* L., in additional to some Micronesian horticultural species grown in gardens. The former species is restricted to cooler tableland districts of south-eastern Australia and the latter is a rare naturalised species in southern NSW and the south-east of SA (Piggin 1977; Forcella *et. al.* 1986; Parsons and Cuthbertson 2001). There are no species of *Echium* native to Australia.

EXPLORATION

Surveys, by Tony Wapshere, initially covered Mediterranean Europe and North Africa; they later involved Jean-François Vayssières and were concentrated in the main evolutionary centre of *Echium* in southern France and Spain, Portugal and Morocco, where the climate is similar to that of Australian areas infested with *E. plantagineum* (Wapshere 1974, 1980).

The surveys listed 80 arthropods and six fungi as natural enemies of the genus *Echium* (Wapshere 1985, earlier unpub. lists). Further surveys and studies by Carey Smith, Guy Forrester and Andy Sheppard from 1987 to 1992 led to the number of insects known to breed on *E. plantagineum* being reduced to eight species of Lepidoptera, eight species of Coleoptera, three Diptera, 12 Hemiptera and a gall mite (C Smith, unpub. CSIRO reports on the European *Echium* research program 1987–89; A Sheppard and G Forrester, unpub. 1992). Not all these species, however, use *E. plantagineum* as the primary or regular host. In addition to the main potential biological control agents discussed below, the cyme feeding moth *Tinagma balteolella* F.R. (Douglasiidae), the pollen beetle *Meligethes tristis* Sturm. (Nitidulidae), the mite *Aceria echii* Canestrini (Eriophyidae) and the gall fly *Contarinia echii* Kieff. (Cecidomyiidae) have regular resident populations on *E. plantagineum* in the native range. The first three were briefly considered as potential agents. Similarly, the pathogens, the leaf blotch fungus *Cercospora echii* Wint. (Moniliales: Dematiaceae) and the downy mildew *Erysiphe horridula* (Wallr.) Lev. (Erysiphales: Erysiphaceae) were considered (the latter because of high observed damage), but both seem to have had too broad a host-range within the Boraginaceae.

The exploration phase spanned 20 years. Prior to the *Wildlife and Countryside Act* 1981 (superseded by the *Environmental Protection and Biodiversity Conservation Act* 1999), risk assessment of biological control agents only had to consider potential impacts on commercial plants, of which there are few in the Boraginaceae. Prior to 1981, therefore, the program considered potential biological control agents which could develop on some native plants, of which there are many in Australia in the Boraginaceae. Several agents that were eventually released had the capacity to develop on native species. The agents finally selected, for which release permits were obtained (except *E. terminella*), and the manner with which they attack the plant are shown in Figure 4.

CANDIDATES

To assist efficacy assessment of potential biological control agents, field exclusion experiments in Europe measured the impact of natural enemies in native populations. Insecticide and fungicide treatments were applied in three field experiments, one in southern France and two in Portugal (G Forrester, C Smith and A Sheppard, unpub. data). Experiments were also undertaken to look at the interaction between natural enemies and interspecific plant competition on Paterson's curse (Sheppard 1996). No effects of natural enemies on *E. plantagineum* were found during these experiments.

Natural enemies

Larvae and adults of the native moth *Utethesia pulchelloides* Hamps. have been observed on *E. plantagineum* throughout NSW and in northern Vic (Common 1990). The native flea beetles *Longitarsus victoriensis* Blackburn and *Longitarsus* sp. 'Victoria' (C Reid, *pers. comm.* 1995) cause shot holes on *E. plantagineum* at many sites throughout NSW and northern Vic and have been confused with introduced flea beetles. The larvae probably feed in the roots and likely spread from native hosts, e.g. *Cynoglossum* spp. on which they are common. One infestation of Paterson's curse near West Wyalong (NSW) in 1992 was killed by this insect (M Smyth and J Moorhouse, *pers. comm.* 1992). Other native insects regularly found developing on the plant include the moths *Amsacta marginata* (Don.), *Spilosoma glatignyi* (Le Guill.) and *Stomopteryx isoscelixantha* (Low), the Rutherglen bug *Nysius vinitor* (Berg) and the red-legged earth mite *Halotydeus destructor* (Tucker). The exotic leaf blotch fungus *C. echii* has also been found in Australia; it was first detected in 1996 (Floyd *et al.* 1996).

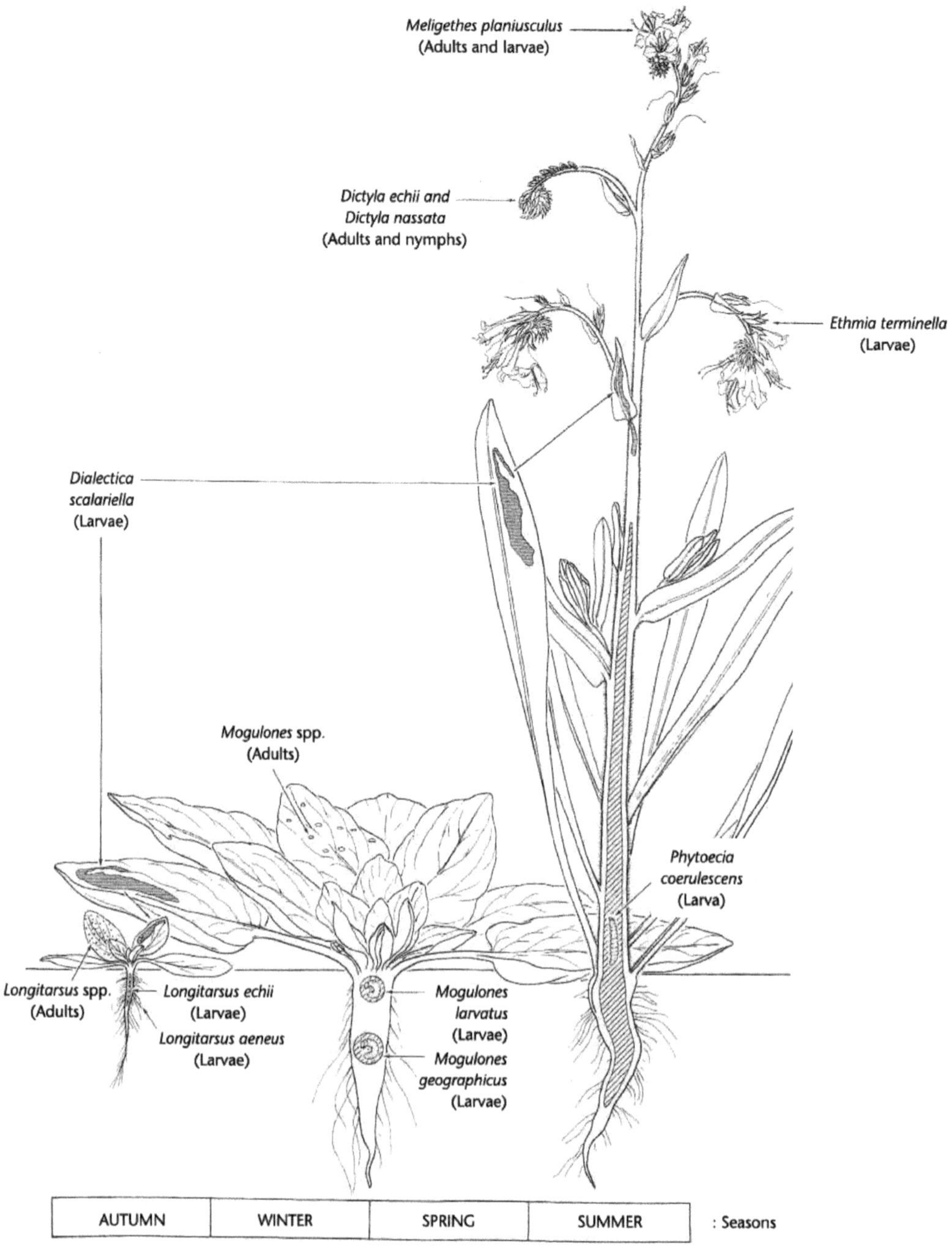

Figure 4: The points of impact of different biological control agents for Paterson's curse, *Echium plantagineum*.

Natural enemies tested but not officially released

Ethmia bipunctella **Fabr (Ethmiidae).** This bivoltine moth feeds externally in spun webs in the shoot and flowers. It was imported into quarantine in Australia in 1990 for host-specificity testing. The tests showed that this insect's host-range includes Australian species in the genera *Cynoglossum*, *Myosotis* and *Trichodesma*. Although successful development on the native Boraginaceae was low (<5% of the controls), a decision was taken not to release this moth. Activities on this insect stopped in 1992 (CSIRO Entomology 1993).

Ethmia terminella **Fletcher (Ethmiidae).** This moth feeds in a web along developing cymes. It proved very difficult to rear, with high levels of parasitism in field-collected material. A detailed comparative field study of this and other flower-feeding insects in Europe in 1994 showed that this insect occurs more abundantly on the large, longer-lived and later-flowering *E. vulgare* and *E. italicum* and performed relatively poorly on *E. plantagineum*. A decision was made to discontinue work on *E. terminella* in 1995 (CSIRO Entomology 1993; CSIRO and VDNRE 1998).

***Dictyla nassata* Puton and *D. echii* Schrank (Tingidae).** These bugs are commonly found on *E. plantagineum* in the native range and, with two to three generations a year, were quickly identified as having biological control potential in limiting seed production (Wapshere 1985; Compton 1988). Testing them against commercially relevant non-target test plants was initiated in the early 1980s (Vayssières 1983); it led to the insects being declared 'agent organisms' against Paterson's curse under the *Biological Control Act* in 1986 (IAC 1985) and therefore permitted for release. By the 1990s, however, testing against Australian native plants was a requirement. Testing in quarantine in 1995 showed these insects had too broad a host-range, including Australian native plants (Shepherd 1995a), so releases were never made.

Natural enemies released

***Dialectica scalariella* Zell. (Gracillariidae).** This is a multivoltine leaf-mining moth with five to seven generations a year, slowing when winter temperatures drop below 7°C (Shepherd and Morley 1992). Host-specificity testing revealed that this moth has quite a broad host-range within the Boraginaceae (Wapshere and Kirk 1977) where host-use was best defined by the presence of large enough leaves to support mines throughout the year. The release permit was issued prior to the *Wildlife and Countryside Act* (1981) in 1979 (IAC 1985). The moth was first released in 1980, but failed to establish (Delfosse *et al.* 1987). These releases instigated a court injunction on the project until 1988, when further successful releases were made throughout temperate Australia (Morley 1990; Shepherd 1990, 1995b), where it now occurs throughout the distribution of *E. plantagineum* (CSIRO 1998). In some upland areas it survives on *E. vulgare* and other broad-leaved herbaceous Boraginaceae. Only under conditions of drought (autumn or spring) has it been observed to reach damage levels necessary to kill Paterson's curse, and then only rarely. Most plants are free from attack in the main winter rosette growth period, and when attacked respond to damage by producing more leaves at a faster rate (Smyth 1993) at no apparent loss to plant reproductive effort (Dodd 1992). Laboratory trials showed that *D. scalariella* damage can significantly reduce plant size and survival, especially in the presence of grass competition. However, these levels of damage could not be reproduced in the field where the lack of a host over summer, and high mortality in winter, severely limit population growth of the moth (Smyth 1993; CSIRO and VDNRE 1998; Sheppard *et al.* 1999). This agent therefore contributes little to the biological control of the target. Its release led to the discovery of a similar native species of *Dialectica* (*Dialectica aemula* (Meyrick)) using *E. plantagineum* (Kumata and Horak 1997). *D. scalariella* has since been found on the west coast of New Zealand (Hoare 2001).

***Mogulones larvatus* Schultze (Curculionidae).** This is a univoltine crown weevil (Fig. 3b) that is a specialist on the genus *Echium*. Adults, which are about 4 mm in length, emerge in spring to early summer to feed on flowering hosts before aestivating over summer. Decreasing autumn day length and rainfall induce the aestivating adults to congregate, feed and become reproductive on the rosette leaves, particularly of *E. plantagineum*. Peak adult emergence post aestivation is in March, often well before autumn rain, and this severely limits the weevil population in drier regions or seasons. The adults feed on the foliage and make regular 5 mm holes. However, adult damage is minor compared to that caused by the larvae. After mating, females lay single eggs in the petioles and leaf blades of *E. plantagineum*. A single female can lay an average of 450 eggs (to >1000 per individual) with oviposition continuing until late spring. Eggs hatch four or five days after being laid (at 25°C) and there are three larval instars (Vayssières 1986). Larvae mine the petiole and crown; attacked plants can be easily identified by the presence of a black necrotic discharge in the crown. Larvae eat the meristem of the rosette and one or two larvae can kill a small rosette, although a large plant can support many larvae (Sheppard *et al.* 2002; Buckley *et al.* 2005). Larvae require a cold period to complete development and leave the plant about 50 days after hatching (at 25°C) once feeding is complete (CSIRO and VDNRE 1998). Pupation occurs in the soil and new adults emerge in spring. Adults feed for several weeks before aestivating over summer in pasture litter.

Testing and studies in Europe showed that this weevil was a sufficient specialist for release in Australia (Vayssières and Wapshere 1983) and a potentially effective agent because meristem damage starts in autumn when the rosette is small (Wapshere 1985). It was declared an 'agent organism' against Paterson's curse under the *Biological Control Act* in 1986 (IAC 1985), which permitted its introduction into Australia for the biological control of *E. plantagineum* in 1990 (Piggin and Sheppard 1995). Releases failed until 1992, at which time the aestivation and cold treatment requirements became understood. Subsequent releases led to very rapid population increases at some initial release sites, causing massive dieback of plants in high-density infestations prior to flowering. *M. larvatus* clearly showed a capacity to eat out its host while simultaneously spreading to new areas. Studies showed:

- the capacity for this agent to avoid competition with *M. geographicus* on the plant (Forrester 1993);
- the levels of potential per plant impact in the field with and without pasture competition in Australia (Sheppard *et al.* 2001);
- oviposition patterns in response to host patch size and density (CSIRO and VDNRE 1998; Shea *et al.* 2000);
- other attack patterns of this weevil (Sheppard *et al.* 1999).

This agent became part of a major redistribution program across all southern states that, to 2005, involved 1760 releases of nearly 1 million individuals (CSIRO Entomology 2006). Overall, these releases led to 66% successful establishment.

M. larvatus caused over 80% mortality of plants prior to flowering at the field site at Yanco, NSW over eight years, albeit with substantial fluctuations in numbers of both plants and agents (Fig. 5b). Percentage of plants attacked or number of larvae per gram plant weight in the field were correlated with within-season plant mortality. Mortality rate was a good predictor of per unit area losses of biomass and seed production (CSIRO Entomology 2006). Drought and autumn rains that arrive later than March limit aestivation survival because adults then emerge in the absence of the host. Thus *M. larvatus* has been less effective in WA, where autumn rains are regularly later than March. *M. larvatus* eggs and larvae are

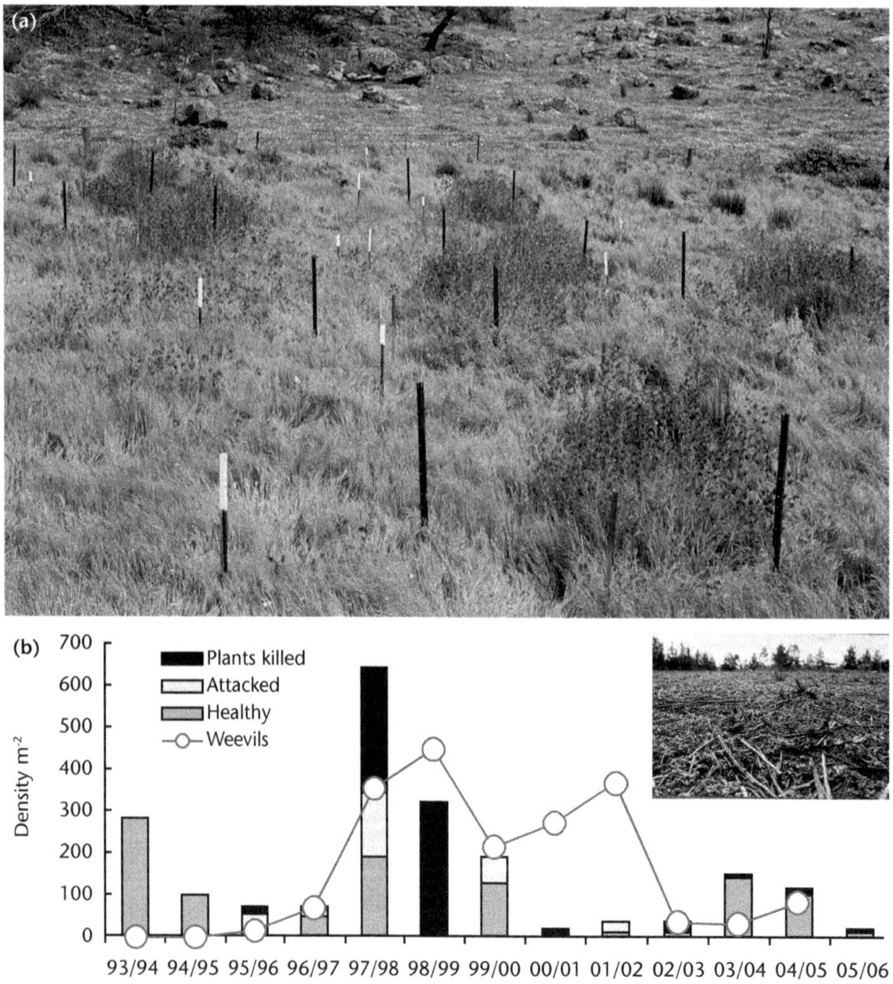

Figure 5: The impacts of biological control agents on Paterson's curse. a) Quadrats treated with insecticide within an infestation of *E. plantagineum* at Euroa, Vic in September 2003, containing *Mogulones larvatus* and *Longitarsus echii*. Photo: T Morley, DPI Vic. b) The impacts and abundance of the crown weevil *Mogulones larvatus* on Paterson's curse over 13 years at the Yanco, NSW release site. Photo: A Sheppard, CSIRO.

located in the above-ground shoot that can be eaten or trampled by grazing animals, especially sheep (Smyth and Sheppard 1996). Nonetheless, up to a third of sites followed from 2000 to 2006 had plant mortalities greater than 50% (CSIRO Entomology 2006) (Fig. 5a). Ecological modelling predicted that *M. larvatus* would be most effective in areas of early (March) autumn rainfall (Smyth and Sheppard 1996; CSIRO and VDNRE 1998; CSIRO Entomology 2006; Smyth *et al.* 2000; Huwer *et al.* 2002, 2005).

Ecological models have since described the population dynamics of the interaction between *M. larvatus* and *E. plantagineum* and how this leads to the suppression of plant density (Sheppard *et al.* 2002; Buckley *et al.* 2005). The role of this agent as part of an integrated weed management strategy has also been explored (Smyth and Sheppard 1996; CSIRO and VDNRE 1998; CSIRO Entomology 2006; Smyth *et al.* 2000; Huwer *et al.* 2002, 2005). *M. larvatus* impact was also used as the basis for a cost:benefit analysis (Nordblom *et al.* 2002a) and the value of further redistribution efforts (Nordblom *et al.* 2002b).

***Mogulones geographicus* Goeze (Curculionidae).** This root weevil (Fig. 3b) has very similar life-cycle and size to those of *M. larvatus* except that the adults emerge from aestivation four to six weeks later in autumn, larvae (Forrester 1993) and require no cold period for pupation (CSIRO and VDNRE 1998). Eggs are laid in the leaf stalks of the larger young rosettes from autumn to spring and females oviposit 250 eggs on average (CSIRO and VDNRE 1998). The first instar larvae burrow down the leaves; later instar larvae cause the most damage in the root cortex (Vayssières and Wapshere 1983; Vayssières 1986; Forrester 1993). In its native range, *M. geographicus* is the most common and damaging root herbivore found on *E. plantagineum*. In 12 of 15 sites where *M. geographicus* was present 45% ± 7.3% plants were attacked and 53% of root cortex of the plants at one site was eaten (CSIRO and VDNRE 1998). Late-stage larval feeding is protected from grazing stock, and the later adult emergence is an advantage in a drought. *M. geographicus* was predicted to be the most damaging agent once released in Australia (CSIRO and VDNRE 1998). Host-specificity testing showed that this weevil was specific enough for release (Vayssières and Wapshere 1983) and it was declared an 'agent organism' against Paterson's curse under the *Biological Control Act* in 1986 (IAC 1985). It was first released in autumn 1993 at Tarcutta, NSW (CSIRO Entomology 1993).

It was harder to confirm establishment of *M. geographicus* than of *M. larvatus* as its presence in the lower root cortex and damage to the plant are harder to assess in the field. In a tub trial in Australia, *M. geographicus* reduced seed production by 30% following 10% root cortex removal. There was no evidence that plants underwent compensatory root growth to replace damaged tissue. In a controlled insecticide experiment in Australia (Sheppard *et al.* 2001), an *M. geographicus* treatment found no direct effect of *M. geographicus* compared to *M. larvatus*. These and later studies, however, assisted in quantifying *M. geographicus* impact in the field, by demonstrating that in general larvae consumption of root tissue leads to a proportional loss in shoot tissue and seed production. *M. geographicus* tends to attack the larger plants, therefore plant size differences between attacked and unattacked plants need to be taken into account in the field when evaluating attack and impact levels. Following releases in Australia, it was found that *M. geographicus* dispersed from release sites before forming dense local populations. No field evidence has yet been found that *M. geographicus* kills plants prior to flowering in the same way *M. larvatus* does. However, high damage levels of attacked plants have been recorded, suggesting that there is a substantial impact at the population level although it is harder to measure (CSIRO Entomology 2006). Under the national redistribution program, to 2006, more than 290 releases were made of more than 135 000 individuals. Releases had similar establishment rates (56%) to those of *M. larvatus*, but the larger numbers of individuals per release increased the likelihood of *M. geographicus* establishment. Attack rates post release remained at 10–20%. Populations of this weevil have increased post release at several locations around Australia; several thousand weevils can be collected in spring. Monitoring has yet to quantify the effectiveness of *M. geographicus* on *E. plantagineum* due to the difficulty in quantifying root damage in the field.

***Longitarsus echii* Koch. (Chrysomelidae).** The life-cycle of this flea beetle (Fig. 3d) is similar to that of the root weevils, except that adults do not emerge in spring. They remain in the soil until winter, when rainfall stimulates their emergence from earthen cells up to 20 cm below ground. *L. echii* is univoltine. Adults, about 3 mm in length, eat shot holes on the foliage of *E. plantagineum* and after one to two weeks lay eggs directly on the tap root. The thin larvae hatch after two to three weeks. They feed on and in the cortex of the tap root and secondary roots, and may attack the undersides of leaf petioles that are prostrate on the ground. When feeding is complete, larvae leave the plant and pupate in the soil. If there are sufficient larvae on a plant the entire root system will be eaten. Feeding can continue into the crown and the elongating flowering stem, killing the plant. Host-specificity testing confirmed that *L. echii* was confined to *Echium* spp. (Wapshere 1982) and this agent was introduced into quarantine in Australia following release approval in 1979 (IAC 1985). Prior to release, it was anticipated that *L. echii* would be well adapted to drier regions and complementary to the weevils elsewhere. First releases of resynchronised quarantine-reared adults were in 1996, then cultures and redistribution activities were set up in all states (CSIRO Entomology 2006). By 2006, 712 releases of over 311 000 individuals had been made across Australia. Establishment rates (84%) and plant attack rates (22–56%) were the highest of any agent at release sites through drought years and were far better than expected, even under heavy grazing and drought. The first site in Australia to record high population levels was at Tarcutta, NSW. *L. echii* was released there in 1996; by 2001 86% of plants were attacked by the larvae. Since 2001 the attack rate of *L. echii* has remained high. The insect population grew rapidly until monitoring stopped in 2005. Forty percent *E. plantagineum* mortality was recorded in 2002 and overall (using a larval number–plant size relationship) plant biomass and seeding were reduced by over 80% per unit area. Intraspecific competition took its toll in that year, however, and the smaller beetle population in 2003 resulted in only an 11% reduction of plant biomass. In 2004 the *L. echii* population had recovered; this led to complete mortality of *E. plantagineum* prior to flowering in both the original 50 ha release paddock and the surrounding 700 ha property. The nearest patch of live *E. plantagineum* was 1 km away, where part of the *L. echii* population took refuge. This allowed the *E. plantagineum* population to reappear in the original release site in 2005, when a 5% biomass reduction was observed. At a second release site, a 200 m wide property border, the beetle population reached 700 larvae per metre and no *E. plantagineum* survived to flowering. Despite the boom–bust cycles observed throughout the drought, the capacity of this agent to spread should ensure sustained year-to-year impact. *L. echii* has proved to be the most damaging agent in many areas, particularly in drier Mediterranean climate areas such as WA. On several farms where this agent has established, *E. plantagineum* density has been suppressed well below economic thresholds. *L. echii* can now be found in large numbers from Tarcutta to Wagga Wagga, NSW. Controlled experimental studies have shown it to be very effective in the presence of competitive pasture grasses (Smyth and Sheppard 2002), despite significant grazing pressure (Smyth *et al.* 2004).

***Longitarsus aeneus* Kutsch (Chrysomelidae).** This flea beetle is the smallest of two univoltine flea beetles, about 2 mm long, which testing showed to be specific enough for release as a biological control agent against *E. plantagineum* (Wapshere 1982). It was first introduced into quarantine in Australia following release approval in 1979. Adults emerge in autumn and eggs are laid in the soil at the base of the rosettes of its host. The thin larvae emerge and migrate down the outside of the root system to the external surface of the finer roots and root hairs, where larval development is completed. Despite several shipments and multiple culturing conditions, attempts to rear *L. aeneus* in quarantine failed. Direct releases of poorly synchronised adults were nonetheless attempted at Yanco, NSW in 1993, but no recoveries were made.

***Phytoecia coerulescens* Scop. (Cerambycidae).** The longicorn beetle (Fig. 3c) is a univoltine stem- and upper root-boring beetle. Adults emerge in spring and lay eggs into the developing flower stems. Larvae bore up the stem before changing direction to bore into the root. This long-horn beetle only survives well on large stems, so *E. vulgare* is often a preferred host (Kirk and Wapshere 1979). Host-specificity studies carried out in Europe demonstrated this insect was sufficiently specific for release. Damage occurred rather too late in the season to be a problem, except perhaps to thinner-stemmed plants (Kirk and Wapshere 1979). A permit was obtained in 1979 prior to the *Wildlife and Countryside Act* 1981. The longer flowering period of *E. plantagineum* and the existence of flowering plants throughout the summer in Australia were thought likely to assist this insect to multiply rapidly and attack a higher proportion of plants than it does in Europe, where incidence is quite low. *P. coerulescens* was reared through three generations in quarantine to resynchronise it then was released into field cages, including on *E. vulgare*, in 1995. This insect successfully established but has never been observed in high numbers. An experiment found no differences in plant performance between attacked and unattacked plants, despite 50% of the stems on attacked plants having been bored (CSIRO 1998). Based on these results, *P. coerulescens* was not included in the nationwide redistribution program. In another study, larvae of *P. coerulescens* were seen in a native species *Cynoglossum australe* R. Br., in the Boraginaceae (M Smyth, *pers. obs.* 2000), but the level of damage was very low and the larvae were small, restricted by the size of the flowering stem.

***Meligethes planiusculus* Heer (Nitidulidae).** The pollen beetle (Fig. 3e) overwinters as an adult before becoming active in early spring, congregating on the elongating and first flowering stems to feed, mate and oviposit. Eggs were mostly laid in the terminal pre-flowering buds of the flowering cyme, usually in protected positions at the base of the sepals or bracts. Upon hatching, larvae mined through the calyx and petals to the inside of the flower bud, where they feed on the anthers, pollen and immature ovules before moving to younger flowers. Once development was complete, the larvae drop from the flowers to pupate in the litter/soil. Adults emerge 10 days later, and feed mostly on pollen and developing seed in open flowers. This species is largely univoltine, but undergoes a partial second generation if newly emerged adults emerge before day length starts to decrease. Overwintering takes place in litter or soil (Compton 1988; CSIRO and VDNRE 1998). Host-specificity testing showed that *M. planiusculus* is restricted to the genus *Echium* and will preferentially oviposit on *E. plantagineum* (Swirepik *et al.* 1996; CSIRO and VDNRE 1998). Impact studies showed it to be the most damaging flower- and seed-feeding insect on *E. plantagineum* in the native range, achieving up to 65% reduced seed production (Swirepik *et al.* 1996). The plant can respond by producing more seeds in unattacked flowers, but it is unlikely that compensation would be complete (CSIRO and VDNRE 1998). This, and the earlier start and extended flowering period in Australia (five versus three months in the native range), suggested that *M. planiusculus* could have a larger second generation in Australia, increasing its effectiveness. A release permit was obtained in 1996 and first releases were made that year. This was the first time that a nitidulid beetle had been used as a biological control agent. Its redistribution became part of the national Paterson's curse biocontrol agent redistribution program through which 174 releases of over 230 000 individuals were made (CSIRO Entomology 2006). Successful establishment was recorded at 45% of release sites. To date, *M. planiusculus* has not reached population densities high enough to limit the seeding of *E. plantagineum* at any known site that can be measured by recording the number of seed per length of flower cyme (Smyth *et al.* 1997). It appears able to spread quite rapidly, but its effectiveness as an agent has so far proved negligible.

DISCUSSION

The Paterson's curse program was the first national biological control program, originally involving government agencies in all southern states, Qld and CSIRO. The program standardised evaluation using three tiers of monitoring to measure establishment, plant attack rates and damage at different spatial scales (CSIRO Entomology 2006). The program is now managed at local levels by landowner groups.

Surveys of farmers in 1996 (n=133) and 2005 (n=46) showed that the majority considered their Paterson's curse problem had remained the same or had decreased (CSIRO Entomology 2006). Loss of production from a heavily infested pasture remained as high as 45%, with a mean area of 114 ha per property costing $5200 each in herbicide alone. There was an average loss of 13 working days in 2005. Despite this, half the respondents in 2005 believed biological control decreased *E. plantagineum* and improved pasture quality. The mean decrease in *E. plantagineum* abundance across all properties was 24%, with 15% of farmers recording a decrease of 70% or greater. They believed there was a potential to reduce *E. plantagineum* by an average of 62%. Reduced *E. plantagineum* had led to a 31% reduction in herbicide use: some farmers with significant historical problems with Paterson's curse stated that they no longer needed to treat *E. plantagineum* on their properties. Importantly, the reductions in *E. plantagineum* abundance were also translating into production gains, with 9% increases in stock condition and 6% increase in stock numbers. On average, this translated to an extra 25 cattle and 90 sheep per property.

Updating the 1985 Industries Commission Analysis, Nordblom *et al.* (2001, 2002a) estimated the discounted (5%) net present value of the benefits during 1972–2015 as $259 million with a benefit:cost ratio of 14, and during 1972–2050 as $916 million with a benefit:cost ratio of 47. These estimates were revised by Page and Lacey (2006) to a cost of $23.1 million and a net present value of $1.2 billion and a benefit:cost ratio of 52 for the same time period.

The program also showed the value of a base in the target's native range. The CSIRO European laboratory acted as a base for much more than exploration. The biology and impact of the agents was studied there, through Masters and PhD studies, in collaboration with Imperial College London and the University of Evora in Portugal. Permanent study plots on the University of Evora field station allowed research into the factors limiting the success of *E. plantagineum* in the native range and comparison with factors promoting invasion in Australia (Grigulis *et al.* 2001). Insecticide and fungicide trials were used to assess the impacts of potential agents in the native range.

These showed that pasture competition was the main limiting factor in the native range, despite the subsequent success of biological control in Australia. The effectiveness of biological control is not therefore dependent on natural enemies being the dominant factor determining abundance of the plant in its native range.

Distribution and establishment of the agents occurred during one of Australia's longest droughts (10 years). It is very likely that the drought conditions favoured *E. plantagineum* over its agents, given the long-lived seed bank. As conditions improve, greater benefits are likely to be observed. The drought revealed difference between agents, with *M. larvatus* proving more effective in areas with an early season break (autumn rains) and *L. echii* being highly effective even through drier areas, notably in Western Australia. This has also shown that effective suppression of *E. plantagineum* has been achieved through the employment of more than one agent (Fig. 5a).

The program had the potential for non-target damage, as several of the agents were released prior to the risk assessment requiring consideration of non-target impact on native plants. *Dialectica scalariella* and *P. coerulescens* were subsequently shown to develop through to adult on Australian native species in the field, but observed damage levels have not been high (M Smyth, *pers. obs.* 2000).

The biological control program led indirectly to the passing of *Biological Control Acts* in Australian states and territories. This was the first legislation anywhere in the world that was designed to manage conflicts of interest around biocontrol programs when potential benefits outweigh costs, and to protect agencies undertaking biological control under such circumstances from common law damage litigation (Cullen and Delfosse 1985). The application and use of these Acts has not significantly changed the practice of biological control in Australia, however, and the legislation has never been tested in court for weed biocontrol.

The lack of a suitable plant pathogen as a biological control agent resulted from lack of specificity more than potential agent efficacy as seen in the native range. For the arthropod agents, predictions made on efficacy before introduction were not confirmed. *M. geographicus* was predicted to be the most effective agent, but it proved to be only the third most effective at attacking the rosette and the least effective agent against the roots.

The program was integrated into a weed management program using sublethal herbicide doses and grazing to maximise agent survival (Smyth and Sheppard 1996; Huwer *et al.* 2002, 2005), but discussing that is beyond the scope of this chapter. By understanding the phenology of the agents, other management practices can be timed to maximise biocontrol effectiveness.

This national biological control program involved 35 years to the point where landholders own the benefits and the further distribution of the agents, with more than $23 million in research investment. The program released seven agents, three of which have paid for the research investment 50-fold, a return which is still increasing. The full long-term impacts and benefits of the program are still to be realised, but the benefits to date are already substantial.

ACKNOWLEDGEMENTS

We gratefully acknowledge the contribution within CSIRO of Jim Cullen, Ernest Delfosse, Guy Forrester, Siraj Hasan, Bobby Lewis, Carey Smith, Anthony Swirepik, Jean-François Vayssières, Tony Wapshere; within Vic, Ros Shepherd, Kerry Roberts, Tom Morley and Raelene Kwong; within NSW, Barry Sampson, Hugh Milvane, John Hosking and Paul Sullivan; within SA, Ken Henry, Susan Ivory; within WA, Paul Wilson and Alan Lord. We also acknowledge the many technical staff who contributed to this biological control program.

REFERENCES

Buckley YM, Rees M, Sheppard AW and Smyth M (2005) Stable coexistence of an invasive plant and biocontrol agent: a coupled plant-herbivore model. *Journal of Applied Ecology* **42**, 70–79.

Common IFB (1990) *Moths of Australia*. Melbourne University Press, Melbourne.

Commonwealth of Australia (1984) *Biological Control Act 1984*. No. 139 of 1984.

Compton JAF (1988) Potential for the biological control of *Echium plantagineum* L. (Boraginaceae) with flower-eating insects, with special reference to *Meligethes planiusculus* Heer (Col.:Nitidulidae). MSc thesis. Imperial College of Science and Technology, University of London.

CSIRO Entomology (1993) *Final Report: Biological Control of Paterson's Curse*. Research report for the Meat Research Corporation, MLA, Locked Bag 991, North Sydney NSW 2059.

CSIRO Entomology (2006) *The Biological Control of Paterson's Curse and Scotch Thistles. A Long-term Investment by Grazing Industries, State and Federal Govern-*

ments. Final report on behalf of CSIRO, WA Dept of Agriculture, Vic Dept of Primary Industry, NSW Dept of Primary Industries and SA Research and Development Institute, Project ECO10D, for Meat and Livestock Australia and Australian Wool Innovation.

CSIRO and VDNRE (1998) *Project CS.209 Biological Control of Paterson's Curse Phase III: Final Report.* Research report for the Meat Research Corporation, MLA, Locked Bag 991, North Sydney NSW 2059.

Cullen JM and Delfosse ES (1985) *Echium plantagineum*: catalyst for conflict and change in Australia. In *Proceedings of the 6th International Symposium on Biological Control of Weeds.* 19–25 August 1984, Vancouver. (Ed ES Delfosse.) pp. 249–292. Agriculture Canada, Ottawa.

Culvenor CCJ (1956) The alkaloids of *Echium plantagineum* L. I. Echiumine and Echimidine. *Australian Journal of Chemistry* **9**, 512–520.

Culvenor CCJ, Jago MV, Peterson JE, Smith LW, Payne AL, Campbell DG, Edgar JA and Frahn JL (1984). Toxicity of *Echium plantagineum* (Paterson's curse). I. Marginal toxic effects in Merino wethers from long-term feeding. *Australian Journal of Agricultural Research* **35**, 293–304.

Delfosse ES and Cullen JM (1981) New activities in biological control of weeds in Australia. 11. *Echium plantagineum:* curse or salvation? In *Proceedings of the 5th International Symposium on Biological Control of Weeds.* 22–29 July 1980, Brisbane. (Ed ES Delfosse) pp. 563–574. CSIRO, Melbourne.

Delfosse ES and Cullen JM (1985) CSIRO Division of Entomology submission to the inquiries into biological control of *Echium plantagineum* L. Paterson's curse/salvation Jane. *Plant Protection Quarterly* **1**, 24–40.

Delfosse ES, Lewis RC and Smith CS (1987) Effect of drought and grasshoppers on establishment of *Dialectica scalariella* (Zeller) (Lepidoptera: Gracillariidae), a potential biological control agent for *Echium plantagineum* L. *Journal of the Australian Entomological Society* **26**, 279–280.

Dodd J (1992) Biological control of Paterson's curse *(Echium plantagineum)*: lessons from the programme using the leaf-mining moth *(Dialectica scalariella)*.In *Proceedings of the 1st International Weed Control Congress.* 17–21 February 1993, Monash University, Melbourne. pp. 155–157. Weed Science Society of Victoria, Melbourne.

Fernandez-Alés R, Laffarga JM and Ortega F (1993) Strategies in Mediterranean grassland annuals in relation to stress and disturbance. *Journal of Vegetation Science* **4**, 313–322.

Floyd RM, Lloyd SG and Shivas RG (1996) First record of *Cercospora echii* in Australia. *Australian Plant Pathology* **25**, 68.

Forcella F, Wood JT and Dillon SP (1986) Characteristics distinguishing invasive weeds within *Echium* (Bugloss). *Weed Research* **26**, 351–364.

Forrester GF (1992) Performance of *Echium plantagineum* (an Australian rangeland weed) in a native habitat: the effects of insects, fungi and vertebrate herbivores. In *Plant-Animal Interactions in Mediterranean-Type Ecosystems. Proceedings of the 6th International Conference on Mediterranean Climate Ecosystems.* 1992, Crete. (Ed CA Thanos) pp. 191–197. University of Athens, Greece.

Forrester GJ (1993) Resource partitioning between two species of *Ceutorhynchus* (Coleoptera: Curculionidae) on *Echium plantagineum* in a Mediterranean habitat. *Bulletin of Entomological Research* **83**, 345–351.

Grain Trade Australia (2010) <http://www.graintrade.org.au/commodity_standards>.

Grigulis K, Sheppard AW, Ash JE and Groves RH (2001) The comparative demography of the pasture weed *Echium plantagineum* between its native and invaded ranges. *Journal of Applied Ecology* **38**, 281–290.

Hoare RJB (2001) Adventive species of Lepidoptera recorded for the first time in New Zealand since 1988. *New Zealand Entomologist* **24**, 23–47.

Huwer RK, Neave MJ, Dowling PM, Lonsdale WM, Sheppard AW, Briese DT and Michalk DL (2002) Integrated weed management (IWM) in perennial pasture using pasture and grazing management, herbicide strategies and biological control. In *Proceedings of the 13th Australian Weeds Conference.* 8–13 September 2002, Perth. (Eds H Spafford Jacob, J Dodd and JH Moore) pp. 727–730. Plant Protection Society of WA, Perth.

Huwer RK, Briese DT, Dowling P, Lonsdale WM, Michalk DL, Kemp D, Neave M, Sheppard AW and Woodburn TL (2005) Can an integrated management approach provide a basis for long-term prevention of weed dominance in Australian pasture systems? *Weed Research* **45**, 175–192.

Industries Assistance Commission (1985) *Report into the Biological Control of Echium Species (Including Paterson's Curse/Salvation Jane).* IAC Report No. 371. Australian Government Publishing Service, Canberra.

Kirk AA and Wapshere AJ (1979) The life history and host specificity of the *Echium* borer, *Phytoecia coerule-*

scens (Col. Cerambycidae). *Entomophaga* **24**, 423–430.

Kumata T and Horak M (1997) The native *Dialectica aemula* (Meyrick) and the introduced *Dialectica scalariella* (Zeller) (Lepidoptera: Gracillariidae) in Australia: characterisation of two closely related species on Boraginaceae. *Australian Journal of Entomology* **36**, 25–35.

Lavorel S, McIntyre S and Grigulis K (1999) Response to grazing in Mediterranean annual pastures: how many functional groups? *Journal of Vegetation Science* **10**, 631–642.

Moore RM (1967) The naturalization of alien plants in Australia. In *Proceedings and Papers: IUCN 10th Technical Meeting, Part III, section 1.* pp. 82–97. Lucerne, Switzerland.

Moore RM (1971) Weeds and weed control in Australia. *Journal of the Australian Institute of Agricultural Science* **37**, 181–191.

Morley T (1990) Rearing the Echium leaf-miner, *Dialectica scalariella.* In *Proceedings of the 9th Australian Weeds Conference.* 6–10 August 1990, Adelaide. (Ed JW Heap) p. 526. Crop Science Society of South Australia, Adelaide.

Nordblom TL, Smyth M, Swirepik A, Sheppard A and Briese D (2001) Benefit-cost analysis for biological control of *Echium* weed species. In *Weed Management Systems: An Impact Assessment.* Technical series No. 6. (Centre for International Economics) pp. 35–43. Weeds CRC, University of Adelaide.

Nordblom TL, Smyth MJ, Swirepik A, Sheppard AW and Briese DT (2002a) Benefit-cost analysis for biological control of *Echium* spp. (Paterson's curse and related species) in Australia, 1972–2050. In *Proceedings of the 13th Australian Weeds Conference.* Perth. (Eds H Spafford Jacob, J Dodd and JH Moore) pp. 753–756. Plant Protection Society of WA, Perth.

Nordblom T, Smyth M, Swirepik A, Sheppard A and Briese D (2002b) Spatial economics of biological control: investing in new releases of insects for earlier limitation of Paterson's curse in Australia. *Agricultural Economics* **27**, 403–424.

Noy-Meir I, Gutman M and Kaplan Y (1989) Responses of Mediterranean grassland plants to grazing and protection. *Journal of Ecology* **77**, 290–310.

Page AR and Lacey KL (2006) *Echium plantagineum* (Paterson's curse). In *Economic Impact Assessment of Australian Weed Biological Control.* Technical series No. 10. pp. 73–74. CRC for Australian Weed Management, Adelaide.

Parsons WT and Cuthbertson EG (2001) *Noxious Weeds of Australia.* 2nd edn. CSIRO Publishing, Melbourne.

Pearce GA (1972) *Paterson's Curse: Its Importance and Control.* WA Dept of Agriculture, Bulletin No. 3855.

Piggin CM (1977) The herbaceous species of *Echium* (Boraginaceae) naturalized in Australia. *Muelleria* **3**, 215–244.

Piggin CM and Sheppard AW (1995). *Echium plantagineum* L. In *The Biology of Australian Weeds.* Vol 1. (Eds RH Groves, RCH Shepherd and RG Richardson) pp. 87–110. RG and FJ Richardson, Melbourne.

Seaman JT, Turvey WS, Ottaway SJ, Dixon RJ and Gilmour AR (1989). Investigations into the toxicity of *Echium plantagineum* in sheep. 1. Field grazing experiments. *Australian Veterinary Journal* **66**, 279–285.

Shea K, Smyth M, Sheppard A, Morton R and Chalimbaud J (2000) Effect of patch size and plant density of Paterson's curse (*Echium plantagineum*) on the oviposition of a specialist weevil, *Mogulones larvatus. Oecologia* **124**, 615–621.

Shepherd RCH (1990) The establishment of the biological control agent for Paterson's curse, *Dialectica scalariella*, in Victoria. In *Proceedings of the 9th Australian Weeds Conference.* 6–10 August 1990, Adelaide. (Ed JW Heap) p. 523. Crop Science Society of South Australia, Adelaide.

Shepherd RCH (1995a) Host specificity of the two lace bugs, *Dictyla echii* and *Dictyla nassata* [Hem.: Tingidae]: potential biological control agents for Paterson's curse (*Echium plantagineum*). *Entomophaga* **40**, 131–143.

Shepherd RCH (1995b) Timing of release of insects, with reference to the Paterson's curse leaf-mining moth, *Dialectica scalariella* (Lepidoptera: Gracillariidae). In *Proceedings of the 8th International Symposium on Biological Control of Weeds.* 2–7 February 1992, Canterbury, NZ. (Eds ES Delfosse and RR Scott) pp. 665–674. DSIR/CSIRO, Melbourne.

Shepherd RCH and Morley TB (1992) The effect of temperature on the establishment of the Paterson's curse leaf-mining moth *Dialectica scalariella* (Zeller) (Lepidoptera: Gracillariidae) in Victoria. In *Proceedings of the 1st International Weed Control Congress.* 17–21 February 1993, Monash University, Melbourne. pp. 474–480. Weed Science Society of Victoria, Melbourne.

Sheppard AW (1996) The interaction between natural enemies and interspecific plant competition in the control of pasture weeds. In *Proceedings of the 9th*

International Symposium on Biological Control of Weeds. 19–26 January, Stellenbosch, South Africa (Eds VC Moran and JH Hoffmann) pp. 47–54. University of Capetown, South Africa.

Sheppard A and Smyth M (2002) Predicting seedbank decay rates: the effects of field conditions on seed longevity and seedling recruitment in *Echium plantagineum* L. In *Proceedings of the 13th Australian Weeds Conference*. 8–13 September 2002, Perth. (Eds H Spafford Jacob, J Dodd and JH Moore) pp. 541–544. Plant Protection Society of WA, Perth.

Sheppard AW, Smyth M and Swirepik A (1999) Impact of the root-crown weevil (*Mogulones larvatus*) and other biological control agents on Paterson's curse in Australia: an update. In *Proceedings of the 12th Australian Weeds Conference*. 12–16 September, Hobart. (Eds AC Bishop, M Boersma and CD Barnes) pp. 343–346. Tasmanian Weed Society, Devonport.

Sheppard AW, Smyth M and Swirepik A (2001) The impact of *Mogulones larvatus* and pasture competition on *Echium plantgineum*. *Journal of Applied Ecology* **38**, 291–300.

Sheppard AW, Rees M, Smyth M, Grigulis K and Buckley Y (2002) Modelling the population interactions between the crown weevil *Mogulones larvatus* and its host plant *Echium plantagineum*. In *Proceedings of the 13th Australian Weeds Conference*. 8–13 September 2002, Perth. (Eds H Spafford Jacob, J Dodd and JH Moore) pp. 248–251. Plant Protection Society of WA, Perth.

Smyth MJ (1993) The effect of *Dialectica scalariella* on *Echium plantagineum*. Graduate Diploma thesis. Australian National University, Canberra.

Smyth M and Sheppard AW (1996) The effect of simulated spray-grazing on the Paterson's curse crown weevil, *Mogulones larvatus*, Schultze. In *Proceedings of the 11th Australian Weeds Conference*. 30 September–3 October 1996, Melbourne. (Ed RHC Shepherd) pp. 291–293. Weed Science Society of Victoria, Melbourne.

Smyth M and Sheppard A (2002) *Longitarsus echii* and its impact on *Echium plantagineum* (Paterson's curse): the insect for the Mediterranean rainfall range of the weed? In *Proceedings of the 13th Australian Weeds Conference*. 8–13 September 2002, Perth. (Eds H Spafford Jacob, J Dodd and JH Moore) pp. 422–425. Plant Protection Society of WA, Perth.

Smyth MJ, Moorhouse J, Sheppard AW and Swirepik A (1992). Factors affecting the dominance of *Echium plantagineum* in annual pastures. *Australian Weeds Research Newsletter* **41**, 11–14.

Smyth MJ, Sheppard AW and Swirepik A (1997) The effect of grazing on seed production in *Echium plantagineum* L. *Weed Research* **37**, 63–70.

Smyth MJ, Sheppard AW, Huwer R and Dowling P (2000) The effect of grazing and herbicides on rosette and root feeding insects for the biological control of Paterson's curse. In *Proceedings of the Wool and Sheepmeat Services Conference*. October 2000, Yanco, Australia. pp. 135–139. NSW Dept of Agriculture, Orange, Australia.

Smyth M, Sheppard A and Huwer R (2004) The population and impact of *Longitarsus echii* Koch (Coleoptera: chrysmelidae), a root-feeding beetle on *Echium plantagineum* L (Boraginaceae) (Paterson's curse) under field grazing conditions. In *Proceedings of the 14th Australian Weeds Conference*. 6–9 September 2004, Wagga Wagga, Australia. (Eds BM Sindel and SB Johnson) pp. 349–352. Weeds Society of NSW, Sydney.

Swirepik A, Sheppard AW and Smyth S (1996) *Meligethes planiusculus* (Nitidulidae): an inflorescence-feeding beetle of good potential for the biological control of Paterson's curse. In *Proceedings of the 11th Australian Weeds Conference*. 30 September–3 October 1996, Melbourne. (Ed RHC Shepherd) pp. 294–297. Weed Science Society of Victoria, Melbourne.

Vayssières JF (1983) Life-histories and host specificities of the *Echium* bugs *Dictyla echii* and *Dictyla nassata* [*Hem.: Tingidae*]. *Entomophaga* **28**, 135–144.

Vayssières JF (1986) The immature stages of two *Ceutorhynchus* sp. (Coleoptera: Curculionidae) found on *Echium* sp. (Boraginaceae). *Nouvelle Revue d'Entomologie* **3**, 325–331.

Vayssières JF and Wapshere AJ (1983) Life-histories and host specificities of *Ceutorhynchus geographicus* (Goeze) and *C. larvatus* Schultze (Coleoptera: Curculionidae), potential biological control agents for *Echium*. *Bulletin of Entomological Research* **73**, 431–440.

Wapshere AJ (1974) Host specificity of phytophagous organisms and the evolutionary centres of plant genera or sub-genera. *Entomophaga* **19**, 301–309.

Wapshere AJ (1980) The biological control of Paterson's curse, *Echium plantagineum*: northern hemisphere studies. In *Proceedings of the 5th International Symposium on Biological Control of Weeds*. 22–29 July 1980, Brisbane. (Ed ES Delfosse) pp. 599–602. CSIRO, Melbourne.

Wapshere AJ (1982) Life histories and host specificities of the *Echium* flea beetles *Longitarsus echii* and *L. aeneus* (Col.: Chrysomelidae). *Entomophaga* **27**, 173–181.

Wapshere AJ (1985) The phytophagous insect guild on *Echium* in the western Mediterranean and the biological control of the weed, *Echium plantagineum*. *Acta Oecologica* **6**, 331–338.

Wapshere AJ and Kirk AA (1977) The biology and host specificity of the *Echium* leaf miner *Dialectica scalariella* (Zeller) (Lepidoptera: Gracillariidae). *Bulletin of Entomological Research* **67**, 627–633.

Eichhornia crassipes (Martius) Solms-Laubach – water hyacinth

Mic Julien

ABSTRACT

Water hyacinth, *Eichhornia crassipes*, a native of Amazonia, was first recorded in Australia in 1894. It is widespread in waterways of the east coast, particularly north of Sydney. Biological control of water hyacinth in Australia began in the 1970s and four agents have been released. *Neochetina bruchi, N. eichhorniae* and *Niphograpta albiguttalis* are widely established, whereas *Xubida infusellus* is known to be established at one site only. Successful control has been achieved at many locations. However, a range of factors such as flooding, drying, cold winters, herbicidal application etc. mitigate against successful biological control at other sites. There is a need to develop integrated management strategies to best utilise the established agents in combination with other control techniques, and there is a need for additional agents, some of which are being studied or have been released overseas. Of particular interest is a plant hopper, *Megamelus scutellaris*, recently released in the USA. Investing in biological control of water hyacinth in Australia and elsewhere, and in integrated management in South Africa, has been extremely cost-effective with benefit:cost estimates up to 27.5:1 in Australia and 124:1 in Benin, and savings of over $US200 per ha compared to herbicidal control in the USA.

Key words: *Neochetina bruchi, Neochetina eichhornia, Niphograpta albiguttalis, Xubida infusellus, Megamelus scutellaris*, integrated management, biological control agents.

INTRODUCTION

Eichhornia crassipes (Mart.) Solms-Laub. (Pontederiaceae), water hyacinth, is indigenous to Brazil, with its centre of origin in Amazonia (Barrett and Forno 1982). Its current distribution is pantropical but it also occurs in many temperate regions, between latitudes 40°N and S. Water hyacinth is a floating macrophyte, with attractive shiny round leaves and mauve flowers. Reproduction is through clonal growth and viable seeds. It forms aggregates of plants, due to stoloniferous development of ramets (daughter plants), and can form very large floating mats (Fig. 1). When crowded, petioles elongate (up to 1 m) (Fig. 2) and when nutrient levels are high, the leaves, which are modified ends of the petioles called pseudolaminae, may reach dinner-plate size. Invading plants and those on the edge of mats in uncrowded situations have short bulbous petioles (to 30 cm) with small kidney-shaped leaves (Fig. 2). Petioles, six to 10, are arranged around the rhizome and the rhizome supports stolon development and fibrous, feathery submerged roots (Julien *et al.* 1999; Coetzee *et al.* 2009). Flowers produce many long-lived (up to 20 years) seeds (Gopal 1987; Coetzee *et al.* 2009; Wright and Purcell 1995). As water hyacinth is a very attractive plant, humans have been the main dispersal vector between catchments, regions and continents. Within catchments, dispersal is mostly by downstream flow or by seed in mud attached to equipment or animals.

Water hyacinth has genetic polymorphism in flower structure, called tristyly (Barrett 1977; Barrett and Forno

Figure 1: Mat of water hyacinth, upper Brisbane River, 2007. Photo: M Julien, CSIRO.

1982). That is, there are three plant forms with different style lengths – short, intermediate and long. The intermediate length style form is prevalent in the invaded ranges and parts of the native range. The long style form also occurs in both native and invaded ranges, including in Australia, but remains less dominant than the other forms. The short styled form is known only from tropical South America. Growth is rapid, doubling in five to 18 days, especially under tropical temperatures and high nutrient availability (Wright and Purcell 1995; Coetzee *et al.* 2009). Growth ceases at around 10°C and prolonged cold, below 5°C, kills plants (Gopal 1987).

Water hyacinth was first recorded in Australia in 1894 in Qld and NSW and was well established by 1900 (Wright and Purcell 1995). Its distribution (Fig. 3) is linked to permanent and semi-permanent waterways, but it can grow on wet mud. On the Gwydir wetlands near Moree, northern NSW, water hyacinth has invaded permanent water impounded behind a natural barrage of fallen timber. In the wet season, vast areas of the surrounding low-lying country may be flooded and vast mats of water hyacinth flourish there. In the dry season, when the flood waters recede, the weed dies. However, since seeds can germinate and plants can mature and set seed in as little as 12 weeks, the floodplains contain a considerable seed bank, some of which germinates during the seasonal flood events. This cycle occurs on many semi-permanent water bodies in Australia. In northern tropical areas, water hyacinth flourishes in waterways and swamps during the dry season and is often flushed to the sea by annual flood events in the wet season.

The rate of growth and its ability to quickly form thick dense mats on water ensures that water hyacinth has negative environmental, social and economical impacts, as documented in Wright and Purcell (1999) and Coetzee *et al.* (2009).

BIOLOGICAL CONTROL HISTORY

Arthropods

Surveys for biological control agents of water hyacinth began when A Silveira-Guido, US Department of Agriculture (USDA), surveyed in Uruguay from 1962 to 1965. In 1962, the USDA set up a laboratory in Buenos Aires, Argentina, for surveys on alligator weed, *Alternanthera philoxeroides* (Martius) Grisebach (Amaranthaceae), but in 1968, under David Perkins and later, continuing under Jack DeLoach after 1971, the research emphasis shifted to water hyacinth, until 1974. In 1968, Fred Bennett and Helmut Zwölfer (CIBC, now CABI) surveyed Guyana, Surinam and Brazil. Around the same time, David

Figure 2: Water hyacinth *Eichhornia crassipes*: short colonising form (left) and the taller mat form (right). Photos: M Julien, CSIRO.

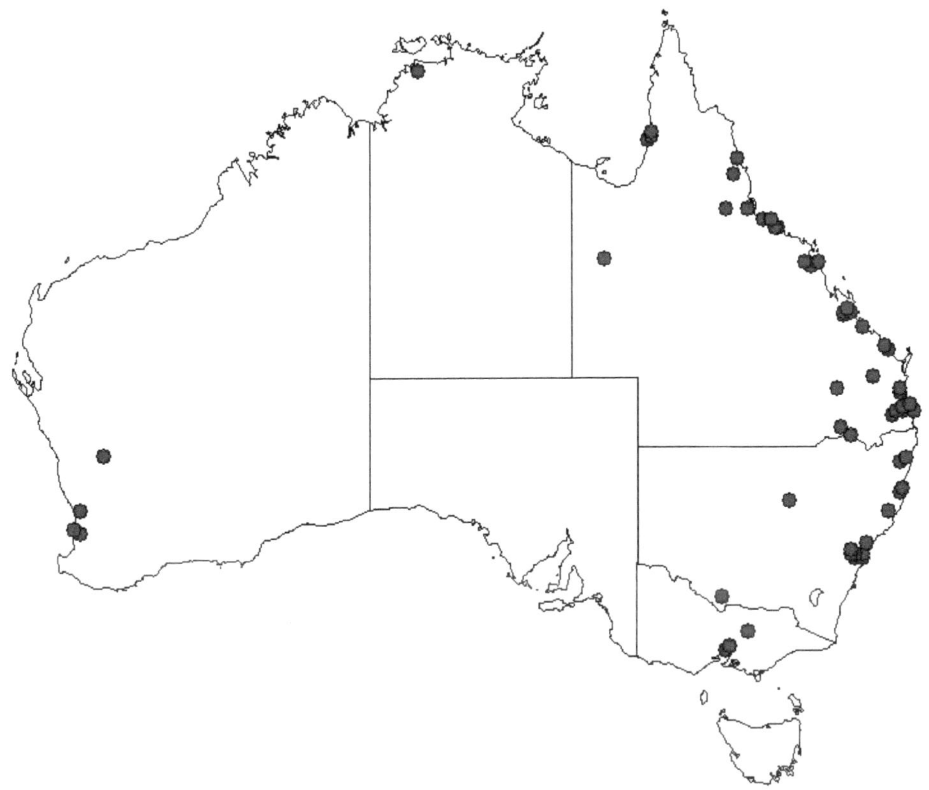

Figure 3: The distribution of *Eichhornia crassipes* in Australia. Specimen data reproduced from Australia's Virtual Herbarium with permission of the Council of Heads of Australasian Herbaria Inc.

Mitchell and Phil Thomas surveyed Argentina, Uruguay, Brazil, Guyana and Trinidad and Fred Bennett (CIBC) surveyed West Indies and Belize, and later Mexico. Various surveys were conducted in the invaded ranges of water hyacinth in the USA and India (Center 1994).

The natural enemies found by these explorers were listed by Center (1994) and Cordo (1999). The important ones, those that have been considered and studied for biological control, are listed here. Most of the important agents were collected by Silviera-Guido from Uruguay. They included the moths *Xubida infusellus* (Walker) (was *Acigona infusella* (Walker) = *Chilo ignitalis* Hampson), the weevil *Neochetina bruchi* Hustache (later separated into two species, *N. bruchi* and *Neochetina eichhorniae* Warner), the grasshopper *Cornops aquaticum* Bruner and the mite *Orthogalumna terebrantis* L. Bennett and Zwölfer added the moths *Niphograpta albiguttalis* Warren (previously *Sameodes albiguttalis* (Warren) and originally *Epipagis albiguttalis* (Warren)), the flies *Thrypticus* spp., a mirid (possibly *Eccritotarsus catariensis* (Carvalho)) and the leaf-hopper *Taosa inexacta* Walker. More recently, the plant-hopper *Megamelus scutellaris* Berg was found (Cordo 1999). Cordo (1999) considered that new natural enemies may be found in areas of the Amazon catchment that have not been surveyed in detail, including the likely centre of origin in the upper Amazon.

Pathogens

Various surveys for fungal pathogens of water hyacinth were conducted in the Americas (1970s to early 2000s), in Australia (early 1980s) and Africa (early 2000s). Of about 70 fungi found on water hyacinth, only a few have been studied and only one as a classical agent. *Cercospora rodmanii* Conway (= *Cersospora piaropi* Tharp), from Florida, was released and established in South Africa during 1991 (Morris *et al.* 1999). Another, *Uredo eichhorniae* Fragosa & Ciferri, may have potential as a classical agent (Charudattan 2001; Morris *et al.* 1999). A number of other species have been studied for their potential as bioherbicides but none are in use, apparently due to lack of commercial interest. The species studied included *Acremonium zonatum* (Sawada) Gams, *Alternaria eichhorniae* Nag Raj & Ponnappa, *C. piaropi* (= *C. rodmanii*), *Myrothecium roridum* Tode ex Fr. and *Rhizoctonia solani* Kühn (Charuadattan 2001). Surveys in Australia from 1982 to 1984 found *C. piaropi* and *A. zonatum*. Studies were conducted on the latter, including interactions with *N. eichhorniae* (Galbraith 1987a, 1987b), but further development did not occur.

From 2000 to 2002, a project that aimed to find appropriate pathogens and to develop mycoherbicides for water hyacinth control in Africa was carried out. Surveys were conducted in a number of African countries as well as in Brazil, Peru and Ecuador. Considerable work concentrated on *A. eichhorniae* and formulations that included low doses of herbicides (Reeder 2002). However, the project was terminated and no useful bioherbicide resulted.

EXPLORATION

No exploration in the native range of water hyacinth for natural enemies has been undertaken for Australia. Australian research utilised organisms already known, although the first assessments of the moth *X. infusellus* were made in Australia.

CANDIDATES

The first water hyacinth project in Australia began in 1974 and terminated in 1991 after testing and releasing four agents. A second project began in mid 1995 and terminated in 1999, re-importing and releasing one of those agents. It also tested a new agent that was found to be unsuitable for release.

Agents tested but not released

***Eccritotarsus catarinensis* (Carvalho) (Hemiptera: Miridae).** This leaf-sucking bug was imported to Australia from South Africa in 1997. It had been collected by Stefan Neser from Santa Catarina Island, Santa Catarina state, Brazil, in 1989. *E. catarinensis* was not released in Australia because it could feed and develop on native *Monochoria vaginalis* (Burman f.) Kunth. (*Pontederiaceae*) (Stanley and Julien 1999). Nor was it released in the USA because of potential damage to *Pontederia cordata* L. (Pontederiaceae). It was released in South Africa in 1996 (established), in Benin (not established), in Malawi and China where its status is not known (Coetzee *et al.* 2009) and in Ghana in 2009, where it established (Akpabey, *pers. comm.* 2009). Its biology was reported by Hill *et al.* (1999) and its contribution to the control of water hyacinth in South Africa by Coetzee *et al.* (2008).

Agents released in Australia

***Neochetina eichhorniae* Warner and *Neochetina bruchi* Hustache (Coleoptera; Curculionidae).** Host-specificity tests of these weevils (Fig. 4) were conducted in Australia and in a number of other countries; the results are listed in Julien *et al.* (1999). Non-target damage was observed in the field in Australia on the introduced *P. cordata*, but only as a spill-over effect when the two plants occurred together.

Neochetina eichhorniae was the first biological control agent released on water hyacinth in Australia, in 1975. The adults are 4–5 mm long, grey coloured and nocturnal. They feed on the young folded leaves and upper petioles, causing characteristic squarish feeding marks (Fig. 4) (Julien *et al.* 1999). Eggs are laid in leaf and petiole tissues and hatch after seven to 10 days. Larvae tunnel inside the petioles and crown of the plants and develop over 30–45 days. Late-stage larvae move to the upper root area and prepare a cocoon of root hairs for pupation. Adults emerge from the cocoon after about seven days and the generation time is about 93 days. Development rates, and therefore rates of population increase, are influenced by temperature and host plant nutrient status (DeLoach and Cordo 1976; Heard and Winterton 2000; Julien *et al.* 1999; Coetzee *et al.* 2009).

Neochetina bruchi was released in Australia in 1990. Its biology and behaviour are fairly similar to those of

Figure 4: a) Neochetina weevils *Neochetina bruchi* (left) and *Neochetina eichhorniae* (right), and b) typical *Neochetina* feeding scars on leaves. Photos: CSIRO.

N. eichhorniae. However, it is more dependent on plants with higher nutrient status (Heard and Winterton 2000), has faster development rates (Center 1994), has different ovipositional site preferences (Coetzee *et al.* 2009) and adults are larger and grey-brown in colour. Morphological and biological differences between these two weevil species are summarised in Julien *et al.* (1999).

The damage to water hyacinth by the two *Neochetina* weevils is similar and, when released together, appears to be complementary. The impacts of adult damage to leaves and of larval tunnelling, which exposes internal tissues to secondary rots and waterlogging, reduces plant vigour, growth rate and productivity. This is observed as markedly reduced flowering and shorter, smaller plants. Under heavy damage, the canopy opens, as there are fewer flowers and fewer, shorter petioles and smaller leaves, which curl and turn brown. Damage and rotting of petiole bases causes petioles to collapse. Eventually, plants die and sink, initially in patches but later as whole sections of mats. The damaged mats are much more susceptible to wind and wave action, which assists their breaking up and sinking, and being flushed downstream (Julien *et al.* 1999; Julien 2001).

These weevils have been released in at least 30 countries and are the main contributors to biological control in 14 countries, including Australia (Julien 2001; Aguilar *et al.* 2003).

***Niphograpta albiguttalis* Warren (Lepidoptera: Pyralidae).** This moth from Argentina was previously referred to as *Samea albiguttalis* (Warren). Adults have a wing-span of 17–25 mm, with variable colours from golden-yellow to charcoal grey, with brown, black and white markings (Fig. 5). They are nocturnal and rest during the day under leaves. Adult females lay eggs in crevices in leaf tissues and the eggs take three to four days to hatch. Larvae initially feed externally on the plant then tunnel into the petioles, often killing leaf buds. After five instars and two to four weeks, pupation occurs inside a petiole. Adults emerge from their silken cocoon after seven to 10 days. In Argentina, adults lived for up to nine days and there were up to five generations per year (DeLoach and Cordo 1978). This moth prefers the young, tender bulbous form of the plant, the form that is typical in uncrowded situations (Julien *et al.* 2001).

A number of countries have conducted host-specificity tests and found that *N. albiguttalis* is specific to Pontedericeae. Observations following releases suggest that it is specific to water hyacinth (Julien *et al.* 2001). *N. albiguttalis* was released in Australia during 1977 and, following natural spread, became widely established. Moth populations were spatially and temporally patchy, therefore water hyacinth was able to tolerate the damage they caused (Center 1984; Wright 1981; Julien *et al.* 2001). However, in combination with the *Neochetina* weevils, *N. albiguttalis* may contribute to management of this weed, especially in preventing growth and reducing spread of the bulbous plant form that grows along the edges of mats of water hyacinth and that is most susceptible to being washed downstream (Julien *et al.* 2001).

Figure 5: Adult and larva of the water hyacinth moth *Niphograpta albiguttalis*. Photos: CSIRO.

***Xubida infusellus* (Walker) (Lepidoptera: Pyralidae).** This moth from South America, previously known as *Acigona infusella*, inflicted considerable damage on *E. crassipes* in its native range (DeLoach 1975). It utilises the tall slender petiole form of water hyacinth. The adult moths are nocturnal, strong fliers, red-brown and 20–25 cm long (Fig. 6). Eggs, in gelatinous masses, are

Figure 6: Adult and larva of *Xubida infusellus*. Photos: J. Stanley, CSIRO.

laid in crevices formed by folded or touching leaves. The newly hatched larvae tunnel inside petioles. As larvae develop and feed, they cause considerable damage to internal plant tissues of the lower petioles and the stolon. Pre-pupal larvae move up to the mid petiole area and tunnel outwards, leaving a thin window of epidermal tissue. The pre-pupal larvae then retreat a short distance and pupation occurs in the tunnel. After emergence, the adult exits through the window. Development from egg to adult takes about two months. A review of the life-history is given in Julien *et al.* (2001).

Host-range studies determined that *X. infusellus* is oligophagous on species within Pontederiaceae. This moth was first imported from Brazil and released in Australia in 1981 (Sands and Kassulke 1983; Julien *et al.* 2001). Although populations were observed for up to nine months after release, they failed to persist, probably due to the development of one site and the drying out of the other. In 1996 it was introduced again, from Argentina, for a more concerted release effort. The moth was also released in Papua New Guinea (PNG) in 1997 and in Thailand in 1999; it established in each country (Julien and Stanley 1999; Julien *et al.* 2001).

Asynchrony in emergence of the adult males and females limits establishment. For this reason, multiple releases at relatively short intervals were made at some sites in Australia and PNG. In Australia, *X. infusellus* was released at 41 sites between the Richmond River area of northern NSW and the vicinity of Cairns, Qld, between September 1996 and April 1999. It was found at one site at Ingham, Qld a year after release, but the site was then destroyed for development. It is only known to be established at one site in Australia, at Loganholme in south-eastern Qld, where releases were made on seven occasions between September 1996 and October 1997. Light trapping at irregular intervals (the last in February 2010) confirmed that the population has persisted, although it is not obviously affecting growth of the weed. As limited post release monitoring was conducted, it may be established elsewhere. A series of releases were made at several sites in PNG, beginning in 1996. Pheromone trapping determined the presence of a persisting low population at Waigani Swamp near Port Moresby in late 1997 (Stanley *et al.* 2000), although its current status is not known.

Other potential agents

***Orthogalumna terebrantis* Wallwork (Acarina: Galumnidae).** This mite is known from Madagascar, south-eastern North America and Central and South America. It was released and has established in India in 1986 and in Zambia in 1971. It spread or was accidentally introduced to other countries (Julien and Griffiths 1998), including South Africa, where evaluations are underway (Coetzee and Hill 2008).

Adults are small (0.5 mm long) and dark brown to black. They lay eggs on the leaves, preferably at adult feeding sites. The eggs hatch in seven to eight days and development of one larval and three nymphal stages is complete after about 15 days (Cordo and DeLoach 1975, 1976). The larvae and nymphs produce characteristic feeding galleries, extending towards the tip of the leaf, between veins. Adults emerge from exit holes at the end of the galleries and live for up to 85 days. Detailed biology of this mite was described by Ganga-Visalakshy and Jayanth (1991). In Argentina, Cordo and DeLoach (1976) found that *O. terebrantis* population numbers varied considerably from year to year, but could produce up to three generations per year. Although they can develop large colonies that are extremely damaging to the weed in South Africa (Hill and Cilliers 1999), other studies indicated that the mite's damage has little impact on growth (Ganga-Visalakshy and Jayanth 1991).

***Megamelus scutellaris* Berg. (Hemiptera: Delphacidae).** This plant-hopper occurs on water hyacinth throughout its native range from Peru to Argentina and is specific to water hyacinth (Sosa *et al.* 2007). Its biology and life-history were described by Sosa *et al.* (2005). It was recently released (May 2010) on water hyacinth in the USA (A Sosa, *pers. comm.* 2010). It has been studied in South Africa, but not released because it does not appear to be better suited to South African conditions (cool winters and euthrophic waters) than agents already established there. South Africans await the outcome of the USA release (J Coetzee, *pers. comm.* 2010).

***Cornops aquaticum* Brüner (Orthoptera: Acrididae).** This grasshopper is very damaging to water hyacinth but it lacks a high level of specificity, accepting other species in Pontederiaceae, *Commelina* sp. (Commelinaceae) and *Canna indica* L. (Cannaceae) (Silveira-Guido and Perkins 1975; Coetzee *et al.* 2009). It was approved for release in South Africa in 2007, but has not been released. Studies on the competitive performance of water hyacinth under attack by *C. aquaticum* concluded that this insect could contribute positively to the management of water hyacinth (Bownes *et al.* 2010). However, the results of current studies on interactions with other agents established in South Africa (Coetzee *et al.* 2009) will contribute to the decision on whether to release this species.

***Thrypticus* spp. (Diptera: Dolichopodidae).** Five species of petiole-mining flies develop in water hyacinth. They include *Thrypticus truncatus* Bickel & Hernandez, *T. sagittatus* Bickel & Hernandez, *T. circularis* Bickel & Hernandez, *T. yanayacu* Bickel & Hernandez and *T. chanophalus* Bickel & Hernandez (Hernández 2007b). The first two species have potential as biological control agents for water hyacinth (Bickel and Hernández 2004) and have been studied in Argentina for possible release in the USA (Cordo *et al.* 2000; Hernández 2008). They have not yet been released, possibly because of lack of evidence that they can contribute to control of the weed.

***Bellura densa* (Walker) (Lepidoptera: Noctuidae).** This moth was considered a potential agent until recently (Wright and Purcell 1995). It is native to North America where its natural host is pickerel weed, *P. cordata*. It is very damaging to water hyacinth but it also attacks taro *Colocasia esculenta* L. (Araceae), an important food plant in the tropics (Centre and Hill 2002), and is therefore unsuitable as an agent.

Other arthropods. Several other insect species may have potential, but have not yet been studied. They include the sucking bugs *Taosa* species (Hemiptera: Dictyopharidae) (De Remes Lenicov and Hernández 2010) and the flower-feeding beetle *Calleida* sp. (Coleoptera: Caribidae), petiole-mining flies *Eugaurax setigena* Sabrosky (Diptera: Chloropidae), *Hydrellia* sp. (Diptera: Ephydridae), *Chironomus falvipilus* Rempel (Diptera: Chironomidae) and the mite *Flechtmannia eichhorniae* Keifer (Acarina: Eriophyidae) (Coetzee *et al.* 2009).

Fungi. As a classical biological control agent, *C. piaropi* from Florida, established in South Africa with no obvious impact (Morris *et al.* 1999). Another fungus, *U. eichhorniae*, may have potential. It was imported several times from Brazil to South Africa in the early 2000s (Charudattan 2001), but the cultures failed to thrive (A Wood, *pers. comm.* 2010).

DISCUSSION

The three most widely utilised water hyacinth agents, *N. bruchi*, *N. eichhorniae* and *N. albuguttalis*, are well established in Australia. These agents have contributed to management of the weed in Australia (Wright 1981; Crutwell McFadyen 2000) and worldwide (Coetzee *et al.* 2009; Julien *et al.* 1999). However, because insect populations require a number of years to increase sufficiently to cause significant damage, the biological control system is often disrupted by seasonal floods that flush the weed, by drying of water bodies that desiccates weed biomass, or by interference from other control methods that cause the demise of the standing crop of water hyacinth. By removing the plants, these factors also remove the insects, but do not remove the seed bank or prevent reinvasion of plant propagules. Subsequent new water hyacinth growth outstrips the ability of the agents to recolonise and, while the agents increase again, the problems caused by the weed are not abated. There are other reasons why control is not satisfactory at specific locations or regions (Julien 2001; Coetzee *et al.* 2009), therefore researchers in other countries have continued to search for new agents, which may in turn be used to assist control in Australia.

The bug *M. scutellaris* should be considered for release in Australia. It is highly specific and causes considerable damage to the weed. It is also capable of developing large populations quickly. This may help it overcome the time lag following perturbations between new growth of the weed and invasion by the agent. Australian *Monochoria* species (Pontederiaceae) are being collected and sent to the USA to determine if *M. scutellaris* will accept them as hosts. If they do not, this bug should be considered for release in Australia.

Of the other potential agents, none are particularly promising. The grasshopper *C. aquaticum* is unlikely to be sufficiently specific. The Australian native *Monochoria* species would be most at risk, although they have yet to be tested on this insect. The *Thrypticus* flies have not been released anywhere because of their low impact on the weed. The mite *O. terebrantis*, although present following releases or natural spread in more than 10 countries (Julien and Griffiths 1998), has not been reported as contributing to control. Current evaluation in South Africa will help determine the value of this mite. Despite considerable effort, no pathogen has been shown to be useful for classical biological control of water hyacinth. Similarly, no useful bioherbicide has been developed for water hyacinth.

The natural or human-induced disruptions to the biological control cycle involving the *Neochetina* weevils has led to the development of *Neochetina* rearing stations, e.g. by Brisbane City Council. Colonies are reared on water hyacinth growing in pools and adults are harvested and released when new infestations are found or when insect numbers in the field are low. This has reduced the resources required to manage water hyacinth in the council jurisdiction (R Wood, *pers. comm.* 2009).

The benefit:cost ratio for biological control of aquatic weeds in Australia, including salvinia, *Salvinia molesta* DS Mitchell (Salviniaceae), water lettuce *Pistia stratiotes* L. (Araceae) and water hyacinth, was 27.5:1 (Page and Lacey 2006). In Benin, the benefit:cost ratio was 124:1 (De Groote *et al.* 2003). A comparison of costs in South Africa for different control techniques indicated that herbicidal control cost $US250 per ha, biological control $US44 per ha and integrated methods $US39 per ha (van Wyk and van Wilgen 2002).

Water hyacinth is a particularly intractable weed because its long-lived seeds survive when conditions are unsuitable for the plant and its rapid growth rate ensures development of large reproductive populations in less than a few months. It was recognised early in the Australian biocontrol project that, in many situations, integrated management would be required to manage water hyacinth (Wright 1984). It is concerning that although some assessment of the effect of herbicides on the quality of water hyacinth as a food source for agents has been studied in Australia (Wright and Bourne 1990) and overseas (Van and Center 1994) and integrated management strategies have been field-tested (Center *et al.* 1999) and implemented successfully (Jones 2001), no adaptive management strategies have been implemented in Australia. These could include judicious use of herbicides or harvesting machines to reduce and maintain weed populations at low levels while biological control agents are permitted to develop. This would require maintenance of refugia for agents, using barriers to hold water hyacinth in place to protect it from herbicide and flushing, and rearing and releasing agents at appropriate times. In South Africa, augmentative releases after a perturbation or during favourable condition for the agents significantly assisted in the management of water hyacinth, even in the most eutrophic conditions (D Conlong, *pers. comm.* 2010). Such strategies would add value to the management capabilities already available and reduce the high levels of resource input currently utilised in non-biological maintenance management programs.

ACKNOWLEDGEMENTS

The initial research project (1974–91) on water hyacinth in Australia was carried out by Ken Harley, Wendy Forno, Tony Wright and Don Sands. The second project (1995–99) was by John Stanley, Andrew White, John Lester and the author. Justine Murray prepared the map (Fig. 3).

REFERENCES

Aguilar JA, Camarena OM, Center TD and Bojórquez G (2003) Biological control of waterhyacinth in Sinaloa, Mexico with the weevils *Neochetina eichhorniae* and *N. bruchi*. *BioControl* **48**, 595–608.

Barrett SCH (1977) Tristyly in *Eichhornia crassipes* (Mart.) Solms (water hyacinth). *Biotropica* **9**, 230–238.

Barrett SCH and Forno IW (1982) Style morph distribution in New World populations of *Eichhornia crassipes* (Mart.) Solms-Laubach (water hyacinth). *Aquatic Botany* **13**, 299–306.

Bickel DJ and Hernández MC (2004) Neotropical *Thrypticus* (Diptera: Dolichopodidae) reared from water hyacinth, *Eichhornia crassipes*, and other Pontederiaceae. *Annals of the Entomological Society of America* **97**, 437–449.

Bownes A, Hill MP and Byrne MJ (2010) Evaluating the impact of herbivory by a grasshopper, *Cornops aquaticum* (Orthoptera: Acrididae), on the competitive performance and biomass accumulation of water hyacinth, *Eichhornia crassipes* (Pontederiaceae). *Biological Control* **53**, 297–303.

Center TD (1984) Dispersal and variation in infestation intensities of the water hyacinth moth, *Sameodes albiguttalis* (Lepidoptera: Pyralidae) populations in Peninsular Florida. *Environmental Entomology* **13**, 482–491.

Center TD (1994) Biological control of weeds: waterhyacinth and waterlettuce. In *Pest Management in the Subtropics: Biological Control – A Florida Perspective.* (Eds D Rosen, FD Bennett and JL Capinera) pp. 481–521. Intercept Publishing, Andover, UK.

Center TD and Hill MP (2002) Field efficacy and predicted host range of the pickerelweed borer, *Bellura densa*, a potential biological control agent of water hyacinth. *BioControl* **47**, 231–243.

Center TD, Dray FA Jr, Jubinsky GP and Grodowitz MJ (1999) Biological control of water hyacinth under conditions of maintenance management: can herbicides and insects be integrated? *Environmental Management* **23**, 241–256.

Charudattan R (2001) Biological control of water hyacinth by using pathogens: opportunities, challenges, and recent developments. In *Proceedings of the 2nd Meeting of the IOBC Global Working for the Biological and Integrated Control of Water Hyacinth.* 9–12 October 2000, Beijing. (Eds MH Julien, TD Center and J Ding) pp. 21–28. ACIAR Proceedings No. 102, Canberra.

Coetzee JA and Hill MP (2008) Biological control of water hyacinth – the South African experience. *OEPP/EPPO Bulletin* **38**, 458–463.

Coetzee JA, Hill MP and Byrne MJ (2008) Ten years after the release of the water hyacinth mirid *Eccritotarsus catariensis* in South Africa: what have we learnt? In *Proceedings of the 12th International Symposium on Biological Control of Weeds.* 22–27 April 2007, La Grande Motte, France. (Eds MH Julien, R Sforza, MC Bon, HC Evans, PE Hatcher, HL Hinz and BG Rector) pp. 512–515. CABI, Wallingford, UK.

Coetzee JA, Hill MP, Julien MH, Center TD and Cordo HA (2009) *Eichhornia crassipes* (Mart.) Solms-Laub. (Pontederiaceae). In *Biological Control of Tropical Weeds Using Arthropods.* (Eds R Muniappan, GVP Reddy and A Raman) pp. 183–210. Cambridge University Press, Cambridge.

Cordo HA (1999) New agents for biological control of waterhyacinth. In *Proceedings of the 1st IOBC Global Working Group Meeting for the Biological and Integrated Control of Water Hyacinth.* 16–19 November 1998, Harare. (Eds MP Hill, MH Julien and TD Center) pp. 68–74. Plant Protection Research Institute, Pretoria.

Cordo HA and DeLoach CJ (1975) Ovipositional specificity and feeding habits of the waterhyacinth mite, *Orthogalumna terebrantis*, in Argentina. *Environmental Entomology* **4**, 561–565.

Cordo HA and DeLoach CJ (1976) Biology of the waterhyacinth mite in Argentina. *Weed Science* **24**, 245–249.

Cordo HA, Sosa AJ and Hernandez MC (2000) The petiole mining fly, *Thrypticus* sp. (Diptera: Dolichopodidae), a new agent for the biological control of water hyacinth (*Eichhornia crassipes*). In *Proceedings of the 10th International Symposium on Biological Control of Weeds.* 4–14 July 1999, Bozeman, Montana. (Ed N Spencer) pp. 315–323. Montana State University, Bozeman, Montana.

Crutwell McFadyen RE (2000) Successes in biological control of weeds. In *Proceedings of the 10th International Symposium on the Biological Control of Weeds.* 4–14 July 1999, Bozeman, Montana. (Ed NR Spencer) pp. 3–14. Montana State University, Bozeman, Montana.

De Groote H, Ajuonu O, Attignon S, Djessou R and Neuenschwander P (2003) Economic impact of biological control of water hyacinth in Southern Benin. *Ecological Economics* **45**, 105–117.

DeLoach CJ (1975) Evaluation of candidate arthropods for biological control of waterhyacinth: studies in Argentina. In *Proceedings of a Symposium on Water Quality Management through Biological Control.* 23–30 January 1975, Gainesville, Florida. (Eds PL Brezonik and JL Fox) pp. 45–50. University of Florida, Florida.

DeLoach CJ and Cordo HA (1976) Life cycle and biology of *Neochetina bruchi*, a weevil attacking water hyacinth in Argentina, with notes on *N. eichhorniae*. *Annals of the Entomological Society of America* **69**, 643–652.

DeLoach CJ and Cordo HA (1978) Life history and ecology of the moth *Sameodes albiguttalis*, a candidate for biological control of water hyacinth. *Environmental Entomology* **7**, 309–321.

De Remes Lenicov AM and Hernández MC (2010) A new species of *Taosa* (Hemiptera: Dictyopharidae) from South America associated with water hyacinth. *Annals of the Entomological Society of America* **103**, 332–340.

Galbraith JC (1987a) Microorganisms on water hyacinth in eastern Australia: a survey in relation to biological control by insects. *Biological Agriculture and Horticulture* **4**, 283–300.

Galbraith JC (1987b) The pathogenicity of an Australian isolate of *Acremonium zonatum* to water hyacinth, and its relationship with the biological control agent, *Neochetina eichhorniae*. *Australian Journal of Agricultural Research* **38**, 219–229.

Ganga-Visalakshy PN and Jayanth KP (1991) Studies on the life history and development of *Orthogalumna terebrantis* Wallwork, an exotic oribatid of *Eichhornia crassipes*. *Entomon* **16**, 53–57.

Gopal B (1987) *Water Hyacinth*. Elsevier, Amsterdam.

Heard TA and Winterton SL (2000) Interactions between nutrient status and weevil herbivory in the biological control of water hyacinth. *Journal of Applied Ecology* **37**, 117–127.

Hernández MC (2007a) Larval feeding behaviour of the *truncates* group of *Thrypticus* (Diptera: Dolichopodidae) that breed in the aerenchyma of Pontederiaceae. *Revista de la Sociedad Entomológica Argentina* **66**, 121–126.

Hernández MC (2007b) Biology of *Thrypticus truncatus* and *Thrypticus sagittatus* (Diptera: Dolichopodidae), petiole miners of water hyacinth, in Argentina, with morphological descriptions of larvae and pupae. *Annals of the Entomological Society of America* **101**, 1041–1049.

Hill MP and Cilliers CJ (1999) A review of the arthropod natural enemies, and factors that influence their efficacy, in the biological control of water hyacinth, *Eichhornia crassipes* (Mart.) Solms-Laubach (Pontederiaceae), in South Africa. In *Biological Control of Weeds in South Africa (1990–1998)*. (Eds T Olckers and MP Hill). *African Entomology*, Memoir No. 1, 103–112.

Hill MP, Cilliers CJ and Neser S (1999) Life history and laboratory host range of *Eccritotarsus catarinensis* (Carvalho) (Heteroptera: Miridae), a new potential natural enemy released on water hyacinth (*Eichhornia crassipes* (Mart.) Solms-Laub.) (Pontederiaceae) in South Africa. *Biological Control* **14**, 127–133.

Jones RW (2001) Integrated control of water hyacinth on the Nseleni/Mposa rivers and Lake Nsezi, Kwa Zulu-Natal, South Africa. In *Biological Control of Water Hyacinth,* Eichhornia crassipes. ACIAR Proceedings 102. (Eds MH Julien, MP Hill, TD Center and J Ding) pp. 123–129. Australian Centre for International Agricultural Research, Canberra.

Julien MH (2001) Biological control of water hyacinth with arthropods: a review to 2000. In *Biological and Integrated Control of Waterhyacinth,* Eichhornia crassipes. (Eds MH Julien, MP Hill, TD Center and J Ding) pp. 8–20. Australian Centre for International Agricultural Research, Canberra.

Julien MH and Griffiths MW (1998) *Biological Control of Weeds: A World Catalogue of Agents and their Target Weeds*. 4th edn. CABI, Wallingford, UK.

Julien MH and Stanley J (1999) Recent research on biological control for water hyacinth in Australia. In *Proceedings of the 1st IOBC Global Working Group Meeting for the Biological and Integrated Control of Water Hyacinth*. 16–19 November 1998, Harare. (Eds MP Hill, MH Julien and TD Center) pp. 52–61. Plant Protection Research Institute, Pretoria.

Julien MH, Griffiths MW and Wright AD (1999) *Biological Control of Water Hyacinth. The Weevils* Neochetina bruchi *and* N. eichhorniae*: Biologies, Host Ranges, and Rearing, Releasing and Monitoring Techniques for Biological Control of* Eichhornia crassipes. Monograph No. 60. Australian Centre for International Agricultural Research, Canberra.

Julien MH, Griffiths MW and Stanley JN (2001) *Biological Control of Water Hyacinth. The Moths* Niphograpta albiguttalis *and* Xubida infusellus*: Biologies, Host Ranges, and Rearing, Releasing and Monitoring Techniques for Biological Control of* Eichhornia crassipes. Monograph No. 79. Australian Centre for International Agricultural Research, Canberra.

Morris MJ, Wood AR and den Breeÿen A (1999) Plant pathogens and biological control of weeds in South Africa: a review of projects and progress during the last decade. *African Entomology*, Memoir No. 1, 129–137.

Page AR and Lacey KL (2006) Economic impact assessment of Australian weed biological control. *CRC for Australian Weed Management Technical Series* **10**, 123–127.

Reeder R (2002) *International Mycoherbicide Programme for* Eichhornia crassipes *Control in Africa (IMPECCA)*. Final report of combined scientific activities (15 February 2000–15 December 2002). CABI Bioscience.

Sands DPA and Kassulke RC (1983) *Acigona infusella* (Walker) (Lepidoptera: Pyralidae), an agent for biological control of waterhyacinth (*Eichhornia crassipes*) in Australia. *Bulletin of Entomological Research* **73**, 625–632.

Silveira-Guido A and Perkins BD (1975) Biology and host specificity of *Cornops aquaticum* (Bruner) (Orthoptera: Acrididae), a potential biological control agent for water hyacinth. *Environmental Entomology* **4**, 400–404.

Sosa AJ, de Remes Lenicov M, Mariani R and Cordo HA (2005) Life history of *Megamelus scutellaris* with description of immature stages (Hemiptera: Delphacidae). *Annals of the Entomological Society of America* **98**, 66–72.

Sosa AJ, Cordo HA and Sacco J (2007) Preliminary evaluation of *Megamelus scutellaris* Berg (Hemiptera: Delphacidae), a candidate for biological control of waterhyacinth. *Biological Control* **42**, 129–138.

Stanley JN and Julien MH (1999) The host range of *Eccritotarsus catarinensis* (Heteroptera: Miridae), a potential agent for the biological control of water hyacinth (*Eichhornia crassipes*). *Biological Control* **14**, 134–140.

Stanley JN, Julien M, Rumbo E and White AJ (2000) Post release monitoring of *Xubida infusellus* (Lep.: Pyralidae): an example of using pheromones for the early detection of establishing populations of biological control agents. In *Proceedings of the 10th International Symposium on the Biological Control of Weeds*. 4–14 July 1999, Bozeman, Montana. (Ed NR Spencer) pp. 753–759. Montana State University, Bozeman, Montana.

Stanley JN, Julien MH and Center TD (2007) Performance and impact of the biological control agent *Xubida infusellus* (Lepidoptera; Pyralidae) on the target weed *Eichhornia crassipes* (waterhyacinth) and on a non-target plant, *Pontederia cordata* (pickerelweed) in two nutrient regimes. *Biological Control* **40**, 298–305.

Van TK and Center TD (1994) Effect of Paclobutrazol and waterhyacinth weevil (*Neochetina eichhorniae*) on plant growth and leaf dynamics of waterhyacinth (*Eichhornia crassipes*). *Weed Science* **42**, 665–672.

Van Wyk E and van Wilgen BW (2002) The cost of water hyacinth control in South Africa: a case study of three options. *African Journal of Aquatic Science* **27**, 141–149.

Wright AD (1981) Biological control of water hyacinth in Australia. In *Proceedings of the 5th International Symposium on Biological Control of Weeds*. 22–29 July 1980, Brisbane. (Ed ES Delfosse) pp. 529–535. CSIRO, Melbourne.

Wright AD (1984) Effect of biological control agents on water hyacinth in Australia. In *Proceedings of the International Conference on Waterhyacinth*. 7–11 February 1983, Hyderabad. (Ed G Thyagarajan) pp. 823–833. UN Environment Program, Nairobi.

Wright AD and Bourne AS (1990) Effects of 2,4-D on the quality of water hyacinth as food for insects. *Plant Protection Quarterly* **5**, 139–140.

Wright AD and Purcell MF (1995) *Eichhornia crassipes* (Mart.) Solms-Laubach. In *The Biology of Australian Weeds*. Vol. 1. (Eds RH Groves, RCH Shepherd and RG Richardson) pp. 111–121. RG and FJ Richardson, Melbourne.

Emex australis Steinheil – doublegee
Emex spinosa (L.) Campdera – lesser jack

Paul B Yeoh, Mic Julien and John K Scott

ABSTRACT

Emex australis and *Emex spinosa* (Polygonaceae) are annual winter-growing herbs. *E. australis*, doublegee, originates from southern Africa and is widespread across southern Australia where it is a weed of crops and pasture. *E. spinosa*, lesser jack, is native to the coastal areas of several countries bordering the Mediterranean Sea. In Australia, distribution of *E. spinosa* has remained restricted to a few locations and it has less impact on crops than *E. australis*. Consequently, Australian biocontrol projects have primarily targeted *E. australis*. A South African weevil, *Perapion antiquum*, was successful in controlling *Emex* in parts of Hawaii. However, when released in Australia it failed to establish at most locations and had no beneficial impact. Other natural enemies found on *E. australis* in South Africa were either not suitably host-specific or were already present as accidental introductions to Australia. An example of the latter is the widely established fungus *Phomopsis emicis* that reduces seed survival in *E. australis*. To expand the list of potential biocontrol agents for *E. australis*, surveys for pathogens and invertebrates were conducted on Mediterranean populations of *E. spinosa*. Two weevils, *Apion frumentarium* and *Lixus linearis*, attack both *Emex* species and were approved for release in Australia, but both failed to establish. A northern hemisphere aphid, *Brachycaudus rumexicolens*, that was not part of the biocontrol program, appeared in Australia in 1985. It spread naturally across southern Australia and under favourable climatic conditions can have a severe effect on *Emex*. Priority organisms for future study include the weevil *Perapion neofallax* because of its ability to diapause over summer, and the pathogen *Cercospora tripolitana*. Both originate from *E. spinosa* in North Africa. Any future biocontrol project will need to take into account the use of *Emex* seeds as a food source by native birds and the risk to native *Rumex* species, as well as an assessment of the continuing importance of *Emex* species due to changing agronomic practices.

Key words: *Apion frumentarium*, *Brachycaudus rumexicolens*, *Cercospora tripolitana*, conflict of interest, *Coniocleonus excoriatus*, cropping weeds, *Dysaphis rumecicola* subsp. *emicis*, *Lixus linearis*, *Lophrotoma analis*, *Microthrix inconspicuella*, *Perapion antiquum*, *Perapion neofallax*, *Perapion violaceum*, *Phomopsis emicis*, plant pathogens, *Rhinoncus australis*, *Rhodometra sacraria*, *Rhytirrhinus inaequalis*, survival over summer, *Uromyces rumicis*.

INTRODUCTION

The genus *Emex* (Polygonaceae) contains only two species – *Emex australis* Steinheil and *Emex spinosa* (L.) Campdera. *E. australis*, doublegee, originates from southern Africa whereas *E. spinosa*, lesser jack, is native to North Africa, the Middle East and Europe, occurring mostly in coastal areas of countries bordering the Mediterranean Sea (Pheloung *et al.* 1996). Plants of both species are autumn–winter germinating annuals typically associated with disturbed environments. *Emex* species only reproduce by seed (Fig. 1). After germination, seedlings quickly develop a rosette and crown with a long and thick tap root. Stems radiate from the crown, branching dichotomously at nodes, and usually remaining prostrate along the surface. Flowers of both *Emex* species are small (~2 mm), green and inconspicuous.They are produced on the crown then on the nodes along the stems. The self-fertile flower produces an achene that contains a single seed. Achenes are dimorphic; those produced on stems have longer spines than those on the crown. The spines aid dispersal of achenes. Plants are capable of producing over 1000 seeds given favourable conditions (Evenari *et al.* 1977; Weiss 1978). All plants die when the winter rains finish and the soil moisture disappears; the species survives over summer as seeds (Weiss 1981).

Emex seeds require an after-ripening period of two to eight months before germination is possible (Panetta and Randall 1993a; Weiss 1980). Germination occurs throughout the whole of the growing season if conditions are favourable, and waves of germination often follow rainfall events. Germination of seeds on the surface is inhibited, with the highest levels of germination occurring at 0.5 cm to 5–6 cm depth (Cheam 1987; Panetta and Randall 1994; Weiss 1980). *E. australis* seeds can remain dormant and viable in the soil for more than eight years (Gilbey 1996) and dormancy rate can be higher in *E. spinosa* (Hagon and Simmons 1978).

E. australis differs from *E. spinosa* in the following ways: when grown under identical conditions, *E. australis* is more prostrate, has fewer seeds and larger achenes (Weiss and Julien 1975); in wheat fields and with high soil-nitrogen levels, *E. spinosa* is less competitive than wheat and *E. australis* (Fig. 2) is more competitive than both *E. spinosa* and wheat. Both species are, however, poor competitors when nitrogen is limiting (Williams *et al.* 1984).

Gilbey and Weiss (1980) state that soon after introduction in 1830 as a salad plant, *E. australis* spread rapidly throughout WA's agricultural area. It appeared in SA in 1870 and in Sydney in 1883. By 1980 it occurred in every WA shire council district with the exception of the Kimberley and was estimated to infest 1 million ha of pasture and 500 000 ha of cereal crops (Gilbey and Weiss 1980). It is most abundant in the grain-growing areas of each state, but only near rivers, towns, homesteads and water holes within pastoral areas (Gilbey and Weiss 1980). It is not common in the tropics. In Tas, it is established only on

Figure 1: Aerial achenes of a) *Emex australis* and b) *Emex spinosa.* Photos: PB Yeoh, CSIRO.

Flinders Island (Fig. 3). It is declared a noxious weed in all states except Qld and ACT (Australian Weeds Committee 2010).

The seed dormancy and longevity ensure survival regardless of cropping system (Cheam 1987) and plant densities can become extreme, with some cereal crops reported as having densities of 460 *E. australis* plants m^{-2}. At these densities, plants produced 16 000 achenes m^{-2}, of

Figure 2: Senescent *Emex australis* in a wheat crop, Mullewa, WA. Photo: PB Yeoh, CSIRO.

which half contained viable seeds (Julien and Hartley 1978). Crop yields and pasture growth decreased 40–50% when *E. australis* was uncontrolled, and additional smaller financial losses due to lameness of sheep and sheep dogs. This resulted in estimated costs in 1980 of more than $20 million annually in WA alone (Gilbey and Weiss 1980).

Today, *E. australis* can be readily controlled in most crops (Panetta and Randall 1993b), however, prior to the mid 1990s, when chemical control options widened (see Commens 1997), herbicide control within the legume pasture phase was difficult because reliable and selective herbicides were not available (Panetta and Randall 1993b).

E. spinosa has been present since at least 1938 in WA, probably introduced as a contaminant of fodder around 1900 as it occurs along the old stock routes that lead from ports to inland pastoral areas (Gilbey 1974). Isozyme studies by Marshall and Weiss (1982) suggested that there have been at least four separate incursions of *E. spinosa* into Australia with the populations in Qld, Vic and Eyre Peninsula, SA all arriving independently from abroad rather than being carried interstate from WA.

Though predicted to become more of a problem than *E. australis* due to its higher seed output and greater potential for contaminating wheat harvests (Weiss and Julien 1975), the distribution of *E. spinosa* (Fig. 3) has remained limited (Gilbey 1974; Pheloung *et al.* 1996; Scott and Yeoh 2004). Poor competitive ability (Williams *et al.* 1984), lower dispersal capacity (Gilbey 1974), low soil suitability (Scott and Yeoh 2004) and the fact that *E. australis* already occupies most of the suitable habitats are factors likely to be restricting the spread of *E. spinosa*. Consequently, *E. spinosa* is a minor weed and is a declared plant only in SA and WA.

BIOLOGICAL CONTROL HISTORY

Outside Australia

Emex species were observed taking over large areas of pasture on Parker Ranch, Hawaii, in about 1940 and were present on other islands of Hawaii (Davis and Krauss 1962; Fullaway 1958). In 1950, the Board of Agriculture and Forestry, Honolulu requested Dr Russo and assistants from Portici Laboratory, Italy to survey *E. spinosa* populations in mainland Italy and Sicily. In 1951 Dr Grillot from the Agricultural Experimentation Service of Rabat, Morocco assisted the Hawaiians with surveys of *E. spinosa* populations in Morocco. Only a few polyphagous insects and a pathogen, *Peronospora rumicis* Corda (Peronosporales: Peronosporaceae), were found to be damaging *E. spinosa*. They were unsuccessful in collecting *P. rumicis* in a follow-up trip and did not study it further (Fullaway 1958).

In 1956, the Hawaiians shifted their attention to *E. australis* in South Africa and NLH Krauss, Board of Agriculture and Forestry, Hawaii surveyed this area (Nakao 1969). From December 1956 to March 1957, Krauss collected *Perapion antiquum* (Gyllenhal) (Coleoptera: Brentidae) (referred to as *Apion antiquum* Gyllenhal) from Durban, South Africa and sent them to Honolulu. First releases were made in January 1957 at Makahalau and nearby Parker Ranch, Hawaii and later on the islands of Lanai, Molokai, Maui and Oahu (Fullaway 1958). On Hawaii, *P. antiquum* effectively controlled *E. australis* and *E. spinosa* at elevations over 1200 m (Simmonds 1967). Within three years of release at Makahalau, *P. antiquum* had reduced a 0.4 ha infestation of *E. australis* by 75% and forage grasses were replacing the weed (Davis and Krauss 1962). No control occurred below 150 m elevation and partial control occurred above 600 m

(a)

(b)

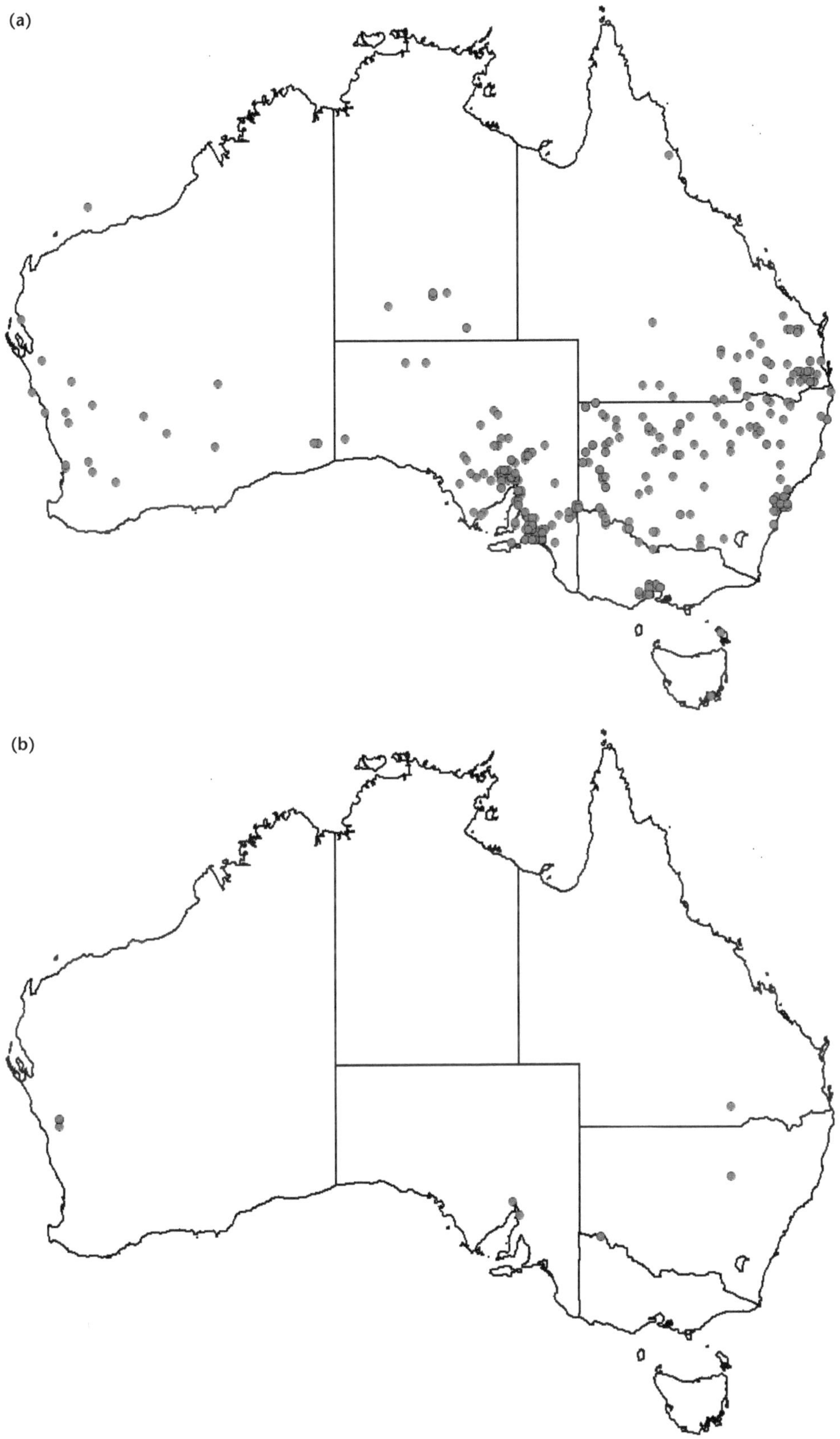

Figure 3: Distribution of a) *Emex australis* and b) *Emex spinosa* in Australia. Specimen data reproduced from Australia's Virtual Herbarium with permission of the Council of Heads of Australasian Herbaria Inc.

(Simmonds 1967). The weevil had a substantial impact on Maui, but provided incomplete control on Oahu and Molokai (Gardner and Davis 1982).

By 1960, Krauss and his collaborators had identified two new potential weevil agents from *E. spinosa* – *Perapion violaceum* (Kirby) (Coleoptera: Brentidae) (referred to as *Apion violaceum* var. *harcyniae* Hubenthal) in Portugal and *Perapion neofallax* (Warner) (Coleoptera: Brentidae) (referred to as *Apion neofallax* Warner) in Morocco (Krauss 1963). In 1962, both species were released to establish control of *Emex* in the drier areas of the Hawaiian islands (Nakao 1969). *P. neofallax* collected in 1960 and 1962 from Rabat, Morocco by Krauss resulted in releases of up to 600 individuals at Omaopio and Waiakoa, Maui and at Molokai Ranch, Molokai during 1962. *P. violaceum* collected from Praia das Macas, Portugal in 1960 and 1962 resulted in releases of 195 and 365 individuals at Omaopio and Waiakoa respectively during 1962 (Davis and Krauss 1963; Krauss 1963). Both species failed to establish (Julien and Griffiths 1998).

Australian biocontrol projects

In 1974, CSIRO imported *P. antiquum* from Hawaii into its quarantine facilities in Brisbane (Harley and Kassulke 1975). *P. antiquum* was subsequently approved for release in Australia and from 1974 to 1978, mass rearing and releases occurred in Qld and all southern states, but most releases failed to establish and there was no impact on the weed. CSIRO also imported and released another weevil, *Lixus linearis* Olivier (Coleoptera: Curculionidae) (= *Lixus cribricollis* Boheman; Fauna Europaea (2010)), but this failed to establish (Julien 1981; Julien *et al.* 1982). Efforts to import and assess new agents for the biocontrol of *Emex* spp. ceased in 1980 (Julien 1981). They recommenced in 1993 when CSIRO in Perth imported and subsequently released a weevil, *Apion frumentarium* (L.) (Coleoptera: Brentidae) from Israel (Scott and Yeoh 2005), but again this failed to establish. Other potential agents both from overseas and within Australia were tested until 2005, but no further agents were released.

EXPLORATION

Surveys in South Africa

In an attempt to find additional agents for *E. australis*, surveys in South Africa were conducted during the 1980s by S Neser and M Morris, Plant Protection Research Institute, Stellenbosch, South Africa, RCH Shepherd, Keith Turnbull Research Institute, Frankston, Vic, RG Shivas, Department of Agriculture, South Perth, WA, JK Scott, CSIRO (then based at the University of Cape Town, Rondebosch, South Africa), and their co-workers. Forty-five insect species and five pathogens were found associated with *E. australis* (Scott and Way 1990; Shepherd 1989; Shivas and Sivasithamparam 1994; Shivas 1995). In addition to *P. antiquum*, three insects, *Microthrix inconspicuella* Ragonot (Lepidoptera: Pyralidae), *Rhodometra sacraria* L. (Lepidoptera: Geometridae) and *Rhytirrhinus inaequalis* (F.) (Coleoptera: Curculionidae), and three pathogens, *Cercospora tripolitana* Sacc. & Trotter (Capnodiales: Mycosphaerellaceae), *Uromyces rumicis* (Schumach.) G. Winter (Pucciniales: Pucciniaceae) and *Phomopsis emicis* R.G. Shivas (Diaporthales: Diaporthaceae), on *E. australis* warranted further investigation.

Surveys in North Africa, Spain and Portugal

Surveys of the insects and pathogens associated with *E. spinosa* were located primarily in Spain, Portugal, Morocco and Israel (Scott and Shivas 1990). Surveys by staff from CSIRO Montpellier (J-F Vayssières, 1978–79, AJ Wapshere, M de Gouvenain and P Rahola, 1977–78 and J-L Sagliocco, 1993–96) of *E. spinosa* in Morocco, Spain and Portugal, identified three curculionids that were subsequently imported into Australian for investigation (*L. linearis*, *P. violaceum* and *Coniocleonus excoriatus* (Gyllenhal) (Coleoptera: Curculionidae)) and observed the pathogens *C. tripolitana* and *P. rumicis* attacking *E. spinosa*.

Surveys in Israel

In 1988, JK Scott found the weevil *A. frumentarium* attacking *E. spinosa* plants in Israel (Scott and Shivas 1990). Also found during this survey were a coreid bug, *Haploprocta sulcicornis* (F.) (Hemiptera: Coreidae) and an aphid, *Dysaphis rumecicola* (Hori) subsp. *emicis* (Mimeur) (Hemiptera: Aphididae), previously named *Dysaphis emicis*; both insects already known from *E. spinosa* in Portugal (Krauss 1963). Benson (1955) also listed the sawfly *Kokujewia ectrapela* Konow (Hymenoptera: Argidae) as occurring on *Emex* in Israel, though it was not found during any of the surveys conducted by Sagliocco or Scott from 1988 to 2006.

Surveys in Australia

Two exotic organisms, *Brachycaudus rumexicolens* (Patch) (Hemiptera: Aphididae) and the fungus *P. emicis*, were discovered on *Emex* in Australia. They are thought to be accidental introductions and were subsequently found to be widespread (see relevant sections in this chapter). Additionally, two insects native to Australia are known to feed on *Emex*: *Lophrotoma analis* (Costa) (Hymenoptera: Pergidae) and *Rhinoncus australis* Oke (Coleoptera: Curculionidae).

CANDIDATES

Agents tested and rejected

***Dysaphis rumecicola* (Hori) subsp. *emicis* (Mimeur) (Hemiptera: Aphididae).** This small aphid was described by Mimeur (1934) from specimens on the roots of *E. spinosa* in Rabat, Morocco. Stekolshchikov (2006) downgraded *D. emicis* to a subspecies of *Dysaphis rumecicola* also known from Sicily, Morocco and India, and from Tajikistan and Japan, areas outside the distribution of *Emex*. A closely related aphid species, *Dysaphis rheicola* Daniyarova ex Narzikulov et Daniyarova, found on rhubarb (Narzikulov and Daniyarova 1971), was also synonymised with *D. rumecicola* (Stekolshchikov 2006). Colonies imported into Australia were collected from five *E. spinosa* sites in Israel (Ashdod to Rehovot) by Scott in February 1997 and sent to quarantine facilities in Perth, WA for study.

Within quarantine, the aphid was found to have a high reproductive rate with apterae on average taking 10 days to mature and producing 39 offspring during their 27-day life-time (25°C day and 15°C night temperatures). They were, however, found to be strongly photophobic, feeding on the crown and stem growing below the surface – a common situation in *E. spinosa* but not in *E. australis*. They were unable to feed or survive upon the roots of *Emex* plants (MJ Majer and PB Yeoh, unpub. data). Rhubarb (*Rheum rhabarbarum* L., Polygonaceae) was also found to be a suitable host, so the quarantine colony was destroyed.

***Microthrix inconspicuella* Ragonot (Lepidoptera: Pyralidae).** This moth was found associated with *E. australis* throughout the plant's distribution in South Africa, causing minor damage (Neser 1979; Scott and Way 1990). However, up to 45% of individuals died from viral infections, braconid parasites killed up to 68% of prepupae and ichneumonid parasites killed 33% in the prepupae/pupae stage. It was considered that, freed of these constraints, *M. inconspicuella* might be a useful biocontrol agent (Neser 1979).

Adult moths lay an average of 175 eggs during their nine-day lifespan. Eggs are mostly deposited singularly on petioles of *E. australis*. The larvae tunnel in the leaves then construct a silken chamber, webbing the leaves together to form a cocoon. The life-cycle takes 26–32 days at 25°C (Harley *et al.* 1979).

Harley *et al.* (1979) imported *M. inconspicuella* into CSIRO's quarantine facilities in Brisbane and found that it could severely damage *E. australis*. However, it completed several generations on *R. crispus* L., *Rheum rhaponticum* L. (European rhubarb) and young apple (cited by Shepherd (1990a) as *Malus domestica* Borkh var. Tropical Beauty).

In subsequent studies in South Africa, Shepherd (1990a, b) found that the moths could feed and develop on apple plants, but mortality was high and multiple generations could not be sustained. No *M. inconspicuella* larvae survived the normal apple orchard spraying regimes and Shepherd (1990b) concluded that the larvae would not survive in an open orchard situation and apples would not be at risk from larval damage. Despite this, *M. inconspicuella* has not been re-imported nor has an application been made for its release in Australia.

***Rhodometra sacraria* L. (Lepidoptera: Geometridae).** This moth was common on *E. australis* throughout the plant's distribution in South Africa (Neser 1979; Scott and Way 1990) where larvae cause extensive damage to the foliage (Shepherd 1989). Generation time is rapid, with the immature stages taking 23–31 days in the laboratory at 25°C and adults living approximately eight to nine days. Females lay on average 193 eggs (Scott and Way 1989b) although infection by a cytoplasmic polyhedrosis virus can significantly lower reproductive output (Shepherd 1990c).

Host-specificity tests found *R. sacraria* to be specific to plants within the Polygonaceae family, with better development on *Polygonum aviculare* L. than on *E. australis* (Scott and Way 1989b). Shepherd (1989) found feeding and development on all four native Australian *Persicaria* species and two of the three native *Rumex* species that she tested. Survival for multiple generations was also possible upon two of the native species of *Persicaria*. This moth was rejected as a biocontrol agent due to its lack of host-specificity.

***Rhytirrhinus inaequalis* (F.) (Coleoptera: Curculionidae).** This weevil was commonly found associated with *E. australis* in South Africa (Scott and Way 1990). The 9 mm long adults are nocturnal and, during the day, hide in or on the soil under *E. australis* plants (Scott and Way 1989a).

Although the adults fed upon *E. australis* stems and petioles under laboratory conditions, Scott and Way (1989a) never observed feeding in the field. Larvae fed upon the roots, making 2–3 mm cavities, sometimes extending further into the crown. The insects appeared to have minimal impact, however, on *E. australis* plants (Scott and Way 1989a). In host-range tests this species attacked plants from the Asteraceae family and so was rejected as a biocontrol agent (Scott and Way 1989a).

Potential agents not released

***Cercospora tripolitana* Sacc. & Trotter (Capnodiales: Mycosphaerellaceae).** This fungus causes circular light brown to reddish brown marks on the leaves and elongated stem lesions. It is widespread in southern and northern Africa on *E. australis* and *E. spinosa* respectively. Despite differences in the morphology of their conidiophores, Morris (1984) considered the southern African and the Moroccan *Cercospora* to be the same.

In South Africa, *C. tripolitana* occurs on *E. australis* plants from areas with a subtropical (Morris 1984) or Mediterranean (Shivas 1995) climate. In preliminary host-specificity tests, South African *C. tripolitana* (of subtropical origin) showed typical symptoms on the stems, petioles and older leaves of inoculated *E. australis* and *Rumex obtusifolius* L. Partial development occurred on *R. crispus*, but the fungus could not develop on species from the other genera within the Polygonaceae, including buckwheat, *Fagopyrum esculentum* Moench, and rhubarb. Under glasshouse conditions, the South African *C. tripolitana* had little impact on the growth of *E. australis* and did not reduce seed production (Morris 1984). In the field, Shivas (1995) observed that the severe attack by this fungus did not normally occur until late in the plant's growing season, after seed production.

Surveys by Wapshere in autumn 1977 and spring 1978 found *C. tripolitana* to be common on *E. spinosa* along the Atlantic coast of Morocco, but less common inland. There was no apparent impact on the plant in any location (AJ Wapshere letter to KLS Harley, 1978). In contrast, in autumn 1999, seedlings and rosettes were being decimated by the fungus in Tunisia (JK Scott, unpub. obs.). Kammoun and Souissi (2009) found that under laboratory conditions the Tunisian strain of *C. tripolitana* was capable of killing *E. spinosa* plants within six weeks of inoculation.

C. tripolitana is easy to culture on artificial media (Shivas and Sivasithamparam 1994). As preliminarily tests showed the species is likely to be host-specific, the possibility of a more virulent form of *C. tripolitana* warrants further investigations.

***Coniocleonus excoriatus* (Gyllenhal) (Coleoptera: Curculionidae).** This 11–14 mm long weevil is found throughout the Mediterranean region (Hoffmann 1950). This univoltine weevil has adults that feed upon the leaves of the *E. spinosa* plants and larval development occurs underground at the junction of the stem and the upper tap root. The larvae pupate in a soil cocoon at the end of spring and adults aestivate within the cocoon. They emerge from the cocoon just after the first autumn rains and the germination of *E. spinosa* seedlings in the field. After feeding on the leaves and stems of the young plants, adults lay eggs on or near the plants; larvae emerge soon after.

In preliminary host-specificity tests, adult *C. excoriatus* fed on *Rumex crispus*, *R. acetosa* L. and rhubarb as much as on *E. spinosa* (J-F Vayssières, unpub. report 1979). Comparatively less feeding by adults occurred on buckwheat, *Rumex conglomeratus* Murray and *R. bucephalophorus* L. and even less on *Polygonum hydropiper* L. and *P. persicaria* L. In the field, Vayssières observed adults on young Malvaceae plants. Trace amounts of feeding by adults were observed on cotton plants in subsequent laboratory trials. It is not known if any of these plants are suitable alternative hosts for *C. excoriatus* as trials using larvae were inconclusive due to high larval mortality on all plants including *E. spinosa* (J-F Vayssières, unpub. report 1979).

Rearing this insect proved very difficult and imports into Australian quarantine facilities in 1993–94 (seven adults, 87 larvae then 74 adults received in shipments from Israel) failed to establish as a laboratory culture. Further progress with *C. excoriatus* awaits successful rearing techniques.

***Perapion neofallax* (Warner) (Coleoptera: Brentidae).** This is a 2.5–3.5 mm long weevil from northern Africa, occurring in Morocco (Krauss 1963) and Tunisia (JK Scott, unpub. obs. 1999). The adults chew small 'shot holes' in the leaves of *E. spinosa*. Eggs are laid singly in pits made in the stem or base of a petiole and larvae complete development while tunnelling inside the stem. Pupation occurs inside the plant. There appears to be only one generation per year (Krauss 1963).

In 1976, larvae of *P. neofallax* within *E. spinosa* stems were collected by R Kassulke (CSIRO, Brisbane) from around Rabat and Casablanca, Morocco and sent to CSIRO's quarantine facility in Brisbane (MH Julien, unpub. data; Julien and Harley 1978). Follow-up importations occurred in 1977 and 1978 when Wapshere collected 549 and 2031 adults together with larvae and pupae in plant material from coastal areas between Skhirat and El Jadida, Morocco and sent them to Qld (MH Julien, unpub. data). The insects sent to Qld remained in their summer diapause state, never became reproductive and did not survive (MH Julien unpub. data; Julien 1981).

The next importations of this species were made in 1994 when Sagliocco sent adults from near Rabat to the quarantine facilities in Perth. The 152 adults that arrived alive appeared to be in a state of diapause. Over the next two years, despite being held in a range of conditions on *E. australis* or *E. spinosa* plants, none became reproductive and all died.

This species' ability to survive long periods of starvation and its apparent tendency to go to ground in search of oversummering sites may make it highly suitable to Australian conditions. Further research with this weevil will depend on developing successful rearing techniques.

***Perapion violaceum* (Kirby) (Coleoptera: Brentidae).** This 2.5–3.5 mm long weevil is closely related to *P. neofallax* and has a similar biology (Krauss 1963). It occurs throughout Europe and around the Mediterranean, where it feeds on many *Rumex* species (Hoffmann 1958). *P. violaceum* was found on *E. spinosa* in Portugal (Krauss 1963) and Spain (J-L Sagliocco, J-F Vayssières and AJ Wapshere, unpub. data). It was recorded from Israel (Melamed-Madjar 1969), although it was not found there during surveys by Scott in February–March 1988, May 1993, March 1994, or February–March 1997.

In preliminary tests using *P. violaceum* collected from *Rumex* near Montpellier, France adults fed upon plants within Polygonaceae, including European *Rumex* spp., an Australian native *Rumex* species, rhubarb, buckwheat and *E. spinosa* (Scott 1985). Eggs laid on the European *Rumex* spp. and *E. spinosa* developed normally. Eggs were also laid into rhubarb, but the larvae died as first instars. The extent to which *P. violaceum* is able to complete a whole generation upon the Australian native *Rumex* spp. is unknown.

In 1976, Kassulke collected 125 adult *P. violaceum* from *Rumex* spp. in southern France and in 1976 he collected larvae in *E. spinosa* stems from Portugal. These were sent to CSIRO in Qld. In 1977, additional larvae in stems together with a single adult collected from Lisbon (MH Julien, unpub. data; Julien and Harley 1978) were again sent to CSIRO in Qld. As with *P. neofallax*, the *P. violaceum* remained in the summer diapause state and did not survive (Julien 1981). In 1996, Sagliocco sent 10 adults to CSIRO in Perth, where they also failed to break diapause. All insects died by February 1997.

Agents tested and released in Australia

***Apion frumentarium* (L.) (Coleoptera: Brentidae).** This red weevil (Fig. 4) from *E. spinosa* was imported under the name *Apion miniatum* Germar, as originally identified by the British Museum. Revision of the genus resulted in *A. miniatum* becoming a junior synonym of *A. frumentarium* (Thompson and Alonso-Zarazaga 1988).

A. frumentarium occurs across Europe, Israel, Syria and the Caucasus (Scott and Yeoh 2005). Adults lay eggs in petioles and stems and feed on the foliage, making small shot holes. The larvae have a long developmental period of 567 degree-days above a lower developmental threshold of 8°C (JK Scott and PB Yeoh, unpub. data), and complete their development within the plant. Newly emerged adults feed on leaves then leave their senescing host plant to seek summer refuge sites, where they aestivate until the following autumn (Yeoh *et al.* 2002, 2004).

The first importations into Australia were of 157 adult *A. frumentarium* collected by Scott in Israel in May 1993. A further 520 adults were collected by Sagliocco and Scott in April–May 1995. All were sent to quarantine in Perth. Host-specificity trials showed that *A. frumentarium* attacks only *Emex* species and some of the European species of *Rumex* that are also weedy in Australia (Scott and Yeoh 2005). Approval for release was obtained in 1997.

In spring 1998 the first releases of 7300 insects were made across 10 sites in WA, but no offspring were found during the following growing season. Subsequent releases were all made in autumn.

Considerable efforts were made to improve release strategies, such as:

- excluding predators (Yeoh *et al.* 1999);
- optimising release numbers per site and determining the influence of different climatic zones (Yeoh *et al.* 2002);

Figure 4: *Apion frumentarium.* Photo: PB Yeoh, CSIRO.

- developing methods of storage of insects over summer to enable spring releases (Yeoh and Woodburn 2004);
- conducting multiple releases at nearby sites and over successive years to reduce Allee effects (PB Yeoh and TL Woodburn, unpub. data);
- discovering the adults' requirement for a carbohydrate source such as the nectar produced by tamarisk plants over summer (Yeoh *et al.* 2004);
- releasing the insect on both species of *Emex* (PB Yeoh, unpub. data).

All failed to assist establishment. Despite nearly 70 000 *A. frumentarium* being released at 66 locations, there has not been a single report of establishment of this insect in Australia.

Lixus linearis **Olivier (Coleoptera: Curculionidae).** Adults of this weevil are 8–14 mm long (Julien *et al.* 1982) and feed on the foliage. They have one or two generations per year, but can live for several years, aestivating during summer. Eggs are deposited in pits chewed into the stem, petiole or mid-rib of the plants. The larvae develop and form pupae inside the stem. Under quarantine laboratory conditions, the adult emerges approximately 39 days after the egg was laid. In non-aestivating adults, there was an 80–90 day pre-oviposition period, but in the quarantine facility in Brisbane the adults began aestivation in November where they remained until June (Julien *et al.* 1982).

The initial colony established in the quarantine facility at CSIRO in Qld came from Morocco in 1976. It was found to be specific to plants within the Polygonaceae family, with *Rumex* spp. and *Emex* spp. susceptible. Rhubarb and buckwheat suffered some degree of adult feeding, but were not capable of supporting larval development (Julien *et al.* 1982). Approval to release the insect for the control of *E. australis*, *E. spinosa* and *R. crispus* was given in August 1979.

During 1980, 670 adults were sent to SA, WA and Vic for mass rearing and field release by the state government agencies. CSIRO released 104 adults over three field sites in south-eastern Qld during that year. In WA, 850 offspring were successfully reared from a starter colony of 15 adults. These were released the following year, being initially contained with *E. australis* for a few weeks in field cages (Richards 1981). *L. linearis* is not known to have established at any of the Australian release sites (Julien and Griffiths 1998).

Perapion antiquum **(Gyllenhal) (Coleoptera: Brentidae).** A native of South Africa, this 3 mm long weevil (Fig. 5) was the first agent to be targeted for release against *Emex* spp. in Australia because of its earlier success in Hawaii. The adults commence feeding immediately after emergence and attack all above-ground parts of the *Emex* plants, sometimes almost defoliating the plants and causing petioles and stems to collapse (Harley and Kassulke 1975).

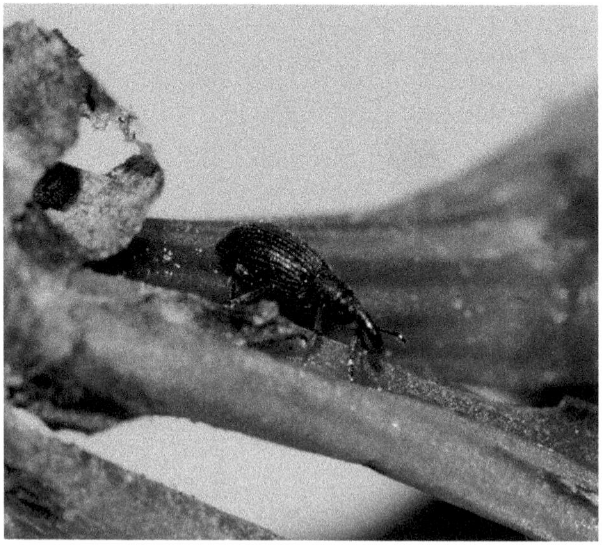

Figure 5: *Perapion antiquum*. Photo: JK Scott, CSIRO.

Like *A. frumentarium* and the other *Perapion* spp., eggs are deposited singly in pits chewed into the stem and petiole of the plants; development, including pupation, occurs entirely within the plant. In the laboratory (~24°C), adults emerge 25 days after eggs are laid with the average life-cycle taking about 33 days (Harley and Kassulke 1975). Lower temperatures greatly decrease the developmental rates, with the life-cycle taking 96 days at 15°C and ceasing below 9.6°C. Adults are without an obligatory aestivation stage and will reproduce continuously under favourable conditions (Julien and Bourne 1983). In South Africa, Scott (1992) found adults aestivating within the old pods of *Acacia cyclops* G. Don (Leguminosae), a shrubby weed introduced from Australia.

In January 1974, H Zimmermann, Plant Protection Research Institute, South Africa collected 32 adult *P. antiquum* from around Franskraal, South Africa and sent them to CSIRO in Qld. Subsequent shipments that summer included 128 adults from Uitenhage and Klipheuwel (MH Julien, unpub. data). Host-specificity testing showed *Emex* and the European *Rumex* to be favourable hosts. *Rumex brownii* Campd., an Australian native, was a suboptimal host plant. Minor feeding occurred on buckwheat and rhubarb. Eggs that were laid into rhubarb hatched, but larvae failed to complete development (Harley and Kassulke 1975). Approval to release the insects in Australia was given mid 1974.

Initially, eight releases of 500 insects were made in August 1974. Establishment rates were poor and so subsequent releases were larger (~2000 individuals) and made over a wider and more diverse area (Julien and Harley 1978). By 1976 over 70 000 adults, reared from colonies originating from the Western and Eastern Cape Province areas, had been released at 64 sites across Australia. Establishment occurred at only two sites; one failed to survive the second summer and the other persisted, but had no impact upon the plants (Julien 1981).

In an attempt to find individuals genetically more suited to Australian habitats, a shipment of 337 adults from Hawaii was imported in March 1974. Additional *P. antiquum* from Hoeko, near Ladysmith, South Africa, a subtropical, hot and arid area (271 adults) and from Grahamstown, South Africa, a warm, temperate and humid area (443 adults) were sent to CSIRO (Qld) in spring 1976 and 1977. From the Hoeko colony, a total of almost 35 000 adults were reared and released over 32 sites and from the Grahamstown colony, 33 500 individuals were reared and released over 30 sites (MH Julien, unpub. data; Julien 1981). Establishment occurred only at two of the sites colonised with weevils from the Hoeko colony. Establishment occurred on Flinders Island, Tas from a 1977 release of stock originating from Franskraal (Julien 1981).

The poor establishment rates in Australia were attributed to the inability of adult *P. antiquum* to survive in the absence of growing host plants over the relatively long summer period. Establishment occurred at sites with irrigation for horticultural crops, which may have assisted out-of-season growth of *Emex*. Even where establishment occurred, the insect's oversummer mortality was too high for them to affect *Emex* populations, as the weevil's population did not reach damaging levels until the plants were already large and producing seed (Julien and Bourne 1983). Follow-up releases of 20 000 insects, in 1987 and 1988, at a site in the central wheat belt area of WA, confirmed that summer-irrigated *E. australis* stands could sustain *P. antiquum* populations if the plants do not perish over the period due to diseases and excessive oviposition. At this site impact was, however, minimal outside the irrigated area as there was only one insect generation during winter (Fisher *et al.* 1990).

CLIMEX models predicted that most of the *Emex*-infested areas of Australia were climatically unsuitable for *P. antiquum* (Scott 1992), with only some coastal areas of Australia favourable for insect development. This includes the coastal areas near Esperance, WA where *A. cyclops* is native and abundant. However, the insect has not been released in this area so far.

Agents already present in Australia

***Brachycaudus rumexicolens* (Patch) (Hemiptera: Aphididae).** Commonly called the dock aphid, this 1.2–1.5 mm aphid is a fortuitous accidental introduction first recorded from Australia in 1985, and found throughout most of the dryland agricultural area of WA three years after the first record. It was known from Europe, Africa, Asia and parts of North America prior to its discovery in Australia (Berlandier and Scott 1993).

The dock aphid feeds upon the growing tip of the stems, flower buds and unfurled leaves of *Emex* spp., causing distorted growth and stunting (Fig. 6). Often, leaves on the infected plant turn yellow and premature death of the plant follows. These symptoms appeared on *E. australis* in the WA grain-growing area in 1990 and became referred to as 'doublegee decline'. It is caused by *B. rumexicolens* rather than a pathogen (Scott *et al.* 1994; Scott and Shivas 1998) or virus (Berlandier *et al.* 1997; Scott *et al.* 1994).

Under heavy attack from the dock aphid, the growing tips of the *E. australis* stems are killed, thus preventing further growth of the stem with its multiple nodes and production of seeds. Under glasshouse conditions, there was an 80% reduction in seed numbers following attack (PB Yeoh and JK Scott, unpub. data). The quality of the seed was also affected, with smaller achenes produced when dock aphid are present (Scott and Shivas 1998). Seed size has been shown to be positively associated with seed viability and dormancy (Scott *et al.* 1994). Published

Figure 6: Leaf distortion to *Emex australis* caused by *Brachycaudus rumexicolens*. Photo: JK Scott, CSIRO.

host records and host-specificity trials (Scott and Yeoh 1998) show that the host-range of the dock aphid is largely restricted to plants within the Polygonaceae family.

In many areas with a Mediterranean climate (such as the south-west of WA), the suitable periods of growth are in autumn and spring, with poor survival predicted over the hot dry summer. If *E. australis* is locally the only suitable host plant, long-range invasion would be required annually within south-western Australia. The aphid is not known to have sexual morphs with their associated resistant life-stage or alternative host cycle (Scott and Yeoh 1999).

The dock aphid has a rapid life-cycle with apterae requiring less than eight days to mature at 20°C and producing up to 45 offspring each. However, its impact upon *Emex* spp. will depend upon how early in the season the insects colonise *Emex* plants. In 1990, for example, rainfall at the end of summer favoured the early establishment of *B. rumexicolens* colonies and a significant impact upon *Emex* spp. could be expected, but in many years conditions are not suitable for aphid populations until autumn. The insect's developmental rate is slow over winter, so the dock aphid will have less influence (Scott and Yeoh 1999). Strategies to increase the effectiveness of the dock aphid as a biocontrol agent for *Emex* spp. by the provision of alternative suitable host plants are discussed by Yeoh and Scott (1996).

***Lophrotoma analis* (Costa) (Hymenoptera: Pergidae).** Adults of this native Australian insect, which occurs in all mainland states of Australia (Schmidt and Smith 2006), are locally known as the dock sawfly. They are large (~2 cm long) with a bright orange abdomen and are poor fliers. They feed on nectar and pollen (Yeoh and Scott 1996) and under laboratory conditions with honey and water they live less than 10 days (Schmidt *et al.* 2006). The female inserts rows of eggs between the upper and lower lamellae along the leaf's edge. The larvae emerge and feed on the leaves, often resulting in total defoliation. The damage is characteristic, as the insects avoid eating the main veins of the leaves as well as the stems. Pupation occurs in the soil and there are several generations per year (Yeoh and Scott 1996; Schmidt *et al.* 2006). In Perth they take two to three months to complete a generation during spring (JK Scott and PB Yeoh, *pers. obs.*).

Larvae feed upon several native and introduced species from within the Polygonaceae (Schmidt and Smith 2006; Yeoh and Scott 1986). At a WA site, Scott and Shivas (1998) measured field densities of 15 larvae m^{-2} from late winter to early spring, declining to 10 larvae m^{-2} in late spring. Adults were present throughout the growing season and eggs were observed during autumn and spring but not in winter, limiting its ability to defoliate *Emex* plants during the main growing season (Scott and Yeoh 1996).

Although the dock sawfly is native to Australia, potential toxicity problems mean further research is needed prior to developing any strategies that may increase this insect's presence within grazed pastures or crops that are infested with *Emex*. Sawflies from the family Pergidae are known to cause sawfly larvae poisoning, when the larvae are accidently and/or actively eaten by mammals including livestock (Oelricks *et al.* 1999).

***Phomopsis emicis* R.G. Shivas (Diaporthales: Diaporthaceae).** This stem blight fungus causes circular or irregular brown to reddish brown leaf spots and elongate stem lesions with a light brown centre and red margins (Fig. 7). Morris (1984) found the pathogen associated with stunted, poorly growing *E. australis* plants in Western Cape, South Africa. Seedlings inoculated with the fungus died within seven days and older plants within three to four weeks. Under natural conditions, 55% fewer seeds germinated when the parent plant was infected with the stem blight (Morris 1984).

Subsequently, *P. emicis* was found to be widespread in Australian *E. australis* populations growing in Mediterranean climatic conditions, but not in subtropical locations. In contrast to South Africa (Morris 1984), only

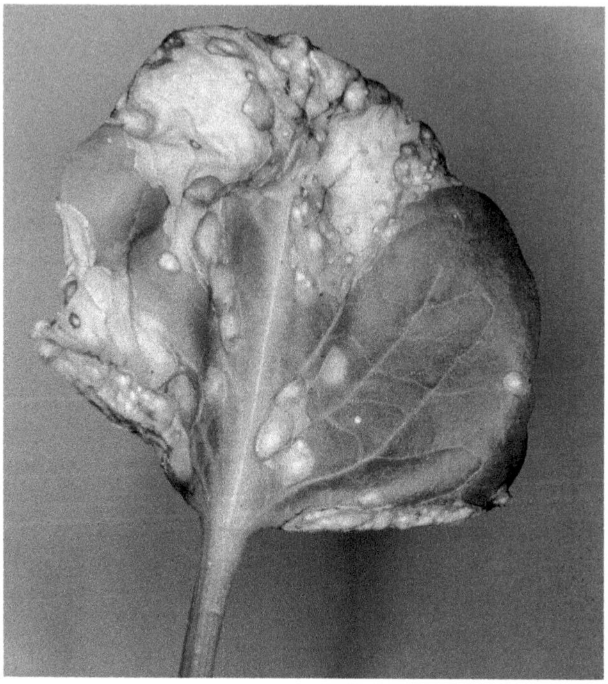

Figure 7: *Phomopsis emicis.* Photo: RG Shivas, DAFWA.

larger and more mature *E. australis* plants show symptoms in Australia (Shivas *et al.* 1994b). Field trials with older plants showed that up to 30% of *E. australis* seeds are killed by the fungus in Australia (Shivas and Scott 1994). Studies on South African *E. australis* showed that young plants were immune to a South African isolate of *P. emicis* (Shivas 1994). Host-specificity tests by Shivas (1994) found that two native species of *Rumex* were also vulnerable to the stem blight fungus.

The importation of strains more virulent against the early stages of *E. australis* and/or the use of this fungus as a mycoherbicide became less feasible when in 1992 it was discovered that *P. emicis* produces large amounts of a mammalian mycotoxin, phomopsin, which is known to cause lupinosis in sheep (Shivas *et al.* 1994a). Of the 900 species of *Phomopsis*, only two are known to produce phomopsin. This discovery halted further attempts to find virulent field-collected strains of the fungi or to use existing strains as mycoherbicides (Shivas and Scott 1994). A non-toxic isolate incapable of reversion would need to be found or developed prior to further consideration of the use of this fungus for the biocontrol of *Emex* in Australia (Shivas and Sivasithamparam 1994).

***Rhinoncus australis* Oke (Coleoptera: Curculionidae).** This ~2.5 mm long, stout weevil is a native of Australia. Julien and Matthews (1980) describe its morphology and biology. Adults feed on all the above-ground vegetative parts of the plant; all immature stages, including pupation, occur inside the plant. In the field, Julien and Matthews (1980) found *R. australis* to be restricted to plants within the Polygonaceae. It has expanded its native host plant range to include the introduced *R. crispus* as well as both species of *Emex*. At some sites, populations of *R. australis* were persisting solely on *E. australis* plants as no other Polygonaceae plants were present in the area.

During two years of field observation at Two Wells, SA and Ma Ma Creek, Qld, *R. australis* was the only insect species found consistently damaging *E. australis* plants (Julien and Matthews 1980). They concluded that it had no significant impact upon field populations of *E. australis* because damage occurred late in the growth season, after the plant had substantially completed development and seed production. A similar situation occurs in WA (PB Yeoh, unpubl. obs.).

***Uromyces rumicis* (Schumach.) G. Winter (Pucciniales: Pucciniaceae).** This rust fungus requires two unrelated hosts to complete its complicated life-cycle, with vegetative spores produced on the primary host and sexual spores on the alternative host (macrocyclic and heteroecious). The first reports of it using *E. australis* as a primary host were from populations of plants in the Western and Eastern Cape provinces, South Africa (Morris 1982). Subsequent surveys by Shivas (1995) found it on *E. australis* in subtropical southern Africa. Morris (1982) considered the fungus to be a strain of the *U. rumicis* found on South African *Rumex crispus*. Preliminary host-specificity tests found that the primary hosts of both strains of *U. rumicis* were *E. australis* and some *Rumex* species from the subgenus *Rumex*. Cross-inoculation studies determined that biotypes of urediniospores collected from one of the host species did not necessary develop normally in the other host.

U. rumicis was not intentionally imported into Australia but has been present here since at least 1986, initially found in a glasshouse in WA on *E. australis*, *R. crispus* and *R. pulcher* L. Surveys conducted during 1990–92 found *U. rumicis* utilising *R. pulcher* plants as a primary host at over half the south-western Australian sites where the plant occurred. It was not, however, found on *E. australis*, *E. spinosa* or any of the other three species of *Rumex* from the subgenus *Rumex* (including *R. crispus*) examined in the field despite several of these species being present at the same sites as *U. rumicis*-infected *R. pulcher* plants (Scott and Shivas 1993).

Ranunculus ficaria L. (Ranunculaceae) is the only species known to be suitable as the alternative host for the sexual phase of *U. rumicis*. Eleven other species from within the genus *Ranunculus*, as well as *R. crispus* and *R. obtusifolius*, were all unsuitable alternative hosts for this stage of the life-cycle (Schubiger *et al.* 1985). Scott and Shivas (1993) did not sample for the sexual stage of *U. rumicis*, as *R. ficaria* was not present at any of the WA sites and is not known to be naturalised in WA. In Rome, where winters are mild, *U. rumicis* can persist without the need of the resistant teliospores, surviving the full year in the vegetative phase upon the living tissue of its *Rumex* spp. primary host (Inman 1970).

Morris (1982) found that, under glasshouse conditions, *U. rumicis* caused significant but modest (<20%) reductions to the size and reproductive output of *E. australis* plants. Field inoculations produced high levels of primary infection, but very little spread of the disease occurred. Shivas and Sivasithamparam (1994) considered that a strain of *U. rumicis* that is virulent on *E. australis* could have potential as a classical biocontrol agent in Australia. Its effectiveness would depend upon it being able to survive the hot summer period in the absence of both growing *Emex* plants and a suitable alternative host plant species. The success of *U. rumicis* as a biocontrol agent of *Emex* may be dependent upon the presence of a co-existing alternative host plant species, such as *R. pulcher*, for the summer periods.

DISCUSSION

As a result of extensive surveys spanning both hemispheres over the past 30 years, 15 potentially host-specific agents for *Emex* from abroad have been identified. Three were released in Australia, but all either failed to establish and/or to have any impact on the weed. Only the dock aphid *B. rumexicolens* and the stem blight fungus *P. emicis*, both thought to be accidental introductions, are widespread and causing significant impact on *Emex* spp. in Australia. Future options include seven potential agents – three weevils that have been difficult to rear in quarantine, a leaf spot fungus *C. tripolitana*, that has been partially studied, the still-to-be-investigated downy mildew fungus *P. rumicis*, a coreid bug *H. sulcicornis* and a sawfly *K. ectrapela*.

Among the insects, the small univoltine weevil *P. neofallax* possibly has the most potential as a biocontrol agent. It is known only from *E. spinosa*, it comes from a region (North Africa) which has hot dry summers similar to *Emex*-infested regions of Australia, and it may have an advantage in Australian conditions by aestivating in the soil; although the last factor needs field confirmation. However, the inability to break aestivation has so far prevented the rearing of this insect within Australia. This problem might be overcome by locating biotypes more attuned to Australian photoperiods (see Dalin *et al.* 2010 for an example of the role of day length in diapause and establishment of biocontrol agents). The other univoltine weevils, *P. violaceum* and *C. excoriatus*, have a host range that includes *Rumex* species, but determining the extent of the host-range and other aspects of the biology requires progress in methods of rearing these insects.

Little is known of the biology and host-range of *H. sulcicornis* other than it occurs around the Mediterranean basin and in central Europe (Linnavuori 1960) and attacks *E. spinosa* in Portugal (Krauss 1963) and Israel (JK Scott, unpub. report). The biology of *K. ectrapela* on *E. spinosa* is also unknown, but it is likely to be similar to that described by Karimpour (2008) on *Rumex* in Urmia, Iran with the exception that it is either quiescent over summer, when *E. spinosa* is not available, instead of winter, or it is not host-specific. Among the insects still to be assessed, we recommend that *P. neofallax* be given priority in a future biological control program.

The known host-range of *P. rumicis* (Hall 1994) indicates that it may be specific to the family Polygonaceae, although there may be pathovars for different host species. It is found in South Africa and New Zealnad and throughout Europe, Africa and Asia, suggesting that it may already be present within Australia. We recommend surveys here before other research is done on this species.

Strains of *C. tripolitana* from Tunisia and *P. rumicis* from Morocco both appear to be virulent against *E. spinosa*, but are currently untested against *E. australis*. *P. rumicis* is an obligate parasite and needs to be reared upon live host plant tissue. As such, it is far harder to rear or test under laboratory conditions than *C. tripolitana*, which will sporulate in culture and for which infection can occur from hyphae. We recommend that *C. tripolitana* be considered first as it appears to have the more restricted host-range.

The knowledge gained during recent years has improved the probability of establishing new importations of *A. frumentarium*. Spring releases of adults upon the trunks of tamarisk plants bordering *Emex* infestations would reduce the effort and space required to maintain rearing colonies over summer, with only a starter colony needed to be held each year or only field collections needed after the first year of releases. The apparent dependence upon neighbouring tamarisk plants is problematic as *Tamarisk* spp. are weeds of national significance, thus there are potential conflicts of interest (Yeoh *et al.* 2004).

Approximately 180 000 *P. antiquum* were released across Australia and yet widespread establishment did not occur. It is hypothesised (Scott 1992) that most of the release areas were climatically unsuitable, as shown by a CLIMEX model. Perhaps, like *A. frumentarium*, an unexpected but critical association is missing from the sites used for the releases in Australia. *Acacia cyclops*, an Australian plant that is now a weed in South Africa, is known to be used as an oversummering site for *P. antiquum* in its native habitat (Scott 1992). Re-importations and releases of *P. antiquum* onto *E. australis* growing adjacent to *A. cyclops*, for example in the Esperance region of WA (Scott 1992), could maximise the probability of establishing a large population of this agent in Australia.

Changing attitudes towards the value of *Emex* species in Australia (Scott *et al.* 2000) now place barriers against the approval for importation and release in Australia of suitable biocontrol agents targeting these weeds. With *E. australis* becoming a major component of large areas denuded of native vegetation, it is not surprising that some native animals now include *E. australis*, especially its seeds, as a food source. For example, two native birds, *Cacatua leadbeateri* (Vigors) (Major Mitchell cockatoo) and *Calyptorhynchus banksii samueli* Mathews (inland subspecies of the red-tailed black cockatoo) eat the seeds of *E. australis*. While the Major Mitchell cockatoo can feed on a range of plants, populations of the red-tailed black cockatoo in grain-growing regions feed almost exclusively on *E. australis* seed. However, the cockatoos eat only about 2% of available *E. australis* seed (Scott *et al.* 2000). The second potential non-target effect involves the risk to native *Rumex* species. It is clear from the host-range tests on organisms associated with *Emex* and *Rumex* that the host-range often includes both *Emex* and the subgenus *Rumex* (*Rumex*), in which the Australian *Rumex* species are included. The potential non-target effects of successful control of *Emex* need to be considered in the context of these two issues – indirect effects on native fauna and direct impact on native *Rumex* species. Further risk assessments and cost:benefit analyses, especially in light of changing agronomic practices which may have lessened the impact of *E. australis* in some situations, may be required prior to any further investigations which aim to import these agents from overseas.

ACKNOWLEDGEMENTS

We thank numerous colleagues in Australia and at the overseas sources of *Emex* and its potential biocontrol agents, especially in France, Hawaii, Israel, Morocco, South Africa and Tunisia, for their support in the quest for *Emex* biocontrol agents.

REFERENCES

Australian Weeds Committee (2010) Noxious weeds list. <http://www.weeds.org.au/noxious.htm> (accessed 12/8/10).

Benson R (1955) The sawflies (Hymen., Symphyta) of Israel. *Bulletin of the Research Council of Israel* **4**, 351–356.

Berlandier FA and Scott JK (1993) Distribution and host records of *Brachycaudus rumexicolens* (Patch) (Hemiptera: Aphididae), a recent arrival in south-west Australia. In *Pest Control and Sustainable Agriculture*. (Eds SA Corey, DJ Dall and WM Milne) pp. 394–396. CSIRO Publishing, Melbourne.

Berlandier FA, Thackray DJ, Jones RAC, Latham LJ and Cartwright L (1997) Determining the relative roles of different aphid species as vectors of cucumber mosaic and bean yellow mosaic viruses in lupins. *Annals of Applied Biology* **131**, 297–314.

Cheam AH (1987) Emergence and survival of buried doublegee (*Emex australis* Steinh.) seeds. *Australian Journal of Experimental Agriculture* **27**, 101–106.

Commens T (1997) Prevent and control doublegee infestations. *Farming Ahead* **69**, 49–51.

Dalin P, Bean DW, Dudley T, Carney VA, Eberts D, Gardner KT, Hebertson E, Jones EN, Kazmer DJ, Michels GJ Jr, O'Meara SA and Thompson DC (2010) Seasonal adaptations to day length in ecotypes of *Diorhabda* spp. (Coleoptera: Chrysomelidae) inform selection of agents against saltcedars (*Tamarix* spp.). *Environmental Entomology* **39**, 1666–1675.

Davis CJ and Krauss NLH (1962) Recent developments in the biological control of weed pests in Hawaii. *Proceedings, Hawaiian Entomological Society* **18**, 65–67.

Davis CJ and Krauss NLH (1963) Recent introductions for biological control in Hawaii–VIII. *Proceedings, Hawaiian Entomological Society* **18**, 245–249.

Evenari M, Kadouri A and Gutterman Y (1977) Eco-physiological investigations on the amphicarpy of *Emex spinosa* (L.) Campd. *Flora* **166**, 223–238.

Fauna Europaea (2010) <http://www.faunaeur.org/full_results.php?id=248123> (accessed 19/12/10).

Fisher K, Fogliani R, Gehradte D and Hutchinson D (1990) Biological control of doublegee (*Emex australis*). In *Entomology Branch Biennial Research Summaries 1988–90*. (Ed S Broughton) pp. 37–38. WA Dept of Agriculture, Perth.

Fullaway DT (1958) Importations of natural enemies of the weed, *Emex spinosa* Campd. *Proceedings, Hawaiian Entomological Society* **16**, 359–360.

Gardner DE and Davis CJ (1982) *The Prospects for Biological Control of Non-native Plants in Hawaiian National Parks*. Cooperative National Park Resources Studies Unit, University of Hawaii at Manoa, Dept of Botany.

Gilbey DJ (1974) *Emex* species in Australia with particular reference to Western Australia. *Journal of the Australian Institute of Agricultural Science* **40**, 114–120.

Gilbey DJ (1996) Control of *Emex* species. *Plant Protection Quarterly* **11**, 156–157.

Gilbey DJ and Weiss PW (1980) The biology of Australian weeds 4. *Emex australis* Steinh. *Journal of the Australian Institute of Agricultural Science* **46**, 221–228.

Hagon MW and Simmons DM (1978) Seed dormancy of *Emex australis* and *E. spinosa*. *Australian Journal of Agricultural Research* **29**, 565–575.

Hall G (1994) IMI descriptions of fungi and bacteria, Set 120, Nos 1191–1200. *Mycopathologia* **126**, 41–63.

Harley KLS and Kassulke RC (1975) *Apion antiquum* (Curculionidae: Apionidae) for biological control of the weed *Emex australis. Journal of the Australian Entomological Society* **14**, 271–276.

Harley KLS, Kassulke RC and Julien MH (1979) Biology and host specificity of *Microthrix inconspicuella* Ragonot (Lepidoptera: Pyralidae), a natural enemy of *Emex australis* in South Africa. *Journal of the Entomological Society of Southern Africa* **42**, 343–348.

Hoffmann A (1950) Coléoptères Curculionides (première partie). *Faune de France* **52**, 1–486.

Hoffmann A (1958) Coléoptères Curculionides (troisième partie). *Faune de France* **62**, 1209–1839.

Inman RE (1970) Observations on biology of Rumex rust Uromyces rumicis (Schum.) Wint. *Botanical Gazette* **131**, 234–241.

Julien MH (1981) A discussion of the limited establishment of *Perapion antiquum* and a review of the current status of biological control of *Emex* spp. in Australia. In *Proceedings of the 5th International Symposium on Biological Control of Weeds.* 22–27 July 1980, Brisbane. (Ed ES Delfosse) pp. 507–514. CSIRO, Melbourne.

Julien MH and Bourne AS (1983) Temperature relations of *Perapion antiquum* (Col., Curculionidae), a weevil introduced to control the weed *Emex australis* in Australia. *Zeitschrift für angewandte Entomologie* **95**, 351–360.

Julien MH and Griffiths MW (1998) *Biological Control of Weeds: A World Catalogue of Agents and their Target Weeds.* 4th edn. CABI, Wallingford, UK.

Julien MH and Harley KLS (1978) The initiation of biological control of *Emex* spp. in Australia. In *Proceedings of the 4th International Symposium on Biological Control of Weeds.* 30 August–2 September 1976, University of Florida, Gainesville. (Ed TE Freeman) pp. 141–144. Center for Environmental Programs, Gainesville, USA.

Julien MH and Matthews NC (1980) The biology of *Rhinoncus australis* Oke (Coleoptera: Curculionidae), a weevil attacking the weed *Emex australis* Steinheil in eastern Australia. *Australian Entomological Magazine* **6**, 87–92.

Julien MH, Kassulke RC and Harley KLS (1982) *Lixus cribricollis* [Col.: Curculionidae] for biological control of the weeds *Emex* spp. and *Rumex crispus* in Australia. *Entomophaga* **27**, 439–446.

Kammoun LG and Souissi T (2009) First report of *Cercospora tripolitana* causing leaf spot of *Emex spinosa* in Tunisia. *Plant Disease* **93**, 763.

Karimpour Y (2008) Host-specificity and potential of *Kokujewia ectrapela* for the control of *Rumex* spp. In *Proceedings of the 12th International Symposium on Biological Control of Weeds.* 22–27 April 2007, La Grande Motte, France. (Eds MH Julien, R Sforza, MC Bon, HC Evans, PE Hatcher, HL Hinz and BG Rector) p. 357. CABI, Wallingford, UK.

Krauss NLH (1963) Biological control investigations on Christmas berry (*Schinus terebinthifolius*) and Emex (*Emex* spp.). *Proceedings, Hawaiian Entomological Society* **18**, 281–287.

Linnavuori R (1960) Hemiptera of Israel. 1. *Annales Zoologici Societatis Zoologicae Botanicae Fennicae 'Vanamo'* **22**, 1–71.

Marshall DR and Weiss PW (1982) Isozyme variation within and among Australian populations of *Emex spinosa* (L.) Campd. *Australian Journal of Biological Sciences* **35**, 327–332.

Melamed-Madjar V (1969) Studies on the phenology of three species of *Apion* (Coleoptera: Curculionidae) occurring on winter leguminous crops in Israel. *Israel Journal of Entomology* **4**, 97–105.

Mimeur JM (1934) Aphididae du Maroc. *Bulletin de la Société des Sciences Naturelles du Maroc* **14**, 178–190.

Morris MJ (1982) *Uromyces rumicis* on *Emex australis* in South Africa. *Phytophylactica* **14**, 13–16.

Morris MJ (1984) Additional diseases of *Emex australis* in South Africa. *Phytophylactica* **16**, 171–175.

Nakao HK (1969) Biological control of weeds in Hawaii. In *Proceedings of the 1st Asian-Pacific Weed Control Interchange.* 12–22 June 1967, Honolulu. (Eds RR Romanowski Jr, DL Plucknett and HF Clay) pp. 93–95. Dept of Agriculture, Hawaii.

Narzikulov MN and Daniyarova MM (1971) A new species of aphid from genus *Dysaphis* (Homoptera, Aphididae) on *Rheum* sp. *Doklady Akademii nauk Tadzhikskoy SSR* **14**, 67–69.

Neser S (1979) Observations in South Africa on *Microthrix inconspicuella* Ragonot (Phycitinae: Pyralidae), a natural enemy of *Emex australis* Steinheil. *Journal of the Entomological Society of Southern Africa* **42**, 337–342.

Oelrichs PB, MacLeod JK, Seawright AA, Moore MR, Ng JC, Dutra F, Riet-Correa F, Mendez MC and Thamsborg SM (1999) Unique toxic peptides isolated from sawfly larvae in three continents. *Toxicon* **37**, 537–544.

Panetta FD and Randall RP (1993a) Variation between *Emex australis* populations in seed dormancy/non-dormancy cycles. *Australian Journal of Ecology* **18**, 275–280.

Panetta FD and Randall RP (1993b) Herbicide performance and the control of *Emex australis* in an annual pasture. *Weed Research* **33**, 345–353.

Panetta FD and Randall RP (1994) An assessment of the colonizing ability of *Emex australis*. *Australian Journal of Ecology* **19**, 76–82.

Pheloung PC, Scott JK and Randall RP (1996) Predicting the distribution of *Emex* in Australia. *Plant Protection Quarterly* **11**, 138–140.

Richards KT (1981) Can insects solve dock problems? *Journal of Agriculture, Western Australia* **22**, 58.

Schmidt S and Smith DR (2006) Pergidae of the world – an online catalogue of the sawfly family Pergidae (Insecta, Hymenoptera, Symphyta). <http://www.pergidae.net/lophyrotoma.htm> (accessed 09/11/10).

Schmidt S, Walter GH, Grigg J and Moore CJ (2006) Sexual communication and host plant associations of Australian pergid sawflies (Hymenoptera: Symphyta: Pergidae). In *Recent Sawfly Research: Synthesis and Prospects*. (Eds SM Blank, S Schmidt and A Taeger) pp. 173–193. Goecke & Evers, Keltern.

Schubiger FX, Defago G, Sedlar L and Kern H (1985) Host range of the haplontic phase of *Uromyces rumicis*. In *Proceedings of the 6th International Symposium on Biological Control of Weeds*. 19–25 August 1984, Vancouver. (Ed ES Delfosse) pp. 1–17. Agriculture Canada, Ottawa.

Scott JK (1985) Candidate insects of the biological control of *Rumex pulcher*. In *Proceedings of the 6th International Symposium on Biological Control of Weeds*. 19–25 August 1984, Vancouver. (Ed ES Delfosse) pp. 829–835. Agriculture Canada, Ottawa.

Scott JK (1992) Biology and climatic requirements of *Perapion antiquum* (Coleoptera: Apionidae) in southern Africa: implications for the biological control of *Emex* spp. in Australia. *Bulletin of Entomological Research* **82**, 399–406.

Scott JK and Shivas RG (1990) Potential biological control agents for *Emex* spp. In *Proceedings of the 9th Australian Weeds Conference*. 6–10 August 1990, Adelaide. (Ed JW Heap) pp. 480–483. Crop Science Society of South Australia, Adelaide.

Scott JK and Shivas RG (1993) Occurrence of the rust fungus *Uromyces rumicis*, a biological control agent of fiddle dock (*Rumex pulcher*) in Western Australia. In *Proceedings of the 10th Australian and 14th Asian-Pacific Weed Conference*. 6–10 September 1993, Brisbane. (Eds JT Swarbrick, CWL Henderson, RJ Jettner, L Streit and SR Walker) pp. 18–19. Weed Society of Queensland, Brisbane.

Scott JK and Shivas RG (1998) Impact of insects and fungi on doublegee (*Emex australis*) in the Western Australian wheatbelt. *Australian Journal of Agricultural Research* **49**, 767–773.

Scott JK and Way MJ (1989a) Biology of *Rhytirrhinus inaequalis* (F.), a weevil associated with *Emex australis* Steinh. (Polygonaceae) in southern Africa. *Journal of the Entomological Society of Southern Africa* **52**, 1–9.

Scott JK and Way MJ (1989b) Host plant specificity and biology of *Rhodometra sacraria* (L.) (Lepidoptera: Geometridae) in South Africa. *Journal of the Entomological Society of Southern Africa* **52**, 245–251.

Scott JK and Way MJ (1990) The phytophagous insects of *Emex australis* Steinheil in southern Africa and their potential for biological control in Australia. *Plant Protection Quarterly* **5**, 43–48.

Scott JK and Yeoh PB (1996) Assessment of potential biological control insects associated with *Emex spinosa*. *Plant Protection Quarterly* **11**, 165–167.

Scott JK and Yeoh PB (1998) Host range of *Brachycaudus rumexicolens* (Patch), an aphid associated with the Polygonaceae. *Biological Control* **13**, 135–142.

Scott JK and Yeoh PB (1999) Bionomics and the predicted distribution of the aphid *Brachycaudus rumexicolens* (Hemiptera: Aphididae). *Bulletin of Entomological Research* **89**, 97–106.

Scott JK and Yeoh PB (2004) *Emex spinosa* (L.) Campd.: a predicted weed threat that has not lived up to expectations? In *Proceedings of the 14th Australian Weeds Conference*. 6–9 September 2004, Wagga Wagga, NSW. (Eds BM Sindel and SB Johnson) pp. 501–504. Weed Society of New South Wales, Sydney.

Scott JK and Yeoh PB (2005) Biology and host specificity of *Apion miniatum* (Coleoptera: Apionidae) from Israel, a potential biological control agent for *Emex australis* and *Emex spinosa* (Polygonaceae) in Australia. *Biological Control* **33**, 20–31.

Scott J, Berlandier F and Hollis K (1994) Doublegee decline and the dock aphid. *Journal of Agriculture, Western Australia* **35**, 69–70.

Scott JK, Yeoh PB and Woodburn TL (2000) *Apion miniatum* (Coleoptera: Apionidae) and the control of *Emex australis* (Polygonaceae): conflicts of interest and non target effects. In *Proceedings of the 10th International Symposium on Biological Control of Weeds.* 4–14 July 1999, Bozeman, Montana. (Ed NR Spencer) pp. 473–485. Montana State University, Bozeman, USA.

Shepherd RCH (1989) Host specificity testing of *Rhodometra sacraria* [Lep.: Geometridae], a possible biological control candidate for *Emex australis* in Australia. *Entomophaga* **34**, 469–476.

Shepherd RCH (1990a) Laboratory tests carried out on *Microthrix inconspicuella* [Lep.: Pyralidae], a biological control agent for *Emex australis* in Australia. *Entomophaga* **35**, 441–447.

Shepherd RCH (1990b) Evaluation of *Microthrix inconspicuella* [Lepidoptera: Pyralidae], a potential biological control agent for *Emex australis* in Australia, carried out in apple orchards in South Africa. *Entomophaga* **35**, 583–588.

Shepherd RCH (1990c) Problems which arise with host-specificity testing of insects. In *Proceedings of the 7th International Symposium on Biological Control of Weeds.* 6–11 March 1988, Rome. (Ed ES Delfosse) pp. 85–92. Istituto Sperimentale per la Patologia Vegetale, Ministero dell'Agricoltura e delle Foreste, Rome.

Shivas RG (1994) Controlled environment studies on the infection of *Emex australis* by *Phomopsis emicis. Plant Pathology* **43**, 547–553.

Shivas RG (1995) Survey for pathogens of *Emex australis* in South Africa. In *Proceedings of the 8th International Symposium on Biological Control of Weeds.* 2–7 February 1992, Canterbury, NZ. (Eds ES Delfosse and RR Scott) pp. 351–354. DSIR/CSIRO, Melbourne.

Shivas R and Scott J (1994) Fungi may be another weapon. *Journal of Agriculture, Western Australia* **35**, 71–72.

Shivas RG and Sivasithamparam K (1994) Pathogens of *Emex australis* Steinheil and their potential for biological control. *Biocontrol News and Information* **15**, 31N–36N.

Shivas RG, Allen JG, Edgar JA, Cockrum PA, Gallagher PF, Ellis Z and Harvey M (1994a) Production of Phomopsin A by a potential mycoherbicide, *Phomopsis emicis*. In *Plant-associated Toxins: Agricultural, Phytochemical and Ecological Aspects.* (Eds SM Colegate and PR Dorling) pp. 161–166. CABI, Wallingford, UK.

Shivas RG, Lewis JC and Groves RH (1994b) Distribution in Australia and host plant specificity of *Phomopsis emicis*, a stem blight pathogen of *Emex australis. Australian Journal of Agricultural Research* **45**, 1025–1034.

Simmonds FJ (1967) Biological control of pests of veterinary importance. *Veterinary Bulletin* **37**, 71–85.

Stekolshchikov A (2006) Aphids of the genus *Dysaphis* Börner (Homoptera, Aphididae) living on plants of the family Polygonaceae. *Entomological Review* **86**, 787–805.

Thompson RT and Alonso-Zarazaga MA (1988) On some weevil species described by Linnaeus (Coleoptera, Curculionoidea). *Entomologica Scandinavica* **19**, 81–86.

Vayssières JF (1979) Observations faites sur *Emex spinosa* en Octobre 1978 et Mars 1979. Unpublished CSIRO internal report, September 1979 (in French). p. 14.

Weiss PW (1978) Reproductive efficiency and growth of *Emex australis* in relation to stress. *Australian Journal of Ecology* **3**, 57–65.

Weiss PW (1980) Germination, reproduction and interference in the amphicarpic annual *Emex spinosa* (L.) Campd. *Oecologia* **45**, 244–251.

Weiss PW (1981) Spatial distribution and dynamics of populations of the introduced annual *Emex australis* in south-eastern Australia. *Journal of Applied Ecology* **18**, 849–864.

Weiss PW and Julien MH (1975) A comparison of two species of spiny emex (*Emex australis* and *E. spinosa*) in north-western Victoria. *Journal of the Australian Institute of Agricultural Science* **41**, 211–213.

Williams JD, Groves RH, Weiss PW and Nicholls AO (1984) Competition between wheat and two *Emex* species. *Australian Journal of Agricultural Research* **35**, 453–461.

Yeoh PB and Scott JK (1996) Utilizing existing natural enemies to enhance the control of *Emex australis* in south-west Australia. *Plant Protection Quarterly* **11**, 162–164.

Yeoh PB and Woodburn TL (2004) Novel techniques for increasing the survival of aestivating biological control insects. In *Proceedings of the 11th International Symposium on Biological Control of Weeds.* 27 April–2 May 2003, Canberra. (Eds JM Cullen, DT Briese, DJ Kriticos, L Lonsdale, L Morin and JK Scott) pp. 464–470. CSIRO Entomology, Canberra.

Yeoh PB, Woodburn TL and Scott JK (1999) The potential of multiplying cages for rearing *Apion miniatum* (a biological control agent of *Emex australis*) and the influence of ant control on production. In *Proceedings of the 7th Australasian Conference on Grassland Inver-*

tebrate Ecology. 4–6 October 1999, Perth. (Ed JN Matthiessen) pp. 211–218. CSIRO Entomology, CSIRO Centre for Mediterranean Agricultural Research, Wembley, Australia.

Yeoh PB, Woodburn TL and Scott JK (2002) Will the red apion (*Apion miniatum*), a potential biocontrol agent for doublegee (*Emex australis*), establish in Australia? In *Proceedings of the 13th Australian Weeds Conference.* 8–13 September 2002, Perth. (Eds H Spafford Jacob, J Dodd and JH Moore) pp. 432–435. Plant Protection Society of Western Australia, Perth.

Yeoh PB, Scott JK and Woodburn TL (2004) Biocontrol of *Emex* at the expense of *Tamarix* incursions? In *Proceedings of the 14th Australian Weeds Conference.* 6–9 September 2004, Sydney. (Eds BM Sindel and SB Johnson) pp. 399–402. Weed Society of New South Wales, Sydney.

Eremophila mitchellii Benth. – budda *Eremophila sturtii* R. Br. – turpentine bush *Dodonaea viscosa* subsp. *angustissima* (DC.) J.G. West – narrow-leaf hopbush *Myoporum deserti* (A. Cunn. Ex Benth.) Chinnock – poison-bush

Mic Julien and David Sparks

ABSTRACT

A native scale, *Pulvinaria* sp. was found to attack two native woody bushes, budda *Eremophila mitchellii* and poison-bush *Myoporum deserti*, in arid western NSW and not to attack other woody species. Similarly, a native eriophyid mite attacks narrow-leaf hopbush, *Dodonaea viscosa* subsp. *angustissima* and not other species. Field trials determined that both established readily when manually distributed and that within 18 months they infested 10% and 62% of their hosts and killed 8% and 17% of them, respectively. Further investigation is needed to determine the potential of these agents to assist in the management of these woody weeds. This with the work on sollya and *Cassinia* spp. is the only research in Australia that involves the study of native organisms to control native plants.

Key words: native woody weeds, native natural enemies, *Pulvinaria* sp., eriophyid mite.

INTRODUCTION

This is a précis of information taken from two reports to the Cuttagoa Catchment Landcare Group written by Sparks and Robinson (1997) and Sparks (2000). No other publications were found that relate to biological control of these woody weeds. The significance of this work is that the researchers attempted to control native weedy species with native natural enemies. Further, these studies were instigated and carried out at a catchment level by a catchment care community group. Since they were working with native organisms there were no requirements for permits. Excepting sollya *Billardiera heterophylla* and *Cassinia* spp. (see this volume), all other biological control of weed research in Australia has involved exotic weed species and required permits to import and to release exotic agents, and the research has been carried out by federal or state government authorities, though often in collaboration with community groups.

Some woody plants that are native to Australia are a major problem on grazing lands (Fig. 1), especially in arid western NSW and Qld. They are a problem in or threaten to become a problem in up to 20 million ha in NSW, almost 25% of the state. They mostly prefer red soils and render the areas invaded unproductive and exposed to erosion. A number of woody species are involved and not

Figure 1: Typical encroachment by *Eremophila mitchellii* and *Eremophila sturtii* at 'Ellerslie', Bourke 2510 NSW. Photo: D Sparks.

all are a problem in the same area. Four species that are not considered beneficial at all were studied for biological control. They were budda *Eremophila mitchellii* Benth., turpentine bush *Eremophila sturtii* R. Br., poison-bush *Myoporum deserti* (A. Cunn. Ex Benth.) Chinnock (Scrophulariaceae) and narrow-leaf hopbush *Dodonaea viscosa* subsp. *angustissima* (DC.) J.G. West (Sapindaceae).

Control methods such as mechanical controls, blitz-grazing with goats or herbicidal control have variable effects and costs depending on plant species and location. Fire management is not very useful as such plants do not carry the hot fires that are required to kill them.

In 1977 Nancy Robinson, grazier, noticed woody weeds being killed by scale insects and determined that the scales could be moved to uninfested plants. Over time she also determined that useful plants such as wilga *Geijera parviflora* Lindley (Rutaceae), mulga *Acacia aneura* F. Muell. ex Benth. (Fabaceae) and gidgee *Acacia* species of arid and semi-arid regions were not suitable host plants. How this was done was not reported.

NATURAL ENEMIES

Over 30 species of insect and mites were found damaging these woody weeds. Ten were causing substantial damage and were selected for further studies, finally focusing on two that showed most promise – a scale *Pulvinaria* sp. (Hemiptera: Coccidae) that attacks budda and poison-bush, and an unidentified mite (Acarina: Eriophyidae) that attacks hopbush.

***Pulvinaria* sp. (Hemiptera: Coccidae).** This scale insect attacks budda *E. mitchellii* and poison bush *M. deserti* and completes its life-cycle in eight to 10 weeks under favourable conditions. The female has a white scale and is very obvious (Fig. 2) whereas the males and immature stages blend better with their environment. Females produce about 200 eggs. The adults are sedentary while the immature stages are mobile. They feed on sap and exude honeydew and consequently are attended by ants. When adverse conditions occur, a protective shield is secreted and the insect diapauses in the third instar until favourable conditions such as rain stimulate development to adult and reproduction. In any particular area the population of scales seem to be synchronised largely as a result of diapausing at the third instar stage in response to poor environmental conditions such as dryness and heat, both regular occurrences in western NSW and Qld. Once conditions improve the scales quickly complete development to the adult stage, mate and begin to reproduce. Early summer is the most likely time of the year for favourable conditions of mild temperatures and rainfall so diapause is broken then, shortly followed by the production of the first instar crawlers.

The scale can readily be spread to new plants by cutting stems at the appropriate time of the year when first instar crawlers are present, and placing the stems in nearby plants. The first instar stage larvae establish more readily than older stages. Populations increase and may kill plants (Fig. 3), but the scale has poor dispersal capabilities so scattered plants are damaged or killed while others remain uninfested under natural conditions. Parasitism of *Pulvinaria* sp. was found to be low, around 5%, and scale populations do not appear to be regulated by natural enemies.

The host-range of this scale insect was assessed through field observation and by starvation trials in which insects were placed on test plants. The details of these trials are not available.

Figure 2: Early-stage infestation of *Pulvinaria* sp. at 'Ellerslie', Bourke 2510 NSW. Photo: D Sparks.

Figure 3: Typical pattern of woody weed mortality caused by *Pulvinaria* sp. at 'Ellerslie', Bourke 2510 NSW. Photo: D Sparks.

Field trials of manually spread scales resulted in establishment at each of six plots in western NSW. Assessment of impact 18 months later indicated an average infestation of 10% and at four of the six sites insect populations were increasing. A maximum of 8% of bushes were severely damaged and it was predicted that they would die. Scales introduced to budda that was regrowing after chaining established and spread so that 34% of trees were infested after 18 months and 6% were killed.

Eriophyid mite, unknown species (Acarina: Eriophyidae. This hopbush mite attacks the growing tip of narrow-leaf hopbush *D. viscosa* subsp. *angustissima*, causing growth distortion. Once all tips are damaged the whole plant dies. It appears to be controlled by natural enemies and can be spread by moving cut stems to new bushes. It is found mostly on old hopbush plants and when spread to adjacent old hopbush plants it quickly establishes and damages them. It is not often found on young hopbush plants. As for the *Pulvinaria* scale, the mites readily established when moved to new bushes.

Field trials at five sites in western NSW showed that this mite spreads more effectively and rapidly than the scale: at one site 62% of bushes were infested and 17% had been killed after 18 months. However, results were variable and infestation rates as low as 2% were observed with no plants killed.

DISCUSSION

The two agents studied show some potential as biological control agents for several species of unwanted woody weeds that are native to Australia. However, to be able to damage the target plants, manual distribution is required. Assessment of impact after distribution to study plots indicated that both agents establish readily and can cause significant impact within 18 months. Further assessment on those plots has not been reported and may not have been carried out. Such an assessment would indicate the real potential of these agents.

REFERENCES

Sparks DR (2000) 'Biological control of woody weeds in western New South Wales and within south western Queensland'. Report to Cuttagoa Catchment Landcare Group.

Sparks DR and Robinson NJ (1997) 'Biological control of woody weeds in western New South Wales and within south west Queensland'. Final report on NRMS Project No. N4080/WN 0140.93. Cuttagoa Catchment Landcare Group.

Euphorbia paralias L. – sea spurge

John K Scott

ABSTRACT

Sea spurge (*Euphorbia paralias*) is an invader of coastal ecosystems of southern Australia. It is implicated in the loss of coastal biodiversity and heritage values, and reduction of human amenity on beaches due to its toxic latex. The weed originates from the Mediterranean region and Atlantic coast of Europe. Initial surveys for biological control agents were made in southern Australia and France in 2009. No potential biocontrol agents have been found so far in Australia. The surveys in France found a rust fungus, *Melampsora euphorbiae*, and other pathogens causing damage to sea spurge. At least three insect species also have potential as biocontrol agents. Future work should focus on the ecology of sea spurge in Australia and the region of origin, and assessment of the potential agents.

Key words: coastal species, environmental weed, dunes, France, rust fungus, *Melampsora euphorbiae*, surveys, potential biocontrol agents, strand species.

INTRODUCTION

Sea spurge (*Euphorbia paralias* L.) (Euphorbiaceae) (Fig. 1) has become a major invader of sandy beaches, dunes and adjacent areas across the southern Australian coastline (Heyligers 2002). This fleshy glabrous herb is found from just north of Perth in WA, around Tas, to Sydney on the NSW coast (Heyligers 2002; Keighery and Dodd 1997; Rudman 2003; Wilcock and Westbrooke 1999a) (Fig. 2). The history of invasion in southern Australia is described

Figure 1: *Euphorbia paralias* at Greens Pool, WA. a) Flowers. b) Growth habit among other foredune plants. Photos: JK Scott, CSIRO.

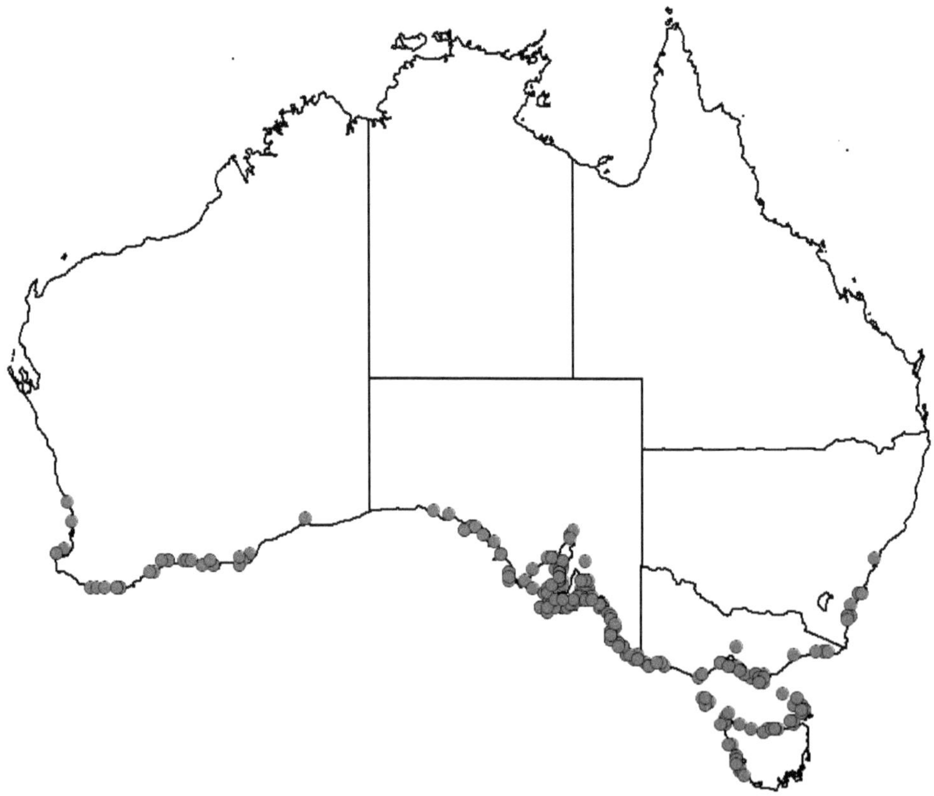

Figure 2: Distribution of sea spurge, *Euphorbia paralias*, in Australia. Specimen data reproduced from Australia's Virtual Herbarium with permission of the Council of Heads of Australasian Herbaria Inc.

in Heyligers (2002), with the first collections made in 1927 at Albany in WA. The weed originates from the Mediterranean region and Atlantic coast of Europe, and was possibly introduced to Australia as a contaminant of shipping ballast.

In its native habitat, sea spurge is found on the foredunes, beaches and islands of all countries contained within the circum-Mediterranean region (Greuter *et al.* 1986). It also grows along the Atlantic coasts of Portugal, Spain, France and Belgium, with its northernmost limits in the south-western part of the Netherlands, southern Britain and Ireland (Bakker 1976; Smith and Tutin 1968). Outside its native range, it is known to be naturalised only in Australia.

Sea spurge is a perennial herb with usually one to 10 stems, 30–35 cm high arising under the sand, from a crown, leading to a long tap root. Smaller roots can extend at least 1 m further into the sand. The stems are leafy, with narrow, overlapping stalkless leaves. The inflorescences are terminal, yellow green, with flowers produced in spring and summer, although flower timing can depend on growing conditions. Individual plants can produce up to 20 000 seeds. The stems die after seed production and are replaced by new shoots from the root crown. The plant's growth enables it to colonise and persist in sand dunes. Long-distance dispersal is by seed which floats in the sea, or by seed accidentally transported on boats (Heyligers 2002; Hussey *et al.* 2007; Keighery and Dodd 1997).

Sea spurge is a risk to public safety and public amenity of beaches due to the toxicity of the sap produced when plants are damaged. Sea spurge sap contains a number of compounds with irritant and cytotoxic properties (Sayed *et al.* 1980). Little research has been targeted on the ecological impacts of sea spurge. However, the dense monocultures established on dune systems have probably displaced native species (Rudman 2003). For example, the nesting habitat of beach-nesting shorebirds is believed to be adversely affected by sea spurge as open ground for nesting is lost (Rudman 2003). Sea spurge establishment is also implicated in transforming the shape of beaches (Heyligers 1985). Sea spurge invasion and damage caused by its removal have the capacity to negatively affect Aboriginal heritage sites and values (Rudman 2003).

Sea spurge it is not currently declared as a weed in any Australian state. As sea spurge occurs almost entirely on public land and is not imported, sold or deliberately transported by people, there is limited utility in its declaration as a weed. Despite its unlisted status, sea spurge is actively managed across its range. The two main methods used by Australian land managers to control sea spurge are physical removal and the application of herbicide (Wilcock and Westbrooke 1999b).

BIOLOGICAL CONTROL HISTORY

Biocontrol of sea spurge has not been attempted elsewhere and in Australia is only in its initial phase – the exploration for potential agents. Surveys by CSIRO were carried out in southern Australia and in southern France during 2009 (Scott *et al.* 2010). Sea spurge was nominated as a target for biocontrol in 2010.

PLANT TAXONOMY

Euphorbia paralias is in the subgenus *Euphorbia (Esula)*, a group of about 500 species with the greatest diversity in the northern hemisphere. Hassall (1977) identified *E. plumerioides* Teysm. & Hassk. as the sole Australian representative in the subgenus *Esula*. However, molecular studies by Steinmann and Porter (2002) and Bruyns *et al.* (2006) showed that this species belongs in Clade 'C' whereas the *Esula* group are placed in Clade 'B' of the genus *Euphorbia*. Subsequently Bruyns *et al.* (2006) defined Clade 'B' as being the subgenus *Euphorbia (Esula)*. A second species, *Euphorbia norfolkiana* Boiss., from Norfolk Island, is mentioned by Hassall (1977) as belonging to subgenus *Esula*, but this has not been confirmed by molecular studies.

EXPLORATION

Surveys were made by CSIRO staff (M Jourdan, L Morin, JK Scott, T Thomann and P Yeoh) during 2008–09 for herbivorous arthropods and plant pathogens associated with sea spurge in its introduced (Australia) and part of its native range (France). Surveys were made of 39 sites on the Mediterranean and Atlantic coasts of France and 36 sites in Australia (Scott *et al.* 2010).

CANDIDATES

Natural enemies

The survey of coastal sites in southern Australia did not identify any arthropods or pathogens specific to sea spurge (Scott *et al.* 2010). In contrast, potential biocontrol agents were found during surveys of plants on the coast of France. These included a rust fungus *Melampsora euphorbiae* (Schub.) Castagne, another very damaging but as yet identified pathogen, and insect species such as the leaf- and stem-feeding tortricid *Acroclita subsequana* (Herrich-Schäffer) (Scott *et al.* 2010).

DISCUSSION

Species from the subgenus *Esula* from Europe (e.g. *E. cyparissias* L., *E. esula* L.) have been successfully controlled by biological means in North America (Faubert and Casagrande 2002; Lesica and Hanna 2009; Lym 2005). Further evidence supporting the suitability of sea spurge as a biocontrol target is the likely absence of native representatives of this subgenus in Australia. The successful biocontrol programs in North America against *Euphorbia* species augers well for a future program against sea spurge in Australia. Research is needed on the host-range of potential biocontrol agents found in Europe, complemented by further surveys in the Mediterranean region to identify other organisms of potential use. Molecular identification of the rust fungus is necessary to help elucidate the host-range in relation to different strains of the rust found on other *Euphorbia* species. The biocontrol work should be accompanied by further study and analysis of the impact on ecosystem diversity and functions, including the role of sea spurge in dune stabilisation in Australia (Pestell *et al.* 2003).

REFERENCES

Bakker JP (1976) Phytogeographical aspects of the vegetation of the outer dunes in the Atlantic province of Europe. *Journal of Biogeography* **3**, 85–104.

Bruyns PV, Mapaya RJ and Hedderson T (2006) A new subgeneric classification for *Euphorbia* (Euphorbiaceae) in southern Africa based on ITS and *psbA-trnH* sequence data. *Taxon* **55**, 397–420.

Faubert H and Casagrande RA (2002) Cypress spurge. In *Biological Control of Invasive Plants in the Eastern United States*. (Eds R Van Driesche, S Lyon, B Blossey, M Hoddle and R Reardon) pp. 195–207. USDA Forest Service Publication FHTET-2002-04.

Greuter W, Burdet HM and Long G (1986) *Med-Checklist. A Critical Inventory of Vascular Plants of the Circum-Mediterranean Countries. 3. Dicotyledones (Convolvulaceae-Labiatae)*. Conservatoire et Jardin Botaniques, Genève.

Hassall DC (1977) The genus *Euphorbia* in Australia. *Australian Journal of Botany* **25**, 429–453.

Heyligers PC (1985) The impact of introduced plants on foredune formation in south-eastern Australia. *Proceedings of the Ecological Society of Australia* **14**, 23–41.

Heyligers PC (2002) The spread of *Euphorbia paralias* (Euphorbiaceae) along the mainland coast of south-eastern Australia. *Cunninghamia* **7**, 563–578.

Hussey BMJ, Keighery GJ, Dodd J, Lloyd SG and Cousens RD (2007) *Western Weeds: A Guide to the Weeds of Western Australia.* 2nd edn. Weeds Society of WA, Perth.

Keighery GJ and Dodd J (1997) Occurrence and spread of sea spurge (*Euphorbia paralias*) along the west coast of Western Australia. *Nuytsia* **11**, 285–286.

Lesica P and Hanna D (2009) Effect of biological control on leafy spurge (*Euphorbia esula*) and diversity of associated grasslands over 14 years. *Invasive Plant Science and Management* **2**, 151–157.

Lym RG (2005) Integration of biological control agents with other weed management technologies: successes from the leafy spurge (*Euphorbia esula*) IPM program. *Biological Control* **35**, 366–375.

Pestell AJ, Raymond CM, Walton H, Petit S and Gibbs J (2003) Abundance of sea spurge (*Euphorbia paralias*) on the foredune at Long Beach, Robe, South Australia, and implications for dune stabilisation. *South Australian Naturalist* **77**, 12–15.

Rudman T (2003) *Tasmanian Beach Weed Strategy for Marram Grass, Sea Spurge, Sea Wheatgrass, Pyp Grass and Beach Daisy.* Nature Conservation Branch, Tas Dept of Primary Industries, Water and the Environment.

Sayed MD, Riszk A, Hammounda FM, El-Missiry MM, Williamson EM and Evans FJ (1980) Constituents of Egyptian Euphorbiaceae. IX. Irritant and cytotoxic ingenane esters from *Euphorbia paralias* L. *Experientia* **36**, 1206–1207.

Scott JK, Jourdan M, Morin L, Thomann T and Yeoh PB (2010) Exploration for potential biological control agents of *Euphorbia paralias*, a major environmental weed of coastal ecosystems in Australia. In *Proceedings of the 17th Australasian Weeds Conference.* 26–30 September 2010, Christchurch, NZ. (Ed SM Zydenbos) pp. 223–226. New Zealand Plant Protection Society, Christchurch.

Smith AR and Tutin TG (1968) *Euphorbia* L. In *Flora Europaea. Vol. 2. Rosaceae to Umbelliferae.* (Eds TG Tutin, VH Heywood, NA Burges, DM Moore, DH Valentine, SM Walters and DA Webb) pp. 213–226. Cambridge University Press, Cambridge.

Steinmann VW and Porter JM (2002) Phylogenetic relationships in Euphorbieae (Euphorbiaceae) based on ITS and *ndh*F sequence data. *Annals of the Missouri Botanical Gardens* **89**, 453–490.

Wilcock PJ and Westbrooke ME (1999a) The distribution and predicted range of *Euphorbia paralias* L. (sea spurge) in Australia. In *Proceedings of the 12th Australian Weeds Conference.* 12–16 September 1999, Hobart. (Eds AC Bishop, M Boersma and CD Barnes) pp. 589–591. Tasmanian Weed Society.

Wilcock PJ and Westbrooke ME (1999b) Aspects of the ecology and control of *Euphorbia paralias* L. (sea spurge) in the Otway National Park, Victoria, Australia. In *Proceedings of the 12th Australian Weeds Conference.* 12–16 September 1999, Hobart. (Eds AC Bishop, M Boersma and CD Barnes) pp. 605–608. Tasmanian Weed Society.

Fumaria spp. – fumitory

John K Scott

ABSTRACT

Some of the seven *Fumaria* species introduced to Australia, especially *F. densiflora*, have become important weeds of cropping in south-eastern Australia. A survey initiated in part of the region of origin of most of the species, France, found two insect species and two plant pathogen species that could be of interest for use in biocontrol. These species were absent in a survey of *Fumaria* species in Australia. Further research is needed on the weed status in Australia, plus additional surveys in the region of origin, and research on the ecology and host-range of potential agents for *Fumaria* species.

Key words: biological control, crop weeds, fumitory, herbicide-resistant weeds, plant pathogens, surveys for phytophagous species.

INTRODUCTION

In Australia, all seven fumitory species (Fumariaceae) (*Fumaria bastardii* Boreau, *F. capreolata* L., *F. densiflora* DC, *F. indica* (Hausskn.) Pugsley, *F. muralis* Sond. ex W.D.J.Koch, *F. officinalis* L., *F. parviflora* Lamarck) are introduced (Walsh and Norton 2007). *Fumaria* species (collectively referred to here as fumitory) were found in all states of Australia (Fig. 1) but are abundant only in southern regions, mostly in NSW, Vic, SA and WA (Walsh and Norton 2007). *Fumaria* species have not been declared noxious weeds as they are mostly widespread and mainly found in cropping situations.

Fumaria species (Fig. 2) are erect or scandent, slender and freely branching herbaceous annuals, 30–70 cm high, with pinnate leaves. Flowers are zygomorphic, pink or white with tips purple, in elongate racemes and are self-fertile. The fruit is a single seeded hairless globular capsule with an elaiosome (Norton 2003). There are up to 22 000 seeds per plant. Seeds are dispersed up to 5 m by ants and as contaminants of sown seed, especially canola and legumes (Norton 2003). Seed can remain dormant for long periods, germinating following soil disturbance. Norton *et al.* (2004) found *F. densiflora* in NSW to have a half-life of 10 years when buried at 15 cm depth. Peak germination occurs in the second year after seed shed, with the main seed loss in the soil in subsequent years due to germinations (Norton *et al.* 2004). Peltzer and Matson (2002) prevented recruitment of seed of *F. muralis* for three years at a study site near Mount Baker in WA and found germination from cultivated soil at densities up to 300 m^{-2} and negligible germination in undisturbed soil. Further details of the biology of *Fumaria* species in Australia can be found in Norton (2003).

Fumaria species are weeds of many parts of the world, mainly in cereal and legume cultivation, vineyards, wastelands and gardens. In Australia, their main impact is as competitors in crops where they reduce crop yields (Taylor and Lill 1986; Wells 1979). They are a problem in canola (Lemerle *et al.* 2001) and pulse crops (particularly chickpeas, faba beans, field peas, lentils and lupins) in SA and southern and central NSW (Holding and Bowcher

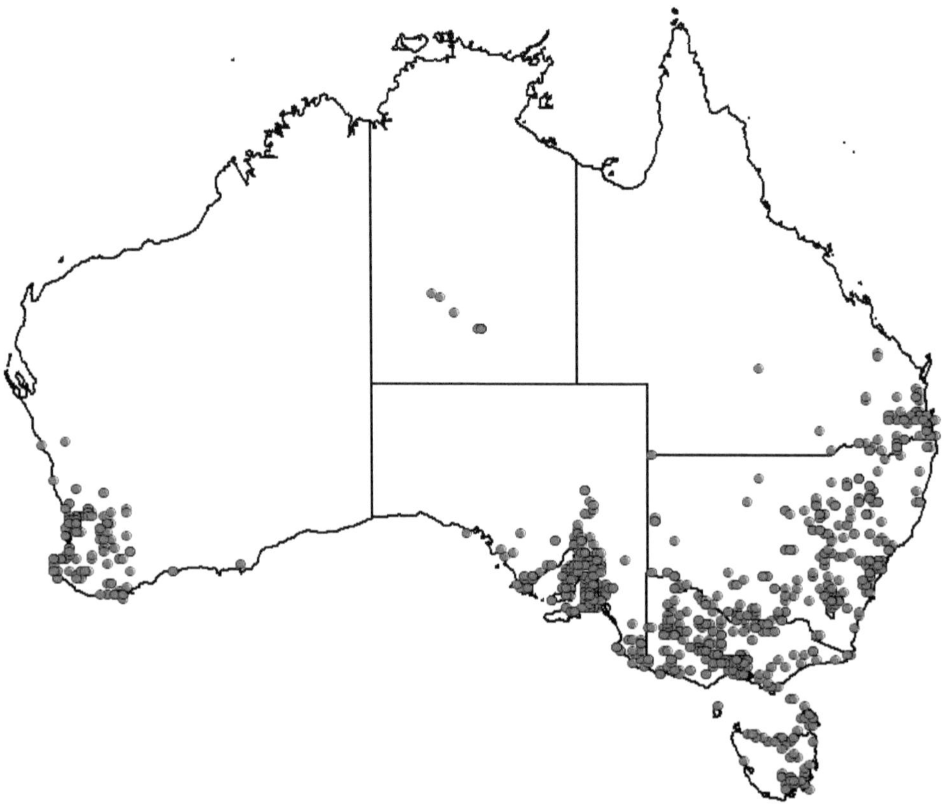

Figure 1: Distribution of the genus *Fumaria* in Australia. The distribution of individual species can be found in Walsh and Norton (2007). Specimen data reproduced from Australia's Virtual Herbarium with permission of the Council of Heads of Australasian Herbaria Inc.

2004). *Fumaria* species were found in 36% of 86 cereal crops in southern NSW during a survey of weeds conducted in spring 1993 (Lemerle *et al.* 1996) and in 58% of 55 canola crops surveyed in NSW, Vic and SA in 1998 (Lemerle *et al.* 2001). Fumitory appears to have become more widespread since earlier surveys. Agronomists and farmers from this region ranked fumitory in the top 10 of 50 species as a potential future problem. Fumitory has been a problem in poppy crops in Tas, where densities up to 256 m^{-2} have been recorded (Baldwin 1977). Fumitory (as *F. muralis*) has more recently been ranked as a weed of increasing importance in cropping in the north-western and northern regions of Tas (Ireson *et al.* 2007). *F. densiflora* first evolved resistance to group K1/3 herbicides (Dinitroanilines and others) in NSW (Heap 2010), underlining that this species is the most important weed of the genus in Australia (Norton *et al.* 2004).

BIOLOGICAL CONTROL HISTORY

Fumitory has not been considered for biological control in any country apart from Australia. The primary aim of the work carried out in part of the native range (France) and in Australia, starting in 2004, was to assess the feasibility of biocontrol. Fumitory has not been subjected to the formal process of being nominated as a target species for biocontrol in Australia.

PLANT TAXONOMY

Six of the seven *Fumaria* species in Australia (*F. bastardii*, *F. capreolata*, *F. densiflora*, *F. muralis*, *F. officinalis* and *F. parviflora*) are introduced from Europe (mostly western Europe) or the Mediterranean region, whereas *F. indica* is from central and south-western Asia. There are no native species in the Fumariaceae nor in the related Papaveraceae in Australia (Kiger 2007; Walsh and Norton 2007).

EXPLORATION

CSIRO, through the CRC for Australian Weed Management, searched for potential classical biocontrol agents for *Fumaria* species in 33 sites in southern France and in 160 sites in southern Australia during 2004 and 2005

Figure 2: *Fumaria* sp. Manjimup, WA. Photo: JK Scott, CSIRO.

(Jourdan *et al.* 2008). The surveys for arthropods and plant pathogens were undertaken by Mireille Jourdan, Janine Vitou and Thierry Thomann in France and by Aaron Maxwell and John K Scott in Australia.

Surveys in France

There are few arthropod species associated with *Fumaria* species in the region of origin (Jourdan *et al.* 2008). Potential biocontrol agents identified by Jourdan *et al.* (2008) during the surveys in France and from the literature included a weevil with stem-mining larvae, *Sirocalodes mixtus* Mulsant and Rey, and a cynipid wasp, *Neaylax versicolor* (Nieves-Aldrey). The larvae of this wasp form galls in the fruits of *Fumaria* species.

Nine plant pathogen species with up to five per site were found in France. Pathogens of interest for biocontrol included the downy mildew, *Peronospora affinis* Rossmann (Peronosporaceae) and the leaf smut, *Entyloma fumariae* J. Schröt. (Entylomataceae) (Jourdan *et al.* 2008; Maxwell *et al.* unpub.).

Surveys in Australia

No fumitory-specific arthropods were found in Australia. Six plant pathogens were identified from diseased *Fumaria* species in Australia, but the two pathogens of interest from the surveys in France were not found (Jourdan *et al.* 2008; Maxwell *et al.* unpub.).

DISCUSSION

Fumaria species have no close relatives in Australia. There are species from 10 genera in the Papaveraceae grown as ornamentals (Jourdan *et al.* 2008) and only one related crop, *Papaver somniferum* L. (opium poppy), which is grown in Tas for production of pharmaceutical chemicals (Kiger 2007; Laughlin *et al.* 1998). The absence of related native species and few related commercial species indicates that weedy *Fumaria* species are ideal targets for biological control. However, being a weed of cropping means that biocontrol will be difficult to achieve because a successful agent will need to respond to the ecological uncertainty created by cultivation, crop rotations and pesticide treatments.

The overseas surveys need to be expanded to include a wider range of regions in western and southern Europe, and a wider range of climates and habitats. Further work is needed on the ecology and host-range of all potential agents, and would be worthwhile given the high likelihood of host-specificity. *Fumaria* species were seen as an increasing problem in cropping (Lemerle *et al.* 2001), and the evolution of herbicide resistance supported this view. Further work is now needed in Australia to establish whether the economic importance of fumitory species has increased sufficiently to justify taking biocontrol further.

REFERENCES

Baldwin BJ (1977) Chemical weed control in oil-seed poppy (*Papaver somniferum*). *Australian Journal of Experimental Agriculture and Animal Husbandry* **17**, 837–841.

Heap I (2010) International survey of herbicide resistant weeds. <http://www.weedscience.org> (accessed 12/8/10).

Holding D and Bowcher A (2004) Weeds in winter pulses – integrated solutions. *CRC Australian Weed Management Technical Series* **9**, 1–11.

Ireson JE, Davies JT, Friend DA, Holloway RJ, Chatterton WS, Van Putten EI and McFadyen REC (2007) Weeds of pastures and field crops in Tasmania: economic impacts and biological control. *CRC for Australian Weed Management Technical Series* **13**, 1–76.

Jourdan M, Vitou J, Thomann T, Maxwell A and Scott JK (2008) Potential biological control agents for fumitory (*Fumaria* spp.) in Australia. In *Proceedings of the 12th International Symposium on Biological Control of Weeds*. 22–27 April 2007, La Grande Motte, France. (Eds MH Julien, R Sforza, MC Bon, HC Evans, PE Hatcher, HL Hinz and BG Rector) pp. 160–164. CABI, Wallingford, UK.

Kiger RW (2007) Papaveraceae. *Flora of Australia* **2**, 386–400.

Laughlin JC, Chung N and Beattie BM (1998) Poppy cultivation in Australia. In *Poppy – The Genus Papaver.* (Ed J Bernath) pp. 249–277. Harwood Academic, Amsterdam.

Lemerle D, Yuan TH, Murray GM and Morris S (1996) Survey of weeds and diseases in cereal crops in the southern wheat belt of New South Wales. *Australian Journal of Experimental Agriculture* **36**, 545–554.

Lemerle D, Blackshaw RE, Smith AB, Potter TD and Marcroft SJ (2001) Comparative survey of weeds surviving in triazine-tolerant and conventional canola crops in south-eastern Australia. *Plant Protection Quarterly* **16**, 37–40.

Norton GM (2003) Understanding the success of fumitory as a weed in Australia. PhD thesis, Charles Sturt University.

Norton GM, Lemerle D and Pratley JE (2004) Persistence of *Fumaria densiflora* DC. seed in the field. In *Proceedings of the 14th Australian Weeds Conference.* 6–9 September 2004, Wagga Wagga, NSW. (Eds BM Sindel and SB Johnson) pp. 519–522. Weed Science Society of New South Wales, Sydney.

Peltzer SC and Matson PT (2002) How fast do the seedbanks of five annual cropping weeds deplete in the absence of weed seed input? In *Proceedings of the 13th Australian Weeds Conference.* 8–13 September 2002, Perth. (Eds H Spafford Jacob, J Dodd and JH Moore) pp. 553–555. Plant Protection Society of Western Australia, Perth.

Taylor AC and Lill WJ (1986) Wheat crop surveys in southern New South Wales. 4. The response by grain yield and other wheat attributes to weeds. *Australian Journal of Experimental Agriculture* **26**, 709–715.

Walsh NG and Norton GM (2007) Fumaria. *Flora of Australia* **2**, 405–412.

Wells GJ (1979) Annual weed competition in wheat crops: the effect of weed density and applied nitrogen. *Weed Research* **19**, 185–191.

Genista monspessulana (L.) L. Johnson – Cape broom

Andy W Sheppard and Ken Henry

ABSTRACT

Cape broom, *Genista monspessulana* (Leguminoseae), is widespread in south-eastern Australia where it forms high-density infestations, affecting native sclerophyll forest, open woodland and grassland, forestry plantations, pastures, roadsides and railways. Since the late 1990s, research has explored the potential for its control using biocontrol mechanisms in a collaborative project with the USA. Surveys of the plant for potential agents were undertaken in the native range, a series of scattered populations from Turkey to Morocco across the northern Mediterranean region of Europe, and via regular monitoring of study sites in north-western Spain and France. A psyllid *Arytinnis hakani* that attacks the young shoots and an apionid weevil *Lepidapion* nr *argentatum* that attacks seeds in pods, were selected as the most promising biocontrol agents and were further investigated. Host-specificity testing, completed only for the psyllid, found that the psyllid would lay eggs on a small number of *Lupinus* species even in field conditions and develop through to adult stage on a subset of these. However, this insect was found in the field in South Australia from 2004, where its impact could be monitored and further field tests against lupins could be completed. Since it was already present in Australia it could not formally be approved as a biocontrol agent, but the risk assessment has led to approval across the affected states for redistribution. Psyllid damage can kill plants, but hot summers reduce survival.

Key words: Montpellier broom, French broom, *Teline monspessulana*, psyllid, *Arytinnis hakani*, apionid weevil, *Lepidapion* nr *argentatum*, Lupinus, Genisteae.

INTRODUCTION

Cape broom (also known as French or Montpellier broom), *Genista monspessulana* (L.) L. Johnson (Leguminoseae), is native to the Mediterranean region of Europe and western North Africa and via movement as a garden plant has become a weed in North America, Hawaii, Chile and forest areas of South Africa (Parsons and Cuthbertson 2001). It is prevalent in Vic (Carr *et al.* 1992), SA, NSW (Leys 1998) and Tas. Populations are also present in the south-west forests and the Margaret River region of WA (G Keighery, *pers. comm.* 2000) (Fig. 1). In south-eastern Australia it forms high-density infestations, affecting native vegetation (e.g. sclerophyll forest, open woodland, grassland), forestry (e.g. pine plantations), grazing (e.g. non-arable pastures) and amenity and infrastructure (e.g. roadsides, railways) (Adams and Simmons 1991). It infests over 600 000 ha of national, state and urban parks (Fig. 2) and fallow land and is proclaimed as a noxious weed in SA and Vic and as a secondary weed under Tas legislation. In 2000, it was ranked 37th in weeds nominated for Weeds of National Significance (SARDI 2008a), but formal approval is pending. Areas invaded can become virtual monocultures, with only mature dominant eucalypts and a few shade-tolerant

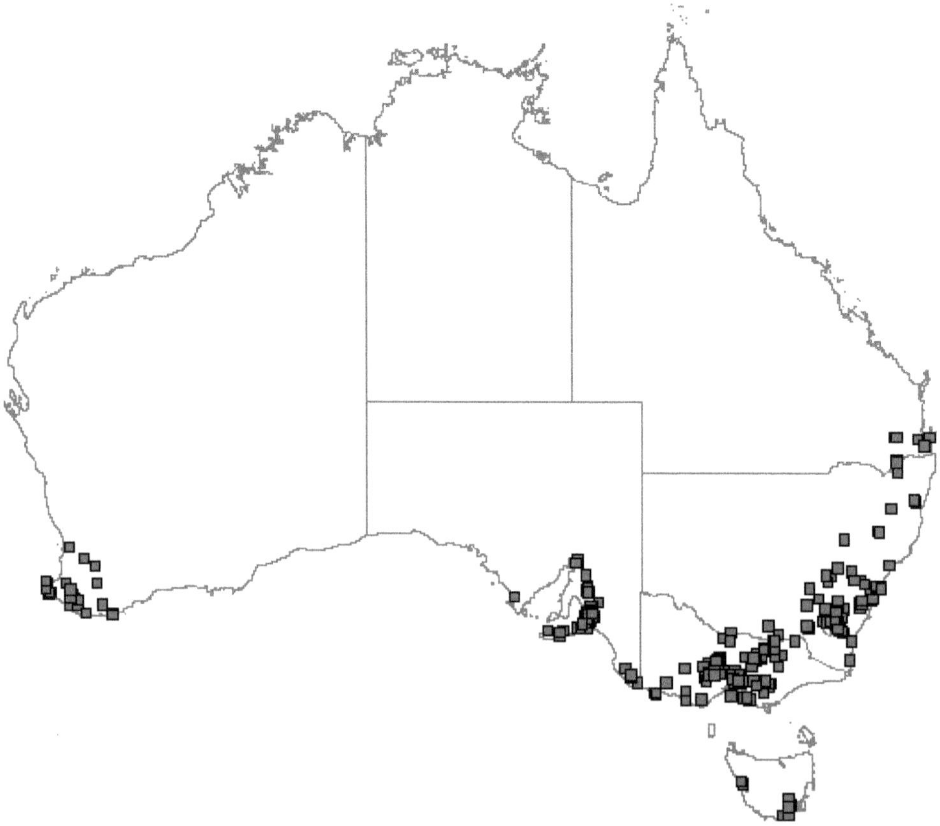

Figure 1: Distribution of Cape broom (*Genista monspessulana*) in Australia. Reproduced from Australia's Virtual Herbarium with permission of the Council of Heads of Australasian Herbaria Inc. Map generated on 1 May 2010.

species interspersed throughout the stand. The soil seed bank is substantial and long-lived. *Genista monspessulana* is invading the south-west forests of WA and several areas of remnant native vegetation in the south and central Mt Lofty Ranges, which contain many rare and endangered species including the only known large populations of the vulnerable orchid, *Pterostylis cucullata* R. Br. As a nitrogen fixer, *G. monspessulana* has the capacity to increase soil N of invaded native habitats, making regeneration of native woodland difficult even after the mature plants are removed.

Taxonomically, Cape broom is one of a number of similar upright spineless species in the genus *Genista* that were previously separated into the genus *Teline*, and have green foliage and bright clusters of yellow flowers that have made them popular garden plants. Recent molecular taxonomy studies have not supported the status of a separate genus (Käss and Wink 1997). In its horticultural history, *G. monspessulana* has been used to produce various floral hybrids, but it is not known whether the invasive form is pure wild type or not.

Cape broom is a perennial shrub growing to 3 m, with ridged green stems covered with short soft hairs, becoming hairless with age. The leaves are shortly stalked with three leaflets with rounded ends (often with a short point) with the upper surface virtually hairless and the lower surface softly haired at least along the midrib (Fig. 2a). The yellow pea-like flowers are 0.8–1.3 cm long with a hairy calyx and the brown or black densely hairy mature oblong ovoid pods (with five to eight brown-black seeds) are 1.5–2.5 cm long and 3–5 mm wide (Fig. 2a).

Cape broom flowers in late winter–spring and sometimes again in late summer–autumn. Seeds are released explosively when ripe and are the only means of reproduction. The plant is an early successional species, found in areas with >600 mm annual rainfall such as oak or pine forest in maquis vegetation within its native range. It produces a long-lived seed bank from which seedlings can germinate whenever there is adequate light, rainfall and temperature. Unlike Scotch broom, *Cytisus scoparius* (L.) Link, which forms large stands that collapse with age, Cape broom stands self-replace, forming persistent monospecific stands that smother native regeneration (Fig. 2b). In the native range, succession and interspecific competition prevents this plant from smothering native vegetation. Flowering can occur in the second to third year of

Figure 2: a) Cape broom *Genista monspessulana* showing leaves, flowers and pods. Photo: J Hosking, NSW Government. b) An infestation of Cape broom in the Adelaide hills. Photo: R Roush, University of Melbourne.

growth. Growth continues whenever temperatures are warm enough. Current control options for *G. monspessulana* are expensive (e.g. herbicides such as triclopyr) or labour-intensive (e.g. pulling or cutting and painting). Mowing or mulching is sometimes a viable option, but stands of the weed are frequently inaccessible to machinery. None of these options deal with the large viable seed bank, unless control is followed up for many years (Parsons and Cuthbertson 2001). Classical biological control was considered the only viable long-term option. Cape broom was successfully nominated as a biological control target through the Australian Weeds Committee in 1998 (APCC 1996).

BIOLOGICAL CONTROL HISTORY

Australia and the USA jointly funded native range biocontrol activities on *G. monspessulana* Cape broom/ French broom from 1999 until funding ended in 2006. Funding from the Californian Department of Forestry and Fire Protection funnelled through the Oregon and then Californian Departments of Agriculture and USDA-ARS and subcontracted to CSIRO to undertake the native range research. The start of work on the biological control of Cape broom led to postgraduate studies being undertaken at the University of Adelaide (Lloyd 2000), Imperial College in the UK (Pareja 2000) and the University of California, Berkeley. The studies focused on the ecology of the plant in its native and exotic ranges. The Cooperative Research Centre of Weed Management Systems provided the support for the biocontrol agent survey work. Survey activities were undertaken in parallel to ongoing studies on *C. scoparius* and gorse, *Ulex europaeus* L., in the native range and were collectively referred to in the USA as the 'international broom initiative'. In 2002 USDA-ARS also initiated a postgraduate scholarship on *G. monspessulana* between California and Montpellier in France, to compare the ecology of the plant and to assist studies on potential agents. USDA-ARS continue this work in quarantine in Albany, California on the psyllid *Arytinnis hakani* (Loginova). This research is ongoing (R Carruthers, *pers. comm.* 2006).

PLANT TAXONOMY

G. monspessulana is one of a number of *Genista* species in Australia, which can be hard to differentiate. *Genista linifolia* L. is also a significant environmental weed in southern Australia, and *Genista stenopetala* Webb & Berthel is increasing in importance as a weed. A number of other species and hybrids of horticultural or botanical interest, including *Genista canariensis* L., *Genista horrida* (Vahl) DC. Lam. & DC., *Genista monosperma* Lam. and *Genista tinctora* L. ssp. *depressa*) (M.Bieb.) PE Gibbs, are also listed as naturalised in Australia (J Hosking, *pers. comm.* 1999).

EXPLORATION

Surveys of the native range of Cape broom in Greece, Italy, France, Spain and Morocco and associated islands were carried out from 1999 to 2005 (CSIRO Entomology 2000, 2002, 2003) and included studies of pathogens found on the plant (CSIRO Entomology 2007). Explorations in Tunisia found no evidence of native populations of Cape broom and in Portugal only adventive stands were found along new road systems, suggesting that this species is not native to those countries (Sheppard and Thomann 2004). The only countries in the native range not explored were the former Yugoslavia, Albania, Turkey and Algeria. Sampling from 50 sites of Cape broom across the visited countries was quantitative and included many related species in the Genisteae that co-occurred to add to the field specificity evidence for any potential agents encountered (Sheppard and Thomann 2004). The diversity of specific natural enemies declined from west to east in the native range and the distribution of *G. monspessulana* was fragmented into quite isolated populations around the Mediterranean. In addition, stands of Cape broom were grown and monitored at CSIRO's European laboratory near Montpellier, southern France. A shortlist of 10 potentially specific agents was reduced to two species for rearing and testing. These were the ones found to be most widespread and consistent in occurrence.

CANDIDATES

Natural enemies

The surveys of *G. monspessulana* in its native range identified 85 species of phytophagous arthropods actively feeding on the plant (CSIRO Entomology 2003). Several fungal pathogens were also identified, including rust specimens like *Uromyces pisi-sativi* (Pers.) Liro, (Pucciniaceae) *Eudarluca caricis* (Biv.) O.E. Erikss (Ascomyceteae) and eight other species identified to genus (CSIRO Entomology 2007). Of the arthropods, 10 species were placed on a shortlist for further investigation based on likely specificity (from field observations, taxonomy and published information) (CSIRO Entomology 2003). Of the selected 10, the psyllid *A. hakani*, the apionid seed weevil *Lepidapion* nr *argentatum* Gerstacker and the buprestid beetle *Agrilus antiquus* Mulsat et Rey were initially considered potential agents. Populations of the broom seed beetle *Bruchidius villosus* F.(Chrysomelidae), already released for the biological control of Scotch broom in Australia (Hosking *et al.*, this volume) were also found on isolated *G. monspessulana* populations (Sheppard *et al.* 2006). A stem-boring fly *Chyliza leptogaster* (Panzer) (Psilidae) was also seen to cause consistently high damage at one site in south-eastern France, but it was difficult to rear and the literature suggested a wide host-range (Chandler 1975; Dengler 1997). Prior to the end of funding for this project, risk assessment research focused almost entirely on *A. hakani*, with only rearing and initial testing work of *L.* nr *argentatum*.

Opportunistic surveys of *G. monspessulana* in Australia and other reports have shown that two moths, the 'tree-lucerne moth' *Uresiphita ornithopteralis* (Guen.) (Common 1990) and the 'lucerne seed web moth' *Etiella behrii* Zeller, both Lepidoptera: Pyralidae (Austin *et al.* 1993), are found on the plant in Australia. The first of these infrequently causes defoliation. In the USA, a *Uromyces* sp. rust fungus is found on *G. monspessulana* in California on the older leaves but does not have much impact (M Pitcairn, *pers. comm.* 1997). Similarly, a rust species has just been found on *G. monspessulana* in Australia, and it has been given the name *Uromyces laburni* (DC.) Otth (=*U. genistae-tinctoriae*) after preliminary identification (L Morin, *pers. comm.* 2010). Both these rusts could be *U. pisi-sativi* identified from Europe, as the taxonomy in this group is complex and a genetic analysis has not been carried out (M Jourdan, *pers comm.* 2010). A European eriophyid mite was also found on the plant in California (Chan and Turner 1998).

Natural enemies tested but not officially released

Arytinnis hakani is a multivoltine psyllid, which attacks young shoots in all life-stages and has only ever been found and recorded on *G. monspessulana* in the native range (Hodkinson and Hollis 1987; Percy 2002) (Fig. 3). Host-specificity testing was carried out using no-choice nymphal development and oviposition tests and choice oviposition tests in the laboratory. Caged and open field experiments were conducted in France (2001–06) and in Adelaide, Australia in quarantine in 2003 and in the field in France in 2007. Host-testing in Australia and the USA was carried out concurrently using an approved test list of 40 test plant species for Australia and a draft list of 60 species for the USA. Critical test plants were representatives of the genus *Lupinus* grown in Australia and native to the USA. Testing showed a capacity for the psyllid to oviposit on and develop through to adult on certain species of *Lupinus*, which are present but not yet grown widely in Australia. A SARDI-run collaborative project funded under the Defeating the Weed Menace program in 2006 was ready to conduct further testing when *A. hakani* was recognised in the field in the Adelaide hills, and stored specimens suggested it had been there at least since 2004. Surveys suggested that the insect occurred on *G. monspessulana* from 40 km south to 80 km north of Adelaide (SARDI 2008a). The project therefore switched focus to completing the risk assessment of the psyllid in the field in Australia. These studies showed that the only *Lupinus* species relevant to Australia that could support psyllid development were annuals and much less preferred than *G. monspessulana*, and so could not support all generations throughout the year. This indicated that only collateral damage was likely in the presence of the target and a completed risk assessment was prepared (SARDI 2008b).

Because *A. hakani* was already in the field in Australia, the completed risk assessment was out of scope for a decision to be made on its release as a biological control agent, so it could not formally be considered such. The risk assessment and the visual damage observed were, however, used in the decisions to redistribute this agent within SA and between Australian states through its acceptance by state-based plant biosecurity officers (SARDI 2008a). *A. hakani* is being redistributed across some states in southern Australia which have approved its release as part of a project funded under the recent Caring for our Country program. The levels of damage it causes to individual plants are being quantified but are generally high, having now been observed killing infestations in the understorey or on protected slopes. It is less effective on exposed north-facing slopes and during particularly hot summers, when observations suggest psyllid populations can crash (SARDI 2008a).

Figure 3: The Cape broom psyllid, *Arytinnis hakani* (female). Photo: K Henry, SARDI.

***Lepidapion* nr *argentatum*.** This apionid seed weevil is univoltine and has only been found on *G. monspessulana* in the native range. Its eggs are laid into immature pods and larvae develop within a single seed. The second generation is passed in stem galls. The taxonomy requires revision as it is certainly an unnamed species that has been morphologically grouped with *L. argentatum*, which develops in the flowers of the related genus *Adenocarpus* (Alonzo-Zarazaga 1985). The *Lepidapion* that feeds on *G. monspessulana* is the first seed-feeder known in the genus, suggesting it is a distinct but as yet unnamed species (Y Jiménez, *pers. comm.* 2000). Three-choice oviposition tests failed to obtain reliable oviposition on pods of the control plants. This appeared to result from a tendency in this weevil to spread eggs thinly. Such testing requires both healthy flowers and young pods to be present on test plant material throughout the duration of the tests, which is logistically difficult.

DISCUSSION

As only one species has been tested as a biological control agent for *G. monspessulana* and it was recently found in the field in Australia, this program is still in its early stages with further development dependent on future funding. *Bruchidius villosus* released against Scotch broom in Australia has not yet been observed on *G. monspessulana*. Further evaluation and redistribution of *A. hakani* should be followed by evaluation of its impact. There have been no activities to develop biological control as part of an integrated control package. The number of potential biological control agents for *G. monspessulana* is not as large as those for the related weeds Scotch broom and gorse, but this may reflect a native abundance at a regional scale that is far lower than for those species. Further work is certainly merited for *L.* nr *argentatum*, including clarification of its taxonomic status.

ACKNOWLEDGEMENTS

We wish to acknowledge the contribution of Thierry Thomann and Mireille Jourdan based at the CSIRO European Laboratory in Montpellier and Susan Ivory and Angela Lush at SARDI for assisting with the research on this program. Thanks also to John Virtue and Rick Roush in Australia and Mike Pitcairn, Walt Decker and Ray Carruthers in the USA for their support for this project.

REFERENCES

Adams R and Simmons D (1991) The invasive potential of *Genista monspessulana* (Montpellier broom) in dry sclerophyll forest in Victoria. *Victorian Naturalist* **108**, 84–89.

Alonzo-Zarazaga MA (1985) *Revisión de los taxones supraespecificos de la familia Apionidae en la region Paleártica (Coleoptera Curculionoidea).* PhD thesis. University de Málaga, Spain.

APCC (Animal & Plant Control Commission of South Australia) (1996) *Nomination of* Genista monspessulana, Genista stenopetala *and* Genista linifolia *for Biological Control.* Adelaide.

Austin AD, White TCR, Maelzer DA and Taylor DG (1993) Biology of *Etiella behrii* Zeller (Lepidoptera: Pyralidae): a pest of seed lucerne in South Australia. *Transactions of the Royal Society of South Australia* **117**, 67–76.

Carr GW, Yugovic JV and Robinson KE (1992) *Environmental Weed Invasions in Victoria: Conservation and Management Implications.* Vic Dept of Conservation and Environment and Ecological Horticulture, Melbourne.

Chan KL and Turner CE (1998) Discovery of the gall mite *Aceria genistae* (Nalepa) (Acarina: Eriophyidae) on gorse and French broom in the United States. *Pan-Pacific Entomologist* **74**, 55–57.

Chandler PJ (1975) Observations on plant associations of the Psilidae (Diptera). *Entomological Record and Journal of Variation* **87**, 13–17.

Common IFB (1990) *Moths of Australia.* Melbourne University Press, Melbourne.

CSIRO Entomology (2000) 'Selection and testing of biological control agents for control of French broom. *Genista monspessulana* (L.) L. Johnson'. Oregon Dept of Agriculture Project 801 GR. Contracted research report.

CSIRO Entomology (2002) 'Foreign exploration and host specificity testing of biological control agents of French broom in California. International Broom Initiative 2001–2002'. Contracted research report.

CSIRO Entomology (2003) 'Foreign exploration and host specificity testing of biological control agents of French broom in California: Phase II'. Contracted research report.

CSIRO Entomology (2007) 'International broom initiative: biological control of French broom: European native-range studies: Phase III'. Contracted research report.

Dengler K (1997) On the life cycle of *Chyliza leptogaster* Panzer and *Chyliza nova* Collin (Psilidae, Diptera) and their importance as producer of phloem-necroses on deciduous trees. *Studia Dipterologica* **4**, 143–164.

Hodkinson ID and Hollis D (1987) The legume feeding psyllids (Homoptera) of the west Palaearctic region. *Bulletin of the British Museum of Natural History (Entomology)* **56**, 1–86.

Käss E and Wink M (1997) Phylogenetic relationships in the Papilionoideae (Family Leguminosae) based on nucleotide sequences of cpDNA (*rbc*L) and ncDNA (ITS1 and 2). *Molecular Phylogenetics and Evolution* **8**, 65–88.

Leys A (1998) 'Survey of distribution and importance of *Genista* brooms in NSW'. Unpublished file report (al290-9). NSW National Parks and Wildlife Service, Sydney.

Lloyd J (2000) Biology and management of *Genista monspessulana* (L.) L. A. S. PhD thesis. University of Adelaide, Australia.

Pareja M (2000) Modelling the biological control of French broom (*Genista monspessulana*). MSc thesis. Dept of Biology, Imperial College, London.

Parsons WT and Cuthbertson EG (2001) *Noxious Weeds of Australia.* 2nd edn. CSIRO Publishing, Melbourne.

Percy DM (2002) Distribution patterns and taxonomy of some legume-feeding psyllids (Hemiptera: Psylloidea) and their hosts from the Iberian Peninsula, Morocco and Macaronesia. *Insect Systematics & Evolution* **33**, 291–310.

SARDI (South Australian Research & Development Institute) (2008a) 'Importation, rearing and field release of the Cape broom psyllid, *Arytinnis hakani*'. K Henry, S Ivory, A Lush and J Kent, contracted research report for Land & Water Australia Defeating the Weed Menace program.

SARDI (South Australian Research & Development Institute) (2008b) 'Application for the approval of the release of the Cape broom psyllid, *Arytinnis hakani*, a potential biological control agent for Cape broom, *Genista monspessulana*'. K Henry, A Sheppard, J Virtue, T Thomann, S Ivory and A Lush, prepared for Biosecurity Australia.

Sheppard AW and Thomann T (2004) Quantitative field surveys for the selection of biological control agents for *Genista monspessulana* based on host range and efficacy assessment. In *Proceedings of the 11th International Symposium on Biological Control of Weeds*. 27 April–2 May 2003, Canberra. (Eds JM Cullen, DT Briese, DJ Kriticos, WM Lonsdale, L Morin and JK Scott) pp. 162–174. CSIRO Entomology, Canberra.

Sheppard AW, Haines M and Thomann T (2006) Native range research assists risk analysis for non-targets in weed biological control: the cautionary tale of the broom seed beetle. *Australian Journal of Entomology* **45**, 292–297.

Harrisia (Eriocereus) martinii (Labour.) Britton – Harrisia cactus *Acanthocereus tetragonus* (L.) Hummelink – sword pear

Rachel Cruttwell McFadyen

ABSTRACT

Harrisia cactus, native to the Chaco area of Argentina and Paraguay, was a major weed in brigalow areas in northern Qld and the Qld–NSW border. Three species occur in Australia, but only one was a major weed. Sword pear is naturalised in a small area of central Qld. Biological control was first investigated in the 1950s, but host-testing was delayed for two decades. Between 1973 and 1976, four insects were tested and released; two established and one, the cactus mealybug *Hypogeococcus festerianus*, rapidly achieved good control in the northern areas but was less effective in the southern range. The cactus mealybug controls all *Harrisia* spp., but is less effective against sword pear. The return on investment from the biological control program has been estimated at $23.5 for each $1 spent, with a net value of $18.6 million. The cactus mealybug has since been introduced to South Africa and New Caledonia.

Key words: *Hypogeococcus*, cactus mealybug, *Alcidion cereicola*, *Cactoblastis doddi*.

INTRODUCTION

Harrisia cactus, *Harrisia martinii* (Labour.) Britton, is a perennial cactus in the subfamily Cactoideae, with long much-branched succulent stems 20–50 mm thick, which have five to six longitudinal ribs bearing spines in areoles, each areole with several short flat spines and one to three sharp spines up to 3 cm long. The large, white funnel-shaped flowers develop in the summer after rain, and arise singly within areoles near stem tips (Fig. 1). Each flower opens for a single night. Fruit are roughly spherical, spiny, red and 35–50 mm diameter when ripe, with up to 1000 small black seeds embedded in a sweetish, crisp white pulp which is attractive to birds. The root system is fibrous and shallow, with extensive deeper tuberous storage roots developing from the crown in larger plants (Mann 1967; McFadyen 1986). In good soil under trees, Harrisia cactus forms dense tangled mats on the ground and clambers 2–3 m up supporting vegetation.

Harrisia cactus is shade-tolerant but will also grow in the open. It prefers deep clay soils and thrives in summer

Figure 1: Flowering plant of Harrisia cactus, *Harrisia martinii*. Photo: R McFadyen, Qld Government.

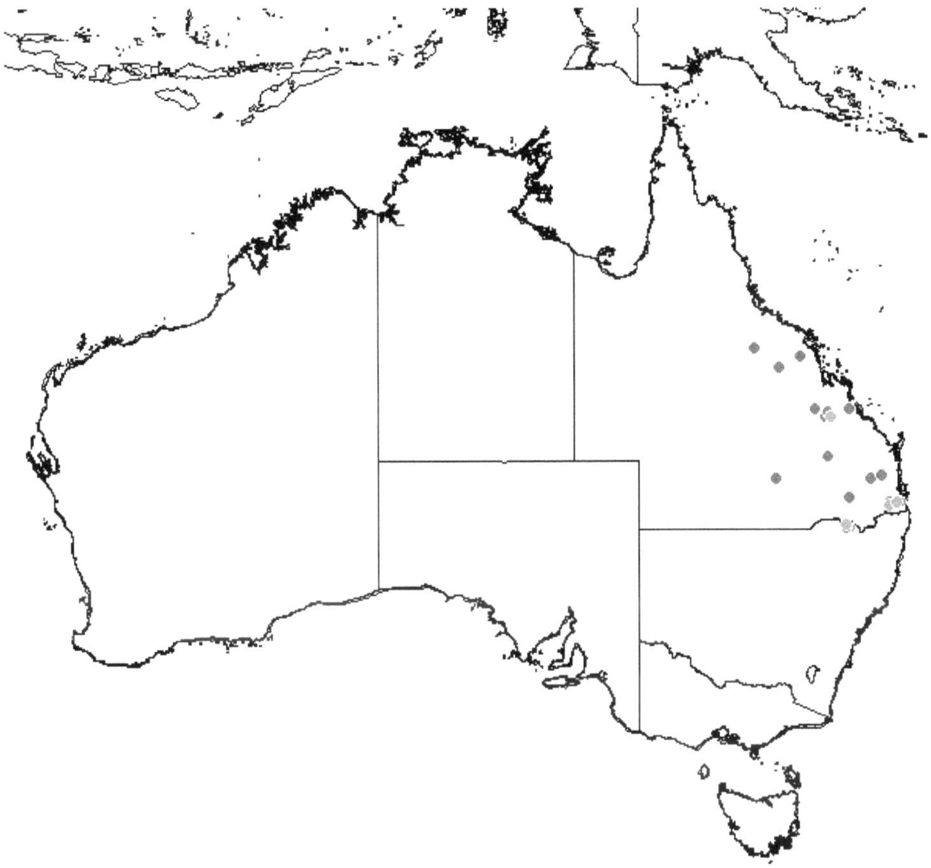

Figure 2: Distribution of Harrisia cactus, *Harrisia martinii*, in Australia. Reproduced from Australia's Virtual Herbarium with permission of the Council of Heads of Australasian Herbaria Inc.

rainfall areas receiving 500–1300 mm annual rainfall. In South America, *H. martinii* is restricted to the Chaco, an extensive sedimentary basin from the sub-Andean hills in the north-west of Argentina to the Parana–Paraguay rivers in north-eastern Argentina and Paraguay (McFadyen 1986). Natural vegetation is tropical deciduous forest in the east, gradually becoming xerophytic thorn-woodlands towards the west. Near the Paraguay River, seasonally flooded open savanna is interspersed with islands of palms and forest. Harrisia cactus, together with many related cactus species, is a minor component of the understorey (McFadyen 1986). Much of the southern Chaco has been cleared and cultivated for summer crops, and Harrisia cactus survives in shelter belts and wasteland. In the drier northern and western Chaco, felling of forest followed by overgrazing has resulted in the development of spiny thorn scrub, where Harrisia is found as an understorey plant in areas subject to flooding interspersed with long droughts. Other *Harrisia* species and related cacti are found from the Chaco west into the drier stony scrub in the foothills of the Andes in northern Argentina.

Harrisia cactus was introduced into Australia as an ornamental about 1900 (McFadyen 1986), and rapidly became a problem in the brigalow scrub country of north and central Qld inland from Bowen and Rockhampton. In suitable conditions, it outcompetes native groundcover plants to form dense impenetrable thickets with up to 140 000 plants ha^{-1} and 80–90% groundcover (McFadyen 1986). By 1935 it was a serious weed in grazing country and the first official attempts to control it, using arsenic pentoxide as a herbicide, started in 1951 (Mann 1967). The above-ground stems were killed relatively easily, but regrowth from root tubers was rapid and effective so that control required repeated treatments. Arsenic pentoxide was poisonous to staff, cattle and other vegetation, and the costs far exceeded the economic value of the land. The cactus continued to spread south, with infestations in Gatton, near Goondiwindi and across the border into NSW (Fig. 2). By 1969, over 77 000 km^2 was infested and the Qld Government was spending \$1 million per year (1982 values) on the control program. In the 1960s, when large bulldozers became available, large-scale clearing, ploughing and cultivation of the forested country reduced infestations in the better soils, as Harrisia does not survive repeated cultivation (McFadyen 1986). However, it continued to spread into new areas and remained a major problem in grazing land, under shade belts and in rocky country where ploughing was not possible.

Sword pear, *Acanthocereus tetragonus* (L.) Hummelink (previously referred to as *A. pentagonus* (L.) Britton & Rose) is a minor weed in a very limited area in central Qld (districts of Leichhardt and Port Curtis) (Mann 1970).

BIOLOGICAL CONTROL HISTORY

The first attempt at biological control was an exploratory survey in southern Brazil, Argentina and Paraguay, undertaken by John Mann of the Qld Department of Lands in December 1958–January 1959. He failed to find any useful insects attacking Harrisia in the Argentine Chaco and reported against prospects for biological control (McFadyen 1990). Continued pressure to consider biological control resulted in CIBC (now CABI Biosciences) being contracted to undertake a second survey in the Chaco in 1972. Several potential agents were found and a CIBC field station was set up in Tucuman, northern Argentina to undertake detailed host-testing and send approved species to quarantine in Brisbane (Bennett 1972). The research station consisted of RE Cruttwell McFadyen as entomologist, a technician plus casual staff as required. It operated from May 1973 to May 1976.

The work proceeded well; the first complete test report on an insect was received in Brisbane in February 1974, and by the time the station closed in May 1976 four insect species had been host-tested and sent to quarantine in Brisbane. Field releases took place between 1974 and 1979; the first indications of success were evident by 1977, with biological control replacing the use of chemical herbicides by 1979 (McFadyen 1990). Widespread collection and redistribution of the cactus mealybug took place throughout the 1980s, including into infestations in border areas of southern Qld and northern NSW. By 1995, the program was regarded as complete and no further work is currently planned.

Prior to the Qld program, there had been no attempts at biological control of Harrisia cactus elsewhere in the world. After the Australian success, the cactus mealybug and the stem-boring beetle were introduced into South Africa for the control of Harrisia cactus and of *Cereus jamacaru* DC; both established and resulted in successful control of the two cacti (Klein 1999). The mealybug was introduced to New Caledonia in 2007 for control of sword pear *Acanthocereus tetragonus* (L.) Hummelink, but with unknown results.

PLANT TAXONOMY

Acanthocereus and *Harrisia* are in different tribes of the subfamily Cactoideae of the family Cactaceae, a purely American family with no species native to Australia (Anderson 2001). *Harrisia* was known as *Eriocereus* for many years (Mann 1970), but has since been returned to the earlier name (Bostock and Holland 2007). All *Harrisia* species are native to northern Argentina and adjacent parts of Paraguay and Brazil, except for *Harrisia adscendens* (Gürke) Britton & Rose, found in the dry scrub country of north-eastern Brazil.

Three *Harrisia* spp. are naturalised in Australia, *H. martinii*, *H. pomanensis* (F.A.C.Weber ex K.Schum.) Britton & Rose (previously misidentified as *H. regelii* (Weing) Borg (Mann 1970)), known only from a single district in Leichhardt, and *H. tortuosa* (J.Forbes ex Otto & A.Dietr.) Britton & Rose, known from five limited infestations in the Darling Downs and Port Curtis regions of Qld and near Denman, Wingen and Grawin in NSW. Unlike the situation in the South American Chaco, where it is native, plants of *H. martinii* in Australia are very uniform, and it is likely that all originate from a single introduction prior to 1900 (McFadyen 1986). Like many cacti, *H. martinii* has been widely grown around the world but is recorded as a weed only in South Africa and Australia.

A. tetragonus (previously referred to as *A. pentagonus* (L.) Britton & Rose) is the only species in the genus naturalised in Australia and occurs in a very limited area in central Qld (districts of Leichhardt and Port Curtis) (Mann 1970; Bostock and Holland 2007). Related plants include *Cereus uruguayanus* R.Kiesling, naturalised in several locations in the Darling Downs, Leichhardt and Maranoa districts of Qld and Grawin and Gunnedah areas in NSW, but not regarded as weedy; *Peniocereus serpentinus* (Lag. & Rodr.) N.P.Taylor (snake cactus), naturalised outside towns throughout Qld (Hosking *et al.* 1988); and *Echinopsis oxygona* (Link) Zucc. ex Pfeiff. & Otto, naturalised at a single site in the Darling Downs. Three other cacti in the same subfamily, also naturalised though not weedy in Australia, *Epiphyllum phyllanthus* (L.) Haw., *Selenicereus macdonaldiae* (Hook.) Britton & Rose, and *Hylocereus undatus* (Haw.) Britton & Rose, are not host plants for the cactus mealybug.

EXPLORATION

In 1972 a survey of the insects on Harrisia in the Chaco was made by FD Bennett of the CIBC and H Zimmermann of South Africa (Bennett 1972; Zimmermann *et al.* 1979). From 1973 to 1976, regular surveys were undertaken by RE Cruttwell McFadyen across northern Argentina from Tucuman in the west to the Parana River in the east and into the Paraguayan Chaco, with surveys of *H. adscendens* in north-eastern Brazil in 1974 and 1975. Surveys were also made south to Cordoba and Mendoza and north to Jujuy in western Argentina. In these surveys, most cacti in the subfamily Cactoideae were routinely examined and all insects causing damage were collected and reared. A cactus garden was maintained in Tucuman. There was no investigation of fungal pathogens or diseases damaging the cacti.

No specific exploration was undertaken for agents against *Acanthocereus*: the genus is native to Central America and the Caribbean, but no exploration was undertaken in this region.

CANDIDATES

Agents not released in Australia

A total of 17 insect species were found attacking *Harrisia* species in South America (McFadyen 1976; Zimmermann *et al.* 1979). Of these, *Tucumania tapiacola* Dyar (Lepidoptera: Phycitidae) and *Diaspis echinocacti* Bouché (Homoptera: Diaspidae) were already present in Australia (Mann 1969), having been introduced during the earlier campaign against *Opuntia* spp.

***Ozamia stigmaferella* Dyar (Lepidoptera: Phycitidae).** This moth was common on *Harrisia* spp. and related cacti in the Chaco. Larvae feed in flowers and fruit which could be destroyed, but damage to plants was generally not significant. Adults would not mate in the laboratory cages used, so no host-testing was undertaken, although it is likely that the species is restricted to *Harrisia* and other cacti in the same subfamily (McFadyen 1976).

***Mimorista pulchellalis* (Dyar) (Lepidoptera: Pyraustidae).** This species was common on *Harrisia* spp. in northern Argentina, but damage caused by feeding of the solitary larvae was minimal and no host-testing was undertaken.

***Asphondylia* sp. indet. (Diptera: Cecidomyiidae).** This midge was common in tips of young stems of *Harrisia* spp. in the Chaco. Damage was not great and the insect was not reared in the laboratory, so no host-testing was undertaken.

***Dasiops bourquini* (Blanchard) (Diptera: Lonchaeidae).** Larvae of this species were common in stems of *Harrisia* spp. in northern Argentina, but were not found in other cactus species and appeared to be restricted to cacti in this genus. The species attacks undamaged cactus, unlike two unnamed *Dasiops* spp. reared from *Harrisia* spp., which attack already damaged and rotting cactus tissue (McFadyen 1976; Zimmerman *et al.* 1979). *D. bourquini* females lay 20–40 eggs per stem tip, each egg being inserted singly into healthy cactus tissue, especially on young stems at or close to ground level. Larvae feed singly initially, tunnelling within plant tissue, but in the third (last) instar their tunnels merge into a single area of decaying cactus in which pupation occurs. In winter (July–August), with day temperatures of 15–20°C, development from egg to adult took two to 2.5 months, with an adult pre-oviposition period of eight to 12 days. Damage was often severe and occurred throughout the year, but mating could not be induced in the laboratory so detailed host-testing was not possible and the species was never introduced to Australia.

Agents released in Australia

***Alcidion cereicola* Fisher (Coleoptera: Cerambycidae).** This stem-boring beetle was the first insect released in Australia. The species occurs widely in northern Argentina and Paraguay, with hosts in several genera of the Cactoideae, including the large columnar *Cereus* and *Stetsonia* spp. The nocturnal adults are 8–15 mm long. They feed on mature green cactus tissue, forming shallow craters a few mm across. After a two-week pre-oviposition period, females lay eggs singly just under the skin surface in mature cactus stems. Larvae feed in the tissue, later moving into the central vascular tissue where they feed and pupate. Diapause occurs as prepupal larvae. There is high mortality of eggs and small larvae in fresh cactus and survival is better in drier stems. Feeding by the larger larvae destroys the vascular tissue leading to stem death, and feeding by 15 to 30 larvae kills even large plants.

Host testing was undertaken on 50 plant species of economic value selected by the Australian quarantine authorities, using multiple-choice tests with ovipositing adults. There was no feeding or oviposition except on Cactaceae and the beetle was approved for release in Australia in 1974 (McFadyen and Fidalgo 1976). First releases were made the same year, and resulted in immediate establishment at most sites. Destruction of large plants was observed within two to three years, but spread was very slow and small plants and regrowth were not suitable for larval development. As a result, the cerambycid was progressively outcompeted by the rapidly spreading cactus mealybug, and is now rare in Australia.

Hypogeococcus festerianus **(Lizer y Trelles) (Homoptera: Pseudococcidae).** The cactus mealybug collected from *Harrisia* spp. and *Cleistocactus baumannii* (Lemaire) in Formosa province, northern Argentina, was identified as *Hypogeococcus festerianus* (Lizer y Trelles) by DJ Williams in 1973 (McFadyen 1979b) and this name has been used in all subsequent publications. A later review redescribed the species sent to Australia as *Hypogeococcus pungens* Granara de Willink (Williams and Granara de Willink 1992); however, the type specimen was collected from *Alternanthera pungens* Kunth (Amaranthaceae) and the host range of *H. pungens* includes *Portulaca* spp. (Portulacaceae). *H. pungens* is now present in the southern USA, where it builds up to damaging populations on *Portulaca* and *Alternanthera* species but does not damage cacti (H Zimmermann, *pers. comm.* 2008). *H. pungens* is also present and continuing to spread in Hawaii, Reunion Island and localities in South Asia where it causes damage to several ornamental plants. As *H. festerianus* in Australia has never been found on any non-cactus plant in the wild, although both *Alternanthera* and *Portulaca* are common and widespread in Australia, the older name is being retained in this and other biocontrol publications.

The cactus mealybug was commonly encountered across northern Argentina in surveys from 1972 to 1976, where it caused local damage to *Harrisia* spp. and other cacti in the Cactoideae with similar growth habits. It proved easy to rear in the laboratory, with a life-cycle of about two months (McFadyen 1979b). All stages feed in protected sites on the cactus, adjacent to spine bases and buds in the areoles and between branches. All stages produce an abundant filamentous wax covering and colonies are covered by the resulting dense cottony waxy mass (Fig. 3). Mature females reach 3.8 mm in size and lay up to 100 eggs. These are laid singly, but immediately hatch into minute (2 mm) active crawlers that disperse over the plant before settling within 24 hours. Winged males are common, though short-lived, and parthenogenesis does not occur.

Feeding by the cactus mealybug does not have an immediate effect on mature stems, but is very damaging to growing tissue. In actively growing stem tips and buds, growth is arrested on the affected side, producing distorted and spiral growth and leading to the development of clumps of distorted stems (Fig. 4). Flower buds are similarly affected and heavily infested plants produce no flowers or fruit (Fig. 5).

Figure 3: Cactus mealybug colony on stem of Harrisia cactus, *Harrisia martinii.* Photo: R McFadyen, CIBC.

Host-testing was undertaken in Argentina on the same list of plants as for the other agents, but using no-choice tests with crawlers. There was complete development from egg to adult on *Portulaca oleracea* L., but it was demonstrated that few crawlers would settle on *P. oleracea* and that adult females confined on the plant did not develop eggs; colonies could not persist on this plant (McFadyen 1979b).

The Australian population was established from a single shipment into quarantine in Brisbane in November 1974, from a colony maintained on *H. martinii* in the laboratory in Tucuman, but originating from insects collected from *H. martinii* and *C. baumannii* in the Argentine Chaco. The Tucuman colony was infested with hymenopterous parasitoids; to eliminate these, a clean colony was established in the Brisbane quarantine using newly hatched crawlers individually placed onto clean plants. It is estimated that the entire subsequent population in Australia originated from six mated females reared in this way in 1975 (A Tomley, *pers. comm.* 1976).

Figure 4: Stem of *Cleistocactus baumannii* distorted by feeding of cactus mealybug. Photo: R McFadyen, CIBC.

First releases were made in northern Qld in 1975: establishment was immediate and by 1978 the cactus mealybug was widely established and causing severe damage.

In Argentina, the cactus mealybug is attacked by five species of hymenopterous parasites, as well as by two species of coccinellid beetles and a predatory cecidomyiid fly. These effectively control mealybug populations and prevent significant damage to the host plants except locally and occasionally (McFadyen 1979b). In Australia, freed from these natural enemies, enormous populations rapidly built up. It was quickly evident that three years of large cactus mealybug populations were sufficient to kill mature plants (Fig. 6). When examined, root tubers were found to be empty and it was clear that plant reserves had been exhausted. By 1978, increasing areas of cactus were dying and in 1979 chemical treatment was ended in favour of intensive rearing and redistribution of the cactus mealybug (McFadyen 1986).

From 1980, the cactus mealybug was released on sword pear, *A. tetragonus*, at Gogango, Qld inland from Rockhampton. Establishment occurred and partial control of this cactus has resulted. Sword pear has fewer spine clusters and is much slower-growing than Harrisia cactus; as a result, mealybug populations are more exposed to predators and do not increase as rapidly. The cactus mealybug has also colonised several other cacti in the Cactoideae occurring in Qld, causing severe damage to many.

Figure 5: Harrisia cactus, *Harrisia martinii*, heavily attacked by cactus mealybug. Photo: R McFadyen, Qld Government.

Figure 6: Harrisia cactus, *Harrisia martinii*, killed by cactus mealybug. Photo: R McFadyen, Qld Government.

Three predators, the coccinellid beetles *Cryptolaemus montrouzieri* Mulsant and *Diomus* sp. and the lacewing fly *Chrysopa ramburi* Schneider, commonly feed on the cactus mealybug in Australia, but do not prevent populations increasing. No parasitoids have been recorded from the species in Australia (McFadyen 1986).

***Eriocereophaga humeridens* O'Brien (Coleoptera: Curculionidae).** This stem-feeding weevil, initially an undescribed genus and species, was found during surveys of insects attacking *H. adscendens* in the caatinga (dry scrub) of north-eastern Brazil, and subsequently described as *Eriocereophaga humeridens* O'Brien. The weevil is only known from north-eastern Brazil, where larvae were found in *H. adscendens* and in small plants of *Cereus jamacaru* and *Pilosocereus (Cephalocereus) gounellei* (F.A.C.Weber) Byles & G.D.Rowle. It was readily reared on *H. martinii* in the laboratory (McFadyen 1979a).

The nocturnal adults are flightless, lacking hind wings, and are robust and solid, 9–12 mm long, spending the long dry season dormant in the soil. After rain (winter in the caatinga), adults feed on fresh young cactus stems. The rather large eggs (1.2–2 mm) are laid singly in protected sites on the cactus stems, and are covered by the female with a 1 mm thick layer of soil and faeces. Larvae feed in cactus tissue, reducing it to a mucilaginous liquid. When mature, they pupate in a cocoon in the damaged stem.

Host-testing was undertaken using multiple-choice tests with adults, on the plant species previously used. There was no attack except on cacti, and the first field releases in Australia were made in 1976 (McFadyen 1979a). Establishment occurred in some sites, but colonies did not increase significantly and died out with the severe reduction in cactus abundance caused by attack by the cactus mealybug (McFadyen 1986). The weevil is now believed to be extinct in Australia.

***Cactoblastis* new species near *doddi* (Lepidoptera: Phycitidae).** Cactoblastis larvae were first collected from *H. martinii* and *Harrisia bonplandii* (Parm. ex Pfeiff) Britton & Rose in the Argentine Chaco in 1972. The species is close to *Cactoblastis doddi* Heinrich, which is common on *Opuntia* species in north-western Argentina, but is not known from Cactoideae or from eastern Argentina (McFadyen 1980, 1985). The life-history of *C.* sp.nr. *doddi* is similar to that of *Cactoblastis cactorum* (Berg); eggs are laid in sticks of 30–60 attached to spines on the host cactus. Newly hatched larvae enter the cactus and feed gregariously inside stems, ejecting frass through the entrance hole. As they grow, larvae enter and feed in new stems. Mature larvae leave stems to pupate singly in soil surface litter. There are two generations a year, a short summer generation and a longer winter generation. Damage to the plant is frequently severe, especially in young small plants.

Host-testing involved use of multiple-choice tests with newly hatched and half-grown larvae, supplemented by no-choice tests on some plants (McFadyen 1980). Full development was possible only in *H. martinii* and *H. bonplandii*; larvae fed in a third species, *H. pomanensis*, but development was slow and the resulting adults were small and infertile. Permission for field release was granted in 1975, but there were problems in laboratory rearing and field releases were delayed, being made from 1978 to 1982. Unfortunately, establishment did not occur at any site. The reason for failure is unknown, but it is possible that *H. martinii* is not in fact a suitable host, as in the field *C.* sp.nr. *doddi* was found only where *H. martinii* grew close to *H. bonplandii*.

EVALUATION

After the release of the first agents in 1974 and 1975, permanent line transects were established in insect release sites in northern Qld. Measurements were taken annually until 1984, during which time the *Harrisia* population fell from 50% cover to less than 0.5%, and plant numbers reduced to fewer than 160 small plants ha^{-1} (Tomley and McFadyen 1985). In 1978 similar transects were established at Goondiwindi in southern Qld, where in the absence of the biocontrol agents the cactus density increased over six years from 15% to 84% cover and from 34 000 to 110 000 plants ha^{-1} (McFadyen 1986). Control was much slower in this cooler area; nevertheless, within 10 years of the mealybug release at the sites, cover was reduced to below 2% and numbers to below 100 plants ha^{-1} (McFadyen, unpub. data).

During 1979, mealybugs from a northern Qld colony (four years after initial releases) were reared at different constant temperatures in the laboratory. The minimum temperature for development was established to be 15.3°C and 424 day-degrees above this were required for a full egg-to-egg generation. Based on this data, there would be seven generations per year in the main cactus areas of northern Qld, two of them during the cooler six months. In Goondiwindi in the south, only four generations would be possible, with only one over the cooler six months (McFadyen, unpub. data).

No evaluation has been undertaken on sword pear, but the main infestation has greatly reduced in extent and control is regarded as adequate.

DISCUSSION

This was an outstandingly successful project, with excellent control of the main infestations of the weed achieved within eight years of the first overseas exploration. Field populations of Harrisia cactus rapidly fell from 50% to 80% groundcover and 140 000 plants ha^{-1} to <0.5% groundcover and <100 plants ha^{-1}. These very low levels have been maintained since the mid 1980s in northern and central Qld, where biological control alone is sufficient and redistribution of the cactus mealybug is not necessary. In southern Qld and adjacent NSW, where temperatures are cooler, the mealybug develops more slowly, especially in shade, and control is slow. Manual distribution of the cactus mealybug may be needed and chemical control may be preferred for a more rapid control. The benefit:cost ratio for the program has been calculated as 23.5 and the investment of $1 million returned a net present value of $18.6 million (Page and Lacey 2006). The partial control of sword pear is an additional benefit.

The program is notable for several factors: success is largely due to an agent which initially was not believed to cause sufficient damage for successful control (McFadyen 1979b); the introduced cactus mealybug population originates from as few as six mated females; field success was achieved within eight years of the first overseas exploration; despite an abundance of native and introduced parasitoids of Pseudococcidae in Qld, none have adopted the cactus mealybug as a host; and two other established agents have subsequently died out due to lack of sufficient host plants.

ACKNOWLEDGEMENTS

The assistance of Helmuth Zimmermann from the start of this project and his continued interest in the identity of the cactus mealybug are gratefully acknowledged. Patricio Fidalgo and Eduardo Willink accompanied me on field surveys in South America and assisted in laboratory rearing and host-testing, and Allan Tomley undertook much of the early insect rearing, releases and field evaluation in Qld; their assistance is gratefully acknowledged.

REFERENCES

Anderson EF (2001) *The Cactus Family.* Timber Press, Portland, Oregon.

Bennett FD (1972) 'Report on a preliminary survey of the insects associated with *Eriocereus* (*Harrisia*) spp. (Cactaceae) in northern Argentina and Paraguay'. Commonwealth Institute of Biological Control, West Indian Station, Curepe, Trinidad & Tobago.

Bostock PD and Holland AE (2007) *Census of the Queensland Flora.* Environment Protection Agency, Qld Government, Brisbane.

Hosking JR, McFadyen RE and Murray ND (1988) Distribution and biological control of cactus species in eastern Australia. *Plant Protection Quarterly* **3**, 115–123.

Klein H (1999) Biological control of three cactaceous weeds, *Pereskia aculeata* Miller, *Harrisia martinii* (Labouret) Britton and *Cereus jamacaru* De Candolle in South Africa. *African Entomology Memoir* **1**, 3–14.

Mann J (1967) *The Harrisia Cactus.* Government Printer, Brisbane.

Mann J (1969) Cactus-feeding insects and mites. *United States National Museum Bulletin* **256**, 1–158.

Mann J (1970) *Cacti Naturalised in Australia and their Control.* Government Printer, Brisbane.

McFadyen RE (1976) 'Final report on investigations of insects attacking *Eriocereus* spp. and related cacti in South America'. Unpublished report, May 1976. Commonwealth Institute of Biological Control, West Indian Station, Curepe, Trinidad & Tobago.

McFadyen RE (1979a) *Eriocereophaga humeridens* (Col.: Curculionidae), a potential agent for the biological control of *Eriocereus martinii* (Cactaceae) in Australia. *Entomophaga* **24**, 49–56.

McFadyen RE (1979b) The cactus mealybug *Hypogeococcus festerianus* (Hem: Pseudococcidae), an agent for the biological control of *Eriocereus martinii* (Cactaceae) in Australia. *Entomophaga* **24**, 281–287.

McFadyen RE (1980) A *Cactoblastis* (Lep.: Phycitidae) for the biological control of *Eriocereus martinii* (Cactaceae) in Australia. *Entomophaga* **25**, 37–42.

McFadyen RE (1985) Larval characteristics of *Cactoblastis* spp. (Lep.: Pyralidae) and the selection of species for biological control of prickly pears (*Opuntia* spp.). *Bulletin of Entomological Research* **75**, 159–168.

McFadyen RE (1986) Harrisia cactus in Queensland. In *The Ecology of Exotic Animals and Plants in Australia.* (Ed. RL Kitching) pp. 240–261. Jacaranda-Wiley, Brisbane.

McFadyen RE (1990) The Harrisia cactus eradication scheme: policy-making by technocrats. MPubAdmin thesis. University of Queensland, Australia.

McFadyen RE and Fidalgo AP (1976) Investigations on *Alcidion cereicola* (Col.: Cerambycidae), a potential agent for the biological control of *Eriocereus martinii* (Cactaceae) in Australia. *Entomophaga* **21**, 103–111.

Page AR and Lacey KL (2006) 'Economic impact assessment of Australian weed biological control'. Technical Series 10. CRC for Australian Weed Management, Adelaide.

Tomley AJ and McFadyen RE (1985) Biological control of Harrisia cactus, *Eriocereus martinii*, in central Queensland by the mealybug *Hypogeococcus festerianus*, nine years after release. In *Proceedings of the 6th International Symposium on Biological Control of Weeds.* 19–25 August 1984, Vancouver. (Ed ES Delfosse) pp. 843–847. Agriculture Canada, Ottawa.

Williams DJ and Granara de Willink MC (1992) *Mealybugs of Central and South America.* CABI, Wallingford, UK.

Zimmermann HG, Erb HE and McFadyen RE (1979) Annotated list of some cactus-feeding insects of South America. *Acta Zoologica Lilloana* **33**, 101–112.

Heliotropium amplexicaule Vahl – blue heliotrope

DT Briese

ABSTRACT

Blue heliotrope (*Heliotropium amplexicaule*), a toxic perennial herbaceous plant from South America, has become an increasingly serious pasture weed in south-eastern Qld and north-eastern NSW since its introduction in the late 19th century. Following preliminary surveys in its native range in the early 1990s, a biological control project was carried out between 1998 and 2004. Five potential control agents (four insects and a pathogen) were identified and a biological control strategy developed based on the target weed population ecology and potential agent impact. Two agents, the leaf-feeding beetle *Deuterocampta quadrijuga* and the root-feeding flea-beetle *Longitarsus* sp., were introduced into Australia for host-specificity testing. Results for *Longitarsus* sp. were inconclusive and it was not approved for release. *D. quadrijuga* was approved, released in 2001 and subsequently established in the field. While observations suggest that *D. quadrijuga* has caused severe local damage to the target weed since release, its impact has not yet been properly evaluated.

Key words: *Deuterocampta quadrijuga*, *Longitarsus* sp., Argentina, biological control.

INTRODUCTION

Heliotropium amplexicaule Vahl (Boraginaceae) (blue heliotrope) is a summer-growing perennial plant, prostrate to 15 cm high and up to 60 cm in diameter. The lanceolate leaves are 2–8 cm long and bear numerous simple and glandular hairs (Fig. 1). Flowers occur on long scorpioid cymes and vary in colour from light blue to lilac (Figs 1 and 2). Plants die back over winter and produce several flushes of new growth and flowers between spring and autumn.

It was introduced into Australia from South America as an ornamental plant in the late 1800s. Over the last 40 years there has been a rapid and continuing expansion of its range and there are now widespread infestations in south-eastern Qld and northern NSW, with scattered colonies extending into Vic and further into SA (Fig. 3) (Parsons and Cuthbertson 1992). Blue heliotrope infests over 110 000 ha in NSW alone (Da Silva 1991), where it is a declared noxious weed in 18 local government areas. Infestations range from subtropical to temperate over a wide range of soil types and habitats. Climatic modelling suggests that the potential range of the weed could extend much further than its current range (Fig. 3).

H. amplexicaule contains pyrrolizidine alkaloids that are toxic to livestock, causing liver damage and stock death (Glover and Ketterer 1987). Production losses occur due to competition by blue heliotrope with more desirable cropping and pasture species and through a decline in animal performance as a result of its toxicity.

The chemicals currently registered for blue heliotrope have had limited success and are not selective. Cultivation encourages spread by stimulation of germination and regeneration from decapitated rootstocks and plant

Figure 1: Inflorescence, leaves and fruit capsule of *Heliotropium amplexicaule*. Illustrations: Royal Botanic Gardens, Sydney.

Figure 2. Blue heliotrope, *Heliotropium amplexicaule*. Photo: CSIRO.

fragments. The increasing rate of spread indicates that current control methods are not successful; Newell (1997) considered *H. amplexicaule* to be on the verge of becoming a much more serious problem for agriculture in eastern Australia. Consequently, *H. amplexicaule* was approved as a target weed for biological control in Australia in 1999.

PLANT TAXONOMY

Heliotropium is a cosmopolitan genus belonging to the subfamily Heliotropioidae within the Boraginaceae (Förther 1998). Originally, it was considered to comprise several sections that evolved in different regions of South America, Europe, Africa and Australia. *H. amplexicaule* is part of the section *Heliotrophytum*, which evolved in South America. Craven (1996) recognised 75 endemic species of *Heliotropium* in Australia, of which 70 belonged to the section *Orthostachys* and the remainder to the section *Pterotropium*.

Recent studies based on molecular markers showed that the vast majority of endemic Australian species form a different phylogenetic clade from other species of *Heliotropium*. Diane *et al.* (2002) described these as a new genus, *Euploca*. Förther (1998) placed the remaining five endemic Australian species in the section *Pterotropium*. Thus, *H. amplexicaule* is less closely related to Australian endemic species than previously thought, a factor useful when considering potential host-range of biological control agents.

BIOLOGICAL CONTROL HISTORY

Australia is the only country in which biological control of this weed has been attempted. Two brief surveys of the weed in its South American native range were made in 1990 and 1991 to investigate the potential for biological control. They identified a number of insects as possible candidate control agents (Wapshere 1991). Climate-matching showed that the areas surveyed for blue heliotrope and its natural enemies were similar to the principal infested areas in Australia (Wapshere 1993), giving optimism about the chance of biological control.

Extensive lobbying by a group of affected landholders, who formed the Blue Heliotrope Action Committee, saw the Rural Industry Research and Development Fund

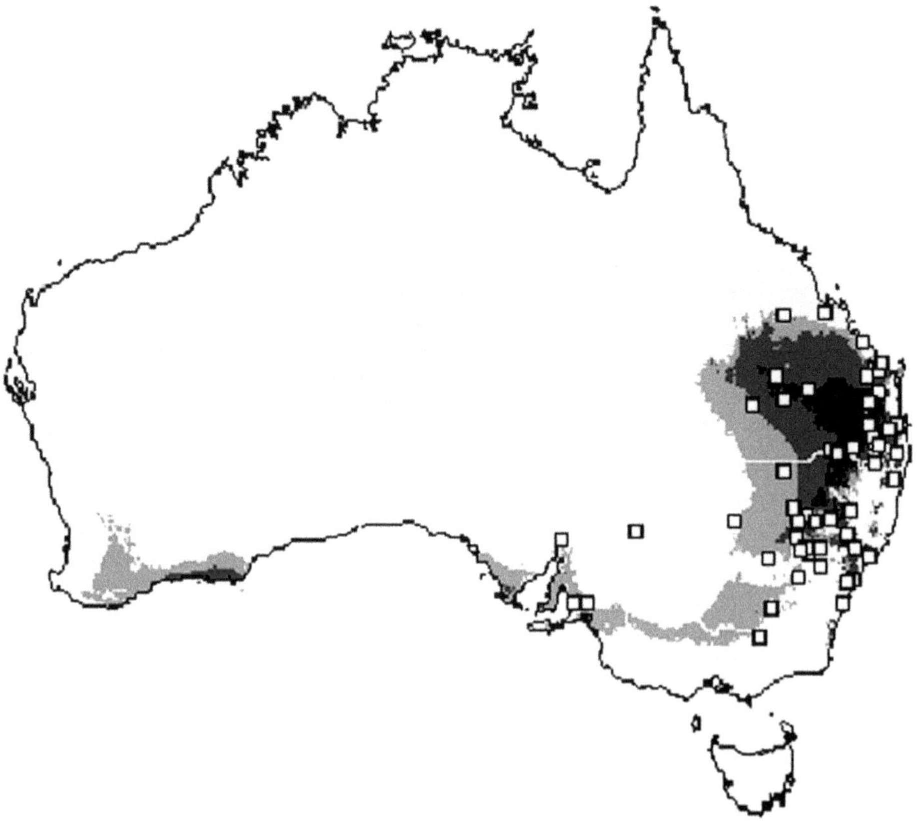

Figure 3: Predicted range of blue heliotrope *Heliotropium amplexicaule* in Australia, based on temperature, rainfall quantity and rainfall seasonality, using Bioclim® (the darker the shading, the more suitable). White squares indicate current distribution records.

support a six-year program by CSIRO Entomology during 1998–2003. During this period more detailed studies were made on the population dynamics of the weed in its native and introduced ranges (Briese and Zapater 2001), extended surveys were carried out in Argentina and experiments were also made there to determine potential agent impact (Briese *et al.* 2000; Briese and Zapater 2001; Zapater *et al.* 2004). These studies enabled Briese and Zapater (2002) to develop a strategy for the biological control of blue heliotrope.

Strategy for the biological control of blue heliotrope

Blue heliotrope is a perennial plant with a deep root system. The plant produces new shoot and leaf growth from spring to autumn, and can flower several times during this period. In Australia, three-quarters of flowering and seed production occurred in the first flush, which finished by late spring (Briese and Zapater 2001). This seed enters a relatively long-lived seed bank. Blue heliotrope reproduces from this seed bank or vegetatively from root buds. In winter the foliage of the plant partially dies back or is killed by frost, to regenerate the following spring from the rootstock. The main problems caused by the weed are an ability to smother desirable summer-growing pasture species and toxicity to livestock due to the presence of pyrrolizidine alkaloids, which can damage the liver.

Biological control therefore had a number of aims. First, any reduction in above-ground biomass would reduce the toxicity problem and reduce photosynthetic capability, rendering the plant less competitive with other pasture species. It was also considered desirable to reduce the production of seed by blue heliotrope and reduce the size of the soil seed reserves, leading to an eventual decline in local blue heliotrope infestation densities and reduction in weed spread into new areas. Finally, any direct damage to the root system would interfere with nutrient uptake from the soil and reduce the plant's capacity to store these nutrients in the rootstock. In its native range in Argentina, blue heliotrope is a coloniser of recently disturbed areas, but in many cases populations do not persist or remain at low levels with individual plants being much shorter-lived than in Australia. This appears to be partially due to the inability of blue heliotrope to compete with later-colonising vegetation, but is also due to continued levels of natural enemy attack, which further reduces its competitiveness.

Two insects, the leaf-beetle *D. quadrijuga* and the root-feeding flea-beetle *Longitarsus* sp., were prioritised for the biological control of *H. amplexicaule* in Australia. *D. quadrijuga* had the potential to build up population levels rapidly, and could quickly and completely defoliate blue heliotrope plants. The strategy was to have a two-pronged attack on the target weed; on the above-ground biomass (photosynthetic tissue), primarily by the leaf beetle with added pressure from the adult flea-beetles, and on the below-ground biomass (root reserves) by the larvae of the flea-beetle. These complementary actions were considered to increase the chances of successful biological control of this toxic weed.

EXPLORATION

H. amplexicaule is native to South America, occurring in northern and central Argentina, southern Bolivia, Uruguay and the extreme south of Brazil (Johnston 1928). Agent exploration was carried out in the southern part of its range in Argentina, as climate-matching showed that these regions were most similar to the principal infested areas in Australia (Wapshere 1993). Four regions were surveyed in particular: the highlands near Cordoba, north of Bahia Blanca, the Pampa north-west of Buenos Aires and the Entre Rios region close to the border with Uruguay. These surveys were accompanied by preliminary testing of host-specificity and field experiments designed to measure the impact of the agents in their native range.

CANDIDATES

Based on exploration and experimental measurement of impact in Argentina during 1998–2000, the candidate agents for biological control were as follows.

***Deuterocampta quadrijuga* Vahl (Coleoptera: Chrysomelidae)**. This is a large (up to 10 mm long) black beetle with an orange stripe along the length of each elytron (Fig. 4). Both larvae and adults feed on the leaves of *H. amplexicaule*. After overwintering as pupae from late May to the end of September, adults emerge to feed, mate and lay eggs. The beetle is very fecund (up to 1400 eggs per female over four months). Eggs are laid on the leaves of *H. amplexicaule* plants and, after emergence and feeding, the larvae leave the plant to pupate in the soil nearby. *D. quadrijuga* can undergo several overlapping generations

Figure 4: The adult of *Deuterocampta quadrijuga*. Photo: CSIRO.

per year in the field (Briese and Zapater 2001). At high densities, *D. quadrijuga* causes complete defoliation of the target weed (Briese and Zapater 2001) and massive populations of the beetle have been recorded defoliating blue heliotrope in its native range (Bruch 1940).

Longitarsus **sp. (Coleoptera: Chrysomelidae)**. This flea-beetle is 2 mm long, dark-brown and halticine. It is active from spring to autumn and there are at least two generations per year. Female *Longitarsus* sp. enter gaps in the soil to lay eggs directly onto the roots; emerging larvae feed on the fine feeder roots that extend from the larger storage roots, impeding the uptake of nutrients and water. After emerging as adults, they feed on the leaf tissue, leaving shot-hole feeding wounds which become infected by saprophytic bacteria. Leaves eventually die and fall from the plant. Damage to the plant accumulates until late autumn when the aerial parts of plants start to senesce and this, coupled with feeding on the root tissue by larvae, severely weakens the plants. Studies in the home range (Zapater *et al.* 2004) found that many attacked plants die before the onset of winter.

***Haplothrips heliotropica* Mound and Zapater (Thysanoptera: Phlaeothripidae).** This 2.2 mm brown thrips feeds on the developing leaf and flower buds and young leaves of *H. amplexicaule*, causing stunting and deformation (Mound and Zapater 2003). Several generations are produced each year and feeding can block stem elongation and subsequent development of flower cymes (Wapshere 1993).

***Dictyla* sp. (Hemiptera: Tingidae)**. This is a small grey-black tingid bug, approximately 4 mm long. Adults and nymphs suck sap from the cells of leaves, causing the damaged area to become translucent and surrounding leaf tissue chlorotic. *Dictyla* sp. seems to be the most widespread of the four insects (Wapshere 1993). It produces several generations per year and caused heavy local damage at several sites (Briese and Zapater 2001).

***Pseudocercosporella* sp. (Deuteromycota: Hyphomycetales).** This fungal pathogen, which caused necrotic lesions on leaf tissue, was found at one site near Entre Rios. While it appeared damaging at this site in Argentina and did not attack a closely related species under quarantine conditions, inoculation of Australian *H. amplexicaule* plants in quarantine did not produce severe symptoms or defoliation, and research on this pathogen for biological control was discontinued (L Morin, *pers. comm.* 2003).

Preliminary host-specificity testing of the insects was carried out using a two-phase open-field test in Argentina (Briese *et al.* 2002). The first phase of the test allowed the insects to choose between target and non-target plants. In the second no-choice phase, these host plants were cut and left on the plot to force natural dispersal of the agents among the remaining non-target plants. Data from these tests suggest that none of the four candidate species attacked plants outside the genus *Heliotropium*, and were most likely limited to a subset of species within that genus.

Results of this preliminary testing, and the observed impacts on the target weed in its native range, confirmed the leaf-beetle *D. quadrijuga* and root-feeding flea-beetle *Longitarsus* sp. to be the best prospects for implementing the biological control strategy. Detailed host-specificity testing of *D. quadrijuga* (Briese and Walker 2002) and *Longitarsus* sp. (Briese and Walker 2008) was conducted in the high-security quarantine at CSIRO Entomology, Canberra. Both choice and non-choice tests were used to test the feeding preferences of adults and larvae and oviposition by adults. Test plant selection was based on the modified centrifugal phylogenetic method, based primarily on degrees of phylogenetic separation rather than taxonomic nomenclature (Briese 2003).

The tests showed that *D. quadrijuga* was able to complete its life-cycle only on plants related to the target weed, *H. amplexicaule*, by one degree of phylogenetic separation (i.e. a group of closely related South American species). *Longitarsus* sp. was slightly less specific, being able to complete its life-cycle on plants related to *H. amplexicaule* by up to two degrees of phylogenetic separation. This included three native Australian species that might overlap the range of the target weed. In both species, exploratory feeding by adults was broader and there was a small risk of localised and temporary collateral damage to non-target plants related to the target weed.

D. quadrijuga was approved as a biological control agent of *H. amplexicaule* by the Australian authorities and released into the field in October 2001. However, the risks of releasing *Longitarsus* sp. were considered to be too great (Briese and Walker 2008). Further studies were recommended to clarify the issues surrounding its risk to non-target native *Heliotropium*, but funding for the project ceased and this agent has not been released.

EVALUATION

Evaluation of the biological control program has been restricted to surveys of the establishment of *D. quadrijuga* following rearing and redistribution programs carried out from 2003 until the present (Day *et al.* 2004). In south-eastern Qld, this was initiated by landholders from the Blue Heliotrope Action Committee (I Crosthwaite, *pers. comm.* 2003). In NSW, redistribution was done under the aegis of the state government (NSW Agriculture), which made more than 240 releases in the seven years since 2003. This was a period of extended drought, which made establishment difficult. As of 2010, there are five confirmed field populations of *D. quadrijuga*; two in the central west slopes, two in the north-western slopes and one in the central tablelands of NSW (R Holtkamp, *pers. comm.* 2010). No formal monitoring of the impact of these beetles has been made, although there is evidence that they have reached population levels able to damage local infestations of *H. amplexicaule*. For example, the newspaper *Agriculture Today* reported in February 2009 that by 2008, in the Capertee Valley of NSW, beetle numbers had increased to the point where they were having a serious impact on the plants, having 'stripped all their leaves, weakening the plants and reducing their ability to flower and produce seed'.

DISCUSSION

This project is a demonstration of grass-roots support for biological control – without strong political lobbying by affected landholders (who formed the Blue Heliotrope Action Committee), research would not have gone further than the preliminary survey phase. Their efforts obtained funding for more detailed exploration and research over a six-year period from 1998 to 2003. The project remained incomplete when external funding ceased and the strategy developed for biological control could not be realised. While one key agent, the leaf-beetle *D. quadrijuga*, was released and established to show promising local impact on the target weed, permission was not given to release the root-feeding beetle *Longitarsus* sp. Although most reviewers considered the agent safe to release, some sought clarification on certain aspects of its host-range and mobility (Briese and Walker 2008). These issues could not be addressed once funding ceased.

From a research rather than a practical viewpoint, the host-testing of these agents demonstrated the utility of two-phased open-field tests for determining host choice in the presence and absence of the primary host (the target weed) (Briese *et al.* 2002), as well as a new approach to host-plant selection proposed by Briese (2003).

Should future funding be obtained, work should concentrate on clarifying the host-specificity of *Longitarsus* sp. Early anecdotal evidence suggest that *D. quadrijuga* might be developing a pattern of boom–bust population cycles, similar to other chrysomelid beetles released for biological control in Australia (Briese 1985). Thus, while it could provide temporal and local relief from the problems caused by *H. amplexicaule*, a complementary species such as *Longitarsus* sp. remains a necessary component of an effective strategy. Sadly, such outcomes remain speculative in the absence of systematic and detailed evaluation of the impact of the established control agent, an element lacking in all such short-term funded projects.

ACKNOWLEDGEMENTS

The efforts of the Blue Heliotrope Action Committee in securing funding, and its subsequent encouragement, are acknowledged.

REFERENCES

Briese DT (1985) A survey to evaluate the long-term relationship between *Chrysolina quadrigemina* and its host weed, St John's wort, in south-eastern Australia. *Proceedings of the 6th International Symposium on Biological Control of Weeds* 19–25 August 1984, Vancouver. (Ed ES Delfosse) pp. 691–708. Agriculture Canada, Ottawa.

Briese DT (2003) The centrifugal phylogenetic method used to select plants for host-specificity testing of weed biological control agents: can and should it be modernised? *Improving the Selection, Testing and Evaluation of Weed Biological Control Agents. CRC Technical Series No. 7*, pp. 23–33.

Briese DT and Walker A (2002) A new perspective on the selection of test plants for evaluating the host specificity of weed biological control agents: the case of *Deuterocampta quadrijuga*, a potential insect control agent of *Heliotropium amplexicaule. Biological Control* **25**, 273–287.

Briese DT and Walker A (2008) Choosing the right plants to test: the host-specificity of *Longitarsus* sp. (Coleoptera: Chrysomelidae) a potential biological control agent of *Heliotropium amplexicaule. Biological Control* **44**, 271–285.

Briese DT and Zapater M (2001) *Biological Control of Blue Heliotrope*. RIRDC Publication No. 01/119.

Briese DT and Zapater M (2002) A strategy for the biological control of blue heliotrope (*Heliotropium amplexicaule*). In *Proceedings of the 13th Australian Weeds Conference.* 8–13 September 2002, Perth. (Eds H Spafford Jacob, J Dodd and JH Moore) pp. 394–397. Plant Protection Society of Western Australia, Perth.

Briese DT, McLaren DA, Pettit W, Zapater M, Anderson F, Delhey R and Distel R (2000) New biological control projects against weeds of South American origin in Australia: blue heliotrope and serrated tussock. In *Proceedings of the 10th Symposium on Biological Control of Weeds.* 4–14 July 1999, Montana State University, Bozeman. (Ed NR Spencer) pp. 215–223.

Briese DT, Zapater M, Andorno A and Perez-Camargo G (2002) A two-phase open-field test to evaluate the host-specificity of candidate biological control agents for *Heliotropium amplexicaule. Biological Control* **25**, 259–272.

Bruch C (1940) Metamorfosis de *Deuterocampta quadrijuga* Stal. *Miscellanea Entomologica* **4**, 200–206.

Craven L (1996) A taxonomic revision of *Heliotropium* (Boraginaceae) in Australia. *Australian Systematic Botany* **9**, 521–657.

Da Silva E (1991) The ecology and control of blue heliotrope (*Heliotropium amplexicaule* Vahl). Unpublished final report to the Wool Research Corporation, Armidale, NSW.

Day MD, Briese DT, Grace BS, Holtkamp RH, Ireson JE, Sheppard AW and Spafford Jacob H (2004) Improving release strategies to increase the establishment rate of weed biocontrol agents. In *Proceedings of the 14th Australian Weeds Conference.* 6–9 September 2004, Wagga Wagga, NSW. (Eds BM Sindel and SB Johnson) pp. 369–373. Weed Science Society of New South Wales, Sydney.

Diane N, Förther H and Hilger HH (2002) A systematic analysis of *Heliotropium*, *Tournefortia* and allied taxa of the Heliotropiaceae (Boraginales) based on ITS1 sequences and morphological data. *American Journal of Botany* **89**, 287–295.

Förther H (1998) Die infragenerische Gliederung der Gattung Heliotropium L. und ihre Stellung innerhalb der subfam. Heliotropioideae (Schrad.) Arn. (Boraginaceae). *Sendtera* **5**, 35–141.

Glover PE and Ketterer PJ (1987) Blue heliotrope kills cattle. *Queensland Agricultural Journal* **113**, 109–110.

Johnston IM (1928) Studies in the Boraginaceae – VII. 1. The South American species of *Heliotropium*. *Contributions of the Gray Herbarium Harvard University* **81**, 3–83.

Jones CD (1971) Blue heliotrope – a difficult weed to control. *Cane Growers Quarterly Bulletin* **34**, 112.

Ketterer PJ, Glover PE and Smith LW (1987) Blue heliotrope (*Heliotropium amplexicaule* Vahl) poisoning in cattle. *Australian Veterinary Journal* **64**, 115–117.

Mound LA and Zapater MC (2003) South American *Haplothrips* species (Thysanoptera, Phlaeothripidae), with a new species of biological control interest to Australia against weedy *Heliotropium amplexicaule* (Boraginaceae). *Neotropical Entomology* **32**, 437–442.

Newell D (1997) Beating blue heliotrope; solutions for long-term control. In *Proceedings of the 9th Biennial Noxious Weeds Conference*. 16–19 September 1997, Dubbo, NSW. pp. 87–92. NSW Agriculture, Goulburn, Australia.

Parsons WT and Cuthbertson EG (1992) *Noxious Weeds of Australia*. Inkata Press, Melbourne.

Wapshere AJ (1991) 'Report on a preliminary survey in Argentina for potential agents for the weed *Heliotropium amplexicaule*'. Final report to the Wool Research and Development Corporation. [add publisher + location]

Wapshere AJ (1993) Climate matching and the prospects for biological control of weeds: the contrasting examples of agents for *Solanum elaegnifolium* and *Heliotropium amplexicaule*. In *Pest Control and Sustainable Agriculture*. (Eds SA Corey, DJ Dall and WM Milne) pp. 198–200. CSIRO Publishing, Melbourne.

Zapater M, Bartoloni N, Perez-Camargo G and Briese DT (2004) Natural impact of the flea-beetle, *Longitarsus* sp., on *Heliotropium amplexicaule* in Argentina and its potential for use as a biological control agent in Australia. In *Proceedings of the 11th International Symposium on Biological Control of Weeds*. 27 April–2 May 2003, Canberra. (Eds JM Cullen, DT Briese, DJ Kriticos, WM Lonsdale, L Morin and JK Scott) pp. 208–214. CSIRO Entomology, Canberra.

Heliotropium europaeum L. – common heliotrope

Andy W Sheppard, Louise Morin and Jim Cullen

ABSTRACT

Common heliotrope is a summer-growing annual weed of crops and pasture, considered native to Mediterranean Europe, Turkey and western North Africa, and present in all states and territories of Australia except Tas. It can be an important component of annual pastures in southern Australia, where it often dominates fallow, disturbed or ploughed land from December to April. When eaten by livestock it causes cumulative liver damage and causes the highest fatality rate from primary pyrrolizidine alkaloid poisoning of any plant. The biological control program targeting common heliotrope started in the 1950s, with a peak of activities from the early 1970s to the early 1990s, and consisted of natural enemy surveys of more than 15 annual *Heliotropium* spp. in the western Mediterranean, eastern Africa and Iran (the centre of origin of the genus). Nearly 40 arthropods and pathogens were identified from common heliotrope out of the 132 species found across *Heliotropium* spp. Host-specificity testing in Europe was undertaken for three arthropod candidate agents, the root- and leaf-feeding flea-beetle *Longitarsus albineus*, the root weevil *Pachycerus segnis* and the flower-cyme feeding moth *Ethmia distigmatella*, and for two plant pathogens, the rust *Uromyces heliotropii* and the leaf-blotch fungus *Cercospora heliotropiicola*. Thousands of flea-beetles were released in Australia from 1979 to the late 1980s following the development of effective laboratory mass-rearing techniques, but the insects never persisted more than a few years at a site and this failure was never adequately explained. The rust fungus was released between 1991 and 1993, widely established and is still found occasionally, but has had no impact on the abundance of the weed. None of the other candidate agents were released. The inability of released agents to reduce weed populations was put down to the difficulty of achieving biological control of an ephemeral summer annual agricultural weed with highly variable between-year dynamics.

Key words: Boraginaceae, *Longitarsus albineus*, *Pachycerus cordiger*, *Ethmia distigmatella*, *Uromyces heliotropii*, *Cercospora heliotropiicola*, *Cercospora taurica*.

INTRODUCTION

Heliotropium europaeum L. (Boraginaceae), common heliotrope, is a summer annual herb considered native to the Mediterranean region and south-western Asia. It is a weed in North America (Parsons and Cuthbertson 2001), but its impacts are far worse in Australia. Its distribution in Australia is given in Figure 1. It is most abundant north and west of the Great Dividing Range in the south-east of the continent, where it occurs in a broad arc from southern Qld, through NSW and northern Vic to SA. It is also abundant in the lower south-west of WA (Parsons and Cuthbertson 2001). This corresponds to a zone of 300–500 mm annual rainfall (Moore 1956). In southern Australia, it germinates from spring to early summer on bare soil with soil moisture once soil temperatures reach 24°C with a growing period from November to May (Moore 1956; Hunt *et al.* 2008). The abundance and distribution in any given year is driven by the seasonal distribution of summer rain (rainfall events generate distinct cohorts of seedlings) and the degree of litter accumulation that provides favourably moist conditions for plant growth. *H. europaeum* is a very ephemeral summer annual, in that few individuals survive more than a few months in a given season and germination requirements rarely occur in the same place twice in consecutive years. Where conditions are right it often dominates fallow, disturbed or ploughed land as little else germinates with it and it can be an important component of annual pastures. When eaten by stock it causes cumulative liver damage and leads to the highest fatality rate from primary pyrrolizidine alkaloid poisoning of any plant. Common heliotrope is six times more toxic than Paterson's curse, which also contains pyrrolizidine alkaloids (O'Dowd and Edgar 1989). Historically, pyrrolizidine alkaloidosis accounts for more than 63% of sheep losses in some years (Edgar *et al.* 1941), although the more recent situation is unclear. Outbreaks of common heliotrope do not occur every year, but in years of high abundance (e.g. 1991–93) farmers generally face economic losses due to weed impact and the costs of implementing largely ineffective chemical or cultural control (Parsons and Cuthbertson 2001). It has been estimated that, in the 1980s, on average, common heliotrope cost the meat and wool industry at least $46.5 million per year ($17.5 million from livestock mortality, $14 million from decreased productivity and $25 million in control costs) (Cullen and Delfosse 1990). In a cropping context, *H. europaeum* is considered to use up valuable soil moisture over summer prior to autumn/winter crop-sowing (Delfosse and Cullen 1980).

Common heliotrope is a hairy, ephemeral summer-annual herb usually 10–30 cm high with a well-developed tap root system and a somewhat offensive odour. It reproduces only by seed. The grey-green erect or semi-prostrate shoot has alternate elliptical to ovate leaves. Flowers are small, white with a yellow throat, in two rows on the flowering coiled cymes, which straighten as seeds ripen. The fruit has four nutlets with brown to black small subglobular seeds (Parsons and Cuthbertson 2001) (Fig. 2a). Plants produce flowers and seeds concurrently with vegetative

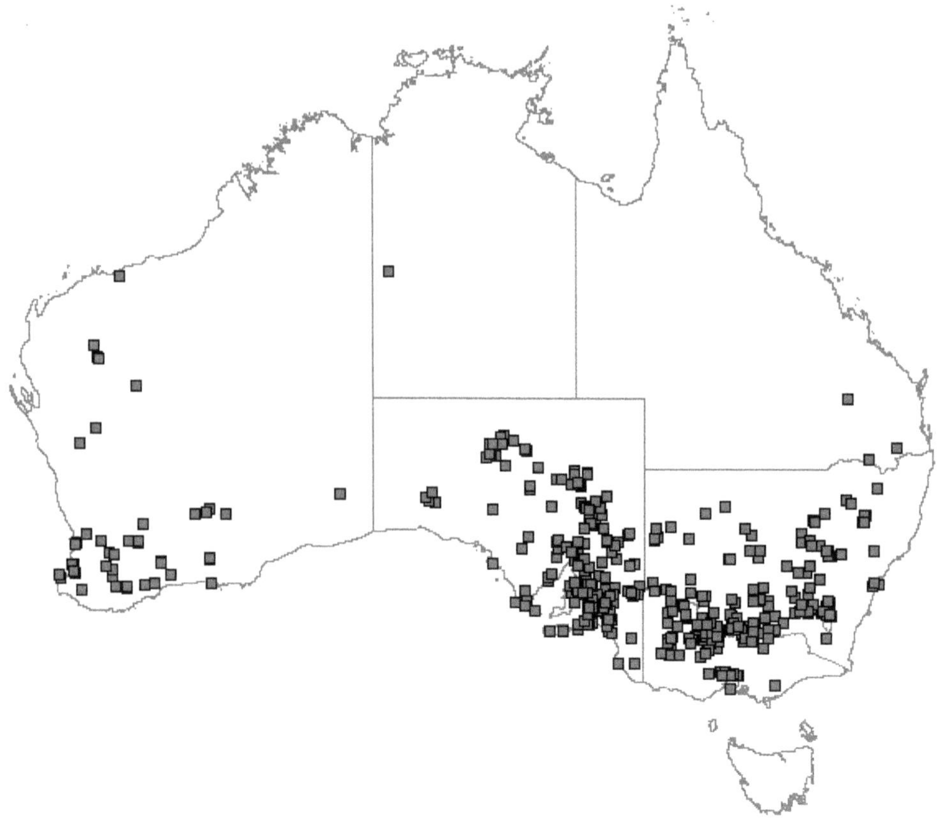

Figure 1: Distribution of common heliotrope, *Heliotropium europaeum*, in Australia. Specimen data reproduced from Australia's Virtual Herbarium with permission of the Council of Heads of Australasian Herbaria Inc. Map generated on 1 May 2010.

Figure 2: Common heliotrope *Heliotropium europaeum*. a) The plant. Photo: CSIRO. b) An infestation near Gundagai, NSW. Photo: L Morin, CSIRO.

growth, soon after germination, thereby creating a persistent seed bank of up to 3 million seeds m^{-2}. Plants senesce in autumn, if they are not already killed by drought or by rain-induced leaf-blight outbreaks over the summer.

In a comparative study of common heliotrope in both its native and exotic ranges, Sheppard *et al.* (1996) found maximum plant densities and basic demography (e.g. seed production versus density) to be similar in both regions although the average infestation size and density were greater and their occurrences more frequent in Australia. The higher summer rainfall and more extensive rotational agriculture in Australia appeared to be the dominant causes of the plant's abundance status. These conditions, however, could not fully explain why the seed banks in Australia were 10 times larger than in the native range.

Control options for common heliotrope are well documented in Parsons and Cuthbertson (2001) and include cultivation, herbicides and crash-grazing by sheep, but each option has significant economic, practical and animal health constraints. The persistent seed bank is the other major hindrance in rotational cropping systems. Classical biological control was identified very early, in the 1950s. When major activity in this biological control program commenced there were no regulatory requirements to nominate a weed as a biological control target or to consider risks that potential arthropod agents might pose to native Australian plants, although this was already an issue for plant pathogens.

BIOLOGICAL CONTROL HISTORY

Australia is the only country to have attempted a biological control program against common heliotrope. The program started with survey work by Frank Wilson from about 1950 in Europe and North Africa, but was deemed 'not especially promising' by him (Delfosse and Cullen 1980) and was shelved until 1971 when exploration was renewed in Europe, the Middle East and North Africa. The early surveys and the work from 1971 to the late 1980s were funded by CSIRO, then jointly by CSIRO and the Meat Research Corporation from 1988 when the program received encouragement from producers, politicians and state agencies with cross-jurisdictional support through the Australian Weeds Committee. Adding biological control to existing management practices was considered the quickest way towards a sustainable control strategy. While the early focus of the program was almost entirely on arthropod agents, one of which was eventually released, the success of the skeleton weed rust in Australia prompted consideration of potential pathogen agents in the 1980s (Hasan 1985, 1992).

PLANT TAXONOMY

Common heliotrope is a member of a complex of closely related *Heliotropium* species in the eastern part of its native range in Eurasia, which has caused confusion on whether certain natural enemies collected during early surveys were indeed found on the target. Only common heliotrope, *H. europaeum*, is considered to be present in Australia, but there are several Australian native *Heliotropium* spp.

EXPLORATION

Frank Wilson's initial exploration included southern France, eastern Spain, coastal Algeria and Tunisia. He then conducted surveys in Kenya, Eritrea and the Sudan. Eventually he concluded that the prospects for biological control of common heliotrope were low (Delfosse and Cullen 1980). There was no further work until 1971, when increasing pressure in Australia to find some solution to the problem, plus the presence of a long-term base in Europe, encouraged renewed exploration. This base allowed more extensive exploration to the east, including the Middle East known to have a high diversity of species in the genus.

Surveys were conducted in Iran in 1971 and 1974 and in Tunisia and Algeria in 1974 mainly by Tony Wapshere, Siraj Hasan and John Huber. Iran was chosen as a primary site because it has a large number of annual *Heliotropium* spp., some of which are similar to *H. europaeum*, and because of the climatic similarity to areas of Australia where common heliotrope is a problem. The surveys yielded a greater diversity of insects on *Heliotropium* spp. in northern Iran than in Mediterranean Europe or North Africa, though the most promising agent, the flea-beetle *Longitarsus albineus* (Foudras) (Coleoptera: Chrysomelidae), was present everywhere.

Surveys by John Huber and Siraj Hasan continued through the 1970s in Mediterranean Europe and Turkey with the promising root-feeding weevil *Pachycerus segnis* Germar (=*Pachycerus cordiger* Germar, =*Pachycerus scabrosus* Brullé) (Coleoptera: Curculionidae) being found in Greece and Turkey as well as in Iran. The rust fungus *Uromyces heliotropii* Sred. (Uredinales: Pucciniaceae) was also first found at this time in Turkey.

A number of later surveys were made in eastern Europe from CSIRO's European laboratory in the 1980s and 1990s by Siraj Hasan, Jim Cullen, Andy Sheppard and Laurent Brun, either to collect live material or to try to understand field host species usage by agents under consideration.

CANDIDATES

Natural enemies

The surveys of *Heliotropium* spp. from 1950 to 1980 in the native range found 132 species of natural enemy, 37 of which were considered to have been found on common heliotrope (list in Delfosse and Cullen 1980). Table 1 lists the species considered to have biological control potential. Diversity of natural enemies on common heliotrope was higher towards the eastern end of its distribution and closer to the presumed evolutionary centre of origin of the genus. Six species were considered to be potentially specific to at least the genus, from the literature at the time (Delfosse and Cullen 1980). The flea-beetle *L. albineus,* the externally feeding moth *Utethesia pulchella* (L.) (Lepidoptera: Arctiideae), the root weevil *P. segnis*, the leaf-blotch fungus *Cercospora heliotropiicola* A.K. Kar & M. Mandal (Moniliales: Dematiaceae) and the rust fungus *U. heliotropii* were eventually found throughout the native range of common heliotrope, while the flower-cyme moth *Ethmia distigmatella* Erschoff (Lepidoptera: Ethmiidae) was found only in the eastern part of the range. Delfosse (1985) discussed strategies for release order, stating that the original proposed order of release had been the flea-beetle then the weevil, then the cyme moth and finally the rust fungus, but that this changed to the rust followed by the moth, then the weevil and finally the flea-beetle, based on the status of the host-specificity testing and agent availability. As it turned out, only the flea-beetle and the rust were ever released.

Four species were found on common heliotrope in Australia prior to the release of exotic agents, a native flea-beetle *Longitarsus* sp. (nr *L. victoriensis* Blackburn) (C Reid, *pers. comm.* 1993), a native arctiid moth *Utethesia pulchelloides* (L.), a native gelechiid moth *Stomopteryx isosceliscantha* (Lower) (Delfosse and Cullen 1980; CSIRO Entomology 1994) and a possibly introduced leaf-blotch fungus *Cercospora taurica* Tanzschel (J Walker, *pers. comm.* 1991; CSIRO Entomology 1994).

Natural enemies tested but not released

***Pachycerus segnis* Germar (Coleoptera: Curculionidae).** This large univoltine curculionid weevil completes its life-cycle on species in the Boraginaceae with larvae developing in an earthen cell on the roots of common heliotrope. The taxonomy has been confusing (Huber and Vayssières 1990), being considered as an eastern European species *P. cordiger* with a sister western European species *P. scabrosus* by two weevil systematists or as one pan-European species later accepted as *P. segnis* by others. Huber and Vayssières (1990) found that adults from eastern Europe (Turkey and Greece) were larger than museum specimens of either species from western Europe, and so tested weevil populations collected from common heliotrope in Greece and Turkey against seven European Boraginaceae and 31 economically important crop species. They found no *Pachycerus* species commonly on common heliotrope in western Europe. They concluded that the tested material was specific to Boraginaceae in the laboratory and to *Heliotropium* spp. in the field, and that therefore should be safe for introduction to Australia. Later concerns that no Australian species in the Boraginaceae had been included in the original tests led to further testing in Europe. Testing of eastern Euro-

Table 1: Species considered with biological control potential and information on where and when observed during native-range surveys of *Heliotropium europaeum*, and current status

Species	Agent type	S. France, E. Spain, Algeria, Tunisia 1950	Iran 1971 1974	Tunisia, Algeria 1974	S Europe, Turkey 1970s	Status
Longitarsus albineus (Chrysomelidae)	Root- and leaf-feeder	X	X	X	X	Released 1979–89, establishment unconfirmed since 1992
Pachycerus segnis (Curculionidae)	Root- and leaf-feeder		X		X	Tested extensively, but not sufficiently host-specific
Neoliturus opacipennis (Leth) (Cicadellidae)	Sap-sucker		X	X*		Not tested
Austroagallia sinuata (M.R.) (Cicadellidae)	Sap-sucker		X	X*		Not tested
Atomoscelis signaticorus Reut. (Miridae)	Sap-sucker		X			Not tested
Polymerus sp. (Miridae)	Sap-sucker		X			Not tested
Ethmia distigmatella (Ethmiidae)	Flower-feeder		X		X	Target proved not the preferred host
Utethesia pulchella (Arctiidae)	Leaf- and shoot-chewer		X		X	Not tested (native in same genus in Australia)
Uromyces heliotropii (Pucciniaceae)	Rust fungus		X			Released 1991, widely established but epidemics never observed
Cercospora heliotropiicola (Dematiaceae)	Leaf-blotch fungus				X	Tested but not released (near-identical congeneric already in Australia)

* Recorded as Neoliturus sp. and Austroagallia sp.

pean populations over four years in the laboratory in France and in the field in Greece showed that no Australian species of *Heliotropium* tested (*H. asperrimum* R.Br., *H. bacciferum* Forsk and *H. crispatum* F. Muell. Ex Benth.) was suitable for larval development, but other Australian Boraginaceae, *Cynoglossum australe* R.Br. and *Myosotis australis* R.Br., supported adult feeding, oviposition and larval development (JM Cullen, unpub.).

Whether or not the phenologies of these Australian species might reduce their exposure to use in the field in Australia by P. segnis was hard to assume and meant the level of risk to them was difficult to assess, thus work on this insect was suspended. A further study discovered a morphologically identical *Pachycerus* species to that from eastern Europe restricted to *H. europaeum* populations in western Europe, adding host-range support that there was one pan-European species (Brun *et al.* 1993), which has since been recognised as *P.* segnis (Alonso-Zarazaga 2010). Given these results and the concerns over specificity, no further application was made to release this species in Australia.

***Ethmia distigmatella* Erschoff (Lepidoptera: Ethmiidae).** This flower-cyme feeding ethmiid moth recorded from common heliotrope in Crete, Turkey and south-western Asia was only once successfully reared in quarantine in Australia on common heliotrope in small numbers, despite several attempts. An initial host-specificity test carried out using freshly emerged adults found no oviposition on common heliotrope or on the Australian native *Heliotropium ovalifolium* Forsk. tested. Using the *Flora of Turkey* (Davis 1984), a survey of the complex of closely related species including *H. europaeum* and the insects on this complex in western Turkey in 1991 failed to confirm whether the moth had ever been recorded from common heliotrope in the native range. Based on this, it was considered that common heliotrope is not the preferred host of this moth (CSIRO Entomology 1994).

Cercospora heliotropiicola **A.K. Kar & M. Mandal (Moniliales: Dematiaceae).** Following rainfall, this leaf-blotch fungus can produce severe disease epidemics on common heliotrope in Europe that can prematurely kill populations prior to much seed production (Brun *et al.* 1995; Hasan *et al.* 1995). However, in the 1980s another *Cercospora* species, *C. taurica*, was observed causing epidemics on common heliotrope in Australia (J Walker, *pers. comm.* 1991). This species was also known from eastern Europe, Russia and western Asia (Chupp 1954). The taxonomic status of the two *Cercospora* spp. attacking common heliotrope was explored morphologically and using isozyme electrophoresis (Brun *et al.* 1996). Results showed consistent differences between isolates classed as *C. heliotropiicola* (from France) or *C. taurica* (from Turkey and Australia) used in the study.

Before host-specificity testing of *C. heliotropiicola* could be justified, it was important to see whether *C. heliotropiicola* (ex. France) was more pathogenic than *C. taurica*, which already occurred in Australia. A comparative demographic study of common heliotrope in Australia and Europe demonstrated that *Cercospora* epidemics were both more common and more damaging in Europe than in Australia, potentially contributing to the observed differences in seed bank sizes between the native and exotic range (Brun *et al.* 1995; Sheppard *et al.* 1996). Brun *et al.* (1996) also showed that *C. heliotropiicola* (ex. France) was more pathogenic on detached leaves of common heliotrope than *C. taurica* (ex. Australia), and that all five Mediterranean *Heliotropium* species tested were susceptible to both species.

C. heliotropiicola was therefore imported from France into quarantine in Australia in 1996 for detailed comparative studies with *C. taurica*. In experiments using whole plants, both *Cercospora* species produced a similar number of leaf lesions on plants exposed to the same temperatures (14°C, 22°C, 30°C) and wetness periods (8 h, 16 h, 24 h, 48 h) during the initial infection phase (Morin and Trueman 1997). Under optimum conditions, the pathogens had a similar impact on plant growth and reduced plant dry weight by 63%. In the first 24 hours, *C. taurica* produced more than five times the number of conidia produced by *C. heliotropiicola* on infected necrotic tissue placed in high-humidity conditions. For both species, production of conidia increased as the period under high humidity increased, although *C. taurica* always produced the highest number of conidia. These results suggested that under the same (controlled) conditions, *C. heliotropiicola* did not appear to have greater value than *C. taurica* for biological control of common heliotrope and consequently was not tested further nor released in Australia.

Natural enemies released

Longitarsus albineus **(Foudras) (Coleoptera: Chrysomelidae).** This multivoltine chrysomelid flea-beetle feeds on the exteriors of fine roots as larvae, pupating in the soil and feeding on foliage as adults (Fig. 3a). In the field in the native range adult feeding was recorded on four annual *Heliotropium* species and typical feeding damage was observed on another 12 species in the genus. A method for observing and rearing the root-feeding larvae was developed in the late 1970s and the beetle was found to have three to four generations a year under field conditions (Huber 1979, 1981). A parasitic nematode (13% of collections) and a hymenopteran parasitoid were recorded from this insect in the native range (Huber 1981). Host-specificity testing was carried out against 41 non-target plants using no-choice adult oviposition and larval starvation tests. The test plant list included only two Australian native species, a *Eucalyptus* sp. and an *Acacia* sp. Adult feeding was observed on 20 of the test plant species, but severity of feeding was less than 10% of that seen on common heliotrope. Oviposition was observed across a wide range of test plants, although the number of eggs laid was a third or less than that found on common heliotrope. Larval development was recorded on *Heliotropium peruvianum* L. and *Myosotis alpestris* F.W.Schmidt, but adults emerged only from the former species. At the time, assessing risks to native Australian species was not a requirement for arthropod agents and therefore Huber (1981) considered that this agent was specific enough for release, despite the test results suggesting that some Australian native species in the Boraginaceae were likely to be within its host-range.

Common heliotrope seedlings were killed by only a few larvae in tests, so significant impact in the field was predicted. Huber (1981) concluded that *L. albineus* was physiologically ill-suited to breeding on non-summer *Heliotropium* or other Boraginaceae, but that its broad climatic adaptability, its host-finding ability and its rapid rate of increase meant that it should have no difficulty establishing in Australia and that survival of overwintering adults would be the main determinant of its effectiveness under Australian conditions.

The flea-beetle was imported into Australia and first released in 1979. Beetles were recovered the following season at most sites, but establishment beyond the first

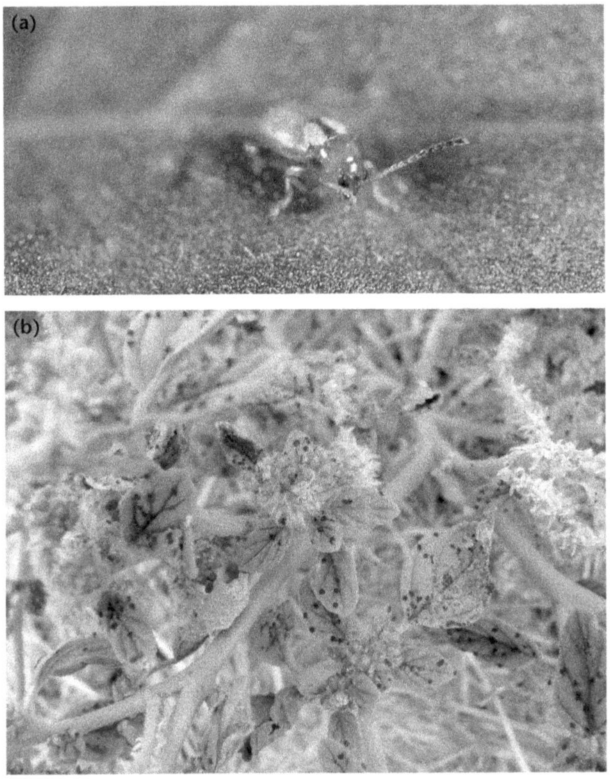

Figure 3: a) The flea-beetle *Longitarsus albineus*. Photo: CSIRO. b) The rust fungus *Uromyces heliotropii*. Photo: L Morin, CSIRO.

season was not confirmed and was put down to a series of drought years (Delfosse 1985). A mass-rearing program was then successfully developed, in Canberra, that produced over 31 000 beetles, which were then released at seven sites across four states (NSW, SA, Vic and WA) from 1987 to 1989. Measures, such as irrigation, were taken to ensure common heliotrope presence in subsequent summers. Successful recoveries of flea-beetles were made at four of these sites in lower numbers in the next year and individuals were recovered at some sites until 1992, but not after. As no flea-beetles persisted at any site longer than three years and it was clear from the search effort that no spread had occurred, monitoring was terminated in 1994. As no further sightings have been made it has been assumed that this agent failed to establish in Australia. Field experiments in the native range suggested that unless the post-release densities of flea-beetles were very high compared to those observed in the native range, this agent would have little impact on common heliotrope abundance, given the already highly variable between-year dynamics of this weed (CSIRO Entomology 1994).

Uromyces heliotropii **Sred. (Uredinales: Pucciniaceae).** This macrocyclic and autoecious rust fungus was originally discovered on common heliotrope in Turkey and was subsequently observed in France, and is generally found, though only sporadically, across its native range (Fig. 3b). Field experiments in the native range suggested that this rust could significantly affect and kill common heliotrope seedlings and could survive winter conditions and reinfect plants the following season (Hasan and Ayres 1990; Hasan and Aracil 1991). Host-specificity tests were carried out on 96 non-target plant species including eight species of *Heliotropium* native to Australia and 16 other native species in the Boraginaceae (Hasan *et al.* 1992). The Australian native *Heliotropium crispatum* F. Muell. Ex Benth. and two exotic *Heliotropium* spp. present in Australia were susceptible to the rust, albeit at a much lower level than common heliotrope (Hasan *et al.* 1992). Any potential threat posed by the rust to *H. crispatum* was discounted as this species is both temporally (winter-growing) and geographically (northern Australia) separated from common heliotrope (Hasan and Delfosse 1995).

An isolate from Turkey was first released in 1991 at a site in NSW, followed by 19 further releases at 13 sites across four states (NSW, SA, Vic and WA) over three years, using sprayed spore suspensions in water, covered by plastic sheeting in the first 24 hours, at each release point. Local spread of the rust up to 500 m from the release plot was observed at 11 sites within the season of release (Sheppard *et al.* 1993) and overwintering teliospores were seen at many sites. Six sites were monitored in the subsequent season and natural reinfection of common heliotrope by the rust was observed at three of them, but on only one or a few plants per site and only at sites where good spread had been observed in the initial year. Infection in the second year never spread nor multiplied the way it did in the initial year, and the rust could be found at only one of the monitored sites as spot infections two years after the releases. While climate appeared limiting in WA (too dry), the main cause of poor establishment of this agent is probably the inability of the overwintering teliospores to synchronise germination with the reappearance of the target in spring (S Hasan, *pers. comm.* 1992). The annual life-cycle and ephemeral nature of common heliotrope may also limit the abundance and therefore the potential impact of the rust, which, following winter without its host, must undergo a sexual cycle on reinfection each spring. Although this agent has had no apparent impact on the target in Australia, its persistence continued to be confirmed through occasional sightings on common heliotrope at sites across south-eastern Australia unrelated to those where original releases were made.

DISCUSSION

As one of Australia's longest-running biological control programs, with very extensive native-range research activities across Europe, North Africa and the Middle East, the program against common heliotrope is most notable by its failure to deliver any kind of effective control of the target. Frank Wilson, the earliest scientist to work on it, did not rate the prospects of biological control very high. His 1950 report has been lost so we cannot be sure of his reasoning, but it is likely that he made this conclusion after his initial efforts did not find many apparently specific candidate agents in the 1950s (listed in Delfosse and Cullen1980). It was only pressure from the agricultural sector for CSIRO to do something, plus the opportunity to explore closer to the centre of origin of the species, that stimulated more work. The limited chance of success of biological control against summer annual cropping weeds is due to a number of factors, including their ephemeral nature (Chaboudez and Sheppard 1995), a factor that was probably not well appreciated at the time. Annual plants are also likely to support a smaller community of natural enemies than longer-lived species because of this.

The only insect to be released, *L. albineus*, was, as Huber (1981) rightly considered, 'a good-looking agent on anyone's books'. That it was so hard to establish, despite a highly successful mass-rearing program, is a significant enigma, unless Australian soil conditions or summer temperatures or overwintering conditions were just too unsuitable in some way for successful beetle survival. As it was, there was every likelihood that this agent would have had to reach very high numbers very quickly within a growing season to have any impact on common heliotrope populations; had it become widespread, it would have also attacked many native *Heliotropium* spp. and possibly other Boraginaceae (CSIRO Entomology 1994).

This program fell at a time when Australia was enthusiastic about plant pathogens as potential agents, following the successful use of a rust fungus for the biological control of skeleton weed in Australia, another cropping weed. Expectations that the two promising pathogens found on common heliotrope, including a rust fungus, could achieve similar levels of success would have been quite high. With hindsight, could the failure of the common heliotrope rust to affect the weed have been predicted from its rarity in the native range? Perhaps, but this program was the first to show that rust fungi have an Achilles heel when their host is so ephemeral. Most of the teliospores, the life-stage of rusts that enables survival overwinter without a host, could have germinated with moisture in the spring when environmental conditions were suitable and produced infective basidiospores, but before temperatures were high enough for germination of the host plant. There may therefore be a problem of phenological synchrony between basidiospore production and host availability, but this has not been explored. In comparison, the *Cercospora* spp. found on common heliotrope cause a much more damaging disease, often resulting in severe epidemics in both the native range and Australia (Brun *et al.* 1995; Sheppard *et al.* 1996). The impact of these pathogens, however, is driven by summer rainfall which provides optimal conditions for infection and associated rain splash necessary for local spread. Keeping the infected litter in the paddocks may be a way of encouraging such epidemics, provided the period until new host foliage is present at the sites is not too long.

This common heliotrope program affected the contemporary over-optimism that successful biological control of cropping weeds was possible. It also provided key lessons on the need to focus on the target weed species rather than the target genus when there are congeneric species native to Australia, in order to avoid wasting time investigating ineffective candidate agents. Since the early 1980s, agents to be released in Australia have had to be more specific to address the legislative requirements of the *Wildlife and Countryside Act* 1981 (superseded by the *Environmental Protection and Biodiversity Conservation Act* 1999).

ACKNOWLEDGEMENTS

We gratefully acknowledge the contribution of Laurent Brun, Ernest Delfosse, Siraj Hasan, John Huber, Bobby Lewis, Carey Smith, Tony Wapshere, Frank Wilson, Tim Woodburn and many technical staff to this biological control program.

REFERENCES

Alonso-Zarazaga MA (2010) Fauna Europaea: *Pachycerus segnis* (Germar 1824). Fauna Europaea version 2.2, <http://www.faunaeur.org>.

Brun LA, Sheppard AW and Carrara A (1993) Host range of *Pachycerus cordiger* Germar (=*P. scabrosus* Brulle) (Col.: Curculionidae). *Entomophaga* **38**, 537–539.

Brun LA, Sheppard AW, Hasan S, Jourdan M and Delmotte E (1995) Development of *Cercospora* blight epidemics and effect on the summer annual weed *Heliotropium europaeum* L. in the field. *Annals of Applied Biology* **127**, 137–150.

Brun LA, Sheppard AW and Lopes V (1996) Variation in pathology of isolates of *Cercospora* spp. attacking *Heliotropium* spp. *Annals of Applied Biology* **128**, 565–570.

Chaboudez P and Sheppard AW (1995) Are particular weeds more amenable to biological control? A reanalysis of mode of reproduction and life history. In *Proceedings of the 8th International Symposium on Biological Control of Weeds.* 2–7 February 1992, Canterbury, NZ. (Eds ES Delfosse and RR Scott) pp. 95–102. DSIR/CSIRO, Melbourne.

Chupp C (1954) *A Monograph of the Fungus Genus* Cercospora. C Chupp, Ithaca, New York.

CSIRO Entomology (1994) 'Project CS.104. Biological control of common heliotrope, *Heliotropium europaeum*'. Contracted research report for Meat Research Corporation, Sydney.

Cullen JM and Delfosse ES (1990) Progress and prospects in biological control of weeds. In *Proceedings of the 9th Australian Weeds Conference.* 6–10 August 1990, Adelaide. (Ed JW Heap) pp. 452–470. Crop Science Society of South Australia, Adelaide.

Davis P (1984) *Flora of Turkey and the East Aegean Islands.* Edinburgh University Press, Edinburgh.

Delfosse ES (1985) Re-evaluation of the biological control program for *Heliotropium europaeum* in Australia. In *Proceedings of the 6th International Symposium on Biological Control of Weeds.* 19–25 August 1984, Vancouver. (Ed ES Delfosse) pp. 735–742. Agriculture Canada, Ottawa.

Delfosse ES and Cullen JM (1980) New activities in biological control of weeds in Australia. I. Common heliotrope, *Heliotropium europaeum*. In *Proceedings of the 5th International Symposium on Biological Control of Weeds.* 22–27 July 1980, Brisbane. (Ed ES Delfosse) pp. 545–561. CSIRO, Melbourne.

Edgar G, Hindmarsh WL, Keast JC and Rose AL (1941) The distribution and incidence of enzootic or toxaemic jaundice in south-eastern Australia with special reference to New South Wales. *Australian Veterinary Journal* **17**, 120–130.

Hasan S (1985) Prospects for biological control of *Heliotropium europaeum* by fungal pathogens. In *Proceedings of the 6th International Symposium on Biological Control of Weeds.* 19–25 August 1984, Vancouver. (Ed ES Delfosse) pp. 617–623. Agriculture Canada, Ottawa.

Hasan S (1992) An update of the biological control of common heliotrope in Australia using plant pathogens. In *Proceedings of the 1st International Weed Control Congress.* Vol. 2. 17–21 February 1992, Melbourne. (Ed RG Richardson) pp. 212–214. Weed Science Society of Victoria, Melbourne.

Hasan S and Aracil E (1991) Biology and effectiveness of *Uromyces heliotropii* Sred., a potential biological control agent of *Heliotropium europaeum* L. *New Phytologist* **118**, 559–563.

Hasan S and Ayres PG (1990) Tansley Review No. 23. The control of weeds through fungi: principles and prospects. *New Phytologist* **115**, 201–222.

Hasan S and Delfosse ES (1995) Susceptibility of the Australian native, *Heliotropium crispatum*, to the rust fungus *Uromyces heliotropii* introduced to control common heliotrope, *Heliotropium europaeum. Biocontrol Science and Technology* **5**, 165–174.

Hasan S, Delfosse ES, Aracil E and Lewis RC (1992) Host-specificity of *Uromyces heliotropii*, a fungal agent for the biological control of common heliotrope (*Heliotropium europaeum*) in Australia. *Annals of Applied Biology* **121**, 697–705.

Hasan S, Jourdan M, Brun L and Sheppard AW (1995) Axenic culture and influence of wetness period and inoculum concentration on infection and development of cercospora blight of *Heliotropium europaeum*. *European Journal of Plant Pathology* **101**, 491–496.

Huber JT (1979) A method for observing and rearing the root-feeding larvae of *Longitarsus albineus* (Foudras) (Col. Chrysomelidae). *Bulletin de la Société Entomologique Suisse* **52**, 431–433.

Huber JT (1981) Observations on the heliotrope flea beetle, *Longitarsus albineus* (Col.: Chrysomelidae) with tests on its host specificity. *Entomophaga* **26**, 265–273.

Huber JT and Vayssières JF (1990) Life cycle and host specificity of the heliotrope weevil, *Pachycerus cordiger* (= *madidus* Auct.) [Col.: Curculionidae]. *Entomophaga* **35**, 475–484.

Hunt JR, Cousens RD and Knights SE (2008) The biology of Australian weeds 51. *Heliotropium europaeum* L. *Plant Protection Quarterly* **23**, 146–152.

Moore CWE (1956) 'Observations on the autoecology of *Heliotropium europaeum* L. in New South Wales and Victoria'. CSIRO Division of Plant Industry Technical Paper No. 7. CSIRO, Melbourne.

Morin L and Trueman HE (1997) Pathogenicity and inoculum potential of two *Cercospora* species on the weed *Heliotropium europaeum*. In *Program and Summaries: 11th Biennial Conference of the Australasian Plant Pathology Society*. 29 September–2 October 1997, Perth. p. 49.

O'Dowd DJ and Edgar JA (1989) Seasonal dynamics in the pyrrolizidine alkaloids of *Heliotropium europaeum*. *Australian Journal of Ecology* **14**, 95–105.

Parsons WT and Cuthbertson EG (2001) *Noxious Weeds of Australia*. 2nd edn. CSIRO Publishing, Melbourne.

Sheppard AW, Lewis RC and Delfosse ES (1993) The establishment of *Uromyces heliotropii* Sred., a biological control agent of *Heliotropium europaeum* L. In *Proceedings of the 10th Australian and 14th Asian-Pacific Weed Conference*. 6–10 September 1993, Brisbane. pp. 89–93. Queensland Weed Society, Brisbane.

Sheppard AW, Brun LA and Lewis RC (1996) A demographic comparison of common heliotrope, *Heliotropium europaeum* L.: southern Australia and southern France. In *Proceedings of the 11th Australian Weeds Conference*. 30 September–3 October 1996, Melbourne. (Ed RCH Shepherd) pp. 286–290. Weed Science Society of Victoria, Melbourne.

Hypericum perforatum L. – St John's wort

DT Briese and Jim Cullen

ABSTRACT

St John's wort is a toxic perennial herb originating from Europe. Since its introduction into Australia in the mid to late 19th century, it has become a serious weed of pastures, poisoning stock and displacing more productive pasture species. It has also spread into natural vegetation beneath open eucalypt woodlands, where it threatens floral diversity. Work on the biological control of St John's wort in Australia first commenced over 80 years ago, and passed through a number of discrete phases. A total of 15 potential control agents were studied in some detail in the native range of the weed in England and southern France and 11 of these were eventually released. Six agents became established, four during earlier phases of work in the 1930s to 1950s and two in the last phase of work from the mid 1970s. Two of the original agents to establish, the defoliating chrysomelid beetles *Chrysolina quadrigemina* and *Chrysolina hyperici*, provide some degree of control in certain situations, but are unable to prevent continued spread of the weed. Later introductions targeted a weakening of the root system, and one, the mite *Aculus hyperici*, became widely established and showed that it could produce local damage on the target weed. Whether further biological control work is required, however, depends on the broader-scale and longer-term impact of *A. hyperici*; careful evaluation of this agent is required. The integration of biological control into management plans, where its impact is reinforced by pasture management and/or more judicious chemical control, is seen as the best option for reducing the costs of this weed to Australian agriculture.

Key words: *Chrysolina quadrigemina*, *Chrysolina hyperici*, *Agrilus hyperici*, *Aculus hyperici*, biological control, France.

INTRODUCTION

St John's wort, *Hypericum perforatum* L. (Clusiaceae), is a perennial herb with two growth forms. In autumn and winter it produces a prostate rosette from a central woody crown, up to 60 cm in diameter with spindly soft stems and a dense mat of bright green elongated leaves (Campbell *et al.* 1995). In spring and summer, one or more woody flowering stems up to 120 cm in height with opposing elongate leaves and clusters of bright yellow five-petalled flowers are produced (Fig. 1). Both the petals and leaves are scattered with translucent oil glands, giving a perforated appearance. Flowering is followed by the formation of reddish-brown seed capsules containing tiny (<1 mm diam.) black seeds. A single plant may produce up to 30 000 seeds and after they fall from the capsules in autumn, the flowering stem senesces. Individual crowns may complete up to eight annual flowering cycles before dying (Briese 1997a). *H. perforatum* is a facultative apomict and fertile seeds may persist for up to 20 years in the soil. It reproduces vegetatively from lateral roots as well as from seed. Infestations of the weed persist mainly by vegetative reproduction, while spread and colonisation

Figure 1: Inflorescence of *Hypericum perforatum*. Photo: CSIRO.

of new sites occur via seed dispersal (Briese 1997a). This can happen by human transport of contaminated materials or by attachment of seeds to stock or native wildlife.

H. perforatum was introduced into Australia from its native range in Europe on several occasions in the mid to late 19th century for ornamental and medicinal purposes (Harris and Gill 1997). It has since spread to infest over 400 000 ha, primarily in south-eastern Australia, but also to some extent in SA and WA (Fig. 2). It grows in a wide range of habitats and soils, from open pastures to beneath open eucalyptus canopies (Clark 1953b). The plant contains the alkaloid hypericin, which causes photosensitisation and liver damage in mammals (Bourke 1997). Losses in animal production in NSW were estimated at $22.5 million by Bourke (1997) and further losses through competition with desirable pasture add to the economic cost of the weed. It is also considered an environmental weed, with over 80% of infestations currently occurring in native vegetation beneath open woodland (Shepherd 1984).

Its use as a natural anti-depressant led to a brief interest in the harvesting of natural infestations of *H. perforatum* in the 1990s. Its costs outweigh its benefits, however, and it remains a declared noxious weed in all states and territories (Campbell *et al.* 1995). It is also considered weedy in other parts of the world where it has been introduced: New Zealand, South Africa, Chile, Canada and the USA.

PLANT TAXONOMY

Hypericum is a large genus within the family Clusiaceae (=Guttiferae), having more than 200 species. Two species, *H. gramineum* J. Forst. and *H. japonicum* Thunb., are native to Australia, and several European species, including *H. androsaemum* L., *H. calycinum* L., *H. crispum* L., *H. tetrapterum* Fries and *H. perforatum*, have become naturalised in Australia. *H. perforatum* is the most widespread of these introduced species. The most common form of this species in Australia is the narrow-leaved var. *angustifolium* DC, though studies by Campbell (1997) demonstrated the presence of broader-leaved varieties that differ from var. *angustifolium* in chemistry and ecology, as well as morphology. Mayo (2004) confirmed its genetic variability, with weak intrapopulation variability and stronger variation between populations, reflecting the clonal nature of reproduction by this species and confirming its multiple introductions into Australia. Mayo also demonstrated genetic variability in susceptibility to attack by the eriophyid mite *Aculus hyperici* (Liro), which demonstrates the importance of plant taxonomy and population genetics to biological control.

BIOLOGICAL CONTROL HISTORY

The biological control project against *H. perforatum*, together with those against prickly pear and lantana, heralded the start of weed biological control in Australia. Its history has been described in some detail by Briese (1997b), who divided the work into four distinct phases.

First mooted in 1917, overseas research commenced in 1928, when CSIR (later CSIRO) set up a laboratory in Buckinghamshire, England (Currie and Garthside 1932). Eight insects were studied in some detail and eventually sent to Australia; five were released in Vic between 1930 and 1937, but with little success (Currie and Fyfe 1938). Eventually one species, the chrysomelid *Chrysolina hyperici* (Forst.), appeared in large numbers in 1939 at the original release site in Bright (Wilson and Campbell 1943).

In the second phase, starting in 1935, operations were based in southern France where climatic conditions were considered closer to those of temperate Australia (Wilson 1943). Work continued until the onset of World War II forced the field station to close in 1940. Two agents, the buprestid root-borer *Agrilus hyperici* (Creutzer) and the chrysomelid defoliator *Chrysolina quadrigemina* (Suffr.), were imported and released in Australia in 1939–40, establishing relatively easily at sites near Bright, Vic and Mudgee, NSW (Wilson and Campbell 1943). The other three agents studied in France (two leaf-feeding moths and a leaf bud-galling cecidomyid) were not considered capable of the same levels of damage as these two species and hence were not released.

A third phase, leading up to 1956, concentrated on the broad-scale redistribution of *C. hyperici* and *C. quadrigemina* from 1940 and 1943, respectively. Agencies such

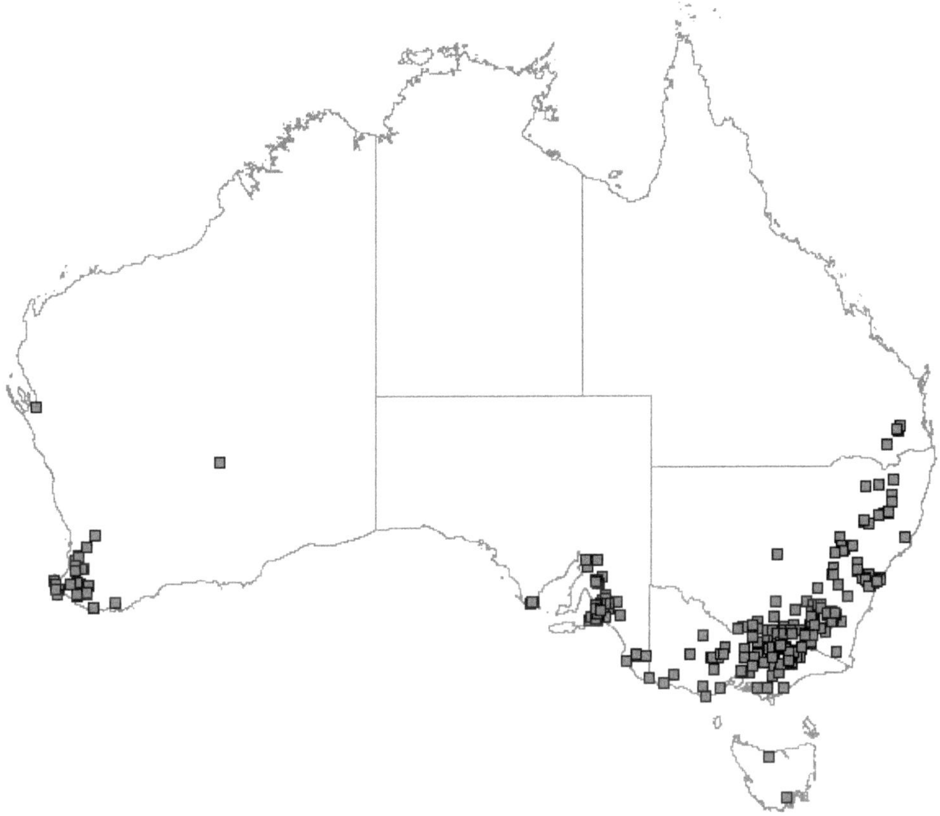

Figure 2: Distribution of *Hypericum perforatum* in Australia. Specimen data and map reproduced from Australia's Virtual Herbarium with permission of the Council of Heads of Australasian Herbaria Inc.

as CSIR and the Vic Department of Lands and Survey distributed over 10 million chrysomelid beetles in Vic (1940–51), NSW (1941–56), Tas (1943), SA (1943–49) and WA (1947–53) (Wilson 1960). By 1951 most redistributions were of *C. quadrigemina*; these have continued intermittently at lower levels until the present day. As well as internal redistribution, *Chrysolina* beetles were sent to NZ in 1943, and have significantly contributed to the control of the weed in some areas (Hancox *et al.* 1986). They were also sent to California from 1945–47, where *C. quadrigemina*, in particular, has been responsible for one of the classical biological control success stories in reducing St John's wort to subeconomic levels over 800 000 ha in 10 years (Huffaker and Kennett 1959).

Despite some substantial success in Australia, it was realised that *Chrysolina* spp. would not solve the St John's wort problem on their own. The cecidomyid leaf-bud galling fly *Zeuxidiplosis giardi* Kieff. was imported from France, released in 1953–54 and established readily (Wilson 1960).

A 20-year hiatus in research followed, until the continuing spread of St John's wort resulted in increased pressure from landholders for further work on biological control. A fourth phase of work, largely funded by industry, commenced in 1978. Research was based in the south of France at the CSIRO Montpellier laboratory, and was aimed at finding agents for timbered situations and areas with summer rainfall, and arthropods, including strains of *Chrysolina* spp., that were better synchronised with the climate of infested areas, able to damage the weed severely enough to prevent summer regrowth and to achieve control in unimproved pasture (Delfosse and Cullen 1981). Following further surveys in Mediterranean Europe, populations of both *Chrysolina* spp. were collected from regions in France with summer rainfall patterns and released at a number of sites in south-eastern Australia in 1980–81 (Delfosse and Cullen 1981). The defoliating moths *Actinotia hyperici* Schiff. and *Anaitis efformata* Guén. were also introduced at this time.

Meanwhile, studies undertaken in Australia indicated that the major factor mitigating against successful control lay more with the ability of the plant to survive defoliation than with the biology of the insects, and that the strong root reserves of established plants seemed the key to this survival. Remaining efforts were concentrated on agents that could exhaust those reserves, such as the previously studied root-borer *Agrilus hyperici*, and an aphid, *Aphis chloris* Koch, both of which were imported from France and released.

During this phase, Wapshere (1984) found that a previously unstudied agent, the eriophyid mite *Aculus hyperici*, was a major controlling agent in southern France, and was not inhibited by shade. This species was subsequently introduced and released.

A strain of the mycoherbicide *Colletotrichum gloeosporioides* (Penzig) Penzig & Saccardo, known to attack *H. perforatum*, was investigated in the 1990s by the Keith Turnbull Research Station, Vic (Shepherd 1995), but was not pursued.

Fifteen agents were studied over the 80-year course of the project; of these, 11 were released and six became established (Briese 1997b). Despite intensive efforts in 1978–96 to find complementary agents, *C. quadrigemina* remains the most effective control agent, providing temporal and localised relief from the problems caused by St John's wort.

EXPLORATION

Initial exploration, from 1928, was based in the UK, a decision almost certainly influenced by political considerations in the early days of CSIR as the then Chief of the Division of Economic Entomology had recommended Montpellier as a base. Following failure of almost all the species imported from England, exploration moved to the climatically more appropriate southern France in 1935, operations being conducted principally in south-eastern France by F Wilson, in the Var region. Based on this work, Wilson (1943) considered *Agrilus hyperici* and *C. quadrigemina* to be the two dominant insects on St John's wort stands in southern France, with the root-borer exerting the greatest impact and being associated with large declines in the density of local patches of the weed.

The resumption of exploration in 1978, this time based in Montpellier, resulted in the importation to Australia of populations of *C. quadrigemina* and *C. hyperici* from summer rainfall regions, and appreciation by Wapshere of the impact of the mite *Aculus hyperici*. Later work by Cullen and Valin (unpub.) demonstrated that *Agrilus hyperici* and *Aculus hyperici* were the two major factors in the decline of *H. perforatum* populations in this region, with *C. quadrigemina* being less important. The exploration program ended in 1996 with the cessation of industry funding.

CANDIDATES

Agents released in Australia

***Chrysolina quadrigemina* (Suffr.) and *C. hyperici* (Forst.) (Coleoptera: Chrysomelidae).** Adults of these two leaf-beetles are very similar in appearance and behaviour and occasionally occur in mixed species populations. *C. hyperici* adults are metallic bronze, oval shaped and 5–6 mm long. *C. quadrigemina* is slightly larger (6–7 mm) and beetles may be metallic blue, green, purple or brown. Both species are active in spring when they feed on flowering stems of their host *H. perforatum* (Fig. 3), followed by a summer aestivation. *C. quadrigemina* resumes activity and oviposition in autumn, several weeks earlier than *C. hyperici*. Females may lay more than 1000 eggs, singly or in small clusters, on the undersides of leaves or leaf buds of the rosettes in autumn. *C. quadrigemina* larvae emerge after about three weeks and overwinter as feeding larvae, whereas *C. hyperici* eggs remain dormant over winter. The emerging larvae are orange, turning pinkish-grey as they mature, and rest concealed in the litter beneath the plant during the day then emerge and feed on the procumbent rosette growth mainly at night. Mature larvae pupate in globular cells in the soil up to 5 cm deep. The pupal stage lasts two to three weeks and adults emerge in late spring. While larvae can defoliate the rosette growth during winter and early spring, adult beetles often form large feeding fronts that defoliate the erect spring plants (Fig. 3).

C. hyperici from England was established from releases made in 1930–34 (Wilson and Campbell 1943). Its greater cold tolerance enables it to contribute to biological control in cooler parts of the range of *H. perforatum* (Clark 1953a). However, *C. quadrigemina* has become the most widespread and effective of the *H. perforatum* biological control agents (Shepherd 1985; Huffaker 1966). This was largely because its earlier autumn activity and population development enabled it to outcompete *C. hyperici* in warmer areas (Clark 1953a). The populations of both species, re-collected from summer rainfall areas of France, were released in 1980–81 (Delfosse and Cullen 1981) to augment control in regions of Australia with high summer rainfall, but they could not be distinguished from the resident populations. As there has been no change in impact by the *Chrysolina* spp., it seems that they did not increase their effectiveness. Two other chrysomelid beetles, *C. varians* and *C. brunsviciensis*, introduced from England in the early 1930s, did not establish.

***Agrilus hyperici* (Creutzer) (Coleoptera: Buprestidae).** Adults of this beetle are elongate, flattened and 4–5 mm

Figure 3: Adults of *Chrysolina quadrigemina* on stem of *Hypericum perforatum*. Photo: CSIRO.

long. They are active in summer, feeding and mating on the foliage of *H. perforatum* plants. Females lay their eggs on the basal part of the flowering stem, protected by a faecal cap. Newly hatched larvae mine down the stem just below the surface, leaving a visible tunnel. The later instars, white and elongate with flat brown heads, tunnel down and feed on root tissue, leaving their frass compressed in the tunnel behind them. Eventually, they turn to tunnel back up and pupate in a chamber just below the crown until the following summer. Several larvae may feed in one root. Damage to the root system reduces shoot growth and may even kill the plant (Briese 1991a).

Agrilus hyperici was collected in southern France and released in Australia in 1939–40. Populations established and began to increase, but were either destroyed by bushfires or overrun by exploding populations of *Chrysolina* spp. Only one colony, restricted to an isolated creek system, survives (Briese 1991b). The efforts made to supplement this colony by further introductions from France in 1984 and 1989, with releases made in NSW and SA, resulted in only one small colony in Tuena, NSW that is known to have persisted.

***Aphis chloris* Koch (Hemiptera: Aphidae).** *Aphis chloris* is a small (1–2 mm), bright green aphid that feeds on the phloem of *H. perforatum*. Adults give birth to live young and both adults and nymphs cluster on the stems, leaves and flowers of the plants to feed. When conditions become crowded, winged females develop within the colony and disperse. Several generations of adults can be produced in a season (Briese 1988). Originally, a population from England was considered for introduction in the 1930s, but it produced overwintering eggs, which did not survive in Australia as cold requirements for hatching were not satisfied (Wilson 1938). In 1986–87, an asexually reproducing strain of *A. chloris* from southern France was released and established (Briese and Jupp 1995). As well as host-testing the aphids, their potential to transmit plant viruses was also studied to ensure that they did not pose a danger to economic plants (Briese 1989), the first study of this type for a biological control agent. Willis *et al.* (1993) found that feeding by *A. chloris* reduced root growth more than shoot growth. *A. chloris* is relatively widespread, but appears to go through seasonal cycles of growth and dispersal without populations building to damaging levels (Briese and Jupp 1995).

***Zeuxidiplosis giardi* (Kieffer) (Diptera: Cecidomyiidae).** The small adult *Z. giardi* (<3 mm) oviposits in the growing terminal and axillary buds of its host. Feeding by the newly hatched larvae induces a loose leafy gall (up to 10 mm diam.) within which the larva or larvae feed, develop and pupate (Wilson 1943). The galls can reduce the vigour and seeding of the plant. *Z. giardi*, originally from France, was shipped from California to Australia and released in 1953–54 (Wilson 1960). It established and is widespread, though not particularly common, and the high numbers required to cause significant damage have not been observed in the field.

***Aculus hyperici* (Liro) (Acari :Eriophyidae).** This is a minute vagrant eriophyid mite from Europe that feeds in the growing points of *H. perforatum*. Adults are cream with a soft, flattened unsegmented body, with two pairs of legs plus chelicerae and pedipalps for feeding (Fig. 4). Females are 0.15 mm long and 0.05 mm wide, and eggs are <0.05 mm in diameter. The nymphs are similar to the adults, but smaller, and there are two nymphal instars. Cast nymphal skins, left on the surface of the plant when they moult, may be the only visible evidence of a mite infestation. The mite has many generations per year and is wind-dispersed, and can rapidly build to damaging levels over a relatively short period. It feeds on the individual cells and the sap, producing no galls but causing shoot dwarfing. Over two to three years it gradually exhausts the root reserves, leading to progressive

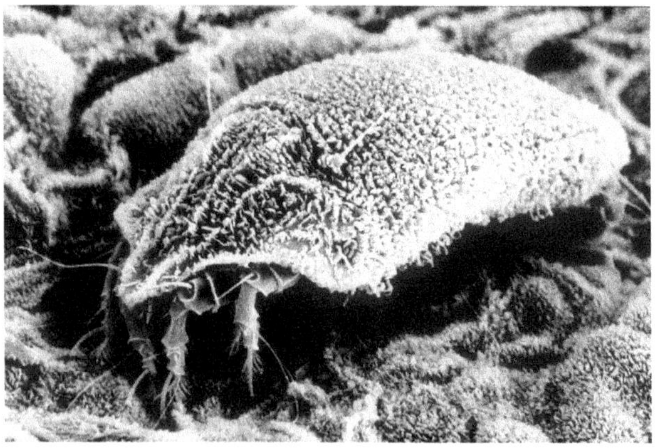

Figure 4: Adult of *Aculus hyperici*. Scanning EM photo: CSIRO.

dwarfing and eventual death of the whole plant (Wapshere 1984; Cullen and Valin, unpub.).

Aculus hyperici is difficult to rear due to its interaction with plant vigour and growth and the cryptic nature of early symptoms of feeding and reproduction. The specificity-testing program was therefore complex, particularly the testing of *Hypericum* spp., and took six years to complete (CSIRO 1991). The mite, from southern France, was first released in Australia in 1990 and is now well established following widespread redistribution (Jupp 1996; Mahr *et al.* 1997). While there is evidence of reduced plant vigour and density at local sites (Mahr *et al.* 1999), its impact has not yet been thoroughly evaluated and some genotypes of *H. perforatum* have been found to be resistant to mite attack (Jupp *et al.* 1997; Mayo and Roush 1997). Despite the damage observed, the impact of *Aculus hyperici* on *H. perforatum* in Australia is less than that observed in France and there may be strains better suited to Australian populations of *H. perforatum*.

***Aplocera efformata* Guenèe (Lepidoptera: Geometridae).** *Aplocera efformata* and the congeneric *A. plagiata* L. are both light grey, bar-winged geometrid moths, with a wingspan up to 4 cm. Females lay several hundred eggs on the leaves of *H. perforatum* and the brown-patterned larvae feed on the foliage before descending to pupate in the soil. They can produce three generations per year (Briese 1986). They were released in Australia in 1936–38 but did not establish, ostensibly due to predation by ants (Currie and Fyfe 1938). In 1981–83, a second unsuccessful attempt was made to release *A. efformata*. This time, it was shown that the exposed larvae and eggs were subject to parasitism and predation by a range of natural enemies, effectively preventing establishment (Briese 1986).

***Actinotia hyperici* Schiff. (Lepidoptera: Noctuidae).** The larvae of *A. hyperici* also feed on the leaves of *H. perforatum*. It was first studied in France during the 1930s, but not introduced. An attempt to establish it in Australia in 1985–86 was unsuccessful (Briese 1997b).

Agents not released

Three other moth species, the gelechiid stem-boring moth *Aristotelia morphochroma* Wals. (Wilson 1943) and two leaf-webbing moths, the oecophorid *Depressaria hypericella* Hubn. and the eucosmid *Lathronympha hypericana* Hubn., were studied, but never released in Australia. The fungus *C. gloeosporioides* was not pursued because of doubts about its host-specificity (McLaren *et al.* 1997).

EVALUATION

Globally, *H. perforatum* has been used as an example of measuring success in biological control (Syrett *et al.* 2000), and evaluation has played a role in shaping the biological control program in Australia from its inception. The shift from England to France for agent exploration in 1935 was the consequence of the belief that climate-matching between native and introduced ranges could be critical to agent success.

Detailed field experiments and observations by the Clarks in the early 1950s and a broader survey by Huffaker (1966) explained the partial control of *H. perforatum* due to *Chrysolina* spp. These beetles controlled the weed in areas with Mediterranean-type climates (Huffaker 1966), such as WA and SA, and in areas of eastern Australia where *H. perforatum* was replaced by competitive pasture species following its reduction by the beetles (Clark 1953a). However, in many areas of eastern Australia this success was limited by several factors. While defoliation by the beetles could kill deep-rooted plants on good soils (Clark 1953b), plants on shallow or stony soils responded to defoliation by producing new shoots vegetatively from lateral roots (Clark and Clark 1952). The boom–bust population cycles of the beetles also provided periods of respite from herbivore pressure (Clark 1953a), which was compounded by occasional asynchrony between the onset of beetle activity after aestivation and growth of the weed (Huffaker 1966). In areas with high summer rainfall, St John's wort infestations were able to recover significantly from winter and spring defoliation by the beetles (Huffaker 1966). Finally, *Chrysolina* spp. were not tolerant of shaded situations and could not defoliate infestations of the weed in timbered areas (Clark 1953a).

Later broad-scale surveys by Shepherd (1984) and Briese (1985) clarified the current status and impact of

Chrysolina spp. These studies showed that, contrary to conventional wisdom, beetles could develop large populations in areas of summer rainfall. However, rainfall after defoliation led to recovery of damaged plants from strong root reserves. This led to a change in strategy to concentrate on agents that could weaken plant roots, either directly, such as the root-borer *Agrilus hyperici*, or indirectly, such as the aphid *Aphis chloris* and the mite *Aculus hyperici*.

Evaluation studies also showed that native natural enemies were detrimental to the establishment of agents with exposed life-stages, such as the larvae of defoliating moths. A field experiment identified generalist predators, parasitoids and pathogens as the primary cause of failure of the moth *Aplocera efformata* (Guenée) to establish in Australia (Briese 1986). Parasitoids also seem to have limited the effectiveness of the gall midge *Z. giardi* in many regions. Other evaluation studies demonstrated how the life-history strategy of agents could affect establishment success, showing how K-selected *Agrilus hyperici* were outcompeted by r-selected chrysomelid beetles at a critical stage in establishment and population growth (Briese 1991a).

In recent years, evaluation extended to examining the risks posed by agents introduced for *H. perforatum* to non-target native *Hypericum* species. Using *Aculus hyperici* as an example, Willis *et al.* (2003) found that, while feeding did occur, there was little risk of damage to non-target plant populations.

Despite the fact that success has been only partial, Page and Lacey (2006) reported an annual benefit of $20.6 million through reduced stock health problems and poisoning, reduced photosensitisation and increased pasture productivity and carrying capacity. They also indicated environmental benefits due to reduced fire hazard risk and decreased incidence of contact rash in humans. The basis for their calculation, however, is not clear and the benefit value should be viewed as an estimate only.

DISCUSSION

As discussed by Briese (1997b), the biological control program against *H. perforatum* in Australia illustrates how the discipline itself has evolved in the past 70 years. Ongoing evaluation of exploration and releases has led to several changes in rationale and priorities for agent introductions. The failures of initial agent introductions from England confirmed the need for matching climates as closely as possible and led to a shift to more Mediterranean-type climates in France. This provided some success, but showed further weaknesses in our understanding of the system. The need to understand agent behaviour became apparent with the demonstrated intolerance of *Chrysolina* beetles to shaded situations. Complexities of interactions between agent biology and weather patterns and competitive interactions between agents themselves were found to have an influence. The overriding importance of understanding the ecology of the target weed, particularly with respect to differences between the introduced and native ranges, was realised. Such assessments have led to a continual refocusing of the characteristics required for an agent to successfully contribute to control of the weed under Australian conditions.

Practically, the program has mirrored the evolution in the way biological control technology is transferred to the end-user; from massive collection and redistribution efforts on selected properties by extension officers to the establishment of community-based networks, where the responsibility for rearing and a level of monitoring resides with the landholder. There has also been a philosophical shift in our view of the role of biological control, away from the idea that natural enemies can by themselves control the weed to the idea that introduced agents form part of a range of control options that need to be integrated in order to develop effective weed management strategies (Briese 1996; Campbell 1997). Modelling activities such as those by Buckley *et al.* (2003), using data from the *H. perforatum* project, could provide a useful tool for this approach.

Finally, Briese (1997b) considered that any need for a further phase of work on the biological control of *H. perforatum* will depend on whether *Aculus hyperici* lives up to its early promise for control of the weed. Unfortunately, there has been no recent evaluation of the impact of this agent and the possible effect of variation in host–mite interaction. It is clear, however, that any future work will need to be directed at a more integrated approach to weed management, building on the impact of biological control agents through the use of pasture management and chemical control techniques that promote rather than interrupt agent activities (Cullen *et al.* 1997).

REFERENCES

Bourke CA (1997) Effects of *Hypericum* spp. (St John's wort) on animal health and production. *Plant Protection Quarterly* **12**, 91–93.

Briese DT (1985) A survey to evaluate the long-term relationship between *Chrysolina quadrigemina* and its host weed, St John's wort, in south-eastern Australia. In *Proceedings of the 6th International Symposium on Biological Control of Weeds.* 19-25 August 1984, Vancouver. (Ed ES Delfosse) pp. 691–708. Agriculture Canada, Ottawa.

Briese DT (1986) Factors affecting the establishment and survival of *Anaitis efformata*, introduced into Australia for the biological control of St John's wort: II. Field trials. *Journal of Applied Ecology* **23**, 821–839.

Briese DT (1988) Bionomics of *Aphis chloris* Koch (Hemiptera: Aphididae), a biological control agent for St John's wort in Australia. *Ecological Entomology* **13**, 365–374.

Briese DT (1989) Host-specificity and virus-vector potential of *Aphis chloris* Koch (Hemiptera: Aphididae), a biological control agent for St John's wort in Australia. *Entomophaga* **34**, 75–92.

Briese DT (1991a) Demographic performance in the root-borer, *Agrilus hyperici* (Coleoptera: Buprestidae), a biological control agent of St John's wort. *Biocontrol Science and Technology* **1**, 195–206.

Briese DT (1991b) Current status of *Agrilus hyperici* (Coleoptera: Buprestidae) released in Australia in 1940 for the biological control of St John's wort: lessons for insect introductions. *Biocontrol Science and Technology* **1**, 207–215.

Briese DT (1996) Biological control of weeds and fire management in protected natural areas: are they compatible strategies? *Biological Conservation* **77**, 135–141.

Briese DT (1997a) Population dynamics of St John's wort in southeastern Australia. *Plant Protection Quarterly* **12**, 59–63.

Briese DT (1997b) Biological control of St John's wort in Australia: past, present and future. *Plant Protection Quarterly* **12**, 73–80.

Briese DT and Jupp PW (1995) Establishment and spread of *Aphis chloris* Koch (Hemiptera: Aphididae), introduced into Australia for the biological control of St John's wort. *Biocontrol Science and Technology* **5**, 271–285.

Buckley YM, Briese DT and Rees M (2003) Demography and management of the invasive plant species *Hypericum perforatum*: 2. Construction and use of an individual-based model to predict invasion dynamics and the effects of management strategies. *Journal of Applied Ecology* **40**, 494–507.

Campbell MH (1997) Control of *Hypericum perforatum* L. (St John's wort) by grazing management. *Plant Protection Quarterly* **12**, 97–99.

Campbell MH, Briese DT and Delfosse ES (1995) *Hypericum perforatum* L. In *The Biology of Australian Weeds.* Vol. 1. (Eds RH Groves, RCH Shepherd and RG Richardson) pp. 149–168. Richardson Publishing, Melbourne.

Clark LR (1953a) The ecology of *Chrysomela gemellata* Rossi and *C. hyperici* Forst., and their effect on St John's wort in the Bright District, Victoria. *Australian Journal of Zoology* **1**, 1–69.

Clark N (1953b) The biology of *Hypericum perforatum* L. var. *angustifolium* DC. (St John's wort) in the Ovens Valley, Victoria, with particular reference to entomological control. *Australian Journal of Botany* **1**, 95–120.

Clark LR and Clark N (1952) A study of the effect of *Chrysomela hyperici* Forst. on St John's wort in the Mannus Valley, NSW. *Australian Journal of Agricultural Research* **3**, 29–59.

Cullen JM, Briese DT and Groves RH (1997) Towards the integration of control methods for St John's wort: workshop summary and recommendations. *Plant Protection Quarterly* **12**, 103–106.

CSIRO (1991) 'The host-specificity of *Aculus hyperici* (Liro) (Acarina: Eriophyidae) in relation to different species in the genus *Hypericum*'. Unpublished report. CSIRO Division of Entomology, Canberra.

Currie GA and Fyfe RV (1938) The fate of certain European insects introduced to Australia for the control of weeds. *Journal of the Council for Scientific and Industrial Research* **11**, 289–301.

Currie GA and Garthside S (1932) 'The possibility of the entomological control of St John's wort in Australia. Progress report'. Pamphlet No. 29. Council for Scientific and Industrial Research, Melbourne.

Delfosse ES and Cullen JM (1981) New activities in the biological control of weeds in Australia. III. St John's wort, *Hypericum perforatum.* In *Proceedings of the 5th International Symposium on Biological Control of Weeds.* 22–29 July 1980, Brisbane. (Ed ES Delfosse) pp. 575–581. CSIRO, Melbourne.

Hancox NG, Syrett P and Scott RR (1986) Biological control of St John's wort (*Hypericum perforatum*) in New Zealand: a review. *Plant Protection Quarterly* **1**, 152–155.

Harris JA and Gill AM (1997) History of the introduction and spread of St John's wort (*Hypericum perforatum* L.) in Australia. *Plant Protection Quarterly* **12**, 52–56.

Huffaker CB (1966) A comparison of the status of biological control of St John's wort in California and Australia. *Mushi* **29** (supplement), 51–73.

Huffaker CB and Kennett CE (1959) A ten-year study of vegetational changes associated with biological control of Klamath weed. *Journal of Range Management* **12**, 69-82.

Jupp PW (1996) The establishment of a distribution network for the mite, *Aculus hyperici*, to control St John's wort, *Hypericum perforatum*, in Australia. In *Proceedings of the 9th International Symposium on Biological Control of Weeds*. 19–26 January 1996, Stellenbosch, South Africa. (Eds VC Moran and JH Hoffman) pp. 451–454. University of Cape Town, South Africa.

Jupp PW, Briese DT and Cullen JM (1997) Evidence for resistance of St John's wort to a biological control agent, the mite *Aculus hyperici*. *Plant Protection Quarterly* **12**, 67–70.

Mahr FA, Kwong RM, McLaren DA and Jupp PW (1997) Redistribution and present status of the mite, *Aculus hyperici*. *Plant Protection Quarterly* **12**, 84–88.

Mahr FA, Mayo G, Ainsworth N and Jupp PW (1999) Monitoring the impact of the biological control agent *Aculus hyperici* on *Hypericum perforatum* across south eastern Australia. In *Proceedings of the 12th Australian Weeds Conference*. 12–16 September 1999, Hobart. (Eds AC Bishop, M Boersma and CD Barnes) pp. 335–338. Tasmanian Weed Society, Hobart.

Mayo GM (2004) Genetic variation in *Hypericum perforatum* L. and resistance to the biological control agent, *Aculus hyperici* (Liro). PhD thesis. University of Adelaide, Australia.

Mayo GM and Roush RT (1997) Genetic variability in *Hypericum perforatum* L. (Clusiaceae) and the detection of resistance to the biological control agent *Aculus hyperici* (Liro) (Eriophyidae). *Plant Protection Quarterly* **12**, 70–72.

McLaren DA, Bruzzese E and Pascoe IG (1997) The potential of fungal pathogens to control *Hypericum* species in Australia. *Plant Protection Quarterly* **12**, 81–83.

Page A and Lacey K (2006) 'Economic impact assessment for Australian weed biological control'. CRC for Australian Weed Management, Technical Series No. 10, CRC Weed Management, Adelaide.

Shepherd RCH (1984) The present status of St John's wort (*Hypericum perforatum* L.) and its biological control agents in Victoria, Australia. *Agriculture, Ecosystems and Environment* **12**, 141–149.

Shepherd RCH (1995) A Canadian isolate of *Colletotrichum gloeosporioides* (Penzig) Penzig & Saccardo as a potential biological control agent for St Johns wort (*Hypericum perforatum* L.) in Australia. *Plant Protection Quarterly* **10**, 148–151.

Syrett P, Briese DT and Hoffmann J (2000) Success in biological control of terrestrial weeds by arthropods. In *Measures of Success in Biological Control*. (Eds SD Wratten and G Gurr) pp. 189–230. Kluwer, Amsterdam.

Wapshere AJ (1984) Recent work in the biological control of *Hypericum perforatum* (Guttiferae) for Australia. *Entomophaga* **29**, 145–156.

Willis AJ, Ash JE and Groves RH (1993) Combined effects of two arthropod herbivores and water stress on growth of *Hypericum* species. *Oecologia* **96**, 517–525.

Willis AJ, Berentson PR and Ash JE (2003) Impacts of a weed biological control agent on recovery from water stress in a target and a non-target *Hypericum* species. *Journal of Applied Ecology* **40**, 320–333.

Wilson F (1938) Some experiments on the influence of environment upon the forms of *Aphis chloris* Koch (Aphididae). *Transactions of the Royal Entomological Society of London* **87**, 165–181.

Wilson F (1943) The entomological control of St John's wort (*Hypericum perforatum* L.) with particular reference to the insect enemies of the weed in southern France. *Bulletin of Council for Scientific and Industrial Research* No. 169, 1–87.

Wilson F (1960) A review of the biological control of insects and weeds in Australia and Australian New Guinea. *Commonwealth Institute of Biological Control Technical Communication* No. 1, 1–102.

Wilson F and Campbell TG (1943) Recent progress in the entomological control of St John's wort. *Journal of the Council for Scientific and Industrial Research* **16**, 45–56.

Hyptis suaveolens (L.) Poit. – hyptis

Mic Julien, Gio Fichera and Ricardo Segura

ABSTRACT

A project on *Hyptis suaveolens*, that lacked continuity and had low support, began in 1979 and was terminated in 2003. Surveys, mostly piggy-backed on other projects, were conducted in Brazil, Venezuela and Mexico and to a lesser extent in Costa Rica, Cuba and Nicaragua. A number of insects and a nematode were found to be insufficiently specific during native range observations, in native range host-range tests and in quarantine studies in Australia. Several insects imported into an Australian quarantine failed to survive on Australian hyptis, suggesting that genotype variation in hyptis may be critical for natural enemy–host plant matching. A range of insect natural enemies are known, including a rust, *Puccinia hyptidis-mutabilis*, that warrant further studies.

Key words: surveys, plant variation, natural enemy–plant genotype matching, potential biocontrol agents.

INTRODUCTION

Hyptis suaveolens (L.) Poit., Lamiaceae (hyptis) (Fig. 1) originates from tropical America from the West Indies south to about the Tropic of Capricorn. It is prevalent in Mexico, but this may not be part of the native range since it was not collected there by Standley (1924) when he recorded other *Hyptis* species. Hyptis was first recorded in Australia at Port Essington, near Darwin, NT in 1845 by Ludwig Leichhardt and it is now widespread in northern Australia (Parsons and Cuthbertson 2001; Wilson 1997). It occurs in north-eastern Qld and in the East Kimberley, West Kimberley and Broome districts of northern WA (Fig. 2). Hyptis is most common along roadsides (Fig. 3), banks and on disturbed areas such as cultivated and overgrazed areas, stockyards and watering points.

Figure 1: *Hyptis suaveolens* in Australia. Photo: C Wilson.

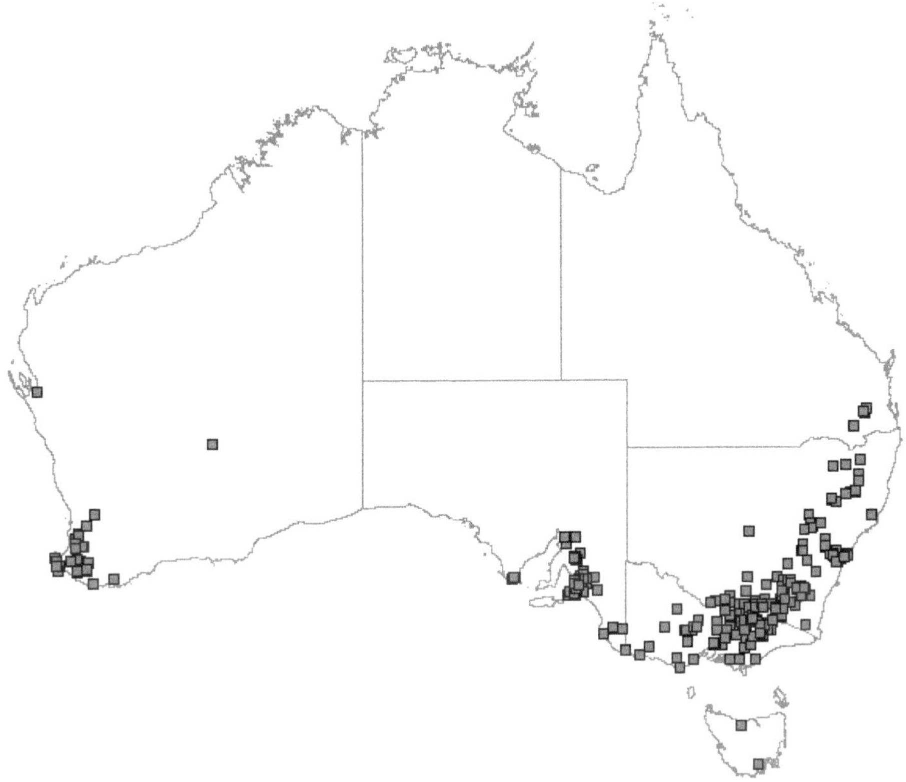

Figure 2: Distribution of *Hyptis suaveolens* in Australia. Specimen data reproduced from Australia's Virtual Herbarium with permission of the Council of Heads of Australasian Herbaria Inc.

Hyptis is an aromatic erect annual, though in favourable conditions it may overwinter and support new season's growth from the base of the plant as a short-lived perennial. Seeds germinate with the onset of the wet season. Flowering occurs in February to May and plants die in the winter (dry season).

The spiny burr is spread by animals, humans and water, in mud on vehicles and machinery, and as a contaminant of hay. The plant is not palatable to stock, thus overgrazing leads to non-productive pastures as hyptis becomes more abundant. It competes for space and nutrients in grain crops and peanuts (Parsons and Cuthbertson 2001). Extracts of hyptis are used worldwide to treat a wide range of ailments and as an insecticide and miticide. The toxic chemicals it contains are probably why, despite its relatively long presence in Australia, only six external feeding herbivores (mostly polyphages) have moved onto hyptis (Wilson 1997).

Figure 3: *Hyptis suaveolens*, NT. Photo: B Lukitsch, NT Government.

BIOLOGICAL CONTROL HISTORY

No other countries have undertaken biological control research on *Hyptis* species.

PLANT TAXONOMY

Plant matching between native and Australian hyptis populations is an issue for surveys and selection of agents. *H. suaveolens* has diverse morphology in the native range and variable chemical composition (Azevedo *et al.* 2002) and may comprise diverse genetic forms. Plant specimens of populations from Venezuela and Australia were deemed by specialists at the Qld Herbarium and Royal Botanic Gardens Herbarium, Kew to be within the parameters that define *H. suaveolens*. They commented that there was great variation within the species and that revision of the taxa is warranted. These plant population differences may be critical for some herbivores, as some insect species from South American hyptis failed to survive when offered Australian hyptis in quarantine conditions.

Preliminary ***random amplification of polymorphic DNA (RAPD)*** analyses of samples from four hyptis populations (three from Venezuela, including populations morphologically similar and dissimilar to Australian hyptis, and Australian hyptis from Mary River, NT) failed to link South American and Australian material. Analyses using a much more comprehensive plant collection are required to determine the origin of the Australian populations and the level of variation within Australia.

EXPLORATION

On behalf of the NT Government, CSIRO initiated research on biological control of hyptis with surveys by John Winder (CSIRO lab, Curitiba, Brazil) for natural enemies in 1979 and 1982 in Paraná, Sáo Paulo, Minas Gerais, southern Goiás and southern Bahia states in Brazil. These areas had climates dissimilar to that of the NT and yielded few potential agents. Consequently brief surveys were conducted in 1981 in Apure state, Venezuela and Guerrero state, Mexico. The Curitiba laboratory closed in 1982. A laboratory at Acapulco, Mexico operated from 1984 to 1986 and during that time John Gillett (CSIRO) surveyed hyptis in the vicinity of Acapulco and in Costa Rica. In 1986 the laboratory was moved to Veracruz, Mexico from where Ricardo Segura (CSIRO) surveyed hyptis, mostly in conjunction with other projects. Regions surveyed between 1985 and 2003 included Veracruz, Guerrero, Quintana Roo, Oaxaca, Jalisco and Morelos states of Mexico; Carabobo, Cojedes, Guárico, Anzoátegui, Bolivar, Barinas and Apure states of Venezuela; Punta Arena state of Costa Rica; Matanzas state of Cuba; and Managua and León states of Nicaragua. At various times other CSIRO staff, including Ken Harley and Wendy Forno in 1981 and 1990 and Mic Julien in 2001, accompanied Winder, Gillett or Segura.

CANDIDATES

In the 1980s four insect species were studied in Mexico and found not to be specific (Julien 2002). They were the two leaf-feeding chrysomelid beetles *Metriona* sp. and *Neocharida* sp. (referred to as *Neocharida bifida* (manuscript name) in CSIRO correspondence dated 1981) collected from Brazil, the agromyzid fly *Calycomyza hyptidis* Spencer (no collection record but presumably from Mexico or Brazil) and the pyralid moth *Pyrausta panopealis* (Walker) collected near Acapulco, Mexico. The identity of the moth used in the host-range tests remains uncertain and there are a number of species of *Pyrausta* moths that utilise hyptis, so further work is warranted. In 1987 a species of root-knot nematode *Meloidogyne* sp. (Nematoda: Heteroderidae) was collected and studied in Veracruz, Mexico. It also was found to be insufficiently specific (Julien 2002).

***Coelocephalapion hyptidis* Kissinger (Coleoptera: Apionidae).** This small black weevil, found in Venezuela in 2000 by Segura, was described by Kissinger (2003). A shipment of 100 field-collected adults was sent to CSIRO Brisbane in November 2000. By February 2001, 300 adult progeny reared on Australian hyptis were in host tests. Feeding and development occurred on plants closely related to hyptis including *Orthosiphon aristatus* (Blume) Miq. (cat's whiskers), *Lavandula pinnata* L. f. (lavender), *Rosmarinus officinalis* L. (rosemary), *Salvia chamaedryoides* Cav. and *Melissa officinalis* L. Feeding but no development occurred on *Hyptis capitata* Jacq. and *Nepeta cataria* L. The weevil colony was destroyed in quarantine.

Seed-feeding beetles (Coleoptera: Bruchidae). *Meibomeus* sp., *Acanthoscelides ramirezi* Johnson and *Sennius rufomaculatus* (Motschulsky) are inflorescence-feeding beetles from Venezuela whose larvae feed in the developing seed of hyptis. They are found in the field from June to March with peak abundance in the peak fruiting period, November to December. *Meibomeus* sp. was abundant in the field, the others much less so. All were imported into Australia but failed to survive on Australian hyptis. While two consignments of *Meibomeus* sp. totalled 771 adults, *A. ramirez* and *S. rufomaculatus* were imported in very low numbers (five and three adults, respectively) and this likely contributed to their failure. Adults (139) of another small brown beetle, *Paratenetus* sp. (Coleoptera: Tenebrionidae), were inadvertently collected and imported with the *Meibomeus* sp.; this species also failed to survive on Australian hyptis. It is not known if it utilises hyptis as a host plant in Venezuela or utilises only the pollen, as a food source.

***Craptus undulatus* (Say) (Coleoptera: Curculionidae).** This stem-galling weevil (Fig. 4) has a life-cycle of about 40 days at field sites near Veracruz and three generations per year, beginning in April. It was collected from sites near Veracruz, Veracruz state, Mexico in

Figure 4: The potential biological control agent *Craptus undulates*. Photo: R Segura, CSIRO.

Figure 5: The potential biological control agent *Omophoita quadrinotata*. Photo: R Segura, CSIRO.

November 2002 and 58 adults and approximately 100 larvae and pupae in stems were imported into Australia.

The imported adults fed on Australian hyptis during their first week in quarantine. Feeding occurred near branch axils and caused swellings, but not like the galls seen in Mexico. None of the swellings contained larvae. Thereafter adults became moribund with little further interest in the plant and many secreted anally a white substance. This secretion was crystalline and did not contain microorganisms. The insect may have been ridding themselves of what was, to them, foreign substances from an unsuitable host plant.

Imported larvae extracted from imported stems accepted diet containing Australian hyptis. They pupated and, after an extended period, emerged as apparently healthy adults but failed to oviposit. Adults that emerged from imported plant stems and were placed on Australian hyptis failed to thrive or oviposit. Later dissections confirmed that both sexes were present.

When moribund adults were placed on hyptis diet they recovered and appeared healthy but when returned to Australian hyptis they quickly became moribund. The diet included hyptis material that had been dried for two days at 70°C. This would have reduced the presence of volatiles and may have denatured oils and chemicals of this aromatic species, thus allowing adults to feed and recover. However, it remained an unsuitable developmental host.

Ten adults were caged for six days onto a four-node stem section of *O. aristatus*, a critical test plant species. Damage and swelling occurred at two nodes but no eggs or larvae were found. Research terminated in April 2003 when the last imported insects died.

***Omophoita quadrinotata* (Fabricius) (Coleoptera: Chrysomeldiae).** This leaf-feeding beetle (Fig. 5) was collected in Venezuela in 1997. Saunders *et al.* (1983) indicated that this insect is an economically important pest of *Zea mays* L. (Poaceae) maize, *Ipomea batatas* (L.) Lam. (Convolvulaceae) sweet potato, *Manihot esculenta* Crantz (Euphorbiaceae) cassava and *Phaseolus vulgaris* L. (Fabaceae) kidney bean, in Costa Rica. Furth and Savini (1996) listed two subspecies in Central America, *Omophoita quadrinotata* subsp. *costaricensis* Bechyné and *O. quadrinotata* subsp. *centralamericana* Bechyné. Preliminary host-range studies are warranted as the insects from hyptis may not be the form recorded as a pest.

***Carmenta lytaea* Druce and *Carmenta* sp. (Lepidoptera: Sesiidae).** *C. lytaea* is a stem-boring moth (Fig. 6) originally collected from Zapote, Guatemala by Druce in 1884 (Duckworth and Eichlin 1978). The only other collections were by Segura from hyptis at Loma Zomorano, near Veracruz in 2001, and Cerro Azul, central east coast, Mexico in 2002. Adults reared from larvae collected from these sites were identified as *C. lytaea* from both sites and as *Carmenta* sp. from Cerro Azul.

Field observations suggest that the young larvae burrow into the softer plant tissue, gradually working downwards into the thicker stems where they tunnel up and down for 5 cm or more. Mature larvae pupate within the stems and possibly the roots, emerging as adults from September to January. Larvae have been found only during the wetter months from July to November. These *Carmenta* species are likely to have only one generation per year, with larvae surviving through the dry season in the plant stem or larger roots. No host-range studies have been conducted.

Figure 6: The potential biological control agent *Carmenta lytaea*. Photo: R Segura, CSIRO.

***Pyrausta* spp. (Lepidoptera: Pyralidae).** In the 1980s the moth *Pyrausta* was collected near Acapulco, Guerrero state, Mexico and identified as *Pyrausta panopealis* (Walker) though there is no record of who identified it and no preserved specimens (CSIRO unpub. report). This insect failed preliminary host-range tests when it fed and developed on *Mentha* x *piperita* L. Lamiaceae (mint) but not on *Origanum vulgare* L. (Lamiaceae) oregano. Another collection was made and identified as *Pyrausta phoenicealis* Hübner. These two species names may be synonyms (Monroe 1976). It is now known that several *Pyrausta* moths utilise hyptis, including *P. panopealis, P. rhodochroia* Dyar collected from Acapulco, México, *P. glirialis* (Herrich-Schaffer) Botys and *P. onythesalis* (Walker), *P. insignatalis* Guenee and *P.* sp.nr *haemapastalis* Hampson collected from Veracruz, Mexico and *P.* sp. probably *rubuscentalis* Hampson from Bolivar state, Venezuela. One of these, *P. panopealis*, occurs in Australia (Wilson 1997). Studies are required to distinguish these species and to identify any that may be specific to hyptis.

***Puccinia hyptidis-mutabilis* Major (Uredinales: Pucciniaceae).** This fungus was collected in Paso del Toro near Boca del Río, Veracruz in 1987 by R Segura and in Yucatan, Mexico in November 1988 by R Segura and Harry Evans (CABI), but it has not been studied.

Coleoptera: Curculionidae. A number of weevils collected from hyptis have been identified. Those that may have potential include *Tyloderma hustachei* Wibmer and *T. pilosellum* (Chev.), both from San Pablo, Apure, Venezuela (collected in 1990), *Apion* sp. (1996) and *Baris* spp. (2000) from El Tigre and *Rhyssomatus* nr *rugulipennis* Ch. (2001) from the Soledad–Carapa road, all in Anzoategui state, Venezuela. Further possible agents include *Apion* sp. (1996) from San Pedro Las Playas, Guerrero state, *Conotrachelus* nr *carolinensis* Schoof. (2001) from Loma Zamorano, Veracruz state and *Lechriops* sp. (2000) from Santa Fe Villarin, Veracruz state, all in Mexico.

Other agents

A moth, *Synanthedon* sp. *nov.* (Sessiidae) was collected from hyptis at Cerro Azul, Veracruz state, Mexico in 2003, a beetle *Epitragus* sp. (Tenebrionidae) near Mantecal, Apure state, Venezuela (2000) and a sap-sucking *Spalacopsis* sp. (Cicadellidae) near Acapulco, Guerrero state, Mexico (2001). Stem-boring beetles (Cerambycidae), *Dorcasta dasycera* (Erichson) were found in hyptis in Anzoategui state during 2000, and *Zepherina* sp. *nov.* in Bolivar state, Venezuela (2001). Unidentified insects collected from hyptis include a pyralid moth, a chrysomelid beetle in the genus *Colaspis*, several other chrysomelid beetles and a bug in the family Tingidae.

In addition, three unidentified stem-borers (Coleoptera: Cerambycidae), found in hyptis in Veracruz (1990), Guerrero state (2000 and 2001) and Jalisco state (2000), in Mexico, were all generalist herbivores, and 17 other insects, identified at least to genus, were recorded as pests, crop disease vectors or generalist feeders. These bring the total natural enemies collected and considered to be using hyptis to approximately 57.

DISCUSSION

The hyptis project has had a long but disjointed duration. It was initially included in an NT Government biological control of weeds funding package arrangement with CSIRO, but later became a lower-priority weed for biological control for the NT and over the years support lacked continuity. Research stopped in 2003 with priority changes by the NT Government.

Should a new hyptis project begin, it would be imperative to assess the variation of hyptis populations in Australia and match them genetically and possibly chemically with populations in the native range. If there is genetic variation in Australia, agents that previously failed on hyptis originating from the Mary River, NT may accept other Australian genotypes. Further native range surveys should be made on populations linked genetically with Australian populations.

The well recorded and studied aromatic, fungicidal and insecticidal properties of hyptis (Azevedo *et al.* 2002) seem to have limited the evolution of a wide diversity in natural enemies. However, a range of insects and a fungus warrant further assessment as potential biological control agents.

ACKNOWLEDGEMENTS

We thank Moisés Martínez and Manuel Juárez for their contributions in the field and laboratory. Our appreciation is extended to taxonomists Carlos Sosa, Harry Evans, Robert Anderson, Alma Solis, Miguel Monné, Shawn Clark, Felipe Noguera, Victor Toledo, Alejandro González, Thomas Eichlin, Natalia Vandenberg, Stuart McKamey, Thomas Hernry, Eric Grissell, John Kingsolver, David Furth and Steven Lingafelter and the Systematic Entomology Laboratory, USDA-ARS for their valuable contribution in the identification of specimens.

REFERENCES

Azevedo NR, Campos IFP, Ferreira HD, Portes TA, Seraphin JC, Realin de Paula J, Santos SC and Ferri PH (2002) Essential oil chemotypes in *Hyptis suaveolens* from Brazilian Cerrado. *Biochemical Systematics and Ecology* **30**, 205–216.

Duckworth WD and Eichlin TD (1978) The type-material of Central and South American clearwing moths (Lepidoptera: Sesiidae). *Smithsonian Contributions to Zoology* **261**, 1–28.

Furth DG and Savini V (1996) Checklist of the Alticinae of Central America, including Mexico (Coleoptera: Chrysolmelidae). *Insecta Mundi* **10**(1–4), March–December.

Julien M (2002) Biological control of tropical weeds with Central and South American origin: current activities by CSIRO Entomology. In *Proceedings of the 13th Australian Weeds Conference.* 8–13 September 2002, Perth. (Eds H Spafford Jacob, J Dodd and JH Moore) pp. 361–365. Plant Protection Society of Western Australia, Perth.

Kissinger DG (2003) A new species of *Coelocephalapion* Wagner (Apionidae) from Venezuela with host *Hyptis suaveolens* Poit. (Lamiaceae). *Coleopterists Bulletin* **57**, 99–104.

Monroe E (1976) Pyraloidea. In *The Moths of America North of Mexico.* Fascicle 13.2B. pp. 111–112. EW Classey, London.

Parsons WT and Cuthbertson EG (2001) *Noxious Weeds of Australia.* 2nd edn. CSIRO Publishing, Melbourne.

Saunders JL, King ABS and Vargas CLS (1983) Plagas de cultivos en América Central – una lista de referencia. Boletín Técnico No. 9. Centro AgronómicoTropical de Investigaciones y Enseñanza, CATIE. Departamento de Producción Vegetal. Turrialba, Costa Rica.

Standley PC (1924) Trees and shrubs of Mexico. *Contributions from the United States Herbarium* **23**, 1273–1276.

Wilson CG (1997) Phytophagous insect fauna of two weeds, *Hyptis suaveolens* (L.) Poit. and *Jatropha gossypifolia* L., in Australia's Northern Territory. *Australian Entomologist* **24**, 55–60.

Jacobaea vulgaris Gaertn. – ragwort

JE Ireson and David McLaren

ABSTRACT

Ragwort, *Jacobaea vulgaris*, is a weed of major economic significance in Vic and Tas due to its detrimental effects on agricultural production and toxicity to livestock. Annual production losses have been conservatively estimated at well over $4 million per year. Biological control programs have resulted in the release of seven agents since 1930. Although five agents have established, the cinnabar moth *Tyria jacobaeae* is confined to only one site in Vic, where its impact on ragwort is not significant, and it has failed to establish in Tas. Two species of flea-beetle, *Longitarsus flavicornis* and *Longitarsus jacobaeae*, have established in both Vic and Tas. In Tas, the root-feeding *L. flavicornis* has become the dominant flea-beetle and is now widespread on ragwort throughout the state. It has had a major impact and is capable of reducing infestations by up to 95% at some sites, resulting in multi-million dollar savings to the dairy and beef industries through increases in production. Biological control now ranges from substantial to complete in many parts of Tas. However, on some properties unfavourable site conditions and incompatible management strategies have restricted its impact. In Vic, *L. flavicornis* has established only in high-rainfall locations above 500 m and has spread slowly. It has not had a significant impact, nor has the less common *L. jacobaeae*. Two other agents, the ragwort stem- and crown-boring moth *Cochylis atricapitana* and the ragwort plume moth *Platyptilia isodactyla* are now well established in Vic and Tas and are continuing to spread naturally. Both species have been observed causing considerable damage to crowns and stems at field sites in these states, with trials in Vic indicating that they are capable of significantly reducing plant vigour and reproductive output. In Tas, these species are also expected to provide significant control, particularly in areas where *L. flavicornis* impact is restricted. Studies in Tas have provided comprehensive information on the biology and behaviour of *L. flavicornis* under field conditions, and its use and effectiveness in integrated management programs including herbicide control. Field studies incorporating *C. atricapitana* and *P. isodactyla* are now required if a more effective integrated control system for ragwort, using a combination of these biological control agents, is to be achieved.

Key words: *Tyria jacobaeae, Botanophila seneciella, Botanophila jacobaeae, Longitarsus flavicornis, Longitarsus jacobaeae, Cochylis atricapitana, Platyptilia isodactyla.*

INTRODUCTION

Ragwort, *Jacobaea vulgaris* Gaertn. (syn. *Senecio jacobaea* L.), is native to Europe and western Asia. It occurs from Scotland and Norway in the north to North Africa in the south and Siberia in the east (Harper and Wood 1957; Parsons and Cuthbertson 2001). Ragwort has now spread into many temperate regions of the world and is found on the east and west coasts of the USA and Canada as well as in South Africa, Argentina, New Zealand and Australia (Harper 1958). In Australia, ragwort is a serious problem

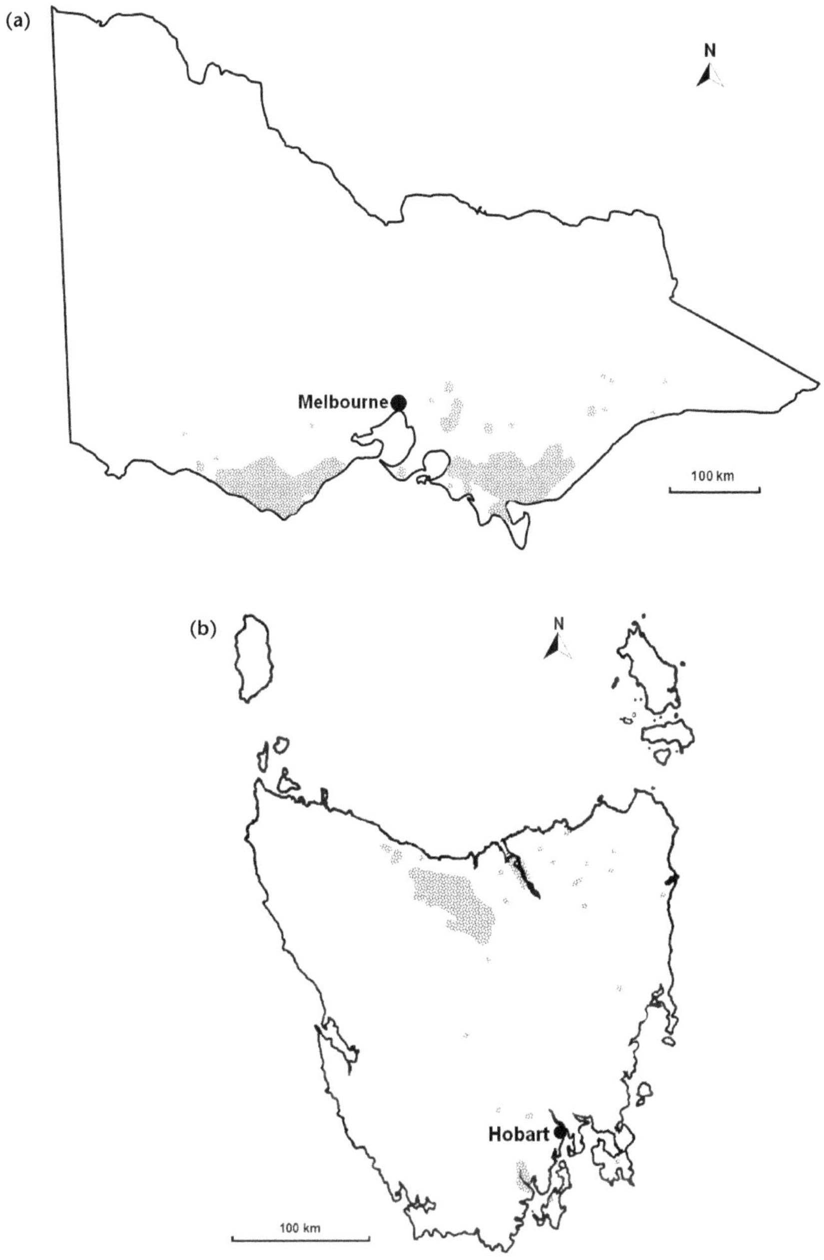

Figure 1: a) Distribution of *Jacobaea vulgaris* in Vic. b) Distribution of *Jacobaea vulgaris* in Tas.

in Tas and Vic. It also occurs in NSW, SA and WA (Richardson *et al.* 2006) but infestations in those states are small and are usually eradicated when located (Parsons and Cuthbertson 2001). The history of its establishment in Australia is not well recorded. Its presence was first noticed in the 1850s but it was not recognised as a serious weed in Australia until the late 19th century when it was first proclaimed under the Vic *Thistle Act 1890* (McLaren *et al.* 2000). The estimated land area now infested with ragwort in Vic is in excess of 820 000 ha, predominately in cattle-grazed pastures and disturbed land in the Otway, Dandenong and Strzelecki ranges of southern Vic (McLaren and Mickan 1997) (Fig. 1a). In Tas, it is believed to have established on the east coast at St Helens in 1905 (Parsons and Cuthbertson 2001) and it is now found over 16 000 ha of cattle-grazed pasture (Ireson *et al.* 2000). The heaviest infestations, ranging from 5–20% of groundcover, usually occur in the higher-rainfall dairying areas of the state where the rainfall exceeds 800 mm (Ireson *et al.* 2000) (Fig. 1b).

Ragwort is a facultative biennial, the biennial life-cycle usually occurring in plants growing on wasteland where plants flower after two years' growth, then die. In pasture situations ragwort normally behaves as a perennial, with plants living for several years and sometimes flowering several times before dying. This perennial life-cycle is promoted by damage to the plant from stock hooves, grazing and cutting (Ireson *et al.* 2007). Ragwort has become a weed of major economic significance due to its detrimental effect on agricultural production and its toxicity to all livestock due to the presence of pyrrolizidine alkaloids (Sharrow *et al.* 1988). The annual cost of ragwort to Australia has been estimated to exceed $4 million; however, this figure was considered conservative as it did not include all the herbicide and labour costs associated with control (McLaren and Mickan 1997). In 1995 and 1996 annual production losses to the Tas dairy and beef industries alone were estimated at about $2.5 million (Ireson 2000).

Apart from herbicides, other direct approaches to control include pulling or grubbing and grazing by sheep, which have a high tolerance to the alkaloids contained in ragwort. Indirect approaches to control include pasture improvement, grazing management, cropping or establishing tree cover. Control by these direct and indirect methods is expensive and needs to be sustained over a long period (>10 years) to be successful. This is primarily due to ragwort infestations developing large seed banks of up to 2000 germinable seeds m^2 (Ireson *et al.* 2007) with seed production ranging from 5000 to 250 000 per plant depending on plant size (McLaren and Mickan 1997; Parsons and Cuthbertson 2001). The main method of dispersal is by seed. These can be dispersed by wind, water and animals as well as by humans via vehicles, clothing, equipment and contaminated hay. Burial of the seed deeper than 2 cm can induce dormancy ranging from 10 to 16 years (McLaren and Mickan 1997). Ragwort's rapid rate of spread, its toxicity to stock and the difficulties in controlling it by chemical and cultural methods led to investigations into the possible use of biological control in Australia.

BIOLOGICAL CONTROL HISTORY

Biological control of ragwort has been attempted in NZ (Syrett 1983), Canada (Harris *et al.* 1984) and the USA (Pemberton and Turner 1990; McEvoy *et al.* 1991) as well as Australia. The agent releases in these countries up to the late 1990s were summarised by Julien and Griffiths (1998). Worldwide, seven biological control agents have been used against ragwort either in Australia or overseas. These are the foliage-feeding cinnabar moth *Tyria jacobaeae* (Linnaeus), two species of seed fly *Botanophila jacobaeae* (Hardy) and *Botanophila seneciella* (Meade), two species of root-feeding flea-beetles *Longitarsus jacobaeae* (Waterhouse) and *Longitarsus flavicornis* (Stephens) and two stem- and crown-boring insects *Cochylis atricapitana* (Stephens) and *Platyptilia isodactyla* (Zeller).

The earliest investigations of potential biological control agents for ragwort were conducted by Cameron (1935) and Newton (1933) on behalf of NZ. Over 60 species were recorded from the plant, although this list was revised and extended to over 80 species by Harper and Wood (1957). Cameron (1935) listed *T. jacobaeae* and *Botanophila* spp. as potential biological control agents; *T. jacobaeae* was the first of four agents to become established in NZ for the biological control of ragwort after it was first released there in 1929 (Miller 1929). Two species of seed fly, *B. jacobaeae* and *B. seneciella*, were introduced to NZ from England in 1936 but only the former species established. In 1983, an Italian biotype of *L. jacobaeae* was imported from the USA and established following its release in 1983. Populations of both *P. isodactyla* and *C. atricapitana* of Spanish origin via Australia were first released in NZ in 2006. Surveys have so far failed to recover *C. atricapitana*, however, *P. isodactyla* has been recovered from 90% of its release sites (H Gourlay, *pers. comm.* 2009). The effectiveness of *B. jacobaeae* in NZ has been negligible, mainly due to the high proportion of seed that escapes predation (Dymock 1987). Similarly, *T. jacobaeae* causes extensive defoliation in some localities but it is still not widespread and its numbers are not high enough to have a major impact. However, *L. jacobaeae* has become widespread and populations have been associated with a significant reduction in ragwort densities throughout many parts of NZ. In areas where *L. jacobaeae* has not managed to control ragwort, its impact is expected to be complemented by the more recent establishment of *P. isodactyla* as there is visual evidence of reductions in ragwort densities at sites where the latter species has been released (H Gourlay, *pers. comm.* 2009).

In Canada, *T. jacobaeae* initially became established on the east coast in Nova Scotia in 1964 and on the west coast in British Columbia in 1965 through the release of stock originating from the border region of France and Switzerland in 1963 and 1964 (Harris *et al.* 1971, 1975). The agent is now present throughout the ragwort-infested regions of Canada (Harris *et al.* 1984). Although there has been a reduction in ragwort biomass due to the effects of annual defoliation, the moth by itself has not reduced ragwort density (Harris *et al.* 1984). A population of *B. seneciella* of Italian origin was first released in 1968, eventually becoming established in British Columbia where it has had little effect on weed density (Harris *et al.* 1971; Harris 2005a). An Italian biotype of *L. jacobaeae* (via California) was first released in 1971, with introductions of this and other biotypes following (Harris *et al.* 1984). The agent established in British Columbia and the Canadian Maritimes; good control of ragwort by *L. jacobaeae* was reported in western British Columbia at elevations below 400 m not subject to flooding (Harris and Crozier 2006). A population of *L. jacobaeae* from south-east England released at Nanaimo on Vancouver Island in 1974 (Harris *et al.* 1984) was later found to consist of small numbers of *L. flavicornis* (P Harris, *pers. comm.* 2010). *L. flavicornis* has had no impact, as it is restricted to only one site on Vancouver Island where it is only about 10% of the *L. jacobaeae* population (Harris and Crozier 2006). Release of *C. atricapitana* in 1990 (ex Spain via Australia) has provided good control in Nova Scotia and has complemented the effects of *L. jacobaeae* in British Columbia (Julien and Griffiths 1998; Harris 2005b).

In the USA, *T. jacobaeae* (ex France), *B. seneciella* (ex France) and *L. jacobaeae* (ex Italy) were successfully established after being first introduced into California in 1959, 1966 and 1969 respectively, then into Oregon and Washington (Frick and Holloway 1964; Pemberton and Turner 1990; McEvoy *et al.* 1991). Studies have shown that a combination of *T. jacobaeae* and *L. jacobaeae* has been responsible for significant reduction in ragwort densities recorded over much of the Pacific North-west (Hawkes and Johnson 1978; Pemberton and Turner 1990; McEvoy *et al.* 1991; James *et al.* 1992). Coombs *et al.* (1996) estimated that the successful biological control of ragwort in Oregon provides an annual benefit of more than $US5 million, with a minimum benefit:cost ratio of 13:1.

The Australian biological control program was initiated by the CSIRO. The first of many unsuccessful attempts to establish *T. jacobaeae* commenced with the importation of stock from England via NZ in 1929 and its release in 1930 (Miller 1929; Currie and Fyfe 1938; Field 1989). Attempts to establish *P. jacobaeae* following releases from 1959 were also unsuccessful (Delfosse and Cullen 1982). Investigations from the late 1970s have resulted in the successful establishment of four agents, *L. flavicornis*, *L. jacobaeae*, *C. atricapitana* and *P. isodactyla*, in Australia (McLaren *et al.* 2000; Ireson *et al.* 2003; Morley and Bonilla 2008).

EXPLORATION

The European distributions of *L. jacobaeae* and *L. flavicornis* differ (Shute 1975). CSIRO imported *L. flavicornis* (originally identified as *L. jacobaeae*, see next section) into quarantine in Australia in 1977 from Annonay in central France (Cullen and Moore 1981), which was near the boundary where the distribution of the two species overlaps. *L. flavicornis* was introduced from Annonay as the climate in that area was considered to be a reasonable climatic match to ragwort-infested areas in Vic and Tas (Cullen and Moore 1981; Field 1989). Permission to introduce the species was based on host-specificity studies already conducted by Frick (1970) that enabled the release of *L. jacobaeae* in the USA, and additional tests conducted by CSIRO (Cullen 1979).

Populations of *L. flavicornis* from several different Spanish localities were imported to Australia between December 1984 and January 1986 following surveys by Vayssières and host testing by CSIRO (Field 1989). Surveys by CSIRO in Portugal, France and Spain (Vayssières 1983, 1984; Vayssières and Rahola 1985) resulted in selection of two additional climatically pre-adapted biological control agents, *C. aticapitana* and *P. isodactyla*, due to their apparent host-specificity and high damage potential. Approval for the release of *C. atricapitana* in Australia was based on host-specificity studies conducted by the CSIRO Biological Control Unit located at Montpellier, France and by the Vic Government at its quarantine facility located at Frankston (McLaren 1992; McLaren *et al.* 1998). Tests on *P. isodactyla* were initially conducted in Montpellier, France by CSIRO (Cullen *et al.* 1985) with additional tests conducted in Australia at Frankston (McLaren *et al.* 2006).

CANDIDATES

Foliage-feeder

***Tyria (= Callimorpha) jacobaeae* (Linnaeus) (Lepidoptera: Arctiidae).** Unsuccessful attempts to establish the cinnabar moth in Vic between 1930 and 1982 were attributed to disease in imported stocks, field predation particularly by the scorpion fly *Harpobittacus nigriceps* (Selys), and the importation of populations from Europe and Canada that were poorly adapted to conditions at the release sites (Bornemissza 1966; Schmidl 1972, 1981; Field

1989). Further attempts to establish *T. jacobaeae* in Tas and Vic from material imported from NZ were conducted during the 1990s (McLaren *et al.* 2000). Failure to establish in Tas was attributed to predation and lack of suitable pupation sites (Ireson *et al.* 1999). In Vic, the species is surviving at one site near Cape Schanck on the Mornington Peninsula following a release in 1994. Here the ragwort is growing on a sandy soil which may be more conducive to pupal survival (McLaren, unpub. data). However, *T. jacobaeae* has not spread from this site since its release. It is unlikely that the moth will become widely established in Vic and it is not expected to play any significant role in ragwort control.

Seed-feeders

***Botanophila seneciella* (Meade) and *B. jacobaeae* (Hardy) (Diptera: Anthomyiidae).** The ragwort seed flies *B. jacobaeae* and *B. seneciella* (both formerly *Pegohylemyia*) were introduced from NZ and England in the 1930s (Wilson 1960; Delfosse and Cullen 1982), but no releases from these importations were made due to rearing difficulties (Field 1989). Further seed fly introductions from NZ were made in 1958 and 1959 (Hoy 1958, 1960). Although releases from these importations have been referred to as *Hylemyia seneciella* (Hoy 1958, 1960; Ireson and Terauds 1982), subsequent examination of voucher specimens suggest that all specimens received from NZ were *B. jacobaeae* (Colless 1982). Releases were conducted in Vic and Tas in 1959 (Field 1989) but correspondence in file records (Tas Dept of Primary Industries) indicates that additional releases were made in Vic in January 1960 and Tas in January 1963 with material supplied through CSIRO. None of the releases resulted in establishment.

Root-feeders

***Longitarsus flavicornis* (Stephens) and *Longitarsus jacobaeae* (Waterhouse) (Coleoptera: Chrysomelidae).** A species of flea-beetle imported from France to Australia in 1977 and first released in Vic and Tas in 1979 was originally identified as *L. jacobaeae* and referred to as that species in some early publications (Cullen and Moore 1981; Delfosse and Cullen 1982; Ireson and Terauds 1982). This species successfully established in Vic and Tas and was subsequently identified as *L. flavicornis* (Field 1989; Ireson *et al.* 1991) (Fig. 2). Populations of *L. flavicornis* introduced from Spain and first released in Vic and Tas in 1985 and 1986 respectively also established (Field 1989; Ireson, unpub. data). *L. flavicornis* is univoltine (Ireson *et al.* 1991) and none of the introduced populations exhibit an adult or egg aestivation stage, as had been recorded for different biotypes of *L. jacobaeae* (Frick and Johnson 1972). Adults emerge from pupae in the soil in late spring and commence laying eggs in summer. Eggs hatch in late summer and larvae develop through three instars in autumn, winter and spring, feeding predominantly on the roots and crown of ragwort (Ireson *et al.* 1991).

Figure 2: Ragwort flea-beetle *Longitarsus flavicornis* and larval damage (inset). Photo: TIAR.

L. flavicornis is now common on ragwort throughout Tas (Ireson *et al.* 2000) and has been a successful control agent. Ragwort densities have been reduced by up to 95% at some sites and biological control ranges from substantial to complete in many parts of Tas (Ireson *et al.* 1991, 2007). However, at some Tas sites the impact of established *L. flavicornis* populations has been restricted (Ireson *et al.* 2000, 2007). Factors thought to be responsible include the frequent pasture flooding and poor drainage that are common at some sites in winter and could cause high larval mortality through drowning (Potter *et al.* 2007), as well as incompatible management strategies (Ireson *et al.* 2007). Studies on the biology and behaviour of *L. flavicornis* (Ireson *et al.* 1991; Potter *et al.* 2004a, 2007) have enabled the impact of other control methods on *L. flavicornis* to be minimised. The key elements of this strategy are detailed by Ireson *et al.* (2007); they include the use of wick-wiped and spot-sprayed herbicides, sheep-grazing and cutting and avoiding heavy stocking of cattle in paddocks prone to waterlogging.

In Vic, *L. flavicornis* has been an effective control agent for ragwort only at high-rainfall sites above 500 m in the Strzelecki Ranges; the beetle has failed to establish in warmer, drier coastal localities (McLaren *et al.* 2000;

Morley and Bonilla 2008). An Italian biotype of *L. jacobaeae* was considered better adapted to drier low-altitude sites in Vic where *L. flavicornis* had failed to establish, as it has an aestivating adult stage. Oviposition commences in autumn and ensures the survival of eggs that are sensitive to desiccation (Frick and Johnson 1973). An Italian biotype (via Oregon) was released in Vic in 1987 (Field 1989). However, its spread and overall impact has been minimal (McLaren *et al.* 2000).

In Tas, establishment of the Italian biotype of *L. jacobaeae* was confirmed at six of 30 sites where releases were conducted between 1988 and 1994 (Ireson, unpub. data). However, its distribution has since been overlapped by *L. flavicornis* spreading from neighbouring release sites. This resulted from mass releases of about 2 million field-collected *L. flavicornis* adults at 879 sites around the state between 1986 and 1999 (Ireson *et al.* 2000). Because of the taxonomic similarity of the two species, a detailed study would be required to distinguish *L. jacobaeae* from *L. flavicornis* and to determine *L. jacobaeae* distribution. It is therefore unlikely that any further work on *L. jacobaeae* will be conducted in Tas, particularly as *L. flavicornis* has performed so well (Ireson *et al.* 1999).

Stem- and crown-borers

***Cochylis atricapitana* (Stephens) and *Platyptilia isodactyla* (Zeller)**. To complement the root-feeding effects of the *Longitarsus* species, stocks of the stem- and crown-boring moth *C. atricapitana* (Fig. 3) were introduced from Salamanca, Spain in 1986. *C. atricapitana* was first released in Vic in 1987 (Field 1989) and in Tas in 1995 (Ireson *et al.* 1999). It is well established in both states and continues to spread naturally. Larvae can cause severe damage to ragwort by boring into leaf and flower buds as well as stems and crowns (McLaren 1992). In Vic, *C. atricapitana* may have three generations annually with larvae diapausing through winter on senescent plant material (McLaren 1992), although it appears that the insect is bivoltine in Tas (Ireson, unpub. data). In Vic, *C. atricapitana* is widespread in the Strzelecki Ranges with efficacy studies showing that it can significantly reduce plant size and survival (McLaren *et al.* 2000; Morley and Bonilla 2008). In Tas, no efficacy studies have been conducted, but the agent has been observed causing significant damage at sites in both the north and south of the state (Ireson, unpub. data) and its impact is probably the same as that indicated by the Victorian efficacy studies.

Introduced to Australia from the province of Lugo in central Spain, the ragwort plume moth *P. isodactyla* (Fig. 4) is the most recent agent introduction (McLaren *et al.* 2000). The moth is well established and spreading in Vic following releases at 53 sites between 1999 and 2004 (T Morley, *pers. comm.* 2009) and in Tas where it was released at 29 sites between 2000 and 2007 (Ireson *et al.* 2003; Ireson, unpub. data). The insect is bivoltine with young larvae overwintering within ragwort plants. Larvae can cause severe damage to ragwort by tunnelling in the petioles, crowns and stems. Field efficacy studies in Vic showed that annual capitulum production was 48–67% lower in plants attacked by *P. isodactyla* (Morley and Bonilla 2008).

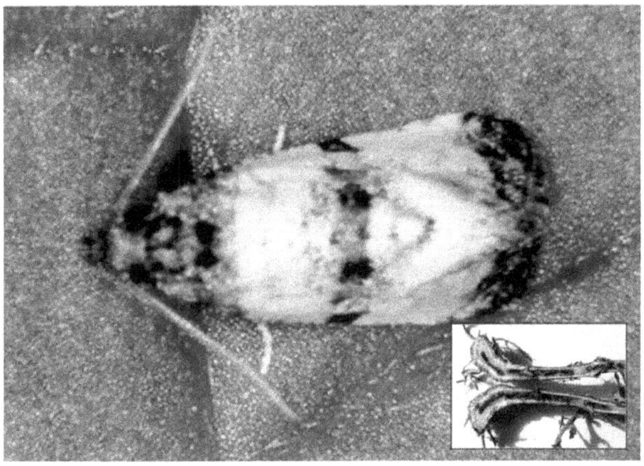

Figure 3: Ragwort stem- and crown-boring moth *Cochylis atricapitana* and larval damage (inset). Photo: R Holloway, TIAR.

DISCUSSION

Evidence that *L. flavicornis* has been a key factor in the gradual but significant decline of the ragwort problem at many sites in Tas since its release in 1979 comes from

Figure 4: Ragwort plume moth *Platyptilia isodactyla* and larval damage (inset). Photo: W Chatterton, TIAR.

long-term efficacy studies, the widespread establishment of *L. flavicornis* populations and photographic records (Ireson *et al.* 1991, 2000, 2007). Many of the sites in Tas where *L. flavicornis* has been less effective are those subject to winter flooding or waterlogging, which can result in high larval mortality (Potter *et al.* 2007). It is unknown why *L. flavicornis* has been less effective in Vic than in Tas, but several factors may be acting in combination to influence *L. flavicornis* population densities such that no single factor or group of factors is dominant (Potter *et al.* 2004b). Efficacy studies suggest that *C. atricapitana* and *P. isodactyla* will have a significant and widespread impact on ragwort in Vic as populations continue to increase and spread naturally (McLaren *et al.* 2000; Morley and Bonilla 2008). Furthermore, as both agents are better adapted to survive waterlogged conditions than *L. flavicornis*, they are expected to provide any additional control needed at Tas sites, particularly those subjected to winter flooding.

Estimates on losses due to ragwort have varied. Although McLaren and Mickan (1997) conservatively estimated the annual cost of ragwort to Australia to exceed $4 million, estimates of annual production losses to the beef and dairy industries in Tas alone ranged from $2.7 million to $9 million when the ragwort problem was still at or near its peak in the mid 1980s to the mid 1990s (Ireson 1998, 2000). Estimates can be affected by variation in stock prices and the area of individual properties affected by ragwort, which can range from 5% to 20%. Page and Lacey (2006) estimated the benefit:cost ratio of the ragwort biocontrol program in Australia at 32:1 primarily as a result of increased production benefits to the dairy and beef industries. It is envisaged that this benefit:cost ratio will increase as populations of *C. atricapitana* and *P. isodactyla* continue to increase and spread in Tas and Vic.

Australian studies have provided comprehensive information on the biology and behaviour of *L. flavicornis* under field conditions and its use and effectiveness in integrated management programs including herbicide control (Ireson *et al.* 1991, 2007; Potter *et al.* 2004a, b, 2010). Additional field studies incorporating *C. atricapitana* and *P. isodactyla* are now required if a more effective integrated control system for ragwort, using a combination of these biological control agents, is to be achieved.

REFERENCES

Bornemissza GF (1966) An attempt to control ragwort in Australia with the cinnabar moth, *Callimorpha jacobaeae* (L.) (Arctiidae: Lepidoptera). *Australian Journal of Zoology* **14**, 201–243.

Cameron E (1935) A study of the natural control of ragwort (*Senecio jacobaea* L.). *Journal of Ecology* **23**, 265–322.

Colless DH (1982) Australian Anthomyiidae (Diptera). *Australian Journal of Zoology* **30**, 81–91.

Coombs EM, Radke H, Isaacson DL and Snyder SP (1996) Economic and regional benefits from the biological control of tansy ragwort, *Senecio jacobaea*, in Oregon. In *Proceedings of the 9th International Symposium on Biological Control of Weeds*. 19–26 January 1996, Stellenbosch. (Eds VC Moran and JH Hoffman) pp. 489–494. University of Cape Town, South Africa.

Cullen JM (1979) 'Results of testing adults and larvae of *Longitarsus jacobaeae* (Waterhouse) against Australian native *Senecio* spp.'. CSIRO unpublished report to Commonwealth Dept of Health (Plant Quarantine).

Cullen JM and Moore AD (1981) Preliminary observations on *Longitarsus jacobaeae* introduced for the control of ragwort in Australia. In *Proceedings of the 5th International Symposium on Biological Control of Weeds*. 22–29 July 1980, Brisbane. (Ed EL Delfosse) pp. 499–505. CSIRO, Melbourne.

Cullen JM, Vayssières JP, Valin A and Rahola P (1985) 'Results of host specificity tests carried out on *Platyptilia isodactyla* (Zeller) (Lepidoptera: Pterphoridae)'. CSIRO unpublished report to Keith Turnbull Research Institute, Frankston, Vic.

Currie GA and Fyfe RV (1938) The fate of certain European insects introduced into Australia for the control of weeds. *Journal of the Council for Scientific and Industrial Research* **11**, 289–301.

Delfosse ES and Cullen JM (1982) Biological control of weeds of Mediterranean origin: a progress report. *Australian Weeds* **2**, 25–30.

Dymock JJ (1987) Population changes of the seedfly, *Pegohylemyia jacobaeae* (Diptera: Anthomyidae) introduced for the biological control of ragwort. *New Zealand Journal of Zoology* **14**, 337–342.

Field RP (1989) Progress towards biological control of ragwort in Australia. In *Proceedings of the 7th International Symposium on Biological Control of Weed*. 6–11 March 1988, Rome. (Ed E Delfosse) pp. 315–322. Instituto Sperimentale per la Patalogia Vegetale (MAF), Italy.

Frick KE (1970) *Longitarsus jacobaeae* (Coleoptera: Chrysomelidae), a flea beetle for the biological control of tansy ragwort. 1. Host plant specificity studies.

Annals of the Entomological Society of America **63**, 284–296.

Frick KE and Holloway JK (1964) Establishment of the cinnabar moth, *Tyria jacobaeae*, on tansy ragwort in the western United States. *Journal of Economic Entomology* **57**, 152–154.

Frick KE and Johnson GR (1972) *Longitarsus jacobaeae* (Coleoptera: Chrsomelidae) a flea beetle for the control of tansy ragwort. 3. Comparison of the biologies of the egg stage of Swiss and Italian biotypes. *Annals of the Entomological Society of America* **65**, 406–410.

Frick KE and Johnson GR (1973) *Longitarsus jacobaeae* (Coleoptera: Chrsomelidae) a flea beetle for the control of tansy ragwort. 4. Life history and adult aestivation of an Italian biotype. *Annals of the Entomological Society of America* **66**, 358–367.

Harper JL (1958) The ecology of ragwort (*Senecio jacobaeae* L.) with special reference to control. *Herbage Abstracts* **28**, 151–157.

Harper JL and Wood WA (1957) Biological flora of the British Isles, *Senecio jacobaeae* L. *Journal of Ecology* **45**, 617–637.

Harris P (2005a) 'Classical biological control of weeds – established biocontrol agent: *Botanophila seneciella* (Meade). Seed-head fly.' Government of Canada, Agriculture and Agri-Food Canada.

Harris P (2005b) 'Classical biological control of weeds – established biocontrol agent: *Cochylis atricapitana* (Stephens) Root-crown feeding moth.' Government of Canada, Agriculture and Agri-Food Canada.

Harris P and Crozier S (2006) 'Classical biological control of weeds – established biocontrol agent *Longitarsus jacobaeae* (Waterhouse) *L. flavicornis* Stephens and *L. succineus* Foudras. Ragwort root beetles.' Government of Canada, Agriculture and Agri-Food Canada.

Harris P, Wilkinson ATS, Neary ME and Thompson LS (1971) *Senecio jacobaea* L., tansy ragwort (Compositae). In *Biological Control Programmes against Insects and Weeds in Canada 1959–1968*. Commonwealth Institute of Biological Control Technical Communication **4**, 97–104.

Harris P, Wilkinson ATS, Neary ME, Thompson LS and Finnamore D (1975) Establishment in Canada of the cinnabar moth, *Tyria jacobaeae* (Lepidoptera: Arctiidae) for controlling the weed *Senecio jacobaea*. *Canadian Entomologist* **107**, 913–917.

Harris P, Wilkinson ATS and Myers JH (1984) *Senecio jacobaea* L., tansy ragwort (Compositae) In *Biological Control Programmes against Insects and Weeds in Canada 1969–1980*. (Eds JS Kelleher and MA Hulme) pp. 195–201. Commonwealth Agricultural Bureau, London.

Hawkes RB and Johnson GR (1978) *Longitarsus jacobaeae* aids moth in the biological control of tansy ragwort. In *Proceedings of the 4th International Symposium on Biological Control of Weeds*. 30 August–2 September 1976, Gainsville. (Ed TE Freeman) pp. 193–196. University of Florida, Gainesville.

Hoy JM (1958) The collection of *Hylemyia seneciella* (Meade) (Diptera: Muscidae) for shipment to Australia. *New Zealand Journal of Science* **1**, 417–422.

Hoy JM (1960) Collection of *Hylemyia seneciella* (Meade) (Diptera: Muscidae) in 1959 season. *New Zealand Journal of Science* **3**, 100–102.

Ireson JE (1998) 'Biological control of ragwort.' Unpublished final report to the Dairy Research and Development Corporation, Project DCN 001. Tasmanian Institute of Agricultural Research/University of Tasmania.

Ireson JE (2000) 'Biological control of ragwort in Tasmania.' Unpublished final report to the Dairy Research and Development Corporation, Project UT 011. Tasmanian Institute of Agricultural Research/University of Tasmania.

Ireson JE and Terauds A (1982) Preliminary observations on the ragwort flea beetle (*Longitarsus jacobaeae*) introduced into Tasmania to control ragwort (*Senecio jacobaeae*). *Australian Weeds* **1**, 3–6.

Ireson JE, Friend DA, Holloway RJ and Paterson S (1991) Biology of *Longitarsus flavicornis* (Stephens) (Coleoptera: Chrysomelidae) and its effectiveness in controlling ragwort (*Senecio jacobaea* L.) in Tasmania. *Journal of the Australian Entomological Society* **30**, 129–141.

Ireson JE, Holloway RJ, Chatterton WS and Leighton SM (1999) An overview of the establishment, distribution and efficacy of biological control agents for ragwort, *Senecio jacobaea* L., in Tasmania. In *Proceedings of the 12th Australian Weeds Conference*. 12–16 September 1999, Hobart. (Eds AC Bishop, M Boersma and CD Barnes) pp. 425–430. Tasmanian Weed Society, Devonport.

Ireson JE, Leighton SM, Holloway R and Chatterton WS (2000) Establishment and redistribution of *Longitarsus flavicornis* (Stephens) (Coleoptera: Chrysomelidae) for the biological control of ragwort (*Senecio jacobaea* L.) in Tasmania. *Australian Journal of Entomology* **39**, 42–46.

Ireson JE, Holloway RJ and Chatterton WS (2003) 'Establishment and dispersal of the ragwort plume moth

Platyptilia isodactyla (Zeller) for the biological control of ragwort in Tasmania.' Unpublished final report to the Dairy Research and Development Corporation, Project UT 11062. Tasmanian Institute of Agricultural Research/University of Tasmania.

Ireson JE, Davies JT, Friend DA, Holloway RJ, Chatterton WS, Van Putten EI and McFadyen RC (2007) 'Weeds of pastures and field crops in Tasmania: economic impacts and biological control'. Technical Series No. 13. CRC for Australian Weed Management, Adelaide.

James RR, McEvoy PB and Cox CS (1992) Combining the cinnabar moth *Tyria jacobaeae* and the ragwort flea beetle *Longitarsus jacobaeae* for control of ragwort *Senecio jacobaea;* an experimental analysis. *Journal of Applied Ecology* **29**, 589–596.

Julien MH and Griffiths MW (1998) *Biological Control of Weeds: A World Catalogue of Agents and their Target Weeds.* 4th edn. CABI, Wallingford, UK.

McEvoy PB, Cox C and Coombs E (1991) Successful biological control of ragwort, *Senecio jacobaea* by introduced insects in Oregon. *Ecological Applications* **1**, 430–442.

McLaren DA (1992) Observations on the life cycle and establishment of *Cochylis atricapitana* (Lep: Cochylidae), a moth for biological control of *Senecio jacobaea* in Australia. *Entomophaga* **37**, 641–648.

McLaren DA and Mickan F (1997) *The Ragwort Management Handbook.* Vic Dept of Natural Resources and Environment, Melbourne.

McLaren DA, Cullen JM and Vayssières JF (1998) 'Host specificity of *Cochylis atricapitana* (Stephens) (Lep: Cochylidae), a biological control agent for ragwort (*Senecio jacobaea* L.) in Australia.' Unpublished report, Vic Dept of Primary Industries, Frankston. pp. 1–17.

McLaren DA, Ireson J and Kwong RM (2000) Biological control of ragwort (*Senecio jacobaea* L.) in Australia. In *Proceedings of the 10th International Symposium on Biological Control of Weeds.* 4–14 July 1999, Bozeman, Montana. (Ed NR Spencer) pp. 67–79. Montana State University, Bozeman.

McLaren DA, Cullen JM, Morley TB and Sagliocco JL (2006) 'Host specificity of *Platyptilia isodactyla* (Lepidoptera: Pterophoridae) a biological control agent for *Senecio jacobaea* (Asteraceae) in Australia.' Unpublished report, Vic Dept of Primary Industries, Frankston.

Miller D (1929) Control of ragwort: experimental work with cinnabar moth. *New Zealand Journal of Science and Technology* **11**, 112–119.

Morley T and Bonilla J (2008) Ragwort plume moth (*Platyptilia isodactyla*) reduces ragwort (*Senecio jacobaea*) plant vigour and makes a useful contribution to biological control of ragwort in Australia. *Australia and New Zealand Biocontrol Conference Abstracts*, p. 38.

Newton HCF (1933) On the biology of some species of *Longitarsus* (Col., Chrysom.) living on ragwort. *Bulletin of Entomological Research* **24**, 511–520.

Page AR and Lacey KL (2006) 'Economic impact assessment of Australian weed biological control.' Technical Series No. 10. CRC for Australian Weed Management, Adelaide.

Parsons WT and Cuthbertson EG (2001) *Noxious Weeds of Australia.* 2nd edn. CSIRO Publishing, Melbourne.

Pemberton RW and Turner CE (1990) Biological control of *Senecio jacobaea* in northern California, an enduring success. *Entomophaga* **35**, 71–77.

Potter KJB, Ireson JE and Allen GR (2004a) Oviposition of the ragwort flea beetle, *Longitarsus flavicornis* (Stephens) (Coleoptera:Chrysomelidae), in relation to the phenology of ragwort, *Senecio jacobaea* L. (Asteraceae). *Biological Control* **30**, 404–409.

Potter KJB, Ireson JE and Allen GR (2004b) Soil characteristics in relation to the long-term efficacy of the biological control agent, the ragwort flea beetle (*Longitarsus flavicornis* (Coleoptera:Chrysomelidae)) in Australia. *Biological Control* **31**, 49–56.

Potter KJB, Ireson JE and Allen GR (2007) Survival of larvae of the ragwort flea beetle, *Longitarsus flavicornis* (Stephens) (Coleoptera:Chrysomelidae), in water-logged soil. *Biocontrol Science and Technology* **17**, 765–770.

Potter KJB, Ireson JE and Allen GR (2010) Desiccation and egg viability of the ragwort flea beetle, *Longitarsus flavicornis* (Coleoptera: Chrysomelidae). *Biocontrol Science and Technology* **20**, 311–316.

Richardson FJ, Richardson RG and Shepherd RCH (2006) *Weeds of the South East – An Identification Guide for Australia.* RG and FJ Richardson, Melbourne.

Schmidl L (1972) Studies on the control of ragwort, *Senecio jacobaeae* L., with the cinnabar moth, *Callimorpha jacobaeae* (L.) (Arctiidae: Lepidoptera), in Victoria. *Weed Research* **12**, 46–57.

Schmidl L (1981) Ragwort, *Senecio jacobaea*, in Victoria and renewed attempts to establish the cinnabar moth, *Tyria jacobaeae*, for its control. In *Proceedings of the 5th International Symposium on Biological Control of Weeds.* 22–29 July 1980, Brisbane. (Ed ES Delfosse) pp. 603–607. CSIRO, Melbourne.

Sharrow SH, Ueckert DN and Johnson AE (1988) Ecology and toxicology of *Senecio* species with special reference to *Senecio jacobaea* and *Senecio longilobus*. In *The Ecology and Economic Impact of Poisonous Plants on Livestock Production*. (Eds LF James, MH Ralphs and DB Nielsen) pp. 181–196. Westview Press, Boulder, Colorado.

Shute SL (1975) *Longitarsus jacobaeae* Waterhouse (Col., Chrysomelidae): identity and distribution. *Entomologist's Monthly Magazine* **111**, 33–40.

Syrett P (1983) Biological control of ragwort in New Zealand: a review. *Australian Weeds* **2**, 96–101.

Vayssières JF (1983) 'Biological control of ragwort (Victoria Funds), December 1982–May 1983'. Unpublished CSIRO Biological Control Unit Report to Keith Turnbull Research Institute, Frankston, Victoria.

Vayssières JF (1984) 'Biological control of ragwort (Victoria Funds), June 1983–December 1983'. Unpublished CSIRO Biological Control Unit Report to Keith Turnbull Research Institute, Frankston, Victoria.

Vayssières JF and Rahola P (1985) 'Investigations in the Iberian Peninsula for insects for the control of *Senecio jacobaea* L. (Compositae)'. CSIRO unpublished report to Keith Turnbull Research Institute, Frankston, Victoria.

Wilson F (1960) 'A review of the biological control of insects and weeds in Australia and Australia New Guinea.' Technical Communication No. 1. CIBC, Ottawa, CAB.

Jatropha gossypiifolia L. – bellyache bush

Tim A Heard, K Dhileepan, Faiz Bebawi, Karen L Bell and Ricardo Segura

ABSTRACT

The shrub *Jatropha gossypiifolia* L. is an invasive and poisonous weed growing in the dry tropics of northern Australia. It has been a target for biological control in Australia since 1996, with a hiatus in the work from 2003 to 2007. *J. gossypiifolia* is now cultivated and invasive in many tropical countries in Asia, Africa and the Pacific. In the Americas, the current geographic range is from USA to Argentina but the native range is probably restricted to the countries around the Caribbean Sea. The plant shows little genetic structuring, with limited genetic differentiation occurring across the native range in the Americas, probably a result of having been moved around by humans. In the search for biocontrol agents against *J. gossypiifolia*, we made over 500 collections in approximately 270 locations in 14 countries. A total of 272 probable herbivore species were collected, however, most of these were rare; only 73 species were collected more than twice. Testing has been conducted on several of these species in Mexico or in Australian quarantine. This chapter documents the significant effort invested in potential agents that led to rejection of those species and summarises the work leading to the release of *Agonosoma trilineatum*, which unfortunately has not established. It also discusses five species, including one fungal pathogen, that are still possible, albeit challenging, candidates.

Key words: Euphorbiaceae, invasive weeds, agent selection.

INTRODUCTION

Jatropha gossypiifolia L. (Euphorbiaceae) is a woody perennial shrub, most commonly known as bellyache bush in Australia. It is a poisonous invasive weed growing in the dry tropics of northern Australia (Bebawi *et al.* 2007b). *J. gossypiifolia* was ranked number 21 of major weeds in Australia (Thorp and Lynch 2000) and has recently been proposed as a new Weed of National Significance by the Australian Weeds Committee. It affects pastoralism by poisoning stock and competing with palatable species. Invasion by this species results in a loss of biodiversity and wildlife habitat, changed fire regimes, increased soil erosion and destabilisation of creek and river banks (Csurhes 1999) (Fig. 1).

J. gossypiifolia is a deciduous shrub. It thrives in wet/dry tropical or subtropical climates. It commonly grows to a height of 2–3 m and has a shallow root system. Leaves vary in colour from green to red to purple with the young leaves particularly displaying non-green colours (Fig. 2). It is characterised by the presence of dense glandular hairs that secrete a sticky resin. It is monoecious and produces more male flowers than female. It produces large quantities of seeds which are explosively ejected from their capsule. The propensity for mass germination of seeds following removal of the adult population makes this plant difficult to manage (Bebawi *et al.* 2007b).

J. gossypiifolia readily colonises disturbed habitats and is frequently found in areas where the natural vege-

Figure 1: Aerial view of an infestation of *Jatropha gossypiifolia* on the banks of the Palmer River, northern Qld. The *J. gossypiifolia* shows as a dense understorey shrub with a red colour to the leaves. Photo: J Vitelli, Qld Government.

tation has been overgrazed or removed by human activity or floodwaters, such as roadsides, around abandoned homesteads and near old mine sites. It also invades lowland habitats such as riparian zones, ephemeral watercourses and pastures. Most naturalised populations of *J. gossypiifolia* in Australia are found in areas receiving 400–1200 mm average annual rainfall (Csurhes 1999). *J. gossypiifolia* is very sensitive to frost (Bebawi *et al.* 2007b).

The distribution of *J. gossypiifolia* in Australia is currently limited to the northern part of Qld, the NT and WA (Fig. 3). Although it is in the relatively early stages of invasion, Lazarides and Hince (1993) indicated that *J. gossypiifolia* has the potential to spread over 75% of the Australian continent, including the entire tropical savannas.

J. gossypiifolia is cultivated and invasive in many tropical countries throughout the world including Asia, Africa and the Pacific (Bebawi *et al.* 2007b). Populations of *J. gossypiifolia* in the invaded range are genetically diverse and appear to have been introduced multiple times from multiple source populations, or alternatively from a representative sample of genetic diversity from a panmictic native range (Prentis *et al.* 2009). The haplotypes observed by Prentis *et al.* (2009) have not been correlated to the

Figure 2: *Jatropha gossypiifolia* growing beside a track in Paraiba state, Brazil showing coloured young leaves and flowers and fruits. Photo: T Heard.

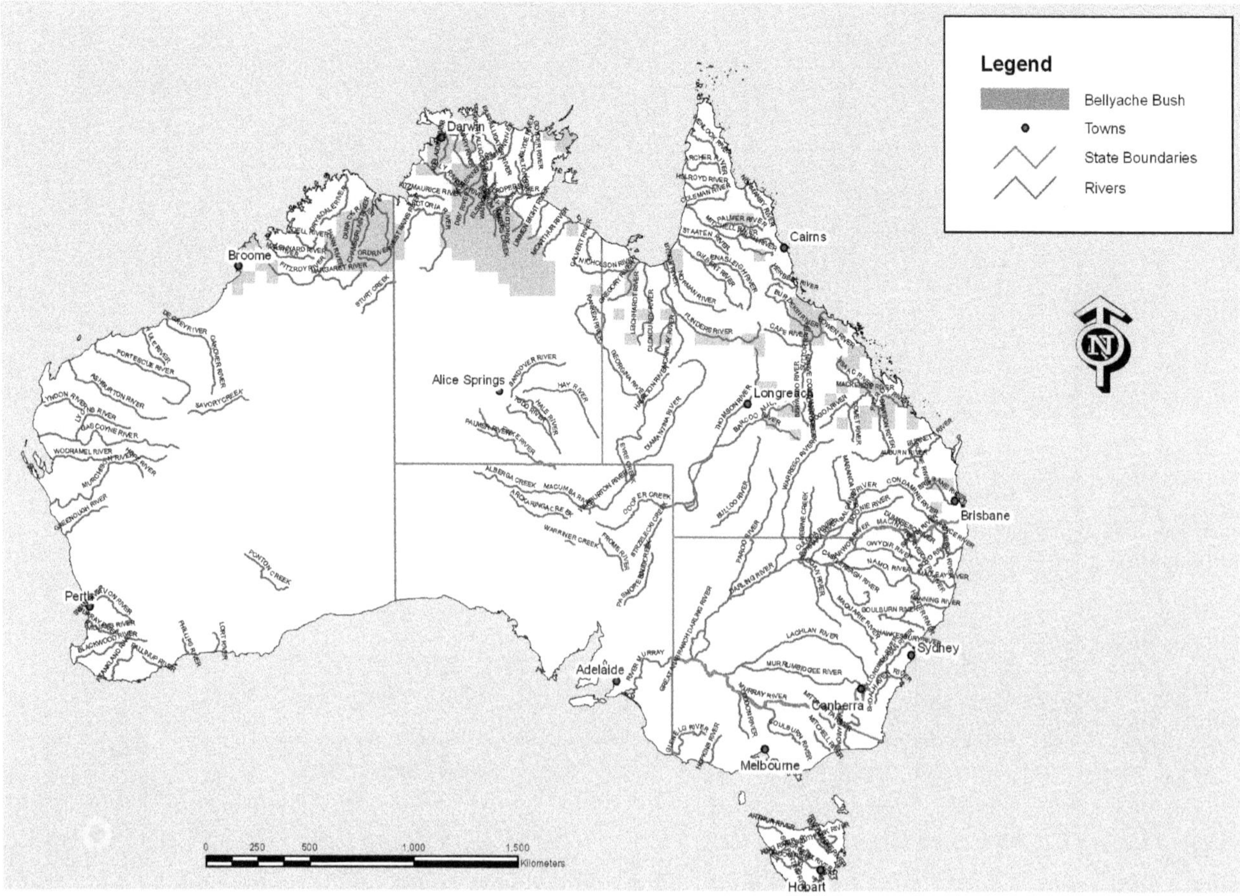

Figure 3: The distribution of *Jatropha gossypiifolia* in Australia. From Bebawi *et al.* (2007b).

morphological types observed by Bebawi *et al.* (2007b) or the varieties of B Deghan (*pers. comm.* 2008).

Another plant in the Euphorbiaceae, *Euphorbia esula* L., leafy spurge, also a rangeland shrub, has been a target for biocontrol and has been successfully controlled over large areas in North America by a group of leaf beetles in the genus *Aphthona* (Cline *et al.* 2008). In Australia, sea spurge, *Euphorbia paralias* L., has become a biocontrol target (Scott *et al.* 2010).

A high level of host-specificity for biological control agents against this weed is not particularly required because no plant species in the tribe Jatropheae are native to, or of economic importance, in Australia. Hence the probability and consequences of non-target impact from biocontrol agents are reduced (Pemberton 2000).

BIOLOGICAL CONTROL HISTORY

J. gossypiifolia has been a target for biological control in Australia since 1996 (Heard *et al.* 2002). The project was led by CSIRO initially in collaboration with the NT Government (1996– 2003) and it later included the Qld Government (1998–2003). Work then ceased until funding from Meat and Livestock Australia and the Qld Government revitalised it in 2007. Despite a significant effort to find, select and test biocontrol agents, only one has been released, the jewel bug *Agonosoma trilineatum* (F.) in 2002. Unfortunately, only a few unlikely prospects remain for further consideration. Biocontrol has not been attempted in any other country.

PLANT TAXONOMY

Three varieties of *J. gossypiifolia* have been described in the Americas: *J. gossypiifolia* var. *elegans* from South America, *J. gossypiifolia* var. *staphysagrifolia* from Mexico to Ecuador and *J. gossypiifolia* var. *jamaicensis* that occurs throughout the Caribbean islands (B Deghan, *pers. comm.* 2008). Some Australian material has been tentatively assigned to these varieties. The Queensland Bronze and Katherine Green were identified from photographs by B Deghan as *J. gossypiifolia* var. *elegans* and *J. gossypiifolia* var. *staphysagrifolia* respectively, but it is possible that the Australian forms are hybrids of the American varieties.

The only other members of the Jatropheae tribe to occur in Australia are exotic ornamental species of *Jatropha*, none of which are economically important and several of which are weeds. The tribe Jatropheae is in the subfamily Crotonoideae with eight other tribes. Many of the approximately 90 members of the subfamily Crotonoideae present in Australia are valued plants, including the potential crop species cassava *Manihot esculenta* Crantz and native species in about 12 genera. These species are well represented on the list of plant species used for host-specificity testing (Heard *et al.* 2009).

PLANT RESPONSE TO HERBIVORY

The response of *J. gossypiifolia* to different types of simulated herbivory (defoliation, shoot damage, root damage and combinations), at varying frequencies (no herbivory, single, two and three events of simulated herbivory) in the glasshouse (Raghu 2004) and in the field (Lockett and Dhileepan, unpub. data) suggest that *J. gossypiifolia* is susceptible to herbivory. Under field conditions, artificial defoliation and shoot damage significantly reduced the shoot length, basal stem diameter, number of branches, root length and above- and below-ground biomass; the reductions were more severe with two or more simulated herbivory events than with a single defoliation event (Lockett and Dhileepan, unpub. data). In contrast, root damage on its own had no negative impact on plant vigour (Raghu 2004). There was no synergy when simulated defoliation and shoot damage were combined. Simulated herbivory studies also suggest that different morphological forms (Queensland Bronze, Kununurra Green, Katherine Green and Darwin Purple) of *J. gossypiifolia* are equally susceptible to defoliation and shoot damage (Pyle 2008). The study highlights the need to prioritise multivoltine leaf herbivores or shoot herbivores, in preference to root herbivores, as potential biological control agents.

EXPLORATION

J. gossypiifolia occurs from Florida, USA south to Argentina. Due to its widespread dispersal by humans, it is difficult to estimate the native range and centre of origin. It thrives in human-disturbed areas, facilitating its spread and further obscuring its origin. It is common in Mexico, Central America and the Caribbean. In South America it occurs predominantly in drier areas around the Amazon basin as far south as Argentina. However, we believe that it is not native to most of South America as there it appears to be strictly associated with human disturbance, with no populations found in natural areas. Furthermore the invertebrate fauna in this region is depauperate, indicating recent establishment. We speculate that its native range corresponds with the countries surrounding the Caribbean Sea, including Mexico, Central America, the northern coast of South American and the Caribbean islands.

Genetic analysis using chloroplast microsatellites (cpSSRs) was used to establish the introduction history and level of genetic diversity present in native and introduced *J. gossypiifolia* populations (Prentis *et al.* 2009). High haplotype diversity was observed in the native range of *J. gossypiifolia*, but limited genetic differentiation occurred across the native range, a fact which was attributed to human translocation. Among the 25 haplotypes found in the native range, nine were also found in Australia. Australian populations of *J. gossypiifolia* were found to be introduced from diverse sources, including three geographically restricted haplotypes from Honduras, Costa Rica and Brazil as well as a number of widely distributed cosmopolitan haplotypes (Prentis *et al.* 2009).

Surveys for natural enemies were primarily conducted by staff at the CSIRO Mexican Field Station, particularly Ricardo Segura, Moisés Martínez, Manuel Juárez, Carlos Pascacio and Quiyari Santiago from 1996 to 2010. Occasionally, Australian-based scientists participated in the surveys, especially Wendy Forno and Tim Heard. In the search for potential agents against *J. gossypiifolia*, we have made over 500 collections in approximately 270 locations in 14 countries (Mexico, Guatemala, Nicaragua, Costa Rica, Dominican Republic, Puerto Rico, Netherlands Antilles, Trinidad, Cuba, Venezuela, Ecuador, Peru, Bolivia and Brazil) (Heard *et al.* unpub.).

Explorations to source climatically suitable biological control agents for *J. gossypiifolia* aimed to target regions in the native range with a climate comparable to that of the infested areas in Australia. Based on the climatic similarity of locations with *J. gossypiifolia* infestations in Australia, locations in Paraguay, Bolivia, Brazil and Venezuela were prioritised for exploration of new biological control agents (Bayliss 2007). Accordingly, further surveys were conducted in Paraguay, Bolivia and Brazil.

CANDIDATES

Natural enemies (exotic range)

In Australia, the leaf-mining moth *Epicephala* sp. (Wilson 1997) and the castor oil looper *Achaea janata* L. cause minor defoliation of *J. gossypiifolia* (Bebawi *et al.* 2007b). A native species of Scutelleridae, *Lampromicra senator* (Fab.), was found feeding on *Jatropha podagrica* Hook. at the Brisbane Botanical Gardens. Laboratory studies show that it can develop on fruits of *J. gossypiifolia* (R Chan, *per. comm.* 2001). Bebawi *et al.* (2007b) listed other species from Australia but many of these are visitors and/or not herbivores. Another Gracillariidae, *Stomphastis thraustica* (Meyrick 1908), is known from *J. gossypiifolia* in Africa and India. Its hosts in India also include *Jatropha curcas* L. and *Sebastaea* sp. (Euphorbiaceae). It was collected in large numbers near Polokwane (Pietersburg) in Limpopo province, South Africa (A Witt, *pers. comm.* 2005).

A soil-borne fungus *Scytalidium dimidiatum* (Penz.) Sutton & Dyko has been observed to cause canker and wilt of *J. gossypiifolia* stems in northern Qld. *S. dimidiatum* (formerly *Torula dimidiata*) is the synanamorph of *Nattrassia mangiferae* (Syd. & P. Syd.) Sutton & Dyko (syn. *Hendersonula toruloidea*). It is a very common soil-borne fungus associated with a wide range of hosts, including eucalypts, crop species (e.g. mangoes in Qld), and causes stem end rot (A Tomley, *pers comm.*2003).

Natural enemies (native range)

Several thousand specimens collected from the surveys for natural enemies have been curated, databased and sent for identification. Species were classified as phytophagous or non-phytophagous based on field observations, rearing from plant material and information in the literature. A total of 272 probable herbivore species or morphospecies have been recorded (Heard, unpub. data). However, most of these are rare with only 73 species collected more than twice. Testing has been conducted on several of the more common species in Mexico or in Australian quarantine.

Agents tested but not released

***Lagocheirus araneiformis* L. (Coleoptera: Cerambycidae).** Shipments of this species took place from 1999 and 2000. To improve the success rate of rearing collected larvae to adults, we developed an artificial diet which also assisted in the elimination of mites from the colonies. Taxonomists identified the individuals from Mexico and Venezuela as *Lagocheirus araneiformis,* a pest of cassava and sugar cane (*Saccharum* spp.) in Florida, Central and South America, Hawaii and the Society Islands (Villegas and Belloti 1985). However, we identified biological differences between individuals from Mexico and Venezuela, which led us to suspect that a species complex may be involved, thus ~800 base pairs of the mitochondrial CO1 gene and ~600 base pairs of the nuclear ribosomal 28S gene were sequenced. The results indicated that the taxa were likely to be different species. Both types were then kept separate for further studies.

Once viable colonies were established, host-testing commenced. The tests were sequential choice-minus-control tests on combined oviposition and larval development. These beetles were eventually rejected as they fed on cassava. Feeding occurred only on cut stems of cassava and very rarely on living plants. However, since cassava is vegetatively propagated by stem cuttings, has potential as a crop in Australia and is currently very important crop in many parts of the world, it was decided to cease testing and to eliminate colonies of these species. If the two types do represent cryptic species, they show no evidence of forming host races as both appear to use cassava.

***Parmenonta valida* Thomson (Coleoptera: Cerambycidae).** *Parmenonta valida* was imported into Australian quarantine in 2001 and 2003 from Mexico, where it is a common stem-borer. The insect was successfully reared on an artificial diet. However, determination of the identity of the species and subsequent literature review showed this species to have a wide host-range.

***Leptostylus* probably *decipiens* Bates (Coleoptera: Cerambycidae).** *Leptostylus* probably *decipiens* was imported from Venezuela into Australian quarantine for host-specificity testing in 2003. A culture was established and host-specificity tests commenced. The test consisted of placing newly emerged larvae onto plant stems. The survival on the target weed was too low and testing was abandoned.

***Leptostylus* sp. (Coleoptera: Cerambycidae).** Two shipments were imported into Australian quarantine in 1998 from the Dominican Republic, but a colony could not be established.

***Styloleptus laticollis* Fisher (Coleoptera: Cerambycidae).** Two shipments were imported into Australian quarantine in 1998 from the Dominican Republic, but a colony could not be established.

***Styloleptus* sp. (Coleoptera: Cerambycidae).** A shipment of *Styloleptus* sp. (referred to as *Styloleptus* sp. 2 in quarantine records), was sent from Venezuela into Australian quarantine in 1999 but a colony could not be established.

***Colaspis musae* Bech. or near (Coleoptera: Chrysomelidae).** This leaf beetle from Venezuela originally appeared promising as heavy adult feeding damage was seen in the field. Observations on surrounding plant species indicated specificity. Data on the host-specificity of this species was obtained from a garden trial in Mexico, from adults collected in Venezuela in 2008. Adult damage was found only on *J. gossypiifolia*, not on other euphorbs, *Cnidoscolus aconitifolius* (Miller), *Ricinus communis* L. and *J. curcas*. One attempt in 1999 at rearing in Australian quarantine by breeding larvae on roots of potted plants was partially successful, with the emergence of three adults from one potted plant, but this success was not repeated with a second shipment in 2001. It was then determined that the species was *Colaspis musae*, a pest of the foliage and fruit of bananas (Briceno and Ramirez 2000; Ostmark 1975), which is thought to breed on roots of grasses (W Flowers, *pers. comm.* 2008). However, there is disagreement on the species in this group and so the possibility remained that it could be a related species. The next step would be to test specificity with young banana leaves, and larval development in grass roots. However, this is a low priority due to the many difficulties with this species, including difficulty in rearing and testing a root-breeding species, taxonomic uncertainty and the expense of obtaining material from Venezuela.

***Cylindrocopturus imbricatus* Champion (Coleoptera: Curculionidae).** *Cylindrocopturus imbricatus* is a damaging stem-borer of *J. gossypiifolia* in its natural habitat. Populations have been found in Mexico, Guatemala and the Dominican Republic. Damage by larvae weakens the stems and causes death of the host plant when the infestation is severe. Given the damage of *C. imbricatus* to *J. gossypiifolia* in the field at Mexico, it was prioritised as a potential biocontrol agent. It was identified first as *C. jatrophae*, but later as *C. imbricatus*.

Eggs are laid in the bark and the emerging larva bore into the stem, making a large tunnel in the fresh living material. The prepupa makes its cocoon with stem fibres in the stem, then pupates and adults emerge from holes chewed through the stem. The adults probably graze on the outer surfaces of plant parts. Adults can be sexed by eye colour: female eyes shine with a golden tone, whereas male eyes are opaque. Adults survive in captivity an average of 14 days (range two to 36). The generation time is ~74 days. Seventy-five individuals of Hymenoptera, most of them Braconidae, possible parasitoids of *C. imbricatus*, emerged from the material collected in January 2009. These represent about 10% of the total of *C. imbricatus* adults emerged from this collection. The relationship of these Hymenoptera to *C. imbricatus* is not conclusively established as they may have been parasitoids of other species of borers present in the stems.

A first attempt to rear this insect at the Mexican Field Station in 2002 was partially successful. Nineteen different rearing systems were tried, including differences in the type of enclosures (white and black bags; cages, containers and pots of different size; soil with grass and other weeds), light condition (full sunlight vs shade), age (young vs old) and traits of plants (number of branches, height, width, presence/absence of flowers and fruits). None of these rearing systems were particularly successful. Overall, more adult insects were added than emerged in the next generation, resulting in a population decline. Often, the same system produced variable results. Rearing of *C. imbricatus* is possible, but a more reliable method is needed.

Five shipments of *C. imbricatus* from Mexico were made to Australia. In 2003, two shipments were sent to CSIRO; in 2008, three shipments were dispatched to the Qld Government facility. All attempts to rear and test the specificity of this insect in Australia failed, as adults did not copulate or lay eggs; the reasons for this have not been determined (Taylor *et al.* 2009). We proved in Mexico that both Green and Bronze varieties of Australian *J. gossypiifolia* are suitable to rear this insect species, thus the failure in Australia is not due to incompatible plant biotypes.

Because of the poor results with rearing this species, host-specificity testing based on larval development did not proceed. Instead, we undertook surveys of natural host plant use in Mexico. In these surveys, a species appearing very similar to *C. imbricatus* emerged from *Xanthium strumarium* L. (Asteraceae), poss. *Simsia* sp. (Asteraceae). However, it was eventually identified as *Cylindrocopturus armatus*. This shows the need for excellent taxonomic support when using field surveys for determination of host-specificity. Otherwise, we found no other evidence of use of alternative plant species by *C. imbricatus*.

***Pachycoris klugii* Burmeister (Hemiptera: Scutelleridae).** Searching the literature on insect fauna of the related plant, *J. curcas*, revealed a potential new agent. The sap-sucking bug *Pachycoris klugii* (Scutelleridae) is recorded as the most damaging herbivore of plantations of *J. curcas* in Nicaragua and is also recorded as being specific to *Jatropha* spp. (Grimm and Maes 1997; Grimm and Somarriba 1998). *P. klugii* was imported into Australian quarantine to determine whether it would develop on *J. gossypiifolia*. Shipments of eggs, nymphs and adults were made in 1999 from Chiapas state, Mexico. Unfortunately, this insect did not develop on *J. gossypiifolia*. All other insect pests of *J. curcas* have been eliminated from consideration as biocontrol agents of *J. gossypiifolia* due to their unacceptably wide host-range. In addition, even though *J. curcas* is a weed in Australia, it has potential here and in many other tropical areas as an important source of biofuel (Gübitz *et al.* 1997).

Agents released

***Agonosoma trilineatum* (F.) (Hemiptera: Scutelleridae).** This jewel bug was first released in the NT in March 2003 and in northern Qld in June 2004 (Heard *et al.* 2009). It feeds on seeds and has the potential to reduce seed production (Bebawi *et al.* 2007a), thereby potentially reducing the rate of spread and recruitment of its host plant. Adults occur in two colour forms, spotted and striped (Fig. 4). To test the host-specificity of *A. trilineatum*, four biological responses to host plant species were determined: development of nymphs, oviposition preferences, adult feeding and frequency of mating. Development of nymphs to adults and adult feeding occurred only on three *Jatropha* spp. These species also supported mating and oogenesis, but only *J. gossypiifolia* was accepted for oviposition. Mating did not occur in the presence of other plant species. The evidence indicates that there is little risk associated with the release of this insect species in Australia and other countries where *J. gossypiifolia* is a problem (Heard *et al.* 2009). The probability of this insect expanding its host-range is low because multiple aspects of the biology would need to change simultaneously. A Climex® model indicated that coastal areas of Qld and the NT would be climatically most suitable for this insect (Heard *et al.* 2009). However, none of the releases made between 2004 and 2007 appear to have resulted in establishment.

Other promising agents

***Phakopsora jatrophicola* Cumm. (Pucciniales).** This rust fungus commonly causes chlorosis and necrosis on leaves of *J. gossypiifolia* in many parts of tropical America. The pathogen is very prolific spore producer. CABI UK and CSIRO MFS staff conducted preliminary studies on this species in 2000 and further work from 2008. Plants of *J. gossypiifolia* were established in the CABI quarantine glasshouse and an accession of the fungus from Mexico was successfully augmented on its host under greenhouse conditions. Studies confirmed that *P. jatrophicola* is pathogenic to all Australian varieties of *J. gossypiifolia* (Seier *et al.* 2009). Host-range testing by CABI UK showed that the rust can also attack and sporulate on *Jatropha multifida* L., *J. integerrima* Jacq. and *J. curcas* (Seier *et al.* 2009). *J. curcas* is invasive in WA, NT and Qld and is an approved target for biological control in Australia (http://www.weeds.org.au/target.htm). However, on *J. curcas* the rust produced limited sporulation (limited number of uredinia and restricted uredinia size), hence this species was classed as a partially susceptible host (Seier *et al.* 2009). No symptoms occurred on *J. podagrica* nor on the euphorb crop species, cassava and rubber (*Hevea brasiliensis* Mull.Arg.). The current focus is to find the most suitable and virulent accession of this rust fungus prior to commencing detailed host-specificity testing. Collections from the Mexican Gulf and Pacific coasts, Nicaragua, Brazil and Trinidad have been tested (Seier and Cortat 2010). Future surveys for the rust will focus in Cuba. Surveys on the extent of the rust on *J. curcas* in the native range are also needed. Work will also be required to elucidate the full life-cycle of the rust to determine whether it is autoecious (can complete its life-cycle on just one host) or heteroecious (needs an alternative host to complete its life-cycle). If heteroecious, it may have other hosts in the sexual stage of its life-cycle and host-specificity testing would be complicated. It would also be useful to test the efficacy of this rust pathogen in the native range to determine whether expenditure on fully testing this fungus is justified.

Figure 4: A striped form of the released biocontrol agent *Agonosoma trilineatum*. Photo: D McClenaghan, CSIRO.

***Euxesta abdominalis* Loew and *Euxesta* aff. *panamena* Curran (Diptera: Ulidiidae).** Larvae of these *Euxesta* spp. mine leaves and stems, causing significant damage to the plant. These species are common with a wide distribution including much of Mexico, but it is difficult to collect adults. Further work on the taxonomic and biological characteristics of these species is needed.

***Pachycoris* prob *fabricii* (Burmeister) (Heteroptera: Scutelleridae).** This fruit-feeding insect was collected in the Dominican Republic. Host-specificity testing was carried out in Mexico on *Cnidoscolus aconitifolius* (Miller) I. M. Johnston and *Ricinus communis* L. but tests were not completed due to lack of fruiting plants.

***Pityophthorus* sp. and *Hypothenemus* sp. (Coleoptera: Scolytidae).** *Pityophthorus* sp. is likely to be a host-specific phytophage capable of killing tips (T Atkinson, *pers. comm.* 2008). *Hypothenemus* sp., the other scolytid found on *J. gossypiifolia*, was rejected due to lack of specificity and saprophagous feeding habits (L Kirkendall, *pers. comm.* 2008). However, *Pityophthorus* sp. is an uncommon insect which was found at only one site in Mexico. Recent collections of 110 dead tips resulted only in *Hypothenemus*; no *Pityophthorus* was found.

***Ormiscus/Eusphyrus* (Coleoptera: Anthribidae).** This insect is locally common at the sites of Mexico, emerging from stems. Larvae have been reared from artificial diet. More specific identification is not yet available. It is possible that the species is not a phytophage, but is a fungus-feeder like most members of its family.

DISCUSSION

Biocontrol of *J. gossypiifolia* has proven a very difficult task indeed. Despite extensive survey work over many years in many countries by experienced field workers, very few prospective agents have been found.

Initial biocontrol prospects seemed good for several reasons. First, there are no other members of the tribe Jatropheae in Australia so a lower level of host-specificity of a biocontrol agent may have been acceptable. Second, the difference in the size of populations and individual plants observed in the native and introduced ranges suggested that the species is under strong herbivore and pathogen pressure in the native range. Third, the high level of pharmacologically active substances in this highly poisonous plant (Das and Das 1994) leads to the prediction, according to co-evolutionary theory, that the suite of natural enemies should be small but host-specific. Fourth, the large geographic range of the plant provided ample opportunities to find climatically suited agents.

Few of these assumptions now appear to be valid. Generally, the level of herbivore pressure on this species in the native range is very low. The suite of natural enemies is not particularly small, at around 270 species. The specificity of these species does not seem to be high; indeed, very few specialists were found. The one specialist found and released, *A. trilineatum*, failed to establish in Australia, possibly because of a poor climate match or non-synchronisation of plant phenology and agent biology.

A few remaining prospective agents are worthy of further attention, but for various reasons none will be easy to develop to the point of release. It is difficult to collect adults of *Euxesta abdominalis* and *E.* aff. *panamena*. *Pachycoris* prob. *fabricii* occurs only on isolated Caribbean islands. *Pityophthorus* sp. is rare. *Ormiscus/Eusphyrus* is difficult to identify and to study, and may not even be a herbivore. *Cylindrocopturus imbricatus* is probably specific but has proven to be very difficult to rear and test. Work on the rust fungus *P. jatrophicola* is active and shows promise, but the agent may attack the potential crop species *J. curcas*, its virulence on the target is not certain and work is required to determine if it is autoecious or heteroecious.

Another possible course of action would be to consider the neoclassical approach (Hokkanen and Pimentel 1989) and search for agents on other species of *Jatropha*. For example, there are various species of *Jatropha* in the Chaco of Argentina and Paraguay that may have associated herbivores and pathogens that could attack *J. gossypiifolia*. It may be productive to look for new associations on *J. gossypiifolia* in the introduced range (South Africa, India etc.) where the plant is known to be subjected to herbivore pressure.

ACKNOWLEDGEMENTS

We thank the following. Tania Yonow (CSIRO) designed the collection database. Marion Seier and Ghislaine Cortat (CABI) did the rust studies. Richard Chan and Karryn Waterworth (CSIRO) reared and tested the agents in Brisbane quarantine. Gio Fichera and Dalio Mira (CSIRO) grew the plant species for quarantine testing. The following taxonomists provided their expert opinion on species mentioned in this chapter: V Toledo and F Noguera (Cerambycidae), F Wills Flowers (Chrysomelidae), H Hespenheide (*Cylindrocopturus*), J Eger (*Pachycoris* and *Agonosoma*), H Evans (*Phakopsora jatrophicola*), V Hernandez (*Euxesta*), Barry Valentine (Anthribidae), Tom Atkinson and Armando Burgos (Scolytidae).

REFERENCES

Bayliss D (2007) *Jatropha gossypiifolia*: prioritising areas for exploration using Climex as a tool. In *Proceedings of the Bellyache Bush Research Workshop.* 17 July 2007, Alan Fletcher Research Station, Brisbane. (Eds K Dhileepan and S Harvey) pp. 36–37. Qld Dept of Primary Industries and Fisheries, Brisbane.

Bebawi FF, Lockett CJ, Davies KM and Lukitsch BV (2007a) Damage potential of the introduced biological control agent (*Agonosoma trilineatum* F.) on bellyache bush (*Jatropha gossypiifolia* L.). *Biological Control* **41**, 415–422.

Bebawi FF, Vitelli JS, Campbell SD, Vogler WD, Lockett CJ, Grace BS, Lukitsch B and Heard TA (2007b) The biology of Australian weeds 47. *Jatropha gossypiifolia* L. *Plant Protection Quarterly* **22**, 42–58.

Briceno Vergara AJ and Ramirez W (2000) Diagnosis of Coleoptera associated with plantain plantations south of Maracaibo Lake-Venezuela. *Revista Forestal Venezolana* **34**, 93–99.

Cline D, Juricek C, Lym RG and Kirby DR (2008) Leafy spurge (*Euphorbia esula*) control with *Aphthona* spp. affects seedbank composition and native grass reestablishment. *Invasive Plant Science and Management* **1**, 120–132.

Csurhes SM (1999) *Bellyache Bush* (Jatropha gossypiifolia) *in Queensland – Pest Status Review Series – Land Protection*. Qld Dept of Natural Resources, Brisbane.

Das B and Das R (1994) Medicinal properties and chemical constituents of *Jatropha gossypiifolia* Linn. *Indian Drugs* **31**, 562–567.

Dhileepan K (2007) Other research: herbivory, haplotypes and hybrids. In *Proceedings of the Bellyache Bush Research Workshop*. 17 July 2007, Alan Fletcher Research Station, Brisbane. (Eds K Dhileepan and S Harvey) pp. 39–40. Qld Dept of Primary Industries and Fisheries, Brisbane.

Grimm C and Maes J-M (1997) Arthropod fauna associated with *Jatropha curcas* L. in Nicaragua: a synopsis of species, their biology and pest status. In *Biofuels and Industrial Products from* Jatropha curcas. (Eds GM Gübitz, M Mittelbach and M Trabi) pp. 31–39. Technische Universität, Graz.

Grimm C and Somarriba A (1998) Lifecycle and rearing of the shield-backed bug *Pachycoris klugii* in Nicaragua (Heteroptera: Scutellidae) *Entomologia Generalis* **22**, 211–221.

Gübitz GM, Mittelbach M and Trabi M (1997) *Biofuels and Industrial Products from* Jatropha curcas. Technische Universität, Graz.

Heard TA, Chan RR and Segura R (2002) Prospects for the biological control of bellyache bush, *Jatropha gossypiifolia*. In *Proceedings of the 13th Australian Weeds Conference*. 8–13 September 2002, Perth. (Eds H Spafford Jacob, J Dodd and JH MooreH) pp. 366–369. Plant Protection Society of WA, Perth.

Heard TA, Chan RR, Wilmot Senaratne KAD, Palmer WA, Lockett C and Lukitsch B (2009) *Agonosoma trilineatum* (Heteroptera: Scutelleridae), a biological control agent of the weed bellyache bush, *Jatropha gossypiifolia* (Euphorbiaceae). *Biological Control* **48**, 196–203.

Hokkanen HMT and Pimentel D (1989) New associations in biological control: theory and practice. *Canadian Entomologist* **121**, 829–840.

Lazarides M and Hince B (1993) *CSIRO Handbook of Economic Plants of Australia*. CSIRO, National Library of Australia.

Ostmark HE (1975) Banana pests in the genus *Colaspis*, including description of a new species (Coleoptera: Chrysomelidae). *Florida Entomologist* **58**, 1–8.

Pemberton RW (2000) Predictable risk to native plants in weed biological control. *Oecologia* **125**, 489–494.

Prentis PJ, Sigg DP, Raghu S, Dhileepan K, Pavasovic A and Lowe AJ (2009) Understanding invasion history: genetic structure and diversity of two globally invasive plants and implications for their management. *Diversity and Distributions* **15**, 822–830.

Pyle K (2008) Ecophysiological responses of different forms of *Jatropha gossypiifolia* ((L.) Euphorbiaceae) to artificial damage. BAppSci Hons thesis. School of Natural Resource Sciences, Queensland University of Technology, Brisbane.

Raghu R (2004) Agent selection for biocontrol of *Jatropha gossypiifolia*. In *Proceedings of the Bellyache Bush Workshop*. 29 November 2004, Tropical Weeds Research Centre, Charters Towers, Qld. (EdW Vogler) pp. 30–35. Qld Dept of Natural Resources and Mines, Charters Towers.

Scott JK, Jourdan M, Morin L, Thomann T and Yeoh PB (2010) Exploration for potential biological control agents of *Euphorbia paralias*, a major environmental weed of coastal ecosystems in Australia. In *Proceedings of the 17th Australasian Weeds Conference*. 26–30 September 2010, Christchurch, NZ. (Ed SM Zydenbos) pp. 223–226. New Zealand Plant Protection Society, Christchurch.

Seier MK and Cortat G (2010) 'Assessment of additional strains of the rust *Phakopsora jatrophicola* ex *Jatropha gossypiifolia* for infectivity and virulence towards *Jatropha curcas* and *J. gossypiifolia;* full host range testing of one selected *Phakopsora jatrophicola* strain'. Interim report for July 2009–June 2010 submitted to the Qld Dept of Employment, Economic Development and Innovation, July 2010.

Seier MK, Cortat G and Evans HC (2009) 'Pathogenicity, life-cycle elucidation and preliminary host-range testing of *Phakopsora jatrophicola* Cummins'. May 2008–

June 2009. Final report submitted to the Qld Dept of Primary Industries and Forestry, June 2009.

Taylor DBJ, Shortus M, Heard TA, Seier MK, Palmer WA and Dhileepan K (2009) Biological control of bellyache bush (*Jatropha gossypiifolia*) – current research and future prospects. In *Proceedings of the 10th Queensland Weed Symposium*. 26–29 July 2009, Yeppoon, Qld. (Eds S Kinnear and T Baker) pp. 102–105. Weed Society of Queensland, Brisbane.

Thorp JR and Lynch R (2000) *The Determination of Weeds of National Significance*. National Weeds Strategy Executive Committee, Launceston.

Villegas GA and Belloti A (1985) Biología, morfología y habitos de *Lagocheirus araneiformis* Linne (Coleoptera: Cerambycidae) barrenador de la yuca en Palmira (Valle del Cauca). *Acta Agronomica* **35**, 56–67.

Wilson CG (1997) Phytophagous insect fauna of two weeds, *Hyptis suaveolens* (L.) Poit. and *Jatropha gossypiifolia* L., in Australia's Northern Territory. *Australian Entomologist* **24**, 55–60.

Lantana camara L. – lantana

Michael Day

ABSTRACT

Lantana is a Weed of National Significance in Australia, being found predominantly throughout coastal and subcoastal eastern Australia. It infests natural, forestry and agricultural ecosystems, displacing native species and reducing biodiversity and productivity. Biological control of lantana in Australia commenced in 1914 and by 1960 eight agents had been introduced, of which six established. Since then, exploration for potential agents by numerous organisations in Australia and overseas, has been conducted throughout most of the native range of lantana, resulting in an additional 23 agents being released in Australia of which 12 established. Six agents, *Aconophora compressa*, *Octotoma scabripennis*, *Uroplata girardi*, *Teleonemia scrupulosa*, *Ophiomyia lantanae* and *Leptobyrsa decora* are regarded as the most damaging, albeit seasonally. Two recently released agents, the rust *Prospodium tuberculatum* and the leaf-mining fly *Ophiomyia camarae*, are becoming more widely distributed but it is too early to determine their impact on lantana. A feature of this program is the strong collaboration between organisations within Australia and overseas, in foreign exploration and in host-specificity testing, supply of agents and plant taxonomy. Despite the large number of agents established in Australia, lantana is not under adequate control. It is often subjected to intense feeding damage seasonally, but plants recover when agent populations wane over winter or during drought. Plant biology and climate are the main factors preventing lantana control. In addition, agents lack adaptation to the wide range of conditions in which the weed can grow, and some agents show preference for some lantana varieties. Current research is focusing on agents that might be better suited and more damaging to lantana. Host-specificity testing of the bud mite *Aceria lantanae* has been completed and an application seeking its release has been submitted. Testing on the rust *Puccinia lantanae* is progressing and it is hoped that these new agents might improve lantana control in Australia.

Key words: climate, phenotypic variation.

INTRODUCTION

Lantana camara L. (Verbenaceae) (lantana) is a pantropical woody perennial shrub originating from tropical America (Swarbrick *et al.* 1998). It was first reported in Australia in 1841, being cultivated as an ornamental in the Botanic Gardens in Adelaide, SA and later at Camden, NSW (34°03'S 150°41'E) in 1843 (Swarbrick 1986). By the late 1850s, it was established outside cultivation and was reported growing along the Brisbane River (27°31'S 152°58'E) in 1861 and along the Hastings River (31°24'S 152°49'E), NSW in the mid 1860s. It was first considered a pest in 1879, and by the 1950s, it was found throughout much of coastal and subcoastal eastern Australia (Swarbrick 1986).

Lantana is now found along coastal and subcoastal eastern Australia from Cape York, Qld south to Eden,

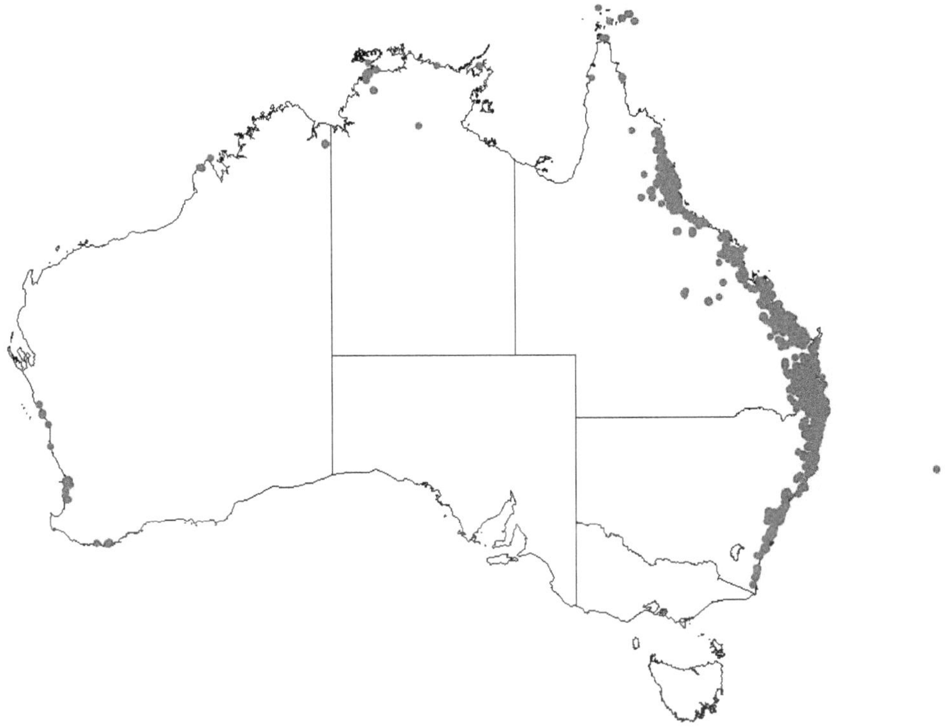

Figure 1: Distribution of *Lantana camara* in Australia. Source: Qld Government.

NSW. It is also present as scattered infestations on some islands in the Torres Strait (9°34'S 143°46'E), around Darwin and Katherine in the NT and Wyndham, Broome, Geraldton, Perth and Albany in WA (Fig. 1). Lantana also occurs on Norfolk Island (29°02'S 167°58'E) and Lord Howe Island (31°31'S 159°03'E) (Day *et al.* 2003a).

Lantana can grow as individual clumps or dense thickets (Fig. 2a, b), displacing desirable species. In disturbed native forests, it can form the dominant understorey, disrupting succession and decreasing biodiversity. Its allelopathic qualities can reduce vigour of plant species nearby and reduce productivity in orchards (Holm *et al.* 1991). In Australia, lantana is classed as a Weed of National Significance, outcompeting native pastures, interfering with the mustering of cattle and causing stock death by poisoning (Swarbrick *et al.* 1998). Costs to the Australian grazing industry alone are over $A100 million per year (AEC Group 2007).

BIOLOGICAL CONTROL HISTORY

Biological control of lantana started in 1902, with the introduction from Mexico into Hawaii of 23 agents of which 14 established. Biocontrol efforts started in Australia in 1914, with the Qld Department of Agriculture and Stock introducing four species that had established in Hawaii (Wilson 1960) (Table 1). CSIR, the forerunner of CSIRO, introduced the sap-sucking bug *Teleonemia scrupulosa* Stål (Hemiptera: Tingidae) to Australia from Fiji in 1936.

From mid 1953 to early 1954, surveys were conducted in Mexico by John Mann, Qld Department of Public Lands and Noel Krauss, Hawaii Department of Agriculture, as part of a collaborative project. Krauss also surveyed British Honduras, Colombia, Costa Rica, Cuba, El Salvador, Florida (USA), Guatemala, Honduras, Nicaragua, Panama, Puerto Rico and Trinidad. Over 150 different insects were found in Mexico but many were not considered for further investigation as they were polyphagous and many were not identified to species level (Mann 1954).

Following these surveys, two agents, the leaf-feeding moths *Neogalea sunia* (Guenée) (Lepidoptera: Noctuidae) (1957) and *Salbia haemorrhoidalis* Guenée (Lepidoptera: Pyralidae) (1958) were introduced into Australia (Wilson 1960) and later established. Another 15 different species were sent to Hawaii for further study (Mann 1954). Of these, the leaf-mining beetle *Octotoma scabripennis* Guérin-Méneville (Coleoptera: Chrysomelidae) (1966) and the stem-boring beetle *Plagiohammus spinipennis* (Thomson)

(a)

(b)

Figure 2: a) *Lantana camara* flowers and fruits. b) *Lantana camara* infestation, Heiffer Creek, Qld. Photos: M Day, Qld Government.

(Coleoptera: Cerambycidae) (1967) were subsequently introduced from Hawaii into Australia. The leaf-feeding moth *Diastema tigris* Guenée (Lepidoptera: Noctuidae) (1965) was introduced into Australia separately.

Several agents, including the leaf-feeding moth *Hypena laceratalis* Walker (Lepidoptera: Noctuidae) (1965), leaf-mining beetle *Octotoma championi* Baly (1975) and stem-boring beetle *Aerenicopsis championi* Bates (Coleoptera: Cerambycidae) (1995), which were also found during these surveys, were introduced into Hawaii then Australia independently some years later.

Other species found during these surveys that were considered to have potential and were introduced into Australia independently, years later, were the leaf-mining beetle *Uroplata girardi* Pic (Coleoptera: Chrysomelidae) (1966), sap-sucking bug *Teleonemia prolixa* (Stål) (1974), leaf-mining beetle *Uroplata fulvopustulata* Baly (1976), stem-sucking bug *Aconophora compressa* Walker (Hemiptera: Membracidae) (1995) and sap-sucking bug *Falconia* sp. (possibly *intermedia*) (2000).

Krauss undertook another survey of Mexico, British Honduras and Belize in 1955. Considerable effort was made to collect and import into Hawaii the two stem-boring beetles *P. spinipennis* and *A. championi*, which were not introduced into Australia until 1967 and 1995 respectively.

Until 1960, seven agents had been introduced deliberately into Australia, with five establishing (Wilson 1960) (Table 1). The flower-feeding moth *Lantanophaga pusillidactyla* (Walker) (Lepidoptera: Pterophoridae), which was introduced into Hawaii in 1902, was thought to be introduced accidentally into Australia on imported potted lantana plants (Day *et al.* 2003a, b).

PLANT TAXONOMY

The genus *Lantana* L. belongs to the family Verbenaceae, of which 10 genera are found in Australia. In Australia, the family consists of mainly introduced species, with only *Phyla nodiflora* (L.) Greene (Julien *et al.* 2004) and three varieties of *Verbena officinalis* L. considered native (Munir 2002). However, the genus *Verbena* is currently under revision and these varieties may be upgraded to species (J Hosking, *pers. comm.* 2010). The genus *Lantana* is divided into four sections, *Camara*, *Calliorheas*, *Sarcolippia* and *Rhytocamara*, consisting of 150 species. The weedy form of lantana is found in the section *Camara* and is commonly referred to as *Lantana camara sensu lato*. It is a composite species, thought to have originated from two or more lantana species from tropical America (Stirton 1977; Sanders 2006). Dutch explorers first introduced the plant into the Netherlands from the West Indies in the 1600s (Stirton 1977) and further introductions from other countries followed. Plants were then hybridised in glasshouses in Europe prior to their introduction as ornamentals into Australia and many other countries (Swarbrick *et al.* 1998), where they became weedy.

Table 1: Origin and status of agents released on *Lantana camara* in Australia pre 1960

Agent	Country	Year released	Organisation	Established	Status
Epinotia lantana	Mexico	1914	QDAS	Yes	b
Eutreta xanthochaeta	Mexico	1914	QDAS	No	
Ophiomyia lantanae	Mexico	1914	QDAS	Yes	a
Strymon bazochii (Godart) (Lepidoptera: Lycaenidae)	Mexico	1914	QDAS	No	
Lantanaphaga pusillidactyla	Mexico	1936	*	Yes	b
Teleonemia scrupulosa	Mexico	1936	CSIR	Yes	a
Neogalea sunia	USA	1957	QDPL	Yes	b
Salbia haemorrhoidalis	Cuba	1958	QDPL	Yes	b

QDAS = Qld Dept of Agriculture and Stock; CSIR = Council for Scientific and Industrial Research; QDPL = Qld Dept of Public Lands.
*: Agent was accidentally introduced, probably via imported plants.
a: Widespread and damaging.
b: Widespread and minimal damage.

There are an estimated 600 varieties of weedy lantana worldwide (Howard 1969) and 29 in Australia (Smith and Smith 1982). Due to the difficulty in identifying each variety, these are commonly sorted into five main groups based on flower colour (pink, pink-edged red, red, white and orange). It is widely recognised among researchers (e.g. Swezey 1924; Sanders 2006) that these weedy forms are morphologically distinct from *Lantana* spp. in their native range and that the term *L. camara* is used as a name of convenience (Sanders 2006). Koebele, when collecting potential agents to send to Hawaii in the early 1900s, remarked that the *Lantana* species in Mexico was quite different from the lantana infesting much of Hawaii (Swezey 1924), and Lindsay Smith could not find any species of lantana in tropical America that corresponded to the weedy lantana in Australia. Harley (1969) stated that the name *L. camara* he used to describe specimens during his surveys may have included other closely related species.

DNA studies in the 1990s suggested that some varieties of weedy lantana were more related to *L. urticifolia* P. Mill. found in Mexico than any other *Lantana* species (Scott *et al.* 2002). However, the sample size of plants from both Australia and overseas was limited. In addition, following a visit to Kew Gardens by Roger Sanders, a review of lantana specimens was proposed, with many *Lantana* species and their subspecies being renamed. Under the proposed new classification, *L. urticifolia* is now a synonym of *L. camara* and the previous incorrect use of *L. camara* is to be replaced by other names, most likely *L. strigocamara* R.W. Sanders or *L. nivea* Vent. subsp. *mutabilis* (Hook.) R.W. Sanders, depending on individual specimens (Sanders 2006).

Recent taxonomic research involving over 280 samples from Australia and 220 from the native range has implicated six *Lantana* species (*L. camara*, *L. depressa* Small via the Callowiana Hybrid Group cultivars, *L. hirsuta* M Martens & Galeotti, *L. horrida* Kunth, *L. nivea* Vent. and *L. strigocamara* RW Sanders, but also possibly *L. cujabensis* Schauer and *L. scabrida* Ait.) in various combinations to describe the weedy lantana in Australia. Research also suggested that the name *L. camara* does not necessarily accurately cover all the varieties and forms found in Australia (Sanders, *pers. comm.* 2010). DNA studies on the same set of samples suggest that the weedy forms in Australia constitute only one highly variable hybrid swarm, with little differentiation and no clear affinity for any species in the native range. They suggest that weedy Australian lantana probably originated from Venezuela or the Dominican Republic (R Watts, *pers. comm.* 2009). Such research is continuing, in an attempt to gain more information on the likely source of Australian *L. camara*.

EXPLORATION

As weedy lantana is considered a hybrid species, it does not possess a native range *per se*. The genus *Lantana* is found in tropical America from southern USA to northern Argentina and some species in the section *Calliorheas* are native to Africa and India, although the actual native range of the species that may be the progenitors of weedy lantana may be much more restricted. The two DNA studies have suggested possible origins of the weedy lantana in Australia, thus limiting the countries in, and *Lantana* species on which, to conduct surveys. However, early biocontrol practitioners did not have this technology or information available when surveys were conducted.

Biocontrol practitioners have probably surveyed most of tropical America searching for lantana agents. In the early 1900s, surveys were conducted mainly in Mexico by Albert Koebele on behalf of the Hawaii Department of Agriculture (Swezey 1924), and in the 1950s numerous countries were surveyed by John Mann and Noel Krauss. Since the 1960s, the main organisations involved in exploration have been Hawaii Department of Agriculture, CSIRO, Qld Department of Public Lands (QDL, now Biosecurity Queensland) and, more recently, CABI and the Plant Protection Research Institute (PPRI), South Africa, with researchers either based overseas for several years or conducting short, highly focused collecting trips. Interestingly, many of the agents subsequently imported from these surveys had been previously listed by Mann and Krauss as warranting further study.

Ken Harley, CSIRO spent from early January to mid April 1969 surveying for potential agents in Brazil, Mexico, Paraguay, Peru, Trinidad and Venezuela. *P. spinipennis* and the sap-sucking bugs *Teleonemia elata* Drake, *T. scrupulosa* and *Leptobyrsa decora* Drake (Hemiptera: Tingidae) were all sent to CSIRO's Long Pocket Laboratories for host-specificity testing and were later released. The leaf-mining fly *Calycomyza lantanae* (Frick) (Diptera: Agromyzidae) was sent to Hawaii (Harley 1969), before being introduced and released in Australia in 1974. Harley also reported the occurrence of several agents, namely *T. prolixa* (1974), the pathogen *Prospodium tuberculatum* (Speg.) Arthur (*Pucciniales: Uropyxidaceae*) (2001) and the herring-bone leaf-mining fly *Ophiomyia camarae* Spencer (Diptera: Agromyzidae) (2007), which were eventually released into Australia. He also found the eriophyid mite *Aceria lantanae* Cook (Acari: Eriophyidae). An application for the release of this species has just been submitted, following testing in South Africa. The pathogen *Puccinia lantanae* Farl. (*Pucciniales: Pucciniaceae*) is undergoing host-specificity testing in the UK.

From 1969 to 1978, John Winder, based in Brazil, conducted surveys on lantana for CSIRO. A total of 345 species were found during this time, of which several had already been released into Australia. Three new agents, *T. prolixa*, the flea-beetle *Alagoasa parana* Samuelson (Coleoptera: Chrysomelidae) and the leaf-mining beetle *Uroplata lantanae* Buzzi and Winder, were introduced but all failed to establish (Winder and Harley 1983; Julien and Griffiths 1998).

Surveys were conducted in Costa Rica by George Diatloff, QDL in 1974–76, resulting in the release of three agents, namely the two hispine beetles *O. championi* and *U. fulvopustulata*, both of which established, and the leaf-feeding moth *Autoplusia illustrata* Guenée (Lepidoptera: Noctuidae) which failed to establish.

From 1988 to 1992, Bill Palmer, QDL conducted surveys in Mexico and southern USA. A total of 261 arthropods and two rust fungi were found on lantana (Palmer and Pullen 1995). From these surveys, *A. compressa* and *A. championi*, which had been reported in earlier surveys by Mann and Krauss, were imported into Australia and later released but only *A. compressa* established. Other species which were thought to show potential were the sap-sucking bug *Adfalconia intermedia* Distant (Hemiptera: Miridae), which was later imported as *Falconia intermedia* via South Africa and released in 2000, *A. lantanae* which is awaiting approval to release, the leaf-mining moth *Cremastobombycia lantanella* Busck (Lepidoptera: Gracillaridae) and the stem-galling fly *Eutreta xanthochaeta* Aldrich (Diptera: Tephritidae) (both successfully introduced into Hawaii in 1902), the root-feeding beetle *Parevander xanthomelas* (Guérin-Melville) (Coleoptera: Cerambycidae), which was introduced unsuccessfully to Hawaii in 1902, and a flower-feeding weevil *Geraeus* sp. One of the rust species *P. tuberculatum*, sighted during these surveys, was later introduced from Brazil and released in Australia in 2000. The rust *P. lantanae* is undergoing host-specificity testing in the UK.

Surveys of fungi associated with lantana were conducted by CABI in southern Brazil in 1988–89 and 1995–96. Ten fungal species were found during the first survey, of which *P. tuberculatum* was introduced into Australia, following host-specificity testing by CABI. *Mycovellosiella lantanae* (Chupp) Deighton (Mycosphaerellaceae) was introduced into South Africa. *P. lantanae*, which is being studied, was also found. Surveys in 1995–96 found an additional six species attacking lantana but no further work has been conducted on these.

Cesar Garcia, contracted by QDL to search for agents on *Parthenium hysterophorus* L. (Asteraceae) in South America, conducted a few surveys in southern Brazil on lantana in the early 1990s, resulting in the introduction of the leaf-feeding beetle *Charidotis pygmaea* Klug (Coleoptera: Chrysomelidae) and the leaf-rolling moth *Ectaga garcia* Becker (Lepidoptera: Oecophoridae), but both failed to establish (see chapter on *Lantana montevidensis* (Spreng.) Briq., this volume).

Various researchers (including Stefan Neser, Fritz Heystek and Alan Urban) from PPRI, South Africa have conducted numerous surveys from the 1980s to the present throughout the range of *Lantana* spp. in Argentina, Belize, Brazil, Costa Rica, Cuba, Dominica, Dominican Republic, Guatemala, Jamaica, Lesser Antilles, Mexico, Paraguay, Puerto Rico, St Christopher, St Lucia, St Vincent, Trinidad, USA (Florida and Texas) and Venezuela (A Urban, *pers. comm.* 2010).

While these surveys were not necessarily conducted on behalf of Australia (although the author accompanied Stefan Neser to Florida, Guatemala and Mexico in 1999), Australia benefited, with the introduction and establishment of *F. intermedia* from Jamaica (2000) and *O. camarae* from Florida (2007), following their establishment in South Africa. The flea-beetle *Alagoasa extrema* Jacoby (Coleoptera: Chrysomelidae) was imported from cultures in South Africa in 2002, but was not released as it lacked specificity. The bud mite *A. lantanae* collected in Mexico, Florida and Jamaica was tested in South Africa, using Australian test plants. Results show the agent is specific to *L. camara* and an application to AQIS and DEWHA has been submitted. Several other agents have subsequently been released in South Africa but their effectiveness is still to be assessed (see below).

Collaboration between countries has always been a strong feature in lantana biocontrol research, with a long history of successful agents being moved from country to country, following their initial introduction (Julien and Griffiths 1998). *Teleonemia scrupulosa* is now present in 32 countries worldwide, following its first introduction into Hawaii in 1902, and *U. girardi* is now present in 24 countries, following its introduction into Australia in 1966 (Day *et al.* 2003b). More recently, Australia has gained from research undertaken by South Africa and CABI in the UK, with the introduction or testing of five species. Thus, while surveys are conducted for particular countries at the time, other countries may benefit directly or indirectly years later.

CANDIDATES

Despite the earlier (pre 1960) efforts, weedy lantana was not considered adequately controlled anywhere within its introduced geographical range. Since 1960 CSIRO, in collaboration with researchers in Hawaii and Brazil, and QDL, in collaboration with South Africa, Brazil and CABI, introduced a further 23 agents into Australia, of which 12 established (Table 2) (Julien and Griffiths 1998; Day *et al.* 2003b). The mealybug *Phenacoccus parvus* Morrison (Hemiptera: Pseudococcidae) was accidentally introduced in 1988 (Swarbrick and Donaldson 1991).

Agents established in Australia

***Octotoma championi* Baly (Coleoptera: Chrysomelidae).** This leaf-mining hispine beetle was released widely in coastal and subcoastal NSW and Qld, but is found only in low numbers and causing minimal damage from Terrigal (33°26'S 151°26'E) to Mt Dromedary (36°17'S 150°05'E) and on Norfolk Island and on the Atherton Tableland in northern Qld. It has not been found at any site between Terrigal and the Atherton Tableland, nor has it established in any other country where it was introduced (Julien and Griffiths 1998; Day *et al.* 2003a, b).

***Octotoma scabripennis* Guérin-Méneville (Coleoptera: Chrysomelidae).** This leaf-mining beetle is one of the most widely distributed and damaging agents present on lantana in Australia. It is found from Shoalwater Bay (22°50'S 150°19'E) in central Qld to Kiama (34°39'S 150°50'E) in NSW. Very small populations have also been found around Hidden Valley (18°58'S 146°01'E) in northern Qld and on the Atherton Tableland. It is most abundant from Kempsey (31°05'S 152°50'E) to Gympie (26°12'S 152°40'E), where it can cause severe defoliation and reduced flowering in late summer and early autumn. Insect numbers decrease in winter when conditions become cool and dry and plants become leafless. While plants can be severely damaged seasonally by *O. scabripennis*, lantana infestations are not necessarily permanently suppressed nor is control achieved (Day *et al.* 2003a).

O. scabripennis has established in six of the 12 countries in which it has been introduced. In those countries, in conjunction with other established agents, it achieves partial to moderate control of lantana (Julien and Griffiths 1998; Day *et al.* 2003b).

***Uroplata fulvopustulata* Baly (Coleoptera: Chrysomelidae).** This is one of three leaf-mining hispine beetles from the genus *Uroplata* to be introduced into Australia (Julien and Griffiths 1998). It was released in northern NSW and on lantana throughout coastal and subcoastal Qld but is currently found only in northern Qld, usually in moderate numbers and causing minimal damage. Larger populations causing moderate seasonal damage have been observed around Tinaroo (17°08'S 145°32'E) on the Atherton Tableland (Day *et al.* 2003a). *U. fulvopustulata* has also been introduced into Fiji and South Africa, but failed to establish in either (Julien and Griffiths 1998).

Table 2: Origin and status of agents released on *Lantana camara* in Australia since 1960

Agent	Country	Year released	Organisation	Established	Status
Diastema tigris	Panama	1965	QDPL	No	
Hypena laceratalis	Kenya	1965	QDPL, CSIRO	Yes	b
Octotoma scabripennis	Mexico	1966	CSIRO	Yes	a
Uroplata girardi	Brazil	1966	CSIRO	Yes	a
Plagiohammus spinipennis	Mexico	1967	CSIRO	No	
Leptobyrsa decora	Colombia	1969	CSIRO	Yes	c
Teleonemia elata	Brazil	1969	CSIRO	No	
Teleonemia harleyi Froeschner	Trinidad	1969	CSIRO	No	
Calycomyza lantanae	Trinidad	1974	CSIRO	Yes	b
Teleonemia prolixa	Brazil	1974	CSIRO	No	
Octotoma championi	Costa Rica	1975	QDL	Yes	d
Autoplusia illustrata	Costa Rica	1976	QDL	No	
Uroplata fulvopustulata	Costa Rica	1976	QDL	Yes	d
Uroplata lantana	Brazil	1977	CSIRO	No	
Alagoasa parana	Brazil	1981, 1998	CSIRO	No	
Phenacoccus parvus		1988	*	Yes	c
Ectaga garcia	Brazil	1993	QDL	No	
Aerenicopsis championi	Mexico	1995	QDL	No	
Charidotis pygmaea	Brazil	1994	QDL	No	
Aconophora compressa	Mexico	1995	QDL	Yes	a
Falconia intermedia		2000	QDNR&M	Yes	c
Prospodium tuberculatum	Brazil	2001	QDNR&M	Yes	e
Ophiomyia camarae	Florida	2007	QDPI&F	Yes	e

CSIRO = Commonwealth Scientific Industrial Research Organisation; QDL = Qld Dept of Lands; QDNR&M = Qld Dept of Natural Resources and Mines; QDPL = Qld Dept of Public Lands; QDPI&F = Qld Dept of Primary Industries and Fisheries.
*: Agent was accidentally introduced.
a: Widespread and damaging.
b: Widespread and minimal damage.
c: Localised and damaging.
d: Localised and minimal damage.
e: Widespread, too early to assess damage.

***Uroplata girardi* Pic (Coleoptera: Chrysomelidae).** This is one of the most widespread and damaging agents present on lantana in Australia (Day *et al.* 2003a). It is found mainly from Cape Tribulation (15°57'S 145°21'E) to Laurieton (31°39'S 152°47'E) in mid NSW, with smaller populations on Norfolk Island and very low numbers around Sydney (33°32'S 150°38'E) (Day *et al.* 2003a). It is common in warm dry areas, particularly in subcoastal regions of southern and central Qld. Here, it can be quite damaging and is often one of a few agents present. In cooler areas, *U. girardi* causes minimal damage to plants. It causes similar damage to that of *O. scabripennis* and the two species are often found together, resulting in substantial defoliation seasonally. Like *O. scabripennis*, populations of *U. girardi* decrease in dry winter months and are not maintained at high enough levels to control lantana (Day *et al.* 2003a). *U. girardi* has established in 24 of the 26 countries in which it has been introduced (Julien and Griffiths 1998; Day *et al.* 2003b) and has been the dominant agent controlling lantana in Guam and the Solomon Islands (Day *et al.* 2003b).

***Calycomyza lantanae* (Frick) (Diptera: Agromyzidae).** This leaf-mining fly is common from Cooktown (15°29'S 145°14'E) to Kempsey. It has also been found around Darwin and Katherine in the NT, although it was never deliberately released in those areas (Day *et al.* 2003a). It is more common in Qld than in NSW as it prefers warm moist areas; populations build quickly over summer following rain. Populations are greatest in the tropics, in late summer and autumn. In temperate areas populations are usually smaller, resulting in little damage to plants (Day *et al.* 2003a). This fly has established in 15 countries worldwide (Julien and Griffiths 1998; Day *et al.* 2003b).

C. lantanae also occurs on two other introduced species, *L. montevidensis* in south-eastern and central coastal and subcoastal Qld and *Lippia alba* (Mill.) N.E.Br. ex Britton & P.Wilson (Verbenaceae). *L. alba* has a limited distribution, being found near Blackwater (23°33'S 148°45'E) in central Qld and in Lakefield National Park (15°15'S 144°11'E) in far-north Qld (Day *et al.* 2003a).

***Ophiomyia camarae* Spencer (Diptera: Agromyzidae).** This leaf-mining fly was introduced into Qld following its release and subsequent establishment in South Africa (Day *et al.* 2009). The fly has been released throughout coastal and subcoastal Qld and has established from Cooktown to Bundaberg (24°52'S 152°28'E). Field releases are continuing in the north, but have been stopped in the south-east where establishment has been poor. CLIMEX models suggest that Bundaberg is about the southern limit of its potential distribution (Day *et al.* 2009). The fly has spread over 30 km from some sites and is causing defoliation around Cooktown, but it is too early to determine its overall impact on lantana. The fly was released in 2001 in South Africa, but it is not as effective as laboratory studies first suggested and is attacked by parasitoids (Simelane and Phenye 2005; A Urban, *pers. comm.* 2010).

***Ophiomyia lantanae* (Froggatt) (Diptera: Agromyzidae).** The seed-fly *O. lantanae* was one of the first agents introduced into Australia, in 1914, and is now present in almost all areas where lantana occurs, from the Torres Strait islands to Eden (37°03'S 149°52'E) in southern NSW. It is also present around Darwin, Perth, Geraldton and on Norfolk Island (Day *et al.* 2003a). The fly is common in warm humid areas where lantana flowers and sets seed readily, but is found only in low numbers in more temperate areas.

The fly can damage up to 85% of fruits in late summer or autumn, but is less common in winter when temperatures are cooler and flowering is reduced. The maggots feed on the fleshy pulp surrounding the seed, but do not damage the embryos (Broughton 1999). Fruit damaged by *O. lantanae* is thought to be less attractive to birds, which are the main vectors for the spread of lantana. Consequently, the presence of *O. lantanae* may decrease the dispersal of lantana (Day *et al.* 2003b). However, due to the enormous quantities of seed produced by lantana plants, there are still many fruits left undamaged by *O. lantanae* even when fly populations are at their highest (Day *et al.* 2003a). *O. lantanae* has established in 23 of the 28 countries in which it was introduced (Julien and Griffiths 1998; Day *et al.* 2003b).

***Aconophora compressa* Walker (Hemiptera: Membracidae).** This stem-sucking bug is widespread along coastal and subcoastal eastern Australia, being found from Gladstone (23°51'S 151°15'E) to Kempsey. Populations are also found on the Atherton Tableland and around Mt Fox (18°50'S 145°53'E) and Sydney (Day *et al.* 2003a). *A. compressa* has the potential to be a useful agent as its feeding habits are independent of whether the plant is in leaf. Adults and nymphs suck the sap, causing branches to die, and there appears to be reduced flowering and seeding compared to uninfested lantana growing nearby. It appears to be a particularly damaging agent, with populations highest in winter and spring. However, high temperatures in summer can cause populations to crash (Day *et al.* 2003a).

A. compressa also attacks *Citharexylum spinosum* L. (fiddlewood) (Verbenaceae), causing significant leaf-drop and a large production of honey-dew, leading to problems for residents who possess these trees. There were reports of *A. compressa* attacking other species, but additional host-range studies suggested that populations cannot be maintained on any species other than lantana and fiddlewood (Day *et al.* 2003a). *A. compressa* has not been released in any other country due to its lack of specificity.

***Falconia intermedia* (Distant) (Hemiptera: Miridae).** This sap-sucking bug was introduced into Australia from South Africa, where it was reported to heavily damage lantana (Day *et al.* 2003a, b). It was released at approximately 50 sites throughout coastal and subcoastal Qld and coastal NSW from Sydney to the Qld border. However, it has established only on the Atherton Tableland where it is spreading and causing extensive damage to lantana and an apparent reduction in flowering, although this has not yet been quantified. It feeds on the intracellular tissue, causing necrosis and premature abscission of leaves. *F. intermedia* has been found only on the pink-edged red flowering variety. It does not attack the common pink flowering variety growing near where it has established (Day *et al.* 2003a).

***Phenacoccus parvus* Morrison (Hemiptera: Pseudococcidae).** This mealybug was reported to have been accidentally introduced to Australia in 1988. It quickly established and spread throughout southern Qld (Marohasy 1997). It has since been found in isolated populations from the Atherton Tableland to Kempsey, as well as around Perth. *P. parvus* is most common in northern NSW and southern Qld, where it can periodically cause severe damage to plants. It has been reported attacking several economically important plants such as tomato and aubergine (eggplant) in the Solanaceae (Marohasy 1997) and is therefore not recommended for redistribution or release in other countries.

Leptobyrsa decora **Drake (Hemiptera: Tingidae).** This sap-sucking bug was released in large numbers throughout lantana infestations in Qld and NSW. However, it is only known to have established on the Atherton Tableland. Populations in open sunny areas reach damaging numbers in late summer and plants attacked by *L. decora* have fewer leaves, flowers and seeds than uninfested plants growing under similar conditions nearby. *L. decora* is not found where lantana grows under a dense forest canopy or in high-rainfall areas (Day *et al.* 2003a). *L. decora* has established in Hawaii but failed to establish in South Africa, Ghana and several countries in the South Pacific (Julien and Griffiths 1998).

Teleonemia scrupulosa **Stål (Hemiptera: Tingidae).** This is one of the most widespread and damaging agents of lantana. It is found from Cooktown to Thirroul (34°18'S 150°55'E), south of Sydney and is present around Darwin and on Norfolk Island. It prefers drier, more exposed regions and is rarely found in areas with high rainfall or where lantana grows under a dense forest canopy. It is more common on north-facing than south-facing slopes. *T. scrupulosa* can cause substantial defoliation of lantana on a seasonal basis, resulting in plants producing fewer new shoots and flowers. It has been particularly damaging in the dry regions of central subcoastal NSW and southern and central subcoastal Qld (Day *et al.* 2003a).

T. scrupulosa has established in 29 of the 31 countries in which it has been introduced and is regarded as one of the most damaging agents on lantana (Julien and Griffiths 1998; Day *et al.* 2003b). It has been reported to feed occasionally on other plant species (Day *et al.* 2003b).

Hypena laceratalis **Walker (Lepidoptera: Noctuidae).** This moth is one of three leaf-feeding Lepidoptera established on lantana in Australia. It is more common in tropical and subtropical regions of Qld, being found from Cow Bay (16°13'S 145°26'E) to Kempsey. *H. laceratalis* has also been found on *L. montevidensis* in moist areas around Brisbane and central subcoastal Qld (Day *et al.* 2003a). It is most common in late summer and autumn, when plant growth is lush and insect populations have had a chance to increase over summer. Damage to lantana is seasonal and localised and, even under favourable conditions, is not sufficient to control lantana. *H. laceratalis* has been introduced and established in 14 countries, causing minimal to moderate levels of damage (Julien and Griffiths 1998; Day *et al.* 2003b).

Neogalea sunia **(Guenée) (Lepidoptera: Noctuidae).** *Neogalea sunia* (previously *Catabena esula*) is now found from Kuranda (16°48'S 145°38'E) to Eden. It is particularly common in drier regions of northern NSW, near Grafton (29°37'S 152°56'E) and throughout subcoastal regions of southern Qld. Together with other biocontrol agents, *N. sunia* can seasonally defoliate plants, reducing development of new shoots (Day *et al.* 2003a). Larvae are well camouflaged, lying longitudinally along the stems during the day and feeding at night. *N. sunia* has been introduced to five other countries but has established only in Hawaii and New Caledonia (Julien and Griffiths 1998).

Lantanophaga pusillidactyla **(Walker) (Lepidoptera: Pterophoridae).** This flower-feeding moth appears to have been accidentally introduced, possibly brought in on imported plants (Julien and Griffiths 1998). It is one of the most widespread biological control agents for lantana in Australia, and is found from Cooktown to Merimbula (36°53'S 149°54'E) in southern NSW and around Darwin and on Norfolk Island (Day *et al.* 2003a). It has also been found on *L. alba* in central Qld and is common on *L. montevidensis* in both naturalised populations and gardens in southern Qld and northern NSW. *L. pusillidactyla* is more common in warmer coastal regions but small numbers have been found in temperate, higher-altitude subcoastal areas. Populations are greatest in late summer, although they have limited impact on lantana as plants still produce many seeds (Day *et al.* 2003a). *L. pusillidactyla* has established in 11 countries and is reported to assist in reducing seed production by up to 80% in Micronesia (Denton *et al.* 1991).

Salbia haemorrhoidalis **Guenée (Lepidoptera: Pyralidae).** This moth is more abundant in warm, moist tropical regions, being found from Cow Bay to Coffs Harbour (30°16'S 153°08'E) in mid coastal NSW. It is also present on Norfolk Island (Day *et al.* 2003a). *S. haemorrhoidalis* causes only moderate damage to lantana and has established in seven of 13 countries in which it has been introduced (Julien and Griffiths 1998; Day *et al.* 2003b).

Epinotia lantana **(Busck) (Lepidoptera: Tortricidae).** This was one of the four original agents imported into Australia in 1914. Larvae of the moth feed on flowers and/or bore into shoot tips or flower peduncles. The moth is found throughout most areas where lantana occurs from Cooktown to Ulladulla (35°20'S 150°28'E) in southern

NSW, as well as around Darwin and on Norfolk Island. However, it is generally more common in warmer coastal districts (Day *et al.* 2003a).

Although *E. lantana* can be seasonally abundant, it does not appear to have a significant impact on the plant and seed production remains high. It has established in all nine countries in which it has been introduced, but impact has been variable (Julien and Griffiths 1998). Damage caused by *E. lantana* appears to be significant in tropical islands in the Pacific, with seed production being reduced by up to 80% in Micronesia (Denton *et al.* 1991).

***Prospodium tuberculatum* (Speg.) Arthur (*Pucciniales: Uropyxidaceae*).** This was the first rust and the 30th agent to be released on lantana in Australia. It has been released throughout coastal and subcoastal Qld and NSW. This autoecious rust is widespread from Maryborough (25°36'S 152°47'E) to Sydney, with populations also on the Atherton Tableland and around Paluma (19°00'S 146°11'E) in northern Qld. Releases are continuing in NSW only. Spores are wind-dispersed and attach to the underside of leaves. The rust causes chlorosis and premature abscission of leaves, but does not seem to have a significant impact on lantana (Day *et al.* 2003a, b). *P. tuberculatum* is highly host-specific, affecting only the common pink flowering lantana variety in Australia (Tomley and Riding 2002).

Agents released but not established

In Australia, 13 agents have been released but failed to establish (Tables 1 and 2). Possible reasons for non-establishment include incompatibility of host plant or climate, or inadequate numbers released due to rearing difficulties (Day *et al.* 2003a, b). More information on each of these agents can be found in Day *et al.* (2003b).

Agents imported but not released

Three potential agents, the stem-boring beetle *Aerencia multipunctata* Le Peletier and Audinet-Serville (Coleoptera: Cerambycidae), the leaf-feeding beetle *A. extrema* and the tingid *Teleonemia validicornis* Stål (Hemiptera: Tingidae) were introduced but not released. The last two were not sufficiently host-specific. More information on these agents can be found in Day *et al.* (2003b).

Agents being tested

***Aceria lantanae* (Cook) (Acari: Eriophyidae).** There are two biotypes of this bud mite, one that attacks the inflorescences and one that attacks the leaves, placing physiological pressures on the plant so that production of new shoots, flowers and seeds is reduced (Craemer and Neser 1996). Populations of the bud mite that attacks the inflorescences have been collected from several countries including the USA (Florida), Jamaica and Mexico by PPRI, South Africa and have been subjected to host-specificity testing. *A. lantanae* appears to be host-specific, inducing galls in only some weedy lantana varieties while not attacking any of the native South African *Lantana* spp. (Urban *et al.* 2001). An application to AQIS and DEHA seeking its field release in Australia has been submitted.

***Puccinia lantanae* Farl. (*Pucciniales: Pucciniaceae*).** This rust fungus is common on lantana in tropical areas of Brazil, but less common on this plant in subtropical regions. Host testing and varietal susceptibility tests have been conducted by CABI and the rust is believed to have a narrow host-range. Symptoms have appeared on *Phyla canescens* (Kunth) Greene (a weed in Australia) and *Verbena africana* (R. Fern. & Verdc.) P.W. Michael (C Ellison, *pers. comm.* 2009). Additional tests are continuing to quantify these results before an application seeking the release of *P. lantanae* is submitted.

Other promising agents

Several other agents have been identified as showing promise as biocontrol agents. Two of these, the petiole-galling beetle *Coelocephalapion camarae* Kissinger (Coleoptera: Brentidae) and the root-feeding weevil *Longitarsus bethae* Savini and Escalona (Coleoptera: Chrysomelidae) have been released in South Africa by PPRI and their impact is being monitored. The stem sap-sucking bug *Barela* sp. was also imported into South Africa, but was not released as it attacked native lantana (A Urban, *pers. comm.* 2009). This insect could be considered for introduction into Australia as no native lantana species exist here. *Eutreta xanthochaeta*, which was introduced into Australia in 1914 and again in the 1970s, but failed to establish, is aiding control of lantana in Hawaii. It is still thought to be a beneficial agent if mass-rearing problems can be overcome, to allow adequate field releases. Details of most of these agents can be found in Day *et al.* (2003b).

DISCUSSION

Biological control of lantana has been in progress, albeit not continuously, since 1902 globally and since 1914 in Australia. Over 40 agents have been released worldwide and 31 in Australia, with 18 agents establishing in Australia. Despite these efforts, lantana is considered under control in only a few countries, including Guam and the

Solomon Islands (Zalucki *et al.* 2007); elsewhere, seasonal damage and defoliation is common but not sustained (Day *et al.* 2003b).

Workshops and reviews have evaluated the Australian lantana program, factors limiting successful biocontrol and the need for possible future research (Broughton 2000; Day and Neser 2000; Day *et al.* 2003a, b; Zalucki *et al.* 2007; Day and Zalucki 2009). A major limiting factor is that lantana has the ability to withstand long periods of defoliation induced by drought, cold temperatures or insect attack. About half the agents that have established feed on the leaves, with populations increasing over summer and autumn, causing plants to defoliate, only to crash over winter when it is too cold or dry. Populations rise again only when temperatures increase and plant health improves, so damage is never sustained. Instead of biocontrol agents suppressing plant health, it appears that plant health is a major driver of agent populations (Day *et al.* 2003b).

Second, lantana grows in a wide geographical and climatic range, from the wet tropics to the NSW–Vic border, as well as around very hot and dry places such as Rockhampton and Charters Towers (Day *et al.* 2003a). Therefore, it is unlikely that any one agent will be adapted to all places where lantana is found.

A third possible factor limiting successful biocontrol is the lantana variety, given that the lantana present in Australia is not the same as any lantana in the native range and that not all agents attack all varieties (Day and Neser 2000). Six of the 13 agents that failed to establish showed varietal preference in the laboratory, and three of the 18 agents that established show preference for one or more varieties in the laboratory and field (Day *et al.* 2003a).

Parasitism was suggested as another limiting factor (Broughton 2000; Day and Neser 2000), but few studies have been conducted (e.g. Broughton 2001). It is suspected that levels may be low, as some agent populations still reach high numbers during summer.

From the 1990s, organisations in Australia and South Africa selected new agents that attack the roots and stems in an effort to improve biocontrol. Although *A. compressa* established and causes severe seasonal damage, populations may crash over summer, following heat waves. *Aerenicopsis championi* showed great potential in the glasshouse but it proved difficult to rear, was released in low numbers and failed to establish (Day *et al.* 2003a). Apart from the two agents (*A. lantanae* and *P. lantanae*) currently undergoing host-specificity testing, consideration will be given to two agents, *L. bethae* and *C. camarae* recently released in South Africa. Field assessments of these agents should provide information on their potential as useful agents. A further two agents, *E. xanthochaeta* and *Barela* sp., may be considered pending the success of the above four agents (Day *et al.* 2003b).

Several studies have identified methods to improve the current levels of control through integrated practices (Day *et al.* 2003b; NRME 2004). The rationale of these studies is to determine methods of control that can be utilised once local populations of biocontrol agents have suppressed or damaged lantana. Such methods include the use of basal bark herbicide applications, fire, stick-raking and manual control when access to the plant is easier. Fundamental to this is the education of land managers, who often rely on biocontrol as the only method of control (NRME 2004).

Although the prospects for complete biocontrol of lantana are limited and possibly unlikely, the opportunities to improve the level of control through use of more effective and adapted agents integrated with other weed management techniques are quite promising. Even minor improvements in overall biocontrol can have big benefits, with economic analyses showing that reducing lantana density by only 5% can produce a nine-fold return on investment (AEC Group 2007).

ACKNOWLEDGEMENTS

The main international agencies involved in exploration and research have been the State Department of Agriculture, Hawaii, Biosecurity Queensland (formerly Qld Department of Public Lands), CSIRO, Plant Protection Research Institute, South Africa and CABI, with Centre of Origin Research and Botanical Research Institute of Texas providing valuable assistance in lantana taxonomy. Within Australia, NSW Department of Industry and Investment, CRC Australian Weed Management, University of Queensland and University of New England have collaborated on research.

REFERENCES

AEC Group (2007) *Economic Impact of Lantana on the Australian Grazing Industry.* Qld Dept of Natural Resources and Water, Brisbane.

Broughton S (1999) Impact of the seed-fly, *Ophiomyia lantanae* (Froggatt) (Diptera: Agromyzidae), on the viability of lantana fruit in south-east Queensland, Australia. *Biological Control* **15**, 168–172.

Broughton S (2000) Review and evaluation of lantana biocontrol programs. *Biological Control* **17**, 272–286.

Broughton S (2001) Parasitism and predation of the lantana leaf mining beetles *Octotoma scabripennis* Guérin-Méneville and *Uroplata girardi* Pic (Coleoptera: Chrysomelidae: Hispinae) in Australia. *Australian Journal of Entomology* **40**, 286–293.

Craemer C and S Neser (1996) Eriophyoid mites (Acari: Eriophyoidea) as possible control agents of introduced plants in South Africa. In *Proceedings of the 9th International Symposium on Biological Control of Weeds.* 19–26 January 1996, Stellenbosch, South Africa. (Eds VC Moran and JH Hoffman) p. 228. University of Cape Town, South Africa.

Day MD and Neser S (2000) Factors influencing the biological control of *Lantana camara* in Australia and South Africa. In *Proceedings of the 10th International Symposium on Biological Control of Weeds.* 4–14 July 1999, Bozeman, Montana. (Ed NR Spencer) pp. 897–908. Montana State University, Bozeman.

Day MD and Zalucki MP (2009) *Lantana camara* Linn. (Verbenaceae). In *Biological Control of Tropical Weeds using Arthropods.* (Eds R Muniappan, GVP Reddy and A Raman) pp. 211–246. Cambridge University Press, Cambridge.

Day MD, Broughton S and Hannan-Jones MA (2003a) Current distribution and status of *Lantana camara* and its biological control agents in Australia, with recommendations for further biocontrol introductions into other countries. *Biocontrol News and Information* **24**, 63N–76N.

Day MD, Wiley CJ, Playford J and Zalucki MP (2003b) *Lantana: Current Management Status and Future Prospects.* ACIAR, Canberra.

Day MD, Riding N and Chamberlain A (2009) Biology and host range of *Ophiomyia camarae* Spencer (Diptera: Agromyzidae), a potential biocontrol agent for *Lantana* spp. (Verbenaceae) in Australia. *Biocontrol Science and Technology* **19**, 627–637.

Denton GRW, Muniappan R and Marutani M (1991) The distribution and biological control of *Lantana camara* in Micronesia. *Micronesica* Suppl. **3**, 71–81.

Harley KLS (1969) 'Report on a survey of South America and Mexico seeking agents for the biological control of *Lantana camara* L.'. CSIRO, Long Pocket Laboratories, Brisbane.

Holm LG, Plucknett DL, Pancho JV and Herberger JP (1991) *A Geographic Atlas of World Weeds.* Krieger Publishing, Malabar, Florida.

Howard RA (1969) A check list of cultivar names used in the genus *Lantana*. *Arnoldia* **29**, 73–109.

Julien MH and Griffiths MW (1998) *Biological Control of Weeds: A World Catalogue of Agents and their Target Weeds.* 4th edn. CABI, Wallingford, UK.

Julien M, Storrie A and McCosker R (2004) Lippia, *Phyla canescens*, an increasing threat to agriculture and the environment. *Proceedings of the 14th Australian Weeds Conference.* 6–9 September 2004, Wagga Wagga, NSW. pp. 476–479.

Mann J (1954) 'Lantana investigations in Mexico July 1953 to January 1954 and a brief summary of the investigations by N.L.H. Krauss in Florida, Mexico and Central America April 1953 to January 1954'. Qld Dept of Public Lands, Biological Section Laboratory, Brisbane.

Marohasy J (1997) Acceptability and suitability of seven plant species for the mealybug *Phenacoccus parvus. Entomologia Experimentalis et Applicata* **84**, 239–246.

Munir AA (2002) A taxonomic revision of the genus *Verbena* L. (Verbenaceae) in Australia. *Journal of the Adelaide Botanical Gardens* **18**, 21–103.

NRME (Dept of Natural Resources, Mines and Energy) (2004) *Lantana Control Manual: Current Management and Control Options for Lantana* (Lantana camara) *in Australia.* Qld Dept of Natural Resources, Mines and Energy, Brisbane.

Palmer WA and Pullen KR (1995) The phytophagous arthropods associated with *Lantana camara, L. hirsuta, L. urticifolia,* and *L. urticoides* (Verbenaceae) in North America. *Biological Control* **5**, 54–72.

Sanders RW (2006) Taxonomy of *Lantana* sect. *Lantana* (Verbenaceae): I. Correct application of *Lantana camara* and associated names. *Sida* **22**, 381–421.

Scott LJ, Hannan-Jones MA and Graham GC (2002) Affinities of *Lantana camara* in the Australia-Pacific region. In *Proceedings of the 13th Australian Weeds Conference.* 8–13 September, 2002, Perth. (Eds H Spafford-Jacob, J Dodd and JH Moore) pp. 471–474. Plant Protection Society of WA, Perth.

Simelane DO and Phenye MS (2005) Suppression of growth and reproductive capacity of the weed *Lantana camara* (Verbenaceae) by *Ophiomyia camarae* (Diptera: Agromyzidae) and *Teleonemia scrupulosa* (Heteroptera: Tingidae). *Biocontrol Science and Technology* **15**, 153–163.

Smith LS and Smith DA (1982) The naturalised *Lantana camara* complex in eastern Australia. *Queensland Botany Bulletin* **1**, 1–26.

Stirton CH (1977) Some thoughts on the polyploidy complex *Lantana camara* L. (Verbenaceae). In *Proceedings of the 2nd National Weeds Conference of South Africa.* 2–4 February 1977, Stellenbosch, South Africa. (Ed DP Annecke) pp. 321–340. AA Balkema, Cape Town.

Swarbrick JT (1986) History of the lantanas in Australia and origins of the weedy biotypes. *Plant Protection Quarterly* **1**, 115–121.

Swarbrick JT and Donaldson JF (1991) Host range studies with the Lantana mealybug (*Phenacoccus parvus* Morrison). *Plant Protection Quarterly* **6**, 68–69.

Swarbrick JT, Willson BW and Hannan-Jones MA (1998) *Lantana camara* L. In *The Biology of Australian Weeds.* (Eds FD Panetta, RH Groves and RCH Shepherd) pp. 119–140. RG and FJ Richardson, Melbourne.

Swezey OH (1924) Present status of lantana and its introduced insect enemies. In *The Introduction into Hawaii of Insects that Attack Lantana.* (Eds RCL Perkins and OH Swezey). *Bulletin of the Experiment Station of the Hawaiian Sugar Planters' Association* **16**, 1–83.

Tomley AJ and Riding N (2002) *Prospodium tuberculatum*, lantana rust, a new agent released for the biocontrol of the woody shrub *Lantana camara.* In *Proceedings of the 13th Australian Weeds Conference.* 8–13 September 2002, Perth. (Eds H Spafford-Jacob, J Dodd and JH Moore) pp. 389–390. Plant Protection Society of WA, Perth.

Urban AJ, Mpedi PF, Neser S and Craemer C (2001) Potential of the flower gall mite, *Aceria lantanae* (Cook) (Acari: Eriophyidae), for biocontrol of the noxious weed, *Lantana camara* L. (Verbenaceae). In *Proceedings of the 13th Entomological Congress.* 2–5 July 2001, Pietermaritzburg, South Africa. (Eds T Olckers and DJ Brothers) pp. 67–68. Entomological Society of Southern Africa, Hatfield.

Wilson F (1960) 'A review of the biological control of insects and weeds in Australia and Australian New Guinea'. Technical Communication No. 1. CAB Farnham Royal, Bucks, UK.

Winder JA and Harley KLS (1983) The phytophagous insects on Lantana in Brazil and their potential for biological control in Australia. *Tropical Pest Management* **29**, 346–362.

Zalucki MP, Day MD and Playford J (2007) Will biological control of *Lantana camara* ever succeed? Patterns, processes and prospects. *Biological Control* **42**, 251–261.

Lantana montevidensis (Spreng.) Briq. – creeping lantana

Michael Day

ABSTRACT

Lantana montevidensis is a low sprawling perennial weed of pastures and disturbed lands, particularly of central and southern Qld. Biological control began in the late 1980s, with surveys in southern Brazil, Uruguay and Argentina. Two agents, the leaf-rolling moth *Ectaga garcia* and the leaf-feeding beetle *Charidotis pygmaea* were released widely in Qld. However, neither established. It is thought that many of the areas where *L. montevidensis* is a problem in Qld are too hot and dry for these agents. Several other agents, including a root-feeding beetle and a rust, were considered worthy of further study. The project ceased in 2001 due to funding constraints.

Key words: varieties, *Ectaga garcia*, *Charidotis pygmaea*.

INTRODUCTION

Lantana montevidensis (Spreng.) Briq. (Verbenaceae), creeping lantana, is native to South America, being found in southern Brazil, Uruguay and Argentina (Moldenke 1983 in Swarbrick 1986; Sanders 2001). The first record of it in Australia was in 1851 (Shepherd 1851 in Swarbrick 1986). It was considered a common weed around Gayndah by 1913 and in other parts of Qld by 1925 (Swarbrick 1986). It is found mainly from Rockhampton, Qld to Casino in northern NSW, as well as around Cairns in northern Qld and Sydney and Nowra in NSW (Munir 1996; Day *et al.* 1998, 1999) (Fig. 1). Horticultural varieties are grown throughout Australia, being found in gardens in most capital cities and many regional towns. While cultivated in many other countries, it has been reported to be invasive in only Hawaii and New Caledonia (Pacific Islands Ecosystems at Risk 2010).

L. montevidensis is a low-growing perennial (Fig. 2) that infests pastures and disturbed land. It has the ability to outcompete preferred species, reducing production and lowering land values, and is toxic to cattle (Parsons and Cuthbertson 1992). Populations in central Qld are particularly serious because they affect the ongoing viability of many farms (O'Donnell 2002). In natural ecosystems, *L. montevidensis* infests open woodlands and dry sclerophyll forests and can become the dominant groundcover species, reducing biodiversity of these areas (Munir 1996; O'Donnell 2002) (Fig. 3).

Local councils and community groups in the Burnett region first wrote to the Qld Department of Lands in the 1960s to request biological control of creeping lantana, as regional infestations were continually spreading and control using conventional means was increasingly difficult.

PLANT TAXONOMY

L. montevidensis belongs to the *Lantana* section *Calliorheas* in Verbenaceae (Sanders 2001). There are several horticultural varieties of *L. montevidensis* which are supposedly sterile, with purple, mauve or white flowers. However, some of these plants have been known to set seed, albeit in low numbers (Neal 1999; M Day, *pers. obs.*). The yellow flowering form, considered by Munir (1996)

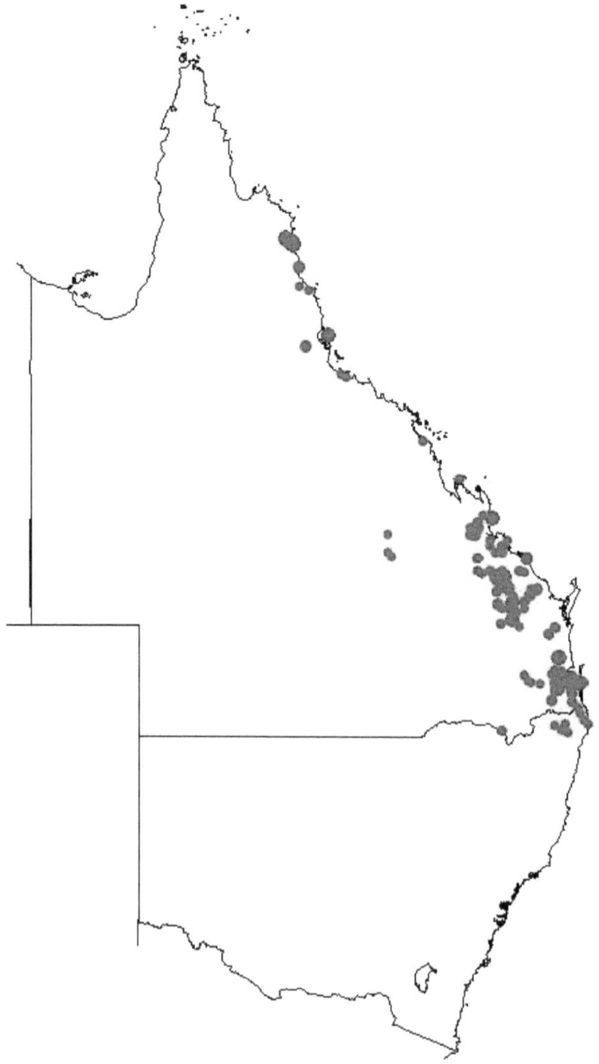

Figure 1: Distribution of *Lantana montevidensis* in Australia. Source: Qld Government and the Australian Virtual Herbarium.

Figure 2: Creeping lantana, *Lantana montevidensis*. Photo: M Day, Qld Government.

to be a *L. montevidensis* derivative, is actually a hybrid of *L. camara* L. and *L. depressa* Small (Sanders 2001). Cytological studies have shown that *L. camara* has a chromosome number of n = 11 while *L. montevidensis* has n = 12, therefore the two cannot hybridise (Sanders 2001). *L. montevidensis* does not have the same taxonomic problems as the weedy lantana, *Lantana camara* L. (section *Camara*), which comprises a complex of horticultural varieties arising from several *Lantana* species.

EXPLORATION

There have been several small dedicated surveys seeking potential agents for the control of *L. montevidensis*. These were conducted in southern Brazil, Uruguay and Argentina from 1988 to 2000 (Garcia 1988, 1989; Winder 1999, 2000). There was initial confusion between the identities of *L. montevidensis* and the closely related *Lantana fucata* Lindl. Thus, some early surveys were conducted on *L. fucata* (C Garcia, *pers. comm.* 1998; J Winder, unpub. report 2000).

Cesar Garcia surveyed in the southern states of Brazil, Soriano province in Uruguay and around Colon, Monte Caseros and Yapeyu along the Argentina–Uruguay and Argentina–Brazil borders in May 1988 and listed over 20 species of natural enemies, representing Diptera, Coleoptera, Hemiptera and Lepidoptera. He found the beetle *Charidotis pygmaea* Klug (Coleoptera: Chrysomelidae) and the moth *Ectaga garcia* Becker (Lepidoptera: Depressariidae), which were later imported and released in Australia. Garcia reported later that both were collected from *L. fucata*, however, not from *L. montevidensis*.

Three surveys were conducted in southern Brazil by John Winder between 1999–2000 in the states of São Paulo, Paraná, Santa Catarina and Rio Grande do Sul. Surveys in February 1999 were later found to have been conducted on *L. fucata* not on *L. montevidensis*. In addition, *L. montevidensis* was subsequently found only in Rio Grande do Sul and not in Curitiba Parana state, where *C. pygmaea* was originally collected, confirming its collection from *L. fucata* (Winder, *pers. comm.* 1999).

About 30 species were collected during the surveys but only a few potential agents were identified, with many thought to be the same as those collected by Garcia. The project ceased in 2001 due to lack of priority for funding.

Figure 3: *Lantana montevidensis* infestation in Kroombit Tops National Park, Qld. Photo: M Day, Qld Government.

CANDIDATES

***Ectaga garcia* Becker (Lepidoptera: Depressariidae).** This leaf-rolling moth was collected from *L. fucata* (section *Calliorheas*) around Curitiba in Parana state in southern Brazil and was introduced to control *L. camara* and *L. montevidensis* in 1993. In laboratory trials, it performed better on *L. camara*, which possessed larger leaves, and so was released mainly on this plant, as adults, pupae and larvae (Day *et al.* 1998). On *L. montevidensis*, it was released around Monto and the northern suburbs of Brisbane; on *L. camara*, it was released widely from Proserpine in northern Qld to southern NSW. It failed to establish on either species (Day *et al.* 2003a), possibly due to not being adapted to the species.

***Charidotis pygmaea* Klug (Coleoptera: Chrysomelidae).** This leaf-feeding beetle was first found on *L. fucata* around Curitiba, Parana state Brazil and later on *L. camara*. It was originally released into Australia to control *L. camara* in 1994. However, in preference and performance trials, the agent preferred and performed better on *L. montevidensis* and subsequent releases were conducted on that species only (Day *et al.* 1999). Over 40 000 beetles were released throughout central and south-eastern Qld but the agent failed to establish at any site. It was proposed that the climate in these regions was not suitable, being too hot and dry (Day and McAndrew 2002). During hot dry periods, plants can be leafless or possess leaves that are brown and tough, thus probably not suitable for young larvae.

Agents introduced to control L. camara and now utilising L. montevidensis

There are several agents introduced to control *L. camara* that also utilise *L. montevidensis* in the field. Although some are quite common and seasonally abundant, they do not provide a significant level of control.

***Calycomyza lantanae* Frick (Diptera: Agromyzidae).** This leaf-mining fly is seasonally common on *L. montevidensis* in all areas where it is a weed. It has also been found on *Lippia alba* (Miller) N. E. Brown ex Britton & P. Wilson. It is more prevalent in late summer and autumn but populations wane over winter or when plants lose their leaves due to dry conditions (Day *et al.* 2003a). It is very common on the horticultural varieties grown in gardens – some home-owners spray the insects to reduce their numbers and damage.

***Lantanophaga pusillidactyla* Walker (Lepidoptera: Pterophoridae).** This flower-feeding moth was thought to be introduced accidentally on imported lantana (Day *et al.* 2003b). It is fairly widespread, occurring on both the weedy and horticultural forms of *L. montevidensis*, even in gardens as far south as Eden, NSW and Adelaide, SA. However, the damage to creeping lantana is minimal and many plants remain able to produce many seeds, even in late summer (Day *et al.* 2003a).

***Hypena laceratalis* Walker (Lepidoptera: Noctuidae).** This leaf-feeding moth tends to be found on *L. montevidensis* growing in warm wet areas, where growth tends to be lush (Day *et al.* 2003a). It was found frequently on *L. montevidensis* growing under shadecloth, but the damage was minimal, and it was never found on *L. montevidensis* growing on exposed dry slopes.

Other potential agents

Several other natural enemies found during the surveys have been reported to show potential. These include a tingid bug, two species of hispine beetles, a cecidomyid fly, several leaf-feeding Lepidoptera and a rust (Garcia 1988; Winder 2000). None were identified to species level before the project terminated.

DISCUSSION

This project had a short life and was largely a spin-off from the *L. camara* project, so was never well resourced. The two agents, *C. pygmaea* and *E. garcia*, that were introduced failed to establish, possibly due to incompatibility with host and to climatic intolerances (Day and McAndrew 2002; Day *et al.* 2003a). A future project on *L. montevidensis* should include thorough surveys of its entire native range in Brazil, Uruguay and Argentina and the identification of new potential agents. A serious constraint will be locating potential agents that can withstand long periods of drought in Australia, during which the plant can become leafless and without flower.

ACKNOWLEDGEMENTS

Cesar Garcia and Dr John Winder conducted surveys and Cesar Garcia supplied agents.

REFERENCES

Day MD and McAndrew TD (2002) Status of *Charidotis pygmaea* (Coleoptera: Chrysomelidae) as a biological control agent of *Lantana montevidensis* (Verbenaceae) in Australia. *Biological Control* **23**, 27–34.

Day MD, Willson BW and Latimer KJ (1998) The life history and host range of *Ectaga garcia*, a biological control agent for *Lantana camara* and *L. montevidensis* in Australia. *BioControl* **43**, 325–338.

Day MD, Willson BW and Nahrung HF (1999) The life history and host range of *Charidotis pygmaea* (Col.: Chrysomelidae), a biological control agent for *Lantana montevidensis* (Verbenaceae). *Biocontrol Science and Technology* **9**, 347–354.

Day MD, Broughton S and Hannan-Jones MA (2003a) Current distribution and status of *Lantana camara* and its biological control agents in Australia, with recommendations for further biocontrol introductions into other countries. *Biocontrol News and Information* **24**, 63N–76N.

Day MD, Wiley CJ, Playford J and Zalucki MP (2003b) *Lantana: Current Management Status and Future Prospects.* ACIAR, Canberra.

Garcia C (1988) 'Report on the insects found attacking *Lantana montevidensis* in Argentina'. Qld Dept of Lands, Brisbane.

Garcia C (1989) 'Report on activities March to June 1989'. Qld Dept of Lands, Brisbane.

Munir AA (1996) A taxonomic review of *Lantana camara* L. and *L. montevidensis* (Spreng.) Briq. (Verbenaceae) in Australia. *Journal of the Adelaide Botanical Gardens* **17**, 1–27.

Neal J (1999) Assessing the sterility of ornamental lantana varieties: are we exacerbating the weed problem? Hons thesis. Department of Botany, University of Queensland, Brisbane.

O'Donnell C (2002) *The Creeping Lantana Handbook. A Guide to Ecology, Control and Management.* Qld Dept of Primary Industries, Natural Resources and Mines, Brisbane.

Pacific Islands Ecosystems at Risk (2010) *Lantana montevidensis* (Spreng.) Briq., Verbenaceae. US Dept of Agriculture, Hawaii. <http://www.hear.org/Pier/species/lantana_montevidensis.htm>.

Parsons WT and Cuthbertson EG (1992). Creeping lantana. In *Noxious Weeds of Australia.* (WT Parsons and EG Cuthbertson) pp. 632–634. CSIRO Publishing, Melbourne.

Sanders RW (2001) The genera of Verbenaceae in the southeastern United States. *Harvard Papers in Botany* **5**, 303–358.

Swarbrick JT (1986) History of the lantanas in Australia and origins of the weedy biotypes. *Plant Protection Quarterly* **1**, 115–121.

Winder JA (1999) 'Report on *Lantana montevidensis* survey in Brazil, November/December, 1999'. Qld Dept of Natural Resources, Brisbane.

Winder JA (2000) 'Report on *Lantana montevidensis* survey in Brazil, March 2000'. Qld Dept of Natural Resources, Brisbane.

Macfadyena unguis-cati (L.) A.H.Gentry – cat's claw creeper

K Dhileepan

ABSTRACT

Cat's claw creeper *Macfadyena unguis-cati* (Bignoniaceae), a perennial woody climbing vine native to tropical America, is a major environmental weed in Qld and NSW, Australia. Biological control of cat's claw creeper was initiated in South Africa in 1996, then in Australia in 2001. Surveys in the native range have identified nine insects, of which six have been prioritised. So far, *Charidotis auroguttata*, *Carvalhotingis visenda*, *Carvalhotingis hollandi*, *Hypocosmia pyrochroma* and *Hylaeogena jureceki* have been imported into Australia for host-specificity tests. The leaf-feeding beetle *C. auroguttata* was not approved for field release due to perceived risk to a native non-target plant. One leaf-feeding tingid, *C. hollandi*, could not be established in quarantine. The second leaf-feeding tingid, *C. visenda* and the leaf-tying moth *H. pyrochroma* were approved for field release in 2007. The tingid has become widely established, but its rate of spread remains very low. There are early indications of the field establishment of the leaf-tying moth. The leaf-feeding jewel beetle *H. jureceki* was imported to Australia in 2009 for host-specificity testing. Though specialist leaf-feeding herbivores could be effective in reducing the existing tuber bank, they may have only limited impact on weed spread arising from seed production. Hence future biological control efforts should focus on reducing seed output by using the specialist pod- and seed-feeding weevil *Apteromechus notatus* from Brazil.

Key words: *Charidotis auroguttata*, *Carvalhotingis visenda*, *Carvalhotingis hollandi*, *Hypocosmia pyrochroma*, *Hylaeogena jureceki*, *Apteromechus notatus*, Qld, NSW, South Africa, Argentina, Paraguay, Brazil.

INTRODUCTION

Cat's claw creeper, *Macfadyena unguis-cati* (L.) Gentry (Bignoniaceae), is a perennial woody climbing vine that is native from Mexico through Central America to tropical South America, including Trinidad and Tobago (Rafter *et al.* 2008). Introduced as an ornamental, the vine became naturalised in several countries in Asia (China, India, Malaysia, Nepal, Sri Lanka and Thailand), Australasia and the South Pacific (Australia, New Zealand, Indonesia, Micronesia, New Caledonia, Hawaii and Cook Islands), Europe (Sicily, Switzerland, Serbia and Montenegro, France and Greece), Africa (Kenya, Mauritius, South Africa, Uganda and Zimbabwe) and North America (southern USA). Cat's claw creeper is now regarded as invasive in Australia, South Africa, India, Mauritius, China, Hawaii and Florida in the USA, New Caledonia, St Helena Island and NZ (King and Dhileepan 2009).

Cat's claw creeper is a high climbing woody vine (Fig. 1), with stems up to 6 cm in diameter and roots becoming elongated and tuberous with age. Upright branches and horizontal runners can develop adventitious roots. Leaves are opposite, compound, with two leaflets and a terminal

Figure 1: Cat's claw creeper, *Macfadyena unguis-cati.* a) Stem. b) Leaf and flower. Photos: M Shortus, Qld Government.

three-forked claw-like tendril that enables the plant to attach itself to tree trunks, other vegetation and artificial structures such as fences. Flowers are yellow and trumpet-shaped, solitary or in small clusters at leaf axils (Fig. 1). The fruit is a linear flat capsule, to 50 cm long, with oblong winged seeds. In Australia, the plant usually has a single annual pulse of flowering in late spring or early summer. The plant can be propagated from seed, and vegetatively from below-ground tubers. Stems trailing along the ground are capable of producing roots at the nodes. Seeds are dispersed by wind and water and the seeds do not remain viable more than a year, suggesting that although the mechanism of spread is through seeds, the mechanism of persistence is through the tuber bank (Osunkoya *et al.* 2009; Vivian-Smith and Panetta 2004).

In Australia, cat's claw creeper is a major environmental weed (Fig. 2) in Qld and NSW and it has the potential to spread throughout eastern Australia (Fig. 3). In Qld and north-eastern NSW, cat's claw creeper is a declared noxious weed. Cat's claw creeper poses a significant threat to biodiversity in riparian areas, rainforest communities, non-agricultural areas and remnant natural vegetation (Vivian-Smith and Panetta 2004; Downey and Turnbull 2007). In densely infested areas, cat's claw creeper covers standing vegetation, including shrubs and large trees up to 30 m tall (Fig. 2), eventually causing canopy collapse (Sparks 1999). In areas without standing vegetation or man-made structures, the vines grow along the forest floor and form dense mats.

The management objectives for cat's claw creeper are focused on reducing the rate of shoot growth to limit the vine's ability to climb and smother native vegetation, as well as reducing tuber biomass. The inaccessibility of root tubers and their ability to regenerate are a major problem for the control of this weed. Chemical control options for managing cat's claw creeper are available, but are often not used due to the sensitive ecosystems in which it occurs. Mechanical control of above-ground growth provides only temporary relief, as regeneration from subterranean tubers continues over many years. As a result, there is a need to treat infested areas with mechanical or chemical control options repeatedly. This severely limits the size of areas that can be treated. Susceptibility of cat's claw creeper to herbivory (Raghu and Dhileepan 2005; Raghu *et al.* 2006; St Pierre 2007) suggests that biological control is the most desirable option to manage this weed.

BIOLOGICAL CONTROL HISTORY

The biological control of *M. unguis-cati* was initiated in South Africa in 1996 by the Agricultural Research Council – Plant Protection Research Institute (ARC-PPRI) (Sparks 1999). Since then, five insects have been screened

Figure 2: *Macfadyena unguis-cati* infestation in Qld. Photo: M Treviño, Qld Government.

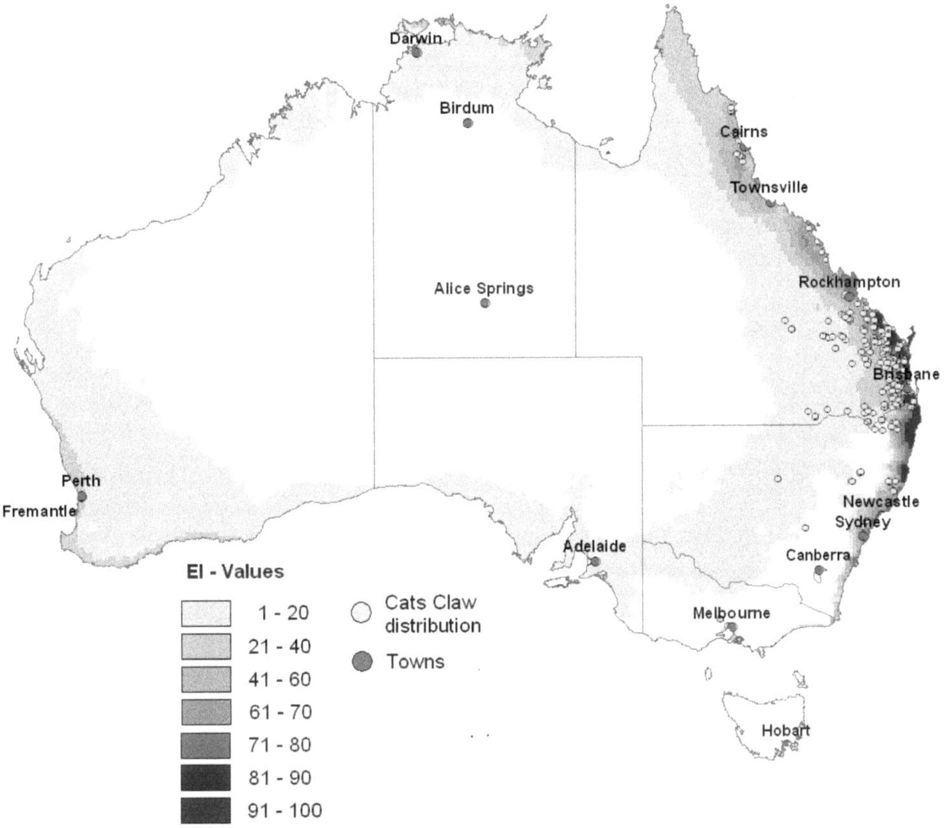

Figure 3: Current and potential distribution of *Macfadyena unguis-cati* in Australia, based on native range distribution (Rafter *et al.* 2008).

and three approved for field release (King and Dhileepan 2009; King *et al.* 2011). The leaf-feeding chrysomelid *Charidotis auroguttata* (Boheman) was the first agent released in South Africa in 1999 (Williams 2002), but its establishment was slow and it failed to survive at several localities. Subsequently, two leaf-feeding tingids *Carvalhotingis visenda* (Drake & Hambleton) and *C. hollandi* (Drake), a leaf-tying pyralid moth *Hypocosmia pyrochroma* Jones and a leaf-mining buprestid beetle *Hylaeogena jureceki* (Obenberger) have been screened (Williams et al. 2008). *C. visenda* and *H. jureceki* have been approved for field release in South Africa, and both agents are showing signs of field establishment (King and Dhileepan 2009; King *et al.* 2011). Biological control of cat's claw creeper in Australia commenced in 2001, in collaboration with ARC-PPRI in South Africa.

PLANT TAXONOMY

M. unguis-cati displays wide genetic diversity throughout its native range (Sigg *et al.* 2006; Prentis *et al.* 2009). In contrast, the genetic diversity in its introduced range is very low (Sigg *et al.* 2006; Prentis *et al.* 2009). Cat's claw creeper populations in most of the introduced ranges, including Australia and South Africa, appear to have originated from Paraguay and are closely genetically related to populations within the Paraguay–Bolivia–Argentina region, representing the southern portion of the species' native range (Sigg *et al.* 2006).

In Australia, two morphologically and genetically distinct populations of *M. unguis-cati* occur (Sigg *et al.* 2006; Shortus and Dhileepan 2011). The more invasive short-pod variety is widespread through Qld and NSW, with a second long-pod variety restricted to a few sites in south-eastern Qld (Shortus and Dhileepan 2011). Both varieties have a yellow trumpet-shaped flower, but the flower of the long-pod variety has a deeper hue of yellow than the short-pod flower. The pods of the short-pod variety mature in late summer to early autumn and the pods of the long-pod variety mature in late winter to early spring. The long-pod variety has significantly larger leaves, larger seed pods and more seeds per pod than the short-pod variety. The short-pod variety has a slightly wider seed pod and thicker leaves than the long-pod variety. Herbarium records at the

Missouri Botanical Gardens suggest that, in the native range, the long-pod variety occurs in Colombia, Costa Rica, Mexico and Nicaragua. The long-pod variety has also been observed in Brazil (S Neser, *pers. comm.* 2009). Only the short-pod variety appears to be present in most countries in the introduced range (Sigg *et al.* 2006), except in Hawaii where the herbarium records suggest that the population is the long-pod variety.

EXPLORATION

Cat's claw creeper is endemic from Mexico through Central America and the Caribbean islands to tropical South America as far south as Chile (Sparks 1999; Rafter *et al.* 2008). Surveys were conducted in Brazil, Argentina, Paraguay, Venezuela and Trinidad in 1996, and in Paraguay and Brazil in 2002 by Stefan Neser of ARC-PPRI, resulting in the identification of around nine insect species as potential biological control agents.

Based on the climatic similarity (using the CLIMEX model) of locations with cat's claw creeper infestations in the introduced range of Australia and South Africa to those in the native range, the areas of central and eastern Argentina, south Brazil, Uruguay and parts of Bolivia and Paraguay were prioritised for exploration for new biological control agents (Rafter *et al.* 2008). Accordingly, further surveys were conducted in northern Paraguay and southern Brazil by S Neser in 2009, and previously tested agents including *C. visenda*, *H. pyrochroma* and a seed-feeding weevil *Apteromechus notatus* (Hustache) were re-collected from climatically suitable areas to enrich the existing colonies of these insects in South Africa and Australia.

CANDIDATES

Natural enemies

There is no major insect damage or disease evident on *M. unguis-cati* in Australia (K Dhileepan, unpub. data). Of the nine insects identified as potential biological control agents in the native range, a leaf-feeding tortoise beetle *C. auroguttata*, two leaf-feeding tingids *C. visenda* and *C. hollandi*, a leaf-tying moth *H. pyrochroma*, a leaf-mining jewel beetle *H. jureceki* and a seed-feeding weevil *A. notatus* have been prioritised for detailed host-specificity testing in quarantine in South Africa and Australia (King and Dhileepan 2009; King *et al.* 2011). A rust fungus *Uropyxis rickiana* Magnus, inducing stem-galls and leaf-spots on cat's claw creeper, has been recorded from Argentina and Brazil (Hernandez and Hennen 2003; S Neser, *pers. comm.* 2009), but its potential as a biological control agent is yet to be ascertained. Other potential agents recorded in the native range, but yet to be identified, include a shoot-tip feeding sawfly from Brazil, a sap-feeding leaf-hopper from Argentina, Brazil and Paraguay, a shoot-galling midge from Brazil and Paraguay and a shoot-feeding mirid bug from Brazil (Sparks 1999).

Agent prioritisation

Insect herbivores from a range of feeding guilds (leaf, shoot and seed), but not subterranean tubers and roots, have been recorded on *M. unguis-cati* in its native range (Sparks 1999; Raghu *et al.* 2006). Since cat's claw creeper is a structural parasite, biological control agents for the liana will need to target its climbing habit. Examining the response of cat's claw creeper to different types of simulated herbivory can help to identify the type of damage to which cat's claw creeper is most vulnerable, so that specialist herbivores likely to cause such damage can be prioritised. Simulated herbivory experiments have shown that defoliation on its own or in combination with shoot damage has the potential to significantly reduce cat's claw creeper growth. Defoliation also reduced the climbing habit of the vine and reduced the rate of subterranean tuber biomass accumulation (Raghu *et al.* 2006). Simulated herbivory studies indicated that repeated defoliations, a minimum of three defoliation events under shade and more than three defoliation events under full sun, are required to stop foliar and tuber growth (St Pierre 2007). In contrast, below-ground damage needs to be avoided, as the plant either tolerates or vigorously compensates for such damage. Specialist herbivores in the leaf-feeding guild have therefore been prioritised as potential biological control agents of this species (Raghu *et al.* 2006). In *M. unguis-cati*, new recruitment is primarily from seeds, hence future biological control efforts need to focus on introducing specialist seed- and pod-feeding insects to reduce seed output, thereby limiting the future tuber bank (Osunkoya *et al.* 2009).

Agents not released

***Charidotis auroguttata* (Boheman) (Chrysomelidae).** This golden-spotted tortoise beetle has a wide geographic native range that includes Argentina, Brazil, Paraguay, Trinidad and Venezuela (Sparks 1999), and was first collected near Las Caracas, Venezuela (Sparks 1999; Williams 2002). It was the first biological control agent for *M. unguis-cati* imported from South Africa to Australia, in 2001 (Dhileepan *et al.* 2005).

The adults are small, 3–5 mm long and wide, and lay eggs singly on both surfaces of the leaves and on the stems. Eggs hatch after nine to 11 days and the larvae feed on both young and old leaves for 21–25 days, passing through five instars. Late fifth instars undergo a non-feeding prepupal period of one to three days, then pupate attached to the lower leaves or on leaf litter around the stem. The adults emerge after eight to 10 days and at first are pale red, turning reddish brown with gold spots after a further five to seven days. The females lay one to three eggs per day, of which ~87% hatch. The short generation time allows several generations per year (Sparks 1999; Williams 2002). Adults are long-lived, and an adult life-span in excess of one year has been recorded. Both adults and larvae feed on the leaves and cause premature leaf abscission and shoot dieback (Sparks 1999).

Host-specificity tests confirmed that *C. auroguttata* was specific to cat's claw creeper (Sparks 1999; Williams 2002). The agent was released in South Africa (Sparks 1999; Williams 2002) and has become established in some release sites there (King and Dhileepan 2009). However, in no-choice trials in Australia approximately 12% of larvae completed development in 59.4 ± 5.2 days on *Myoporum boninense australe* Chinnock (Myoporaceae), a non-target native plant, as against 95% larval development in 35.2 ± 0.43 days on *M. unguis-cati* (Dhileepan *et al.* 2005). In the no-choice demography trials, adults laid eggs from the second week after emergence on cat's claw creeper, with an average of 0.286 eggs/female/day, resulting in an 18-fold increase in the adult population over 16 weeks. Adult survival on *M. boninense australe* remained high, but oviposition commenced only from the 10th week after emergence, with an average of 0.023 eggs/female/day, and none of the eggs hatched. In the choice demography trials, oviposition on *M. unguis-cati* was evident from the fourth week onwards, while on the non-target *M. boninense australe* oviposition commenced only from the 14th week. Only 10% of total adults and 11.3% of total eggs were on the non-target plant, and none of these eggs hatched. These tests demonstrated that although the beetle can spill-over from the target weed to the non-target native plant and cause feeding damage, the non-target plant could not sustain a viable insect population on its own. However, owing to the perceived risk to non-target species, this agent was not approved for field release in Australia (Dhileepan *et al.* 2005; Raghu *et al.* 2007).

***Carvalhotingis hollandi* (Drake) (Tingidae).** This leaf-feeding tingid was collected on cat's claw creeper from Curitiba in Brazil and from Posadas in Argentina in April 2002 (Williams *et al.* 2008). A laboratory colony of the tingid was established in quarantine at ARC-PPRI, Pretoria. Females lay eggs along the vein on the upper (axial) side of the leaves and the emerging nymphs feed as a group, causing chlorosis. Studies in South Africa confirmed that *C. hollandi* is highly host-specific, hence it was approved for field release (Williams *et al.* 2008). The tingid was imported into Australia in 2004 but, due to difficulties in establishing a colony under high-security conditions, it could not be cultured in quarantine and host-specificity tests were not conducted.

Agents released

***Carvalhotingis visenda* (Drake & Hambleton) (Tingidae).** This tingid (Fig. 4) was collected on cat's claw creeper from Curitiba in Brazil and from Posadas in Argentina in April 2002 (Williams *et al.* 2008), and imported from South Africa to Australia in 2004 (Treviño *et al.* 2006). The tingid developed and reproduced throughout the year and there was no evidence of winter diapause. The females lived longer (48.0 ± 7.3 days) than males (24.4 ± 7.9 days) and laid 82.2 ± 13.2 eggs in their life-time, with an average of 6.8 ± 0.8 eggs per day after a pre-oviposition period of 3.4 ± 0.7 days (Dhileepan *et al.* 2010b). The eggs were laid in groups along the main vein on the undersides of leaves. The emerging nymphs fed as a group on the underside of leaves, sucking out the cell contents of leaves and causing chlorosis. The nymphs developed through five nymphal instars in 16.9 ± 1.4 days. Temperatures between 20°C and 30°C were the most favourable for adult survival, oviposition, egg hatching and nymphal development (Dhileepan *et al.* 2010a).

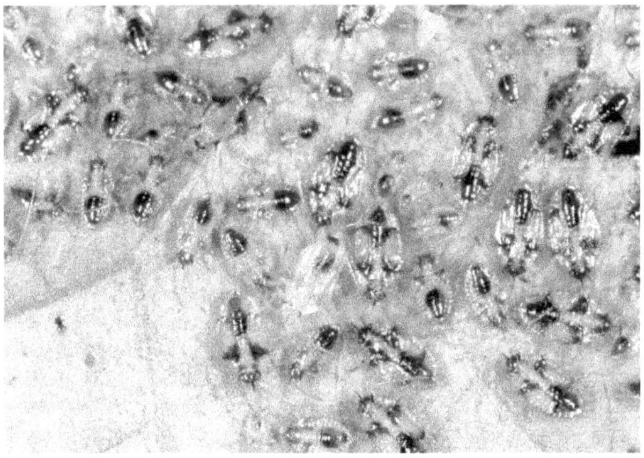

Figure 4: *Carvalhotingis visenda* adult and nymphs. Photo: J McCarthy, Qld Government.

The tingid fed and reproduced on both long- and short-pod varieties of *M. unguis-cati*. Pre-release evaluations confirmed that feeding by the tingid significantly affected the growth and productivity of cat's claw creeper seedlings (Conrad and Dhileepan 2007) by reducing the chlorophyll contents of damaged leaves by 5–17%, resulting in a 31–49% reduction in photosynthesis (Bayliss 2006). The pre-release evaluations also indicated that even a single tingid generation could significantly reduce above-ground shoot and subterranean tuber growth (Conrad and Dhileepan 2007; Bayliss 2006).

The host-specificity trials in Australia confirmed that the tingid is host-specific and does not pose risk to any non-target plants (Dhileepan *et al.* 2007b). Climatic modelling (Rafter *et al.* 2008) and thermal tolerance (Dhileepan *et al.* 2010a) studies suggest that majority of the cat's claw creeper-infested areas in Qld and NSW are climatically suitable for *C. visenda*. The potential number of generations it can complete in a year in Australia ranged from three to eight, with more generations in Qld than in NSW (Dhileepan *et al.* 2010a). To increase the virulence of the existing lab colony, a fresh colony of *C. visenda* was imported from Paraguay in 2007 and mixed with the existing colony in Australia; progeny from the mixed colony are currently being field released. The tingid was approved for field release in 2007 and since then more than half a million adults and nymphs have been released at 72 sites in Qld and northern NSW in partnership with local community groups (Dhileepan *et al.* 2010b). Field establishment of *C. visenda* was evident on both varieties of *M. unguis-cati* in 80% of the release sites, but the agent appears to spread slowly in the field (Dhileepan *et al.* 2010b). The spread was mostly horizontal along the ground-level infestations, less often vertically on the plants climbing on trees. Field establishment of *C. visenda* has also been reported from South Africa (King and Dhileepan 2009).

***Hypocosmia pyrochroma* Jones (Lepidoptera: Pyralidae).** A laboratory colony of this leaf-tying moth was first established in Pretoria, South Africa, from material collected on cat's claw creeper from sites near Curitiba in Brazil and Posadas in Argentina in April 2002 (Williams 2003a). The moth (Fig. 5) was imported to Australia in 2005 after tests in South Africa proved it to be highly host-specific. Host-specificity tests in Australia involving 38 plant species in 10 families confirmed that *H. pyrochroma* is a highly specific biological control agent that does not pose risk to non-target plants (Dhileepan *et al.* 2007a).

The moth lays eggs singly on the undersides of leaves and in the crevices of woody stems. Eggs hatch in two

Figure 5: *Hypocosmia pyrochroma* adult. Photo: J McCarthy, Qld Government.

weeks. The larvae feed destructively by tying leaves together by silk webs, which create silken tunnels. Larvae go though six instars in four weeks. Fully grown larvae pupate in the soil, 2–3 cm below the soil surface, and the majority of the adults emerge after four weeks. Pupae undergo diapause from late autumn (April–May) to early spring (September–October). Larvae reared under a higher temperature (25–30°C) and longer light regime (14 h) in the laboratory did not undergo pupal diapause. Females lived for around 10 days and laid up to 120 eggs after a two-day pre-oviposition period (Williams 2003a).

The female moth laid eggs and the emerging larvae developed equally well on both long- and short-pod varieties of cat's claw creeper. Feeding and leaf-tying by *H. pyrochroma* larvae severely damage foliage (Williams 2003b) and result in reduced plant growth and tuber production (Snow *et al.* 2006). Under quarantine conditions, *H. pyrochroma* larval feeding resulted in death of cat's claw creeper seedlings as leaves were fed upon and turned into silken tunnels. It is anticipated that the moth will establish along the coastal regions of south-eastern Qld and northern NSW, and has the potential to complete at least three generations in a year (October to April).

The leaf-tying moth was approved for field release in 2008 and since then field releases of larvae and adults have been made across 17 sites in south-eastern Qld and northern NSW. An improved laboratory rearing method, replacing potted plants with corrugated paper for oviposition, cut foliage for larval feeding and sterilised sand for pupation in temperature-controlled rearing cages, has resulted in the field release of large numbers of larvae and adults. Though larvae have been recovered from some of

Figure 6: Leaf-mining by *Hylaeogena jureceki* larvae. a) With pupal cells (P). b) The larva. Photo: D Taylor, Qld Government.

the release sites, it is too early to determine its field establishment status in Australia.

Other promising agents

***Hylaeogena jureceki* (Obenberger) (Coleoptera: Buprestidae).** This leaf-mining jewel beetle native to Argentina, Brazil, Paraguay and Trinidad was collected from Argentina and Brazil in April 2002 (Williams 2003b). The adults are small and blackish in colour with a slight metallic sheen and live for more than five months. Both adults and larvae cause damage; adults feed on leaves, while the larvae mine the leaves (Fig. 6). Females lay one to two eggs per day singly on the leaves, and the eggs hatch in 10–12 days. The emerging larvae go though three larval instars in 10–13 days. Mature larvae pupate on the leaf (Fig. 6) and adults emerge after 12–16 days. Host-specificity tests in South Africa indicated that the jewel beetle is specific to cat's claw creeper and hence it was approved for field release there (King *et al.* 2011). The jewel beetle was imported to Australia in 2009 and host-specificity tests have been completed and the agent is awaiting approval for field release.

***Apteromechus notatus* (Hustache) (Coleoptera: Curculionidae).** In its native range, *A. notatus* has been observed in large numbers destroying up to 80% of the seeds found within developing pods, thereby reducing seed rain (King and Dhileepan 2009). Adults are long-lived and are thought to lay eggs on green or immature pods. Thereafter, hatching larvae burrow into the pod and feed on numerous seeds before pupating. After overwintering as either pupae or newly emerged adults within the pod, the next generation of adults emerges in spring to coincide with flowering and early pod production. This life-history, however, presents a significant challenge to the rearing and specificity testing of the weevil in quarantine.

DISCUSSION

Simulated herbivory studies indicated that cat's claw creeper is susceptible to leaf herbivory. Hence, leaf-feeding agents have been prioritised and imported for host-specificity tests. However, for the leaf herbivory to be effective, multiple defoliation events would be required to slow the climbing habit of the liana and reduce its tuber resources (St Pierre 2007). As a result, five agents, all leaf-feeding, have been imported for host-specificity testing. Among them, only two agents have been field released so far, while the third agent is undergoing host-specificity tests in quarantine.

There is no evidence of any competition between the two leaf-feeding insects being field released, *C. visenda* and *H. pyrochroma*, as both agents show distinct preference for different niches within the plant in laboratory studies. The tingid prefers to feed and oviposit on mature and older leaves (Bayliss 2006; Conrad and Dhileepan 2007), while the leaf-tying moth preferentially oviposits on shoot tips and young leaves (Snow *et al.* 2006). This is further confirmed by field observations, where the tingid (*C. visenda*) infestation was more evident on mature and older leaves at ground level. In contrast, larvae of the leaf-tying moth (*H. pyrochroma*) were often recovered from foliage on the tree stems.

Climate-matching and plant genotypic studies directed surveys to areas around the southern parts of the native distribution of cat's claw creeper, in particular Paraguay, the southern reaches of Brazil and north-eastern Argentina. Biological control agents sourced from these regions with similar climatic conditions and matching plant genotypes are more likely to be successful. A freshly field-collected culture of the leaf-tying moth and the tingid from Paraguay was mixed with the existing lab colony that originated from field-collected materials in 2002 to enhance the virulence and genetic diversity of the lab colonies of both agents.

The recruitment of cat's claw creeper in the field is primarily from seeds, not from vegetative propagation as previously thought (Osunkoya *et al.* 2009). Though the specialist leaf-herbivores, through reduction of photosynthetic leaf surfaces and hence assimilates, should be effective in reducing the existing tuber bank (Conrad and Dhileepan 2007; Bayliss 2006; St Pierre 2007), they may have only limited impact on the spread or establishment of new tubers or populations from new seed inputs. Hence, future biological control efforts should focus equally on reducing seed production and spread, by targeting fruit pods or mature seeds using a specialist pod- and seed-feeding weevil (e.g. *A. notatus*) from Brazil.

ACKNOWLEDGEMENTS

We thank Dr Stefan Neser, Hester Williams and Anthony King of ARC-PPRI, South Africa, for supplying the biological control agents; Mariano Treviño, Jayd McCarthy, Mathew Shortus, Deanna Bayliss and Di Taylor for technical support; and Bill Palmer and Dane Panetta for comments on the manuscript.

REFERENCES

Bayliss D (2006) The pre-release evaluation of a potential biological control agent (*Carvalhotingis visenda*) for cat's claw creeper (*Macfadyena unguis-cati*) – plant growth and physiological response to herbivory. BSc Hons thesis. School of Integrative Biology, University of Queensland, Australia.

Conrad KA and Dhileepan K (2007) Pre-release evaluation of the efficacy of the leaf-sucking bug *Carvalhotingis visenda* (Heteroptera: Tingidae) as a biological control agent for cat's claw creeper *Macfadyena unguis-cati* (Bignoniaceae). *Biocontrol Science and Technology* **17**, 303–311.

Dhileepan K, Treviño M, Donnelly GP and Raghu S (2005) Risk to non-target plants from *Charidotis auroguttata* (Chrysomelidae: Coleoptera), a potential biocontrol agent for cat's claw creeper *Macfadyena unguis-cati* (Bignoniaceae) in Australia. *Biological Control* **32**, 450–460.

Dhileepan K, Snow EL, Rafter MA, McCarthy J, Treviño M and Wilmot Senaratne KAD (2007a) Leaf-tying moth *Hypocosmia pyrochroma* (Lepidoptera: Pyralidae), a host specific biological control agent for cat's claw creeper *Macfadyena unguis-cati* (Bignoniaceae) in Australia. *Journal of Applied Entomology* **131**, 564–568.

Dhileepan K, Treviño M and Snow EL (2007b) Specificity of *Carvalhotingis visenda* (Hemiptera: Tingidae) as a biocontrol agent for cat's claw creeper *Macfadyena unguis-cati* (Bignoniaceae) in Australia. *Biological Control* **41**, 282–290.

Dhileepan K, Bayliss D and Treviño M (2010a) Thermal tolerance and potential distribution of *Carvalhotingis visenda* (Hemiptera: Tingidae), a biological control agent for cat's claw creeper *Macfadyena unguis-cati* (Bignoniaceae). *Bulletin of Entomological Research* **100**, 159–166.

Dhileepan K, Treviño M, Bayliss D, Saunders M, McCarthy J, Shortus M, Snow EL and Walter GH (2010b) Introduction and establishment of *Carvalhotingis visenda* (Hemiptera: Tingidae) as a biological control agent for cat's claw creeper *Macfadyena unguis-cati* (Bignoniaceae) in Australia. *Biological Control* **55**, 58–62.

Downey PA and Turnbull I (2007) The biology of Australian weeds. 48. *Macfadyena unguis-cati* (L.) A.H. Gentry. *Plant Protection Quarterly* **22**, 82–91.

Hernandez JR and Hennen JF (2003) Rust fungi causing galls, witches' brooms, and other abnormal plant growths in northwestern Argentina. *Mycologia* **95**, 728–755.

King AM and Dhileepan K (2009) Clinging on: a review on the biological control of cat's claw creeper. *Biocontrol News and Information* **30**, 53N–56N.

King AM, Williams HE and Madire LG (2011) Biological control of cat's claw creeper. *Macfadyena unguis-cati* (L.) A.H.Gentry (Bignoniaceae), in South Africa. *African Entomology* **19**, 366–377.

Osunkoya OO, Pyle K, Scharaschkin T and Dhileepan K (2009) What lies beneath? The pattern and abundance of the subterranean tuber bank of the invasive liana cat's claw creeper, *Macfadyena unguis-cati* (Bignoniaceae). *Australian Journal of Botany* **57**, 132–138.

Prentis PJ, Sigg DP, Raghu S, Dhileepan K and Lowe AJ (2009) Worldwide phylogeography of two globally invasive plants reveals contrasting introduction histories and genetic dynamics. *Diversity and Distributions* **15**, 822–830.

Rafter MA, Wilson AJ, Wilmot Senaratne KAD and Dhileepan K (2008) Modelling climatic similarities in native and introduced ranges of cat's claw creeper *Macfadyena unguis-cati* (Bignoniaceae) to prioritise areas for exploration and release of biological control agents. *Biological Control* **44**, 169–179.

Raghu S and Dhileepan K (2005) The value of simulating herbivory in selecting effective weed biological control agents. *Biological Control* **34**, 265–273.

Raghu S, Dhileepan K and Treviño M (2006) Response of an invasive liana to simulated herbivory: implications for its biological control. *Acta Oecologica* **29**, 335–345.

Raghu S, Dhileepan K and Scanlan J (2007) Predicting the risk to non-target species in classical biological control: a systems modeling approach. *Ecological Modelling* **208**, 247–262.

Shortus M and Dhileepan K (2011) Two varieties of the invasive liana, cat's claw creeper, *Macfadyena unguis-cati* (Bignoniaceae) in Queensland, Australia. *Proceedings of the Royal Society of Queensland* **116**, 13–20.

Sigg DP, Lowe A, Raghu S and Dhileepan K (2006) 'Using genetic tools to assess the provenance of an invasive liana, *Macfadyena unguis-cati*, in Australia'. Report to the Qld Dept of Natural Resources, Mines and Water, July 2006.

Snow EL, McCarthy J, Treviño M, Senaratne KADW and Dhileepan K (2006) Biological control of cat's claw creeper (*Macfadyena unguis-cati*): the potential of the leaf-tying pyralid moth *Hypocosmia pyrochroma* Jones. In *Proceedings of the 15th Australian Weeds Conference*, 24–28 September, Adelaide. (Eds C Preston, JH Watts and ND Crossman) pp. 599–600. Weed Management Society of South Australia, Adelaide.

Sparks HE (1999) The initiation of a biological control programme against *Macfadyena unguis-cati* (L.) Gentry (Bignoniaceae) in South Africa. *African Entomology Memoir No. 1*, 153–157.

St Pierre S (2007) The response of an invasive liana to defoliation in sun and shade and the effects of plant light history on behaviour of a potential biocontrol species. BSc Hons thesis. School of Integrative Biology, University of Queensland, Australia.

Treviño M, Snow EL, Senaratne KADW, Conrad C and Dhileepan K (2006) Leaf-sucking tingid (*Carvalhotingis visenda*): a potential biological control agent for cat's claw creeper (*Macfadyena unguis-cati*). In *Proceedings of the 15th Australian Weeds Conference*. 24–28 September, Adelaide. (Eds C Preston, JH Watts and ND Crossman) pp. 605–606. Weed Management Society of South Australia, Adelaide.

Vivian-Smith G and Panetta FD (2004) Seedbank ecology of the invasive vine, cat's claw creeper (*Macfadyena unguis-cati* (L.) Gentry). In *Proceedings of the 14th Australian Weeds Conference*. 6–9 September, Wagga Wagga, NSW. (Ed SB Johnson) pp. 531–537. Weed Society of New South Wales, Wagga Wagga.

Williams HE (2002) Life history and laboratory host range of *Charidotis auroguttata* (Boheman) (Coleoptera: Chrysomelidae), the first natural enemy released against *Macfadyena unguis-cati* (L.) Gentry (Bignoniaceae) in South Africa. *Coleopterists Bulletin* **56**, 299–307.

Williams HE (2003a) 'Host specificity report to the National Department of Agriculture: Directorate of Plant Health and Quality in relation to an application for permission to undertake the release of the cat's claw creeper leaf-tying moth, *Hypocosmia pyrochroma*, a potential biocontrol agent for *Macfadyena unguis-cati* in South Africa'. Agricultural Research Council, Plant Protection Research Institute, Pretoria, South Africa.

Williams HE (2003b) 'Host specificity report to the National Department of Agriculture: Directorate of Plant Health and Quality in relation to an application for permission to undertake the release of *Hylaeogena (Hedwigiella) jureceki* Obenberger, a biological control agent for *Macfadyena unguis-cati* L. Gentry (Bignoniaceae) in South Africa'. Agricultural Research Council, Plant Protection Research Institute, Pretoria, South Africa.

Williams HE, Neser S and Madire S (2008) Candidates for biocontrol of *Macfadyena unguis-cati* in South Africa: biology, host ranges and potential impact of *Carvalhotingis visenda* and *Carvalhotingis hollandi* under quarantine. *BioControl* **53**, 945–956.

Marrubium vulgare L. – horehound

John Weiss and Jean-Louis Sagliocco

ABSTRACT

Horehound is a widespread weed in southern Australia. In 1990, a 10-year biocontrol program (six years exploration and host-specificity testing and four years mass-rearing and releasing) was initiated. Two moths were found to be host-specific and released across the south-eastern states. The defoliator horehound plume moth is now widely distributed at over 100 sites and in moderate-to high-rainfall areas is having a suppressing effect on the weed and reducing seed production. The root-boring horehound clearwing moth has been released at fewer sites in Vic, SA and NSW. It appears to be well established and is increasing the mortality of plants at those sites. Two other insects have been identified as potential agents.

Key words: plume moth, *Wheeleria spilodactylus*, clearwing moth, *Chamaesphecia mysiniformis*.

INTRODUCTION

Horehound, *Marrubium vulgare* L. (Lamiaceae), is an erect perennial forb, or spreading, bushy, aromatic, multistemmed perennial herb usually 0.30–0.75 m high and wide and with up to 200 individual stems (Fig. 1). It is native to temperate Eurasia, Europe, the Middle East and the Mediterranean region including North Africa. It has become a weed in southern USA (including California and Texas), in South America (Argentina, Chile, Peru, Uruguay), New Zealand and Australia. Horehound was possibly first introduced into Australia from Europe via a shipment of botanical plants sent by Sir Joseph Banks in 1798 (Frost 1993).

Horehound has dense clusters of small white flowers and the resulting fruits, when dry, form brown burrs with

Figure 1: Horehound, *Marrubium vulgare*. Photo: J Weiss, Vic Government.

Figure 2: Horehound infestation in Wyperfeld National Park, January 1994. Photo: J Weiss, Vic Government.

small hooked spines. Each burr contains up to four but more often three, small (1–2 mm long) spear-shaped seeds. Horehound reproduces from seed and, while some germination can occur throughout winter and spring whenever sufficient water is available, most germination occurs in response to autumn rainfall (Lippai *et al.* 1996). Horehound is drought-tolerant, occurring in areas that receive a minimum of 200 mm annual rainfall. It is frost-resistant, but is susceptible to fire and waterlogging. It grows generally on alkaline and very poor soils and is an early coloniser of eroded areas, roadsides, channel banks, sheep camps, rabbit warrens and other disturbed sites, from which it encroaches into bushland and adjoining farmland. In natural ecosystems, it invades open bushland, open grassland and rocky outcrops, especially if the areas have been disturbed or overgrazed (Fig. 2). Horehound does not appear to be invasive in undisturbed native vegetation. Horehound, as with most members of the Lamiaceae (mint) family, is primarily bee-pollinated. There are no native plants closely related to *M. vulgare* in Australia. Overseas, *M. vulgare* can hybridise with *M. supinum* L., which is sometimes found in herb gardens in Australia. Also present in Australia and belonging to the same subfamily Lamioideae are the introduced genera *Ballota*, *Lamium*, *Leonotis*, *Leucas*, *Molucella* and *Stachys*.

Horehound is adapted to southern Australia's Mediterranean-type climate and the cooler climates of adjoining areas and it now occurs in both arid and high-rainfall zones. In southern Australia, an estimated 26 million ha, primarily in SA and Vic, were infested by horehound (Lane *et al.* 1980). Horehound also extends into Qld and WA but is either restricted or not important there (Weiss *et al.* 2000). Horehound is found throughout rural agricultural Tas but is most prevalent and troublesome in the grazing areas of the midlands (Ireson *et al.* 2000) (Fig. 3).

Grazing animals avoid *M. vulgare* if alternative food is available because of the bitter alkaloid, marrubin, contained in the leaves (Everist 1981). *M. vulgare* decreases pasture productivity and reduces the value of wool fleece contaminated with the hooked burrs. The cost of horehound in the contamination of wool has been estimated at 1.5–3.4%, thereby reducing the price of the wool by 20.3 cents per kilogram compared to clean wool – an annual loss in SA of up to \$52 700 per year (Carter, *pers. comm.* 1998). The overall cost of *M. vulgare* to Australian wool producers in 1985–86 was estimated to be \$A680 000 per year (Sloan *et al.* 1988). Grazing animals contribute to seed dispersal via the movement of burrs and seeds attached to their fleece. In certain situations, *M. vulgare* directly threatens native plants. In Vic, the endangered marble daisy bush, *Olearia astroloba* Lander & N.G. Walsh (Asteraceae) is at risk from horehound invasion (Carter and Walsh 2006).

BIOLOGICAL CONTROL HISTORY

Australia is the only country to have investigated biological control of horehound. July 1990 saw the start of a six-year research project on the biocontrol of *M. vulgare* with work done simultaneously at Frankston, Vic and at the CSIRO laboratory in France (Sagliocco and Weiss 1996; Ziegler *et al.* 1993).

EXPLORATION

The native range of *M. vulgare* includes Europe, western and central Asia and North Africa (Parsons and Cuthbertson 2001). In Mediterranean Europe, *M. vulgare* is generally found on wastelands with well-drained calcareous soils. Populations persist there due to sheep or horse grazing suppressing neighbouring plant competition and dispersing horehound seeds.

Between 1991 and 1996, the arthropod fauna associated with *M. vulgare* and its ecology was studied in

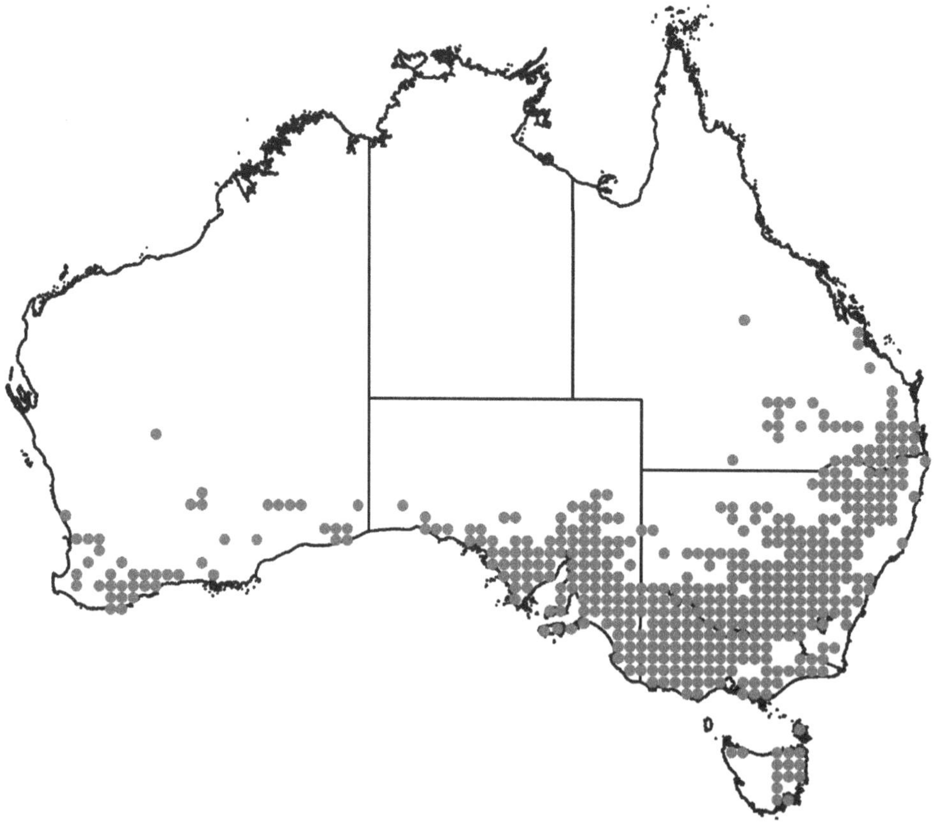

Figure 3: Distribution of horehound in Australia. Adapted from Parsons and Cuthbertson (2001), Lane *et al.* (1980) and the Australian Virtual Herbarium.

southern France at two locations near Montpellier and at one site near Zaragoza, Spain. During the same period extensive surveys for associated organisms were conducted in Spain, Portugal, Italy, the former Yugoslavia and Morocco (Fig. 4). A total of 55 horehound populations were surveyed, all arthropods observed feeding on horehound plants were collected and preserved for identification, and the type of damage was recorded. A list of arthropods associated with horehound can be found in Sagliocco (2000).

Figure 4: Areas of Mediterranean Europe and Africa surveyed for horehound insects in 1991–96.

CANDIDATES

Based on surveys and ecological studies conducted in Europe, four agents were initially selected for further host-specificity testing. *Wheeleria spilodactylus* Curtis (Lepidoptera: Pterophoridae), *Carcharodus boeticus* (Rambur) (Lepidoptera: Hesperiidae) and *Meligethes rotroui* Easton (Coleoptera: Nitidulidae) were imported to Australia for host-specificity studies while host testing of *Chamaesphecia mysiniformis* (Boisduval) (Lepidoptera: Sesiidae) was conducted at CSIRO European Laboratory, France.

Agents released in Australia

***Wheeleria spilodactylus* (Lepidoptera: Pterophoridae).** The agent most frequently found in Europe was the multivoltine horehound plume moth *W. spilodactylus* (previously referred to as *Pterophorus spilodactylus*), whose larvae feed on leaves, shoot tips and stems (Fig. 5). It was found across the study range. Its distribution includes western and continental Europe, the British islands, the Mediterranean region, Asia Minor and Northern Africa. It was the first candidate insect to be considered for the biological control of horehound. It was abundant at most sites in France and Spain, and had been recorded feeding on *M. vulgare* (Gielis 1996) and *Ballota nigra* (Bigot and Picard 1983). Females lay eggs underneath the leaves where the green larvae later feed. They also feed on the stem's soft tissues and shoot apices, often causing abortion of buds, thus affecting plant vigour and reducing the number of flowers and seeds produced. Pupae are found on the upper sides of leaves. A first consignment of a French ecotype collected at Cape d'Agde, France was sent to the quarantine facility at Frankston, Vic in 1991. Host-specificity tests conducted during 1991–92 showed *W. spilodactylus* to be host-specific to horehound (Weiss and Sagliocco, unpub. data). It was approved for release in late 1993 and by early 1995 the insect had been released at a number of sites in Vic (Weiss 1996a) and later in SA, ACT, NSW and Tas. In 1996, an ecotype originating from dry regions of Spain with long hot summers was shipped to Frankston. In 1997, rearing programs resulted in the release of both ecotypes at over 100 sites in the four south-eastern Australian states (Ireson *et al.* 2000; Wills 2000). By 1998 the plume moth had established at 90% of sites and the insect had dispersed over 20 km, and 41% of the sites were used for redistribution programs (Kwong 2004).

Figure 5: The adult horehound plume moth, *Wheeleria spilodactylus*. Photo: J Weiss, Vic Government.

The Spanish ecotype of the moth, obtained from an arid region, was considered to be better suited to the inland and mallee conditions of SA and Vic. The original French population was found established at fewer field sites in SA and had a slower rate of population increase compared to the Spanish or a more recently imported French ecotype (Clarke *et al.* 2000). However, results from a laboratory study showed that the original French ecotype was suffering from inbreeding depression which had limited its establishment and rate of growth (Clarke 2001).

Larval feeding at high densities reduced plant biomass and life-span. Once present for more than four years at a particular location, densities of at least one (in areas with <400 mm rainfall) or two (in areas with >400 mm rainfall) larvae per shoot indicated that the plume moth was having or would have some impact on the infestation (Weiss *et al.* 2000).

Studies on the plume moth include metapopulation structure (Menéndez and Thomas 2000, 2006), life-history (Weiss and Sagliocco, unpub. data), courtship and mating (Leyson 1999), dispersal (Traeger 2000), establishment (Baker 2002) and colonisation (Clarke 2001).

***Chamaesphecia mysiniformis* (Boisduval) (Lepidoptera: Sesiidae).** In Spain, the horehound clearwing moth *C. mysiniformis* (Fig. 6) was observed to inflict significant damage and often cause plant death. *C. mysiniformis* was present at eight of the 21 horehound populations surveyed in Spain, and at only one of the 15 populations in southern France. The species was also reported from Portugal and Italy (Laštůvka and Laštůvka 2001). In Spain, large populations of the insect were found around Zaragoza (northern Spain) which has a continental climate with summer maximum temperatures commonly reaching 40–45°C and at one location near Alicante (eastern Spain) with a Mediterranean-type climate. *C. mysiniformis* is univoltine and adults emerge from roots in spring and feed on nectar of horehound flowers and other species. Females emerge in the morning and mate during the first day of their adult life. Soon after mating they lay eggs at the base of plants over a period of seven to 12 days, with an average 96 eggs per female (Sagliocco and Coupland 1995). Larvae develop within the roots over 10 months.

Host-specificity testing was completed in France in 1994 and several shipments of eggs were sent to Frankston, Vic between 1995 and 1998. The culture was synchronised to the southern hemisphere season while in quarantine. In March 1997, 146 eggs were released at Wyperfeld National Park, Vic. Small releases of eggs were also made at Cambrai and near Murray Bridge, SA. Further releases

Figure 6: The adult horehound clearwing moth, *Chamaesphecia mysiniformis*. Photo: J Weiss, Vic Government.

were made at Wyperfeld in February and December 1998 (Wills 2000). Establishment was confirmed at Wyperfeld in 2001 (Faithfull and Sagliocco 2002) and by 2003, 50.3% of plants at the original release site were attacked (Sagliocco and Weiss 2004). By 2004 larvae were present in roots of 46% of live plants dissected from 10 sites in the national park. Larvae were also found in plants up to 4 km from the original release site. In 2006, larvae were found in 29.2% of plants sampled at 19 sites over a more extensive area (Table 1).

In 1998 a colony, imported from Spain, was briefly established at the quarantine facilities at the Waite Campus of the University of Adelaide. During 1999, 5830 eggs (with 61% fertility) were released at three sites, Narrung, Monarto Zoo and Wilpena in SA (Clarke *et al.* 2000).

In December 2004, over 4600 eggs obtained from a colony collected at Wyperfeld, Vic in October 2004 were sent to NSW. The Department of Primary Industries NSW conducted six releases on horehound infestations located on the north and central western slopes of NSW. In 2005, DPI NSW reared over 7700 eggs and made 11 releases in the same areas. The insect appears to have established at some of these release sites (P Sullivan, *pers. comm.* 2010).

Table 1: *Chamaesphecia mysiniformis* attack levels on *Marrubium vulgare* at Wyperfeld National Park, November, various years

	2001	2003	2004	2006
Number of locations sampled	1	3	11	19
Total number of plants sampled	141	162	1190	1067
Percentage of plants attacked	29.8	49.4	45.9	29.2

Source: Sagliocco and Weiss (unpub.).

Releases of eggs from a colony collected from Monarto Zoo, near Murray Bridge, SA were made on a north-facing horehound infestation near Melton Mowbray, southern Tas in 2008. A pupa was recovered in December 2009, suggesting that this insect is able to complete a life-cycle under the climatic conditions prevailing in that part of Tas (J Ireson, *pers. comm.* 2010).

Agents imported but not tested

***Carcharodus boeticus* Rambur (Lepidoptera: Hesperiidae).** The horehound skipper butterfly, *C. boeticus*, was common in France and Spain where it was found at low densities at most sites. It was absent from Morocco. Larvae are nocturnal and feed on horehound leaves, hiding during the day in a nest made of rolled leaves spun with silk. In Europe, adults mate in full sun while flying and when temperature is favourable. In 1994 and 1995 consignments of *C. boeticus* larvae from France were sent to quarantine at Frankston, Vic where mating did not occur and the colonies died. No further importations of this insect have been made.

***Meligethes rotroui* Easton (Coleoptera: Nitidulidae).** Of very common occurrence on horehound in Morocco and Spain were adults of the *Meligethes* group (Coleoptera: Nitidulidae) (Easton 1952). They included *M. villosus* Brisout, *M. tropicus* Reitter, *M. rotroui* Easton and a minor species, *M. nanus* Erichson, which occurs in France and Spain only. As they feed on ovaries and seeds they may play an important role by reducing horehound seed output. The horehound pollen beetle *M. rotroui* was common on horehound populations throughout Morocco and Tunisia. Like the other *Meligethes* species recorded on horehound (Easton 1952), adult *M. rotroui* overwinter in the ground. In Morocco, they emerge when the first horehound flowers are produced in March and feed on anthers, pollen, ovaries and newly formed seeds. As they become sexually mature, females lay eggs among flower clusters. The emerging larvae feed in the calyxes on nectar and pollen, moving between flowers and, when fully developed, drop to the base of plants to pupate in the litter. In May, the new adults emerge to build up fat reserves by feeding on the last flowers produced then quickly return to the ground to diapause until the following spring. These emerging post-diapause adults live for only a few weeks. Attempts to obtain colonies from Morocco were made on three occasions between March 1995 and March 1997 but, due to difficulties in obtaining appropriate permits, certificates and maintaining the colonies during transportation, none were successful. No further collection has been attempted.

Other potential agents

***Phytoecia melanocephala* Fabricius (Coleoptera: Cerambycidae).** *Phytoecia melanocephala* F. (Coleoptera: Cerambycidae) is a univoltine root-boring insect reported from Morocco, Algeria, Tunisia and Sicily (Villiers 1946) but whose host plant was unknown until these horehound surveys. In northern Morocco, it was present at six of the 11 populations surveyed (Sagliocco 2000; Sagliocco and Weiss 1996). Random sampling near Meknes in 1994 showed that 40% of horehound roots contained larvae. A large horehound population discovered in 1994 near Mohammedia, Morocco appeared to have been almost completely destroyed by *P. melanocephala* by 1996. The same year, about 300 adults were obtained at CSIRO, France from 180 horehound roots collected near Meknes. Unfortunately, the short-lived adults did not reproduce in the laboratory and this agent was not investigated further.

Natural enemies in Australia

The native brightly coloured orange and black horehound bug *Agonoscelis rutila*, Fabricius (Hemiptera: Pentatomidae) has often been seen in great numbers on horehound but has not provided any worthwhile control. *A. rutila* has been found associated with *Citrus* spp., *Eucalyptus* spp., *Hibiscus sabdariffa* L. (Malvaceae) and *Leptospermum coriaceum* (F.Muell. ex Miq.) Cheel (*Myrtaceae*) and is an occasional pest of grain legumes and fruit trees (Field *et al.* 1998). Polyphagous insects such as grasshoppers cause occasional minor damage to horehound leaves.

DISCUSSION

Integrated control with herbicides

Horehound is one of the few weed species in Australia for which integration of chemical and biological control methods has been researched (Ainsworth 1999) Several herbicides are registered for use on horehound (Carter 1990) and recommended rates had no significant effect on *W. spilodactylus* eggs or larvae. However, adult moths showed large reductions in longevity and fecundity in response to the subsequent loss of flowers as a food source. Female moths avoided herbicide-treated plants when ovipositing. This has implications for integrated control, as oviposition and subsequent larval feeding would be concentrated on plants least affected by the herbicide treatment. Overall, it appeared that herbicides and application rates which caused rapid death of shoots were less suitable for integration with *W. spilodactylus* than herbicides or rates that had slower effects (Ainsworth 1999). Preserving *W. spilodactylus* populations would require keeping some horehound areas unsprayed. An alternative would be to devise herbicide treatment that allows longer survival of some plant parts (flowers) so that the moth populations could persist and continue suppressing the weed before the majority of plants die (Ainsworth 1999).

Autumn burning is an effective method of horehound suppression and seed bank reduction (Weiss 1996b; Weiss and Wills 2000). Burning is not compatible, however, with the preservation of biological control agents, unless relatively large reserves of the weed and agents are left untouched to allow later colonisation of new infestations.

Evaluation

Two agents, the horehound plume moth and the horehound clearwing moth, have been released to control *M. vulgare* in Australia. The plume moth is now very common in the field and appears to be performing well in areas where horehound plants rarely experience extremes of summer temperature and therefore remain green throughout the year. In the field during summer, development of the first instar larvae halted during very hot weather and/or when host plants became severely heat-stressed. It has not been determined whether this halt in development is an aestival diapause or quiescence (Clarke 2001). As a guideline, it appears that the plume moth performs better at sites with an annual rainfall exceeding 450 mm per year and with summer maximum temperatures rarely above 35°C (Weiss 1996a).

Regular monitoring has shown a steady increase of *C. mysiniformis* densities at Wyperfeld National Park, Vic resulting in plant mortality. Tas and NSW have implemented redistribution programs for this insect. However, impact assessment is required to measure both agents' actual level of control of horehound.

The most promising potential agent remains *M. rotroui*. Attempts to export colonies from its native range failed. Maintaining breeding colonies in the country of origin under an agreement with a local research organisation to conduct host-specificity tests and observations on its biology may be a feasible solution. Similar remarks apply to the horehound skipper *C. boeticus*, as mating may be achieved in larger outside enclosures in the native range. Further work could be attempted on the root-boring agent *P. melanocephala*. However, these potential agents may not be required if it can be shown that *C. mysiniformis* is providing a satisfactory level of control.

ACKNOWLEDGEMENTS

We thank the field staff in the many states that assisted the project, especially David Clifford (SA), Barry Sampson (NSW) and the rangers at Wyperfeld National Park (Vic).

REFERENCES

Ainsworth N (1999) Potential for integrated control of horehound *Marrubium vulgare* L. by the plume moth *Wheeleria spilodactylus* (Curtis) and 2,4-D. *Biocontrol Science and Technology* **9**, 355–364.

Baker J (2002) Factors affecting the establishment of a classical biological control agent, the horehound plume moth (*Wheeleria spilodactylus*) in South Australia. PhD thesis. University of Adelaide, Australia.

Bigot L and Picard J (1983) Lépidoptères Pterophorides du département des Bouches-du-Rhône et de la région de la Sainte-Baume (supplément au catalogue de P. Siépi). *Bulletin du Museum d'Histoire Naturelle, Marseille* **43**, 53–69.

Carter RJ (1990) Biology and control of horehound, *Marrubium vulgare*. In *Proceedings of the 9th Australian Weeds Conference*. 6–10 August 1990, Adelaide. (Ed JW Heap) pp. 382–386. Crop Sciences Society of South Australia, Adelaide.

Carter O and Walsh N (2006) National recovery plan for the marble daisy-bush *Olearia astroloba*. Vic Dept of Sustainability and Environment, Melbourne.

Clarke CR (2001) Climate matching in the colonisation of biological control agents against *Chrysanthemoides monilifera* and *Marrubium vulgare*. PhD thesis. University of Adelaide, Australia.

Clarke CR, Baker J, Keller MA and Roush RT (2000) Biological control of horehound: lessons from South Australia. *Plant Protection Quarterly* **15**, 29–32.

Easton AM (1952) A review of the *Meligethes* (Col. Nitidulidae) associated with the Labiate *Marrubium vulgare* L. *Bulletin de la Société des Sciences Naturelles du Maroc* **32**, 181–192.

Everist SL (1981) *Poisonous Plants of Australia*. Angus and Robertson, London.

Faithfull I and Sagliocco J-L (2002) Horehound clearwing moth survives at Wyperfeld. *Under Control, Pest Plants and Animals News* **19**, 9–10.

Field SA, Keller MA and Austin AD (1998) Field ecology and behaviour of the egg parasitoid *Trissolcus basalis* (Wollaston) Hymenoptera: Scelionidae). *Transactions of the Royal Society of South Australia* **122**, 65–71.

Frost A (1993) *Sir Joseph Banks and the Transfer of Plants to and from the South Pacific 1786–1798*. Colony Press, Melbourne.

Gielis C (1996) *Pterophoridae*. Vol. 1. *Microlepidoptera of Europe*. (Eds P Huemer, O Karsholt and L Lyneborg). Apollo Books, Stenstrup.

Ireson JE, Holloway RJ and Chatterton WS (2000) Progress on the rearing, release and establishment of the horehound plume moth, *Wheeleria spilodactylus* (Curtis), for the biological control of horehound in Tasmania. *Plant Protection Quarterly* **15**, 33–35.

Kwong R (2004) Biological control of weeds in Victoria. *Under Control, Pest Plants and Animals News, Supplement* **27**, 13.

Lane D, Riches K and Combellack H (1980) A survey of the distribution of noxious weeds in Victoria. Vic Dept of Conservation and Natural Resources, Melbourne.

Laštůvka Z and Laštůvka A (2001) *The Sesiidae of Europe*. Apollo Books, Stenstrup.

Leyson M (1999) Courtship and mating of the horehound plume moth, *Wheeleria spilodactylus*. BSc Hons thesis. University of Adelaide, Australia.

Lippai A, Smith PA, Price TV, Weiss J and Lloyd CJ (1996) Effects of temperature and water potential on germination of horehound (*Marrubium vulgare*) seeds from two Australian localities. *Weed Science* **44**, 91–99.

Menéndez R and Thomas C (2000) Metapopulation structure depends on spatial scale in the host-specific moth *Wheeleria spilodactylus* (Lepidoptera: Pterophoridae). *Journal of Animal Ecology* **69**, 935–951.

Menéndez R and Thomas C (2006) Can occupancy patterns be used to predict distributions in widely separated geographic regions? *Oecologia* **149**, 396–405.

Parsons WT and Cuthbertson EG (2001) *Noxious Weeds of Australia*. CSIRO Publishing, Melbourne.

Sagliocco J-L (2000) The insect fauna associated with horehound (*Marrubium vulgare* L.) in western Mediterranean Europe and Morocco: potential for biological control in Australia. *Plant Protection Quarterly* **15**, 21–25.

Sagliocco J-L and Coupland JB (1995) Biology and host specificity of *Chamaesphecia mysiniformis* (Lepidoptera: Sesiidae), a potential biological control agent of *Marrubium vulgare* (Lamiaceae) in Australia. *Biocontrol Science and Technology* **5**, 509–516.

Sagliocco J-L and Weiss J (1996) 'Biocontrol of horehound (*Marrubium vulgare* L.) work in Europe 1990–1996'. Final report on project DCV14 to International Wool Secretariat, by CSIRO and Vic Dept of Natural Resources and Environment, Melbourne.

Sagliocco J-L and Weiss J (2004) Importation, establishment and preliminary impact assessment of *Chamaesphecia mysiniformis* (Lepidoptera: Sesiidae) for the biological control of horehound in Australia. In *Proceedings of the 14th Australian Weeds Conference.* 6–9 September 2004, Wagga Wagga, NSW. (Eds BM Sindel and SB Johnson) pp. 388–390. Weed Society of New South Wales, Sydney.

Sloan, Cook and King Pty Ltd (1988) 'The economic impact of pasture weeds, pests and diseases on the Australian wool industry'. Report to the Australian Wool Commission. Australian Wool Commission, Sydney.

Traeger AM (2000) The dispersal of horehound plume moth (*Wheelaria spilodactylus*). BAgSci Hons thesis. University of Adelaide, Australia.

Villiers A (1946) *Coleoptères Cerambycides de l'Afrique du Nord*. Editions du Museum, Paris.

Weiss J (1996a) 'The biological control of horehound in southern Australia'. Progress report to the SA Animal and Plant Control Commission. Dept of Natural Resources and Environment, Melbourne.

Weiss J (1996b) Control of horehound, *Marrubium vulgare* L., in Wyperfeld National Park, Victoria. In *Proceedings of the 11th Australian Weeds Conference.* 30 September–3 October 1996, Melbourne. (Ed RCH Sheppard) pp. 417–419. Weed Science Society of Victoria, Melbourne.

Weiss J and Wills E (2000) Integrated management of horehound (*Marrubium vulgare* L.) in Wyperfeld National Park. *Plant Protection Quarterly* **15**, 40–42.

Weiss J, Ainsworth N and Faithfull I (2000) Best practice management guide. 8. Horehound, *Marrubium vulgare*. CRC for Weed Management Systems, University of Adelaide, Australia.

Wills E (2000) The release and establishment of two biological control agents of horehound (*Marrubium vulgare* L.) in south-eastern Australia. *Plant Protection Quarterly* **15**, 26–28.

Ziegler CR, Weiss J, Sagliocco J-L and Walker A (1993) 'The biological control of horehound in southern Australia'. Final report on project DCV003 to Wool Research and Development Corporation, by CSIRO and Vic Dept of Conservation, Forest and Lands, Melbourne.

Mikania micrantha Kunth – mile-a-minute

Michael Day

ABSTRACT

Mikania micrantha or mile-a-minute is a Class 1 weed in Qld and the target of a national cost-share eradication program. It is a fast-growing vine native to tropical America and a major weed in Asia and the Pacific, affecting subsistence agriculture and plantations. Biocontrol worldwide began in 1978, with surveys for insects conducted by CIBC (now CABI) in Central and South America. Following host-specificity testing, *Liothrips mikaniae* was introduced unsuccessfully into the Solomon Islands in 1988 and Malaysia in 1990. Surveys targeting pathogens by CABI between 1988–89 and 1996–99 resulted in the successful release of the rust *Puccinia spegazzinii* in Taiwan in 2008 and its unsuccessful release in India in 2005 and China in 2006. In 2008, the rust was released and established in Papua New Guinea and Fiji. In PNG, the rust is spreading and reducing stands of *M. micrantha* in some areas. The butterflies *Actinote anteas* and *A. thalia pyrrha* were released in Indonesia in 1999, as part of an ACIAR-funded project to control *Chromolaena odorata.* They are also aiding the control of *M. micrantha* in Sumatra. Effective agents used elsewhere will be introduced into Australia if the current eradication program fails and the emphasis shifts to containing the weed.

Key words: *Liothrips mikaniae*, *Puccinia spegazzinii*, *Actinote* spp.

INTRODUCTION

Mikania micrantha Kunth (Asteraceae), mile-a-minute, is a fast-growing vine, native to Central and South America (Waterhouse and Norris 1987). It has a pantropical distribution, being found throughout Asia and many Pacific island countries and territories (Waterhouse and Norris 1987; Waterhouse 1994). *M. micrantha* was first reported in Australia in 1998 at Bingil Bay in northern Qld, where it had been introduced illegally for medicinal purposes, and near Ingham (pathway unknown). In 2001, *M. micrantha* was found near Speewah on the Atherton Tableland, where it had been introduced as a contaminant of packaging material (DEEDI 2009). *M. micrantha* is found in several areas in northern Qld, including Speewah, Mission Beach, Bingal Bay and Ingham (Fig. 1). It is also present on Christmas Island, an Australian Territory in the Indian Ocean.

M. micrantha grows about 1 m per month (Day *et al.*, unpubl. data) and it flowers prolifically (Fig. 2). Seeds are spread by wind and machinery as well as by water along watercourses. Sections of plant such as broken stems and leaves can easily produce roots and develop into new plants (Waterhouse and Norris 1987; A Kawi, *pers. obs.* 2009).

In many Pacific island countries such as Papua New Guinea (PNG), Samoa and Fiji, *M. micrantha* is a major problem in food gardens and young plantations, smothering crops such as taro, bananas, papaw and sweet potato (Pene *et al.* 2007) and reducing yields of oil palm and cocoa. In south-east Asia, it is a problem in teak, rubber and tea plantations (Cock 1982a). Cost of control of *M. micrantha* in oil palm and rubber plantations in Malaysia exceeds $10 million per year (Cock *et al.* 2000). In such areas, control by hand-pulling vines or slashing is very labour-intensive and time-consuming.

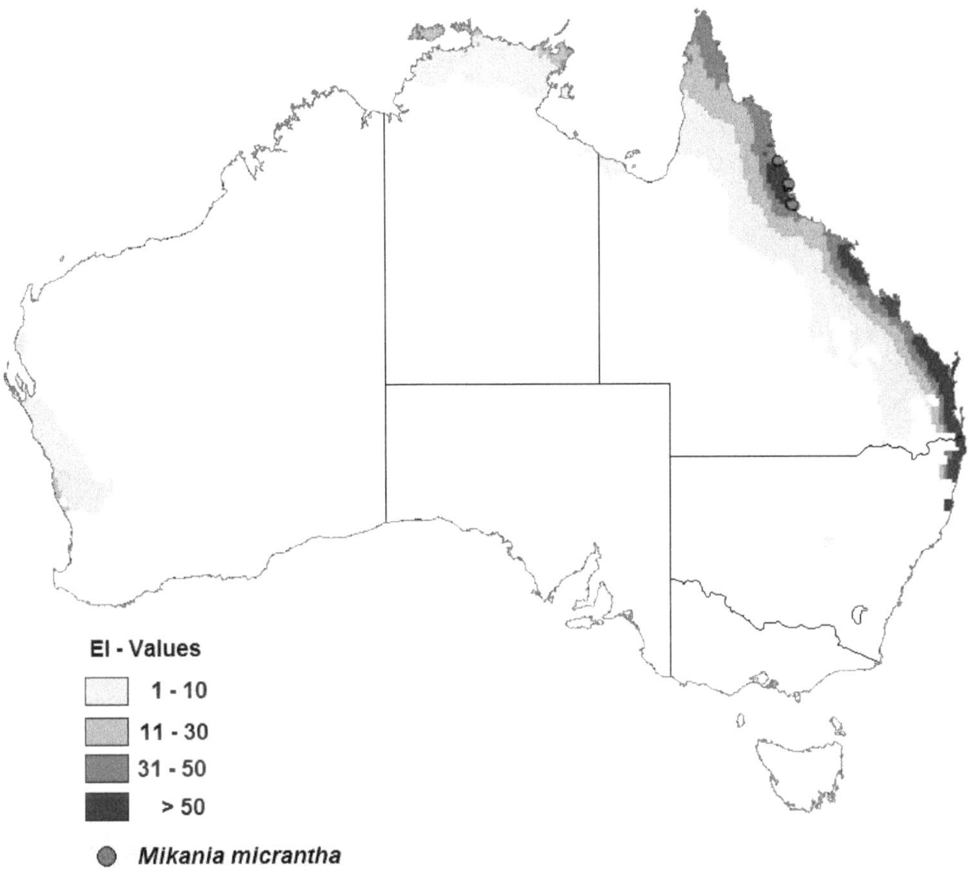

Figure 1: The current distribution of *Mikania micrantha* in Qld and its potential distribution in Australia based on a CLIMEX model. Source: Qld Government.

M. micrantha is a Class 1 weed in Qld and is one of six targets of a national cost-share eradication program, costing over $A400 000 per year, with the *M. micrantha* component accounting for about $A72 000 per year (DEEDI 2009). Therefore, biocontrol is currently not a consideration. If left uncontrolled in northern Qld, *M. micrantha* may infest agricultural areas and reduce productivity, and also invade natural ecosystems where it may smother rainforest species and reduce biodiversity. It has potential to spread west into the NT and south down the Qld coast (Fig. 1). Should the eradication program fail, biocontrol could be initiated as part of a containment program and effective agents utilised elsewhere could be introduced into Australia (Day and Panetta 2010).

Figure 2: Flowering *Mikania micrantha*. Photo: M Day, Qld Government.

BIOLOGICAL CONTROL HISTORY

Biocontrol of *M. micrantha* began in 1978 when Matthew Cock from the Commonwealth Institute of Biological Control (CIBC, now CABI) funded by the Overseas Development Administration of the UK, conducted surveys on *M. micrantha* in Trinidad, Colombia, Costa Rica, Ecuador, Panama, Peru and Venezuela. These surveys were primarily for insects and resulted in 29 species being recorded, of which nine were considered for further studies (Cock 1982a). Due to limited funding, only *Liothrips mikaniae* (Priesner) (Thysanoptera: Phlaeothripidae) was host-tested (Cock *et al.* 2000). *Liothrips mikaniae* was subsequently released in the Solomon Islands in 1988 and Malaysia in 1990, but failed to establish in either country (Cock *et al.* 2000). A consignment of insects was also sent to PNG, but the colony died in quarantine (Cock *et al.* 2000; Pene *et al.* 2007).

In 1988–89, surveys in southern Brazil for pathogens on *M. micrantha* were conducted by Robert Barreto. Of the nine species found, *Mycosphaerella mikania-micranthae* RW Barreto (*Capnodiales: Mycosphaerellaceae*) and *Puccinia spegazzinii* De Toni (*Pucciniales: Pucciniaceae*) were considered the most promising, due to their wide climatic range and potential damage (Barreto and Evans 1995).

Additional surveys by CABI for potential pathogens were conducted between 1996–99 throughout the range of *M. micrantha* in tropical America, in a collaborative project between Kerala Forestry Research, Assam Agricultural University (AAU), India and Viçosa University, Brazil. Three species, including *P. spegazzinii*, which was found throughout the native range of *M. micrantha*, were collected for further study (Evans and Ellison 2005). Host-specificity testing was subsequently conducted on *P. spegazzinii* and the agent was released unsuccessfully at Assam, India in 2005 and China in 2006 (Ellison *et al.* 2008; J Ding, *pers. comm.* 2011). The rust was released successfully in Taiwan in 2008 (C Ellison, *pers. comm.* 2010) but its impact has not been assessed.

In 1999, two butterflies *Actinote anteas* (Doubleday and Hewitson) (Lepidoptera: Arctiidae) and *A. thalia pyrrha* Fabr. were released in Indonesia as part of the ACIAR-funded biocontrol of *Chromolaena odorata* (L.) RM King and H Robinson (Asteraceae) project, administered by the Qld Government (Desmier de Chenon *et al.* 2002). The butterflies also attacked *M. micrantha* and readily established on both species in the field. In Sumatra, the butterflies have controlled *M. micrantha* in some lowlands areas (R Desmier de Chenon, *pers. comm.* 2006).

In 2006, an ACIAR-funded biocontrol of *M. micrantha* project, administered by the Qld Government, was initiated to control *M. micrantha* in Fiji and PNG (Pene *et al.* 2007) and thereby reduce the risk of weed spread into Australia (Day and Panetta 2010). Both butterfly species were introduced into Fiji from Indonesia in 2006, following their success there, but both died in quarantine and were never field released. After additional host-specificity testing by CABI, *P. spegazzinii* was introduced into Fiji and PNG in 2008. This rust is currently being released in both countries, where it has established and is spreading.

PLANT TAXONOMY

The genus *Mikania* contains about 250 species and belongs to the tribe Eupatorieae, within the family Asteraceae. There are no crops or ornamentals in this tribe (Waterhouse 1994). There are 20 other species in the Eupatorieae in Australia but only *Adenostemma lavenia* (L.) Kuntze and *A. macrophyllum* (Blume) DC. are native. *Ageratina adenophora* (Spreng.) RM King & H Rob. (crofton weed), *Ageratina riparia* (Regel) RM King & H Rob. (mistflower) and *C. odorata* have been targeted for biocontrol elsewhere.

M. micrantha has been previously confused in early literature with *M. cordata* (Burm. f.), which also has a pantropical distribution and is native to south-east Asia and PNG (Cock 1982a; Waterhouse and Norris 1987).

CANDIDATES

No agent has yet been introduced or studied in Australia for the biocontrol of *M. micrantha*. However, several agents that have been utilised elsewhere, have the potential to control the weed and could be introduced if biocontrol is warranted.

Promising agents

***Puccinia spegazzinii* De Toni (*Pucciniales: Pucciniaceae*).** *Puccinia spegazzinii* is an autoecious, microcyclic rust. Chlorotic spots first appear six to nine days after inoculation and teliospores appear after 10–14 days. Basidiospores, which infect new plants, are released from the mature teliospores under conditions of high humidity. Infection occurs when temperatures are between 15–25°C and there are about 10 hours of dew (Ellison *et al.* 2008). Sixty plant species, including 33 from the family Asteraceae, of which 17 were from the tribe Eupatorieae, including 11 species of *Mikania*, were tested prior to its introduction into India (Ellison *et al.* 2008) and a further nine species, including three from Asteraceae were tested on behalf of PNG and Fiji.

In PNG, the rust has been released in 15 of the 16 provinces where *M. micrantha* has been reported. It has established at over 70 sites, and has spread up to 37 km from some sites. At several sites, percent plant cover is being reduced as rust populations increase. In laboratory studies, growth of plants infected with the rust was significantly reduced (Day *et al.*, unpubl. data). In Fiji, the rust has been released on the four main islands, and it has established at 20 sites, on Viti Levu and Vanua Levu (Day *et al.*, unpubl. data).

***Actinote* spp.** *Actinote anteas* and *A. thalia pyrrha* were released in Indonesia in 1999 under an ACIAR-funded project to control chromolaena (see *C. odorata* chapter, this volume). *Actinote* spp. were able to complete their life-cycle on both chromolaena and *M. micrantha* and were subsequently approved for release (Desmier de Chenon *et al.* 2002). Both butterflies established in the field and are causing significant damage to *M. micrantha*, especially in the lowlands of Sumatra (R Desmier de Chenon, *pers. comm.* 2006).

***Liothrips mikaniae* (Priesner) (Thysanoptera: Phlaeothripidae).** This thrips was found in Colombia, Costa Rica, Peru, Surinam, Trinidad and Venezuela, causing substantial damage to young growth of *M. micrantha* (Cock 1982a). Adults and nymphs feed on the underside of leaves, causing small lesions. Eggs are deposited on the underside of leaves and take 12–14 days to hatch. Larvae feed in groups and pupation occurs in the leaf litter. Emerging adults feed on fresh new growth and the pre-oviposition phase is about 10–14 days (Cock 1982b). Total life-cycle is about 35 days. Host-specificity was conducted on 34 species representing 16 families, including 15 species in the family Asteraceae. Development was restricted to *M. micrantha*, although some adult feeding was observed on *M. cordata* and *M. vitifolia* DC. In the field, *L. mikaniae* was collected only from *M. micrantha* (Cock 1982a, b).

DISCUSSION

In Qld, *M. micrantha* is the target of a national eradication program and as such, biocontrol is not considered an option. However, work overseas suggests that several agents could be utilised if the program shifts from eradication to containment. *P. spegazzinii* would probably be the highest priority, due to its high specificity, ease of culturing and early results of its impact on *M. micrantha* in PNG. For Australia, extra testing should be minimal, involving the two native *Adenostemma* species and possibly the two weeds *A. adenophora* and *Ageratum conyzoides* L. subsp. *conyzoides*. Field releases would not require large stands of *M. micrantha* to achieve establishment.

L. mikaniae should also be considered for similar reasons. However, there are concerns that predation may affect its establishment and levels of control (Cock *et al.* 2000). *Actinote* spp., while causing damage to *M. micrantha* in Indonesia, is labour-intensive to rear. It requires large stands of *M. micrantha* to achieve establishment and so may not establish in the small and scattered infestations that currently occur in Qld. *Actinote* spp. would have the added advantage, however, of attacking chromolaena that is also a weed in northern Qld (see *C. odorata* chapter).

If one or more of the above agents are established in Australia to control *M. micrantha*, the benefit:cost ratio is likely to be higher than for many other successful biocontrol projects which would normally have an exploration phase and detailed host-specificity testing included. Similar principles have been used previously to justify overseas biocontrol work on *C. odorata*.

ACKNOWLEDGEMENTS

The author gratefully acknowledges the research conducted by staff at the Secretariat of the Pacific Community, National Agricultural Research Institute (PNG), Cocoa and Coconut Institute (PNG), PNG Oil Palm Research Association, Fiji Ministry of Primary Industries, CABI Europe-UK and Dr Roch Desmier de Chenon, France.

REFERENCES

Barreto RW and Evans HC (1995) The mycobiota of the weed *Mikania micrantha* in southern Brazil with particular reference to fungal pathogens for biological control. *Mycological Research* **99**, 343–352.

Cock MJW (1982a) Potential biological-control agents for *Mikania-micrantha* HBK from the neotropical region. *Tropical Pest Management* **28**, 242–254.

Cock MJW (1982b) The biology and host specificity of *Liothrips-mikaniae* (Priesner) (Thysanoptera, Phlaeothripidae), a potential biological-control agent of *Mikania-micrantha* (Compositae). *Bulletin of Entomological Research* **72**, 523–533.

Cock MJW, Ellison CA, Evans HC and Ooi PAC (2000) Can failure be turned into success for biological control of mile-a-minute weed (*Mikania micrantha*)? In *Proceedings of the 10th International Symposium on Biological Control of Weeds.* 4–14 July 1999, Bozeman, Montana. (Ed NR Spencer) pp. 155–167. Montana State University, Bozeman.

Day MD and Panetta FD (2010) Preventing and managing incursions of Class 1 weeds in Queensland. In *Proceedings of the Global Biosecurity Conference.* 28 February–3 March 2010, Brisbane. pp. 69–70. CRC National Plant Biosecurity.

Department of Employment, Economic Development and Innovation (DEEDI) (2009) 'National Four Tropical Weeds Eradication Program: annual report'. July 2008–June 2009. Brisbane.

Desmier de Chenon R, Sipayung A and Sudharto P (2002) A new biological agent, *Actinote anteas*, introduced into Indonesia from South America for the control of *Chromolaena odorata*. In *Proceedings of the 5th International Workshop on Biological Control and Management of* Chromolaena odorata. Durban, South Africa. (Eds C Zachariades, R Muniappan and LW Strathie) pp. 170–176. ARC-PPRI.

Ellison CA, Evans HC, Djeddour DH and Thomas SE (2008) Biology and host range of the rust fungus *Puccinia spegazzinii*: a new classical biological control agent for the invasive, alien weed *Mikania micrantha* in Asia. *Biological Control* **45**, 133–145.

Evans HC and Ellison CA (2005) The biology and taxonomy of rust fungi associated with the neotropical vine *Mikania micrantha*, a major invasive weed in Asia. *Mycologia* **97**, 935–947.

Pene S, Orapa W and Day M (2007) First fungal pathogen to be utilized for weed biocontrol in Fiji and Papua New Guinea. *Biocontrol News and Information* **28**, 55N–56N.

Waterhouse DF (1994) *Biological Control of Weeds: Southeast Asian Prospects*. ACIAR, Canberra.

Waterhouse DF and Norris KR (1987) *Biological Control: Pacific Propects*. Inkata Press, Melbourne.

Mimosa diplotricha C.Wright ex Sauvalle – giant sensitive plant

Rachel Cruttwell McFadyen

ABSTRACT

Mimosa diplotricha, native to the tropical Americas, has become a serious weed of the wet tropics in south and east Asia and the Pacific, and of sugarcane and grazing lands in northern Qld. Exploration in southern Brazil resulted in host-testing of three insects and release of two, but only the psyllid *Heteropsylla spinulosa* became established. Feeding by this minute insect has successfully controlled the weed in Qld, Papua New Guinea and several Pacific island countries.

Key words: *Mimosa invisa*, *Heteropsylla spinulosa*.

INTRODUCTION

Mimosa diplotricha C.Wright ex Sauvalle (Fabaceae), previously known as *Mimosa invisa* Mart., is a sprawling prickly annual plant native to tropical South and Central America (Waterhouse 1994). It is in the family Mimosaceae and shares many characteristics with its weedy congeners *Mimosa pudica* L. and *Mimosa pigra* L. Leaves are compound and bipinnate, consisting of many leaflets on a long central vein, which fold up at night or when touched. Stems are covered with numerous short recurved thorns. Flowers are pink on short stalks and seed pods are covered with stiff bristles and break up when mature into single-seeded segments. These float readily and also adhere to animal fur and human clothing, leading to rapid spread. *M. diplotricha* grows quickly, clambering over and smothering supporting vegetation (Fig. 1). It has become a serious weed in many tropical countries across Africa and south and south-east Asia including Papua New Guinea and Pacific island countries (Waterhouse 1994; Orapa 2009).

M. diplotricha has been present in far northern Qld since the 1920s, probably introduced as a contaminant in seed of tropical legumes, and subsequently spread to infest thousands of hectares of pastures, sugarcane fields and horticultural plantations from the Daintree to Mackay. By the 1990s it was the most serious weed of grazing and sugarcane lands in northern Qld. The sharp prickles caused problems for field workers and the dense growth choked sugarcane harvesting machinery. Mechanical and

Figure 1: Infestation of *Mimosa diplotricha* in northern Qld. Photo: M Ablin, Qld Government.

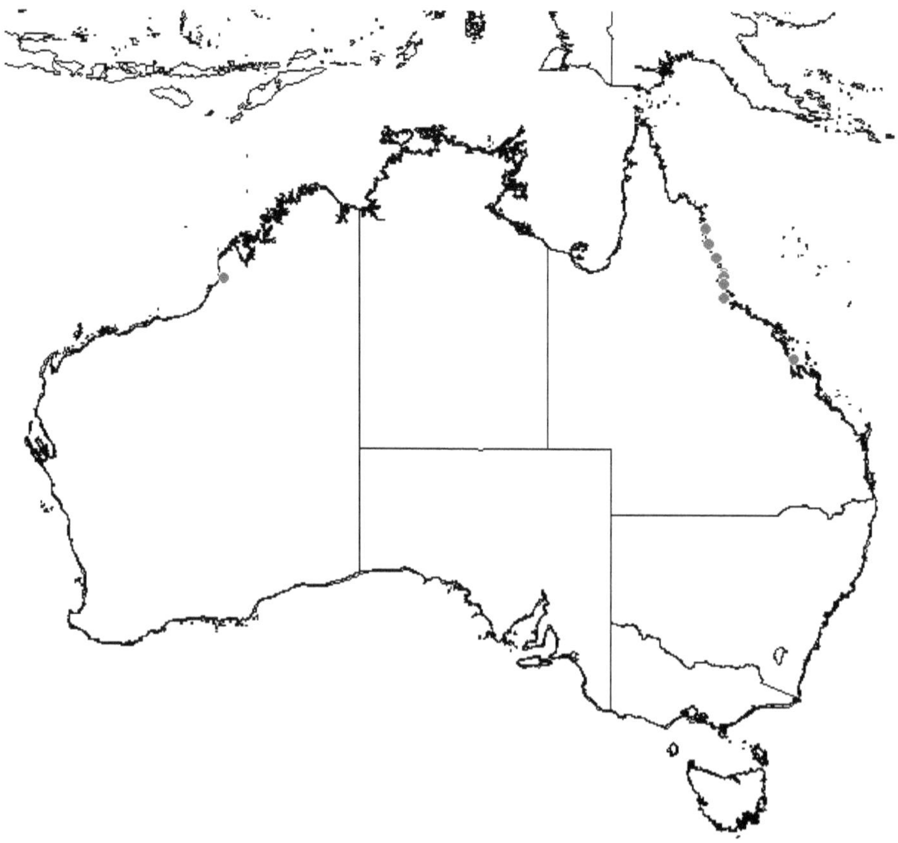

Figure 2: Distribution of *Mimosa diplotricha* in Australia. Specimen data reproduced from Australia's Virtual Herbarium with permission of the Council of Heads of Australasian Herbaria Inc.

cultural control methods were not effective and use of herbicides was not economically sustainable (Parsons and Cuthbertson 2001).

Although common in the adjacent areas of East and West Timor, there is only a single record of the plant from the wet tropical areas of the NT and WA which would be suitable habitats (Fig. 2).

BIOLOGICAL CONTROL HISTORY

The possibilities for biological control were first investigated by the Qld Government in the early 1980s. A field station had been maintained in southern Brazil from 1973 as part of a biocontrol program against groundsel bush *Baccharis halimifolia* L. (see separate chapter, this volume). Surveys were undertaken from this station from 1982 onwards and nearly 70 species of insects were found to attack *M. diplotricha* in Brazil: a complete list of these is published in Waterhouse (1994). Three insects were brought into quarantine in Brisbane for detailed host-testing.

One of these, the psyllid, *Heteropsylla spinulosa* (Muddiman, Hodkinson and Hollis), was successful in controlling *M. diplotricha* in northern Qld. It has since been released in several Pacific island countries (Waterhouse 1994; Kuniata and Korowi 2004; Orapa 2009), and recently in East Timor (M Day, *pers. comm.* 2010).

EXPLORATION

Exploration was undertaken by Cesar Garcia, an entomologist employed by the Qld Government, and was limited to southern Brazil, with occasional surveys in Colombia. Most of the insects discovered were unable to be identified and are probably undescribed species (Waterhouse 1994). Investigations ceased when C Garcia moved to northern Argentina in 1987 to undertake work on another weed. No surveys of pathogens were undertaken.

CANDIDATES

***Scamurius* sp. (Hemiptera: Coreidae).** This large (2.2 cm), undescribed sucking bug is native to southern Brazil where both the nymphs and adults feed on the shoots of *M. diplotricha*. Eggs are laid in groups on the shoots and development from egg to adult takes about eight weeks. Adults live for up to six months and adult feeding causes death of the shoots. Detailed host-testing in quarantine in Brisbane demonstrated that development was only possible on the weedy species of *Mimosa*. Releases were made from 1987 to 1990 in northern Qld, but there was no establishment. Releases were also made in Western Samoa in 1989, but again establishment did not occur.

***Heteropsylla spinulosa* Muddiman, Hodkinson and Hollis (Hemiptera: Psyllidae).** This psyllid is native to southern Brazil and was an undescribed species when found in the initial surveys. It has a very similar life-cycle and impact to the leucaena psyllid *Heteropsylla cubana* Crawford which in the 1980s spread rapidly throughout South Asia and the Pacific, including northern Australia, as a pest of the valuable fodder tree *Leucaena leucocephala* (Lamk.) de Wit. Adults are green, 2.5 mm long, and feed on the leaflets and shoots. Eggs are attached by a pedicel to the plant between the overlapping leaflets, and nymphs feed on the leaflets (Fig. 3). High populations cause severe stunting and distortion of the leaves and growing tips, preventing flowering and production of seed (Fig. 4). The

Figure 3: Adult and nymphs of *Heteropsylla spinulosa*. Photos: M Ablin, Qld Government.

Figure 4: Damage caused by *Heteropsylla spinulosa* feeding. Photo: M Ablin, Qld Government.

development period from egg to adult is about 28 days, so populations can increase rapidly after the plant germinates with the first rainfall of the wet season.

Extensive host-specificity testing was undertaken in quarantine in Brisbane and development found to occur only on *M. diplotricha* (Willson and Garcia 1992). First releases were made in 1988 and establishment occurred immediately. The psyllid spread rapidly and the impact on the weed was quickly evident (Ablin 1993).

In Brazil the psyllid is attacked by several parasitoids and predators, but there are no data on attack levels in Australia. During the 1980s and 1990s, several parasitoids and predators were introduced from the Caribbean into Hawaii, Thailand and other countries for the control of the leucaena psyllid *H. cubana*. Of these, only the parasitoid *Psyllaephagus yaseeni* Noyes (Hymenoptera: Encyrtidae) and the predator *Curinus coeruleus* Mulsant (Coleoptera: Coccinellidae) have become widely established in Pacific island countries and in south and south-east Asia (Bapompeth 1993). However, their impact is not great (Geiger and Gutierrez 2000) and they are not yet present in Australia or PNG (M Day, *pers. comm.* 2010).

***Psigida walkeri* (Grote) (Lepidoptera: Citheroniidae).** This moth occurs widely in Brazil and Colombia, with host plants in several genera of Mimosaceae. Adult wing-span is up to 5 cm and the large green larvae feed on the leaves, flowerbuds, young seed pods and young stems, causing severe damage and preventing both flowering and seed production. The life-cycle takes about two months, with a pupal diapause of up to four months in southern Brazil. Detailed host-testing was undertaken in quarantine in Brisbane, where it was found that several Australian native plants were suitable hosts for oviposition and larval development, at least in laboratory conditions. The risk was considered unacceptable and the quarantine colony destroyed without any releases being made (Vitelli *et al.* 2001). The moth was released in the Cook Islands, where these non-target plants do not occur, but establishment did not occur.

***Corynespora cassiicola* (Burke and Curtis) Wei.** *Corynespora cassiicola* is a pest stem-spot fungus, indigenous to Australia as well as other countries, which causes disease on a wide variety of plants including vegetable and fruit crops. A strain of this fungus isolated from northern Qld appears to be specific to *M. diplotricha*. The disease causes defoliation and dieback in very hot humid conditions. Initially older leaflets are shed, then small dark oval spots develop along the stems. As the spotted area increases, the growing tip dies. Damage in the field can be severe, but artificial spread has not been used as the disease is already widespread on *M. diplotricha* in Qld (Biosecurity Queensland 2007).

EVALUATION

The impacts of the psyllid were quickly evident in the field in northern Qld. Insecticide exclusion studies demonstrated that seed production was reduced by 80% and growing tip elongation reduced by 77% in plants exposed to psyllid damage compared with plants kept free of the psyllid (Ablin 1993). Evaluation in Papua New Guinea demonstrated similar impacts; reductions of groundcover by *M. diplotricha* from 100% to <5% in two years, and 80% reduction in seed production (Kuniata and Korowi 2004). By 2000, *M. diplotricha* was reduced to being a very minor weed of tropical eastern Australia (Vitelli *et al.* 2001). Based on these data, Page and Lacey (2006) estimated the net present value of the program to be $19.7 million in 2005 values, a benefit:cost ratio of 17.6 on expenditure of $1.7 million.

DISCUSSION

This was an outstandingly successful project, with excellent control of the weed achieved in most of the affected areas. The control continues to be effective 20 years later and giant sensitive plant is no longer a serious weed for the sugar and other agricultural industries in northern Qld. This achievement has been little recognised because the infestations were geographically remote and the results not adequately publicised at the time. The successful agent, the psyllid *H. spinulosa*, has since been transferred to Papua New Guinea (Kuniata and Korowi 2004), Fiji, Pohnpei, Western Samoa (Waterhouse 1994), Solomon Islands, Cook Islands, Vanuatu and Niue and has spread unaided to Palau, resulting in successful control in most areas (Orapa 2009). It has recently been successfully established in East Timor (M Day, *pers. comm.* 2010), from where it is expected to spread naturally throughout the eastern Indonesian islands. There have been no reports of significant negative interactions with the parasitoid and coccinellid introduced for the control of the leucaena psyllid (Bapompeth 1993) though no proper evaluations have been undertaken.

ACKNOWLEDGEMENTS

I wish to acknowledge the efforts of Doug Waterhouse (1916–2000) who brought together much unpublished information from government reports in his 1994 book, thus making it available to a wider audience. John Hosking kindly obtained the map for Figure 2.

REFERENCES

Ablin MP (1993) Insecticide exclusion with carbofuran demonstrates the effectiveness of *Heteropsylla spinulosa* as a biological control agent for *Mimosa invisa* in North Queensland (abstract only). In *Proceedings of the 10th Australian and 14th Asian-Pacific Weed Conference*. 6–10 September, Brisbane. p. 128. Queensland Weed Society, Brisbane.

Bapompeth N (1993) Leucaena psyllid in the Asia-Pacific region: implications for its management in Africa. <http://www.fao.org/docrep/008/v5020e/V5020E06.htm>.

Biosecurity Queensland (2007) 'Giant sensitive plant *Mimosa diplotricha* (= *Mimosa invisa*).' Fact Sheet PP27. Qld Dept of Primary Industries and Fisheries, Brisbane.

Geiger CA and Gutierrez AP (2000) Ecology of *Heteropsylla cubana* (Homoptera: Psyllidae): psyllid damage, tree phenology, thermal relations, and parasitism in the field. *Environmental Entomology* **29**, 76–86.

Kuniata LS and Korowi KT (2004) Bugs offer sustainable control of *Mimosa invisa* and *Sida* spp. in the Markham Valley, Papua New Guinea. In *Proceedings of the 11th International Symposium on Biological Control of Weeds*. 27 April–2 May 2003, Canberra. (Eds JM Cullen, DT Briese, DJ Kriticos, WM Lonsdale, L Morin and JK Scott) pp. 567–573. CSIRO Entomology, Canberra.

Orapa W (2009) 'History of weed biological control in the Pacific Islands'. Presentation to Biological Control Workshop, Auckland, November 2009. <www.issg.org>.

Page AR and Lacey KL (2006) 'Economic impact assessment of Australian weed biological control'. Technical Series 10. CRC for Australian Weed Management, Adelaide.

Parsons WT and Cuthbertson EG (2001) *Noxious Weeds of Australia*. 2nd edn. CSIRO Publishing, Melbourne.

Vitelli MP, Garcia C, Lockett CJ, West GM and Willson BW (2001) Host specificity and biology of the moth *Psigida walkeri* (Lepidoptera: Citheroniidae), a potential biological control agent for *Mimosa diplotricha* in Australia and the South Pacific. *Biological Control* **22**, 1–8.

Waterhouse DF (1994) *Biological Control of Weeds: Southeast Asian Prospects*. Australian Centre for Agricultural Research, Canberra.

Willson BW and Garcia CA (1992) Host specificity and biology of *Heteropsylla spinulosa* (Hom.: Psyllidae) introduced into Australia and Western Samoa for the biological control of *Mimosa invisa*. *Entomophaga* **37**, 293–299.

Mimosa pigra L. – mimosa

Tim A Heard

ABSTRACT

Mimosa pigra is a serious weed of agriculture and the environment in northern Australia, particularly the NT. It is a Weed of National Significance and is listed in the Global Invasive Species Database as one of the *One Hundred of the World's Worst Invasive Alien Species*. It is a prickly, perennial, woody shrub which forms monocultures on floodplains and around wetlands. *M. pigra* has been the subject of a biological control program in Australia from 1979. Projects were also conducted in several south-east Asian countries. A large suite of agents were assessed but rejected for various reasons, and a total of 13 insects and two fungi were released in Australia to control *M. pigra* of which 10 insects and one fungus are known to have established. Two agents, *Carmenta mimosa* and *Neurostrota gunniella*, are currently inflicting severe damage, reducing both seed production and seed banks, and defoliating plants, which favours competing vegetation and leads to lower seedling survival and increasingly senescent stands. The impact of more recently released agents, particularly *Malacorhinus irregularis* and *Macaria pallidata*, appears to be substantial but has not been measured. Some agents were also released in Asia but their fate has barely been evaluated. The biological control program against this weed is characterised by the selection of challenging agents that might have been rejected for practical reasons in a less well-funded project. The experience gained in overcoming these challenges, for the flower-feeding agents and the green seed-feeders in particular, provides a model for the use of these types of agents in future programs. The *M. pigra* program also provides a leading example of how to measure the impact of biocontrol agents.

Keywords: woody shrub, *Carmenta mimosa*, *Neurostrota gunniella*, *Malacorhinus irregularis*, *Macaria pallidata*, flower-feeders, impact measurement.

INTRODUCTION

Mimosa pigra L. is one of Australia's 20 Weeds of National Significance (Thorpe and Lynch 2000) and is listed in the Global Invasive Species Database (2010). Infestations form dense thickets (Fig. 1) that affect both conservation areas and pastoral land. These reduce the diversity of plants and animals and impact negatively on agriculture by competing with pasture species, hindering mustering of livestock and restricting access to water by humans and livestock (Braithwaite *et al.* 1989; Lonsdale *et al.* 1989).

M. pigra is a prickly, perennial, woody shrub (Fig. 2) which grows to a height of 6 m in Australia (Fig. 3) but rarely more than 2 m in the native range of the American tropics (Fig. 4). It forms monocultures on floodplains and around wetlands, wherever seasonal inundation occurs (Lonsdale *et al.* 1989). Untreated infestations may double in area each year (Lonsdale 1993) and spread by seeds which float on the water surface (Lonsdale *et al.* 1989).

In Australia, *M. pigra* is found mainly in the NT with major infestations on the floodplains of the Adelaide, Daly,

Figure 1: Aerial view of a monoculture of *Mimosa pigra* on a floodplain in the NT. Photo: CSIRO.

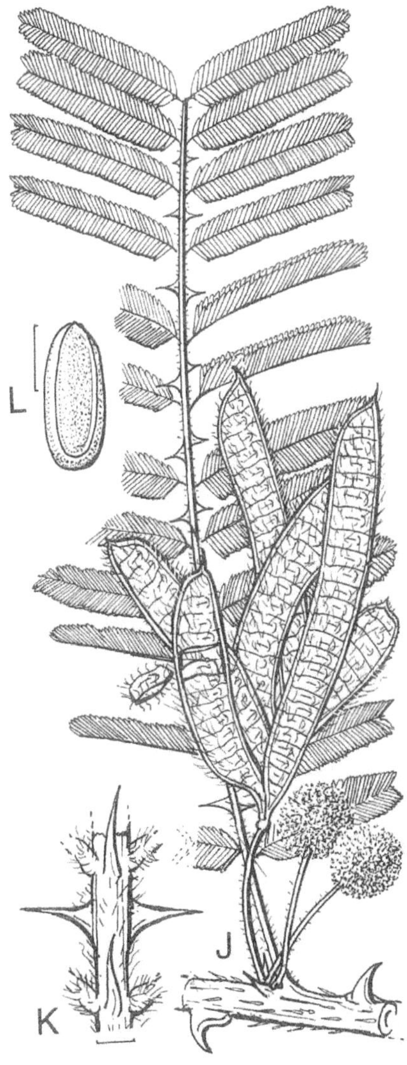

Figure 2: *Mimosa pigra*. J) Node with leaf, inflorescence and pod cluster. K) Section of leaf rachis with prickles. L) Seed. Drawn by Milton Andrews. Reproduced by permission from *Flora of Australia* vol. 12 (1998).

Figure 3: A tall and dense invasion of *Mimosa pigra* on a floodplain in the NT. Photo: CSIRO.

Finniss, Mary and East Alligator River systems (Fig. 5). In Qld an infestation was discovered in 2001 at Peter Faust Dam, near Proserpine, but it is under eradication (Vitelli *et al.* 2006). In 2009 a small infestation of approximately 1 ha was discovered at Ivanhoe Station, near Kununurra in WA and it is also the subject of an eradication plan.

Other than Australia, *M. pigra* has been introduced to many African and Asian countries and Papua New Guinea. In many of these countries it is considered a serious weed. In Africa, it has been reported from Angola, Benin, Botswana, Burkina Faso, Burundi, Cameroon, Central African Republic, Chad, Egypt, Ethiopia, Gambia, Ghana, Guinea, Guinea Bissau, Ivory Coast, Kenya, Liberia, Malawi, Mali, Mauritania, Mozambique, Namibia, Niger, Nigeria, Rwanda, Senegal, Sierra Leone,

Figure 4: A typical population of *Mimosa pigra* in Mexico. Note the low-growing plants and competing vegetation. Photo: CSIRO.

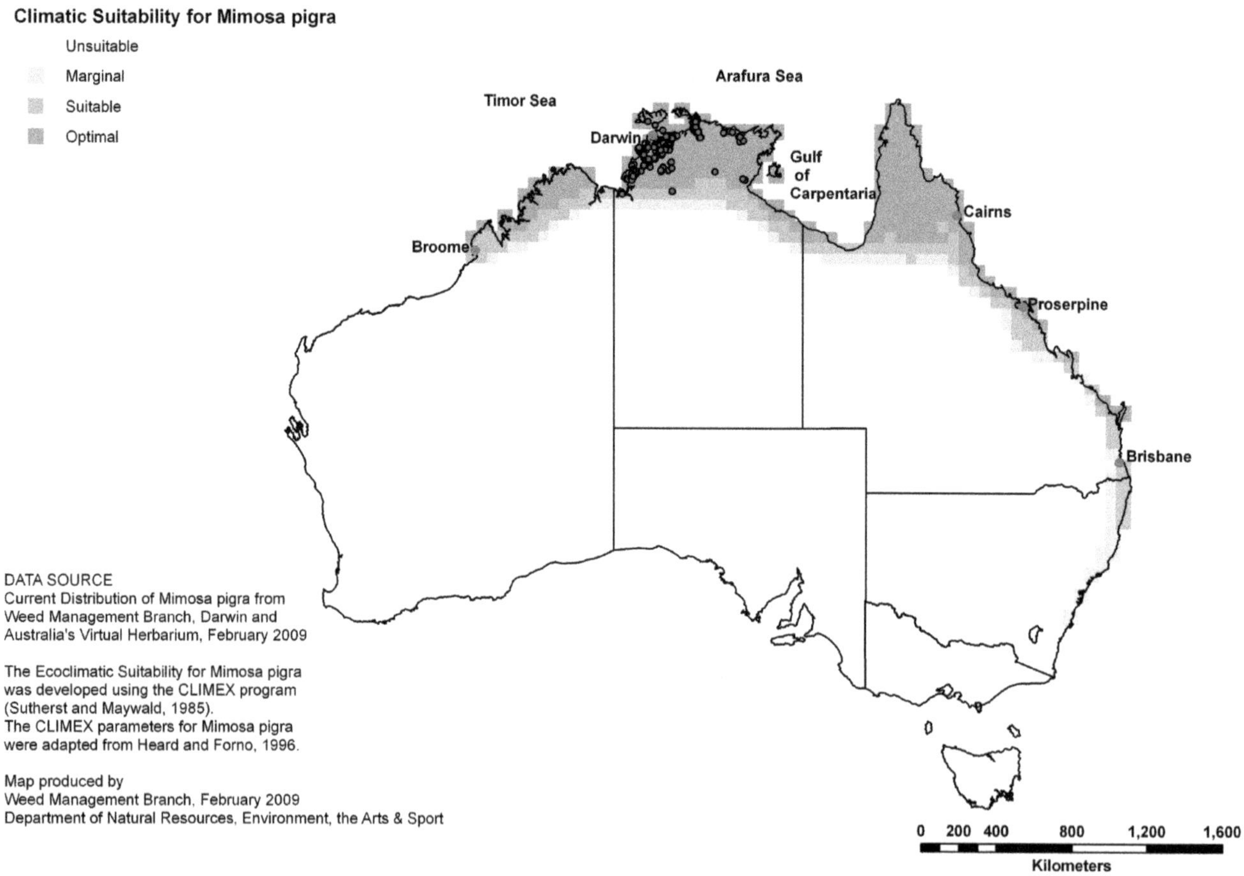

Figure 5: Distribution in 2009 (red dots) and potential distribution, based on climatic suitability, of *Mimosa pigra* in Australia. Image: Vinod Chejara, NT Government Weed Management Branch.

Somalia, South Africa, Sudan, Tanzania, Togo, Uganda, Zaire, Zambia and Zimbabwe. It is widespread in Asia, e.g. Thailand, Myanmar, Laos, Cambodia, Vietnam, Malaysia, Indonesia, India and Sri Lanka (Global Invasive Species Database 2010). It is also found in Papua New Guinea and in the Indian Ocean islands of Madagascar and Mauritius (Heard, unpub. data).

BIOLOGICAL CONTROL HISTORY

A biological control project in Australia against *M. pigra* commenced in 1979 with the first exploration for natural enemies taking place in Brazil in 1980. No previous biocontrol work had been done on this species by any other country. The project has involved the close collaboration of CSIRO and the NT Weeds Management Branch and has received long-term funding from the Australian Government's Department of Environment and ACIAR. The ACIAR financed two projects: the first project from 1984–91 involved Thailand only and resulted in the release of five agents in that country, and the second from 1995–97 involved Malaysia, Vietnam and Indonesia, and resulted in the release of one to four agents in each of those countries (Heard and Paynter 2009).

A total of 15 agents (13 insect species and two pathogenic fungi) have been released against *M. pigra* in Australia, of which 11 (10 insects and one fungus) are known to have established (Table 1). Two agents, *Carmenta mimosa* Eichlin & Passoa and *Neurostrota gunniella* (Busck), are currently inflicting severe damage, reducing both seed production and seed banks, and defoliating plants, which favours competing vegetation and leads to lower seedling survival and increasingly senescent stands. The impact of more recently released agents has not been measured, but appears to be substantial (Heard and Paynter 2009).

PLANT TAXONOMY

M. pigra is here referred to in the strict sense, which excludes *Mimosa asperata* L., which some authors have lumped with *M. pigra* as one species (Heard 2004). *Mimosa* is one of the largest genera of mimosoid legumes with approximately 500 species (Barneby 1991).

Table 1: Status of agents released for biological control of *Mimosa pigra* (in chronological order of release in Australia)

Agent	Plant part attacked	When released	Establishment and impact
Acanthoscelides puniceus	Mature hard seeds	1983	Established, 0–10% seed kill
Acanthoscelides quadridentatus	Mature hard seeds	1983	Initially established but did not persist
Chlamisus mimosae	Leaves, stems	1985	Established, but causes no significant damage
Neurostrota gunniella	Bores in pinnae and small stems	1989	Established, very abundant, substantial impact
Carmenta mimosa	Bores in large stems	1989	Established, very damaging
Coelocephalapion aculeatum	Flower buds	1992	Did not establish or did not persist
Coelocephalapion pigrae	Leaves and flower buds	1994	Established, destroys 10% of flowers
Phloeospora mimosae-pigrae	Leaves, stems and pods	1995	Did not establish or did not persist
Diabole cubensis	Leaves	1996	Establishment confirmed in 2011
Chalcodermus serripes	Mature green seeds	1996	Establishment confirmed in 2008
Sibinia fastigiata	Young green seeds	1997	No evidence of establishment
Malacorhinus irregularis	Roots and leaves	2000	Established at several sites, impact unknown
Macaria pallidata	Leaves	2002	Widespread establishment confirmed
Leuciris fimbriaria	Leaves	2004	Limited establishment confirmed
Nesaecrepida infuscata	Roots and leaves	2007	Early signs of establishment

See text for species summaries.

EXPLORATION

The native range of *M. pigra* extends from Mexico to northern Argentina and includes many Caribbean countries and all mainland countries except Chile (Barneby 1991). *M. pigra* is the most widely dispersed and most common of all shrubby *Mimosa* species in the region, being abundant on stream banks and seasonally flooded shores, along ditches and in seasonally wet savanna, and sometimes as a pest of irrigated land (Barneby 1991). The centre of origin of the genus is probably the moist region in central South America (Polhill and Raven 1981). The major centre of diversification is central Brazil, which today has very high diversity and endemism of *Mimosa* species (Simon and Proença 2000). Whether *M. pigra* migrated to North America across the land bridge or originated in North America and migrated south is unknown. It is of interest that its closest relative, *M. asperata*, occurs only in North America and therefore probably evolved there.

Surveys for natural enemies were conducted in seven countries of the native range from 1980 until 2009. By 1995 approximately 420 species of insects were found attacking *M. pigra* (Harley *et al.* 1995). Several more species were revealed in later surveys. Heard and Pettit (2005) showed that while large areas of the natural distribution were not surveyed, most natural enemies occur over the majority of the range of the host, including North and South America. Therefore, although many areas were not explored, the surveys were sufficiently thorough. Indeed, most species eventually released in Australia were discovered early in the exploration phase.

Many workers were involved in the exploratory surveys, all employed by CSIRO unless otherwise stated. Initially John Winder and Jair Gabriel da Silva conducted the searches in Brazil from 1980 until 1981. Mexican-based staff Ricardo Segura, John Gillett, Kim Pullen, Hugo Miranda, Jose Garcia and Moises Martinez then conducted searches over an extended period in Mexico from 1984 until 2009, and in Central America and northern South America. Australian-based staff, in particular Ken Harley, Wendy Forno, Richard Kassulke and Tim Heard, also participated in surveys. Harry Evans and Marion Seier (CABI, UK), Robert Barreto (Universidade de Viçosa, Brazil) and Gloria Carrión (Instituto de Ecologia, Mexico) performed surveys for fungal pathogens. Later, other field workers including Manuel Juarez, Paty Juarez, Carlos Pascacio and Quiyari Santiago contributed to open field tests, native range biology studies and collections of specific agents for transport to Australia.

CANDIDATES

Natural enemies (exotic range)

More than 100 native species have been recorded on *M. pigra* in Australia (Wilson *et al.* 1990). Some of these inflict conspicuous damage, although their population level impact is unknown. *Platyomopsis humeralis* White (Coleoptera: Cerambycidae) occasionally causes visible damage by girdling stems (Flanagan *et al.* 1990) and efforts were made to redistribute this insect. *Mictis profana* (F.) (Hemiptera: Coreidae) breeds on *M. pigra*, causing stem tip dieback and is considered to make a positive if small contribution to biological control (Flanagan 1994). Larvae of the moth *Maroga setiotricha* Meyrick (Lepidoptera: Xyloryctidae) feed in stems and cause a lot of damage per insect (Flanagan *et al.* 1990). The widespread tropical plant pathogenic fungus *Lasiodiplodia theobromae* (Pat.) Griffon & Maubl. induces dieback of *M. pigra* in the dry season, especially on plants stressed by drought and attacked by the introduced insect *N. gunniella* (Wilson and Pitkethley 1992).

Risbecoma pigrae Rasplus (Hymenoptera: Eurytomidae) is a common seed-feeding wasp on *M. pigra* in Africa, first described from the Ivory Coast of Africa (Rasplus 1988), and since collected from north-eastern Namibia, north-eastern South Africa and Uganda (S Neser, *pers. comm.* 2000). A total of 24 collections have been made over many years in Mexico and Central America of pods of both *M. pigra* and *M. asperata*. Insects were reared from pods and the Hymenoptera sent to O Neser (PPRI, South Africa) for identification. Although over 700 Hymenoptera have been recovered, none were *R. pigrae*, suggesting this species originated in Africa, not the Americas, and therefore from a host plant other than *M. pigra*. Nevertheless, as this type of seed-feeding chalcoid is generally host-specific it has potential for biological control purposes elsewhere.

Napompeth (1982) listed the insects associated with *M. pigra* in Thailand and Indonesia. Cam *et al.* (1997) listed the insects on *M. pigra* in Vietnam. None was considered sufficiently damaging or safe for release outside their native range.

Natural enemies (native range)

Harley *et al.* (1995) recorded approximately 420 species of insects that attack *M. pigra* and shortlisted 25 species regarded as having the greatest potential. This list evolved, as is typical for lists of potential biocontrol agents, as the project progressed; some species were tested and released, others were removed from or added to the list as more was learnt about them.

Agents assessed but not released

The following is a summary of work on agents assessed but not released (Table 2). These species were either imported into an Australian quarantine or studied in the native range.

***Cryptocephalus* sp. nr. *miserablis* Suffrian (Coleoptera: Chrysomelidae).** This species was readily cultured in quarantine. Adult and/or larval feeding was confined to 11 of the 75 species tested in choice trials. Six of these species supported larval development to adults, with subsequent oviposition of viable eggs on four of them. Performance (as inferred from longevity, developmental times, survival and fecundity) was compared across these six species in a series of experiments. *Cryptocephalus* sp. nr. *miserabilis* performed equally well, and sometimes better, on *Acacia irrorata* Sieber ex Spreng. when compared with *M. pigra*. Performance indicators were also relatively high on *Acacia spectabilis* A.Cunn. ex Benth. in most experiments. *Cryptocephalus* sp. was therefore insufficiently specific for release as a biocontrol agent in Australia. In any case, it is doubtful how effective an agent this species would be, as adults cause only minor damage to growing *M. pigra* plants, and larvae little or none as they preferentially feed on dead leaves.

***Lexiphanes* sp. nr. *guerini* (Perbosc) (Coleoptera: Chrysomelidae).** *Lexiphanes* sp. nr. *guerini* feed on foliage as larvae and feed and oviposit on foliage as adults. Host-range choice tests, using larvae on cut plants and adults on cut and whole plants, were conducted for ~56 species. Starvation trials using unfed adults were conducted on the seven species fed upon in the choice trials. *Lexiphanes* sp. nr. *guerini* has an exceptionally broad adult and larval host-range that spans unrelated plant families. Larvae were considerably more specific than adults (15 species as compared with 39 species fed on as adults).

***Diplacaspis* sp. nr. *prosternalis* (Schaeffer) (Coleoptera: Chrysomelidae).** This species was introduced into Australian quarantine but proved difficult to culture. Forno and Harley (1992) stated that it was tested and proved to be not specific.

***Syphrea* sp., *Genaphthona* sp. and *Paria* sp. (Coleoptera: Chrysomelidae).** These three leaf-beetles, along with *Nesaecrepida infuscata* (Schaeffer), are united by their life-history of larval development on roots, which made laboratory breeding and host-testing more difficult. To assist with prioritisation within this group, we designed a new method to determine the host-specificity of adult feeding (Heard and van Klinken 2004). Using this method, one species only, *N. infuscata* (= *Syphrea bibiana* (Bechyne)), was selected for further assessment of

Table 2: Natural enemies of *Mimosa pigra* assessed for biological control potential but not released

Agent (in order of families)	Plant part attacked	Status
Cryptocephalus nr *miserablis*	Leaves	Studied in Australia in 1981. Rejected – not specific
Lexiphanes nr *guerini*	Young leaves	Studied in Australia in 1985–86. Rejected – not specific
Diplacaspis nr *prosternalis*	Stem and leaves	Studied in Australia in 1985–86. Rejected – not specific
Syphrea sp. (*Syphrea flavipes* group nr. *S. cardiaca)*	Leaves and roots	Studied in Australia in 1995–97. Rejected – not specific
Genaphthona nr. *transversicollis*	Leaves and roots	Studied in Australia in 1995–97. Rejected – not specific
Paria sp.	Leaves and roots	Studied in Australia in 1995–97. Rejected – not specific
Coelocephalapion spretissimum	Flowers	Accidentally introduced in quarantine as contaminant with *C. aculeatum*
Chalcodermus persimilis	Mature green seed	Studied in Australia in 1994–95. Rejected – not specific
Sibinia seminicola	Young green seed	Studied in Mexico in 1993–94. Rejected, normal host is *Mimosa asperata* but see text
Sibinia ochreosa	Flowers	Studied in Mexico in 1997–99. Host testing 15% complete
Sibinia peruana	Flowers	Studied in Mexico in 1997–99. Host testing 15% complete
Temncerus debilis	Tips	Studied in Australia in 2003–09. Host testing completed but showed potential impact on *Neptunia* spp.
Marmara sp.	Tunnels under surface of stems	Studied in Mexico, c. 1995. Rejected – not sufficiently damaging
nr. *Aroga* sp. = *Aristotelia* cf *howardi* = *Gelechia benitella = new genus, new sp.*	Leaves and stems	Studied in 1984–85. Rejected – not specific
Aristotelia sp. nr. *dasypoda*	Leaves	Studied in Mexico and Australia in 1997–2001. Rejected- unable to determine specificity
Pococera gelidalis	Leaves	Studied in Mexico and Australia in 1997–2001. Rejected – not specific
Apotoforma rotundipennis	Leaves	Studied in Mexico and Australia in 1997–2001. Rejected – unable to determine specificity
Ithome sp.	Pods	Studied in Mexico in 1997–2000. Rejected – rare and not available when pods occur in Australia
Morpheis pyracmon	Stems	Attempted to collect in Brazil in 1995–96 but could not find
Spilococcus prosopidis	Young leaves	Attempted to collect in Brazil[?] in 1995–96 but could not find
Other pseudococcids	Young leaves	Studied in Mexico in 1995–96. Rejected – see text
Cicadellidae: several spp.	Leaves	Studied in Mexico in 1996–98. Rejected – not specific, difficult to rear, nymphs not found
Scamurius sp.	Leaf tips	Released 1988, did not establish. Normal host is *Mimosa diplotricha*
Risbecoma pigrae	Seeds	Attempted to collect in various countries in the American tropics but not found; see section 'Other promising agents'
Microstroma ruizii-belinii	Leaves	Studied in Mexico. Rejected – not damaging in native range
Mycosphaerella mimosicola	Leaves	Studied in Mexico. Rejected – only on *M. asperata*

See text for species summaries

the larval feeding host-range (see below). The remaining three species lack sufficiently narrow host-specificity of adult feeding for further consideration.

***Coelocephalapion spretissimum* (Sharp) (Coleoptera: Curculionidae).** This apionine from Mexico was accidentally introduced in quarantine as a contaminant in shipments of *Coelocephalapion aculeatum* (Fall). Although Harley *et al.* (1995) listed this species as having potential, it was considered at the time to be a lower priority than the other *Coelocephalapion* species, probably because it was less abundant, and was not further assessed.

***Chalcodermus persimilis* O'Brien (Coleoptera: Curculionidae).** This species was accidentally imported into an Australian quarantine from Mexico in a shipment believed to be *Chalcodermus serripes*. Differences in morphology and behaviour were observed, which led to its description as a new species (Heard *et al.* 1998). Oviposition preferences, adult feeding preferences and larval development of *C. persimilis* were determined for selected legume plant species. Adults fed and oviposited on several species of Mimosaceae. Larvae developed into normal adults on several of these species. The host-range of this species was therefore too broad for it to be considered as a biological control agent of *M. pigra* (Heard *et al.* 1998).

***Sibinia seminicola* Clark (Coleoptera: Curculionidae).** This species was considered of great potential by Harley *et al.* (1995). To determine field host preferences, we collected large numbers of mature pods of *M. pigra* and the closely related *M. asperata*, from many plants growing at several sites in Mexico in 1993, and held them for adult emergence. The results showed that the normal host for *S. seminicola* in Mexico is *M. asperata* (Heard and Segura 2004). The earlier belief that *S. seminicola* was a natural enemy of *M. pigra* may have originated from the previous lumping of *M. pigra* and *M. asperata* together under the one species *M. pigra* (Turner 1959). On the basis of this data, *S. seminicola* was dropped from the list, as it was unlikely that an insect with such a low preference for, or suitability to, *M. pigra* would be an effective agent. However, *S. seminicola* is often collected on *M. pigra* in Brazil where *M. asperata* does not occur (Heard and Segura 2004). It is therefore possible that biotypes of *S. seminicola* from Brazil may be adapted to *M. pigra* and worthy of further assessment.

***Sibinia ochreosa* Casey and *Sibinia peruana* Pierce (Coleoptera: Curculionidae).** Huge populations of these small flower-feeding weevils are present in the native range. Both species co-exist in the field, apparently occupying the same niche. Host-specificity tests on this insect in Mexico in 1998 and 1999 showed that only *M. pigra* and *M. asperata* were acceptable for oviposition. However, work on these species ceased for three reasons: the insects cannot be bred in the laboratory; considerable effort is required to sort adults into species before tests can be conducted; and the effectiveness of a flower-feeding insect on a plant that produces a super-abundance of flowers was doubtful (Heard and Segura 2004).

***Temnocerus debilis* (Sharp) (Coleoptera: Rhynchitidae).** *Temnocerus debilis* is a host-specific natural enemy of *M. pigra* from Central America. The adults feed and oviposit into the tips of the host plant, killing them. The larvae feed and develop in the dying and dead tips, potentially slowing or even arresting the growth of plants if high field densities are attained. It was expected to complement previously released agents and other control methods. Host-specificity studies were conducted in Australian quarantine and in the field in Mexico. A series of four laboratory tests and one field test confirmed that the host-range is limited to *M. pigra*, with possible occasional use of two Australian native species of *Neptunia*. It was predicted that the use of and impact on these *Neptunia* species would be minor as laboratory tests showed that they were inferior hosts and much less preferred. Furthermore, in the native range, Neotropical species of *Neptunia* are not used by this insect. In light of these results, the level of risk associated with releasing *T. debilis* was deemed acceptable and it was believed that it would be a potentially effective biological control agent. Approval to release this agent (Heard *et al.* 2009) was refused, however, due to potential non-target impacts on native species of *Neptunia*.

***Marmara* sp. (Lepidoptera: Gracillariidae).** This moth, known from Mexico, Brazil and Venezuela was considered a high priority by Harley *et al.* (1995) as the larvae feed extensively under the epidermis of the stem. In subsequent preliminary studies, the direct damage of *Marmara* sp. was reassessed and considered to be insignificant, and the species dropped from the priority list.

nr *Aroga* sp. = *Aristotelia* cf *howardi* = *Gelechia benitella* = new genus, new sp. (Lepidoptera: Gelechiidae). Given the proliferation of names, there has clearly been taxonomic uncertainty surrounding this species. Specimens collected in Mexico were sent to an Australian quarantine in 1984–85. Larvae live in a tunnel bored into the stem from which they emerge to feed on leaves, tied together as a temporary shelter. A rearing technique was developed for this leaf-tying species and host-specificity tests were conducted against 49 legume species and 31 species from 17 other families using both cut tips and whole plants. Feeding occurred on most plant species and development to adult occurred on three legume species, including two species of the important genus *Acacia* (G Maynard, *pers. comm.* 1995). On this basis, this candidate agent was not pursued further.

***Aristotelia* sp. nr *dasypoda* (Lepidoptera: Gelechiidae), *Pococera gelidalis* (Walker) (Lepidoptera: Pyralidae), *Apotoforma rotundipennis* (Walsingham) (Lepidoptera: Tortricidae).** These three species of moths feed on leaves and tie leaves together to form a protective shelter, causing heavy damage in the native range. All species were imported into Australian quarantine to evaluate their host-range. Parts of their natural host selection behaviour were not expressed in cages, resulting in indiscriminate oviposition on many plant species. As oviposition preferences could not be tested in these circumstances, the innate developmental host-range of larvae was tested. Although *M. pigra* was the superior host for all three species, development to adult occurred on several other plant species. Open field tests in semi-natural conditions were then conducted in Mexico. Four individuals of 26 test plant species of Australian and Mexican origin were grown in a field plot and successive cohorts of laboratory-reared adults of the three species introduced. All resulting leaf ties were bagged to capture emerging adults. The numbers of adults reared on the 48 *M. pigra* plants was disappointingly low, being 11, 17 and 103 for *A. rotundipennis*, *Aristotelia* sp. and *P. gelidalis*, respectively. The original numbers of adults introduced to the plot were 437, 295 and 150, respectively. No adults of *Aristotelia* sp. were reared from the test plants, but two adults of *A. rotundipennis* were reared from the closely related *M. asperata*. However, not much value was placed on these results as the poor return of reared adults to released adults on *M. pigra* suggested that the oviposition pressure on plant species was too low. One useful result was that adults of *P. gelidalis* were reared from one test plant species, *Desmanthus virgatus* (L.) Willd., confirming that this insect is not sufficiently specific to release in Australia. However, the other two species remain as possible, but difficult to test, candidates (Heard *et al.* 2004).

***Ithome* sp. (Lepidoptera: Cosmopterigidae).** This small moth was considered a potential agent by Harley *et al.* (1995). Larvae primarily feed on the pods of *M. pigra*, but can also eat leaves, flowers and stems. Little is known about its biology. Larvae tunnel in pods, eating the seeds and pod wall, often tying two pods together with silk. The taxonomy is not clear: more than one species may occur and other genera such as Obithome may be present in collections. Specimens of *Ithome* sp. were common in collections, and 34 individuals from Veracruz were reared in 1994 from field-collected pods of both *M. asperata* and *M. pigra*. This species is currently not highly rated as a prospective agent because it is difficult to rear, as adults are very small, delicate and difficult to handle, and it is not very damaging in the field. Also, it is not available in the field in the native range when pods are present in Australia (March–May). More work is needed, e.g. to find the preferred feeding site of larvae. It is possibly not compatible with other seed-feeding insects because *Ithome* damage the pod wall, seeds or perhaps the other species of larvae directly.

***Morpheis pyracmon* (Cramer) (Lepidoptera: Cossidae).** This moth was considered a potential agent by Harley *et al.* (1995). However, no specimens are present in any later collections. Specific attempts to find this insect in Brazil in 1995 and 1996 failed to locate any specimens.

***Spilococcus prosopidis* (Cockerell) (or nr) (Hemiptera: Pseudococcidae).** This mealybug was considered a potential agent by Harley *et al.* (1995) who recorded it from Tabasco state, Mexico. According to the literature, *S. prosopidis* is known from the USA on *Prosopis velutina* Wooton and from *Poradendron* sp. in Mexico (Ben Dov 1994). Attempts to find this species in Tabasco in 1995 located several other species of pseudococcids (identified by G Watson, Natural History Museum, UK). The species not recorded in Harley *et al.* (1995) are *Dysmicoccus grassii* (Leonardi), *Nipaecoccus neogaeus* Williams & Granara de Willink, *Paracoccus lycopersici* Ezzat & McConnell and *Pseudococcus cryptus* / calceolariae / comstocki complex. All members of this group have wide host-ranges and were dropped from further consideration.

Cicadellidae (Hemiptera: Cicadellidae). Harley *et al.* (1995) listed about 20 species of leaf-hoppers collected from *M. pigra* which were known to have a wide host-range and so were rated as having no potential, or whose host-range was unknown and thus rated as having unknown potential. New identifications by Paul Freytag in 1997 of species from Mexico and Venezuela showed that several species of leaf-hoppers occur on *M. pigra*, including *Kunzella pseudomarginella* (Cardwell), *Zygina* sp. ceonothana group, *Kunzeana flavella* Ruppel & DeLong, *Kunzeana vomerella* Ruppel & DeLong, and *Kunzeana* spp. The most abundant and widespread species is *K. pseudomarginella*, occurring in Mexico and Venezuela. Species from this family are abundant: approximately 30 adults per hour per person can be collected with a sweep net at most sites in Mexico. The feeding damage consists of white, star-shaped patches, 0.5 mm in diameter, radiating from a central insertion point. Several problems have limited research on this taxon. First, only adults were found on *M. pigra* in the field, never nymphs or eggs, thus *M. pigra* may not be a breeding host but be used only for adult feeding. Second, it is difficult to transport adults alive from the field to the laboratory. Third, preliminary tests showed that adult feeding was not host-specific. Finally, their taxonomy was confused. Further work was therefore discontinued and preliminary findings are recorded in Heard and Segura (2004).

***Microstroma ruizii-belinii* Evans & Carrión (Fungus: Basidiomycotina).** This fungus was observed on leaf pinnae of *M. pigra* on the Pacific coast of Mexico, but not on the Caribbean coast (Evans *et al.* 1995). It was not rated as having promise as a classical biocontrol agent.

***Mycosphaerella mimosicola* Henn. (Fungus: Ascomycotina).** This species was reported from *M. pigra* (Evan *et al.* 1995), but apparently from when *M. pigra* and *M. asperata* were considered a single species. Examination of field and herbarium material led to the conclusion that this species has a disparate range: in North America on *M. asperata* and in Amazonia on possibly *Mimosa orinocöensis* Barneby (Evan *et al.* 1995).

Agents released in Australia

The following account of species released in Australia is arranged in approximate order of year of release. Photographs of many of these species are shown in Heard and Segura (2004).

***Acanthoscelides puniceus* Johnson and *Acanthoscelides quadridentatus* (Schaeffer) (Coleoptera: Bruchidae).** Both *A. puniceus* and *A. quadridentatus* were imported into Australian quarantine from Mexico, although the latter insect is widely distributed throughout the American tropics (Heard and Pettit 2005). Seed-feeders were chosen to target the dispersal of seed (Kassulke *et al.* 1990) associated with rapid expansion of infestations (Lonsdale 1993). In oviposition tests, conducted on a mixed colony later shown to be only 10% *A. puniceus*, eggs were laid on 16 species but the resulting larvae developed only on *M. pigra* (Kassulke *et al.* 1990). Both *Acanthoscelides* spp. were released in Australia in 1983 using rearing and release methods described by Wilson *et al.* (1992).

Both species initially established in Australia although *A. puniceus* dominated from an early stage (Wilson and Flanagan 1991) and only *A. puniceus* has been recovered in recent years (Paynter 2004a). Ostermeyer and Grace (2007) showed that overall less than 2% of seeds were attacked and that the abundance of *A. puniceus* declined between 1997 and 2004, both in number of sites where found and the abundance at those sites. Their surveys, however, were done at times that were not ideal for recovering this species. Paynter (2005) found rates of attack up to 10%. While very high levels of seed predation are often assumed to be necessary for a seed-feeder to control established infestations, if establishment of seedlings at the edge is seed limited then even the relatively low proportion of seeds consumed by *A. puniceus* may have an impact on the rate of invasion of *M. pigra* (Paynter 2006). *Acanthoscelides* spp. may prove more effective in regions where seed production is less strongly seasonal (Paynter 2005). Both species were introduced in at least four south-east Asian countries and are now widely distributed in that region (Heard and Paynter 2009).

***Chlamisus mimosae* Karren (Coleoptera: Chrysomelidae).** This species was described from insects collected in Brazil in 1981 (Karren 1989). Both larvae and adults feed on the epidermis of leaves and stems. Host-specificity was tested using choice and no-choice tests of adult and first instar larvae feeding. Only minor feeding damage by adults was detected and no larval feeding or development beyond first instar occurred on any of the non-target species tested (Kassulke and Harley 1987).

The first release in Australia was made in 1985. This species established in the Finniss River catchment of the NT, where it has since occasionally reached high numbers and caused moderate damage. Despite considerable efforts to redistribute it, *C. mimosae* has not persisted in other river systems (Ostermeyer and Grace 2007). Even in the Finniss River system, *C. mimosae* numbers were low throughout the survey period 1997–2004. Ostermeyer and Grace (2007) suggested that it is unlikely to have a significant effect on *M. pigra* at such densities. Its long life-cycle (three to four months) makes it susceptible to predation, for example by green tree ants (Heard, unpub. data).

***Carmenta mimosa* Eichlin & Passoa (Lepidoptera: Sesiidae).** *C. mimosa* is known from southern Mexico, Honduras and Nicaragua (Eichlin and Passoa 1983). Eggs are laid on the plant and larvae enter the stem at the node and bore within the stem. Large larvae sometimes girdle the stem and kill it (Fig. 6). Pupation occurs within the stem. Adults live for six to 12 days and females lay an average of 295 eggs (Eichlin and Passoa 1983). Host testing was confined to no-choice tests of larval development using transferred eggs. Larvae developed to adults on *M. pigra* only.

C. mimosa was first released in Australia in 1989 (Forno *et al.* 1991). The natural dispersal of *C. mimosa* is not rapid, presumably due to the long life-cycle (mean: 98 days) and low rates of dispersal by adult moths. Two populations in the NT increased by five and 10 times in density and three and eight times in distribution in two years, indicating a rate of spread of about 2 km per year (Ostermeyer 2000). Redistribution efforts were therefore continued and by 2004 *C. mimosa* was present on all river catchments with major *M. pigra* infestations and abundance at all sites was still increasing (Ostermeyer and

Figure 6: Plants of *Mimosa pigra* on a floodplain in the NT heavily damaged by larval feeding of *Carmenta mimosa*. Photo: CSIRO.

Grace 2007). Wilson *et al.* (1992) gave rearing and release methods for this agent. Ostermeyer *et al.* (2004) described a project for involving community groups and school children in the rearing and release of *C. mimosa* and *Macaria pallidata* (Warren).

C. mimosa was predicted to be a useful agent (Forno *et al.* 1991). In Australia, this prediction has proved to be correct: Paynter (2005) showed that in three years, four out of eight *M. pigra* stands where *C. mimosa* was absent expanded and none contracted. In contrast, none of nine stands where *C. mimosa* was present expanded and three contracted. Analysis of the age structure of *M. pigra* stands indicated that contracting stands were typically devoid of seedlings. *M. pigra* seed rain was negatively correlated with *C. mimosa* damage and declined by more than 90% at the highest *C. mimosa* densities. Seed banks also declined with *C. mimosa* damage (Paynter 2005). Furthermore, percentage cover of competing vegetation, mainly grasses and sedges, was significantly higher under stands defoliated by *C. mimosa*, probably because the dead branches allowed light penetration to areas that would otherwise be heavily shaded (Fig. 6). Competing vegetation inhibited *M. pigra* seedling establishment (Lonsdale and Farrell 1998) and apparently increased the susceptibility of *M. pigra* to fire, by increasing fuel loads beneath stands (Paynter 2005). Although Paynter (2005) subsequently predicted that *C. mimosa* should cause widespread reductions in *M. pigra* populations, modelling of the impact of biological control on *M. pigra* indicated that it might take decades to clear the huge existing patches of the weed (Paynter and Flanagan 2002; Buckley *et al.* 2004).

***Neurostrota gunniella* (Busck) (Lepidoptera, Gracillariidae).** *N. gunniella* is widespread from southern Texas to Costa Rica and Cuba where adults are frequently bred from *M. pigra* and *M. asperata* and also from *Neptunia plena* (L.) Benth., another legume native to the Neotropics (Davis *et al.* 1991). Eggs are laid singly on leaves, the first and second instar larvae mine the pinnules, third instar larvae enter the primary rachis and tunnel towards the stem where the remaining instars complete their development. The eighth (final) instar exits the stem, spins a cocoon between pinnules and then pupates. The total development time (egg–adult) is ~30 days (at ~24°C). Adult females lay a mean of 86 eggs (Davis *et al.* 1991). Wilson *et al.* (1992) have described the rearing and release methods for *N. gunniella*.

No-choice host-specificity tests in the laboratory examined the oviposition preferences of adults and the ability of larvae to develop on test species. Adults accepted two species of *Mimosa* (both weeds) and the four native Australian species of *Neptunia* for oviposition and the larvae completed development on all these species. The duration of larval development did not differ greatly among plant species but larval mortality on *Neptunia* spp. was higher than on *M. pigra*. Damage to *M. pigra* plants was much greater than to *Neptunia* species (Davis *et al.* 1991). On the basis of these data and predictions, a decision was made to release *N. gunniella* in Australia. However, Thailand did not approve release of this insect as laboratory trials showed that *N. gunniella* could reproduce on the introduced aquatic vegetable *Neptunia oleracea* Lour. (Forno *et al.* 2000). It was predicted that, in Australia, *Neptunia* spp. would probably not support high populations and that non-target attack to most *Neptunia* spp. was unlikely because *Neptunia major* (Benth.) Windler is the only species occurring in the same regions or habitats as *M. pigra*. However, this risk may have been understated because adult *N. gunniella* disperse rapidly (Wilson and Flanagan 1990) and *M. pigra* has continued to spread to new areas. Nevertheless, Taylor *et al.* (2007) found that *N. major* suffered only about 10% of the level of attack experienced by *M. pigra* in the field, and only plants growing in close proximity to *M. pigra* were affected. At the highest level of attack recorded on *N. major* in the field, tip death increased by 54% and plant height decreased by 6%. The effect of *N. gunniella* on the reproductive output of *N. major* could not be clearly resolved, but is likely to be small. Therefore, Taylor *et al.* (2007) concluded that the impact of *N. gunniella* on *N. major* is minor, confirming the predictions of the pre-release studies.

In contrast, *N. gunniella* damage significantly reduced both growth (14% reduction in radial growth in one season) and seed output (such that seed output was 60% lower than normal at the highest densities of the insect) of *M. pigra* plants (Lonsdale and Farrell 1998), although these authors considered the damage insufficient to control the weed. By contrast, Paynter (2005) found that *N. gunniella* abundance was not correlated with seed set. Paynter (2005) confined sampling to within dense *M. pigra* thickets and found that *N. gunniella* attack was greater on plants on the edge of stands compared to within stands (Smith and Wilson 1995). Paynter (2006) hypothesised that a plant's capacity for compensatory growth (i.e. a density-dependent increase in growth and flowering of stems that escape attack) should be greater within stands, where *N. gunniella* attack is relatively low, compared with those at stand edges, where every stem may be heavily infested. Therefore, these contrasting results may indicate that spatial variation in *N. gunniella* abundance results in significant impacts (a greater than 50% reduction in seed production) on plants growing at the edges of *M. pigra* stands but not within dense stands.

***Coelocephalapion aculeatum* (Fall) and *Coelocephalapion pigrae* Kissinger (Coleoptera: Curculionidae).** *C. aculeatum* was sourced from Mexico (it also occurs in Texas (Kissinger 1992)) and first released in Australia in 1992. *C. pigrae*, native to Venezuela and Brazil, was discovered after the release of *C. aculeatum* and was first released in 1994 in Australia (Heard and Forno 1996). Adults of both species feed on and oviposit into unopened flower buds, although *C. pigrae* also feeds on *M. pigra* leaves (Fig. 7). Larvae of both species complete their development on the flowers. Host-specificity studies showed that both *C. aculeatum* and *C. pigrae* larvae could complete their development only on *M. pigra*, except for low levels of development on *Neptunia dimorphantha* Domin. (Forno *et al.* 1994; Heard and Forno 1996).

Although initially reported to have established in Australia, *C. aculeatum* has not been found in surveys for several years (Paynter 2004a). *C. aculeatum* was released in Thailand in 1991, where it also failed to establish (Suasa-ard *et al.* 2004). In contrast, *C. pigrae* is now widespread in Australia and is still spreading to isolated infestations (Ostermeyer and Grace 2007).

Although *C. pigrae* prefers perfect flowers (i.e. containing female and male parts) over the male flowers, potentially allowing a greater impact on seed set (Heard, unpub. data), Paynter (2005, 2006) found no correlation between *C. pigrae* abundance and seed production in Australia. At one site in Australia, Paynter (2006) found *C. pigrae* abundance lagged behind sharp seasonal peaks

Figure 7: Damage to leaves of *Mimosa pigra* in Australia caused by adult feeding by *Coelocephalapion pigrae*. Photo: CSIRO.

in flower production and, while 55% of inflorescences were infested with *C. pigrae* and an average of 2.7 beetles were reared per inflorescence, this level of infestation was estimated to destroy only 11% of flowers.

The contrasting fortunes of *C. aculeatum* and *C. pigrae* in Australia are possibly related to their ability to survive the dry season. Flowering almost ceases during the dry season in northern Australia (Lonsdale 1988). *C. aculeatum* is an obligate flower-feeder, whereas the adult *C. pigrae* also feed on *M. pigra* leaves, allowing them to survive in the absence of flowers. Both *Coelocephalapion* species may prove more effective in regions where flower production occurs throughout the year.

***Phloeospora mimosae-pigrae* H.C. Evans & G. Carrión (Fungus: Ascomycotina).** The wet-season coelomycete pathogen *P. mimosae-pigrae* was introduced into Australia in 1995. The anamorph (asexual form) of *P. mimosae-pigrae* was successfully mass-cultured in liquid medium and a standardised culturing and application protocol was developed and tested for large-scale field applications during the wet seasons from 1996 to 1998. The fungal pathogen established temporarily in the field and caused a considerable reduction in growth of *M. pigra* seedlings. The pathogen was never observed to survive beyond the wet season of its release. Mature plants were less affected by *P. mimosae-pigrae* due to premature leaf-drop of diseased plant tissue. In addition, the fungal pathogen failed to develop its sexual form (teleomorph) in Australia, which is assumed to be vital for its survival (Hennecke 2004).

***Diabole cubensis* (Arthur & J.R. Johnston) (Fungus: Basidiomycotina).** The dry-season rust fungus *D. cubensis* (Fig. 8) was introduced into Australia in 1996. Pre-release evaluation and observation in Mexico, Central America

Figure 8: *Diabole cubensis* pustules on an *M. pigra* leaf. Photo: CSIRO.

Figure 9: *Chalcodermus serripes* attacking seeds of *Mimosa pigra*. Note the oviposition scars on most seeds. Photo: CSIRO.

and South America, the centre of origin of *D. cubensis*, revealed a range of characteristics that suggested the rust had great potential to exert some control over *M. pigra* in Australia (Evans *et al.* 1995). This rust was cultured and field inoculations were carried out during the northern Australian dry season when ambient conditions were suitable for infection. Post-release assessments showed that *M. pigra* developed symptoms of rust infection in the field but that the rust apparently failed to persist or spread at release sites and therefore did not seem to establish in Australia (Hennecke 2006). However, in 2011, this pathogen was detected for the first time in Australia, on indigenous land on the Finniss River, NT (Anon 2011).

Chalcodermus serripes **Fahraeus (Coleoptera: Curculionidae).** *C. serripes* is a common seed predator through the entire native range of *M. pigra*. Adults fed on growing tips, flowers and pods of the host plant and lay eggs on seeds that are still green though close to full size (Fig. 9). Females avoided laying more than one or at most two eggs per seed even in the absence of other oviposition sites. Larvae developed inside the pod, each on a single seed. One larva developed per seed and took approximately 14 days to completely consume the seed. They then changed to prepupae which, in the field, stayed in their pod segments until the following wet season when they left the segment and pupated in the soil before emerging as adults. In the laboratory, few of these prepupae completed development to adults despite being provided with a range of environmental conditions to break diapause. Host-specificity testing focused on the oviposition and feeding preferences of adults, which proved to be entirely specific (Heard *et al.* 1999).

Mass rearing this insect for release was difficult. Initially, permission was obtained to release field-collected adults following an intensive regimen process including holding the insects for two weeks, inspection for correct taxonomic identity and absence of parasites, washing in sodium hypochlorite solutions to eliminate adhering micro-organisms, and internal inspection of a sample (5%) of beetles for pathogens. Eventually a method for mass rearing was developed and approximately 9300 field-collected or laboratory-reared adults were released between 1996 and 2000 (N Graham, *pers. comm.* 2002). Evidence was lacking that this species had persisted (Paynter 2004a; Ostermeyer and Grace 2007) until 2008 when 11 adults were found in a beat sample of 40 tips at a site on the Finniss River (N Burrows and B Lukitsch, *pers. comm.* 2008). Large populations of this species, causing heavy damage to flowers and seeds, have recently been observed in several river systems (L. Elliot, *pers. comm.* 2010). This species is clearly slowly increasing in population and hopes are high that it will become a useful agent.

Sibinia fastigiata **Clark (Coleoptera: Curculionidae).** *S. fastigiata* is a common seed predator in Mexico, Central and South America (Clark 1984). Adults oviposit on small green seeds. Larvae develop inside the pod, each on a single seed. The first two larval instars feed on the perimeter of the seed, allowing it to continue growing. Host-specificity testing focused on the oviposition preferences of adults, which proved to be highly specific (Heard *et al.* 1997). Adults feed non-destructively on pollen and possibly on nectar from open flowers.

Australian Quarantine and Inspection Service (AQIS) normally requires that imported insects be reared through one generation in quarantine before field release. Because *S. fastigiata* could not be reared in quarantine (as seeds are not produced in the low light intensities), an exception was made by AQIS to release field-collected adults following an intensive regimen similar to that practised with *C. serripes*. Approximately 25 importations from both Mexico and Brazil (1997–2002) resulted in the release of ~2450 field-collected adults. However, there is no evidence that *S. fastigiata* has established in Australia (Paynter 2004a; Ostermeyer and Grace 2007).

***Malacorhinus irregularis* Jacoby (Coleoptera: Chrysomelidae).** *M. irregularis* is known only from Mexico. The adults feed on the leaves (Fig. 10) and the larvae develop in the soil, feeding on small seedlings, imbibed seeds, roots and root nodules. Larval feeding can kill seedlings (McIntyre *et al.* 2007). Larvae prefer nitrogen-fixing root nodules to other plant parts and *M. pigra* relies on these nodules for nitrogen supply in low-nutrient conditions, suggesting that below-ground herbivory should affect *M. pigra* vigour (McIntyre *et al.* 2007). The larvae complete development from egg to adult in about 36 days and adults can live for up to six months (Heard *et al.* 2005). Host-specificity tests were conducted to determine the suitability of seedlings and leaves for larval development, and suitability of leaves for adult feeding. No larval survival occurred on any plant species other than *M. pigra*. The extent of adult feeding on the test plants was negligible, being less than 1% of that on *M. pigra* (Heard *et al.* 2005).

This insect is not common in Mexico, possibly due to the low abundance of *M. pigra* seedlings. It was predicted that, due to the availability of large numbers of seedlings, it could become common in Australia and damage seedling and defoliate mature plants, thereby complementing the impacts of *C. mimosa* and *N. gunniella*. The first release occurred in 2000. Establishment was confirmed in the Adelaide River system (Heard *et al.* 2005), from where adults were collected and redistributed until 2005, with the release of a total of 35 500 beetles (Ostermeyer and Grace 2007). Light traps have been used to confirm establishment at other sites in other river systems including one site where considerable damage was visible (Routley and Wirf 2006). Abundance of this species has so far been variable in space and time.

Figure 10: Plants of *Mimosa pigra* on a floodplain in the NT heavily damaged by adult feeding of *Malacorhinus irregularis*. Photo: Bron Routley, NT Government Weed Management Branch.

***Macaria pallidata* (Warren) (Lepidoptera: Geometridae).** *M. pallidata* is a common and damaging species that is widespread across tropical America. Material for testing and eventually release into Australia was obtained from Mexico. Adults are short-lived, nocturnal and feed on nectar. Females deposit eggs directly onto leaves and stems. Larvae feed at night, initially on the top surface of the leaf and subsequently on young and mature foliage, completely stripping plants when larval densities are high. During the day, larvae rest on the tips of leaves in positions that imitate stems. They move very little from their original position until forced to look for more food after almost completely consuming pinnules. There are five larval instars. Pupation is in the soil or among damaged plant tissue. Time from egg to adult under laboratory conditions at 25–27°C is 25 days and there are several generations per year (Heard *et al.* 2010).

The host-specificity of *M. pallidata* was tested using laboratory larval development tests on 70 test plant species. Development to adult occurred on six species other than *M. pigra*. However, survival rates were so low that these plants were considered unable to sustain a population of this insect species (Heard *et al.* 2010).

Released in 2002, *M. pallidata* was found to have spread widely by 2004. It causes conspicuous larval feeding damage, and adults can be highly visible in the undergrowth around *M. pigra* populations (Routley and Wirf 2006). However, populations experience severe fluctuations and densities were low for most of the year at all sites surveyed (Heard *et al.* 2010). Wirf (2006) demonstrated significant reductions in the growth of potted plants artificially infested with *M. pallidata* larvae, but field experiments have not yet quantified the impact of the agent under natural conditions. When populations are high, damage appears to be significant in the field, contrary to predictions made before its release: Harley *et al.* (1995) listed *M. pallidata* as having no potential 'because ectophagous larvae of Lepidoptera are often subjected to high levels of parasitism and predation'. Grace (2005) showed that *M. pallidata* larvae are indeed subject to moderate levels of predation, and larval survival increased when ants were excluded. Larvae drop on a silken thread when disturbed (Heard *et al.* 2010), perhaps providing a useful protection from predators. It will be interesting to see whether this species attracts more predators and parasitoids with time and, if so, whether they reduce its impact.

***Leuciris fimbriaria* (Stoll) (Lepidoptera: Geometridae).** *L. fimbriaria* is an abundant and damaging defoliator of *M. pigra* in Central and South America. Larvae feed on leaves of all ages. Adults are non-feeding, short-lived moths. Generation times are short and fecundity is high, allowing rapid population increase. Culturing is easy, allowing large releases to be made with minimal resources (Heard *et al.* 2010).

The host-specificity of this species was tested on colonies introduced from Mexico, using laboratory larval development tests on 69 plant species. Development to adult did not occur on any species other than *M. pigra* and *M. asperata*. Releases were made from 2004 and field recoveries indicate that it is widespread but not abundant (Heard *et al.* 2010).

***Nesaecrepida infuscata* (Schaeffer) (Coleoptera: Chrysomelidae).** *N. infuscata* is among the most common insects on *M. pigra* in Mexico, Central America and Venezuela. It was known as *Syphrea bibiana* Bechyne in earlier references, until *N. infuscata* was shown to be a senior synonym (Furth 2006). The larvae feed on roots while the adults feed on leaves. Development time from egg to adult is 29 days, with adults living a mean of 86 days and laying 4.4 eggs per day. As both roots and leaves of *M. pigra* are relatively undamaged in Australia, this species has potential to limit the growth and seed production through both leaf-feeding and root-feeding damage. Furthermore, direct feeding on seedlings may increase their mortality. In host-specificity tests, larvae of *N. infuscata* did not develop on any plant species other than *M. pigra* and adult feeding on other plant species was minimal (Heard *et al.* 2006). Permission for its release in Australia was gained in 2006. Colonies were sent to the NT for mass rearing and release in 2007. The first signs of establishment were confirmed in 2010 when good numbers of adults were collected at a site on the Adelaide River more than 20 weeks after they were released there, suggesting that they are breeding in the field (S Casanova, *pers. comm.* 2010).

***Scamurius* sp. (Hemiptera: Coreidae).** Qld Government workers introduced *Scamurius* sp. into Australia from Brazil for biological control of *Mimosa diplotricha* Sauvalle (= *Mimosa invisa*) (see separate chapter, this volume). Host-specificity tests in the laboratory showed that *M. pigra* is also acceptable for adult feeding and suitable for larval development, although native range observations indicated that it would not transfer to *M. pigra* (Wild 1986). In 1988, as an example of the neoclassical approach to biological control (Hokkanen and Pimentel 1984), NT Government workers released this tip-feeding bug in *M. pigra* infestations in the hope that it may prove to be an effective agent. No evidence of establishment was ever found (C Wilson, *pers. comm.* 2002).

DISCUSSION

Assessment of level of control

Although it has proved difficult to separate the impacts of individual agents, it is clear that by 2003 both *C. mimosa* and *N. gunniella* were having a major impact on *M. pigra*, resulting in reduced seed rain, reduced seedling regeneration and lower seed banks (Table 3). Subsequent surveys, conducted after the release and establishment of *M. pallidata*, have indicated that *M. pigra* seed banks have continued to decline (Routley and Wirf 2006) and are now approximately 10% of the pre-biological control levels recorded by Lonsdale *et al.* (1988). However, they have to fall even further to reach the levels measured in the native range (Lonsdale and Segura 1987).

Benefit:cost analysis

Heard and Pettit (2005) estimated the biological control and integrated management components of the *M. pigra*

Table 3: Mean soil seed banks (seeds m^{-2}) of *Mimosa pigra* under dense canopy of the plant in Australia (introduced range) and Mexico (native range)

Year sampled	Mean soil seed bank	Site information	Reference
1986	8500–12 000	Australia, before biological control	Lonsdale *et al.* (1988)
2002	3710 (SE=755)	Australia, after biological control	Paynter (2005)
2005	991	Australia, after biological control	Routley and Wirf (2006)
1985	117	Mexico, natural conditions	Lonsdale and Segura (1987)

project cost $A27 million over 24 years, with the 14 species released by 2004 costing an average of about $A500 000 per agent released (including exploration, testing, rearing and release but excluding evaluation). These figures underline that biological control work is expensive at the beginning. Indeed, while Page and Lacey (2006) estimated the benefits of *M. pigra* biological control to be an annual saving of $A1.5 million (from decreased costs of chemical and mechanical control), the resulting benefit:cost ratio was just 0.8:1. This is low compared to outstanding examples such as 312:1 for prickly pear and 109:1 for rubbervine. However, Page and Lacey (2006) considered their *M. pigra* figure to be an underestimate as productivity gains and environmental, recreational, tourism and cultural benefits were not considered. Furthermore, benefits should continue to increase, as the total impact of recently established agents improves the overall benefits from biological control.

Integrated control

Miller *et al.* (1992) suggested that an integrated approach combining biological control with other methods should provide the most effective management strategy for *M. pigra* in Australia. To investigate whether biological control could be integrated with other management options, a large-scale experiment was performed to measure the impact of single and repeated applications of herbicide and crushing by bulldozer, either alone or in combination, on both *M. pigra* and the five biological control agents that were abundant at the site (Paynter and Flanagan 2004). Used individually, herbicide, bulldozing and fire were not effective, but several combinations of these techniques cleared *M. pigra* thickets and promoted establishment of competing vegetation that inhibited *M. pigra* regeneration from the seed bank. Some treatment combinations found to be effective were predicted by Buckley *et al.* (2004) to succeed only in combination with biological control. *N. gunniella* increased dramatically in response to all treatments due to the increased number of stand edges as a result of cutting tracks into the initial dense stands, which led to more attack by this species. By reducing *M. pigra* populations from monocultures to smaller patches or individual plants, all control treatments increased the ratio of 'edge' plants to 'thicket' plants and therefore the proportion of plants susceptible to *N. gunniella* attack. *C. mimosa* declined dramatically following the fire. Although *C. mimosa* is also most abundant at stand edges (Paynter 2006), fire temporarily reduced the proportion of plants large enough to support *C. mimosa* larvae. Paynter and Flanagan (2004) concluded that integrating control techniques can successfully control dense *M. pigra* thickets and that biological control integrates well with other control options. Paynter (2003) demonstrated that biological control could be compatible with herbicidal control because *N. gunniella* larvae were able to successfully complete their development before the death of plants treated with herbicide.

Buckley *et al.* (2007) outlined the important role of disturbance in the invasion of *M. pigra* into unoccupied areas. Firn *et al.* (2008) suggested that management efforts should focus on future populations by attempting to reduce the size of the *M. pigra* seed bank, if there is a high probability that native vegetation will be disturbed. Reducing a seed bank requires intensive actions which are likely to also negatively affect the seed bank of native species. If, however, the probability of disturbance of native sites can be reduced, managers can shift control from future populations to the current population, which is more straightforward to implement and monitor.

Revegetation, using native grasses and sedges, allows recovery of native vegetation and prevents *M. pigra* regeneration from seed following control (Paynter 2004b). Stolons of the native perennial grass *Hymenachne acutigluma* (Steud) Gilliland (Poaceae) established well but spread rather slowly, hence revegetation must be combined with the diligent control of *M. pigra* seedlings regenerating following control of thickets.

Native wetland vegetation was just as effective in preventing regrowth of *M. pigra*, as was the introduced pasture species para grass *Urochloa mutica* (Forssk.), and managing grazing pressure was more important in controlling regrowth of *M. pigra* than the introduction of exotic pasture species (B Grace and Q Paynter, *pers. comm.* 2008). Grazing trials showed that, after dense stands of *M. pigra* have been cleared, fewer *M. pigra* plants will regenerate in plots that had reduced grazing pressure, and there was no difference between plots with reduced grazing pressure and ungrazed plots (B Grace and Q Paynter, *pers. comm.* 2008).

Future work

At this stage, there is no proposal for the introduction of further agents for *M. pigra* into Australia. Several prospective agents are available but further investigations would not necessarily be straightforward: some would require considerable resources (e.g. *R. pigrae*) or be difficult to test (e.g. *A. rotundipennis*, *Aristotelia* sp. nr *dasypoda*), show variation in field host plant use (biotype of *S. seminicola* from Brazil) or are perceived as insufficiently effective (*S. ochreosa, S. peruana, C. spretissimum*).

The future holds many opportunities for the introduction, re-introduction and redistribution in Asian and African countries of existing agents found in the native range of *M. pigra* and tested in Australia. This is especially true for those agents released in Australia after 1997 when the projects in Asia terminated. A related activity would be to assess the establishment of the agents already released in Asia and to redistribute those seen to be effective. Agents successfully established in Australia, such as *M. pallidata* and *M. irregularis*, will be the most promising candidates, but those that did not establish in Australia should also be considered. In particular, the green seed-feeders and pathogens may offer potential if introduced to other regions. It may be that establishment did not occur in Australia for reasons that will not limit establishment in other countries. For example, the dry season in Australia is more severe than in other areas of introduction, and these agents might establish in climates with a less extreme dry season.

Clearly it would be useful to determine the size of the soil seed bank in northern Australia given that it has not been measured since 2005, at which time it was showing a downward trajectory. Further detailed study of the current reproduction and spread of this invader, with an effort to determine the agents responsible for any change, is needed.

ACKNOWLEDGEMENTS

The following people are gratefully acknowledged for providing technical assistance in quarantine: Areli Mira, Richard Chan, Jenny Burcher, Ryan Zonneveld, John Carew, Gio Fichera, Dalio Mira, Shaun Winterton, Sounthi Subaaharan, Leonie Wittenberg and Toby Burgess. The following provided technical input in the field in Australia: Merrilyn Paskins, Lynne Burrell, Bruce Hitchins, Bert Lukitsch, Nadine Graham, Natasha Burrows, Tim Schatz, Suzanne Gall, Magen Pettit, Suzanne Casanova and others. Richard Milner and Caroline Hauxwell identified insect pathogens. The following taxonomists assisted with insect identifications: Otilie Neser, Robert Hamilton, David Kissinger, Charlie O'Brien, Wayne Clark, Malcolm Scoble, Gillian Watson, Paul Freytag, David Furth, Alma Solis, Gerhard Scherer, Wills Flowers, David Agassiz, Klaus Sattler and Józef Razowski. The many people involved in native range exploration and research are acknowledged in the Exploration section. I am grateful to them all.

REFERENCES

Anonymous (2011) Biological control project with Bulgul indigenous rangers. *Weed All About It.* NT Dept of Natural Resources, Environment, the Arts and Sport (NRETAS) **13**, 5.

Barneby RC (1991) Sensitivae Censitae, a description of the genus *Mimosa* Linnaeus (Mimosaceae) in the New World. *Memoirs of the New York Botanical Garden* **65**, 1–835.

Ben Dov Y (1994) *A Systematic Catalogue of the Mealybugs of the World.* Intercept, Andover.

Braithwaite RW, Lonsdale WM and Estbergs JA (1989) Alien vegetation and native biota in tropical Australia: the spread and impact of *Mimosa pigra. Biological Conservation* **48**, 189–210.

Buckley YM, Paynter Q and Lonsdale M (2004) Modelling integrated weed management of an invasive shrub in tropical Australia. *Journal of Applied Ecology* **41**, 547–560.

Buckley YM, Bolker BM and Rees M (2007) Disturbance, invasion and re-invasion: managing the weed-shaped hole in disturbed ecosystems. *Ecology Letters* **10**, 809–817.

Cam NV, Lam PV, Dien HC, Huong TT and Hien NT (1997) 'Biological control of giant sensitive plant *Mimosa pigra* L. in Vietnam'. Report of ACIAR Project No. 9313. VN Biological Control Research Center, National Institute for Plant Protection, Hanoi.

Clark WE (1984) Species of *Sibinia* Germar (Coleoptera: Curculionidae) associated with *Mimosa pigra* L. *Proceedings of the Entomological Society of Washington* **86**, 358–368.

Davis DR, Kassulke RC, Gillett JD and Harley KLS (1991) Systematics, morphology, biology and host specificity of *Neurostrota gunniella* (Buesk) (Lepidoptera: Gracillaridae), an agent for biological control of *Mimosa pigra* in Australia. *Proceedings of the Entomological Society of Washington* **93**, 16–44.

Eichlin TD and Passoa S (1983) A new clearwing moth (Sesiidae) from Central America: a stem borer in *Mimosa pigra. Journal of the Lepidopterists Society* **37**, 193–206.

Evans HC, Carrión G and Ruiz-Belin F (1995) Mycobiota of the giant sensitive plant, *Mimosa pigra sensu lato* in the Neotropics. *Mycological Research* **99**, 420–428.

Firn J, Rout T, Possingham H and Buckley YM (2008) Managing beyond the invader: manipulating disturbance of natives simplifies control efforts. *Journal of Applied Ecology* **45**, 1143–1151.

Flanagan GJ (1994) The Australian distribution of *Mictis profana* and its life cycle on *Mimosa pigra*. *Journal of the Australian Entomological Society* **33**, 111–114.

Flanagan GJ, Wilson CG and Gillett JD (1990) The abundance of native insects on the introduced weed *Mimosa pigra* in Northern Australia. *Journal of Tropical Ecology* **6**, 219–230.

Forno IW and Harley KLS (1992) Host testing of biological control agents for *Mimosa pigra*. In *A Guide to the Management of Mimosa pigra*. (Ed KLS Harley) pp. 43–48. CSIRO, Canberra.

Forno IW, Kassulke RC and Day MD (1991) Life-cycle and host testing procedures for *Carmenta mimosa* Eichlin and Passoa (Lepidoptera: Sesiidae), a biological control agent for *Mimosa pigra* L. (Mimosaceae) in Australia. *Biological Control* **1**, 309–315.

Forno IW, Heard TA and Day MD (1994) Host specificity and aspects of the biology of *Coelocephalapion aculeatum* (Coleoptera: Apionidae), a potential biological control agent of *Mimosa pigra* (Mimosaceae). *Environmental Entomology* **23**, 147–153.

Forno IW, Fichera J and Prior S (2000) Assessing the risk to *Neptunia oleracea* Lour. by the moth *Neurostrota gunniella* (Busck), a biological control agent for *Mimosa pigra* L. In *Proceedings of the 10th International Symposium on Biological Control of Weeds*. 4–14 July 1999, Bozeman, Montana. (Ed NR Spencer) pp. 449–457. Montana State University, Bozeman.

Furth DG (2006) The current status of knowledge of the Alticinae of Mexico (Coleoptera: Chrysomelidae). *Bonner Zoologische Beitrage* **54,** 209–237.

Global Invasive Species Database (2010) *One Hundred of the World's Worst Invasive Alien Species*. <http://www.issg.org/database/species/search.asp?st=100ss>.

Grace B (2005) Do predators affect the survival of *Macaria pallidata* larvae? Implications for the biological control of *Mimosa pigra* in the NT. *Northern Territory Naturalist* **18**, 8–13.

Harley KLS, Gillett JD, Winder J, Forno IW, Segura R, Miranda H and Kassulke RC (1995) Natural enemies of *Mimosa pigra* and *M. berlandieri* (Mimosaceae) and prospects for biological control of *M. pigra*. *Environmental Entomology* **24**, 1664–1678.

Heard TA (2004) The taxonomy of *Mimosa pigra*. In *Research and Management of* Mimosa pigra. (Eds M Julien, G Flanagan, T Heard, B Hennecke, Q Paynter and C Wilson) p. 10. CSIRO Entomology, Canberra.

Heard TA and Forno IW (1996) Host selection and host range of the flower feeding weevil, *Coelocephalapion pigrae*, a potential biological control agent against *Mimosa pigra*. *Biological Control* **6**, 83–95.

Heard TA and Paynter Q (2009) *Mimosa pigra* (Leguminosae). In *Biological Control of Tropical Weeds Using Arthropods*. (Eds R Muniappan, GVP Reddy and A Raman) pp. 256–273. Cambridge University Press, Cambridge.

Heard TA and Pettit W (2005) A review of the surveys for natural enemies of *Mimosa pigra*: what does it tell us about surveys for broadly distributed hosts? *Biological Control* **34**, 247–254.

Heard TA and Segura R (2004) Agents for biological control of *Mimosa pigra* in Australia: review and future prospects. In *Research and Management of* Mimosa pigra. (Eds M Julien, G Flanagan, T Heard, B Hennecke, Q Paynter and C Wilson) pp. 126–140. CSIRO Entomology, Canberra.

Heard TA and van Klinken RD (2004) Rapid preliminary characterisation of host specificity of leaf-beetles (Coleoptera: Chrysomelidae). *BioControl Science and Technology* **14**, 499–511.

Heard TA, Segura R, Martinez M and Forno IW (1997) Biology and host range of the green seed weevil, *Sibinia fastigiata*, for biological control of *Mimosa pigra*. *Biocontrol Science and Technology* **7**, 631–644.

Heard TA, O'Brien CW, Forno IW and Burcher JA (1998) *Chalcodermus persimilis* O'Brien n. sp. (Coleoptera: Curculionidae): description, biology, host range, and suitability for biological control of *Mimosa pigra* L. *Transactions of the American Entomological Society* **124**, 1–11.

Heard TA, Forno IW and Burcher J (1999) *Chalcodermus serripes* Coleoptera: Curculionidae for biological control of *Mimosa pigra*: host relations and life cycle. *Biological Control* **15**, 1–9.

Heard TA, Segura R, Zonneveld R and Martinez M (2004) Limited success of open field tests to clarify the host range of three species of Lepidoptera of *Mimosa pigra*. In *Proceedings of the 11th International Symposium on Biological Control of Weeds*. 27 April–2 May 2003, Canberra. (Eds JM Cullen, DT Briese, DJ Kriticos, WM Lonsdale, L Morin and JK Scott) pp. 277–282. CSIRO Entomology, Canberra.

Heard TA, Paynter Q, Chan R and Mira A (2005) *Malacorhinus irregularis* for biological control of *Mimosa pigra*: host specificity, life cycle, and establishment in Australia. *Biological Control* **32**, 252–262.

Heard TA, Mira A, Chan R and Fichera G (2006) 'Application to release *Nesaecrepida infuscata* (Coleoptera: Chrysomelidae: Alticinae) into Australia for biological control of the weed *Mimosa pigra*'. Report submitted to AQIS.

Heard TA, Mira A, Chan R and Fichera GA (2009) 'Application to release the tip weevil *Temnocerus debilis*, a potential biological control agent of the weed *Mimosa pigra*'. Report submitted to AQIS.

Heard TA, Elliott L, Zonneveld R, Mira A, Chan R, Fichera GA, Routley B, Wirf L, Burrows N and Segura R (2010) Biology, host specificity, release and establishment of *Macaria pallidata* and *Leuciris fimbriaria* (Lepidoptera: Geometridae), potential biocontrol agents of *Mimosa pigra*. *Biological Control* **55**, 248–255.

Hennecke B (2004) The prospect of biological control of *Mimosa pigra* with fungal pathogens in Australia. In *Research and Management of* Mimosa pigra. (Eds M Julien, G Flanagan, T Heard, B Hennecke, Q Paynter and C Wilson) pp. 117–121. CSIRO Entomology, Canberra.

Hennecke BR (2006) Failure of *Diabole cubensis*, a promising classical biological control agent, to establish in Australia. *Biological Control* **39**, 121–127.

Hokkanen H and Pimentel D (1984) New approach for selecting biological-control agents. *Canadian Entomologist* **116**, 1109–1121.

Kassulke RC and Harley KLS (1987) *Chlamisus* sp. nr. *sidae* Monrós (Coleoptera: Chrysomelidae: Chlamisinae) an agent for the biological control of *Mimosa pigra* L. in Australia. Unpublished ms. CSIRO Entomology, Brisbane.

Kassulke RC, Harley KLS and Maynard GV (1990) Host specificity of *Acanthoscelides quadridentatus* and *A. puniceus* (Col.: Bruchidae) for biological control of *Mimosa pigra* (with preliminary data on their biology). *Entomophaga* **35**, 85–96.

Karren JB (1989) *Chlamisus mimosae*, sp. nov. (Coleoptera: Chrysomelidae: Chlamisinae) from Brazil and imported into Australia and Thailand. *Coleopterists Bulletin* **43**, 355–358.

Kissinger DG (1992) Apionidae from North and Central America. Part 4. Generic classification and introduction to the genus *Coelocephalapion Wagner*, with new species from Mexico and Venezuela (Coleoptera). *Insecta Mundi* **6**, 65–77.

Lonsdale WM (1988) Litterfall in an Australian population of *Mimosa pigra*, an invasive tropical shrub. *Journal of Tropical Ecology* **4**, 381–392.

Lonsdale WM (1993) Rates of spread of an invading species: *Mimosa pigra* in northern Australia. *Journal of Ecology* **81**, 513–521.

Lonsdale WM and Farrell GS (1998) Testing the effects on *Mimosa pigra* of a biological control agent *Neurostrota gunniella* (Lepidoptera: Gracillaridae), plant competition and fungi under field conditions. *Biocontrol Science and Technology* **8**, 485–500.

Lonsdale WM and Segura R (1987) A demographic comparison of native and introduced populations of *Mimosa pigra*. In *Proceedings of the 8th Australian Weeds Conference*, 21–25 September 1987, Sydney. (Eds D Lemerle and AR Leys) pp. 163–166. Weeds Society of New South Wales, Sydney.

Lonsdale WM, Harley KLS and Gillett JD (1988) Seed bank dynamics of *Mimosa pigra*, an invasive tropical shrub. *Journal of Applied Ecology* **25**, 963–976.

Lonsdale WM, Miller IL and Forno IW (1989) The biology of Australian Weeds 20. *Mimosa pigra* L. *Plant Protection Quarterly* **4**, 119–131.

McIntyre V, Grace B and Schmidt S (2007) Impacts of the biocontrol agent *Malacorhinus irregularis* (Coleoptera, Chrysomelidae) on *Mimosa pigra* seedlings and the importance of root nodules. *Biocontrol Science and Technology* **17**, 365–374.

Miller IL, Napompeth B, Forno IW and Siriworakul M (1992) Strategies for the integrated management of *Mimosa pigra*. In *A Guide to the Management of* Mimosa pigra. (Ed KLS Harley) pp. 110–115. CSIRO, Canberra.

Napompeth B (1982) Preliminary screening of insects for biological control of *Mimosa pigra* L. in Thailand. In Mimosa pigra *Management, Proceedings of an International Symposium*. 22–26 February 1982, Chiang Mai, Thailand. (Eds GL Robert and DH Habeck) pp. 121–128. International Plant Protection Center, Corvallis, Oregon.

Ostermeyer N (2000) Population density and distribution of the biological control agent *Carmenta mimosa* on *Mimosa pigra* in the Adelaide and Finniss River catchments of the Northern Territory. *Plant Protection Quarterly* **15**, 46–49.

Ostermeyer N and Grace BS (2007) Establishment, distribution and abundance of *Mimosa pigra* biological control agents in northern Australia: implications for biological control. *BioControl* **52**, 703–720.

Ostermeyer N, Grace B, Paskins M, McIntyre V and Routley B (2004) Biological control in the Top End: getting community involved. In *Proceedings of the 14th Australian Weeds Conference.* 6–9 September 2004, Wagga Wagga, NSW. (Eds BM Sindel and SB Johnson) pp. 337–340. Weeds Society of New South Wales, Sydney.

Page AR and Lacey KL (2006) 'Economic impact assessment of Australian weed biological control'. CRC for Australian Weed Management, Technical Series 10. CRC for Australian Weed Management, Adelaide.

Paynter Q (2003) Integrated weed management: effect of herbicide choice and timing of application on the survival of a biological control agent of the tropical wetland weed, *Mimosa pigra. Biological Control* **26**, 162–167.

Paynter Q (2004a) Evaluating *Mimosa pigra* biological control in Australia. In *Research and Management of* Mimosa pigra. (Eds M Julien, G Flanagan, T Heard, B Hennecke, Q Paynter and C Wilson) pp. 141–148. CSIRO Entomology, Canberra.

Paynter Q (2004b) Revegetation of a wetland following control of the invasive woody weed, *Mimosa pigra*, in the Northern Territory, Australia. *Ecological Management and Restoration* **5**, 191–198.

Paynter Q (2005) Evaluating the impact of biological control against *Mimosa pigra* in Australia. *Journal of Applied Ecology* **42**, 1054–1062.

Paynter Q (2006) Evaluating the impact of biological control against *Mimosa pigra* in Australia: comparing litterfall before and after the introduction of biological control agents. *Biological Control* **38**, 166–173.

Paynter Q and Flanagan GJ (2002) Integrated management of *Mimosa pigra*. In *Proceedings of the 13th Australian Weeds Conference*, 8–13 September 2002, Perth. (Eds H Spafford Jacob, J Dodd and JH Moore) pp. 165–168. Plant Protection Society of Western Australia, Perth.

Paynter Q and Flanagan GJ (2004) Integrating herbicide and mechanical control treatments with fire and biological control to manage an invasive wetland shrub, *Mimosa pigra. Journal of Applied Ecology* **41**, 615–629.

Pollhill RM and Raven PH (1981) *Advances in Legume Systematics Part 1.* Royal Botanic Gardens, Kew.

Rasplus JY (1988) Nouvelles espèces d'Eurytomidae principalment parasites de coleoptères seminivores de legumineuses en Côte d'Ivoire (Lamto) (Hymenoptera Chalcidoidea). *Bollettino di Zoologia Agraria e di Bachicoltura* **20**(Ser.ii), 89–114.

Routley BM and Wirf LA (2006) Advancements in biocontrol of *Mimosa pigra* in the Northern Territory. In *Proceedings of the 15th Australian Weeds Conference.* 24–28 September 2006, Adelaide. (Eds C Preston, JH Watts and ND Crossman) pp. 561–564. Weed Management Society of South Australia, Adelaide.

Simon MF and Proença C (2000) Phytogeographic patterns of *Mimosa* (Mimosoideae, Leguminosae) in the Cerrado biome of Brazil: an indicator genus of high-altitude centers of endemism? *Biological Conservation* **96**, 279–296.

Smith CS and Wilson CG (1995) Close to the edge: microhabitat selection by *Neurostrota gunniella* Busck (Lepidoptera: Gracillariidae), a biological control agent for *Mimosa pigra* L. in Australia. *Journal of the Australian Entomological Society* **34**, 177–180.

Suasa-ard W, Sommartya P and Jaitui S (2004) Evaluation of seed feeding bruchids, *Acanthoscelides* species, as biological control agents for *Mimosa pigra* in Thailand. In *Research and Management of* Mimosa pigra. (Eds M Julien, G Flanagan, T Heard, B Hennecke, Q Paynter and C Wilson) pp. 122–125. CSIRO Entomology, Canberra.

Taylor DBJ, Heard TA, Paynter Q and Spafford Jacob H (2007) Non-target impact of a weed biological control agent on a native plant in northern Australia. *Biological Control* **42**, 25–33.

Thorp JR and Lynch R (2000) *The Determination of Weeds of National Significance.* National Weeds Strategy Executive Committee, Launceston.

Turner BL (1959) *The Legumes of Texas.* University of Texas Press, Austin.

Vitelli JS, Madigan BA and Worsley KJ (2006) *Mimosa pigra* in Queensland. In *Proceedings of the 15th Australian Weeds Conference.* 24–28 September, Adelaide. (Eds C Preston, JH Watts and ND Crossman) pp. 251–254. Weed Management Society of South Australia, Adelaide.

Wild CH (1986) Host specificity report on *Scamurius* sp. (Hemiptera: Heteroptera: Coreidae), an agent for the biological control of giant sensitive plant (*Mimosa invisa* Mart.: Mimosaceae) in Queensland. Qld Dept of Lands, Alan Fletcher Research Station. Unpublished report.

Wilson CG and Flanagan GJ (1990) Establishment and spread of *Neurostrota gunniella* on *Mimosa pigra* in the Northern Territory. In *Proceedings of the 9th Australian Weeds Conference.* 6–10 August 1990, Adelaide. (Ed JW Heap) pp. 505–507. Crop Science Society of South Australia, Adelaide.

Wilson CG and Flanagan GJ (1991) Establishment of *Acanthoscelides quadridentatus* (Schaffer) and *A. puni-*

ceus Johnson (Coleoptera: Bruchidae) on *Mimosa pigra* L. *Journal of the Australian Entomological Society* **30**, 279–280.

Wilson CG and Pitkethley RN (1992) *Botryodiplodia* dieback of *Mimosa pigra*, a noxious weed in northern Australia. *Plant Pathology* **41**, 777–779.

Wilson CG, Flanagan GJ and Gillett JD (1990) The phytophagous insect fauna of the introduced shrub, *Mimosa pigra* in northern Australia and its relevance to biological control. *Environmental Entomology* **19**, 776–784.

Wilson CG, Forno IW, Smith CS and Napompeth B (1992) Rearing and release methods for biological control agents. In *A Guide to the Management of* Mimosa pigra. (Ed KLS Harley) pp. 49–62. CSIRO, Canberra.

Wirf LA (2006) The effect of manual defoliation and *Macaria pallidata* (Geometridae) herbivory on *Mimosa pigra*: implications for biological control. *Biological Control* **37**, 346–353.

Moraea flaccida Sweet – one-leaf Cape tulip
Moraea miniata Andrews – two-leaf Cape tulip

John K Scott and Louise Morin

ABSTRACT

In 1999, the Iridaceae species one- and two-leaf Cape tulips (*Moraea flaccida* and *Moraea miniata*, previously in the genus *Homeria*), were the first in this family to be approved as target weeds for biocontrol in Australia. Cape tulips invade natural ecosystems and pasture and are toxic to livestock across southern Australia, with the greatest infestations recorded in the south-west of WA. Surveys for natural enemies of Cape tulips in South Africa, the region of origin of these weeds, identified the rust fungus *Puccinia moraeae* as the most promising candidate biocontrol agent. Insects feeding on the corms and seeds were also found, but basic information on their biology is needed before their biocontrol potential can be assessed. Thirty-nine accessions of the rust collected across the Cape tulips' range in South Africa were recently imported into quarantine in Australia, but only one accession from two-leaf Cape tulip was successfully cultured on Australian plants. Preliminary host-specificity tests with this rust accession were performed on a few selected non-target species within the Iridaceae; none developed any symptoms. However, more collection and testing is required to select the rust isolates most pathogenic on Australian Cape tulips prior to undertaking comprehensive host-specificity testing. Identification of putative genotypes comprised in both Cape tulip species in Australia would facilitate this selection process.

Key words: *Cryptolarynx* spp., environmental weed, *Homeria* spp., pasture weed, rust fungus *Puccinia moraeae*, *Urodontellus* spp.

INTRODUCTION

One-leaf Cape tulip (*Moraea flaccida* Sweet) and two-leaf Cape tulip (*Moraea miniata* Andrews) (Iridaceae) (Fig. 1) have become major environmental and pasture weeds of southern Australia (Parsons and Cuthbertson 2001). Cape tulips were originally cultivated for their attractive flowers and this, along with their hardiness, was probably the reason for their introduction into Australia. Today, Cape tulips are not cultivated to any extent, although vegetative propagation material is occasionally offered for sale despite these species being declared noxious in most southern Australian states – NSW, SA, Tas, Vic and WA.

The genus *Moraea* (including species previously placed in the genus *Homeria*) is native to southern Africa (Goldblatt *et al.* 2008) and defined in Goldblatt (1998). Most species of *Moraea*, including one- and two-leaf Cape tulips, are native to the south-west and western regions of the Western Cape province, South Africa (Goldblatt 1981). In their introduced range in Australia, one- and two-leaf Cape tulips have a southern distribution, and are primarily found in WA, SA and Vic (Cooke 1986; Parsons and Cuthbertson 2001) (Fig. 2). Chromosome studies

Figure 1: a) *Moraea flaccida* flowers and b) *Moraea* species in pasture in WA. Photos: JK Scott, CSIRO.

indicated that Australian one-leaf Cape tulip accessions corresponded to the hexaploid form of this species that is naturally found on the west coast of the Western Cape province and that Australian two-leaf Cape tulip accessions corresponded to the triploid form of this species found near Stellenbosch in that province (Goldblatt 1981; Morrison and Scott 1995). A recent pilot genetic study of one-leaf Cape tulip using amplified fragment length polymorphisms showed that there was significant genotypic variation between populations from SA and WA (Blackman *et al.* 2006). In SA, there was limited evidence of genetic divergence between populations and low genetic diversity within populations. In contrast, the two WA populations studied had distinctly different genotypes, suggesting that multiple introductions of the plant occurred in that state.

Cape tulips produce shoots during winter, flower in early spring and senesce in late spring. One-leaf Cape tulip reproduces by seed and regeneration of corms (swollen underground plant stem that serves as a storage organ). Two-leaf Cape tulip does not produce seed and reproduces only by corms and cormils (small corms), which form around the main corm and in the axil of leaves. Corms and cormils have a dormancy period, which is broken in some, but not all, of them in any one year when soil temperature and moisture are adequate for germination, usually after the first autumn rains.

Cape tulips are unpalatable to grazing animals and are known to be toxic in some instances. The toxin identified in other *Moraea* species is a cardiac glycoside, bufadienolide, 1a, 2a-epoxyscillirosidin (Naudé and Potgieter 1971). Poisoning and death of cattle, sheep, horses, goats and donkeys due to Cape tulips has been reported, but most losses have been among cattle (Roberts *et al.* 1988). Usually this is due to animals being introduced to the weed for the first time. In addition, subclinical toxicity affects animal performance (notably sheep) by causing scouring, and lower body weights and wool yields. The unpalatability and weeping habit of the leaves probably confers a competitive advantage to Cape tulips in pasture and it is estimated that Cape tulips can reduce the carrying capacity of infested areas by up to 50% (Meadly 1954). One-leaf Cape tulip is also recognised as a major environmental weed of Banksia woodlands on the Spearwood dune system near Perth, WA (Keighery 1989) and Cape tulips are considered to severely affect the growth and regeneration of indigenous ground flora in south-east Australia (Muyt 2001).

Cape tulips can be effectively controlled with herbicide during the crop phase of the agricultural rotation system typical of south-west Australia (Parsons and Cuthbertson 2001). However, the dormant corms and cormils that remain ensure the weed regenerates in the subsequent pasture phase. Chemical control in pasture is difficult as many herbicides have the potential to seriously damage the annual legume component. For this reason there is demand for more appropriate herbicides for use in pastures and for biocontrol agents in the 500–800 mm rainfall annual pasture region of southern Australia. Cape tulips can also be controlled by cultivation, if it is carried out when the old corms have shrivelled considerably and the new corms have only partially developed (Parsons and Cuthbertson 2001). The use of this control method is limited, however, because paddocks are often too wet for machinery to operate at the time when cultivation should occur for effective control.

The WA Agricultural Protection Board's control program against Cape tulips was subjected to a cost:benefit analysis in the 1980s (Roberts *et al.* 1988). The net cost of Cape tulips, if the herbicide control program had not been implemented in the 13 most affected shires in WA

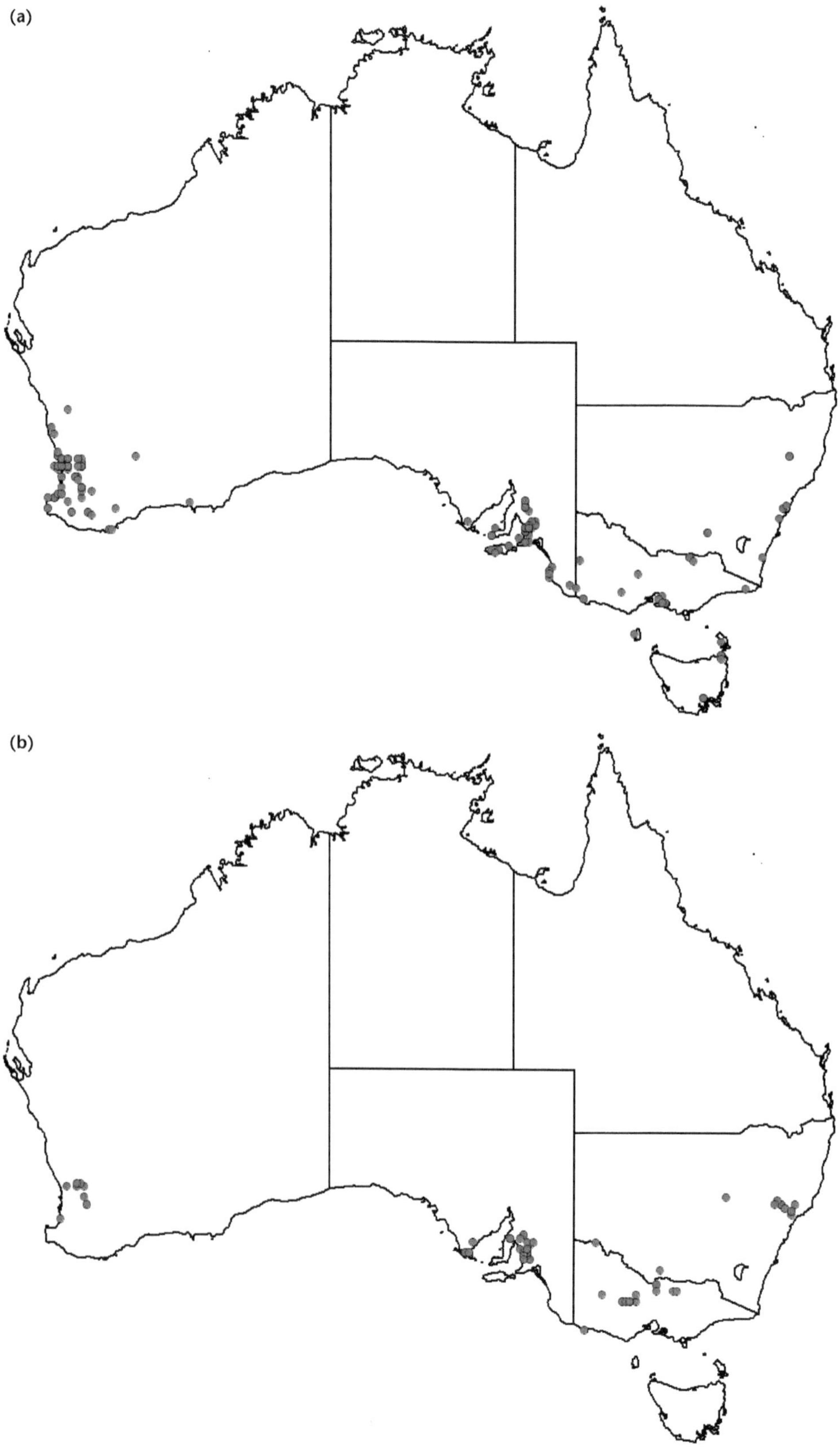

Figure 2: Distribution of a) one-leaf Cape tulip (*Moraea flaccida*) and b) two-leaf Cape tulip (*Moraea miniata*) in Australia. Specimen data reproduced from Australia's Virtual Herbarium with permission of the Council of Heads of Australasian Herbaria Inc.

over 30 years (calculated at the net present value) was estimated at $108 million. The industries affected were wool and meat. The cost in SA and Vic has not been calculated, but would be of similar magnitude given the area involved.

BIOLOGICAL CONTROL HISTORY

So far, there have been no attempts to establish biocontrol programs for Cape tulips or any other Iridaceae (Julien and Griffiths 1998) in any countries other than Australia. Cape tulips (as *Homeria flaccida* (Sweet) Steud. and *H. miniata* (Andrews) Sweet) were proposed as target species for biocontrol in Australia by WA and were approved in 1999 by the Standing Committee on Agriculture and Resource Management, Forests and Lands (equivalent to the current Natural Resource Management Standing Committee) (Australian Weeds Committee 2010). The CSIRO has made three short surveys in South Africa since 1995 to identify potential agents. In 2009, one such agent was imported for initial assessment in quarantine in Australia.

EXPLORATION

In the mid 1990s, the CSIRO began explorations for natural enemies of Cape tulips in South Africa, largely on an ad hoc basis. Surveys were made in the Western Cape province by John K Scott and Stefan Neser, July 1995, Petra Müller, August 2000 to June 2001 and John K Scott and Aaron Maxwell, June–July 2004. In the 2001 growing season, 12 study sites were selected and Cape tulips were monitored weekly for natural enemy damage (P Müller, unpub. data). In August–September 2009, Louise Morin and Petra Müller made several collections of one of the promising candidate agents across the range of Cape tulips in the Western Cape prior to importation into an Australian quarantine facility.

CANDIDATES

Natural enemies

Very little leaf damage and no corm, cormil or seed damage has been observed during surveys in WA (PB Yeoh and JK Scott, unpub. data 2001, 2009) nor reported from elsewhere in Australia. There are no specimens of diseased Cape tulips collected in Australia that have been lodged in any of the Australian mycological herbaria.

Exit holes on damaged corms and cormils of Cape tulips were frequently observed during surveys in South Africa. The damage was possibly made by weevils from the genus *Cryptolarynx*, but little is known of the biology of these species (Oberprieler *et al.* 2007). Seed-feeding weevils, *Urodontellus* sp. (formerly included in the genus *Urodon*), were also found on one-leaf Cape tulips in South Africa. This genus is associated with *Moraea* species where the larvae feed on immature seeds in the capsules (Louw 1993; JK Scott, *pers. obs.*). The rust fungus *Puccinia moraeae* Henn. was the sole natural enemy observed attacking the foliage of Cape tulips in South Africa.

Puccinia moraeae **Henn. (Uredinales: Pucciniaceae).** This rust fungus (Fig. 3) infects leaves of Cape tulips and has been recorded from South Africa only on *Moraea* species including *Moraea collina* Thunberg (syn. *Homeria breyniana* (L.) Lewis *sensu* Lewis), *M. fugax* (Roche) Jacq. subsp. *fugax* (syn. *M. edulis* (L.f.) Ker), *M. flaccida* (syn. *H. collina* var. *aurantiaca* (Zuccagni) Baker *sensu* Baker), *M. miniata* and *M. pallida* (Baker) Goldblatt (syn. *Homeria humilis* N.E. Brown) (Farr and Rossman 2010, species names based on Goldblatt 1981, 1998). Recent surveys showed that the rust fungus is widespread throughout the range of Cape tulips in South Africa, and is most prevalent on two-leaf Cape tulip. It is not known to occur on any other native South African species belonging to the Iridaceae family (e.g. *Iris*, *Gladiolus* and *Freesia* spp.) or exotic Iridaceae species cultivated in South Africa (Glen 2002), including *Patersonia glabrata*

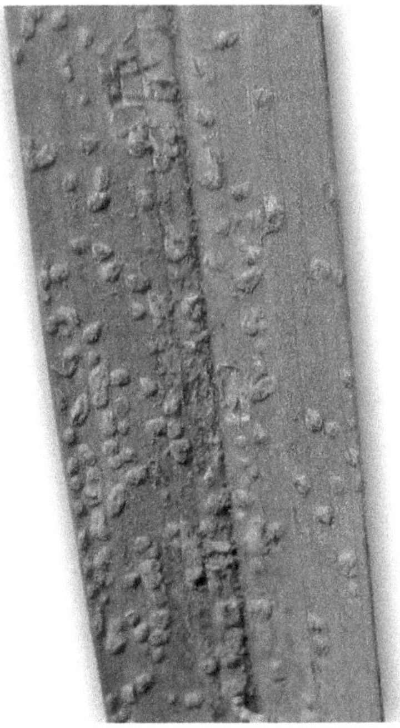

Figure 3: *Puccinia moraeae* on leaves of *Moraea* sp. Photo: L Morin, CSIRO.

R.Br. that is indigenous to eastern Australia (Farr and Rossman 2010). There are no indigenous Australian species in the tribe Irideae to which the *Moraea* genus belongs (Goldblatt *et al.* 2008).

In 2001, initial data on the impact of the rust fungus on two-leaf Cape tulip were gathered from three sites in the Western Cape where severe infection had been observed the previous year (P Müller, unpub. data). At each site, half the selected plants were sprayed monthly with a fungicide to prevent rust infection, while the remaining plants were left intact. Significant stunting and reduced flowering of rust-infected plants was observed by the end of the season, but the impact on corms and cormils was not measured.

Several accessions of the rust fungus were collected in late winter 2009 in the Western Cape and were imported into quarantine in Australia: six accessions from one-leaf Cape tulip, 31 from two-leaf Cape tulip and two from a mixture of one- and two-leaf Cape tulips (L Morin, unpub. data). One- and two-leaf Cape tulip plants that had been dug up from different sites in WA, SA and Vic were inoculated with each of the accessions. Although plants initially appeared in good health, many of them died during the experiments, most likely due to transplantation shock. While one or two uredinia or telia developed on both Cape tulip species inoculated with some of the rust accessions, only one accession from a two-leaf Cape tulip site (ex. road from Wellington to Porterville) was successfully cultured, albeit with great difficulties due to the poor quality of plants available during spring and summer, and the tendency of the rust accession to produce mostly telia and only a few uredinia from which limited amount of spores could be recovered to inoculate new plants. Some of the Australian two-leaf Cape tulip accessions inoculated with this rust accession were found to be resistant, indicating that intraspecific variation in plant resistance exists in this host–pathogen system. A preliminary host-specificity test was also performed with the rust accession using a few key non-target plant species closely related to Cape tulips within the Iridaceae. None of the species inoculated, except the two-leaf Cape tulip plants, developed any uredinia or telia.

DISCUSSION

Basic information on the biology of the insects found on Cape tulips in South Africa is needed before their biocontrol potential can be assessed. Both corm- and seed-feeding weevils appear to have annual life-cycles, and this has made them difficult to study. The seed-feeding weevils are relevant only to one-leaf Cape tulip because it is the only species that produces seed. Even so, the relative ecological role of seeds, cormils (functionally like seeds) and corms in the persistence of these weed species requires study to help determine the organs most susceptible to biocontrol targeting.

The initial screening of rust accessions on Australian Cape tulip plants faced major hurdles due to the poor quality of plants available at the time. Dormancy in Cape tulip corms and cormils is the main barrier to a continuous supply of vigorous Cape tulip plants for testing. Nonetheless, a culture of one of the rust accessions collected from two-leaf Cape tulip in South Africa was established in quarantine and tested on a range of two-leaf Cape tulip accessions from Australia and on a few non-target plant species within the Iridaceae. More collection and testing to select the rust isolates most pathogenic on Australian Cape tulips is required prior to undertaking comprehensive host-specificity testing. There are indications from previous genetic work (Blackman *et al.* 2006) and from field observations in South Africa (L Morin, *pers. obs.*) that Cape tulip species may comprise different genotypes that vary in susceptibility to different rust genotypes. This could mean that more than a single rust isolate may be required to attack all Cape tulip genotypes in Australia. *A priori* identification of the putative genotypes of both species in Australia would greatly facilitate future searches for pathogenic rust isolates. Indeed, the establishment of an outdoor garden in South Africa planted with the different Australian Cape tulip genotypes could be an effective approach to consider, rather than random field-collecting, to source rust isolates that will have the greatest impact in Australia.

Despite its long standing as a weed, there is little published on the ecology of the plant in its pasture environment. This, and its role as an increasingly important invader of natural ecosystems in WA, points to the need for baseline and impact studies in conjunction with the assessment of biocontrol agents.

ACKNOWLEDGEMENTS

We gratefully acknowledge the WA Department of Agriculture and Food and Prof. John Hoffmann of the University of Cape Town for their support of studies relating to Cape tulip biocontrol. Survey and quarantine work in South Africa and Australia was supported by Ruth Aveyard, Dr Aaron Maxwell, Petra Müller, Dr Stefan Neser and Paul Yeoh.

REFERENCES

Australian Weeds Committee (2010) Target species for biological control. <http://www.weeds.org.au/management.htm> (accessed 23.3.10).

Blackman D, Baker J and Preston C (2006) What does genetic variation of one-leaf Cape tulip (*Moraea flaccida*) populations in Australia tell us: implications for classical biological control? In *Proceedings of the 15th Australian Weeds Conference*. 24–28 September 2004, Adelaide. (Eds C Preston, JH Watts and ND Crossman) p. 295. Weed Management Society of South Australia, Adelaide.

Cooke DA (1986) Iridaceae. *Flora of Australia* **46**, 1–66.

Farr DF and Rossman AY (2010) Fungal databases, systematic mycology and microbiology laboratory. ARS, USDA. <http://nt.ars-grin.gov/fungaldatabases/> (accessed 27.3.10).

Glen HF (2002) *Cultivated Plants of Southern Africa*. Jacana, Johannesburg.

Goldblatt P (1981) Systematics and biology of *Homeria* (Iridaceae). *Annals of the Missouri Botanical Garden* **68**, 413–503.

Goldblatt P (1998) Reduction of *Barnardiella*, *Galaxia*, *Gynandriris*, *Hexaglottis*, *Homeria*, and *Roggeveldia* in *Moraea* (Iridaceae: Irideae). *Novon* **8**, 371–377.

Goldblatt P, Rodriguez A, Powell MP, Davies TJ, Manning JC, van der Bank M and Savolainen V (2008) Iridaceae 'out of Australasia'? Phylogeny, biogeography, and divergence time based on plastid DNA sequences. *Systematic Botany* **33**, 495–508.

Julien MH and Griffiths MW (1998) *Biological Control of Weeds: A World Catalogue of Agents and their Target Weeds*. CABI, Wallingford, UK.

Keighery GJ (1989) Banksia woodland weeds. *Journal of the Royal Society of Western Australia* **71**, 111–112.

Louw S (1993) Systematics of the Urodontidae (Coleoptera: Curculionoidea) of southern Africa. *South Africa Department of Agriculture Entomology Memoir* **87**, 1–92.

Meadly GRW (1954) Cape tulip (*Homeria breyniana* (L.) Lewis and *H. miniata* Sweet). *Journal of Agriculture of Western Australia Third Series* **3**, 289–296.

Morrison SM and Scott JK (1995) Chromosome numbers of Cape tulips (*Homeria* species) in South Australia and Western Australia. *Plant Protection Quarterly* **10**, 96–98.

Muyt A (2001) *Bush Invaders of South-east Australia. A Guide to the Identification and Control of Environmental Weeds Found in South-East Australia*. RG & FJ Richardson, Melbourne.

Naudé TW and Potgieter DJJ (1971) Studies on South African cardiac glycosides. 1. Isolation of toxic principles of *Homeria glauca* and observations on their chemical and pharmacological properties. *Onderstepoort Journal of Veterinary Research* **38**, 255–277.

Oberprieler RG, Marvaldi AE and Anderson RS (2007) Weevils, weevils, weevils everywhere. *Zootaxa* **1668**, 491–520.

Parsons WT and Cuthbertson EG (2001) *Noxious Weeds of Australia*. 2nd edn. CSIRO Publishing, Melbourne.

Roberts EJ, Hector JM, Bunny CJ and Carter GD (1988) 'The economics of Cape tulip control in Western Australia: a cost-benefit analysis'. WA Agricultural Protection Board Discussion Paper 3/88.

Nassella trichotoma (Nees) Hack ex Arechav. – serrated tussock *Nassella neesiana* (Trin. & Rupr.) Barkworth – Chilean needle grass

David McLaren, Freda E Anderson and Jane Barton

ABSTRACT

Nassella trichotoma (serrated tussock) and *Nassella neesiana* (Chilean needle grass) (Poaceae) are significant agricultural and environmental weeds in Australia and New Zealand. A biological control program was initiated in 1999 with detailed field exploration in Argentina and study of pathogens on both species. The potential biological control candidates identified for serrated tussock were either not host-specific (*Puccinia nassellae*), not sufficiently pathogenic to Australian accessions of the weed (*P. nassellae*, *Tranzscheliella* spp.) or their biology and life-cycle could not be fully determined (*P. nassellae*, *Tranzscheliella* spp., *Corticiaceae* sp.), precluding further work. Concurrently, *P. nassellae* ex Chilean needle grass and two other rust fungi, *Puccinia graminella* and *Uromyces pencanus*, were identified as potential biological control candidates for Chilean needle grass. *P. nassellae* ex Chilean needle grass causes premature senescence of foliage and can be very damaging to this grass in the field. It has proven to be extremely host-specific, to a point where only three out of seven Australian Chilean needle grass accessions have been susceptible to one of the various isolates with which they have been inoculated. Mass rearing and storage of this rust have proven problematic and consequently it has not undergone comprehensive host-specificity testing. *P. graminella* was found to be damaging to Chilean needle grass and was common and widespread during the initial surveys. However, it later became scarce, probably due to prevailing drought conditions. A large number of Chilean needle grass plants within each accession from both Australia and Argentina were found to be resistant to this rust, thus complicating mass culturing for host-specificity testing. To date, the most promising candidate agent for Chilean needle grass is *U. pencanus*: it was host-specific during preliminary testing; it is easy to mass culture; its spores can be frozen for later use; and it can be very damaging to Chilean needle grass populations in the field. It has now been approved for release in NZ, which is a historical decision as this is the first pathogen biological control agent in the world approved for release on a grass. The next phase of this project will be to seek permission to release this fungus in Australia.

Key words: serrated tussock, Chilean needle grass, *Puccinia nassellae*, *Tranzscheliella* spp., *Puccinia graminella*, Corticiaceae, *Uromyces pencanus*, host-specificity.

INTRODUCTION

Nassella trichotoma (Nees) Hack ex Arechav. (serrated tussock, Poaceae) is a perennial, drought-resistant tussock grass species that is native to the pampas grasslands of Argentina, Uruguay, Chile and Peru (Parodi 1930; Rosengurtt *et al.* 1970) and has also been reported from Bolivia (Walsh and Entwisle 1994). It has now naturalised in Australia (Fig. 1), NZ and South Africa and small infestations also occur in England and Italy (Campbell and Vere 1995). In south-eastern Australia, serrated

Figure 1: Distribution of *Nassella trichotoma* in Australia. Specimen data reproduced from Australia's Virtual Herbarium with permission of the Council of Heads of Australasian Herbaria Inc.

tussock is widespread and covers more than 2 million ha of land (Osmond *et al.* 2008). The potential distribution of serrated tussock in Australia has been estimated at 32 million ha, with substantial areas of NSW, Vic and Tas at risk of invasion (McLaren *et al.* 1998).

Serrated tussock (Fig. 2) is a C3 grass that produces active growth during the winter, spring and early summer months and is dormant over summer (Osmond *et al.* 2008). Flowering and seeding occurs during spring and early summer and a single serrated tussock plant can produce more than 100 000 seeds. Seed dispersal can occur by wind (up to 20 km), vehicles and machinery, animals, stock feed and produce, soil and waterways (Osmond *et al.* 2008). It is an insidious weed as it can be easily confused with other pasture grasses, meaning that land managers may not realise the extent of the problem until serrated tussock dominates.

It has been described as potentially causing greater reductions in carrying capacity than any other pasture weed in Australia (Parsons and Cuthbertson 1992). It has been estimated to cost land managers $40 million and $5 million in control and lost production costs annually in NSW (Jones and Vere 1998) and Vic (Nicholson *et al.* 1997), respectively.

The only registered herbicides for control of serrated tussock in pastures are flupropanate, glyphosate and 2,2-DPA. Serrated tussock resistance to flupropanate has been identified at two locations in Vic and one location in NSW (McLaren *et al.* 2008). Osmond *et al.* (2008) described integrated management strategies for serrated tussock including use of chemical control, cultural control, competition and techniques for prevention of seed spread. Carr *et al.* (1992) classified serrated tussock as a very serious environmental weed and McLaren *et al.* (1998) provided examples of vegetation formations invaded by serrated tussock that include dry coastal vegetation, lowland grassland, grassy woodland, sclerophyll forest and woodland and rocky outcrop vegetation.

Nassella neesiana (Trin. and Rupr.) Barkworth (Chilean needle grass, Poaceae) is a long-lived perennial

Figure 2: *Nassella trichotoma* seeds. Illustration: E Mayfield.

tussock-forming grass that is indigenous to Argentina, Bolivia, Chile, Ecuador, southern Brazil and Uruguay (Rosengurtt *et al.* 1970). In Australia it is both a serious environmental weed (Carr *et al.* 1992; McLaren *et al.* 1998) and a problem weed of agriculture (Grech 2007) and has been determined a Weed of National Significance (WONS) as a consequence (Thorp and Lynch 2000). *N. neesiana* is a serious weed in NZ (Bourdôt and Hurrell 1992). It has been recorded from France, Spain, Italy, Portugal and England in Europe (Verloove 2005; Stace 1997), South Africa (Gibbs Russell *et al.* 1985; Wells *et al.* 1986) and the USA (USDA ARS 2006). It is now widespread in Vic, NSW and the ACT with recent outbreaks identified in Qld, SA and Tas (Fig. 3) (Snell *et al.* 2007).

Once reproductive, Chilean needle grass becomes unpalatable and even at stocking rates as high as 300 dryland sheep equivalents (DSE) sheep will not graze it (Grech 2007). Rotational grazing of Chilean needle grass is useful as a management strategy to reduce Chilean needle grass seed production only if the infestation can be

Figure 3: Distribution of *Nassella neesiana* in Australia. Specimen data reproduced from Australia's Virtual Herbarium with permission of the Council of Heads of Australasian Herbaria Inc.

completely grazed in the two-week window between seedhead development and panicle emergence (Grech 2007). In particular, Chilean needle grass threatens the wool industry through contamination of wool, reductions in animal condition and physical damage from the sharp-pointed seeds (Fig. 4) penetrating the fleece, skins and eyes of livestock (Slay 2002). It has been estimated that Vic would save $82 million over 30 years in control costs and lost production if it could contain Chilean needle grass infestations to 2002 levels (estimated at 815 ha) (Morfe *et al.* 2003). Chilean needle grass is difficult to control and, according to Gardener and Sindel (1998) there is 'overwhelming evidence' that it is 'almost impossible to eradicate' because of the difficulty of killing mature plants, the size and longevity of the soil seed bank and the production of basal cleistogenes.

Chilean needle grass is also a significant environmental weed. It rated as the most significant weed threat to temperate grassland biodiversity in Australia (McLaren *et al.* 1998; Groves and Whalley 2002) and as 'the worst environmental weed threatening native grasslands' (Snell *et al.* 2007). However, Faithfull *et al.* (2009) found that major losses of native plant biodiversity in areas occupied by Chilean needle grass probably often preceded its invasion and may have been caused by other degrading processes such as *Themeda triandra* (R. Br) Stapf. dieback, overgrazing and major soil disturbance. Both serrated tussock and Chilean needle grass are WONS because of their significant agricultural and environmental impacts (Thorp and Lynch 2000).

Weedy grasses have only recently been targeted for biological control (Witt and McConnachie 2004). Previously they had not been considered good targets for biological control for a number of reasons, including: the great economic and ecological importance of related species; the simple chemical composition and morphology of grasses that may preclude any great degree of pathogen and herbivore speciation; and the great adaptability of grasses to grazing and harvesting (i.e. it was thought they would be able to recover from damage inflicted by biological control agents) (Palmer *et al.* 2008). To address these issues: host-specificity testing will be extensive to ensure safety of valued grasses; pathogens have been found which are specific; and pathogens have been found which are thought to be sufficiently damaging to reduce populations of the weed. Broad surveys of the host-range of insects have indicated that grass-feeding species are less specific than those of non-grass hosts and this has led to a recommendation that classical biological control using insects should not be used for grass weeds (Berneys 1985). However, more recently host-specific insects such as the gall forming wasp *Tetrames romana* Walker have been identified as potential biological agents for the grass *Arundo donax* (Goolsby and Moran 2009), and this wasp has now been released in the United States of America. The rapid rate of spread and unpalatability to stock of both serrated tussock and Chilean needle grass, and the difficulties in controlling them by chemical and cultural methods, has led to investigations into the possible use of pathogens for biological control of these two grasses in Australia and NZ.

BIOLOGICAL CONTROL HISTORY

Previous studies by South African scientists (Wells 1976) on serrated tussock in its native South American range indicated that insects and pathogens were able to suppress this plant, although the insects were found to be insufficiently specific for use for classical biological control. The Australian biological control program therefore decided to assess only pathogens as potential biological control agents of serrated tussock and Chilean needle grass. A preliminary survey was commissioned by CSIRO in 1998.

During three brief visits to Argentina, Harry Evans (CABI Bioscience) found a number of pathogens on serrated tussock, including some related to fungi that had been used successfully for biological control of other weed species (Briese and Evans 1998). His conclusions were that serrated tussock had a rich fungal flora in its native range and there was cause for optimism regarding their potential for biological control. However, given the difficulties of studying the pathogens under laboratory conditions outside their native range, he recommended that a long-term *in situ* study was required. This would provide a full inventory of the pathogens on the weed, determine their pathogenicity and epidemiology in the field, and clarify the biology and life-cycles of those showing sufficient specificity to warrant further investigation. A project in South America was developed with a research base identified in Bahía Blanca, Argentina during 1999.

Three pathogens looked promising for biological control of serrated tussock. These were the rust fungus *Puccinia nassellae*, a smut fungus *Tranzscheliella* sp. and a crown- and root-rotting fungus, *Corticiaceae* sp. Unfortunately these were either not host-specific, not virulent or too difficult to mass rear (see below for details). As a consequence, this project is now concentrating on biological control of Chilean needle grass. The rust *Uromyces pencanus* Arth. & Holw. (Anderson *et al.* 2010a) has been prioritised for assessment because preliminary investigations suggested it is virulent and host-specific.

Biological control has not been attempted against these grasses in any other country.

EXPLORATION

The evolutionary centre of the stipoid grass genus *Nassella* is South America, where it has diversified widely. Of the 98 species of *Nassella*, 90 can be found in South America (Torres 1997). Serrated tussock is a common pasture component in the Argentine pampas, an area of fertile soils in the central eastern part of the country used extensively for cropping. In the drier regions to the west and the south of the pampas grasslands, the distribution of serrated tussock becomes patchy, as it does in the more northern and damper areas (Fig. 5). The distribution of Chilean needle grass in South America is less restricted. As well as the pampas, it is found in the west from Tucuman province in northern Argentina to Chilean Patagonia. Therefore, South America was the logical place to begin a search for highly specific pathogens that had evolved with their host grasses, and the chances of finding such organisms were thought to be promising (Briese and Evans 1998). Detailed surveys for pathogens were carried out between 1994 and 2009 throughout a large part of the range of the two target weeds in Argentina by HC Evans and CA Ellison in 1994, HC Evans in 1996 and 1999, B Petitt and F Anderson from 1999 to 2001, F Anderson and P Martín from 2002 to 2005, F Anderson, M Díaz and A Flemmer from 2005 to 2006, F Anderson, J Barton and A Flemmer in 2006, F Anderson, A Flemmer and G Roth from 2007 to 2008 and F Anderson and G Roth in 2009.

During the surveys, serrated tussock was found to be most common in the region of Sierra de la Ventana, an outcrop of low mountains about 100 km north of Bahía Blanca, while large populations of Chilean needle grass were found to occur commonly around the Sierras de Córdoba in central western Argentina (Fig. 5).

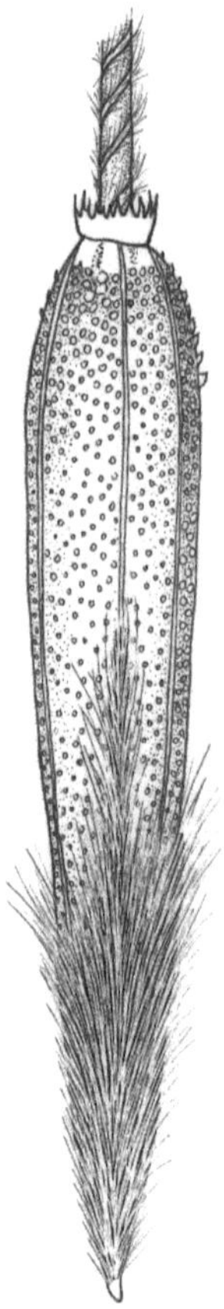

Figure 4: *Nassella neesiana* seed. Illustration: E Mayfield.

CANDIDATES

Puccinia nassellae **Arth. & Holw.** This rust has been recorded on both serrated tussock and Chilean needle grass. Levels of infection of *P. nassellae* in the field are highly dependent on environmental conditions, ranging from hardly detectable after prolonged dry periods to severe outbreaks that kill plants under favourable wet conditions. On serrated tussock, the most severe levels of infection have been recorded at shaded sites, where plants would likely experience longer dew periods than they would out in the open. On Chilean needle grass, levels of infection seem to be more uniform across plants located in shade and the open. Serrated tussock plants infected with *P. nassellae* show chlorotic bands on leaves, sometimes with necrotic centres. Uredinia on the adaxial side of blades commonly remain unexposed inside the tightly rolled leaves. In contrast, on Chilean needle grass, uredinia are obvious on the wide-open leaf blades.

After thorough field searches carried out for almost three years, covering most of serrated tussock's geographical range in Argentina, only the uredinial state of *P. nassellae* was found on that grass. However, both uredinia and telia are commonly formed on Chilean needle grass. Field observations did not provide conclusive evidence of how the rust completes its life-cycle in nature, but they indicated that it cycles as urediniospores on both hosts. No experimental work to elucidate the life-cycle was possible with the isolates collected from serrated tussock, as no telia were ever found on this host. For isolates from Chilean needle grass, teliospores were induced to form basidiospores in the laboratory. However, these basidiospores failed to infect Chilean needle grass or the two identified potential alternate hosts found in association with Chilean needle grass, *Clematis montevidensis* Spreng. and an unidentified *Morrenia* species, so that it was not possible to determine the role of basidiospores in the life-cycle of the rust. It is possible that *P. nassellae* behaves differently and completes its life-cycle in other habitats and/or on other hosts than the ones studied, but it would appear that on serrated tussock it has lost its telial stage and that on Chilean needle grass the role of telia remains unknown (Anderson *et al.* 2011).

P. nassellae ex serrated tussock has proved very difficult to mass rear. Inoculation procedures are very time-consuming as spores need to be applied directly on the adaxial side of leaves by unrolling the blades under the microscope. Other, more traditional inoculation methods (e.g. application of spores by misting) rendered very poor results.

Evidence indicates that the rust is highly specialised. The strains that attack serrated tussock are apparently different from those that attack Chilean needle grass: cross-inoculations of *P. nassellae* isolates between the two grass species failed (Anderson *et al.* 2002). Moreover, in artificial inoculation trials the *P. nassellae* strains from Chilean needle grass did not infect all tested accessions of Chilean needle grass, while all the tested strains that came from serrated tussock did attack all available accessions of serrated tussock.

In initial host-range tests, several grasses were inoculated with urediniospores collected from naturally infected serrated tussock plants from different sites in Argentina. The plants tested included several species of stipoid grasses native to Argentina (Chilean needle grass *Nassella clarazii* (Ball) Barkworth, *Nassella gynerioides* (Phil.) Barkworth, *Piptochaetium napostaense* (Speg.) Hack.) and the Australian species *Austrostipa scabra* (Lindl.) S.W.L. Jacobs & J. Everett. None of these showed signs of infection by *P. nassellae*, while all four accessions of serrated tussock tested became infected. However, the other Australian species tested, *Austrostipa aristiglumis* (F. Muell.) S.W.L. Jacobs & J. Everett, was found to be moderately susceptible to the two isolates tested of *P. nassellae* ex serrated tussock (11% of young plants and 8% of older ones tested developed a single normal pustule per plant compared to up to 12 pustules per plant on 62% of inoculated serrated tussock plants) (Anderson *et al.* 2002). In contrast, neither of the two Australian *Austrostipa* species became infected with either of the two tested isolates of *P. nassellae* ex Chilean needle grass, nor did any of the species tested that are native to Argentina (serrated tussock, *Nassella tenuissima* (Trin.) Barkworth, *Stipa brachychaeta* Godr., *Stipa clarazii* and *Poa ligularis* Nees ex Steud.) (Anderson *et al.* 2004, 2005). In another test, plants from six Australian accessions of Chilean needle grass were inoculated with three *P. nassellae* isolates ex Chilean needle grass. Plants from the ACT, Laverton (Vic) and Truganina (Vic) were found to be susceptible to one of these isolates originating from Bahía Blanca in Argentina (PN 27). No infection was achieved with the other two tested isolates.

Ultimately, the *P. nassellae* isolates from serrated tussock were rejected as potential agents for that weed because they were so difficult to manipulate in the laboratory, they lacked specificity and they required very specific environmental conditions in order to severely damage plants in the field. No final decision has been made with respect to the *P. nassellae* isolates from Chilean needle grass, but these may prove too specific to be effective.

***Tranzscheliella* spp.** Both serrated tussock and Chilean needle grass have been found infected by two distinct species of smut, whose identity remains to be confirmed. The smut on Chilean needle grass is most likely *Tranzscheliella hypodytes* (Schltdl.) Vánky & McKenzie (formerly *Ustilago hypodytes sensu lato* (Vánky and McKenzie 2002)), but the one on serrated tussock would appear to be a new, hitherto undescribed species of *Tranzscheliella* (Marcin Piatek, *pers. comm.* 2006).

The disease symptoms caused by the smut fungi in the field are different on each host. Serrated tussock has its inflorescences completely replaced by smutted heads. In Chilean needle grass, smut spores cover upper internodes of culms, occasionally allowing the formation of some seed.

It has been proved that infection occurs at the time of germination of the host plants, though infection rates under experimental conditions have been extremely low (see below). Field observations indicate that the incidence of disease on both serrated tussock and Chilean needle grass tends to be low. For serrated tussock, although the smut occurred at most sites surveyed, it was usually present on only a few isolated plants at each site. Exceptions were two sites, both located near Villa Ventana around 100 km north-east of Bahía Blanca, where large sections of serrated tussock populations showed high levels of infection and varying levels of inflorescence replacement. At these two sites, many plants failed to produce inflorescences bearing viable seed. For Chilean needle grass, high levels of infection and inflorescence replacement were observed at only one site close to Los Reartes, in Córdoba province. At this same site, a large population of neighbouring serrated tussock plants exhibited no signs of smut infection whatsoever, suggesting that it was not susceptible to the smut isolate infecting the Chilean needle grass.

Several species of stipoid grasses were tested for their susceptibility to different isolates of the smut collected from naturally infected serrated tussock plants. Plants tested were Australian and Argentinean accessions of serrated tussock and Chilean needle grass, *N. tenuis* (Phil.) Barkworth, *N. tenuissima*, *P. napostaense*, *S. clarazii* and *S. gynerioides* from Argentina and *A. scabra* from Australia. Of these, only *N. tenuis*, *N. tenuissima* and an Australian accession of serrated tussock were found to be susceptible. However, positive control plants of serrated tussock included in these tests showed either no or little infection, so these results are by no means conclusive (Anderson *et al.* 2002, 2004). Large-scale inoculation experiments in which *Tranzscheliella* sp. was applied to serrated tussock yielded very low rates of infection (<1%). This result precludes it as a potential biological control agent, at least until more effective inoculation techniques can be developed. The *Tranzscheliella* sp. that occurs on Chilean needle grass has not been examined for virulence or host-specificity.

Corticiaceae. Some patches of serrated tussock were found showing severe dieback leading to desiccation and death of affected tussocks (surveys by HC Evans and CA Ellison in 1995 and HC Evans in 1996). When pulled from the ground, infected plants showed root and crown necrosis and a white mycelial mat at the ground level on dead leaves. Fine hyphae, ~2 μm thick with clamp connections (typical of Basidiomycota fungi), were sometimes associated with much thicker (~5 μm) hyaline *Rhizoctonia*-like hyphae on which no clamp connections could be found. Whether both types of mycelia correspond to the same fungus remains to be clarified. White-brownish crust-like structures on which basidiospores were formed were associated with these mycelia. All attempts to isolate the Basidiomycete fungus on artificial media have failed. Instead, *Fusarium* sp. has been repeatedly isolated from affected plants. When the fungus was found by Evans and Ellison in 1995 it was tentatively placed in the family Corticiaceae because at that time, this family was characterised by the presence of resupinate basidiomata and a distinctive anatomical feature in the form of clamp connections on hyphae. Since then there have been changes in the taxonomy of this group of fungi; the Corticiaceae have been redefined (Cannon and Kirk 2007) and restricted to a group in which clamp connections are mostly absent. So, according to modern taxonomy, this fungus may not in fact belong in that group.

To attempt to reproduce symptoms observed in the field on serrated tussock and determine if non-target grass species could also become infected, trials were set up in both the field and the glasshouse (Anderson *et al.* 2002). In the field, healthy plants of several different grass species, including serrated tussock, were transplanted into a patch of naturally infected plants. In the glasshouse, infected plants from the field were transplanted into a large container planted with healthy plants of serrated tussock and other non-target grass species (Anderson *et al.* 2002). Results of both trials were assessed after six months. None of the tested non-target grass species became infected, but nor did any of the *N. trichotoma* plants included as positive controls become infected under the given conditions. Thus it was not possible to draw any conclusions on the specificity of this pathogen. This, combined with the fact that it could not be identified even to the family level, makes it unsuitable as a potential biological control agent.

***Uromyces pencanus* Arth. & Holw.** This rust fungus was observed during field surveys causing severe infections and killing foliage of Chilean needle grass, especially under hot dry conditions (Figs 5 and 6). Seed production was greatly reduced in rust-infected plants, compared with healthy plants at the same site. This was noticed both at sites where only *U. pencanus* was present and when mixed infections with *P. graminella* (see below) occurred (Anderson *et al.* 2010a). In an experiment carried out in the laboratory under controlled conditions, all Chilean needle grass plants grown from seed collected at site 27 in Bahía Blanca, Argentina and 70% of Australian plants (ex ACT) were infected by an accession of *U. pencanus* from site 27 (Giordano *et al.* 2009). Inoculated plants from Argentina grew significantly less than uninoculated plants of the same accession and Australian plants produced fewer leaves and had a lower dry weight of living material than did uninoculated plants.

Methods have been successfully developed to maintain and multiply a pure culture of *U. pencanus* in the glasshouse and to store urediniospores (Anderson *et al.* 2010a). Storage in the freezer at –70°C proved the best method to keep spores viable in the long term, with high survival for at least 12 months.

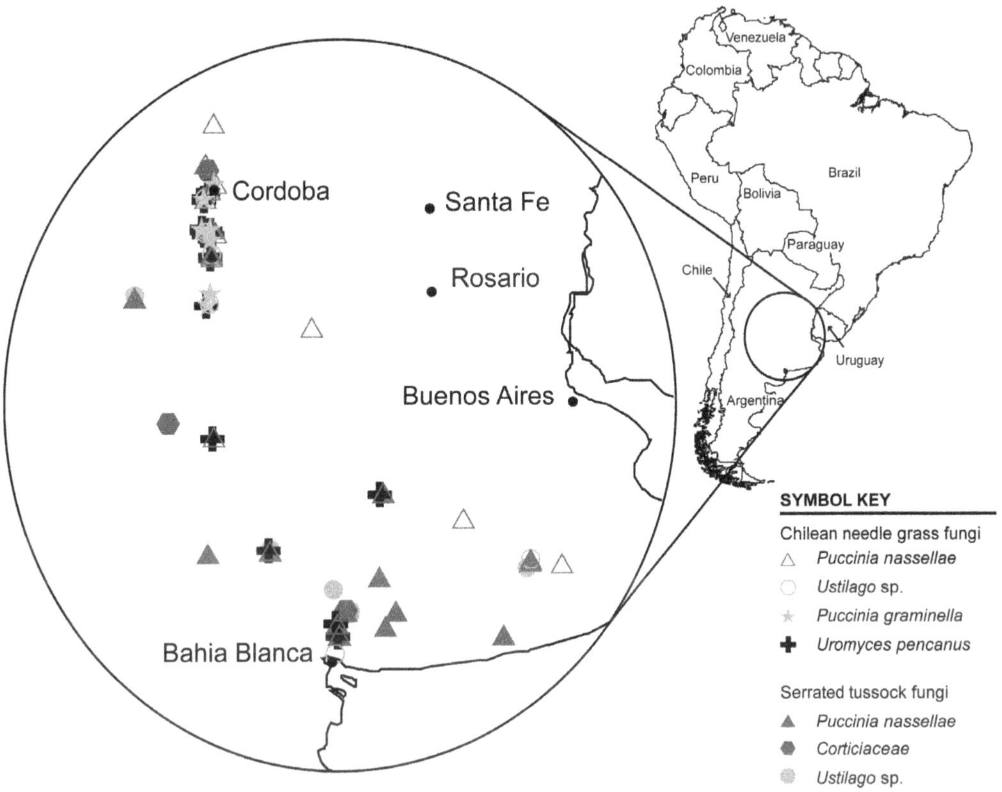

Figure 5: Locations of *Nassella trichotoma* and *Nassella neesiana* fungi from surveys in Argentina.

U. pencanus is autoecious according to the literature (Arthur 1925; Greene and Cummins 1958; Cummins 1971; Lindquist 1982), with aecia, uredinia and telia on its grass host. However, we have not observed aecia on Chilean needle grass at any of the sites where *U. pencanus* was the only rust present. The only sites where aecia were observed were those where another rust, *P. graminella*, was also found to be present. In addition, teliospores of *U. pencanus* did not germinate and produce basidiospores under any of the treatments tested in the laboratory to promote germination (Anderson *et al.* 2011). It thus appears that the rust is capable of continuous cycling in the form of urediniospores on Chilean needle grass. The role of teliospores in the completion of the life-cycle remains unclear. The infection process of *U. pencanus* urediniospores on Chilean needle grass was investigated and found to be similar to that of most other rusts (Flemmer *et al.* 2010).

Based on the literature, *U. pencanus* appears to have a narrow host-range, confined to the genus *Nassella* (Arthur 1925; Cummins 1971; Lindquist 1982; Barkworth and Torres 2001). Inoculation of many plant species within the Poaceae is under way, to test their susceptibility to *U. pencanus* isolate UP27 collected at Bahía Blanca, Argentina.

Figure 6: The rust fungus *Uromyces pencanus* telia infecting *Nassella neesiana*. Photo: P Martin, Universidad Nacional del Sur, Argentina.

Isolate UP27 has been selected on the basis of its pathogenicity towards most Australian accessions of Chilean needle grass tested. The results so far are promising, in that no uredinia have developed on any test species other than the target species Chilean needle grass. Development of intercellular mycelium and a few haustoria have been observed within leaves of several *Austrostipa* species and *Piptatherum miliaceum*. However, signs of plant resistance (e.g. thickening of cell walls upon hyphal contact) were also observed within sections of the same samples, probably explaining why the rust did not develop further to form uredinia on these species (Anderson *et al.* 2010b).

Puccinia graminella **Diet. & Holw.** This rust occurred only on Chilean needle grass and was seldom encountered alone during the surveys, being mostly in mixed infections with other rusts, namely *U. pencanus* and *P. nassellae*. Notwithstanding, on the few occasions it was found alone, it was observed severely damaging individual plants by causing premature senescence of foliage. For unknown reasons, this rust used to be more common around Bahía Blanca not only on Chilean needle grass but also on *N. tenuis* and *N. hyalina* (Nees) Barkworth (Briese *et al.* 2000), but has recently become rare in the region (F Anderson, unpub. data).

Both aecia and telia of *P. graminella* have been recorded on Chilean needle grass, their relative abundance varying with the time of year. It has been shown that infection by aeciospores of this rust produce new aecia and not uredinia as in other rusts. Therefore, aeciospores should be referred to as aecidioid urediniospores (Kirk *et al.* 2008). It was also found that teliospores do not have a dormant period and germinate readily after formation, producing typical basidiospores, which in turn also germinate readily on water agar. It was expected that these basidiospores would infect Chilean needle grass plants and give rise to new aecia, but no infection by basidiospores was achieved under our experimental conditions. With the evidence collected to date, it appears *P. graminella* cycles as aecidioid urediniospores; the roles of the teliospores and basidiospores in its life-cycle remain unknown (Anderson *et al.* 2011).

According to the literature, *P. graminella* infects *Nassella*, *Stipa* and *Piptochaetium* species (Cummins 1971; Lindquist 1982; Mujica and Vergara 1945). The occurrence of high numbers of resistant Chilean needle grass individuals among plant accessions used in inoculation experiments (Anderson *et al.* 2008, 2011) did not allow the establishment of a pure culture in the laboratory. This precluded the performance of comprehensive host-specificity tests. Nevertheless, preliminary tests on Chilean needle grass plants from Australia (ACT accession) revealed that only a small percentage (15%) became infected.

DISCUSSION

To be an effective biological control agent, an organism must possess two key characteristics: it must be able to damage the target weed sufficiently to cause population level effects, either alone or in combination with other agents; and it must possess sufficient host-specificity to preclude any unacceptable risk to non-target plant species. Unfortunately, the potential biological control candidates identified for serrated tussock were either not host-specific (*P. nassellae*) or not sufficiently pathogenic to Australian accessions of the weed (*P. nassellae*, *Tranzscheliella* spp.), or their biology and life-cycle could not be fully determined (*P. nassellae*, *Tranzscheliella* spp., *Corticiaceae* sp.), precluding further work.

The apparent inability of strains of *P. nassellae* from serrated tussock and Chilean needle grass to cross-infect each other's host plant suggest they could be considered pseudo separate species. Although work on the *P. nassellae* strains infecting Chilean needle grass has been less intense to date, these isolates appear to be more damaging and specific than those on serrated tussock. Moreover, the formation of telia on this host has allowed the study of the biology of the rust to proceed further. However, the confirmation of its life-cycle remains elusive. Inoculation trials using basidiospores as inocula have failed to provide conclusive results. While further host-specificity tests are needed, evidence collected to date indicates *P. nassellae* strains from Chilean needle grass are specific to their host. Indeed, they may be too specific: results in Argentina vary according to the geographic origin of the isolates and the test plants. An isolate has yet to be identified (and may not exist) that is capable of infecting most, if not all, Australian and NZ accessions of the weed.

Unfortunately, experimentation involving *P. graminella* had to be interrupted – for unknown reasons, all cultures of this rust were lost in the laboratory and recent drought conditions have resulted in the rust not being found again in the field in the region around Bahía Blanca.

U. pencanus appears to be an excellent candidate for the biological control of Chilean needle grass. It has been found causing severe damage to Chilean needle grass in the field, its spores can be frozen and stored for long durations without significant loss of viability and to date it has proved to be very host-specific. Authorities in Australia have previously approved the release of biocontrol agents despite an incomplete knowledge of their life-cycle, as it was concluded the benefits of such a release outweighed the potential risks (Ellison *et al.* 2006; Evans 1993).

In conclusion, to date it has not been possible to find a pathogen with good potential as a biocontrol agent for serrated tussock, but the rust *U. pencanus* appears to be a very promising agent for Chilean needle grass. An isolate of *U. pencanus* is presently undergoing host-range testing on many species in the Poaceae. NZ has fewer native stipoid grass species than Australia and has therefore been successful in obtaining permission to release *U. pencanus.* Upon completion of host-specificity testing, relevant authorities in Australia will have to decide whether the potential benefits of such a release outweigh the slight risk that the rust could complete its life-cycle by means of a hypothetical alternate host in the new environment. To our knowledge, this is the first pathogen biological control agent in the world approved for release on a grass.

ACKNOWLEDGEMENTS

CERZOS-CONICET Bahía Blanca is acknowledged for hosting the project in Argentina. Eduardo Botto is warmly thanked for his assistance and support during work performed at INTA Castelar in Buenos Aires.

REFERENCES

Agrios GN (1997) *Plant Pathology.* 4th edn. Academic Press, California.

Anderson F, Pettit W, Briese DT and McLaren DA (2002) Biological control of serrated tussock and Chilean needlegrass. *Plant Protection Quarterly* **17**, 104–111.

Anderson F, Pettit W, Morin L, Briese DT and McLaren DA (2004) Pathogens for the biological control of weedy stipoid grasses in Australia: completion of investigations in Argentina. In *Proceedings of the 11th International Symposium on Biological Control of Weeds.* 27 April–2 May 2003, Canberra. (Eds JM Cullen, DT Briese, DJ Kriticos, WM Lonsdale, L Morin and JK Scott) pp. 69–74. CSIRO Entomology, Canberra.

Anderson FE, Diaz ML and McLaren DA (2005) Biological control agents for *Nassella neesiana* (Poaceae) invading Australian native grasslands. In *20th International Grassland Congress: Offered Papers.* 26 June–1 July 2005, Dublin. (Eds FP O'Mara, RJ Wilkins, LT Mannetje, DK Lovett, PAM Rogers and TM Boland) p. 411. Wageningen Academic, Ireland and UK.

Anderson FE, Flemmer AC, Hansen PV, Barton J, Díaz ML and McLaren DA (2008) An update on progress towards potential biological control of *Nassella neesiana* in Australia and New Zealand. In *Proceedings of the 16th Australian Weeds Conference.* 18–22 May 2008, Cairns. (Eds RD Van Klinken, VA Osten, FD Panetta and JC Scanlan) pp. 237–239. Queensland Weeds Society, Brisbane.

Anderson FE, Barton J and McLaren DA (2010a) Studies to assess the suitability of *Uromyces pencanus* as a biological control agent for *Nassella neesiana* (Poaceae) in Australia and New Zealand. *Australasian Plant Pathology* **39**, 69–78.

Anderson FE, Gallego L, Roth G, Botto E, McLaren DA and Barton J (2010b) Investigations into biological control of *Nassella neesiana* in Australia and New Zealand. *Proceedings of the 17th Australasian Weeds Conference.* 26–30 September 2010, Christchurch. (Ed SM Zydenbos) pp. 215–218. New Zealand Plant Protection Society, Christchurch.

Anderson FE, Díaz L, Barton J, Flemmer AC, Hansen PV and McLaren DA (2011) Exploring the life cycles of three South American rusts that have potential as biocontrol agents of the stipoid grass *Nassella neesiana* in Australasia. *Fungal Biology* **115**, 370–380.

Arthur JC (1925) The grass rusts of South America; based on the Holway collections. *Proceedings of the American Philosophical Society* **64**, 131–223.

Barkworth ME and Torres MA (2001) Distribution and diagnostic characters of *Nassella* (*Poaceae:Stipeae*). *Taxon* **50**, 439–468.

Berneys EA (1985) Arthropods for weed control in IPM systems. In *Biological Control in Agricultural IPM Systems.* (Eds MA Hoy and DC Herzog) pp. 373–391. Academic Press, California.

Bourdôt GW and Hurrell GA (1992) Aspects of the ecology of *Stipa neesiana* Trin. and Rupr. seeds. *New Zealand Journal of Agricultural Research* **35**, 101–108.

Briese DT and Evans HC (1998) Biological control of serrated tussock (*Nassella trichotoma*): is it worth pursuing? *Plant Protection Quarterly* **13**, 94–97.

Briese DT, McLaren DA, Pettit W, Zapater M, Anderson F, Delhey R and Distel R (2000) New biological control initiatives against weeds of South American origin in Australia: blue heliotrope and serrated tussock. In *Proceedings of the 10th International Symposium on Biological Control of Weeds.* 4–14 July, Bozeman, Montana. (Ed NR Spencer) pp. 215–223. US Dept of Agriculture, Agricultural Research Services, Sidney, MT and Montana State University, Bozeman.

Campbell MH and Vere DT (1995) *Nassella trichotoma* (Nees) Arech. In *The Biology of Australian Weeds. Vol. 1.* (Eds RH Groves, RCH Shepherd and RG Richardson) pp. 189–202. RG and FJ Richardson, Melbourne.

Cannon PF and Kirk PM (2007) *Fungal Families of the World.* CABI, Wallingford, UK.

Carr GW, Yugovic JV and Robinson KE (1992) *Environmental Weed Invasion in Victoria. Conservation and Management Implications.* Vic Dept of Conservation and Environment/Ecological Horticulture Pty Ltd, Melbourne.

Cummins GB (1971) *The Rust Fungi of Cereals, Grasses and Bamboos.* Springer-Verlag, Berlin.

Ellison CA, Pereira JM, Thomas SE, Barreto RW and Evans HC (2006) Studies on the rust *Prospodium tuberculatum*, a new classical biological control agent released against the invasive alien weed *Lantana camara* in Australia. 1. Life-cycle and infection parameters. *Australasian Plant Pathology* **35**, 309–319.

Evans HC (1993) Studies on the rust *Maravalia cryptostegiae*, a potential biological control agent for rubber-vine weed (*Cryptostegia grandiflora*, Asclepiadaceae: Periplocoideae) in Australia, I: Life cycle. *Mycopathologia* **124**, 164–174.

Faithfull I, Hocking C and McLaren DA (2009) Impacts of Chilean needle grass *Nassella neesiana* on native plant diversity in Australia's indigenous grasslands. In *Proceedings of the 4th Victorian Weeds Conference: Plants Behaving Badly: In Agriculture and the Environment.* pp. 16–25. Weed Society of Victoria, Melbourne.

Flemmer A, Anderson FE, Hansen PV and McLaren DA (2010) Microscopic observations of a compatible host/pathogen interaction between a potential biocontrol agent (*Uromyces pencanus*) and its target weed (*Nassella neesiana*). *Mycoscience* **51**, 396–400.

Gardener MR and Sindel BM (1998) The biology of *Nassella* and *Achnatherum* species naturalized in Australia and the implications for management on conservation lands. *Plant Protection Quarterly* **13**, 76–79.

Gibbs Russell GE, Reid C, Van Rooy J and Smook L (1985) List of species of Southern African plants. Edition 2. Recent literature and synonyms. Part 1. Cryptogams, Gymnosperms, Monocotyledons. *Memoirs of the Botanical Survey of South Africa* **51**, 1–152.

Giordano L, Anderson FE and McLaren DA (2009) Efectos de la infección por la roya *Uromyces pencanus* sobre el vigor de plantas de *Nassella neesiana* (*Poaceae*). *Resúmenes de las XIII Jornadas Fitosanitarias Argentinas.* Termas de Río Hondo, Santiago del Estero, 30 de septiembre–2 de Octubre. E045.

Goolsby JA and Moran P (2009) Host range of *Tetramesa romana* Walker (Hymenoptera: Eurytomidae), a potential biological control of giant reed, *Arundo donax* L. in North America. *Biocontrol* **49**, 160–168.

Grech C (2007) Grazing management for the long term utilisation and control of Chilean needle grass (*Nassella neesiana*). PhD thesis. University of New England, Armidale, NSW.

Greene HC and Cummins GB (1958) A synopsis of the uredinales which parasitize grasses of the genera *Stipa* and *Nassella*. *Mycologia* **50**, 6–35.

Groves RH and Whalley RDB (2002) Grass and grassland ecology in Australia. In *Flora of Australia. Vol. 43. Poaceae 1. Introduction and Atlas.* (Eds K Mallett and AE Orchard) pp. 157–182. Australian Biological Resources Survey/CSIRO, Melbourne.

Jones RE and Vere DT (1998). The economics of serrated tussock in New South Wales. *Plant Protection Quarterly* **13**, 70–76.

Kirk PM, Cannon PF, Minter DW and Stalpers JA (Eds) (2008) *Dictionary of the Fungi.* 10th edn. CABI, Wallingford, UK.

Lindquist JC (1982) *Royas de la República Argentina y zonas limítrofes.* INTA, Buenos Aires.

McLaren DA, Stajsic V and Gardener MR (1998) The distribution and impact of South/North American stipoid grasses (Poaceae: Stipeae) in Australia. *Plant Protection Quarterly* **13**, 62–70.

McLaren DA, Merton E, Pritchard GS, Grech C and Bonilla J (2008) Evaluating the extent of serrated tussock (*Nassella trichotoma*) resistance to the herbicide, flupropanate in Australia. In *Proceedings of the 16th Australian Weeds Conference.* (Eds RD van Klinken, VA Osten, FD Panetta and JC Scanlan) pp. 94–96, Queensland Weeds Society, Brisbane.

Morfe TA, McLaren DA and Weiss J (2003) Treatment of Chilean needle grass (*Nassella neesiana*) in Victoria: a benefit-cost analysis. In *Proceedings II. 19th Asian-Pacific Weed Science Society Conference.* 17–21 March, Manila. pp. 875–881. Weed Science Society of the Philippines.

Mujica F and Vergara C (1945) *Flora fungosa chilena. Indice preliminar de los huéspedes de los hongos chilenos y sus referencias bibliográficas.* Imprenta, Stanley.

Nicholson C, Patterson A and Miller L (1997) 'The cost of *Nassella trichotoma* control in central western Victoria'. Unpublished report prepared for the Victorian Serrated Tussock Working Group.

Osmond R, Veebeek M, McLaren DA, Michelmore M, Wicks B, Grech CJ and Fullerton P (2008) *Serrated Tussock – National Best Practice Manual.* Vic Dept of Primary Industries, Melbourne.

Palmer WA, Yobo KS and Witt ABR (2008) Prospects for the biological control of weedy Sporobolus grasses in Australia. In *Proceedings of the 16th Australian Weeds Conference.* 18–22 May 2008, Cairns. (Eds RD van Klinken, VA Osten, FD Panetta and JC Scanlan) pp. 246–266. Queensland Weeds Society, Brisbane.

Parodi LR (1930) Ensayo fitogeográfico sobre el partido de Pergamino. Revista de la Facultad de Agronomía y Veterinaria. *Entrega* **1**, 65–289.

Parsons WT and Cuthbertson EG (1992) *Noxious Weeds of Australia.* Inkata Press, Melbourne.

Rosengurtt B, Arrillaga De Maffei BR and Izaguirre De Artucio P (1970) *Gramineas Uruguayas.* Universidad de la Republica. Departmento de publicaciones, coleccion ciencias 5. Montevideo.

Slay M (2002) *Chilean Needle Grass. A Guide to Identification and Management in Hawke's Bay.* Hawke's Bay Regional Council, Cliff Press Printers.

Snell K, Grech C and Davies J (2007) *National Best Practice Management Manual: Chilean Needle Grass. Defeating the Weed Menace.* Vic Dept of Primary Industries/ National Chilean Needle Grass Taskforce, Melbourne.

Stace CA (1997) *New Flora of the British Isles.* 2nd edn. Cambridge UniversityPress, Cambridge.

Thorp JR and Lynch R (2000) *The Determination of Weeds of National Significance.* National Weeds Strategy Executive Committee, Launceston.

Torres MA (1997) '*Nassella (Gramineae) del noroeste de la Argentina. Stipa (Gramineae) del noroeste de la Argentina. Nicoraella (Gramineae) un nuevo género para América del Sur.*' Comisión de Investigaciones Científicas, Monografia 13, La Plata., Ministerio de la Producción y el Empleo, Provincia de Buenos Aires.

USDA ARS (US Dept of Agriculture, Agricultural Research Service) (2006) Germplasm Resources Information Network (GRIN) [online database]. National Germplasm Resources Laboratory, Beltsville, Maryland. <http://www.ars-grin.gov/cgi-bin/npgs/html/taxon.pl?402089>.

Vánky K and McKenzie EHC (2002) *Smut Fungi of New Zealand. Fungi of New Zealand. Vol. 2.* Fungal Diversity Press, Hong Kong.

Verloove F (2005) A synopsis of *Jarava* Ruiz and Pav. and *Nassella* E. Desv. (*Stipa* L. *s.l.*) (Poaceae: Stipeae) in southwestern Europe. *Candollea* **60**, 97–117.

Walsh NG and Entwisle TJ (1994) *Flora of Victoria. Vol. 2. Ferns and Allied Plants, Conifers and Monocotyledons.* Inkata Press, Melbourne.

Wells MJ (1976) Progress with research on *Nassella* tussock. In *Proceedings of the 2nd National Weeds Conference of South Africa*, pp. 48–55. Balkema, Cape Town.

Wells MJ, Balsinhas AA, Joffe H, Engelbrecht VM, Harding G and Stirton CH (1986) A catalogue of problem plants in southern Africa incorporating the National Weed List of Southern Africa. *Memoirs of the Botanical Survey of South Africa* **53.**

Witt ABR and McConnachie AJ (2004) The potential for classical biological control of invasive grass species with special reference to invasive *Sporobolus* spp. (Poaceae) in Australia. In *Proceedings of the 11th International Symposium on Biological Control of Weeds.* 27 April–2 May 2003, Canberra. (Eds JM Cullen, DT Briese, DJ Kriticos, WM Lonsdale, L Morin and JK Scott) pp. 198–202. CSIRO Entomology, Canberra.

Onopordum acanthium L. – Scotch thistle *Onopordum illyricum* L. – Illyrian thistle hybrids

DT Briese

ABSTRACT

Onopordum acanthium and *Onopordum illyricum* are monocarpic biennial or facultative perennial thistles originating in Europe. Since their introduction in the 19th century, these two species and their hybrids have become serious pasture weeds of south-eastern Australia. This chapter describes a biological control program carried out against them from 1987 to 2006. Studies were carried out on the population dynamics of *Onopordum* spp. in Australia and Europe, agent exploration undertaken in the Mediterranean region of Europe and experiments carried out in Europe to determine the potential impact of candidate agents. From these, a strategy for biological control was developed, based on target weed and agent ecology. This led to the release of seven control agents in Australia and the establishment of four of these. Two agents, the seed-weevil *Larinus latus* and the stem-boring weevil *Lixus cardui*, are now widely established and making substantial measurable reductions to seed production and, in the latter case, plant vigour of the target weeds. The rosette-attacking crown weevil *Trichosirocalus briesei* and crown moth *Eublemma amoena* are established but remain much more restricted in range. Damage to individual plants has been observed, but agent impact at the population level remains unknown. While economic assessments predict a high return on investment from this biological control project, further field evaluation will be needed to determine the extent to which this has been achieved.

Key words: biological control, Mediterranean Europe, *Larinus latus*, *Lixus cardui*, *Trichosirocalus briesei*, *Eublemma amoena*, *Tephritis postica*, *Botanophila spinosa*, *Tettigometra sulphurea*, *Urophora terebrans*.

INTRODUCTION

Thistles of the genus *Onopordum* (Asteraceae: Cardueae) were introduced into Australia during the 19th century as ornamental plants (Parsons and Cuthbertson 1992). Scotch thistle *Onopordum acanthium* L. (Fig. 1), native to continental Europe, is a monocarpic biennial which germinates to produce a flat rosette of deeply indented spine-tipped leaves covered with fine cottony hairs. Rosettes require vernalisation in winter, before producing a bolting spiny-winged stem, up to 2 m high, in spring. The purple-flowering globular capitula (Fig. 2) are borne on multibranched side branches in summer and are protected by elongate spined bracts. After flowering and seed-set, the plant dies. Illyrian thistle *Onopordum illyricum* L. (Fig. 1), native to the Mediterranean basin, is also monocarpic, but is a facultative perennial that has a minimum size requirement in addition to vernalisation to induce bolting and flowering. It is distinguished from *O. acanthium* by reduced leaf indentation and hairiness, reduced branching, larger capitula and broader recurved capitulum bracts. Hybrids of these species, which are more common in Australia than the parental forms, have intermediate characteristics, both morphological and ecological (O'Hanlon *et al.* 1999).

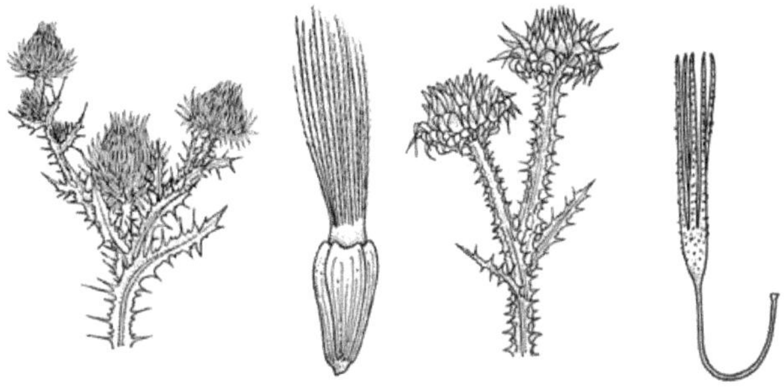

Figure 1: Capitula and floret of *Onopordum illyricum* (left) and capitula and seed of *Onopordum illyricum* (right). Illustrations: Royal Botanic Gardens, Sydney.

Onopordum thistles soon became naturalised, but remained localised until the onset of widespread pasture improvement programs in the latter half of the 20th century. The resultant increase in soil fertility led to an expansion in the range of these nitrophilous plants to become some of the most serious broad-leaf weeds of higher-rainfall grazing lands in parts of south-eastern Australia (Figs 3 and 4). Infestations cover close to 1 million ha (Briese *et al.* 1990). Chemical herbicides have generally not succeeded in reducing *Onopordum* infestations and need to be used on a continuing basis (Briese 1996a).

Figure 2: Flowering plant of *Onopordum acanthium x illyricum* hybrid. Photo: CSIRO.

Figure 3: An infestation of *Onopordum* sp. Photo: CSIRO.

PLANT TAXONOMY

The genus *Onopordum* belongs to the subtribe Carduinae, the largest subtribe of the *Cardueae* (Family Asteraceae). The Carduinae comprise about 1600 species, many of which have ecological and economic significance, especially as weeds (Häffner 2000). *Onopordum* has a Eurasian distribution; its centre of origin seems to be central-western Asia, where it shows the greatest diversity, but the range of *Onopordum* species extends west into Mediterranean Europe and North Africa, north into European Russia and east as far as Mongolia and north-western China. There are some 50 species, of which eight have become weedy following introduction into other parts of the world. Four of these, *O. acanthium*, *O. illyricum*, *Onopordum acaulon* L. and *Onopordum tauricum* Willd., occur in Australia. Michael (1996) considered the taxonomy of these introduced species, particularly *O. acanthium* and *O. illyricum*, to be quite complex owing to the existence of intermediate forms. Pure *O. acanthium* populations occur in southern NSW, Vic and Tas, while there are only two recorded localities with pure populations of *O. illyricum*, one in central Vic and one in northern NSW. Recently, molecular studies by O'Hanlon *et al.* (1999) confirmed that most field populations in the northern part of their range are of hybrid origin, with varying levels of genetic contribution from the two parent species. For biological control purposes, these thistles are therefore best considered as a hybrid complex.

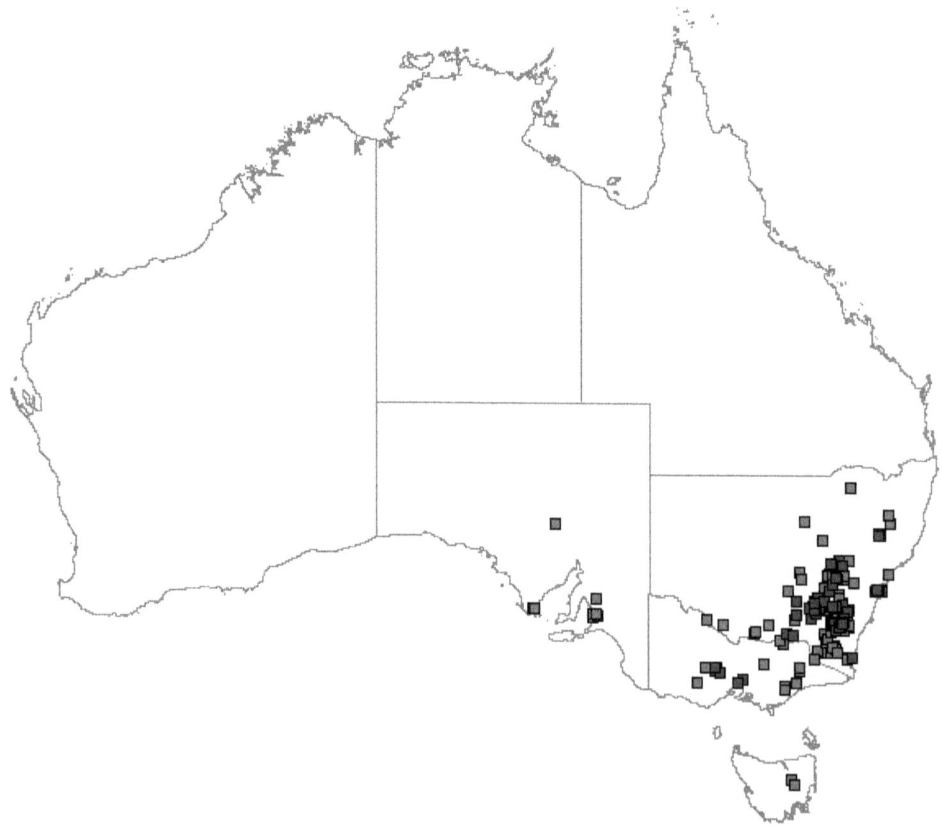

Figure 4: Distribution of *Onopordum acanthium* (red dots), *Onopordum illyricum* and hybrid forms (blue dots) in Australia. Specimen data and map with permission of the Council of Heads of Australasian Herbaria Inc.

BIOLOGICAL CONTROL HISTORY

CSIRO Entomology first looked at the possibility of biological control in 1985 as a possible follow-up to existing projects on other species of carduine thistle in Australia. The distribution of infestations was mapped (Briese *et al.* 1990) and a survey of insects that had colonised the thistles in their Australian range was carried out (Briese 1989). The latter study revealed an impoverished herbivore fauna associated with *Onopordum* spp. in Australia with distinct gaps in feeding niches, compared to that known for these thistles in their native range (Zwölfer 1965).

Research on the biological control of these thistles commenced in 1987, based at the CSIRO laboratory in Montpellier, France (Briese 1990). Southern Europe, in particular the Mediterranean basin, comprises that part of the native range of *Onopordum* thistles most similar to areas infested in Australia. From 1988 to 1992, surveys were made of populations of the two target weeds and several other *Onopordum* species in selected parts of Spain, France, Italy and Greece (Briese *et al.* 1994).

Field and cage experiments were conducted on agents of interest to measure their potential impact on the target weeds during the latter stages of that period, and continued until 1997 (Briese *et al.* 2002a; Briese 2006). These experiments were coupled with preliminary host-specificity studies (Briese *et al.* 1995, 2002b) and molecular studies to accurately identify candidate agents (Michalakis *et al.* 1992; Briese *et al.* 1996).

Parallel studies on the demography of *Onopordum* spp. were begun in Australia in 1991 (Pettit *et al.* 1996) and continued for 10 years, to complement earlier studies of *Onopordum* population dynamics (Cavers *et al.* 1995). Knowledge of the ecology of the target weeds in Australia was used in conjunction with the information obtained on potential agent impact to develop a strategy for *Onopordum* biological control (see below).

Formal host-specificity testing was carried out in quarantine in Canberra from 1990 to 2000, leading to eight candidate agents being approved for release (Briese *et al.* 2002c). Seven of these insects were released between 1991 and 2002 and four established (Briese 2006). Redistribution of those four agents was carried out until 2006, in collaboration with community and landholder groups, under a pasture biocontrol agent program within the

Cooperative Research Centre for Australian Weed Management (Swirepik *et al.* 2004).

Some field evaluation work was done at the same time as redistribution, which suggested that at least two of the agents were responsible for a decrease in plant vigour and for over 80% reduction in seed production in *Onopordum* infestations (Swirepik *et al.* 2008). Anecdotal reports also suggest that landholders consider the agents to have reduced the problems associated with *Onopordum* thistles. *Onopordum* thistles were included in small-scale experiments which looked at integrating biological control with herbicide use and other forms of pasture weed management (Huwer *et al.* 2005), but the results were inconclusive. Whether the longer-term studies required to demonstrate impact on a regional scale and to quantify the contribution of biological control can be done is problematic, as the project funding was terminated in 2006, following the completion of the redistribution work.

Strategy for the biological control of Onopordum spp.

Studies of *Onopordum* population dynamics (Cavers *et al.* 1995; Pettit *et al.* 1996) showed that infestations were characterised by a very large, long-lived soil seed reserve, which was maintained by the absence of insects that could reduce seed production (Briese 1989; Leigh *et al.* 1993). The studies also identified the transitional stages in the life-cycle of the weed that were most vulnerable to agent impact and a strategy was developed to target these stages (Briese *et al.* 2002b).

The strategy had both long- and short-term aims; in the longer term it was considered essential to reduce soil seed reserves to levels at which emergent plant densities were not economically damaging, and in the shorter term it aimed to minimise the impact of plants that did emerge during the period of seed bank decline. From more than 120 insect species found to feed on *Onopordum* spp. in their native European range, six were selected to form a complementary herbivore guild, targeting the small rosette to large rosette, stem elongation to flowering, and seed production to soil seed reserve transitions of their host plant.

EXPLORATION AND NATIVE RANGE STUDIES

Surveys carried out in the Mediterranean basin of Europe in 1987–92 identified 129 insect species that fed on *Onopordum* spp. (Briese *et al.* 1994). Over half of these insects were polyphagous, one-third restricted to the tribe Carduae and 5% restricted to the genus *Onopordum* (Briese *et al.* 1994). The last insects seemed to be genus-specific rather than monophagous.

Almost half of the 74 insect species that bred on *Onopordum* spp. were external leaf-chewing or sap-sucking insects. The remaining 40 species were endophages, predominantly species that fed within the capitula and rosette crown/root area (Briese *et al.* 1994). A number of these attacked the identified target transitional stages of their host weeds. To ensure the complementarity of the selected guild of agents, attention was given to the manner in which agents attacking the same transitional stage partitioned other niche dimensions, such as time or microsite utilisation.

Measurements of natural damage levels in the field (e.g. Briese 2000) and cage experiments, in which agent densities were augmented (e.g. Briese 1996b; Briese *et al.* 2002a, 2003), indicated whether these agents had the potential to make an impact on the life-cycle transitions of *Onopordum* spp. in Australia. Population genetic studies were also carried out to clarify agent host-use patterns (Briese *et al.* 1996). By the time the exploration phase finished in 1992, eight insects had been selected as candidates for the biological control of *Onopordum* spp. on the basis of their potential to affect transitional stages of the target weed.

CANDIDATES

Agents targeting the seed production to soil seed reserve transition

***Larinus latus* Herbst.** This large (adults up to 25 mm long) univoltine weevil (Fig. 5) from Greece becomes active in spring just before *Onopordum* plants produce flowering stems. Adults chew holes in the leaves and stems of thistle plants and feed on the pollen in their flowers. Females lay their eggs into the bracts or florets of *Onopordum* capitula in late spring to early summer. The eggs are laid into a small hole chewed into the surface by the female then are covered with a faecal deposit for protection. Each female may lay up to 70 eggs (Pettit and Briese 2000). Oviposition is spread over the period from the green bud stage until the flowers have dried off. Newly hatched larvae bore into the capitulum to feed on receptacle tissue and developing seeds (Briese 2000) and, depending on capitulum size and number of developing larvae, can destroy all the seed. Following pupation in the capitula, adults emerge in late summer and seek protected

Figure 5: *Larinus latus* adult. Photo: CSIRO.

overwintering sites. *L. latus* is the dominant seed-feeding insect on *Onopordum* in its native range. It is not uncommon to find populations of the thistle with 80–100% of capitula containing eggs, and total seed loss in Greece was strongly correlated with the density of *L. latus* (Briese 2000). *L. latus* was released in 1992 and is now well established at many sites. It was also released against *O. acaulon* in SA and WA, but its fate is unknown.

Tephritis postica **Loew.** This univoltine tephritid fly occurs throughout the range of *Onopordum* spp. in Europe. After becoming active in spring, courtship and mating occur on *Onopordum* host plants, after which females lay their eggs about the immature floral tubes. Larvae feed on developing seeds and the surrounding receptacle tissue. After several weeks, the mature larvae pupate within the capitulum. Adults emerge in late summer and remain dormant in protected sites until the following spring. In most field surveys in Mediterranean Europe, attack rates were relatively low, but on occasion 50% seed destruction was observed (Briese *et al.* 2002b). *T. postica* was released in three consecutive years from 1995–97, but failed to establish. Mating, oviposition and larval development took place normally in the field, but the emerging generation of adults did not survive. Mortality of overwintering adults, coupled with overdispersal of any survivors, could have led to failure of starter colonies. In the native range, dispersing flies would always reach areas with existing populations of *T. postica*; during initial colonisation in Australia, the opposite would be true.

Urophora terebrans **(Loew).** The seed-galling fly *U. terebrans* was selected to replace *T. postica* in the release strategy, to avoid the above problem. This species diapauses as a larva within woody galls in the capitula of *Onopordum* spp., hence new adults emerge at a similar time in the one place. *U. terebrans* was collected from *Onopordum* spp. in central Italy, where it was the dominant capitulum insect attacking *Onopordum* spp. It becomes active in spring and lays eggs directly into unopened buds. The feeding of *U. terebrans* larvae induces formation of gall tissue, which destroys a number of surrounding seeds. It was hoped that the adult emergence behaviour of this species would increase the probability of survival, but there was no establishment following releases during 2000–01. The insect also proved difficult to rear under laboratory conditions and was discarded as a control agent following the widespread establishment of *L. latus*.

Agents targeting the stem elongation to flowering transition

Lixus cardui **Olivier.** This univoltine weevil occurs throughout the range of *Onopordum* spp. in Mediterranean Europe (Briese *et al.* 1994). Adults of this elongate weevil (adults up to 15 mm long) emerge in early spring and feed voraciously on the leaves of *Onopordum* rosettes. As soon as the thistles begin to produce flowering stems, females lay their eggs into the growing stems, protecting the eggs with a plug of plant frass. Newly hatched larvae bore deeper into the stem and feed on structural tissue. Adults continue to feed on leaf tissue and heavy adult and larval attack can reduce plant growth and subsequent seed production. Many larvae (over 100) may develop in a single plant stem, depending on its size. New adults remain in the dead stems over summer and winter, where they hibernate until the following spring. In a cage experiment in France, plant size was reduced by ~50% and subsequent seed production by over 70% (Briese 1996b). Effects on plant growth are evident before there is significant larval development, indicating that adult feeding is an important contributor to plant damage. *L. cardui* was released in 1993 and is the most widely established of the *Onopordum* biological control agents.

Tettigometra sulphurea **(Mulsant and Rey).** This small green bivoltine plant-hopper feeds gregariously on the roots, crown, stem and flowering parts of *Onopordum* and was considered for biological control. It appeared to be specific to *Onopordum* and experiments on potted plants show that heavy feeding by *T. sulphurea* can cause dieback of bolting plants and abortion of seed heads (Boettcher 1993). However, one aspect of concern was the fact that this plant-hopper attracts large numbers of attendant ant species (Boettcher 1993). While this may be advantageous for the survival of *T. sulphurea*, there was a risk of the ants interfering with other biological control agents or preying on their eggs, so it was not released.

Agents targeting the small rosette to large rosette transition

***Trichosirocalus briesei*.** A crown weevil observed during surveys of *Onopordum* spp. in Spain was identified as a new species, *T. briesei* (Alonso-Zarazaga and Sanchez-Ruiz 2002). This small univoltine weevil (Fig. 6) becomes active in autumn, and eggs are laid from autumn to spring in pockets in the midrib on the underside of thistle rosette leaves. Females may each lay several hundred eggs over this period. Upon hatching, the larvae mine down through the midrib to the crown of the plant, where they feed in the base of the petiole and on meristem tissue. There are three larval instars. Fully fed third instar larvae migrate from the crown to the soil, where they pupate inside earthen cells. The adults emerge in late spring, feed on foliage then aestivate until the following autumn. In the field in Spain, the proportion of rosettes attacked by *T. briesei* ranged from 34% to 90%, with up to 84 larvae per rosette (Briese *et al.* 2002a). Field-cage experiments in France showed that this level of attack could kill small rosettes and reduce the growth of larger ones (Briese *et al.* 2002a). *T. briesei* was released in 1997 and is now established at a number of sites in south-eastern NSW.

Figure 6: *Trichosirocalus briesei* adult. Photo: CSIRO.

***Botanophila spinosa* (Rondani).** This crown fly attacks *Onopordum* spp. rosettes throughout the range in Mediterranean Europe, but is active later than *T. briesei*. Female flies lay up to 78 eggs each on young rosette leaves near the meristem from early to late spring (Vitou *et al.* 2001). Newly hatched larvae feed on the meristem, leading to gradual necrosis of plant tissue. Developing larvae can also be found at leaf axils along the elongating stem as a result of late oviposition on plant rosettes. Pupation may occur on or near the plant and adults emerge in early summer. Most males appear to mate after emergence, then die (Vitou *et al.* 2001). Females pass the winter in sheltered sites. In France, 10–33% of rosettes were attacked in different field populations of *Onopordum*, with up to 29 larvae per plant (Vitou *et al.* 2001). Cage experiments showed that, while late-season attack had no effect, attack early in the season could reduce plant size by over 20% and seed production by over 50% (Briese *et al.* 2003). *B. spinosa* was released in Australia in 1999 and 2002, but did not establish in the field.

***Eublemma amoena* (Hübner).** The noctuid moth *E. amoena* was collected in southern France where it has three generations per year, one in early spring, one in summer and an overwintering one. It overwinters as a late instar larva inside a hard cocoon. Female moths lay bluish-green eggs individually on the leaves of *Onopordum* plants in mid to late spring. Newly hatched larvae feed within the leaf petiole and cause the leaf to become deformed and die. The summer generation can feed in the leaves of flowering stems, as well as those of perennating rosettes. If larvae bore into the crown tissue and root of rosettes, this can kill smaller plants or lead to the production of several smaller crowns. Attack rates observed in the field in Europe range from 0% to 25%, with one to four larvae per plant (Briese *et al.* 2002a), levels which can severely damage smaller rosettes. Second-generation moths lay eggs on the leaves of bolting stems, leading to similar leaf deformation and death. *E. amoena* was released in 1998 and populations have become established at a number of sites in south-eastern NSW.

EVALUATION

Evaluation of the *Onopordum* biological control project followed the three-tiered approach set up by the Cooperative Research Centre for Australian Weed Management (Swirepik and Smyth 2003). The lowest tier provided information on the establishment rates of released populations of the agent (Swirepik *et al.* 2004). Chudleigh and

Simpson (2008) showed that there were substantial economic returns in speeding up the redistribution of pasture weed agents, including those established on *Onopordum*. They reported a potential benefit:cost ratio of 37.8, and found that even if benefits were as low as 10% of those predicted the return on investment would still be positive (benefit:cost ratio of 3.8).

Tier two evaluation involved monitoring plant density and agent population size at seven sites in southern NSW from 1997 to 2003 to get a more detailed picture of agent–weed dynamics (Swirepik *et al.* 2008). Tier three relied on more detailed observations of post-release agent population dynamics (Pettit and Briese 2000) and experiment-based studies to provide better understanding of the observed dynamics. These included the determination of agent density–plant damage relationships (Briese *et al.* 2004). Ideally, the data collected from these three tiers should be integrated to provide a regional estimation of agent impact on *Onopordum*, but this had not been done at the time the project ceased.

An economic assessment of this project estimated that it cost \$3.7 million, resulting in a net present value of \$18 million and a benefit:cost ratio of 9.6 at a discount rate of 8.0% (Page and Lacey 2006). It was calculated that by 2017 (should control be fully effective), biological control would reduce productivity losses by approximately \$7.4 million per year, through increased carrying capacity of land and reduced vegetable fault in wool (Page and Lacey 2006). It should be noted, however, that until accurate field evaluation of agent impact is undertaken these represent potential rather than actual benefits.

DISCUSSION

This project demonstrated that it is possible to develop an *a priori* biological control strategy population based on the dynamics of the target weed. *The* life-cycles of the target weeds formed a template for developing initial strategy aims and for all subsequent research. Even though the information available at the beginning of the project was incomplete, parallel studies on *Onopordum* populations in both native and introduced ranges helped to refine the strategy as the study progressed.

While only four of the seven released agents established, they covered the three key transitional stages of the target weeds, *L. latus* affecting seed production, *L. cardui* affecting the rosette to flowering stem transition and *T. briesei* and *E. amoena* affecting rosette growth and survival, in complementary ways. It was also demonstrated that the impact of the two most common agents, *L. latus* and *L. cardui*, was cumulative rather than a single agent being primarily responsible for impact. This augurs well for when the rosette-feeding agents begin to contribute to overall control.

Having an explicit control strategy provides a mechanism for ongoing evaluation, both at the practical control level and to further refine the process of agent selection. For example, evaluation suggested that agent selection could have been more parsimonious if the requirements for establishment, dispersal and overwintering vulnerability had been weighted more heavily, that the order of agent release can be critical to establishment and impact, that timing of agent attack is probably more critical than the time over which such attack occurs, and that agent-induced changes to population dynamics of the target weed may change the rules of the game. The issues are discussed more fully in Briese (2006).

While the indicators are that this project will contribute to a reduction in the problems caused by and cost of *Onopordum* thistles, funding ceased before a thorough evaluation of agent impact was undertaken. Such an evaluation would be best conducted when the rosette-feeding agents reach their full potential. Simple surveys could be undertaken every few years to monitor changes in agent density and distribution. Once agent populations have stabilised, a more detailed assessment would be needed to inform both the scientific and economic evaluations of impact.

REFERENCES

Alonso-Zarazaga MA and Sanchez-Ruiz M (2002) A revision of the *Trichosirocalus horridus* species complex, with description of two new species (Coleoptera, Curculionidae, Ceutorhynchinae). *Australian Journal of Entomology* **41**, 199–208.

Boettcher S (1993) *Tettigometra sulphurea* (Mulsant and Rey): life-cycle, host selection, and impact on the host plant. Diplomarbeit, Universität Bayreuth.

Briese DT (1989) Natural enemies of carduine thistles in New South Wales. *Journal of the Australian Entomological Society* **28**, 125–134.

Briese DT (1990) A new biological control programme against thistles of the genus *Onopordum* in Australia. In *Proceedings of the 7th International Symposium on Biological Control of Weeds*. 6–11 March 1988, Rome. (Ed ES Delfosse) pp. 151–160. Instituto Sperimentale per la Patologia Vegetale – MAF, Rome.

Briese DT (1996a) Landholder attitudes to *Onopordum* thistles and their control: a preliminary view. *Plant Protection Quarterly* **11**, 281–284.

Briese DT (1996b) Potential impact of the stem-boring weevil *Lixus cardui* on the growth and reproductive capacity of *Onopordum* spp. thistles. *Biocontrol Science and Technology* **6**, 251–261.

Briese DT (2000) Impact of the *Onopordum* capitulum weevil *Larinus latus* on seed production by its host plant. *Journal of Applied Ecology* **37**, 238–246.

Briese DT (2006) Can an *a priori* strategy be developed for biological control? The case of *Onopordum* spp. thistles in Australia. *Australian Journal of Entomology* **45**, 317–323.

Briese DT, Lane D, Hyde-Wyatt BH, Crocker J and Diver RG (1990) Distribution of thistles of the genus *Onopordum* in Australia. *Plant Protection Quarterly* **5**, 23–27.

Briese DT, Sheppard AW, Zwölfer H and Boldt PE (1994) Structure of the phytophagous insect fauna of *Onopordum* thistles in the northern Mediterranean basin. *Biological Journal of the Linnaean Society* **53**, 231–253.

Briese DT, Sheppard AW and Reifenberg JM (1995) Open-field host-specificity testing for potential biological control agents of *Onopordum* thistles. *Biological Control* **5**, 158–166.

Briese DT, Espiau C and Pouchot-Lermans A (1996) Micro-evolution in the weevil genus *Larinus*: host-race formation and speciation. *Molecular Ecology* **5**, 531–545.

Briese DT, Thomann T and Vitou J (2002a) Studies on the ecology of the crown weevil, *Trichosirocalus briesei*, and its impact on the growth and reproduction of its host *Onopordum* spp. thistles. *Journal of Applied Ecology* **39**, 688–698.

Briese DT, Pettit W, Swirepik A and Walker A (2002b) A strategy for the biological control of *Onopordum* spp. thistles in south-eastern Australia. *Biocontrol Science and Technology* **12**, 121–137.

Briese DT, Walker A, Pettit W and Sagliocco J-L (2002c) Host-specificity of candidate agents for *Onopordum* spp. thistles in Australia: an assessment of testing procedures. *Biocontrol Science and Technology* **12**, 149–163.

Briese DT, Vitou J, Boettcher S and Thomann T (2003) Prevalence and impact of the crown fly, *Botanophila spinosa*, on its host thistle, *Onopordum acanthium*, in southern France. *Acta Oecologia* **24**, 77–86.

Briese DT, Pettit W and Walker A (2004) Evaluation of the biological control agent, *Lixus cardui*, on *Onopordum* thistles: experimental studies on agent demography and impact. *Biological Control* **31**, 165–171.

Cavers PB, Groves RH and Kaye PE (1995) Seed population dynamics in *Onopordum* over one year in southern New South Wales. *Journal of Applied Ecology* **32**, 425–433.

Chudleigh P and Simpson S (2008) 'Final report: measuring and communicating the value of R&D'. Meat and Livestock Australia, Sydney.

Häffner E (2000) On the phylogeny of the subtribe Carduinae (Cardueae, Compositae). *Englera* **21**, 1–208.

Huwer RK, Briese DT, Dowling PM, Kemp DR, Lonsdale WM, Michalk DL, Neave MJ, Sheppard AW and Woodburn TL (2005) Can an integrated management approach provide a basis for long-term prevention of weed dominance in Australian pasture systems? *Weed Research* **45**, 175–192.

Leigh JH, Briese DT, Holgate MD and Slee AV (1993) Seed production by thistles, cut, goat-grazed or attacked by insects. In *Pests of Pastures. Weed, Invertebrate and Disease Pests of Australian Sheep Pastures.* (Ed ES Delfosse) pp. 163–171. WRDC, Melbourne.

Michael PW (1996) Necessary background for studies in the taxonomy of *Onopordum* in Australia. *Plant Protection Quarterly* **11**, 239–241.

Michalakis Y, Briese DT and Sheppard AW (1992) Taxonomic status of *Larinus cynarae* F. and *Larinus latus* Herbst (Coleoptera: Curculionidae), and its implication for the biological control of *Onopordum* (Asteraceae: Cardueae) in Australia. *Biocontrol Science and Technology* **2**, 275–280.

O'Hanlon PC, Peakall R and Briese DT (1999) AFLP reveals introgression in weedy *Onopordum* thistles: hybridisation and invasion. *Molecular Ecology* **8**, 1239–1246.

Page A and Lacey K (2006) 'Economic impact assessment for Australian weed biological control'. Report to CRC for Australian Weed Management by AEC Group. Technical Series No. 10. CRC for Australian Weed Management, Adelaide.

Parsons WT and Cuthbertson EG (1992) *Noxious Weeds of Australia*. Inkata Press, Melbourne.

Pettit WJ and Briese DT (2000) The demographic performance of the capitulum weevil, *Larinus latus*, on *Onopordum* thistles in its native and introduced ranges. In *Proceedings of the 10th International Symposium on Biological Control of Weeds.* 4–14 July 1999, Bozeman, Montana. (Ed NR Spencer) pp. 739–745. Montana State University, Bozeman.

Pettit WJ, Briese DT and Walker A (1996) Aspects of thistle population dynamics with reference to *Onopordum*. *Plant Protection Quarterly* **11**, 232–235.

Swirepik AE and Smyth MJ (2003) 'Evaluating biological control at the regional scale. Improving the selection, testing and evaluation of weed biological control agents'. CRC Technical Series No. 7, pp. 23–33. CRC for Australian Weed Management, Adelaide.

Swirepik AE, Smyth MJ and Briese DT (2004) Delivering pasture weed biological control through community networks in temperate Australia. In *Proceedings of the 11th International Symposium on Biological Control of Weeds*. 27 April–2 May 2003, Canberra. (Eds JM Cullen, DT Briese, DJ Kriticos, WM Lonsdale, L Morin and JK Scott) pp. 451–456. CSIRO Entomology, Canberra.

Swirepik A, Turner P and Briese DT (2008) Evaluation of the biological control agent, *Lixus cardui*, on *Onopordum* thistles: establishment and initial field impact. *Biological Control* **47**, 108–114.

Vitou J, Briese DT, Sheppard AW and Thomann T (2001) Comparative biology of two rosette crown-feeding flies of the genus *Botanophila* (Diptera: Anthomyiidae) with potential for biological control of their thistle hosts. *Journal of Applied Entomology* **125**, 85–95.

Zwölfer H (1965) Preliminary list of phytophagous insects attacking wild Cynarae (Compositae) in Europe. *Technical Bulletin of the Commonwealth Institute for Biological Control* **6**, 81–154.

Opuntia robusta H. L. Wendl. ex Pfeiff. – wheel cactus

J Baker

ABSTRACT

Cactoblastis cactorum and *Dactylopius* spp., historically released as biocontrol agents for *Opuntia* in Australia, have limited efficacy against *Opuntia robusta* and current methods of control are expensive, requiring long-term commitment to be successful. *O. robusta* has potential to infest up to 6.2 million square kilometres of Australia and, while control in prime agricultural regions may be feasible, controlling *O. robusta* in semi-arid pastoral and conservation regions will require the development of biological control options that can be incorporated into integrated management programs. There is a high level of genetic diversity in populations in SA with evidence of gene flow over 300 km. Determining which previously introduced *Dactylopius* spp. remain in Australia has not progressed beyond the initial experimental stages and current regional control programs are questioning the usefulness of conducting additional efficacy and redistribution trials using field-sourced *C. cactorum* or *Dactylopius* spp. Development of successful biological control will involve locating and testing new potential agents or investigating those not followed up during historical Australian *Opuntia* research efforts. A National Invasive Cacti Group has been formed to assist development of improved management programs.

Key words: biocontrol efficacy, benefit:cost ratio, *Cactoblastis cactorum*, *Dactylopius*, National Invasive Cacti Group.

INTRODUCTION

Opuntia robusta H. L. Wendl. ex Pfeiff. (Cactaceae), wheel cactus, is a significant environmental and pastoral weed in southern Australia, particularly in the Flinders Ranges and the Murray-Darling Basin in SA, and north-western/central Vic, with scattered infestations across the border into NSW. It is an erect shrub to small tree growing to 6 m high. Its native range is northern and central Mexico (Coahuila, Durango, Nuevo Leon, San Luis Potosi, Tamaulipas, Aguascalientes, Guanajuato, Guerrero, Hidalgo, Mexico, Queretaro provinces). *O. robusta* has expanded into North and South America and has been introduced into New Zealand and the Mediterranean region of southern Europe, South Africa and Australia (Global Biodiversity Information Facility).

The exact origin of invasive *O. robusta* present in Australia is unknown, but it is likely that populations have established from multiple introductions from central or northern Mexico. *O. robusta* is particularly well adapted to arid areas, is known to survive in areas with annual rainfall greater than 150 mm and is resistant to the cold (Figs 1 and 2) (Parsons and Cuthbertson 2001; Chaouche and Abdul-Hussain 2008). Hybridisation between *Opuntia* species has been reported (Barrientos-Perez *et al.* 1993; Hoffman 2004). It is suspected that *O. robusta* may have hybridised with other *Opuntia* species in Australia, but this has not been confirmed. If proven, hybridisation between *O. robusta* and other species could alter the

Figure 1: Wheel cactus, *Opuntia robusta*. Photo: L Edmunds.

efficacy of biological control agents introduced for *O. robusta*.

O. robusta has been widely used in ornamental cultivation in dry areas and is still traded on the internet. It is a proclaimed plant for enforced control in all parts of SA, NSW, Qld, NT and WA. *O. robusta* is proclaimed as a separate species for the whole of Vic except the Melbourne metropolitan area. Many cactus and succulent plant societies still promote *O. robusta* as 'a useful weed' (e.g. Ballarat Cactus and Succulent Society). This indicates that there will be problems in controlling future reinfestations from garden waste or escapes.

Figure 2: Wheel cactus, *Opuntia robusta*, infestation in the Flinders Ranges, SA. Photo: L Edmunds.

Parsons and Cuthbertson (2001) concluded that areas receiving annual rainfall of less than 150 mm were unlikely to be suitable for the development of dense infestations of *O. robusta*; however, sporadic rain and flood events and microhabitat can lead to the establishment of individual plants. Based on climate matching, *O. robusta* has the potential to establish across most of the continent if other species-specific resources are available (Fig. 3) (Bomford 2003; Bomford *et al.* 2009a, b, 2010; Brown *et al.* 2006; Crombie *et al.* 2008). It is also accepted that there will be a general shift southwards for most pest plant species in Australia as predicted climate changes occur (Scott *et al.* 2008; Baker 2009). Using the CSIRO Mk3.5 2020 climate change dataset and Climatch software (www.brs.gov.au/climatch) this trend appears true for *O. robusta*, with most of the continent, including prime agricultural areas, remaining climatically suitable for establishment into the 2020s.

O. robusta monocultures impact across economic, environmental and social amenities. Calculating the cost of these impacts is difficult, with economic costs influenced by a broad range of variables. For example, the cost of conventional control (including labour, herbicides and equipment) has been estimated to be $4.50 per plant in the Flinders Ranges region of SA (Edmunds 2006), while in the neighbouring Northern and Yorke NRM pastoral region, Mortimer and Patrick (2005) estimated $1.15 per plant for control. Where ongoing control costs exceed the annual production value of the land, control is likely to remain opportunistic and sporadic. Rapid recruitment following control has been observed (Lethbridge 2005; Edmunds 2006). The only option for control in areas where cost of control exceeds the land value will be biocontrol.

BIOLOGICAL CONTROL HISTORY

There are no records of any investigation of biological control of *O. robusta* outside Australia although it is known that *Cactoblastis cactorum* Berg (Lepidoptera: Pyralidae) and *Dactylopius* spp. (Hemiptera: Dactylopiidae) can cause damage to *O. robusta* plants (Hernandez *et al.* 1988; Zimmerman *et al.* 2000). The plant has a thick cuticle that reduces the penetration of herbicides and

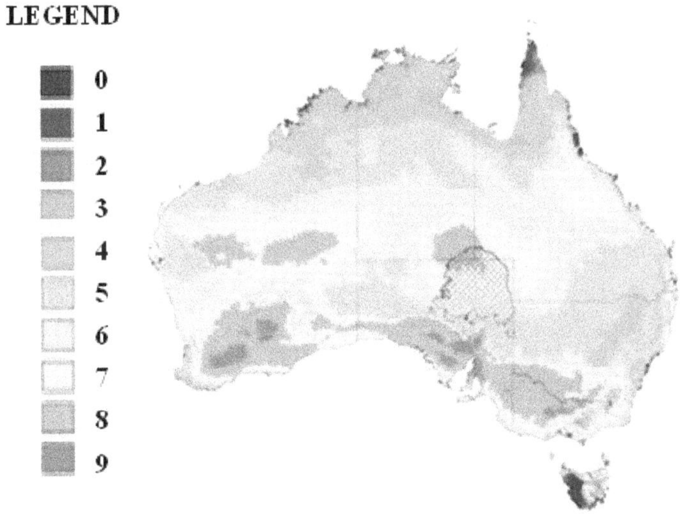

Figure 3: Climate match for *Opuntia robusta* in Australia using current climate data and Global Biodiversity Information Facility records of current distribution. Hatched areas indicate areas where rainfall is <150 mm per year. Colour legend = climate match class, where class match 9 = closest climate match to climate for GBIF records and 0 = poorest climate match. Class match <5 represents regions where *O. robusta* is unlikely to establish (Climatch www.brs.gov.au/climatch; GBIF http://data.gbif.org/occurrences/search.htm?c[0].s=20&c[0].p=0&c[0].o=15307022, accessed January 2009).

limits attack by *Cactoblastis* caterpillars. *Dactylopius* has been observed feeding on *O. robusta* (Fig. 4), but not all individuals are damaged by it and it is rare that plants are killed outright. Records of historical searches for biocontrol agents to control *Opuntia* and related species in Australia do not specifically mention *O. robusta* or its synonyms (Mann 1970). Therefore, it must be assumed that agents currently in Australia were introduced primarily to control other species.

Similarly, there are no records of additional agents being brought in following the early work of the Commonwealth Prickly Pear Board. Redistribution of *C. cactorum* and *Dactylopius* spp. has been carried out in attempts to control *O. robusta*, but this has failed to be effective. A report by Lewis *et al.* (1978) noted that various *Opuntia* species became established over wide areas of the north-east pastoral country of SA. In an attempt to control these infestations, *Dactylopius opuntiae* (Cockerell) and *C. cactorum* were released in this area in 1972. *Dactylopius* established on and reduced infestations of *O. imbricata*, while *Cactoblastis* failed to establish on any of the *Opuntia* species present. Between 1972 and 2005, control in Australia appears to have been sporadic and localised, with a focus on the use of conventional control.

O. robusta remains one of the most difficult to control *Opuntia* species in Australia.

Figure 4: *Dactylopius* spp. damage on *Opuntia robusta* plant in the Flinders Ranges region, SA. Photo: L Edmunds.

Genetic diversity

Alarmed by a dramatic increase in the distribution and density of *Opuntia* species across the south Australian semi-arid lands, a group of pastoralists formed the Blinman/Parachilna Pest Plant Control Group (BPPPCG) in 2004. They obtained funding to implement a regional pest plant control project, including additional scientific research and redistribution of existing biological control agents.

Investigations into the genetic diversity of *O. robusta* and *Dactylopius* spp. were initiated at the University of Adelaide, with additional funding from the University and the Cooperative Research Centre for Australian Weed Management. The results revealed high levels of

Table 1: Redistribution efforts for *Cactoblastis cactorum* and *Dactylopius* spp. on Gum Creek pastoral station, Flinders Ranges region, SA, May 2007–November 2009

Biocontrol agent	Obs. 1	Obs. 2	Obs. 3	Obs. 4	Conclusion
C. cactorum	9 plants infected (May 2007)	22 plants infested, 21 surviving (Jun 2007)	29 plants infested, 28 surviving (Sep 2009)	Plants destroyed by foraging *Macropus robustus* Gould (Marsupialia : Macropodidae) (euros)	Ineffective
Dactylopius spp. (Oratunga strain nursery)	2 plants infected (Jul 2008)	60% of 1 plant infested, 2 surviving (Aug 2008)	110 plants infested over 0.5 km^{-2}, 110 surviving (Nov 2009)		Virulent, decreased fruit production, but otherwise ineffective
Dactylopius spp. (sourced from *C. imbricata*, Warcoo station, Lake Victoria, NSW)	3 plants infected (Jul 2008)	No plants infested (Sep 2008)	No plants infested (Nov 2009)		Ineffective

gene flow, or relocation of vegetative parts. For example, different genotypes occurred in the same population, while similar genotypes were recorded 300 km apart. This suggests a high level of genetic diversity in Australian *O. robusta* (Baker *et al.* 2008).

Initial investigation into the genetic diversity of *Dactylopius* spp. present on *Opuntia* species using BARcoding to sequence part of the cytochrome c oxidase subunit 1 gene (Herbert *et al.* 2003) was not successful. However, an alternate method, utilising existing information for two species of *Dactylopius* (*Dactylopius austrinus* (De Lotto) and *Dactylopius confusus* (Cockerell)) and seeking primers to sequence a similar region from DNA extracted from existing samples (Gullen and Cook 1999) might clarify the number of *Dactylopius* biotypes in Australia.

Redistribution efforts

The BPPPCG undertook opportunistic redistribution work using *C. cactorum* and unknown *Dactylopius* spp. occurring in the Flinders Ranges region between 2007 and 2009. Overall, these efforts were unsuccessful (Table 1); only one trial, using the 'Oratunga' strain of *Dactylopius* (named after the location where it was first sourced), produced positive results. Fruit production was reduced, but there were no lethal effects on the infested plants. The BPPPCG was still maintaining a nursery of this biotype in December 2009.

The BPPPCG redistribution work cost about six contractor days per year and the value of the work is being questioned. It may be discontinued until the most effective strain, or strains, of *Dactylopius* present in Australia can be identified or effective new agents imported.

DISCUSSION

Cactoblastis and *Dactylopius* spp. previously introduced into Australia for control of other *Opuntia* spp. rarely kill *O. robusta* plants, although *Dactylopius* may reduce fruit output. Limited research suggests a high level of genetic diversity and gene flow in *O. robusta.* We do not know how many of the *Dactylopius* biotypes introduced historically survive today. Further work is required to identify *Opuntia* agents, or their strains, that may be useful for controlling *O. robusta.*

The annual cost of existing control programs is high, around \$566 km^{-2} based on densities of 200 plants per square metre (Mortimer and Patrick 2005; Edmunds 2006), and follow-up work would be required for a minimum of 25 years because seed can persist for up to 20 years. Considering the potential for spread to about 80% of Australia, with most areas being remote, inaccessible and with low economic productivity but high conservation value, a specific biological control program for *O. robusta* is required.

There are no native species from the Cactaceae family in Australia, although in some areas *Opuntia ficus-indica* and *Hylocereus undatus* are being grown commercially for their edible fruit. This commercial production could complicate the search for an effective biological control agent. Some of the promising biological control agents identified by the Commonwealth Prickly Pear Board were never released or if released may have died out. These may be more effective against *O. robusta* than current strains of existing *Opuntia* agents and they should be reassessed.

Australian surveys in the 1930s revealed several American fungal and bacterial diseases present, including *Gloeosporium lunatum* Ellis and Everh., *Phyllostica concave* Seaver and *Bacillus cacticidis* Johnston and Hitchcock (Evans 2002). These micro-organisms are often associated with the rapid decay of *Opuntia* segments following attack by *Cactoblastis* (Dodd 1936; Mann 1970). Understanding the ecology and association of such micro-organisms with *Cactoblastis* in semi-arid regions may also provide options for increasing the efficacy of biological control in those regions.

There is potential to tap into research investigating pests affecting commercial *Opuntia* operations overseas to identify new biological control agents. The National Invasive Cacti Group, convened in December 2009, brought together representatives of state biosecurity agencies, scientists, taxonomists and natural resource and community groups working on the control of invasive cacti. This group may be a useful resource through which to run coordinated research and management programs.

Factors contributing to the success of a biocontrol program will include adequate resourcing, availability of suitable candidates as agents, testing to meet release requirements, survival in Australia and the incorporation of successful agents into current integrated control methods. However, doing nothing may well be a more expensive option.

ACKNOWLEDGEMENTS

The assistance of the Blinman/Parachilna Pest Plant Control Group, especially Lorraine Edmunds, and the technical work of Justin Williams and Jo Kent are gratefully acknowledged.

REFERENCES

Baker J (2009) Watching the sleepers. *Weedscene* July 2009. Weed Society of Victoria, Meredith, Vic.

Baker J, Keller M and Preston C (2008) Genetic variability of wheel cactus (*Opuntia robusta* Wendl.) in southern Australia: implications for weed spread and biological control. In *Proceedings of the 16th Australian Weeds Conference*. 18–22 May 2008, Cairns. (Eds RD van Klinken, VA Osten, FD Panetta and JC Scanlan) p. 153. Queensland Weeds Society, Brisbane.

Barrientos-Perez F, Muratalla-Lua A and Barrientos-Priego AF (1993) New hybrids of *Opuntia*. In *Biotechnology for Aridland Plants*. (Eds TJ Mabry, HT Ngyuen, RA Dixon and MS Bounness) pp. 243–251. IC2 Institute, University of Texas, Austin.

Bomford M (2003) 'Risk assessment for the establishment of exotic vertebrate in Australia: recalibration and refinement of models'. Bureau of Rural Sciences, Canberra.

Bomford M, Darbyshire RO and Randall L (2009a) Determinants of establishment success for introduced mammals. *Wildlife Research* **36**, 192–202.

Bomford M, Kraus F, Barry SC and Lawrence E (2009b) Predicting establishment success for alien reptiles and amphibians: a role for climate matching. *Biological Invasions* **11**, 713–724.

Bomford M, Barry S and Lawrence E (2010) Predicting establishment success for introduced freshwater fishes: a role for climate matching. *Biological Invasions 1387–3547* (Print) *1573-1464* (Online). doi:10.1007/s10530-009-9665-3.

Brown L, Barry S, Cunningham D and Bomford M (2006) Current practice in applying CLIMATE for weed risk assessment in Australia. In *Proceedings of the 15th Australian Weeds Conference*. 24–28 September 2006, Adelaide. (Eds C Preston, JH Watts and ND Crossman) pp. 703–706. Weed Management Society of South Australia, Adelaide.

Chaouche FZ and Abdul-Hussain MS (2008) Contribution to the study of opuntia and prospects for improvement in the steppe region. *Agricultura – Revista de Stiinta si Practica Agricola* **17**, 24–31.

Crombie J, Brown L, Lizzio J and Hood G (2008) *Climatch User Manual*. Bureau of Rural Sciences, Canberra. <http://adl.brs.gov.au:8080/Climatch/docs/climatch_manual.doc> (accessed April 2010).

Dodd AP (1936) The control and eradication of prickly pear in Australia. *Bulletin of Entomological Research* **27**, 503–517.

Edmunds LJ (2006) A community approach to pest plant control in South Australia's rangelands. In *Proceedings of the 15th Australian Weeds Conference*. 24–28 September 2006, Adelaide. (Eds C Preston, JH Watts and ND Crossman) pp. 757–760. Weed Management Society of South Australia, Adelaide.

Evans HC (2002) Biocontrol of weeds. In *The Mycota: A Comprehensive Treatise on Fungi as Experimental Systems for Basic and Applied Research. Agricultural Applications VXI.* (Eds K Esser and F Kempken) pp. 135–152. Springer-Verlag, Berlin.

Global Biodiversity Information Facility (2009) <http://data.gbif.org/welcome.htm>.

Gullen PJ and Cook LG (1999) Are cochineal insects eriococcids? *Entomologica*, Bari, **33**, 91–99.

Herbert PDN, Ratnasingham S and deWaard JR (2003) Barcoding animal life: cytochrome c oxidase subunit 1 divergences among closely related species. *Proceedings of the Royal Society of London. Series B, Biological Sciences. Royal Society (Great Britain)* **7**, 270 (Suppl. 1), S96–S99.

Hernandez R, Pisaca SI and Saavedra DO (1988) The cochineal, a singular insect. *Agricultura* **57**, 677, 914–919.

Hoffman JH (2004) Biotypes, hybrids and biological control: lessons from cochineal insects on *Opuntia* weeds. In *Proceedings of the 11th International Symposium on Biological Control of Weeds.* 27 April–2 May 2003, Canberra. (Eds JM Cullen, DT Briese, DJ Kriticos, WM Lonsdale, L Morin and JK Scott) pp. 283–286. CSIRO Entomology, Canberra.

Lethbridge M (2005) Modelling wheel cactus: problems and possibilities. Presentation for the Weed Management Society of South Australia. 7 April 2005, University of Adelaide, SA. Unpublished.

Lewis AW, O'Neil JM and Tideman AF (1978) 'A review of weed control in the arid zone of South Australia outside of local government areas – 1945 to 1976'. Agronomy Branch Report No. 81, SA Dept of Agriculture and Fisheries, Adelaide.

Mann J (1970) *Cacti Naturalised in Australia and their Control.* Government Printer, Brisbane.

Mortimer R and Patrick G (2005) 'Report on wheel cactus'. Upper North Animal and Plant Control Board, SA.

Parsons WT and Cuthbertson EG (2001) *Noxious Weeds of Australia.* 2nd edn. CSIRO Publishing, Melbourne.

Scott JK, Batchelor K, Ota N and Yeoh PB (2008) 'Modelling climate change on sleeper and alert weeds'. Final report for Defeating the Weed Menace, Land and Water, Canberra.

Zimmerman HG, Moran VC and Hoffman JH (2000) The renowned cactus moth, *Cactoblastis cactorum*: its natural history and threat to native *Opuntia* floras on Mexico and the United States of America. *Diversity and Distributions* **6**, 259–269.

Opuntia spp.

John R Hosking

ABSTRACT

Opuntia spp. were major weeds in Qld and NSW. The biological control program that ran from 1911 to 1939 resulted in the release of 20 biological control agents, with 14 reported as established, some of which appear to have subsequently died out. *Cactoblastis cactorum* and *Dactylopius* spp. are considered to be the main agents that controlled *Opuntia* spp. following their release, and still do so today. This biological control program is considered to be the most successful biological control program in Australia. No *Opuntia* spp. in Qld or NSW is now considered to be a major problem. In 2005 dollar terms, a conservative estimate of the return on investment was $147.4 for each $1 spent, with a net present value of $3110.3 million.

Key words: *Opuntia stricta, Opuntia aurantiaca, Opuntia tomentosa, Opuntia monocantha, Opuntia elata, Opuntia humifusa, Opuntia streptacantha*, prickly pear, *Cactoblastis cactorum*, cactoblastis, *Dactylopius*, cochineal.

INTRODUCTION

Opuntia spp. naturalised in Australia vary from low shrubs to trees up to 8 m high. They are often referred to as prickly pears, but this term has been used to cover differing numbers of cactus species in various parts of Australia so will not be used in this chapter. *Opuntia* spp. stems are succulent, consisting of a number of joints, often incorrectly termed leaves, which are flattened or almost cylindrical. Areoles form at the base of small leaves that are shed shortly after new joints emerge. These areoles contain glochids (very small spines found only in this genus and genera that were previously included in *Opuntia*) adjacent to longer spines, although long spines may be absent in some species. Long spines are not covered by a detachable sheath as is the case in *Cylindropuntia* spp. Flowers and fruit form within or near areoles on upper stem joints. Flattened stamens of flowers resemble sepals and petals; these vary from yellow to orange on species found in Australia. Fruit form seed for most of the problem species, but not for *Opuntia aurantiaca* Lindl. All species spread via movement of detached stem joints. These stem joints will root if in contact with the ground, particularly in wet conditions.

Opuntia spp. are native to the Americas and the West Indies. Nineteen *Opuntia* spp. are considered to have naturalised to some extent in Australia. This does not include the eight *Cylindropuntia* spp. naturalised in Australia. *Cylindropuntia* is often considered to be part of *Opuntia* as in publications by Dodd (1940), Mann (1970) and Hosking *et al.* (1988). Only those *Opuntia* spp. considered to be a problem are covered in this chapter, these being *Opuntia stricta* (Haw.) Haw., *O. aurantiaca*, *Opuntia elata* Link & Otto ex Salm-Dyck (referred to, based on incorrect identification, as *Opuntia paraguayensis* K.Schum. in Hosking *et al.* 1988), *Opuntia humifusa* (Raf.) Raf. (*Opuntia opuntia*, an invalid name, was used for this species in Dodd 1940 and Mann 1970), *Opuntia*

Figure 1: *Opuntia stricta* in flower and fruit, between Gunnedah and Tamworth, NSW. Photo: JR Hosking, NSW Government.

monocantha Haw. (referred to as *Opuntia vulgaris* in Hosking *et al.* 1988), *Opuntia streptacantha* Lem. and *Opuntia tomentosa* Salm-Dyck. Other naturalised cacti are of minor importance, some of which are dealt with in Hosking *et al.* (1988). *Opuntia robusta* H.L.Wendl. ex Pfeiff. and *Cylindropuntia* spp. are the subject of current biological control programs and are discussed elsewhere in this book.

O. stricta (Fig. 1) was the main cactus problem when the biological control of cactus campaign began in Australia. At that time it was divided into *Opuntia inermis* (DC.) DC. and *O. stricta*, both now considered to be part of the natural variation of *O. stricta*. In 1910 it was estimated that *O. stricta* occupied 4 million ha; this had increased to greater than 24 million ha in 1925, with half the latter area being unusable (Dodd 1940). The current distribution of *O. stricta* in Australia is shown in Figure 2.

There are no accurate figures for areas infested by other *Opuntia* spp. in Australia. In NSW it was estimated that close to 200 000 ha were infested with *O. aurantiaca* in 1988. Although not determined, the area in Qld was probably similar. *O. aurantiaca* is present, but much less common in other states.

The area covered by other *Opuntia* spp. is smaller. For distribution of these species, see Hosking *et al.* (1988) for eastern Australia and AVH (2010) for all species for all of

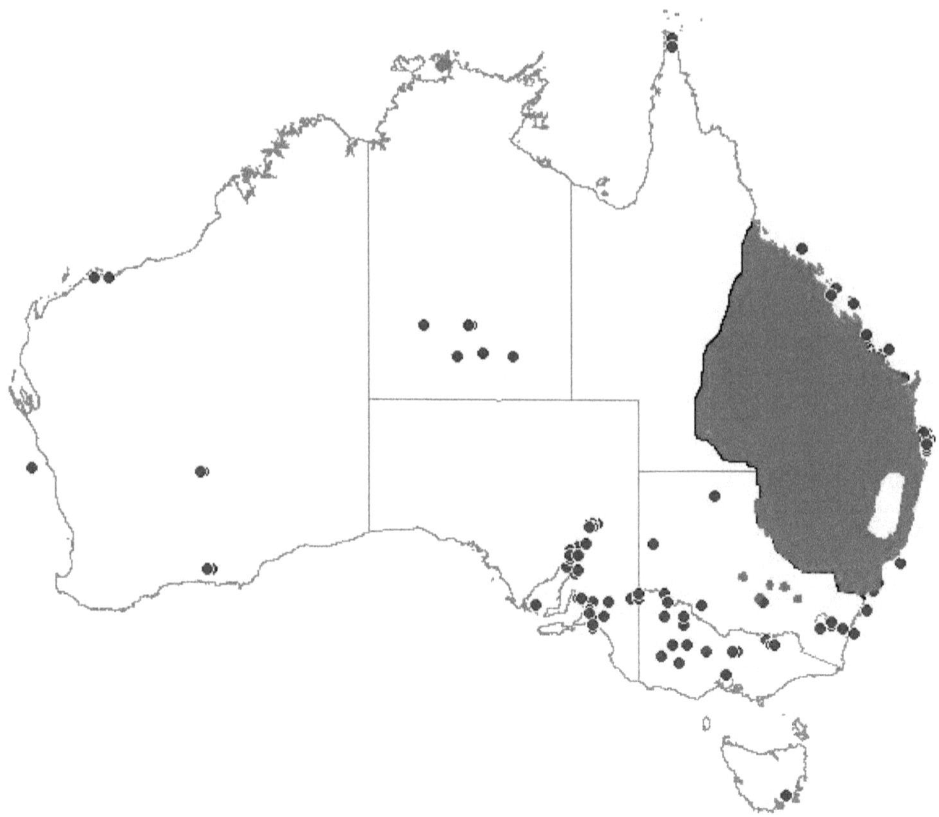

Figure 2: Distribution of *Opuntia stricta* in Australia based on data in Australia's Virtual Herbarium (2010), with additional points and filling in of continuous *O. stricta* areas based on Hosking *et al.* (1988) and knowledge of the author. Specimen data were reproduced from Australia's Virtual Herbarium with permission of the Council of Heads of Australasian Herbaria Inc.

Australia. However, records in AVH (2010) are partial because cactus is rarely collected due to difficulties with specimen preservation.

A short summary of the biological control program is given here: more detail can be found in Dodd (1940), Mann (1970) and Hosking *et al.* (1988).

BIOLOGICAL CONTROL HISTORY

The first deliberate use of insects to control a weed involved deliberate distribution of a cochineal on *Opuntia* sp. in India in 1863 (Tryon 1910). This resulted in control of the common *Opuntia* sp. at that time, but the *Opuntia* sp. and the cochineal involved could not be determined in 1910 (Tryon 1910). The same cochineal was introduced to control an *Opuntia* sp. in Sri Lanka prior to 1865 and again in that year, again resulting in control of this cactus (Tryon 1910). Later studies suggest that the cochineal and cactus involved in India were *Dactylopius ceylonicus* (Green) and *O. monacantha* respectively (De Lotto 1974).

Cactus biological control began in Australia in 1903 with an unsuccessful attempt to establish a cochineal, *Dactylopius* sp., from India and Sri Lanka (Tryon 1910). In 1912 the Queensland Prickly Pear Travelling Commission was established. It sent a cochineal, *D. ceylonicus*, to Qld from Sri Lanka in 1913, for control of *O. monocantha*. The first shipment did not survive, but insects from the second shipment were successfully reared and released in 1914. Further work was interrupted by World War I.

The Commonwealth Prickly Pear Board was appointed in 1920 by the Commonwealth Government and the state governments of Qld and NSW (Dodd 1940). The Board's charter included a search for prickly pear insects in the Americas, investigation of host restriction of prickly pear insects, selection and transport to Australia of suitable insects plus their large-scale rearing and later distribution, and study of the progress of the established insects (Dodd 1940). There was also to be a survey of plant disease organisms.

EXPLORATION

The Queensland Prickly Pear Travelling Commission undertook survey work from 1912 to 1914. It sent insects to Qld from India, Sri Lanka, South Africa and South America (Mann 1970). From 1920 to 1937 the Commonwealth Prickly Pear Board had entomologists based in North and/or South America investigating potential biological control agents (Dodd 1940).

CANDIDATES

Many insects for control of *Opuntia* spp. were tested, many were released and several established. Only those having some impact are discussed below. For more details on all biological control agents considered, released and their impact see Dodd (1940), Mann (1970), Hosking *et al.* (1988) and Julien and Griffiths (1988).

A native insect, *Helicoverpa puntigera* (Wallengren) (Lepidoptera: Noctuidae), causes death of some flower buds and fruit, and minor damage to young joints of *O. aurantiaca* and *O. stricta* (Hosking *et al.* 1988) and probably to other *Opuntia* spp. However, damage is generally minor.

Cactus-feeding moths

***Cactoblastis cactorum* (Bergroth) (Lepidoptera: Pyralidae).** *C. cactorum*, a South American insect (Fig. 3) originating from Argentina, was released in 1926 and was the main agent responsible for the biological control of *O. stricta*, the most serious of the *Opuntia* spp. naturalised in Australia. It still successfully controls *O. stricta* over most of the range of that cactus (Fig. 4). It is not able to control *O. stricta* in cooler areas where the moth cannot complete two generations a year, although in such areas *D. opuntiae* is able to control this cactus (Hosking *et al.* 1994). *C. cactorum* has also proved unsuitable in areas where it is hot most of the time and *O. stricta* segments are often dehydrated. Larvae of this moth also feed on *O. aurantiaca, O. elata, O. ficus-indica, O. humifusa, O. monocantha, O streptacantha* and *O. tomentosa*, but damage in most cases is not as spectacular as for *O. stricta*. *C. cactorum* is the only agent damaging *O. humifusa* and damage is sufficient for control of this species in the Hunter Valley, NSW.

***Tucumania tapiacola* Dyar (Lepidoptera: Pyralidae).** This insect was released in 1935 and is locally damaging to *O. aurantiaca*. Like *C. cactorum*, the larvae hollow out segments. It has been particularly damaging in the Hunter Valley near Ravensworth, NSW in some years, but is generally not common on *O. aurantiaca* or other cactus species (Hosking, unpub. data).

Cochineal insects *Dactylopius* spp. (Hemiptera: Dactylopiidae)

***Dactylopius austrinus* De Lotto.** Released in 1933, this cochineal is the most successful biological control agent for *O. aurantiaca*. In Australia, *O. aurantiaca* is its only host and is successfully controlled in all areas where *D. austrinus* has been used (Figs 5 and 6). *D. austrinus* is

Figure 3: *Cactoblastis cactorum* larvae hollow out cactus segments at Wellington, NSW. Photo: GR Sainty, consultant.

Figure 5: *Opuntia aurantiaca* with *Dactylopius austrinus*, Tamworth, NSW. One segment has recently been destroyed by this cochineal. Photo: JR Hosking, NSW Government.

most damaging in hot dry times; in hot wet seasons the cactus often grows more rapidly than the rate at which the insect is capable of destroying segments (Hosking, unpub. data).

***Dactylopius ceylonicus* (Green).** This cochineal was released in 1914 and successfully controlled large patches of *O. monocantha* in Qld such that by 1923 'occasional plants only have been reported' (Mann 1970). It still controls *O. monocantha* in Australia, although control is slower in NSW than in Qld, apparently as a result of lower temperatures. This insect also causes damage to *O. elata*

Figure 4: Destruction of *Opuntia stricta* by *Cactoblastis cactorum* near Gunnedah, NSW. Upper image taken on 5 January 1982, lower image taken on 23 August 1982. Photos: JR Hosking, NSW Government.

Figure 6: Destruction of *Opuntia aurantiaca* by *Dactylopius austrinus* near Gunnedah, NSW. Upper image taken on 5 January 1982, lower image taken on 10 January 1984. Photos: JR Hosking, NSW Government.

and this, with damage caused by *C. cactorum*, is probably sufficient for control of *O. monocantha*.

***Dactylopius opuntiae* (Cockerell).** Released in 1921, this species had some initial success and appeared more suited to control of *O. stricta* growing in scrub country where plant foliage protected cochineal from weather effects (Dodd 1940). Today *D. opuntiae* is used for biological control of *O. stricta* in cooler areas where cactoblastis cannot complete two generations per year (Hosking *et al.* 1994) and in hot dry areas where segments are often dehydrated. The species is also damaging to tree pears such as *O. streptacantha* and *O. tomentosa*, with damage being greater if plants are felled.

Diseases

These were not investigated in any detail by the Commonwealth Prickly Pear Board as they were considered to be of little value. Many were already present in Australia, and it would be difficult to prove that disease organisms could not transfer to other plants (Dodd 1940).

DISCUSSION

This remains the most successful biological control program in Australia and was the stimulus for greatly increased national and international interest in this approach. *C. cactorum* was released in 1926 and by 1932 the second (regrowth) wave of *O. stricta* was being destroyed by larvae of this moth. Land previously abandoned to this cactus was brought back into production. So successful was the biological control program, that the people of Boonarga (200 km NW of Brisbane) in Qld built and named their local hall the Boonarga Cactoblastis Memorial Hall in honour of the cactoblastis moth. Construction of the hall reflected the strength of the relief felt by farmers when cactoblastis succeeded against *O. stricta*. Eighty years later this moth still controls *O. stricta*. Biological control of *O. aurantiaca* by cochineal is also considered to be adequate. No other *Opuntia* in Qld or NSW is now considered to be a major problem. The benefit:cost ratio has been calculated as 312.3 and, in 2005 dollar terms, the investment of $21.1 million returned a net present value of $3110.3 million (Page and Lacey 2006). As a result of the success of the biological control program for *Opuntia* spp. in Australia, *C. cactorum* was successfully transferred to other countries, including those in the West Indies. It reportedly had a significant impact in the West Indies, particularly on *Opuntia triacanthos* (Willd.) Sweet and *O. stricta* (Julien and Griffith 1988). The moth unfortunately spread from the West Indies to North and Central America and is causing concern as it is damaging native *Opuntia* spp. in those areas (Zimmermann and Pérez Sandi Cuen 2006; Mafokoane *et al.* 2007). Cactoblastis, cochineal species and several other cactus biological control agents were also transferred to South Africa, following their success in Australia, and resulted in generally successful biological control of *Opuntia* spp. there (Moran and Zimmermann 1991a, b; Zimmermann and Moran 1991; Julien and Griffiths 1998; Hoffmann and Moran 2008).

REFERENCES

AVH (2010) *Australia's Virtual Herbarium*. <http://www.chah.gov.au/avh/public_query.jsp> (accessed 10.3.10).

De Lotto G (1974) On the status and identity of the cochineal insects (Homoptera: Coccoidea: Dactylopiidae). *Journal of the Entomological Society of Southern Africa* **37**, 167–193.

Dodd AP (1940) 'The biological campaign against prickly-pear'. Commonwealth Prickly Pear Board Bulletin. Government Printer, Brisbane.

Hoffmann JH and Moran VC (2008) Assigning success in biological weed control: what do we really mean? In *Proceedings of the 12th International Symposium on Biological Control of Weeds*. 22–27 April 2007, La Grande Motte, France. (Eds M Julien, R Sforza, MC Bon, HC Evans, PE Hatcher, HL Hinz and BG Rector) pp. 685–690. CABI, Wallingford, UK.

Hosking JR, McFadyen RE and Murray ND (1988) Distribution and biological control of cactus species in eastern Australia. *Plant Protection Quarterly* **3**, 115–123.

Hosking JR, Sullivan PR and Welsby SM (1994) Biological control of *Opuntia stricta* (Haw.) Haw. var. *stricta* using *Dactylopius opuntiae* in an area of New South Wales, Australia, where *Cactoblastis cactorum* (Berg) is not a successful biological control agent. *Agriculture, Ecosystems and Environment* **48**, 241–255.

Julien MH and Griffiths MW (1998) *Biological Control of Weeds: A World Catalogue of Agents and their Target Weeds*. 4th edn. CABI, Wallingford, UK.

Mafokoane LD, Zimmermann HG and Hill MP (2007) Development of *Cactoblastis cactorum* (Berg) (Lepidoptera: Pyralidae) on six North American *Opuntia* species. *African Entomology* **15**, 295–299.

Mann J (1970) *Cacti Naturalised in Australia and their Control*. Government Printer, Brisbane.

Moran VC and Zimmermann HG (1991a) Biological control of jointed cactus, *Opuntia aurantiaca* (Cactaceae), in South Africa. *Agriculture, Ecosystems and Environment* **37**, 5–27.

Moran VC and Zimmermann HG (1991b) Biological control of cactus weeds of minor importance in South Africa. *Agriculture, Ecosystems and Environment* **37**, 37–55.

Page AR and Lacey KL (2006) 'Economic impact assessment of Australian weed biological control'. Technical Series 10. CRC for Australian Weed Management, Adelaide.

Tryon H (1910) The 'Wild cochineal,' with reference to its injurious action on prickly pear (*Opuntia* spp.) in India, etc., and to its availability for the subjugation of this plant in Queensland and elsewhere. *Queensland Agriculture Journal* **25**, 188–197.

Zimmermann HG and Moran VC (1991) Biological control of prickly pear, *Opuntia ficus-indica* (Cactaceae), in South Africa. *Agriculture, Ecosystems and Environment* **37**, 29–35.

Zimmermann H and Pérez Sandi Cuen M (2006) 'The consequences of introducing the cactus moth *Cactoblastis cactorum* to the Caribbean, and beyond'. IAEA/FAO report. PRONATURA, FMCN and USAID, Mexico City, Mexico.

Parkinsonia aculeata L. – parkinsonia

Rieks D van Klinken and Tim A Heard

ABSTRACT

Parkinsonia aculeata (Leguminosae: Caesalpinioideae) is a shrub or tree from the Americas that can form dense thorn thickets that negatively affect both the environment and the pastoral industry in rangeland Australia. A biological control program was initiated by Australia in 1983, with extensive surveys conducted in North America. This resulted in the release of two agents into Australia in the 1990s, a seed-feeding bruchid *Mimosestes ulkei* and a sap-sucking mirid *Rhinacloa callicrates*. In addition, a bruchid *Penthobruchus germaini* previously found in Argentina, was tested and released in this period. Only *P. germaini* became widespread and common, but it is not causing population-level impacts. Native range surveys recommenced in 1999. They initially focused on Central America and Venezuela, before expanding into the rest of South America. The broadening geographic focus was the result of improved understanding of the extent of parkinsonia's native range, from which approximately 150 species of herbivorous insects and one pathogen have now been recorded. However, most are rare (~53% represented by only one or two specimens) and few appear to be both damaging and host-specific. Nonetheless, several potential agents have been shortlisted for more detailed study on the basis of field host-range, preliminary host-testing and likely impact. The biological control program has also provided more general contributions to biological control theory and practice, including the use of seed-predators, the role of genetics, and the use of biogeography and new analytical tools to optimise native range survey design.

Key words: Caesalpinioideae, rangelands, Bruchidae.

INTRODUCTION

Parkinsonia aculeata L. (Leguminosae: Caesalpinioideae) is a perennial thorny shrub or small tree that is native to the Americas (Hawkins *et al.* 2007). It was intentionally spread around the world, principally as an ornamental tree owing to its brilliant yellow flowers, but also because of its potential value for hedging, fodder and land rehabilitation (Stewart *et al.* 1992; Hawkins 2001). *P. aculeata* now has a pan-tropical distribution (van Klinken *et al.* 2009a). Significant negative impacts are largely restricted to Australia, although it chokes irrigation channels in northern Mexico, extensive naturalised populations occur near Nakuru in Kenya (Hawkins 2001) and it was considered as a biological control target in its native range in the USA (DeLoach 1981). Its greatest environmental impact is probably through exclusion of the herbaceous layer (van Klinken 2006). Thicket formation interferes with stock management, but existing infestations are generally too patchy to cause substantial and widespread reductions in carrying capacities (van Klinken 2006).

P. aculeata was probably already naturalised in Australia by the late 1800s, either from Venezuela or from genetically similar Meso-American populations (Hawkins *et al.* 2007). It is now widely naturalised in diverse

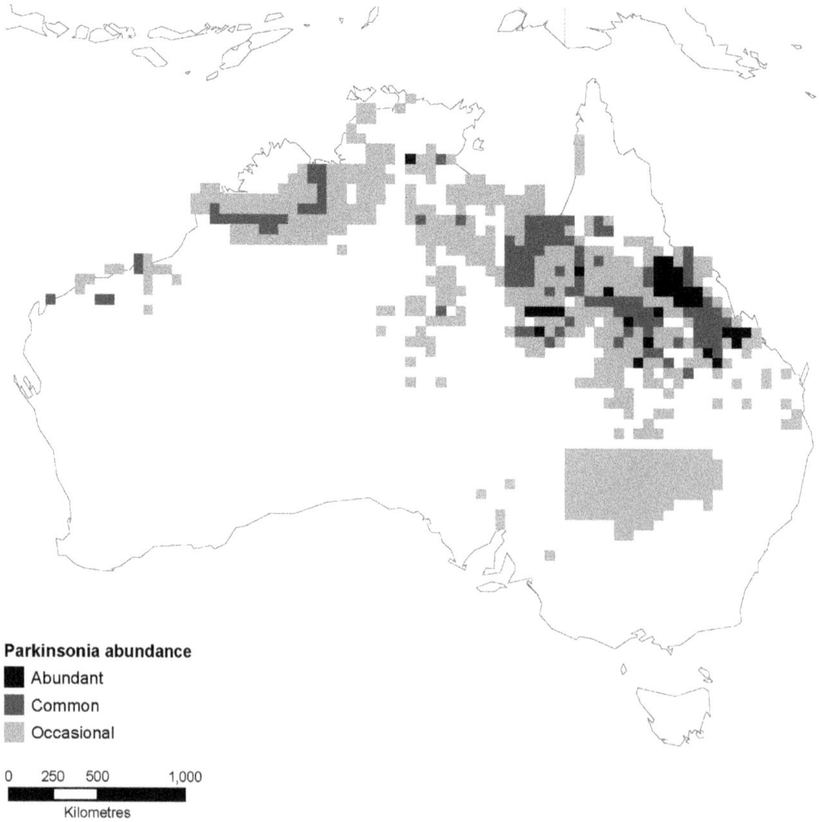

Figure 1: Current distribution and abundance of *Parkinsonia aculeata* in Australia. Source: Queensland Biosecurity.

climates, from arid to wet-dry tropics (Fig. 1) and habitats including seasonally flooded wetlands, riparian corridors and uplands. However, it is typically most invasive in more mesic habitats in semi-arid and semi-humid regions (van Klinken *et al.* 2009a, b). It is recognised as one of the 20 worst weeds in Australia (Thorp and Lynch 2000) and has been declared in all states and territories. Considerable resources have been spent on its management. For example, over $2 million of Australian Government funding alone was spent on parkinsonia-related projects between 2001 and 2005 (Martin and van Klinken 2006).

P. aculeata is readily identified in Australia by its smooth green bark, very distinctive pendulous leaves with minute, easily shed pinnules, bright yellow five-petalled flowers, and pods which contain one to 11 seeds and are straw-coloured when mature (Fig. 2). Adults typically grow to 5–7 m tall and wide, and probably rarely live more than 20–30 years (van Klinken *et al.* 2009a). They can produce large numbers of seeds, which are mostly dispersed by floodwaters within floating pods, or become incorporated into the seed bank under or adjacent to parent trees. Seeds are hard-seeded and are released from dormancy by 'wet heat' (van Klinken and Flack 2005; van Klinken *et al.* 2006, 2008). Populations are typically very dynamic as a result of often rare major recruitment events and a wide range of mortality factors, including dieback caused by a suite of soil-borne pathogens (Toh *et al.* 2008; Diplock *et al.* 2006, 2008; Toh 2009; van Klinken *et al.* 2009a), severe frosts, fires, and browsing by macropods or sheep (van Klinken *et al.* 2009a). In fact, most of the 23 initially healthy populations monitored across Australia since 1999–2000 have declined in adult density, and local extinctions are probably common (van Klinken *et al.* 2009a; JB Pichancourt and RD van Klinken, unpub. data). Browsing by sheep, goats and other livestock (generally not cattle) is likely to be an important factor preventing invasions in other countries.

BIOLOGICAL CONTROL HISTORY

The possibility of biological control of *P. aculeata* in its native range in North America by the introduction of natural enemies from South America was suggested by DeLoach (1981) following preliminary surveys in South

Figure 2: *Parkinsonia aculeata* in Australia. a) Leaves (pinnae and pinnules) and thorns. b) Flowers. c) Mature pods. d) Adult plant in flower. e) A large infestation in wetlands of the Gulf Region, Qld. Source: Nathan March, Biosecurity Queensland.

America, although he concluded that 'the conflict of interest may be too great to introduce control agents'. Surveys for natural enemies were conducted in Argentina and Paraguay between 1975 and 1978 (from notes and collections of Hugo Cordo). However, Australia is the only country where *P. aculeata* has been formally targeted for biological control.

The Australian biological control program commenced in 1983 as a joint project between government agricultural agencies in WA, Qld and the NT. It was approved as a target for biological control in 1983 (Donnelly 2000) and native range surveys were conducted by William Woods from 1983 to 1987, principally in the USA and northern Mexico. Two insects were subsequently tested in quarantine by the Qld Government, the sap-sucking mirid *Rhinacloa callicrates* Herring and the seed-feeding bruchid *Mimosestes ulkei* (Horn). Both were released from 1993. A third insect, the seed-feeding bruchid *Penthobruchus germaini* Pic (Fig. 3) from Argentina and Chile, was identified as a potential agent from the literature (Kingsolver 1973) and in collections made by Hugo Cordo in 1977. *P. germaini* was sent to Australia from 1990 to 1993 following field and laboratory studies

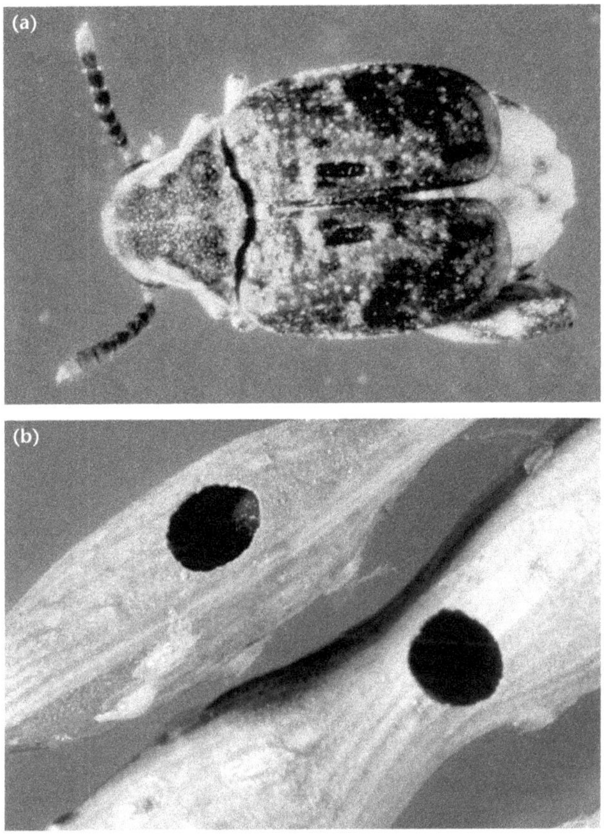

Figure 3: *Penthobruchus germaini.* a) Adult. b) Eggs and adult emergence hole on a mature pod. Source: Biosecurity Queensland.

at the USDA-ARS South American Biological Control Laboratory in Argentina (SABCL) from 1986 to 1989. It was tested in quarantine by the Qld Government and subsequently released from 1995 (Briano *et al.* 2002).

The biological control program was reinitiated by CSIRO in 1999 and is ongoing (Heard 2006; van Klinken *et al.* 2009a; Heard *et al.*, unpub. data). Its main focus is to conduct surveys in previously unexplored or underexplored parts of the native range across Central and South America. It has included preliminary host-range testing of several potential agents at the CSIRO Mexican field station in Veracruz, as well as intensive sampling on *P. aculeata* and other leguminous species at a site in Mexico to determine field host-ranges during 2007–10. From 2008, the SABCL conducted further field surveys and laboratory studies in Argentina on behalf of CSIRO. The first agents have been imported into Australian quarantine for host-specificity testing, including a geometrid moth *Eueupithecia cisplatensis* Prout and a tortricid moth *Ofatulena luminosa* Heinrich.

PLANT TAXONOMY

P. aculeata readily hybridises with other *Parkinsonia* species in the Americas, but there is no evidence of hybrid populations in Australia. In fact, Australian populations appear to be morphologically and genetically very uniform, suggesting a single translocation event from Meso-America, perhaps Venezuela (Hawkins *et al.* 2007).

EXPLORATION

There has been considerable controversy and confusion regarding the native range of *P. aculeata* (Hawkins *et al.* 2007; van Klinken *et al* 2009a), resulting in the changing focus of successive exploration programs for biological control agents. This confusion was partly the result of *P. aculeata* apparently being largely restricted across much of its American distribution to disturbed habitats, such as irrigation ditches, heavily grazed farms and roadsides.

The first surveys were carried out by the USDA in Argentina from 1975 to 1978. Australia became involved later (April 1983 until January 1987) with exploration in Texas and Arizona (USA) and Sonora and Baja California (northern Mexico), and with brief surveys in Costa Rica and Guerrero (southern Mexico) (Woods 1986, 1988, 1992). A special focus was placed on the Sonoran Desert as it was thought to be the centre of origin for the genus and few insects were found elsewhere (Woods 1988). As a result, North America (USA and northern Mexico) were relatively thoroughly surveyed.

Recent genetic studies support the conclusion that *P. aculeata* has an exceptionally large native range, taking in semi-arid and wet-dry tropical regions from southern USA through to Argentina (Hawkins *et al.* 2007). Marked genetic divergence was found between *P. aculeata* populations in northern and western Mexico and south-western USA; central and southern Mexico and Texas; Venezuela; Central America; and Argentina. A molecular clock analysis concluded that the split between North and South America predated the emergence of the Panama Isthmus 3.5–5 million years ago (Hawkins *et al.* 2007). This biogeographic understanding of *P. aculeata* is likely to improve further with ongoing genetic work sampling populations in South America more thoroughly (J. Hawkins, *pers. comm.*).

Surveys by CSIRO from 1999 initially focused on Central America and Venezuela, in part because limited sampling by Woods suggested a relatively rich natural enemy fauna in Costa Rica, and observations by Hughes (Hughes and Styles 1984; Hughes 1986) of extensive natural populations in sunken volcanic lake beds in Central America. The Central American populations were surveyed in visits to southern Mexico, Guatemala, Nicaragua and Costa Rica between 1999 and 2008. Venezuela was also surveyed during that time, not because there was evidence it was part of the native range, but opportunistically with other project trips. A promising site was found in eastern Mexico (near the city of Tampico) in 2006 and it became the focus of activity until 2009 due to its accessibility, extensive parkinsonia populations and abundance of natural enemies. Following the interpretation of Hawkins *et al.* (2007) of the native status of the Argentinean populations, brief surveys were made in other South American sites in Ecuador, Peru, Brazil and Paraguay between 2006 and 2008. Also, the curated collection of parkinsonia natural enemies held by SABCL (Argentina) was re-examined and the Caribbean island nations of Dominican Republic and Puerto Rico were surveyed in 2007 and 2008 respectively. However, several areas remain understudied; Argentina, Colombia and Brazil in particular would benefit from further surveys (Bell *et al.*, unpub. data).

CANDIDATES

Natural enemies (Australia)

The most important natural enemies in Australia are soil-borne pathogens that cause dieback of *P. aculeata* populations across much of its Australian distribution (Toh *et al.* 2008; Diplock *et al.* 2006, 2008; Toh 2009; van Klinken *et al.* 2009a; N Diplock, unpub. data). This dieback is likely to be a primary cause of population declines that are commonly observed across Australia (van Klinken *et al.* 2009a; JB Pichancourt and RD van Klinken, unpub. data), and is the likely reason why *P. aculeata* has not become a serious problem in some regions such as the Victoria River district and parts of Qld (van Klinken *et al.* 2009a). Experimental work in the laboratory, glasshouse and field implicates several generalist cosmopolitan species of Botrysphaeriaceae, a family of soil-borne pathogens that are commonly implicated in dieback in natural and agricultural systems elsewhere in the world (de Wit *et al.* 2008). It is still not known what triggers the dieback phenomenon, why other co-existing native plant species do not appear to be affected and why it apparently does not affect *P. aculeata* in its native range.

Several insect species can locally damage *P. aculeata* in Australia, although they often appear to be most abundant on plants that are already stressed (van Klinken *et al.* 2009a). The most damaging is the giant termite *Masto-*

termes darwiniensis Froggatt (Mastotermitidae), which is capable of killing mature plants. An unidentified coccid (mealybug) has also been observed at high densities in the Victoria River district, where it can cause gnarling of growth tips.

Natural enemies (Americas)

The total phytophagous fauna of *P. aculeata* can be estimated from the results of the unpublished USDA surveys by Hugo Cordo in Argentina and Paraguay, the surveys of Woods (1988) in USA, Mexico and Costa Rica, and more recent CSIRO surveys (Heard *et al.*, unpub. data). A review of the specimens collected in the USDA Argentinean collection reveals 36 insect species in Argentina, Bill Woods collected 142 insects and two mites (Woods 1985, 1986, 1988, 1992) and CSIRO collected approximately 250 insect and one fungus species (Heard *et al.*, unpub. data). After removing the species common to these collection efforts, the total identified fauna is approximately 350 species. However, most species (~53%) were rare on parkinsonia, with only one or two specimens collected. This rarity is reflected by field observations that most parkinsonia plants surveyed had few natural enemies present, and there was rarely heavy damage to any plant parts. Analyses using species accumulation curves suggest that only 48% of all natural enemies on parkinsonia have been collected, but 90% of the common species have been collected and these are the ones most likely to be useful biocontrol agents (K Bell *et al.*, unpub. data).

Of the 145 arthropod species for which we had reliable data on host-specificity, seven (4.8%) were considered specific to the genus *Parkinsonia*, lower than on other host plants examined (Bell *et al.*, unpub. data). Of those, three species have already been released in Australia (see below).

Only one pathogen, *Septoria* sp., has been reported from *P. aculeata* collected in five localities in Nicaragua and Venezuela (H Evans, *pers. comm.*). It was reported to cause significant rachis and branch cankering (H Evans, *pers. comm.*), but does not appear to cause widespread damage.

Agents tested and rejected

Most potential agents have been rejected on the basis of field data, rarity or literature records. However, three species were studied at the Mexican field station before being rejected subsequently.

Atrypanius irrorellus **Bates (Coleoptera: Cerambycidae).** This seed-feeding beetle is common in the gulf coast of Mexico. Its host-specificity was tested in the field and laboratory garden in Mexico and it was found to develop on various legume species (R Segura and TA Heard, unpub. data).

Iridopsis aglauros **Shaus (Lepidoptera: Geometridae).** This leaf-feeder is relatively abundant in Mexico and Central America. There were no previous host-plant records in the literature, so field host-specificity tests were carried out in Mexico. These found that the insect could complete its life-cycle on several other legume species (*Caesalpinia pulcherrima*, *Leucaena leucocephala*, *Acacia cornigera*, *Mimosa asperata*, *Acacia farnesiana*, *Desmanthus virgatus* and *Prosopis tamaulipana*), making it unsuitable as a biocontrol agent (R Segura and TA Heard, unpub. data).

Tolype nanus **(Druce) (Lepidoptera: Lasiocampidae)** This leaf-feeding species is relatively abundant in Mexico. It was tested at the field station garden and found to develop on a range of legume species (*Mimosa asperata*, *Tamarindus indica* and *Delonix regia*) (R Segura and TA Heard, unpub. data).

Agents released

Three agents have been released, two seed-predators and a sap-sucking mirid.

Rhinacloa callicrates **Herring (Hemiptera: Miridae).** This species is a North American sap-sucking bug that has been observed to cause distortion of growth and cessation of flowering at very high population levels in its native range (Woods 1988, 1992). Cigar-shaped eggs are oviposited singly under the epidermis, mainly of the leaf rachis (Donnelly 2000). Eggs take an average of 12.3 days to hatch and nymphal development takes approximately 19 days in an insectary (22–37°C) (Donnelly 2000). Adults and nymphs feed on green foliage and on softer terminal sections of growing tips (Donnelly 2000). In its native range adults fly readily and are attracted to light, and can be observed swarming around host plants at night (Donnelly 2000). Initial host-testing in Arizona showed that it would lay eggs and feed only on *P. aculeata* (Woods 1986), and only very minor oviposition and feeding were recorded on non-targets during extensive testing in Australia (Donnelly 2000). Releases were made between 1989 and 1994, with approximately 500 000 insects released in Qld (80 sites) and additional insects released in the NT and WA (East Kimberley) (Flanagan *et al.* 1996; Julien and Griffiths 1998; Donnelly 2000).

Mimosestes ulkei **(Horn) (Coleoptera: Bruchidae).** This species is multivoltine, oviposits on almost-mature or mature pods on both the tree and the ground (Woods

1985) and is common and widespread on *P. aculeata* in the USA and Mexico (Woods 1988). Yellow cigar-shaped eggs are cemented firmly to the pod surface, or through open sutures, cracks and insect entry and emergence holes, in which case they are usually attached by a peduncle at one end of the egg (G Donnelly, unpub. report, February 1991). First instar larvae can wander freely within and between pods. Host-testing predicted that it would complete development only on *P. aculeata* in Australia (G Donnelly, unpub. report, February 1991), although it can attack several *Parkinsonia* species in its native range (Woods 1986). It was released on *P. aculeata* between 1995 and 1999 throughout much of its range in Qld and the NT, and in the Kimberley (Donnelly 1998; Lockett *et al.* 1999; Lukitsch and Wilson 1999; van Klinken *et al.* 2009a).

Penthobruchus germaini **(Pic.) (Coleoptera: Bruchidae).** This multivoltine seed-feeder oviposits on mature pods, primarily on the tree, and is relatively common and widespread in Argentina (Briano *et al.* 2002; H Cordo, *pers. comm.*). The life-cycle is completed in 35–40 days at a constant 30°C, but individuals can overwinter as larvae (Briano *et al.* 2002). Eggs are individually glued onto the surface of the pod or free seed and covered by a characteristic thin membrane. Larvae destroy seeds by consuming approximately 90–100% of the cotyledons (Briano *et al.* 2002). Host-testing showed that it would attack only *P. aculeata* in Australia, with the possible exception of *Delonix regia* (Hook.) Raf. (poinciana) (G Donnelly, unpub. report, November 1994). A single adult emerged from *D. regia* in laboratory trials, but pods were expected to deter oviposition. Releases were made between 1995 and 1999 in Qld and the NT, and in the Kimberley in WA (Donnelly 1998; Lockett *et al.* 1999; Lukitsch and Wilson 1999; van Klinken 2005; van Klinken and Flack 2008; van Klinken *et al.* 2009a).

Other promising agents

Several species have been identified as having some potential as biological control agents based on current knowledge of their suspected host-specificity and/or the damage they cause (Table 1). Work on most of these candidates is ongoing, and is summarised below. There may be other potential agents, but very little is known about them at this stage.

Ofatulena luminosa **Heinrich (Lepidoptera: Tortricidae).** This multivoltine species is consistently common in the Tampico delta area of the Mexican Gulf and is also known from the USA (California, Arizona and Texas) (Brown *et al.* 2010). Larvae bore in growing tips, mature green stems and green seeds. Up to five larvae have been dissected from the distal 25 cm of a *P. aculeata* stem. A single larva may also develop in a green seed, eating out the seed and killing it. Larvae of *O. luminosa* are heavily parasitised by Hymenoptera (Woods 1992). No success has yet been achieved with rearing this species for a full generation, either in Mexico or in Australia, as adults are fragile and short-lived (several days) and larvae may need stems of a specific stage. In a field survey of potential host plants, this species was not collected from any other legume (>20 species surveyed) except from pods of *Parkinsonia florida* (Benth. Ex A. Gray). No damage was caused to other legumes in no-choice host-specificity tests conducted in garden plots (Brown *et al.* 2010).

Rudenia leguminana **complex sp. B (Lepidoptera: Tortricidae).** This species is widely distributed from the USA to Venezuela and has been reared in large numbers from *P. aculeata* in Guatemala, Mexico, Nicaragua and Venezuela (Brown *et al.* 2010). First instar larvae feed inside the rachis before making a tunnel in the axil, in which they hide during the day. Larvae leave the tunnel at night to feed on the pinnules and rachis of leaves. Larvae can also develop in flowers and occasionally pods. It could potentially be a species complex, as it has an unusually broad geographic range for Neotropical tortricids, and analyses of two molecular markers strongly suggest that the individuals examined belong to more than one species (Brown *et al.* 2010). Host-specificity studies at the Mexican field station lent evidence to the hypothesis that a host-specific cryptic species may be included in the currently defined species, because in open-field trials it successfully developed only on *P. aculeata*. Records in the literature of the use of various legumes may therefore refer to other species in this complex.

Calosima **sp. (Lepidoptera: Coleophoridae).** At least one undescribed species of this genus was found on *P. aculeata* at several sites in Central America and northern South America. It is a stem-borer with a similar habit to *O. luminosa*, although adults are larger. Formal taxonomic description of these species is being conducted by David Adamski (USDA-ARS Systematic Entomology Laboratory). It is possibly host-specific, but may not be sufficiently common to establish cultures for host-specificity testing.

Eueupithecia cisplatensis **Prout (Lepidoptera: Geometridae).** This multivoltine looper is abundant and widespread in Argentina. Adults are relatively short-lived (females survived for up to seven days). Larvae tend to be evenly dispersed in the field and so spectacular damage is

Table 1: Potential biological control agents that are still under consideration

Species	Feeding	Origin	Notes	Current status
Lepidoptera: Tortricidae				
Ofatulena luminosa	Stem- and seed-boring moth	Mexico, USA	May be host-specific. Not culturable yet	Testing in Australia
Rudenia leguminana complex sp. B	A defoliating/flower bud-feeding moth that makes a shelter by boring into tips	USA to Venezuela	May be host-specific. Can be lab-reared	Testing in Mexico
Lepidoptera: Geometridae				
Eueupithecia cisplatensis	Defoliating looper caterpillar	Argentina	Likely to be host-specific. Easily cultured	Testing almost completed, application for release in preparation
Lepidoptera: Coloephoridae				
Calosima sp.	Stem-boring moth	Central and northern South America	Rare in the field. May be host-specific. Culturing not attempted yet	Potential agent
Diptera: Cecidomyiidae				
Neolasioptera n. sp.	Galler on growing tips	Argentina, Paraguay	Difficult to culture, may be host-specific	Testing in Argentina and attempted in Australian quarantine
Coleoptera: Chrysomelidae				
Glyptoscelis sonorensis	Defoliating leaf-beetle, larvae may feed on roots	Mexico	Culturing not attempted yet	Potential agent
Myochrous austrinus		Mexico, Panama, Colombia		Potential agent
Myochrous melancholicus		Mexico		Potential agent
Coleoptera: Buprestidae				
Agrilus parkinsoniae	Stem-borer	Mexico		Potential agent

not observed but, free of its natural enemies, it may cause heavy damage to both pinnules and rachises. Both field and laboratory studies by USDA collaborators in Argentina have provided convincing evidence of the host-specificity of this species (F McKay, unpub. data). A total of 28 legume species have been tested in replicated experiments; these showed that *E. cisplatensis* is incapable of developing on any species other than *P. aculeata* including *Parkinsonia praecox* (Ruiz & Pav.) Hawkins. Field surveys of host plant use confirmed that the species only occurs naturally on the target weed. This insect was imported into Australian quarantine for final testing in 2010 and at August 2011, testing was almost complete with no evidence of any damage to any other plant species. An application for release is in preparation. A new cryptic species of *Eueupithecia* has recently been diagnosed by Axel Hausmann (Bavarian State Collection of Zoology, Munich). It appears to also be a specialist of *P. aculeata* with potential for biocontrol.

***Neolasioptera aculeatae* Gagné (Diptera: Cecidomyiidae)**. This gall midge causes damage to growing tips of *P. aculeata* in Argentina and Paraguay, where it can be very abundant. *Neolasioptera* species are generally specific and tied to host biology. A formal description of this species has recently been completed by Dr Raymond Gagné (USDA-ARS Systematic Entomology Laboratory) (Gagné *et al.* 2011). A shipment has been received into Australian quarantine but no success with culturing has yet been achieved.

Leaf-beetles (Coleoptera: Chrysomelidae). Three species appear to have potential as biological control agents although their host-range is not yet known. *Glyptoscelis sonorensis* Blake is widespread in Mexico and locally abundant, *Myochrous melancholicus* Jacoby is known from southern Mexico (Oaxaca) and *Myochrous austrinus* Blake (Coleoptera: Chrysomelidae: Eumolpinae) is known from southern Mexico (Oaxaca), Panama and Colombia. All three species have been recorded only on *P. aculeata*. Larvae of all three species may be root-feeders as they all belong to the subfamily Eumolpinae which typically feed on plant roots, and larvae have never been found on above-ground parts. However, larval feeding habits would need to be understood before their potential to significantly damage *P. aculeata*, which has an extensive root system, can be assessed.

***Agrilus parkinsoniae* Hespenheide (Coleoptera: Buprestidae).** Adults of this species are relatively common on *P. aculeata* in Oaxaca state, Mexico. Although only adults have so far been collected, taxonomy suggests that it is a stem-borer (Hespenheide and Bellamy 2009) and may therefore have potential as a biological control agent. This species has no other recorded host plants. However, at this stage its host-range has not been examined thoroughly.

EVALUATION

Objective measurable criteria to assess the success of a *P. aculeata* biocontrol program were developed by van Klinken (2006). Three sets of criteria were used, ecological (whether the desired population-level changes are achieved), geographical (whether control is being achieved where parkinsonia poses the greatest threat) and value to integrated management in the event that the ecological and geographical objectives cannot be met by biological control alone. Criteria included:

- reducing patch density (to <30% cover) and size (to <0.1 ha);
- reducing rates of spread and in-fill (by reducing seed production and understorey seed bank densities to <100 viable seeds m^{-2});
- explicitly targeting regions most at risk from parkinsonia impacts (the Barkly Tablelands and central Qld to the Gulf of Carpentaria to reduce both pastoral and environmental threats, and the Pilbara to reduce environmental threats);
- reducing growth and recruitment (by 50%) and increasing time to reproduction (by at least one year, currently two to three years) to reduce the cost of control.

Currently, none of those objectives have been met.

Field inspections in Qld over seven years following the release of *R. callicatres* indicated that it had established at 30 sites (37.5% of release sites) and had spread at least 20 km from one site, but no plant damage was found (e.g. shoot distortion or necrosis) (Donnelly 2000). Establishment has not been confirmed in either the NT or WA (Donnelly 2000). The reason for its apparent failure to establish or to reach damaging densities is not known.

Post-release evaluation of the seed predators was conducted by CSIRO (van Klinken 2005; van Klinken and Flack 2008) and state research agencies (Donnelly 1998; Lockett *et al.* 1999; Lukitsch and Wilson 1999). *M. ulkei* was reported to kill up to 5.3% of seeds in the NT (Donnelly 1998; Locket *et al.* 1999), but has not been reported in the past 10 years despite intensive and widespread sampling across Australia between 2000 and 2004 (van Klinken 2005; van Klinken and Flack 2008). *P. germaini* established readily and rapidly dispersed, even reaching isolated *P. aculeata* populations in the Alice Springs region, several hundred kilometres from the nearest known release sites (van Klinken and Flack 2008). It is now rare to find a *P. aculeata* population that is free of *P. germaini*. The main exception is the Pilbara region where systematic releases have not yet been made. However, *P. germaini* is unlikely to be causing population-level impacts (van Klinken 2005; van Klinken and Flack 2008). Seed predation rates of more than 30% were rare, and generally peaked before or after the period when most seeds were available for predation. Several factors combined to limit predation rates: the inability of *P. germaini* to track sudden seasonal fluctuations in pod supply which resulted in a lag-phase between seed availability and beetle numbers, high egg parasitism rates (10–70%), high unexplained larval mortality within seeds (a constant 62%) and slightly aggregated (negative binomial) oviposition. Larval mortality and oviposition behaviour alone were sufficient to ensure that oviposition rates would need to be unrealistically high (an average of >8.5 eggs per seed) to achieve 80% seed mortality (van Klinken and Flack 2008).

DISCUSSION

Classical biological control is currently having little or no population-level impacts on *P. aculeata* in Australia. However, several potential agents have been identified for further study. In addition, the biological control program on *P. aculeata* has proved to be a useful system for addressing several important areas in biological control, including agent selection (van Klinken 2006), the use of seed predators (van Klinken 2005; van Klinken and Flack 2008), genetics of the host (Hawkins *et al.* 2007) and the design and implementation of native range surveys (Bell *et al.*, unpub. data). Some of these developments are summarised below.

Seed predators are commonly used as biological control agents, in part because they are easily cultured and tested, and because they are frequently host-specific. However, with a few spectacular exceptions, most seed predators used in biological control have resulted in limited population-level control (van Klinken and Flack 2008). Detailed continent-scale evaluation of seed predation on *P. aculeata* showed that seed predation rates were fundamentally constrained, even though the main seed predator *P. germaini* was typically abundant. The most important constraints were the inability of seed predator populations to respond rapidly enough to seasonal peaks in seed availability (temporal inertia), egg parasitism, within-seed larval mortality and oviposition behaviour (van Klinken 2005; van Klinken and Flack 2008). Although these constraints are likely to be common among seed predators, they can be examined *a priori* to help prioritise agents based on likely efficacy. Results suggest that univoltine insects are less likely to have these constraints. Indeed, the most effective seed-predating biocontrol agents are univoltine (van Klinken and Flack 2008).

The value of understanding the genetic structure of the target weeds for biological control is being increasingly appreciated (Goolsby *et al.* 2006) and has been instrumental in the development of the parkinsonia biological control program. Genetic structuring in the native range helped confirm the extent of its native range, and helped direct exploration work aimed at comprehensively documenting the native range natural enemies on this species. The genetic work also showed that Australian populations are likely to represent a single genotype from Venezuela (Hawkins *et al.* 2007). This may be an important consideration when host-testing potential biological control agents, but at this stage there is no indication that any natural enemies on *P. aculeata* discriminate between hosts at the genotype level.

Finally, *P. aculeata* has proved to be a valuable model system for improving native range survey methodology, especially with respect to optimising sampling strategies across a geographically extensive and biogeographically diverse area (Bell *et al.*, unpub. data). Analyses of native range data suggested that biogeographic province and geographic distance were the best predictors of natural-enemy dissimilarity, and in this case are therefore the most important parameters to consider when designing survey strategies. The predominance of rare natural enemies means that ongoing surveys at even the best-sampled sites are expected to reveal more species. However, gap analyses have helped identify regions where return for effort is likely to be greatest (Bell *et al.*, unpub. data). The CSIRO Mexican and the USDA Argentinean field stations have been fundamental to this project, as they have allowed surveys to be conducted across the large geographic range of this plant and allowed preliminary testing of potential agents.

Future priorities within the parkinsonia biological control program are to complete native range exploration and to conduct detailed host-specificity testing of the most promising agents. Final native range exploration is targeting the few remaining undersampled regions, specifically Argentina, Colombia and Brazil (Bell *et al.*, unpub. data). One agent, *E. cisplatensis*, has been subjected to detailed study in Australian quarantine following work in the native range to develop culturing techniques and to show that it is likely to be host-specific. The other potential agents, particularly *Neolasioptera* sp. and *O. luminosa*, have also been brought into quarantine for detailed host-specificity testing, but need further work to develop culturing techniques and/or to confirm through preliminary host-testing that they are likely to be host-specific.

ACKNOWLEDGEMENTS

Field work was ably conducted by Ricardo Segura, Moises Martinez, Carlos Pascacio, Quiyari Santiago (CSIRO Mexico), Hugo Cordo, Fernando Mckay, Alejandro Sosa and Marcelo Parisi (USDA Argentina). We thank Andrew White, Gio Fichera and Areli Mira for the laboratory host-specificity testing of new potential agents.

REFERENCES

Briano JA, Cordo HA and DeLoach CJ (2002) Biology and field observations of *Penthobruchus germaini* (Coleoptera: Bruchidae), a biological control agent for *Parkinsonia aculeata* (Leguminosae: Caesalpinioideae). *Biological Control* **24**, 292–299.

Brown JW, Segura R, Heard TA, Rota J and Heard TA (2010) Tortricid moths reared from the invasive weed Mexican palo verde, *Parkinsonia aculeata*, with comments on their host specificity, biology, geographic distribution, and systematics. *Journal of Insect Science* **11**, 1–17.

DeLoach CJ (1981) Prognosis for biological control of weeds of southwestern US rangelands. In *Proceedings of the 5th International Symposium on Biological Control of Weeds*. 22–27 July 1980, Brisbane. (Ed ES Delfosse) pp. 175–199. CSIRO, Melbourne.

de Wit J, Slippers B, Preisig O, Wingfield BD and Wingfield MJ (2008) Phylogeny of Botryosphaeriaceae reveals patterns of host association. *Molecular Phylogenetics and Evolution* **46**, 116–126.

Diplock N, Galea V, van Klinken RD and Wearing A (2006) A preliminary investigation of dieback on *Parkinsonia aculeata*. In *Papers and Proceedings of the 15th Australian Weeds Conference*. 24–28 September 2006, Adelaide. (Eds C Preston, JH Watts and ND Crossman) pp. 585–587. Weed Management Society of South Australia, Adelaide.

Diplock N, Galea V and van Klinken RD (2008) Movement of dieback front through a stand of parkinsonia – a time series study. In *Proceedings of the 16th Australian Weeds Conference*. 18–22 May 2008, Cairns. (Eds RD van Klinken, VA Osten, FD Panetta and JC Scanlan) p. 186. Queensland Weeds Society, Brisbane.

Donnelly GP (1998) Levels of attack and destruction of *Parkinsonia aculeata* seeds by bruchid biological control agents. In *Proceedings of the 6th Australasian Applied Entomological Research Conference. Vol. 2*. 29 September–2 October 1998, Brisbane. (Eds MP Zalucki, RAI Drew and GG White) p. 310. University of Queensland Press, Brisbane.

Donnelly GP (2000) Biology and host specificity of *Rhinacloa callicrates* Herring (Hemiptera: Miridae) and its introduction and establishment as a biological control agent of *Parkinsonia aculeata* L. (Caesalpiniaceae) in Australia. *Australian Journal of Entomology* **39**, 89–94.

Flanagan GJ, van Rangelrooy DS and Kerin S (1996) Integrated management of *Parkinsonia aculeata* on the Roper River, Northern Territory, Australia. In *Proceedings of the 9th International Symposium on Biological Control of Weeds*. 19–26 January 1996, Cape Town. (Eds VC Moran and JH Hoffmann) pp. 441–443. University of Cape Town, South Africa.

Gagné RJ, Mc Kay F and Heard TA (2011) A new species of *Neolasioptera* (Diptera: Cecidomyiidae) from *Parkinsonia aculeata* (Leguminosae) in Argentina for possible use in biological control in Australia, with a key to Neotropical species of *Neolasioptera*. *ZooTaxa* **2866**, 61–68.

Goolsby JA, van Klinken RD and Palmer WA (2006) Maximizing the contribution of native-range studies towards the identification and prioritization of weed biocontrol agents. *Australian Journal of Entomology* **45**, 276–286.

Hawkins JA (2001) *Parkinsonia aculeata* (Mexican paloverde). In *Forestry Compendium*. CABI, Wallingford, UK.

Hawkins JA, Boutaoui N, Cheung KY, van Klinken RD and Hughes CE (2007) Intercontinental dispersal prior to human translocations revealed in a cryptogenic invasive plant. *New Phytologist* **175**, 575–587.

Heard TA (2006) *Parkinsonia aculeata*: surveys for natural enemies, native range ecological studies and prospects for biological control. In *Proceedings of the 15th Australian Weeds Conference*. 24–28 September 2006, Adelaide. (Eds C Preston, JH Watts and ND Crossman) pp. 581–584. Weed Management Society of South Australia, Adelaide.

Hespenheide HA and Bellamy CL (2009) New species, taxonomic notes, and records for *Agrilus curtis* (Coleoptera: Buprestidae) of Mexico and the United States. *Zootaxa* **2084**, 50–68.

Hughes CE (1986) Progress in collection and evalution of NFT germplasm from Central America. *Nitrogen Fixing Tree Research Reports* **4**, 49–51.

Hughes CE and Styles BT (1984) Exploration and seed collection of multiple-purpose dry zone trees in central America. *International Tree Crops Journal* **3**, 1–31.

Julien JH and Griffiths MW (1998) *Biological Control of Weeds: A World Catalogue of Agents and their Target Weeds*. 4th edn. CABI, Wallingford, UK.

Kingsolver JM (1973) Description of a new genus and a new species of Bruchidae from South America (Coleoptera). *Journal of the Washington Academy of Science* **63**, 142–146.

Lockett C, Gray E and Donnelly G (1999) The prevalence of seed damage by the introduced bruchid, *Penthobruchus germaini* (Coleoptera: Bruchidae), a biological control agent for parkinsonia (*Parkinsonia aculeata*), in far north and central Queensland. In *Proceedings of the 12th Australian Weeds Conference*. 12–16 September 1999, Hobart. (Eds AC Bishop, M Boersma and CD Barnes) pp. 431–435. Tasmanian Weeds Society, Hobart.

Lukitsch B and Wilson A (1999) Distribution and impact of the mature seed feeding bruchid, *Penthobruchus germaini* on *Parkinsonia aculeata* in northern Australia. In *Proceedings of the 12th Australian Weeds Conference*. 12–16 September 1999, Hobart. (Eds AC

Bishop, M Boersma and CD Barnes) pp. 436–440. Tasmanian Weeds Society, Hobart.

Martin T and van Klinken RD (2006) Value for money? Investment in weed management in Australian rangelands. *Rangeland Journal* **28**, 63–75.

Stewart JL, Dunsdon AJ, Hellin JJ and Hughes CE (1992) *Wood Biomass Estimation of Central American Dry Zone Trees*. Tropical Forestry Papers No. 26. Oxford Forestry Institute, Oxford.

Thorp JR and Lynch R (2000) *The Determination of Weeds of National Significance*. National Weeds Strategy Executive Committee, Launceston.

Toh R (2009) Investigation of fungi pathogenic towards seedlings of *Parkinsonia aculeata* – their potential use as mycoherbicides. PhM thesis. University of Queensland, Australia.

Toh R, Galea V, Diplock N and van Klinken RD (2008) Evaluation of fungal isolates for potential use as mycoherbicides for seedbank reduction of *Parkinsonia aculeata*. In *Proceedings of the 16th Australian Weeds Conference*. 18–22 May 2008, Cairns. (Eds RD van Klinken, VA Osten, FD Panetta and JC Scanlan) pp. 284–286. Queensland Weeds Society, Brisbane.

van Klinken RD (2005) Total annual seed loss on a perennial legume through predation by insects: the importance of within-season seed and seed-feeder dynamics. *Austral Ecology* **30**, 414–425.

van Klinken RD (2006) Parkinsonia biocontrol: what are we trying to achieve? *Australian Journal of Entomology* **45**, 268–271.

van Klinken RD and Flack L (2005) The relationship between wet heat and hard-seeded dormancy and germination. *Weed Science* **53**, 663–669.

van Klinken RD and Flack L (2008) What limits predation rates by a specialist seed feeder? *Journal of Applied Ecology* **45**, 1600–1611.

van Klinken RD, Flack L and Pettit W (2006) Wet-season dormancy release in seed banks of a tropical leguminous shrub is determined by wet heat. *Annals of Botany* **98**, 875–883.

van Klinken RD, Lukitsch B and Cook C (2008) Interaction between seed dormancy-release mechanism, environment and seed bank strategy for a widely distributed perennial legume (*Parkinsonia aculeata*, Caesalpinaceae). *Annals of Botany* **102**, 255–264.

van Klinken RD, Campbell SD, Heard TA, McKenzie J and March N (2009a) The biology of Australian weeds. *Parkinsonia aculeata* L. *Plant Protection Quarterly* **24**, 100–117.

van Klinken RD, Lawson BE and Zalucki MP (2009b) Predicting invasions in Australia by a Neotropical shrub under climate change: the challenge of novel climates and parameter estimation. *Global Ecology and Biogeography* **18**, 688–700.

Woods W (1985) Bruchid seed beetles for control of *Parkinsonia aculeata* in Australia. In *Proceedings of the 6th International Symposium on Biological Control of Weeds*. 19–25 August 1984, Vancouver. (Ed E Delfosse) pp. 855–862. Agriculture Canada, Ottawa.

Woods W (1986) Biological control of parkinsonia. *Western Australian Journal of Agriculture* **27**, 80–83.

Woods W (1988) The potential for biological control of *Parkinsonia aculeata* L.: phytophagous insects collected from the U.S.A., Mexico and Costa Rica and the effect of insect damage on the growth and survival of the plant. MSc thesis. University of Western Australia, Australia.

Woods W (1992) Phytophagous insects collected from *Parkinsonia aculeata* (Leguminoseae: Caesalpiniaceae) in the Sonoran Desert Region of the south western United States and Mexico. *Entomophaga* **37**, 465–474.

Parthenium hysterophorus L. – parthenium

K Dhileepan and Rachel Cruttwell McFadyen

ABSTRACT

Parthenium hysterophorus, an annual herbaceous plant native to the tropical Americas, is now a major crop and pasture weed in Australia as well as many parts of Africa and Asia. It is also a human health hazard, causing severe allergic dermatitis and respiratory problems. A major biological control program against parthenium in Qld started in 1977 with scientists based in Mexico and Brazil, and continued until 2002. Over 144 species of herbivorous insects and several pathogenic fungi were identified, of which nine insect species and two rust fungi were introduced into Qld. Seven species of insects and two rust fungi have established on parthenium and are widespread. Key agents are the stem-galling moth *Epiblema strenuana*, the leaf-feeding beetle *Zygogramma bicolorata*, the stem- and root-feeding weevil *Listronotus setosipennis*, the seed-feeding weevil *Smicronyx lutulentus* and the summer rust *Puccinia xanthii* var. *parthenii-hysterophorae*. The combined impact of these agents has resulted in significant reductions in the abundance and impact of parthenium in most situations and seasons, though serious infestations can still occur. Some of these agents have been successfully introduced into India, Sri Lanka and China, and are being considered for use in eastern and southern Africa.

Key words: *Bucculatrix parthenica, Carmenta ithacae, Epiblema strenuana, Listronotus setosipennis, Parthenium hysterophorus, Platphalonidia mystica, Puccinia abrupta* var. *partheniicola, Puccinia xanthii* var. *parthenii-hysterophorae; Smicronyx lutulentus, Stobaera concinna, Zygogramma bicolorata.*

INTRODUCTION

Parthenium hysterophorus L. (Asteraceae), commonly known as parthenium, is a weed of global significance affecting many countries, including Australia, where it is a Weed of National Significance (WONS). Parthenium was first identified in Australia in 1955, and proclaimed as a noxious plant in 1975 (Auld *et al.* 1982–83). There were two separate introductions into Australia; a less invasive biotype introduced through US military bases north of Brisbane in 1942–45, and a second, highly invasive biotype introduced as a contaminant of pasture seed from the USA into grazing country in central Qld in the 1960s (Graham and Lange 1998, Navie *et al.* 1996). This second biotype spread very rapidly after introduction (Auld *et al.* 1982–83) and is now found throughout large areas of Qld and to a limited extent in NSW.

Parthenium is an annual herb (Fig. 1) with a deeply penetrating tap root system and an erect shoot system. Young plants form a rosette of leaves close to the soil surface. As it matures, the plant develops many branches on its upper half, and may eventually grow up to 2 m (McFadyen 1992). Parthenium has the ability to germinate and establish at any time of year, given good rainfall. Flowering usually commences six to eight weeks after germination, although in winter plants can remain

Figure 1: Parthenium, *Parthenium hysterophorus*. Photo: Qld Government.

at the non-flowering rosette stage for some weeks. Parthenium is a prolific seed producer and a fully grown plant can produce more than 15 000 seeds in its lifetime. Seeds persist and remain viable in the soil for years, with a seed bank half-life of approximately six years (Navie *et al.* 1998a).

Parthenium can grow in a wide range of landscapes, including degraded and disturbed lands, degraded pastures, crops, orchards, forests, along railway tracks and roadsides, and streams and rivers (Navie *et al.* 1996). In Australia, parthenium mainly occurs in Qld (Fig. 2), affecting 170 000 km^2 of prime grazing country (Chippendale and Panetta 1994; McFadyen 1992). Parthenium causes human (e.g. McFadyen 1995) and animal health (e.g. Tudor *et al.* 1982) problems, agricultural losses (e.g. Navie *et al.* 1996) and serious environmental problems (Chippendale and Panetta 1994). Parthenium is a serious problem in perennial grasslands in central Qld, where it reduces beef production by as much as \$A16.5 million per year (Chippendale and Panetta 1994). Parthenium also acts as a reservoir host for insect pests and plant pathogens of crop plants (Dhileepan and Strathie 2009).

Since the 1950s, parthenium has spread to most tropical and subtropical areas of the world. Initial transcontinental spread is usually in grains from the US delivered as food aid and subsequent intracountry and intraregion spread is via agricultural machinery and vehicles. Parthenium is now a major weed in India, Pakistan, Taiwan, Vietnam, Pacific island countries, and countries of eastern and southern Africa, including Madagascar (Dhileepan and Strathie 2009). It is predominantly a weed of agriculture in areas with a distinct wet/dry season climate and is not as severe a problem in the wet tropics.

BIOLOGICAL CONTROL HISTORY

Biological control of parthenium was first initiated in Australia in 1977. Nine species of insects and two rust fungi have been released, of which seven insect species and two rust fungi are known to be established in the field. Field releases of biological control agents ceased in 2002, and recent research efforts have concentrated on monitoring the establishment and impact of the released agents. Based on the Australian program, biological control efforts have been underway in India from 1983, in Sri Lanka from 2003, in Papua New Guinea in 2004, and in South Africa and Ethiopia since 2005, with the importation of successful biological control agents from Australia (Dhileepan and Strathie 2009).

PLANT TAXONOMY

Parthenium is in the subtribe Ambrosiinae of the tribe Heliantheae, family Asteraceae. The genus *Parthenium* includes 17 species, all native to tropical America; one species, *Parthenium argentatum* Gray (Guayule), has potential economic value. Other weedy genera in the same subtribe Ambrosiinae are *Xanthium*, *Ambrosia*, *Iva*, *Hymenoclea*, *Dicoria*, *Parthenice* and *Euphrosyne*, all native to the Americas. Crop plants of economic importance within the tribe Heliantheae include sunflower (*Helianthus annus* L.), Jerusalem artichoke (*Helianthus tuberosus* L.) and niger seed (*Guizotia abyssinica* Cass.).

Several biotypes or strains of *P. hysterophorus* have been identified. The main invasive biotype of *P. hysterophorus* is native to the Gulf of Mexico (North American population) and has subsequently spread throughout the southern USA, the Caribbean and southern Brazil (Navie *et al.* 1996). A morphologically distinct biotype with

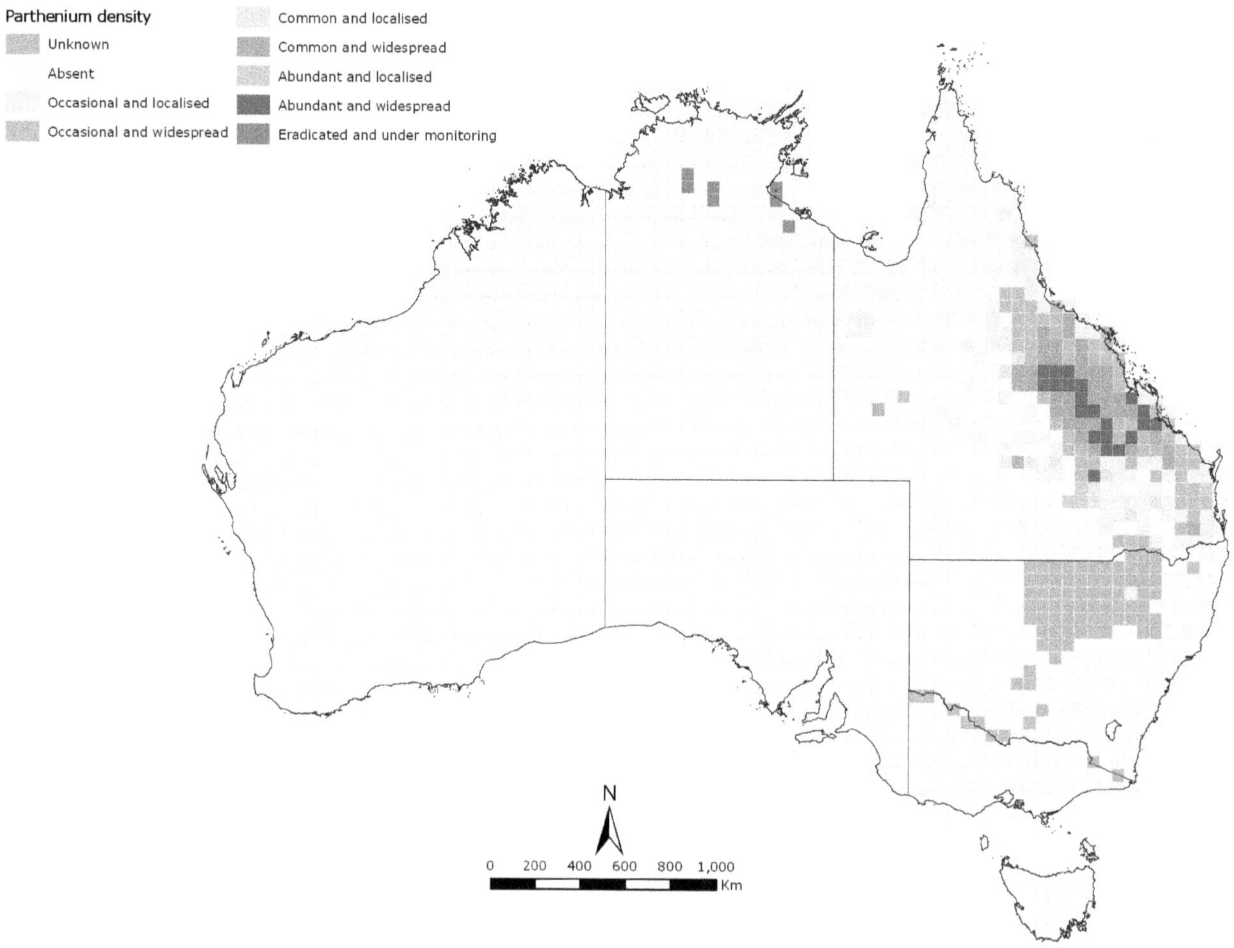

Figure 2: Current and recorded *Parthenium hysterophorus* infestations in Australia. Source: Qld Government.

yellow flowers and different chemical constituents occurs in Argentina, Bolivia, Chile, Paraguay, Peru and Uruguay (South American population) (Dale 1981). Occurrence of distinct North and South American *P. hysterophorus* populations has been confirmed by genetic studies, which also confirmed that *P. hysterophorus* genotypes in Australia originated from southern Texas (Graham and Lange 1998). Different strains are also present within the white-flowered form in the USA (Dale 1981).

EXPLORATION

On the basis of preliminary surveys conducted in Mexico and southern USA in North America, Brazil and Argentina in South America, and the Caribbean islands (Bennett 1976; McFadyen 1979), the Qld Government undertook investigations in both North and South America. A McClay (CIBC) under a contract with Qld was based in Monterrey, Mexico from 1977 to 1982, working full-time on parthenium, conducting surveys throughout Mexico and collecting suitable agents for further study (McClay 1980; McClay *et al.* 1995). Parallel exploration was conducted in central South America, where C Wild (Qld) was based in Londrina, southern Brazil from 1977 to 1980, working full-time on parthenium and surveying into Argentina and Bolivia as well as southern Brazil. From 1982 to 1994 surveys were conducted by W Palmer (Qld), based at Temple, Texas and assisted by K Pullen (Qld) based in Cuernavaca, Mexico (McClay *et al.* 1995). From 1995 to 1997, pathogens on parthenium were investigated by M Seier (CIBC), assisted by A Romero based in Xalapa, Mexico and the CSIRO field station in Veracruz, Mexico (Evans 1997). From 1987 to 1992 C Garcia (Qld) was based in Tucuman, Argentina, also full-time on parthenium, surveying and collecting agents throughout western Argentina.

In summary, from 1977 to 1992 there was a full-time scientist on this program almost continuously in both North and South America, with a total of 13 full-time scientist years plus another six part-time scientist years, based in the countries of origin of the weed, one of the largest exploration programs undertaken by Australia. The result was a comprehensive knowledge of the insects and pathogens attacking parthenium in its native range (McClay *et al.* 1995).

CANDIDATES

Potential agents found

Around 262 phytophagous arthropod species and several fungal pathogens have been catalogued from parthenium in Mexico and southern USA (Evans 1997; McClay 1980; McClay *et al.* 1995). At least 144 were found to feed on parthenium at some stage of their life-cycle, of which 13 species were restricted to Ambrosiinae. During surveys in Brazil and Argentina, around 100 insect species were recorded (McFadyen 1979), of which only three were prioritised for further host-specificity tests. None of the 15 species of insects recorded on parthenium in the Caribbean islands (Bennett 1976) was selected for host-specificity tests.

Some agents known to be host-specific in the native range were not reared and host-tested. The parthenium aphid *Uroleucon* sp. nr *ambrosiae* (Thomas) was tested in Mexico, where indications were that it was restricted to parthenium, but concerns regarding its relationship to similar aphids feeding on sunflower meant that further testing was not undertaken. The gelechiid moths *Gnorimoschaema saphirinella* (Chambers) and *Helcystogramma chambersella* (Murtfeldt) and the noctuid *Spragueia guttata* (Hodges), also from North America, might warrant further investigations.

Agents tested but not released

***Thecesternus hirsutus* Pierce (Coleoptera: Curculionidae).** This Mexican weevil (McClay and Anderson 1985) was imported into Australia for host-specificity tests in 1982 (McFadyen 1992) and 1997. The large adults feed on leaves, the eggs are laid onto the soil surface and the larvae feed on the roots, inducing development of a superficial gall. On both occasions no further progress was made as the insect could not be reared in quarantine and there was a suspicion that the primary host was in fact *Parthenium confertum* A. Gray, on which it was commonly encountered in Mexico (McClay and Anderson 1985; McFadyen 1992).

Agents released

A total of nine species of insects and two rust fungi have been host-tested and released in Australia. All have become established in the field or recovered in at least some localities, although their abundance and impact varies greatly.

***Zygogramma bicolorata* Pallister (Coleoptera: Chrysomelidae).** This leaf-feeding beetle (Fig. 3) was introduced from Mexico into Australia and released in 1980 after host-testing in Mexico and Brisbane showed the beetle could develop on only one or two species in the genera *Parthenium* and *Ambrosia* (McFadyen and McClay 1981). Both adults and larvae feed on parthenium leaves, preferentially on younger leaves. Adults lay eggs singly or in groups on the leaves, flower heads, stems and terminal and axillary buds. Newly hatched larvae feed voraciously on young leaves and fully grown larvae burrow into the soil to pupate inside an earthen cell, the pupal stage lasting two weeks. The life-cycle is completed in six to eight weeks and there are up to four generations per year depending on rainfall and food availability. In Australia, adult beetles enter diapause in the soil in autumn (April–May), triggered by short days and cool temperatures, the adults emerging in spring (September–November) in response to rainfall, increased temperature and longer days. Adult beetles can spend as much as six months in diapause in the soil in autumn and winter.

The insect became established on annual ragweed (*Ambrosia artemisiifolia* L.) in south-eastern Qld and northern NSW within two years of first releases (McFadyen 1992), but establishment on *P. hysterophorus* was much slower. Between December 1980 and May 1983 at least 84 800 beetles were released at 20 sites from south of Charters Towers and Collinsville in northern Qld to south of Rolleston in central Qld (McFadyen, unpubl. data), with several releases of >7000 beetles at a time. Early summer 1982 was exceptionally dry in central Qld, which severely limited the availability of green parthenium for beetle releases. As a result, between October 1982 and April 1983 a total of 23 000 beetles were released

Figure 3: *Zygogramma bicolorata* adult. Photo: Qld Government.

at two sites about 20 km apart on a creek 40 km south of Rolleston. This was the only site where any beetles were seen more than 12 months after releases; at all other sites, although beetles were sometimes seen in the following summer, there was no evidence of survival beyond 12 months. Along the creek, large numbers of adults and larvae were seen 12 months later over a distance of several kilometres, and in subsequent years small numbers of beetles continued to be seen on parthenium growing along the creeks. Large numbers were seen in the area in January 1990, and in 1993 a large outbreak occurred along another creek in the same catchment but 10 km north of the release site (Dhileepan and McFadyen 1997). That summer, outbreaks of *Z. bicolorata* resulting in complete defoliation of small patches of *P. hysterophorus* were reported from within an area of 200 km^2 centred on these creeks near Rolleston in central Qld. Outbreak populations therefore took 10 years to develop, and it is likely that some degree of selection or adaptation to the seasonally very dry conditions was occurring. *Z. bicolorata* is known to show variability in dormancy patterns (e.g. Jayanth and Bali 1993) and it is likely that selection in this character was critical to survival.

Since then, due to both natural spread by the beetle and deliberate spread by farmers, the area with *Z. bicolorata* defoliation has increased to about 12 000 km^2, covering more than 50 properties in central Qld where outbreaks of *Z. bicolorata* periodically cause complete defoliation (Fig. 4), resulting in reductions in plant vigour, reproductive output, weed density and soil seed banks (Dhileepan *et al.* 2000, Dhileepan 2003b). To date (2010), however, there is no indication of field establishment in northern Qld.

***Smicronyx lutulentus* Dietz (Coleoptera: Curculionidae).** This seed-feeding weevil (Fig. 5) occurs commonly on *P. hysterophorus* in Mexico and Texas (Bennett 1976). It can also survive and reproduce on *P. confertum*, but its incidence was less than on *P. hysterophorus*. Host-specificity studies in Brisbane confirmed that *S. lutulentus* is restricted to a few species within the genus *Parthenium*, with *P. hysterophorus* the preferred host, and it was approved for field release in 1980 (McFadyen and McClay 1981). A total of 15 100 adults were released between January 1981 and February 1983 at 11 release sites ranging from Collinsville in northern Qld to south of Rolleston in central Qld. All releases used young adults reared in laboratory cages in Brisbane (McFadyen, unpub. data). The weevil was initially thought to have failed to establish, as no insects were seen at release sites in the field (McFadyen 1992). However, a large population of adults and larvae were found south of Rolleston in December 1995; the weevil has subsequently been found in many areas including northern Qld (Dhileepan and McFadyen 1997).

Adults are 1.5–2.0 mm long, black/grey in colour (Fig. 5) and feed on flower buds and tender leaves. They live for up to three months, and females lay an average of 237 eggs. Oviposition occurs in buds or freshly opened capitula, and emerging larvae feed within the seeds, with a single larva per capitulum. The attacked seed swells greatly and other seeds in the head do not develop. When the mature flower head dehisces and falls to the ground, larvae leave the seeds and burrow into the soil to pupate in an earthen cell 2–6 cm below the surface. An extended prepupal stage of seven to eight weeks occurs in the soil (McFadyen and McClay 1981). Emergence of adults is stimulated by rainfall; in the absence of rainfall, prepupae can survive for months.

Adults feed on young leaves, but the damage is negligible. Larval feeding causes significant reductions in seed

Figure 4: Defoliation by *Zygogramma bicolorata* in central Qld. Photo: Qld Government.

Figure 5: *Smicronyx lutulentus* adults. Photo: Qld Government.

output. In Mexico, *S. lutulentus* appears to have two generations in a year, coinciding with autumn and summer rains, and can cause up to 30% seed destruction (McClay 1985). In Australia, the incidence of *S. lutulentus* is sporadic though very large populations can occur after rain and may have a comparable impact on seed production (Dhileepan and McFadyen, unpub. data 2010).

***Epiblema strenuana* Walker (Lepidoptera: Tortricidae).** This stem-galling moth is widely distributed in North America, including Mexico, Virgin Islands and Antigua in the Caribbean, and uses *Ambrosia* spp. and *Xanthium* spp. in the northern parts and *P. hysterophorus* in the southern parts of the range (Bennett 1976). The moth imported from *P. hysterophorus* in Mexico was approved for field release in 1982 after host-specificity tests in Brisbane (McClay 1987; McFadyen 1982). Field releases were made at 64 sites from December 1982 to February 1984, using 700–1500 pupae at each site, placed in absorbent paper in small wooden boxes protected from ant attack and with an exit hole for the adult moths. Subsequent field releases were made using galls collected from the initial establishment sites. The moth became established on *P. hysterophorus* from the first releases. It also occurs on annual ragweed (*A. artemisiifolia*) and Noogoora burr (*X. occidentale*) and is now widespread in all areas in Qld and NSW where host plants occur.

Eggs are laid singly on the leaves and the emerging larvae feed initially on the leaf, then bore into the stem (Fig. 6) through the terminal or axillary meristem. Once inside the stem, larval feeding induces a hollow fusiform gall in which the larvae feed and pupate. Four to six weeks later, the moth emerges through a hole in the gall, leaving the pupal skin extruded from the exit hole. Gall development can occur in all stages of plant growth and usually there is one larva per gall. In summer, the life-cycle is around five weeks and in central and northern Qld there are up to seven generations per year (McFadyen 1987). Females lay 400–500 eggs, with very large females laying >1000 eggs: as a result, very rapid increases can occur in favourable conditions. In autumn, larvae enter prepupal diapause to overwinter within the galls. Emergence occurs in spring in response to rising temperature, and the first spring generation can suffer very high mortality if, due to seasonally dry conditions, there is little or no green parthenium available for oviposition (McFadyen 1992).

E. strenuana causes visible symptoms on parthenium with gall damage evident on both rosette and flowering stages throughout the growing season (McFadyen 1992). The impact becomes significant when young plants are attacked (Dhileepan 2001, 2003b; Dhileepan and McFadyen 2001) and in the presence of pasture competition (Navie *et al.* 1998b). Several parasitoids (*Antrocephalus* sp., *Bracon* sp., *Brachymeria* sp., *Chelonus curvimaculatus* Cameron, *Elasmus* sp., *Euplemus* sp., *Glabridorsum* sp., *Goniozus* sp. and *Xanthopimpla* sp.) have been recorded in Australia, but parasitism is generally low (Dhileepan *et al.* 2005).

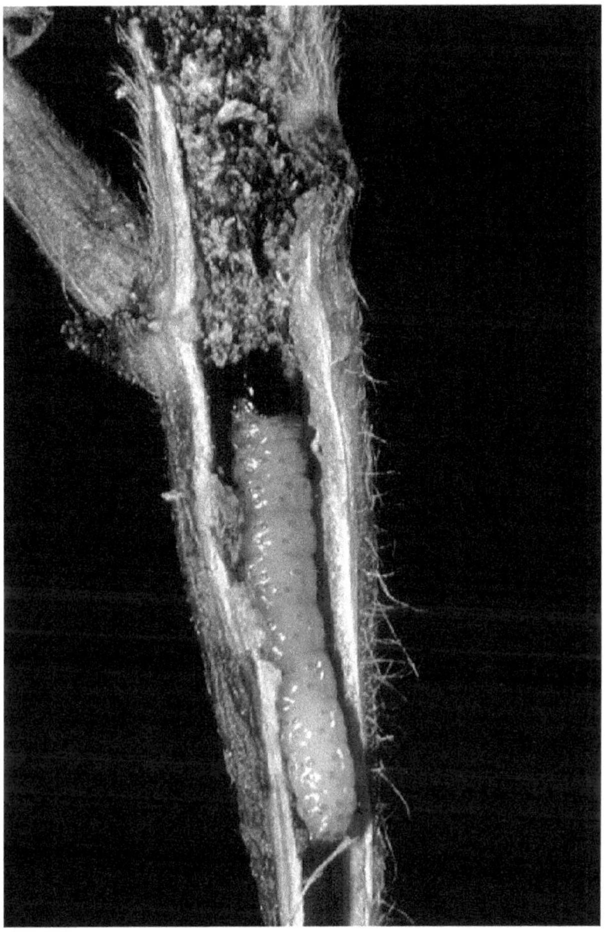

Figure 6: *Epiblema strenuana* larva. Photo: Qld Government.

***Listronotus setosipennis* (Hustache) (Coleoptera: Curculionidae).** This stem-boring weevil (Fig. 7) has a wide distribution in its native range in northern Argentina and southern Brazil, where the only known host plant is *P. hysterophorus*. Permission for release was granted in 1982, after host-specificity tests in Brazil and Australia (Wild *et al.* 1992). All releases were of adult weevils reared in the laboratory colony in Brisbane, established from weevils collected in Brazil in 1980. A total of 23 800 adult weevils was released (1000–3000 weevils per site per release) between December 1982 and March 1985 at 21 sites in central Qld, from south of Rolleston to north of Clermont. Field establishment was confirmed in 1983 (Wild *et al.* 1992), and subsequent spread was rapid. In 1990 the weevil was re-collected from north-western Argentina between Tucuman and Santiago del Este, a much drier area with hot summers and cool to cold winters, though frosts are very rare. The second collection was sent to Australia and field releases were made in northern Qld near Charters Towers from December 1991 to May 1992. As the weevil was already widespread by the time the second collection was made, there is no way of determining its contribution, if any, to the field population. The weevil is now widespread, but populations remain patchy and seasonal.

Adult weevils are 5 mm long, nocturnal and feed on parthenium leaves and flowers. Adults are winged, but are not known to fly. The female weevil lays eggs singly in holes chewed in flowers, leaf bases or stem surfaces that are then covered by frass, and the newly emerged larvae tunnel into the stem. Five larval instars occur. Mature larvae may leave the stem and move to the root, where they feed prior to pupation. Pupation occurs in a cell in the soil and the life-cycle takes about seven weeks in summer, though adult emergence is triggered by rain and will be delayed in dry conditions. The adults are long-lived, up to eight months in the laboratory, and pass the winter or dry season buried in the soil in diapause (Wild *et al.* 1992).

Adult feeding and oviposition damage is negligible. Larval feeding can kill or prevent further development of parthenium seedlings and seriously damage larger plants, especially if several larvae feed within the crown (Wild *et al.* 1992). In Australia, *L. setosipennis* was recorded in 48% of the parthenium-infested sites sampled in 1996–98, but only 16% of the sites showed more than one larva per plant (Dhileepan 2003a). *L. setosipennis* was more prevalent in plants growing on alluvial and black soils than on clay and sandy soils, possibly due to the impact of soil quality on the survival of pupation chambers. As diapause occurs in the soil, with emergence triggered by rainfall, this weevil is particularly suitable for regions with prolonged dry periods and an erratic rainfall pattern.

Figure 7: *Listronotus setosipennis* adult. Photo: Qld Government.

Stobaera concinna **(Stäl) (Homoptera: Delphacidae).** In its native range in Mexico, this sap-feeding plant-hopper was found only on *P. hysterophorus*. At sites in Nuevo Leon, Tamaulipas and San Luis Potosi in Mexico, *S. concinna* was generally rare, and dense populations were never found in the field (McClay 1983). In the southern USA and in the West Indies, *S. concinna* feeds and reproduces primarily on *Ambrosia* spp. (Calvert *et al.* 1987). Based on the preliminary host-specificity tests in Mexico (McClay 1983), *S. concinna* was imported to Australia in 1982, where detailed specificity tests under quarantine conditions confirmed that *S. concinna* could feed and reproduce only on *P. hysterophorus* and *Ambrosia* spp. (McClay 1983). The plant-hopper was approved for field release in 1983 and releases of several thousand individuals (5000–10 000 adults and nymphs per site per release) were made at 26 sites, from Townsville in northern Qld to Pentland in the west, and west of Brisbane in south-eastern Qld, from October 1983 to April 1986.

Adults are 3–4 mm long, with males smaller and more strongly pigmented than females. The eggs are inserted singly in stems and the eggs hatch in two weeks. There are five nymphal instars and development from egg to adult takes 30–56 days. Nymphs feed on leaves and new shoots, while adults feed on main and axillary shoots. At high population densities in laboratory cages, *S. concinna* caused yellowing of the leaves and spindly plant growth. This agent did not establish on *P. hysterophorus*, though it has become established and is present at low levels on annual ragweed (*A. artemisiifolia*) in south-eastern Qld (McFadyen 1992). It is not known why establishment failed in central Qld nor what limits the populations in the south, but the absence of a diapause stage may be the main factor, as the long dry season in parthenium-infested areas results in long periods with no green host plant available. All stages are also very vulnerable to predators such as spiders, ants and predatory mites.

Bucculatrix parthenica **Bradley (Lepidoptera: Bucculatricidae).** This leaf-mining moth is native to Mexico where it is rare and causes no major damage to the plant, probably due to specialist natural enemies (McClay *et al.* 1990). After host-testing in Brisbane, releases were made in seven sites in central Qld from February 1984 to September 1985, in three sites in northern Qld during

August–September 1985, and in one site each in western (Charleville) and south-eastern (Toogoolawah) Qld. Releases were of plant material infested with pupal cocoons, as well as adult moths. Field establishment in central Qld was confirmed in 1987 (McClay *et al.* 1990).

Adults are 3–4 mm long, and eggs are laid singly on the leaves. The first and second instar larvae are leaf-miners and later instars feed externally on the leaves. Pupation takes place within a characteristic ribbed silk cocoon usually attached to the leaf midrib. Adults emerge in seven to 11 days and survive for up to 14 days. The life-cycle is completed in about 25 days under field conditions. Larval feeding is evident on all growth stages of parthenium, but seldom causes significant damage. The moth is now widespread on parthenium in Qld, but is abundant only late in the summer and does not appear to have a major impact on the weed populations. No parasitoids have been reared from larvae in Qld.

Conotrachelus albocinereus **Fiedler (Coleoptera: Curculionidae).** This stem-galling weevil is native to South America, where it has been found in Colombia, Argentina and Brazil feeding on *P. hysterophorus* and *Ambrosia elatior* L. There are three generations a year in northern Argentina and the adults overwinter in the leaf litter or the soil. A colony was established in quarantine in Australia from adults collected near Tucuman, Argentina. During host-specificity trials, no adult feeding or oviposition was evident on any of the non-target plants tested except for the closely related weed *A. artemisiifolia* (McFadyen 2000). The agent was approved for field release in 1995, and since then over 15 000 weevils have been released at several sites (40–800 weevils per site per release) in central Qld between 1995 and 2000. A single release of 900 weevils was made at Greenvale in northern Qld in 1999.

Adults are 4–5 mm long, nocturnal and live for up to three months (McFadyen 2000). They feed on the stem tips and leaves and lay eggs singly on the stem, mainly in the leaf axils. Eggs hatch in 10–14 days. Newly emerged larvae feed on epidermal cells and burrow vertically into the stem to feed on the nutrient-rich parenchyma cells. Larval feeding induces elliptical galls on the main shoot axes. Larvae remain within the galls for about two months. The final larval instar chews a hole at the end of the tunnel, through which it exits to pupate in an earth cell in the soil. The adult emerges after 30 days: in the field, adult emergence is triggered by rain.

Adult feeding damage causes only minor damage to young leaves (McFadyen 2000). Damage is mainly due to larval feeding, which results in the fracturing of the vertical continuity of vascular tissues, thereby disrupting the host plant's overall metabolism (Florentine *et al.* 2002). Gall induction usually destroys axillary shoots, but the main stem remains unaffected.

Though recovered from a few release sites in central Qld, there is no evidence of widespread field establishment of this agent. No releases were made on *A. artemisiifolia* in Australia.

Platphalonidia mystica **(Razowski & Becker) (Lepidoptera: Tortricidae).** This stem-boring moth is native to northern Argentina, Bolivia and southern Brazil, where the only known host plant is *P. hysterophorus*. Host-specificity testing took place in both Argentina and Australia. Larval feeding was evident on members of the subtribe Ambrosiinae and two members of the subtribe Helianthineae, *Dahlia* and *Helianthus*. However, risk of damage to sunflower crops under field conditions was considered negligible and the moth was approved for field release (Griffiths and McFadyen 1993). Field releases in Australia were made as potted parthenium plants infested with *P. mystica* larvae and pupae (20–80 potted plants per sites per release) in central Qld from 1992 to 1996. Occasionally, newly emerged adults were also field released.

Adults are light grey to white, 5–6 mm long and partly diurnal, with oviposition occurring in the daytime. Newly emerged larvae feed on the leaf surface initially then enter the main stem via axillary shoots. Larvae generally feed close to the surface of the phloem tissue of the stem and pupate within the stem. Prior to pupation, the larvae cut emergence holes to which the exuvium remains attached following adult emergence. The life-cycle from egg to adult takes eight to 10 weeks. Larval feeding results in the death of shoot tips and weakening of the main stem, and high larval densities can kill the plant (Griffiths and McFadyen 1993).

Releases were made using stems containing large larvae and pupae from the laboratory colony, which were placed into field infestations of parthenium. Therefore, exact numbers of adults released are not known. Monitoring was complicated because of the universal presence of the stem-galling moth *E. strenuana*, and the difficulty of distinguishing larvae in the field. However, adults of *P. mystica* have been reared from field-collected stems on several occasions and the species is believed to be established, though at a low level. Seven unidentified species of parasitoids were recorded from larvae in Argentina, but none have been reared so far from Australia (Griffiths and McFadyen 1993).

***Carmenta* sp. nr. *ithacae* (Beutenmüller) (Lepidoptera: Sesiidae).** This root-feeding clearwing moth (Fig. 8) has a wide native range distribution in the USA (Arizona, Colorado, Kansas, Missouri, North Carolina, northern Florida, New Mexico, New York, Mississippi, Texas and Wisconsin), where larvae have been recorded feeding in *Helenium autumnale* L. and *Heliopsis helianthoides* (L.) Sweet (Asteraceae). The only records from *P. hysterophorus* are from Mexico (McClay *et al.* 1995; Withers *et al.* 1999). The population of *C. ithacae* from *P. hysterophorus* in Mexico may be a different species from the US species, and hence is referred to as *C.* sp. nr. *ithacae* (Withers *et al.* 1999). Studies in Australia confirmed that *C.* sp. nr. *ithacae* collected from *P. hysterophorus* in Mexico is highly host-specific and does not feed in *Helenium* or *Heliopsis* or any other species outside the subtribe Ambrosiinae (Withers *et al.* 1999). The moth was approved for field release in 1998; field releases were made by planting or placing potted parthenium plants infested with *C.* nr. *ithacae* larvae and pupae in the ground at 23 sites in central Qld (40–513 plants per site) and at nine sites in northern Qld (four to 234 plants per site) from 1998 to 2002. Occasionally, newly emerged adults were also released directly in the field.

Moths are 10 mm long, strongly patterned in black and yellow (Fig. 8) and diurnal. Emergence occurs soon after sunrise, mating between 0800 h and 1100 h, and oviposition after 1100 h. Females lay the hemispherical ribbed brown eggs on any part of the plant, and the eggs hatch in 10–14 days. Larvae migrate to the stem where they feed just below the outer layer. As they grow they migrate down the stem to feed in the cortical tissue of the tap root and crown. After five to six weeks, the fully developed larvae pupate in the root or stem base in a silk cocoon. Adults emerge after 10–12 days from the clearly visible silk tube protruding from the plant at or above the soil level. Adults live for three to 12 days and females lay 64–235 eggs in their life-time. Larvae are found on all growth stages of parthenium, and heavily infested plants often die.

Figure 8: *Carmenta* sp. nr. *ithacae* female moth. Photo: Qld Government.

This moth is now present in the field in central Qld both at release sites and at non-release sites nearby. It is neither widespread nor abundant and there is no evidence of field establishment in parthenium-infested areas of northern Qld.

***Puccinia abrupta* Diet. & Holw. var. *partheniicola* (Jackson) Parmelee (Uredinales).** The parthenium winter rust occurs naturally in Argentina, Bolivia, Brazil and Central America (Evans 1997) and was collected from semi-arid upland regions of Mexico (Evans 1987). It is highly host-specific to *P. hysterophorus* (Parker *et al.* 1994) and was the first pathogen to be approved for release against *P. hysterophorus* in Australia. Field releases were made at 50 sites in central Qld from 1991 to 1996.

The winter rust is a macrocyclic and autoecious rust, having all spore stages on parthenium. It forms brown powdery pustules (uredinia) on the upper leaf surface and stems throughout the year, and black non-powdery pustules on leaves, stems and inflorescences (telia) seasonally (Parker *et al.* 1994). The teliospores are not dispersed and remain on the host plant. The rust appears to overwinter in the urediniospore stage, the teliospores being non-functional under the conditions observed. The rust requires surface moisture for spore germination and infection, and the highest rate of infection occurs at 15°C with a dew period of eight to 10 hours. Spore germination is optimal at 15°C, with a steady decrease in the rate of germination with increasing temperature and almost no germination at 30°C (Fauzi *et al.* 1999). The rust infects the leaves and (rarely) stems, forming brown 'rust' patches on the upper surface of the leaves.

The winter rust established easily in southern Qld, but in central Qld it established only in a few localised areas with wetter winters and cooler temperatures (Dhileepan and McFadyen 1997). Even there, its impact on parthenium appears to be minor (Dhileepan 2003b). The spread and effectiveness of this pathogen has been hindered by frequent drought over the last 20 years. The rust has not established in northern Qld, where conditions are hotter and drier.

***Puccinia xanthii* var. *parthenii-hysterophorae* Seier, Evans & Romero var. nov. (Uredinales).** The parthenium summer rust (Fig. 9), previously known as *Puccinia melampodii* Dietel & Holway, is native to Central America and

Figure 9: *Puccinia xanthii* var. *parthenii-hysterophorae* infection in *Parthenium hysterophorus* leaf. Photo: Qld Government.

Mexico (Evans 1997; Seier *et al.* 2009). Collected from low-altitude regions of Mexico and Texas (Evans 1997; Seier *et al.* 1997), it is highly host-specific, virulent and adapted to areas with high temperatures and limited periods of humidity (Seier 1999). However, its incidence in Mexico was highest during the wet season. Field release of parthenium summer rust in Qld commenced in January 2000 and since then releases have been made at more than 50 sites in northern and central Qld (Dhileepan *et al.* 2006).

The parthenium summer rust is a microcyclic rust species which cycles exclusively through the telial stage. It therefore completes its reduced life-cycle on only the one host. It forms telia predominantly on the lower leaf surface as well as on the stems, initially as distinct sori, which constitute aggregates of individual telia that coalesce over time (Seier 1999). Sporulation can occur over the entire leaf surface (Fig. 9), leading to necrosis and eventual dieback of affected leaves. Successive infection cycles can cause severe stunting and premature plant death (Seier 1999).

Field establishment of parthenium summer rust was evident in most release sites soon after release (Dhileepan *et al.* 2006). The prevalence and intensity of parthenium summer rust was higher in northern than in central Qld; however, dry summers in the 2000s resulted in low rust populations (Dhileepan *et al.* 2006). It is believed that the pathogen maintains sufficient residual inoculum on the infected plant during dry periods (Seier 1999).

EVALUATION

From the first releases in 1980, rearing and releases were undertaken from the Alan Fletcher Research Station (AFRS) at Brisbane, 1000–1500 km from the release sites. Travel to these sites included many stretches of unsealed road and/or low creek crossings which became impassable after rain. The local human population was very small, typically one or two families on a 30 000 ha property, sometimes only five or six people on properties as large as 80 000 ha. Towns were small, with fewer than 1000 inhabitants, and 100–500 km apart. Under these circumstances, regular post-release monitoring or evaluation was not possible. Most release sites were visited only two or three times a year, and not always by the same people. Consequently, highly visible insects such as adults of *Z. bicolorata* were seen and recorded, but small inconspicuous species such as *S. lutulentus*, *S. concinna* and *B. parthenica* could pass unnoticed until seen by one of the few experienced entomological staff. The highly irregular emergence of some species, tied to local rainfall rather than to temperature or month, also meant that *S. lutulentus* in particular went undetected for years. Spring rainfall in central Qld generally falls as storms; these can be heavy but often affect areas of only a few square kilometres, which means that weevils might be visible and feeding at one site but absent from all others even on the same grazing property.

The only detailed evaluation of the impact of biological control on parthenium was undertaken at two properties with contrasting climates (Mt Panorama in central Qld and Plain Creek in northern Qld) during 1996–2000, and was based on an exclusion experiment using insecticides (monocrotophos and carbofuran) and a fungicide (mancozeb) (Dhileepan 2001, 2003b). At Mt Panorama, *Z. bicolorata* and *E. strenuana* were the prominent agents. At Plain Creek, *E. strenuana* was the predominant agent until the summer of 1999–2000, when the summer rust *P. xanthii* var. *parthenii-hysterophorae* became established. These biological control agents reduced plant height and flower production in 1996–97, especially at Mt Panorama. They reduced weed density by 90% at Mt Panorama in 1996–97 and prevented 28% and 18% of plants from producing flowers at Plain Creek in 1996–97 and 1997–98 respectively. However, biological control had only limited impact on host population density at Plain Creek, due to below-average summer rainfall producing non-synchrony between parthenium germination and *E. strenuana* emergence. It also resulted in low *Z. bicolorata* and *E. strenuana* activity at Mt Panorama. There was a 70% reduction in the soil seed bank at Mt Panorama over the four-year period due to biological control, but the soil seed bank reduction at Plain Creek was not statistically significant (Dhileepan 2003b).

A significant increase in grass production due to biological control of parthenium was observed at both sites, but in only one out of four years at Mt Panorama and two out of four years at Plain Creek (Dhileepan 2007). At Mt Panorama there was a 40% increase in grass production in 1997 due to 96% defoliation by *Z. bicolorata* and galling by *E. strenuana* in 100% of the parthenium plants. At Plain Creek, grass production increased by 52% in 1998 due to reduced parthenium seedling emergence, and by 45% in 2000 (Dhileepan 2007) due to the combined effects of galling by *E. strenuana* and the incidence of summer rust *P. xanthii* var. *parthenii-hysterophorae* in 72% of the plants.

DISCUSSION

Seven species of insects and two rust fungi have been successfully established as biological control agents against parthenium in Australia and two other insect species, *S. concinna* and *P. mystica*, are recovered in the field, though only occasionally. Among the established agents, *Z. bicolorata*, *C. ithacae*, *C. albocinereus* and *P. abrupta* var. *partheniicola* continue to have restricted distributions which limits their overall impact on the target weed. *S. lutulentus*, *B. parthenica*, *L. setosipennis* and *P. xanthii* var. *parthenii-hysterophorae* are widespread, but only seasonally present in large numbers causing significant damage. Only the stem-galling moth *E. strenuana* is both widespread and damaging, occurring in all parthenium-infested areas and frequently at high population levels.

The time taken for establishment of the different biological control agents varied widely, possibly due to climatic differences between the countries of origin and introduction. In Australia, *Z. bicolorata* became established on *A. artemisiifolia* in south-eastern Qld immediately and was visibly abundant within two years of introduction (McFadyen 1992), but the same beetle took more than 12 years to become abundant on parthenium in central Qld (Dhileepan and McFadyen 1997). The seed-feeding weevil *S. lutulentus* also took more than 14 years to become abundant in Australia (Dhileepan and McFadyen 1997). The longer time taken for the two agents to become abundant in central Qld is possibly due to the need for adaptation to extreme local climatic conditions. In contrast, *E. strenuana*, *L. setosipennis* and *P. xanthii* var. *parthenii-hysterophorae* all became established almost immediately on field release. Significant problems with field monitoring also meant that field establishment often went undetected until populations were very large.

The distribution and abundance of biological control agents and their interactions with their host plants are influenced by a range of abiotic factors. Rainfall (either the total summer rainfall in central Qld or the timing of the onset of summer rainfall in northern Qld) is the major factor affecting the abundance and therefore effectiveness of parthenium biological control agents (Dhileepan 2003b). Parthenium is a fast-growing opportunistic plant with a large seed bank and long-lived seed. Plant populations develop very rapidly in response to local rainfall events, and plants flower and senesce equally rapidly as moisture dries up. Lack of suitable host plant material is therefore the main constraint on population increase of the biological control agents, and this is in turn driven by the extremely variable local rainfall. In central Qld, where early summer rainfall is more reliable, *Z. bicolorata*, *E. strenuana* and *L. setosipennis* are widespread and effective. In northern Qld, *E. strenuana* and *P. xanthii* var. *parthenii-hysterophorae* are the prominent agents.

Cumulatively, biological control has had a significant negative impact on parthenium weed and its soil seed banks in Qld, but the impact varies widely between years and between locations (Dhileepan 2003b). Overall, the impact was greater in central Qld than in northern Qld, and a significant increase in grass biomass production due to biological control was observed more often in central Qld than in northern Qld (Dhileepan 2007). In economic terms, the benefit from increased grass production due to biological control was estimated to be \$1.25 per ha per year for buffel grass *Cenchrus ciliaris* L. in central Qld, and $1.19 per ha per year for the Qld blue grass *Dichanthium sericeus* (R. Br.) in northern Qld (Adamson and Bray 1999). As these values were based on cattle prices of $1.00 per kilo at the market at Roma, Qld where prices are now (2010) around $2.00 per kilo, current economic benefits from biological control are roughly double those published amounts. In addition, biological control resulted in the saving of $8 million per year in health costs in treating allergic dermatitis and asthma in workers in infested areas, giving an overall benefit:cost ratio for the program of 7.2 and a net present value (in 2005) of $33.3 million for a total cost of $11.0 million (Page and Lacey 2006). Biological control in Qld has also greatly reduced seed production, which should be reducing the problem of seed transported in machinery, trucks and grain shipments (mostly feed sorghum) south into NSW and Vic. However, no data are available.

Parthenium continues to be controlled using herbicides in high-value crops, along some roads in central Qld and adjacent to homesteads if allergies are an issue. In pastures, parthenium is now managed through grazing management, adjusting stocking rates to maintain good pasture cover especially in spring and early summer. With the seed bank greatly reduced by the impact of the biological control agents, and the reduced growth of seedlings and earlier senescence of established plants due to this impact, grazing management is now sufficient to control parthenium, which was not possible prior to the establishment of these agents. This approach is also suitable for managing parthenium in protected forests and nature reserves (Dhileepan 2009). In general, landholders are no longer making field releases, redistributing agents or taking any active steps to encourage biological control, as most agents are already widespread and population size and impact are dependent on rainfall.

Not all potential agents have been properly investigated, and there are still possibilities for further introductions. Detailed surveys in lowland Bolivia might lead to discovery of additional agents (White 1994). However, at this stage there is no funding for further work.

The continued spread of parthenium in Asia and Africa, where chemicals are often too expensive and much weed control relies on hand weeding, has resulted in serious crop losses and major human health problems. There is increasing interest in biological control in these countries, with the possibility of using some of the agents known to be successful in Australia. The best agent, *E. strenuana*, has been released in Sri Lanka (Jayasuriya 2005) but has been rejected in India due to feeding on sunflower (*H. annuus*) and niger seed (*G. abyssinica*) in laboratory experiments (Jayanth 1987), though no attack has ever occurred on sunflower in the field in Qld. It is likely that *L. setosipennis* will also be rejected in those countries for the same reason. In India, *Z. bicolorata* has been released; it is now widespread and effective in parts of India and Pakistan, but is probably less effective in very dry climates. There were reports of widespread feeding of *Z. bicolorata* on sunflower crops in India, but so far there is no evidence of economic loss to sunflower crops there (Dhileepan and Strathie 2009). The seed weevil *S. lutulentus* is very host-specific and is well adapted to seasonally dry climates, and the summer rust *P. xanthii* var. *parthenii-hysterophorae* is also specific and able to persist in hot dry climates. The root-boring clearwing moth *C.* sp. nr. *ithacae* is in the early stages of field establishment and dispersal. It might be a very effective agent when not competing with *E. strenuana*, and is entirely restricted to parthenium. Several of these agents are now under investigation for countries in eastern and southern Africa (A McConnachie and L Strathie, *pers. comm.* 2010).

ACKNOWLEDGEMENTS

We thank Calvert Moya for the parthenium incidence map (Fig. 2). Some of this work was carried out under the auspices of the Cooperative Research Centre for Tropical Pest Management, and the Cooperative Research Centre for Weed Management.

REFERENCES

Adamson DC and Bray S (1999) The economic benefit from investing in insect biological control of parthenium weed (*Parthenium hysterophorus*). RDE Connections, NRSM, University of Queensland, Brisbane.

Auld BA, Hosking J and McFadyen RE (1982–83) Analysis of the spread of tiger pear and parthenium weed in Australia. *Australian Weeds* **2**, 56–60.

Bennett FD (1976) 'A preliminary survey of the insects and diseases attacking *Parthenium hysterophorus* L. (Compositae) in Mexico and the USA to evaluate the possibilities of its biological control in Australia.' Mimeographical report. Commonwealth Institute of Biological Control.

Calvert PD, Wilson SW and Tsai JH (1987) *Stobaera concinna* (Homoptera: Delphacidae): field biology laboratory rearing and descriptions of immature stages.

Journal of the New York Entomological Society **95**, 91–98.

Chippendale JF and Panetta FD (1994) The cost of parthenium weed to the Qld cattle industry. *Plant Protection Quarterly* **9**, 73–76.

Dale IJ (1981) Parthenium weed in the Americas: a report on the ecology of *Parthenium hysterophorus* in South, Central and North America. *Australian Weeds* **1**, 8–14.

Dhileepan K (2001) Effectiveness of introduced biocontrol insects on the weed *Parthenium hysterophorus* (Asteraceae) in Australia. *Bulletin of Entomological Research* **91**, 167–176.

Dhileepan K (2003a) Current status of the stem-boring weevil *Listronotus setosipennis* (Coleoptera: Curculionidae) introduced against the weed *Parthenium hysterophorus* (Asteraceae) in Australia. *Biocontrol Science and Technology* **13**, 3–12.

Dhileepan K (2003b) Seasonal variation in the effectiveness of leaf-feeding beetle *Zygogramma bicolorata* (Coleoptera: Curculionidae) and stem-galling moth *Epiblema strenuana* (Lepidoptera: Totricidae) as biocontrol agents on the weed *Parthenium hysterophorus* (Asteraceae). *Bulletin of Entomological Research* **93**, 393–401.

Dhileepan K (2007) Biological control of parthenium (*Parthenium hysterophorus*) in Australian rangeland translates to improved grass production. *Weed Science* **55**, 497–501.

Dhileepan K (2009) Managing *Parthenium hysterophorus* across landscapes: limitations and prospects. In *Management of Invasive Weeds.* (Ed S Inderjit) pp. 227–260. Invading Nature – Springer Series in Invasion Ecology Vol. 5. Springer Science + Business Media, Dordrecht.

Dhileepan K and McFadyen RE (1997) Biological control of parthenium in Australia – progress and prospects. In *1st International Conference on Parthenium Management. Vol. I.* 6–8 October 1997, Dharwad, Karnataka, India. (Eds M Mahadevappa and VC Patil) pp. 40–44. University of Agricultural Sciences, Dharwad, India.

Dhileepan K and McFadyen RE (2001) Effects of gall damage by the introduced biocontrol agent *Epiblema strenuana* (Lepidoptera: Tortricidae) on the weed *Parthenium hysterophorus* (Asteraceae). *Journal of Applied Entomology* **124**, 1–8.

Dhileepan K and Strathie L (2009) 20. *Parthenium hysterophorus.* In *Weed Biological Control with Arthropods in the Tropics: Towards Sustainability.* (Eds R Muniappan, DVP Reddy and A Raman), pp. 272–316. Cambridge University Press, Cambridge.

Dhileepan K, Setter S and McFadyen RE (2000) Impact of defoliation by the introduced biocontrol agent *Zygogramma bicolorata* (Coleoptera: Chrysomelidae) on parthenium weed in Australia. *BioControl* **45**, 501–512.

Dhileepan K, Lockett CJ and McFadyen RE (2005) Larval parasitism by native insects on the introduced stem-galling moth *Epiblema strenuana* Walker (Lepidoptera: Tortricidae) and its implications for biological control of *Parthenium hysterophorus* (Asteraceae). *Australian Journal of Entomology* **44**, 83–88.

Dhileepan K, Florentine SK and Lockett CJ (2006) Establishment, initial impact and persistence of parthenium summer rust *Puccinia melampodii* in north Qld. In *Proceedings of the 15th Australian Weeds Conference.* 24–28 September 2006, Adelaide. (Eds C Preston, JH Watts and ND Crossman) pp. 577–580. Weed Management Society of South Australia, Adelaide.

Evans HC (1987) Life cycle of *Puccinia abrupta* var. *partheniicola*, a potential biological control agent of *Parthenium hysterophorus. Transactions of the British Mycological Society* **88**, 105–111.

Evans HC (1997) The potential of neotropical fungal pathogens as classical biological control agents for management of *Parthenium hysterophorus* L. In *1st International Conference on Parthenium Management.* (Eds M Mahadeveppa and VC Patil) pp. 55–62. University of Agricultural Sciences, Dharwad, India.

Fauzi MT, Tomley AJ, Dart PJ, Ogle HJ and Adkins SW (1999) The rust *Puccinia abrupta* var. *partheniicola*, a potential biocontrol agent of parthenium weed: environmental requirements for disease progress. *Biological Control* **14**, 141–145.

Florentine SK, Raman A and Dhileepan K (2002) Response of the weed *Parthenium hysterophorus* (Asteraceae) to the stem gall-inducing weevil *Conotrachelus albocinereus* (Coleoptera: Curculionidae). *Entomologia Generalis* **26**, 195–206.

Graham GC and Lange CL (1998) 'Genetic analysis of relationship of parthenium occurrences in Australia and indications of its origins.' Internal report. Cooperative Research Centre for Tropical Pest Management, University of Queensland, Brisbane.

Griffiths MW and McFadyen RE (1993) Biology and host-specificity of *Platphalonidia mystica* (Lep, Cochylidae) introduced into Qld to biologically control *Parthenium hysterophorus* (Asteraceae). *Entomophaga* **38**, 131–137.

Jayanth KP (1987) Investigations on the host-specificity of *Epiblema strenuana* (Walker) (Lepidoptera: Tortricidae), introduced for biological control trials against *Parthenium hysterophorus* in India. *Journal of Biological Control* **1**, 133–137.

Jayanth KP and Bali G (1993) Diapause behaviour of *Zygogramma bicolorata* (Coleoptera: Chrysomelidae), a biological control agent for *Parthenium hysterophorus* (Asteraceae) in Bangalore, India. *Bulletin of Entomological Research* **83**, 383–388.

Jayasuriya AHM (2005) Parthenium weed – status and management in Sri Lanka. In *2nd International Conference on Parthenium Management.* (Eds TV Ramachandra Prasad, HV Nanjappa, R Devendra, A Manjunath, SC Subramanya, Chandrashekar, VK Kiran Kuman, KA Jayaram and TK Prabhakara Setty) pp. 36–43. University of Agricultural Sciences, Bangalore, India.

McClay AS (1980) Studies of some potential biocontrol agents for *Parthenium hysterophorus* in Mexico. In *Proceedings of the 5th International Symposium on Biocontrol of Weeds.* 22–29 July 1980, Brisbane. (Ed ES Delfosse) pp. 471–482. CSIRO, Canberra.

McClay AS (1983) Biology and host-specificity of *Stobaera concinna* (Stal) (Homoptera: Delphacidae), a potential biocontrol agent for *Parthenium hysterophorus* L. (Compositae). *Folia Entomologica Mexicana* **56**, 21–30.

McClay AS (1985) Biocontrol agents for *Parthenium hysterophorus* from Mexico. In *Proceedings of the 6th International Symposium on Biological Control of Weeds.* 19–25 August 1984, Vancouver. (Ed ES Delfosse), pp. 771–778. Agriculture Canada, Vancouver.

McClay AS (1987) Observations on the biology and host specificity of *Epiblema strenuana* (Lepidoptera, Tortricidae), a potential biocontrol agent for *Parthenium hysterophorus* (Compositae). *Entomophaga* **32**, 23–34.

McClay AS and Anderson DM (1985) Biology and immature stages of *Thecesternus hirsutus* Pierce (Coleoptera: Curculionidae) in north-eastern Mexico. *Proceedings of the Entomological Society of Washington* **87**, 207–215.

McClay AS, McFadyen RE and Bradley JD (1990) Biology of *Bucculatrix parthenica* Bradley sp. n. (Lepidoptera: Bucculatricidae) and its establishment in Australia as a biological control agent for *Parthenium hysterophorus* (Asteraceae). *Bulletin of Entomological Research* **80**, 427–432.

McClay AS, Palmer WA, Bennett FD and Pullen KR (1995) Phytophagous arthropods associated with *Parthenium hysterophorus* (Asteraceae) in North America. *Environmental Entomology* **24**, 796–809.

McFadyen PJ (1979) A survey of insects attacking *Parthenium hysterophorus* L. (Compositae) in Argentina and Brazil. *Dusenia* **11**, 42–45.

McFadyen RE (1982) 'Host specificity of parthenium stem galling moth *Epiblema strenuana* Walker from Mexico, a potential biocontrol agent against *Parthenium hysterophorus*'. Internal report. Alan Fletcher Research Station, Brisbane.

McFadyen RE (1987) The effect of climate on the stem-galling moth, *Epiblema strenuana*, in eastern Australia. In *Proceedings of the 8th Australian Weeds Conference.* 21–25 September 1987, Sydney. (Eds D Lemerle and AR Leys) pp. 97–99. Weed Society of New South Wales, Sydney.

McFadyen RE (1992) Biological control against parthenium weed in Australia. *Crop Protection* **24**, 400–407.

McFadyen RE (1995) Parthenium weed and human health in Qld. *Australian Family Physician* **24**, 1455–1459.

McFadyen RE (2000) Biology and host specificity of the stem galling weevil *Conotrachelus albocinereus* Fielder (Col: Curculionidae), a potential biocontrol agent for parthenium weed *Parthenium hysterophorus* L. (Asteraceae) in Qld, Australia. *Biocontrol Science and Technology* **10**, 195–200.

McFadyen RE and McClay AR (1981) Two new insects for the biological control of parthenium weed in Qld. In *Proceedings of the 6th Australian Weeds Conference.* 13–18 September 1981, Gold Coast, Qld. (Eds BJ Wilson and JD Swarbrick) pp. 145–149. Weed Science Society of Queensland, Brisbane.

Navie SC, McFadyen RE, Panetta FD and Adkins SW (1996) The biology of Australian weeds 27. *Parthenium hysterophorus* L. *Plant Protection Quarterly* **11**, 76–88.

Navie SC, Panetta FD, McFadyen RE and Adkins SW (1998a) Behaviour of buried and surface-lying seeds of parthenium weed (*Parthenium hysterophorus* L.). *Weed Research* **38**, 338–341.

Navie SC, Priest TE, McFadyen RE and Adkins SW (1998b) Efficacy of the stem-galling moth *Epiblema strenuana* Walk. (Lepidoptera: Tortricidae) as a biological control agent for the ragweed parthenium (*Parthenium hysterophorus* L.). *Biological Control* **13**, 1–8.

Page AR and Lacey KL (2006) 'Economic impact assessment of Australian weed biological control.' Technical Series No. 10. CRC for Australian Weed Management, Adelaide.

Parker A, Holden ANG and Tomley AJ (1994) Host specificity testing and assessment of the pathogenicity of the rust, *Puccinia abrupta* var. *partheniicola*, as a biological control agent of parthenium weed (*Parthenium hysterophorus*). *Plant Pathology* **43**, 1–16.

Seier MK (1999) 'Studies on the rust *Puccinia melampodii* Diet. & Holw., a potential biological control agent for parthenium weed (*Parthenium hysterophorus* L.) in Australia'. CABI Bioscience UK Centre, Silwood Park, UK. Report submitted to the Qld Dept of Natural Resources and Mines.

Seier MK, Harvey JL, Romero A and Kinnersley RP (1997) Safety testing of the rust *Puccinia melampodii* as a potential biocontrol agent of *Parthenium hysterophorus* L. In *1st International Conference on Parthenium Management. Vol. I.* 6–8 October 1997, Dharwad, Karnataka, India. (Eds M Mahadeveppa and VC Patil) pp. 63–69. University of Agricultural Sciences, Dharwad, India.

Seier MK, Morin L, Van der Merwe M, Evans HC and Romero A (2009) Are the microcyclic rust species *Puccinia melampodii* and *Puccinia xanthii* conspecific? *Mycological Research* **113**, 1271–1282.

Tudor GD, Ford AL, Armstrong TR and Bromage EK (1982) Taints in meat from sheep grazing *Parthenium hysterophorus. Australian Journal of Experimental Agriculture and Animal Husbandry* **22**, 43–46.

White GG (1994) 'Workshop report: parthenium weed'. Cooperative Research Centre for Tropical Pest Management, Brisbane.

Wild CH, McFadyen RE, Tomley AJ and Wilson BW (1992) The biology and host specificity of the stem-boring weevil *Listronotus setosipennis* (Coleoptera: Curculionidae), a potential biocontrol agent for *Parthenium hysterophorus* (Asteraceae). *Entomophaga* **37**, 591–598.

Withers TM, McFadyen RE and Marohasy J (1999) Importation protocols and risk assessment of weed biological control agents in Australia: the example of *Carmenta* nr. *ithacae*. In *Nontarget Effects of Biological Control.* (Eds PA Follett and JJ Duan) pp. 195–214. Kluwer, Boston.

Phyla canescens (Kunth) Greene – lippia

Mic Julien, Alejandro Sosa and Guadalupe Traversa

ABSTRACT

Lippia, *Phyla canescens*, originating from South America, is widespread in south-eastern Australia, especially in the Murray-Darling Basin. It continues to spread and is increasing in density. Lippia is having very significant impacts on agricultural production and the environment. Surveys for potential biological control agents, both arthropods and pathogens, began in 2005 in Argentina and continued through to 2010. Most surveys were during an extended drought that may have limited collection of natural enemies. The leaf-feeding beetle *Kuschelina bergi* is recommended for importation to Australia for host-specificity testing. Four other insects, a mite and two fungi are under study in Argentina and a small group of other insects maybe considered in the future. Project funding ended in June 2010, which may have caused the premature termination of the lippia biological control research.

Key words: biological control, surveys, Argentina, *Phyla nodiflora*, arthropods, pathogens, native range studies.

INTRODUCTION

Phyla canescens (Kunth) Greene (Verbenaceae), lippia, is a serious weed of riparian and floodplain production and conservation areas including RAMSAR sites, roadsides and stock routes with considerable impact on rural production, land values and ecosystem services (Julien *et al.* 2004; Earl 2003). The greatest invasions and impacts occur in the Murray-Darling Basin's northern catchments, but it exists throughout the Basin (occurring on 5.3 million ha therein) and elsewhere in Australia (Fig. 1) and is spreading and increasing in density. In the Murray-Darling Basin, lippia costs the grazing industry a conservatively estimated \$38 million per year and environmental costs of \$1.8 billion per year (Earl 2003).

Lippia is a perennial forb with a prostrate habit and creeping stems that can root at each node (Fig. 2). It is competitive with other plant species, invades bare patches, is not palatable and therefore invades overgrazed pastures. It tolerates inundation and stems fragment with submergence and float downstream in flood events. Copious seeds are produced with flowering from spring to autumn and the persistent seed bank is estimated to have 2500–7500 seeds m^2 (McCosker 1994; Macdonald 2008). It is native to South America, with the centre of origin probably in central and northern Argentina.

BIOLOGICAL CONTROL HISTORY

The research reported here is the only biological control research attempted on lippia.

PLANT TAXONOMY

In Australia

Only two species in the genus *Phyla*, *P. canescens* and *Phyla nodiflora* (L.) Greene, occur in Australia. Recently *P. nodiflora* was confirmed to be native to Australia

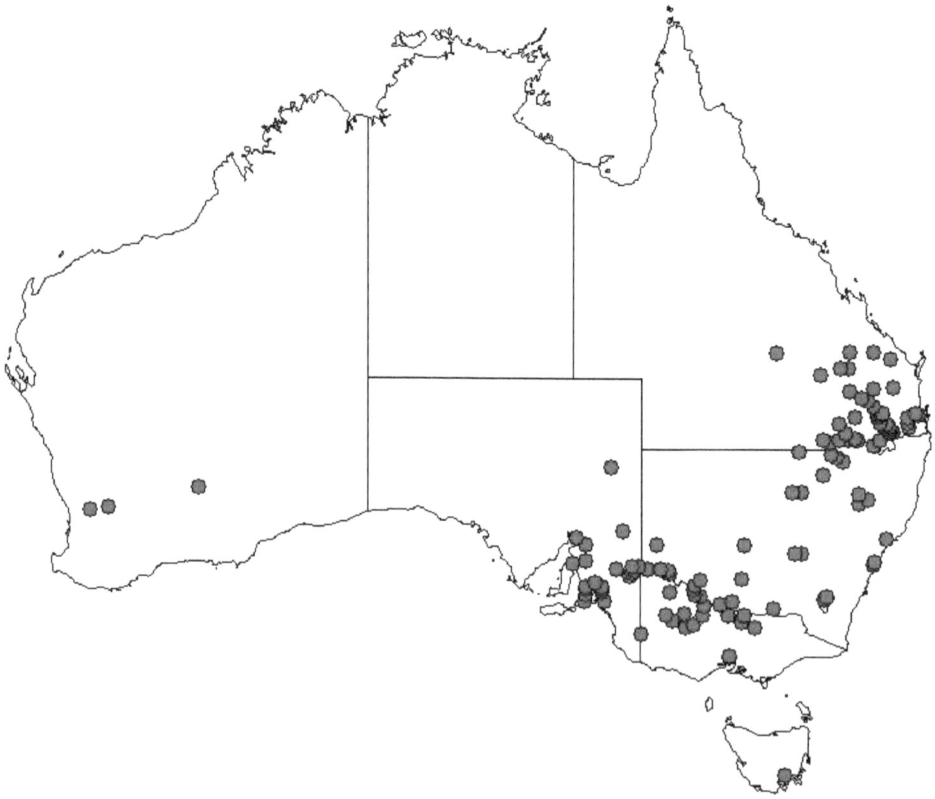

Figure 1: Distribution of *Phyla canescens* in Australia. Map from Australian Virtual Herbarium. Specimen data reproduced from Australia's Virtual Herbarium with permission of the Council of Heads of Australasian Herbaria Inc and from Earl (2003) and Macdonald (2008).

(Fatemi *et al.*, unpub.) although it has a worldwide distribution, and this has implications for the selection of biological control agents. In addition, there has been confusion in the binomial name used for lippia, which until recently was referred to in Australia as *P. nodiflora* and *Lippia nodiflora*. Munir (1993) separated *P. nodiflora* and *P. canescens*, showing that the former has a northern distribution and the later, lippia, a southern distribution.

Figure 2: Lippia, *Phyla canescens*. Photo: A White, CSIRO.

Recent genetic assessment of populations of lippia from the native range and Australia indicate that multiple introductions of lippia have occurred followed by admixture of genes, making it difficult to pinpoint the origins of many Australian populations, though all appear to have come from Argentina and possibly Chile (Fatemi *et al.*, unpub). In addition, irregular meiosis and low pollen viability suggest that some populations in Australia are infertile hybrids (Greizerstein *et al.*, unpub.).

In the native range, Argentina

There are considered to be three *Phyla* entities in Argentina, *P. canescens*, *Phyla fruticosa* (Mill.) Kennedy and *P. nodiflora* based on a morphological revision of *Phyla* that has been submitted for publication (N O'Leary, *pers. comm.*). However, recent genetic studies suggested that *P. nodiflora*, as represented by samples from around the world, does not occur in Argentina but that such populations are included in *P. canescens sensu latu* (M Fatemi *et al.*, unpub.). *P. fruticosa* has been referred to as *P. reptans* (Kunth) Greene.

EXPLORATION

Surveys to delimit the native range of lippia in Argentina and nearby countries and to seek natural enemies and identify potential biological control agents began in late 2005 (Sosa *et al.* 2008). CSIRO subcontracted the arthropod survey work to the USDA-ARS South American Biological Control Laboratory (SABCL) at Hurlingham, Buenos Aires, and the plant pathogen survey work to the Universidad Nacional del Sur (UNS), Bahía Blanca, Argentina, with scientific management by CSIRO.

Surveys were conducted by Alejandro Sosa, SABCL, Guadalupe Traversa, UNS and occasionally Mic Julien, CSIRO. Populations of *Phyla* species were located in all provinces in Argentina north of San Antonio Oeste, Rio Negro province (40°45'S) (Fig. 3), with exceptions of the Andean provinces of Neuquén and Catamarca and the north-eastern tropical wet province of Misiones. *P. canescens* occurs throughout those provinces, but north of about 30°S another species, *P. fruticosa*, also occurs.

Collections of natural enemies (insects and pathogens) were made throughout the range in Argentina on numerous occasions each growing season from December 2005 until late summer 2010. A single survey was conducted in Chile in 2007 from Copiapó (27°21'S) to Valdivia (~40°S) and another in Bolivia during August 2007 near Santa Cruz de La Sierra, Santa Cruz province and Coroico, La Paz province.

CANDIDATES

Surveys found relatively few (about 20) natural arthropod enemies of *Phyla* species in Argentina and about 17 fungal species. Damage typical of leaf-feeding beetles (Chrysomelidae) and a leaf-mining fly was found in Chile and similar leaf-feeding damage was observed in Bolivia. Species that may have potential as biological control agents are included in Table 1. The others were eliminated as generalist feeders or non-obligate fungi. There is little information about the last eight arthropods included in the table. They have been found in low numbers and at low frequencies so it has not been possible to work with them other than to submit specimens for identification. The rest, six arthropods and two fungi (Table 1), are considered to have prospects and are discussed below.

Table 1: Natural enemies of *Phyla* species found in Argentina

Natural enemy			
Order/Family	**Genus/species**	**Distribution**	**Guild**
Natural enemies that may have potential as biological control agents			
Chrysomlidae (beetle)	*Kuschelina bergi* Harold	Rio Negro, Buenos Aires	Leaf-feeder
Chrysomelidae (beetle)	*Longitarsus* sp. 2	Rio Negro, San Luis, Santiago del Estero, Jujuy	Leaf-feeder
Acari: Eriophyidae (mite)	Unidentified	Buenos Aires	Leaf-feeder
Stem-galler 1	Unidentified	Rio Negro, Jujuy	Stem-galler
Stem-galler 2	Unidentified	Formosa	Stem-galler
Diptera: Cecidomyidae (fly)	*Neolasioptera* new sp.	Formosa	Stem-galler
Basidiomycetes: Uredinales (fungus)	*Puccinia lantanae* Farl.	Salta, Entre Rios, Jujuy, Formosa	Leaf disease
Hyphomycetes (fungus)	*Cercospora lippiae* Ellis & Everh.	Widespread on *Phyla*	Leaf disease
Natural enemies with unknown potential as biological control agents			
Curculionidae (weevil)	Unidentified	Formosa	Tip-galler?
Agromyzidae? leaf-mining fly	Unidentified	San Luis, Salta, Santiago del Estero	Leaf-miner
Thysanoptera 1 (thrips)	Unidentified	Buenos Aires	Sap-feeder
Thysanoptera 2 (thrips)	Unidentified	Corrientes	Sap-feeder
Lepidoptera 1 (hairy caterpillar)	Unidentified	Jujuy	Leaf-feeder
Lepidoptera 2 (black caterpillar)	Unidentified	Jujuy	Leaf-feeder
Lepidoptera 3 (micro moth)	Unidentified	Jujuy	Unknown
Lepidoptera 4 (micro moth)	Unidentified	Jujuy	Unknown

Figure 3: Provinces of Argentina. Source: Wikipedia.

Kuschelina bergi **Harold (Coleoptera: Chrysomelidae).** Adults and larvae of this beetle eat leaves, making characteristic holes. Scraping marks are made by first instar larvae. Eggs are laid in batches in the soil. Pupation occurs in the substrate and adults, 5 mm long (Fig. 4), are invariably found in the litter under lippia plants. The species has been found at locations in southern Buenos Aires and La Pampa provinces, only on *P. canescens*.

Figure 4: *Kuschelina bergi.* a) Adults. b) Eggs. Photos: A Sosa, USDA.

Adults field-collected from Tres Arroyos, Buenos Aires province in January 2009 were set up in non-replicated no-choice feeding tests on *P. canescens*, *Lantana megapotamica* (Spreng.) Tronc (subfamily Lantaneae), *Glandularia platensis* (Spreng.) Schnack & Covas, *Glandularia tenera* (Spreng.) Cabrera (subfamily Verbeneae) and *Duranta* sp. (subfamily Citharexyleae) during a field trip. After two days feeding had occurred only on *P. canescens*, suggested that this insect was specific at least to *Phyla*.

Replicated laboratory tests, conducted at SABCL, confirmed that *K. bergi* is specific to *Phyla*. In no-choice feeding tests using the following Verbenaceae, *P. canescens*, *P. fruticosa* (Mill.) Kennedy, *P. nodiflora* (as determined by N O'Leary Darwinion Herbarium), *L. megapotamica*, *G. platensis*, *G. tenera* and *Duranta* sp., feeding occurred only on *Phyla* species after one week of exposure and there was a distinct preference for *P. canescens* (A Sosa, unpub.). In a no-choice development test, larval development occurred only on *Phyla* species with no preferences apparent, though in a separate experiment survivorship from early larvae to adults was greater on lippia than on other *Phyla* species (A Sosa, unpub.). This species is recommended for importation to Australia for testing on Australian *P. nodiflora*. If no significant damage or development occurs on Australian populations of *P. nodiflora*, detailed host-specificity testing should be conducted.

Longitarsus **sp. 1 and sp. 2 (Coleoptera: Chrysomelidae).** These are small brown leaf-feeding beetles with adults <2 mm long. Larvae have not been observed in the field and it is likely that they feed on the roots. Adults are cryptic although their small, numerous feeding holes in leaves are characteristic and very common. They can be readily collected using vacuum techniques.

These may be two new species (V Savini, *pers. comm.* 2008) as they appear to be distinguished by colour. The northern species, designated *Longitarsus* sp. 1, is lighter in colour than the southern *Longitarsus* sp. 2. Recent genetic analysis using mitochondrial (12S rDNA) and nuclear (28S rDNA) DNA sequence data from material from Laguna Yema and Bazán, Formosa province (*Longitarsus* sp. 1) and from Tres Arroyos, Buenos Aires province (*Longitarsus* sp. 2) shows that there has been no recent gene flow between populations from the two provinces and that they likely represent two different species (L Cook, *pers. comm.* 2010).

Longitarsus sp. 1 has been collected from Entre Rios, Jujuy, Salta, Formosa, Santiago del Estero, Buenos Aires, Chaco and Santa Fé provinces; *Longitarsus* sp. 2 has been

Figure 5: Gall clusters found on *Phyla canescens* in Argentina. Photo: A Sosa, USDA.

collected from Rio Negro, La Pampa, Mendoza, Buenos Aires, Cordoba, San Luis and Formosa provinces (Fig. 3). Both have been collected from *P. canescens* and *P. fruticosa*. *Longitarsus* sp. 2 was also collected from *P. fruticosa* at Abapó, Santa Cruz province, Bolivia in August 2007.

***Longitarsus* sp. 1.** Attempts to laboratory rear *Longitarsus* sp. 1 have not been successful. In replicated no-choice tests conducted at SABCL using field-collected adults from Monte Redondo, Santiago del Estero province, no feeding occurred on *Duranta* sp., but feeding occurred on all other test plant species – *P. canescens*, *P. fruticosa*, *P. nodiflora*, *L. megapotamica*, *G. platensis* and *G. tenera*. There was significant preference based on consumption rate for *P. canescens* and *Glandularia* (both species), but significantly more feeding marks occurred on *P. canescens* and *P. fruticosa* than on the other species. We concluded that this insect is not sufficiently specific to warrant further studies.

***Longitarsus* sp. 2.** In a non-replicated no-choice feeding trial, conducted during a survey trip in Argentina in January 2009, five to seven field-collected adults from Choele Choele, Rio Negro province, were sleeved onto potted plants of *P. canescens*, *L. megapotamica*, *G. platensis*, *G. tenera* and *Duranta* sp. and held in the back of a vehicle. After 2.5 days feeding damage was assessed; the results suggest that this insect is not host-specific. However, replicated lab tests are required to confirm lack of specificity.

Mite, new species (Acari: Eriophyidae). This tiny mite, that requires a binocular microscope to be seen, has been found throughout the range of lippia and on *P. fruticosa* in Argentina. Collections preserved in alcohol have been sent to Prof. S Ozman, Ondokuz Mayis University, Turkey for identification and taxonomic description, to determine if there is more than one species.

Stem-galler 1. Galls were found on stems of *P. canescens* from Santiago del Estero, Buenos Aires, Salta, Rio Negro, Formosa, Jujuy and Tucuman provinces of Argentina and from Arauco, Chile, and from *P. fruticosa* from Jujuy province. Multiple galls occurred per stem (Fig. 5), mostly associated with older nodes and therefore found next to or in the surface substrate. The causal agent is likely a fly but remains unconfirmed.

Stem-galler 2. Galls, larger than those formed by stem-galler 1, were found on stems of *P. fruticosa* from Pichanal, Salta province, Argentina. Stem-galler 1 and 2 may be due to the same causal agent and the gall size difference due to plant responses. The causal agent is unknown.

***Neolasioptera* new species (Diptera: Cecidomyidae).** Young galls caused by this fly were found midway on erect stems, one gall per stem (Fig. 6). Older galls with emergence holes were found lower in the canopy. Often the stem above a younger gall was dead and multiple misshapen new branches grew from the galls. Collections were made at a site in western Formosa province, northwestern Argentina, where *P. nodiflora* (determined by N O'Leary Darwinion Herbarium) or *P. canescens sensu latu* (Fatemi *et al.*, unpub. genetic tree) and *P. fruticosa* grow together. No galls were found on *P. fruticosa*. A cecid fly and several parasitic wasps emerged from galls held in the lab. The fly was identified by R Gagné as *Neolasioptera* new species. Attempts to rear this insect are currently underway at SABCL.

***Puccinia lantanae* Farl. (Basidiomycetes: Uredinales).** This microcyclic rust (Fig. 7) was recorded by Spegazzini (1909) on *P. canescens* from Salta province. From our collections from Pichanal, Salta province, the host is *P. fruticosa* (after N O'Leary) or *P. canescens sensu latu* (after Fatemin *et al.*). It has been recorded on two other *Phyla* species – *Phyla dulcis* (Trevir.) Moldenke (as *Phyla scaberrima* (Juss.) Moldenke) in Cuba (Schmiedeknecht 1984) and *Phyla stoechadifolia* (L.) Small (as *Lippia stoechadifolia* (L.) Kunth) in both North and South America (Farr *et al.* 1989; Viégas 1961).

Figure 6: Single stem galls found on *Phyla canescens* in northern Argentina. Photo: M Julien, CSIRO.

In Argentina, rust was collected from *P. fruticosa* at Pichanal and near San Salvador Mazza in Salta province, at Pre-Delta in Entre Rios province, at Calilegua in Jujuy province, and from *P. nodiflora* or *P. canescens sensu latu* (see comments above) in western Formosa province.

Cross-inoculation glasshouse trials at UNS using rust spores from *P. fruticosa*, from Pichanal, caused considerable damage through local and semi-systemic infection on *P. fruticosa* plants from Pichanal and Calilegua, but only localised infections on *P. canescens* plants from Bahía Blanca, Buenos Aires (G Traversa, unpub.). In another trial using the rust from Formosa, *P. canescens* from Bahía Blanca remained uninfected and *P. fruticosa* from Calilegua developed pustules on leaves. Further trials with rust from Pre-Delta were not successful, i.e. the host plants (*P. canescens* from Bahía Blanca and *P. fruticosa* from Rio Bermejo, Chaco province) failed to develop pustules though necrosis occurred on basal leaves (G Traversa, unpub.). Further field collections are underway so that additional cross-inoculation trials can be conducted. These trials will include a range of *Phyla* forms from Argentina that include the three forms according to O'Leary and representatives of the clades of *P. canescens sensu latu* according to Fatemi *et al.* The aim is to determine if there is a *formae speciales* occurring on *P. canescens* or if the *formae speciales* from other *Phyla* can infect and cause significant damage to forms of *P. canescens*. Host-range tests on selected Verbenaceae are also planned.

Figure 7: *Puccinia lantanae* infecting *Phyla fruticosa* in Argentina. Photo: G Traversa, Universidad del Sur, Bahía Blanca.

Cercospora lippiae **Ellis & Everh. (Hyphomycetes).** This widespread pathogen occurs throughout the north–south distribution (Jujuy to Rio Negro provinces) of *Phyla* in Argentina but not in the central western provinces of La Rioja, San Juan, Cordoba, Mendoza, San Luis, La Pampa and Neuquen, or in the northern provinces of Formosa and Corrientes. It has been collected from the three *Phyla* forms (N O'Leary, *pers. comm.* 2007) that occur in Argentina. Symptoms are angular vein-delimited, irregular leaf spots and leaf necrosis (Fig. 8). Such symptoms were commonly observed on *Phyla* throughout Chile but samples were not collected. Isolation techniques have recently been developed and inoculation trials using plant forms and species described above for *P. lantanae* are planned for this fungus.

Other natural enemies of interest

There are various other natural enemies about which little is known, but which continue to be of interest (Table 1). These include a leaf-miner (Diptera) found on all *Phyla* forms in Buenos Aires, San Luis, Entre Rios, Salta and Jujuy provinces and at Arauco, Chile; four species of moth (Lepidoptera) that together occur throughout the *P. canescens* range; a weevil (Curculionidae) on *P. canescens* and *P. fruticosa* from Jujuy, Chaco, Formosa and Buenos Aires provinces and a thrips (Thysanoptera) on *Phyla* species found in Formosa and Santiago del Estero. The leaf-miner, moths and weevil were found at just a few sites and in very low numbers, so further studies have not been possible. The thrips was found recently.

Figure 8: *Cercospora lippiae* on *Phyla canescens* in Argentina. Photo: G Traversa. Universidad del Sur, Bahía Blanca.

In addition, leaf-hoppers (Cicadellidae) are widespread and relatively common on *Phyla* species, but appear not to be specific. Four *Colletotrichum* fungi have been found; one may be a new species. Though these are of no interest for classical biological control, they may be useful for development of a mycoherbicide if they are also found to occur in the lippia-invaded regions of Australia.

DISCUSSION

We had considered that after 4.5 years of surveys we were unlikely to find additional natural enemies. However, in early 2010 we found a new stem-galler. We also found the rust and *C. lippiae* at one and seven survey sites, respectively, for the first time. All the survey work prior to the summer 2009–10 was conducted during an extended drought period in Argentina. As it may require several non-drought seasons for some natural enemies to recover sufficiently for them to be found during surveys, it may be sensible to continue surveying for the next few summer seasons if normal rainfall conditions occur.

Lippia is competitive with other plants in Australia due to clonal subsidisation (Xu *et al.* 2010). Spread and density increase is due to vegetative growth, movement of stem fragments and seed production. Most populations of lippia are subjected to seasonal inundation, which can cause stem fragmentation and downstream movement. Best management of lippia would therefore be achieved by reducing production of propagules, i.e. nodes and seeds. Stem- and crown-borers and rust fungi would likely have direct impacts on such growth; defoliators and root feeders may have similar effects, but indirectly. Selected agents should be able to survive flooding and extended hot dry periods, so agents that have internal feeding stages may be more protected than external feeders. Surprisingly, lippia has very few insects that feed internally other than the Dipteran node-gallers. Weevils and moths that may have internal feeding stages have been found on lippia, but in such low numbers that we have not been able to develop colonies for study.

As a consequence, the current main activities in this project involve assessing the available potential agents in Argentina, developing rearing techniques, conducting preliminary host-range studies and recording biology so that agents that warrant importation to an Australian quarantine, in addition to *K. bergi*, can be identified.

The uncertainty of the status of *P. nodiflora* in Argentina and the lack of permission to import Australian clones of *P. nodiflora* into Argentina for host-range tests means that this critical test plant species should be included in the first tests conducted on all potential agents imported to Australia. If the agents do not cause significant damage to *P. nodiflora*, further host-specificity testing should be made.

The funding for Australian and overseas research ended in June 2010. Continued uncertainty around future weed research funding threatens to terminate this project prematurely.

ACKNOWLEDGEMENT

We thank Justine Murray for preparing the distribution map (Fig. 1).

REFERENCES

Earl J (2003) 'The distribution and impacts of Lippia (*Phyla canescens*) in the Murray Darling system'. Agriculture Information & Monitoring Services, Guyra, Australia.

Farr DF, Bills GF, Chamuris GP and Rossman AY (1989) *Fungi on Plants and Plant Products in the United States.* APS Press, St Paul.

Julien M, Storrie A and McCosker R (2004) Lippia, *Phyla canescens*, an increasing threat to agriculture and the environment. In *Proceedings of the 14th Australian Weeds Conference.* 6–9 September 2004, Wagga Wagga, NSW. (Eds BM Sindel and SB Johnson) pp. 476–479. Weed Society of New South Wales, Sydney.

Macdonald MJ (2008) Ecology of *Phyla canescens* (Verbenaceae) in Australia. PhD thesis. University of New England, Australia.

McCosker R (1994) 'Lippia (*Phyla nodiflora*): an invasive plant of floodplain ecosystems in the Murray-Darling Basin'. Report on the distribution and ecology of lippia in the lower Gwydir Valley and the Murray-Darling Basin prepared for the Gingham Watercourse Landcare Group. Dept of Ecosystem Management, University of New England, Australia.

Munir AA (1993) A taxonomic revision of the genus *Phyla* Lour. (Verbenaceae) in Australia. *Journal of the Adelaide Botanic Gardens* **15**, 109–128.

Schmiedeknecht M (1984) Uredinales aus Kuba. *Feddes Repertorium* **94**, 683–708.

Sosa AJ, Traversa MG, Delhey R, Kiehr M, Cardo MV and Julien MH (2008) Biological control of lippia (*Phyla canescens*): surveys for the plant and its natural enemies in Argentina. In *Proceedings of the 12th International Symposium on Biological Control of Weeds.* 22–27 April 2007, La Grande Motte, France. (Eds MH Julien, R Sforza, MC Bon, HC Evans, PE Hatcher, HL Hinz and BG Rector) pp. 211–215. CABI, Wallingford, UK.

Spegazzini C (1909) Mycetes argentinenses. (Series IV). *Anales del Museo Nacional de Buenos Aires* **19**, Ser. III, tomo XII, 257–458.

Viégas AP (1961) Índice de fungos da América do Sul. Instituto Agronomico, Campinas, Brazil.

Xu C, Schooler SS and van Klinken RD (2010) Effect of clonal integration and light availability on the growth and physiology of two invasive herbs. *Journal of Ecology* **98**, 833–844.

Pistia stratiotes L. – water lettuce

Michael Day

ABSTRACT

Pistia stratiotes, water lettuce, is a free-floating aquatic weed that has naturalised in Qld and NSW. It is also found in the NT, where it is considered to be native. Water lettuce is fast-growing and has the ability to form dense mats, blocking waterways and interfering with access to water, fishing and hydro-electric flows. Biocontrol began in Australia in 1982, with the opportunistic release of the weevil *Neohydronomus affinis* from Brazil. This insect quickly established in most parts of Qld where it was released. Complete control in northern Qld was achieved in 18 months, while control in southern Qld was achieved in over two years. However, water lettuce infestations in south-western Qld are also actively treated mechanically or with herbicide, or are flushed out during floods, and this inhibits the maintenance of weevil populations. In NSW, the weevil was not released, as the weed is a target for eradication and is being actively controlled.

Key words: *Neohydronomus affinis, Samea multiplicalis, Spodoptera pectinicornis, Argentinorhynchus* spp.

INTRODUCTION

Pistia stratiotes L. (Araceae), water lettuce, is a free-floating perennial weed native to tropical regions of America, Asia, Malesia and the NT (Gillett *et al.* 1988; Waterhouse 1994). The large number of herbivores collected from *P. stratiotes* suggests that its centre of origin is South America (Waterhouse 1994). However, recent studies using chloroplast- and mitochondrial-DNA sequences suggest that it is Palearctic in origin (Renner and Zhang 2004), but has become extinct in that region.

Water lettuce is considered a weed in Africa, Asia, parts of Australia, parts of North America, the Caribbean and the Pacific (Holm *et al.* 1991; Waterhouse 1994). It was first recorded in Australia in the NT in 1946 and in Qld in 1967 (Kleinschmidt and Johnson 1977). It is now found from Townsville to central coastal NSW, as well as in the NT (Fig. 1). However, specific locations of the weed, especially in NSW and southern Qld, vary seasonally due to active control programs or seasonal flushing of waterways due to flooding.

Water lettuce (Fig. 2) is a fast-growing weed which has the ability to form dense mats, completely covering dams and waterways, restricting access to and movement on the water and interfering with fisheries and hydro-electric flows (Waterhouse 1994). Infestations are known to harbour mosquitoes and other pests that transmit diseases (Thompson and Habeck 1989; Waterhouse 1994). Water lettuce reproduces by seed (Harley *et al.* 1990) and vegetatively through the production of daughter plants connected to the parent plants by stolons (Waterhouse 1994). It grows best at 22–30°C and in waters with pH 4 (Holm *et al.* 1991; Neuenschwander *et al.* 2009). Plants are free-floating rosettes, similar to lettuce or young cabbage, and have spongy leaves covered in water-repelling hairs. Good descriptions of the plant are given in Waterhouse (1994) and Neuenschwander *et al.* (2009).

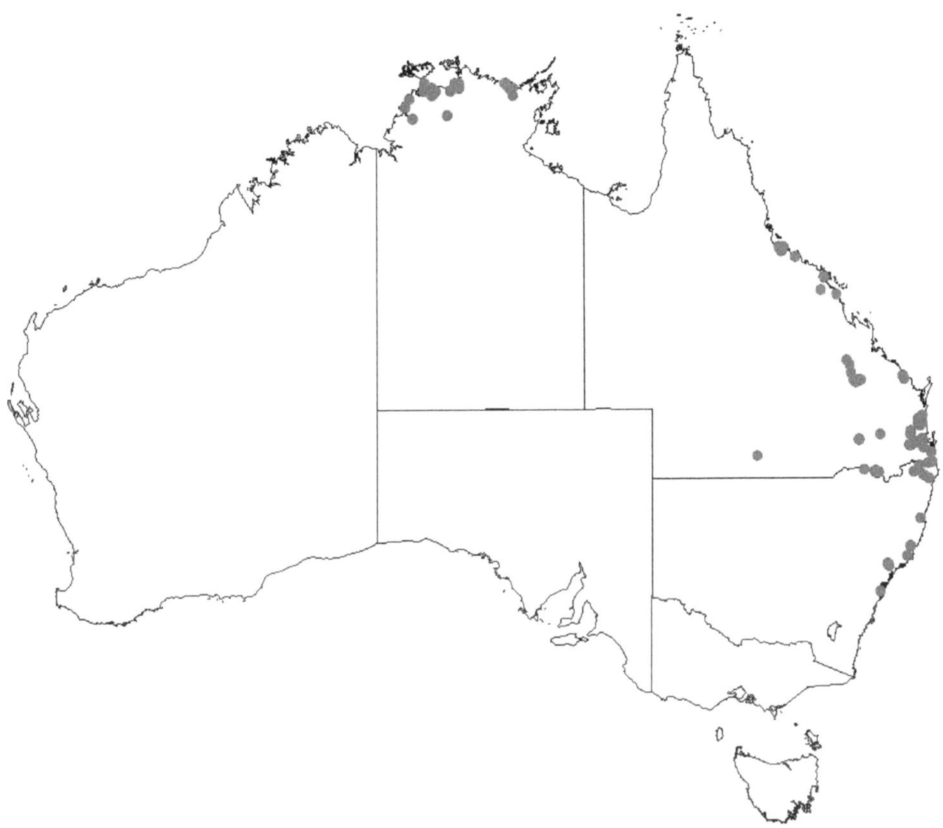

Figure 1: Recorded locations of water lettuce, *Pistia stratiotes*, in Australia. Source: AVH and Qld Government.

BIOLOGICAL CONTROL HISTORY

Biological control of water lettuce began in 1973, with initial surveys for potential biocontrol candidates conducted in Argentina, Brazil and Paraguay by Bennet (CIBC) (Bennett 1975) and DeLoach *et al.* (USDA) (DeLoach *et al.* 1976). The weevil *Neohydronomus affinis* Hustache (Coleoptera: Curculionidae) was assessed for its ability to damage water lettuce and for its host-specificity (DeLoach *et al.* 1976). It was subsequently introduced into Australia in 1981 from Brazil. Following further host-specificity testing, the first field releases were conducted near Brisbane in 1982 (Harley *et al.* 1984). The weevil established quickly, resulting in control of water lettuce on several dams (Harley *et al.* 1984, 1990). It has since been released in many countries, especially in Africa where water lettuce is also a problem (Julien and Griffiths 1998).

Figure 2: *Pistia stratiotes* plants. Photo: CSIRO.

Apart from these initial surveys, there have been only a few additional surveys by Cordo *et al.* (1978) and Dray *et al.* (1993) in tropical America. Some surveys have been conducted in Asia and Africa, but these resulted in fewer phytophagous insects being found than in tropical America. Surveys conducted in Florida in 1985 and 1986 by researchers at the USDA Aquatic Plant Management Laboratory, Fort Lauderdale found 18 phytophagous species, of which only six are known to attack water lettuce, although not exclusively (Dray *et al.* 1993). Apart from *Samea multiplicalis* Guenée (Lepidoptera: Pyralidae), which also attacks salvinia, *Salvinia molesta* D.S. Mitchell (see chapter on salvinia, this volume), the others were not considered for further study.

The moth *Spodoptera pectinicornis* (Hampson) (Lepidoptera: Noctuidae) was found attacking water lettuce in parts of Asia (Waterhouse 1994) and was subsequently introduced to the USA from Thailand. However, it failed to persist (Dray and Center 2002).

A total of 46 organisms have now been found on water lettuce. Of these, 25 species were collected from tropical America, 13 from Asia and eight from Africa. However, of these, only 14, of which 12 are from tropical America, are considered monophagous (Cordo and Sosa 2000). A total of 23 species are deemed polyphagous and therefore unsuitable as biocontrol agents. The remaining nine are oligophagous.

Figure 3: *Neohydronomus affinis*. Photo: CSIRO.

EXPLORATION

Australian researchers did not undertake any dedicated surveys on water lettuce. *N. affinis* was collected in Brazil opportunistically by Wendy Forno in 1981 while she was based in South America surveying potential agents for biocontrol of salvinia and water hyacinth. Following release and establishment of *N. affinis* in Australia, no further work has been conducted by Australians in the native range.

CANDIDATES

Natural enemies released

***Neohydronomus affinis* Hustache (Coleoptera: Curculionidae).** *N. affinis* (at one time incorrectly referred to as *Neohydronomus pulchellus* Hustache (Waterhouse 1994; Julien and Griffiths 1998)) is a small (~1.8 mm long) brown-grey weevil that feeds on the leaves, creating tiny shot-holes that may go through the leaf (Fig. 3). Eggs are laid singly on the upper surface and hatch in three to four days (DeLoach *et al.* 1976). Larvae feed within the leaf, but have not been observed to feed on the roots. Adults emerge after 15–19 days, with the total life-cycle taking 25–30 days (Harley *et al.* 1990). High populations (~130 adults per plant) of *N. affinis* can severely damage and even kill plants (DeLoach *et al.* 1976).

N. affinis was first released in Australia in 1982, around the outskirts of Brisbane. By the mid 1980s, it had been released throughout Qld. Control of water lettuce was more rapid and more complete in the tropics and subtropics (north of 22°S), with control occurring within 18 months (Harley *et al.* 1990) (Figs 4 and 5). In south-eastern Qld, control took over two years, with results confounded by localised flood events.

N. affinis has not been deliberately released in NSW, as the plant is under eradication there and infestations are being actively treated (R Ensbey, *pers. comm.* 2010). In south-western Qld, especially around Boggabilla and Cunnamulla, water lettuce populations vary due to use of herbicides or mechanical control, or flushing by floods. Consequently, weevil populations vary widely and *N. affinis* has often been reintroduced following reinfestation of waterways.

***Samea multiplicalis* Guenée (Lepidoptera: Pyralidae).** *Samea multiplicalis* was introduced into Australia to control salvinia (see chapter on salvinia), but in its native range was also found to damage growing buds of water lettuce and to cause severe damage on occasions. It has been reported to feed on *Lemna* sp. (Araceae), water hyacinth *Eichhornia crassipes* (Mart.) Solms-Laub. (Pontederiaceae) and *Azolla* spp. (Azollaceae) (DeLoach *et al.* 1979; Waterhouse 1994). *S. multiplicalis* has rarely been found on water lettuce in the field in Australia (Waterhouse 1994).

Figure 4: *Pistia stratiotes* infestation at Park Ridge, Qld in 1982. Photo: CSIRO.

Figure 5: *Pistia stratiotes* under control at Park Ridge, Qld in 1984. Same location as in Figure 4. Photo: CSIRO.

Other promising agents

Argentinorhynchus bruchi **(Hustache) (Coleoptera: Curculionidae).** This weevil was found on water lettuce in Argentina and Paraguay and was thought to be very damaging. Adults feed on the leaves, producing tiny holes in the leaf surface. Eggs are laid on the leaf surface and larvae tunnel down to feed on the upper part of the stolon (Cordo and Sosa 2000). In laboratory trials, *A. bruchi* was found to be highly host-specific. However, there were difficulties rearing it in the laboratory due to mortality in the pupal stage. Its usefulness as an agent was considered limited, as it was thought that the exposed eggs would be subject to high predation in the field (Cordo *et al.* 1978).

Argentinorhynchus breyeri **Brethes.** The weevil was recorded on water lettuce in eight countries, including Argentina, Brazil, Bolivia, Paraguay and Venezuela (Cordo and Sosa 2000). Adults feed at night and females lay eggs on the lower leaf surface. Larvae feed within the leaf and tunnel towards the centre of the rosette, feeding on the meristematic tissue. Pupation occurs in the substrate (Cordo and Sosa 2000).

Spodoptera pectinicornis **(Hampson) (Lepidoptera: Noctuidae).** The moth is found in several Asian countries and is reported to cause extensive damage to water lettuce, controlling infestations in some waterways in Thailand within six to eight weeks (Waterhouse 1994). It was subsequently introduced into the USA, where it was thought to have established in Florida (Julien and Griffiths 1998), but did not persist (Dray and Center 2002).

DISCUSSION

Only one agent (*N. affinis*) has been actively introduced and released on water lettuce in Australia. Water lettuce now appears to be under control in areas north of 22°S, with weed infestations disappearing within 18 months of the weevils' release (Harley *et al.* 1990). Weevil impact in southern Qld is limited, with control at some sites being achieved after two to three years (Harley *et al.* 1990). However, in most southern Qld sites, water lettuce populations fluctuate widely. In south-western Qld (e.g. near St George and Cunnamulla), river systems can be choked with water lettuce one season and free of it the next due to flushing by floodwaters. Elsewhere, water lettuce populations are being actively controlled, as cooler temperatures mean that populations of *N. affinis* are not maintained at sufficient levels to exert adequate control. In NSW, where water lettuce is under eradication, spraying and manual removal are the preferred and most effective control options.

The biological control program for water lettuce has never been evaluated but it should be considered at least a partial success, as the weed is under control in the tropics. Overall, the cost of the project was relatively low, as *N. affinis* was originally collected by DeLoach and tested in Florida on behalf of the USA. Following these tests, which showed that the agent was host-specific and damaging (DeLoach *et al.* 1976), the agent was imported into Australia opportunistically during work on salvinia and water hyacinth in South America.

At this stage, there is no intention to import other agents which have been introduced elsewhere or studied previously. Water lettuce is under eradication in NSW and there are only occasional outbreaks of the weed in southern Qld. It is no longer considered a problem in northern Qld.

REFERENCES

Bennett FD (1975) Insects and plant pathogens for the control of *Salvinia* and *Pistia*. In *Proceedings of the Symposium for Water Quality Management through Biological Control*. 29–31 January 1975, Gainesville, Florida. (Ed TE Freeman) pp. 28–35. EPA, Gainesville, Florida.

Cordo HA and Sosa A (2000) The weevils *Argentinorhynchus breyeri*, *A. bruchi* and *A. squamosus* (Coleoptera: Curculionidae), candidates for biological control of waterlettuce *Pistia stratiotes*. In *Proceedings of the 10th International Symposium on Biological Control of Weeds*. 4–14 July 1999, Bozeman, Montana. (Ed NR Spencer) pp. 325–335. Montana State University, Bozeman.

Cordo HA, DeLoach JC, Runnacles J and Ferrer R (1978) *Argentinorhynchus bruchi*, a weevil from *Pistia strati-*

otes in Argentina: biological studies. *Environmental Entomology* 7, 329–333.

DeLoach CJ, DeLoach AD and Cordo HA (1976) *Neohydronomus pulchellus*, a weevil attacking *Pistia stratiotes* in South America: biology and host specificity. *Annals of the Entomological Society of America* **69**, 830–834.

DeLoach CJ, Deloach DJ and Cordo HA (1979) Observations on the biology of the moth, *Samea multiplicalis*, on water lettuce in Argentina. *Journal of Aquatic Plant Management* **17**, 42–44.

Dray FA and Center TD (2002) Waterlettuce. In *Biological Control of Invasive Plants in the Eastern United States.* (Technical coordinators R van Driesche, S Lyon, B Blossey, M Hoddle and R Reardon) pp. 65–78. USDA Forest Service. FHTET-2002-04.

Dray FA, Center TD and Habeck DH (1993) Phytophagous insects associated with *Pistia stratiotes* in Florida. *Environmental Entomology* **22**, 1146–1154.

Gillett JD, Dunlop CR and Miller IL (1988) Occurrence, origin, weed status and control of water lettuce (*Pistia stratiotes* L.) in the Northern Territory. *Plant Protection Quarterly* **3**, 144–148.

Harley KLS, Forno IW, Kassulke RC and Sands DPA (1984) Biological control of water lettuce. *Journal of Aquatic Plant Management* **22**, 101–102.

Harley KLS, Kassulke RC, Sands DPA and Day MD (1990) Biological control of water lettuce *Pistia stratiotes* (Araceae) by *Neohydronomus affinis* (Coleoptera: Curculionidae). *Entomophaga* **35**, 363–374.

Holm LG, Plucknett DL, Pancho JV and Herberger JP (1991) *The World's Worst Weeds: Distribution and Biology.* Kreiger Publishing, Malabar, Florida.

Julien MH and Griffiths MW (1998) *Biological Control of Weeds: A World Catalogue of Agents and their Target Weeds.* 4th edn. CABI, Wallingford, UK.

Kleinschmidt HE and Johnson RW (1977) *Weeds of Queensland.* Qld Dept of Primary Industries, Brisbane.

Neuenschwander P, Julien MH, Center TD and Hill MP (2009) *Pistia stratiotes* L. (Araceae). In *Biological Control of Tropical Weeds Using Arthropods.* (Eds R Muniappan, GVP Reddy and A Raman) pp. 332–352. Cambridge University Press, Cambridge.

Renner SS and Zhang LB (2004) Biogeography of the *Pistia* clade (Araceae): based on chloroplast and mitochondrial DNA sequences and Bayesian divergence time inference. *Systematic Biology* **53**, 422–432.

Thompson CR and Habeck DH (1989) Host specificity and biology of the weevil *Neohydronomus affinis* (Coleoptera: Curculionidae), a biological control agent of *Pistia stratiotes*. *Entomophaga* **34**, 299–306.

Waterhouse DF (1994) *Biological Control of Weeds: Southeast Asian Prospects.* Monograph 26. ACIAR, Canberra.

Prosopis spp. – mesquite

Rieks D van Klinken

ABSTRACT

Prosopis species (Leguminoseae) and their hybrids (together referred to as 'mesquite') are long-lived shrubs or trees that have intentionally been spread around the world and subsequently naturalised. In Australia they are recognised as one of 20 Weeds of National Significance due largely to their potential impacts on pastoral production in hot, semi-arid and arid rangelands. Mesquite was first targeted for biological control in Australia in 1994 following the release of seed predators in South Africa. As a consequence, releases were made of two seed predators, *Algorobius prosopis* and *Algorobius bottemeri* (Bruchidae) from the USA, and a leaf-tying moth *Evippe*. sp. #1 (Gelechiidae) and a sap-sucking insect *Prosopidopsylla flava* (Psyllidae) from Argentina. Only the leaf-tying moth and one seed predator (*A. prosopis*) have become widely established, and only the leaf-tying moth is causing population-level impacts. The moth has caused continuous high-level defoliation of the largest mesquite infestation in Australia (located in the Pilbara region of WA) since its release in 1998, resulting in seed production and growth rates approaching zero. This is already greatly facilitating integrated control work and population modelling predicts a long-term reduction in population growth rates. Population-level impacts of the moth are expected to be negligible in south-western Qld and intermediate in northern Qld. There are likely to be many more potential agents for mesquite as it has a rich and diverse native range fauna.

Key words: Gelechiidae, Psyllidae, Leguminoseae.

INTRODUCTION

Prosopis (Leguminoseae) is a genus of woody shrubs or trees native to the Americas (40 species) and south-west Asia and Africa (four species) (Burkart 1976). Several species within section Algarobia (together referred to as 'mesquite') have been intentionally introduced into hot, semi-arid and arid habitats around the world as multipurpose plants. Mesquite can provide a range of services including timber, highly nutritious and palatable pods (for livestock and human food), honey, shade and restoration of degraded rangelands (Pasiecznik *et al.* 2001, 2004). However, most of these utilised mesquite species can also be highly invasive, resulting in dense thorn forests which cause wide-ranging economic, social and environmental impacts.

In Australia at least three species, *Prosopis pallida* (Humb. & Bonpl. Ex Willd.) Kunth, *Prosopis velutina* Wooton and *Prosopis glandulosa* Torr., together with their hybrids, are widely naturalised, and a fourth species, *Prosopis juliflora* (Sw.) DC., has been recorded (van Klinken and Campbell 2009). All four species belong to the New World Section Algarobia which includes about 26 species (Burkart 1976). Together these species and their hybrids are ranked as one of 20 Weeds of National Significance (WONS) on the basis of their actual and potential impacts to the pastoral industry and the

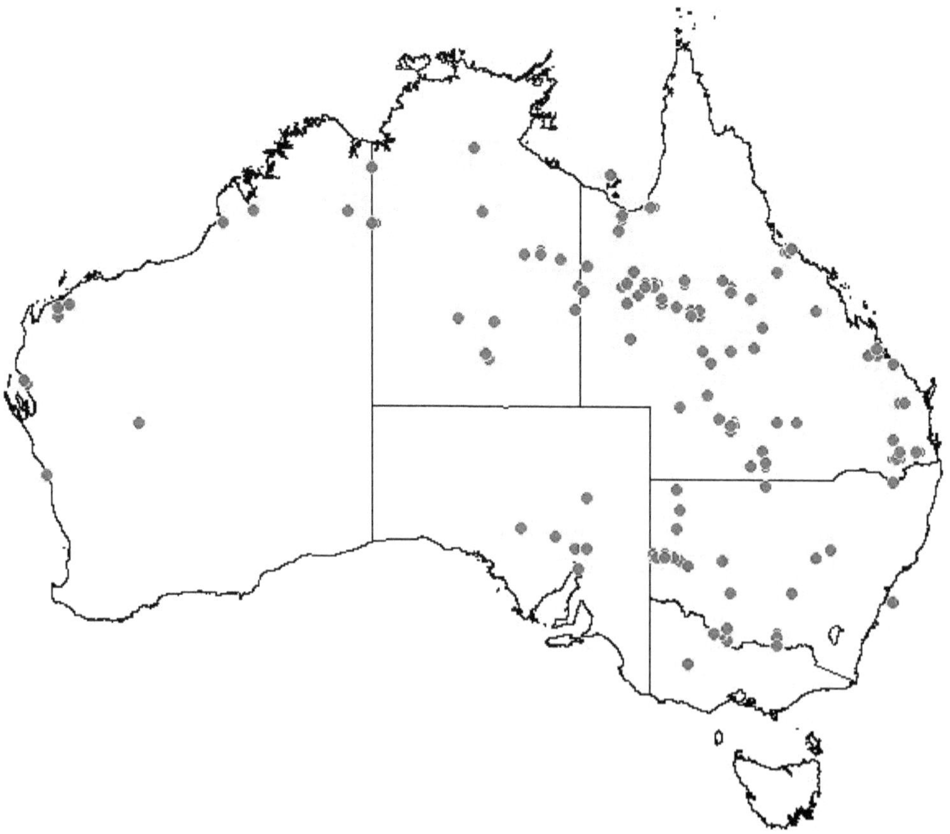

Figure 1: Mesquite distribution based on herbarium records. Specimen data reproduced from Australia's Virtual Herbarium with permission of the Council of Heads of Australasian Herbaria Inc. See van Klinken and Campbell (2009) for a more detailed description of the distribution and density of the different mesquite taxa in Australia.

environment (Thorp and Lynch 2000). Mesquite is now naturalised in all mainland Australian states and the NT (Fig. 1) (van Klinken and Campbell 2009). Active management programs have greatly reduced the density and extent of most major infestations in Australia, and large dense populations are now restricted to two core infestations, in the Pilbara region of WA (200 800 ha with 32 500 of dense infestation) (van Klinken *et al.* 2007) and in northern Qld (~20 000 ha of mesquite around Hughenden) (N March, *pers. comm.* 2010).

The mesquite species in Australia all have bipinnate leaves with one to four pairs of pinnae, and have characteristic 'lambs-tail' flowers and fleshy pods (Fig. 2a, b). They can grow to 3–15 m tall, either with a single stem and spreading canopy or as a multistemmed shrub. Mortality is very low after the seedling stage. They are long-lived species, so age-related deaths in Australia, where most populations are less than 90 years old, are probably still rare. Seed production can be prodigious, especially following wet years, with most seeds subsequently being dispersed through the gut of vertebrate herbivores (Fig. 2c). Seeds have hard-seeded dormancy and can therefore be long-lived, although most probably lose dormancy within a year or two under typical field conditions (van Klinken and Campbell 2009).

BIOLOGICAL CONTROL HISTORY

Research into mesquite biological control first commenced in the USA (DeLoach 1988) with surveys of natural enemies in North and South America by the US Department of Agriculture (Cordo and DeLoach 1987). There had long been an interest in the natural enemies of *Prosopis* species owing to it being an important group in its native range ecologically, and also at times economically and socially (Ward *et al.* 1977). Many studies were conducted on the life-histories, ecology and host-specificity of several of the most abundant or damaging natural enemies present on mesquite in the USA (DeLoach 1982, 1983a, b, 1994; Cuda *et al.* 1990; DeLoach and Cuda 1994; Cuda and DeLoach 1998). However, no formal biocontrol program was conducted owing to concerns regarding the use of classical biological control to target native species in the USA.

Figure 2: a) Flowers and leaves (with pinnules) of *Prosopis velutina* × *P. glandulosa* var. *torreyana* near Port Augusta, SA. b) Large thorns on hybrid mesquite in the Pilbara, WA. Photos: R van Klinken, CSIRO. c) Mature pods and seeds in emu dung at Comongin Station, Qld. Photo: Qld Government.

Mesquite was first targeted for biological control in South Africa from 1985 (Zimmermann 1991; Moran *et al.* 1993; Impson *et al.* 1999; van Klinken *et al.* 2009a). Testing was restricted to flower- and pod-feeders to avoid impacting on any commercial value of the plant. Initially three seed-feeding bruchids were tested and released – *Algarobius prosopis* (Le Conte), *Algarobius bottimeri* Kingsolver and *Neltumius arizonensis* (Schaeffer). All three became established, although *A. bottimeri* has not been recovered for several years and may no longer be present in South Africa (Roberts 2006). Fungal isolates were also isolated from green pods in North and South America and from diseased plant material in South Africa and Namibia from 1994 to 2003. Nine isolates from the Americas showed promise for classical biological control but none were found suitable following extensive testing in 2000–02 (Zimmermann *et al.* 2006; van Klinken *et al.* 2009a). In addition, in South Africa permission is being sought to release *Coelocephalapion gandolfoi* Kissinger (Curculionidae), which feeds on green pods, and testing of the flower bud-galler *Asphondylia prosopidis* Cockerell is ongoing (van Klinken *et al.* 2009a).

A biological control program targeting mesquite commenced in Australia in the early 1990s, by the Qld Government. The two *Algarobius* species previously released in South Africa were retested and subsequently released from 1997, and post-release evaluations have recently been done (van Klinken *et al.* 2009a). CSIRO commenced a biocontrol program in 1995 to test new potential agents. Five insect species were studied in quarantine between 1996 and 1998 (van Klinken 1999, 2000; van Klinken and Heard 2000; van Klinken *et al.* 2009a; van Klinken and Campbell 2009). Two new agents, *Evippe* sp. #1 (Gelechiidae) and *Prosopidopsylla flava* (Psyllidae), were released between 1998 and 2000 (van Klinken *et al.* 2003). *Mozena obtusa* (Coreidae) was tested and found not to be sufficiently host-specific (van Klinken 1999), *Oncideris rhodostica* was partially tested with results proving inconclusive (RD van Klinken, unpub.) and a second *Evippe* species was reared (van Klinken *et al.* 2009a).

There have been efforts to introduce the seed-feeding biological control agents from South Africa into other countries including Ascension Island (British overseas territory) and Kenya (van Klinken *et al.* 2009a). Host-specificity testing of *A. prosopis* has been conducted in Kenya and it was found to be safe to release. Permission to release it there is currently being sought (van Klinken *et al.* 2009a). The seed-feeding bruchids, especially *A. prosopis*, are likely to have been moved around unintentionally in northern Africa and the Middle East with seed used in reforestation (Delobel and Fédière 2002; van Klinken *et al.* 2009a).

PLANT TAXONOMY

Prosopis is a taxonomically difficult genus owing to the paucity of morphological characters, considerable phenotypic variation and overlap in morphological characters, and widespread hybridisation and introgression between a wide range of species, in both their native and introduced ranges. There has been little taxonomic study since the genus-level monograph of Burkart (1976), although recent genetic studies are helping to resolve species-level problems (Landeras *et al.* 2006; Burghardt and Espert 2007).

At least three species are naturalised in Australia, *P. velutina* and *P. glandulosa* (including both varieties, *glandulosa* and *torreyana*) from North America, and *P. pallida* from Central America and Peru (van Klinken *et al.* 2009a). Australian populations variously include true species or hybrids of two or all three of those species. For example, the core infestation at Hughenden (northern Qld) is predominantly or entirely *P. pallida* while the core infestation in the Pilbara region appears to be a hybrid swarm of all three species. A fourth species, *P. juliflora*, has also been reported from a few sites, but may no longer be present (van Klinken and Campbell 2009). Genetic studies are currently underway to confirm the taxonomic status of Australian populations (R Watts and RD van Klinken, unpub.).

EXPLORATION

The Australian mesquite biological control programs have relied entirely on exploratory work previously conducted by others (Ward *et al.* 1977; Cordo and DeLoach 1987; DeLoach 1988), including unpublished expert opinion of local specialists in America (J DeLoach) and Argentina (H Cordo) (T Heard, *pers. comm.* 2010). Biological control agents were required that could attack multiple mesquite species. There was little evidence that natural enemies specialised on particular species within the section Algarobia. Potential agents were therefore also considered from *Prosopis* species that were native to Argentina, even though they did not occur in Australia (i.e. new associations).

CANDIDATES

Natural enemies (native range)

Prosopis species have a rich and diverse natural enemy fauna in their native range, with over 945 species attacking all plant parts recorded (Swenson 1969; Ward *et al.* 1977; Cordo and DeLoach 1987; Silva 1988; Paziecznik *et al.* 2001; van Klinken *et al.* 2009a). Survey work in Argentina and Paraguay was relatively systematic (Cordo and DeLoach 1987). In contrast, survey work in the USA was more ad hoc, and has been mostly summarised from disparate sources by Ward *et al.* (1977). Many of these natural enemies may not be candidates for biological control because they are likely to be transients on *Prosopis*, have broad host-ranges or are not particularly damaging.

Natural enemy diversity is nonetheless likely to be seriously underreported on *Prosopis* species. Much of the Americas have not been systematically surveyed, including northern South America (Peru, Ecuador, Colombia and Venezuela), Central America and the Caribbean. Also, with a few exceptions, pathogens, stem-borers, root-feeders and natural enemies that preferentially attack immature mesquite plants have been largely overlooked (van Klinken *et al.* 2009a).

Natural enemies (exotic range)

Natural enemies of mesquite in its introduced range have not been systematically surveyed, with the exception of fungal isolates in South Africa and Namibia (Zimmermann *et al.* 2006). Nonetheless, native phytophages and pathogens have been observed to cause significant damage to mesquite plants in India and Pakistan (Tewari *et al.* 1998; Yousuf and Gaur 1998). Several native natural enemies damage mesquite in Australia, including the crusader bug *Mictis profana* (Fabricius), a seed-feeding bruchid *Caryedon serratus* (Olivier) and an unidentified stem-boring cerambycid that can kill seedlings (van Klinken and Campbell 2009), but none are likely to cause significant population-level impacts. The exception might be vertebrate herbivores feeding on pods. Although consumption of pods is the principal means of seed dispersal, seed survival through the gut varies greatly between herbivore species (van Klinken *et al.* 2009a). For example, around 60% of seeds survive passage through cattle while only a few percent may survive through animals that masticate, such as sheep and macropods. Slower invasion rates may therefore be expected when pods are predominantly consumed by masticators, although the largest infestation in Australia is on a property that had sheep until 2000 (Robinson *et al.* 2007).

Agents tested but not released (in Australia)

***Heteropsylla texana* Crawford.** This psyllid was tested in 1991 by the Qld Government (Donnelly 1995, 2002). Complete development did occur on two of the 45 non-target species tested, *Acacia bidwillii* Benth. and *Dichrostachys spicata* (F. Muell. Domin), but oviposition was much lower, fewer adults emerged and development times were longer than on *Prosopis* species. Results therefore suggest that *H. texana* may be safe to release in Australia, but more intensive testing on at-risk plant species would be required to confirm this.

***Mozena obtusa* Uhler (Coreidae).** This large coreid is one of the more common natural enemies on mesquite in the USA. Testing in quarantine showed that it could complete development on Australian species in the closely related genus *Neptunia* (van Klinken 1999). However, nymphal performance was highly variable between replicates, possibly because variation in plant nitrogen composition affected plant quality. The relationship between nitrogen composition, plant quality and nymphal development would need to be resolved (Cuda *et al.* 1995; van Klinken 1999), and the degree of oviposition specificity under field conditions determined, before this insect could be definitively assessed as a biocontrol agent.

Oncideres rhodosticta **Bates.** *O. rhodosticta* is a common stem-boring and stem-girdling beetle on mesquite in the USA and Mexico, and its field host-range may be restricted to *P. glandulosa* (D Ueckert, *pers. comm*). Preliminary host-testing was conducted using adults that emerged from two shipments received from Texas in 1996 (>220 adults) and 1997 (15 adults) but proved inconclusive (RD van Klinken, unpub.). Adults did feed on non-target plant species, especially in the absence of *Prosopis*, but their performance was relatively poor if only non-target hosts were available. Development of culturing techniques, possibly on artificial diets, would be a priority for further work on this species.

Agents released in Australia

Algarobius prosopis **(LeConte) (1996) and** ***Algarobius bottimeri*** **Kingsolver (1997).** These two seed-feeding bruchid beetles were released in Australia following host-specificity testing in quarantine additional to that conducted in South Africa (Donnelly *et al.* 1997). Both species were widely released in WA and Qld (G Donnelly, *pers. comm.*), although only *A. prosopis* has been found in recent surveys in both states (van Klinken and White 2009). It is unlikely to be causing population-level impacts as seed predation rates were low, almost always less than 20% (RD van Klinken and A White, unpub.).

Prosopidopsylla flava **and** ***Evippe*** **sp. #1.** These were both released in large numbers throughout Australia (NSW, Qld, the NT and WA) between 1998 and 2000 (van Klinken *et al.* 2003). In laboratory trials both species did equally well on all Australian mesquite taxa (van Klinken 2000; van Klinken and Heard 2000), although *Evippe* sp. #1 does not perform as well on *P. pallida* in the field (van Klinken *et al.* 2009a). *P. flava* only became tenuously established in north-western NSW. Its failure to establish elsewhere despite multiple large releases at some locations (up to 31 releases totalling more than 49 000 individuals), may be due to climatic factors (with hot conditions limiting populations), predation from ants, or in-breeding in the laboratory (releases were sourced from a population that had been in culture for ~11–23 generations). In contrast, *Evippe* sp. #1 established everywhere (although it remains rare in NSW). It thrives in hot conditions, doing best in the Pilbara region where conditions are already hot when the insect emerges from obligatory diapause about August.

Evippe sp. #1 rapidly dispersed from release sites, ~115 km from one release site in three years (van Klinken *et al.* 2003). Although it attracted a wide range of larval and pupal parasitoids, overall parasitism rates have remained low (averaging 1.8%) (van Klinken and Burwell 2005). In the Pilbara it caused over 90% defoliation within 12 months of release and has since maintained similarly high rates of defoliation, despite high interyear variation in rainfall (van Klinken *et al.* 2003, 2009a; Anderson *et al.* 2006; van Klinken and White 2009; RD van Klinken, unpub.). As a result, mesquite plants now flush asynchronously and rarely produce much foliage (Fig. 3). Repeated defoliation has contributed to the very low seed production and growth rates observed since 2000–01, and tree death is becoming apparent (van Klinken *et al.* 2009a).

Other promising agents

There are likely to be numerous other potential biological control agents for mesquite in Australia. Available evidence suggests that most natural enemies do not discriminate greatly between *Prosopis* species within Section Algarobia, so potential agents could be sourced from any Algarobia species and are likely to be similarly effective

Figure 3: a) A mesquite infestation in the Pilbara region viewed from the air. b) Viewed from the ground, following prolonged defoliation by the leaf-tying moth *Evippe*. sp. #1. Photos: a) R Parr. b) R van Klinken, CSIRO.

on all Australian taxa, including hybrids. In fact, the potential for biocontrol agents among the natural enemy fauna on mesquite remains largely untapped. This is especially true for organisms that attack plant parts other than flowers and pods (which were the focus for South Africans), pathogens (which have not been systematically surveyed) and for natural enemies outside the USA.

South Africa has had an ongoing biological control program and therefore is the most ready source for new agents (van Klinken *et al.* 2009a). However, its program has been restricted to potential agents that attack only flowers and seeds. The most exciting agent there is the weevil *Coelocephalapion gandolfi* Kissinger, which attacks green pods and is host-specific to the Algarobia section. A release application for South Africa is being prepared (C Zachariades, *pers. comm.* 2010). The multivoltine galler of flower buds, *A. prosopidis* Cockerell (Cecidomyiidae), also shows potential, although it is proving difficult to culture and test. Genetic work suggests it may be a complex of at least four 'species' (D Thompson, *pers. comm.* 2010). To be effective, both species will require a mechanism to synchronise activity with mesquite reproduction, which typically peaks over a relatively short period (months) in Australian populations.

EVALUATION

Measurable criteria for successful biological control are important as they help guide the selection of potential biological control agents and serve as an objective measure against which the success of a biological control program can be assessed (van Klinken and Raghu 2006). For this reason, success criteria were recently developed for mesquite, combining impact reduction and contribution to integrated management (van Klinken *et al.* 2009a). Complete success was defined as when the canopy cover of populations at equilibrium is reduced below 10% and when the long-term removal of all adult mesquite plants is feasible using other management strategies. Partial success was defined as when a canopy cover target of 10% could be realistically met through a combination of biological control and other control techniques.

Biological control of mesquite has been partially successful, as defined above, in the Pilbara region of WA where the largest infestation in Australia occurs. It is also likely to be causing significant impacts in northern Qld. Most evaluation work has been conducted at Mardie Station in the Pilbara. Population modelling based on this field work predicts that the population will stabilise or reduce to an equilibrium of sparse to moderately dense canopy cover over the next century (Fig. 4) (J Pichancourt and RD van Klinken, unpub.). The exception is in parts of the landscape that act as seed sinks, such as around watering points where cattle congregate, where continued increases in population densities are expected. These predictions are contingent on leaf-tier activity remaining as high as it has been in the past 10 years, most importantly that it is sufficient to continue to suppress seed production even following a series of high-rainfall years.

Most mesquite management in Australia relies on a combination of mechanical control (especially dozing, which severs trees at least 20 cm under the ground), chemical control (mostly basal barking, which involves ring-barking trees with herbicide) and dispersal management (especially controls on stock movement). Repeat applications of control are required to kill plants that survive or are missed by the primary treatment, or that establish from the seed bank. Kill rates using these techniques typically approach 100% if properly applied, however, a high proportion of seedlings and juveniles can be missed by spray teams, especially when applying herbicide over large areas. Leaf-tier damage in the Pilbara region has greatly facilitated existing integrated management practices by reducing seed bank densities, reducing dispersal risks and increasing time to maturity of regenerating or newly established plants following the primary control. The excellent dispersal ability of the moth *Evippe* sp. #1, and its rapid rate of increase, means that it will quickly re-establish on surviving or regenerating plants following a control program.

DISCUSSION

The mesquite biocontrol program in Australia has provided some important insights of general value. First, it provided insights into why biocontrol agents will not necessarily perform equally well across the distribution of the target, and showed that predicting where the agents will do best may not be possible if only distribution data from the native range is used, especially when novel climates are encountered (in this case, year-round hot conditions in the Pilbara) (van Klinken *et al.* 2003). A range of data sources is likely to be required for reliable predictions of where agents will do best (van Klinken *et al.* 2009b). Second, considerable attention is often devoted to locating potential biocontrol agents from the native range of target weed genotypes (Goolsby *et al.* 2006). Mesquite biocontrol provides a counter example where tested natu-

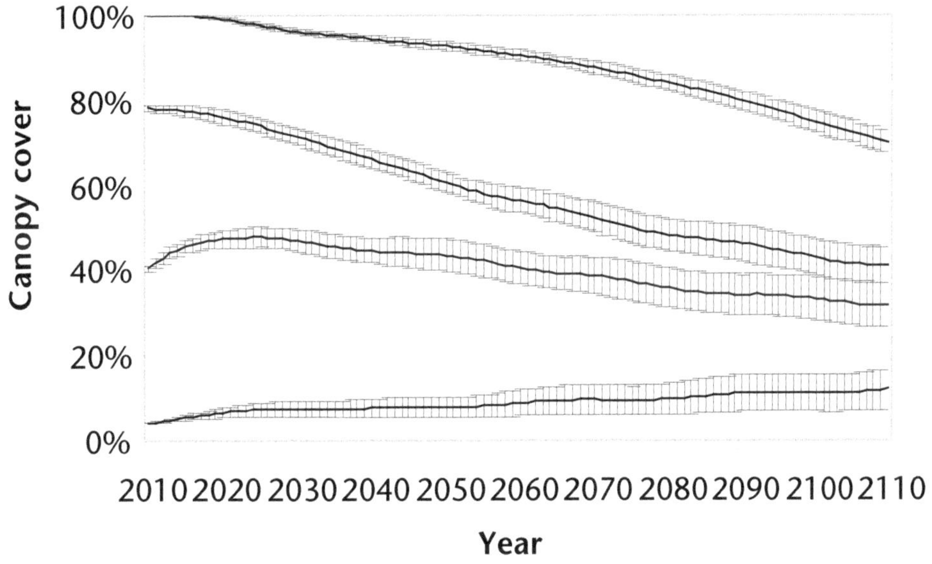

Figure 4: Predicted population dynamics of mesquite in the Pilbara region, assuming no anthropogenic climate change and no net seed dispersal (J Pichancourt and RD van Klinken, unpub.). The population structure for representative starting populations were based on field observations (van Klinken *et al.* 2006) and the matrix model used to generate the predictions was parameterised using data obtained since the release of the leaf-tying moth in 1998.

ral enemies do not appear to differentiate between hosts at the species level, at least not in Section Algarobia. This means that suitable biocontrol agents can be found from novel hosts, and that agents should be equally effective against hybrids. Third, the program demonstrated again that defoliators can be highly effective, even against a particularly hardy and long-lived perennial shrub such as mesquite. However, impact on mesquite was contingent on the leaf-tying moth *Evippe* sp. #1 maintaining very high densities for long periods. This was assisted by low parasitism rates in the field despite immature stages attracting a diverse range of generalist parasitoids (van Klinken and Burwell 2005). Finally, seed predators are commonly used as biocontrol agents because they are often easy to obtain, culture and test. They also circumvent potential conflict with economic exploitation of the target plant, as in South Africa. In the case of mesquite, seed predation rates are fundamentally constrained because pods are typically consumed by vertebrate herbivores, which means that most of the seed crop may be available for seed predation only for brief periods. Post-dispersal seed predation is unlikely to be sufficient to overcome this in Australia (RD van Klinken and A White, unpub.) despite observations to the contrary in South Africa (van Klinken *et al.* 2009a).

In Australia, further biocontrol agents are not currently considered a priority by the relevant national expert body, although potential agents that could be introduced into Australia with relatively little testing are likely to be considered (National Prickle Bush Management Group; N March, *pers. comm.* 2010). Existing management tools (largely mechanical and basally applied herbicide) have been hugely successful in managing mesquite where sufficient resources have been available and *Evippe* sp. #1 is currently highly effective in suppressing the main population in the Pilbara. Nonetheless, sparse populations remain across vast areas of suitable habitat, representing the nascent foci for new core infestations. For example, in a recent telephone survey of land managers on 109 pastoral properties in four shires in north-western Qld mesquite was described as common and abundant on 5% of properties but 'occasional and localised' on a further 34% (March 2010). These nascent populations represent the biggest challenge for management as they all have the potential to become new core infestations if the right climate conditions (e.g. several successive high-rainfall years) combine with good seed dispersal and ineffective management. If biocontrol becomes a priority again, there should be several new potential biocontrol agents. Potential agents may be sourced from South African work, from known natural enemies in the Americas, from 'pests' identified in areas where mesquite is utilised, and from new native-range survey work directed at areas and guilds that have been undersampled. Attention would be required to clearly define the objectives for biocontrol (e.g. with respect to target regions and overall management objectives) (van Klinken and Raghu 2006; van Klinken *et al.* 2009a) and to pick the agents that are most likely to be effective, from the large number of available natural enemies.

REFERENCES

Anderson LJ, van Klinken RD, Parr RJ, Climas R and Barton D (2006) Integrated management of hybrid mesquite: a collaborative fight against one of Australia's worst woody weeds. In *Papers and Proceedings of the 15th Australian Weeds Conference*. 24–28 September 2006, Adelaide. (Eds C Preston, JH Watts and ND Crossman) pp. 239–242. Weed Management Society of South Australia, Adelaide.

Burghardt AD and Espert SM (2007) Phylogeny of *Prosopis* (Leguminosae) as shown by morphological and biochemical evidence. *Australian Systematic Botany* **20**, 332–339.

Burkart A (1976) A monograph of the genus *Prosopis* (Leguminosae subf. Mimosoideae). *Journal of the Arnold Arboretum* **57**, 219–249, 450–525.

Cordo HA and DeLoach J (1987) 'Insects that attack mesquite (*Prosopis* spp.) in Argentina and Paraguay: their possible use for biological control in the United States.' US Dept of Agriculture, ARS62.

Cuda JP and DeLoach CJ (1998) Biology of *Mozena obtusa* (Hemiptera: Coreidae), a candidate for the biological control of mesquite, *Prosopis* spp. (Fabaceae). *Biological Control* **13**, 101–110.

Cuda JP, DeLoach CJ and Robbins TO (1990) Population dynamics of *Melipotis indomita* (Lepidoptera: Noctuidae), an indigenous natural enemy of mesquite, *Prosopis* spp. *Environmental Entomology* **19**, 415–422.

Cuda JP, Johnson HB and Tischler CR (1995) Trophic interactions between *Mozena*, mesquite and a microbe: implications for host-specificity testing of insects of leguminous weeds. In *Proceedings of the 8th International Symposium on Biological Control of Weeds*. 2–7 February 1992, Canterbury, NZ. (Eds ES Delfosse and RR Scott) p. 575. DSIR/CSIRO, Melbourne.

DeLoach CJ (1982) Biological studies on a mesquite leaf tier, *Tetralopha euphemella*, in central Texas. *Environmental Entomology* **11**, 261–267.

DeLoach CJ (1983a) Life history of a mesquite looper, *Semiothisa cyda* (Lepidoptera: Geometridae), in central Texas. *Annals of the Entomological Society of America* **76**, 83–86.

DeLoach CJ (1983b) Field biology and host range of a mesquite looper, *Semiothisa cyda*, (Lepidoptera: Geometridae), in central Texas. *Annals of the Entomological Society of America* **76**, 87–93.

DeLoach CJ (1988) 'Mesquite: a weed of southwestern rangelands: a review of its taxonomy, biology, harmful and beneficial values, and its potential for biological control'. A petition to the technical advisory group on introduction of biological control agents of weeds, USDA, Animal and Plant Health Inspection Service.

DeLoach CJ (1994) Feeding behaviour of *Melipotis indomita* (Lepidoptera: Noctuidae), a herbivore of mesquite (*Prosopis* spp.). *Environmental Entomology* **23**, 161–166.

DeLoach CJ and Cuda JP (1994) Host range of the mesquite cutworm, *Melipotis indomita* (Lepidoptera: Noctuidae), a potential biological control agent for mesquite (*Prosopis* spp.). *Biological Control* **4**, 38–44.

Delobel A and Fédière G (2002) First report in Egypt of two seed-beetles (Coleoptera: Bruchidae) on *Prosopis* spp. *Bulletin of Faculty of Agriculture, Cairo University* **53**, 129–139.

Donnelly G (1995) Biological control of mesquite. In *Exotic Woody Weeds and their Control in North West Queensland*. (Ed N March) pp. 32–33. Isa Printing Service, Mt Isa, Qld.

Donnelly GP (2002) The host range and biology of the mesquite psyllid *Heteropsylla texana*. *Biocontrol* **47**, 363–371.

Donnelly G, Dodd J and Addison S (1997) Biological control of mesquite with insects. In *Annual Report on Weed and Pest Animal Research, Queensland Department of Natural Resources, Technical Highlights 1996–7*, pp. 10–11.

Goolsby JA, van Klinken RD and Palmer WA (2006) Maximizing the contribution of native-range studies towards the identification and prioritization of weed biocontrol agents. *Australian Journal of Entomology* **45**, 276–286.

Impson FAC, Moran VC and Hoffmann JH (1999) A review of the effectiveness of seed-feeding bruchid beetles in the biological control of mesquite, *Prosopis* species (Fabaceae), in South Africa. In *Biological Control of Weeds in South Africa (1990–1998)*. (Eds T Olckers and MP Hill) pp. 81–88. African Entomology Memoir No. 1.

Landeras G, Alfonso M, Pasiecznik NM, Harris PJC and Ramirez L (2006) Identification of *Prosopis juliflora* and *Prosopis pallida* accessions using molecular markers. *Biodiversity and Conservation* **15**, 1829–1844.

March N (2010) Assessing property level awareness and best practice adoption for prickly acacia and mesquite management in western Queensland. Qld Dept of Employment, Economic Development and Innovation. Accessed from ipcu@dpi.qld.gov.au.

Moran VC, Hoffmann JH and Zimmermann HG (1993) Objectives, constraints and tactics in the biological control of mesquite weeds (*Prosopis*) in South Africa. *Biological Control* **3**, 80–83.

Pasiecznik NM, Harris PJC, Harsh LN, Cruz G, Tewari JC, Cadoret K and Maldonado LJ (2001) *The* Prosopis juliflora-Prosopis pallida *Complex: A Monograph.* HDRA Publishing, Coventry, UK.

Pasiecznik NM, Harris PJC and Smith SJ (2004) *Identifying Tropical* Prosopis *Species: A Field Guide.* HDRA Publishing, Coventry, UK.

Roberts A (2006) Biological control of alien invasive mesquite species (*Prosopis*) in South Africa: the role of introduced seed-feeding bruchids. PhD thesis. University of Cape Town, South Africa.

Robinson TP, van Klinken RD and Metternicht G (2007) Spatial and temporal rates and patterns of mesquite (*Prosopis* species) encroachment in Western Australia. *Journal of Arid Environment* **72**, 175–188.

Silva S (1988) *Prosopis juliflora* (Sw) DC. in Brazil. In *The Current State of Knowledge on Prosopis juliflora.* (Eds MA Habit and JC Saavedra) pp. 29–55. FAO UN, Recife, Brazil.

Swenson WH (1969) Natural enemies of mesquite. In *Literature on the Mesquite (Prosopis L.) of North America: An Annotated Bibliography.* (Ed JL Schuster) pp. 23–24. Texas Tech University, Texas.

Tewari JC, Pasiecznik NM, Harsh LN and Harris PJC (Eds) (1998) *Prosopis Species in the Arid and Semi-Arid Zones of India. Prosopis* Society of India and Henry Doubleday Research Association, Jodhpur.

Thorp JR and Lynch R (2000) *The Determination of Weeds of National Significance.* National Weeds Strategy Executive Committee, Launceston.

van Klinken RD (1999) Developmental host-specificity of *Mozena obtusa* (Heteroptera: Coreidae), a potential biocontrol agent for mesquite (*Prosopis* species). *Biological Control* **16**, 283–290.

van Klinken RD (2000) Host-specificity constrains evolutionary host change in the psyllid *Prosopidopsylla flava. Ecological Entomology* **25**, 413–422.

van Klinken RD and Burwell C (2005) Evidence from a gelechiid leaf-tier on mesquite (Mimosaceae: *Prosopis*) that semi-concealed Lepidopteran biological control agents may not be at risk from parasitism in Australian rangelands. *Biological Control* **32**, 121–129.

van Klinken RD and Campbell S (2009) Australian weeds series: *Prosopis* species. *The Biology of Australian Weeds. Vol. 3.* (Ed FD Panetta), pp. 238–273. RG and FJ Richardson, Melbourne.

van Klinken RD and Heard TA (2000) Estimating fundamental host range: a host-specificity study of a biocontrol agent for *Prosopis* species (Leguminosae). *Biocontrol Science and Technology* **10**, 331–342.

van Klinken RD and Raghu S (2006) A scientific approach to agent selection. *Australian Journal of Entomology* **45**, 253–258.

van Klinken RD and White A (2009) 'Optimising management of core mesquite infestations in Australia. Land and Water Australia, Final Report.' pp. 1–39. <http://lwa.gov.au/products/pn22407>.

van Klinken RD, Fichera G and Cordo H (2003) Targeting biological control across diverse landscapes: the release, establishment and early success of two insects on mesquite (*Prosopis*) in rangeland Australia. *Biological Control* **26**, 8–20.

van Klinken RD, Shepherd D, Parr R, Robinson TP and Anderson L (2007) Mapping mesquite (*Prosopis*) distribution and density using visual aerial surveys. *Rangeland Management and Ecology* **60**, 408–416.

van Klinken RD, Hoffmann JH, Zimmermann HG and Roberts AP (2009a) *Prosopis* species (Leguminosae). In *Biological Control of Tropical Weeds Using Arthropods.* (Eds R Muniappan, GVP Reddy and A Raman) pp. 353–377. Cambridge University Press, Cambridge.

van Klinken RD, Lawson BE and Zalucki MP (2009b) Predicting invasions in Australia by a Neotropical shrub under climate change: the challenge of novel climates and parameter estimation. *Global Ecology and Biogeography* **18**, 688–700.

Ward CR, O'Brien CW, O'Brien LB, Foster DE and Huddleston EW (1977) 'Annotated checklist of New World insects associated with *Prosopis* (mesquite).' US Dept of Agriculture Research Service Bulletin 1557.

Yousuf M and Gaur M (1998) Some noteworthy insect pests of *Prosopis juliflora* from Rajasthan. In *Prosopis Species in the Arid and Semi-arid Zones of India.* (Eds JC Tewari, NM Pasiecznik, LN Harsh and PJC Harris) pp. 91–94. *Prosopis* Society of India and Henry Doubleday Research Association, Jodhpur.

Zimmermann HG (1991) Biological control of mesquite, *Prosopis* spp. (Fabaceae), in South Africa. *Agriculture, Ecosystems and Environment* **37**, 175–186.

Zimmermann HG, Hoffmann JH and Witt ABR (2006) A South African perspective on *Prosopis. Biocontrol News and Information* **27**, 6–10.

Raphanus raphanistrum L. – wild radish

John K Scott

ABSTRACT

Raphanus raphanistrum (Brassicaceae), wild radish, is an annual winter-growing herb that is widespread in southern Australia where it is one of the economically most important weeds of annual cropping. Wild radish was surveyed for potential agents, insects and pathogens in part of the region of origin, southern Europe and North Africa. Potential agents, pathogens and the gall midge *Gephyraulus raphanistri*, were tested to assess their host-range, but none were sufficiently host-specific. Complementary surveys of pathogens, including soil bacteria, have been made in Australia to determine which organisms are already present and could be used in development of bioherbicides. Ants are an important cause of seed predation in both the regions of origin and introduction. Wild radish is very closely related to edible radish *Raphanus sativus* and the important crop, canola. This limits the possible range of biocontrol agents due to lack of host-specificity to the weed. Priorities for future research include examining low-dose herbicide–pathogen interactions, examining the species boundaries and host-specificity of the downy mildew *Hyaloperonospora parasitica*, developing pathogens already in Australia as mycoherbicides, and promoting beneficial pathogens and rhizobacteria through stubble and soil management.

Key words: biological control, conflict of interest, *Gephyraulus raphanistri*, plant pathogens, surveys for phytophagous species, *Urodon* species.

INTRODUCTION

Raphanus raphanistrum L. (Brassicaceae), wild radish (Fig. 1) is an annual or biennial herb growing to 1.5 m high. Stems are erect, flexuous and slightly angled. Leaves are initially formed into a rosette, and are variable in size (up to 30 cm long) and shape. The main root can be up to 160 cm long and it produces a mass of fibrous lateral roots in the upper 20 cm of soil (Parsons and Cuthbertson 2001). Flowers vary in colour from white to purple with light or dark veins. Wild radish is self-incompatible and pollination is ensured by insects. Reproduction is entirely by seed, which can be produced as early as seven weeks after sowing (Cheam 1986). The fruit is a silique or pod, comprised on average of four to seven (range one to 12) segments each containing a single seed (Vitou and Scott 2002). Reeves *et al.* (1981) recorded up to 17 275 seed m^{-2} in a wheat crop. Dispersal occurs via movement of segments containing seeds by humans, animals, wind and water (Cheam and Code 1995). Most seed germinates in autumn and winter, but seedlings can emerge at any time if soil moisture is sufficient (Cheam and Code 1995). Several germinations can occur in one season, with early germinations growing into plants that contribute most to the seed bank (Cheam 1986). Some seed remains viable either on the surface or buried in soil for up to six years (Cheam and Code 1995). More detailed description of the plant can be found in Cheam and Code (1995), Parsons and Cuthbertson (2001) and Warwick and Francis (2005).

Figure 1: a) Wild radish *Raphanus raphanistrum* flowers, WA. b) A dense *Raphanus raphanistrum* infestation, WA. Photos: JK Scott, CSIRO.

The phylogeny and origins of wild radish were investigated (Jolivet 2002; J-F Martin and JK Scott, unpub.). Seed collections were made in Australia and the Mediterranean region, including edible (commercial) radish *Raphanus sativus* L. and two subspecies of wild radish, *R. r. raphanistrum* and *R. r. landra* (Moretti ex DC.) Bonnier & Layens. The study used two techniques, gene sequencing and AFLPs. Gene sequencing did not distinguish between wild and edible radish nor between the wild radish subspecies. In contrast, the AFLP results showed that edible radish formed a group separate from wild radish. The results for the two subspecies, *raphanistrum* and *landra*, indicated considerable gene flow between the subspecies. Snow and Campbell (2005) and Ellstrand *et al.* (2010) noted that wild radish was separated from the edible radish by a chromosomal translocation and a suite of morphological characters. Gene flow and hybridisation also occurs in Californian wild and edible species of radish (Ellstrand *et al.* 2010; Hegde *et al.* 2006). In the AFLP analysis wild radish from Australia was grouped with populations from the Mediterranean region, indicating that this was an appropriate region for surveys for biocontrol agents. The evidence from hybridisation studies and gene sequencing showed the two radish species were very closely related, which indicated that it will be difficult to find biocontrol agents that are specific to wild radish.

Wild radish is widespread in Australia (Fig. 2) except in the north. It is most abundant in areas for crop production in southern Australia. Wild radish has noxious weed status in NSW and WA (Australian Weeds Committee 2010).

Wild radish is found across a range of soil types, although it has a preference for acidic soils. It is ranked among the top three weed species of winter cereal and grain legume crops in southern Australia (Alemseged *et al.* 2001; Poole and Gill 1987). The third highest cost component due to weeds has been attributed to wild radish and is estimated to be at least $71.8 million per year (Jones *et al.* 2005). *R. raphanistrum* was less frequent in pasture and undisturbed areas and generally absent from native vegetation (Cheam and Code 1995). It can usually be controlled by a combination of herbicide, crop and pasture rotation and management (Cheam and Code 1995). While a range of herbicides can control wild radish in cereal crops, the herbicide options are limited in pulse or canola crops (Cheam and Lee 2006). One result of current management has been the evolution of herbicide resistance, with populations of wild radish in southern Australia resistant across four different modes of herbicide action (Walsh *et al.* 2004).

BIOLOGICAL CONTROL HISTORY

Wild radish has not been considered for biocontrol in any country apart from Australia. Even then, the primary aim of the work carried out in the native distribution, the circum-Mediterranean region, starting in 1997 and with subsequent research in Australia, was to assess the feasibility of applying biocontrol. Wild radish has not been subjected to the formal process of being nominated as a target species for biocontrol in Australia.

EXPLORATION

CSIRO, through the Cooperative Research Centre for Australian Weed Management, searched for potential

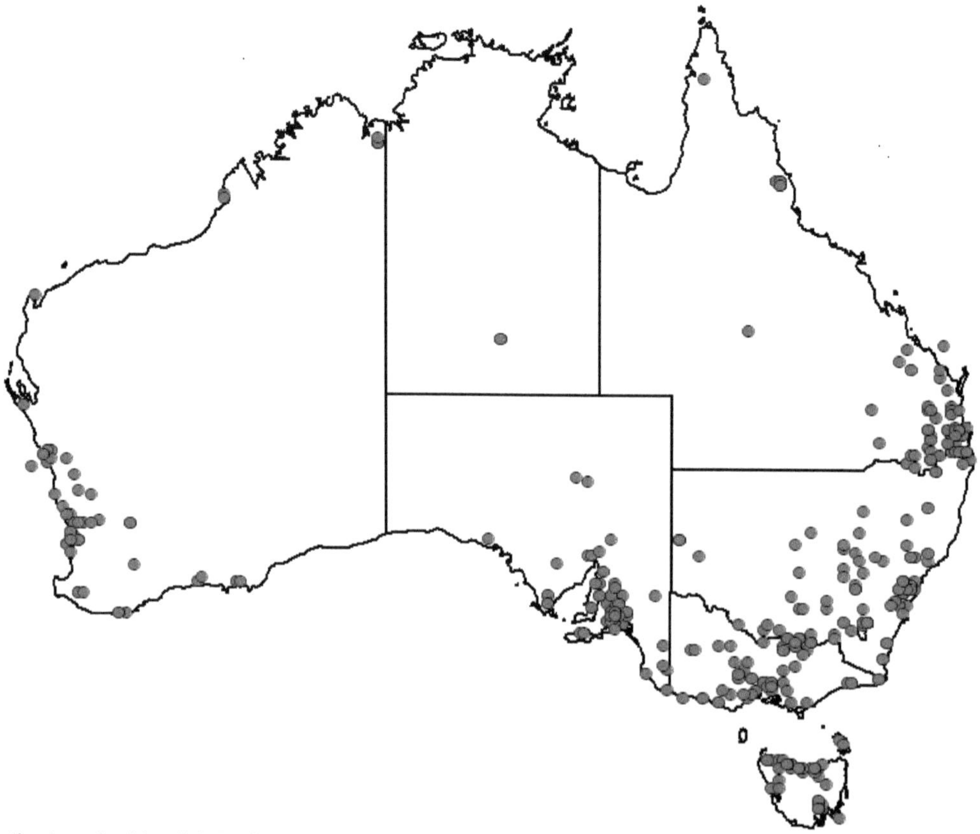

Figure 2: Distribution of wild radish *Raphanus raphanistrum* in Australia. Specimen data reproduced from Australia's Virtual Herbarium with permission of the Council of Heads of Australasian Herbaria Inc.

classical biocontrol agents in southern Europe and the circum-Mediterranean region concentrating on southern Portugal, northern Tunisia, the Mediterranean coast of France and southern Greece (Scott *et al.* 2002) from 1997 to 2003. Wild radish is a common weed of vineyards, cereal crops and legumes in this region. The surveys were undertaken by Mireille Jourdan, Deirdre Lemerle, Mark Lonsdale, David Luckett, Jean-Louis Sagliocco, John Scott and Janine Vitou. Plant pathogens and insects including seed-feeders and ants were examined on wild radish. The surveys focused in particular on northern Tunisia, in collaboration with Naceur Djébali and Thouraya Souissi from the Institut National Agronomique de Tunisie. This region has rain-fed cropping similar to that in the wild radish-infested regions of Australia. Subsequent surveys by Aaron Maxwell and John Scott were made during 2004 in WA for organisms that could be used in a mycoherbicide approach. This work in the area of introduction was complemented by other studies in WA on soil bacteria (Flores-Vargas and O'Hara 2002, 2006) and the impact of seed-eating ants (Spafford Jacob *et al.* 2006).

Insects in the region of origin

Many of the ~50 insect species found feeding on wild radish in the native distribution are specialists on the family Brassicaceae. However, most species were not sufficiently host-specific to exclude off-target risk to canola. One species and three species groups were identified by Scott *et al.* (2002) for closer study, the gall midge *Gephyraulus raphanistri* Kieffer (Cecidomyiidae), sawflies *Elinora* spp. (Tenthredinidae), seed weevils *Urodon* spp. (Urodontidae) and the weevils *Ceutorhynchus* spp. (Curculionidae). Aside from the gall-midge, further investigation of the host-range was limited by the availability of sufficient numbers of insects to undertake experimental work.

Plant pathogens in the region of origin

Wild radish from 107 populations from the circum-Mediterranean regions of Greece, Spain, Portugal, Tunisia and France were examined for pathogens from 1997 to 2001 (Scott *et al.* 2002). The results of surveys in Tunisia are reported in Djébali *et al.* (2009).

To estimate the effect of fungi in Tunisia, the weed was grown in randomised blocks and was treated with and without a fungicide. There was a two-fold decrease in number and weight of pods in the untreated blocks. Thus, it seems that proliferation of wild radish is naturally reduced by spontaneous fungi attacks although the causal agents were not specifically identified (Djébali *et al.* 2009). Similar results were obtained from a fungicide and insecticide exclusion experiment in France where there was twice the biomass and seed production in plots with treated plants compared to untreated controls (J Vitou, unpub. data). The importance of seed predation by ants was also assessed in France (Vitou *et al.* 2004) (see below).

Of 15 species of pathogens found on wild radish, 12 were tested for host-specificity and all were shown to attack canola (Djébali *et al.* 2009; M Jourdan, unpub. data); consequently these are not being studied further. Two biotrophic pathogens, *Hyaloperonospora parasitica* (Pers. ex Fr.) Constant. (downy mildew) (Oomycetes: Peronosporaceae) (discussed further below) and *Albugo candida* (Pers. ex. Chev.) Kuntze (white rust) (Oomycetes: Albuginaceae) have also been recorded from Australia and were not tested for host-range in this study, although *A. candida* associated with wild radish is not likely to be host-specific (Morris and Knox-Davies 1980, Petkowski *et al.* 2010).

Pathogens in the region of introduction

Twelve fungal species were isolated from diseased wild radish during a survey of south-western Australia during 2004. These included a previously undescribed species of *Mycosphaerella* (Capnodiales: Mycosphaerellaceae) and *H. parasitica* (previously not recorded on wild radish in WA). Further options for conservation or augmentative approaches were reviewed by Maxwell and Scott (2004, 2006, 2008). Potentially useful fungal pathogens included *Alternaria alternata* (Fr.) Keissler, *Alternaria brassicae* (Berk.) Sacc., *Alternaria brassicicola* (Schw.) Wiltshire, *Alternaria japonica* Yoshii (Dothideales: Pleosporaceae), *Athelia rolfsii* (Curzi) Tu and Kimbrough (Atheliales: Atheliaceae), *Colletotrichum coccodes* (Wallr.) Hughes (Phyllachorales: Phyllachoraceae), *Fusarium oxysporum* Schlechi. emend Snyder and Hansen (Hypocreales: Nectriaceae) and *H. parasitica*.

CANDIDATES

***Gephyraulus raphanistri* (Kieffer) (Diptera: Cecidomyiidae).** This flower-gall midge was considered a potential agent because recent host records showed restriction to wild radish throughout Europe and it has never been confirmed from commercially grown *Brassica* crops. The galls have a direct impact on seed production. It was found in surveys on wild radish throughout the Mediterranean in a range of habitats, including wheat fields. Field and experimental studies (Vitou *et al.* 2008) manipulated host-plant phenology and showed that this insect could develop on canola as long as wild radish and canola flowered at the same time. Thus it is unsuitable for use as a biological control agent because there would be high risks of transfer to canola in Australia.

***Hyaloperonospora parasitica* (Pers. ex Fr.) Constant. (Oomycetes: Peronosporaceae).** Downy mildew was the earliest pathogen to show disease symptoms on field populations of wild radish in the circum-Mediterranean region (Scott *et al.* 2002). Similar disease was not evident on neighbouring Brassicaceae plants, making this species the highest priority for further study. Downy mildew was the most damaging pathogen of wild radish in WA (Maxwell and Scott 2008). Phylogenetic analysis showed that the *H. parasitica* isolates from wild radish are genetically distinct from the isolates found on *Brassica* species including canola. This finding opens up the possibility of developing a conservation or augmentative approach to the biological control of wild radish using *H. parasitica* (Maxwell and Scott 2005, 2008).

Seed-harvesting ants

Seed-harvesting ants caused the greatest seed losses of maturing seeds in southern France (11–91% seed loss per unit area) (Vitou *et al.* 2004). The ants consumed fresh seed by removing green siliquas from plants in spring, and harvested mature segments over summer. While ants cannot be considered for introduction to Australia as potential biological control agents because they have a wide host-range, their overriding effect relative to other seed predators in the native range suggests associations between ants and other generalist seed predators on wild radish in Australia may be providing some form of natural control (Vitou *et al.* 2004).

Spafford Jacob *et al.* (2006) reported on the loss of seeds due to post-dispersal weed seed predation by rodents, ants or other small invertebrates during the summer fallow period in south-west Australian cropping regions. Seed removal was very variable, but in general highest near the edges and lowest in the centre of fields. Annual ryegrass *Lolium rigidum* Gaudin (Poaceae) seeds were preferred over wild oat *Avena fatua* L. (Poaceae) seeds, followed by wild radish pod segments. Similar to the situation in France (Vitou *et al.* 2004), the actions of seed harvesters, in particular ants, may play an important role in the ecology of wild radish in Australia.

Soil bacteria

Flores-Vargas and O'Hara (2002, 2006) investigated the potential of deleterious rhizosphere inhabiting bacteria (DRB) for control of weed species including wild radish found in vineyards in WA. Three strains, two of *Pseudomonas fluorescens* (Bacteria: Pseudomonadaceae) and one of *Alcaligenes xylosoxidans* (Bacteria: Alcaligenaceae), were identified from mature wild radish and found to inhibit growth of wild radish seedlings but to have no significant deleterious effects on either grapevine or subterranean clover. Peltzer and Ash (2005) proposed the research idea that the addition of a DRB to a competitive cereal crop at high seedling rates could have a synergistic effect causing reduction in weed growth.

DISCUSSION

Wild radish is not an obvious target for biocontrol for two reasons. First, wild radish is very closely related to edible radish (*R. sativus*) and is closely related to an important crop, canola, which implies that a particularly high level of host-specificity will be needed in any biocontrol agent. An initial strategy for biocontrol was to focus on seed-feeding insects with host-specificity to or within the genus *Raphanus*. Of the two options under this strategy, *G. raphanistri* was shown to include canola as a potential host (Vitou *et al.* 2008). The second option was the highly specialised *Urodon* species. These seed-feeding weevils were difficult to find as adults and may be very rare, as very few larvae were found in an extensive x-ray analysis of wild radish seed from the Mediterranean region (J Vitou and JK Scott, unpub. data).

Another strategy for classical biocontrol was to assess plant pathogens in the hope of finding host-specificity restricted to *R. raphanistrum*. Of the field observations in the region of origin, a form of *H. parasitica* appeared the most promising as a potential agent. However, it is likely that this form was the one found subsequently in WA (Maxwell and Scott 2008), thus negating further consideration of this pathogen as an exotic biocontrol agent.

The second reason for wild radish being a difficult target is that any biocontrol agent will need to be adapted to the annual life-cycle of the weed and may need a dormant stage to pass summer when the weed oversummers as seed, as occurs in southern Australia. The weed is found in highly disturbed habitats typical of cropping systems. This implies that the biological control agent will need to be highly mobile or have resistant life-stages so it can survive cultivation and rapidly colonise new infestations; that is, it must have characteristics similar to those of pest species. This implies that the best approach is to focus on the organisms that have successfully colonised the region of interest and to investigate a conservation or mycoherbicide approach. This includes encouraging seed-feeding ants (Spafford Jacob *et al.* 2006), investigating the use of plant pathogens as mycoherbicides (Maxwell and Scott 2006, 2008) or using deleterious rhizobacteria (Peltzer and Ash 2005).

Aspects of the biological control potential of wild radish using plant pathogens were reviewed and research priorities proposed in a workshop in 2005 (Maxwell and Scott 2005). Thirty-one ideas were assessed for feasibility, potential impact on control of wild radish and time required to achieve a research output. The highest priorities for future work included examining low-dose herbicide–pathogen interactions, examining the species boundaries and host-specificity of *H. parasitica*, developing pathogens already in Australia as mycoherbicides, and promoting beneficial pathogens and rhizobacteria through stubble and soil management. The considerable economic impact and the continuing and developing incidence of herbicide resistance in wild radish points to the need to explore these alternative means of control for this very important crop weed.

ACKNOWLEDGEMENTS

I thank Janine Vitou and Mireille Jourdan for help with unpublished data from the work in France.

REFERENCES

Alemseged Y, Jones R and Medd R (2001) A farmer survey of weed management and herbicide resistance problems of winter crops in Australia. *Plant Protection Quarterly* **16**, 21–25.

Australian Weeds Committee (2010) Noxious weeds list. <http://www.weeds.org.au/noxious.htm> (accessed 19.7.10).

Cheam AH (1986) Seed production and seed dormancy in wild radish (*Raphanus raphanistrum* L.) and some possibilities for improving control. *Weed Research* **26**, 405–413.

Cheam AH and Code GR (1995) The biology of Australian weeds. 24. *Raphanus raphanistrum* L. *Plant Protection Quarterly* **10**, 2–13.

Cheam A and Lee S (2006) Why is wild radish the worst cruciferous weed? In *Proceedings of the Wild Radish and other Cruciferous Weeds Symposium*. 11–12 July 2006, Perth. (Ed AH Cheam) pp. 11–19. WA Dept of Agriculture and Food, Perth.

Djébali N, Scott JK, Jourdan M and Souissi T (2009) Fungi pathogenic on wild radish (*Raphanus raphanistrum* L.) in northern Tunisia as potential biocontrol agents. *Phytopathologia Mediterranea* **48**, 205–213.

Ellstrand NC, Heredia SM, Leak-Garcia JA, Heraty JM, Burger JC, Yao L, Nohzadeh-Malakshah S and Ridley CE (2010) Crops gone wild: evolution of weeds and invasives from domesticated ancestors. *Evolutionary Applications* **3**, 494–504.

Flores-Vargas RD and O'Hara GW (2006) Isolation and characterization of rhizosphere bacteria with potential for biological control of weeds in vineyards. *Journal of Applied Microbiology* **100**, 946–954.

Flores-Vargas RD and O'Hara GW (2002) Towards ecologically based weed management systems in vineyards. In *Proceedings of the 13th Australian Weeds Conference*. 8–13 September 2002, Perth. (Eds H Spafford Jacob, J Dodd and JH Moore) pp. 228–231. Plant Protection Society of Western Australia, Perth.

Hegde SG, Nason JD, Clegg JM and Ellstrand NC (2006) The evolution of California's wild radish has resulted in the extinction of its progenitors. *Evolution* **60**, 1187–1197.

Jolivet C (2002) Phylogénie du genre *Raphanus* et lutte biologique contre la ravenelle (*Raphanus raphanistrum raphanistrum*). Diplôme d'Etudes Approfondies. Ecole Nationale Supérieure Agronomique, Montpellier.

Jones RE, Vere DT, Alemseged Y and Medd RW (2005) Estimating the economic cost of weeds in Australian annual winter crops. *Agricultural Economics* **32**, 253–265.

Maxwell A and Scott JK (2004) The potential for using mycoherbicides to control weeds from the family Brassicaceae in Australia. In *Proceedings of the 14th Australian Weeds Conference*. 6–9 September 2004, Wagga Wagga, NSW. (Eds BM Sindel and SB Johnson) pp. 333–336. Weed Society of New South Wales, Sydney.

Maxwell A and Scott JK (2005) Report on research priorities identified from the workshop on biological control options for wild radish. In *Biological Control of Wild Radish: Proceedings of a Workshop on Pathogen-based Biological Control Options for Wild Radish (Raphanus raphanistrum)*. 8 March 2005, Floreat, WA. (Eds A Maxwell and JK Scott) pp. 3–8. CRC for Australian Weed Management, Adelaide.

Maxwell A and Scott JK (2006) Opportunities for the control of Brassicaceae weeds in cropping systems using mycoherbicides. *Advances in Botanical Research* **43**, 143–174.

Maxwell A and Scott JK (2008) Pathogens on wild radish, *Raphanus raphanistrum* (Brassicaceae), in south-western Australia – implications for biological control. *Australasian Plant Pathology* **37**, 523–533.

Morris MJ and Knox-Davies PS (1980) *Raphanus raphanistrum* as a weed host of pathogens of cultivated cruciferae in the Western Cape Province of South Africa. *Phytophylactica* **12**, 53–55.

Parsons WT and Cuthbertson EG (2001) *Noxious Weeds of Australia*. 2nd edn. CSIRO Publishing, Melbourne.

Peltzer S and Ash G (2005) Deleterious rhizobacteria to control wild radish? In *Biological Control of Wild Radish: Proceedings of a Workshop on Pathogen-based Biological Control Options for Wild Radish (Raphanus raphanistrum)*. 8 March 2005, Floreat, WA. (Eds A Maxwell and JK Scott) p. 18. CRC for Australian Weed Management, Adelaide.

Petkowski JE, Cunnington JH, Minchinton EJ and Cahill DM (2010) Molecular phylogenetic relationships between *Albugo candida* collections on the Brassicaceae in Australia. *Plant Pathology* **59**, 282–288.

Poole ML and Gill GS (1987) Competition between crops and weeds in southern Australia. *Plant Protection Quarterly* **2**, 86–96.

Reeves TG, Code GR and Piggin CM (1981) Seed production and longevity, seasonal emergence and phenology of wild radish (*Raphanus raphanistrum* L.). *Australian Journal of Experimental Agriculture and Animal Husbandry* **21**, 524–530.

Scott JK, Vitou J and Jourdan M (2002) Review of the potential for biological control of wild radish (*Raphanus raphanistrum*) based on surveys in the Mediterranean region. In *Proceedings of the 13th Australian Weeds Conference*. 8–13 September 2002, Perth. (Eds H Spafford Jacob, J Dodd and JH Moore) pp. 377–380. Plant Protection Society of Western Australia, Perth.

Snow AA and Campbell LG (2005) Can feral radishes become weeds? In *Crop Ferality and Volunteerism*. (Ed J Gressel) pp. 193–207. CRC Press, Boca Raton.

Spafford Jacob H, Minkey DM, Gallagher RS and Borger CP (2006) Variation in postdispersal weed seed predation in a crop field. *Weed Science* **54**, 148–155.

Vitou J and Scott JK (2002) Seed quality and germination in wild radish (*Raphanus raphanistrum* L.). In *Proceedings of the 13th Australian Weeds Conference*. 8–13 September 2002, Perth. (Eds H Spafford Jacob, J Dodd and JH Moore) pp. 570–573. Plant Protection Society of Western Australia, Perth.

Vitou J, Scott JK and Sheppard AW (2004) The patterns of harvester ant removal of wild radish seeds in the native range: the importance of generalist seed predators to weed management. In *Proceedings of the 11th International Symposium on Biological Control of Weeds*. 27 April–3 May 2003, Canberra. (Eds JM Cullen, DT Briese, DJ Kriticos, WM Lonsdale, L Morin and JK Scott) pp. 188–192. CSIRO, Canberra.

Vitou J, Skuhrava M, Skuhravy V, Scott JK and Sheppard AW (2008) The role of plant phenology in the host specificity of *Gephyraulus raphanistri* (Diptera: Cecidomyiidae) associated with *Raphanus* spp. (Brassicaceae). *European Journal of Entomology* **105**, 113–119.

Walsh MJ, Powles SB, Beard BR, Parkin BT and Porter SA (2004) Multiple-herbicide resistance across four modes of action in wild radish (*Raphanus raphanistrum*). *Weed Science* **52**, 8–13.

Warwick SI and Francis A (2005) The biology of Canadian weeds. 132. *Raphanus raphanistrum*. L. *Canadian Journal of Plant Science* **85**, 709–733.

Reseda lutea L. – cutleaf mignonette

Peter Bailey

ABSTRACT

Reseda lutea is a weed in parts of Australian cropping lands that have a Mediterranean-type climate and calcareous soils. In its native Mediterranean range it is uncommon and is not reported as a weed in cultivated crops. Different agronomic practices partially explain this difference but *R. lutea* also has a number of natural enemies identified during field surveys of Mediterranean countries during 1994–96. Promising agents included the weevil *Baris picicornis* which tunnels into stems and roots, the flower- and seed-eating beetle *Bruchela suturalis* and the leaf-spot fungus *Cercospora resedae*. Both species of insects were imported to Australia during 1998–99 for specificity testing in quarantine but neither was released. Further work may provide data to support the release of these promising agents. *C. resedae* is unlikely to provide useful control of *R. lutea*.

Key words: *Reseda luteola*, *Baris picicornis*, *Bruchela suturalis*, *Bruchela rufipes*, *Cercospora resedae*, Mediterranean surveys.

INTRODUCTION

In Australia, *Reseda lutea* L. (Resedaceae) is a weed of crops grown on calcareous soils that coincide with the cereal belt covering Eyre Peninsula, Yorke Peninsula, the mid-north of SA and the Murray Mallee of SA and Vic (Fig. 1). *R. lutea* tolerates rainfall in the range 100–625 mm (Heap *et al.* 1987).

R. lutea competes with crops and its small seeds contaminate grain (Figs 2 and 3). Crop competition studies (Heap 1993) indicate the cost to SA grain growers was $3 million in 2001 (I Black, *pers. comm.* 1994). This weed is difficult to control with herbicides, which burn top growth but leave the deep tap roots to regenerate. It is unpalatable to stock; when ingested, it taints the flesh and milk of livestock.

In its native Mediterranean and European habitats, *R. lutea* grows on calcareous soils in newly disturbed habitats such as road cuttings and forest clearings. After some years in a location, *R. lutea* is replaced by other plants in the succession. Immature *R. lutea* plants are particularly sensitive to competition (Heap *et al.* 1987). In SA cropping areas, the habitat is frequently disturbed. Shallow cultivation (50–75 mm deep) produces cuttings which may be spread by cultivation machinery, leaving the parent plant to reshoot. Spread within a crop is mainly by vegetative propagation (Heap 1997b). Propagation by seed within a crop is relatively unimportant but it may be important in new areas (Heap 1997a), especially where seed has passed through livestock (Heap and Honan 1993).

In Mediterranean Europe *R. lutea* rarely invades crops. This may be partly attributable to the use of deeper cultivation than that practised by SA farmers, as deep cultivation appears to destroy the deep tap root of the plant. A second factor may be the higher sowing rate used in southern Spain (150–200 kg per ha), that produces denser crops than is normal in SA where the sowing rate is

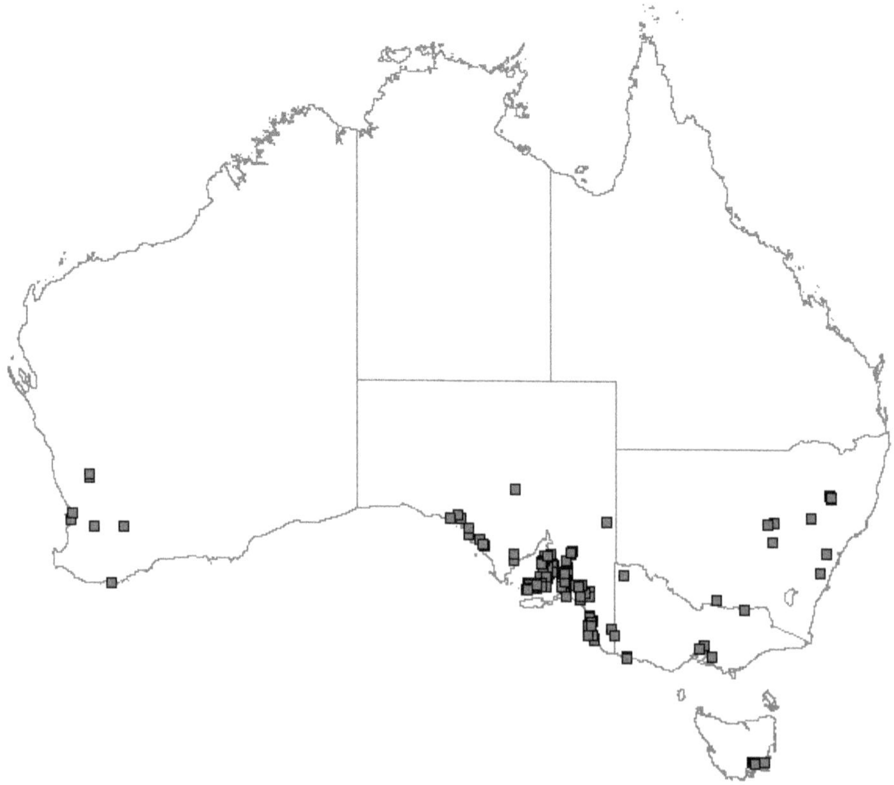

Figure 1: Distribution of *Reseda lutea* in Australia. Specimen data reproduced from Australia's Virtual Herbarium with permission of the Council of Heads of Australasian Herbaria Inc.

50–100 kg per ha. Seedlings of *R. lutea* appear to be out-competed by densely planted crops. In addition to these agronomic factors, a number of natural control agents have been identified which may contribute to the benign status of *R. lutea* in Mediterranean lands.

BIOLOGICAL CONTROL HISTORY

No past attempts at biological control of *R. lutea* have been recorded. *R. lutea* was a recorded host of several insects (Bailey *et al.* 2002) including *Baris picicornis* Marsham (Coleoptera: Curculionidae), *Bruchela rufipes* (Schőnherr) and *Bruchela suturalis* (Fabricius) (Coleoptera: Urodontidae) and species of flea-beetles including *Phyllotreta corrugata* (Fabricius), *Phyllotreta nigripes* (Fabricius) and *Phyllotreta procera* (Redtenbacher) (Coleoptera: Chrysomelidae).

The present project spanned six years, comprising surveys in the Mediterranean region between 1994 and 1996 followed by importation, culturing and specificity testing during 1997 to 2000.

Figure 2: Field of *Reseda lutea* on Yorke Peninsula during summer fallow. Photo: P Bailey, SARDI.

PLANT TAXONOMY

There are no native members of the family Resedaceae in Australia and the introduced members, *Reseda alba* L., *R. lutea*, *Reseda luteola* and *Reseda phyteuma* L. are weeds while *Reseda odorata* L. is a minor ornamental with weedy tendency. The family Resedaceae is closely related to Cruciferae.

EXPLORATION

A survey for biological control agents was made during April and May 1994 where species diversity was greatest,

Figure 3: *Reseda lutea* competes with cereals. Photo: P Bailey, SARDI.

the Iberian Peninsula and Morocco (Fig. 4). A further survey during May 1995 in Turkey and Israel aimed to collect agents from the eastern range of *R. lutea* (Fig. 4). In addition to these surveys, plots of *R. lutea*, *R. luteola* and *Brassica oleracea* L. (Brassicaceae) were established at Elvas in Portugal, Montpellier in France and Pozantl in Turkey to act as source areas for importation of biological control agents to Australia and for field host-specificity observations.

CANDIDATES

Natural enemies

Fungal pathogens and insect leaf-, stem- and root-feeding insects found during Mediterranean surveys are listed in Table 1. These organisms were collected almost exclusively from mature plants as immature plants were rarely encountered. Few dead or dying plants were seen.

Agents tested but not released

Baris picicornis. 120 adults of this weevil, sourced from Montpellier, France which has a climate similar to that of Yorke Peninsula, SA, were imported into quarantine in Adelaide in November 1997. A second batch of 19 adult weevils collected at Pozantl, Turkey (1100 m altitude) was introduced at the same time. These two batches were the only weevils imported during this project. Culture methods and specificity testing were described by Bailey *et al.* (2002). About 50% of the French adults were infected with the nematode *Beddingia ?barisii* Laumond & Bonifassi (Nematoda: Sphaerularioidae) (K Davies, *pers. comm.* 1998). Subsequent laboratory generations were cleaned of these parasites. During culture, it was noted that potted *R. lutea* were killed by attack from eight or more weevil larvae per plant (Fig. 5).

For laboratory specificity studies, *B. picicornis* was introduced to the following plants in no-choice specificity

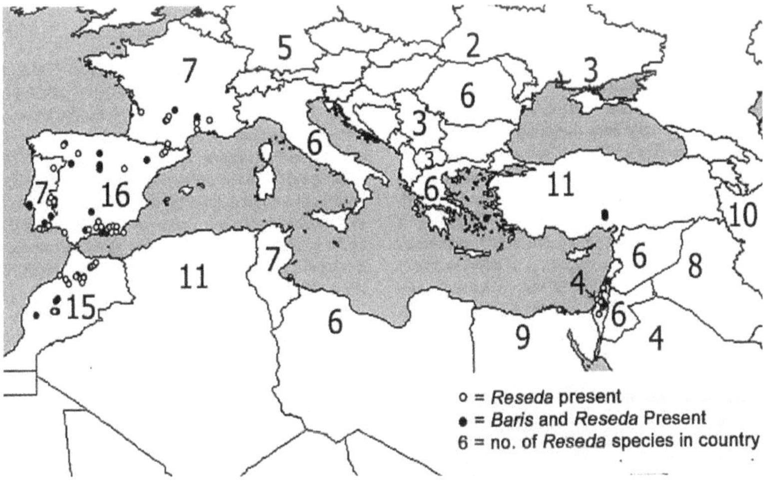

Figure 4: Mediterranean countries, showing number of *Reseda* spp. recorded by Abdallah and de Wit (1978). Dots indicate sites sampled during the present study (Bailey *et al.* 2002).

Table 1: Putative natural enemies recorded on mature *Reseda lutea* plants during field surveys of Mediterranean Europe and North Africa

Agent	Plant part attacked	Observed field occurrence
Fungi		
Alternaria tenuissima Wiltshire (Pleosporaceae)	Leaf	Very uncommon, minor damage
Cercospora resedae Fuckel (Dermatiaceae))	Leaf	<1% of mature plants symptomatic, minor damage. Main host was *R. luteola*
Cladosporium sp. (Ascomycota)	Leaf	Very uncommon, minor damage
Peronospora crispula Fuckel (Peronosporaceae)	Leaf	Very uncommon, minor damage
Phoma resedicola Garb. (Pleosporaceae)	Stem	Very uncommon, some stem damage
Insects		
Baris morio (Boheman)	Adults on flowers and leaves	Uncommon, no damage observed on *R. lutea*. Main host was *R. luteola*
Baris picicornis	Adults eat leaves and flowers, larvae channel stems and roots	Present at 34 of 99 sites
Bruchela rufipes and *B. suturalis*	Adults abort flowers, larvae eat seeds	Very common on mature plants at all sites. Flower abortion and seed damage commonly observed
Phyllotreta corrugata, *P. nigripes* and *P. procera*	Adults common on flowers and leaves	Damage not obvious

tests: *Moringia oleifera* Lam. (Moringaceae), *Cleome viscosa* L. (Capparaceae), *Arabidella trisecta* (F. Muell), *Brassica chinensis* L., *Brassica oleracea* L., *Brassica napus* L., *Brassica nigra* (L.), *Brassica rapa* L., *Brassica rapa pekinensis* (Lour.), *Lepidium pseudohyssopifolium* Hewson O.E.Schulz, *Stenopetalum sphaerocarpum* F. Muell. (Brassicaceae) and *Cardamine paucijuga* Turez. (Arabidae). The testing procedure was detailed in Bailey *et al.* (2002). Adult weevils fed on leaves of *B. nigra*, *B. rapa chinensis* (choy sum) and *B. rapa pekinensis* (Chinese cabbage) and on leaves of the *Lepidium* species. Eggs were laid only on *B. rapa chinensis* and *B. rapa pekinensis*. Some developed to third instar larvae, indicating the life-cycle could be completed on both these subspecies. In a subsequent series of choice experiments when adult weevils were given the choice of laying on *R. lutea* and these plants, weevils laid most eggs on *R. lutea* but one laid on *B. rapa chinensis* and the larva developed to third instar. No larvae of *B. picicornis* were found during searches of the roots of cabbages grown in field plots contiguous with plots of *R. lutea* in France, Portugal or Turkey (unpub. data).

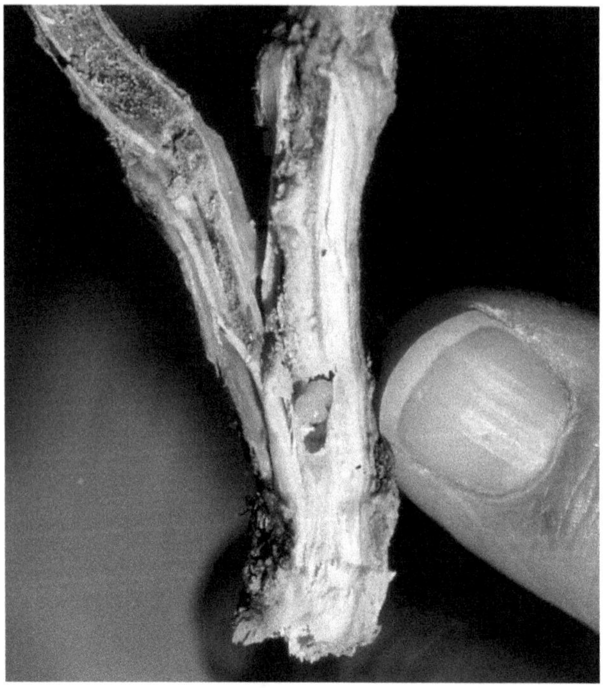

Figure 5: Larvae of the weevil *Baris picicornis* tunnelling the main stem and root of *Reseda lutea*. Photo: P Bailey, SARDI.

Bruchela suturalis. Several hundred pupae cultured from larvae collected from *R. lutea* at la Blaquererie, France were imported as a single batch into quarantine at Adelaide in September 1999 (Bailey *et al.* 2002) (Fig. 6). No further importations were made. By manipulation of temperature and daylight, adult emergence was synchronised with the southern hemisphere phenology of its host. However, adults from this shipment did not develop

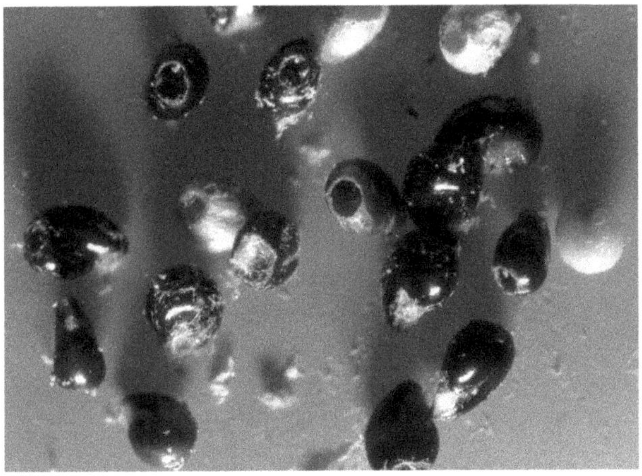

Figure 6: Seeds of *Reseda lutea* eaten by *Bruchela suturalis*. Photo: J-L Sagliocco, CSIRO.

mature ovaries and the culture died out. With more time and further shipments, it is likely that rearing difficulties could be overcome, but the project had reached its end-point and continuation funding was not available.

Fungi. Some fungal pathogens identified in field surveys (Table 1) were cultured and tested for pathogenicity at the School of Plant Sciences, University of Reading, UK (Mutinda 1998; Giles *et al.* 2002). The results indicated that isolates of the most common leaf pathogen, *Cercospora resedae*, collected from various Mediterranean sites were less damaging to *R. lutea* than Australian isolates derived from *R. luteola*. Under test conditions, four- to six-week-old seedlings of *R. lutea* treated with spore suspensions from Australian isolates suffered up to 55% of leaf area diseased at 22°C and over 90% of leaf area at 27°C (Giles *et al.* 2002). It was concluded that that *C. resedae* was likely to be more damaging to seedlings than to established *R. lutea*, and that the pathogen was more adapted to *R. luteola*.

Pathogenicity testing of *Alternaria tenuissima*, *Cladosporum* sp. and the downy mildew *Peronospora crispula* did not demonstrate damage of more than 5% of *R. lutea* leaf surface.

These results did not provide justification for importation of fungal pathogens into Australia.

Agents released

None of the exotic agents were released in Australia as a result of this project. However, *C. resedae* isolated from *R. luteola* in a 630 mm average annual rainfall site in Australia was transferred to aggregations of *R. lutea* in 360 mm cereal-growing site; it did not induce disease outbreaks.

DISCUSSION

The weevil *B. picicornis* was the most promising agent for control of established populations of *R. lutea* in cereal-growing areas of Australia. Tunnelling by larvae into the stem and tap root of the weed during summer would likely stress the plant, if not killing it outright. Freed of the nematodes that infect larvae and adults in Europe, *B. picicornis* would be expected to be more effective at limiting the weed in Australia than in Europe. The life-cycle of the weevil appears to be compatible with cultivation and cropping practices in Australian winter cereal areas.

This program terminated when funding ran out and before further specificity testing of *B. rapa chinensis* and *B. rapa pekinensis* under field conditions in Europe could be undertaken. For future work, it may be possible to demonstrate low likelihood of damage to these vegetable crops in the field, especially since these crops are normally treated with insecticides and are geographically separated in Australia from habitats where *R. lutea* grows. Adult weevils appear to be flightless, and no larvae of *B. picicornis* were found during searches of the roots of cabbages grown in field plots contiguous with plots of *R. lutea* in France, Portugal or Turkey.

A future introduction of the seed-destroying *B. suturalis* could be facilitated by developing methods for laboratory culture for specificity testing. Destruction of flowers and seeds by this insect in the field would likely reduce the risk of contamination of harvested grain and the risk of weed spread to new areas via seed dispersal. The potential of a second seed-feeder, *B. rufipes*, also warrants investigation.

ACKNOWLEDGEMENTS

Jean-Louis Sagliocco and Janine Vitou (CSIRO Entomology), David Cooke (SA Animal and Plant Control Commission), Irene Giles and Roland Fox (University of Reading) and Robin Coles and Trevor Wicks (SARDI) participated in this project. We acknowledge CSIRO Publishing for permission to use Figures 2 to 6, previously published in *Australian Journal of Experimental Agriculture*.

REFERENCES

Abdallah MS and de Wit HCD (1978) The Resedaceae – a taxonomical revision of the family. *Mededelingen Landbouwhogeschool Wageningen* **14**, 99–416.

Bailey P, Sagliocco J-L, Vitou J and Cooke D (2002) Prospects for biological control of cutleaf mignonette,

Reseda lutea (Resedaceae), by *Baris picicornis* and *Bruchela* spp. in Australia. *Australian Journal of Experimental Agriculture* **42**, 185–194.

Giles I, Bailey P, Fox R, Coles R and Wicks T (2002) Prospects for the control of cutleaf mignonette, *Reseda lutea* (Resedaceae) by *Cercospora resedae* and other pathogens. *Australian Journal of Experimental Agriculture* **42**, 37–41.

Heap JW (1993) Biology and control of *Reseda lutea* L. (cutleaf mignonette). MSc thesis. University of Adelaide, Australia.

Heap JW (1997a) Biology and control of *Reseda lutea* L. 1. Seed biology and seedling growth. *Australian Journal of Agricultural Research* **48**, 511–515.

Heap JW (1997b) Biology and control of *Reseda lutea* L. 2. Life cycle: seedling emergence, recruitment and vegetative reproduction. *Australian Journal of Agricultural Research* **48**, 517–524.

Heap JW and Honan I (1993) Weed excretion by sheep – temporal patterns of germinability. In *Proceedings of the 10th Australian and 14th Asia-Pacific Weed Conference.* 6–10 September 1993, Brisbane. (Ed JT Swarbrick) pp. 431–434. Weed Society of Queensland, Brisbane.

Heap JW, Wilcocks MC and Kloot PM (1987) The biology of Australian weeds. 17. *Reseda lutea* L. *Plant Protection Quarterly* **2**, 178–185.

Mutinda I (1998) Biological control of mignonette weeds. PhD thesis. Dept of Horticulture and Landscape, School of Plant Sciences, University of Reading, UK.

Rubus fruticosus L. aggregate – European blackberry

Louise Morin and Katherine J Evans

ABSTRACT

European blackberry, a group of closely related species belonging to the *Rubus fruticosus* aggregate (Rosaceae), is an introduced invasive shrub of agriculture, forestry and natural ecosystems in Australia. This chapter describes the various Australian biological control projects targeting European blackberry carried out over the years and summarises similar efforts undertaken in other countries. Surveys in Europe in the 1970s identified three potential agents for European blackberry, the stem-boring sawfly *Hartigia albomaculata*, purple blotch fungus *Septocyta ruborum* and leaf-rust fungus *Phragmidium violaceum*. The purple blotch fungus was not investigated at the time and preliminary host-specificity tests with the sawfly showed that larvae were able to feed on a number of cultivated blackberry and rose varieties. In 1984, while efforts were underway in Europe to select isolates of the leaf-rust fungus, it was found in Australia after an unauthorised introduction. Nonetheless, one of the selected and tested isolates (F15) was officially released in Australia in the early 1990s. Subsequent research addressed knowledge gaps considered critical to the biological control success of European blackberry, including a taxonomic revision of the *R. fruticosus* aggregate in Australia, climate modelling to predict impact of the rust fungus, and studies of pathogenicity and population genetics of the fungus to explain variation in disease intensity across the landscape. In 2000, eight additional and genetically diverse isolates of the rust fungus were sourced using a trap garden established in Europe that comprised blackberry clones from Australia. Following demonstration of host-specificity and approval for release in 2004, a large-scale release program, including re-release of F15, was established and the fate of the additional isolates monitored using molecular markers. Concurrently, surveys for other potential candidate agents were performed in Europe and as a result the purple blotch fungus is being investigated in more detail.

Key words: *Rubus* systematics, leaf-rust fungus *Phragmidium violaceum*, sawfly *Hartigia albomaculata*, purple blotch fungus *Septocyta ruborum*.

INTRODUCTION

European blackberry is a group of closely related species belonging to the *Rubus fruticosus* L. aggregate (agg.) (Rosaceae), hereafter named 'blackberry'. Blackberry is an introduced perennial, woody and thorny invasive shrub of agriculture, forestry and natural ecosystems (Amor *et al.* 1998; Parsons and Cuthbertson 2001). It occupies nearly 9 million ha of Australia, an area of land greater than the size of Tas, and is one of the 20 Weeds of National Significance (Thorp and Lynch 2000). It invades disturbed areas and is found mainly in temperate Australia south of the Qld–NSW border in areas where annual rainfall exceeds 700 mm (Amor *et al.* 1998) (Fig. 1). The ecology and distribution of the most widespread species within the group, *Rubus anglocandicans* A. Newton, is presented by Evans

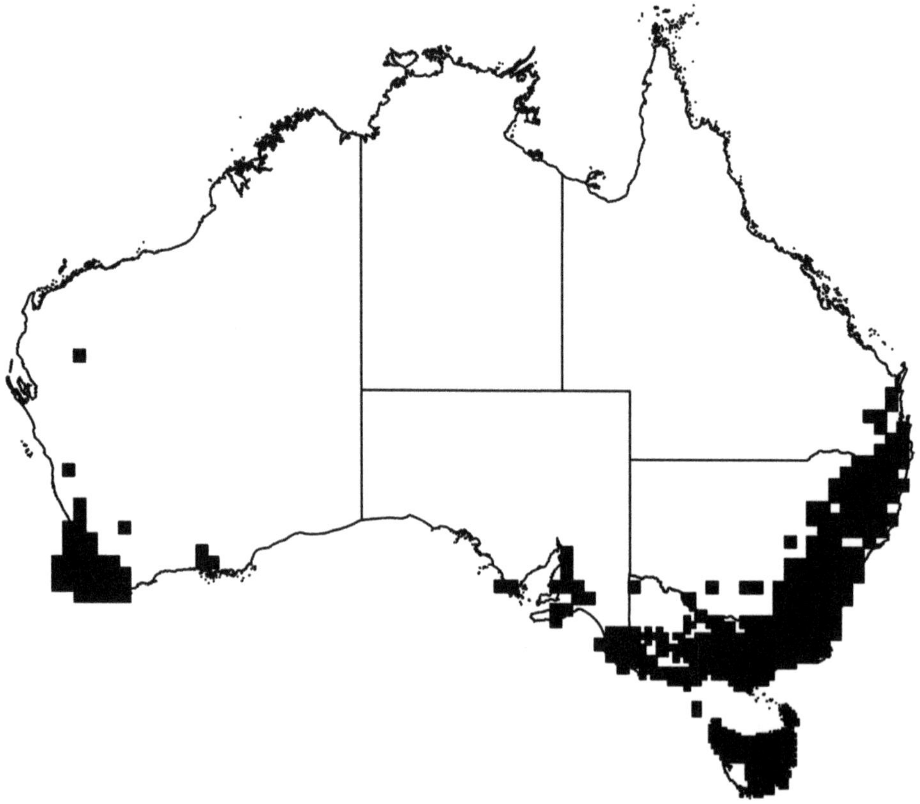

Figure 1: Distribution of European blackberry, *Rubus fruticosus* agg., in Australia. Map drawn by the Bureau of Rural Sciences using data presented in Thorp and Lynch (2000). Distribution shown as 0.5, 0.25 or 0.125 square degrees depending on the state or territory.

and Weber (2003). Evans *et al.* (2007) provide distribution maps for the other known species of the *R. fruticosus* agg. It is noteworthy that several varieties of commercially grown blackberry have species of *R. fruticosus* agg. in their breeding pedigree.

Using data from the early 1980s, the net cost of blackberry infestations to the Australian community was determined to be $41.5 million per year in 1984 (Amor *et al.* 1998). This was based on the estimated losses from primary production and the costs of control conducted at that time. Page and Lacey (2006) revised this cost estimate to $95–103 million per year. The cost of negative impacts such as loss of biodiversity, reduced water quality and recreational values is incalculable at this time (James and Lockwood 1998). Blackberry can have a major impact on the values of natural ecosystems. It can reduce diversity of native flora and fauna, provide a food source for competitive or predatory fauna (e.g. starlings, blackbirds) and harbour vermin such as foxes and feral cats (Davies 1998; James and Lockwood 1998). Blackberry infestations on river banks (Fig. 2a) can prevent access to the water by native fauna. The plant is reported to contribute significantly to the decline of a number of rare and threatened plant species, such as ballantinia *Ballantinia antipoda* (F.Muell.) E.A.Shaw, Brassicaceae, McIvor spider-orchid *Caladenia audasii* R.S. Rogers, Orchidaceae and Buxton gum *Eucalyptus crenulata* Blakely & Beuzev., Myrtaceae (Davies 1998).

The root system of blackberry is perennial and a crown develops where canes grow from the roots. Plants are semi-deciduous and canes are produced, mature and flower over two growing seasons, dying back to the crown in the winter following flower production (Amor *et al.* 1998). First-year canes (primocanes) are vegetative and can produce roots and a resting bud when the tip contacts the ground (Fig. 2b), usually in autumn. In late winter, a new cane may develop from this bud, creating a daughter plant at the same time as other canes develop from the mother plant's crown. Flowering canes (floricanes) (Fig. 2c) develop from the leaf axils of the previous year's primocane. Germination of seed occurs mainly in spring, but shading results in a low rate of seedling survival. Local vegetative spread is by production of daughter plants and occasional production of suckers from lateral roots.

Figure 2: a) Typical infestation of European blackberry, *Rubus fruticosus* agg. b) Roots and resting bud produced at the tip of a primocane that has come into contact with the ground; the shoot that will develop from the bud completes formation of the daughter plant. c) Fruit-bearing flowering canes (floricanes) developed from the leaf axils of the previous year's primocane. d) The leaf-rust fungus *Phragmidium violaceum*. Photos: L Morin, CSIRO.

Herbicides have been extensively used for control of blackberry (NSW Dept of Primary Industries Weed Management Unit 2009). Grubbing, mowing, cultivation, grazing management, fire or combinations of these methods are also used. In several instances, inaccessibility, the extensive areas infested and the risk of damage to native vegetation make control by herbicides or cultural means difficult or impossible (Bruzzese and Lane 1996; Amor *et al.* 1998). Biological control in these situations has been considered the only management option.

BIOLOGICAL CONTROL HISTORY

Biological control in other countries

A biological control project for the *R. fruticosus* agg. was initiated in New Zealand in 1925, with explorations in North America and Europe (Hill 1989). Of the extensive list of phytophagous insects and diseases found, only eight species were imported from 1925 to 1931 for further evaluation in NZ. Host-specificity tests involving raspberry, *Rubus idaeus* L., and in some instances a few other non-target plant species were performed with the moth *Thyatira batis* L., leaf midge *Dasineura* (=*Perrisia*) *plicatrix* H.Loew, gall midge *Lasioptera rubi* Heeger and buprestid beetle *Coraebus rubi* (L.) (Hill 1989). All these insects demonstrated insufficient specificity in preliminary tests and were thus abandoned as potential biological control agents.

In 1973, the leaf-rust fungus *Phragmidium violaceum* (Schultz) Winter (Fig. 2d), which occurs throughout Europe and the Middle East (Laundon and Rainbow 1969), was introduced to Chile from Germany for the biological control of two blackberry species belonging to the *R. fruticosus* agg.: *Rubus constrictus* Lefèvre & P.J.Müll. and *Rubus ulmifolius* Schott (Oehrens 1977). The rust fungus established and spread readily, and *R. constrictus* was observed to be more susceptible than *R. ulmifolius*. Oehrens (1977) reported that rust infections hastened defoliation by several months and severe attacks reduced seed production by 45%. Plants infected over successive years were reduced in height from 2.5 m to 1 m and became considerably less competitive, allowing colonisation by other species.

A biological control project based on the leaf-rust fungus is in the early stages of development in the USA to target Himalayan blackberry (tentatively identified as *Rubus armeniacus* Focke), another member of the *R. fruticosus* agg. (WL Bruckart, *pers. comm.* 2009). Recently, the rust fungus was discovered in the USA on this invasive blackberry species as well as on feral and commercially grown cutleaf or evergreen blackberry, *Rubus laciniatus* Willd. (Osterbauer *et al.* 2005). Disease management strategies have since been developed for control of the rust fungus in commercial production areas (Johnson and Mahaffee 2010). Although the fungus has been damaging on Himalayan blackberry in certain locations, initial research showed considerable variation in the susceptibility of different US accessions of this species to isolates of the fungus recovered in the USA (WL Bruckart, *pers. comm.* 2009). The proposed project is modelled after the most recent Australian approach (see below): it will further characterise variability in both the host and rust fungus in terms of disease susceptibility then acquire pathogenic rust isolates from Europe to target the resistant blackberry clones in the USA.

Australian biological control projects

The inaccessibility of more than 40% of blackberry infestations to chemical and cultural control (Amor and Harris 1977) is one of the reasons why the Vic Government and CSIRO initiated a biological control project in the late 1970s, primarily based on the leaf-rust fungus *P. violaceum* (Bruzzese and Hasan 1986a, b, c). The sawfly *Hartigia albomaculata* (Stein), which was identified as a potential candidate agent during initial surveys, was also investigated (Bruzzese 1982a). *R. fruticosus* agg. was proposed as a target for biological control by the Vic Department of Conservation, Forests and Lands and approved under the *Biological Control Act 1984* (Commonwealth) in July 1985 (Australian Weeds Committee 2011). The Act, however, was never implemented to resolve the conflict of interest that arose between people who considered blackberry useful and others who wanted its biological control when a submission was made in 1986 to introduce the first agent in Australia.

In the late 1990s, research on blackberry biological control was resumed by the Cooperative Research Centre (CRC) for Weed Management Systems, based on new knowledge of the taxonomy of the *R. fruticosus* agg. (Evans *et al.* 2007) and pathogenicity studies confirming the presence of blackberry clones resistant to several of the leaf-rust fungus isolates recovered in Australia (Evans *et al.* 2005, 2011). The overall aim of the project was to enhance the rust's biological control effectiveness by introducing additional isolates from the native range, after molecular genetic studies had revealed that the fungus was considerably less diverse in Australasia than in its European native range (Gomez *et al.* 2006). Population-genetic studies also provided evidence of sexual recombination and genetic differentiation among rust populations in Australia, indicating the potential for natural selection of pathogenic genotypes, including recombinants emerging after introduction of additional isolates (Gomez *et al.* 2008). The underpinning hypothesis was that by increasing the genetic diversity of the leaf-rust fungus it may be possible to increase its chances to evolve and adapt to the variable environmental conditions and hosts encountered in Australia.

In the mid 2000s, the Vic Department of Primary Industries renewed efforts to find new biological control agents for blackberry, with a series of surveys in Europe. A review of the literature had identified a number of organisms with biological control potential that had not been found during surveys in the 1970s or investigated in detail at the time (Sagliocco and Bruzzese 2004). One of these candidate agents, the purple blotch fungal pathogen *Septocyta ruborum* (Lib.) Petr., which causes cane dieback of blackberry, has been the focus of a recent study to fully assess its potential for biological control (Bagnant *et al.* 2008).

PLANT TAXONOMY

Most blackberry species belonging to the *R. fruticosus* agg. are polyploid, facultatively apomictic and pseudogamous (Nybom 1988). Even though low levels of genetic variability occur within apomictic species in Europe (Kraft *et al.* 1996), interspecific hybridisation and high phenotypic plasticity (Nybom 1987) are thought to contribute significantly to the taxonomic complexity of blackberry. The consequence is that a taxonomy based solely on morphological characters has resulted in the description of numerous species, which differ from one another to varying degrees. Weber (1996) named blackberry species if a morphological variant occupied a diameter of distribution greater than 20 km, resulting in the description of some 300 species of the *R. fruticosus* agg. in central Europe alone. A single European taxonomic account is unavailable, and not surprisingly, a consensus in the taxonomic treatment of blackberry in Europe and elsewhere has not been reached.

Species of the *R. fruticosus* agg. have been introduced to Australia on multiple occasions: records begin in 1842, although earlier importation is likely (Parsons and Cuthbertson 2001). Evans *et al.* (2007) listed 15 naturalised species within the *R. fruticosus* agg. in Australia. This is a small subset of the total genetic diversity present in the native range of Europe and the Middle East (western Asia). *Rubus anglocandicans* is the most widespread species in Australia and appears to exist as a single clonal lineage (Evans and Weber 2003).

Evans *et al.* (2007) identified 50 M13/*Hae*III–DNA banding-pattern phenotypes among 198 collections of blackberry clones putatively belonging to the *R. fruticosus* agg. across Australia. Thirty-five DNA phenotypes were correlated to 15 species of the *R. fruticosus* agg. after morphological examination by *Rubus* experts in Europe; the remaining 15 DNA phenotypes correlated poorly with these species or could not be named. Consequently, a significant number of the clones collected remain unnamed. Species of the *R. fruticosus* agg. have had over 160 years to evolve in Australia and new clones may have arisen by hybridisation and/or somatic mutation. It is also possible that some of the clones that remain unnamed (Evans *et al.* 2007) have not yet been recognised and characterised as species in Europe. When problems were encountered naming a blackberry clone, those sharing the same M13/*Hae*III–DNA phenotype were considered members of the same species (Evans *et al.* 2007). This form of DNA typing, which is analogous to barcoding items in a grocery store, allowed blackberry clones to be identified with certainty during recent research. An interactive taxonomic key, the LucID tool and CD-ROM *Blackberry* (Barker and Barker 2005), has been developed from the recent taxonomic update (Evans *et al.* 2007) to aid land managers, community groups and researchers distinguish and identify cultivated, native and invasive *Rubus* species occurring in Australia.

EXPLORATION

The first surveys of blackberry's natural enemies were carried out in the mid 1970s by the Commonwealth Institute of Biological Control in countries across continental Europe, mostly in central and southern Europe, from Spain in the west to Turkey in the east (Bruzzese 1982b). A total of 38 arthropods and 15 pathogens restricted to *Rubus* spp. were found, but most of them were not investigated further because they also attacked cultivated *Rubus* spp. or did not cause significant damage to invasive blackberry species. Based on field observations, only three organisms were considered to have some potential for biological control, the leaf-rust fungus *P. violaceum*, the stem-boring sawfly *H. albomaculata* and the purple blotch fungus *S. ruborum* (Bruzzese 1982b) (see below for detail).

From 1978 to 1980, additional surveys specifically focusing on the leaf-rust fungus occurring on blackberry species were undertaken in Europe by E Bruzzese (Bruzzese and Hasan 1986a). An ad hoc sampling strategy was adopted whereby urediniospores were collected from diseased leaves at 48 sites across Spain, France, Italy, former Yugoslavia, Greece, Bulgaria and Turkey, and stored for subsequent testing.

A different approach was adopted 20 years later to source additional isolates of the leaf-rust fungus. Twenty-one blackberry clones, representing 19 DNA phenotypes, were collected across Australia, based on the taxonomic update of Evans *et al.* (2007), and a trap garden of these clones was established at the CSIRO European Laboratory near Montpellier, France (Scott *et al.* 2002; Morin *et al.* 2011). The objective was to expose Australian blackberry clones to the airspora in the native range for direct selection of pathogenic isolates of the rust fungus. This approach was chosen since some of the blackberry clones in Australia may now be extinct in Europe or have evolved from clones originally introduced in Australia. The clones were shipped to Montpellier as pest- and disease-free micropropagated plantlets to satisfy French quarantine authorities, and were planted in the garden in 2000 and 2001. Plants were pruned back severely during the winter and all tips of flowering canes were removed, dried and burnt to prevent gene flow to other blackberry plants close by. Urediniospores were collected from the various clones as they became infected during the growing season, and stored for subsequent use.

A climate-matching approach was also taken at that time in an attempt to source isolates that would have greater impact than the existing population of the leaf-rust fungus in south-western Australia where blackberry is an increasing problem. Scott *et al.* (2002) identified various areas of Portugal that climatically matched Manjimup in WA, using the CLIMEX program and geographic information system (GIS) techniques, and collected rust isolates in those areas. Australian blackberry clones inoculated with these isolates, however, were found to be moderately resistant or immune, highlighting that compatible host–pathogen interactions in this case were more critical than climate-matching in sourcing isolates of the leaf-rust fungus.

More surveys specifically targeting previously overlooked crown- and root-feeding organisms were carried out in France, Italy and Spain in the mid 2000s (Adair and Bruzzese 2006; F Mahr, *pers. comm.* 2010). Crowns and roots from more than 30 sites were brought back to the laboratory for dissection. Other natural enemies attacking above-ground parts were collected but not investigated further.

CANDIDATES

Natural enemies

Blackberry is known to be associated with a wide range of natural enemies in its native and introduced ranges, most of which are generalists or oligophagous feeders (Bruzzese 1980a, b; Amor *et al.* 1998; Pennycook 1998; Sagliocco and Bruzzese 2004; Scott *et al.* 2008). Three fungal pathogens were found to be common on invasive, native and cultivated *Rubus* spp. in Vic in the late 1970s, the rust *Kuehneola uredinis* (Link) Arthur, leaf-spot *Coryneopsis rubi* (Westend.) Grove (previously *Septoria rubi* Westend.) and *Elsinoë veneta* (Burk.) Jenkins (anthracnose) (Bruzzese 1980a). The sawfly *Priophorus morio* (Lepeletier), which was first recorded in Tas in 1959 (Julien and Griffiths 1998), was the only insect found in the late 1970s to be restricted to *Rubus* spp. (Bruzzese 1980b).

Agents tested but not released

Throughout Mediterranean Europe, the sawfly *H. albomaculata* oviposits into succulent primocanes and larvae tunnel into the pith, leading to cane collapse and dieback when levels of attack are high (Bruzzese 1982b). Preliminary host-specificity tests under laboratory conditions carried out in the late 1970s on 35 non-target plant species showed that larvae were able to feed on a number of cultivated blackberry and rose varieties but not on raspberry (Bruzzese 1982a). In light of these results, this potential agent was not pursued further. Recent surveys carried out in southern France confirmed that the sawfly is common on *R. ulmifolius* (Sagliocco and Bruzzese 2004), but it was never found on the closely related species *Rubus caesius* L., *Rosa canina* L. and *Rosa rubiginosa* L. present at sites where *R. ulmifolius* was attacked. This raises the possibility that results from previous preliminary host-specificity tests (Bruzzese 1982b) had been influenced by the laboratory procedure used. Additional testing under natural conditions in the native range would be necessary to fully assess its host-range.

Agent released

The only agent released to date for blackberry biological control is the leaf-rust fungus *P. violaceum*, a biotrophic pathogen, or obligate parasite, which obtains nutrients from living plant cells. It infects leaves of blackberry and occasionally infects green stems, petioles, the green tissues of the flowers and unripe fruits. Through its absorption or diversion of assimilates from the host plant, it reduces plant development and reproduction. The fungus also destroys leaf tissue by producing fruiting bodies, thus reducing the photosynthetic surface of the plant. The leaf-rust fungus is autoecious, completing its life-cycle on a single host (blackberry), and macrocyclic, producing all five spore states – urediniospores, teliospores, basidiospores, spermatia and aeciospores. The wind-dispersed urediniospores correspond to the repetitive stage of the rust that is responsible for epidemics during the growing season (Evans and Bruzzese 2003). The rust overwinters on infected plant debris in the form of resistant teliospores that are produced at the end of the growing season.

The leaf-rust fungus was first reported in Vic in 1984, following an unauthorised introduction, and soon became widespread in blackberry infestations across southern Australia (Marks *et al.* 1984; Bruzzese 1985; Bruzzese and Field 1985). Inoculation trials confirmed that this rust accession was pathogenic on all blackberry species found in Vic, except *Rubus cissburiensis* W.C. Barton & Ridd., *Rubus erythrops* Edees & A. Newton (previously referred to as *Rubus rosaceus* Weihe (Evans *et al.* 2007)) and some *R. ulmifolius* hybrids (probably corresponding to *Rubus leucostachys* Schleich. ex Sm. of Evans *et al.* (2007)) (Bruzzese and Field 1985). It allegedly travelled on wind currents across the Tasman Sea and was first recorded in NZ in 1990 (Pennycook 1998).

The incursion of the leaf-rust fungus in Australia did not stop research underway in Europe to select host-specific and highly pathogenic isolates of the rust for biological control in Australia (Bruzzese and Hasan 1986a, b, c). The main criterion for selecting a rust isolate from Europe for release in Australia was that it produced a relatively high density of uredinia on more than one of four widespread blackberry species in Vic (*R. anglocandicans* (previously referred to as *R. procerus* (Evans *et al.* 2007)), *R. ulmifolius*, *Rubus polyanthemus* Lindeb. and *R. laciniatus*) (Bruzzese and Hasan 1986a). Relative to 14 other isolates tested, isolate F15 produced, on average, a high density of uredinia on three of the widespread blackberry species; another isolate was needed to cause significant rust on the fourth species, *R. laciniatus*. The founding population of the fungus in Australia, from the unauthorised introduction, was reported to attack *R. laciniatus*. Therefore, following extensive host-specificity testing (Bruzzese and Hasan 1986b, c) only isolate F15 was released in 1991 and 1992, in all southern states of mainland Australia (Bruzzese and Lane 1996). It is noteworthy that *Rubus* spp. of North American origin, such as *Rubus laudatus* A.Berger, which is well established in WA (Evans *et al.* 2007) and possibly elsewhere, are not susceptible to the leaf-rust fungus (Evans *et al.* 2011).

Molecular characterisation of rust isolates collected in Australia between 1997 and 1999, based on M13/*Hae*III digestion and selective amplification of microsatellite polymorphic loci (SAMPL), revealed that all genotypes identified were different from that of isolate F15 (Evans *et al.* 2000; Gomez *et al.* 2006). Only three isolates collected in WA in 1998 shared DNA restriction fragments with isolate F15 (Evans *et al.* 2000). This result suggests that genetic material of F15 may have been introduced into the rust population in WA through sexual recombination (Gomez *et al.* 2008), and that F15's genotype may have failed to persist following establishment of this isolate in that state.

Eight additional isolates of the leaf-rust fungus, sourced from the trap garden in France, were released in April 2004 by CSIRO at experimental sites in NSW and WA (Morin *et al.* 2006) and in 2005 at other sites in those states and in Vic by the Vic Department of Primary Industries (R Adair, *pers. comm.* 2005). While the rust fungus developed on all 21 blackberry clones planted in the garden over the two-year period (Scott *et al.* 2002; Morin *et al.* 2011), pressure from stakeholders to accelerate the biological control project meant that evaluation of isolates was restricted to those collected and purified in 2000 from a subset of blackberry clones. Prior to their release, the eight isolates were bioassayed across a range of blackberry clones, each with a different DNA phenotype and previously used for characterising physiologic forms of the rust for pathogenicity (Evans *et al.* 2011; Morin *et al.* 2011). Host-specificity testing was also performed on a representative number of non-target plant species related to blackberry (Morin *et al.* 2011). Genetic characterisation of these eight additional isolates indicated that they were genetically distinct from isolates collected in Australia before 2004 (Gomez *et al.* 2006; Morin *et al.* 2011).

A large-scale, intensive national project was initiated in 2006 to coordinate the release of the additional isolates at numerous sites across the distribution of blackberry in Australia over a three-year period (Morin *et al.* 2008). The project involved the participation of land managers and was designed to accommodate the temporal and spatial asynchrony previously observed in the population dynamics of the rust in Australia (Gomez *et al.* 2008), thereby increasing chances of establishment. An important consideration was that inoculum of each isolate being released was extremely small compared to that present from the existing rust population. Consequently, a 'swamping' effect could easily have occurred at the time of release and prevented establishment. Polymorphic co-dominant loci from simple sequence repeats (SSRs or microsatellites) were developed to monitor the fate of the additional isolates (Molecular Ecology Resources Primer Development Consortium 2010). Unlike SAMPL used previously, SSRs enabled determination of heterozygosity and tracking of alleles in populations in which sexual reproduction is occurring. Screening of rust samples, recovered from four sites two years after release, showed that alleles of the additional isolates had been incorporated into the existing population of the rust (DR Gomez, L Morin and KJ Evans, unpub. data).

Other promising agents

Based on an extensive review of the literature and field observations made in the 1970s, Sagliocco and Bruzzese (2004) listed several natural enemies, in addition to the sawfly *H. albomaculata*, that may have potential for blackberry biological control. Five other sawfly species have been reported in the literature to feed exclusively on *Rubus* spp., while the eriophyid mite *Eriophyes rubicolens* (Canestrini) and the moth *Ectoedemia erythrogenella* (de Joannis) (Nepticulidae) have been recorded only on the *R. fruticosus* agg. The moth, however, has only one generation a year and consequently would be unlikely to cause sufficient damage to suppress blackberry. The purple blotch fungus *S. ruborum* was listed as having potential for biological control based on the significant dieback it can cause to primocanes (Punithalingam 1980), although it is recorded on some of the commercial blackberry cultivars grown in Europe (Nordskog *et al.* 2003). The fungus is currently being investigated to determine whether pathotypes with different physiologic specialisation exist and to assess the susceptibility of Australian native *Rubus* spp. and commercial blackberry varieties (Bagnant *et al.* 2008). The biological control potential of the buprestid-type beetle larvae found associated with root-feeding damage during recent surveys was not assessed because larvae could not be reared to the adult stage and thus were not formally identified and investigated further (F Mahr, *pers. comm.* 2010).

DISCUSSION

Infection by the leaf-rust fungus *P. violaceum* rarely results in more than 40% of leaves on a blackberry shoot being severely diseased, yet biological control with this rust has been considered to be successful in some locations where the nutrient drain from infection and defoliation over multiple growing seasons has slowed the vegetative spread of blackberry and/or reduced it to a non-competitive understorey species (Bruzzese and Lane 1996; Mahr and Bruzzese 1998). Mahr and Bruzzese (1998) monitored the impact of the rust at two sites in Vic, one with *R. polyanthemus* and the other with *R. ulmifolius*, over a 12-year period and reported reduction in total biomass and daughter plant production of more than 56% and 95%, respectively. By 2004–05, a 2.5:1 benefit:cost ratio was estimated for the blackberry biological control program in Australia; this did not take into account any benefits to biodiversity and ecosystems services (Page and Lacey 2006).

Climate and weather can explain regional and local site differences in the impact of the rust observed across the distribution of blackberry in Australia. A rust impact model developed by Pigott *et al.* (2003) predicted that the fungus would have the highest impact in localities of Vic where the annual rainfall was greater than 750 mm, where summer rainfall occurred or relative humidity was high, and when the average maximum daily temperatures in January were close to 20°C. In addition to climate and weather, pathogenicity studies identified both isolate-specific and age-related disease resistance as host factors that might limit the severity of rust in some blackberry infestations (Evans *et al.* 2005, 2011). Within local infestations, compatible host–pathogen interactions may not occur in a timely manner even when the weather is highly favourable for disease development. Studies of the population genetics of the rust fungus revealed that stochastic migration of genotypes across the landscape was most likely a factor affecting temporal and spatial variation in disease intensity in blackberry infestations (Gomez *et al.* 2008). These are the main reasons why further exploration for additional candidate agents was undertaken in the early 2000s. To date, the only additional agents released in Australia have been further isolates of the leaf-rust fungus, but work is continuing on the purple blotch fungus *S. ruborum* and could be resumed on the other candidate insect agents identified if funding becomes available. Expectations should be realistic, however, as it will remain a challenge to find agents that are highly specific and do not pose a threat to native and cultivated varieties of *Rubus* spp.

Researchers require a useful taxonomy of plant species in order to evaluate the responses of host and non-target species to biological control agents. The incomplete taxonomic treatment of *Rubus* in Europe, let alone Australia, has presented significant difficulties in the selection of rust isolates that can potentially cause severe disease in all clones of blackberry present in Australia. Phenotypic plasticity in the *R. fruticosus* agg. is high but DNA markers can identify blackberry clones with certainty. These markers were employed to characterise plant material used in pathogenicity studies with rust isolates and as plants in the trap garden in Europe to source additional isolates (Evans *et al.* 2000, 2007; Gomez *et al.* 2006). The trap garden of Australian blackberry clones proved useful, albeit labour-intensive, for directly sourcing genetically diverse and pathogenic isolates of the rust fungus (Morin *et al.* 2011). Such an approach should be considered in other biological control projects involving rust fungi because they often comprise several pathotypes specific to different forms or genotypes of their hosts.

The population genetic studies of Gomez *et al.* (2008) highlight some potential pitfalls in releasing a rust pathogen for biological control of a genetically diverse weed when done in an ad hoc or restricted manner, influenced by regulation, funding and/or unauthorised importation of the pathogen, as happened with the leaf-rust fungus more than two decades ago. While many factors affect the success of establishment of an exotic organism, the effort expended to introduce a biological control agent is likely to be key to successful establishment (Hopper and Roush 1993). The amount of inoculum of each additional isolate of the leaf-rust fungus released was small compared with the large existing population of the pathogen. The 'numbers game' involved in recruitment of additional isolates meant their persistence depended on the introgression of their alleles into the existing population through sexual recombination in the season following the release. While there is evidence that such introgression happened (DR Gomez, L Morin and KJ Evans, unpub. data), there is no guarantee that the novel rust genotypes will disperse to other locations to become well established there, despite the apparent capacity for long-distance dispersal (Gomez *et al.* 2008). These potential limitations justified the large-scale releases of the additional isolates, using intensive methods that ensured field inoculations resulted in disease development (Morin *et al.* 2008). Engagement from multiple agencies and individuals at the national level has been crucial to the successful implementation of such a large-scale release program.

Over time, it is hoped that introduction of genetic material from the additional rust isolates in Australia will contribute to the co-evolution of the leaf-rust fungus with its blackberry host and to the enhancement of biological control. While the recovery of large areas of productive pasture now infested with impenetrable blackberry thickets across southern Australia would be the ultimate success, land managers would be satisfied with any significant decrease in spread and density of blackberry. At present, biological control is best used on large inaccessible infestations with low priority for other control methods (Bruzzese and Lane 1996; NSW Dept of Primary Industries Weed Management Unit 2010). It is also worth recognising that naturalised *Rubus* spp. and stabilised hybrid cultivars (e.g. loganberry *Rubus loganobaccus* L.H. Bailey) of North American origin growing adjacent to patches of European blackberry could replace the latter if biological control were successful. Integration of control methods will remain an essential element for the management of all invasive *Rubus* spp. at the landscape scale.

ACKNOWLEDGEMENTS

We thank the community of land managers and weed scientists in Australia who supported this research in many ways. Special thanks go to our collaborators and technical staff, including all our cited co-authors, for their significant contributions. We also thank the Bureau of Rural Sciences, and Dr Lucy Randall in particular, for the distribution map (Fig. 1).

REFERENCES

Adair RJ and Bruzzese E (2006) Blackberry: treading a prickly path to effective biological control in Australia. In *Proceedings of the 15th Australian Weeds Conference*. 24–28 September 2006, Adelaide. (Eds C Preston, JH Watts and ND Crossman) pp. 557–560. Weed Management Society of South Australia, Adelaide.

Amor RL and Harris RV (1977) Survey of the distribution of blackberry (*Rubus fruticosus* L. agg) in Victoria and the use and effectiveness of chemical control measures. *Journal of the Australian Institute of Agricultural Science* **45**, 260–263.

Amor RL, Richardson RG, Pritchard GH and Bruzzese E (1998) *Rubus fruticosus* L. agg. In *The Biology of Australian Weeds. Vol. 2*. (Eds FD Panetta, RH Groves and RCH Shepherd) pp. 225–246. RG and FJ Richardson, Melbourne.

Australian Weeds Committee (2011) Target species for biological control. <http://www.weeds.org.au/management.htm> (accessed 3.9.11).

Baguant J, Adair RJ and Lawrie AC (2008) Infection process of *Septocyta ruborum*, a coelomycetous fungus with potential for biological control of European blackberry in Australia. In *Proceedings of the 16th Australian Weeds Conference*. 18–22 May 2008, Cairns. (Eds RD Van Klinken, VA Osten, FD Panetta and JC Scanlan) pp. 251–253. Queensland Weeds Society, Brisbane.

Barker RM and Barker WR (2005) Blackberry. An identification tool to introduced and native *Rubus* in Australia. SA Dept for Environment and Heritage, Adelaide. <http://www.cbit.uq.edu.au/software/blackberry/> (accessed 3.9.11).

Bruzzese E (1980a) A check list of the fungal pathogens of *Rubus* spp. (Rosaceae) in Victoria. *Australasian Plant Pathology* **9**, 117–120.

Bruzzese E (1980b) The phytophagous insect fauna of *Rubus* spp. (Rosaceae) in Victoria, a study on the biological control of blackberry (*Rubus fruticosus* L. agg.). *Journal of the Australian Entomological Society* **19**, 1–6.

Bruzzese E (1982a) The host specificity of *Hartigia albomaculatus* (Hymenoptera: Cephidae) and its potential effectiveness in the biological control of European blackberry. *Entomophaga* **27**, 335–342.

Bruzzese E (1982b) Recent work on the assessment of the biological control agents of European blackberry (*Rubus fruticosus*). In *Proceedings of the Brighton Crop Protection Conference – Weeds. Vol. 3*. 22–25 November 1982, Brighton, UK. pp. 787–790. British Crop Protection Council, Brighton, UK.

Bruzzese E (1985) Spread of blackberry rust. An update. *Plant Protection Quarterly* **1**, 46.

Bruzzese E and Field RP (1985) Occurrence and spread of *Phragmidium violaceum* on blackberry (*Rubus fruticosus*) in Victoria, Australia. In *Proceedings of the 6th International Symposium on Biological Control of Weeds*. 19–25 August 1984, Vancouver. (Ed ES Delfosse) pp. 607–612. Agriculture Canada, Vancouver.

Bruzzese E and Hasan S (1986a) The collection and selection in Europe of isolates of *Phragmidium violaceum* (Uredinales) pathogenic to species of European blackberry naturalized in Australia. *Annals of Applied Biology* **108**, 527–533.

Bruzzese E and Hasan S (1986b) Host specificity of the rust *Phragmidium violaceum*, a potential biological control agent of European blackberry. *Annals of Applied Biology* **108**, 585–596.

Bruzzese E and Hasan S (1986c) Infection of Australian and New Zealand *Rubus* subgenera *Dalibarda* and

Lampobatus by the European blackberry rust fungus *Phragmidium violaceum*. *Plant Pathology* **35**, 413–416.

Bruzzese E and Lane M (1996) *The Blackberry Management Handbook*. Keith Turnbull Research Institute, Vic Dept of Conservation and Natural Resources, Melbourne.

Davies RJ-P (1998) Regeneration of blackberry-infested native vegetation. *Plant Protection Quarterly* **13**, 189–195.

Evans KJ and Bruzzese E (2003) Life history of *Phragmidium violaceum* in relation to its effectiveness as a biological control agent of European blackberry. *Australasian Plant Pathology* **32**, 231–239.

Evans KJ and Weber HE (2003) *Rubus anglocandicans* (Rosaceae) is the most widespread taxon of European blackberry in Australia. *Australian Systematic Botany* **16**, 527–537.

Evans KJ, Jones MK, Mahr FA and Roush RT (2000) DNA phenotypes of the blackberry biological control agent, *Phragmidium violaceum*, in Australia. *Australasian Plant Pathology* **29**, 249–254.

Evans KJ, Jones MK and Roush RT (2005) Susceptibility of invasive taxa of European blackberry to rust disease caused by the uredinial stage of *Phragmidium violaceum* under field conditions in Australia. *Plant Pathology* **54**, 275–286.

Evans KJ, Symon DE, Whalen MA, Hosking JR, Barker RM and Oliver JA (2007) Systematics of the *Rubus fruticosus* aggregate (Rosaceae) and other exotic *Rubus* taxa in Australia. *Australian Systematic Botany* **20**, 187–251.

Evans KJ, Gomez DR, Jones MK, Oakey H and Roush RT (2011) Pathogenicity of *Phragmidium violaceum* isolates on European blackberry clones and on leaves of different ages. *Plant Pathology* **60**, 532–544.

Gomez DR, Evans KJ, Harvey PR, Baker J, Barton J, Jourdan M, Morin L, Pennycook SR and Scott ES (2006) Genetic diversity in the blackberry rust pathogen, *Phragmidium violaceum*, in Europe and Australasia as revealed by analysis of SAMPL. *Mycological Research* **110**, 423–430.

Gomez DR, Evans KJ, Baker J, Harvey PR and Scott ES (2008) Dynamics of introduced populations of *Phragmidium violaceum* and implications for biological control of European blackberry in Australia. *Applied and Environmental Microbiology* **74**, 5504–5510.

Hill RL (1989) *Rubus fruticosus* L. agg., blackberry (Rosaceae). In *A Review of Biological Control of Invertebrate Pests and Weeds in New Zealand 1874 to 1987.* (Eds PJ Cameron, RL Hill, J Bain and WP Thomas) pp. 357–360. CABI, Wallingford, UK.

Hopper KR and RT Roush (1993) Mate finding, dispersal, number released and the success of biological control introductions. *Ecological Entomology* **18**, 321–331.

James R and Lockwood M (1998) Economics of blackberries: current data and rapid valuation techniques. *Plant Protection Quarterly* **13**, 175–179.

Johnson KB and Mahaffee WF (2010) Factors influencing epidemiology and management of blackberry rust in cultivated *Rubus laciniatus*. *Plant Disease* **94**, 581–588.

Julien MH and Griffith MW (1998) *Biological Control of Weeds: A World Catalogue of Agents and their Target Weeds*. 4th edn. CABI, Wallingford, UK.

Kraft T, Nybom H and Werlemark G (1996) DNA fingerprint variation in some blackberry species (*Rubus* subg. *Rubus, Rosaceae*). *Plant Systematics and Evolution* **199**, 93–108.

Laundon GF and Rainbow AF (1969) *Phragmidium violaceum* (Schultz) Wint. In 'Descriptions of Pathogenic Fungi and Bacteria' (No. 209). Commonwealth Mycological Institute, Kew, UK.

Mahr FA and Bruzzese E (1998) The effect of *Phragmidium violaceum* (Shultz) Winter (Uredinales) on *Rubus fruticosus* L. agg. in south-eastern Victoria. *Plant Protection Quarterly* **13**, 182–185.

Marks GC, Pascoe IG and Bruzzese E (1984) First record of *Phragmidium violaceum* on blackberry in Victoria. *Australasian Plant Pathology* **13**, 12–13.

Molecular Ecology Resources Primer Development Consortium (2010) Permanent genetic resources added to Molecular Ecology Resources Database 1 December 2009–31 January 2010. *Molecular Ecology Resources* **10**, 576–579.

Morin L, Aveyard R, Batchelor KL, Evans KJ, Hartley D and Jourdan M (2006) Additional isolates of *Phragmidium violaceum* released for the biological control of blackberry. In *Proceedings of the 15th Australian Weeds Conference.* 24–28 September 2006, Adelaide. (Eds C Preston, JH Watts and ND Crossman) pp. 565–568. Weed Management Society of South Australia, Adelaide.

Morin L, Adair R, Aveyard R, Evans K, Gomez D, Lester J and Yeoh P (2008) National blackberry biological control program in partnership with the community. In *Proceedings of the 16th Australian Weeds Conference.* 18–22 May 2008, Cairns. (Eds RD Van Klinken, VA Osten, FD Panetta and JC Scanlan) pp. 344–346. Queensland Weeds Society, Brisbane.

Morin L, Evans KJ, Jourdan M, Gomez DR and Scott JK (2011) Use of a trap garden to find additional genetically distinct isolates of the rust fungus *Phragmidium violaceum* to enhance biological control of European blackberry in Australia. *European Journal of Plant Pathology* **131**, 289–303.

Nordskog B, Stensvand A and Heiberg N (2003) Fungi occurring on aerial constituents of cultivated blackberry (*Rubus fruticosus* L.) in Norway. **Acta** *Agriculturae Scandinavica, Section B: Soil and Plant Science* **53**, 21–28.

NSW Dept of Primary Industries Weed Management Unit (2009) 'Blackberry control manual: management and control options for blackberry (*Rubus* spp.) in Australia.' Dept of Primary Industries, Victoria. <http://www.weeds.org.au/WoNS/blackberry/docs/blackberry-control-manual-intro.pdf> (accessed 3.9.11).

Nybom H (1987) A demographic study of the apomictic blackberry, *Rubus nessensis* (Rosaceae). *Nordic Journal of Botany* **7**, 365–372.

Nybom H (1988) Apomixis versus sexuality in blackberries (*Rubus* subgen. *Rubus*, Rosaceae). *Plant Systematics and Evolution* **160**, 207–218.

Oehrens E (1977) Biological control of the blackberry through the introduction of rust, *Phragmidium violaceum*, in Chile. *FAO Plant Protection Bulletin* **25**, 26–28.

Osterbauer N, Trippe A, French K, Butler T, Aime MC, McKemy J, Bruckart WL, Peerbolt T and Kaufman D (2005) First report of *Phragmidium violaceum* infecting Himalaya and evergreen blackberries in North America. *Plant Health Progress*. Online. doi:10.1094/PHP-2005-0923-01-BR.

Page AR and Lacey KL (2006) 'Economic impact assessment of Australian weed biological control'. CRC for Australian Weed Management, Technical Series No. 10, Adelaide. <http://www.southwestnrm.org.au/sites/default/files/uploads/ihub/page-ar-lacey-kl-2006-economic-impact-assessment-australian-weed-biological-control.pdf> (accessed 3.9.11).

Parsons WT and Cuthbertson EG (2001) *Noxious Weeds of Australia*. 2nd edn. CSIRO Publishing, Melbourne.

Pennycook SR (1998) Blackberry in New Zealand. *Plant Protection Quarterly* **13**, 163–174.

Pigott JP, Weiss J, Evans KJ and Mahr FA (2003) Predicting the potential distribution of the biological control agent blackberry leaf rust (*Phragmidium violaceum*) and its impact on European blackberry (*Rubus fruticosus*) in Victoria, Australia. In *Plant Invasions: Ecological Threats and Management Solutions*. (Eds L Child, JH Brock, G Brundu, K Prach, P Pysek, PM Wade and M Williamson) pp. 373–381. Backhuys Publishers, Leiden.

Punithalingam E (1980) *Septocyta ruborum*. In 'Descriptions of pathogenic fungi and bacteria' (No. 667). Commonwealth Mycological Institute, Kew, UK.

Sagliocco J-L and Bruzzese E (2004) Biological control of *Rubus fruticosus* agg. (blackberry): is the leaf rust the only option for Australia? In *Proceedings of the 11th International Symposium on Biological Control of Weeds*. 27 April–2 May 2003, Canberra. (Eds JM Cullen, DT Briese, DJ Kriticos, WM Lonsdale, L Morin and JK Scott) pp. 141–144. CSIRO Entomology, Canberra.

Scott JK, Jourdan M and Evans KJ (2002) Biological control of blackberry: progress towards finding additional isolates of the rust fungus, *Phragmidium violaceum*. In *Proceedings of the 13th Australian Weeds Conference*. 8–13 September 2002, Perth. (Eds H Spafford Jacob, J Dodd and JH Moore) pp. 418–421. Plant Protection Society of WA, Perth.

Scott JK, Yeoh PB and Knihinicki K (2008) Redberry mite, *Acalitus essigi* (Hassan) (Acari: Eriophyidae), an additional biological control agent for *Rubus* species (blackberry) (Rosaceae) in Australia. *Australian Journal of Entomology* **47**, 261–264.

Thorp JR and Lynch R (2000) *The Determination of Weeds of National Significance*. National Weeds Strategy Executive Committee, Launceston.

Weber HE (1996) Former and modern taxonomic treatment of the apomictic *Rubus* complex. *Folia Geobotanica et Phytotaxonomica* **31**, 373–380.

Rumex spp. – docks

GR Strickland, R Fogliani and John K Scott

ABSTRACT

Several species of *Rumex* are important weeds of pasture in many temperate climate regions of the world. In Australia, increasing economic losses attributed to the exotic species *Rumex pulcher* and *Rumex crispus* in the 1970s led to an investigation of options for the biological control of these weeds. In 1982, the WA Department of Agriculture and CSIRO commenced a search for biological control agents of *Rumex* in France and northern Africa, considered to be the likely origin of the weedy species of *Rumex*. Two moths were recommended for release in Australia, *Pyropteron doryliformis* and *Pyropteron chrysidiformis* (Sesiidae), but only *P. doryliformis* was successfully synchronised to a southern hemisphere life-cycle. A novel mass rearing and distribution methodology was developed; more than 31 million eggs were produced over the period 1994–99 to facilitate landowners inoculating their own properties with starter colonies of *P. doryliformis*. This insect established at over 70% of release sites and was capable of dispersing at least 4 km in seven years and up to 1 km in a single generation. *P. doryliformis* is well established in Australia and having a significant impact on the densities of *Rumex* populations. More research is required to determine whether the introduction of *P. chrysidiformis* would be likely to improve the biological control of *Rumex* in colder and wetter regions, where it may be better adapted than *P. doryliformis*.

Key words: biological control, *Pyropteron chrysidiformis*, *Pyropteron doryliformis*, *Rumex conglomeratus*, *Rumex crispus*, *Rumex obtusifolius*, *Rumex pulcher*, Sesiidae.

INTRODUCTION

The genus *Rumex* (Polygonaceae) comprises about 200 species of docks and sorrels that are common throughout the northern hemisphere. Several species have spread around the world and become weeds, particularly in high-rainfall areas and irrigated pastures in areas with Mediterranean or temperate climates. Introduced *Rumex* species that have developed into serious weeds in Australia include *Rumex conglomeratus* Murray, *Rumex crispus* L., *Rumex obtusifolius* L. and *Rumex pulcher* L. Their distributions are shown in Figure 1. These species are perennial herbs with large tap roots that enable the plants to survive long dry summer seasons and regenerate rapidly with the onset of autumn rains. Broad rosettes of leaves and stored nutrients within the tap roots provide docks with a competitive advantage over other pasture species that typically rely on seeds to re-establish after summer.

In the 1970s, primary producers in WA became concerned about the increasing dominance of *Rumex* in pastures, especially *R. pulcher.* In a 1972/73 Agriculture WA survey, Allen (1974) reported that dock was in moderate to high densities in more than 60 000 ha of pastures. The high-rainfall area surveyed covered approximately 1 million ha, 42% of which was sown to pastures. More alarming was a later report by Allen (1975), which estimated

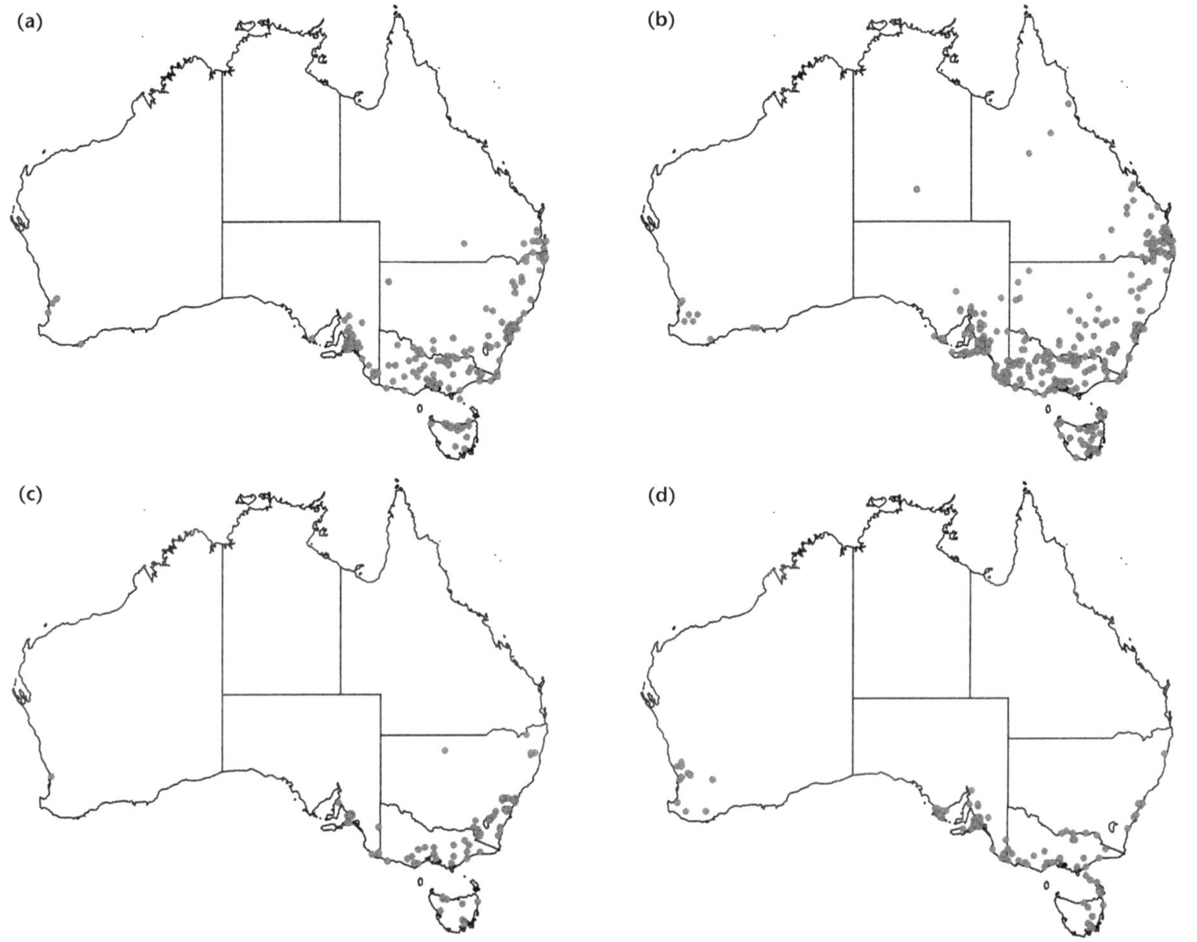

Figure 1: Distribution of *Rumex* species in Australia that are targeted by biocontrol. a) *Rumex conglomeratus*. b) *Rumex crispus*. c) *Rumex obtusifolius*. d) *Rumex pulcher*. Specimen data reproduced from Australia's Virtual Herbarium with permission of the Council of Heads of Australasian Herbaria Inc.

that dock had expanded its range to about 100 000 ha. The reasons for the rapid expansion of dock were not clearly identified, but it was widely accepted that the shift from sheep to cattle and dairy production was an important factor. Sheep will graze docks, including the flowering stems in spring if stocking rates are high, but cattle avoid the plants, which they find unpalatable.

Allen (1975) estimated that *Rumex* species cost farmers in WA about \$A400 000 per year in lost production (\$A2.97 million in 2010 values) (Australian Bureau of Statistics 2010). Controlling dock with herbicides in pastures is problematic because the effective herbicides are non-selective and seriously affect the desirable broad-leaf species, particularly clovers.

While *Rumex* spp. in Australia are conspicuous weeds in high-rainfall agricultural areas, especially permanent pastures, there is no evidence of *Rumex* being an invasive species in undisturbed or natural plant communities. The WA Department of Agriculture successfully nominated the weedy *Rumex* species as targets for biological control in 1983 (Australian Weeds Committee 2010).

BIOLOGICAL CONTROL HISTORY

Detailed studies of the rust fungus *Uromyces rumicis* (Schum.) Wint. (Pucciniaceae) were conducted by Inman (1970, 1971) with a view to importing the rust into the USA from Italy for biocontrol of *R. crispus*. While importation did not occur, the work on pathogens was followed by studies of insects associated with *Rumex* species. Biocontrol studies on *R. obtusifolius* were conducted in Spain by Garcia-Baudin and Santiago-Alvarez (1976) with emphasis on *Gastrophysa unicolor* Marsham (Coleoptera: Chrysomelidae). In Japan, Miyazaki (1979) investigated the complex of phytophagous insects on *R. obtusifolius* and listed a total of 65 species, 18 of which were considered oligophagous. Spencer (1981) produced a list of insects and pathogens attacking *R. crispus* in Italy and the pathogens attacking the genus *Rumex* were listed by Sedlar *et al.* (1983). Despite this activity, no importations occurred and biocontrol of *Rumex* for the USA was discontinued.

More recent interest in biocontrol of *Rumex* has been sponsored by the UK, Switzerland and Norway (Hatcher *et al.* 2008), driven by the needs of organic agriculture in Europe (Grossrieder and Keary 2004), where alternatives to herbicides are sought for intensive cropping. This has included investigations into the development of mycoherbicides (Huber-Meinicke *et al.* 1989).

EXPLORATION

Despite the considerable research into the biological control of *Rumex* spp., it appears that a program of classical biocontrol has been undertaken only in Australia. Cattle and dairy producers in WA initiated the demand for a project in 1980 to investigate the prospects of controlling dock biologically. In 1982, Agriculture WA and Meat and Livestock Australia (then the Meat Research Corporation) commenced a partnership focused on the biocontrol of dock in southern Australian pastures.

R. pulcher and *R. crispus* were the main targets for biocontrol, although *R. conglomeratus* and *R. obtusifolius* were also common weeds. Their origin was considered to be in the western Mediterranean region; the WA Department of Agriculture commenced exploration there from the CSIRO base in Montpellier, France (Scott 1985). The search for phytophagous insects focused on southern France, including Corsica, Italy including Sardinia and Sicily, Spain, Portugal and Morocco and resulted in 37 species of phytophagous insects being recorded feeding on various *Rumex* plant parts. Of these, nine species were considered worthy of detailed study after taking into account published host plant records (Scott 1985). Some of the potential candidates were also included in the assessment of biocontrol of *Emex* species (see chapter on *Emex* species), as these are closely related to *Rumex* and many organisms are associated with both genera. Of the agents considered for *Emex*, *Lixus* (*Dilixellus*) *linearis* Olivier (= *Lixus cribricollis* Boheman, = *Lixus ferrugatus* Olivier) (Coleoptera: Curculionidae) (Fauna Europaea 2010) included *R. crispus* as a target when it was released in Australia (Julien *et al.* 1982).

The following organisms have been assessed or partly assessed for host-specificity to *Rumex*: *Perapion violaceum* (Kirby) (Coleoptera: Brentidae) (Scott 1985), *Apion frumentarium* (L.) (Coleoptera: Brentidae) (= *Apion miniatum* Germar) (Scott and Yeoh 2005), *Lixomorphus algirus* (L.) (= *Lixomorphus ocularis* (F.)) (Coleoptera: Curculionidae) (Campobasso and Murano 1988) and *Pegomya solennis* (Meigen) (= *Pegomya nigritarsis* Zett.) (Diptera: Anthomyiidae) (Scott and Sagliocco 1989). Of these species, only *A. frumentarium* has been released in Australia, where it failed to survive (see chapter on *Emex*). The biology and host-range of the following species have not been assessed: *Perapion hydrolapathi* Marsham (Coleoptera: Brentidae) and *Pegomya bicolor* Wied. (Diptera: Anthomyiidae). Detailed host-range studies and importations to Australia were made on *Pyropteron chrysidiformis* (Esper) (= *Bembecia chrysidiformis* (Esper)) (Lepidoptera: Sesiidae) and *Pyropteron doryliformis* (Ochsenheimer) (= *Chamaesphecia doryliformis* (Ochsenheimer) (Lepidoptera: Sesiidae); these are discussed in the following section.

CANDIDATES

***Lixus* (*Dilixellus*) *linearis* Olivier(= *Lixus cribricollis* Boheman) (Coleoptera: Curculionidae).** The description of this weevil and the biocontrol activities against *Emex* and *Rumex* are outlined in the chapter on *Emex*. It is not known to have established in Australia.

***Pyropteron chrysidiformis* (Esper) (Lepidoptera: Sesiidae).** This clearwing moth has larvae that tunnel inside the larger roots of *Rumex* species including the tap root, feeding over summer to complete development, then entering a diapause period until the following spring. The short-lived adults emerge from the roots, mate and lay eggs on the dried seed-bearing stems of its host plant. Eggs hatch after about two weeks and the first instar larvae enter the root at about soil level. Host-range studies carried out on insects from Rome, Montpellier and Evora (Portugal) showed a host-range restricted to *Rumex* species and closely related Polygonaceae (Scott and Sagliocco 1991a). *P. chrysidiformis* was approved for release in Australia in 1989.

P. chrysidiformis was imported from France into quarantine at South Perth, WA in 1987. Seven consignments of eggs were imported; four consignments in 1988 and three consignments in 1990. In all cases the imported eggs hatched in sufficient numbers to inoculate over 1000 potted dock roots. However, by the end of 1991, only seven moths had emerged, during late August to December. The very short adult life-span of these insects makes mating difficult unless they emerge within a day or so of each other. Only one pairing was achieved, and the eggs were infertile.

During 1990, *P. chrysidiformis* larvae were subjected to a variety of manipulations to find the cues for larvae to commit to pupation and emergence. The success was limited to a few emergences, but they were not contiguous enough for mating. Despite two years of effort it was not possible to synchronise its one-year life-cycle from northern hemisphere to southern hemisphere conditions and the insects did not survive (Fisher *et al.* 1996).

***Pyropteron doryliformis* (Ochsenheimer) (Lepidoptera: Sesiidae).** The dock moth (Fig. 2) has a life-cycle similar to that of *P. chrysidiformis* and a host-range restricted to *Rumex* and closely related Polygonaceae (Scott and Sagliocco 1991b). *P. doryliformis* larvae feed extensively inside the tap roots of *Rumex*, often preventing normal plant regeneration with autumn rain. Several larvae may infest a single large tap root and conduct most of their feeding during the hot dry summer months when the host plant is effectively dormant (Spafford *et al.* 2008). Mature larvae enter a quiescent phase during the cold winter months. During late spring and early summer, the larvae become active and create openings from the tap root to facilitate moth emergence following pupation. The adults are day-active and feed on the nectar of flowering plants. Eggs are laid on the mature flowering stems of *Rumex* plants.

In 1987, the insect was imported from France into quarantine at South Perth, WA, the population originally sourced from insects from Morocco. Two years were required to synchronise the dock moth's one-year life-cycle from northern hemisphere to southern hemisphere conditions and to rear a generation of moths that could be safely released from quarantine.

Figure 2: *Pyropteron doryliformis* (male) emerging from pupal case. Photo: WA Dept of Agriculture and Food.

After quarantine release approval in 1989, *P. doryliformis* was released at 23 sites in the south-west of WA. The inoculation of plants was achieved by painstakingly placing delicate neonate larvae onto the flowering stems of target species. While this technique successfully established populations of *P. doryliformis* in the field, it was clear that new mass rearing and release methodologies would be needed to achieve widespread releases.

Between 1991 and 1994, research focused on developing mass rearing and field release techniques (Fisher 1992). A novel mass rearing and distribution system was developed by the WA Department of Agriculture (described in detail in Fogliani and Strickland 2000). Dock roots were collected from the field and grown in vermiculite in tote boxes to provide food plants for rearing the dock moth. The development rate of *P. doryliformis* was manipulated by controlling the temperature at which it was reared in large temperature-controlled laboratories. It was found that temperatures below 15°C arrested larval development. The moth's one-year life-cycle was shortened to just a few months by exposure to simulated constant 'springtime' temperatures and day length. This overcame the winter dormancy that characterised the moth's life-cycle in the field. Thus, temperature-controlled conditions enabled multiple generations in a 12-month period and facilitated the production of the millions of eggs required for the release season from October to March.

Colony health was maintained through vigorous quality control protocols and supplementation with naturally reared moths. The field release period of the dock moth is restricted to spring and early summer, but the laboratory colony could be manipulated to produce moths over this season by subtle increases in temperature. Annual production of *P. doryliformis* averaged just over 40 000 moths, the most that could be produced in the facility without compromising quality. About 17% of moths were produced in tunnel houses in natural environmental conditions while the bulk was reared in large temperature-controlled laboratories.

The second essential component of the field release process for the dock moth was the development of an efficient inoculation protocol that enabled landowners to establish dock moth populations on their own properties. The efficacy of moth, larval and egg releases into the field were assessed; the final system utilised the eggs, which are robust, hard-shelled, poppyseed-like, simple to collect and handle, and able to be transported and stored for suitable periods of time.

Figure 3: Mass rearing *Pyropteron doryliformis* using an automated eggstick machine to glue dock moth eggs onto toothpicks.

In 1994, an 'eggstick machine' was developed (Fig. 3). This enabled harvested moth eggs to be mechanically glued onto bamboo toothpicks, which could then be dispatched to landholders to be placed in the field. The machine consisted of a conveyor belt assembly with dispensers for dilute PVA glue and moth eggs. The toothpicks moved along the conveyor belt and received a drop of glue, followed by a sprinkling of moth eggs, before being unloaded at the end of the conveyor and packaged ready for dispatch. The use of pre-prepared eggstick 'combs' (polystyrene strips with 25 evenly spaced toothpicks inserted along one edge) allowed for increased production efficiency. An experienced operator was able to produce 1000–1500 eggsticks per hour, enough to inoculate one to two release sites. To avoid storage problems, eggsticks were prepared as closely as possible to the time of dispatch, usually within 24 hours, and no more than five days prior to the anticipated hatch date to allow enough time for them to be placed in the field. During the five years 1994–99, the total production of dock moth eggs was almost 31 million, with the maximum annual production of almost 7 million eggs achieved during the 1995–96 summer (Fogliani and Strickland 2000).

The release program for *P. doryliformis* focused on WA until production and release protocols were refined, then it expanded nationally between 1994 and 1999. WA remained the eggstick production centre and distributed eggsticks to farmer collaborators who inoculated and monitored their own sites. Collaborators from state agencies in SA, Vic and NSW were also strongly involved. The state collaborators functioned as coordinators for their respective jurisdictions, and were responsible for recruiting, training and monitoring the performance of landowners who were ultimately responsible for selecting, preparing and inoculating sites with *Rumex* species on their properties. Landowners were assisted by field days, publications and a Freecall helpline to ensure that reliable site inoculations were conducted.

Combs with enough eggsticks to inoculate 950–1000 dock plants per site were supplied to each participant, to be placed into the release area within two days of delivery. Eggsticks were inserted securely into the central pith of the cut dock stems, at the rate of one eggstick per plant. Every mature dock plant in the release area was inoculated, until all the eggsticks were inserted. The greatest number of dispatches was to farmers in WA (56.0%), followed by SA (16.1%), Vic (15.3%), NSW (12.3%) and Tas (0.3%). The total number of dispatches between 1989 and 1999 was 716.

Widespread field establishment at approximately 70% of sites was confirmed in WA, SA, Vic and NSW (Fogliani and Strickland 2000). Reports of establishment in the Tamar Valley in Tas were noted, but not confirmed.

Other agents present in Australia

The dock sawfly *Lophrotoma analis* (Costa) (Hymenoptera: Pergidae) is occasionally reported to cause extensive defoliation of *Rumex* and *Emex* species. Another insect associated with *Rumex* species is the weevil *Rhinoncus australis* (Oke) (Coleoptera: Curculionidae) (see *Emex* chapter for more details on these insects). During the collection of *Rumex* tap roots for insect-rearing purposes, some minor chewing damage to roots was observed, caused by larvae of the vegetable weevil *Listroderes difficilis* Germain (Coleoptera: Curculionidae). Adult weevils also chewed on leaves, but damage was minor and appeared inconsequential to plant growth.

A natural enemy of *Rumex* that established in Australia by unknown means was the aphid *Brachycaudus rumexicolens* (Patch) (Hemiptera: Aphididae). Surveys in WA found the aphid to be widespread on *Rumex* species (see *Emex* chapter for details). The rust fungus *Uromyces rumicis*, which had been considered for biological control, appeared on *R. pulcher* in WA in the late 1980s (Shivas 1987). Surveys conducted during 1990–92 in the agricultural regions of south-western WA found the rust only on *R. pulcher* at 14 sites even when other *Rumex* and *Emex* species were present (Scott and Shivas 1993). A later arrival (1993) is the dock sawfly from Europe, *Ametastegia glabrata* (Fallén) (Hymenoptera: Tenthredinidae) (Malipatil *et al.* 1995). This species is recorded from *R. obtusifolius* in Vic. The larvae feed on *Rumex* leaves but pupae have been found in a range of plants, thus giving it pest status.

DISCUSSION

The classical biological control of *Rumex* spp. in Australia was a 20-year program, from initiation through to final field releases. The program is a clear success story and an example of the long-term commitment that is often required to realise the potential benefits from biological control. In this case, a sound scientific base was strongly supported by industry funds through Meat and Livestock Australia and by high levels of collaboration with landowners and state-based research partners. Challenges to program success were met with novel innovations and persistence, which led to the mass production and establishment of the previously almost unknown sesiid *P. doryliformis*.

Surveys of the release sites in each state indicated an overall establishment success rate of more than 70%, with the moth persisting in the field years after release (Fogliani and Strickland 2000). In WA, surveys along transects extending from established sites showed that the moth was capable of dispersing at least 4 km in seven years, and up to 1 km in a single generation. Dock densities in the sites were estimated around the time of release and again one to six years later. The data show considerable reduction in the number of dock plants at each site, from 37.4% two years after release, to 100% after five and six years. The percentage reduction in dock density at release sites was shown to be related to the duration of time post-release. The reduction in dock density was consistently highest in those sites surveyed one year after release. This was explained by the fact that all the plants in the release site were inoculated with eggsticks, and a high kill rate would be expected. For sites surveyed one year after release the mean percentage reduction in dock each year was 67.2%, after two years it was 30.3%, four years 22.9%, five years 14.4% and six years 14.7%. A visually impressive decrease in the abundance of dock elsewhere in the paddocks where the release sites were established was also observed at some of the older sites.

In Vic, surveys in the 2003–04 summer showed that the dock moth was very abundant; it was recovered from 77% of sites in northern Vic (Morley *et al.* 2004). The host was predominantly *R. crispus*, with an average of 48% plants sampled being infested or damaged by dock moth. *P. doryliformis* was not recovered from *R. brownii* or *R. conglomeratus* and only a small proportion of *R. obtusifolius* and *R. pulcher* showed evidence of attack.

In southern Vic, where the climate is considerably colder and wetter, the establishment and dispersal of *P. doryliformis* was much lower than in the north. The host species also differed with *R. obtusifolius* being the most common host, followed by *R. pulcher*. However, infestation rates were low for both species, 4% and 2% respectively.

The results from Vic are consistent with the observations of Scott (1985) who reported that *P. doryliformis* was more abundant and presumably better adapted to the hotter drier regions of northern Africa and Europe, while the related species *P. chrysidiformis* dominated in the colder and wetter areas of Europe. Further consideration of *P. chrysidiformis*, an agent already approved for release, is warranted to enhance the biological control of *Rumex* species in the colder regions of Australia, such as Vic and Tas. This would require research into its life-cycle to enable it to be synchronised to southern hemisphere conditions.

Scott (1985) and Scott and Sagliocco (1991a, b) concluded that all the potential agent insects would have host-ranges within the genus *Rumex*, and therefore some degree of attack on native Australian *Rumex* could occur if the biological program were to proceed. Taylor *et al.* (2008) provided some evidence to support that conclusion, although no larvae or pupae of *P. doryliformis* were found in native *Rumex* in field surveys.

The potential risk to native *Rumex* species in Australia was a major consideration in recommending the importation and release of the two sesiids *P. chrysidiformis* and *P. doryliformis*. Both species attack the host late in the growing season, after the plants have set seed. For this reason, even if native *Rumex* species were attacked, propagation from seed would still occur and consequently the viability of the species would not be threatened. This hypothesis still needs to be tested, as does the associated hypothesis expressed in Scott and Sagliocco (1991a, b) that the attack on mature, post-reproductive weedy *Rumex* plants changes the plant demography in a way that leads to improved weed control.

The dock moth *P. doryliformis* has established and spread in Australian pastures, and is having a significant impact on the *Rumex* population. Its effect in reducing dock to below economic thresholds will deliver benefits to producers in the form of improved pasture quality and utilisation, as well as reduced herbicide use. New surveys and experiments to measure the changed pasture species composition are recommended, to quantify the economic benefits from the *Rumex* biological control program.

ACKNOWLEDGEMENTS

Interstate collaborators include Ian Faithful and Tom Morley (Vic), Dennis Hopkins, Ken Henry and John Heap (SA) and Barry Sampson, Bob Thurling, Peter Proctor and Paul Sullivan (NSW). The assistance of WA Dept of Agriculture staff is gratefully acknowledged, in particular, technical officers Dane Gehradte, Dawn Hutchinson, Shelley McDavitt and Mark Pleysier. Kingsley Fisher managed early projects and was fundamental to the design and concepts utilised in the program. Special mention is made of Bill Russell, Bill Smart, Ron McTaggart, Mike Grimm, Ernie Haggett, Phil Barrett-Lennard and Tom Sweeny for assistance in the rural regions of WA. J-L Sagliocco provided valuable assistance in Europe and North Africa.

REFERENCES

Allen JM (1974) Preliminary observations and investigations on docks (*Rumex* spp.) in Western Australia. *Western Australian Department of Agriculture Technical Bulletin* **23**, 1–5.

Allen JM (1975) Docks in Western Australia. *Journal of Agriculture Western Australia* **16**, 67–71.

Australian Bureau of Statistics (2010) 6401.0 – Consumer Price Index, Australia, September 2010 Canberra, Australia. <http://www.abs.gov.au/ausstats/abs@.nsf/mf/6401.0> (accessed 19.12.10).

Australian Weeds Committee (2010) Target species for biological control. <http://www.weeds.org.au/management.htm> (accessed 19.7.10).

Campobasso G and Murano F (1988) Laboratory and field biology of *Lixomorphus ocularis* (Fabricius) (Coleoptera, Curculionidae). *Fragmenta Entomologica* **10**, 309–316.

Fauna Europaea (2010) Database. <http://www.faunaeur.org/full_results.php?id=248123> (accessed 19.12.10).

Fisher K (1992) Clearwing moths are key to dock control. *Journal of Agriculture Western Australia* **33**, 152–155.

Fisher KT, Fogliani RG and Strickland GR (1996) 'Final report to the Meat Research Corporation Project DAW.045 1991–1994. Biological control of dock: field augmentation'. WA Dept of Agriculture, Perth.

Fogliani RG and Strickland GR (2000) 'Final report to Meat and Livestock Australia Project DAW.057 1994–2000. Biological control of dock: enhanced distribution of the dock moth.' WA Dept of Agriculture, Perth.

Garcia-Baudin JM and Santiago-Alvarez C (1976) Biological control of *Rumex obtusifolius* L. I. Preliminary considerations on the species *Gastroidea* (=*Gastrophysa*) *unicolor* Marsham (Coleoptera Chrysomelidae). *Anales del Instituto Nacional de Investigaciones Agrarias. Serie: Proteccion Vegetal* **6**, 127–139.

Grossrieder M and Keary IP (2004) The potential for the biological control of *Rumex obtusifolius* and *Rumex crispus* using insects in organic farming, with particular reference to Switzerland. *Biocontrol News and Information* **25**, 65N–79N.

Hatcher PE, Brandsaeter LO, Davies G, Luscher A, Hinz HL, Eschen R and Schaffner U (2008) Biological control of *Rumex* species in Europe: opportunities and constraints. In *Proceedings of the 12th International Symposium on Biological Control of Weeds*. 22–27 April 2007, La Grande Motte, France. (Eds MH Julien, R Sforza, MC Bon, HC Evans, PE Hatcher, HL Hinz and BG Rector) pp. 470–475. CABI, Wallingford, UK.

Huber-Meinicke G, Defago G and Sedlar L (1989) *Ramularia rubella* (Bon.) Nannf. as a potential mycoherbicide against *Rumex* weeds. *Botanica Helvetica* **99**, 81–89.

Inman RE (1970) Observations on the biology of *Rumex* rust *Uromyces rumicis* (Schum.) Wint. *Botanical Gazette* **131**, 234–241.

Inman RE (1971) A preliminary evaluation of *Rumex* rust as a biological control agent for curly dock. *Phytopathology* **61**, 102–107.

Julien MH, Kassulke RC and Harley KLS (1982) *Lixus cribricollis* (Col.: Curculionidae) for biological control of the weeds *Emex* spp. and *Rumex crispus* in Australia. *Entomophaga* **27**, 439–446.

Malipatil MB, Baumann ID and Williams DG (1995) First record of dock sawfly *Ametastegia glabrata* (Fallén) in Australia (Hymenoptera: Tenthredinidae). *Journal of the Australian Entomological Society* **34**, 95–96.

Miyazaki M (1979) An annotated list of insects feeding on *Rumex obtusifolius* L. (Polygonaceae) in Japan. *Bulletin of the National Grassland Research Institute* **14**, 124–139.

Morley TB, Faulkner S and Faithfull IG (2004) Establishment and dispersal of dock moth *Pyropteron doryliformis* (Ochsenheimer) (Lepidoptera: Sesiidae) in Victoria. In *Proceedings of the 14th Australian Weeds Conference.* 6–9 September 2004, Wagga Wagga, NSW. (Eds BM Sindel and SB Johnson) pp. 381–384. Weed Society of New South Wales, Sydney.

Scott JK (1985) Candidate insects of the biological control of *Rumex pulcher.* In *Proceedings of the 6th International Symposium on Biological Control of Weeds.* 19–25 August 1984, Vancouver. (Ed ES Delfosse) pp. 829–835. Agriculture Canada, Ottawa.

Scott JK and Sagliocco JL (1989) Biology and host-specificity of *Pegomya solennis* (Diptera; Anthomyiidae), a possible biological control agent for *Rumex* spp. in Australia. *Acta Oecologica, Oecologia Applicata* **10**, 157–163.

Scott JK and Sagliocco JL (1991a) Host-specificity of a root borer, *Bembecia chrysidiformis* (Lep.: Sesiidae), a potential control agent for *Rumex* spp. (Polygonaceae) in Australia. *Entomophaga* **36**, 235–244.

Scott JK and Sagliocco JL (1991b) *Chamaesphecia doryliformis* (Lep.: Sesiidae), a second root borer for the control of *Rumex* spp. (Polygonaceae) in Australia. *Entomophaga* **36**, 245–251.

Scott JK and Shivas RG (1993) Occurrence of the rust fungus *Uromyces rumicis*, a biological control agent of fiddle dock (*Rumex pulcher*) in Western Australia. In *Proceedings of the 10th Australian and 14th Asian-Pacific Weed Conference.* 6–10 September 1993, Brisbane. (Eds JT Swarbrick, CWL Henderson, RJ Jettner, L Streit and SR Walker) pp. 18–19. Weed Society of Queensland, Brisbane.

Scott JK and Yeoh PB (2005) Biology and host specificity of *Apion miniatum* (Coleoptera: Apionidae) from Israel, a potential biological control agent for *Emex australis* and *Emex spinosa* (Polygonaceae) in Australia. *Biological Control* **33**, 20–31.

Sedlar L, Vonmoos R, Defago G and Kern H (1983) *Fungi occurring on Selected Weeds: A Compilation from Literature Data.* Institut fur Phytomedizin, ETH-Zentrum, Zurich, Switzerland.

Shivas RG (1987) Rumex rust (*Uromyces rumicis*) in Western Australia. *Australasian Plant Pathology* **16**, 40–41.

Spafford H, Hawley J and Strickland G (2008) Survival of dock moth larvae, *Pyropteron doryliformis* (Lepidoptera: Sesiidae), in tubers of fiddle dock (*Rumex pulcher*). In *Proceedings of the 16th Australian Weeds Conference.* 18–22 May 2008, Cairns. (Eds RD van Klinken, VA Osten, FD Panetta and JC Scanlan) pp. 272–274. Queensland Weeds Society, Brisbane.

Spencer NR (1981) Exploration for biotic agents for the control of *Rumex crispus.* In *Proceedings of the 5th International Symposium on Biological Control of Weeds.* 22–27 July 1980, Brisbane. (Ed ES Delfosse) pp. 125–151. CSIRO, Melbourne.

Taylor DBJ, Spafford H and Heard TA (2008) Native *Rumex* spp. (Polygonaceae) and their utilisation by the dock moth *Pyropteron doryliformis* (Ochsenheimer) (Lepidoptera: Sesiidae) in Australia. In *Proceedings of the 16th Australian Weeds Conference.* 18–22 May 2008, Cairns. (Eds RD van Klinken, VA Osten, FD Panetta and JC Scanlan) p. 212. Queensland Weeds Society, Brisbane.

Salvinia molesta D.S. Mitchell – salvinia

Mic Julien

ABSTRACT

Salvinia molesta, salvinia, is a sterile floating fern that has spread to most tropical and subtropical countries. Rapid growth rates allow it to blanket still or slow-moving water bodies very quickly, causing environmental, sociological and economic problems. A small black weevil, *Cyrtobagous salviniae*, from Brazil, which is specialised on *Salvinia* species, was released in Australia in 1980 on Lake Moondarra, Mt Isa and provided extraordinary control of the weed within 15 months. Thereafter, releases elsewhere in Australia and in Papua New Guinea, Sri Lanka and many other countries resulted in extremely high levels of control, mostly within three years. There are situations where the weevil is not effective, such as on multi-layered salvinia mats or when it grows as an understorey plant, but integrated management strategies that include biological control can overcome these restrictions. Recent studies have confirmed the value of this agent in temperate regions. Other control methods for salvinia are impractical, ineffective and costly, whereas biological control has provided long-term sustainable management with very positive benefit:cost ratio estimates from 11 to 53:1. The control of salvinia by the salvinia weevil is a classic contemporary example of the benefits of biological control of weeds.

Key words: biological control, *Cyrtobagous salviniae*, *Samea multiplicalis*, *Paulinia acuminata*, benefit:cost.

INTRODUCTION

Salvinia molesta D.S. Mitchell (Salviniaceae), salvinia (Fig. 1), is a floating aquatic fern. It has a pair of floating leaves at each node and a submerged leaf that is modified to act as a root. Among the filamentous roots hang sporocarps, however, salvinia is a sterile pentaploid (chromosome number of 45) and no viable spores are produced. Population increase is by growth and fragmentation. Dispersal is mediated by downstream flow and wind, and by the movement on birds and animals of viable fragments, i.e. nodes. It has been spread by man around the world as an ornamental and aquarium plant or as a contaminant of other plant shipments (Julien *et al.* 2009; Room and Julien 1995).

Figure 1: *Salvinia molesta*, salvinia, tertiary form. Photo: M Julien, CSIRO.

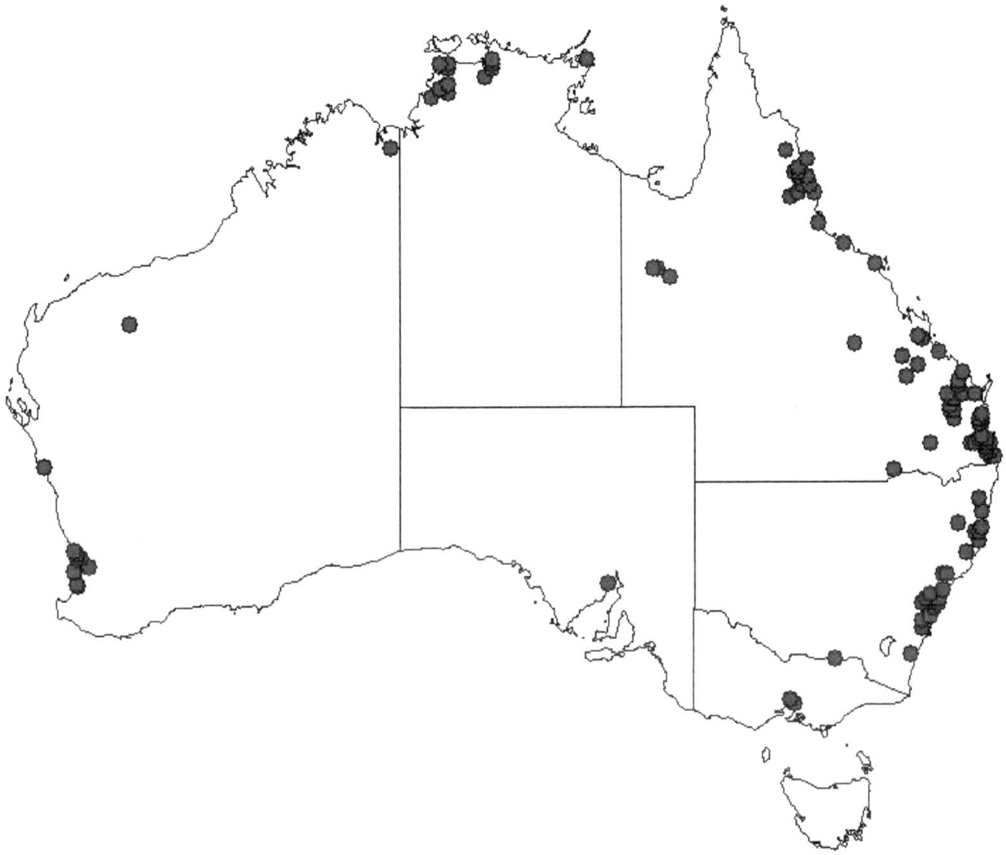

Figure 2: Distribution of *Salvinia molesta* in Australia. Specimen data reproduced from Australia's Virtual Herbarium with permission of the Council of Heads of Australasian Herbaria Inc.

There is considerable morphological variation determined by age, crowding and available nutrients. Three forms were described (Mitchell and Tur 1975), but there is continuity of growth between them. The primary and secondary forms occur during early colonisation; they have leaves nearly flat on the water's surface and space between leaves. The tertiary form (Fig. 1) occurs when plants are crowded and the leaves become folded and upright (Julien *et al.* 2009; Room and Julien 1995).

Salvinia is native to south-eastern Brazil (Forno and Harley 1979), is widespread throughout the tropical world and was first recorded in Australia at Ludenham, NSW in 1952 and Bulimba Creek, Brisbane in 1953. The history of its distribution, worldwide and in Australia, was outlined in Room and Julien (1995). By the 1970s it had become a serious widespread weed in Australia and it now occurs in waterways from the NT (rivers near Darwin, Kakadu National Park, Nhulunbuy), Qld (Cairns to Brisbane including Atherton Tablelands and Lake Moondarra near Mt Isa) and NSW south to Sydney (Fig. 2). It has been recorded from many locations south of Sydney in NSW, Vic and southern WA, but because of low temperatures it is much less problematic and populations may not persist.

Salvinia grows best in still or slow-moving water. It can survive low nutrient levels but rapidly takes up nutrients when they become available, for example during increased flows. It has very rapid growth rates (doubling its number and dry weight in less than three days) under optimum conditions (tropical temperatures and unlimited nutrients) (Room 1986) and forms dense mats (Fig. 3) that can be several layers of plants deep, blanketing the surface of water. In Kakadu National Park, NT dry weights doubled in five to 30 days (Storrs and Julien 1996) and on Lake Kariba, Zimbabwe ramet numbers doubled in nine to 17 days (Mitchell and Tur 1975). The range in doubling time was likely due to nutrient availability and temperature changes over time. Salvinia has been used as a biological weapon to damage fisheries in Papua New Guinea (Gewertz 1983).

Details of its taxonomic relationships, biology, ecology, distribution, pest status and impacts can be found in Julien *et al.* (2009), Room and Julien (1995) and van Oosterhout *et al.* (2006) and references cited therein.

Figure 3: *Salvinia molesta* covering a creek at Kaban, Qld. Photo: W Forno, CSIRO.

BIOLOGICAL CONTROL HISTORY

The first surveys for biological control agents for salvinia were conducted in the native range of *Salvinia auriculata*, including Trinidad, British Guiana and Brazil, from 1961 to 1963 (Bennett 1966, 1975). Of 25 phytophagous insect species found on the *S. auriculata* complex (Bennett 1975) three species, a curculionid weevil *Cyrtobagous singularis* Hustache, a pyralid moth *Samea multiplicalis* (Guenée) and a grasshopper *Paulinia acuminata* (De Geer), were selected for further study. Host-specificity studies were conducted at Belem, Brazil and Curepe, Trinidad during 1964–65 (Bennett 1966).

C. singularis, a small black weevil 2 mm long, has been collected from species in the *S. auriculata* complex from northern Argentina, Brazil, Guyana, Paraguay and Trinidad and from *Salvinia oblongata* Mart. in Brazil (Bennet 1966, 1975). It was released in 1971 in Botswana and Zambia and established (Julien and Griffiths 1998), and spread to the Caprivi Strip of Namibia and to Zimbabwe. It was released in Fiji (1979) and established there (Julien *et al.* 2009). Its biology is described by Bennett (1966) and its feeding behaviour by Sands and Schotz (1985). It has not contributed to control of *S. molesta*.

The moth *S. multiplicalis* has been collected from species in the *S. auriculata* complex in Argentina, Brazil, Florida, Guyana, Panama, Trinidad and Uruguay (Bennett 1966; Forno 1981). It was released in Zambia during 1970 and Botswana in 1972, but did not establish in either. Releases in Fiji in 1976 resulted in establishment but no control of the weed. Though common in the USA, it was not deliberately released there and it provided no control of salvinia (Julien *et al.* 2009).

The grasshopper *P. acuminata* occurs widely in South America and following releases it established in Zimbabwe (1969), Zambia (1970), Botswana (1975), Sri Lanka (1973), India (1974) and Fiji (1975), but failed to establish in Kenya after releases in 1970. It has since spread to Mozambique, downstream from Zambia and Zimbabwe. With the possible exception of the decline of salvinia on Lake Kariba in 1973, *P. acuminata* was not considered an effective biological control agent (Julien *et al.* 2009). Salvinia declined markedly four years after *P. acuminata* was released on Lake Kariba. Other factors contributed to that decline and it was not clear how much could be attributed to damage caused by the grasshopper (Mitchell and Rose 1979; Marshall and Junor 1981).

EXPLORATION

Before 1978 *S. molesta* was known only from outside South America. Since it was sterile and had the egg-beater shaped hairs on the upper leaf surfaces (see Julien *et al.* 2009) characteristic of species in the *S. auriculata* complex, it was considered a hybrid of species from that complex, but without a native range (Mitchell 1972, 1978). Wendy Forno (CSIRO), based in Curitiba, Brazil from January 1978 to March 1982 to survey for potential biological control agents for salvinia, found the native range of *S. molesta* within the latitudes 24°05'S and 32°05'S in south-eastern Brazil (Forno and Harley 1979). This permitted targeted surveys on salvinia in its native range, which led to the discovery of the salvinia weevil *Cyrtobagous salviniae* Calder & Sands that has been so important in the control of salvinia worldwide.

Surveys by Forno on species in the *S. auriculata* complex were conducted in Argentina, Brazil, Guyana, Trinidad, Uruguay, Venezuela and Paraguay. The natural enemies found – 31 insect and mite species, including those found by Bennett (1966) during earlier surveys – were listed by Forno and Bourne (1984). Three of these were imported into Australia for study in quarantine at CSIRO, Brisbane – the weevil *C. salviniae* (imported as *C. singularis*), the moth *S. multiplicalis* and the grasshopper *P. acuminata*. The first two were subsequently released in Australia.

CANDIDATES

Agents released in Australia

***Cyrtobagous salviniae* Calder & Sands (Coleoptera: Curculionidae).** Adults of this small (2 mm) black long-snouted weevil (Fig. 4) were very similar in appearance to adults of *C. singularis*. Calder and Sands (1985) described

Figure 4: *Cyrtobagous salviniae*, the salvinia weevil. Photo: M Julien, CSIRO.

the adult and May and Sands (1986) described the larvae. When first collected from salvinia in south-eastern Brazil it was thought to be a biotype of *C. singularis* adapted to *S. molesta* (Forno *et al.* 1983). It was only after releases in Australia that detailed comparative studies determined that it was a new undescribed species, later to be named *C. salviniae* (Calder and Sands 1985). The comparisons identified differences that suggest why this species, which fed on buds and internal tissues and had higher rates of increase, became an effective agent while *C. singularis*, which fed externally and not on buds, did not (Sands and Schotz 1985; Sands *et al.* 1986).

C. salviniae was found on all species in the *S. auriculata* complex in South America. The material introduced into Australia was collected from salvinia at a site (26°30'S, 48°42'W) near Joinville, Santa Catarina province, Brazil. Host testing determined that it was specific to species in the *S. auriculata* complex (Forno *et al.* 1983). The biology and ecology of *C. salvinia* were presented in Forno *et al.* (1983), Sands *et al.* (1983, 1986), Forno and Bourne (1984), Julien *et al.* (1987), Room and Thomas (1985), Schotz and Sands (1988) and Hennecke and Postle (2006) and summarised in Julien *et al.* (2009). Some literature that was published before *C. salviniae* was recognised as a new species uses the name *C. singularis* when the information actually refers to *C. salviniae* (e.g. Forno *et al.* 1983; Room *et al.* 1981).

The first releases of the salvinia weevil as a biological control agent were at Lake Moondarra, Mt Isa, Qld in June 1980. Lake Moondarra had a 400 ha mat of salvinia, weighing >50 000 t fresh weight. The damage caused by the huge population of weevils that developed following release resulted in spectacular destruction of the mat within 15 months, reducing it to less than 1 t (Room *et al.* 1981) (Fig. 5). Similar successes were repeated on impoundments, billabongs, rivers, creeks and dams throughout tropical and subtropical Australia (Forno 1987; Room and Julien 1995; Storrs and Julien 1996). Biological control of salvinia in temperate areas of Australia took longer than in tropical areas, but was successful in coastal catchments in central NSW (Julien *et al.* 2009; Sullivan *et al.* 2011). *C. salviniae* has been released on salvinia in at least 20 other countries with good to excellent control occurring in one to five years in the 17 countries where the outcomes were known (Julien *et al.* 2009). The Australian research that led to these successes was recognised with awards including the 1985 UNESCO Science Prize (team), the AIDAB (now AusAid) 1988 Award for Excellence in Overseas Development (team) and the 1991 CSIRO Chairman's Medal (team leader Peter Room).

***Samea multiplicalus* (Guenée) (Lepidoptera: Pyralidae).** This moth has tan colouration with darker marking on the wings (Fig. 6) and forewing length of 6.5–10.5 mm (Sands and Kassulke 1984). Eggs are laid singly among leaf hairs. Larvae have five to seven, usually six, instars

Figure 5: Lake Moondara, Mt Isa, Qld before and 15 months after release of the salvinia weevil *C. salviniae*. Photos: P Room, CSIRO.

that feed on leaves under a canopy of silk and leaf hairs. When older larvae tunnel through tertiary form salvinia they cause a 'shot hole' effect. When larvae are numerous they destroy most leaves and the salvinia looks very ragged. Pupation occurs in a silk cocoon attached to a leaf. The life-cycle takes 50–60 days and development is influenced by host plant nutrition (Taylor and Sands 1986; Taylor 1984). The biology and host-range of this moth is described by Bennett (1966), Knopf and Habeck (1976), DeLoach *et al.* (1979) and Sands and Kassulke (1984). Its main hosts include *S. molesta*, *Salvinia minima* Baker, *Pistia stratiotes* L. (Araceae), *Azolla caroliniana* Willd. and *Azolla pinnata* R. Br. (Azollaceae). It fed but could not complete development on *Lemna* species (Araceae) and *Eichhornia crassipes* (Mart.) Solms-Laub. (Pontederiaceae). It was heavily parasitised and predated upon in its native range (Bennett 1966) and has attracted parasites in Australia (Semple and Forno 1987). Though widely established on salvinia, it has rarely been found on *P. stratiotes* in the field in Australia (Waterhouse 1994).

The material imported, tested and released in Australia originated from Joinville, Santa Catarina province, Brazil. The first releases were at Mt Isa and Ingham, Qld in January and February 1981, where it established. Dispersal was rapid and it caused considerable but patchy defoliation of salvinia. Although defoliation could be severe, buds were not damaged and so plants continued to grow (Julien and Bourne 1988). *S. multiplicalis* adults sought undamaged plants for oviposition (Taylor and Forno 1987), so there was no follow-up damage until the plants regenerated. *S. multiplicalis* has not contributed to control of salvinia in Australia.

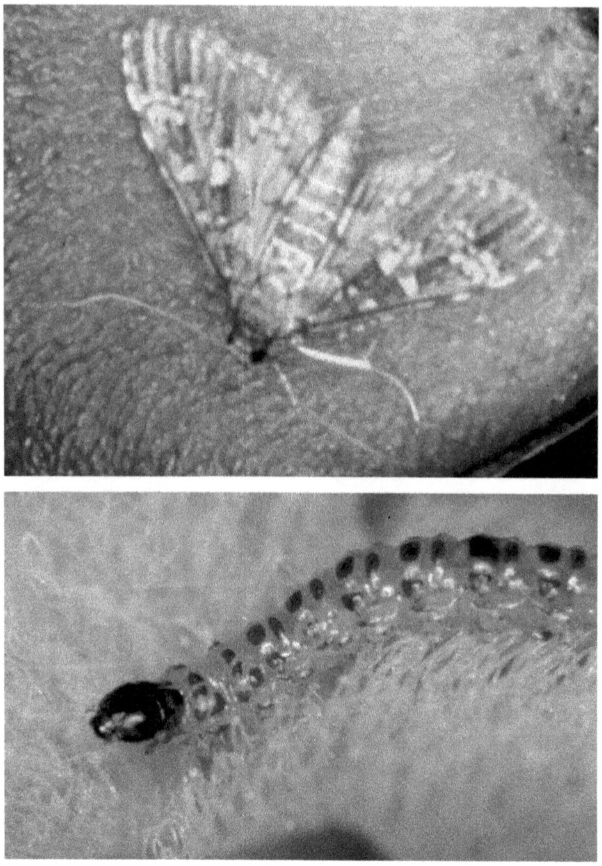

Figure 6: *Samea multiplicalis* adult and larva. Photos: P Room, CSIRO.

Agents studied but not released in Australia

***Paulina acuminata* (De Geer) (Orthoptera: Pauliniidae).** The adults of this semi-aquatic, green and brown grasshopper are 3–4 cm long and there are five or six nymphal stages. Population survival is dependent on high humidity. The biology and development were described by Sands and Kassulke (1986), Vieira and Adis (2000) and Thomas (1980). This grasshopper preferred to eat the leaves of the tertiary form of salvinia. Egg sacs were attached to leaves and life-cycle development took 70–90 days. It completed development on a range of plants in the families Salviniaceae, Araceae and Hydrocharitaceae (Bennett 1966; Sands and Kassulke 1986; Vieira and Adis 2000).

A population of this grasshopper was imported from Rio Guarguacu, Praia de Leste, Parana state, Brazil into quarantine at Brisbane in March 1981 where its biology and host-specificity were studied (Sands and Kassulke 1986). The decision to release it was put off until the impacts of *C. salviniae* and *S. multiplicalis* were known. Following the success of the weevil in controlling salvinia, no further consideration has been given to the release of *P. acuminata* in Australia.

DISCUSSION

The success of the salvinia weevil *C. salviniae* is a contemporary classic for biological control of weeds because of the extraordinarily rapid (one to five years) and high level (up to 95% reduction in cover or biomass) control of salvinia. It is a good example of the importance of knowing the native range of the weed. In this case new associations (biological control agents selected from other *Salvinia* species) were not useful, whereas a useful agent was found once the native range had been defined. It is also an example of the importance of understanding the biology of each biological control agent. The useful agent, assumed to be *C. singularis*, was found to be a new species and named *C. salviniae*; it had very important feeding and behaviour differences that led to its controlling salvinia. Both of these taxonomic issues were solved by sound applied research as part of exploration in the case of the native range of the weed (Forno and Harley 1979), and as part of post-release studies in relation to insect taxonomy and behaviour (Sands and Schotz 1985; Sands *et al.* 1986). The salvinia weevil system as a model system to study contributed to improved understanding of plant–insect–environment interactions (Julien *et al.* 1987; Room 1990; Room and Thomas 1985; Room *et al.* 1989).

Although the salvinia weevil had been successful in controlling salvinia in cooler climates in Australia in the 1980s, the technology was not taken up there (Julien *et al.* 2009). Salvinia infestations in temperate Australia at that time were small, scattered and largely limited to small impoundments. It was easier to apply herbicides than to manage weevils and wait several years for control. It was not until a significant infestation occurred on the Hawkesbury River, north-west of Sydney, that biological control was successfully implemented (Sullivan *et al.* 2011), as part of a long-term management strategy to maintain salvinia at low levels after huge amounts of biomass had been mechanically removed from the river (van Oosterhout *et al.* 2006). Concurrent studies showed that oviposition by the salvinia weevil can adapt to cooler temperatures (Hennecke and Postle 2006).

The salvinia weevil could not develop large populations on multi-layered mats of salvinia (Storrs and Julien 1996). The reasons are likely to relate to the effects of low host plant quality on weevil population development (Room *et al.* 1984; Room and Thomas 1985). The adult weevils ate buds while larvae tunnelled in the stem; the amount of nitrogen available in these food sources regulated weevil population growth rates (Sands *et al.* 1986). Multi-layered salvinia tended to have low nitrogen content, therefore low growth rates and few buds, and could support only low weevil numbers. Populations of the weevil increased most rapidly when numerous high-quality buds were available. This occurred when salvinia grew rapidly in uncrowded situations and movement of assimilates concentrated nitrogen in the tips of the plants for bud development. Any activity, such as strip application of herbicide, physical control or downstream flushing, that removed part of the multi-layered mat stimulated rapid growth of salvinia and therefore population increase of *C. salviniae*, that led to control. Integrated strategies using herbicide and biological control have been used successfully in temperate regions (Sullivan *et al.* 2011; G Popple, *pers. comm.* 2009) and high-altitude tropical areas (Forno 1987). They have also been suggested for Magela Creek, Kakadu National Park in the NT, a tropical monsoon area where local conditions can lead to very high biomass build-up on billabongs (Storrs and Julien 1996).

When salvinia grew as an understorey plant the salvinia weevil maintained populations but failed to control the weed (Storrs and Julien 1996). Emergent plants on the fringes of water bodies or plants that used salvinia as a substrate may have helped to contain salvinia and so maintain weevil populations when all other salvinia was washed away. Those weevils could move to and control new salvinia growth on adjacent open water. Alternatively, non-biological techniques could be used to remove the overstorey plants to assist in the management of salvinia.

Following seasonal flushing of salvinia from water bodies, recolonisation may occur due to vector dispersal of propagules. The rapid growth of salvinia and the relatively low dispersal rate of the weevil between catchments has meant that mats of salvinia have developed quickly and covered water surfaces. To counteract this, local and state government agencies have established rearing facilities to provide *C. salviniae* for use in their jurisdictions and in others on request (van Oosterhout *et al.* 2006). This has added considerable value to the biological control of salvinia while reducing the use of chemical and physical controls, methods that do not provide long-term solutions.

A study in Zimbabwe found that physical and chemical control of salvinia was less effective and more expensive than biological control, which provided a benefit:cost ratio of 10.6:1 (Chikwenhere and Keswani 1997). The *S. molesta* project in Australia cost $A4.2 million over 11 years, beginning in 1978. An assessment of the combined cost of biological control of the three water weeds *S. molesta*, *E. crassipes* and *P. stratiotes* was $A5.1 million (1974–93), while the estimated combined benefit:cost ratio was 27.5:1 (Page and Lacey 2006). The value, in environmental restoration and protection, social and economic benefits, of transferring the salvinia weevil to many other countries continues to be enormous. In Sri Lanka the benefit:cost ratios for successful biological control of salvinia in the early 1980s were estimated to be 53:1 in financial terms and 1671:1 in labour terms (Doeleman 1989).

ACKNOWLEDGEMENTS

The salvinia research team, led by Peter Room, included Wendy Forno, Don Sands, Martin Taylor and the author. Thanks to the many technical staff, students and research collaborators who contributed.

REFERENCES

Bennett FD (1966) Investigations on the insects attacking the aquatic ferns, *Salvinia* spp. in Trinidad and northern South America. In *Proceedings of the Southern Weed Conference* **19**, 497–504.

Bennett FD (1975) Insects and plant pathogens for the control of *Salvinia* and *Pistia*. In *Proceedings of the Symposium on Water Quality Management through Biological Control.* January 1975, University of Florida, Gainesville. (Eds PL Brezonik and JL Fox) pp. 28–35.

Calder AA and Sands DPA (1985) A new Brazilian *Cyrtobagous* Hustache (Coleoptera: Curculionidae) introduced into Australia to control salvinia. *Journal of the Australian Entomological Society* **24**, 57–64.

Chikwenhere GP and Keswani CL (1997) Economics of biological control of Kariba weed (*Salvinia molesta* Mitchell) at Tengwe in north-western Zimbabwe – a case study. *International Journal of Pest Management* **43**, 109–112.

DeLoach CJ, DeLoach DJ and Cordo HA (1979) Observations on the biology of the moth, *Samea multiplicalis*, on waterlettuce in Argentina. *Journal of Aquatic Plant Management* **17**, 42–44.

Doeleman JA (1989) 'Biological control of *Salvinia molesta* in Sri Lanka: an assessment of costs and benefits.' Australian Centre for International Agriculture Research. Economic Assessment Series 1. Canberra.

Forno IW (1981) Progress in the exploration for biological control agents for *Salvinia molesta*. In *Proceedings of the 5th International Symposium on Biological Control of Weeds*. 22–29 July 1980, Brisbane. (Ed ES Delfosse) pp. 167–173. CSIRO, Canberra.

Forno IW (1987) Biological control of the floating fern *Salvinia molesta* in north-eastern Australia: plant-herbivore interactions. *Bulletin of Entomological Research* **77**, 9–17.

Forno IW and Bourne AS (1984) Studies in South America of arthropods on the *Salvinia auriculata* complex of floating ferns and their effects on *S. molesta*. *Bulletin of Entomological Research* **74**, 609–621.

Forno IW and Harley KLS (1979) The occurrence of *Salvinia molesta* in Brazil. *Aquatic Botany* **6**, 185–187.

Forno IW, Sands DPA and Sexton W (1983) Distribution, biology and host specificity of *Cyrtobagous singularis* Hustache (Coleoptera: Curculionidae) for the biological control of *Salvinia molesta*. *Bulletin of Entomological Research* **73**, 85–95.

Gewertz DB (1983) *Salvinia molesta*: the destruction of an ecosystem. In *Sepik River Societies: A Historical Ethnography of the Chambri and their Neighbours*. (Ed DB Gewertz) pp. 196–217. Yale University Press, New Haven.

Hennecke B and Postle L (2006) The key to success: an investigation into oviposition of the salvinia weevil in cool climate regions. In *Proceedings of the 15th Australian Weeds Conference*. 24–28 September 2006, Adelaide. (Eds C Preston, JH Watts and ND Crossman) pp. 780–783. Weed Management Society of South Australia, Adelaide.

Julien MH and Bourne AS (1988) Effects of leaf-feeding by the larvae of the moth *Samea multiplicalis* Guen. (Lep., Pyralidae) on the floating weed *Salvinia molesta*. *Journal of Applied Entomology* **106**, 518–526.

Julien MH and Griffiths MW (1998) *Biological Control: A World Catalogue of Agents and their Target Weeds*. 4th edn. CABI, Wallingford, UK.

Julien MH, Chan RR and Bourne AS (1987) Effects of adult and larval *Cyrtobagous salviniae* on the floating weed *Salvinia molesta*. *Journal of Applied Ecology* **24**, 935–944.

Julien MH, Hill MP and Tipping PW (2009) *Salvinia molesta* D. S. Mitchell Salviniaceae). In *Biological Control of Tropical Weeds using Arthropods*. (Eds R Muniappan, GV Reddy and A Raman) pp. 378–407. Cambridge University Press, Cambridge.

Knopf KW and Habeck DH (1976) Life history and biology of *Samea multiplicalis*. *Environmental Entomology* **5**, 539–542.

Marshall BE and Junor FJR (1981) The decline of *Salvinia molesta* on Lake Kariba. *Hydrobiologia* **83**, 477–484.

May BM and Sands DPA (1986) Descriptions of larvae and biology of *Cyrtobagous* (Coleoptera: Curculionidae): agents for biological control of salvinia. *Proceedings of the Entomology Society of Washington* **88**, 303–312.

Mitchell DS (1972) The Kariba weed: *Salvinia molesta*. *British Fern Gazette* **10**, 251–252.

Mitchell DS (1978) *Aquatic Weeds in Australian Waters*. Australian Government Publishing Service, Canberra.

Mitchell DS and Rose DJW (1979) Factors affecting fluctuations in extent of *Salvinia molesta* on Lake Kariba. *PANS* **25**, 171–177.

Mitchell DS and Tur NM (1975) The rate of growth of *Salvinia molesta* (*S. auriculata* Auct.) in laboratory and natural conditions. *Journal of Applied Ecology* **12**, 213–225.

Page AR and Lacey KL (2006) 'Economic impact assessment of Australian weed biological control'. CRC for

Australian Weed Management. Technical Series No. 10, pp. 123–127.

Room PM (1986) Equations relating growth and uptake of nitrogen by *Salvinia molesta* to temperature and the availability of nitrogen. *Aquatic Botany* **24**, 43–59.

Room PM (1990) Ecology of a simple plant–herbivore system: biological control of *Salvinia*. *Trends in Ecology & Evolution* **5**, 74–79.

Room PM and Julien MH (1995) *Salvinia molesta* D.S. Mitchell. In *Biology of Australian Weeds. Vol. 1.* (Eds RH Groves, RCH Shepherd and RG Richardson) pp. 217–230. RG Richardson and JF Richardson, Melbourne.

Room PM and Thomas PA (1985) Nitrogen and establishment of a beetle for biological control of the floating weed salvinia in Papua New Guinea. *Journal of Applied Ecology* **22**, 139–156.

Room PM, Harley KLS, Forno IW and Sands DPA (1981) Successful biological control of the floating weed salvinia. *Nature* **294**, 78–80.

Room PM, Forno IW and Taylor MFJ (1984) Establishment in Australia of insects for biological control of the floating weed *Salvinia molesta* Mitchell. *Bulletin of Entomological Research* **74**, 505–516.

Room PM, Julien MH and Forno IW (1989) Vigorous plants suffer most from herbivores: latitude, nitrogen and biological control of the weed *Salvinia molesta*. *OIKOS* **54**, 92–100.

Sands DPA and Kassulke RC (1984) *Samea multiplicalis [Lep. Pyralidae]*, for biological control of two water weeds, *Salvinia molesta* and *Pistia stratiotes* in Australia. *Entomophaga* **29**, 267–273.

Sands DPA and Kassulke RC (1986) Assessment of *Paulinia acuminata [Orthoptera: Acrididae]* for the biological control of *Salvinia molesta* in Australia. *Entomophaga* **31**, 11–17.

Sands DPA and Schotz M (1985) Control or no control: a comparison of the feeding strategies of two salvinia weevils. In *Proceedings of the 6th International Symposium on Biological Control of Weeds.* 19–25 August 1984, Vancouver. (Ed ES Delfosse) pp. 551–556. Agriculture Canada, Vancouver.

Sands DPA, Schotz M and Bourne AS (1983) The feeding characteristics and development of larvae of a salvinia weevil *Cyrtobagous* sp. *Entomologia Experimentalis et Applicata* **34**, 291–296.

Sands DPA, Schotz M and Bourne AS (1986) A comparative study of the intrinsic rates of increase of *Cyrtobagous singularis* and *C. salviniae* on the water weed *Salvinia molesta*. *Entomologia Experimentalis et Applicata* **42**, 231–237.

Schotz M and Sands DPA (1988) Diel pattern of feeding and oviposition by *Cyrtobagous salviniae* Calder and Sands (Coleoptera: Curculionidae). *Australian Entomological Magazine* **15**, 31–32.

Semple JL and Forno IW (1987) Native parasitoids and pathogens attacking *Samea multiplicalis* Guenée (Lepidoptera: Pyralidae) in Queensland. *Journal of the Australian Entomological Society* **26**, 365–366.

Storrs MJ and Julien MH (1996) '*Salvinia*. A handbook for the integrated control of *Salvinia molesta* in Kakadu National Park'. Northern Landscapes Occasional Papers No. 1. Australian Nature Conservation Agency, Darwin.

Sullivan PR, Postle LA and Julien M (2011) Biological control of *Salvinia molesta* by *Cyrtobagous salviniae* in temperate Australia. *Biological Control* **57**, 222–228.

Taylor MFJ (1984) The dependence of development and fecundity of *Samea multiplicalis* on early larval nitrogen intake. *Journal of Insect Physiology* **30**, 779–785.

Taylor MFJ and Forno IW (1987) Oviposition preferences of the salvinia moth *Samea multiplicalis* Guenée (Lep., Pyralidae) in relation to host plant quality and damage. *Zeitschrift für angewandte Entomologie* **104**, 73–78.

Taylor MFJ and Sands DPA (1986) Effects of ageing and nutrition on the reproductive system of *Samea multiplicalis* Guenée (Lepidopteran: Pyralidae). *Bulletin of Entomological Research* **76**, 513–517.

Thomas PA (1980) Life-cycle studies on *Paulinia acuminata* (DeGeer) (Orthoptera: Pauliniidae) with particular reference to the effects of constant temperature. *Bulletin of Entomological Research* **70**, 381–389.

van Oosterhout E, Coventry R and Julien M (2006) 'Salvinia control manual. Management and control options for salvinia (*Salvinia molesta*) in Australia.' NSW Dept of Primary Industries, Sydney.

Vieira MF and Adis J (2000) Aspectos da biologia e etologia de *Paulinia acuminata* (De Geer), 1773 (Orthoptera: Pauliniidae), um gafanhoto semi-aquático, na Amazônia Central. *Acta Amazonica* **30**, 333–346.

Waterhouse DF (1994) *Biological Control of Weeds: Southeast Asian Prospects.* Monograph 26. ACIAR, Canberra.

Senecio madagascariensis Poir. – fireweed

Rachel Cruttwell McFadyen and Louise Morin

ABSTRACT

Senecio madagascariensis, fireweed, is one of the worst weeds of coastal pastures of south-eastern Qld and NSW. Originating in south-eastern Africa, it was accidentally introduced to the Hunter Valley in Australia prior to 1918 and has since spread to other coastal areas of NSW and southern Qld and is continuing to spread. It is also invasive in Japan, Hawaii, Argentina and Uruguay. A biocontrol program started in 1987, but only two insects were tested and neither was released. Investigations of a rust pathogen from South Africa also failed to result in any releases. Biocontrol continues to be promoted as a solution for fireweed, but the challenge remains of finding agents that would not attack the many Australian native *Senecio* spp. closely related to fireweed.

Key words: pasture weeds, native species, South Africa, *Puccinia*, *Lobesia*, *Phycitodes*.

INTRODUCTION

Fireweed, *Senecio madagascariensis* Poir. (Asteraceae), is a short-lived perennial herb, with many stems reaching 70 cm tall from a single branched tap root up to 20 cm deep. It produces numerous bright yellow flower heads which make it conspicuous when in flower. Cypselas (fruits that look like seeds, each containing a single seed, often called achenes) are 1–3 mm long with a pappus and are spread by wind, animals and machinery. Seeds germinate readily after rain from autumn to spring and seedlings can establish even in well-managed pastures. Fireweed is a serious invasive weed in temperate and subtropical pastures along the east coast of Australia. Containing pyrrolizidine alkaloids, it is poisonous to livestock, particularly cattle. However, once cattle are familiar with the weed, they tend to avoid it, which allows it to compete more vigorously with pasture species and reduce their productivity (Fig. 1). Poisoning is more likely to occur where other feed is limited, when plants are young and not easily differentiated from the rest of the pasture, when contaminated hay is consumed or when stock are newly introduced to the weed (Sindel *et al.* 1998). Herbicides are available that effectively kill fireweed, but year-long management is difficult and uneconomic because of the weed's ability to germinate and flower throughout much of the year (Sindel *et al.* 1998).

In Australia, fireweed was introduced to the Hunter Valley some time before 1918, probably as a contaminant of hay used in shipping horses. However, for many years it was thought to be a form of the native *Senecio lautus* G.Forst. ex Willd. (now *Senecio pinnatifolius* A.Rich.) complex and was not correctly identified until 1980 (Michael 1981). By the 1980s the weed had spread north and south in coastal NSW and southern Qld in climatic regions similar to those where it originated in southern Africa (Sindel and Michael 1992). It is continuing to spread in NSW, including the Bega area where it previously occurred only in isolated patches, the Monaro region in the Southern Tablelands and the Coffs

Figure 1: Fireweed *Senecio madagascariensis* infestation in pasture at Bega, NSW, September 2006. Photo: R McFadyen, Weeds CRC.

Harbour and Dorrigo areas on the Northern Tablelands. In 2007 the weed was also found growing on the Atherton Tablelands in northern Qld, in line with predictions by Sindel and Michael (1992). It is therefore clear that it has not yet reached its potential distribution in Australia. Climate change is likely to affect the continued spread of fireweed and its potential distribution: movement into cooler highland areas and further south may increase, while continued spread north into the tropics is likely to slow. The current distribution based on specimens in Australia's Virtual Herbarium is shown in Figure 2.

BIOLOGICAL CONTROL HISTORY

Prior to 1988, biocontrol programs against *S. madagascariensis* had been suggested for both Argentina and Hawaii, but no work undertaken. A program began in Hawaii in the late 1990s with the importation and testing of potential insect agents and a rust fungus (Culliney *et al.* 2003). Introduction of the arctiid moth *Secusio extensa* (Butler) has been proposed, but no releases have yet been made (Hawaii Dept of Agriculture 2008).

As a result of two short surveys in 1987 and 1988 of the insects attacking *S. madagascariensis* in southern Madagascar, a five-year project to import into quarantine, test and release suitable insects started in July 1990.

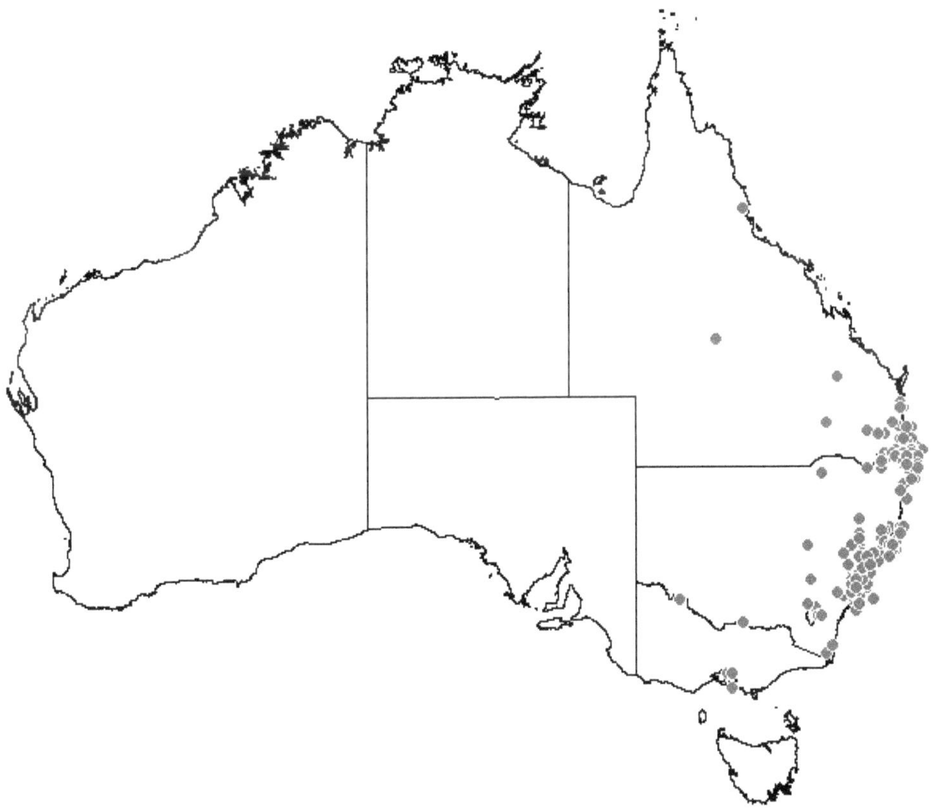

Figure 2: *Senecio madagascariensis* distribution in Australia. Specimen data reproduced from Australia's Virtual Herbarium with permission of the Council of Heads of Australasian Herbaria Inc, June 2010.

The taxonomy of *S. madagascariensis* continued to be questioned and it became clear that, despite its name, *S. madagascariensis* was in fact native to the KwaZulu-Natal region of South Africa. A further survey in South Africa was therefore undertaken in 1991.

Because of the lack of specificity of the agents tested, and the fact that Australia has a rich flora of native *Senecio* species, including species very close to *S. madagascariensis*, the prospects for successful biocontrol were thought to be small and the program was discontinued in 1995. However, as fireweed continues to spread in the high-value coastal pastures of NSW and southern Qld, there has been continued stakeholder and political pressure for further biocontrol research against fireweed. Further surveys in South Africa were therefore undertaken in 2002 and a rust pathogen was investigated, but was found to have limited potential. Additional South African surveys, particularly focusing on invertebrates, are planned for the future.

PLANT TAXONOMY

S. madagascariensis was first described by Poiret in 1817 from specimens collected in Madagascar. In Australia it was initially confused with a similar native *Senecio* species (then called *S. lautus*), and was first recognised as a southern African plant and correctly named as *S. madagascariensis* in 1980 (Michael 1981). Subsequent investigations have shown that Australian *S. madagascariensis* is most similar to plants from the KwaZulu-Natal province in South Africa, and that it is from this region that our fireweed originated (Radford *et al.* 2000; Lafuma *et al.* 2003).

Continuing confusion over the identification of this group led to detailed taxonomic work on the *S. lautus* complex, as a result of which many Australian native plants previously confused with *S. madagascariensis* have now been re-assigned to *S. pinnatifolius* and other *Senecio* spp. (Thompson 2005). The native species *Senecio brigalowensis* I.Thomps., which is common and has become increasingly abundant and weedy in central Qld from Roma to Rockhampton, will also cause cattle poisoning (Noble *et al.* 1994) and is easily confused with *S. madagascariensis*. Unfortunately there are still difficulties in the identification of some Australian native *Senecio* species, with possible hybridisation occurring (Prentis *et al.* 2007). There are also difficulties in the identification of *S. madagascariensis* and related species in South Africa such as *Senecio inaequidens* DC., which is a major weed in Europe (Lafuma *et al.* 2003). A recent molecular phylogeny of the tribe Senecioneae has resulted in a new delimitation of the genus *Senecio* (Pelser *et al.* 2007), but additional molecular studies targeting the Australian native species are required to clarify interspecific relationships.

EXPLORATION

Fireweed was originally thought to originate from Madagascar and initial surveys were therefore concentrated in south-eastern Madagascar, where *S. madagascariensis* is known to occur and the climate is most similar to that of affected areas of Australia. Initial surveys were undertaken in January 1987 and June 1988 by Jennifer Marohasy (then Turnour) who was based in south-western Madagascar studying potential biocontrol agents against rubbervine *Cryptostegia grandiflora* R.Br. (see chapter, this volume). Both surveys were restricted to the Fort Dauphin area in the extreme south of Madagascar, where *S. madagascariensis* is present in disturbed land along tracks and in clearings (Marohasy 1989). In August 1991 she made a three-week survey in South Africa, searching on related *Senecio* species in KwaZulu-Natal and the Drakensberg regions (Marohasy 1991) (Table 1). She discovered several new insects feeding on *S. madagascariensis* and recommended that further surveys be undertaken in the summer season. This was not followed up, however, due to lack of financial support. She also recommended further study and host-specificity testing on the stem- and flower-feeder *Platyptilia* sp. nr. *molopias* Meyrick (Lep.: Pterophoridae), the flower-feeder *Homoeosoma stenotea* Hampson (Lep.: Pyralidae) and the stem-borer *Melanagromyza* sp. (Dipt.: Agromyzidae).

In 1996/97, *Senecio* specimens were collected in KwaZulu-Natal for a taxonomic study (Radford *et al.* 2000) and a rust fungus was discovered attacking the leaves (P. Müller, *pers. comm.* 1997). A second field trip was made in September 2002 specifically to collect this rust for study in quarantine in Australia (Morin 2003; Morin *et al.* 2009). However, no general survey of the pathogens attacking *S. madagascariensis* in KwaZulu-Natal has been undertaken.

CANDIDATES

Pathogens

While an extensive range of pathogens, from rusts and mildews to facultative parasitic fungi, have been recorded on *Senecio* species in South Africa (Farr and Rossman 2010), no fungi had been recorded on *S. madagascariensis* prior to the 2002 survey in KwaZulu-Natal. In that survey, disease symptoms caused by what was thought to be an *Aecidium* sp. rust fungus and the cosmopolitan biotrophic fungus *Albugo tragopogonis* (Pers.) Gray were found on *S. madagascariensis* (L Morin, unpub. data 2002). Since *A. tragopogonis* is known for its wide host-range within the Asteraceae and already occurs on fireweed in Australia (Plant Health Australia 2001), only the rust fungus was imported into quarantine in Canberra for detailed studies (Morin 2003). This rust fungus was initially referred to as an *Aecidium* sp. because only the aecial stage was present on specimens collected. DNA sequencing and RFLP analyses revealed that one of the imported rust isolates was similar to Australian isolates of *Puccinia lagenophorae* Cooke (Morin *et al.* 2009), a fungus known to attack fireweed and other *Senecio* species in Australia and several other countries (Holtkamp and Hosking 1993; Farr and Rossman 2010). In contrast, all other South African rust isolates investigated were found to be putative hybrids between *P. lagenophorae* and another unknown rust species. In pathogenicity tests, none of the three South African isolates tested (including the non-hybrid isolate) produced more damage on Australian fireweed accessions than did Australian *P. lagenophorae* isolates (Morin 2003). In concurrent tests, the native *S. pinnatifolius* did not develop any disease symptoms following inoculation with two of the isolates from South Africa and two from Australia. While these initial results suggest that the South African rust isolates may not pose a threat to Australian native *Senecio* spp., it is unlikely that their introduction into Australia would lead to improved control of fireweed.

Insects

In her surveys in Madagascar and South Africa, J Marohasy found several potentially useful insects, nine species from Madagascar and 18 from South Africa (Marohasy 1989, 1991). The two most easily collected species from Madagascar, the flower-feeding pyralid *Phycitodes* new sp. and the stem-feeding tortricid moth *Lobesia* new sp., were imported into quarantine in Brisbane in July 1990 for detailed host-testing. There were considerable logistical problems, as J Marohasy was based in Kenya at the time and went to Madagascar to make the collections, where she had to use the facilities of a Canadian mining firm operating in Fort Dauphin in southern Madagascar. Shipments were sent via Mauritius and Singapore and took several days but about 70% of the larvae survived, sufficient to establish quarantine colonies. Repeat collections in 1991 were not possible due to rioting in the capital and major towns, but in July 1992 Marohasy was able to collect and send a shipment of several hundred larvae.

In 1992 permits were obtained to import two of the recommended species from South Africa, the flower-feeding pyralid moth *Homoeosoma stenotea* and the stem-boring agromyzid fly *Melanagromyza* sp. However, problems in finding collaborators to collect the insects meant that no shipments were received before the funding finished in 1995 (McFadyen 1995) and no further investigations have been made on these species. In addition, both species have close relatives in Australia that attack fireweed and it was potentially unlikely that the introduction of similar African species would improve overall biocontrol. *H. stenotea* is closely related to the blue stem-borer moth *Patagoniodes (Homoeosoma) farinaria* (Turner), native to Australia and New Zealand where it bores stems and roots of a small group of native and introduced *Senecio* species (McQuillan and Ireson 1987). *Melanagromyza* is a large genus of mainly host-specific species, with the native Australian species *M. seneciophila* Spencer known to attack *S. madagascariensis* as well as other *Senecio* spp. (Holtkamp and Hosking 1993).

***Phycitodes* new sp. (Lepidoptera: Pyralidae).** Adults of this flower-feeding moth were reared from larvae feeding in seed heads of *S. madagascariensis* in southern Madagascar and it was identified by the Commonwealth Institute of Entomology (CIE) as a new species (Marohasy 1989). There is little published information on the biology of other species in the genus; *Phycitodes binaevella* (Hübner) has been recorded from the flower heads of *Cirsium vulgare* (Savi) Ten. in Holland and larvae of an undetermined species of *Phycitodes* were recorded from flower heads of *Ptilostemon (Cirsium) gnaphaloides* (Cirillo) Soják in Crete, both Asteraceae (McFadyen 1995).

Larvae from Madagascar were readily reared on fireweed flowers in quarantine. The larvae fed primarily on the developing achenes in the flower heads, tying the damaged flower heads together with a few strands of silk and moving to adjacent flower heads as they were consumed. Once all flower heads were consumed, larvae moved down to bore into leaf axils and fed in the stem, killing the stem portion above their feeding site. Feeding by several larvae killed plants. Full-grown larvae left the plant to pupate in a silk cocoon in leaf litter, dry leaves or attached to dry bark. Adults, 10–12 mm long, with narrow grey-brown wings held rolled over the back, emerged in a few days and lived for 10–18 days. They were nocturnal, and fed on nectar. Females laid only on flower heads

Table 1: Consolidated list of insects collected on *S. madagascariensis* in Madagascar and South Africa

Insect species		Country	Plant part attacked	Other information
Lepidoptera	Noctuidae:			
	Thysanoplusia orichalcea (Fab.)	Madagascar	Leaves	Known polyphagous pest species
	Condica conducta Walker	Madagascar	Leaves	Known generalist on Asteraceae
	Geometridae:			
	Indet. genus and sp.	Madagascar	Leaves	No other information
	Pyralidae:			
	Homoeosoma stenotea Hampson	South Africa	Flowers	Damaging – potential agent
	Phycitodes new sp.	Madagascar	Flowers and stems	Imported and tested in quarantine
	Crambidae:			
	Udea ferrugalis (Hübner)	South Africa	Leaves	
	Tortricidae:			
	Indet. genus and sp.	South Africa	Stem-boring in pith	Not very damaging
	Epichorestodes acerbella Walker	South Africa	Leaves	
	Lobesia new sp.	Madagascar	Stems	Imported and tested in quarantine
	Platyptilia ?molopias Meyrick	South Africa	Stems and flowers	Damaging – potential agent
Diptera	Tephritidae:			
	Sphenella marginata (Fallen)	Madagascar	Flowers	Cosmopolitan species present in Australia
	Cryptophorellia peringueyi (Bezzi)	South Africa	Flowers	Also from other *Senecio* spp.
	Agromyzidae:			
	Liriomyza trifolii (Burgess)	Madagascar	Leaves	Known polyphagous pest species
	Melanagromyza sp.	South Africa	Stem-boring	Damaging but similar to *M. seneciophila* in Australia
	Cecidomyiidae:			
	Indet. genus and sp.	South Africa	Flowers	Host-range unknown
	Sciaridae:			
	Indet. genus and sp.	South Africa	Boring in pith of stem and root	Host-range unknown
Homoptera	Pseudococcidae:			
	Tylococcus ?harongae Mamet	Madagascar	Phloem-feeding	Host-range unknown
	Coccidae:			
	Pulvinaria sp.	South Africa	Phloem-feeding	Host-range unknown

Insect species		Country	Plant part attacked	Other information
	Cicadellidae: Indet. genus and sp.	South Africa	Phloem-feeding	Host-range unknown
	Tettigometridae: *Hilda* sp.	South Africa	Phloem-feeding	Host-range unknown
	Aphididae: *Aphis fabae* subsp. *solanella* Theobald	South Africa	Phloem-feeding	Known polyphagous pest species
	Aphis gossypii Glover	Madagascar	Phloem-feeding	Known polyphagous pest species
	Brachycaudus helichrysi (Kaltenbach)	South Africa	Phloem-feeding	Known polyphagous pest species
	Macrosiphum euphorbiae (Thomas)	South Africa	Phloem-feeding	Known polyphagous pest species
	Myzus ornatus Laing	South Africa	Phloem-feeding	Known polyphagous pest species
Heteroptera	Lygaeidae: *Nysius albipennis* Distant	Madagascar	Seed heads	Host-range unknown; other *Nysius* spp. attacking *Senecio* spp. in Australia
	Nysius ericae (Schilling)	Madagascar and South Africa	Seed heads	Host-range unknown
	Nysius ?graminicola (Kolenati)	South Africa	Seed heads	Host-range unknown
	Nysius ?senecionis Schilling	South Africa	Seed heads	Host-range unknown
	Rhopalidae: *Stictopleurus scutellaris coquerelli* (Signoret)	Madagascar	Phloem-feeding	Host-range unknown
	Miridae: *Ellenia obscuricornis* (Poppius)	South Africa	Flowers	Host-range unknown
Hymenoptera	Eucoilidae: *Diglyphosema* sp.	South Africa	Stem-boring	Host-range unknown; may be parasite of agromyzid
Coleoptera	Curculionidae: *? Throgonius* sp.	Madagascar	Adults at flowers	Larval host unknown
	Gasteroclisus tricostalis (Thunberg)	South Africa	Boring in stem pith	Common but not very damaging
Thysanoptera	Phlaeothripidae: *Haplothrips nigricornis* (Bagnall)	Madagascar	Seed heads	Pest of sunflower?

Source: Marohasy (1989,1991); identifications from CIE.

and the cylindrical eggs were laid singly, carefully inserted into a concealed position between bracts and florets on the flower heads. In summer temperatures (daily range 20–30°C), the life-cycle from egg to adult was completed in 21 days (McFadyen and Sparks 1996).

In quarantine, *Phycitodes* new sp. was extremely destructive to fireweed plants, with 20 adult moths on 12 well-grown fireweed plants resulting in the complete destruction of plants by the time the larvae had reached maturity in two weeks. However, the colony died out suddenly in January 1991, and in January 1993 the new colony died out in the same way. Fireweed flowers less often in summer in Australia, but flowering plants were always provided for adult oviposition and eggs were laid as usual. However, on both occasions the resulting larvae did not survive and the colonies died out. The reason for this failure to survive is not known; no disease could be found and temperatures were not unusually high. It is unlikely that it was due to Australian *S. madagascariensis* being an unsuitable host, as there was not a gradual decline in each quarantine generation, but rather in both years a sudden complete failure of survival of larvae from eggs laid in December and January.

Lobesia **new sp. (Lepidoptera: Tortricidae).** Adults of this stem- and root-boring moth reared from larvae feeding in *S. madagascariensis* in Madagascar were identified by CIE as a new species of *Lobesia*. Other species in the genus have been recorded from plants in several plant families in the Old World. Some are damaging pest species, but all are specific to one or a small group of host plants. For example, *Lobesia euphorbiana* (Freyer) was introduced into Canada and the USA for the biocontrol of its host plant leafy spurge *Euphorbia esula* L. (Julien and Griffiths 1998).

Adults of the *Lobesia* sp. from Madagascar are 4–8 mm long, grey with wings at rest folded across the body. Although largely nocturnal, oviposition in quarantine occurred during the day with a peak at dawn (D Sparks, *pers. comm.* 1995). Adults fed on nectar and lived for eight to 14 days. Flattened, circular, translucent white eggs were laid singly on leaves, petioles or stem tips. Newly hatched larvae wandered over the leaf to the axil, where they bored into and fed down the stems as far as the crown or even roots. Frass was ejected through holes in the stems and pupation occurred in the stem. Feeding destroyed the stems above the feeding site. The complete life-cycle took about 28 days at temperatures of 20°C (night) to 30°C (day) (McFadyen and Sparks 1996).

HOST TESTING

Plants tested

The list of plants to be tested, approved in 1989, was largely restricted to the family Asteraceae with only five species from other families (McFadyen and Sparks 1996). Asteraceae is a natural family with well-established limits, and the many biocontrol programs against Asteraceae weeds have demonstrated that suitable biocontrol agents, i.e. insects specific to a small group of species within the family, do not damage plants outside the family. The list (Table 2) included all 18 native *Senecio* species then recorded from coastal eastern NSW and south-eastern Qld, including the *S. lautus* (now *S. pinnatifolius*) complex, plus four ornamental *Senecio* species. Because of confusion in the taxonomy of *Senecio* in eastern Australia prior to the revision by Thompson (2006), it is not possible to assign many of the plants used in the tests in 1989 to currently recognised species. The test list also included species (native and introduced ornamentals) from other genera in the tribe Senecioneae present in Australia. Native species and/or species of economic importance from other tribes within the Asteraceae, which occur in Australia, were included. A representative species from other families of the order Asterales was included, selecting species with a similar habit to fireweed. In addition, for testing of the moth *Lobesia* new sp., the known host plants of other species of *Lobesia* were added.

Methods used

For both species, the host-selection stage is the adult female, as larvae do not normally move from plant to plant. Therefore all tests used adult moths, usually in standard cages 50 × 80 × 100 cm with metal bases and gauze sides and top. Test plants were grown in pots. For tests, the cage base rested on the pot rims with the upper parts of the plants entering the cage through narrow holes in the cage base. Moths thus had free access to the plants but not to soil. Honey and water in cotton wicks were supplied, and moths of both species were seen to feed on these.

In initial tests, two test plants and four fireweed plants were used per test. In later tests, two or more test plants were used per cage, with separate control cages containing fireweed plants. Ten female and 10 male moths were released into each cage. After two weeks (by which time all moths were dead) the plants were removed, examined then held in separate cages for larval development. Damage from larval feeding was recorded on a four-point assessment scale and any adult emergence recorded.

Table 2: Plants tested with *Lobesia* sp. and *Phycitodes* sp.

Family Asteraceae, Tribe Senecioneae	***Senecio:* 21 native species or subspecies, 5 also tested with *Phycitodes* (determination of actual species is not possible). *S. cineraria* DC., *S. confusus* Britten, *S. macroglossus* DC., *S. tamoides* DC.**
	Bedfordia arborescens Hochr.[a]
	Brachyglottis greyi (Hook.f.) B.Nord.
	Crassocephalum crepidioides (Benth.) S Moore
	Delairea odorata Lem.
	Emilia sonchifolia (L.) DC[a]
	Erechtites valerianifolius (Wolf) DC.
	Euryops chrysanthemoides (DC.) B.Nord.[b]
	Gynura drymophila (F.Muell.) FG Davies[a,b]
	Ligularia sp.
	Roldana (*Senecio*) *petasitis* (Sims) H.Rob. & Bretell.
Tribe Calenduleae	*Calendula* sp.[b]
Tribe Heliantheae	*Dahlia* sp.[b]
	Flaveria australasica Hook.[a]
	Gaillardia sp.
	Helianthus annuus L. (two commercial varieties)[b]
Tribe Inuleae	*Cassinia laevis* R.Br.[a,b]
	Euchiton sphaericus (Willd.) Holub[a]
	Xerochrysum bracteatum (Vent.) Tzvelev[a]
Tribe Anthemideae	*Chrysanthemum* sp.[b]
	Cotula australis (Sieber ex Spreng.) Hook.f.[a,b]
	Soliva anthemifolia (Juss.) Sweet
	Soliva sessilis Ruiz & Pav.
Tribe Astereae	*Aster subulatus* Michx.[c]
	Callistephus chinensis (L .) Nees[b]
	Olearia nernstii (F.Muell.) Benth.[a,b]
	Vittadinia sulcata N.T.Burb.[a]
Tribe Eupatoriaea	*Adenostemma lavenia* (L.) Kuntze[a,b]
Tribe Vernonieae	*Vernonia cinerea* (L.) Less.
Tribe Lactuceae	*Cichorium intybus* L.
	Lactuca sativa L.[b]
Tribe Mutiseae	*Gerbera* sp.
Tribe Cynareae	*Carthamus tinctorius* L.
Family Campanulaceae	*Pratia purpurascens* (R.Br.) E.Wimm.[a]
	Isotoma anethifolia Summerh.
Family Goodeniaceae	*Brunonia australis* Sm. ex R.Br.[a]
	Scaevola humilis R.Br.[a]
Family Stylidiaceae	*Stylidium graminifolium* Sw.[a,b]
Family Euphorbiaceae	*Euphorbia* sp.
Family Rubiaceae	*Coffea arabica* L.
Family Vitaceae	*Vitis vinifera* L.
Family Pinaceae	*Larix europaea* DC

a: Australia native.
b: Also tested with *Phycitodes*.
c: Tested only with *Phycitodes*.

For *Lobesia* sp., additional tests in a larger cage 3 × 3 × 2.5 m were conducted with those test plants attacked in the smaller cages. Four plants of each test species plus four plants of fireweed were distributed in the cage and 15 male and 15 female moths released into the cage. Subsequent procedures were as above.

Results

***Phycitodes* sp.:** some larval feeding occurred on the ornamentals chrysanthemum (*Chrysanthemum indicum* L.), dahlia (*Dahlia* x hybrid), and *Calendula* sp., in other tribes within the family Asteraceae, and on the native plants *Euryops chrysanthemoides* (DC.) B.Nord., *Callistephus chinensis* (L.) Nees, and *Olearia nernstii* (F.Muell.) in the Asteraceae, and *Stylidium graminifolium* Sw. in the Stylidiaceae. Development through to the adult occurred only on chrysanthemum and dahlia, and numbers were small and development abnormally slow. On plants within the tribe Senecioneae, development to the adult occurred on fireweed and on the native species *S. pinnatifolius* var. *pinnatifolius* and *S. quadridentatus* Labill. On both these species, the number of adults reared was 50% or less of those on fireweed, but this may not be significant due to differences in the size of the plants. There was no attack on the other species tested including the natives *S. pinnatifolius* var. *serratus* I.Thomps., *S. bipinnatisectus* Belcher, *S. minimus* Poir. and *Gynura drymophila* (F.Muell.) F.G.Davies. Only six species of *Senecio* were tested and no large cage tests had been undertaken by January 1993 when the colony died out.

***Lobesia* sp.:** in the small cage tests, 23 *Senecio* species were tested (Table 2) and larval feeding and normal development through to adult occurred on four of these (*S. pinnatifolius* var. *pinnatifolius*, *S. pinnatifolius* var. *serratus*, *S. quadridentatus* and *S. anethifolius* A. Cunn. ex DC.), with a few adults also developing on the ornamental *Senecio confusus* Britten. Thirty-seven other species were tested (Table 2) and comparable numbers of adults were reared from *Calendula* sp., *Dahlia* sp. and *Flaveria australasica* Hook. from other tribes within the Asteraceae. Small numbers of adults were reared from *Crassocephalum crepidioides* (Benth.) S. Moore, *Emilia sonchifolia* (L.) DC., *Erechtites valerianifolius* (Wolf) DC., *Cotula australis* (Sieber ex Spreng.) Hook.f. and *Soliva anthemifolia* (Juss.) Sweet.

Results from the large cage tests were very different; of the five *Senecio* species tested, larval feeding and normal adult development occurred only on *S. pinnatifolius* var. *pinnatifolius*, and feeding, but greatly reduced numbers of adults (10% of controls), on *S. pinnatifolius* var. *serratus*. Of the eight species in other tribes of Asteraceae tested, there was no feeding in *E. sonchifolia*, *Dahlia* sp., *F. australasica*, *C. australis* and *S. anthemifolia*, and feeding, but greatly reduced numbers of adults, from *C. crepidioides* (50% of controls), *E. valerianifolius* (10% of controls) and in one of five replicates of *Calendula* sp. (50% of controls). However, two generations of apparently normal moths were reared on *Calendula* sp. In the light of these results, which implied a potential host-range across two or more tribes within the family Asteraceae, host-testing was terminated and the colony destroyed (McFadyen and Sparks 1996).

DISCUSSION

The Australian biocontrol program against fireweed has been plagued throughout by the taxonomic problems with *Senecio*. The taxonomy of *Senecio* in southern Africa, the native range of *S. madagascariensis*, was confused; Australian fireweed was initially thought to originate from Madagascar, but subsequent molecular studies indicated that its origin was KwaZulu-Natal, South Africa (Radford *et al.* 2000; Lafuma *et al.* 2003). Similarly, there are many native Australian plants in the genus *Senecio*, some very difficult to distinguish from each other and from fireweed. Given the large number of native Australian *Senecio* species that are very similar to fireweed, both chemically and in habit and habitat, it is clear that only a very specific agent will attack fireweed without also causing damage to at least some of these native species. Yet one of the native species, *S. brigalowensis*, is known to cause cattle deaths and economic losses in grazing properties in central Qld (Noble *et al.* 1994) and has greatly increased in abundance due to clearing of native forest for grazing in central Qld (McFadyen, unpub. data 2008). It would seem appropriate to undertake a cost:benefit analysis of the advantages and disadvantages of the release of agents that might attack these native species as well as fireweed; this is possible under the *Biological Control Act 1984*. However, other native Australian *Senecio* species are listed as rare and threatened (Briggs and Leigh 1996), and there is little likelihood of support for the release of any agents that might damage any of these species.

Host-specificity testing of the two moths from Madagascar was never fully completed. With *Lobesia* sp. a decision was made that the host-range was unacceptably wide. The difference between the small- and large-cage test results demonstrated that female moths exhibited a significantly narrower choice in their egg-laying when given more natural conditions. It is therefore likely that their behaviour in the wild would be even more selective (McFadyen *et al.* 2002). Nevertheless, the apparently normal development of the insect on plants in other tribes within the Asteraceae was judged to indicate an unacceptably wide host-range, and it is unlikely that further testing would reverse this decision.

For *Phycitodes* sp. the situation is somewhat different. Even in small cages, oviposition and larval development occurred only on two of six native *Senecio* spp. tested, and it is likely that even fewer species would have been attacked if tested in larger cages (and hence by extrapolation in the wild) (McFadyen *et al.* 2002). Further, the *Phycitodes* sp. tested was collected in Madagascar from a *Senecio* which is now known not to be the same as that found in Australia. If a similar *Phycitodes* sp. were to be found in South Africa it might prove to be specific to the Australian fireweed.

There is already a significant suite of native insects and at least one damaging pathogen *P. lagenophorae* which attack fireweed in the field in Australia (Holtkamp and Hosking 1993). This does not provide satisfactory control of fireweed in most grazing situations. A successful agent would therefore need to attack plants earlier in the season, i.e. during the seedling stage, or achieve much higher densities than the native insects by being, for example, less susceptible to native parasitoids and predators (McQuillan and Ireson 1987).

As fireweed continues to spread in eastern Australia and conventional control measures remain ineffective or uneconomic, there is continued pressure to seek a biocontrol solution. A thorough and detailed investigation of the insects and pathogens attacking *S. madagascariensis* and related *Senecio* species in KwaZulu-Natal, together with their specificity and potential efficacy undertaken by expert(s) resident in the area over several seasons, will be essential to identify any new potential biocontrol agents.

ACKNOWLEDGEMENTS

We thank John Hosking and Royce Holtkamp for their very helpful additions to and comments on earlier drafts of this chapter, and especially John for his help with the taxonomic confusion and for the distribution map sourced from the Australian Virtual Herbarium. Quarantine rearing and host-testing of the two insect species was carried out by Tanya Girle and David Sparks.

REFERENCES

Briggs J and Leigh J (1996) *Rare or Threatened Australian Plants*. CSIRO Publishing, Melbourne.

Culliney TW, Nagamine WT and Teramoto KK (2003) Introductions for biological control in Hawaii 1997–2001. *Proceedings of the Hawaiian Entomological Society* **36**, 145–153.

Farr DF and Rossman AY (2010) Fungal databases, systematic mycology and microbiology laboratory. ARS-USDA. <http://nt.ars-grin.gov/fungaldatabases/ (accessed 11.6.10).

Hawaii Dept of Agriculture (2008) Field release of *Secusio extensa* (Butler) (Lepidoptera: Arctiidae), for biological control of fireweed, *Senecio madagascariensis* Poiret (Asterales: Asteraceae), in Hawaii – draft environmental assessment. <http://www.hear.org/species/senecio_madagascariensis/> (accessed 1.7.10).

Holtkamp RH and Hosking JR (1993) Insects and diseases of fireweed *Senecio madagascariensis*, and the closely related *Senecio lautus* complex. In *Proceedings of the 10th Australian and 14th Asian-Pacific Weed Conference*. 6–10 September 1993, Brisbane. (Eds JT Swarbrick, CWL Henderson, RJ Jettner, L Streit and SR Walker) pp. 104–106. Weed Society of Queensland, Brisbane.

Julien MH and Griffiths MW (1998) *Biological Control of Weeds: A World Catalogue of Agents and their Target Weeds*. 4th edn. CABI, Wallingford, UK.

Lafuma L, Balkwill K, Imbert E, Verlaque R and Maurice S (2003) Ploidy level and origin of the European invasive weed *Senecio inaequidens* (Asteraceae). *Plant Systematics and Evolution* **243**, 59–72.

Marohasy JJ (1989) A survey of fireweed (*Senecio madagascariensis* Poir) and its natural enemies in Madagascar with a view to biological control in Australia. *Plant Protection Quarterly* **4**, 139–140.

Marohasy JJ (1991) 'A survey of the insect fauna of the *Senecio madagascariensis* complex (Compositae) in South Africa'. Unpublished report. Qld Government, Brisbane.

McFadyen RE (1995) 'MRC project QDL 003 "Biological control of fireweed" final report'. Unpublished report, December 1995. Qld Government, Brisbane.

McFadyen RE and Sparks D (1996) Biological control of fireweed. In *Proceedings of the 11th Australian Weeds Conference*. 30 September–3 October 1996, Melbourne. (Ed RCH Shepherd) pp. 305–308. Weed Science Society of Victoria, Melbourne.

McFadyen REC, Vitelli M and Setter C (2002) Host specificity of the rubber vine moth, *Euclasta whalleyi* Popescu-Gorj and Constantinescu (Lepidoptera: Crambidae: Pyraustinae): field host-range compared to that predicted by laboratory tests. *Australian Journal of Entomology* **41**, 321–323.

McQuillan PB and Ireson JE (1987) The identity and immature stages of *Patagoniodes farinaria* (Turner) comb. n. (Lepidoptera: Pyralidae: Phycitinae) in Australia and New Zealand. *Journal of the Australian Entomological Society* **26**, 239–247.

Michael PW (1981) Alien plants. In *Australian Vegetation.* (Ed RH Groves) pp. 44–64. Cambridge University Press, Cambridge.

Morin L (2003) 'Potential for fireweed control'. Unpublished report, May 2003. CSIRO, Canberra.

Morin L, van der Merwe M, Hartley D and Muller P (2009) Putative natural hybrid between *Puccinia lagenophorae* and an unknown rust fungus on *Senecio madagascariensis* in KwaZulu-Natal, South Africa. *Mycological Research* **113**, 725–736.

Noble JW, Crossley J, Hill BD, Pierce RJ, McKenzie RA, Debritz M and Morley AA (1994) Pyrrolizidine alkaloidosis of cattle associated with *Senecio lautus. Australian Veterinary Journal* **71**, 196–200.

Pelser PB, Nordenstam B, Kadereit JW and Watson LE (2007) An ITS phylogeny of tribe Senecioneae (Asteraceae) and a new delimitation of *Senecio* L. *Taxon* **56**, 1077–1104.

Plant Health Australia (2001) Australian plant pest database. Online database. <www.planthealthaustralia.com.au/appd> (accessed 16.6.10).

Prentis PJ, White EM, Radford IJ, Lowe AJ and Clarke AR (2007) Can hybridization cause local extinction: a case for demographic swamping of the Australian native *Senecio pinnatifolius* by the invasive *Senecio madagascariensis. New Phytologist* **176**, 902–912.

Radford IJ, Muller P, Fiffer S and Michael PW (2000) Genetic relationships between Australian fireweed and South African and Madagascan populations of *Senecio madagascariensis* Poir. and closely related *Senecio* species. *Australian Systematic Botany* **13**, 409–423.

Sindel BM and Michael PW (1992) Spread and potential distribution of *Senecio madagascariensis* Poir. (fireweed) in Australia. *Australian Journal of Ecology* **17**, 21–26.

Sindel BM, Radford IJ, Holtkamp RH and Michael PW (1998) *Senecio madagascariensis* Poir. In *The Biology of Australian Weeds. Vol. 2.* (Eds FD Panetta, RH Groves and RCH Shepherd) pp. 247–267. RG and FJ Richardson, Melbourne.

Thompson IR (2005) Taxonomic studies of Australian *Senecio* (Asteraceae): 5. The *S. pinnatifolius /S. lautus* complex. *Muelleria* **21**, 23–76.

Thompson IR (2006) A taxonomic treatment of tribe Senecioneae (Asteraceae) in Australia. *Muelleria* **24**, 51–110.

Senna obtusifolia (L.) Irwin and Barneby – sicklepod

Bill Palmer

ABSTRACT

Sicklepod *Senna obtusifolia* and its congener foetid senna *Senna tora* are serious weeds of field and pasture in many tropical and subtropical regions such as Australia, the USA and the Pacific islands. In Australia, sicklepod infests over 60 000 ha of the tropical eastern coast of Qld from Cape York to Mackay. However eco-climatic modelling suggests a larger potential range.

Classical biological control was considered by the Commonwealth Institute of Biological Control for several island nations while American scientists researched mycoherbicides for use in cropping situations along the Gulf Coast of the USA. Brazilian scientists also investigated possibilities for Brazil. Surveys were then conducted in Mexico and Central America for Qld from 1991 to 1993 and 108 species representing six insect orders and 40 families were found.

Two insects were studied in quarantine in Australia. The psyllid *Mitrapsylla albalineata* was distributed widely in Mexico and infested terminal leaves. However, it utilised several native Australian plant species, bred successfully through three generations on nine *Senna* spp., and was considered too problematic for release in Australia. The second, a weevil *Conotrachelus* sp., entered a reproductive diapause after being reared for five generations. It ceased oviposition and the colony was lost before host testing could be undertaken. Mixed populations of four bruchid seed-feeding beetles *Sennius fallax*, *Sennius lebasi*, *Sennius morosus* and *Sennius rufomaculatus* were commonly found in Mexico. They are not monophagous, but have host-ranges mostly confined to *Senna* spp. Although it is unlikely that they would be approved for release in Australia under present guidelines, they may warrant further study. The hesperiid *Typhedanus undulatus*, the pierid *Eurema salome* and the pyralid *Anabasis ochrodesma*, all from Mexico, may warrant further study.

The project was terminated prematurely because of funding issues. No agent has been released and sicklepod remains a serious concern in northern Qld.

Key words: Caesalpiniaceae, biological control, Mexico, *Mitrapsylla albalineata*, *Conotrachelus*.

INTRODUCTION

Senna (= *Cassia*) *obtusifolia* (L.) Irwin and Barneby (Caesalpiniaceae), known as sicklepod, is a serious weed of field and pasture in many tropical and subtropical regions such as Australia, the USA and the Pacific islands. Sicklepod is a vigorously growing, annual or perennial, competitive woody shrub growing to 2.5 m tall. It has hairless compound leaves with two to three pairs of obovate leaflets, yellow one- to two-flowered inflorescences and very characteristic sickle-shaped pods containing five to 10 seeds (Fig. 1). *S. obtusifolia* is thought to be neotropical in origin (Irwin and Barneby 1982), but is now almost circumtropical and also occurs in various warm temperate regions. It is found in abundance in Mexico, Central America, Colombia, the Brazilian Planalto and north-western Argentina. In its native range it is usually found along lake shores and river banks, but in introduced areas of the tropical and subtropical world it has become a significant weed of pastures, cropping systems, orchards, environmental areas and waste places. It is a serious agricultural weed in the southern USA, particularly of leguminous crops (e.g. soy bean) since herbicide use is limited because of plant similarity.

Little is known about the introduction of sicklepod into Australia, but it may have been introduced during World War II around Darwin (Mackey *et al.* 1997). Sicklepod is now found along the tropical eastern coast of Qld from Cape York to Mackay (Fig. 2) and often grows in association with a similar and related species, foetid senna (cassia) *Senna tora* (L.) Roxb. (Mackey *et al.* 1997). The two species are not commonly distinguished in weedy situations and field-based information on sicklepod in Australia probably includes observations on foetid senna. Sicklepod infests more than 600 000 ha in Qld. Eco-climatic modelling suggests that it could spread along the entire eastern coast of Qld (Fig. 2) and that foetid senna could grow across the base of Cape York and around the Gulf of Carpentaria.

In Australia, sicklepod is a weed of pastures and sugar cane but it has the potential to be a major weed of several other crops (Mackey *et al.* 1997). In pasture, it is an aggressive invader and can completely dominate grass species, preventing pasture growth and excluding stock. Carrying capacities can be greatly reduced and it is conceivable that properties could become completely unproductive (Fig. 3). Sicklepod is generally unpalatable, and is toxic to cattle if eaten. Infestations in sugar cane crops decrease yields and cause machinery breakdowns at sugar mills. It is an environmental weed in some native ecosys-

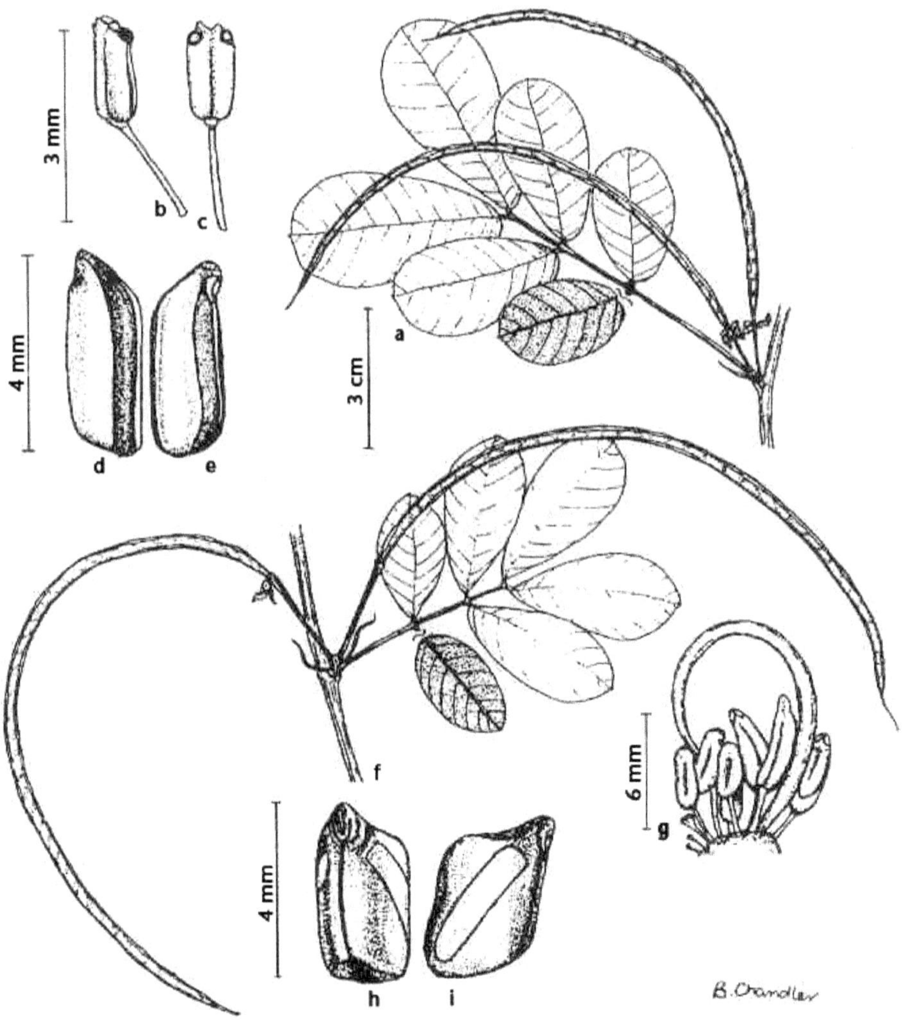

Figure 1: *Senna obtusifolia* in the bottom half of illustration with *Senna tora* (top half) showing pods and compound leaves, anthers and seeds with areoles. Modified after Randell (1988) from illustrations prepared by B Chandler.

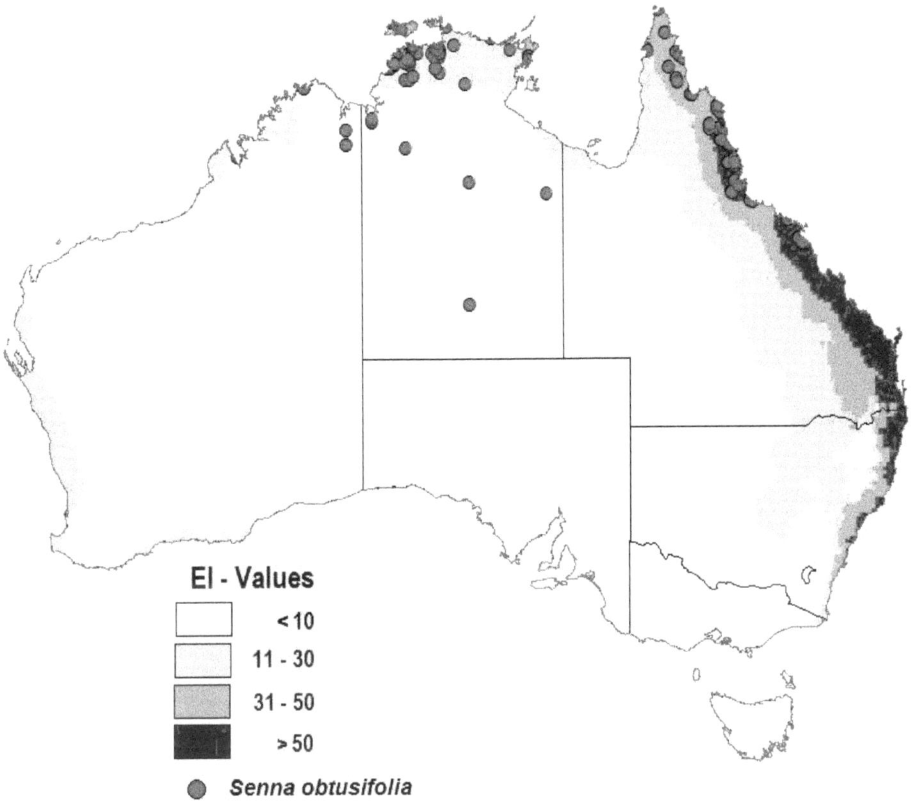

Figure 2: Distribution of *Senna obtusifolia* in Australia based on herbarium records from the Australian Virtual Herbarium and its potential distribution predicted by the climate modelling software CLIMEX. EI values >30 are considered favourable, while values ≤10 are considered climatically unfavourable.

tems in far northern Qld. Only preliminary economic assessments have been made, but these indicate that the weed costs Australia more than $1 million per year (Mackey *et al.* 1997).

Sicklepod is dispersed by water (stream flow and floods), in mud on machinery and stock, in mulch and by stock which ingest the pods (Mackey *et al.* 1997). Seed production is profuse (8000 seeds per plant) and, as the seed is long-lived, a large seed bank can develop in the soil. Scarification of the seed is required for germination, and therefore fires can cause mass emergence of seedlings. The plant grows on a variety of soil types and is usually an annual, although it can perennate.

Figure 3: A severe infestation of *Senna obtusifolia* extending as a monoculture over hundreds of hectares on the eastern coast of Cape York. Photo: B Palmer, Qld Government.

BIOLOGICAL CONTROL HISTORY

Biocontrol of sicklepod and foetid senna has been considered at various times, particularly by international agencies. CABI (then the Commonwealth Institute of Biological Control) reviewed the possibilities for the biological control of both species (Cock and Evans 1984). Scientists in Brazil, where sicklepod may be native, also became interested in its biological control (Fontes *et al.* 1995; Sujii *et al.* 1996), and in the USA there has been a long-standing interest in developing mycoherbicides for the control of sicklepod in cropping systems (Walker and Tilley 1997; Pitelli *et al.* 1998).

The first serious effort to investigate classical biological control was made by the then Qld Department of Lands through its Alan Fletcher Research Station in Brisbane and its North American Field Station in Temple, Texas. This project was undertaken from 1992 to 2002 and involved exploration in the weed's native range, host testing of promising agents in quarantine facilities and obtaining baseline data to assess the effects of any agents released.

PLANT TAXONOMY

A complete revision of the American Cassiinae was undertaken by Irwin and Barneby (1982). Previously known as *Cassia obtusifolia* L., sicklepod was placed in the genus *Senna* in this revision, and foetid senna was placed in the same genus after a revision of Australian species (Randell 1988; Randell and Barlow 1998). The correct names for sicklepod and foetid senna are thus *S. obtusifolia* (L.) Irwin and Barneby and *S. tora* (L.) Roxb. respectively (Randell and Barlow 1998). Interestingly, some two decades after this revision, sicklepod was still appearing in the literature as *C. obtusifolia*.

The genus *Senna* (Bauhin) Miller has over 300 species. It is well represented in South and Central America, southern Africa and Australia, but has fewer representatives in North America, northern Africa and Asia (Irwin and Barneby 1982; Randell and Barlow 1998). Within the genus *Senna*, *S. obtusifolia* and *S. tora* have been placed within the section *Chamaefistula* (Colladon) Irwin & Barneby, a section represented in the Americas by 143 described species (Irwin and Barneby 1982). Section *Chamaefistula* is not well represented in Australia, where there are nine exotic and three endemic species (Randell 1988). *S. tora* is morphologically very similar to *S. obtusifolia*, from which it is thought to have evolved in Asia (Randell 1995).

From the biological control perspective, the facts that there are over 50 native Australian *Senna* spp. (Ross 1998) and that three of these are in section *Chamaefistula* present difficulties for finding sufficiently host-specific agents for Australia. Any prospective agent would need to be very nearly host-specific to be approved for release.

EXPLORATION

Various insect species and pathogens were known from sicklepod and foetid senna before the Australian project commenced. A literature review (Cock and Evans 1984) summarised existing knowledge and indicated several potential agents.

Brazilian investigations revealed that at least 21 insect species were associated with *S. obtusifolia* in that country (Sujii *et al.* 1996). However, most of these species were not identified beyond family level.

The Australian exploration was conducted from Temple and Cuernavaca (Morelos, Mexico) between September 1991 and December 1993 (Palmer and Pullen 2001). The area of search in Mexico consisted of the Gulf Coast states of Veracruz and Tamaulipas, the Pacific Coast states of Guerrero, Oaxaca, Michoacán and Jalisco, the central state of Morelos and the Yucatan Peninsula states of Yucatan and Quintana Roo. One collecting trip was also made to Honduras in December 1993. Each species was recorded in a computer database after its identification (Palmer 1995), along with details gathered at the time of collection and relevant literature.

Some 108 species representing six orders and 40 families were found. The orders represented were Orthoptera (three species, 3% of species), Thysanoptera (three, 3%), Hemiptera (49, 45%), Coleoptera (45, 42%), Diptera (one, 1%) and Lepidoptera (seven, 7%).

The number of species found in this survey was less than has been found in surveys of comparable plants (Palmer and Pullen 2001). Furthermore, the insect fauna was relatively depauperate in Lepidoptera, which usually comprise about 10–15% of a total phytophagous fauna (Palmer and Pullen 2001). The important families Geometridae, Noctuidae and Tortricidae, usually well represented in surveys, were completely absent.

Inherent attributes of the host plant such as biochemistry and morphology may be factors in the relatively low number of arthropod species found in this survey. Although surveys were conducted in Mexico for more than two years, the study was terminated earlier than intended before sufficient searches could be conducted in Central America. As *S. obtusifolia* was found to be abundant in Honduras, further work in Central America is desirable.

Interestingly, only three species found in North America (Palmer and Pullen 2001) were also reported in Brazil (Sujii *et al.* 1996). Although many insects found in Brazil were not identified to species, it was apparent that different insect faunas exist on *S. obtusifolia* in North and South America.

CANDIDATES

Several insect species were investigated further as they initially appeared to have potential as biocontrol agents for Australia.

***Mitrapsylla albalineata* Crawford (Hemiptera: Psyllidae).** This psyllid was frequently found infesting terminal leaves and appeared to be distributed widely in Mexico. The nymphs of this rather large psyllid (2–3 mm in length) extrude a white waxy filament, giving infested parts of the plant a 'cottony' appearance. Although its host-range was not known (Crawford 1914), it was thought narrowly stenophagous. It was readily reared in the laboratory in Cuernavaca. *M. albalineata* was therefore imported into quarantine in Australia. There it was found to be narrowly stenophagous, but able to utilise several native Australian species. In continuation trials, *M. albalineata* bred successfully through three generations on nine *Senna* spp. including the natives *Senna gaudichaudii* (Hook. & Arn.) Irwin & Barneby, *Senna magnifolia* (F. Muell.) Randell, *Senna costata* (J. F. Bailey & C. T. White) Randell and *Senna barclayana* (Sweet) Randell (Anon 2000). This host-range was considered too problematic for release in Australia and the insect was studied no further.

***Conotrachelus cruciferae* Champion and *Conotrachelus* sp. 'Morelos' (Coleoptera: Curculionidae)**. Mexican populations of the curculionid *Conotrachelus* were reported (Palmer and Pullen 2001) as two species, *C. cruciferae* Champion from near the coast in Guerrero and *Conotrachelus* sp. 'Morelos' from around Lake Tequesquitengo in Morelos. However, the two may be conspecific (ML Cox, *pers. comm.* 1995). Both species produce a bulbous stem gall and are considered potential biocontrol agents. *Conotrachelus* sp. 'Morelos' was imported into a quarantine facility in Australia. The insect was reared through five generations, but then entered a reproductive diapause and ceased oviposition (Anon 2000). The diapause was not broken and the colony was ultimately lost (Anon 2002). It may warrant further study in the future.

***Sennius fallax* (Boheman), *Sennius lebasi* (Fahraeus), *Sennius morosus* (Sharp) and *Sennius rufomaculatus* (Motschulsky) (Coleoptera: Chrysomelidae).** Mixed populations of four bruchid seed-feeding beetles *S. fallax* (Boheman), *S. lebasi* (Fahraeus), *S. morosus* (Sharp) and *S. rufomaculatus* (Motschulsky) were commonly found in the native range as plants matured and produced pods (Palmer and Pullen 2001). On occasions, adults were very abundant and the pods were covered with many eggs. Similarly, large numbers of adults emerged in the laboratory from collections of maturing pods. All four species are regarded as continuous breeders (Johnson and Slobodchikoff 1979). None of these four *Sennius* spp. is host-specific, all having eight to 13 recorded host species, all within *Senna* spp. (CD Johnson, *pers. comm.* 1993) except for a record of *S. fallax* from *Cassia hintoni* Sandwith. Although these species are clearly not monophagous they do have host-ranges basically confined to *Senna*. As they also have the potential to attack a high proportion of seed, they warrant further study, although it is likely that they would attack native *Senna* spp. and be difficult to approve for release in Australia under present guidelines.

***Anabasis ochrodesma* (Zeller) (Lepidoptera: Pyralidae).** *A. ochrodesma* was found throughout the survey area and throughout the growing season (Palmer and Pullen 2001). The larvae were commonly found feeding within webbed terminal leaves which they killed, within the terminal stem or within pods where they fed on the immature seed. This species is known from other *Senna* spp. and also from *Cassia javanica* L. (Heinrich 1956). Because it is narrowly stenophagous and is also damaging to the plant, it could be considered a potential biocontrol agent.

***Typhedanus undulatus* (Hewitson) (Lepidoptera: Hesperiidae).** Nominated by Sujii *et al.* (1996) as a potential agent for Brazil, this was found in low numbers along both Mexican coasts (Palmer and Pullen 2001). Its host-range is unknown although oviposition has been observed on *Senna bicapsularis* (L.) Roxb. (RO Kendall, *pers. comm.* 1990). It warrants further study. Similarly, *Eurema salome* Felder (Lepidoptera: Pieridae) could be studied. It was also found in low numbers and its host-range is unknown.

***Aphis craccivora* Koch (Hemiptera: Aphididae) and *Caliothrips phaseoli* (Hood) (Thysanoptera: Thripidae)**. These were found at many sites, often causing appreciable damage to *S. obtusifolia* (Palmer and Pullen 2001). However, both are well documented cosmopolitan pests of many legumes (Singh 1990).

***Deuteronoda suturalis* (Lefevre) (Coleoptera: Chrysomelidae)**. This eumolpine insect was found in all the areas of study (Palmer and Pullen 2001). Typically only a few adults were collected on any occasion and they were never found in large damaging numbers. Adults were found feeding on leaves and in flowers but were also collected on *Senna alata* (L.) Roxb., *Senna uniflora* (Miller) Irwin and Barneby and an unidentified Caesalpiniaceae. Larvae were not found, but those of the very closely related genus *Colaspis* are root-feeders (RE White, *pers. comm.* 1992) and the larvae often have a broader host-range than the adult. For this reason, this species is unlikely to be useful for biological control.

***Prepops latipennis* (Stål) (Hemiptera: Miridae).** *P. latipennis* was found along the Pacific coast and the Yucatan Peninsula of Mexico, sometimes in appreciable numbers of both nymphs and adults (Palmer and Pullen 2001). Infested plants showed characteristic speckling of the leaves. However, the wide host-range of this bug, which includes sweet potato, beans and cowpea (King and Saunders 1984), eliminated it from consideration for biological control.

DISCUSSION

Sicklepod remains a major problem in coastal areas of far northern Qld and a serious concern to affected stakeholders. It is consistently towards the top of prioritised weed lists generated in those areas. Further, because it grows on low-value broad-area holdings, biological control would be a very desirable solution. There is also the issue that sicklepod may not have yet occupied its favourable climatic range and that areas to the south of presently infested areas may, in time, be seriously affected.

The project was terminated prematurely when it was necessary to prioritise the various projects because of funding constraints. Although Mexico was quite thoroughly searched, further exploration in Central America would have been desirable. Similarly, some of the identified prospects have not yet been fully studied and so remain potential agents. At this time *M. albalineata* is the only species to have been properly studied.

ACKNOWLEDGEMENTS

The services of the Systematic Entomology Laboratory, Agricultural Research Service, USDA, and the associated Department of Entomology, National Museum of Natural History, Smithsonian Institution in identifying the collected specimens are acknowledged. The Universidad Autónoma del Estado de Morelos, Cuernavaca made facilities available for the Mexican studies. KADW Senaratne developed the CLIMEX model and prepared Figure 2. Colleagues at the AFRS provided valuable comments on the manuscript.

REFERENCES

Anon (2000) '1999/2000 technical highlights – annual report on weed and pest animal research'. Qld Dept of Natural Resources, Brisbane.

Anon (2002) '2000/01 technical highlights – annual report on weed and pest animal research'. Qld Dept of Natural Resources and Mines, Brisbane.

Cock MJW and Evans HC (1984) Possibilities for biological control of *Cassia tora* and *C. obtusifolia*. *Tropical Pest Management* **30**, 339–350.

Crawford DL (1914) *A Monograph of the Jumping Plant Lice or Psyllidae of the New World*. Smithsonian Institution, Washington, DC.

Fontes EG, Teixeira CSS and Sujii ER (1995) Current status of the biological control of weeds in Brazil. In *Proceedings of the 8th International Symposium on Biological Control of Weeds*. 2–7 February 1992, Canterbury, NZ. (Eds ES Delfosse and RR Scott) p. 693. DSIR/CSIRO, Melbourne.

Heinrich C (1956) *American Moths of the Subfamily Phycitinae*. Smithsonian Institution, Washington, DC.

Irwin HS and Barneby RC (1982) The American Cassiinae. A synoptic revision of Leguminosae tribe Cassieae subtribe Cassiinae in the New World. *Memoirs of the New York Botanical Garden* **35**, 1–918.

Johnson CD and Slobodchikoff CN (1979) Coevolution of *Cassia* (Leguminoseae) and its seed beetle predators (Bruchidae). *Environmental Entomology* **8**, 1059–1064.

King ABS and Saunders JL (1984) *The Invertebrate Pests of Annual Food Crops in Central America*. London Overseas Development Administration, London.

Mackey AP, Miller EN and Palmer WA (1997) *Sicklepod* (Senna obtusifolia) *in Queensland*. Qld Dept of Natural Resources, Brisbane.

Palmer WA (1995) The use of computer databases during the foreign exploration phase of a biological control programme. In *Proceedings of the 8th International Symposium on Biological Control of Weeds*. 2–7 February 1992, CanterburyNZ. (Eds ES Delfosse and RR Scott) pp. 705–708. DSIR/CSIRO, Melbourne.

Palmer WA and Pullen KR (2001) The phytophagous arthropods associated with *Senna obtusifolia* (Caesalpiniaceae) in Mexico and Honduras and their prospects for utilization for biological control. *Biological Control* **20**, 76–83.

Pitelli RA, Charudattan R and Devalerio JT (1998) Effect of *Alternaria cassiae*, *Pseudocercospora nigricans*, and soybean (*Glycine max*) planting density on the biological control of sicklepod (*Senna obtusifolia*). *Weed Technology* **12**, 37–40.

Randell BR (1988) Revision of the Cassiinae in Australia. 1. *Senna* Miller sect. *Chamaefistula* (Colladon) Irwin and Barnaby. *Journal of Adelaide Botanic Gardens* **11**, 19–49.

Randell BR (1995) Taxonomy and evolution of *Senna obtusifolia* and *S. tora*. *Journal of Adelaide Botanic Gardens* **16**, 55–58.

Randell BR and Barlow BA (1998) *Senna*. *Flora of Australia* **12**, 89–138.

Ross JH (1998) Caesalpiniaceae. *Flora of Australia* **12**, 50–55.

Singh SR (1990) *Insect Pests of Tropical Food Legumes.* John Wiley & Sons, Chichester.

Sujii ER, Fontes EG, Pires CSS and Teixeira CAD (1996) Application of multivariate analysis for the selection of candidates for biological control agents. *Biological Control* 7, 288–292.

Walker HL and Tilley AM (1997) Evaluation of an isolate of *Myrothecium verrucaria* from sicklepod (*Senna obtusifolia*) as a potential mycoherbicide agent. *Biological Control* **10**, 104–112.

Sida acuta Burm.f – spinyhead sida
Sida rhombifolia L. – paddy's lucerne
Sida cordifolia L. – flannel weed

Tim A Heard and Michael Day

ABSTRACT

Sida acuta, *Sida rhombifolia* and *Sida cordifolia* are invasive weeds in northern Australian rangelands. Surveys for biocontrol agents of *S. acuta* and *S. rhombifolia* were conducted in Mexico from 1984. Sixty-two species of phytophagous arthropods were found and 12 were considered to have potential as biocontrol agents. Six were imported into Australian quarantine and three were approved for release in Australia. The first agent *Calligrapha pantherina* was released in 1989 and was widely established by 1997. This beetle defoliates *S. acuta* plants and reduces seed production. After several years of defoliation, populations of *S. acuta* were replaced by desirable native and exotic pasture species. *C. pantherina* has its greatest impact on *S. acuta* in coastal and subcoastal areas. The fate of the other two agents released, two species of *Eutinobothrus*, has not been investigated, although they are known to have survived the first dry season.

Key words: *Calligrapha pantherina*, *Eutinobothrus*, Malvaceae.

INTRODUCTION

Sida acuta Burm.f, *Sida rhombifolia* L. and *Sida cordifolia* L. (Malvaceae) are weeds in many parts of the world, including Australia. They are native to the Americas, although *S. rhombifolia* and *S. cordifolia* may be native to all tropical regions of the world, including Australia. They can tolerate drought as well as high-rainfall conditions. All species are erect perennial shrubs up to 1.5 m in height, occurring on a wide range of soil types. All have yellow flowers and reproduce by seed. They are weeds in disturbed and cultivated areas (Holm *et al.* 1991; Parsons and Cuthbertson 1992).

S. acuta is common from the north-eastern corner of WA through the NT and south, particularly along the coastal areas of Qld to Gladstone. *S. rhombifolia* occurs in tropical to warm temperate areas, extending further south than *S. acuta* to southern NSW. *S. cordifolia* is also found throughout the tropical and subtropical north of the continent (Parsons and Cuthbertson 1992) (Fig. 1).

BIOLOGICAL CONTROL HISTORY

Biocontrol of *Sida* spp. began in 1984 when the NT Government provided funds for exploratory studies on *S. acuta* and *S. rhombifolia* in their native range. Some work was also conducted opportunistically on *S. cordifolia*. Surveys were conducted out of the CSIRO's Mexican field station at Acapulco between 1984 and 1986. Collection of agents for export to Australia and less intensive opportunistic surveys continued until 1997 from Veracruz to where the field station was relocated. From 1985 until 1997 the highest-priority agents were imported into the quarantine facility at CSIRO's Long Pocket Laboratories in Brisbane, where host-specificity testing was conducted. Three agents, *Calligrapha pantherina* Stål (Coleoptera: Chrysomelidae), *Eutinobothrus* sp. (Coleoptera: Curculio-

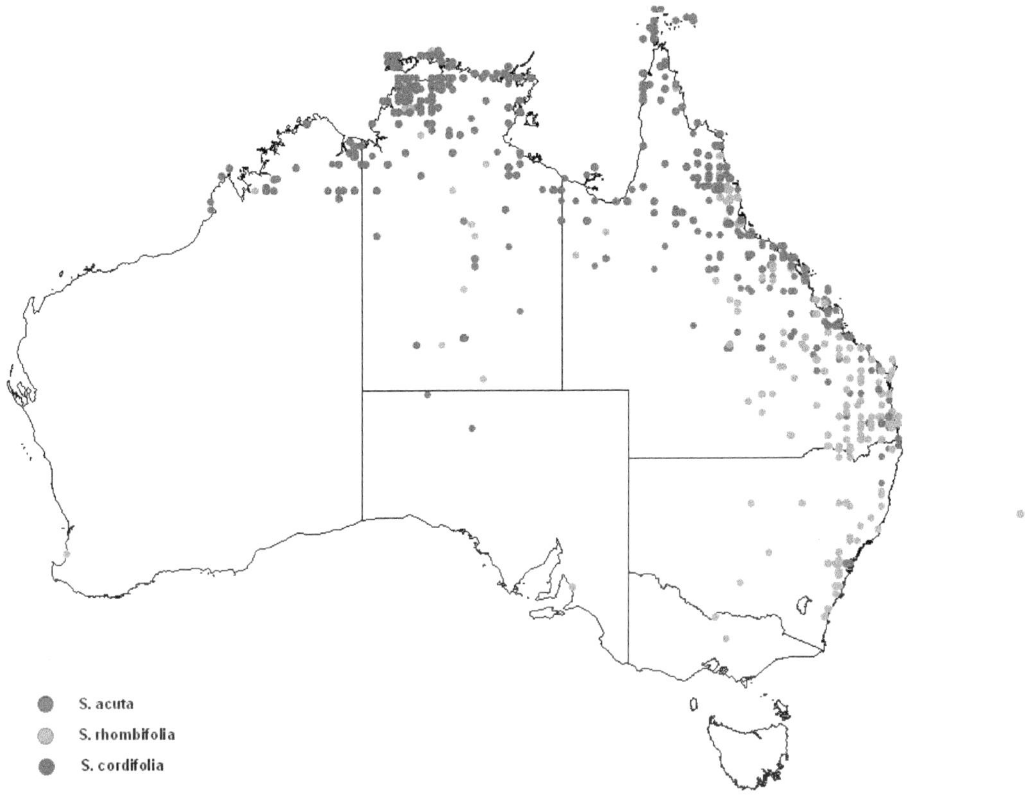

Figure 1: Distribution of *Sida acuta, Sida rhombifolia* and *Sida cordifolia* in Australia. Source: Specimen data reproduced from Australia's Virtual Herbarium with permission of the Council of Heads of Australasian Herbaria Inc.

nidae) and *Eutinobothrus pilosellus* Faust, were approved for field release in 1989, 1994 and 1997 respectively. The majority of releases were conducted in the NT, where the most severe infestations were found. Some field releases of all three species were also conducted in northern Qld. Mass rearing and release of *Eutinobothrus* spp. continued until 1999, but it is not known if establishment occurred. Several other agents were also imported, but were not released as they were not specific to the target plants. *C. pantherina* established quickly and offers good control of *Sida* spp. in the NT. It was subsequently released in Papua New Guinea (Kuniata and Korowi 2004), Fiji and Vanuatu, where it is offering some level of control against *Sida* spp. (W Orapa, *pers. comm.* 2009).

PLANT TAXONOMY

The genus *Sida* L. is a large, poorly circumscribed group of species of pantropical distribution belonging to the tribe Malveae in the family Malvaceae (Fuertes Aguilar *et al.* 2003). The three *Sida* species considered in this chapter can be distinguished in the following ways. *S. cordifolia* is a taller plant than *S. acuta* and *S. rhombifolia*. Its stems and leaves have a felt-like appearance due to the dense stellate hairs. The flowers are produced in clusters and the seed capsules divide into 10 segments. The two sharp points or awns on the end of each seed segment are longer than those of *S. acuta*. *S. rhombifolia* has leaves in which the undersurface is paler than the upper surface, whereas the leaves of *S. acuta* are dark green on both sides (Fig. 2). *S. rhombifolia* produces pale yellow flowers which grow singly on stalks 10–35 mm long. The flowers of *S. acuta* are on stalks 3–8 mm long. The seed capsules of *S. rhombifolia* divide into 10 segments, each segment having two blunt points. The seed capsules of *S. acuta* divide into five to eight segments, each segment having two sharp points (Pitt 1992; Parsons and Cuthbertson 1992).

EXPLORATION

Native range

The Acapulco area of Mexico was chosen as the base for exploration due to a close climate match with Darwin. In particular, the dry season on the south Pacific coast of Mexico is long, potentially providing a source of agents capable of surviving the severe northern Australian dry

Figure 2: A flowering and fruiting tip of *Sida acuta*. Photo: CSIRO.

season (Gillett *et al.* 1991). From 1984 to 1986, 90 collections were made by John Gillett, Hugo Miranda, Ricardo Segura, Kim Pullen and Jose Garcia at 19 locations, all in Mexico. Moises Martinez contributed to later collections, investigations and shipments of agents to Australia. Preliminary host-specificity testing was conducted in the native range (Gillett *et al.* 1991).

Exotic range

The phytophagous insect fauna of *S. acuta* and *S. cordifolia* was investigated in the NT (Wilson and Flanagan 1990). This study attempted to reveal any vacant niches which might be exploited by introduced biocontrol agents. Most of the 20 species on *S. acuta* and 23 species on *S. cordifolia* were rarely encountered. *S. cordifolia* was more heavily utilised and supported six species that damaged the plant heavily. The only species found in large numbers on *S. acuta* was the seed-sucking bug *Oxycarenus luctuosus* Montrouzier and Signoret. It was concluded that vacant plant niches occurred on both plant species for introduced biocontrol agents (Wilson and Flanagan 1990).

CANDIDATES

Natural enemies (native range)

A total of 62 phytophagous arthropod species were collected from Mexico; 35 only on *S. acuta*, 11 only on *S. rhombifolia* and 16 on both plant species. Data were obtained on the host-specificity of 38 herbivorous species, which indicated that five were restricted to *Sida* spp. (Gillett *et al.* 1991). Species were ranked according to their perceived potential for control. Twelve species were selected for further consideration. Six species were imported into Australian quarantine and are treated in detail below.

Agents tested but not released

***Meskea horor* Dyar (Lepidoptera: Thyrididae).** This species was imported into Australian quarantine for study between 1988 and 1990. Females lay eggs on the underside of leaves and the emerging active larvae enter the stem between the nodes. The larvae feed in stems, inducing plants to form galls. Pupation occurs in the galls and adults emerge through a clear window on the side of the gall. Development from egg to adult is about 190 days. During the testing stage, females preferred to oviposit on *S. rhombifolia* rather than *S. acuta* and there was a higher development rate on *S. rhombifolia*. In host-specificity trials, oviposition occurred and development was completed on several other species, including *Abutilon oxycarpum* (F. Muell.) F. Muell. ex Benth. var. *acutatum*, *Gossypium hirsutum* L., *Hibiscus tiliaceus* L. and *Sida subspicata* F. Mueller ex Bentham (all in the family Malvaceae). Consequently, its field release was not sought and the colony was destroyed (Day *et al.* 1997).

***Zygogramma* nr. *bigenera* (Coleoptera: Chrysomelidae).** A colony of the leaf-feeding chrysomelid *Z.* nr. *bigenera* was studied in Australia during 1994. Its life-cycle and damage to *S. acuta* and *S. rhombifolia* are similar to those of *C. pantherina*. However, attack on some native species of *Sida* was substantial and it was judged unsuitable as a biocontrol agent in Australia. The colony was destroyed.

***Acanthoscelides brevipes* (Sharp) (Coleoptera: Bruchidae).** This species was imported into Australian quarantine for study between 1988 and 1990. This small black beetle feeds on mature seeds of *Sida* spp. Adults drink nectar from the flowers or water. The pre-oviposition period is about two weeks and females lay eggs between the seed and the calyx. Eggs can also be laid between two seeds within the infructescence. Upon emerging, larvae tunnel into the seed and feed for about six to seven weeks. Pupation occurs within the seed (M Day, unpub.).

During host-specificity testing, adults emerged from 10 species, of which seven were *Sida* spp., including *S. acuta*. Apart from *S. acuta*, the main species supporting *A. brevipes* were *Sida cunninghamii* White and *Sida fibulifera* Lindley. Of the non-*Sida* spp., nine adults emerged from *Abutilon oxycarpum* var *acutatum* and one adult each from *Abutilon oxycarpum* var *subsagittatum* and *Triumfetta rhomboidea* Jacquin (M Day, unpub.). In no-choice trials, adults emerged from eight species, of which six were *Sida* spp., including *S. acuta*. The main species supporting development were *Sida brachypoda* F. Mueller ex Holland and Reynier, *Sida fibulifera* and *Sida spinosa* L. A total of 16 adults emerged from *A. oxycarpum* var *acutatum* while one adult emerged from *A. oxycarpum* var *subsagittatum* (M Day, unpub.). Due to the number of species that supported development, including several native species, the agent was not considered for release and the colony was destroyed.

***Brachycoryna pumila* Guerin or near (Coleoptera: Chrysomelidae).** Adults of the chrysomelid *B. pumila* were screened against several *Sida* species in Mexico and close relatives in the Malvaceae. Adults damaged and reproduced on a number of non-target *Sida* species and so this insect was considered unsuitable to import and further test in quarantine in Australia (W Forno, *pers. comm.* 1995).

Agents released

***Calligrapha pantherina* Stal (Coleoptera: Chrysomelidae).** *Calligrapha pantherina* feeds on leaves of *S. acuta* and *S. rhombifolia* in Mexico. It is only known to occur in the states of Guerrero, Oaxaca, Veracruz and Chiapas. Adults and larvae both feed on leaves of their host plants (Gillett *et al.* 1991). Eggs are laid on the underside of mature leaves. Larvae feed gregariously for the first three instars (Fig. 3) and singly during the fourth (final) instar. Older instar larvae may also feed on the flowers and fruits. Pupation occurs in the leaf litter or in fine soil and lasts for six to 12 days. The total time from egg to adult is about 24 days, with a pre-oviposition period of about 18 days. Newly emerged adults are dull brown but change to iridescent green with black markings, and become receptive to mating after about seven days. The adults live for an average of 18 weeks and lay two batches of about 50 eggs each per week (Forno *et al.* 1992).

Figure 3: Larvae and leaf damage of *Calligrapha pantherina*. Photo: CSIRO.

Host-specificity studies showed that only *S. acuta*, *S. rhombifolia* and *S. spinosa* (another exotic weed) were good hosts for *C. pantherina*. The feeding, survival and oviposition on other native hosts was sufficiently poor for this agent to be acceptable for release in Australia (Forno *et al.* 1992).

Following release in 1989, *C. pantherina* quickly established and caused severe defoliation of *S. acuta* near Darwin in the NT (Wilson and Lonsdale 1992). Figure 4 shows an infestation before the release of *C. pantherina* and Figure 5 shows a similar infestation after damage by this insect. Populations of *C. pantherina* move as a front through stands of *S. acuta* and provide good control with maximum impact on coastal and subcoastal stands of *S. acuta* in the NT (Flanagan *et al.* 2000). *C. pantherina* reduces annual seed production by an order of magnitude and reduces plant density (Lonsdale *et al.* 1995). Beetle survival during the severe dry season in northern Australia can be poor, and it often needs to be reintroduced to areas where high densities were present the previous dry season (Lonsdale *et al.* 1995). Many landowners value the insect and take measures to ensure its survival over the dry season (Grace *et al.* 2006).

Following its success in Australia, *C. pantherina* was sent to Papua New Guinea in 1999 and soon provided effective control of *Sida* spp. (Kuniata and Korowi 2004). It has also been released in Fiji and Vanuatu, where it is offering some level of control against *S. acuta*, *S. rhombifolia* and *Senna retusa* (W Orapa, *pers. comm.* 2009).

C. pantherina failed to establish on *S. rhombifolia* near Brisbane in south-eastern Qld and only limited

Figure 4: An undamaged stand of *Sida acuta* at Katherine Rural College in December 1987 before the release of *Calligrapha pantherina*. Photo: Colin Wilson.

establishment has occurred on *S. rhombifolia* near Townsville in northern Qld. Both *S. acuta* and *S. rhombifolia* are nutritionally similar hosts for *C. pantherina* (Heard and Gardner 1994). Climate matching shows that the location where the insect has established (near Darwin) is very similar to Acapulco (close to the epicentre of the native range). The location near Brisbane where the insect did not establish is not climatically similar to Acapulco, while the location near Townsville where limited establishment has occurred has an intermediate level of similarity (Heard and Gardner 1994). Hence, it appears likely that climatic factors influence the observed distribution of *C. pantherina* in Australia.

***Eutinobothrus* sp. (Coleoptera: Curculionidae).** *Eutinobothrus* sp. was first introduced into Australia in 1989. Following host-specificity testing, it was released in May 1994. Adults graze on the thicker stems and eggs are laid in feeding scars or in nodes on the stems. Eggs hatch in about nine days and larvae tunnel into the stem, at the site of oviposition. Larvae feed for 34–56 days. Pupation occurs in the stems and adults emerge after 13 days and can live for about nine months (Day *et al.* 1995). *Eutinobothrus* sp. survived the first dry season in the NT at sites not too distant from Darwin. A total of 40 670 adults were reared at the CSIRO Long Pocket Laboratories, sent to the NT, WA and northern Qld and released at many widely dispersed sites from May 1994 to April 1999. Early surveys revealed establishment, continued survival and spread at most release sites that had not been disturbed by spraying or slashing (G Fichera, *pers. comm.* 2010).

Figure 5: A stand of *Sida acuta* at Marlow Lagoon, NT in January 1991 defoliated by *Calligrapha pantherina*. Photo: Colin Wilson.

***Eutinobothrus pilosellus* Faust (Coleoptera: Curculionidae).** Adults of this beetle were imported accidentally with shipments of *Eutinobothrus* sp. These beetles differ from *Eutinobothrus* sp., being slightly smaller and more elongate. Adults of *E. pilosellus* tend to feed on leaves whereas those of *Eutinobothrus* sp. tend to feed on the stems. A separate colony of *E. pilosellus* was established and host-specificity testing was conducted once numbers were adequate. Approval for release of *E. pilosellus* was gained in March 1997 and from then until March 1999, 20 690 adults were reared at the CSIRO Long Pocket Laboratories and released in the NT. Approximately 4000 were also released at sites in coastal Qld, from Bowen to Gordonvale in 1998 (G Fichera, *pers. comm.* 2010).

Other promising agents

Calligrapha polyspila heavily defoliates and even kills plants of *S. rhombifolia* in pine plantations in Argentina (Lopez *et al.* 2009). It is used to control *Sida* in Brazil (Garcia 1991). Hugo Cordo (USDA-ARS South American Biological Control Laboratory) studied the host-range of *C. polyspila* for the control of *S. spinosa* in USA and concluded that it was specific to *S. rhombifolia*. It completed one generation on *S. spinosa* but not a second and hence was not released in the USA (H Cordo, *pers. comm.* 2010). This insect may be a useful agent for *S. rhombifolia* in cooler areas of Australia where *C. pantherina* does not thrive.

In addition to the six agents imported into Australia for testing, Gillett *et al.* (1991) ranked the following species as having potential: *Calligrapha felina* Stal (Chrysomelidae), *Calycomyza* sp. poss. *C. sidae* Spencer or *C. ipomaeae* (Frost) (Agromyzidae), *Asphondylia* sp. poss. *sidae* Mohn (Cecidomyiidae), *Neolasioptera* sp. prob. *sidae* Mohn (Cecidomyiidae), *Neoptilia biramosa* (Klug) (Argidae), *Stegasta albocapitella* (F.) (Gelechiidae),

Pyrgus adepta Ploetz (Hesperiidae) and *Bucculatrix* sp. (Lyonetiidae). Further work is needed to assess the potential of these agents.

DISCUSSION

Of the 62 species of phytophagous arthropods collected in three years of surveying in Mexico, 12 were considered to have potential as biocontrol agents (Gillett *et al.* 1991). Six were imported into Australian quarantine and three were approved for release in Australia. One of these, *C. pantherina*, is severely affecting the weed (Flanagan *et al.* 2000; Lonsdale *et al.* 1995). The other released agents, two species of *Eutinobothrus*, are known to have survived the first dry season following their release but their current status is unknown.

On the basis of the number of phytophagous species reared on *S. acuta* and *S. rhombifolia* in Mexico, Gillett *et al.* (1991) concluded that Mexico is part of the native range of both these species. However, the presence of *C. polyspila*, an insect specific to *S. rhombifolia* in Argentina (H Cordo, *pers.comm.* 2010), indicates that the native range of *S. rhombifolia* may be very broad through the American continents. Indeed, it has been suggested that both *S. rhombifolia* and *S. cordifolia* are pantropical species (Parsons and Cuthbertson 1992).

Page and Lacey (2006) estimated the present value costs of the biocontrol program against *S. acuta* and *S. rhombifolia* to outweigh the benefits, resulting in a negative net present value and a benefit:cost ratio of 0.5:1. This implies that for every dollar invested, a benefit of only $0.50 was generated. However, this was with a high discount rate of 8% and the analysis was considered to underestimate the benefits, as it only included benefits received in the Darwin area.

ACKNOWLEDGEMENTS

We thank Wendy Forno for useful comments on a draft.

REFERENCES

Day MD, Forno IW, Segura R and Martinez M (1995) Life cycle and host specificity of *Eutinobothrus* sp. (Col: Curculionidae) an agent for biological control of *Sida acuta* (Malvaceae) in the Northern Territory, Australia. *Entomophaga* **40**, 345–355.

Day MD, Segura R and Martinez M (1997) Life cycle and host range of the gall-forming moth, *Meskea horor* (Lep.: Thyrididae), and its suitability as a biological control agent for *Sida acuta* and *S. rhombifolia* (Malvaceae). *Entomophaga* **42**, 393–403.

Flanagan GJ, Hills LA and Wilson CG (2000) The successful biological control of spinyhead sida, *Sida acuta* (Malvaceae), by *Calligrapha pantherina* (Col: Chrysomelidae) in Australia's Northern Territory. In *Proceedings of the 10th International Symposium on Biological Control of Weeds*. 4–14 July, Bozeman, Montana. (Ed NR Spencer) pp. 35–41. Montana State University, Bozeman.

Forno IW, Kassulke RC and Harley KLS (1992) Host specificity and aspects of the biology of *Calligrapha pantherina* (Col.: Chrysomelidae), a biological control agent of *Sida acuta* [Malvaceae] and *S. rhombifolia* in Australia. *Entomophaga* **37**, 409–417.

Fuertes Aguilar J, Fryxell PA and Jansen RK (2003) Phylogenetic relationships and classification of the *Sida* generic alliance (Malvaceae) based on nrDNA ITS evidence. *Systematic Botany* **28**, 352–364.

Garcia MA (1991) Arthropods in a tropical corn field: effects of weeds and insecticides on community composition. In *Plant-Animal Interactions: Evolutionary Ecology in Tropical and Temperate Regions*. (Eds PW Price, TM Lewinsohn, GW Fernandes and WW Benson) pp. 619–634. Wiley, New York.

Gillett JD, Harley KLS, Kassulke RC and Miranda HJ (1991) Natural enemies of *Sida acuta* NL Burman and *S. rhombifolia* L. (Malvaceae) in Mexico and their potential for biological control of these weeds in Australia. *Environmental Entomology* **20**, 882–888.

Grace B, Paynter J, van Klinken R and Wirf L (2006) Biocontrol of weeds in the Northern Territory: what has it achieved? In *Proceedings of the 15th Australian Weeds Conference. Managing Weeds in a Changing Climate*. 24–26 September 2006, Adelaide. (Eds C Preston, JH Watts and ND Crossman) pp. 553–556. Weed Management Society of South Australia, Adelaide.

Heard TA and Gardner MG (1994) Comparative studies of development and host utilisation by *Calligrapha pantherina* on *Sida acuta* and *S. rhombifolia*. *Biological Control* **4**, 336–340.

Holm LG, Plucknett DL, Pancho JV and Herberger JP (1991) *The World's Worst Weeds: Distribution and Biology*. Kreiger Publishing, Malabar, Florida.

Kuniata LS and Korowi KT (2004) Bugs offer sustainable control of *Mimosa invisa* and *Sida* spp. in the Markham Valley, Papua New Guinea. In *Proceedings of the 11th International Symposium on Biological Control of*

Weeds. 27 April–2 May 2003, Canberra. (Eds JM Cullen, DT Briese, DJ Kriticos, WM Lonsdale, L Morin and JK Scott) pp. 567–573. CSIRO Entomology, Canberra.

Lonsdale WM, Farrell GS and Wilson CG (1995) Biological control of a tropical weed: a population model and experiment for *Sida acuta*. *Journal of Applied Ecology* **32**, 391–399.

Lopez A, Demaestri M, Garcia J and Crenna C (2009) Control biológico de *Sida rhombifolia* por *Calligrapha polyspila* en forestaciones del Valle de Calamuchita-Argentina. In *Abstracts of the 13th Congreso Forestal Mundial* 2009. Buenos Aires, Argentina.

Page AR and Lacey KL (2006) 'Economic impact assessment of Australian weed biological control'. Technical Series No. 10. CRC for Australian Weed Management, Adelaide.

Parsons WT and Cuthbertson EG (1992) *Noxious Weeds of Australia*. Inkata Press, Melbourne.

Pitt JL (1992) 'Spinyhead sida (*Sida acuta*)'. DPIF Agnote No. 496.

Wilson CG and Flanagan GJ (1990) The phytophagous insect fauna of the introduced shrubs *Sida acuta* Burm. F. and *Sida cordifolia* L. in the Northern Territory, Australia. *Australian Entomological Magazine* **17**, 7–15.

Wilson CG and Lonsdale WM (1992) Biological control of *Sida acuta* in Australia's Northern Territory. *Proceedings of the 8th International Symposium on the Biological Control of Weeds*, 2–7 February 1992, Canterbury, NZ. (Eds ES Delfosse and RR Scott) p. 451. CSIRO, Canberra.

Silybum marianum (L.) Gaertner – variegated thistle

Tom Morley

ABSTRACT

Silybum marianum is one of the most widespread exotic weedy carduine thistles in Australia. In Vic it infests more than 4.7 million ha and vast tracts of NSW are also affected. The only natural enemy released as a biological control agent against it in Australia is the thistle receptacle weevil *Rhinocyllus conicus*. This weevil is probably established in Vic but this requires confirmation. Two rust species may warrant further investigation as potential agents. If more agents are required, exploration should be focused in south-western Spain and north-western Africa.

Key words: *Rhinocyllus conicus*.

INTRODUCTION

Silybum marianum (L.) Gaertner (Asteraceae: Cardueae) (variegated thistle) is a winter annual, or occasionally summer annual or biennial, that occurs in all states except the NT and is one of the most widespread thistles in temperate eastern Australia (Fig. 1). It produces a rosette up to 1 m diameter and may attain 2.5 m height when in flower (Fig. 2). Its importance is mainly as a weed of pasture, although cereal crops are also affected. It grows best on high-fertility soils such as alluvial flats, sheep camps, stockyards and any other areas with elevated soil nitrogen levels (Parsons and Cuthberson 2001). Seedling establishment is enhanced on bare or disturbed soil and under favourable conditions dense stands may exclude almost all other vegetation. Lane *et al.* (1980) estimated that in Vic 4.76 million ha were infested.

Variegated thistle is a serious threat to several indigenous vegetation formations in Vic (Carr *et al.* 1992). However, it is also regarded as a beneficial plant and has been utilised for a variety of purposes including food, medicine and oil production (Parsons and Cuthberson 2001). Goeden (1976) speculated that its relict habitats likely lay either along the Algerian coast or in the north-western African range of its only congener *Silybum eburneum* Coss. and Durieu and that, aided by man, it spread eastwards and northwards from there across Africa and into southern Europe, the British Isles and Asia. *S. marianum* is also a serious weed in North and South America and New Zealand.

BIOLOGICAL CONTROL HISTORY

A coordinated program to develop classical biological control for *Carduus* L. and *Silybum* L. thistles in North America was initiated in the early 1960s by the Agricultural Research Service, US Department of Agriculture, the Canada Department of Agriculture and the Commonwealth Institute of Biological Control–Switzerland (CIBC) (Dunn 1978). Goeden (University of California) (1976) documented the following surveys for *S. marianum* natural enemies: H Zwölfer (CIBC) in southern France (Zwölfer 1965), R Goeden in southern Italy and southern Greece and M Hafez in Egypt, Lebanon and

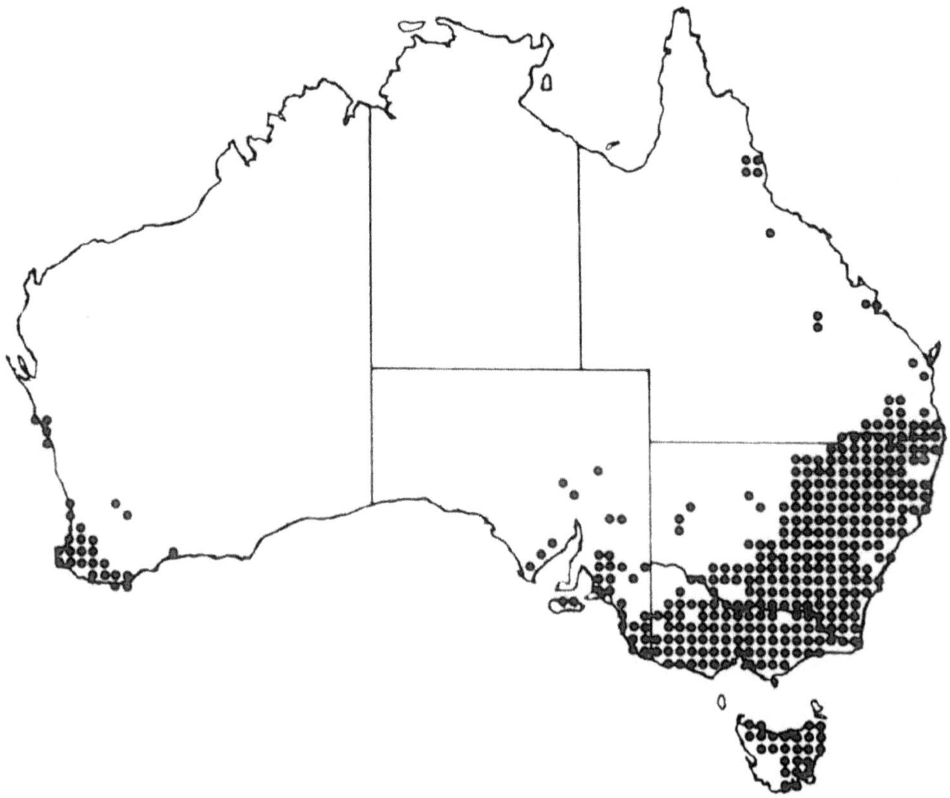

Figure 1: Distribution of *Silybum marianum* in Australia. Reproduced from Parsons and Cuthberson (2001) with kind permission of CSIRO Publishing.

Syria in 1971–72, and MA Ghani in Pakistan in 1972–73. More than 130 insects were associated with *S. marianum* in these surveys and Goeden recorded 115 in his, apparently the most comprehensive. Further outcomes of the program included the discovery of host biotypes of the thistle receptacle weevil *Rhinocyllus conicus* Frölich (Coleoptera: Curculionidae), the realisation that *S. marianum*-adapted populations of the weevil seemed to be the only arthropod suitable for use in classical biological of this weed, and the establishment of an Italian *S. marianum* biotype of the weevil in North America (Goeden 1976; Goeden and Ricker 1977; Goeden *et al.* 1985; Zwölfer and Preiss 1983), where it caused significant damage to *S. marianum* flower heads but did not appear to be reducing *S. marianum* populations (Boldt and DeLoach 1985).

In 1985 the CSIRO Division of Entomology initiated a program to develop biological control for thistles in Australia, where the main collaborator for *S. marianum* research was the Vic Government's Keith Turnbull Research Institute (KTRI). Aeschlimann (1999) improved the level of knowledge and understanding of *R. conicus* biotypes in Europe and exported to KTRI a French *S. marianum* biotype for mass rearing and releases in Australia. The first releases of this biotype were made in December 1988 at Omeo, Vic and a second importation and release in Vic of the same biotype was made in 1994 (Bruzzese 1996). Although *R. conicus* was released and is established on *Carduus nutans* L. in New Zealand (Kelly and McCallum 1995) the weevil has never been released specifically against nor recovered from *S. marianum* (R Groenteman, *pers. comm.* 2010).

CANDIDATES

***Rhinocyllus conicus* Frölich (Coleoptera: Curculionidae).** *R. conicus* is a natural enemy of carduine thistles in northern Africa and Europe and is the only enemy of *S. marianum* in its native range to have been tested and deployed for classical biological control of this weed anywhere in the world. The weevil has multiple biotypes and two subspecies that can be discriminated on the bases of preferred host, oviposition period, morphometrics and allozymes (Aeschlimann 1999; Klein and Seitz 1994). Its potential as a biological control agent is due to its larvae feeding exclusively in the receptacle tissue of the develop-

Figure 2: a) *Silybum marianum* in flower. b) *Silybum marianum* infestation. Photos: a) M Imhof, Vic Government. b) T Morley, Vic Government.

ing capitulum and consequent suppression of seed production. There is only one generation per year. Eggs are deposited singly in spring on the exterior of bracts of young thistle capitula and covered with chewed plant material. Larvae burrow through the bracts to the receptacle to feed. In mid to late summer, the next generation of adults emerges and moves to overwintering sites in the ground litter. Biology and host-specificity of *R. conicus* are documented in Zwölfer and Harris (1984).

Twenty releases of *R. conicus* were made on *S. marianum* at 13 sites in Vic (mostly western Vic) from 1988 to 1999 using weevils derived from *S. marianum* at St Martin de Crau, Provence, France (B Roberts, *pers. comm.* 2001; Bruzzese 1996). These were the only releases of any *S. marianum* biotype of *R. conicus* to have been made in Australia. By 2000, the weevil appeared to have established at at least five sites, with dispersal observed at some. However, no decline in the *S. marianum* infestations that could be attributed to *R. conicus* has been observed (B Roberts, *pers. comm.* 2001). No other observations or assessments of *R. conicus* on *S. marianum* have been attempted or made.

Other potential agents

Two defoliating rust fungi utilise *S. marianum* in Europe but neither has been developed for biological control. Although Bruzzese (1996) discounted the rust *Puccinia mariana* Sacc. as a potential agent, Aeschlimann (1999) considered it worth investigating. Bruzzese (1996) considered that the rust *Puccinia cruchetiana* Mayor might be worth investigating.

DISCUSSION

Knowledge of the current status of biological control of *S. marianum* in Australia is very poor and should be rectified. It seems likely that *R. conicus* is established but this is unconfirmed and there is no current knowledge of dispersal from release sites, distribution or impacts. The impacts of *R. conicus* on *S. marianum* in the USA are not encouraging. Bruzzese (1996) related that while the weevil caused significant damage to flower heads in Texas, it did not appear to have reduced the weed population and would probably not achieve successful biological control alone. Dodd's (1989) prognosis was similar: the restricted oviposition period of *R. conicus* and the low density of larvae per capitulum might limit the weevil's impact on *S. marianum* in Australia.

If other biological control agents are sought for *S. marianum*, Goeden (1976) concluded that further exploration should be conducted in south-western Spain and north-western Africa, as these are likely closer to the weed's centre of evolution than the localities of earlier surveys.

ACKNOWLEDGEMENTS

Jean-Paul Aeschlimann (Agropolis, Montpellier, France) collected *R. conicus* in France for shipment to Australia. Brad Roberts, Peter Stevens and Karen Green (Vic Government) reared and released the weevils in Australia. Eligio Bruzzese (Vic Government) was the principal investigator in Australia and Raelene Kwong (Vic Government) was project leader in the later parts of the program. The author did not participate in the program.

REFERENCES

Aeschlimann JP (1999) Specificity and bionomics of south-western Palaearctic biotypes of *Rhinocyllus conicus* Frölich (Col., Curculionidae), a biological control agent of Palaearctic thistles (Asteraceae) accidentally introduced to Australia. *Mitteilungen der Schweizerischen Entomologischen Gesellschaft* **72**, 11–22.

Boldt PE and DeLoach CJ (1985) Evaluating *Rhinocyllus conicus* (Coleoptera: Curculionidae) on *Silybum marianum* (Compositae) in Texas. In *Proceedings of the 6th International Symposium on Biological Control of Weeds.* 19–25 August 1984, Vancouver. (Ed ES Delfosse) pp. 417–422. Agriculture Canada, Vancouver.

Bruzzese E (1996) Ecology of *Cirsium vulgare* and *Silybum marianum* in relation to biological control. *Plant Protection Quarterly* **11**, 245–249.

Carr WG, Yugovic JV and Robinson KE (1992) *Environmental Weed Invasions in Victoria.* Vic Dept of Conservation and Environment, Melbourne.

Dodd J (1989) Phenology and seed production of variegated thistle, *Silybum marianum* (L.) Gaertn., in Australia in relation to mechanical and biological control. *Weed Research* **29**, 255–263.

Dunn PH (1978) History of the biological control of musk thistle in North America and studies with the flea beetle *Psylliodes chalcomera.* In *Biological Control of Thistles in the Genus* Carduus *in the United States.* (Ed KE Frick) pp. 1–6. Science Education Administration, US Dept of Agriculture, Washington, DC.

Goeden RD (1976) The palearctic insect fauna of milk thistle, *Silybum marianum*, as a source of biological control agents for California. *Environmental Entomology* **5**, 345–353.

Goeden RD and Ricker DW (1977) Establishment of *Rhinocyllus conicus* on milk thistle in southern Calfornia. *Weed Science* **25**, 288–292.

Goeden RD, Ricker DW and Hawkins BA (1985) Ethological and genetic differences among three biotypes of *Rhinocyllus conicus* (Coleoptera: Curculionidae) introduced to North America for the biological control of Asteraceous thistles. In *Proceedings of the 6th International Symposium on Biological Control of Weeds.* 19–25 August 1984, Vancouver (Ed ES Delfosse) pp. 181–189. Agriculture Canada, Vancouver.

Kelly D and McCallum K (1995) Evaluating the impact of *Rhinocyllus conicus* on *Carduus nutans* in New Zealand. In *Proceedings of the 8th International Symposium on Biological Control of Weeds.* 2–7 February 1992, Canterbury, NZ. (Eds ES Delfosse and RR Scott) pp. 205–211. CSIRO, Canberra.

Klein M and Seitz A (1994) Geographic differentiation between populations of *Rhinocyllus conicus* Frölich (Coleoptera: Curculionidae): concordance of allozyme and morphometric analysis. *Zoological Journal of the Linnean Society* **110**, 181–191.

Lane D, Riches K and Combellack H (1980) 'A survey of the distribution of the noxious weeds of Victoria'. Keith Turnbull Research Institute, Vic Dept of Crown Lands and Survey, Frankston.

Parsons WT and Cuthberson EG (2001) *Noxious Weeds of Australia.* 2nd edn. CSIRO Publishing, Melbourne.

Zwölfer H (1965) Preliminary list of phytophagous insects attacking wild Cynareae (Compositae) in Europe. *Commonwealth Institute of Biological Control Technical Bulletin* **6**, 81–154.

Zwölfer H and Harris P (1984) Biology and host specificity of *Rhinocyllus conicus* (Froel.) (Col., Curculionidae), a successful agent for biocontrol of the thistle, *Carduus nutans* L.1. *Zeitschrift für Angewandte Entomologie* **97**, 36–62.

Zwölfer H and Preiss M (1983) Host selection and oviposition behavior in west-European ecotypes of *Rhinocillus conicus* Froel. (Col., Curculionidae). *Zeitschrift für Angewandte Entomologie* **95**, 113–122.

Solanum elaeagnifolium Cav. – silverleaf nightshade

Raelene M Kwong and Jean-Louis Sagliocco

ABSTRACT

Solanum elaeagnifolium, silverleaf nightshade, is a significant weed in the cropping/pasture and perennial pasture zones of temperate Australia as well as many other countries with Mediterranean-type climates. The extensive and deep perennial root system makes this plant extremely competitive, and the limited control techniques currently available are uneconomical for the treatment of large dense infestations. Despite the large guild of phytophagous fauna associated with silverleaf nightshade in the Americas, few appear to have narrow host-ranges, limiting prospects for introduction into Australia. This chapter provides an overview of natural enemy surveys, biocontrol attempts in South Africa and outcomes of the Australian program. Recommendations for future research are also provided.

Key words: natural enemy surveys, climate matching, host-range.

INTRODUCTION

Solanum elaeagnifolium Cav. (Solanaceae), silverleaf nightshade, is a summer-growing perennial herb (Fig. 1) that has become a major agronomic weed in warm temperate regions throughout the world (Parsons and Cuthbertson 2001).

S. elaeagnifolium is native to the Americas with geographically separate distributions occurring in south-western USA/Central America and South America. According to Boyd *et al.* (1984), the species' most likely centre of geographic origin is the southern USA or northern Mexico; this is in contrast to M Nee's (Solanaceae specialist, New York Botanical Garden, *pers. comm.* 2006) suggestion that the region of origin is Argentina and Chile.

Outside its regions of origin, silverleaf nightshade is invasive in Australia, India, Egypt, Israel, Greece, Sicily, Spain, Morocco, Tunisia, Syria, South Africa and Zimbabwe (Boyd *et al.* 1984; Bouhache and Tanji 1985; Mekki 2007). In Australia, silverleaf nightshade is estimated to infest approximately 140 000 ha and is particularly problematic in SA, NSW and Vic, with isolated infestations in Qld and WA (Fig. 2). A weed risk assessment analysis indicated that silverleaf nightshade is a highly invasive weed (scoring 0.668 out of a maximum potential score of 1) with the potential to invade up to 398 million ha across Australia (Kwong *et al.* 2008).

The extensive and deep root system enables silverleaf nightshade to compete strongly with other vegetation for water and nutrients. Cereal yield reductions of up to 77% have been recorded in SA (Heap and Carter 1999). In pastures, silverleaf nightshade competes with summer-growing pastures such as lucerne, lowering production and leading to reduced carrying capacity. On average, silverleaf nightshade costs affected farmers $1730 per year in control costs and $7786 per year in production losses (McLaren *et al.* 2004). In SA alone, silverleaf nightshade has been estimated to cost producers more that $10 million per year (I Honan, *pers. comm.* 2006).

Figure 1: a) Silverleaf nightshade, *Solanum elaeagnifolium* flowers. b) Dense patch of *Solanum elaeagnifolium* in southern NSW. Photos: a) R Kwong. b) R Stanton.

Considerable research into herbicide control of silverleaf nightshade has been conducted in many countries since the 1930s (Heap and Carter 1999). Some herbicides are effective against seedlings and as spot-spray treatments for isolated plants, but as yet no effective and affordable treatments for the control of large infestations are available.

Cultivation is ineffective because most of the roots are below the depth of cultivation and new plants may reshoot from fragments. Under dry conditions, deep cultivation may reduce infestations (Parsons and Cuthbertson 2001).

BIOLOGICAL CONTROL HISTORY

Several surveys have been conducted since the 1960s to identify the arthropod fauna associated with silverleaf nightshade populations in the Americas. Between 1966 and 1968, Goeden (1971) surveyed Colorado, southern California, Arizona, Texas and New Mexico. During the 1970s, South African scientist Dr Helmuth Zimmerman surveyed populations in Argentina (1971 and 1972) and south-western USA (1973), listed associated organisms and ranked potential agents to be tested in South Africa (Zimmermann 1974). In the 1980s, Australian scientist Dr Tony Wapshere surveyed silverleaf nightshade populations in Texas and around Monterrey, Mexico. On the basis of its greatest arthropod diversity, Wapshere suggested that the Monterrey region was the centre of origin and evolution of silverleaf nightshade (Wapshere 1988). Despite reporting on the biocontrol potential of a number of organisms, Wapshere concluded that herbivores sourced from the Monterrey region would not be sufficiently climatically adapted to summer-drought, cereal-growing areas of southern Australia. Organisms associated with silverleaf nightshade with potential for biological control, including biological control agents introduced into other countries, are summarised in Table 1.

Host testing: South Africa

During the 1970s and 1980s, South African scientists imported and tested a number of candidate agents (Zimmermann 1974; Neser *et al.* 1989). Some were later rejected due to various degrees of attack and survival on eggplant *Solanum melongena* L. and native *Solanum* species.

Olckers (1996) highlighted the difficulty of interpreting host-specificity test results, especially after results led to the rejection of most of the agents considered, including the most promising leaf-feeding beetles, *Leptinotarsa* species (Table 2).

Analysis suggests that the risks posed to eggplant cultivations were minimal and were mitigated by the existing cultural and insecticidal control measures. This led to the release of *Leptinatarsa texana* Schaeffer and *Leptinatarsa defecta* Stål in South Africa in 1992. Both insects have established, with *L. texana* populations having increased more than those of *L. defecta*. The impact of *L. texana* on silverleaf nightshade is substantial (Hoffmann *et al.* 1998), and in summer-rainfall areas the weed is no longer a problem (J Hoffmann, *pers. comm.* 2006). Despite the lack of quantitative data, it appears that the beetles reduce the regenerative capacity of the plants, possibly due to the insect attack on the aerial parts depleting root reserves.

Host testing: Australia

Following approval as a target for biological control in 1985, host testing of the leaf-galling nematode *Ditylenchus phyllobius* (Thorne) Filip'ev from Texas commenced in 1987 in the quarantine facilities of the Keith Turnbull Research Institute in Vic (Field *et al.* 2009). A total of 118 plant taxa were screened and only species in the genus

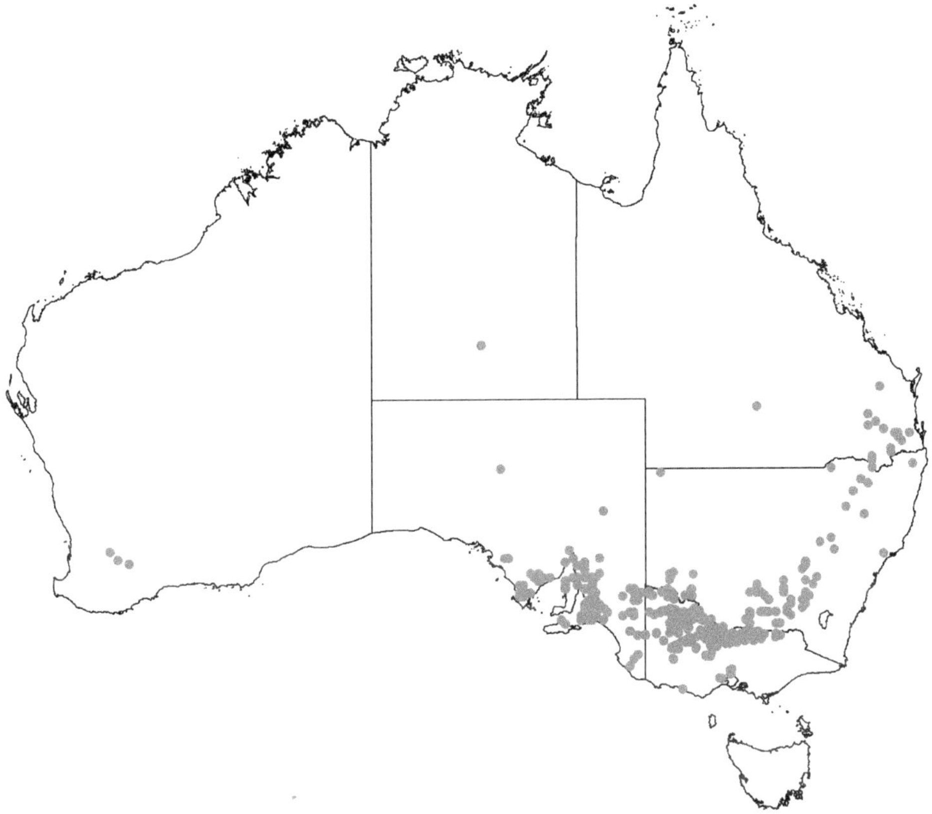

Figure 2: *Solanum elaeagnifolium* distribution in Australia. Location data sourced from Australia's Virtual Herbarium (http://www.ersa.edu.au/avh/ (verified 25.3.10)) and Integrated Pest Management System (2006), Vic Dept of Primary Industries.

Solanum and from the Sections Oliganthes and Melongena appeared to be suitable hosts for the nematode. *D. phyllobius* galls were produced on 13 native *Solanum* species and on 13 of 15 *S. melongena* cultivars tested. Therefore, *D. phyllobius* was considered an unsuitable agent for release in Australia. The biological control program ceased at this point due to concerns that there were limited prospects of finding climatically adapted, host-specific agents for Australia.

DISCUSSION

The success of the leaf-feeding beetle *L. texana* in South Africa has renewed interest in the biological control of silverleaf nightshade in other parts of the world. In recent years, two studies have been undertaken to assess the feasibility of biological control in the Mediterranean basin (Sforza and Jones 2007) and Australia (Kwong *et al.* 2008).

Despite the extensive guild of natural enemies associated with silverleaf nightshade in the Americas, few have proved host-specific when subject to rigorous host-range screening tests (Olckers 1996; Field *et al.* 2009). With 117 species within the *Solanum* genus present in Australia (Haegi *et al.* 1982), including many important agricultural crops such as eggplant, potato, capsicum and tobacco, finding sufficiently host-specific natural enemies may be challenging. Hence, priority for agent selection should be directed to typically highly host-specific organisms such as fungal pathogens and eriophyid mites. It is interesting to note that no pathogens or mites were identified in the natural enemy surveys conducted by Goeden, Zimmermann or Wapshere. Trotter (1900) described a leaf mite *Aceria bicornis* Trotter (Acarina: Eriophyidae) found in abundance on *S. elaeagnifolium* in Argentina, and the leaf-spot fungus *Pseudocercospora atromarginalis* (Atk.) Deighton (Dothideomycetidae: Mycosphaerellaceae) was believed to have a narrow host-range including *S. elaeagnifolium* in the USA (Farr 2006).

Future directions

Kwong *et al.* (2008) made two recommendations for future biological control research in Australia. First, the genetic variation of silverleaf nightshade in Australia should be studied and possible sources of origin from the

Table 1: Agents associated with *Solanum elaeagnifolium* with potential for biological control

Order: Family Genus Species	Plant association	Host-range	Distribution	References
Nematoda: Anguinidae				
Orrina phyllobia (Thorne) (= *Ditylenchus phyllobius* (Thorne) Filipjev (=*Nothanguina phyllobia* Thorne)	Leaf-galling nematode	*S. elaeagnifolium*, *S. viarum*, *S. tampicense*	Mexico, Texas, Arizona	Esser and Orr (1979); Field *et al.* (2009)
Acarina: Eriophyidae				
Getrapodili sp.	Leaf galls	Unknown	Argentina	Zimmermann (1974)
Eriophyes bicornis Trotter	Leaf erineum	Possibly restricted to *S. elaeagnifolium*	Argentina	Trotter (1900)
Eriophyes sp.	Deformed leaves	Possibly restricted to *S. elaeagnifolium*	Argentina	Kieffer and Jörgensen (1910); Amrine (1994)
Hemiptera: Tingidae				
Gargaphia arizonica Drake and Carvalho	Leaves, cell-sucking	*S. elaeagnifolium*	Mexico	Wapshere (1988)
Lepidoptera: Gelechiidae				
Frumenta nephelomicta Meyrick as *Asapharca nephelomicta* Meyrick	Fruits and seeds larval feeding	*S. elaeagnifolium*	Mexico	Wapshere (1988); Zimmermann (1974)
Frumenta (sp. A)	Fruits and seeds	*S. elaeagnifolium*	Texas	Olckers (1996)
Frumenta solanophaga Adamski and Brown	Fruits and seeds	*S. elaeagnifolium* only host known	Mexico	Adamski and Brown (2002)
Symmetrischema ardeola (Meyr)	Flowers and flower buds, stamens and pistils	*S. elaeagnifolium*	Argentina (Tucuman)	Zimmermann (1974)
Keiferia sp.	Leaf-miner	Unknown	Argentina	Zimmermann (1974)
Gnorimoschema sp.	Stem galls	Unknown	Argentina	Zimmermann (1974)
Lepidoptera: Carposinidae				
Unidentified species	Fruits and seeds	Unknown	USA	Goeden (1971)
Coleoptera: Nitidulidae				
Carpophilus sp.	Flowers and flower buds	*S. elaeagnifolium*	Argentina	Zimmermann (1974)
Coleoptera: Chrysomelidae				
Gratiana pallidula (Boh.)	Leaves	*Solanum* spp.	Mexico, USA	Wapshere (1988); Goeden (1971)
Gratiana lutescens (Boh.)	Leaves, petioles and flower buds	*S. elaeagnifolium* and possibly *S. melongena*	Argentina	Zimmermann (1974); Siebert (1975); Hill (1999)

Order: Family Genus Species	Plant association	Host-range	Distribution	References
Leptinotarsa defecta Stahl	Leaves	*S. elaeagnifolium* and *S. dimidatium*	Mexico, Texas	Olckers and Zimmermann (1994)
Leptinotarsa texana (Schaeffer)	Leaves	*S. elaeagnifolium*, possibly *S. rostratum*	Mexico, Texas	Olckers and Hulley (1994); Olckers *et al.* (1999)
Leptinotarsa decimlineata (Say)	Leaves	*S. elaeagnifolium*, possibly *S. rostratum*	USA	Goeden (1971)
Coleoptera: Curculionidae				
Trichobaris texana LeConte	Stem-borer	*S. elaeagnifolium* and three closely related *Solanum* spp.	Mexico	Wapshere (1988); Zimmermann (1974); Cuda and Burke (1985)
Diptera: Cecidomyiidae				
Asphondylia sp.	Flowers	Unknown	Texas	Provisional checklist of the Cecidomyiidae (Diptera) (gall midges) of Brackenridge Field Laboratory, compiled by RW Patrock (24.8.89); Anon (2005) <http://www.utexas.edu/research/bfl/species/cecido.html>
Unknown species	Stem galls	Unknown	Mexico	Zimmermann (1974)
Unnamed species	Stems, distortion by galling	*S. elaeagnifolium*	Mexico	Wapshere (1988)
Diptera: Agromyzidae				
Haplomyza sp.	Leaves	Unknown	Argentina	Zimmermann (1974)
Diptera: Tephritidae				
Zonosemata vittigera (Coquillett)	Fruits and seeds	*S. elaeagnifolium*	Mexico; Arizona	Wapshere (1988); Goeden and Ricker (1971)

Table 2: Status of candidate agents evaluated for the biological control of ***Solanum elaeagnifolium*** in South Africa

Agent	Year	Origin	Damage	Status
Gratiana lutescens (Boh.) (Coleoptera: Chrysomelidae)	1973	Argentina, Texas	Defoliator	Rejected
Arvelius albopunctatus (DeGeer) (Hemiptera: Pentatomidae)	1974	Argentina	Seed-feeder	Rejected
Conotrachelus bisignatus (Boh.) (Coleoptera: Curculionidae)	1974	Argentina	Seed-feeder	Rejected
Frumenta nephelomicta Meyrick (Lepidoptera: Gelechiidae)	1976	Mexico	Fruit-galler	Released; not established
Ditylenchus phyllobius (Thorne) Filip'ev (Nematoda)	1984	Texas	Leaf-galler	Rejected
Leptinotarsa texana Schaeffer (Coleoptera: Chrysomelidae)	1985–89	Texas	Defoliator	Released; established
Leptinotarsa defecta Stål (Coleoptera: Chrysomelidae)	1985–87	Texas	Defoliator	Released; established
Frumenta sp. nov. (Lepidoptera: Gelechiidae)	1989	Texas	Fruit- and stem-galler	Released locally; not established

Source: Olckers (1996).

native range determined. Despite morphological differences between the North and South American populations, the taxonomic relationships and genetic variation within and between these populations are not well understood. Nor do we know from where the Australian populations originated. As highly host-specific agents are required, potential agents should be sourced from populations closely related to those occurring in Australia.

A genetic study is underway to unravel the origins of silverleaf nightshade from a number of countries, including Australia. The study, which is being led by USDA-ARS European Biological Control Laboratory, France will compare accessions from invasive populations with silverleaf nightshade from North and South America.

Further natural enemy surveys are needed and should be conducted in areas with comparable climates to temperate Australia. A climate analysis comparing the distribution of silverleaf nightshade in south-eastern Australia with the native ranges of the weed indicated that the most comparable climates are in the central regions of Argentina and Chile (Kwong *et al.* 2008). These regions have not been surveyed; Zimmerman's surveys in 1971–72 included only the northern parts of Argentina.

Natural enemy prospects for Australia

Of the 11 phytophagous species originating from Argentina, the literature reveals very little information on their biology, host-range and impact on silverleaf nightshade, making it difficult to adequately assess their biological control potential. However, the gelechiid moth *Symmetrischema ardeola* (Meyr) warrants further investigation as preliminary tests by Zimmermann (1974) in Argentina suggest that it might be specific to silverleaf nightshade. The larvae of *S. ardeola* feed on flower buds, stamens and pistils and thus could reduce seed production.

REFERENCES

Adamski D and Brown JW (2002) A new species of *Frumenta* Busk (Lepidoptera: Gelichiidae: Gnorimoschemini) from Mexico: a potential biocontrol agent against *Solanum elaeagnifolium* (Solanaceae). In *Proceedings of the Entomological Society of Washington* **104**, 1029–1035.

Amrine JW and Stasny TA (1994) *Catalogue of the Eriophyoidea (Acarina: Prostigmata) of the World.* Indira Publishing, West Bloomfield, Michigan.

Anon (2005) Provisional checklist of the Cecidomyidae (Diptera) (gall midges) of Brackenridge Field Laboratory. <http://www.utexas.edu/research/bfl/species/cecido.html>.

Boyd JW, Murray DS and Tyrl RJ (1984) Silverleaf nightshade, *Solanum elaeagnifolium*, origin, distribution and relation to man. *Economic Botany* **38**, 210–216.

Bouhache M and Tanji A (1985) Evaluation du stock en semences de la morelle jaune (*Solanum elaeagnifolium* Cav.) dans le sol du Tadla (Maroc). *Weed Research* **25**, 11–14.

Cuda JP and Burke HR (1985) Biology and impact of *Trichobaris texana* (Coleoptera: Curculionidae) on silverleaf nightshade, *Solanum elaeagnifolium* in central Texas. In *Proceedings of the 6th International Symposium on Biological Control of Weeds.* 19–25 August 1985, Vancouver. (Ed ES Delfosse) pp. 721–734. Minister of Supply and Services, Canada.

Esser RP and Orr CC (1979) '*Nothanguina phyllobia* a nematode pest of a noxious weed *Solanum elaeagnifolium*'. Circular 51. Florida Department of Agriculture and Consumer Services, Division of Plant Industry, Gainesville, Florida.

Farr DF, Rossman AY, Palm ME and McCray EB (2006) Fungal databases, systematic botany and mycology laboratory. ARS-USDA. <http://nt.ars-grin.gov/fungaldatabases.htm> (retrieved 31.1.06).

Field RP, Kwong RM and Sagliocco J-L (2009) Host specificity of *Ditylenchus phyllobius*, a potential biological control agent of silver-leaf nightshade (*Solanum elaeagnifolium* Cav.) in Australia. *Plant Protection Quarterly* **24**, 141–144.

Goeden RD (1971) Insect ecology of silverleaf nightshade. *Weed Science* **19**, 45–51.

Goeden RD and Ricker DW (1971) Biology of *Zonosemata vittigera* relative to silverleaf nightshade. *Journal of Economic Entomology* **64**, 417–421.

Haegi L, Purdie RW and Symon DE (1982) *Flora of Australia. Vol. 29 Solanaceae.* Australian Government Publishing Service, Canberra.

Heap JW and Carter RJ (1999) The biology of Australian weeds. 35. *Solanum elaeagnifolium* Cav. *Plant Protection Quarterly* **14**, 2–12.

Hill MP (1999) *Gratiana lutescens* (Boheman) (Coleoptera: Chrysomelidae: Cassidinae) reconsidered as a natural enemy of satansbos, *Solanum elaeagnifolium* Cavanilles (Solanaceae), in South Africa. *African Entomology* **7**, 177–181.

Hoffman JH, Moran VC and Impson FAC (1998) Promising results from the first biological programme against a solanaceous weed (*Solanum elaeagnifolium*). *Agriculture, Ecosystems and Environment* **170**, 145–150.

Kieffer J-J and Jörgensen P (1910) Gallen und Gallentiere aus Argentinien. *Zentralblatt fuer Bakteriologie, Parasitenkunde un Infektionskrankheiten* **27**, 435.

Kwong RM, Sagliocco JL, Weiss J, Hunt T and Morfe D (2008) Feasibility of biological control of silverleaf nightshade in Australia. In *Proceedings of the 16th Australian Weeds Conference.* 18–22 May 2008, Cairns. (Eds RD van Klinken, VA Osten, FD Panetta and JC Scanlan) pp. 256–258. Queensland Weeds Society, Brisbane.

McLaren DA, Morfe TA, Honan I and Holtkamp R (2004) Distribution, economic impact and attitudes towards silverleaf nightshade (*Solanum elaeagnifolium* Cav.) in Australia. In *Proceedings of the 14th Australian Weeds Conference.* 6–9 September 2004, Wagga Wagga, NSW. (Eds BM Sinden and SB Johnson) p. 701. Weed Society of New South Wales, Sydney.

Mekki M (2007) Biology, distribution and impacts of silverleaf nightshade (*Solanum elaeagnifolium* Cav.). *EPPO Bulletin* **37**, 114–118.

Neser S, Zimmermann HG, Erb HE and Hoffman JH (1989) Progress and prospects for the biological control of two *Solanum* weeds in South Africa. In *Proceedings of the 7th International Symposium on Biological Control of Weeds.* 6–11 March 1988, Rome. (Ed ES Delfosse) pp. 371–381. MAF, Rome.

Olckers T (1996) Improved prospects for biological control of three solanum weeds in South Africa. In *Proceedings of the 9th International Symposium on Biological Control of Weeds.* 19–26 January 1996, Stellenbosch, South Africa. (Ed VC Moran and JH Hoffmann) pp. 307–312. University of Cape Town, South Africa.

Olckers T and Hulley PE (1994) Resolving ambiguous results of host-specificity tests: the case of two *Leptinotarsa* species (Coleoptera: Chrysomelidae) for biological control of *Solanum elaeagnifolium* (Solanaceae) in South Africa. *African Entomology* **2**, 137–144.

Olckers T and Zimmermann HG (1994) Host specificity tests on leaf-feeding insects: aberrations from the use of excised leaves. *African Entomology* **2**, 68–70.

Olckers T, Hoffmann JH, Moran VC, Impson FAC and Hill MP (1999) The initiation of biological control programmes against *Solanum elaeagnifolium* Cavanilles and *S. sisymbriifolium* Lamarck (Solanaceae) in South Africa. *African Entomology Memoir* **1**, 55–63.

Parsons WT and Cuthbertson EG (2001) *Noxious Weeds of Australia.* 2nd edn. CSIRO Publishing, Melbourne.

Sforza R and Jones WA (2007) Potential for classical biocontrol of silverleaf nightshade in the Mediterranean Basin. *Bulletin OEPP/EPPO Bulletin* **37**, 156–162.

Siebert MW (1975) Candidates for the biological control of *Solanum elaeagnifolium* Cav. (Solanaceae) in South Africa. 1. Laboratory studies on the biology of *Gratiana lutescens* (Boh.) and *Gratiana pallidula* (Boh.) (Coleoptera: Cassidae). *Journal of the Entomological Society of South Africa* **38**, 297–304.

Trotter A (1900) Description d'une espèce nouvelle d'*Eriophyes* (Acar.) de l'Amérique du Sud. *Bulletin de la Société entomologique de France* **11**, 224–226.

Wapshere AJ (1988) Prospects for the biological control of silver-leaf nightshade, *Solanum elaegnifolium*, in Australia. *Australian Journal of Agricultural Research* **39**, 187–197.

Zimmermann HG (1974) Preliminary work on the biological control of satansbos, *Solanum elaeagnifolium* Cav. Unpublished papers. *1st Medicinal Weeds Conference.* 13–14 August 1974, Pretoria, South Africa. pp. 151–168. Weeds Research Laboratory, Plant Protection Research Institute.

Sonchus oleraceus L. – sowthistle

John K Scott and Kathryn McCarren

ABSTRACT

Sowthistles, *Sonchus* spp. (Asteraceae), are widely known as serious weeds affecting crop production in many regions of the world. Australia has three species, the weedy introduced *Sonchus oleraceus* and *Sonchus asper* and the native *Sonchus hydrophilus*. *S. oleraceus* is considered a major weed of the south-east Qld and northern NSW cropping region, whereas *S. asper* has relatively minor importance in Australia. Surveys of *S. oleraceus* identified several potential biocontrol agents already present in Australia including *Miyagia pseudosphaeria*, *Aceria thalgi* and the potential mycoherbicide pathogen *Aschochyta sonchi*. In Australia, *A. thalgi* and *M. pseudosphaeria* are widespread and could be used in a conservative approach to biological control of exotic *Sonchus* spp. Potential agents identified in preliminary surveys in the native distribution in France could be considered for a classical biocontrol approach. Host-range studies would need to take into account the close relationship between introduced and native *Sonchus* in Australia.

Key words: biological control, crop weeds, sowthistle, herbicide-resistant weeds, plant pathogens, surveys for phytophagous species.

INTRODUCTION

Sowthistle, *Sonchus oleraceus* L. (Asteraceae) (Fig. 1) is an erect annual and occasionally biennial herb of Eurasian origins. Capitula production (flowering) occurs throughout the year. Flowers are self-fertile and reproduction is entirely by seed with up to 8000 seeds produced per plant. Widderick *et al.* (2004) measured seed densities up to 39 340 seeds m^{-2}. Achenes contain a single seed and have a pappus of bristles for dispersal by the wind. Very few seeds are dormant; after two years, less than 2% of seeds on the surface and up to 12% of seeds at 10 cm soil depth are viable (Walker *et al.* 2006; Widderick *et al.* 2010). Germination is mainly in autumn and winter although it can occur year-round (Widderick *et al.* 2004) in response to a wide range of temperatures (Chauhan *et al.* 2006). The seedling forms a rosette and develops a strong tap root 0.5–1.5 cm in diameter, 10–35 cm long. The stem bolts after about 640 degree days (base temperature for development 5°C) (Cici

Figure 1: *Sonchus oleraceus*. Floreat, WA. Photo: JK Scott, CSIRO.

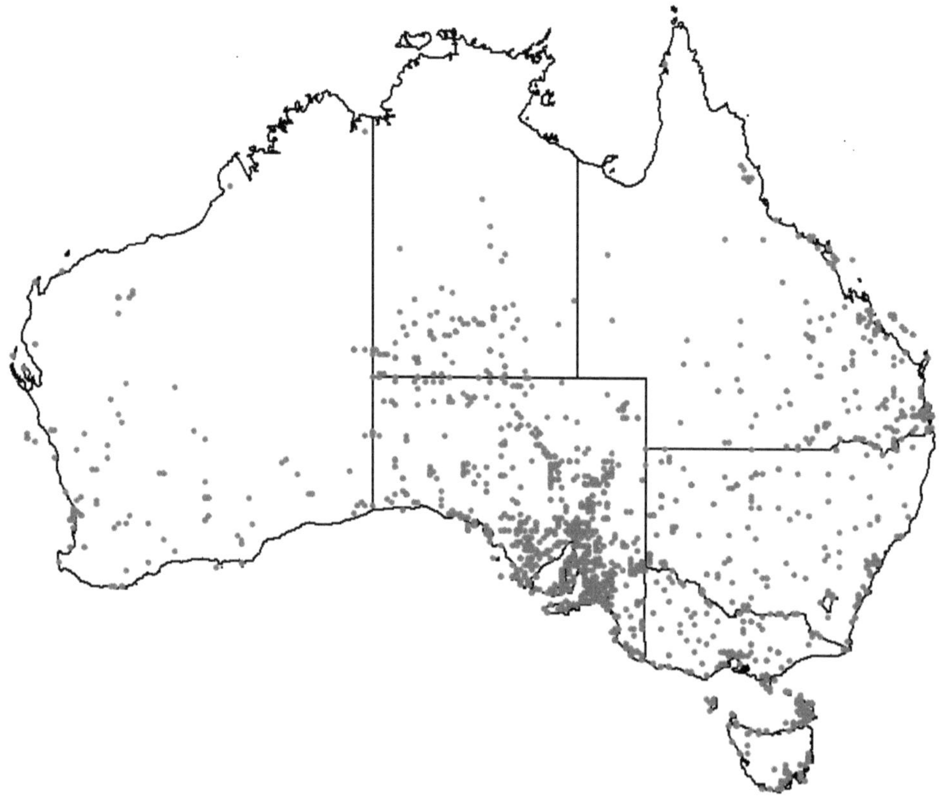

Figure 2: Distribution of *Sonchus oleraceus* in Australia. Specimen data reproduced from Australia's Virtual Herbarium with permission of the Council of Heads of Australasian Herbaria Inc.

et al. 2009) and can grow up to 140 cm high (Boulos 1973). Further details on the biology and growth of *S. oleraceus* can be found in Cici *et al.* (2009), Hutchinson *et al.* (1984) and Widderick *et al.* (2010).

Australia has three species of *Sonchus*, the introduced *Sonchus asper* (L.) Hill and *S. oleraceus* and a native species *Sonchus hydrophilus* Boulos (Thompson 2007). These species are phylogenetically closely related (Kim *et al.* 2007). The only other related native species is *Actites megalocarpa* (Hook. f.) Lander (Kim *et al.* 2004; Lander 1976) and there are no plants of economic importance in the subtribe Sonchinae (Asteraceae: Cichorieae).

Sowthistle is widespread in Australia (Fig. 2) and a common weed of disturbed ecosystems. *S. oleraceus* is also a major weed of Australian cropping systems, especially those of south-east Qld and northern NSW, with regional grower surveys labelling it one of the main weeds of the area (Chauhan *et al.* 2006; Osten *et al.* 2007; Walker *et al.* 2005). Its incidence is increasing (Widderick *et al.* 2010) largely due to growers adopting conservation tillage and the evolving resistance of the weed to chemical controls such as chlorsulfuron and atrazine (Adkins *et al.* 1997; St John-Sweeting *et al.* 2008). In contrast, *S. asper* has relatively minor importance in Australia.

BIOLOGICAL CONTROL HISTORY

The biocontrol of sowthistles (primarily targeting *Sonchus arvensis* L.) was investigated in the 1970s for Canada (Peschken 1984; Zollinger and Parker 1999). Surveys for potential biocontrol agents in central Europe are reported in Schroeder (1974) and for Kursk, Russia by Shurovenkov (1983). The tephritid fly *Tephritis dilacerata* (Loew), that has flower head feeding larvae, was released in Canada but it failed to establish. The agromyzid fly *Liriomyza sonchi* Hendel and the cecidomyid fly *Cystiphora sonchi* (Bremi) established in Canada, the former without impact over a limited distribution. The cecidomyid is more widespread, but generally not effective except in unmown sites (Julien and Griffiths 1998; Peschken 1984). Classical biocontrol has also been proposed for *S. arvensis* in Brazil (Ellison and Barreto 2004).

More recent work in Europe has focused on developing mycoherbicides and the identification of phytotoxic metabolites for *Sonchus* species and other weeds based on *Alternaria sonchi* J.J.Davis (Pleosporales: Pleosporaceae) (Evidente *et al.* 2009), *Ascochyta sonchi* (Sacc.) Grove (Ascomycetes) (= *Phoma exigua* var. *exigua*) (Cimmino *et al.* 2008) and *Ascochyta tussilaginis* Oudem. (Berestetski and Parkhomenko 2004). Mycoherbicides based on cultures of *Ascochyta sonchi* have been tested in the glasshouse and the field, but it was determined that more aggressive isolates were needed (Gasich *et al.* 2007).

The primary aim of the work carried out for Australia since 2004 was to determine the potential for biocontrol of *S. oleraceus*. Sowthistle has not been nominated as a target species for biocontrol in Australia.

EXPLORATION

As the plant's origins are thought to lie in Europe, CSIRO, through the CRC for Australian Weed Management, searched for potential classical biocontrol agents for *S. oleraceus* in France, to complement northern European surveys made by other researchers mentioned above. Surveys were also made in Australia to determine the presence of potential agents.

Preliminary surveys were made by Mireille Jourdan, Thierry Thomann and Janine Vitou for biological control agents from April 2004 to May 2005. Fourteen species of plant pathogens were found on *S. asper* and *S. oleraceus* in southern France (M Jourdan, *pers. comm.* 2005). Seven insect species were recorded during these surveys, the most frequent being *Cystiphora sonchi* (J Vitou, *pers. comm.* 2005).

Surveys by Kathryn McCarren and John K Scott of *Sonchus* species in Australia, primarily in south-western WA, south-eastern Qld and coastal SA and Vic, examined 137 sites from November 2006 to February 2008, mainly for the presence of the mite *Aceria thalgi* Knihinicki and the rust *Miyagia pseudosphaeria* (Mont.) Jorst. (McCarren and Scott 2008). The survey was not intended to be a comprehensive listing of all plant pathogens because there were expected to be many species (Farr and Rossman 2010), and the time and resources available permitted only major diseases or potential biocontrol organisms to be examined. For example, *Ascochyta sonchi*, a species with mycoherbicide potential, was isolated from diseased plants from WA and the identification confirmed by its DNA sequence (Genbank Accession Number HQ201951).

CANDIDATES

***Miyagia pseudosphaeria* (Mont.) Jorst. (Uredinales: Pucciniaceae).** The rust fungus *M. pseudosphaeria* (Fig. 3) has a worldwide distribution and probably originated from Europe or north Asia. It was found at six sites, all Mediterranean coastal sites, out of the 73 surveyed (M Jourdan, *pers. comm.* 2005). It has spread from Eurasia to New Zealand (1919) and Australia (1953) (McKenzie and Johnston 2004) and more recently North America (2002) (Hernández *et al.* 2003). *M. pseudosphaeria* was the most frequently encountered organism found in our surveys of *S. oleraceus* (83 of 137 sites) and was widespread across southern and eastern Australia (K McCarren and JK Scott, unpub.). It has been hypothesised that the recent spread may be explained by the appearance of a more virulent strain of the fungus (McKenzie and Johnston 2004).

M. pseudosphaeria is an autoecious rust. Uredospores and aeciospores were common on stems and leaf adaxial and abaxial surfaces. Teliospores were rare on field material from Australia examined by microscopy (K McCarren and JK Scott, unpub.). Field observations indicated that *M. pseudosphaeria* was capable of infecting *S. asper*, *S. oleraceus* and *S. hydrophilus*; however, infection of *S. hydrophilus* was rare and when infected, damage was minimal. However, in glasshouse experiments, *M. pseudosphaeria* would infect *S. oleraceus* and *S. hydrophilus* to the same extent (K McCarren and JK Scott, unpub.). McKenzie and Johnston (2004) reported that the rust also infects the

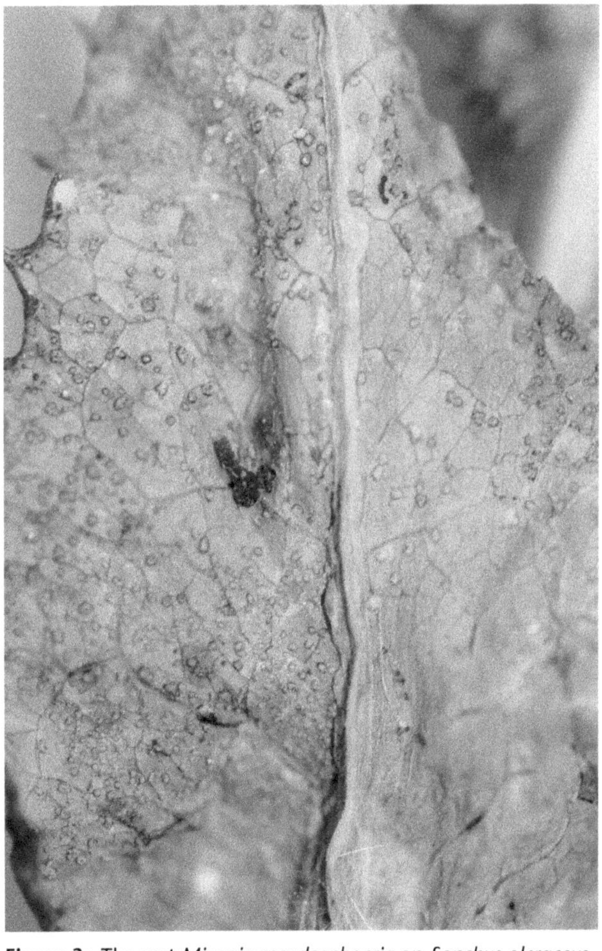

Figure 3: The rust *Miyagia pseudosphaeria* on *Sonchus oleraceus*. Photo: K McCarren, CSIRO.

endangered NZ endemic *Sonchus kirkii* Hamlin. Kim *et al.* (2007) showed that these four species are very closely related, so some overlap in the host-range might be expected. In glasshouse host no-choice specificity tests, *M. pseudosphaeria* urediospores were unable to infect *Actites megalocarpus* (Hook. f.) Lander, (a native which shares the Sonchinae subtribe), *Reichardia tingitana* (L.) Roth (Sonchinae), *Cichorium endiva* L. (Hyoseridinae), *Lactuca sativa* L. (Lactucinae), *Lactuca serriola* L. (Lactucinae) and *Taraxacum officinale* F.H. Wigg (Crepidinae) while readily infecting *S. asper*, *S. oleraceus* and *S. hydrophilus* (K McCarren and JK Scott, unpub.).

***Aceria thalgi* Knihinicki (Acari: Eriophyidae).** *A. thalgi* is a mite that causes severe curling and rolling of *S. oleraceus* leaves in Australia (Fig. 4). It also affects the introduced *S. asper* and the native *S. hydrophilus*, to a lesser degree (Knihinicki *et al.* 2009). Glasshouse pathogenicity trials of this eriophyid mite showed a significant growth reduction in infested plants of both *S. oleraceus* and *S. hydrophilus*, although age of plant when infection occurred played an important role in the extent of damage. *S. hydrophilus* was not affected to the same extent as *S. oleraceus* (K McCarren and JK Scott, unpub). The mite is only known from Australia, and possibly NZ (Knihinicki *et al.* 2009). It is widespread in Australia and it is possible that the native *Sonchus* species may be the original host. Preliminary host-specificity testing and studies of the interrelationship between the mite and host in both the glasshouse and field environments showed that the mite is likely to be specific to *Sonchus* species (K McCarren and JK Scott, unpub.).

Figure 4: Curling of leaf edges due to *Aceria thalgi* on *Sonchus oleraceus*. Photo: P Yeoh, CSIRO.

DISCUSSION

The Sonchinae subtribe is widely distributed throughout the world and includes many successful weeds and several island endemics (Kim *et al.* 2007). Phylogenetic evidence (Kim *et al.* 2007) shows a close relationship between *S. oleraceus* and Australian and NZ endemic *Sonchus* species and *Actites*, *Kirkianella* and *Embergeria* genera. In addition, the evidence from pollen deposits indicates the presence of *S. oleraceus* on Norfolk Island before Polynesian colonisation (MacPhail *et al.* 2001). These two factors raise the possibility that *S. oleraceus* might be a prehistoric colonist of Australia. Despite this, the presence of associated fauna and pathogens points to a Eurasian origin and introduced status in Australia.

There are no economic plants in the subtribe Sonchinae in Australia, which supports the suitability of *S. oleraceus* as a target for biocontrol. However, being a weed of cropping means that biocontrol will be difficult because a successful agent will need to respond to the ecological uncertainty created by cultivation, crop rotations and pesticide treatments.

There are several options – mycoherbicide, conservative, inoculative or classical approaches (Wapshere *et al.* 1989) – that could be further developed for the biocontrol of *S. oleraceus* in Australia. *Ascochyta sonchi*, now known to occur in Australia, is being developed as a mycoherbicide overseas (Cimmino *et al.* 2008) and could be considered locally. The two agents that are widespread in Australia, *M. pseudosphaeria* and *A. thalgi*, could be considered for the conservative approach to biocontrol. Both would require further understanding of their biology and dispersal to enable management options that maximise the agents' damage on the weedy hosts.

Studies of the genetic variation in *M. pseudosphaeria* will aid our understanding of its origins and whether importation of further strains are warranted in a classical approach to biocontrol. Also requiring investigation are the mechanisms of dispersal and survival in the absence of the host. Although *M. pseudosphaeria* was seen to produce teliospores which are long-term survival structures for rust fungi, they were observed only rarely. It is likely that the pathogen relies on wind dispersal for its distribution, rather than alternative *Sonchus* hosts or soil survival mechanisms for infection of germinating *S. oleraceus*. The search for a more virulent strain would need to take into context the dispersal mechanism and threat to *S. kirkii*, an NZ species undergoing gradual decline, especially as trans-Tasman Sea spread is possible for rusts (McKenzie and Johnston 2004).

Further to the classical approach would be to expand overseas surveys to include a wider range of regions in western and southern Europe and in Asia, covering a wider range of climates and habitats. For example, another eriophyid mite that forms galls on leaves, *Aceria sonchi* (Nalepa), is known from *Sonchus* species in the Mediterranean region but has not been found in recent surveys (Knihinicki *et al.* 2009). Further work is needed on all potential agents, such as the cecidomyid fly *C. sonchi* to determine the host-specificity between the weedy and native *Sonchus* species. Lastly, *A. thalgi* could be considered as a 'new association' biocontrol agent for Canada, where sowthistles are major weeds of agriculture.

ACKNOWLEDGEMENTS

We thank Mireille Jourdan, Thierry Thomann and Janine Vitou for information about surveys in France.

REFERENCES

Adkins SW, Wills D, Boersma M, Walker SR, Robinson G, McLeod RJ and Einam JP (1997) Weeds resistant to chlorsulfuron and atrazine from the north-east grain region of Australia. *Weed Research* **37**, 343–349.

Berestetski AO and Parkhomenko NV (2004) Effect of temperature, light and nutrients on morphological and cultural characteristics of *Ascochyta tussilaginis*. *Mikologiya I Fitopatologiya* **38**, 78–88.

Boulos L (1973) Révision systématique du genre *Sonchus* L. s.l. IV. Sous-genre 1. Sonchus. *Botaniska Notiser* **126**, 155–196.

Chauhan BS, Gill G and Preston C (2006) Factors affecting seed germination of annual sowthistle (*Sonchus oleraceus*) in southern Australia. *Weed Science* **54**, 854–860.

Cici SZH, Adkins S and Hanan J (2009) Modelling the morphogenesis of annual sowthistle, a common weed in crops. *Computers and Electronics in Agriculture* **69**, 40–45.

Cimmino A, Andolfi A, Berestetskiy A and Evidente A (2008) Production of phytotoxins by *Phoma exigua* var. *exigua*, a potential mycoherbicide against perennial thistles. *Journal of Agricultural and Food Chemistry* **56**, 6304–6309.

Ellison CA and Barreto RW (2004) Prospects for the management of invasive alien weeds using co-evolved fungal pathogens: a Latin American perspective. *Biological Invasions* **6**, 23–45.

Evidente A, Punzo B, Andolfi A, Berestetskiy A and Motta A (2009) Alternethanoxins A and B, polycyclic ethanones produced by *Alternaria sonchi*, potential mycoherbicides for *Sonchus arvensis* biocontrol. *Journal of Agricultural and Food Chemistry* **57**, 6656–6660.

Farr DF and Rossman AY (2010) Fungal databases, systematic mycology and microbiology laboratory. ARS-USDA. <http://nt.ars-grin.gov/fungaldatabases/> (accessed 17.8.10).

Gasich EL, Khlopunova LB and Levitin MM (2007) Evaluation of the fungus *Ascochyta sonchi* for biological control of *Cirsium arvense* and *Sonchus arvensis*. In *Proceedings of the 14th European Weed Research Society Symposium*. 17–21 June 2007, Hamar, Norway. (Ed E Fløistad) p. 29. European Weed Research Society, Doorwerth, Netherlands.

Hernández JR, Koike ST, Scheck HJ and Palm Hernandez ME (2003) First report of the rust *Miyagia pseudosphaeria* on *Sonchus oleraceus* in the Americas. *Plant Disease* **87**, 752.

Hutchinson I, Colosi J and Lewin RA (1984) The biology of Canadian weeds. 63. *Sonchus asper* (L.) Hill and *Sonchus oleraceus* L. *Canadian Journal of Plant Science* **64**, 731–744.

Julien MH and Griffiths MW (1998) *Biological Control of Weeds: A World Catalogue of Agents and their Target Weeds*. 4th edn. CABI, Wallingford, UK.

Kim SC, Lu CT and Lepschi BJ (2004) Phylogenetic positions of *Actites megalocarpa* and *Sonchus hydrophilus* (Sonchinae: Asteraceae) based on ITS and chloroplast non-coding DNA sequences. *Australian Systematic Botany* **17**, 73–81.

Kim SC, Chunghee L and Mejías JA (2007) Phylogenetic analysis of chloroplast DNA matK gene and ITS of nrDNA sequences reveals polyphyly of the genus *Sonchus* and new relationships among the subtribe Sonchinae (Asteraceae: Cichorieae). *Molecular Phylogenetics and Evolution* **44**, 578–597.

Knihinicki DK, McCarren KL and Scott JK (2009) A new species of *Aceria* (Acari: Eriophyidae) damaging sow-thistles, *Sonchus* spp. (Asteraceae), in Australia with notes on *Aceria sonchi* (Nalepa, 1902). *Zootaxa* **2119**, 23–38.

Lander NS (1976) *Actites*, a new genus of Compositae from Australia. *Telopea* **1**, 129–135.

MacPhail MK, Hope GS and Anderson A (2001) Polynesian plant introductions in the Southwest Pacific: initial pollen evidence from Norfolk Island. *Records of the Australian Museum, Supplement* **27**, 123–134.

McCarren KL and Scott JK (2008) Two biological control options for *Sonchus oleraceus* in Australia. In *Proceedings of the 16th Australian Weeds Conference*. 18–22 May 2008, Cairns. (Eds RD van Klinken, VA Osten, FD Panetta and JC Scanlan) pp. 259–261. Queensland Weeds Society, Brisbane.

McKenzie EHC and Johnston PR (2004) *Puccinia embergeriae* sp. nov. on Chatham Islands sow thistle (*Embergeria grandifolia*) and a note on *Miyagia pseudosphaeria* on sow thistles (*Sonchus* spp.) in New Zealand. *New Zealand Journal of Botany* **42**, 657–661.

Osten VA, Walker SR, Storrie A, Widderick M, Moylan P, Robinson GR and Galea K (2007) Survey of weed flora and management relative to cropping practices in the north-eastern grain region of Australia. *Australian Journal of Experimental Agriculture* **47**, 57–70.

Peschken DP (1984) *Sonchus arvensis* L., perennial sow-thistle, *S. oleraceus* L., annual sow-thistle, and *S. asper* (L.) Hill, spiny annual sow-thistle (Compositae). In *Biological Control Programmes against Insects and Weeds in Canada 1969–1980*. (Eds JS Kelleher and MA Hulme) pp. 205–209. Commonwealth Agriculture Bureau, Slough, UK.

Schroeder D (1974) The phytophagous insects attacking *Sonchus* spp. (Compositae) in Europe. In *Proceedings of the 3rd International Symposium on Biological Control of Weeds*. 10–14 September 1973, Montpellier, France. (Ed AJ Wapshere) pp. 89–96. Commonwealth Agricultural Bureau, Slough, UK.

Shurovenkov BG (1983) Phytophages of yellow sowthistle. *Zashchita Rastenii (Moscow)* **11**, 22–23.

St John-Sweeting RS, Preston C, Baker J, Walker S and Widderick M (2008) Gene movement in herbicide resistant sowthistle (*Sonchus oleraceus* L.). In *Proceedings of the 16th Australian Weeds Conference*. 18–22 May 2008, Cairns. (Eds RD van Klinken, VA Osten, FD Panetta and JC Scanlan) pp. 113–115. Queensland Weeds Society, Brisbane.

Thompson IR (2007) A taxonomic treatment of tribe Lactuceae (Asteraceae) in Australia. *Muelleria* **25**, 59–100.

Walker SR, Taylor IN, Milne G, Osten VA, Hoque Z and Farquharson RJ (2005) A survey of management and economic impact of weeds in dryland cotton cropping systems of subtropical Australia. *Australian Journal of Experimental Agriculture* **45**, 79–91.

Walker S, Wilson B, Wu H, Widderick M and Taylor I (2006) Weed seed persistence with changing farming practices in southern Queensland. In *Proceedings of the 15th Australian Weeds Conference*. 24–28 September 2006, Adelaide. (Eds C Preston, JH Watts and ND Crossman) pp. 343–346. Weed Management Society of South Australia, Adelaide.

Wapshere AJ, Delfosse ES and Cullen JM (1989) Recent developments in biological control of weeds. *Crop Protection* **8**, 227–250.

Widderick M, Walker S and Sindel B (2004) Better management of *Sonchus oleraceus* L. (common sowthistle) based on the weed's ecology. In *Proceedings of the 14th Australian Weeds Conference*. 6–9 September 2004, Wagga Wagga, NSW. (Eds BM Sindel and SB Johnson) pp. 535–537. Weed Society of New South Wales, Sydney.

Widderick MJ, Walker SR, Sindel BM and Bell KL (2010) Germination, emergence, and persistence of *Sonchus oleraceus*, a major crop weed in subtropical Australia. *Weed Biology and Management* **10**, 102–112.

Zollinger RK and Parker R (1999) Sowthistles. In *Biology and Management of Noxious Rangeland Weeds*. (Eds RL Sheley and JK Petroff) pp. 336–349. Oregon State University Press, Corvallis, Oregon.

Sporobolus spp. – weedy sporobolus grasses

Bill Palmer

ABSTRACT

Five exotic *Sporobolus* spp., collectively known as the weedy sporobolus grasses, have become serious weeds along the entire eastern seaboard of Australia. The grasses are of extremely low palatability and high tensile strength such that cattle cannot utilise them. They are also invasive and easily spread. A project to develop biological control commenced in 2000 when surveys were undertaken in southern Africa, where *Sporobolus pyramidalis*, *Sporobolus natalensis* and *Sporobolus africanus* originate. Some 70 phytophagous insect species and 23 plant pathogens were found, but only two organisms were considered potential biocontrol agents; the leaf smut *Ustilago sporoboli-indici* and the stem wasp *Tetramesa* sp. Techniques to culture the smut were developed and it was found to be infective for Australian populations of four of the target species, but not the American *Sporobolus jacquemontii*. However, it was also infective on four Australian native *Sporobolus* spp. and was therefore rejected. Attempts to rear the stem wasp failed, negating further study. Although other areas such as Asia and North America could be searched, the prospects for biological control of these grasses are not good.

Key words: *Sporobolus africanus*, *Sporobolus fertilis*, *Sporobolus jacquemontii*, *Sporobolus natalensis*, *Sporobolus pyramidalis*, *Ustilago sporoboli-indici*, *Tetramesa*.

INTRODUCTION

Five grasses (Poaceae), giant rat's tail grass *Sporobolus natalensis* (Steud.) Dur. & Schinz. and *Sporobolus pyramidalis* P. Beauv., giant Parramatta grass *Sporobolus fertilis* (Steud.) Clayton, Parramatta grass *Sporobolus africanus* (Poir.) Robyns & Tournay and American rat's tail grass *Sporobolus jacquemontii* Kunth, collectively known as the weedy sporobolus grasses, are serious pastoral weeds in Australia, affecting productivity, property management and land values. The detrimental effects of these grasses are such that the potential annual losses to beef production in northern Australia, if weedy sporobolus grasses spread to their limits, have been estimated at $60 million per year (Walton 2001; Bray and Officer 2007).

Carr *et al.* (1992) considered *S. africanus* a serious threat to vegetation formations in Vic and recorded its invading dry coastal vegetation, heathland and healthy woodland, lowland grassland and grassy woodland, dry sclerophyll forest and woodland, riparian vegetation and rock outcrop vegetation. In addition, a survey of Australian herbaria has shown that the weedy *Sporobolus* species invade grasslands (*S. africanus*, *S. pyramidalis* nd *S. fertilis*), grassy woodlands (*S. africanus*, *S. pyramidalis* and *S. fertilis*), bushland (*S. africanus*), coastal vegetation (*S. africanus*), riparian vegetation (*S. africanus*, *S. fertilis* and *S. pyramidalis*), roadside reserves (*S. africanus*) and crops (*S. pyramidalis*) (DA McLaren, *pers. comm.* 2010). The variety of vegetation associations invaded by this weedy *Sporobolus* complex highlights their invasiveness and the serious threat they pose as both agricultural and environmental weeds.

The five grasses have different origins, with giant rat's tail grass (both species) and Parramatta grass being of African origin, giant Parramatta grass coming from Asia and American rat's tail grass from North America (Simon and Jacobs 1999). These species arrived in Australia from the early 1800s (Parramatta grass) to the 1960s (giant rat's tail grass). Little is known of the methods of introduction, though contaminated pasture seed importation has been implicated, at least for some species (Walton 2001).

The weedy sporobolus grasses are all tussock grasses. They are very difficult to tell apart and to distinguish from some of the native species. One distinguishing feature is that the weedy species have hairless leaf margins. *S. pyramidalis* and *S. africanus* are illustrated in Figure 1. The grasses are all highly invasive and are very easily spread by machinery, vehicles, livestock and native animals. The entire eastern coast of Australia is climatically favourable for one or more species and it is most unlikely that they have reached their ultimate geographic range (Fig. 2).

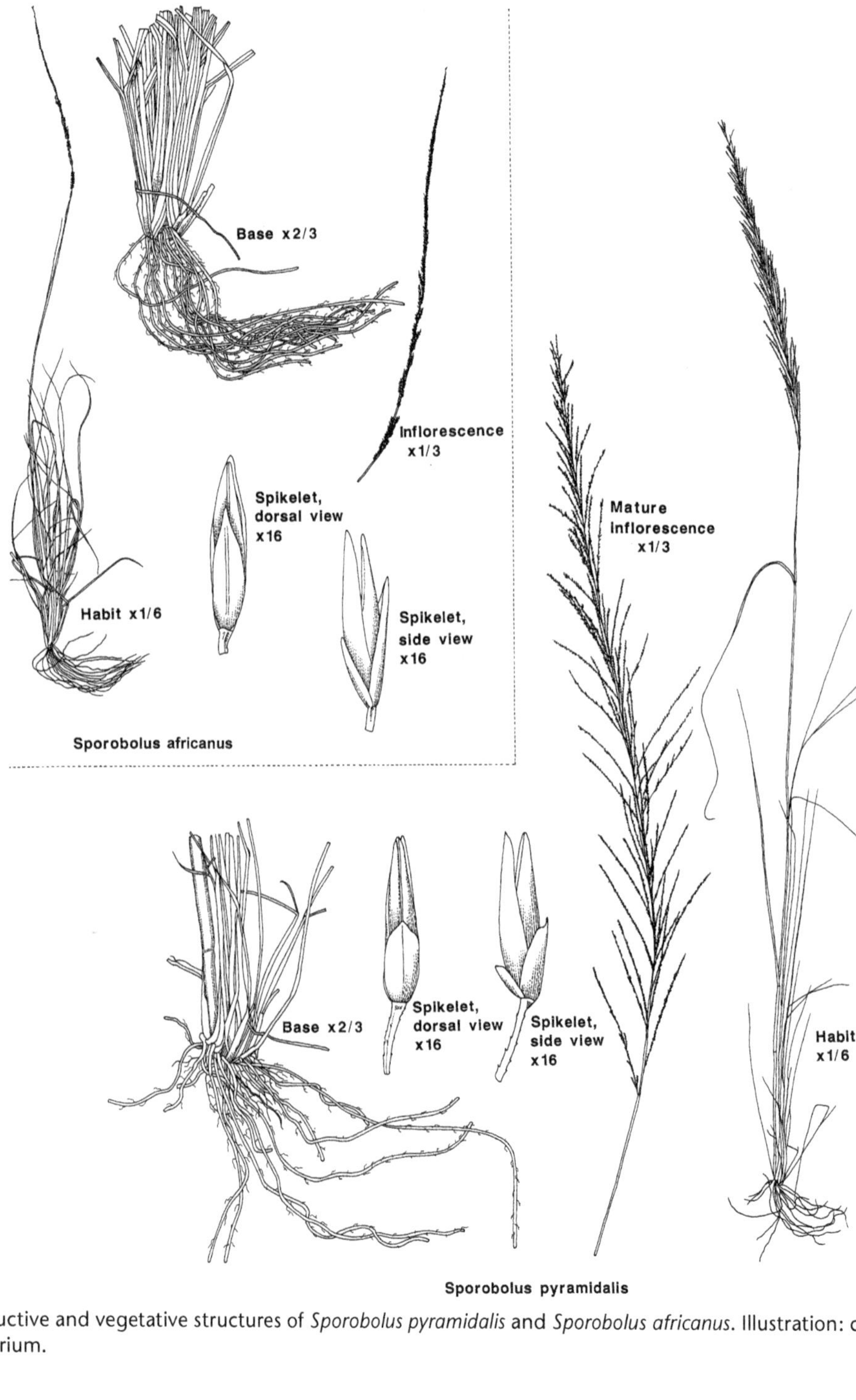

Figure 1: Reproductive and vegetative structures of *Sporobolus pyramidalis* and *Sporobolus africanus*. Illustration: courtesy of B Simon, Qld Herbarium.

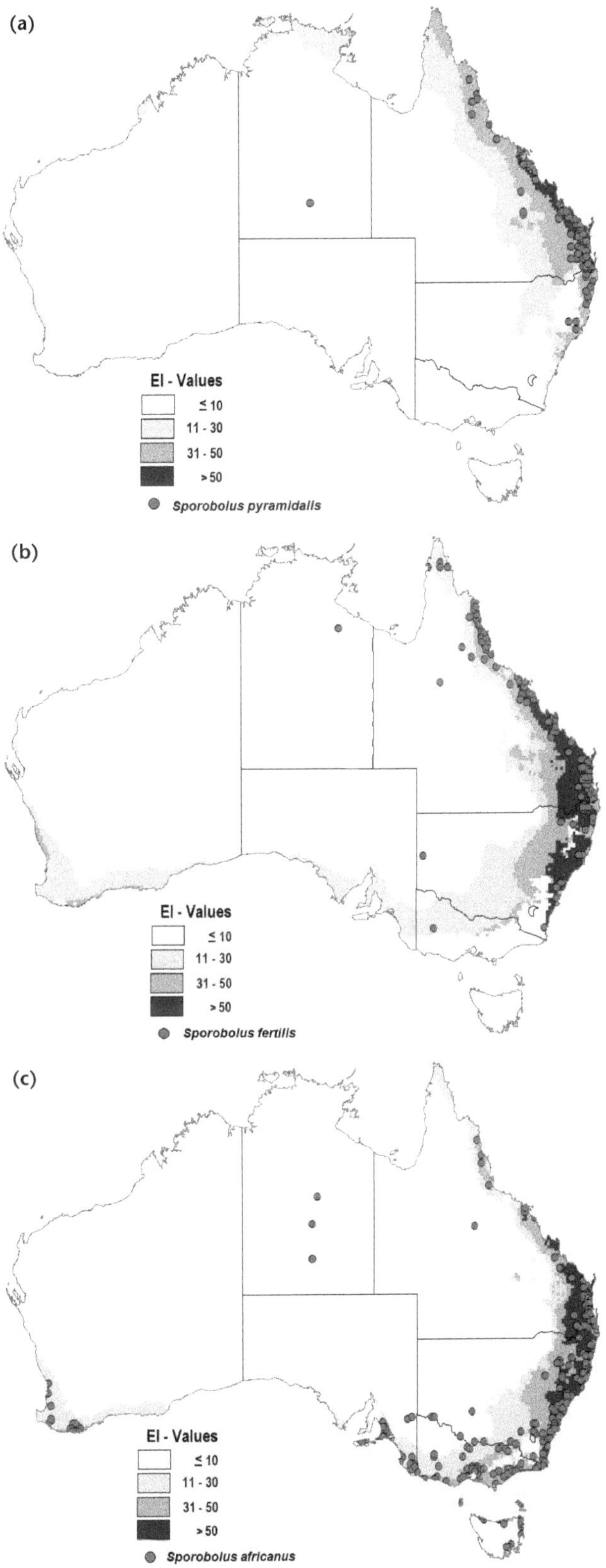

Figure 2: Distribution of a) *Sporobolus pyramidalis*, b) *Sporobolus fertilis* and c) *Sporobolus africanus* infestations in Australia based on herbarium records from the Australian Virtual Herbarium and their potential distributions (Walton 2001) predicted using the climate modelling software CLIMEX (Sutherst *et al.* 2007). EI values >30 are considered favourable while values ≤10 are considered climatically unfavourable.

Figure 3: A beast in very poor condition in a paddock of *Sporobolus pyramidalis* – starving in the midst of plenty. Photo: Qld Government.

The major problem with these grasses is that after germination they rapidly become useless for animal production. Not only do they become unpalatable, but they become so tough that cattle cannot physically break the leaves from the tussock. Consequently, animals in paddocks with a high proportion of a weedy sporobolus grass may face starvation (Fig. 3).

The weedy sporobolus grasses are generally regarded as weeds although there is anecdotal evidence that some landholders in NSW may regard *S. fertilis* as a useful drought fodder. They are all declared class 2 weeds (requiring landholders to control them) in Qld. In NSW, both *S. fertilis* and *S. pyramidalis* are declared as category W2 (landholder required to fully and continuously suppress and destroy the weed) or W3 weeds (landholder required to prevent the weed from spreading and to reduce its numbers), depending upon region.

BIOLOGICAL CONTROL HISTORY

Although the weedy sporobolus grasses are not regarded favourably anywhere outside their native ranges, they have only become serious weeds in Australia, particularly in recent decades. No other country has shown any interest in their biological control.

Concern about the weedy sporobolus grasses increased in the 1990s, resulting in requests to find a biological control and for the development of a national weedy sporobolus grasses strategy (Walton 2001). The five species were all approved as targets for biological control in Australia by the Natural Resource Management Standing Committee in 2007.

Three of the five weedy species (*S. africanus*, *S. natalensis* and *S. pyramidalis*) originated in southern Africa and therefore searches for biological control agents were initiated in South Africa during 2000 using the existing Qld Government's South African Field Station. ABR Witt conducted the faunal surveys, supported by plant pathologists I Rong, RG Shivas and K Vanky. Following the selection of the leaf smut *Ustilago sporoboli-indici* L. Ling (Ustilaginomycetes) as a potential agent, biology and host-specificity studies were conducted at the University of Kwa-Zulu Natal under the direction of MD Laing and KS Yobo. The *Ustilago* smuts found on the *Sporobolus* grasses worldwide have been described (Vanky 2003).

Investigations of pathogens already present in Australia have been undertaken (Hetherington 1997). A fungus identified as *Nigrospora oryzae* (Berk & Broome) Petch (Sordariomycetes) (Ramasamy *et al.* 2008) has been observed causing stunted growth, yellowing of leaves and rot of tillers of *S. fertilis* around Grafton, NSW since 2003 (D Officer, *pers. comm.* 2008). Investigations into using *N. oryzae* as a bioherbicide are underway (S Ramasamy, *pers. comm.* 2010).

PLANT TAXONOMY

The weedy sporobolus grasses are all exotic and belong to a section of the *Sporobolus* genus known as the *indicus* complex (Simon and Jacobs 1999). The species included in this complex are morphologically very similar and it is quite likely that the species will be redefined should appropriate molecular studies be conducted. The *indicus* complex is presently represented in Australia by 11 species, including six native species, and these species form a monophyletic group (Shrestha *et al.* 2003). A further 13 species outside this complex complete the 24 *Sporobolus* spp. found in Australia. Because the five weedy sporobolus grasses are so morphologically similar and difficult to distinguish from each other and from some native species, an identification tool based on random amplified polymorphic DNA (RAPD) profiles was developed to identify these five from all others (Shrestha *et al.* 2005).

From a biocontrol point of view, the fact that there are six native species within this monophyletic intragenus group reduces the probabilities of finding agents sufficiently specific to be approved for release in Australia

EXPLORATION

Exploration for phytophagous arthropod fauna and pathogens on *S. pyramidalis*, *S. natalensis* and *S. africanus* in southern Africa was undertaken during 2001–03. It was not possible to visit some countries, e.g. Zimbabwe, because of political and safety issues, but South Africa, Botswana and Swaziland were surveyed. Identification of the individual species of *Sporobolus* presented some difficulty, as they are morphologically very similar and interbreed (van Oudtshoorn 1999). *S. pyramidalis* and *S. natalensis* did not occur in the Western Cape, whereas *S. africanus* was quite abundant in pastures in that region. All three species co-occurred in areas further north and were particularly abundant in disturbed sites.

An arthropod fauna of at least 70 species was found on the three weedy sporobolus grasses (Witt and McConnachie 2004). Many were only partially determined (usually to genus) as they belonged to groups that have not yet been properly described in southern Africa. Most species had a casual association with the plant rather than utilising the grass as a host plant.

The only insect seen as a prospective biological control agent was the eurytomid wasp *Tetramesa* sp. (Hymenoptera: Eurytomidae). Larvae of this wasp feed in the culm, which results in malformation of the inflorescence and hence significant damage. The wasp was found at many localities throughout the survey area, often at high levels of infestation.

Twenty-three pathogens, including five primary pathogens, were found on the *Sporobolus* spp. (Witt and McConnachie 2004). Only the leaf smut *U. sporoboli-indici* was thought to be sufficiently host-specific to be considered further as a potential biological agent for Australia.

CANDIDATES

***Ustilago sporoboli-indici* (Ustilaginomycetes).** This leaf smut was found quite commonly throughout South Africa (Yobo *et al.* 2007). Preliminary studies of its biology were made in Germany (K Vanky, unpub.) and a phylogenetic comparison of *U. sporoboli-indici* with other *Ustilago* species revealed strong support for an evolutionary radiation of *Ustilago* spp. infecting the Choridoideae and Pooideae, of which *S. sporoboli-indici* forms a major lineage (Cunnington and Shivas 2006).

The pathogenicity of the smut fungus was tested against Australian populations of all five weedy sporobolus grasses (Yobo *et al.* 2009). Four species (*S. pyramidalis*, *S. africanus*, *S. fertilis* and *S. natalensis*) were hosts for the leaf smut but *S. jacquemontii* was a non-host.

The smut was then tested against 10 species of *Sporobolus* native to Australia (Yobo *et al.* 2009). These were *Sporobolus actinocladus* (F. Muell.) F. Muell., *Sporobolus australasicus* Domin, *Sporobolus caroli* Mez, *Sporobolus contiguous* Blake, *Sporobolus coromandelianus* (Retz.) Knuth, *Sporobolus creber* De Nardi, *Sporobolus disjunctus* R. Mills ex B. Simon, *Sporobolus laxus* Simon, *Sporobolus scabridus* Blake and *Sporobolus sessilis* Simon. *S. creber*, *S. elongatus*, *S. sessilis* and *S. scabridus* developed typical symptoms of infection though those on *S. sessilis* and *S. scabridus* were minimal, did not spread, and remained localised.

This result was sent to about 20 stakeholders (agronomists, regulators, botanists, plant pathologists and industry representatives) to determine whether the weedy sporobolus grasses problem warranted considering a biocontrol agent that might also attack a native plant. A consensus to progress the study did not emerge, and it was decided to terminate work with the leaf smut (Palmer 2008).

***Tetramesa* sp. (Hymenoptera: Eurytomidae).** Studies in South Africa failed to develop a successful rearing technique for the stem wasp *Tetramesa* sp. (ABR Witt, unpub.), so further work was not possible. The wasp might have a winter diapause and may have less effect on the plant than originally thought (Palmer 2008; Palmer *et al.* 2008).

DISCUSSION

Grasses have generally not been considered good candidates for biological control (Wapshere 1990a, b). There are various reasons for this, including the perception that they lack specific herbivores because they have simple morphology and lack secondary compounds (Witt and McConnachie 2004). Other reasons include the great economic and ecological importance of related species and the great adaptability of grasses to grazing and harvesting (Palmer *et al.* 2008). However, in recent years programs against *Nassella* spp. (McLaren *et al.* 2002), *Spartina alterniflora* Loisel (Grevstad *et al.* 2003) and *Arundo donax* L. (Jones and Sforza 2007) have been initiated, as well as those against the weedy sporobolus grasses.

The relationship between the five weedy sporobolus grasses and the native Australian congeners was always recognised as the major impediment to finding biological control agents. Indeed, one criterion recommended by Wapshere (1990a) was that the weedy grass not be related generically to valued grasses. A second observation (Wapshere 1990a) was that only a few types of agents, notably gall-makers and smut fungi, were likely to be sufficiently specific for introduction.

When the leaf smut *U. sporoboli-indici* did not infect *S. jacquemontii* there were high hopes that there might be sufficient difference between the natives and the exotic weeds to allow safe agents to be selected. That was not to be: the experimental evidence, albeit from the laboratory, was that at least two native Australian *Sporobolus* spp. could be infected similarly to the exotics. This leaf smut has a known narrow host-range (Vanky 2003) and the fact that it can infect both exotics and some natives does not bode well for finding suitable agents elsewhere in the world. In light of these results, the project was terminated.

It is unfortunate that this project was unsuccessful, because the weedy sporobolus grasses are of increasing concern in Australia and successful biological control would have offered the best hope of controlling their abundance.

ACKNOWLEDGEMENTS

The Biosystematics Division of the ARC-Plant Protection Research Institute for identifying the collected specimens and the ARC-Plant Protection Research Institute for hosting the South African Field Station are acknowledged. KADW Senaratne prepared Figure 2. Valuable comments on the manuscript were provided by colleagues at the Alan Fletcher Research Station.

REFERENCES

Bray S and Officer D (Eds) (2007) *Weedy Sporobolus Grasses Best Practice Manual*. 3rd edn. Qld Dept of Primary Industries and Fisheries, Brisbane.

Carr GW, Yugovic JV and Robinson KE (1992) 'Environmental weed invasions in Victoria – conservation and management implications'. Vic Dept of Conservation and Environment and Ecological Horticulture Pty Ltd, Melbourne.

Cunnington JH and Shivas RG (2006) Phylogenetic considerations for predicting the host range of *Ustilago sporoboli-indici*, a potential biological control agent for *Sporobolus* species in Australia. *Plant Protection Quarterly* **21**, 7–9.

Grevstad FS, Strong DR, Garcia-Rossi D, Switzer RW and Wecker MS (2003) Biological control of *Spartina alterniflora* in Willapa Bay, Washington using the planthopper *Prokelisia marginata*: agent specificity and early results. *Biological Control* **27**, 32–42.

Hetherington S (1997) Evaluation of endemic fungal pathogens as biological control agents of *Sporobolus* spp. PhD thesis. University of Queensland, Brisbane.

Jones WA and Sforza R (2007) The European Biological Control Laboratory: an existing infrastructure for biological control of weeds in Europe. *EPPO Bulletin* **37**, 163–165.

McLaren DA, Morfe TA and Weiss J (2002) Distribution, economic impact and attitudes towards Chilean needlegrass (*Nassella neesiana* (Trin. & Rupr.) Barkworth) in Australia. In *Proceedings of the 13th Australian Weeds Conference*. 8–13 September 2002, Perth. (Eds JH Spafford, J Dodd and JH Moore) pp. 749–752. Plant Protection Society of Western Australia, Perth.

Palmer WA (2008) 'Biological control of weedy sporobolus grasses by two host specific agents'. Meat & Livestock Australia, Sydney.

Palmer WA, Yobo KS and Witt ABR (2008) Prospects for the biological control of the weedy sporobolus grasses in Australia. In *Proceedings of the 16th Australian Weeds Conference*. 18–22 May 2008, Cairns. (Eds RD van Klinken, VA Osten, FD Panetta and JC Scanlan) pp. 264–266. Queensland Weeds Society, Brisbane.

Ramasamy S, Officer D, Lawrie AC and McLaren DA (2008) *Nigrospora oryzae*, a potential bio-control agent for giant Parramatta grass (*Sporobolus fertilis*) in Australia. In *Proceedings of the 12th International Symposium on Biological Control of Weeds*. 22–27 April 2007, La Grande Motte, France. (Eds MH Julien, R Sforza, MC Bon, HC Evans, PE Hatcher, HL Hinz and BG Rector) p. 130. CABI, Wallingford, UK.

Shrestha S, Adkins SW, Graham GC and Loch DS (2003) Phylogeny of the *Sporobolus indicus* complex, based on internal transcribed spacer (ITS) sequences. *Australian Systematic Botany* **16**, 165–176.

Shrestha S, Adkins SW, Graham GC and Loch DS (2005) An identification tool for the Australian weedy *Sporobolus* species based on random amplified polymorphic DNA (RAPD) profiles. *Australian Journal of Agricultural Research* **56**, 157–167.

Simon BK and Jacobs SWL (1999) Revision of the genus *Sporobolus* (Poaceae, Chloridoidae) in Australia. *Australian Systematic Botany* **12**, 375–448.

Sutherst RW, Maywald GF and Kriticos D (2007) *CLIMEX Version 3 User's Guide*. Hearne Scientific Software, Melbourne.

van Oudtshoorn F (1999) *Guide to Grasses of Southern Africa*. Briza Publications, Pretoria.

Vanky K (2003) The smut fungi (Ustilaginomycetes) of *Sporobolus* (Poaceae). *Fungal Diversity* **14**, 205–241.

Walton C (Ed) (2001) 'Weedy sporobolus grasses strategy'. Qld Dept of Natural Resources and Mines, Cooparoo.

Wapshere AJ (1990a) Biological control of grass weeds in Australia: an appraisal. *Plant Protection Quarterly* **5**, 62–75.

Wapshere AJ (1990b) Biological control options for some grass weeds in Australia. *FRI Bulletin* **155**, 80–84.

Witt ABR and McConnachie AJ (2004) The potential for classical biological control of invasive grass species with special reference to invasive *Sporobolus* spp. (Poaceae) in Australia. In *Proceedings of the 11th International Symposium on Biological Control of Weeds*. 27 April–2 May 2003, Canberra. (Eds JM Cullen, DT Briese, DJ Kriticos, WM Lonsdale, L Morin and JK Scott) pp. 198–202. CSIRO Entomology, Canberra.

Yobo KS, Laing MD, Palmer WA and Shivas RG (2007) Potential of *Ustilago sporoboli-indici* for biological control of five invasive *Sporobolus* grasses in Australia. In *Proceedings of the 12th International Symposium on Biological Control of Weeds*. 22–27 April 2007, La Grande Motte, France. (Eds MH Julien, R Sforza, MC Bon, HC Evans, PE Hatcher, HL Hinz and BG Rector) p. 365. CABI, Wallingford, UK.

Yobo KS, Laing MD, Palmer WA and Shivas RG (2009) Evaluation of *Ustilago sporoboli-indici* as a classical biological control agent for invasive *Sporobolus* grasses in Australia. *Biological Control* **50**, 7–12.

Tribulus terrestris L. – caltrop

John K Scott

ABSTRACT

Tribulus terrestris, caltrop, is an annual summer-growing prostrate herb that is widespread in Australia. Its burrs contaminate agricultural produce and the plant is toxic to livestock. It is an approved target for biocontrol, but no introductions of biocontrol agents have been made into Australia. Two organisms, the seed-feeding moth *Ephysteris subdiminutella ferritincta* and an unidentified eriophyid mite similar to *Aceria tribuli* causing galls on leaves and stems, are already present in Australia and damaging on the plant. The main biocontrol agents used overseas, *Microlarinus* weevils, have a history of spreading between islands and could potentially arrive in Australia from New Guinea where at least one species is established. Issues that will need to be resolved before the application of biocontrol include determining the native or introduced status of *T. terrestris*, the host-specificity of potential agents in the context of the risk to the diverse flora of native *Tribulus* species, and the lack of economic assessment showing the costs associated with the weed.

Key words: biological control, conflict of interest, *Microlarinus lareynii*, *Microlarinus lypriformis*.

INTRODUCTION

Tribulus terrestris L. (Zygophyllaceae), caltrop or puncture vine (Fig. 1), is a prostrate annual herb. Stems grow up to 3 m long, radiating from a central woody tap root. The flower consists of five yellow petals 5–6 mm long. Reproduction is solely by seed, which can be produced as soon as five weeks after germination (Parsons and Cuthbertson 2001). The fruit consists of five wedge-shaped schizocarps (cocci or burrs), each with one or two pairs of large spines. There are usually three (range one to five) seeds per burr (Scott and Morrison 1996). Under optimum conditions a single plant can produce up to 400 burrs m^{-2} (Squires 1969). Dispersal occurs via tyres, footwear, animals and movement of soil. The seeds germinate rapidly after late spring and summer rains in warm conditions (24–27°C). Several germination events can occur in one season. The seeds can remain viable for three to six years (Ernst and Tolsma 1988), and thus after successive generations a large reservoir of seeds can accumulate (Squires 1979; Parsons and Cuthbertson 2001). More detailed description of the plant can be found in Parsons and Cuthbertson (2001) and Squires (1979).

The genus *Tribulus* and the close relative *Tribulopis* have around 17 and 10 species respectively in Australia (Barker 1998a, b; Sheahan and Chase 2000). Most are regarded as native, which makes Australia a region of diversity for this genus and relatives.

There is a long-standing question about the origins, native or introduced, of forms of *T. terrestris* in Australia (Squires 1969, 1979). A study of the morphology, chromosome numbers and isozyme variation in *T. terrestris* from Australia and overseas attempted to identify the weed's origins (Morrison and Scott 1996a, b; Scott and Morrison

Figure 1: Caltrop, *Tribulus terrestris*, showing flowers and burrs. Photo: JK Scott, CSIRO.

1996). Populations from northern Australia were possibly a separate Australian species. Other populations from Australia had a high genetic similarity to burrs obtained from the Mediterranean region, west Asia, South Africa, Namibia and the USA, the last being an incontrovertible introduced population. Barker (1998a) identified that style length could be an important identifier for the purported introduced populations of *T. terrestris* from Australia. Both studies underline that a much more detailed worldwide study using current molecular techniques is needed to identify *Tribulus* species and their origins.

T. terrestris sensu lato is widespread in Australia (Fig. 2) except that it is not established in Tas (Tas Dept of Primary Industries, Parks, Water and Environment 2010). *T. terrestris* has a noxious weed status in all or parts of the NT, SA, Tas, Vic and WA (Australian Weeds Committee 2010a).

T. terrestris grows best where there has been soil disturbance, low competition from other plants, high temperatures, adequate soil moisture and low atmospheric humidity, i.e. conditions that occur with irregular frequency in desert regions. The weed is regularly found associated with human activities in dry regions. Consequently the plant can be a nuisance for tourists, especially in camping grounds, and cyclists due to the damage to tyres (hence the common name, puncture vine). Contamination of dried fruit by spiny burrs, including *T. terrestris*, has been a problem in the Australian dried fruit industry (Johnstone 1990) and is managed by crop hygiene, cultural and herbicide methods (MacGregor 1990).

The alkaloids in *T. terrestris* cause a locomotor disorder of sheep in Australia (Bourke *et al.* 1992; Bourke

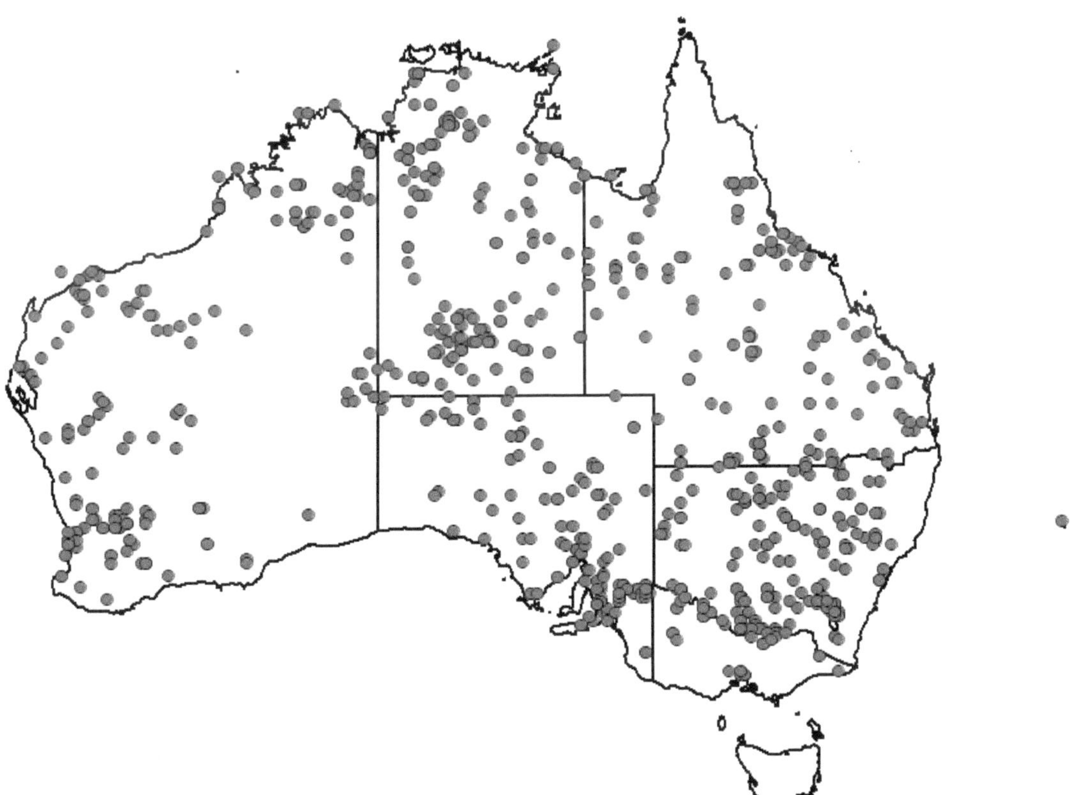

Figure 2: Distribution of *Tribulus terrestris* in Australia. Specimen data reproduced from Australia's Virtual Herbarium with permission of the Council of Heads of Australasian Herbaria Inc.

2006). *T. terrestris* is also associated with a photosensitisation disease of livestock, called geeldikkop in South Africa (Kellerman and Coetzer 1984), with similar symptoms reported from Australia (Glastonbury *et al.* 1984).

BIOLOGICAL CONTROL HISTORY

Surveys for potential biological control agents have been undertaken in areas of the Rajasthan desert and near Bangalore and New Delhi, India, coastal and interior areas of southern France and Italy (Andres and Angalet 1963), Israel (Gerling and Kugler 1973a, b), South Africa and Namibia (Kluge 1975) and Bangalore and nearby areas in South India (Sankaran and Ramaseshiah 1981).

***Microlarinus* spp. (Coleoptera: Curculionidae).** The first biocontrol introductions were the weevils *Microlarinus lareynii* (Jacquelin du Val) and *Microlarinus lypriformis* (Wollaston) from Italy, which were released in 1961 in the USA. From there they were distributed to Hawaii in 1962 and Canada in 1986 (Julien and Griffiths 1998). Twenty years after the first releases in USA, the weevils were shown to significantly reduce seed production and growth of *T. terrestris* (Huffaker *et al.* 1983).

The seed weevil *M. lareynii* and the stem weevil *M. lypriformis* were released in New Guinea in the 1960s for the control of *Tribulus*. Only the stem weevil established near Port Moresby (Young 1982). There is evidence that these weevils can spread between islands. An adult of *M. lareynii* was captured on La Réunion Island in 2001, where the species had not been deliberately released (Perrin *et al.* 2007). Apparently seeds of *Tribulus* species have not been examined on the island to see if it is established. *M. lypriformis* has spread from its initial release in western USA to Florida and throughout the Caribbean, to Mexico and Venezuela (Bennett 1989; Julien and Griffiths 1998). The host species, *Tribulus cistoides* L., is a coastal species present throughout the north of Australia and islands of the Pacific. The proximity of established populations in New Guinea, the evidence of spread and a suitable host plant indicates that there is a possibility that this insect will arrive in Australia. While it might provide useful control of weedy *Tribulus* species, it is also a predictable risk to native *Tribulus* species (Pemberton 2000).

Only *M. lareynii* and *M. lypriformis* have been extensively tested for host-specificity to *T. terrestris*. Other *Microlarinus* species have been collected from *Tribulus* but little is known of their biology, although it seems that the genus is restricted to the Zygophyllaceae.

***Grammodes stolida* (F.) (Noctuidae) and *Deroplax* sp. (Scutelleridae).** Kluge (1975) carried out preliminary tests on two insect species associated with *T. terrestris* in South Africa, the moth *G. stolida* and the bug *Deroplax* sp. that feeds on green and dry burrs. Neither species has been assessed further, although *G. stolida* appears to have a wide distribution and host-range, and is present in Australia without being recorded on *Tribulus*.

***Hyaloperonospora tribulina* (Pass.) Constant. (= *Peronospora tribulina* Pass.).** Little is known of the host-range of the pathogen species recorded from *Tribulus* (Farr and Rossman 2010). The downy mildew *H. tribulina* (Constantinescu and Fatehi 2002) is known only from *Tribulus* and has not been reported from Australia (L Morin, *pers. comm.* 2010). It may have potential as a classical biocontrol agent.

T. terrestris was approved as a biocontrol target in 1989 (Australian Weeds Committee 2010b). So far, there have been no attempts to release biocontrol agents for caltrop in Australia. Scott (1990) and Shepherd (1990) reviewed the potential of biocontrol against *T. terrestris* in Australia and concluded that the question of the weed origins should be resolved first.

ARTHROPODS PRESENT IN AUSTRALIA

***Ephysteris subdiminutella* (Stainton) *ferritincta* (Turner) (Lepidoptera: Gelechiidae).** This moth is the Australian subspecies of a widespread European and African species (Common 1990; Povolny 1977). The Australian subspecies feeds on developing burrs of *T. terrestris*. While the insect is widespread in Australia (Povolny 1977) and the level of infestation can be high, Squires (1965, 1979 (as *Aristotelia* sp.)) considered the overall impact as low, although there are no detailed studies of this species.

***Aceria* spp. *cf tribuli* (Keifer) (Acari: Eriophyidae).** This mite is found on native and introduced *Tribulus* species in Australia. Typical symptoms of damage were found on herbarium specimens of 10 native species of *Tribulus* and *T. terrestris* (JR Hosking and JK Scott, unpub. data). Evidence of damage was found in herbarium specimens from all mainland states and was present on a *T. terrestris* specimen collected by Dallachy and Goodwin in 1858 from the Darling River. Mites from Australia resemble *Aceria tribuli* (D Knihinicki, *pers. comm.* 1992). This species was first described from Sudan on the stems and leaves of *T. terrestris* (Keifer 1974) and was found in Bangalore and nearby areas in India on

Figure 3: *Tribulus terrestris* showing distorted leaves due to infestation by *Aceria* sp. *cf tribuli*. Photo: JK Scott, CSIRO.

T. terrestris plants surveyed by Sankaran and Ramaseshiah (1981). It causes hair-like growth on stems and leaves (Fig. 3). At present it is unclear whether the damage on *Tribulus* species in Australia is caused by one or a number of mite species.

DISCUSSION

The level of host-specificity of potential biocontrol organisms is likely to be a problem since Australia is a centre of species diversity in the genus *Tribulus* and many native species may be subject to damage by biocontrol agents. The native species are mostly restricted to the central and north-west of the country and are largely absent from regions of Mediterranean climate. This suggests that biocontrol could be implemented by selecting only those agents restricted to Mediterranean climates, even though their host-range might include species of *Tribulus* other than *T. terrestris*. Examples are *H. tribulina* or *Microlarinus humeralis* (Tournier). The latter has been reported from Israel, where it attacks both stems and seeds of *T. terrestris* (Gerling and Kugler 1973a, b). Careful assessment of its ecophysiology will be needed to show, with a high level of confidence, that the potential distribution (fundamental niche) in Australia will be as restricted as that in the region of origin. Alternatively, known successful agents could be released in Australia with the acceptance of the likelihood of damage to native species. Further information on the costs associated with introduced *Tribulus* species will help authorities in Australia decide between the options for control.

ACKNOWLEDGEMENTS

I thank John Hosking and Danuta Knihinicki for advice on *Tribulus* and its associated mite.

REFERENCES

Andres LA and Angalet GW (1963) Notes on ecology and host specificity of *Microlarinus lareynii* and *M. lypriformis* (Coleoptera: Curculionidae) and the biological control of puncture vine, *Tribulus terrestris*. *Journal of Economic Entomology* **56**, 333–340.

Australian Weeds Committee (2010a) Noxious weeds list. <http://www.weeds.org.au/noxious.htm> (accessed 19.7.10).

Australian Weeds Committee (2010b) Target species for biological control. <http://www.weeds.org.au/management.htm> (accessed 19.7.10).

Barker RM (1998a) A trial key and notes on *Tribulus* (Zygophyllaceae) in Australia, including one new species and validation of *Tribulus suberosus*. *Nuytsia* **12**, 9–35.

Barker RM (1998b) Notes on the genus *Tribulopis* (Zygophyllaceae) in Australia. *Journal of the Adelaide Botanic Garden* **18**, 77–93.

Bennett FP (1989) *Microlarinus lypriformis* (Coleoptera: Curculionidae) in Curacao, Venezuela, and Puerto Rico: new distribution records. *Coleopterists Bulletin* **43**, 390–391.

Bourke CA (2006) Abnormal turning behaviour, GABAergic inhibition and the degeneration of astrocytes in ovine *Tribulus terrestris* motor neuron disease. *Australian Veterinary Journal* **84**, 53–58.

Bourke CA, Stevens GR and Carrigan MJ (1992) Locomotor effects in sheep of alkaloids identified in Australian *Tribulus terrestris*. *Australian Veterinary Journal* **69**, 163–165.

Common IFB (1990) *Moths of Australia*. Melbourne University Press, Melbourne.

Constantinescu O and Fatehi J (2002) *Peronospora*-like fungi (Chromista, Peronosporales) parasitic on Brassicaceae and related hosts. *Nova Hedwigia* **74**, 291–338.

Ernst WHO and Tolsma DJ (1988) Dormancy and germination of semi-arid annual plant species, *Tragus berteronianus* and *Tribulus terrestris*. *Flora* **181**, 243–251.

Farr DF and Rossman AY (2010) Fungal databases, systematic mycology and microbiology laboratory. ARS-USDA. <http://nt.ars-grin.gov/fungaldatabases/> (accessed 18.7.10).

Gerling D and Kugler J (1973a) An examination of the possibilities for biological control of some weeds in Israel. *Phytoparasitica* **1**, 80.

Gerling D and Kugler J (1973b) 'Final technical report, evaluation of enemies of noxious plants in Israel as potential agents for the biological control of weeds'. Tel Aviv University, Israel.

Glastonbury JRW, Doughty FR, Whitaker SJ and Sergeant E (1984) A syndrome of hepatogenous photosensitization, resembling geeldikkop, in sheep grazing *Tribulus terrestris*. *Australian Veterinary Journal* **61**, 314–316.

Huffaker CB, Hamai J and Nowierski RM (1983) Biological control of puncturevine, *Tribulus terrestris* in California after twenty years of activity of introduced weevils. *Entomophaga* **28**, 387–400.

Johnstone RB (1990) Opening address: the problems caused by weed seeds to the dried vine fruit industry. *Plant Protection Quarterly* **5**, 84.

Julien MH and Griffiths MW (1998) *Biological Control of Weeds: A World Catalogue of Agents and their Target Weeds*. CABI, Wallingford, UK.

Keifer HH (1974) Eriophyid studies C-9. *Eriophyid Studies (Series C)* **9**, 1–24.

Kellerman TS and Coetzer JAW (1984) Hepatogenous photosensitivity diseases in South Africa. *Technical Communication, Department of Agriculture, South Africa* **193**, 1–23.

Kluge RL (1975) Observations on insects associated with *Tribulus terrestris* L. in southern Africa. MSc thesis. University of Pretoria, South Africa.

MacGregor A (1990) The potential cultural control of *Tribulus*, *Cenchrus* and *Emex* in Sunraysia vineyards. *Plant Protection Quarterly* **5**, 116–119.

Morrison SM and Scott JK (1996a) Variation in populations of *Tribulus terrestris* (Zygophyllaceae). 2. Chromosome numbers. *Australian Journal of Botany* **44**, 191–199.

Morrison SM and Scott JK (1996b) Variation in populations of *Tribulus terrestris* (Zygophyllaceae). 3. Isozyme analysis. *Australian Journal of Botany* **44**, 201–212.

Parsons WT and Cuthbertson EG (2001) *Noxious Weeds of Australia*. 2nd edn. CSIRO Publishing, Melbourne.

Pemberton RW (2000) Predictable risk to native plants in weed biological control. *Oecologia* **125**, 489–494.

Perrin H, Poussereau J, Quilici S and Voisin J-F (2007) Trois espèces de Curculionides nouvelles pour l'île de la Réunion (Coleoptera). *Bulletin de la Société entomologique de France* **112**, 127–130.

Povolny D (1977) Notes on Gnorimoschemini of Australia and New Zealand (Lepidoptera, Gelechiidae). *Acta Entomologica Musei Nationalis Pragae* **39**, 403–443.

Sankaran T and Ramaseshiah G (1981) Studies on some natural enemies of puncturevine *Tribulus terrestris* occurring in Karnataka state, India. In *Proceedings of the 5th International Symposium on Biological Control of Weeds*. 22–29 July 1980, Brisbane. (Ed ES Delfosse) pp. 153–160. CSIRO, Melbourne.

Scott JK (1990) *Tribulus terrestris* L. (Zygophyllaceae) in southern Africa: an outline of biological and potential biological control agents for Australia. *Plant Protection Quarterly* **5**, 103–106.

Scott JK and Morrison SM (1996) Variation in populations of *Tribulus terrestris* (Zygophyllaceae). 1. Burr morphology. *Australian Journal of Botany* **44**, 175–190.

Sheahan MC and Chase MW (2000) Phylogenetic relationships within Zygophyllaceae based on DNA sequences of three plastid regions, with special emphasis on Zygophylloideae. *Systematic Botany* **25**, 371–384.

Shepherd RCH (1990) Past Victorian work on *Emex australis* Stenheil and *Tribulus terrestris* L. *Plant Protection Quarterly* **5**, 100–102.

Squires VR (1969) Distribution and polymorphism of *Tribulus terrestris sens. lat.* in Australia. *Victorian Naturalist* **86**, 328–334.

Squires VR (1979) Biology of Australian weeds. 1. *Tribulus terrestris* L. *Journal of the Australian Institute of Agricultural Science* **45**, 75–82.

Tas Dept of Primary Industries, Parks, Water and Environment (2010) Caltrop (*Tribulus terrestris*). Status of caltrop in Tasmania. <http://www.dpiw.tas.gov.au/inter,nsf/WebPages/BVIS-87A4LM?open> (accessed 19.7.10).

Young GR (1982) Recent work on biological control in Papua New Guinea and some suggestions for the future. *Tropical Pest Management* **28**, 107–114.

Ulex europaeus L. – gorse

JE Ireson and JT Davies

ABSTRACT

Gorse, *Ulex europaeus*, is one of the most invasive weeds in south-eastern Australia and is difficult to control by traditional methods. The annual cost of gorse management to agricultural industries alone has been estimated at $7 million. Biological control is being investigated as a cheaper and long-term control option to reduce the spread and impact of gorse, particularly in areas where it is widespread or inaccessible. Three foliage-feeding agents and a seed-feeder have now been released for its biological control in Australia. The gorse seed weevil *Exapion ulicis*, released in 1939, and the gorse spider mite *Tetranychus lintearius*, released in 1998, are now widespread across south-eastern Australia. The gorse thrips *Sericothrips staphylinus*, released in 2001, is now established and although initially slow to spread, populations have started to disperse exponentially in Tas. Establishment of the gorse soft shoot moth *Agonopterix umbellana* was confirmed in Tas in 2010 following its release in 2007. These four agents are expected to reduce plant vigour and seed output in the long term. Release of a second seed-feeding agent, the gorse pod moth *Cydia succedana*, will be dependent on the outcome of host-specificity tests. In combination with the gorse seed weevil, this species has the potential to reduce gorse seed banks below critical replacement levels in some areas. European surveys have found no new agents worth pursuing. Integrated control techniques offer the best prospects for long-term control in areas where gorse is actively managed, but the extent to which biological control will play a role will only be determined by future research once the complete guild of available agents is established.

Key words: biological control, *Exapion ulicis*, *Tetranychus lintearius*, *Sericothrips staphylinus*, *Agonopterix umbellana*, *Cydia succedana*.

INTRODUCTION

Gorse, *Ulex europaeus* L. (Fabaceae), is a perennial, spiny, woody shrub native to western Europe that now occurs in most temperate regions of the world. It is a major problem in the Pacific Coast states of North America, Hawaii, Chile, New Zealand and Australia. Gorse was deliberately introduced to Australia from Europe during the early 19th century for aesthetic reasons and was widely promoted as a hedge plant and a fodder source in Tas (Parsons and Cuthbertson 2001). Gorse, listed as a Weed of National Significance (Thorp and Lynch 2000), occurs in SA, NSW, WA and the ACT, with the heaviest infestations occurring in Vic and Tas (Fig. 1).

Gorse is common in agricultural and urban areas. It can form dense impenetrable thickets that compete strongly with pasture and other desirable plant species, significantly reducing pasture and animal productivity and providing habitat and shelter for vertebrate pests. In forestry plantations it reduces tree establishment and growth. Gorse also invades riparian environments and native bushland, thereby reducing access and conservation values and

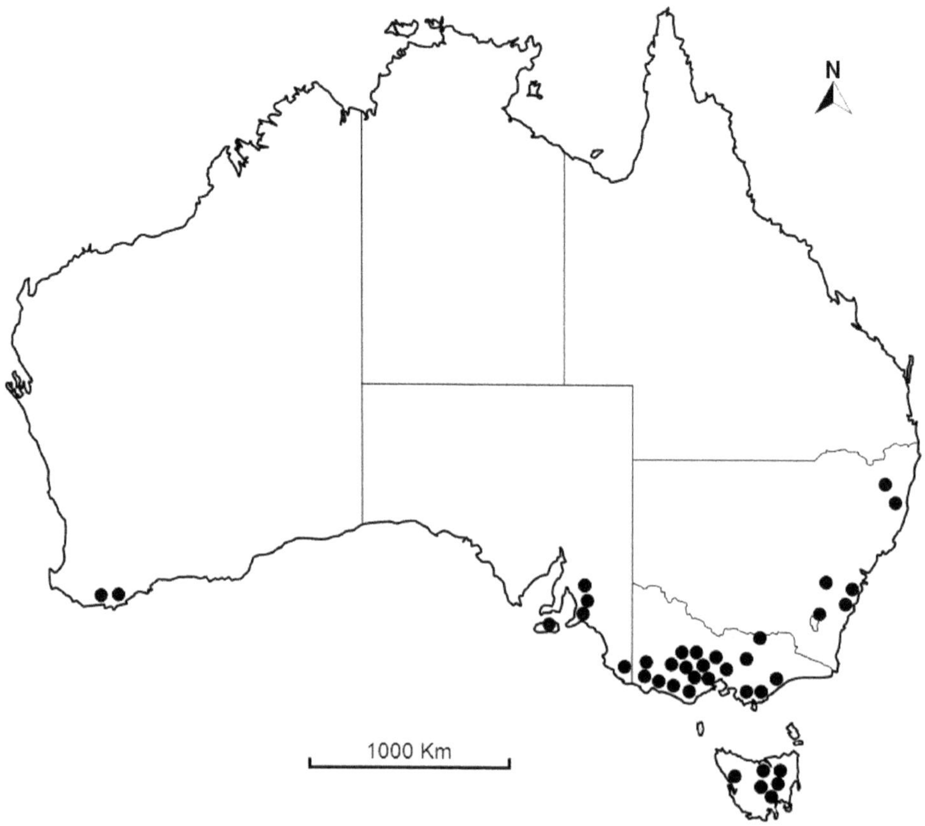

Figure 1: Distribution of *Ulex europaeus* in Australia.

threatening the survival of rare and endangered plants and plant communities (Richardson and Hill 1998). It is also a significant fire hazard. The annual cost of gorse management to Australian agricultural and forest industries was estimated at $7 million in 2000 (Anon 2006).

Gorse can live to 29 years of age and grow up to 7 m tall (Lee *et al.* 1986), more commonly growing to 4 m in height and ranging from 1–3 m in diameter (Parsons and Cuthbertson 2001). Gorse can flower in autumn, winter and spring, but flowering phenology and pod production of gorse vary considerably not only between sites but also on individual bushes within sites (Tarayre *et al.* 2007; Davies *et al.* 2008), influenced by environmental conditions and a high level of genetic variability (Tarayre *et al.* 2007). Seed is produced in small pods mainly in late spring and summer. Annual seed fall densities have been measured at a rate of ~600 seeds m^{-2} per year (Hill *et al.* 1996). These seeds can remain viable in the soil for more than 25 years (Moss 1959). Seed dispersal is mainly via pod dehiscence, but also via animals, man and water (Moss 1959). When pods dehisce, most seeds are dispersed less than 2.5 m from bushes; the maximum distance measured is about 5 m (Hill *et al.* 1996).

Gorse can be particularly difficult to control because it is a competitive pioneering species with a long-lived seed bank that readily invades disturbed ground (Rees and Hill 2001). The cost of control can be highly variable, depending on the height and density of the infestation. For instance, the cost of ground-spraying with herbicide can be $300–1600 ha (Anon 2006). Because of the difficulty and expense of controlling gorse by traditional methods such as herbicides, mechanical clearing and cultivation, biological control is being investigated as a cheaper and long-term control option.

BIOLOGICAL CONTROL HISTORY

Gorse was originally targeted for biological control in NZ and Hawaii using the gorse seed weevil *Exapion ulicis* (Forster). This species was first introduced to Hawaii from England in 1926 and released on Maui but failed to establish, despite several subsequent attempts, until populations imported from France to Hawaii in 1955 and 1956 did eventually establish (Markin and Yoshioka 1989). Selection of *E. ulicis* for NZ resulted from surveys of phytophagous insects on gorse (Chater 1931) and

Table 1: Status of agents investigated for the biological control of *Ulex europaeus* in Australia

Species	Taxonomic group	Origin	Year of first release[b]	Main part of plant attacked	Status
Exapion ulicis (Forster)	Coleoptera: Brentidae	Ex England via NZ	1939	Seeds in pods	Widespread in Tas, Vic, SA, NSW
Tetranychus lintearius Dufour	Acari: Tetranychidae	Ex England, Portugal and north-west Spain via NZ	1998	Foliage	Widespread in Tas, Vic, SA, NSW; present in WA
Sericothrips staphylinus Haliday	Thysanoptera: Thripidae	Ex England via NZ	2001	Young foliage and seedlings	Established in Tas and Vic
		Ex Portugal via Hawaii via NZ	2003		Established in Tas, Vic, SA and NSW
Agonopterix umbellana (Fabricius)[a]	Lepidoptera: Oecophoridae	Ex England via NZ	2007	Young foliage	Established in Tas; releases continuing
Cydia succedana (Denis & Schiffermüller)	Lepidoptera: Tortricidae	Ex England	-	Seeds in pod	Undergoing further host-specificity testing

a: *Agonopterix umbellana* (Fabricius 1794) has been referred to in earlier publications by its junior synonym, *Agonopterix ulicetella* (Stainton 1849).
b: Host testing of *T. lintearius*, *S. staphylinus*, *A. umbellana* and *C. succedana* on Australian plant species was conducted in NZ by Landcare Research New Zealand under contract from the Tas Institute of Agricultural Research. Some additional host testing on Australian plant species was conducted on *A. umbellana* and *C. succedana* by Vic Dept of Primary Industries at Frankston.

investigations into its biology and host-specificity by Davies (1928) in England. *E. ulicis* established successfully in NZ following its introduction from England in 1931 (Miller 1947) and in Tas following its importation from NZ in 1939 (Evans 1942).

Additional surveys for biological control agents of gorse were conducted in France on behalf of NZ by CAB International (CABI) in France in 1962–65. Results from these surveys were reported by Zwölfer (1962, 1963, 1964), Schröder and Zwölfer (1970) and Girling (1977). A survey of the fauna in north-western Spain and Portugal was also conducted on behalf of NZ in 1986 by CABI (O'Donnell 1986). Information obtained from these surveys on the identity, life-history and host-range of the gorse fauna, together with the work conducted by Hill (1982a), enabled the assembly of a list of potential host-specific biological control agents. These were investigated for introduction to NZ (Hill 1983) and later for Australia (Ireson *et al.* 2004). In NZ, the new program resulted in the host testing and release of the following six agents between 1989 and 1996: the gorse spider mite *Tetranychus lintearius* Dufour, the gorse thrips *Sericothrips staphylinus* Haliday, the gorse soft shoot moth *Agonopterix umbellana* (Fabricius), the gorse pod moth *Cydia succedana* (Denis and Schiffermüller), the colonial hard shoot moth *Pempelia genistella* Duponchel and the gorse hard shoot moth *Scythris grandipennis* (Hayworth) (Hill *et al.* 2008). Of these, only *S. grandipennis* failed to establish (Hill *et al.* 2008). The NZ work provided the basis for the new Australian program that was initiated after gorse was declared a target for biological control in Australia in 1995. Since then, four of the agents now established in NZ, *T. lintearius*, *S. staphylinus*, *A. umbellana* and *C. succedana*, have been investigated for release in Australia (Table 1). A summary of the status of biological control agents released worldwide against gorse is given by Hill *et al.* (2008).

EXPLORATION

Additional European surveys for the Australian program were conducted by CSIRO based at Montpellier, France on behalf of the Tas Institute of Agricultural Research in 2005, 2006 and 2007. These complemented the earlier surveys conducted in the native range of gorse by CABI in the 1960s and 1986 and a later survey in Spain and Portugal by CSIRO in 2003. The latter survey was conducted to search for potential autumn-active seed- and root-feeding insects, but no suitable candidates were found active at this time of year (Sheppard 2004). In 2005, a spring/summer survey for gorse root-feeders in France, Portugal and Spain between May and July found a lack of specialised root-feeding agents and very low levels of damage across the native range. It was therefore concluded that root-boring insects were unlikely to be useful in the biological control of gorse in its exotic range (Sheppard and Thomann 2005). Surveys for fungal pathogens were conducted in south-western France, north-western Spain and northern Portugal in November 2006 and June 2007 (Jourdan 2009). Analysis of diseased gorse specimens enabled 32 isolates belonging to at least 15 different species to be tested for their pathogenicity. Of these, three isolates of *Fusarium* sp., *Thielaviopsis basicola* and two unidentified strains of a rust (probably *Uromyces pisi* f. subsp. *europaei*) were the only ones found to be pathogenic to gorse. However, none are sufficiently host-specific to be considered as biological control agents (Jourdan 2009). The results of these extensive surveys indicate it is unlikely that any new host-specific invertebrate species or fungal pathogens with potential as gorse biological control agents will be found in Europe.

CANDIDATES

Foliage-feeders

Tetranychus lintearius **Dufour**. The gorse spider mite (Fig. 2) lives on gorse bushes in large webbed colonies often consisting of hundreds of thousands of mites which move around the host plant, feeding en masse and web-spinning as they go. The period for egg to adult development is ~32 days at 20°C and populations can increase rapidly at warm temperatures (Stone 1986). The mites use their mouth parts to pierce cell walls and extract the contents, giving the foliage a bleached appearance. They are dispersed rapidly by wind currents, feed on mature foliage as well as new season's growth and are capable of causing severe damage.

The suitability of *T. lintearius* for release in Australia was based on host-specificity tests conducted on over 130 plant species (Hill and O'Donnell 1991; Ireson *et al.* 2003). Colonies of *T. lintearius* were first released in Tas in December 1998 (Ireson *et al.* 1999) and the mite is now widespread on gorse in Tas, Vic, SA, NSW and WA. Many of the early releases were conducted by lodging 15 cm gorse cuttings infested with ~200 mites into individual gorse bushes, with up to 10 cuttings distributed between two to five bushes in close proximity at each site (Ireson *et al.* 2003). In Tas, Davies *et al.* (2007) showed that the presence of mite colonies on gorse bushes over a period of 2.5 years from the time of release reduced foliage dry weight by ~36%. However, surveys have indicated that predation by the introduced Chilean predatory mite *Phyoseiulus*

Figure 2: Colony of gorse spider mite *Tetranychus lintearius* Dufour. Photo: W Chatterton, TIAR.

persimilis Athias-Henriot and native predatory ladybird beetles *Stethorus* spp. is widespread in *T. lintearius* colonies in south-eastern Australia (Ireson *et al.* 2003; Ireson, unpub. data). Studies have shown that these predators play a major role in suppressing *T. lintearius* populations (Peterson *et al.* 1994; Pratt *et al.* 2003; Davies *et al.* 2009) and are restricting its usefulness as a biological control agent in Australia.

Sericothrips staphylinus **Haliday**. The gorse thrips (Fig. 3) is bivoltine. Adults enter a reproductive diapause once gorse shoots have matured and hardened by the end of the summer and commence egg laying towards the end of the winter. Eggs start hatching in spring, at which time succulent new shoot growth is available for the larvae (Ireson *et al.* 2008b). Development from egg to adult takes ~32 days at 20°C (Ireson *et al.* 2008b).

Release of *S. staphylinus* was approved following tests on 38 Australian plant species or cultivars (Ireson *et al.* 2008a) and on the basis of tests conducted previously on

Figure 3: Gorse thrips *Sericothrips staphylinus* Haliday. Photo: W Chatterton, TIAR.

80 plant species that enabled the release of *S. staphylinus* in NZ and Hawaii (Hill *et al.* 2001). *S. staphylinus* was first released in Tas in January 2001 (Ireson *et al.* 2008a). Tasmanian field studies showed that population growth, reproduction rate and the number of *S. staphylinus* recovered 14 months after release can be non-linear functions of release size and that establishment could be achieved with as few as 10 thrips. However, as *S. staphylinus* is easily cultured, ~250 thrips were chosen as the minimum number for release because this release size produced close to the maximum population growth (Ireson *et al.* 2008a). By March 2009, releases had been made at 466 sites in Tas, 146 sites in Vic, 46 sites in SA and 15 sites in NSW in an attempt to accelerate the slow natural dispersal of *S. staphylinus* (Ireson, unpub. data). Surveys in early 2007 recovered *S. staphylinus* from 80% of 30 sites surveyed in Tas, the post release period ranging from one to six years. However, densities were low (<1 thrips per cm of tip growth), with no evidence of visible plant damage and a maximum dispersal range estimated at 180–250 m (Ireson *et al.* 2008a). Repeat surveys at the same sites in early 2009 indicated that *S. staphylinus* populations were starting to disperse exponentially at some sites. The maximum dispersal was recorded at 4000 m six years after release and a maximum density of approximately four thrips per cm of tip growth was recorded four years after release, too low to cause observable field damage (Ireson, unpub. data). A glasshouse study (Davies *et al.* 2005) showed that a combination of *S. staphylinus*, ryegrass competition and simulated grazing resulted in gorse seedling mortality of 93% compared to no mortality in the untreated control. Individual and combined treatments significantly reduced the shoot dry weight of seedlings; the greatest reduction was observed when all treatments were combined. The experiment therefore indicated the potential of *S. staphylinus* in an integrated control program if field populations are able to increase to sufficient densities.

***Agonopterix umbellana* (Fabricius).** This univoltine moth (Fig. 4) overwinters and becomes active in spring to mate and lay eggs near buds in the axils of spines and on stems. Newly hatched larvae migrate to young buds and spin a silken tube, feeding on the developing spines of the apical tip. As the larvae develop they spin larger tubes and feed on the lateral spines of the growing shoot. Larvae start to pupate within the silken tube from early summer (Hill *et al.* 1995).

A. umbellana was first released at one site each in Vic and Tas in October 2007 following tests on 37 Australian plant species or cultivars (Ireson and Gourlay 2003; Sagliocco *et al.* 2004). The Australian host-specificity test list was approved on the basis of tests previously conducted on over 200 plant species and cultivars in Hawaii (Markin and Brown 1988), NZ (Hill *et al.* 1995) and Chile (Martinez *et al.* 2000). *A. umbellana* had been released at eight sites in Tas by the end of May 2010. Although the only Vic site was destroyed by fire, establishment of *A. umbellana* was confirmed at the first Tas release site in December 2010. In NZ, large populations are now being recorded 15 years after the release of the moth but gorse plants appear to compensate for the severe foliar damage by initiating new shoots and the effect on plant growth is unknown (Hill *et al.* 2008).

Figure 4: Larva of gorse soft shoot moth *Agonopterix umbellana* (Fabricius). Photo: R Holloway, TIAR.

Seed-feeders

***Exapion ulicis* (Forster).** The gorse seed weevil (Fig. 5) is univoltine, but the adults live for up to 12 months and can be found on gorse bushes all year feeding on foliage and flowers. Adults overwinter then mate and commence egg laying in spring. The females bore a hole in the developing seed pods through which they insert an average of nine eggs per pod, although the number of eggs laid can vary considerably (Davies *et al.* 2008). Eggs are laid during spring and summer with peak egg production occurring in mid to late spring. Larval numbers per pod vary considerably, averaging around six per pod. Larvae feed on the seeds and are usually active only from spring through summer, with maximum numbers occurring in late spring to early summer (Davies *et al.* 2008). Pupation occurs inside the pod. Depending on site conditions, some larvae start to pupate from late spring, with others pupating in summer. Adults emerge when the mature pods dehisce (Davies *et al.* 2008).

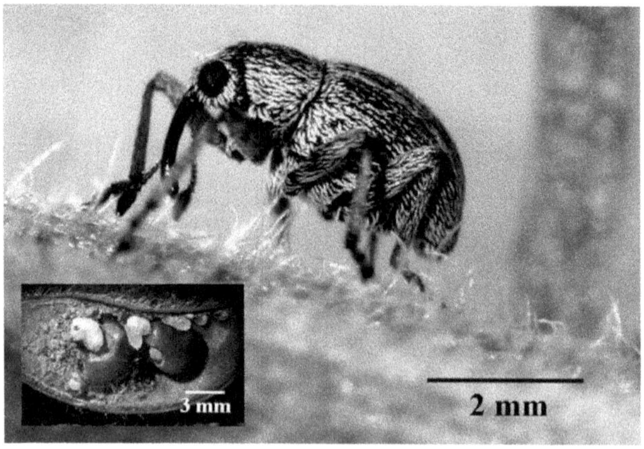

Figure 5: Gorse seed weevil *Exapion ulicis* (Forster) and seed damaged by larvae (inset). Photo: W Chatterton, TIAR.

Released in 1939, *E. ulicis* is now common on gorse throughout Tas, Vic, SA and NSW (Ireson, unpub. data). A Tas study showed that 12–55% of mature seeds are destroyed annually by the weevil (Davies *et al.* 2008), levels not high enough to significantly affect plant densities. The larvae feed on only a proportion of seed produced in spring and summer; as they are not present during the autumn/winter period, a significant proportion of the annual seed crop escapes attack (Fig. 6).

Cydia succedana **(Denis & Schiffermüller).** The first generation of adults of this bivoltine moth become active in spring and produce larvae that feed on developing seed in gorse pods. The second late summer/autumn generation feed on the related gorse species *Ulex minor* Roth and *Ulex gallii* Planch, as these species produce green pods before autumn/winter-flowering gorse begins to bloom (Paynter *et al.* 2008). Neither *U. minor* nor *U. gallii* occurs in Australia.

Release of *C. succedana* in Australia was approved in 2001 following host-specificity studies on 35 species or cultivars of Australian plants conducted in NZ (Ireson and Gourlay 2001) and earlier tests on 44 species of plants that enabled its release in NZ (Hill and Gourlay 2002). However, the Australian release was postponed when field surveys in NZ revealed that the moth could exploit some exotic species of Genisteae, including the tree lupin *Lupinus arboreus* Sims, as well as *Lotus pedunculatus* Cav. and *Lotus corniculatus* L. (Loteae) (Withers *et al.* 2008; Paynter *et al.* 2008). The original host-range tests for NZ were conducted on moths collected in southern England, but the offspring of Portuguese moths were also released in NZ, so the populations used for the Australian host testing were probably mixed. Paynter *et al.* (2008) found that the Portuguese population of *C. succedana* had a different host-range from the English population. Tests on moths sourced from England concurred with the original tests, indicating that these populations would be unlikely to exploit non-target species and would be safe to release in Australia. Paynter *et al.* (2008) concluded that the release of untested moths from Portugal, coupled with asynchrony between the flight period of the gorse pod moth and gorse flowering, explained the unanticipated non-target attack in NZ.

Additional tests conducted in quarantine by the Vic Department of Primary Industries have provided further evidence that moths sourced from England may be safe to release in Australia (Ireson *et al.* 2011), but tests are still being conducted to confirm this. The moth is widely established in NZ and, in combination with *E. ulicis*, has been found to result in seed destruction by up to 92% (TR Partridge, unpub. data cited by Hill and Gourlay 2002). Modelling studies (Rees and Hill 2001) have indicated this level of seed destruction should be high enough to reduce the recruitment of gorse below replacement levels if management practices that reduce seedling survival are also used. If *C. succedana* is ever released in Australia and becomes established, it has the potential to play a significant role in the biological control of gorse.

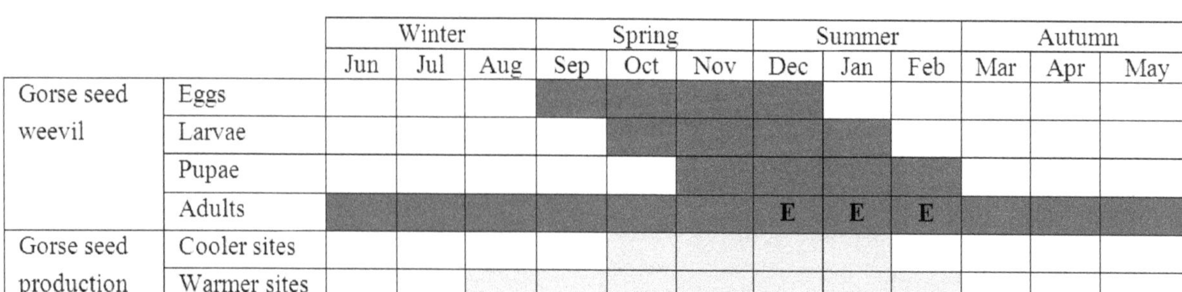

Figure 6: Main periods of activity in the life-cycle of *Exapion ulicis* in relation to gorse seed production patterns at cooler and warmer sites in Tas. The main period of larval activity is October to January, so a large proportion of seed produced annually is not attacked. Cooler sites are generally at high altitude and/or inland, warmer sites are generally at low altitude and/or near the coast; minor seed production on some plants can occur outside the periods indicated. E = emergence of adult weevils from pods.

DISCUSSION

Of the four agents released for the biological control of gorse in Australia, only *E. ulicis* and *T. lintearius* are widespread throughout the range of gorse in most of south-eastern Australia. However, it is not uncommon for biological control agents such as *S. staphylinus* to exist at low levels for several generations following their introduction, after which populations start to increase exponentially to high densities and start to disperse (Hopper and Roush 1993). Indications that *S. staphylinus* is starting to increase exponentially at some sites in Tas suggest the agent will become widespread there during the next two to three years. The rate at which this occurs in other states will depend on the number of new sites established through continuing field release programs.

The establishment of *A. umbellana* in Tas is a promising development but release of the agent will also need to be ongoing to accelerate widespread field establishment and dispersal, particularly in mainland states. Sites in north-western, south-eastern or western Tas will be optimal for initial nursery site establishment as they occur in areas less likely to be subject to drought stress. Outbreak populations of *A. umbellana* began to appear at NZ sites 15 years after the moth was first released (Hill *et al.* 2008), indicating a low initial rate of natural increase, as has also been recorded for *S. staphylinus*. Lags in population explosions are often inherent because of the nature of population growth and range expansion, and may be prolonged by environmental and genetic factors (Crooks and Soule 1999). A low intrinsic rate of natural increase was discussed by Lawton and McNeill (1979) and Price (2000) as being characteristic of the invertebrate fauna associated with perennial, stress-tolerant woody plants such as gorse. The majority of invertebrate species on gorse in its native habitat are characteristically confined to one or a few closely related species of plant, with many being very sedentary (Hill 1982b; McNeill and Prestidge 1982).

Efficacy studies have shown that an established combination of three foliage-feeders and one seed-feeding agent will contribute to the biological control of gorse in Australia (Davies *et al.* 2005, 2007, 2008), but these agents are constrained by predation and the phenology and seasonality of gorse (Ireson *et al.* 2003, 2008a, 2008b; Davies *et al.* 2009). However, release of a second seed-feeding agent (*C. succedana*) in Australia is a possibility if the host-specificity of the population from southern England, as indicated by the tests of Paynter *et al.* (2008), is confirmed.

Although surveys have now shown that it is unlikely that additional invertebrate species or fungal pathogens with potential as biological control agents will be found in Europe (Sheppard 2004; Sheppard and Thomann 2005; Jourdan 2009), the guild of agents that are available may have a long-term impact through sublethal effects on maximum plant age and increasing susceptibility to fungal attack. Seed-feeding insects could play the major role in reducing seed banks below critical replacement levels, but foliage-feeding agents could also cause a decline in annual seed production through a reduction in plant vigour. Integrated control techniques (herbicides, mechanical clearing, cultivation and grazing) offer the best prospects for long-term control in areas where gorse is actively managed, but the extent to which biological control will play a role will only be determined by future research once the complete guild of available agents is established (Hill *et al.* 2008).

REFERENCES

Anon (2006) *Gorse National Best Practice Manual: Managing Gorse* (Ulex europaeus *L.*) *in Australia*. National Gorse Taskforce. Tas Dept of Primary Industries and Water.

Chater EH (1931) A contribution to the study of the natural control of gorse. *Bulletin of Entomological Research* **22**, 225–235.

Crooks JA and Soule ME (1999) Lag times in population explosions of invasive species: causes and implications. In *Invasive Species and Biodiversity Management*. (Eds OT Sandlund, PJ Schei and A Viken) pp. 103–125. Kluwer Academic, Dordrecht.

Davies WM (1928) The bionomics of *Apion ulicis* F. (gorse weevil), with special reference to its role in the control of *U. europaeus* in New Zealand. *Annals of Applied Biology* **15**, 263–286.

Davies JT, Ireson JE and Allen GR (2005) The impact of gorse thrips, ryegrass competition and simulated grazing on gorse seedling performance in a controlled environment. *Biological Control* **32**, 280–286.

Davies JT, Ireson JE and Allen GR (2007) The impact of the gorse spider mite, *Tetranychus lintearius*, on the growth and development of gorse, *Ulex europaeus*. *Biological Control* **41**, 86–93.

Davies JT, Ireson JE and Allen GR (2008) The phenology and impact of the gorse seed weevil, *Exapion ulicis*, on gorse, *Ulex europaeus*, in Tasmania. *Biological Control* **45**, 85–92.

Davies JT, Ireson JE and Allen GR (2009) Pre-adult development of *Phytoseiulus persimilis* on diets of *Tetranychus urticae* and *Tetranychus lintearius*: implications for the biological control of *Ulex europaeus*. *Experimental and Applied Acarology* **47**, 133–145.

Evans JW (1942) The gorse weevil. *Tasmanian Journal of Agriculture* **13**, 15–18.

Girling DJ (1977) 'Report on the possibilities of finding natural enemies to control gorse, *Ulex europaeus* L., in New Zealand'. Report of the Commonwealth Institute of Biological Control.

Hill RL (1982a) The phytophagous fauna of gorse and host plant quality. PhD dissertation. Imperial College, University of London, UK.

Hill RL (1982b) Seasonal patterns of phytophage activity on gorse (*Ulex europaeus*), and host plant quality. In *Proceedings of the 5th International Symposium on Insect-Plant Relationships*. 1–4 March 1982, Wageningen, Netherlands. (Eds JH Visser and AK Minks) pp. 237–242. Pudoc, Wageningen.

Hill RL (1983) Prospects for the biological control of gorse. *Proceedings of the 36th New Zealand Weed and Pest Control Conference*, 9–11 August 1983, Hastings. (Ed MJ Hartley) pp. 56–58. New Zealand Weed and Pest Control Society, NZ.

Hill RL and Gourlay AH (2002) Host-range testing, introduction, and establishment of *Cydia succedana* (Lepidoptera: Tortricidae) for biological control of gorse in New Zealand. *Biological Control* **25**, 173–186.

Hill RL and O'Donnell DJ (1991) The host range of *Tetranychus lintearius* (Acarina: Tetranychidae). *Experimental and Applied Acarology* **11**, 253–269.

Hill RL, O'Donnell DJ, Gourlay AH and Speed CB (1995) Suitability of *Agonopterix ulicetella* (Lepidoptera: Oecophoridae) as a control for *Ulex europaeus* (Fabaceae: Genisteae) in New Zealand. *Biocontrol Science and Technology* **5**, 3–10.

Hill RL, Gourlay AH, Lee WG and Wilson JB (1996) Dispersal of seeds under isolated gorse plants and the impact of seed feeding insects. In *Proceedings of the 49th New Zealand Plant Protection Conference*, 13–15 August 1996, Nelson. (Ed M O'Callaghan) pp. 114–118. New Zealand Plant Protection Society, NZ.

Hill RL, Markin GP, Gourlay AH, Fowler SV and Yoshioka E (2001) Host range, release and establishment of *Sericothrips staphylinus* Haliday (Thysanoptera: Thripidae) as a biological control agent for gorse, *Ulex europaeus* L. (Fabaceae), in New Zealand and Hawaii. *Biological Control* **21**, 63–74.

Hill RL, Ireson J, Sheppard AW, Gourlay AH, Norambuena H, Markin GP, Kwong R and Coombes EM (2008) A global view of the future for biological control of gorse, *Ulex europaeus* L. In *Proceedings of the 12th International Symposium on Biological Control of Weeds*. 22–27 April 2007, La Grande Motte, France. (Eds MH Julien, R Sforza, MC Bon, HC Evans, PE Hatcher, HL Hinz and BG Rector) pp. 680–686. CABI, Wallingford, UK.

Hopper KR and Roush RT (1993) Mate finding, dispersal, number released, and the success of biological control introductions. *Ecological Entomology* **18**, 321–331.

Ireson JE and Gourlay AH (2001) 'Application for approval of the release of gorse pod moth, *Cydia succedana* (Denis and Schiffermüller), a potential biological control agent for gorse, *Ulex europaeus* L'. Unpublished report. Tas Institute of Agricultural Research.

Ireson JE and Gourlay AH (2003) 'Application for approval of the release of gorse soft shoot moth, *Agonopterix ulicetella* (Stainton), a potential biological control agent for gorse, *Ulex europaeus* L'. Unpublished report. Tas Institute of Agricultural Research.

Ireson JE, Gourlay AH, Kwong RM, Holloway RJ and Chatterton WS (1999) Progress on the rearing, release and establishment of the gorse spider mite, *Tetranychus lintearius* Dufour, for the biological control of gorse in Australia. In *Proceedings of the 12th Australian Weeds Conference*. 12–16 September 1999, Hobart. (Eds AC Bishop, M Boersma and CD Barnes) pp. 320–324. Tasmanian Weed Society, Devonport.

Ireson JE, Gourlay AH, Kwong RM, Holloway RJ and Chatterton WS (2003) Host specificity, release and establishment of the gorse spider mite, *Tetranychus lintearius* Dufour (Acarina: Tetranychidae), for the biological control of gorse, *Ulex europaeus* L. (Fabaceae), in Australia. *Biological Control* **26**, 117–127.

Ireson JE, Kwong RM, Gourlay AH, Davies JT, Chatterton WS and Holloway RJ (2004) Progress on the biological control of gorse (*Ulex europaeus*) in Australia. In *Proceedings of the 11th International Symposium on Biological Control of Weeds*. 27 April–2 May 2003, Canberra. (Eds JM Cullen, DT Briese, DJ Kriticos, WM Lonsdale, L Morin and JK Scott) pp. 415–418. CSIRO Entomology, Canberra.

Ireson JE, Gourlay AH, Holloway RJ, Chatterton WS, Foster SD and Kwong RM (2008a) Host specificity, establishment and dispersal of the gorse thrips, *Sericothrips staphylinus* Haliday (Thysanoptera: Thripidae), a biological control agent for gorse, *Ulex europaeus* L. (Fabaceae) in Australia. *Biological Control* **45**, 460–471.

Ireson JE, Holloway RJ and Chatterton WS (2008b) Phenology and development of the gorse thrips, *Sericothrips staphylinus* Haliday (Thysanoptera: Thripidae), a biological control agent for gorse, *Ulex europaeus* L. (Fabaceae), in Tasmania. *Biological Control* **45**, 64–71.

Ireson J, Relf M, Sagliocco J-L, Kwong R, Bruzzese A and Chatterton W (2011) 'Biological control of gorse in Australia: host testing of the gorse pod moth'. Publication No. 11/030. Rural Industries Research and Development Corporation.

Jourdan M (2009) 'Final report on European surveys for potential plant pathogen biological control agents of *Ulex europaeus* L. (Fabaceae) (gorse) in Australia'. CSIRO Entomology contracted research report for Tas Institute of Agricultural Research.

Lawton JH and McNeill S (1979) Between the devil and the deep blue sea: on the problems of being a herbivore. *Symposium of the British Ecological Society* **20**, 223–244.

Lee WG, Allen RB and Johnson PN (1986) Succession and dynamics of gorse (*Ulex europaeus* L.) communities in the Dunedin Ecological District South Island, New Zealand. *New Zealand Journal of Botany* **24**, 279–292.

Markin GP and Brown RE (1988) 'Petition for introduction and release in the United States of the moth, *Agonopterix ulicetella* (Stainton) (Lepidoptera: Oecophoridae), for a biological control of gorse (*Ulex europaeus* L., Leguminoseae)'. Unpublished report. US Dept of Agriculture/Oregon Dept of Agriculture.

Markin GP and Yoshioka ER (1989) Present status of biological control of the weed gorse (*Ulex europaeus* L.) in Hawaii. In *Proceedings of the 7th International Symposium on Biological Control of Weeds.* 6–11 March 1988, Rome. (Ed ES Delfosse) pp. 357–362. Instituto Sperimentale per la Patalogia Vegetale, Rome.

Martinez GG, Norambuena HM, Carrillo LL and Rodriguez FA (2000) Estudio de especificidad de la polilla del espinillo *Agonopterix ulicetella* (Stainton) para el control biologico del espinillo (*Ulex europaeus* L.). *Agro sur* **28**, 133–151.

McNeill S and Prestidge RA (1982) Plant nutritional strategies and insect herbivore community dynamics. In *Proceedings of the 5th International Symposium on Insect-Plant Relationships.* 1–4 March 1982, Wageningen, Netherlands. (Eds JH Visser and AK Minks) pp. 225–235. Pudoc, Wageningen.

Miller D (1947) Control of gorse by seed weevil. *New Zealand Journal of Agriculture* **75**, 341–344.

Moss GR (1959) The gorse seed problem. In *Proceedings of the 12th New Zealand Weed Control Conference.* 7–9 July 1959, Tauranga, NZ. pp. 59–64. New Zealand Weed Control Conference.

O'Donnell D (1986) 'A survey of natural enemies of gorse (*Ulex* spp.) in northern Spain and Portugal'. Unpublished report. CIBC, Imperial College, Silwood Park, Ascot, UK.

Parsons WT and Cuthbertson EG (2001) *Noxious Weeds of Australia.* 2nd edn. CSIRO Publishing, Melbourne.

Paynter Q, Gourlay AH, Oboyski PT, Fowler SV, Hill RL, Withers TM, Parish H and Hona S (2008) Why did specificity testing fail to predict the field host range of the gorse pod moth in New Zealand? *Biological Control* **46**, 453–462.

Peterson PG, McGregor PG and Springett BP (1994) Development of *Stethorus bifidus* in relation to temperature: implications for regulation of gorse mite populations. *Proceedings of the 47th New Zealand Plant Protection Conference.* 9–11 August 1994, Waitangi, NZ. (Ed AJ Popay) pp. 103–106. New Zealand Weed and Pest Control Society, NZ.

Pratt PD, Coombs EM and Croft BA (2003) Predation by Phytoseiid mites on *Tetranychus lintearius*, an established weed biological control agent of gorse (*Ulex europaeus*). *Biological Control* **26**, 40–47.

Price PW (2000) Host plant resource quality, insect herbivores and biological control. In *Proceedings of the 10th International Symposium on Biological Control of Weeds.* 4–14 July 1999, Bozeman, Montana. (Ed NR Spencer) pp. 583–590. Montana State University, Bozeman, Montana.

Rees M and Hill RL (2001) Large-scale disturbances, biological control and the dynamics of gorse populations. *Journal of Applied Ecology* **38**, 364–377.

Richardson RG and Hill RL (1998) The biology of Australian weeds 34. *Ulex europaeus* L. *Plant Protection Quarterly* **13**, 46–58.

Sagliocco J-L, Bruzzese A and Kwong R (2004) 'Report on the rearing and additional host-specificity tests with *Agonopterix ulicetella* (Stainton) for the biological control of gorse (*Ulex europaeus*)'. Unpublished report. Vic Dept of Primary Industries, Frankston, Vic.

Schröder D and Zwölfer H (1970) Studies on insects associated with gorse, *Ulex europaeus* L. In *Proceedings of the 1st International Symposium on Biological Control of Weeds.* 6–8 March 1969, Delémont, Switzerland. (Ed FJ Simmonds) pp. 55–58. CIBC, Wallingford, UK.

Sheppard A (2004) 'A search in Spain and Portugal for potential biocontrol agents for gorse (*Ulex europaeus*

europaeus L.) in Hawaii'. CSIRO Entomology, contracted research report for Parker Ranch, Hawaii.

Sheppard A and Thomann T (2005) 'European surveys for potential root-feeding biological control agents of gorse (*Ulex europaeus europaeus* L.)'. CSIRO Entomology, contracted research report no. 87 for Tas Institute of Agricultural Research.

Stone C (1986) An investigation into the morphology and biology of *Tetranychus lintearius* Dufour (Acari: Tetanychidae). *Experimental and Applied Acarology* **2**, 173–186.

Tarayre M, Bowman G, Schermann-Legionnet A, Barat M and Atlan A (2007) Flowering phenology of *Ulex europaeus*: ecological consequences of variation within and among populations. *Evolutionary Ecology* **21**, 395–409.

Thorp JR and Lynch R (2000) *The Determination of Weeds of National Significance*. National Weeds Strategy Executive Committee, Launceston.

Withers TM, Hill RL, Paynter QP, Fowler SV and Gourlay AH (2008) Post-release investigations into the field range of the gorse pod moth *Cydia succedana* Denis and Schiffermüller (Lepidoptera: Tortricidae) in New Zealand. *New Zealand Entomologist* **31**, 67–76.

Zwölfer H (1962) '*Ulex europaeus* project: European investigations for New Zealand'. Unpublished report no. 1. Commonwealth Institute of Biological Control, Delémont, Switzerland.

Zwölfer H (1963) '*Ulex europaeus* project: European investigations for New Zealand'. Unpublished report no. 2. Commonwealth Institute of Biological Control, Delémont, Switzerland.

Zwölfer H (1964) '*Ulex europaeus* project: European investigations for New Zealand'. Unpublished report no. 3. Commonwealth Institute of Biological Control, Delémont, Switzerland.

Xanthium occidentale Bertol. – Noogoora burr

Rieks D van Klinken and Louise Morin

ABSTRACT

The annual herb, *Xanthium occidentale*, Noogoora burr, was one of the first weeds targeted by biocontrol worldwide. In the 1950s it was considered one of the worst weeds in Australia, owing largely to its impact on the wool industry. In Australia it was the focus of major efforts both prior to and following World War II; these resulted in the release of three biocontrol agents, two stem-boring beetles and a seed-feeding fly. In 1974 the rust fungus *Puccinia xanthii* was introduced through unknown means, and in 1982 the stem-galling moth *Epiblema strenuana* was released against *Parthenium hysterophorus* but it also attacked Noogoora burr. The rust caused almost complete control across much of Australia, except in more arid regions and in the wet-dry tropics. The moth may also have provided some control, but the remaining three insects had no impact. Recent work has focused on finding pathogenic rust isolates that are better adapted climatically to conditions in northern Australia, but it has so far not been successful. Nonetheless, the benefits of the biocontrol program (including the rust) were estimated to exceed \$A23 million in 2004–05, and Noogoora burr is now considered a relatively minor problem nationally.

Key words: Noogoora burr complex, rust fungus *Puccinia xanthii*, stem-galling moth *Epiblema strenuana*.

INTRODUCTION

Xanthium occidentale Bertol. (Asteraceae), Noogoora burr, is an annual herb from the Neotropics. It is listed as a serious weed of 11 major agricultural crops in 28 countries, and as one of the 10 most noxious weeds in the USA (Hocking and Liddle 1995). It was first introduced into Australia (Qld) in the mid 1800s in imported cotton seed, probably from the USA, and subsequently spread over approximately 2 million km^2 in Qld, NSW, SA, WA and the NT (Liddle 1986). At least until the 1970s, it was one of the most important weeds in Australia, affecting the sheep and grazing industries. It was declared a noxious weed in Australia in 1926, was considered the second most important weed in NSW and Qld in 1929 after prickly pear (*Opuntia stricta* (Haworth) Haworth), and was regarded as one of Australia's worst weeds by 1953 (Chippendale 1995). It lowered wool values due to fleece contamination by spiny fruits, reduced carrying capacities of infested lands, poisoned and discomforted livestock, was costly to control with traditional methods and resulted in interstate sale embargoes on livestock. These far-reaching impacts triggered the initiation of a biocontrol program in 1929, making Noogoora burr one of the first weeds to be systematically targeted for biocontrol in Australia and indeed worldwide (van Klinken and Julien 2003).

Noogoora burr is now much less important in Australia and was ranked 28 out of 71 species assessed as part of a national weed ranking process in 1999 (Thorp and Lynch 2000). Several reasons have been put forward to explain this change in importance status: a drastic reduction in densities in wetter parts of its range where the rust fungus *Puccinia xanthii* Schw. has had a severe impact, sheep farming has contracted from much of northern Australia for a wide range of reasons and there have been improvements in wool-cleaning procedures such that contamination with Noogoora burr fruits can be handled more easily (Chippendale 1995; van Klinken and Julien 2003).

The densest Noogoora burr infestations are now largely restricted to riparian areas and floodplains in semi-arid parts of NSW, Qld and NT (although only following exceptional rainfall events) and in the wet-dry tropics around the Gulf of Carpentaria (Qld and NT), the Darwin region and Victoria River district (NT) and the Kimberley (WA). An eradication program was terminated in the Kimberley in the early 1990s; the remaining populations have been contained within a quarantine zone maintained until recently to prevent spread of the weed to other parts of WA. The high cost of such containment, and the continued presence of Noogoora burr across the wet–dry tropics, renewed interest in the early 2000s in finding a biocontrol solution particularly adapted for this region.

BIOLOGICAL CONTROL HISTORY

Australia and Fiji are the only two countries that established a biocontrol program for Noogoora burr. Fiji did not undertake any exploration; it took advantage of the Australian biocontrol program, which led to the introduction into Fiji of two of the insect agents that were released in Australia (Julien and Griffiths 1998).

There were two major phases in the Australian program: 1929–40 and 1953–75 (see van Klinken and Julien (2003) for details). The first phase (led by CSIR, now CSIRO) included extensive surveys in the USA, the putative origin of Noogoora burr plants that invaded Australia, as well as cursory searches in Central and South America and southern Asia. The second phase commenced in 1953 following renewed interest by CSIRO and the Qld Government. The problem was sufficiently important that a conference of specialists, with the support of the then Prime Minister Robert Menzies, formulated a national response. Recommencement of biocontrol resulted in the re-evaluation and subsequent release of two previously rejected candidate agents. In addition, more comprehensive surveys for natural enemies were conducted in Pakistan by the Commonwealth Institute of Biological Control but they yielded no new candidate agents.

Opportunistic studies were conducted on the rust fungus in Europe in the late 1960s following the discovery of an outbreak on *Xanthium strumarium* L. *sensu lato* in Po Valley, Italy (Hasan 1974a). It was suggested as a possible candidate agent pending host-specificity testing (Hasan 1974 a, b), but a naturalised population was found in a field near Brisbane in 1975 (Alcorn 1975). The pathway of introduction into Australia has never been identified.

In the early 2000s, interest in revisiting biocontrol of Noogoora burr was renewed, particularly for the persisting infestations in the wet-dry tropics of WA and NT. Climate matching had indicated that the hot conditions that prevail during the Noogoora burr growing season were not conducive to high levels of infection by the rust fungus (Morin *et al.* 1996). It was proposed that overseas searches for rust isolates climatically adapted to tropical conditions could be a fruitful avenue to enhance biocontrol in the region (van Klinken and Julien 2003).

PLANT TAXONOMY

There are four species recognised in Australia within the Noogoora burr complex (*X. occidentale*, *Xanthium italicum* Moretti, *Xanthium orientale* L. and *Xanthium cavanillesii* Schouw), all of which are often referred to as *X. strumarium s. l.* in Europe and America (Hocking and Liddle 1995). Noogoora burr (*X. occidentale*) is the most widespread species of the complex (Fig. 1) and has been the target for biocontrol. It can generally be easily differentiated from the other species in Australia using leaf shape and fruit characteristics (Fig. 2) (Hocking and Liddle 1995), although two morphological forms have been recognised near Rockhampton in central Qld (P Michael, *pers. comm.* 2000). Putative hybrids between Noogoora burr and Hunter burr (*X. italicum*) have been reported where their geographical distributions coincide (McMillan 1973).

Identification of Noogoora burr, including delimitation of its distribution, is much more difficult in the Americas where there is extensive morphological variation within species lumped under *X. strumarium s. l.*, possibly due to extensive hybridisation (Weaver and Lechowicz 1982; Hocking and Liddle 1995). A recent comparison of morphological characteristics between species in the Noogoora burr complex and *Xanthium* accessions from Mexico and the Dominican Republic supported conjecture that the Dominican Republic and Mexican plants examined were putative hybrids between closely related species (L Morin, unpub. data). The leaf morphology of Dominican Republic *Xanthium* herbarium specimens was similar to that of Noogoora burr but fruits (when present) had beaks similar in morphology to those of Hunter burr. In contrast, Mexican *Xanthium*

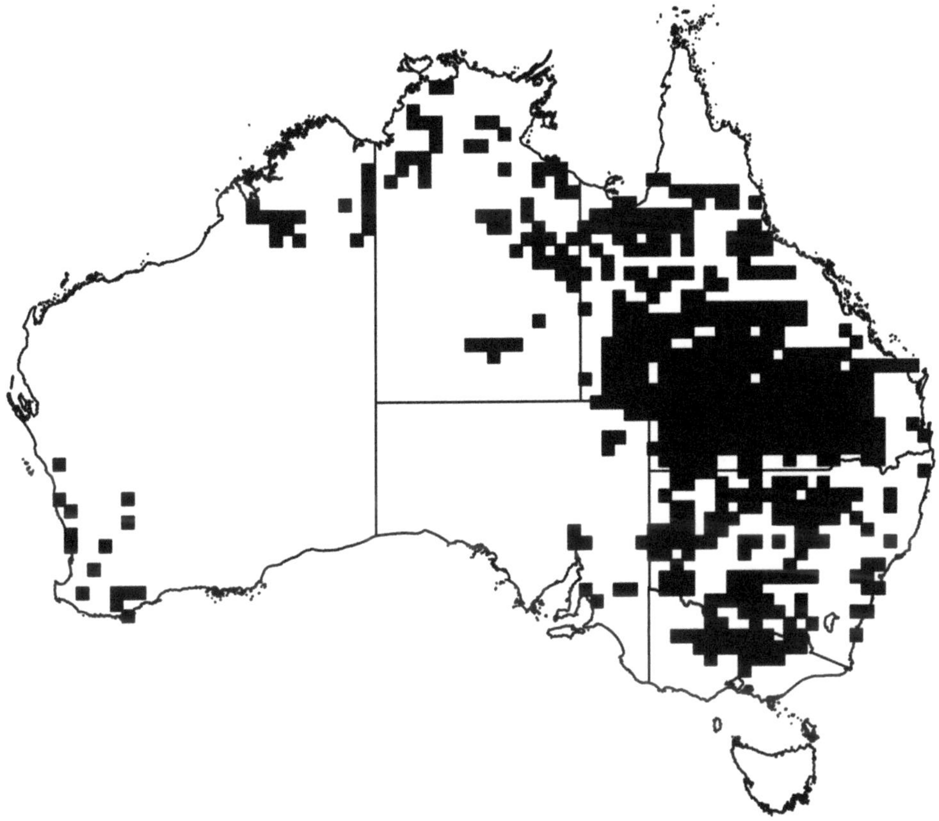

Figure 1: Distribution of Noogoora burr, *Xanthium occidentale*, in Australia. Map drawn by the Bureau of Rural Sciences using data presented in Thorp and Lynch (2000). Distribution shown as 0.5, 0.25 or 0.125 square degrees depending on the state or territory.

herbarium specimens had leaf and fruit morphology that resembled those of South American burr *X. cavanillesii.*

EXPLORATION

Noogoora burr is probably native to southern USA, Mexico and the West Indies, although its southern boundary in Central or South America, which remains poorly documented, is unclear (van Klinken and Julien 2003). The common Australian 'form' of Noogoora burr occurs in south-eastern USA (including Louisiana) west to Texas, according to P Michael (*pers. comm.* 2000) who examined herbarium records in the USA. However, closely related species occur across Central and South America (Love and Dansereau 1959). This means that biocontrol agents specific to the *Xanthium* genus could potentially be sourced from across the Americas, as there are no at-risk native *Xanthium* species in Australia. The Americas beyond the USA and Mexico remain largely underexplored, despite being considered the centre of origin for the genus (Love and Dansereau 1959). In addition, species of the Noogoora burr complex have been introduced into Canada, Trinidad, Europe, Africa, Central Asia, south Asia, south-east Asia, New Zealand, Australian, Melanesia and a few other Pacific islands (Hocking and Liddle 1995). Some of these introduced ranges have been searched for potential biocontrol agents.

Field surveys in the USA were initially conducted in Kansas (1929–35) because of its climatic similarity to Australian regions where Noogoora burr was a problem, but they later moved to Texas (1936–40) (van Klinken and Julien 2003). Cursory surveys were conducted during that time in other parts of the Americas, including southern states from Florida to California, and in Mexico and Argentina. Considerable additional ecological research was conducted in the 1970s on the natural enemies of Ambrosiinae, including *X. strumarium s. l.* (Hilgendorf and Goeden 1983; Goeden and Palmer 1995; references therein). This research was conducted independently of the biocontrol program but provided a better understanding of the natural enemies. Faunal surveys were conducted in India in 1938–40 and Pakistan in 1960–70 (van Klinken and Julien 2003). The surveys in Pakistan were more extensive and involved some experimental work.

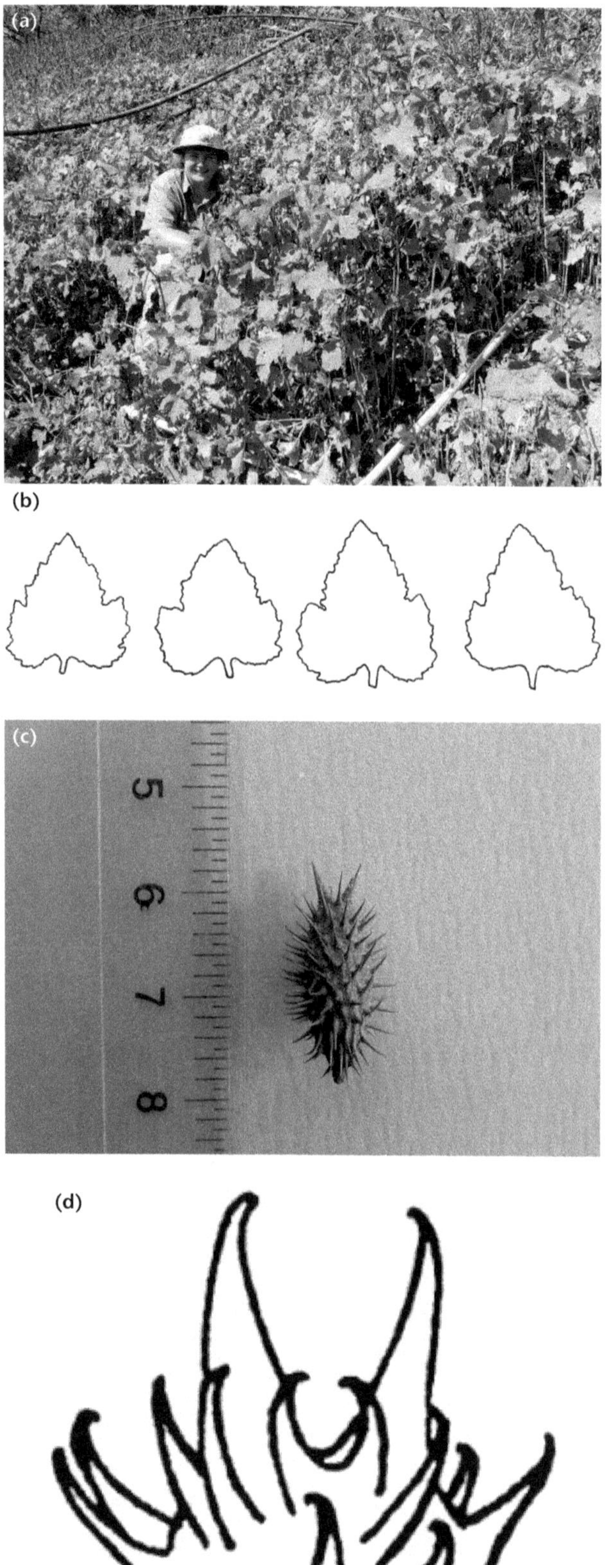

Figure 2: a) Noogoora burr, *Xanthium occidentale*, infestation at a study site on the Daly River, NT. b) Examples of leaf shapes. c) Typical fruit ('burr'). d) Noogoora burr fruit beaks morphology. Photos: a) R van Klinken, CSIRO. c) L Morin, CSIRO.

Searches for isolates of the rust fungus *P. xanthii* that are better adapted to the climatic conditions in northern Australia were made in tropical America from August to November 2007 (L Morin, unpub. data). Field surveys were undertaken in Venezuela, Mexico and the Dominican Republic, areas that climatically match those of northern Australia where the rust fungus has not been highly effective (van Klinken and Julien 2003).

CANDIDATES

Natural enemies

Surveys of natural enemies on Noogoora burr were conducted in Australia between 1984 and 1987 and recorded over 50 insect species feeding on Noogoora burr (Wilson and Flanagan 1993). However, none were causing significant impact, most were rare and not known to breed on Noogoora burr, and all 26 species for which host-range data were known were polyphagous.

About 80 insect species were found attacking *X. strumarium s. l.* in the USA during the 1920s and 1930s surveys, but the species list could not be found in recent archive searches (van Klinken and Julien 2003). However, at least 182 phytophagous insect species have been recorded on this host in surveys conducted by RD Goeden and colleagues in the USA and Mexico since the 1970s (Hilgendorf and Goeden 1982, 1983; Goeden and Palmer 1995). Only six of these species were considered to be sufficiently host-specific for consideration as candidate biocontrol agents based on field host-range records (Goeden and Palmer 1995). Of these, three had been released in Australia in the early 1930s and mid 1960s and two were considered realistic prospects based on damage caused, but were not investigated further (Hilgendorf and Goeden 1983) (see below).

Detailed records of the opportunistic field surveys performed in India in the late 1930s could not be relocated, but few natural enemy species were found, although one of the agents eventually released in Australia was sourced from there (Table 1). More than 90 species were collected from *X. strumarium s. l.* in Pakistan in the 1960s, but many were known as pests of economic plants. Only 17 appeared to have a definite association with *Xanthium* and had no other known hosts (van Klinken and Julien 2003). Six of these species were shortlisted as having potential for biocontrol.

Pathogens of *X. strumarium s. l.* have not been systematically surveyed in the native range. However, more than 50 fungi, one virus and some nematodes and bacteria

Table 1: Biocontrol agents released in Australia against Noogoora burr, *Xanthium occidentale*

Order, family and species	First released	Origin	Damage	Status
Diptera: Tephritidae				
Euaresta aequalis (Loew)	1932	USA (California, Kansas, Texas)	Seed-feeder	Established at low levels. No control
Coleoptera: Cerambycidae				
Mecas saturnina LeConte	1963	USA (Texas)	Stem-girdler and borer	Previously established at one site. No recent recoveries
Nupserha vexator (Pascoe)	1964	India	Stem-borer	Established widely. Provides little control
Lepidoptera: Tortricidae				
Epiblema strenuana (Walker)	1982 (primarily for parthenium)	Mexico	Stem-galler	Widely established. Possibly reducing weed vigour and competitiveness
Fungus: Uredinales: Pucciniaceae				
Puccinia xanthii Schweinitz	1975	Accidental or illegal introduction	Infects cells of vegetative aerial parts	Widespread throughout eastern Australia. Excellent control in north-eastern NSW and parts of Qld through progressive reduction in seed bank. Little or no control elsewhere

Source: After Julien and Griffiths (1998).

have been recorded from this species, from both the native and introduced ranges (Morin 1993; Morin *et al.* 1994; Farr and Rossman 2010). Although their potential for biocontrol has not been systematically assessed, none appear to be obvious candidates.

Natural enemies tested but not released

Surveys conducted during the first exploration phase shortlisted a seed-feeding fly, six stem-boring and two root-breeding weevil species from the USA and one stem-boring insect from southern Asia as being sufficiently damaging and potentially host-specific to warrant detailed study (van Klinken and Julien 2003). All but one of these, a stem-boring weevil, was imported into an Australian quarantine facility for further study between 1930 and 1940. All species, except the seed-feeding fly, were assessed as unsuitable for release due to poor host-specificity. Three of the stem-borers from the USA and the stem-borer from southern Asia were re-evaluated as part of the second phase of the biocontrol program, and two of the stem-borers from the USA (both *Dectes* species) were again rejected.

Of the six insects shortlisted as potential agents in Pakistan, four were studied in more detail and only one stem-borer (sourced from India) was eventually released in Australia (see below) (van Klinken and Julien 2003). Two moths, the leaf-feeder *Anania verbascalis* (Schiff.) (Pyralidae) and *Eublemma parva* (Hubner) (Noctuidae), which attacks buds and tender burrs, were rejected because they were not sufficiently host-specific. The sap-sucker *Coptosoma* sp. nr *cardoni* Montan (Heteroptera: Plataspidae) was considered to have potential and appeared host-specific in initial tests but the damage it caused to plants was limited unless they were exposed to very high densities of the insect. The remaining two shortlisted species, the leaf-miner *Liriomyza compositella* Spencer and stem-borer *Melanagromyza* sp. (Diptera: Agromyzidae), were not studied.

Agents released

Three insects have been released against Noogoora burr in Australia since 1932: the fly *Euaresta aequalis* (Loew) and the stem-boring beetles *Mecas saturnine* LeConte and *Nupserha vexator* (Pascoe) (Table 1). The stem-galling moth *Epiblema strenuana* (Walker) released against parthenium weed *Parthenium hystephorus* L. (Asteraceae) in Australia in 1982 also attacks Noogoora burr. The rust fungus *P. xanthii* arrived in Australia as the result of an unauthorised introduction in the mid 1970s (Alcorn 1975) and quickly controlled Noogoora burr across large parts of its range.

***Euaresta aequalis* (Loew) (Diptera: Tephritidae).** This univoltine seed-feeding fly (ex Kansas and Texas) was released in southern and central Qld between 1932 and 1940. It established in coastal regions of eastern Qld, where it kills up to 15% of the seed, but has not been reported west of the Great Dividing Range (Wapshere 1974; Haseler 1981). Females oviposit into green fully formed fruits and larvae consume one of the two seeds contained therein before pupating in the hollowed-out fruit. Both seeds are rarely attacked within a fruit, possibly because of oviposition deterrence pheromones, thereby limiting the impact of this species (van Klinken and Julien 2003).

***Mecas saturnina* LeConte and *Nupserha vexator* (Pascoe) (Coleoptera: Cerambycidae).** The univoltine stem-boring beetle *M. saturnina* (ex Texas) was released in 1963. A year later the other stem-boring beetle *N. vexator* (ex India) was released. Releases of these agents were approved at the time, even though they were capable of causing some damage on sunflower *Helianthus annuus* L., Jerusalem artichoke *Helianthus tuberosus* L. and related Asteraceae, because they were not recorded as pests in their native ranges and the Qld Government was willing to accept the risk of damage to sunflowers (van Klinken and Julien 2003). *M. saturnina* was recorded as established at one site near Roma, Qld (Haseler 1981), but has not been recovered since. In contrast, *N. vexator* established widely and is sometimes quite common in midsummer, but provides little control (Haseler 1981). Various hypotheses have been suggested as to why these insects have failed to either establish or to reach high densities, including poor synchronisation with the host, Noogoora burr being a marginal host (*M. saturnina* appeared to perform better on naturalised sunflower near Roma) and poor adaptation to low-rainfall climates (van Klinken and Julien 2003). To have any population-level impact the densities of these stem-borers would need to be very high, as Noogoora burr has a strong compensatory mechanism and because larvae generally reach the roots of tall plants too late to prevent seeding (Wapshere 1974).

***Epiblema strenuana* (Walker) (Lepidoptera: Tortricidae).** The multivoltine stem-galling moth (ex Mexico) was initially released against parthenium weed, although it had also been identified as a potential agent for Noogoora burr in 1930 (Kelly 1931). Larvae form fusiform galls within green stems of parthenium weed, Noogoora burr and annual ragweed *Ambrosia artemisiifolia* L. It is widely established across Australia, with the possible exception of the western Kimberley, and galls can be common at times (van Klinken and Julien 2003) but generally not at densities sufficient to severely affect adult plants (30–100 galls per plant; McFadyen 1987). For example, during surveys of Noogoora burr populations at sites on the Daly and Victoria rivers in the NT in 2005–07, an average of 49% of Noogoora burr stems showed some damage from this biocontrol agent but there were rarely more than 20 galls per plant (B Grace *et al.*, unpub. data). The lack of hosts (Noogoora burr, parthenium weed or annual ragweed) when adults emerge from winter diapause has been suggested as one regulating factor, especially in northern Australia where summer, germinating rains can arrive quite late (McFadyen 1987).

***Puccinia xanthii* Schw. (Uredinales: Pucciniaceae).** This fungus (Fig. 3) is an obligate parasite of *Xanthium* and *Ambrosia* species that affects leaves and sometimes stems of its hosts and that overwinters in dead plant material as spores. It is of American origin (Seier *et al.* 2009) but is widely naturalised in Europe and elsewhere where its hosts have been introduced. Following its introduction in Australia, the rust rapidly spread throughout much of the Australian distribution of Noogoora burr, both through natural means and human assistance (Morin *et al.* 1994, 1996). It is now responsible for the complete control of Noogoora burr across most of eastern Australia. *P. xanthii* was found on sunflower in the field soon after its introduction (Alcorn and Kochman 1976), the first record of this rust fungus attacking that species, but the limited damage caused has never been an issue for the industry. Other species such as Calendula, various cultivars of sunflower and all species within the Noogoora burr complex in Australia were subsequently found to be susceptible to Australian accessions of *P. xanthii* that infect Noogoora burr (Alcorn 1976; Morin *et al.* 1993). Extensive studies on the biology and epidemiology of the rust fungus were conducted in the early 1990s to fill knowledge gaps on this agent, which by then was having a significant impact on populations of Noogoora burr (Morin 1993; Morin *et al.* 1993, 1994, 1996). All microcyclic *Puccinia* species that infect Asteraceae and exhibit the same morphology (including the rust fungus released for the biocontrol of parthenium weed in Australia and previously referred to as *Puccinia melampodii* Diet. & Holw. (Uredinales: Pucciniaceae)) are now considered to belong to the single morphospecies name P. xanthii, which comprises distinct host-specific genotypes (Seier *et al.* 2009).

Other promising agents

Several other insect species have been identified as having potential for biocontrol, but they are unlikely to

Figure 3: Fruiting bodies of the rust fungus *Puccinia xanthii* on the undersurface of a Noogoora burr, *Xanthium occidentale*, leaf. Photo: L Morin, CSIRO.

be effective agents (van Klinken and Julien 2003). Two species from the USA that were not evaluated as part of the two phases of the program were considered worthy of additional study: the seed-feeding moth *Phaneta imbridiana* (Fenrald) (Tortricidae) and the root-breeding fly *Oedoopa* sp. nr *capito* Loew (Otitidae) (Hilgendorf and Goeden 1983). Both may be host-specific. The moth damages up to 28% of seeds in New York (mean <10%), but damage by the fly has not been quantified.

The greatest opportunity to enhance biocontrol of Noogoora burr in regions of northern Australia where the current agents have not had a major impact may be to locate, in the native range, isolates of the rust fungus that are better adapted to environmental conditions in those regions (van Klinken and Julien 2003). Characterisation of rust isolates from Australia and elsewhere using microsatellite markers revealed that the Australian population comprises only a fraction of the genetic variation in the species (L Morin and D Gomez, unpub. data), further justifying the search for additional isolates to release. Climate matching identified the Caribbean islands and coastal parts of South and Central America as priority regions for surveys, as they meet the climatic criteria of northern Australia most closely (van Klinken and Julien 2003).

X. strumarium s. l. plants infected by *P. xanthii* were found at 12 sites in Mexico and the Dominican Republic, and infected material was imported into the high-security CSIRO quarantine in Canberra for testing. However, no infection of Australian Noogoora burr accessions was obtained with the imported rust material. In contrast, plants inoculated with two rust isolates recovered from Australian populations developed severe symptoms, confirming that experimental conditions were suitable for infection. This came as a major surprise, considering previous research showed that an Australian isolate of the rust fungus was capable of infecting the four different species found in the Noogoora burr complex in Australia (Morin *et al.* 1993). As a result, it was not possible to establish cultures from the imported material and perform host-specificity tests (L Morin, unpub. data). Although *P. xanthii* is already present in Australia, host-specificity testing of additional isolates, albeit not as extensive as for a new species, would be required as part of the risk assessment process prior to introduction because specificity can vary among isolates (Alcorn 1976; Morin *et al.* 1993; Seier *et al.* 2009).

Genetic differences between Australian Noogoora burr accessions and *Xanthium* plants from which rust-infected material was collected in Mexico and the Dominican Republic most likely explain why no infection occurred with the imported rust material. Analysis of plant sequences of non-coding DNA regions from the chloroplast and nuclear ribosomal genomes revealed that Noogoora burr accessions from northern Australia (ex Daly and Victoria rivers) are closely related but well separated from three of the Dominican Republic *Xanthium* accessions and genetically distant from the Mexican and the other Dominican Republic accessions sequenced (L Morin and D Gomez, unpub. data).

DISCUSSION

Intentionally released biocontrol insect agents have caused little damage to Noogoora burr populations in Australia, with the possible exception of the stem-galling moth for which the population-level impacts may be underappreciated. The moth was the only multivoltine insect released against Noogoora burr, and is therefore likely to be better equipped to respond to the high inter-annual variability in host availability. Damage caused by the rust fungus *P. xanthii* is dramatic and has resulted in complete control over large parts of Noogoora burr's range. Extensive infestations still occur, especially along river banks in parts of the wet-dry tropics. However, they are typically patchy and are mostly not in areas where they could seriously affect the sheep industry, although environmental impacts and limited impacts on the cattle industry are possible.

An economic analysis of the Noogoora burr biocontrol program was done by Chippendale (1995). The research program was estimated to have cost \$A7.2 million (in 1991 dollars), although this probably excluded pre-World War II costs. The rust fungus, an unauthorised introduction, was responsible for saving over \$A16.75 million from losses resulting from contaminated wool alone. Updated estimates in 2004–05, based largely on these figures, correspond to a present value benefit of \$23.3 million at a 5% discount rate (Page and Lacey 2006). The ongoing economic impacts of Noogoora burr in Australia were not calculated in these estimates, but anecdotally are likely to be relatively modest. Noogoora burr was identified in a recent review of weeds of significance to the pastoral industry as a low research priority for rangelands in temperate regions, and not a priority in the five other climate regions considered (Grice 2002). Noogoora burr is not identified as a funding priority by the principal R&D provider for the wool industry in Australia, Australia Wool Innovation, and a 1989 mail survey that covered 131 local government authorities in Qld indicated that Noogoora burr was considered of nil to low importance throughout most of the state (Chippendale 1995).

The relatively well resourced and long-term nature of the biocontrol program against Noogoora burr means that it provides both historical perspectives into the development of biocontrol as a discipline and insights that are relevant for future biocontrol efforts (van Klinken and Julien 2003). Considerable post-release studies have helped identify, and in some cases test, why the rust fungus *P. xanthii* works where it does and why most agents have failed to have an impact. This has helped direct the latest recommencement of the biocontrol program and will guide future efforts targeting similar weeds (annual herbs) in comparable environments (arid to tropical). Clearly, agents require strategies to overcome the typically high interyear variation in resource availability (in terms of quantity, quality and timing) that is typical of such systems. Lack of synchronisation of insect life-histories with host availability has been particularly important for the three univoltine agents. In addition, the strong compensatory ability of Noogoora burr to overcome damage, at least when growing in good habitat and when attacked by stem-borers (Wapshere 1974), means that agents will need to reach sufficient densities to surmount such plant responses. However, some agents may not be able to cause significant population-level impacts even at high densities. For example, the seed-feeding fly *E. aequalis* usually attacks only one of the two seeds in each fruit, and damage by stem-boring beetle larvae is typically too late to prevent seeding (Wapshere 1974). These results demonstrate the importance of considering potential efficacy as well as host-specificity when prioritising potential biocontrol agents. They illustrate that, at least in these cases, poor potential efficacy could have been demonstrated prior to their release.

Future work

Additional candidate agents could be sourced from across the Americas, as the regions beyond the USA and Mexico remain largely underexplored despite being considered the centre of origin for the genus (Love and Dansereau 1959). Searches could also be extended to other *Xanthium* species as there are no native *Xanthium* species at risk in Australia. Any new agent would have more chance of being effective in controlling Noogoora burr if it is climatically adapted to the wet-dry tropics, able to synchronise its life-cycle with Noogoora burr availability and cope with extreme events such as flooding (van Klinken and Julien 2003). However, we believe that the most promising approach for managing dense Noogoora burr across northern Australia involves locating rust isolates that are better adapted climatically. This search has been greatly complicated by recent findings of unexpected host-specialisation within the *Xanthium* genus. It may not be possible to find isolates in regions that are climatically matched to northern Australia, because of mismatches between the pathogen and Australian plants. The establishment of trap gardens consisting of northern Australian accessions of Noogoora burr in tropical America would greatly assist in sourcing pathogenic rust isolates.

ACKNOWLEDGEMENTS

We gratefully acknowledge our collaborators from DAFWA and NRETA who gathered baseline data on Noogoora burr populations in northern Australia, as part of the most recent biocontrol project targeting Noogoora burr. We also thank Ricardo Segura, Carlos Pascacio and Moises Martinez (CSIRO Ecosystem Sciences, Mexico) for conducting the most recent surveys for new rust isolates in the native range, and Melissa Piper (CSIRO Ecosystem Sciences, Canberra) for technical assistance with the quarantine work.

We thank the Bureau of Rural Sciences, Dr Lucy Randall in particular, for the distribution map (Fig. 1).

REFERENCES

Alcorn JL (1975) A new disease of Noogoora burr. *Queensland Agricultural Journal* **101**, 162.

Alcorn JL (1976) Host range of *Puccinia xanthii*. *Transactions of the British Mycological Society* **66**, 365–367.

Alcorn JL and Kochman JK (1976) A field record of *Puccinia xanthii* on sunflower. *Australian Plant Pathological Society Newsletter* **5**, 33.

Chippendale JF (1995) The biological control of Noogoora burr (*Xanthium occidentale*) in Queensland: an economic perspective. In *Proceedings of the 8th International Symposium on Biological Control of Weeds*. 2–7 February 1992, Canterbury, NZ. (Eds ES Delfosse and RR Scott) pp. 185–192. DSIR/CSIRO, Melbourne.

Farr DF and Rossman AY (2010) Fungal databases, systematic mycology and microbiology laboratory. ARS-USDA. <http://nt.ars-grin.gov/fungaldatabases/> (retrieved 8.3.10).

Goeden RD and Palmer WA (1995) Lessons learned from studies of the insects associated with Ambrosiinae in North America in relation to the biological control of weedy members of this group. In *Proceedings of the 8th International Symposium on Biological Control of Weeds*. 2–7 February 1992, Canterbury, NZ. (Eds ES Delfosse and RR Scott) pp. 565–573. DSIR/CSIRO, Melbourne.

Grice AC (2002) 'Weeds of significance to the grazing industries of Australia'. Final report. CRC for Weed Management, Adelaide.

Hasan S (1974a) Xanthium rust as a possible biological control agent of Bathurst and Noogoora burrs in Australia. In *Proceedings of the 3rd International Symposium on Biological Control of Weeds*. 10–14 September 1973, Montpellier, France. (Ed AJ Wapshere) pp. 137–140. Commonwealth Institute of Biological Control, Miscellaneous Publication 8.

Hasan S (1974b) Recent advances in the use of plant pathogens as biological control agents of weeds. *PANS* **20**, 437–443.

Haseler WH (1981) Lessons from early attempts at biological control of weeds in Queensland. In *Proceedings of the 5th International Symposium on Biological Control of Weeds*. 22–29 July 1980, Brisbane. (Ed ES Delfosse) pp. 3–9. CSIRO, Canberra.

Hilgendorf JH and Goeden RD (1982) Phytophagous insects reported worldwide form the noxious weeds, spiny clotbur, *Xanthium spinosum*, and cocklebur, *Xanthium strumarium*. *Bulletin of the Entomological Society of America* **28**, 147–152.

Hilgendorf JH and Goeden RD (1983) Phytophagous insect fauna of spiny clotbur, *Xanthium spinosum*, and cocklebur, *Xanthium strumarium*, in southern California. *Environmental Entomology* **12**, 404–411.

Hocking PJ and Liddle MJ (1995) *Xanthium occidentale* Bertol. Complex and *X. spinosum* L. *The Biology of Australian Weeds. Vol. 1.* (Eds RH Groves, RCH Shepherd and RG Richardson) pp. 241–302. RG and FJ Richardson, Melbourne.

Julien MH and Griffiths MW (1998) *Biological Control of Weeds: A World Catalogue of Agents and their Target Weeds*. 4th edn. CABI, Wallingford, UK.

Kelly SG (1931) The control of Noogoora burr and Bathurst burr by insects. *Journal of the Council of Scientific and Industrial Research* **4**, 161–172.

Liddle MJ (1986) Noogoora burr – a successful suite of weeds. In *The Ecology of Exotic Animals and Plants*. (Ed RL Kitching) pp. 188–220. John Wiley and Sons, Brisbane.

Love D and Dansereau P (1959) Biosystematic studies on *Xanthium*: taxonomic appraisal and ecological status. *Canadian Journal of Botany* **37**, 173–208.

McFadyen RE (1987) The effect of climate on the stem galling moth *Epiblema strenuana*, in eastern Australia. In *Proceedings of the 8th Australian Weeds Conference*. 21–25 September, Sydney. (Eds D Lermie and AR Leys) pp. 97–99. Weed Society of New South Wales, Sydney.

McMillan C (1973) Photoperiodic adaptation in *Xanthium strumarium* populations introduced to Australia. *Canadian Journal of Botany* **51**, 221–229.

Morin L (1993) The prospects for biological control of the Noogoora burr complex with naturally occurring fungi in Australia. PhD thesis. University of New England, Australia.

Morin L, Auld BA and Brown JF (1993) Host-range of *Puccinia xanthii* and post-penetration development on *Xanthium occidentale*. *Canadian Journal of Botany* **71**, 959–965.

Morin L, Auld BA, Brown JF and Cholil MA (1994) Pathogenic fungi occurring on the Noogoora burr complex in Australia. *Proceedings of the Linnean Society NSW* **114**, 133–148.

Morin L, Auld BA and Smith HE (1996) Rust epidemics, climate and control of *Xanthium occidentale*. In *Proceedings of the 9th International Symposium on Biological Control of Weeds*. 19–26 January 1996, Stellenbosch, South Africa. (Eds VC Moran and JH

Hoffmann) pp. 385–391. University of Cape Town, South Africa.

Page AR and Lacey KL (2006) 'Economic impact assessment of Australian weed biological control'. Technical Series No. 10. CRC for Australian Weed Management, Adelaide. <http://www.southwestnrm.org.au/sites/default/files/uploads/ihub/page-ar-lacey-kl-2006-economic-impact-assessment-australian-weed-biological-control.pdf> (accessed 3.9.11).

Seier MK, Morin L, van der Merwe M, Evans HC and Romero A (2009) Are the microcyclic rust species *Puccinia melampodii* and *Puccinia xanthii* conspecific? *Mycological Research* **113**, 1271–1282.

Thorp JR and Lynch R (2000) The determination of Weeds of National Significance. National Weeds Strategy Executive Committee, Launceston. <http://www.weeds.org.au/docs/WONS/> (accessed 3.9.11).

van Klinken RD and Julien MH (2003) Learning from past attempts: does classical biological control of Noogoora burr (Asteraceae: *Xanthium occidentale*) have a promising future? *Biocontrol Science and Technology* **13**, 139–153.

Wapshere A J (1974) An ecological study of an attempt at biological control of Noogoora burr (*Xanthium strumarium*). *Australian Journal of Agricultural Research* **25**, 275–292.

Weaver SE and Lechowicz MJ (1982) The biology of Canadian weeds 56. *Xanthium strumarium* L. *Canadian Journal of Plant Science* **63**, 211–225.

Wilson GC and Flanagan GJ (1993) Phytophagous insect fauna of the introduced noxious weed *Xanthium occidentale* in northern Australia and its relevance to biological control. *Environmental Entomology* **22**, 254–261.

Xanthium spinosum L. – Bathurst burr

Louise Morin and Bruce A Auld

ABSTRACT

Bathurst burr, *Xanthium spinosum* (Asteraceae), is a widespread summer annual weed in rangeland, pastures and crops in eastern Australia. Possibilities for both the inundative (bioherbicide) and classical biocontrol of Bathurst burr have been explored. In the 1980s–90s, the anthracnose fungus *Colletotrichum orbiculare* that naturally occurs on Bathurst burr in Australia was extensively investigated to develop a bioherbicide for use in infested crops and pastures. Despite initial optimism, commercial development of the potential bioherbicide did not go ahead. Prospects for classical biocontrol of Bathurst burr were first explored in the 1930s then again in the 1990s with surveys in the USA and South America, the putative centre of origin. A range of arthropod natural enemies was identified; only one, a tetranychid webbing mite, was investigated further and found not to be sufficiently specific for biocontrol. Three of the fungal pathogens most frequently recovered during the 1990s surveys, the anthracnose fungus, the leaf spot fungus *Cercospora xanthicola* and the powdery mildew *Golovinomyces cichoracearum*, were studied in more detail. Isolates of the anthracnose fungus collected in semi-arid and arid areas of Argentina were found not to be superior, in terms of dew requirement for infection, to the Australian isolate of the fungus used in the bioherbicide research. Bathurst burr and three of the eight sunflower cultivars tested with Argentinian isolates of the powdery mildew became heavily infected. Isolates of the leaf spot fungus caused severe necrosis on Bathurst burr leaves, but did not infect stems. They were also shown in preliminary tests to cause small necrotic lesions on lucerne and five sunflower cultivars. These fungi were not investigated further as possible biocontrol agents due to lack of specificity.

Key words: anthracnose fungus *Colletotrichum orbiculare*, powdery mildew fungus *Golovinomyces cichoracearum*, leaf spot fungus *Cercospora xanthicola*, bioherbicide.

INTRODUCTION

Bathurst burr (*Xanthium spinosum* L.) (Asteraceae) is a widespread summer annual weed in rangeland, pastures and crops in eastern Australia, although it does occur in all states and territories (Hocking and Liddle 1986) (Fig. 1). It was ranked 40th out of 71 species during a prioritisation process to determine the Weeds of National Significance in the late 1990s (Thorp and Lynch 2000). It occurs predominantly in temperate climates and high-fertility disturbed soils, especially along watercourses and around dams, but it also grows well in drier conditions. It is an erect branched herb that can grow as tall as 1 m depending on the habitat (Hocking and Liddle 1986) (Fig. 2a). It has characteristic long (1.5–2.5 cm) whitish-yellow spines at the base of leaf stalks and produces spiny fruits (burrs) that contain two seeds, although fruits with only one seed are common (Auld 1993) (Fig. 2b).

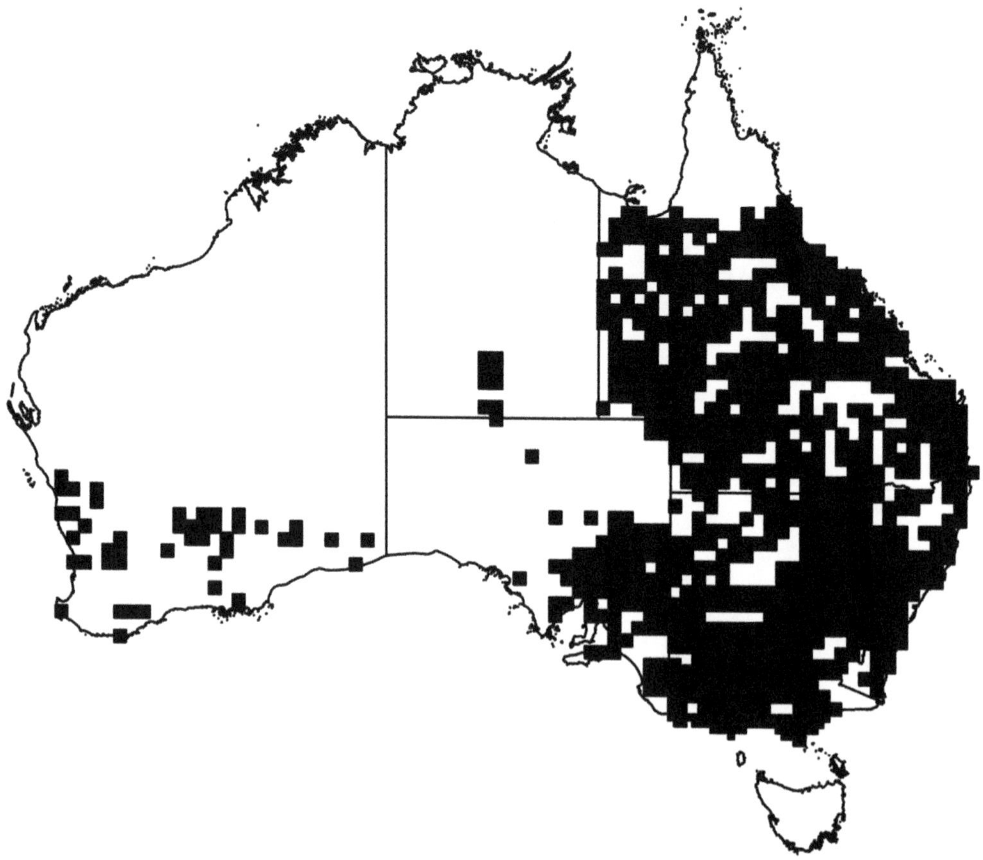

Figure 1: Distribution of Bathurst burr, *Xanthium spinosum*, in Australia. Map drawn by the Bureau of Rural Sciences using data presented in Thorp and Lynch (2000). Distribution shown as 0.5, 0.25 or 0.125 square degrees depending on the state or territory.

Contamination of fleeces of grazing sheep by the spiny fruits, which causes significant wool fault, is the main nuisance from Bathurst burr (Martin and Carnahan 1982). Further, seedlings are reputedly poisonous to stock and mature plants are unpalatable (Hocking and Liddle 1986). Dense infestations can hinder stock movement and adversely affect pasture growth. Bathurst burr is also prevalent in irrigated crops such as sunflower, soybean, corn and cotton (Andrews 1993; Clarke and Porter 1993), where it can reduce yields through competition (Fig. 2c). It has been observed to cause allergic contact dermatitis in some people (Fisher *et al.* 1988).

The wool industry estimates the cost of controlling Bathurst burr to be over $A82 million per year (Australian Wool Innovation 2009). Herbicides are highly effective at killing it, but are uneconomical for use over the huge area of low-value pastoral land infested. Herbicides are recommended for Bathurst burr control in some crops (Andrews 1993), but selective control can be difficult in some cases. In these situations, hand-hoeing of young plants before flowering is recommended (Hocking and Liddle 1986). The impracticality of chemical control in rangeland and pastures has been the impetus to explore the possibility for classical biocontrol of Bathurst burr. Potential crop damage from herbicide application stimulated extensive research into the development of an inundative biocontrol approach (bioherbicide).

BIOLOGICAL CONTROL HISTORY

Australia is the only country that has investigated options for the biocontrol of Bathurst burr. Prospects for the classical biocontrol of Bathurst burr were first explored in the 1930s by CSIR (now CSIRO) in tandem with investigations targeting Noogoora burr (*Xanthium occidentale* Bertol.) (Wilson 1960). Interest in Bathurst burr classical biocontrol was renewed six decades later with surveys in the native range, initially performed by CSIRO (Wapshere *et al.* 1995) and subsequently by the NSW Department of Primary Industries (NSW DPI), with the assistance of the International Institute of Biological Control (now CABI Bioscience) (B Auld, H Evans and C

Figure 2. a) Bathurst burr, *Xanthium spinosum*, plant. b) Close-up of a stem with spines and fruits. c) Typical infestation of Bathurst burr in an irrigated soybean crop. Photos: B Auld, NSW Government.

Ellison, unpub. data). In the field, a mite appeared to be the most promising of the arthropod natural enemies found. The mite and three of the fungal pathogens most frequently recovered during the 1990s surveys were imported into quarantine facilities in the UK and in Australia and studied in more detail (Morin *et al.* 2001; B Auld, H Evans and C Ellison, unpub. data). All these candidate agents either attacked some of the non-target plant species tested or did not produce symptoms sufficiently damaging to Bathurst burr plants.

In the 1980s–90s, a team based at NSW DPI extensively investigated the anthracnose fungus *Colletotrichum orbiculare* (Berk. et Mont.) v. Arx that naturally occurs on Bathurst burr in Australia (Butler 1951; Walker *et al.* 1991), to develop a bioherbicide to control the weed in crops and pastures. The project focused on a highly pathogenic isolate of the fungus capable of producing girdling stem lesions and killing Bathurst burr plants. The research undertaken:

- defined optimum temperature and moisture conditions and inoculum dosage for severe disease development (Fig. 3);
- determined the fungus' host-range;
- explored mass production, formulation and application techniques;
- tested the efficacy of the prototype bioherbicide in the field (Auld *et al.* 1988, 1989, 1990a, b, 2003; McRae and Auld 1988; McRae *et al.* 1988; McRae and Stevens 1990; Klein and Auld 1995, 1996; Klein *et al.* 1995; Chittick and Auld 2001).

Figure 3: Effect of the anthracnose fungus *Colletotrichum orbiculare* on Bathurst burr, *Xanthium spinosum*, 15 days after application of increasing densities of spores (10^4–10^7 spores mL^{-1}) under ideal laboratory conditions (Auld *et al.* 1988). Photo: B Auld, NSW Government.

A patent was granted for the novel use of the fungus (Auld 1988) and a commercial partner (Sandoz Corporation, Switzerland) produced a prototype product. Despite initial optimism, commercial development of the potential bioherbicide did not go ahead mainly because of inconsistent results in the field. This suggested a need for an improved formulation of the bioherbicide. Further investment in this area was declined by the commercial partner because of potentially high production costs and limited market for the product.

EXPLORATION

Bathurst burr's centre of origin lies to the east of the Andes in South America, probably in Argentina (Wapshere *et al.* 1995). It also occurs widely in Chile, Uruguay and the southern parts of Brazil, Paraguay and Bolivia. It has naturalised in most warm temperate and semi-arid regions of the world, from the Mediterranean region of

Europe to western Asia and Australasia (Hocking and Liddle 1986).

The earliest surveys of arthropod natural enemies of Bathurst burr were conducted in the USA, Chile and Argentina in 1936–40 (Wilson 1960). Only four insect species were found, most likely due to the limited exploration in South America. None were studied or deliberately introduced to Australia. Another survey targeting arthropods and fungi attacking Bathurst burr in Chile and Argentina was carried out in 1991 (Wapshere *et al.* 1995). Results of this preliminary survey gave cause for optimism that a damaging fungal pathogen could be found. Additional surveys specifically for fungal pathogens of Bathurst burr were thus undertaken in Argentina in March and November 1995 and in April 1996 (B Auld, H Evans and C Ellison, unpub. data). The first of these surveys followed the trajectory of Wapshere's 1991 survey (Wapshere *et al.* 1995) and covered a range of habitats in 14 provinces of Argentina. Subsequent surveys concentrated in the area south-west and north-east of San Miguel de Tucumán, embracing the Argentinian provinces of Catamarca, Salta, Tucumán, San Luis and La Rioja.

CANDIDATES

Natural enemies present in Australia

Only four species of insects are known to attack Bathurst burr in Australia. The seed fly *Euaresta bullans* (Wiedemann), discovered in NSW in 1928, is reported to destroy large numbers of seeds in Qld and NSW, though apparently without effect on the abundance of the weed (Wilson 1960). Three native beetle stem-borers (*Zygrita diva* Thoms., *Corrhenes paulla* Germ. and *Lixus tasmanicus* Pascoe) are also known to attack Bathurst burr, with *C. paulla* observed to cause greatest mortality in some local situations (Wilson 1960). The anthracnose fungus *C. orbiculare* is commonly found on Bathurst burr in eastern Australia and is most severe following prolonged periods of cool wet weather in late summer and autumn (Butler 1951; Walker *et al.* 1991). Other fungi such as *Phomopsis* sp. and *Sclerotinia sclerotiorum* (Lib.) de Bary, and to a lesser extent *Macrophomina phaseolina* (Tassi) Goidanich and *Verticillium dahliae* Klebahn, have also been found associated with diseased Bathurst burr in NSW (Nikandrow *et al.* 1990). *Sclerotinia minor* Jagger, which causes stem rot of Bathurst burr, was reported in a sunflower (*Helianthus annuus* L.) crop in Vic (Clarke and Porter 1993). The rust fungus *Puccinia xanthii* Schw., which is common on Noogoora burr in eastern Australia (see separate chapter, this volume), has also been recorded on Bathurst burr (Cook and Dubé 1989), although it was shown in glasshouse trials not to be capable of developing mature fruiting bodies on this host (Morin *et al.* 1993).

Natural enemies identified overseas

Wilson (1960) reported the finding, in the 1930s, of the following insects on Bathurst burr in its native range: the seed fly *E. bullans* in Chile, two unidentified curculionid weevils in Argentina and the weevil *Baris callida* Casey in the USA. In a worldwide literature review of recorded phytophagous insects of *Xanthium* spp., five additional species attacking Bathurst burr in the USA were listed: the leafhoppers *Colladonus geminatus* (Van Duzee), *Colladonus montanus* (Van Duzee), *Macrosteles fascifrons* (Stal) and *Texanamus latipex* DeLong and the weevil *Cylindrocopturus adspersus* LeConte (Hilgendorf and Goeden 1982). Subsequently, Hilgendorf and Goeden (1983) reported that the phytophagous insect fauna of Bathurst burr in southern California comprised 21 species representing six orders and 14 families. They concluded, however, that none of the insect species were suitable as biocontrol agents. A more recent survey in South America found several insects and a mite attacking Bathurst burr, but none were observed to kill plants (Wapshere *et al.* 1995) (Table 1).

A wide range of fungi has been recorded from Bathurst burr in its introduced range outside Australia, but limited information is available (Farr and Rossman 2010). The mycoflora of Bathurst burr in its South American native range was poorly documented until the 1990s surveys. Wapshere *et al.* (1995) observed disease symptoms on Bathurst burr resembling those caused by the anthracnose fungus and the *Phomopsis* sp. in Australia (Nikandrow *et al.* 1990) throughout Argentina. They also recorded seven other fungi attacking foliage, fruits and/or roots of Bathurst burr: an unidentified root rot fungus, various *Alternaria* spp. and *Cladosporium* spp. and the powdery mildew *Golovinomyces cichoracearum* (DC.) V.P. Heluta. The powdery mildew, the anthracnose fungus and *Phomopsis* sp. were found in subsequent surveys in Argentina (B Auld, H Evans and C Ellison, unpub. data). Disease symptoms caused by the leaf spot fungus *Cercospora xanthicola* Heald & Wolf were also observed at several sites in the mid 1990s. Several other fungal species from genera such as *Fusarium*, *Phialiphora*, *Verticilium* and *Sclerotinia* were isolated from dieback and wilt symptoms.

Table 1: Arthropods found on *Xanthium spinosum* during a survey in the native range in 1991

Country	Species
Chile	Agromyzid stem-miner larva
	Leaf-feeding chrysomelid *Systena argentiniensis* Jacoby
	Leaf-webbing tetranychid mite
	Seed flies *Euaresta* spp.
Argentina	Stem-borer long-horned beetle *Emphytoecia versicolor* (Boh.)
	Tip- and stem-mining weevil larvae: two *Rhyssomatus* spp. and one other species
	Stem-boring mordellid larvae (two species)
	Tip- and stem-mining moth larvae: *Platphalonidia* sp., an oecophorid and another lepidopteran
	Agromyzid stem-miner larvae (at least four species)
	Leaf-webbing lepidopteran larva
	Lace bug *Gargaphia* sp.
	Seed flies *Euaresta* spp.

Source: Wapshere *et al.* (1995).

Natural enemies tested but not released

None of the arthropod natural enemies found during surveys, except the leaf-webbing tetranychid mite, were investigated further. Active populations of the mite were collected in the Andean foothills in Argentina in November 1995. In host-specificity tests in quarantine in the UK, the mite was found to attack sunflower and Noogoora burr (H Evans, *pers. comm.*).

***Colletotrichum orbiculare* (Berk. et Mont.) v. Arx** Isolates of the anthracnose fungus collected from diseased Bathurst burr from a wide range of environments from semi-arid, arid and moist areas of Argentina were imported into an Australian quarantine facility (Auld and Say 1999). The aim of the study was to determine if isolates from the drier areas of the native range would be capable of producing significantly greater disease at dew periods of 8 h or less than the Australian isolate used in previous bioherbicide research (see above). Results from experiments comparing pathogenicity of the different isolates under a range of dew periods as well as vegetative growth rate and sporulation on artificial media showed that none of the exotic isolates were superior to the Australian isolate (Auld and Say 1999).

***Golovinomyces cichoracearum* (DC.) V.P. Heluta** (previously referred to as *Erysiphe cichoracearum* DC. or *Oidium erysiphoides* Fr.; Braun *et al.* 2001). Three isolates of this powdery mildew collected in Argentina in April 1996 were imported into quarantine in the UK for further investigations (C Ellison and R Reeder, unpub. data). Preliminary host-specificity testing of the isolates involved a range of non-target plant species belonging to the Asteraceae, including several cultivars of sunflower (Morin *et al.* 2001). Severe infection and heavy sporulation developed on Bathurst burr, while the other *Xanthium* spp. developed only mild symptoms or none. All other Asteraceae tested were immune or resistant to the fungus, apart from three of the eight sunflower cultivars tested, which became heavily infected. Similar results were obtained in subsequent trials using a wind tunnel set up to simulate natural infection and consequently comprehensive host-specificity testing was not undertaken.

***Cercospora xanthicola* Heald & Wolf.** Four isolates of this leaf spot fungus, also collected in Argentina in 1996, were imported into quarantine in the UK. No infection of Bathurst burr was obtained with any of the isolates in the single test performed and no further work was carried out on this fungus until the isolates were imported into an Australian quarantine facility two years later.

Before any testing was undertaken in Australia, an efficient method was devised to produce abundant spores of the leaf spot fungus. Bathurst burr and six cultivars of sunflower were then inoculated with a dense suspension of spores of each isolate. Only a few necrotic lesions developed on leaves of Bathurst burr, while none of the sunflower cultivars developed any disease symptoms. Five additional 'fresh' isolates were collected in Argentina in May 2000 from sites where severe disease symptoms were observed (F Anderson and W Pettit, unpub. data). Experiments with two of these isolates showed that they were capable of causing severe necrosis on inoculated leaves if adequate conditions were present during the infection period (minimum of two days exposure to high humidity conditions and a 12 h photoperiod) (Morin *et al.* 2001). However, no stem lesions developed in any of the experiments performed and Bathurst burr plants recovered rapidly from the disease symptoms on leaves. In further tests on non-target plant species using one of the selected isolates, the fungus produced a few small necrotic lesions on lucerne *Medicago sativa* L. and five of the sunflower cultivars tested and developed only a few chlorotic flecks on the remaining nine cultivars.

Other promising agents

The stem-borer and stem-miner insects recorded in different climatic regions of Argentina and Chile during Wapshere *et al.*'s (1995) survey (Table 1) warrant further investigation as they were the most damaging on Bathurst burr, although they were never seen killing plants.

DISCUSSION

Until the mid 1990s, most of the research into Bathurst burr biocontrol had focused on the development of a bioherbicide. The anthracnose fungus *C. orbiculare*, the active ingredient of the potential bioherbicide, requires several hours of available water following application of spores onto Bathurst burr for infection to occur (McRae and Auld 1988). Isolates of the fungus from dry areas of western Argentina were investigated in the hope that they would be capable of infecting plants under drier conditions, but it was not the case (Auld and Say 1999). In light of these findings, the development of a formulation to overcome the fungus' dew requirement was perceived as the only solution. Consequently, work was undertaken to develop liquid formulations that retain moisture around the spores after spray applications to the weed, with variable results (Klein *et al.* 1995; Chittick and Auld 2001). Although subsequent research that developed a complex water-in-oil-in-water emulsion formulation was promising (Auld *et al.* 2003), the cost of the process was too expensive for further consideration for use in a broadscale agricultural spray. The expected cost of industrial production of the bioherbicide was considered prohibitive by the commercial partner. An augmentative biocontrol approach may be a possible solution to commercial limitations on use of this fungus for Bathurst burr control. The fungus could be mass produced and stored at a regional level using low-tech methods on inexpensive solid substrates, such as wheat chaff. The material colonised by the fungus would then be distributed manually or mechanically in infested fields when environmental conditions are suitable for infection of Bathurst burr.

The impact of the powdery mildew *G. cichoracearum* on Bathurst burr observed during native range field surveys and initial glasshouse experiments demonstrated exceptional promise from a biocontrol point of view (Morin *et al.* 2001). The leaf spot fungus *C. xanthicola* showed less potential, requiring long exposure to humid conditions to infect plants and then infecting only leaves, not stems, of Bathurst burr. The ability of both fungi, as well as the webbing tetranychid mite, to attack sunflower, however, militates against their possible use as biocontrol agents for Bathurst burr in Australia.

It is not clear why the insects found attacking Bathurst burr during Wapshere *et al.*'s (1995) surveys in the native range were never investigated further, but it was possibly due to the fact that none were observed killing plants. Moreover, damaging insect species were different in the various regions surveyed, suggesting that several agents would need to be introduced, to have an impact across Bathurst burr's range in Australia. Finding a natural enemy that affects only Bathurst burr and not the closely related crop sunflower remains a challenge; we will not know if any of the insects found in the native range have potential for biocontrol until they are identified and preliminary host-specificity tests are performed.

Bathurst burr is widespread in Argentina but rarely occurs there in high density as it commonly does in Australia, even though individual plants may be equally large in both countries (B Auld, H Evans and C Ellison, unpub. data). A long-term demographic study of Bathurst burr in Argentina should reveal factors limiting its population sizes and may lead to further possibilities for biocontrol in Australia.

ACKNOWLEDGEMENTS

We thank our colleagues, Harry Evans and Carol Ellison (UK) and Freda Anderson and Bill Pettit (Argentina), for their enthusiastic collaboration. We also thank the Bureau of Rural Sciences, Dr Lucy Randall in particular, for the distribution map (Fig. 1).

REFERENCES

Andrews JA (1993) Control of Bathurst burr (*Xanthium spinosum*) in irrigated soybeans in southern New South Wales. *Plant Protection Quarterly* **8**, 15–18.

Auld BA (1988) (Inventor) Mycoherbicide. Australian Patent 602223. 14 pp. with 6 figures. Official Journal 4 October 1990. Letters Patent 5 February 1991.

Auld BA (1993) Emergence and flowering in *Xanthium spinosum*. *Agriculture, Ecosystems & Environment* **47**, 215–222.

Auld BA and Say MM (1999) Comparison of isolates of *Colletotrichum orbiculare* from Argentina and Australia as potential bioherbicides for *Xanthium spinosum* in Australia. *Agriculture, Ecosystems & Environment* **72**, 53–58.

Auld BA, McRae CF and Say MM (1988) Possible control of *Xanthium spinosum* by a fungus. *Agriculture, Ecosystems & Environment* **21**, 219–223.

Auld BA, McRae CF and Say MM (1989) Potential for *Xanthium spinosum* control by *Colletotrichum orbiculare* as a mycoherbicide. In *Proceedings of the 7th International Symposium on Biological Control of Weeds*. 6–11 March 1988, Rome. (Ed ES Delfosse) pp. 435–443. Instituto Sperimentale per la Patologia Vegetale, Rome.

Auld BA, Say MM and Millar GD (1990a) Influence of potential stress factors on anthracnose development on *Xanthium spinosum*. *Journal of Applied Ecology* **27**, 513–519.

Auld BA, Say MM, Ridings HI and Andrews J (1990b) Field applications of *Colletotrichum orbiculare* to control *Xanthium spinosum*. *Agriculture, Ecosystems & Environment* **23**, 315–323.

Auld BA, Hetherington SD and Smith HE (2003) Advances in bioherbicide formulation. *Weed Biology and Management* **3**, 61–67.

Australian Wool Innovation (2009) Bathurst burr (*Xanthium spinosum*). <http://www.wool.com/Grow_Pastures-and-Nutrition_Pasture-Pests-and-Weeds_Bathurst-Burr.htm> (accessed 3.9.11).

Braun U, Kiehr M and Delhey R (2001) Some new records of powdery mildew fungi from Argentina. *Sydowia* **53**, 34–43.

Butler FC (1951) Anthracnose and seedling blight of Bathurst burr caused by *Colletotrichum xanthii* Halst. *Australian Journal of Agricultural Research* **2**, 401–410.

Chittick AT and Auld BA (2001) Polymers in bioherbicide formulation: *Xanthium spinosum* and *Colletotrichum orbiculare* as a model system. *Biocontrol Science and Technology* **11**, 691–702.

Clarke RG and Porter IJ (1993) *Sclerotinia minor* – a pathogen of Bathurst burr (*Xanthium spinosum*). *Australasian Plant Pathology* **22**, 98–99.

Cook RP and Dubé AJ (1989) *Host-pathogen Index of Plant Diseases in South Australia*. SA Dept of Agriculture, Adelaide.

Farr DF and Rossman AY (2010) Fungal databases, systematic mycology and microbiology laboratory. ARS-USDA. <http://nt.ars-grin.gov/fungaldatabases/> (accessed 31.3.10).

Fisher G, Gow E and Freeman S (1988) Allergic contact dermatitis to Noogoora burr and Bathurst burr. *Australasian Journal of Dermatology* **29**, 155–159.

Hilgendorf JH and Goeden RD (1982) Phytophagous insects reported worldwide from the noxious weeds spiny clotbur, *Xanthium spinosum*, and cocklebur, *X. strumarium*. *Bulletin of the Entomological Society of America* **28**, 147–152.

Hilgendorf JH and Goeden RD (1983) Phytophagous insect faunas of spiny clotbur, *Xanthium spinosum* & cocklebur, *Xanthium stumarium* in southern California. *Environmental Entomology* **12**, 404–411.

Hocking PJ and Liddle MJ (1986) The biology of Australian weeds: 15. *Xanthium occidentale* Bertol. complex and *Xanthium spinosum* L. *Journal of the Australian Institute of Agricultural Science* **52**, 191–221.

Klein TA and Auld BA (1995) Influence of spore dose and water volume on a mycoherbicide's efficacy in field trials. *Biological Control* **5**, 173–178.

Klein TA and Auld BA (1996) Wounding can improve efficacy of *Colletotrichum orbiculare* as a mycoherbicide for Bathurst burr. *Australian Journal of Experimental Agriculture* **36**, 185–187.

Klein TA, Auld BA and Fang W (1995) Evaluation of oil suspension emulsions of *Colletotrichum orbiculare* as a mycoherbicide in field trials. *Crop Protection* **14**, 193–197.

Martin RJ and Carnahan JA (1982) Distribution and importance of Noogoora and Bathurst burrs in eastern Australia. *Australian Weeds* **2**, 27–32.

McRae CF and Auld BA (1988) The influence of environmental factors on anthracnose of *Xanthium spinosum*. *Phytopathology* **78**, 1182–1186.

McRae CF and GR Stevens (1990) Role of conidial matrix of *Colletotrichum orbiculare* in pathogenesis of *Xanthium spinosum*. *Mycological Research* **94**, 890–896.

McRae CF, Ridings HI and Auld BA (1988) Anthracnose of *Xanthium spinosum* – quantitative disease assessment and analysis. *Australasian Plant Pathology* **17**, 11–13.

Morin L, Auld BA and Brown JF (1993) Host-range of *Puccinia xanthii* and post-penetration development on

Xanthium occidentale. *Canadian Journal of Botany* **71**, 959–965.

Morin L, Auld B, Evans H, Ellison C, Reeder R, Anderson F and Pettit B (2001) Argentinian fungi fall at the first hurdle as potential biocontrol agents for Bathurst burr. In *Conference Handbook: 13th Biennial Conference of the Australasian Plant Pathology Society*. 24–27 September 2001, Cairns. p. 347. Centre for Tropical Agriculture, Mareeba, Qld.

Nikandrow A, Weidemann GJ and Auld BA (1990) Incidence and pathogenicity of *Colletotrichum orbiculare* and a *Phomopsis* sp. on *Xanthium* spp. *Plant Disease* **74**, 796–799.

Thorp JR and Lynch R (2000) *The Determination of Weeds of National Significance*. National Weeds Strategy Executive Committee, Launceston.

Walker J, Nikandrow A and Millar GD (1991) Species of *Colletotrichum* on *Xanthium* (Asteraceae) with comments on some taxonomic and nomenclatural problems in *Colletotrichum*. *Mycological Research* **95**, 1175–1193.

Wapshere AJ, Erb H and Bunster L (1995) Implications of a preliminary survey on the biological control prospects for the pasture and rangeland weed Bathurst burr, *Xanthium spinosum* in Australia. In *Proceedings of the 8th International Symposium on the Biological Control of Weeds*. 2–7 February 1992, Canterbury, NZ. (Eds ES Delfosse and RR Scott) pp. 355–359. DSIR/CSIRO, Melbourne.

Wilson F (1960) *A Review of the Biological Control of Insects and Weeds in Australia and Australian-New Guinea*. Commonwealth Agricultural Bureau, Bucks., UK.

Zantedeschia aethiopica (L.) Spreng. – arum lily

John K Scott

ABSTRACT

Arum lily *Zantedeschia aethiopica* is an invader of native wetlands and wet pastures of the lower south-west of southern Australia that has proved difficult to control by conventional means. The flowers are harvested from weed populations and the plant is used as an ornamental although it is declared noxious in WA. The weed originates from southern Africa, where initial surveys for biocontrol agents focused on seed-feeding insects because of the potential conflict of interest with the plant's use in horticulture. Suitable organisms for use in biocontrol were not found. Surveys were then made in WA for disease-causing organisms that could be used in an inundative approach to biocontrol. Potential plant pathogens were found or identified from the literature, but have not been studied further. Future work should focus on completing surveys for pathogens in Australia and investigating the potential for mycoherbicide development, especially for organisms associated with the rhizome, the main perennating organ for arum lily.

Key words: conflict of interest, environmental weed, mycoherbicides, southern Africa, surveys for potential biocontrol agents.

INTRODUCTION

Arum lily *Zantedeschia aethiopica* (Araceae) (Fig. 1) is naturalised across southern Australia and has become a major invader of wetter areas of south-western WA (Keighery 1997). It is native to southern Africa where it is found in the Western Cape, Eastern Cape and Kwazulu-Natal, inland mountains of Mpumalanga and Limpopo provinces of South Africa and in Lesotho and Swaziland (Scott 1997a). It was probably an early introduction into Australia as a garden plant because of its large white 'flowers' and attractive green foliage.

Arum lily is a perennial herb with large leaves and fleshy rhizomes. In summer dry areas such as south-western Australia, plants regrow from the rhizome in autumn, flower in winter and senesce in summer. Otherwise, the plant is evergreen in its native habitat (Scott and Neser 1996). The 'flower' consists of a fleshy yellow axis (spadix) enclosed in a large, leaf-like white bract, the spathe (Letty 1973; Singh *et al.* 1996a). It develops through five phases: pre-female, female, male, fruit development and fruit ripening (Singh *et al.* 1996a). This protogyny ensures the plant is self-incompatible and pollination is probably ensured by beetles, mainly Scarabaeidae (Singh *et al.* 1996a). The fruits consist of green berries that turn orange, soft and mucilaginous when ripe (Singh *et al.* 1996a). In south-western Australia, ripe fruits are found in January (Panetta 1988). On average 4.4 seeds are produced per berry and 38 berries per fruit (or infructescence) (Panetta 1988; Scott and Neser 1996). Seed dispersal is by water, or by birds and possibly mammals consuming the ripe fruits. Seeds have a low temperature requirement for germination, indicating germination in autumn and winter (Panetta 1988). Seeds live less than a

Figure 1: Arum lily *Zantedeschia aethiopica*. a) Flowers. b) Growth habit in pasture at Busselton, WA. Photos: JK Scott, CSIRO.

year and do not form a persistent seed bank (Panetta 1988). Perenniation is via the rhizomes. Daughter rhizomes or offsets are formed, ensuring the regrowth of clumps of plants should the main rhizome be killed, e.g. by herbicide. In horticulture, 25–30, and up to 55, offsets are produced per plant (Plummer 1997).

The genus *Zantedeschia* (Araceae) consists of eight species, all of African origin (Letty 1973; Singh 2003; Singh *et al.* 2007). The Araceae have few genera and species native to Australia and their distributions, being mainly in tropical regions, do not overlap that of *Z. aethiopica* in southern Australia except for outliers along the east coast of Australia (Fig. 2; Scott 1997a).

In WA, arum lily is found in fringing vegetation of freshwater wetlands, creek lines and pasture in wetter areas of the south-west. It is a major understorey weed of dunes and swales on Garden and Eclipse islands (Keighery 1997). Some major infestations, such as on Garden Island in WA, appear to be of relatively recent origin (1950s) (Wykes 1997). The weed was first recorded on the neighbouring Carnac Island in 1966, where it is considered a potential threat to the indigenous vegetation (Abbott *et al.* 2000). Elsewhere in Australia it is found in urban rubbish dumps, gullies, irrigation ditches and poorly drained swampy pastures (Parsons and Cuthbertson 1992). It is sparingly naturalised in widely separated locations around the world (Scott 1997a), but nowhere in the abundance observed in Australia. In its native habitat in South Africa it is found in marshy depressions, roadsides and wetlands (Letty 1973; Scott and Neser 1996). Arum lily is comparatively difficult to find and has an apparent lower abundance in southern Africa than in the south-west of Australia (Scott and Neser 1996).

Little targeted research has been conducted on the ecological impacts of arum lily. However, the dense monocultures established on dune and swale systems of Garden Island have led to it being categorised as an 'ecological disaster' (Wykes 1997). No other ecological assessments have been made.

Arum lily is only declared noxious in WA (Australian Weeds Committee 2010; Dodd 1997). Multiple treatments over at least three years by cultivation or herbicides are required to control plants (Moore 1997). Some herbicides provide effective control of above-ground growth but have little effect on the rhizome, and repeated applications are necessary for long-term control (Moore and Hoskins 1997). A second reaction is for rhizome dormancy to be associated with herbicide application, resulting in regrowth in later years (Moore and Hoskins 1997). This effect, of plant recovery two to three years later, was observed in a major spraying program on a 350 ha infestation on Garden Island (Wykes 1997). In contrast, Rayner and Peirce (2002) used a blanket wiper to apply paraquat, which reduced rhizome density by 82%.

BIOLOGICAL CONTROL HISTORY

Biocontrol of arum lily has not been attempted elsewhere and is only in its initial phase in Australia – the exploration for potential agents. The taxonomic and geographical separation of arum lily in Australia from other Araceae support the use of a biocontrol strategy (Scott 1997b). The difficulties of control by cultural and herbicide methods also point to the need to assess biocontrol. An application for nomination as a target for biocontrol has not been attempted, in part because of the commercial importance of the plant in horticulture and the collection of 'flowers' from weed populations (Scott 1997b). This conflict of interest points to a strategy of either controlling seed production, i.e. reducing dispersal by using

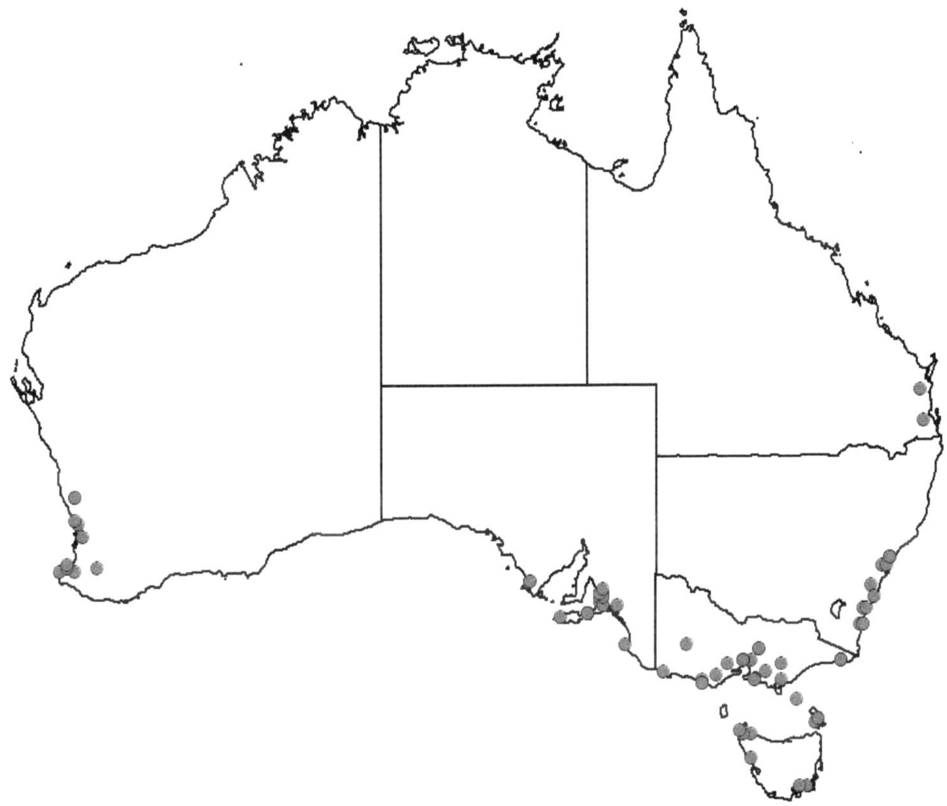

Figure 2: Distribution of arum lily *Zantedeschia aethiopica* in Australia. Specimen data reproduced from Australia's Virtual Herbarium with permission of the Council of Heads of Australasian Herbaria Inc.

exotic biocontrol agents, and/or using organisms that already exist in Australia in an inundative biocontrol approach (Scott 1997b).

EXPLORATION

Fruits of arum lily were surveyed in South Africa for potential biocontrol agents in 1996 (Scott and Neser 1996). While up to 31% of seeds were damaged by insects, only polyphagous species were found, none of which would be suitable for release in Australia (Scott and Neser 1996). Plant pathogens associated with arum lily were surveyed in WA in 1997 (Maxwell 1997a, b). A diverse range of pathogens was found, including some new associations such as *Fusarium oxysporum* Schlecht. (Hypocreaceae), a pathogen that is known to form host-specific pathovars. Overall, the survey revealed a low level of endemic disease (Maxwell 1997a, b).

DISCUSSION

Research priorities for arum lily were assessed at a workshop in 1997 (Scott and Wykes 1997) which concluded that the highest priority for research is the interaction between herbicides and rhizomes. Aside from work by Rayner and Peirce (2002), little progress has been made on this issue. Likewise, there has been little progress towards biocontrol, which was also rated a high priority for research. Arum lily continues to be a major weed in south-western Australia (Lloyd and Dodd 2006). However, the use of the plant in horticulture and for flower collection makes it not feasible to use most exotic agents. Reducing seed set is one possibility, but Scott and Neser (1996) did not find suitable seed-feeding species in South Africa. This points to the main biocontrol option being the use of organisms already resident in Australia; some initial work has been done in this direction through surveys in south-western Australia (Maxwell 1997b).

Further surveys are required on diseases associated with arum lily in other regions of Australia. The organism *Phytophthora richardiae* Buisman (Pythiaceae) (Erwin and Ribeiro 1996), recorded from NSW (Hall 1991), stands out as having potential for further study. This root rot of *Z. aethiopica* has been confirmed as a separate species (Blair *et al.* 2008), has a wide distribution (Hall 1991) and is known only from plants of the genus *Zantedeschia* (Erwin and Ribeiro 1996), allowing for some likely erroneous records in the older literature. A

Phytophthora species has been used as a mycoherbicide in the past (Kenney 1986; Ridings *et al.* 1978), indicating a potential direction for development of *P. richardiae*. Little is known about the biology and host-range of this species, but an assessment of its potential for use in biocontrol would be a highly suitable place to start if it can be rediscovered in Australia.

REFERENCES

Abbott I, Marchant N and Cranfield R (2000) Long-term change in the floristic composition and vegetation structure of Carnac Island, Western Australia. *Journal of Biogeography* **27**, 333–346.

Australian Weeds Committee (2010) Noxious weeds list. <http://www.weeds.org.au/noxious.htm> (accessed 7.7.10).

Blair JE, Coffey MD, Park SY, Geiser DM and Kang S (2008) A multi-locus phylogeny for *Phytophthora* utilizing markers derived from complete genome sequences. *Fungal Genetics and Biology* **45**, 266–277.

Dodd J (1997) Arum lily – implications of its declared plant status in Western Australia. In *Proceedings of a Workshop on Arum Lily* (*Zantedeschia aethiopica*). 7 August 1997, HMAS Stirling, Garden Island, WA. (Eds JK Scott and BJ Wykes) pp. 38–39. CRC for Weed Management Systems, Adelaide.

Erwin DC and Ribeiro OK (1996) *Phytophthora Diseases Worldwide*. APS Press, St Paul, Minnesota.

Hall G (1991) IMI descriptions of fungi and bacteria No. 1066 *Phytophthora richardiae*. *Mycopathologia* **115**, 231–232.

Keighery G (1997) Arum lily as an environmental weed in Western Australia – overview of distribution and threat to natural systems. In *Proceedings of a Workshop on Arum Lily* (*Zantedeschia aethiopica*). 7 August 1997, HMAS Stirling, Garden Island, WA. (Eds JK Scott and BJ Wykes) pp. 17–19. CRC for Weed Management Systems, Adelaide.

Kenney DS (1986) DeVine® – the way it was developed – an industralist's view. *Weed Science* **34** (Supplement 1), 15–16.

Letty C (1973) The genus *Zantedeschia*. *Bothalia* **11**, 5–26.

Lloyd S and Dodd J (2006) 'Garden plants that become bushland weeds. Western Australia Department of Agriculture and Food Garden Note 185'. <http://www.agric.wa.gov.au/objtwr/imported_assets/content/pw/gard/gn2006_gweeds_slloyd.pdf> (accessed 8.7.10).

Maxwell A (1997a) The potential for biological control of *Zantedeschia aethiopica*: a survey of south-western Australia for pathogens. In *Proceedings of a Workshop on Arum Lily* (*Zantedeschia aethiopica*). 7 August 1997, HMAS Stirling, Garden Island, WA. (Eds JK Scott and BJ Wykes) p. 40. CRC for Weed Management Systems, Adelaide.

Maxwell A (1997b) The potential for bio-control of *Zantedeschia aethiopica*: a survey of south-western Australia for pathogens of this noxious weed. BSc (Hons) thesis. Murdoch University, Australia.

Moore JH (1997) Arum lily: a general description. In *Proceedings of a Workshop on Arum Lily* (*Zantedeschia aethiopica*). 7 August 1997, HMAS Stirling, Garden Island, WA. (Eds JK Scott and BJ Wykes) pp. 4–6. CRC for Weed Management Systems, Adelaide.

Moore JH and Hoskins EB (1997) Arum lily (*Zantedeschia aethiopica*) control in the south west of Western Australia. In *Proceedings of a Workshop on Arum Lily* (*Zantedeschia aethiopica*). 7 August 1997, HMAS Stirling, Garden Island, WA. (Eds JK Scott and BJ Wykes) pp. 31–37. CRC for Weed Management Systems, Adelaide.

Panetta FD (1988) Studies on the seed biology of arum lily (*Zantedeschia aethiopica* (L.) Spreng.). *Plant Protection Quarterly* **3**, 169–171.

Parsons WT and Cuthbertson EG (1992) *Noxious Weeds of Australia*. Inkata Press, Melbourne.

Plummer JA (1997) Reproductive biology of *Zantedeschia aethiopica*. In *Proceedings of a Workshop on Arum Lily* (*Zantedeschia aethiopica*). 7 August 1997, HMAS Stirling, Garden Island, WA. (Eds JK Scott and BJ Wykes) pp. 7–9. CRC for Weed Management Systems, Adelaide.

Rayner BJ and Peirce JR (2002) Control of weeds using an experimental blanket wiper. In *Proceedings of the 13th Australian Weeds Conference*. 8–13 September 2002, Perth. (Eds H Spafford Jacob, J Dodd and JH Moore) pp. 158–160. Plant Protection Society of Western Australia, Perth.

Ridings WH, Mitchell DJ, Schoulties CL and El-Gholl NE (1978) Biological control of milkweed vine in Florida citrus groves with a pathotype of *Phytophthora citrophthora*. In *Proceedings of the 4th International Symposium on Biological Control of Weeds*. 30 August–2 September 1976, University of Florida, Gainesville. (Ed TE Freeman) pp. 224–240. University of Florida, Gainesville.

Scott JK (1997a) The distribution of arum lily and its invasive potential. In *Proceedings of a Workshop on Arum Lily* (*Zantedeschia aethiopica*). 7 August 1997, HMAS Stirling, Garden Island, WA. (Eds JK Scott and BJ Wykes) pp. 10–16. CRC for Weed Management Systems, Adelaide.

Scott JK (1997b) The potential for classical biological control of arum lily (*Zantedeschia aethiopica*) in Australia. In *Proceedings of a Workshop on Arum Lily* (*Zantedeschia aethiopica*). 7 August 1997, HMAS Stirling, Garden Island, WA. (Eds JK Scott and BJ Wykes) pp. 41–48. CRC for Weed Management Systems, Adelaide.

Scott JK and Neser S (1996) Prospects for the biological control of the environmental weed, *Zantedeschia aethiopica* (arum lily). In *Proceedings of the 11th Australian Weeds Conference*. 30 September–3 October 1996, Melbourne. (Ed RCH Shepherd) pp. 413–416. Weed Science Society of Victoria, Melbourne.

Scott JK and Wykes BJ (1997) Report on research priorities for arum lily (*Zantedeschia aethiopica*). In *Proceedings of a Workshop on Arum Lily* (*Zantedeschia aethiopica*). 7 August 1997, HMAS Stirling, Garden Island, WA. (Eds JK Scott and BJ Wykes) pp. 1–3. CRC for Weed Management Systems, Adelaide.

Singh Y (2003) Araceae. *Strelitzia* **14**, 972–973.

Singh Y, van Wyk AE and Baijnath H (1996a) Floral biology of *Zantedeschia aethiopica* (L.) Spreng. (Araceae). *South African Journal of Botany* **62**, 146–150.

Singh Y, van Wyk AE and Baijnath H (1996b) Taxonomic notes on the genus *Zantedeschia* Spreng. (Araceae) in southern Africa. *South African Journal of Botany* **62**, 321–324.

Singh Y, Craib C and Condy G (2007) *Zantedeschia jucunda* Araceae. *Flowering Plants of Africa* **60**, 2–7.

Wykes BJ (1997) HMAS Stirling arum lily management program. In *Proceedings of a Workshop on Arum Lily* (*Zantedeschia aethiopica*). 7 August 1997, HMAS Stirling, Garden Island, WA. (Eds JK Scott and BJ Wykes) pp. 20–26. CRC for Weed Management Systems, Adelaide.

Index: Scientific names of weeds

Index: Common names of weeds

Index: Key words

www.ingramcontent.com/pod-product-compliance
Lightning Source LLC
LaVergne TN
LVHW071343210926
848486LV00003B/215